BASIC ALGEBRA FORMULAS

Arithmetic Operations

$$a(b + c) = ab + ac, \qquad \frac{a}{b} \cdot \frac{c}{d} = \frac{ac}{bd}$$

$$\frac{a}{b} + \frac{c}{d} = \frac{ad + bc}{bd}, \qquad \frac{a/b}{c/d} = \frac{a}{b} \cdot \frac{d}{c}$$

Laws of Signs

$$-(-a) = a, \qquad \frac{-a}{b} = -\frac{a}{b} = \frac{a}{-b}$$

Zero   Division by zero is not defined.

$$\text{If } a \neq 0: \frac{0}{a} = 0, \quad a^0 = 1, \quad 0^a = 0$$

$$\text{For any number } a: \ a \cdot 0 = 0 \cdot a = 0$$

Laws of Exponents

$$a^m a^n = a^{m+n}, \qquad (ab)^m = a^m b^m, \qquad (a^m)^n = a^{mn}, \qquad a^{m/n} = \sqrt[n]{a^m} = \left(\sqrt[n]{a}\right)^m$$

If $a \neq 0$,

$$\frac{a^m}{a^n} = a^{m-n}, \qquad a^0 = 1, \qquad a^{-m} = \frac{1}{a^m}.$$

The Binomial Theorem   For any positive integer n,

$$(a + b)^n = a^n + na^{n-1}b + \frac{n(n-1)}{1 \cdot 2} a^{n-2}b^2$$

$$+ \frac{n(n-1)(n-2)}{1 \cdot 2 \cdot 3} a^{n-3}b^3 + \cdots + nab^{n-1} + b^n.$$

For instance,

$$(a + b)^2 = a^2 + 2ab + b^2, \qquad\qquad (a - b)^2 = a^2 - 2ab + b^2$$

$$(a + b)^3 = a^3 + 3a^2b + 3ab^2 + b^3, \qquad (a - b)^3 = a^3 - 3a^2b + 3ab^2 - b^3.$$

Factoring the Difference of Like Integer Powers, $n > 1$

$$a^n - b^n = (a - b)(a^{n-1} + a^{n-2}b + a^{n-3}b^2 + \cdots + ab^{n-2} + b^{n-1})$$

For instance,

$$a^2 - b^2 = (a - b)(a + b),$$

$$a^3 - b^3 = (a - b)(a^2 + ab + b^2),$$

$$a^4 - b^4 = (a - b)(a^3 + a^2b + ab^2 + b^3).$$

Completing the Square   If $a \neq 0$,

$$ax^2 + bx + c = au^2 + C \qquad \left(u = x + (b/2a), C = c - \frac{b^2}{4a}\right)$$

The Quadratic Formula   If $a \neq 0$ and $ax^2 + bx + c = 0$, then

$$x = \frac{-b \pm \sqrt{b^2 - 4ac}}{2a}.$$

GEOMETRY FORMULAS

A = area, B = area of base, C = circumference, S = lateral area or surface area, V = volume

Triangle

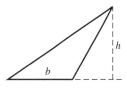

$$A = \tfrac{1}{2}bh$$

Similar Triangles

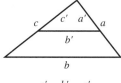

$$\frac{a'}{a} = \frac{b'}{b} = \frac{c'}{c}$$

Pythagorean Theorem

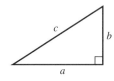

$$a^2 + b^2 = c^2$$

Parallelogram

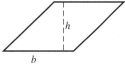

$$A = bh$$

Trapezoid

$$A = \tfrac{1}{2}(a + b)h$$

Circle

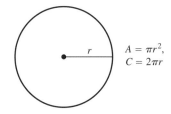

$$A = \pi r^2,$$
$$C = 2\pi r$$

Any Cylinder or Prism with Parallel Bases

$$V = Bh$$

Right Circular Cylinder

$$V = \pi r^2 h$$
$$S = 2\pi rh = \text{Area of side}$$

Any Cone or Pyramid

$$V = \tfrac{1}{3}Bh$$

Right Circular Cone

$$V = \tfrac{1}{3}\pi r^2 h$$
$$S = \pi rs = \text{Area of side}$$

Sphere

$$V = \tfrac{4}{3}\pi r^3, \; S = 4\pi r^2$$

THOMAS'
CALCULUS
EARLY TRANSCENDENTALS

Twelfth Edition

Based on the original work by
George B. Thomas, Jr.
Massachusetts Institute of Technology

as revised by

Maurice D. Weir
Naval Postgraduate School

Joel Hass
University of California, Davis

Addison-Wesley

Boston   Columbus   Indianapolis   New York   San Francisco   Upper Saddle River
Amsterdam   Cape Town   Dubai   London   Madrid   Milan   Munich   Paris   Montréal   Toronto
Delhi   Mexico City   São Paulo   Sydney   Hong Kong   Seoul   Singapore   Taipei   Tokyo

Editor-in-Chief: Deirdre Lynch
Senior Acquisitions Editor: William Hoffman
Senior Project Editor: Rachel S. Reeve
Associate Editor: Caroline Celano
Associate Project Editor: Leah Goldberg
Senior Managing Editor: Karen Wernholm
Senior Production Project Manager: Sheila Spinney
Senior Design Supervisor: Andrea Nix
Digital Assets Manager: Marianne Groth
Media Producer: Lin Mahoney
Software Development: Mary Durnwald and Bob Carroll
Executive Marketing Manager: Jeff Weidenaar
Marketing Assistant: Kendra Bassi
Senior Author Support/Technology Specialist: Joe Vetere
Senior Prepress Supervisor: Caroline Fell
Manufacturing Manager: Evelyn Beaton
Production Coordinator: Kathy Diamond
Composition: Nesbitt Graphics, Inc.
Illustrations: Karen Heyt, IllustraTech
Cover Design: Rokusek Design

Cover image: Forest Edge, Hokuto, Hokkaido, Japan 2004 © Michael Kenna

About the cover: The cover image of a tree line on a snow-swept landscape, by the photographer Michael Kenna, was taken in Hokkaido, Japan. The artist was not thinking of calculus when he composed the image, but rather, of a visual haiku consisting of a few elements that would spark the viewer's imagination. Similarly, the minimal design of this text allows the central ideas of calculus developed in this book to unfold to ignite the learner's imagination.

Library of Congress Cataloging-in-Publication Data

Weir, Maurice D.
   Thomas' calculus : early transcendentals / Maurice D. Weir, Joel Hass, George B. Thomas.—12th ed.
      p. cm
   Includes index.
   ISBN 978-0-321-58876-0
   1. Calculus—Textbooks.   2. Geometry, Analytic—Textbooks.   I. Hass, Joel.   II. Thomas, George B. (George Brinton), 1914–2006.   III. Title   IV. Title: Calculus.

QA303.2.W45 2009
515—dc22   2009023070

3 4 5 6 7 8 9 10—CRK—12 11

Addison-Wesley
is an imprint of

www.pearsoned.com

ISBN-10:   0-321-58876-2
ISBN-13: 978-0-321-58876-0

CONTENTS

Preface   ix

1   Functions   1

1.1   Functions and Their Graphs   1
1.2   Combining Functions; Shifting and Scaling Graphs   14
1.3   Trigonometric Functions   22
1.4   Graphing with Calculators and Computers   30
1.5   Exponential Functions   34
1.6   Inverse Functions and Logarithms   40
    QUESTIONS TO GUIDE YOUR REVIEW   52
    PRACTICE EXERCISES   53
    ADDITIONAL AND ADVANCED EXERCISES   55

2   Limits and Continuity   58

2.1   Rates of Change and Tangents to Curves   58
2.2   Limit of a Function and Limit Laws   65
2.3   The Precise Definition of a Limit   76
2.4   One-Sided Limits   85
2.5   Continuity   92
2.6   Limits Involving Infinity; Asymptotes of Graphs   103
    QUESTIONS TO GUIDE YOUR REVIEW   116
    PRACTICE EXERCISES   117
    ADDITIONAL AND ADVANCED EXERCISES   119

3   Differentiation   122

3.1   Tangents and the Derivative at a Point   122
3.2   The Derivative as a Function   126

3.3   Differentiation Rules   135
3.4   The Derivative as a Rate of Change   145
3.5   Derivatives of Trigonometric Functions   155
3.6   The Chain Rule   162
3.7   Implicit Differentiation   170
3.8   Derivatives of Inverse Functions and Logarithms   176
3.9   Inverse Trigonometric Functions   186
3.10   Related Rates   192
3.11   Linearization and Differentials   201
        QUESTIONS TO GUIDE YOUR REVIEW   212
        PRACTICE EXERCISES   213
        ADDITIONAL AND ADVANCED EXERCISES   218

4   Applications of Derivatives   222

4.1   Extreme Values of Functions   222
4.2   The Mean Value Theorem   230
4.3   Monotonic Functions and the First Derivative Test   238
4.4   Concavity and Curve Sketching   243
4.5   Indeterminate Forms and L'Hôpital's Rule   254
4.6   Applied Optimization   263
4.7   Newton's Method   274
4.8   Antiderivatives   279
        QUESTIONS TO GUIDE YOUR REVIEW   289
        PRACTICE EXERCISES   289
        ADDITIONAL AND ADVANCED EXERCISES   293

5   Integration   297

5.1   Area and Estimating with Finite Sums   297
5.2   Sigma Notation and Limits of Finite Sums   307
5.3   The Definite Integral   313
5.4   The Fundamental Theorem of Calculus   325
5.5   Indefinite Integrals and the Substitution Method   336
5.6   Substitution and Area Between Curves   344
        QUESTIONS TO GUIDE YOUR REVIEW   354
        PRACTICE EXERCISES   354
        ADDITIONAL AND ADVANCED EXERCISES   358

6   Applications of Definite Integrals   363

6.1   Volumes Using Cross-Sections   363
6.2   Volumes Using Cylindrical Shells   374
6.3   Arc Length   382

6.4   Areas of Surfaces of Revolution   388
6.5   Work and Fluid Forces   393
6.6   Moments and Centers of Mass   402
QUESTIONS TO GUIDE YOUR REVIEW   413
PRACTICE EXERCISES   413
ADDITIONAL AND ADVANCED EXERCISES   415

7   Integrals and Transcendental Functions   417

7.1   The Logarithm Defined as an Integral   417
7.2   Exponential Change and Separable Differential Equations   427
7.3   Hyperbolic Functions   436
7.4   Relative Rates of Growth   444
QUESTIONS TO GUIDE YOUR REVIEW   450
PRACTICE EXERCISES   450
ADDITIONAL AND ADVANCED EXERCISES   451

8   Techniques of Integration   453

8.1   Integration by Parts   454
8.2   Trigonometric Integrals   462
8.3   Trigonometric Substitutions   467
8.4   Integration of Rational Functions by Partial Fractions   471
8.5   Integral Tables and Computer Algebra Systems   481
8.6   Numerical Integration   486
8.7   Improper Integrals   496
QUESTIONS TO GUIDE YOUR REVIEW   507
PRACTICE EXERCISES   507
ADDITIONAL AND ADVANCED EXERCISES   509

9   First-Order Differential Equations   514

9.1   Solutions, Slope Fields, and Euler's Method   514
9.2   First-Order Linear Equations   522
9.3   Applications   528
9.4   Graphical Solutions of Autonomous Equations   534
9.5   Systems of Equations and Phase Planes   541
QUESTIONS TO GUIDE YOUR REVIEW   547
PRACTICE EXERCISES   547
ADDITIONAL AND ADVANCED EXERCISES   548

10   Infinite Sequences and Series   550

10.1   Sequences   550
10.2   Infinite Series   562

10.3   The Integral Test   571
10.4   Comparison Tests   576
10.5   The Ratio and Root Tests   581
10.6   Alternating Series, Absolute and Conditional Convergence   586
10.7   Power Series   593
10.8   Taylor and Maclaurin Series   602
10.9   Convergence of Taylor Series   607
10.10 The Binomial Series and Applications of Taylor Series   614
       QUESTIONS TO GUIDE YOUR REVIEW   623
       PRACTICE EXERCISES   623
       ADDITIONAL AND ADVANCED EXERCISES   625

11   Parametric Equations and Polar Coordinates   628

11.1   Parametrizations of Plane Curves   628
11.2   Calculus with Parametric Curves   636
11.3   Polar Coordinates   645
11.4   Graphing in Polar Coordinates   649
11.5   Areas and Lengths in Polar Coordinates   653
11.6   Conic Sections   657
11.7   Conics in Polar Coordinates   666
       QUESTIONS TO GUIDE YOUR REVIEW   672
       PRACTICE EXERCISES   673
       ADDITIONAL AND ADVANCED EXERCISES   675

12   Vectors and the Geometry of Space   678

12.1   Three-Dimensional Coordinate Systems   678
12.2   Vectors   683
12.3   The Dot Product   692
12.4   The Cross Product   700
12.5   Lines and Planes in Space   706
12.6   Cylinders and Quadric Surfaces   714
       QUESTIONS TO GUIDE YOUR REVIEW   719
       PRACTICE EXERCISES   720
       ADDITIONAL AND ADVANCED EXERCISES   722

13   Vector-Valued Functions and Motion in Space   725

13.1   Curves in Space and Their Tangents   725
13.2   Integrals of Vector Functions; Projectile Motion   733
13.3   Arc Length in Space   742
13.4   Curvature and Normal Vectors of a Curve   746
13.5   Tangential and Normal Components of Acceleration   752
13.6   Velocity and Acceleration in Polar Coordinates   757

QUESTIONS TO GUIDE YOUR REVIEW   760
PRACTICE EXERCISES   761
ADDITIONAL AND ADVANCED EXERCISES   763

14 | Partial Derivatives   765

14.1   Functions of Several Variables   765
14.2   Limits and Continuity in Higher Dimensions   773
14.3   Partial Derivatives   782
14.4   The Chain Rule   793
14.5   Directional Derivatives and Gradient Vectors   802
14.6   Tangent Planes and Differentials   809
14.7   Extreme Values and Saddle Points   820
14.8   Lagrange Multipliers   829
14.9   Taylor's Formula for Two Variables   838
14.10 Partial Derivatives with Constrained Variables   842
       QUESTIONS TO GUIDE YOUR REVIEW   847
       PRACTICE EXERCISES   847
       ADDITIONAL AND ADVANCED EXERCISES   851

15 | Multiple Integrals   854

15.1   Double and Iterated Integrals over Rectangles   854
15.2   Double Integrals over General Regions   859
15.3   Area by Double Integration   868
15.4   Double Integrals in Polar Form   871
15.5   Triple Integrals in Rectangular Coordinates   877
15.6   Moments and Centers of Mass   886
15.7   Triple Integrals in Cylindrical and Spherical Coordinates   893
15.8   Substitutions in Multiple Integrals   905
       QUESTIONS TO GUIDE YOUR REVIEW   914
       PRACTICE EXERCISES   914
       ADDITIONAL AND ADVANCED EXERCISES   916

16 | Integration in Vector Fields   919

16.1   Line Integrals   919
16.2   Vector Fields and Line Integrals: Work, Circulation, and Flux   925
16.3   Path Independence, Conservative Fields, and Potential Functions   938
16.4   Green's Theorem in the Plane   949
16.5   Surfaces and Area   961
16.6   Surface Integrals   971

16.7   Stokes' Theorem   980
16.8   The Divergence Theorem and a Unified Theory   990
       QUESTIONS TO GUIDE YOUR REVIEW   1001
       PRACTICE EXERCISES   1001
       ADDITIONAL AND ADVANCED EXERCISES   1004

17 | Second-Order Differential Equations

online

17.1   Second-Order Linear Equations
17.2   Nonhomogeneous Linear Equations
17.3   Applications
17.4   Euler Equations
17.5   Power Series Solutions

Appendices

AP-1

A.1   Real Numbers and the Real Line   AP-1
A.2   Mathematical Induction   AP-6
A.3   Lines, Circles, and Parabolas   AP-10
A.4   Proofs of Limit Theorems   AP-18
A.5   Commonly Occurring Limits   AP-21
A.6   Theory of the Real Numbers   AP-23
A.7   Complex Numbers   AP-25
A.8   The Distributive Law for Vector Cross Products   AP-35
A.9   The Mixed Derivative Theorem and the Increment Theorem   AP-36

Answers to Odd-Numbered Exercises

A-1

Index

I-1

Credits

C-1

A Brief Table of Integrals

T-1

PREFACE

We have significantly revised this edition of *Thomas' Calculus: Early Transcendentals* to meet the changing needs of today's instructors and students. The result is a book with more examples, more mid-level exercises, more figures, better conceptual flow, and increased clarity and precision. As with previous editions, this new edition provides a modern introduction to calculus that supports conceptual understanding but retains the essential elements of a traditional course. These enhancements are closely tied to an expanded version of MyMathLab® for this text (discussed further on), providing additional support for students and flexibility for instructors.

In this twelfth edition early transcendentals version, we introduce the basic transcendental functions in Chapter 1. After reviewing the basic trigonometric functions, we present the family of exponential functions using an algebraic and graphical approach, with the natural exponential described as a particular member of this family. Logarithms are then defined as the inverse functions of the exponentials, and we also discuss briefly the inverse trigonometric functions. We fully incorporate these functions throughout our developments of limits, derivatives, and integrals in the next five chapters of the book, including the examples and exercises. This approach gives students the opportunity to work early with exponential and logarithmic functions in combinations with polynomials, rational and algebraic functions, and trigonometric functions as they learn the concepts, operations, and applications of single-variable calculus. Later, in Chapter 7, we revisit the definition of transcendental functions, now giving a more rigorous presentation. Here we define the natural logarithm function as an integral with the natural exponential as its inverse.

Many of our students were exposed to the terminology and computational aspects of calculus during high school. Despite this familiarity, students' algebra and trigonometry skills often hinder their success in the college calculus sequence. With this text, we have sought to balance the students' prior experience with calculus with the algebraic skill development they may still need, all without undermining or derailing their confidence. We have taken care to provide enough review material, fully stepped-out solutions, and exercises to support complete understanding for students of all levels.

We encourage students to think beyond memorizing formulas and to generalize concepts as they are introduced. Our hope is that after taking calculus, students will be confident in their problem-solving and reasoning abilities. Mastering a beautiful subject with practical applications to the world is its own reward, but the real gift is the ability to think and generalize. We intend this book to provide support and encouragement for both.

Changes for the Twelfth Edition

CONTENT In preparing this edition we have maintained the basic structure of the Table of Contents from the eleventh edition, yet we have paid attention to requests by current users and reviewers to postpone the introduction of parametric equations until we present polar coordinates. We have made numerous revisions to most of the chapters, detailed as follows:

- **Functions** We condensed this chapter to focus on reviewing function concepts and introducing the transcendental functions. Prerequisite material covering real numbers, intervals, increments, straight lines, distances, circles, and parabolas is presented in Appendices 1–3.

- **Limits** To improve the flow of this chapter, we combined the ideas of limits involving infinity and their associations with asymptotes to the graphs of functions, placing them together in the final section of Chapter 3.

- **Differentiation** While we use rates of change and tangents to curves as motivation for studying the limit concept, we now merge the derivative concept into a single chapter. We reorganized and increased the number of related rates examples, and we added new examples and exercises on graphing rational functions. L'Hôpital's Rule is presented as an application section, consistent with our early coverage of the transcendental functions.

- **Antiderivatives and Integration** We maintain the organization of the eleventh edition in placing antiderivatives as the final topic of Chapter 4, covering applications of derivatives. Our focus is on "recovering a function from its derivative" as the solution to the simplest type of first-order differential equation. Integrals, as "limits of Riemann sums," motivated primarily by the problem of finding the areas of general regions with curved boundaries, are a new topic forming the substance of Chapter 5. After carefully developing the integral *concept*, we turn our attention to its evaluation and connection to antiderivatives captured in the Fundamental Theorem of Calculus. The ensuing applications then *define* the various geometric ideas of area, volume, lengths of paths, and centroids, all as limits of Riemann sums giving definite integrals, which can be evaluated by finding an antiderivative of the integrand. We return later to the topic of solving more complicated first-order differential equations.

- **Differential Equations** Some universities prefer that this subject be treated in a course separate from calculus. Although we do cover solutions to separable differential equations when treating exponential growth and decay applications in Chapter 7 on integrals and transcendental functions, we organize the bulk of our material into two chapters (which may be omitted for the calculus sequence). We give an introductory treatment of first-order differential equations in Chapter 9, including a new section on systems and phase planes, with applications to the competitive-hunter and predator-prey models. We present an introduction to second-order differential equations in Chapter 17, which is included in MyMathLab as well as the *Thomas' Calculus: Early Transcendentals* Web site, **www.pearsonhighered.com/thomas**.

- **Series** We retain the organizational structure and content of the eleventh edition for the topics of sequences and series. We have added several new figures and exercises to the various sections, and we revised some of the proofs related to convergence of power series in order to improve the accessibility of the material for students. The request stated by one of our users as, "anything you can do to make this material easier for students will be welcomed by our faculty," drove our thinking for revisions to this chapter.

- **Parametric Equations** Several users requested that we move this topic into Chapter 11, where we also cover polar coordinates and conic sections. We have done this, realizing that many departments choose to cover these topics at the beginning of Calculus III, in preparation for their coverage of vectors and multivariable calculus.

- **Vector-Valued Functions** We streamlined the topics in this chapter to place more emphasis on the conceptual ideas supporting the later material on partial derivatives, the gradient vector, and line integrals. We condensed the discussions of the Frenet frame and Kepler's three laws of planetary motion.

- **Multivariable Calculus** We have further enhanced the art in these chapters, and we have added many new figures, examples, and exercises. We reorganized the opening material on double integrals, and we combined the applications of double and triple integrals to masses and moments into a single section covering both two- and three-dimensional cases. This reorganization allows for better flow of the key mathematical concepts, together with their properties and computational aspects. As with the eleventh edition, we continue to make the connections of multivariable ideas with their single-variable analogues studied earlier in the book.

- **Vector Fields** We devoted considerable effort to improving the clarity and mathematical precision of our treatment of vector integral calculus, including many additional examples, figures, and exercises. Important theorems and results are stated more clearly and completely, together with enhanced explanations of their hypotheses and mathematical consequences. The area of a surface is now organized into a single section, and surfaces defined implicitly or explicitly are treated as special cases of the more general parametric representation. Surface integrals and their applications then follow as a separate section. Stokes' Theorem and the Divergence Theorem are still presented as generalizations of Green's Theorem to three dimensions.

EXERCISES AND EXAMPLES We know that the exercises and examples are critical components in learning calculus. Because of this importance, we have updated, improved, and increased the number of exercises in nearly every section of the book. There are over 700 new exercises in this edition. We continue our organization and grouping of exercises by topic as in earlier editions, progressing from computational problems to applied and theoretical problems. Exercises requiring the use of computer software systems (such as *Maple*® or *Mathematica*®) are placed at the end of each exercise section, labeled **Computer Explorations**. Most of the applied exercises have a subheading to indicate the kind of application addressed in the problem.

Many sections include new examples to clarify or deepen the meaning of the topic being discussed and to help students understand its mathematical consequences or applications to science and engineering. At the same time, we have removed examples that were a repetition of material already presented.

ART Because of their importance to learning calculus, we have continued to improve existing figures in *Thomas' Calculus: Early Transcendentals*, and we have created a significant number of new ones. We continue to use color consistently and pedagogically to enhance the conceptual idea that is being illustrated. We have also taken a fresh look at all of the figure captions, paying considerable attention to clarity and precision in short statements.

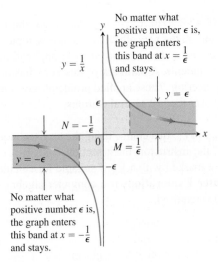

FIGURE 2.50, page 104 The geometric explanation of a finite limit as $x \to \pm\infty$.

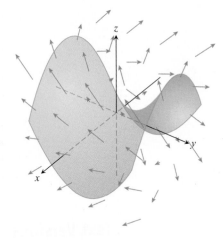

FIGURE 16.9, page 926 A surface in a space occupied by a moving fluid.

MYMATHLAB AND MATHXL The increasing use of and demand for online homework systems has driven the changes to MyMathLab and MathXL® for *Thomas' Calculus:*

Early Transcendentals. The **MyMathLab** course now includes significantly more exercises of all types.

Continuing Features

RIGOR   The level of rigor is consistent with that of earlier editions. We continue to distinguish between formal and informal discussions and to point out their differences. We think starting with a more intuitive, less formal, approach helps students understand a new or difficult concept so they can then appreciate its full mathematical precision and outcomes. We pay attention to defining ideas carefully and to proving theorems appropriate for calculus students, while mentioning deeper or subtler issues they would study in a more advanced course. Our organization and distinctions between informal and formal discussions give the instructor a degree of flexibility in the amount and depth of coverage of the various topics. For example, while we do not prove the Intermediate Value Theorem or the Extreme Value Theorem for continuous functions on $a \leq x \leq b$, we do state these theorems precisely, illustrate their meanings in numerous examples, and use them to prove other important results. Furthermore, for those instructors who desire greater depth of coverage, in Appendix 6 we discuss the reliance of the validity of these theorems on the completeness of the real numbers.

WRITING EXERCISES   Writing exercises placed throughout the text ask students to explore and explain a variety of calculus concepts and applications. In addition, the end of each chapter contains a list of questions for students to review and summarize what they have learned. Many of these exercises make good writing assignments.

END-OF-CHAPTER REVIEWS AND PROJECTS   In addition to problems appearing after each section, each chapter culminates with review questions, practice exercises covering the entire chapter, and a series of Additional and Advanced Exercises serving to include more challenging or synthesizing problems. Most chapters also include descriptions of several **Technology Application Projects** that can be worked by individual students or groups of students over a longer period of time. These projects require the use of a computer running *Mathematica* or *Maple* and additional material that is available over the Internet at **www.pearsonhighered.com/thomas** and in MyMathLab.

WRITING AND APPLICATIONS   As always, this text continues to be easy to read, conversational, and mathematically rich. Each new topic is motivated by clear, easy-to-understand examples and is then reinforced by its application to real-world problems of immediate interest to students. A hallmark of this book has been the application of calculus to science and engineering. These applied problems have been updated, improved, and extended continually over the last several editions.

TECHNOLOGY   In a course using the text, technology can be incorporated according to the taste of the instructor. Each section contains exercises requiring the use of technology; these are marked with a **T** if suitable for calculator or computer use, or they are labeled **Computer Explorations** if a computer algebra system (CAS, such as *Maple* or *Mathematica*) is required.

Text Versions

THOMAS' CALCULUS: EARLY TRANSCENDENTALS, Twelfth Edition
Complete (Chapters 1–16), ISBN 0-321-58876-2 | 978-0-321-58876-0
Single Variable Calculus (Chapters 1–11), 0-321-62883-7 | 978-0-321-62883-1
Multivariable Calculus (Chapters 10–16), ISBN 0-321-64369-0 | 978-0-321-64369-8

The early transcendentals version of *Thomas' Calculus* introduces and integrates transcendental functions (such as inverse trigonometric, exponential, and logarithmic functions) into the exposition, examples, and exercises of the early chapters alongside the algebraic functions. The Multivariable book for *Thomas' Calculus: Early Transcendentals* is the same text as *Thomas' Calculus, Multivariable*.

THOMAS' CALCULUS, Twelfth Edition

Complete (Chapters 1–16), ISBN 0-321-58799-5 | 978-0-321-58799-2
Single Variable Calculus (Chapters 1–11), ISBN 0-321-63742-9 | 978-0-321-63742-0
Multivariable Calculus (Chapters 10–16), ISBN 0-321-64369-0 | 978-0-321-64369-8

Instructor's Editions

Thomas' Calculus: Early Transcendentals, ISBN 0-321-62718-0 | 978-0-321-62718-6
Thomas' Calculus, ISBN 0-321-60075-4 | 978-0-321-60075-2
In addition to including all of the answers present in the student editions, the *Instructor's Editions* include even-numbered answers for Chapters 1–6.

University Calculus (Early Transcendentals)
University Calculus: Alternate Edition (Late Transcendentals)
University Calculus: Elements with Early Transcendentals

The *University Calculus* texts are based on *Thomas' Calculus* and feature a streamlined presentation of the contents of the calculus course. For more information about these titles, visit **www.pearsonhighered.com**.

Print Supplements

INSTRUCTOR'S SOLUTIONS MANUAL

Single Variable Calculus (Chapters 1–11), ISBN 0-321-62717-2 | 978-0-321-62717-9
Multivariable Calculus (Chapters 10–16), ISBN 0-321-60072-X | 978-0-321-60072-1
The *Instructor's Solutions Manual* by William Ardis, Collin County Community College, contains complete worked-out solutions to all of the exercises in *Thomas' Calculus: Early Transcendentals*.

STUDENT'S SOLUTIONS MANUAL

Single Variable Calculus (Chapters 1–11), ISBN 0-321-65692-X | 978-0-321-65692-6
Multivariable Calculus (Chapters 10–16), ISBN 0-321-60071-1 | 978-0-321-60071-4
The *Student's Solutions Manual* by William Ardis, Collin County Community College, is designed for the student and contains carefully worked-out solutions to all the odd-numbered exercises in *Thomas' Calculus: Early Transcendentals*.

JUST-IN-TIME ALGEBRA AND TRIGONOMETRY FOR EARLY TRANSCENDENTALS CALCULUS, Third Edition

ISBN 0-321-32050-6 | 978-0-321-32050-6
Sharp algebra and trigonometry skills are critical to mastering calculus, and *Just-in-Time Algebra and Trigonometry for Early Transcendentals Calculus* by Guntram Mueller and Ronald I. Brent is designed to bolster these skills while students study calculus. As students make their way through calculus, this text is with them every step of the way, showing them the necessary algebra or trigonometry topics and pointing out potential problem spots. The easy-to-use table of contents has algebra and trigonometry topics arranged in the order in which students will need them as they study calculus.

CALCULUS REVIEW CARDS

The Calculus Review Cards (one for Single Variable and another for Multivariable) are a student resource containing important formulas, functions, definitions, and theorems that correspond precisely to the *Thomas' Calculus* series. These cards can work as a reference for completing homework assignments or as an aid in studying, and are available bundled with a new text. Contact your Pearson sales representative for more information.

Media and Online Supplements

TECHNOLOGY RESOURCE MANUALS

Maple Manual by James Stapleton, North Carolina State University
Mathematica Manual by Marie Vanisko, Carroll College
TI-Graphing Calculator Manual by Elaine McDonald-Newman, Sonoma State University
These manuals cover *Maple* 13, *Mathematica* 7, and the TI-83 Plus/TI-84 Plus and TI-89, respectively. Each manual provides detailed guidance for integrating a specific software package or graphing calculator throughout the course, including syntax and commands. These manuals are available to qualified instructors through the *Thomas' Calculus: Early Transcendentals* Web site, **www.pearsonhighered.com/thomas**, and MyMathLab.

WEB SITE www.pearsonhighered.com/thomas

The *Thomas' Calculus: Early Transcendentals* Web site contains the chapter on Second-Order Differential Equations, including odd-numbered answers, and provides the expanded historical biographies and essays referenced in the text. Also available is a collection of *Maple* and *Mathematica* modules, the Technology Resource Manuals, and the **Technology Application Projects**, which can be used as projects by individual students or groups of students.

MyMathLab Online Course (access code required)

MyMathLab is a text-specific, easily customizable online course that integrates interactive multimedia instruction with textbook content. MyMathLab gives you the tools you need to deliver all or a portion of your course online, whether your students are in a lab setting or working from home.

- **Interactive homework exercises**, correlated to your textbook at the objective level, are algorithmically generated for unlimited practice and mastery. Most exercises are free-response and provide guided solutions, sample problems, and learning aids for extra help.

- **"Getting Ready" chapter** includes hundreds of exercises that address prerequisite skills in algebra and trigonometry. Each student can receive remediation for just those skills he or she needs help with.

- **Personalized Study Plan**, generated when students complete a test or quiz, indicates which topics have been mastered and links to tutorial exercises for topics students have not mastered.

- **Multimedia learning aids**, such as video lectures, Java applets, animations, and a complete multimedia textbook, help students independently improve their understanding and performance.

- **Assessment Manager** lets you create online homework, quizzes, and tests that are automatically graded. Select just the right mix of questions from the MyMathLab exercise bank and instructor-created custom exercises.

- **Gradebook**, designed specifically for mathematics and statistics, automatically tracks students' results and gives you control over how to calculate final grades. You can also add offline (paper-and-pencil) grades to the gradebook.

- **MathXL Exercise Builder** allows you to create static and algorithmic exercises for your online assignments. You can use the library of sample exercises as an easy starting point.

- **Pearson Tutor Center** (**www.pearsontutorservices.com**) access is automatically included with MyMathLab. The Tutor Center is staffed by qualified math instructors who provide textbook-specific tutoring for students via toll-free phone, fax, email, and interactive Web sessions.

MyMathLab is powered by CourseCompass™, Pearson Education's online teaching and learning environment, and by MathXL, our online homework, tutorial, and assessment system. MyMathLab is available to qualified adopters. For more information, visit **www.mymathlab.com** or contact your Pearson sales representative.

Video Lectures with Optional Captioning

The Video Lectures with Optional Captioning feature an engaging team of mathematics instructors who present comprehensive coverage of topics in the text. The lecturers' presentations include examples and exercises from the text and support an approach that emphasizes visualization and problem solving. Available only through MyMathLab and MathXL.

MathXL Online Course (access code required)

MathXL is an online homework, tutorial, and assessment system that accompanies Pearson's textbooks in mathematics or statistics.

- **Interactive homework exercises,** correlated to your textbook at the objective level, are algorithmically generated for unlimited practice and mastery. Most exercises are free-response and provide guided solutions, sample problems, and learning aids for extra help.

- **"Getting Ready" chapter** includes hundreds of exercises that address prerequisite skills in algebra and trigonometry. Each student can receive remediation for just those skills he or she needs help with.

- **Personalized Study Plan,** generated when students complete a test or quiz, indicates which topics have been mastered and links to tutorial exercises for topics students have not mastered.

- **Multimedia learning aids,** such as video lectures, Java applets, and animations, help students independently improve their understanding and performance.

- **Gradebook,** designed specifically for mathematics and statistics, automatically tracks students' results and gives you control over how to calculate final grades.

- **MathXL Exercise Builder** allows you to create static and algorithmic exercises for your online assignments. You can use the library of sample exercises as an easy starting point.

- **Assessment Manager** lets you create online homework, quizzes, and tests that are automatically graded. Select just the right mix of questions from the MathXL exercise bank, or instructor-created custom exercises.

MathXL is available to qualified adopters. For more information, visit our Web site at **www.mathxl.com**, or contact your Pearson sales representative.

TestGen®

TestGen (**www.pearsonhighered.com/testgen**) enables instructors to build, edit, print, and administer tests using a computerized bank of questions developed to cover all the objectives of the text. TestGen is algorithmically based, allowing instructors to create multiple but equivalent versions of the same question or test with the click of a button. Instructors can also modify test bank questions or add new questions. Tests can be printed or administered online. The software and testbank are available for download from Pearson Education's online catalog.

PowerPoint® Lecture Slides

These classroom presentation slides are geared specifically to the sequence and philosophy of the *Thomas' Calculus* series. Key graphics from the book are included to help bring the concepts alive in the classroom. These files are available to qualified instructors through the Pearson Instructor Resource Center, **www.pearsonhighered/irc**, and MyMathLab.

Acknowledgments

We would like to express our thanks to the people who made many valuable contributions to this edition as it developed through its various stages:

Accuracy Checkers

Blaise DeSesa

Paul Lorczak

Kathleen Pellissier

Lauri Semarne

Sarah Streett

Holly Zullo

Reviewers for the Twelfth Edition

Meighan Dillon, *Southern Polytechnic State University*

Anne Dougherty, *University of Colorado*

Said Fariabi, *San Antonio College*

Klaus Fischer, *George Mason University*

Tim Flood, *Pittsburg State University*

Rick Ford, *California State University—Chico*

Robert Gardner, *East Tennessee State University*

Christopher Heil, *Georgia Institute of Technology*

Joshua Brandon Holden, *Rose-Hulman Institute of Technology*

Alexander Hulpke, *Colorado State University*

Jacqueline Jensen, *Sam Houston State University*

Jennifer M. Johnson, *Princeton University*

Hideaki Kaneko, *Old Dominion University*

Przemo Kranz, *University of Mississippi*

Xin Li, *University of Central Florida*

Maura Mast, *University of Massachusetts—Boston*

Val Mohanakumar, *Hillsborough Community College—Dale Mabry Campus*

Aaron Montgomery, *Central Washington University*

Christopher M. Pavone, *California State University at Chico*

Cynthia Piez, *University of Idaho*

Brooke Quinlan, *Hillsborough Community College—Dale Mabry Campus*

Rebecca A. Segal, *Virginia Commonwealth University*

Andrew V. Sills, *Georgia Southern University*

Alex Smith, *University of Wisconsin—Eau Claire*

Mark A. Smith, *Miami University*

Donald Solomon, *University of Wisconsin—Milwaukee*

John Sullivan, *Black Hawk College*

Maria Terrell, *Cornell University*

Blake Thornton, *Washington University in St. Louis*

David Walnut, *George Mason University*

Adrian Wilson, *University of Montevallo*

Bobby Winters, *Pittsburg State University*

Dennis Wortman, *University of Massachusetts—Boston*

1

FUNCTIONS

OVERVIEW Functions are fundamental to the study of calculus. In this chapter we review what functions are and how they are pictured as graphs, how they are combined and transformed, and ways they can be classified. We review the trigonometric functions, and we discuss misrepresentations that can occur when using calculators and computers to obtain a function's graph. We also discuss inverse, exponential, and logarithmic functions. The real number system, Cartesian coordinates, straight lines, parabolas, and circles are reviewed in the Appendices.

1.1 | Functions and Their Graphs

Functions are a tool for describing the real world in mathematical terms. A function can be represented by an equation, a graph, a numerical table, or a verbal description; we will use all four representations throughout this book. This section reviews these function ideas.

Functions; Domain and Range

The temperature at which water boils depends on the elevation above sea level (the boiling point drops as you ascend). The interest paid on a cash investment depends on the length of time the investment is held. The area of a circle depends on the radius of the circle. The distance an object travels at constant speed along a straight-line path depends on the elapsed time.

In each case, the value of one variable quantity, say y, depends on the value of another variable quantity, which we might call x. We say that "y is a function of x" and write this symbolically as

$$y = f(x) \qquad \text{("y equals f of x").}$$

In this notation, the symbol f represents the function, the letter x is the **independent variable** representing the input value of f, and y is the **dependent variable** or output value of f at x.

> **DEFINITION** A **function** f from a set D to a set Y is a rule that assigns a *unique* (single) element $f(x) \in Y$ to each element $x \in D$.

The set D of all possible input values is called the **domain** of the function. The set of all values of $f(x)$ as x varies throughout D is called the **range** of the function. The range may not include every element in the set Y. The domain and range of a function can be any sets of objects, but often in calculus they are sets of real numbers interpreted as points of a coordinate line. (In Chapters 13–16, we will encounter functions for which the elements of the sets are points in the coordinate plane or in space.)

Often a function is given by a formula that describes how to calculate the output value from the input variable. For instance, the equation $A = \pi r^2$ is a rule that calculates the area A of a circle from its radius r (so r, interpreted as a length, can only be positive in this formula). When we define a function $y = f(x)$ with a formula and the domain is not stated explicitly or restricted by context, the domain is assumed to be the largest set of real x-values for which the formula gives real y-values, the so-called **natural domain**. If we want to restrict the domain in some way, we must say so. The domain of $y = x^2$ is the entire set of real numbers. To restrict the domain of the function to, say, positive values of x, we would write "$y = x^2, x > 0$."

Changing the domain to which we apply a formula usually changes the range as well. The range of $y = x^2$ is $[0, \infty)$. The range of $y = x^2, x \geq 2$, is the set of all numbers obtained by squaring numbers greater than or equal to 2. In set notation (see Appendix 1), the range is $\{x^2 \mid x \geq 2\}$ or $\{y \mid y \geq 4\}$ or $[4, \infty)$.

When the range of a function is a set of real numbers, the function is said to be **real-valued**. The domains and ranges of many real-valued functions of a real variable are intervals or combinations of intervals. The intervals may be open, closed, or half open, and may be finite or infinite. The range of a function is not always easy to find.

A function f is like a machine that produces an output value $f(x)$ in its range whenever we feed it an input value x from its domain (Figure 1.1). The function keys on a calculator give an example of a function as a machine. For instance, the $\sqrt{x}$ key on a calculator gives an output value (the square root) whenever you enter a nonnegative number x and press the $\sqrt{x}$ key.

A function can also be pictured as an **arrow diagram** (Figure 1.2). Each arrow associates an element of the domain D with a unique or single element in the set Y. In Figure 1.2, the arrows indicate that $f(a)$ is associated with a, $f(x)$ is associated with x, and so on. Notice that a function can have the same *value* at two different input elements in the domain (as occurs with $f(a)$ in Figure 1.2), but each input element x is assigned a *single* output value $f(x)$.

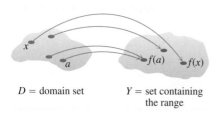

FIGURE 1.1 A diagram showing a function as a kind of machine.

D = domain set Y = set containing the range

FIGURE 1.2 A function from a set D to a set Y assigns a unique element of Y to each element in D.

EXAMPLE 1   Let's verify the natural domains and associated ranges of some simple functions. The domains in each case are the values of x for which the formula makes sense.

Function	Domain (x)	Range (y)
$y = x^2$	$(-\infty, \infty)$	$[0, \infty)$
$y = 1/x$	$(-\infty, 0) \cup (0, \infty)$	$(-\infty, 0) \cup (0, \infty)$
$y = \sqrt{x}$	$[0, \infty)$	$[0, \infty)$
$y = \sqrt{4 - x}$	$(-\infty, 4]$	$[0, \infty)$
$y = \sqrt{1 - x^2}$	$[-1, 1]$	$[0, 1]$

Solution   The formula $y = x^2$ gives a real y-value for any real number x, so the domain is $(-\infty, \infty)$. The range of $y = x^2$ is $[0, \infty)$ because the square of any real number is nonnegative and every nonnegative number y is the square of its own square root, $y = \left(\sqrt{y}\right)^2$ for $y \geq 0$.

The formula $y = 1/x$ gives a real y-value for every x except $x = 0$. For consistency in the rules of arithmetic, *we cannot divide any number by zero*. The range of $y = 1/x$, the set of reciprocals of all nonzero real numbers, is the set of all nonzero real numbers, since $y = 1/(1/y)$. That is, for $y \neq 0$ the number $x = 1/y$ is the input assigned to the output value y.

The formula $y = \sqrt{x}$ gives a real y-value only if $x \geq 0$. The range of $y = \sqrt{x}$ is $[0, \infty)$ because every nonnegative number is some number's square root (namely, it is the square root of its own square).

In $y = \sqrt{4 - x}$, the quantity $4 - x$ cannot be negative. That is, $4 - x \geq 0$, or $x \leq 4$. The formula gives real y-values for all $x \leq 4$. The range of $\sqrt{4 - x}$ is $[0, \infty)$, the set of all nonnegative numbers.

Logarithmic Functions   These are the functions $f(x) = \log_a x$, where the base $a \neq 1$ is a positive constant. They are the *inverse functions* of the exponential functions, and we discuss these functions in Section 1.6. Figure 1.23 shows the graphs of four logarithmic functions with various bases. In each case the domain is $(0, \infty)$ and the range is $(-\infty, \infty)$.

FIGURE 1.23   Graphs of four logarithmic functions.

FIGURE 1.24   Graph of a catenary or hanging cable. (The Latin word *catena* means "chain.")

Transcendental Functions   These are functions that are not algebraic. They include the trigonometric, inverse trigonometric, exponential, and logarithmic functions, and many other functions as well. A particular example of a transcendental function is a **catenary**. Its graph has the shape of a cable, like a telephone line or electric cable, strung from one support to another and hanging freely under its own weight (Figure 1.24). The function defining the graph is discussed in Section 7.3.

Exercises 1.1

Functions

In Exercises 1–6, find the domain and range of each function.

1. $f(x) = 1 + x^2$

2. $f(x) = 1 - \sqrt{x}$

3. $F(x) = \sqrt{5x + 10}$

4. $g(x) = \sqrt{x^2 - 3x}$

5. $f(t) = \dfrac{4}{3 - t}$

6. $G(t) = \dfrac{2}{t^2 - 16}$

In Exercises 7 and 8, which of the graphs are graphs of functions of x, and which are not? Give reasons for your answers.

7. a.

b.

8. a.

b.

Finding Formulas for Functions

9. Express the area and perimeter of an equilateral triangle as a function of the triangle's side length x.

10. Express the side length of a square as a function of the length d of the square's diagonal. Then express the area as a function of the diagonal length.

11. Express the edge length of a cube as a function of the cube's diagonal length d. Then express the surface area and volume of the cube as a function of the diagonal length.

12. A point P in the first quadrant lies on the graph of the function $f(x) = \sqrt{x}$. Express the coordinates of P as functions of the slope of the line joining P to the origin.

13. Consider the point (x, y) lying on the graph of the line $2x + 4y = 5$. Let L be the distance from the point (x, y) to the origin $(0, 0)$. Write L as a function of x.

14. Consider the point (x, y) lying on the graph of $y = \sqrt{x - 3}$. Let L be the distance between the points (x, y) and $(4, 0)$. Write L as a function of y.

Functions and Graphs

Find the domain and graph the functions in Exercises 15–20.

15. $f(x) = 5 - 2x$

16. $f(x) = 1 - 2x - x^2$

17. $g(x) = \sqrt{|x|}$

18. $g(x) = \sqrt{-x}$

19. $F(t) = t/|t|$

20. $G(t) = 1/|t|$

21. Find the domain of $y = \dfrac{x + 3}{4 - \sqrt{x^2 - 9}}$.

22. Find the range of $y = 2 + \dfrac{x^2}{x^2 + 4}$.

23. Graph the following equations and explain why they are not graphs of functions of x.

  a. $|y| = x$ **b.** $y^2 = x^2$

24. Graph the following equations and explain why they are not graphs of functions of x.

  a. $|x| + |y| = 1$ **b.** $|x + y| = 1$

Piecewise-Defined Functions

Graph the functions in Exercises 25–28.

25. $f(x) = \begin{cases} x, & 0 \le x \le 1 \\ 2 - x, & 1 < x \le 2 \end{cases}$

26. $g(x) = \begin{cases} 1 - x, & 0 \le x \le 1 \\ 2 - x, & 1 < x \le 2 \end{cases}$

27. $F(x) = \begin{cases} 4 - x^2, & x \le 1 \\ x^2 + 2x, & x > 1 \end{cases}$

28. $G(x) = \begin{cases} 1/x, & x < 0 \\ x, & 0 \le x \end{cases}$

Find a formula for each function graphed in Exercises 29–32.

29. a.

  b.

30. a.

  b.

31. a.

  b.

32. a.

  b.

The Greatest and Least Integer Functions

33. For what values of x is

  a. $\lfloor x \rfloor = 0$?   **b.** $\lceil x \rceil = 0$?

34. What real numbers x satisfy the equation $\lfloor x \rfloor = \lceil x \rceil$?

35. Does $\lceil -x \rceil = -\lfloor x \rfloor$ for all real x? Give reasons for your answer.

36. Graph the function

$$f(x) = \begin{cases} \lfloor x \rfloor, & x \ge 0 \\ \lceil x \rceil, & x < 0. \end{cases}$$

Why is $f(x)$ called the *integer part* of x?

Increasing and Decreasing Functions

Graph the functions in Exercises 37–46. What symmetries, if any, do the graphs have? Specify the intervals over which the function is increasing and the intervals where it is decreasing.

37. $y = -x^3$

38. $y = -\dfrac{1}{x^2}$

39. $y = -\dfrac{1}{x}$

40. $y = \dfrac{1}{|x|}$

41. $y = \sqrt{|x|}$

42. $y = \sqrt{-x}$

43. $y = x^3/8$

44. $y = -4\sqrt{x}$

45. $y = -x^{3/2}$

46. $y = (-x)^{2/3}$

Even and Odd Functions

In Exercises 47–58, say whether the function is even, odd, or neither. Give reasons for your answer.

47. $f(x) = 3$

48. $f(x) = x^{-5}$

49. $f(x) = x^2 + 1$

50. $f(x) = x^2 + x$

51. $g(x) = x^3 + x$

52. $g(x) = x^4 + 3x^2 - 1$

53. $g(x) = \dfrac{1}{x^2 - 1}$

54. $g(x) = \dfrac{x}{x^2 - 1}$

55. $h(t) = \dfrac{1}{t - 1}$

56. $h(t) = |t^3|$

57. $h(t) = 2t + 1$

58. $h(t) = 2|t| + 1$

Theory and Examples

59. The variable s is proportional to t, and $s = 25$ when $t = 75$. Determine t when $s = 60$.

60. Kinetic energy   The kinetic energy K of a mass is proportional to the square of its velocity v. If $K = 12{,}960$ joules when $v = 18$ m/sec, what is K when $v = 10$ m/sec?

61. The variables r and s are inversely proportional, and $r = 6$ when $s = 4$. Determine s when $r = 10$.

62. Boyle's Law   Boyle's Law says that the volume V of a gas at constant temperature increases whenever the pressure P decreases, so that V and P are inversely proportional. If $P = 14.7$ lbs/in^2 when $V = 1000$ in^3, then what is V when $P = 23.4$ lbs/in^2?

63. A box with an open top is to be constructed from a rectangular piece of cardboard with dimensions 14 in. by 22 in. by cutting out equal squares of side x at each corner and then folding up the sides as in the figure. Express the volume V of the box as a function of x.

64. The accompanying figure shows a rectangle inscribed in an isosceles right triangle whose hypotenuse is 2 units long.

  a. Express the y-coordinate of P in terms of x. (You might start by writing an equation for the line AB.)

  b. Express the area of the rectangle in terms of x.

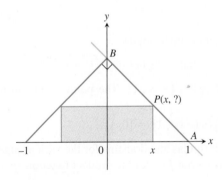

In Exercises 65 and 66, match each equation with its graph. Do not use a graphing device, and give reasons for your answer.

65. a. $y = x^4$ **b.** $y = x^7$ **c.** $y = x^{10}$

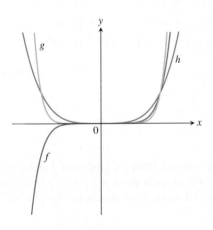

66. a. $y = 5x$ **b.** $y = 5^x$ **c.** $y = x^5$

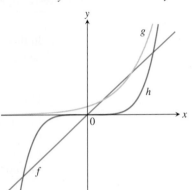

T 67. a. Graph the functions $f(x) = x/2$ and $g(x) = 1 + (4/x)$ together to identify the values of x for which
$$\frac{x}{2} > 1 + \frac{4}{x}.$$

  b. Confirm your findings in part (a) algebraically.

T 68. a. Graph the functions $f(x) = 3/(x - 1)$ and $g(x) = 2/(x + 1)$ together to identify the values of x for which
$$\frac{3}{x - 1} < \frac{2}{x + 1}.$$

  b. Confirm your findings in part (a) algebraically.

69. For a curve to be *symmetric about the x-axis*, the point (x, y) must lie on the curve if and only if the point $(x, -y)$ lies on the curve. Explain why a curve that is symmetric about the x-axis is not the graph of a function, unless the function is $y = 0$.

70. Three hundred books sell for $40 each, resulting in a revenue of $(300)(\$40) = \$12{,}000$. For each $5 increase in the price, 25 fewer books are sold. Write the revenue R as a function of the number x of $5 increases.

71. A pen in the shape of an isosceles right triangle with legs of length x ft and hypotenuse of length h ft is to be built. If fencing costs $5/ft for the legs and $10/ft for the hypotenuse, write the total cost C of construction as a function of h.

72. Industrial costs   A power plant sits next to a river where the river is 800 ft wide. To lay a new cable from the plant to a location in the city 2 mi downstream on the opposite side costs $180 per foot across the river and $100 per foot along the land.

  a. Suppose that the cable goes from the plant to a point Q on the opposite side that is x ft from the point P directly opposite the plant. Write a function $C(x)$ that gives the cost of laying the cable in terms of the distance x.

  b. Generate a table of values to determine if the least expensive location for point Q is less than 2000 ft or greater than 2000 ft from point P.

1.2 | Combining Functions; Shifting and Scaling Graphs

In this section we look at the main ways functions are combined or transformed to form new functions.

Sums, Differences, Products, and Quotients

Like numbers, functions can be added, subtracted, multiplied, and divided (except where the denominator is zero) to produce new functions. If f and g are functions, then for every x that belongs to the domains of both f and g (that is, for $x \in D(f) \cap D(g)$), we define functions $f + g$, $f - g$, and fg by the formulas

$$(f + g)(x) = f(x) + g(x).$$
$$(f - g)(x) = f(x) - g(x).$$
$$(fg)(x) = f(x)g(x).$$

Notice that the $+$ sign on the left-hand side of the first equation represents the operation of addition of *functions*, whereas the $+$ on the right-hand side of the equation means addition of the real numbers $f(x)$ and $g(x)$.

At any point of $D(f) \cap D(g)$ at which $g(x) \neq 0$, we can also define the function f/g by the formula

$$\left(\frac{f}{g}\right)(x) = \frac{f(x)}{g(x)} \qquad (\text{where } g(x) \neq 0).$$

Functions can also be multiplied by constants: If c is a real number, then the function cf is defined for all x in the domain of f by

$$(cf)(x) = cf(x).$$

EXAMPLE 1   The functions defined by the formulas

$$f(x) = \sqrt{x} \qquad \text{and} \qquad g(x) = \sqrt{1 - x}$$

have domains $D(f) = [0, \infty)$ and $D(g) = (-\infty, 1]$. The points common to these domains are the points

$$[0, \infty) \cap (-\infty, 1] = [0, 1].$$

The following table summarizes the formulas and domains for the various algebraic combinations of the two functions. We also write $f \cdot g$ for the product function fg.

Function	Formula	Domain
$f + g$	$(f + g)(x) = \sqrt{x} + \sqrt{1 - x}$	$[0, 1] = D(f) \cap D(g)$
$f - g$	$(f - g)(x) = \sqrt{x} - \sqrt{1 - x}$	$[0, 1]$
$g - f$	$(g - f)(x) = \sqrt{1 - x} - \sqrt{x}$	$[0, 1]$
$f \cdot g$	$(f \cdot g)(x) = f(x)g(x) = \sqrt{x(1 - x)}$	$[0, 1]$
f/g	$\dfrac{f}{g}(x) = \dfrac{f(x)}{g(x)} = \sqrt{\dfrac{x}{1 - x}}$	$[0, 1) \; (x = 1 \text{ excluded})$
g/f	$\dfrac{g}{f}(x) = \dfrac{g(x)}{f(x)} = \sqrt{\dfrac{1 - x}{x}}$	$(0, 1] \; (x = 0 \text{ excluded})$

■

The graph of the function $f + g$ is obtained from the graphs of f and g by adding the corresponding y-coordinates $f(x)$ and $g(x)$ at each point $x \in D(f) \cap D(g)$, as in Figure 1.25. The graphs of $f + g$ and $f \cdot g$ from Example 1 are shown in Figure 1.26.

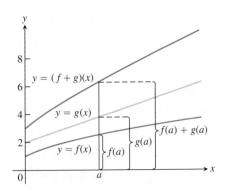

FIGURE 1.25 Graphical addition of two functions.

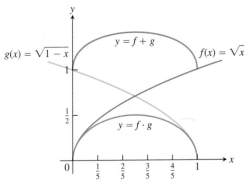

FIGURE 1.26 The domain of the function $f + g$ is the intersection of the domains of f and g, the interval $[0, 1]$ on the x-axis where these domains overlap. This interval is also the domain of the function $f \cdot g$ (Example 1).

Composite Functions

Composition is another method for combining functions.

DEFINITION If f and g are functions, the **composite** function $f \circ g$ ("f composed with g") is defined by

$$(f \circ g)(x) = f(g(x)).$$

The domain of $f \circ g$ consists of the numbers x in the domain of g for which $g(x)$ lies in the domain of f.

The definition implies that $f \circ g$ can be formed when the range of g lies in the domain of f. To find $(f \circ g)(x)$, *first* find $g(x)$ and *second* find $f(g(x))$. Figure 1.27 pictures $f \circ g$ as a machine diagram and Figure 1.28 shows the composite as an arrow diagram.

FIGURE 1.27 Two functions can be composed at x whenever the value of one function at x lies in the domain of the other. The composite is denoted by $f \circ g$.

FIGURE 1.28 Arrow diagram for $f \circ g$.

To evaluate the composite function $g \circ f$ (when defined), we find $f(x)$ first and then $g(f(x))$. The domain of $g \circ f$ is the set of numbers x in the domain of f such that $f(x)$ lies in the domain of g.

The functions $f \circ g$ and $g \circ f$ are usually quite different.

EXAMPLE 2   If $f(x) = \sqrt{x}$ and $g(x) = x + 1$, find

(a) $(f \circ g)(x)$ **(b)** $(g \circ f)(x)$ **(c)** $(f \circ f)(x)$ **(d)** $(g \circ g)(x)$.

Solution

Composite	**Domain**
(a) $(f \circ g)(x) = f(g(x)) = \sqrt{g(x)} = \sqrt{x + 1}$	$[-1, \infty)$
(b) $(g \circ f)(x) = g(f(x)) = f(x) + 1 = \sqrt{x} + 1$	$[0, \infty)$
(c) $(f \circ f)(x) = f(f(x)) = \sqrt{f(x)} = \sqrt{\sqrt{x}} = x^{1/4}$	$[0, \infty)$
(d) $(g \circ g)(x) = g(g(x)) = g(x) + 1 = (x + 1) + 1 = x + 2$	$(-\infty, \infty)$

To see why the domain of $f \circ g$ is $[-1, \infty)$, notice that $g(x) = x + 1$ is defined for all real x but belongs to the domain of f only if $x + 1 \geq 0$, that is to say, when $x \geq -1$. ∎

Notice that if $f(x) = x^2$ and $g(x) = \sqrt{x}$, then $(f \circ g)(x) = \left(\sqrt{x}\right)^2 = x$. However, the domain of $f \circ g$ is $[0, \infty)$, not $(-\infty, \infty)$, since $\sqrt{x}$ requires $x \geq 0$.

Shifting a Graph of a Function

A common way to obtain a new function from an existing one is by adding a constant to each output of the existing function, or to its input variable. The graph of the new function is the graph of the original function shifted vertically or horizontally, as follows.

Shift Formulas

Vertical Shifts

$y = f(x) + k$ Shifts the graph of f *up* k units if $k > 0$

Shifts it *down* $|k|$ units if $k < 0$

Horizontal Shifts

$y = f(x + h)$ Shifts the graph of f *left* h units if $h > 0$

Shifts it *right* $|h|$ units if $h < 0$

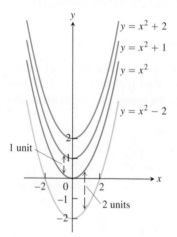

FIGURE 1.29   To shift the graph of $f(x) = x^2$ up (or down), we add positive (or negative) constants to the formula for f (Examples 3a and b).

EXAMPLE 3

(a) Adding 1 to the right-hand side of the formula $y = x^2$ to get $y = x^2 + 1$ shifts the graph up 1 unit (Figure 1.29).

(b) Adding -2 to the right-hand side of the formula $y = x^2$ to get $y = x^2 - 2$ shifts the graph down 2 units (Figure 1.29).

(c) Adding 3 to x in $y = x^2$ to get $y = (x + 3)^2$ shifts the graph 3 units to the left (Figure 1.30).

(d) Adding -2 to x in $y = |x|$, and then adding -1 to the result, gives $y = |x - 2| - 1$ and shifts the graph 2 units to the right and 1 unit down (Figure 1.31). ∎

Scaling and Reflecting a Graph of a Function

To scale the graph of a function $y = f(x)$ is to stretch or compress it, vertically or horizontally. This is accomplished by multiplying the function f, or the independent variable x, by an appropriate constant c. Reflections across the coordinate axes are special cases where $c = -1$.

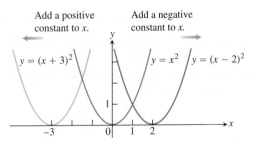

FIGURE 1.30 To shift the graph of $y = x^2$ to the left, we add a positive constant to x (Example 3c). To shift the graph to the right, we add a negative constant to x.

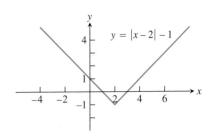

FIGURE 1.31 Shifting the graph of $y = |x|$ 2 units to the right and 1 unit down (Example 3d).

Vertical and Horizontal Scaling and Reflecting Formulas

For $c > 1$, the graph is scaled:

$y = cf(x)$ Stretches the graph of f vertically by a factor of c.

$y = \dfrac{1}{c} f(x)$ Compresses the graph of f vertically by a factor of c.

$y = f(cx)$ Compresses the graph of f horizontally by a factor of c.

$y = f(x/c)$ Stretches the graph of f horizontally by a factor of c.

For $c = -1$, the graph is reflected:

$y = -f(x)$ Reflects the graph of f across the x-axis.

$y = f(-x)$ Reflects the graph of f across the y-axis.

EXAMPLE 4   Here we scale and reflect the graph of $y = \sqrt{x}$.

(a) Vertical: Multiplying the right-hand side of $y = \sqrt{x}$ by 3 to get $y = 3\sqrt{x}$ stretches the graph vertically by a factor of 3, whereas multiplying by $1/3$ compresses the graph by a factor of 3 (Figure 1.32).

(b) Horizontal: The graph of $y = \sqrt{3x}$ is a horizontal compression of the graph of $y = \sqrt{x}$ by a factor of 3, and $y = \sqrt{x/3}$ is a horizontal stretching by a factor of 3 (Figure 1.33). Note that $y = \sqrt{3x} = \sqrt{3}\sqrt{x}$ so a horizontal compression *may* correspond to a vertical stretching by a different scaling factor. Likewise, a horizontal stretching may correspond to a vertical compression by a different scaling factor.

(c) Reflection: The graph of $y = -\sqrt{x}$ is a reflection of $y = \sqrt{x}$ across the x-axis, and $y = \sqrt{-x}$ is a reflection across the y-axis (Figure 1.34). ∎

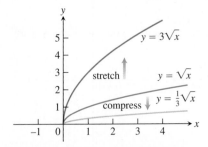

FIGURE 1.32 Vertically stretching and compressing the graph $y = \sqrt{x}$ by a factor of 3 (Example 4a).

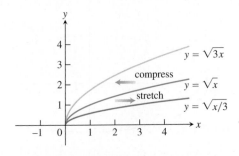

FIGURE 1.33 Horizontally stretching and compressing the graph $y = \sqrt{x}$ by a factor of 3 (Example 4b).

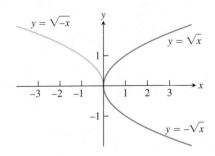

FIGURE 1.34 Reflections of the graph $y = \sqrt{x}$ across the coordinate axes (Example 4c).

EXAMPLE 5   Given the function $f(x) = x^4 - 4x^3 + 10$ (Figure 1.35a), find formulas to

(a) compress the graph horizontally by a factor of 2 followed by a reflection across the y-axis (Figure 1.35b).

(b) compress the graph vertically by a factor of 2 followed by a reflection across the x-axis (Figure 1.35c).

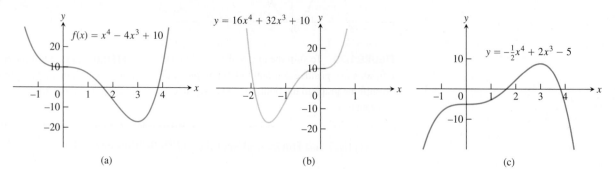

FIGURE 1.35   (a) The original graph of f. (b) The horizontal compression of $y = f(x)$ in part (a) by a factor of 2, followed by a reflection across the y-axis. (c) The vertical compression of $y = f(x)$ in part (a) by a factor of 2, followed by a reflection across the x-axis (Example 5).

Solution

(a) We multiply x by 2 to get the horizontal compression, and by -1 to give reflection across the y-axis. The formula is obtained by substituting $-2x$ for x in the right-hand side of the equation for f:

$$y = f(-2x) = (-2x)^4 - 4(-2x)^3 + 10$$
$$= 16x^4 + 32x^3 + 10.$$

(b) The formula is

$$y = -\frac{1}{2}f(x) = -\frac{1}{2}x^4 + 2x^3 - 5.$$ ∎

Ellipses

Although they are not the graphs of functions, circles can be stretched horizontally or vertically in the same way as the graphs of functions. The standard equation for a circle of radius r centered at the origin is

$$x^2 + y^2 = r^2.$$

Substituting cx for x in the standard equation for a circle (Figure 1.36a) gives

$$c^2x^2 + y^2 = r^2. \tag{1}$$

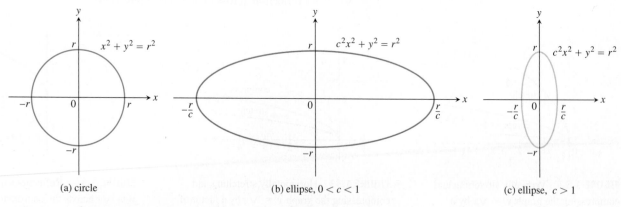

FIGURE 1.36   Horizontal stretching or compression of a circle produces graphs of ellipses.

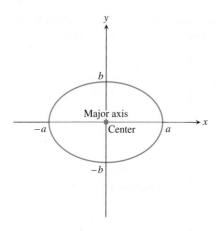

FIGURE 1.37 Graph of the ellipse $\dfrac{x^2}{a^2} + \dfrac{y^2}{b^2} = 1$, $a > b$, where the major axis is horizontal.

If $0 < c < 1$, the graph of Equation (1) horizontally stretches the circle; if $c > 1$ the circle is compressed horizontally. In either case, the graph of Equation (1) is an ellipse (Figure 1.36). Notice in Figure 1.36 that the y-intercepts of all three graphs are always $-r$ and r. In Figure 1.36b, the line segment joining the points $(\pm r/c, 0)$ is called the **major axis** of the ellipse; the **minor axis** is the line segment joining $(0, \pm r)$. The axes of the ellipse are reversed in Figure 1.36c: The major axis is the line segment joining the points $(0, \pm r)$, and the minor axis is the line segment joining the points $(\pm r/c, 0)$. In both cases, the major axis is the longer line segment.

If we divide both sides of Equation (1) by r^2, we obtain

$$\frac{x^2}{a^2} + \frac{y^2}{b^2} = 1 \tag{2}$$

where $a = r/c$ and $b = r$. If $a > b$, the major axis is horizontal; if $a < b$, the major axis is vertical. The **center** of the ellipse given by Equation (2) is the origin (Figure 1.37).

Substituting $x - h$ for x, and $y - k$ for y, in Equation (2) results in

$$\frac{(x - h)^2}{a^2} + \frac{(y - k)^2}{b^2} = 1. \tag{3}$$

Equation (3) is the **standard equation of an ellipse** with center at (h, k). The geometric definition and properties of ellipses are reviewed in Section 11.6.

Exercises 1.2

Algebraic Combinations

In Exercises 1 and 2, find the domains and ranges of $f, g, f + g$, and $f \cdot g$.

1. $f(x) = x$, $\quad g(x) = \sqrt{x - 1}$

2. $f(x) = \sqrt{x + 1}$, $\quad g(x) = \sqrt{x - 1}$

In Exercises 3 and 4, find the domains and ranges of $f, g, f/g$, and g/f.

3. $f(x) = 2$, $\quad g(x) = x^2 + 1$

4. $f(x) = 1$, $\quad g(x) = 1 + \sqrt{x}$

Composites of Functions

5. If $f(x) = x + 5$ and $g(x) = x^2 - 3$, find the following.

  a. $f(g(0))$ **b.** $g(f(0))$

  c. $f(g(x))$ **d.** $g(f(x))$

  e. $f(f(-5))$ **f.** $g(g(2))$

  g. $f(f(x))$ **h.** $g(g(x))$

6. If $f(x) = x - 1$ and $g(x) = 1/(x + 1)$, find the following.

  a. $f(g(1/2))$ **b.** $g(f(1/2))$

  c. $f(g(x))$ **d.** $g(f(x))$

  e. $f(f(2))$ **f.** $g(g(2))$

  g. $f(f(x))$ **h.** $g(g(x))$

In Exercises 7–10, write a formula for $f \circ g \circ h$.

7. $f(x) = x + 1$, $\quad g(x) = 3x$, $\quad h(x) = 4 - x$

8. $f(x) = 3x + 4$, $\quad g(x) = 2x - 1$, $\quad h(x) = x^2$

9. $f(x) = \sqrt{x + 1}$, $\quad g(x) = \dfrac{1}{x + 4}$, $\quad h(x) = \dfrac{1}{x}$

10. $f(x) = \dfrac{x + 2}{3 - x}$, $\quad g(x) = \dfrac{x^2}{x^2 + 1}$, $\quad h(x) = \sqrt{2 - x}$

Let $f(x) = x - 3$, $\quad g(x) = \sqrt{x}$, $\quad h(x) = x^3$, and $j(x) = 2x$. Express each of the functions in Exercises 11 and 12 as a composite involving one or more of f, g, h, and j.

11. a. $y = \sqrt{x} - 3$ **b.** $y = 2\sqrt{x}$

    c. $y = x^{1/4}$ **d.** $y = 4x$

    e. $y = \sqrt{(x - 3)^3}$ **f.** $y = (2x - 6)^3$

12. a. $y = 2x - 3$ **b.** $y = x^{3/2}$

    c. $y = x^9$ **d.** $y = x - 6$

    e. $y = 2\sqrt{x - 3}$ **f.** $y = \sqrt{x^3 - 3}$

13. Copy and complete the following table.

$g(x)$	$f(x)$	$(f \circ g)(x)$
a. $x - 7$	$\sqrt{x}$	?
b. $x + 2$	$3x$	?
c. ?	$\sqrt{x - 5}$	$\sqrt{x^2 - 5}$
d. $\dfrac{x}{x - 1}$	$\dfrac{x}{x - 1}$	?
e. ?	$1 + \dfrac{1}{x}$	x
f. $\dfrac{1}{x}$	?	x

14. Copy and complete the following table.

$g(x)$	$f(x)$	$(f \circ g)(x)$
a. $\dfrac{1}{x-1}$	$|x|$	?
b. ?	$\dfrac{x-1}{x}$	$\dfrac{x}{x+1}$
c. ?	$\sqrt{x}$	$|x|$
d. $\sqrt{x}$	?	$|x|$

15. Evaluate each expression using the given table of values

x	-2	-1	0	1	2
$f(x)$	1	0	-2	1	2
$g(x)$	2	1	0	-1	0

a. $f(g(-1))$ **b.** $g(f(0))$ **c.** $f(f(-1))$
d. $g(g(2))$ **e.** $g(f(-2))$ **f.** $f(g(1))$

16. Evaluate each expression using the functions

$$f(x) = 2 - x, \quad g(x) = \begin{cases} -x, & -2 \le x < 0 \\ x - 1, & 0 \le x \le 2. \end{cases}$$

a. $f(g(0))$ **b.** $g(f(3))$ **c.** $g(g(-1))$
d. $f(f(2))$ **e.** $g(f(0))$ **f.** $f(g(1/2))$

In Exercises 17 and 18, **(a)** write formulas for $f \circ g$ and $g \circ f$ and find the **(b)** domain and **(c)** range of each.

17. $f(x) = \sqrt{x + 1}, \; g(x) = \dfrac{1}{x}$

18. $f(x) = x^2, \; g(x) = 1 - \sqrt{x}$

19. Let $f(x) = \dfrac{x}{x - 2}$. Find a function $y = g(x)$ so that
$(f \circ g)(x) = x$.

20. Let $f(x) = 2x^3 - 4$. Find a function $y = g(x)$ so that
$(f \circ g)(x) = x + 2$.

Shifting Graphs

21. The accompanying figure shows the graph of $y = -x^2$ shifted to two new positions. Write equations for the new graphs.

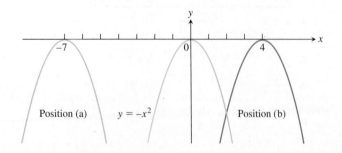

22. The accompanying figure shows the graph of $y = x^2$ shifted to two new positions. Write equations for the new graphs.

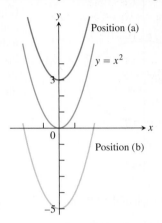

23. Match the equations listed in parts (a)–(d) to the graphs in the accompanying figure.

a. $y = (x - 1)^2 - 4$ **b.** $y = (x - 2)^2 + 2$
c. $y = (x + 2)^2 + 2$ **d.** $y = (x + 3)^2 - 2$

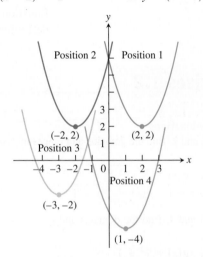

24. The accompanying figure shows the graph of $y = -x^2$ shifted to four new positions. Write an equation for each new graph.

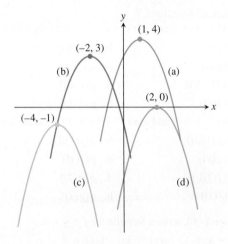

Exercises 25–34 tell how many units and in what directions the graphs of the given equations are to be shifted. Give an equation for the shifted graph. Then sketch the original and shifted graphs together, labeling each graph with its equation.

25. $x^2 + y^2 = 49$ Down 3, left 2

26. $x^2 + y^2 = 25$ Up 3, left 4

27. $y = x^3$ Left 1, down 1

28. $y = x^{2/3}$ Right 1, down 1

29. $y = \sqrt{x}$ Left 0.81

30. $y = -\sqrt{x}$ Right 3

31. $y = 2x - 7$ Up 7

32. $y = \frac{1}{2}(x + 1) + 5$ Down 5, right 1

33. $y = 1/x$ Up 1, right 1

34. $y = 1/x^2$ Left 2, down 1

Graph the functions in Exercises 35–54.

35. $y = \sqrt{x + 4}$

36. $y = \sqrt{9 - x}$

37. $y = |x - 2|$

38. $y = |1 - x| - 1$

39. $y = 1 + \sqrt{x - 1}$

40. $y = 1 - \sqrt{x}$

41. $y = (x + 1)^{2/3}$

42. $y = (x - 8)^{2/3}$

43. $y = 1 - x^{2/3}$

44. $y + 4 = x^{2/3}$

45. $y = \sqrt[3]{x - 1} - 1$

46. $y = (x + 2)^{3/2} + 1$

47. $y = \dfrac{1}{x - 2}$

48. $y = \dfrac{1}{x} - 2$

49. $y = \dfrac{1}{x} + 2$

50. $y = \dfrac{1}{x + 2}$

51. $y = \dfrac{1}{(x - 1)^2}$

52. $y = \dfrac{1}{x^2} - 1$

53. $y = \dfrac{1}{x^2} + 1$

54. $y = \dfrac{1}{(x + 1)^2}$

55. The accompanying figure shows the graph of a function $f(x)$ with domain $[0, 2]$ and range $[0, 1]$. Find the domains and ranges of the following functions, and sketch their graphs.

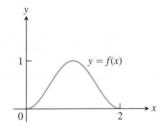

a. $f(x) + 2$ **b.** $f(x) - 1$

c. $2f(x)$ **d.** $-f(x)$

e. $f(x + 2)$ **f.** $f(x - 1)$

g. $f(-x)$ **h.** $-f(x + 1) + 1$

56. The accompanying figure shows the graph of a function $g(t)$ with domain $[-4, 0]$ and range $[-3, 0]$. Find the domains and ranges of the following functions, and sketch their graphs.

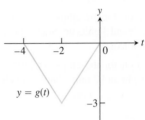

a. $g(-t)$ **b.** $-g(t)$

c. $g(t) + 3$ **d.** $1 - g(t)$

e. $g(-t + 2)$ **f.** $g(t - 2)$

g. $g(1 - t)$ **h.** $-g(t - 4)$

Vertical and Horizontal Scaling

Exercises 57–66 tell by what factor and direction the graphs of the given functions are to be stretched or compressed. Give an equation for the stretched or compressed graph.

57. $y = x^2 - 1$,   stretched vertically by a factor of 3

58. $y = x^2 - 1$,   compressed horizontally by a factor of 2

59. $y = 1 + \dfrac{1}{x^2}$,   compressed vertically by a factor of 2

60. $y = 1 + \dfrac{1}{x^2}$,   stretched horizontally by a factor of 3

61. $y = \sqrt{x + 1}$,   compressed horizontally by a factor of 4

62. $y = \sqrt{x + 1}$,   stretched vertically by a factor of 3

63. $y = \sqrt{4 - x^2}$,   stretched horizontally by a factor of 2

64. $y = \sqrt{4 - x^2}$,   compressed vertically by a factor of 3

65. $y = 1 - x^3$,   compressed horizontally by a factor of 3

66. $y = 1 - x^3$,   stretched horizontally by a factor of 2

Graphing

In Exercises 67–74, graph each function, not by plotting points, but by starting with the graph of one of the standard functions presented in Figures 1.14–1.17 and applying an appropriate transformation.

67. $y = -\sqrt{2x + 1}$

68. $y = \sqrt{1 - \dfrac{x}{2}}$

69. $y = (x - 1)^3 + 2$

70. $y = (1 - x)^3 + 2$

71. $y = \dfrac{1}{2x} - 1$

72. $y = \dfrac{2}{x^2} + 1$

73. $y = -\sqrt[3]{x}$

74. $y = (-2x)^{2/3}$

75. Graph the function $y = |x^2 - 1|$.

76. Graph the function $y = \sqrt{|x|}$.

Ellipses

Exercises 77–82 give equations of ellipses. Put each equation in standard form and sketch the ellipse.

77. $9x^2 + 25y^2 = 225$

78. $16x^2 + 7y^2 = 112$

79. $3x^2 + (y - 2)^2 = 3$

80. $(x + 1)^2 + 2y^2 = 4$

81. $3(x - 1)^2 + 2(y + 2)^2 = 6$

82. $6\left(x + \dfrac{3}{2}\right)^2 + 9\left(y - \dfrac{1}{2}\right)^2 = 54$

83. Write an equation for the ellipse $(x^2/16) + (y^2/9) = 1$ shifted 4 units to the left and 3 units up. Sketch the ellipse and identify its center and major axis.

84. Write an equation for the ellipse $(x^2/4) + (y^2/25) = 1$ shifted 3 units to the right and 2 units down. Sketch the ellipse and identify its center and major axis.

Combining Functions

85. Assume that f is an even function, g is an odd function, and both f and g are defined on the entire real line $\mathbb{R}$. Which of the following (where defined) are even? odd?

a. fg	**b.** f/g	**c.** g/f
d. $f^2 = ff$	**e.** $g^2 = gg$	**f.** $f \circ g$
g. $g \circ f$	**h.** $f \circ f$	**i.** $g \circ g$

86. Can a function be both even and odd? Give reasons for your answer.

T 87. (*Continuation of Example 1.*) Graph the functions $f(x) = \sqrt{x}$ and $g(x) = \sqrt{1 - x}$ together with their (a) sum, (b) product, (c) two differences, (d) two quotients.

T 88. Let $f(x) = x - 7$ and $g(x) = x^2$. Graph f and g together with $f \circ g$ and $g \circ f$.

1.3 | Trigonometric Functions

This section reviews radian measure and the basic trigonometric functions.

Angles

Angles are measured in degrees or radians. The number of **radians** in the central angle $A'CB'$ within a circle of radius r is defined as the number of "radius units" contained in the arc s subtended by that central angle. If we denote this central angle by θ when measured in radians, this means that $\theta = s/r$ (Figure 1.38), or

$$s = r\theta \qquad (\theta \text{ in radians}). \qquad (1)$$

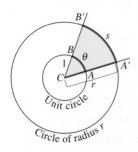

FIGURE 1.38 The radian measure of the central angle $A'CB'$ is the number $\theta = s/r$. For a unit circle of radius $r = 1$, θ is the length of arc AB that central angle ACB cuts from the unit circle.

If the circle is a unit circle having radius $r = 1$, then from Figure 1.38 and Equation (1), we see that the central angle θ measured in radians is just the length of the arc that the angle cuts from the unit circle. Since one complete revolution of the unit circle is $360°$ or 2π radians, we have

$$\pi \text{ radians} = 180° \qquad (2)$$

and

$$1 \text{ radian} = \frac{180}{\pi} (\approx 57.3) \text{ degrees} \qquad \text{or} \qquad 1 \text{ degree} = \frac{\pi}{180} (\approx 0.017) \text{ radians}.$$

Table 1.2 shows the equivalence between degree and radian measures for some basic angles.

TABLE 1.2 Angles measured in degrees and radians

Degrees	−180	−135	−90	−45	0	30	45	60	90	120	135	150	180	270	360
θ (radians)	$-\pi$	$\dfrac{-3\pi}{4}$	$\dfrac{-\pi}{2}$	$\dfrac{-\pi}{4}$	0	$\dfrac{\pi}{6}$	$\dfrac{\pi}{4}$	$\dfrac{\pi}{3}$	$\dfrac{\pi}{2}$	$\dfrac{2\pi}{3}$	$\dfrac{3\pi}{4}$	$\dfrac{5\pi}{6}$	π	$\dfrac{3\pi}{2}$	2π

An angle in the *xy*-plane is said to be in **standard position** if its vertex lies at the origin and its initial ray lies along the positive *x*-axis (Figure 1.39). Angles measured counterclockwise from the positive *x*-axis are assigned positive measures; angles measured clockwise are assigned negative measures.

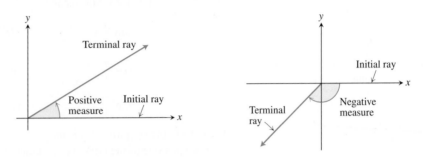

FIGURE 1.39 Angles in standard position in the *xy*-plane.

Angles describing counterclockwise rotations can go arbitrarily far beyond 2π radians or 360°. Similarly, angles describing clockwise rotations can have negative measures of all sizes (Figure 1.40).

FIGURE 1.40 Nonzero radian measures can be positive or negative and can go beyond 2π.

Angle Convention: Use Radians From now on, in this book it is assumed that all angles are measured in radians unless degrees or some other unit is stated explicitly. When we talk about the angle $\pi/3$, we mean $\pi/3$ radians (which is 60°), not $\pi/3$ degrees. We use radians because it simplifies many of the operations in calculus, and some results we will obtain involving the trigonometric functions are not true when angles are measured in degrees.

The Six Basic Trigonometric Functions

You are probably familiar with defining the trigonometric functions of an acute angle in terms of the sides of a right triangle (Figure 1.41). We extend this definition to obtuse and negative angles by first placing the angle in standard position in a circle of radius r. We then define the trigonometric functions in terms of the coordinates of the point $P(x, y)$ where the angle's terminal ray intersects the circle (Figure 1.42).

$$\textbf{sine:}\quad \sin\theta = \frac{y}{r} \qquad\qquad \textbf{cosecant:}\quad \csc\theta = \frac{r}{y}$$

$$\textbf{cosine:}\quad \cos\theta = \frac{x}{r} \qquad\qquad \textbf{secant:}\quad \sec\theta = \frac{r}{x}$$

$$\textbf{tangent:}\quad \tan\theta = \frac{y}{x} \qquad\qquad \textbf{cotangent:}\quad \cot\theta = \frac{x}{y}$$

These extended definitions agree with the right-triangle definitions when the angle is acute. Notice also that whenever the quotients are defined,

$$\tan\theta = \frac{\sin\theta}{\cos\theta} \qquad\qquad \cot\theta = \frac{1}{\tan\theta}$$

$$\sec\theta = \frac{1}{\cos\theta} \qquad\qquad \csc\theta = \frac{1}{\sin\theta}$$

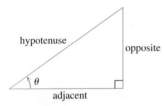

$$\sin\theta = \frac{\text{opp}}{\text{hyp}} \qquad \csc\theta = \frac{\text{hyp}}{\text{opp}}$$

$$\cos\theta = \frac{\text{adj}}{\text{hyp}} \qquad \sec\theta = \frac{\text{hyp}}{\text{adj}}$$

$$\tan\theta = \frac{\text{opp}}{\text{adj}} \qquad \cot\theta = \frac{\text{adj}}{\text{opp}}$$

FIGURE 1.41 Trigonometric ratios of an acute angle.

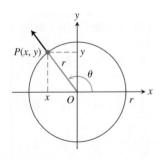

FIGURE 1.42 The trigonometric functions of a general angle θ are defined in terms of x, y, and r.

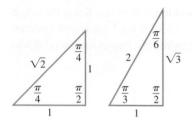

FIGURE 1.43 Radian angles and side lengths of two common triangles.

As you can see, $\tan \theta$ and $\sec \theta$ are not defined if $x = \cos \theta = 0$. This means they are not defined if θ is $\pm \pi/2$, $\pm 3\pi/2, \ldots$. Similarly, $\cot \theta$ and $\csc \theta$ are not defined for values of θ for which $y = 0$, namely $\theta = 0$, $\pm \pi$, $\pm 2\pi, \ldots$.

The exact values of these trigonometric ratios for some angles can be read from the triangles in Figure 1.43. For instance,

$$\sin \frac{\pi}{4} = \frac{1}{\sqrt{2}} \qquad \sin \frac{\pi}{6} = \frac{1}{2} \qquad \sin \frac{\pi}{3} = \frac{\sqrt{3}}{2}$$

$$\cos \frac{\pi}{4} = \frac{1}{\sqrt{2}} \qquad \cos \frac{\pi}{6} = \frac{\sqrt{3}}{2} \qquad \cos \frac{\pi}{3} = \frac{1}{2}$$

$$\tan \frac{\pi}{4} = 1 \qquad \tan \frac{\pi}{6} = \frac{1}{\sqrt{3}} \qquad \tan \frac{\pi}{3} = \sqrt{3}$$

The CAST rule (Figure 1.44) is useful for remembering when the basic trigonometric functions are positive or negative. For instance, from the triangle in Figure 1.45, we see that

$$\sin \frac{2\pi}{3} = \frac{\sqrt{3}}{2}, \qquad \cos \frac{2\pi}{3} = -\frac{1}{2}, \qquad \tan \frac{2\pi}{3} = -\sqrt{3}.$$

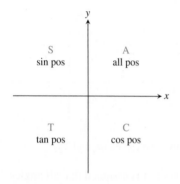

FIGURE 1.44 The CAST rule, remembered by the statement "Calculus Activates Student Thinking," tells which trigonometric functions are positive in each quadrant.

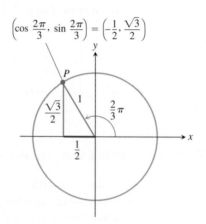

FIGURE 1.45 The triangle for calculating the sine and cosine of $2\pi/3$ radians. The side lengths come from the geometry of right triangles.

Using a similar method we determined the values of $\sin \theta$, $\cos \theta$, and $\tan \theta$ shown in Table 1.3.

TABLE 1.3 Values of $\sin \theta$, $\cos \theta$, and $\tan \theta$ for selected values of θ

Degrees	−180	−135	−90	−45	0	30	45	60	90	120	135	150	180	270	360
θ (radians)	$-\pi$	$\dfrac{-3\pi}{4}$	$\dfrac{-\pi}{2}$	$\dfrac{-\pi}{4}$	0	$\dfrac{\pi}{6}$	$\dfrac{\pi}{4}$	$\dfrac{\pi}{3}$	$\dfrac{\pi}{2}$	$\dfrac{2\pi}{3}$	$\dfrac{3\pi}{4}$	$\dfrac{5\pi}{6}$	π	$\dfrac{3\pi}{2}$	2π
$\sin \theta$	0	$\dfrac{-\sqrt{2}}{2}$	−1	$\dfrac{-\sqrt{2}}{2}$	0	$\dfrac{1}{2}$	$\dfrac{\sqrt{2}}{2}$	$\dfrac{\sqrt{3}}{2}$	1	$\dfrac{\sqrt{3}}{2}$	$\dfrac{\sqrt{2}}{2}$	$\dfrac{1}{2}$	0	−1	0
$\cos \theta$	−1	$\dfrac{-\sqrt{2}}{2}$	0	$\dfrac{\sqrt{2}}{2}$	1	$\dfrac{\sqrt{3}}{2}$	$\dfrac{\sqrt{2}}{2}$	$\dfrac{1}{2}$	0	$-\dfrac{1}{2}$	$\dfrac{-\sqrt{2}}{2}$	$\dfrac{-\sqrt{3}}{2}$	−1	0	1
$\tan \theta$	0	1		−1	0	$\dfrac{\sqrt{3}}{3}$	1	$\sqrt{3}$		$-\sqrt{3}$	−1	$\dfrac{-\sqrt{3}}{3}$	0		0

21. $\sin\left(x - \dfrac{\pi}{4}\right) + 1$ **22.** $\cos\left(x + \dfrac{2\pi}{3}\right) - 2$

Graph the functions in Exercises 23–26 in the *ts*-plane (*t*-axis horizontal, *s*-axis vertical). What is the period of each function? What symmetries do the graphs have?

23. $s = \cot 2t$ **24.** $s = -\tan \pi t$

25. $s = \sec\left(\dfrac{\pi t}{2}\right)$ **26.** $s = \csc\left(\dfrac{t}{2}\right)$

T **27. a.** Graph $y = \cos x$ and $y = \sec x$ together for $-3\pi/2 \le x \le 3\pi/2$. Comment on the behavior of $\sec x$ in relation to the signs and values of $\cos x$.

   b. Graph $y = \sin x$ and $y = \csc x$ together for $-\pi \le x \le 2\pi$. Comment on the behavior of $\csc x$ in relation to the signs and values of $\sin x$.

T **28.** Graph $y = \tan x$ and $y = \cot x$ together for $-7 \le x \le 7$. Comment on the behavior of $\cot x$ in relation to the signs and values of $\tan x$.

29. Graph $y = \sin x$ and $y = \lfloor \sin x \rfloor$ together. What are the domain and range of $\lfloor \sin x \rfloor$?

30. Graph $y = \sin x$ and $y = \lceil \sin x \rceil$ together. What are the domain and range of $\lceil \sin x \rceil$?

Using the Addition Formulas

Use the addition formulas to derive the identities in Exercises 31–36.

31. $\cos\left(x - \dfrac{\pi}{2}\right) = \sin x$ **32.** $\cos\left(x + \dfrac{\pi}{2}\right) = -\sin x$

33. $\sin\left(x + \dfrac{\pi}{2}\right) = \cos x$ **34.** $\sin\left(x - \dfrac{\pi}{2}\right) = -\cos x$

35. $\cos(A - B) = \cos A \cos B + \sin A \sin B$ (Exercise 57 provides a different derivation.)

36. $\sin(A - B) = \sin A \cos B - \cos A \sin B$

37. What happens if you take $B = A$ in the trigonometric identity $\cos(A - B) = \cos A \cos B + \sin A \sin B$? Does the result agree with something you already know?

38. What happens if you take $B = 2\pi$ in the addition formulas? Do the results agree with something you already know?

In Exercises 39–42, express the given quantity in terms of $\sin x$ and $\cos x$.

39. $\cos(\pi + x)$ **40.** $\sin(2\pi - x)$

41. $\sin\left(\dfrac{3\pi}{2} - x\right)$ **42.** $\cos\left(\dfrac{3\pi}{2} + x\right)$

43. Evaluate $\sin\dfrac{7\pi}{12}$ as $\sin\left(\dfrac{\pi}{4} + \dfrac{\pi}{3}\right)$.

44. Evaluate $\cos\dfrac{11\pi}{12}$ as $\cos\left(\dfrac{\pi}{4} + \dfrac{2\pi}{3}\right)$.

45. Evaluate $\cos\dfrac{\pi}{12}$.   **46.** Evaluate $\sin\dfrac{5\pi}{12}$.

Using the Half-Angle Formulas

Find the function values in Exercises 47–50.

47. $\cos^2\dfrac{\pi}{8}$ **48.** $\cos^2\dfrac{5\pi}{12}$

49. $\sin^2\dfrac{\pi}{12}$ **50.** $\sin^2\dfrac{3\pi}{8}$

Solving Trigonometric Equations

For Exercises 51–54, solve for the angle θ, where $0 \le \theta \le 2\pi$.

51. $\sin^2\theta = \dfrac{3}{4}$ **52.** $\sin^2\theta = \cos^2\theta$

53. $\sin 2\theta - \cos\theta = 0$ **54.** $\cos 2\theta + \cos\theta = 0$

Theory and Examples

55. The tangent sum formula The standard formula for the tangent of the sum of two angles is

$$\tan(A + B) = \frac{\tan A + \tan B}{1 - \tan A \tan B}.$$

Derive the formula.

56. (*Continuation of Exercise 55.*) Derive a formula for $\tan(A - B)$.

57. Apply the law of cosines to the triangle in the accompanying figure to derive the formula for $\cos(A - B)$.

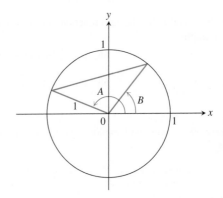

58. a. Apply the formula for $\cos(A - B)$ to the identity $\sin\theta = \cos\left(\dfrac{\pi}{2} - \theta\right)$ to obtain the addition formula for $\sin(A + B)$.

   b. Derive the formula for $\cos(A + B)$ by substituting $-B$ for B in the formula for $\cos(A - B)$ from Exercise 35.

59. A triangle has sides $a = 2$ and $b = 3$ and angle $C = 60°$. Find the length of side c.

60. A triangle has sides $a = 2$ and $b = 3$ and angle $C = 40°$. Find the length of side c.

61. The law of sines *The law of sines* says that if a, b, and c are the sides opposite the angles A, B, and C in a triangle, then

$$\frac{\sin A}{a} = \frac{\sin B}{b} = \frac{\sin C}{c}.$$

Use the accompanying figures and the identity $\sin(\pi - \theta) = \sin\theta$, if required, to derive the law.

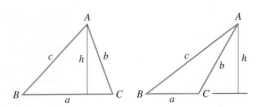

62. A triangle has sides $a = 2$ and $b = 3$ and angle $C = 60°$ (as in Exercise 59). Find the sine of angle B using the law of sines.

63. A triangle has side $c = 2$ and angles $A = \pi/4$ and $B = \pi/3$. Find the length a of the side opposite A.

T 64. The approximation $\sin x \approx x$ It is often useful to know that, when x is measured in radians, $\sin x \approx x$ for numerically small values of x. In Section 3.11, we will see why the approximation holds. The approximation error is less than 1 in 5000 if $|x| < 0.1$.

  a. With your grapher in radian mode, graph $y = \sin x$ and $y = x$ together in a viewing window about the origin. What do you see happening as x nears the origin?

  b. With your grapher in degree mode, graph $y = \sin x$ and $y = x$ together about the origin again. How is the picture different from the one obtained with radian mode?

General Sine Curves

For

$$f(x) = A \sin\left(\frac{2\pi}{B}(x - C)\right) + D,$$

identify A, B, C, and D for the sine functions in Exercises 65–68 and sketch their graphs.

65. $y = 2 \sin(x + \pi) - 1$ **66.** $y = \dfrac{1}{2} \sin(\pi x - \pi) + \dfrac{1}{2}$

67. $y = -\dfrac{2}{\pi} \sin\left(\dfrac{\pi}{2} t\right) + \dfrac{1}{\pi}$ **68.** $y = \dfrac{L}{2\pi} \sin \dfrac{2\pi t}{L}, \quad L > 0$

COMPUTER EXPLORATIONS

In Exercises 69–72, you will explore graphically the general sine function

$$f(x) = A \sin\left(\frac{2\pi}{B}(x - C)\right) + D$$

as you change the values of the constants A, B, C, and D. Use a CAS or computer grapher to perform the steps in the exercises.

69. The period B Set the constants $A = 3, C = D = 0$.

  a. Plot $f(x)$ for the values $B = 1, 3, 2\pi, 5\pi$ over the interval $-4\pi \le x \le 4\pi$. Describe what happens to the graph of the general sine function as the period increases.

  b. What happens to the graph for negative values of B? Try it with $B = -3$ and $B = -2\pi$.

70. The horizontal shift C Set the constants $A = 3, B = 6, D = 0$.

  a. Plot $f(x)$ for the values $C = 0, 1$, and 2 over the interval $-4\pi \le x \le 4\pi$. Describe what happens to the graph of the general sine function as C increases through positive values.

  b. What happens to the graph for negative values of C?

  c. What smallest positive value should be assigned to C so the graph exhibits no horizontal shift? Confirm your answer with a plot.

71. The vertical shift D Set the constants $A = 3, B = 6, C = 0$.

  a. Plot $f(x)$ for the values $D = 0, 1$, and 3 over the interval $-4\pi \le x \le 4\pi$. Describe what happens to the graph of the general sine function as D increases through positive values.

  b. What happens to the graph for negative values of D?

72. The amplitude A Set the constants $B = 6, C = D = 0$.

  a. Describe what happens to the graph of the general sine function as A increases through positive values. Confirm your answer by plotting $f(x)$ for the values $A = 1, 5$, and 9.

  b. What happens to the graph for negative values of A?

1.4 Graphing with Calculators and Computers

A graphing calculator or a computer with graphing software enables us to graph very complicated functions with high precision. Many of these functions could not otherwise be easily graphed. However, care must be taken when using such devices for graphing purposes, and in this section we address some of the issues involved. In Chapter 4 we will see how calculus helps us determine that we are accurately viewing all the important features of a function's graph.

Graphing Windows

When using a graphing calculator or computer as a graphing tool, a portion of the graph is displayed in a rectangular **display** or **viewing window**. Often the default window gives an incomplete or misleading picture of the graph. We use the term *square window* when the units or scales on both axes are the same. This term does not mean that the display window itself is square (usually it is rectangular), but instead it means that the x-unit is the same as the y-unit.

When a graph is displayed in the default window, the x-unit may differ from the y-unit of scaling in order to fit the graph in the window. The viewing window is set by specifying an interval $[a, b]$ for the x-values and an interval $[c, d]$ for the y-values. The machine selects equally spaced x-values in $[a, b]$ and then plots the points $(x, f(x))$. A point is plotted if and only if x lies in the domain of the function and $f(x)$ lies within the interval $[c, d]$. A short line segment is then drawn between each plotted point and its next neighboring point. We now give illustrative examples of some common problems that may occur with this procedure.

EXAMPLE 1   Graph the function $f(x) = x^3 - 7x^2 + 28$ in each of the following display or viewing windows:

(a) $[-10, 10]$ by $[-10, 10]$ **(b)** $[-4, 4]$ by $[-50, 10]$ **(c)** $[-4, 10]$ by $[-60, 60]$

Solution

(a) We select $a = -10$, $b = 10$, $c = -10$, and $d = 10$ to specify the interval of x-values and the range of y-values for the window. The resulting graph is shown in Figure 1.50a. It appears that the window is cutting off the bottom part of the graph and that the interval of x-values is too large. Let's try the next window.

    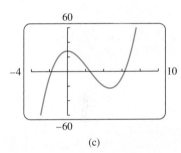

(a)   (b)   (c)

FIGURE 1.50   The graph of $f(x) = x^3 - 7x^2 + 28$ in different viewing windows. Selecting a window that gives a clear picture of a graph is often a trial-and-error process (Example 1).

(b) Now we see more features of the graph (Figure 1.50b), but the top is missing and we need to view more to the right of $x = 4$ as well. The next window should help.

(c) Figure 1.50c shows the graph in this new viewing window. Observe that we get a more complete picture of the graph in this window, and it is a reasonable graph of a third-degree polynomial.   ■

EXAMPLE 2   When a graph is displayed, the x-unit may differ from the y-unit, as in the graphs shown in Figures 1.50b and 1.50c. The result is distortion in the picture, which may be misleading. The display window can be made square by compressing or stretching the units on one axis to match the scale on the other, giving the true graph. Many systems have built-in functions to make the window "square." If yours does not, you will have to do some calculations and set the window size manually to get a square window, or bring to your viewing some foreknowledge of the true picture.

Figure 1.51a shows the graphs of the perpendicular lines $y = x$ and $y = -x + 3\sqrt{2}$, together with the semicircle $y = \sqrt{9 - x^2}$, in a nonsquare $[-4, 4]$ by $[-6, 8]$ display window. Notice the distortion. The lines do not appear to be perpendicular, and the semicircle appears to be elliptical in shape.

Figure 1.51b shows the graphs of the same functions in a square window in which the x-units are scaled to be the same as the y-units. Notice that the scaling on the x-axis for Figure 1.51a has been compressed in Figure 1.51b to make the window square. Figure 1.51c gives an enlarged view of Figure 1.51b with a square $[-3, 3]$ by $[0, 4]$ window.   ■

If the denominator of a rational function is zero at some x-value within the viewing window, a calculator or graphing computer software may produce a steep near-vertical line segment from the top to the bottom of the window. Here is an example.

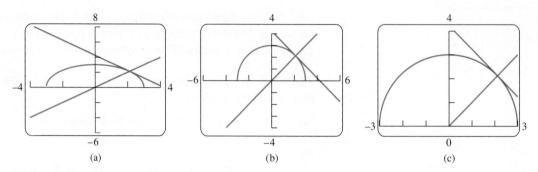

FIGURE 1.51 Graphs of the perpendicular lines $y = x$ and $y = -x + 3\sqrt{2}$, and the semicircle $y = \sqrt{9 - x^2}$ appear distorted (a) in a nonsquare window, but clear (b) and (c) in square windows (Example 2).

EXAMPLE 3   Graph the function $y = \dfrac{1}{2 - x}$.

Solution   Figure 1.52a shows the graph in the $[-10, 10]$ by $[-10, 10]$ default square window with our computer graphing software. Notice the near-vertical line segment at $x = 2$. It is not truly a part of the graph and $x = 2$ does not belong to the domain of the function. By trial and error we can eliminate the line by changing the viewing window to the smaller $[-6, 6]$ by $[-4, 4]$ view, revealing a better graph (Figure 1.52b).   ∎

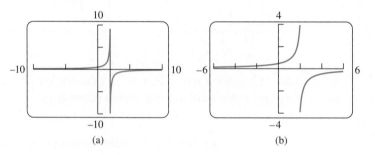

FIGURE 1.52 Graphs of the function $y = \dfrac{1}{2 - x}$. A vertical line may appear without a careful choice of the viewing window (Example 3).

Sometimes the graph of a trigonometric function oscillates very rapidly. When a calculator or computer software plots the points of the graph and connects them, many of the maximum and minimum points are actually missed. The resulting graph is then very misleading.

EXAMPLE 4   Graph the function $f(x) = \sin 100x$.

Solution   Figure 1.53a shows the graph of f in the viewing window $[-12, 12]$ by $[-1, 1]$. We see that the graph looks very strange because the sine curve should oscillate periodically between -1 and 1. This behavior is not exhibited in Figure 1.53a. We might

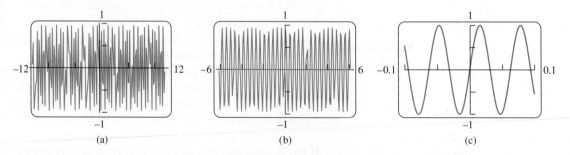

FIGURE 1.53 Graphs of the function $y = \sin 100x$ in three viewing windows. Because the period is $2\pi/100 \approx 0.063$, the smaller window in (c) best displays the true aspects of this rapidly oscillating function (Example 4).

experiment with a smaller viewing window, say $[-6, 6]$ by $[-1, 1]$, but the graph is not better (Figure 1.53b). The difficulty is that the period of the trigonometric function $y = \sin 100x$ is very small ($2\pi/100 \approx 0.063$). If we choose the much smaller viewing window $[-0.1, 0.1]$ by $[-1, 1]$ we get the graph shown in Figure 1.53c. This graph reveals the expected oscillations of a sine curve. ∎

EXAMPLE 5   Graph the function $y = \cos x + \dfrac{1}{50}\sin 50x$.

Solution   In the viewing window $[-6, 6]$ by $[-1, 1]$ the graph appears much like the cosine function with some small sharp wiggles on it (Figure 1.54a). We get a better look when we significantly reduce the window to $[-0.6, 0.6]$ by $[0.8, 1.02]$, obtaining the graph in Figure 1.54b. We now see the small but rapid oscillations of the second term, $1/50 \sin 50x$, added to the comparatively larger values of the cosine curve. ∎

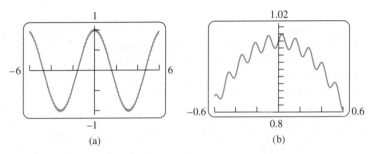

FIGURE 1.54   In (b) we see a close-up view of the function $y = \cos x + \dfrac{1}{50}\sin 50x$ graphed in (a). The term $\cos x$ clearly dominates the second term, $\dfrac{1}{50}\sin 50x$, which produces the rapid oscillations along the cosine curve. Both views are needed for a clear idea of the graph (Example 5).

Obtaining a Complete Graph

Some graphing devices will not display the portion of a graph for $f(x)$ when $x < 0$. Usually that happens because of the procedure the device is using to calculate the function values. Sometimes we can obtain the complete graph by defining the formula for the function in a different way.

EXAMPLE 6   Graph the function $y = x^{1/3}$.

Solution   Some graphing devices display the graph shown in Figure 1.55a. When we compare it with the graph of $y = x^{1/3} = \sqrt[3]{x}$ in Figure 1.17, we see that the left branch for

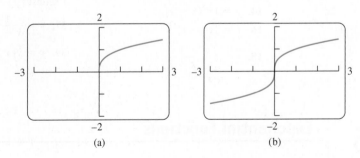

FIGURE 1.55   The graph of $y = x^{1/3}$ is missing the left branch in (a). In (b) we graph the function $f(x) = \dfrac{x}{|x|} \cdot |x|^{1/3}$, obtaining both branches. (See Example 6.)

$x < 0$ is missing. The reason the graphs differ is that many calculators and computer software programs calculate $x^{1/3}$ as $e^{(1/3)\ln x}$. Since the logarithmic function is not defined for negative values of x, the computing device can produce only the right branch, where $x > 0$. (Logarithmic and exponential functions are introduced in the next two sections.) To obtain the full picture showing both branches, we can graph the function

$$f(x) = \frac{x}{|x|} \cdot |x|^{1/3}.$$

This function equals $x^{1/3}$ except at $x = 0$ (where f is undefined, although $0^{1/3} = 0$). The graph of f is shown in Figure 1.55b. ∎

Exercises 1.4

Choosing a Viewing Window

T In Exercises 1–4, use a graphing calculator or computer to determine which of the given viewing windows displays the most appropriate graph of the specified function.

1. $f(x) = x^4 - 7x^2 + 6x$

    a. $[-1, 1]$ by $[-1, 1]$ **b.** $[-2, 2]$ by $[-5, 5]$

    c. $[-10, 10]$ by $[-10, 10]$ **d.** $[-5, 5]$ by $[-25, 15]$

2. $f(x) = x^3 - 4x^2 - 4x + 16$

    a. $[-1, 1]$ by $[-5, 5]$ **b.** $[-3, 3]$ by $[-10, 10]$

    c. $[-5, 5]$ by $[-10, 20]$ **d.** $[-20, 20]$ by $[-100, 100]$

3. $f(x) = 5 + 12x - x^3$

    a. $[-1, 1]$ by $[-1, 1]$ **b.** $[-5, 5]$ by $[-10, 10]$

    c. $[-4, 4]$ by $[-20, 20]$ **d.** $[-4, 5]$ by $[-15, 25]$

4. $f(x) = \sqrt{5 + 4x - x^2}$

    a. $[-2, 2]$ by $[-2, 2]$ **b.** $[-2, 6]$ by $[-1, 4]$

    c. $[-3, 7]$ by $[0, 10]$ **d.** $[-10, 10]$ by $[-10, 10]$

Finding a Viewing Window

T In Exercises 5–30, find an appropriate viewing window for the given function and use it to display its graph.

5. $f(x) = x^4 - 4x^3 + 15$ **6.** $f(x) = \dfrac{x^3}{3} - \dfrac{x^2}{2} - 2x + 1$

7. $f(x) = x^5 - 5x^4 + 10$ **8.** $f(x) = 4x^3 - x^4$

9. $f(x) = x\sqrt{9 - x^2}$ **10.** $f(x) = x^2(6 - x^3)$

11. $y = 2x - 3x^{2/3}$ **12.** $y = x^{1/3}(x^2 - 8)$

13. $y = 5x^{2/5} - 2x$ **14.** $y = x^{2/3}(5 - x)$

15. $y = |x^2 - 1|$ **16.** $y = |x^2 - x|$

17. $y = \dfrac{x + 3}{x + 2}$ **18.** $y = 1 - \dfrac{1}{x + 3}$

19. $f(x) = \dfrac{x^2 + 2}{x^2 + 1}$ **20.** $f(x) = \dfrac{x^2 - 1}{x^2 + 1}$

21. $f(x) = \dfrac{x - 1}{x^2 - x - 6}$ **22.** $f(x) = \dfrac{8}{x^2 - 9}$

23. $f(x) = \dfrac{6x^2 - 15x + 6}{4x^2 - 10x}$ **24.** $f(x) = \dfrac{x^2 - 3}{x - 2}$

25. $y = \sin 250x$ **26.** $y = 3 \cos 60x$

27. $y = \cos\left(\dfrac{x}{50}\right)$ **28.** $y = \dfrac{1}{10} \sin\left(\dfrac{x}{10}\right)$

29. $y = x + \dfrac{1}{10} \sin 30x$ **30.** $y = x^2 + \dfrac{1}{50} \cos 100x$

31. Graph the lower half of the circle defined by the equation $x^2 + 2x = 4 + 4y - y^2$.

32. Graph the upper branch of the hyperbola $y^2 - 16x^2 = 1$.

33. Graph four periods of the function $f(x) = -\tan 2x$.

34. Graph two periods of the function $f(x) = 3 \cot \dfrac{x}{2} + 1$.

35. Graph the function $f(x) = \sin 2x + \cos 3x$.

36. Graph the function $f(x) = \sin^3 x$.

Graphing in Dot Mode

T Another way to avoid incorrect connections when using a graphing device is through the use of a "dot mode," which plots only the points. If your graphing utility allows that mode, use it to plot the functions in Exercises 37–40.

37. $y = \dfrac{1}{x - 3}$ **38.** $y = \sin \dfrac{1}{x}$

39. $y = x\lfloor x \rfloor$ **40.** $y = \dfrac{x^3 - 1}{x^2 - 1}$

1.5 Exponential Functions

Exponential functions are among the most important in mathematics and occur in a wide variety of applications, including interest rates, radioactive decay, population growth, the spread of a disease, consumption of natural resources, the earth's atmospheric pressure, temperature change of a heated object placed in a cooler environment, and the dating of fossils. In this section we introduce these functions informally, using an intuitive approach. We give a rigorous development of them in Chapter 7, based on important calculus ideas and results.

Exponential Behavior

When a positive quantity P doubles, it increases by a factor of 2 and the quantity becomes $2P$. If it doubles again, it becomes $2(2P) = 2^2P$, and a third doubling gives $2(2^2P) = 2^3P$. Continuing to double in this fashion leads us to the consideration of the function $f(x) = 2^x$. We call this an *exponential* function because the variable x appears in the exponent of 2^x. Functions such as $g(x) = 10^x$ and $h(x) = (1/2)^x$ are other examples of exponential functions. In general, if $a \neq 1$ is a positive constant, the function

$$f(x) = a^x$$

is the **exponential function with base a**.

Don't confuse 2^x with the power x^2, where the variable x is the base, not the exponent.

EXAMPLE 1   In 2000, \$100 is invested in a savings account, where it grows by accruing interest that is compounded annually (once a year) at an interest rate of 5.5%. Assuming no additional funds are deposited to the account and no money is withdrawn, give a formula for a function describing the amount A in the account after x years have elapsed.

Solution   If $P = 100$, at the end of the first year the amount in the account is the original amount plus the interest accrued, or

$$P + \left(\frac{5.5}{100}\right)P = (1 + 0.055)P = (1.055)P.$$

At the end of the second year the account earns interest again and grows to

$$(1 + 0.055) \cdot (1.055P) = (1.055)^2P = 100 \cdot (1.055)^2. \qquad P = 100$$

Continuing this process, after x years the value of the account is

$$A = 100 \cdot (1.055)^x.$$

This is a multiple of the exponential function with base 1.055. Table 1.4 shows the amounts accrued over the first four years. Notice that the amount in the account each year is always 1.055 times its value in the previous year.

TABLE 1.4   Savings account growth

Year	Amount (dollars)	Increase (dollars)
2000	100	
2001	$100(1.055) = 105.50$	5.50
2002	$100(1.055)^2 = 111.30$	5.80
2003	$100(1.055)^3 = 117.42$	6.12
2004	$100(1.055)^4 = 123.88$	6.46

In general, the amount after x years is given by $P(1 + r)^x$, where r is the interest rate (expressed as a decimal). ∎

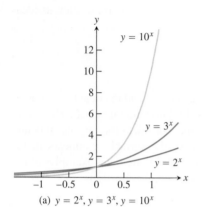

(a) $y = 2^x$, $y = 3^x$, $y = 10^x$

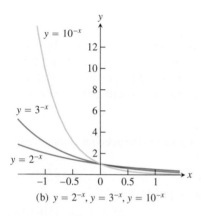

(b) $y = 2^{-x}$, $y = 3^{-x}$, $y = 10^{-x}$

FIGURE 1.56   Graphs of exponential functions.

TABLE 1.5   Values of $2^{\sqrt{3}}$ for rational r closer and closer to $\sqrt{3}$

r	2^r
1.0	2.000000000
1.7	3.249009585
1.73	3.317278183
1.732	3.321880096
1.7320	3.321880096
1.73205	3.321995226
1.732050	3.321995226
1.7320508	3.321997068
1.73205080	3.321997068
1.732050808	3.321997086

For integer and rational exponents, the value of an exponential function $f(x) = a^x$ is obtained arithmetically as follows. If $x = n$ is a positive integer, the number a^n is given by multiplying a by itself n times:

$$a^n = \underbrace{a \cdot a \cdot \; \cdots \; \cdot a}_{n \text{ factors}}.$$

If $x = 0$, then $a^0 = 1$, and if $x = -n$ for some positive integer n, then

$$a^{-n} = \frac{1}{a^n} = \left(\frac{1}{a}\right)^n.$$

If $x = 1/n$ for some positive integer n, then

$$a^{1/n} = \sqrt[n]{a},$$

which is the positive number that when multiplied by itself n times gives a. If $x = p/q$ is any rational number, then

$$a^{p/q} = \sqrt[q]{a^p} = \left(\sqrt[q]{a}\right)^p.$$

If x is *irrational*, the meaning of a^x is not so clear, but its value can be defined by considering values for rational numbers that get closer and closer to x. This informal approach is based on the graph of the exponential function. In Chapter 7 we define the meaning in a rigorous way.

We displayed the graphs of several exponential functions in Section 1.1, and show them again here in Figure 1.56. These graphs describe the values of the exponential functions for all real inputs x. The value at an irrational number x is chosen so that the graph of a^x has no "holes" or "jumps." Of course, these words are not mathematical terms, but they do convey the informal idea. We mean that the value of a^x, when x is irrational, is chosen so that the function $f(x) = a^x$ is *continuous*, a notion that will be carefully explored in the next chapter. This choice ensures the graph retains its increasing behavior when $a > 1$, or decreasing behavior when $0 < a < 1$ (see Figure 1.56).

Arithmetically, the graphical idea can be described in the following way, using the exponential $f(x) = 2^x$ as an illustration. Any particular irrational number, say $x = \sqrt{3}$, has a decimal expansion

$$\sqrt{3} = 1.732050808 \ldots .$$

We then consider the list of numbers, given as follows in the order of taking more and more digits in the decimal expansion,

$$2^1, 2^{1.7}, 2^{1.73}, 2^{1.732}, 2^{1.7320}, 2^{1.73205}, \ldots . \qquad (1)$$

We know the meaning of each number in list (1) because the successive decimal approximations to $\sqrt{3}$ given by 1, 1.7, 1.73, 1.732, and so on, are all *rational* numbers. As these decimal approximations get closer and closer to $\sqrt{3}$, it seems reasonable that the list of numbers in (1) gets closer and closer to some fixed number, which we specify to be $2^{\sqrt{3}}$.

Table 1.5 illustrates how taking better approximations to $\sqrt{3}$ gives better approximations to the number $2^{\sqrt{3}} \approx 3.321997086$. It is the *completeness property* of the real numbers (discussed briefly in Appendix 6) which guarantees that this procedure gives a single number we define to be $2^{\sqrt{3}}$ (although it is beyond the scope of this text to give a proof). In a similar way, we can identify the number 2^x (or a^x, $a > 0$) for any irrational x. By identifying the number a^x for both rational and irrational x, we eliminate any "holes" or "gaps" in the graph of a^x. In practice you can use a calculator to find the number a^x for irrational x, taking successive decimal approximations to x and creating a table similar to Table 1.5.

Exponential functions obey the familiar rules of exponents listed on the next page. It is easy to check these rules using algebra when the exponents are integers or rational numbers. We prove them for all real exponents in Chapters 4 and 7.

Rules for Exponents

If $a > 0$ and $b > 0$, the following rules hold true for all real numbers x and y.

1. $a^x \cdot a^y = a^{x+y}$ **2.** $\dfrac{a^x}{a^y} = a^{x-y}$

3. $(a^x)^y = (a^y)^x = a^{xy}$ **4.** $a^x \cdot b^x = (ab)^x$

5. $\dfrac{a^x}{b^x} = \left(\dfrac{a}{b}\right)^x$

EXAMPLE 2   We illustrate using the rules for exponents.

1. $3^{1.1} \cdot 3^{0.7} = 3^{1.1+0.7} = 3^{1.8}$

2. $\dfrac{\left(\sqrt{10}\right)^3}{\sqrt{10}} = \left(\sqrt{10}\right)^{3-1} = \left(\sqrt{10}\right)^2 = 10$

3. $\left(5^{\sqrt{2}}\right)^{\sqrt{2}} = 5^{\sqrt{2}\cdot\sqrt{2}} = 5^2 = 25$

4. $7^\pi \cdot 8^\pi = (56)^\pi$

5. $\left(\dfrac{4}{9}\right)^{1/2} = \dfrac{4^{1/2}}{9^{1/2}} = \dfrac{2}{3}$ ∎

The Natural Exponential Function e^x

The most important exponential function used for modeling natural, physical, and economic phenomena is the **natural exponential function**, whose base is the special number e. The number e is irrational, and its value is 2.718281828 to nine decimal places. It might seem strange that we would use this number for a base rather than a simple number like 2 or 10. The advantage in using e as a base is that it simplifies many of the calculations in calculus.

If you look at Figure 1.56a you can see that the graphs of the exponential functions $y = a^x$ get steeper as the base a gets larger. This idea of steepness is conveyed by the slope of the tangent line to the graph at a point. Tangent lines to graphs of functions are defined precisely in the next chapter, but intuitively the tangent line to the graph at a point is a line that just touches the graph at the point, like a tangent to a circle. Figure 1.57 shows the slope of the graph of $y = a^x$ as it crosses the y-axis for several values of a. Notice that the slope is exactly equal to 1 when a equals the number e. The slope is smaller than 1 if $a < e$, and larger than 1 if $a > e$. This is the property that makes the number e so useful in calculus: **The graph of $y = e^x$ has slope 1 when it crosses the y-axis**.

FIGURE 1.57 Among the exponential functions, the graph of $y = e^x$ has the property that the slope m of the tangent line to the graph is exactly 1 when it crosses the y-axis. The slope is smaller for a base less than e, such as 2^x, and larger for a base greater than e, such as 3^x.

In Chapter 3 we use that slope property to prove e is the number the quantity $(1 + 1/x)^x$ approaches as x becomes large without bound. That result provides one way to compute the value of e, at least approximately. The graph and table in Figure 1.58 show the behavior of this expression and how it gets closer and closer to the line $y = e \approx 2.718281828$ as x gets larger and larger. (This *limit* idea is made precise in the next chapter.) A more complete discussion of e is given in Chapter 7.

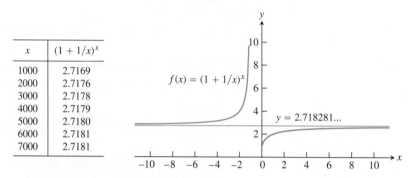

x	$(1 + 1/x)^x$
1000	2.7169
2000	2.7176
3000	2.7178
4000	2.7179
5000	2.7180
6000	2.7181
7000	2.7181

FIGURE 1.58 A graph and table of values for $f(x) = (1 + 1/x)^x$ both suggest that as x gets larger and larger, $f(x)$ gets closer and closer to $e \approx 2.7182818 \ldots$.

Exponential Growth and Decay

The exponential functions $y = e^{kx}$, where k is a nonzero constant, are frequently used for modeling exponential growth or decay. The function $y = y_0 e^{kx}$ is a model for **exponential growth** if $k > 0$ and a model for **exponential decay** if $k < 0$. Here y_0 represents a constant. An example of exponential growth occurs when computing interest **compounded continuously** modeled by $y = P \cdot e^{rt}$, where P is the initial investment, r is the interest rate as a decimal, and t is time in units consistent with r. An example of exponential decay is the model $y = A \cdot e^{-1.2 \times 10^{-4} t}$, which represents how the radioactive element carbon-14 decays over time. Here A is the original amount of carbon-14 and t is the time in years. Carbon-14 decay is used to date the remains of dead organisms such as shells, seeds, and wooden artifacts. Figure 1.59 shows graphs of exponential growth and exponential decay.

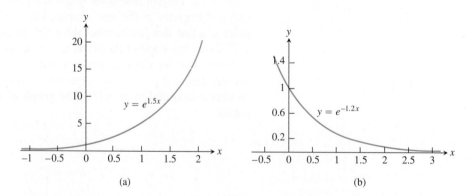

(a)

(b)

FIGURE 1.59 Graphs of (a) exponential growth, $k = 1.5 > 0$, and (b) exponential decay, $k = -1.2 < 0$.

EXAMPLE 3 Investment companies often use the model $y = Pe^{rt}$ in calculating the growth of an investment. Use this model to track the growth of \$100 invested in 2000 at an annual interest rate of 5.5%.

Solution Let $t = 0$ represent 2000, $t = 1$ represent 2001, and so on. Then the exponential growth model is $y(t) = Pe^{rt}$, where $P = 100$ (the initial investment), $r = 0.055$ (the annual interest rate expressed as a decimal), and t is time in years. To predict the amount in the account in 2004, after four years have elapsed, we take $t = 4$ and calculate

$$y(4) = 100e^{0.055(4)}$$

$$= 100e^{0.22}$$

$$= 124.61. \qquad \text{Nearest cent using calculator}$$

This compares with \$123.88 in the account when the interest is compounded annually from Example 1. ∎

EXAMPLE 4 Laboratory experiments indicate that some atoms emit a part of their mass as radiation, with the remainder of the atom re-forming to make an atom of some new element. For example, radioactive carbon-14 decays into nitrogen; radium eventually decays into lead. If y_0 is the number of radioactive nuclei present at time zero, the number still present at any later time t will be

$$y = y_0 \, e^{-rt}, \qquad r > 0.$$

The number r is called the **decay rate** of the radioactive substance. (We will see how this formula is obtained in Section 7.2.) For carbon-14, the decay rate has been determined experimentally to be about $r = 1.2 \times 10^{-4}$ when t is measured in years. Predict the percent of carbon-14 present after 866 years have elapsed.

Solution If we start with an amount y_0 of carbon-14 nuclei, after 866 years we are left with the amount

$$y(866) = y_0 \, e^{(-1.2 \times 10^{-4})(866)}$$

$$\approx (0.901)y_0. \qquad \text{Calculator evaluation}$$

That is, after 866 years, we are left with about 90% of the original amount of carbon-14, so about 10% of the original nuclei have decayed. In Example 7 in the next section, you will see how to find the number of years required for half of the radioactive nuclei present in a sample to decay (called the *half-life* of the substance). ∎

You may wonder why we use the family of functions $y = e^{kx}$ for different values of the constant k instead of the general exponential functions $y = a^x$. In the next section, we show that the exponential function a^x is equal to e^{kx} for an appropriate value of k. So the formula $y = e^{kx}$ covers the entire range of possibilities, and we will see that it is easier to use.

Exercises 1.5

Sketching Exponential Curves

In Exercises 1–6, sketch the given curves together in the appropriate coordinate plane and label each curve with its equation.

1. $y = 2^x, y = 4^x, y = 3^{-x}, y = (1/5)^x$
2. $y = 3^x, y = 8^x, y = 2^{-x}, y = (1/4)^x$
3. $y = 2^{-t}$ and $y = -2^t$ 4. $y = 3^{-t}$ and $y = -3^t$
5. $y = e^x$ and $y = 1/e^x$ 6. $y = -e^x$ and $y = -e^{-x}$

In each of Exercises 7–10, sketch the shifted exponential curves.

7. $y = 2^x - 1$ and $y = 2^{-x} - 1$
8. $y = 3^x + 2$ and $y = 3^{-x} + 2$
9. $y = 1 - e^x$ and $y = 1 - e^{-x}$
10. $y = -1 - e^x$ and $y = -1 - e^{-x}$

Applying the Laws of Exponents

Use the laws of exponents to simplify the expressions in Exercises 11–20.

11. $16^2 \cdot 16^{-1.75}$
12. $9^{1/3} \cdot 9^{1/6}$
13. $\dfrac{4^{4.2}}{4^{3.7}}$
14. $\dfrac{3^{5/3}}{3^{2/3}}$
15. $\left(25^{1/8}\right)^4$
16. $\left(13^{\sqrt{2}}\right)^{\sqrt{2}/2}$
17. $2^{\sqrt{3}} \cdot 7^{\sqrt{3}}$
18. $\left(\sqrt{3}\right)^{1/2} \cdot \left(\sqrt{12}\right)^{1/2}$
19. $\left(\dfrac{2}{\sqrt{2}}\right)^4$
20. $\left(\dfrac{\sqrt{6}}{3}\right)^2$

Composites Involving Exponential Functions

Find the domain and range for each of the functions in Exercises 21–24.

21. $f(x) = \dfrac{1}{2 + e^x}$ **22.** $g(t) = \cos(e^{-t})$

23. $g(t) = \sqrt{1 + 3^{-t}}$ **24.** $f(x) = \dfrac{3}{1 - e^{2x}}$

Applications

[T] In Exercises 25–28, use graphs to find approximate solutions.

25. $2^x = 5$ **26.** $e^x = 4$

27. $3^x - 0.5 = 0$ **28.** $3 - 2^{-x} = 0$

[T] In Exercises 29–36, use an exponential model and a graphing calculator to estimate the answer in each problem.

29. Population growth The population of Knoxville is 500,000 and is increasing at the rate of 3.75% each year. Approximately when will the population reach 1 million?

30. Population growth The population of Silver Run in the year 1890 was 6250. Assume the population increased at a rate of 2.75% per year.

    a. Estimate the population in 1915 and 1940.

    b. Approximately when did the population reach 50,000?

31. Radioactive decay The half-life of phosphorus-32 is about 14 days. There are 6.6 grams present initially.

    a. Express the amount of phosphorus-32 remaining as a function of time t.

    b. When will there be 1 gram remaining?

32. If John invests $2300 in a savings account with a 6% interest rate compounded annually, how long will it take until John's account has a balance of $4150?

33. Doubling your money Determine how much time is required for an investment to double in value if interest is earned at the rate of 6.25% compounded annually.

34. Tripling your money Determine how much time is required for an investment to triple in value if interest is earned at the rate of 5.75% compounded continuously.

35. Cholera bacteria Suppose that a colony of bacteria starts with 1 bacterium and doubles in number every half hour. How many bacteria will the colony contain at the end of 24 hr?

36. Eliminating a disease Suppose that in any given year the number of cases of a disease is reduced by 20%. If there are 10,000 cases today, how many years will it take

    a. to reduce the number of cases to 1000?

    b. to eliminate the disease; that is, to reduce the number of cases to less than 1?

1.6 | Inverse Functions and Logarithms

A function that undoes, or inverts, the effect of a function f is called the *inverse* of f. Many common functions, though not all, are paired with an inverse. In this section we present the natural logarithmic function $y = \ln x$ as the inverse of the exponential function $y = e^x$, and we also give examples of several inverse trigonometric functions.

One-to-One Functions

A function is a rule that assigns a value from its range to each element in its domain. Some functions assign the same range value to more than one element in the domain. The function $f(x) = x^2$ assigns the same value, 1, to both of the numbers -1 and $+1$; the sines of $\pi/3$ and $2\pi/3$ are both $\sqrt{3}/2$. Other functions assume each value in their range no more than once. The square roots and cubes of different numbers are always different. A function that has distinct values at distinct elements in its domain is called one-to-one. These functions take on any one value in their range exactly once.

> **DEFINITION**   A function $f(x)$ is **one-to-one** on a domain D if $f(x_1) \neq f(x_2)$ whenever $x_1 \neq x_2$ in D.

EXAMPLE 1 Some functions are one-to-one on their entire natural domain. Other functions are not one-to-one on their entire domain, but by restricting the function to a smaller domain we can create a function that is one-to-one. The original and restricted functions are not the same functions, because they have different domains. However, the two functions have the same values on the smaller domain, so the original function is an extension of the restricted function from its smaller domain to the larger domain.

(a) One-to-one: Graph meets each horizontal line at most once.

(b) Not one-to-one: Graph meets one or more horizontal lines more than once.

FIGURE 1.60 (a) $y = x^3$ and $y = \sqrt{x}$ are one-to-one on their domains $(-\infty, \infty)$ and $[0, \infty)$. (b) $y = x^2$ and $y = \sin x$ are not one-to-one on their domains $(-\infty, \infty)$.

(a) $f(x) = \sqrt{x}$ is one-to-one on any domain of nonnegative numbers because $\sqrt{x_1} \neq \sqrt{x_2}$ whenever $x_1 \neq x_2$.

(b) $g(x) = \sin x$ is *not* one-to-one on the interval $[0, \pi]$ because $\sin(\pi/6) = \sin(5\pi/6)$. In fact, for each element x_1 in the subinterval $[0, \pi/2)$ there is a corresponding element x_2 in the subinterval $(\pi/2, \pi]$ satisfying $\sin x_1 = \sin x_2$, so distinct elements in the domain are assigned to the same value in the range. The sine function *is* one-to-one on $[0, \pi/2]$, however, because it is an increasing function on $[0, \pi/2]$ giving distinct outputs for distinct inputs. ∎

The graph of a one-to-one function $y = f(x)$ can intersect a given horizontal line at most once. If the function intersects the line more than once, it assumes the same y-value for at least two different x-values and is therefore not one-to-one (Figure 1.60).

The Horizontal Line Test for One-to-One Functions

A function $y = f(x)$ is one-to-one if and only if its graph intersects each horizontal line at most once.

Inverse Functions

Since each output of a one-to-one function comes from just one input, the effect of the function can be inverted to send an output back to the input from which it came.

DEFINITION Suppose that f is a one-to-one function on a domain D with range R. The **inverse function** f^{-1} is defined by

$$f^{-1}(b) = a \quad \text{if} \quad f(a) = b.$$

The domain of f^{-1} is R and the range of f^{-1} is D.

The symbol f^{-1} for the inverse of f is read "f inverse." The "-1" in f^{-1} is *not* an exponent; $f^{-1}(x)$ does not mean $1/f(x)$. Notice that the domains and ranges of f and f^{-1} are interchanged.

EXAMPLE 2 Suppose a one-to-one function $y = f(x)$ is given by a table of values

x	1	2	3	4	5	6	7	8
$f(x)$	3	4.5	7	10.5	15	20.5	27	34.5

A table for the values of $x = f^{-1}(y)$ can then be obtained by simply interchanging the values in the columns (or rows) of the table for f:

y	3	4.5	7	10.5	15	20.5	27	34.5
$f^{-1}(y)$	1	2	3	4	5	6	7	8

∎

If we apply f to send an input x to the output $f(x)$ and follow by applying f^{-1} to $f(x)$ we get right back to x, just where we started. Similarly, if we take some number y in the range of f, apply f^{-1} to it, and then apply f to the resulting value $f^{-1}(y)$, we get back the value y with which we began. Composing a function and its inverse has the same effect as doing nothing.

$$(f^{-1} \circ f)(x) = x, \quad \text{for all } x \text{ in the domain of } f$$

$$(f \circ f^{-1})(y) = y, \quad \text{for all } y \text{ in the domain of } f^{-1} \text{ (or range of } f)$$

Only a one-to-one function can have an inverse. The reason is that if $f(x_1) = y$ and $f(x_2) = y$ for two distinct inputs x_1 and x_2, then there is no way to assign a value to $f^{-1}(y)$ that satisfies both $f^{-1}(f(x_1)) = x_1$ and $f^{-1}(f(x_2)) = x_2$.

A function that is increasing on an interval so it satisfies the inequality $f(x_2) > f(x_1)$ when $x_2 > x_1$ is one-to-one and has an inverse. Decreasing functions also have an inverse. Functions that are neither increasing nor decreasing may still be one-to-one and have an inverse, as with the function $f(x) = 1/x$ for $x \neq 0$ and $f(0) = 0$, defined on $(-\infty, \infty)$ and passing the horizontal line test.

Finding Inverses

The graphs of a function and its inverse are closely related. To read the value of a function from its graph, we start at a point x on the x-axis, go vertically to the graph, and then move horizontally to the y-axis to read the value of y. The inverse function can be read from the graph by reversing this process. Start with a point y on the y-axis, go horizontally to the graph of $y = f(x)$, and then move vertically to the x-axis to read the value of $x = f^{-1}(y)$ (Figure 1.61).

(a) To find the value of f at x, we start at x, go up to the curve, and then over to the y-axis.

(b) The graph of f^{-1} is the graph of f, but with x and y interchanged. To find the x that gave y, we start at y and go over to the curve and down to the x-axis. The domain of f^{-1} is the range of f. The range of f^{-1} is the domain of f.

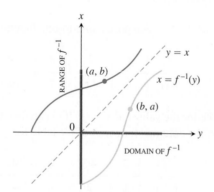

(c) To draw the graph of f^{-1} in the more usual way, we reflect the system across the line $y = x$.

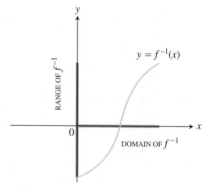

(d) Then we interchange the letters x and y. We now have a normal-looking graph of f^{-1} as a function of x.

FIGURE 1.61 Determining the graph of $y = f^{-1}(x)$ from the graph of $y = f(x)$. The graph of f^{-1} is obtained by reflecting the graph of f about the line $y = x$.

We want to set up the graph of f^{-1} so that its input values lie along the x-axis, as is usually done for functions, rather than on the y-axis. To achieve this we interchange the x

and y axes by reflecting across the 45° line $y = x$. After this reflection we have a new graph that represents f^{-1}. The value of $f^{-1}(x)$ can now be read from the graph in the usual way, by starting with a point x on the x-axis, going vertically to the graph, and then horizontally to the y-axis to get the value of $f^{-1}(x)$. Figure 1.61 indicates the relationship between the graphs of f and f^{-1}. The graphs are interchanged by reflection through the line $y = x$.

The process of passing from f to f^{-1} can be summarized as a two-step procedure.

1. Solve the equation $y = f(x)$ for x. This gives a formula $x = f^{-1}(y)$ where x is expressed as a function of y.

2. Interchange x and y, obtaining a formula $y = f^{-1}(x)$ where f^{-1} is expressed in the conventional format with x as the independent variable and y as the dependent variable.

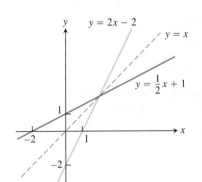

FIGURE 1.62 Graphing $f(x) = (1/2)x + 1$ and $f^{-1}(x) = 2x - 2$ together shows the graphs' symmetry with respect to the line $y = x$ (Example 3).

EXAMPLE 3   Find the inverse of $y = \dfrac{1}{2}x + 1$, expressed as a function of x.

Solution

1. *Solve for x in terms of y:* $y = \dfrac{1}{2}x + 1$

$$2y = x + 2$$
$$x = 2y - 2.$$

2. *Interchange x and y:* $y = 2x - 2.$

The inverse of the function $f(x) = (1/2)x + 1$ is the function $f^{-1}(x) = 2x - 2$. (See Figure 1.62.) To check, we verify that both composites give the identity function:

$$f^{-1}(f(x)) = 2\left(\frac{1}{2}x + 1\right) - 2 = x + 2 - 2 = x$$

$$f(f^{-1}(x)) = \frac{1}{2}(2x - 2) + 1 = x - 1 + 1 = x. \qquad \blacksquare$$

FIGURE 1.63 The functions $y = \sqrt{x}$ and $y = x^2, x \ge 0$, are inverses of one another (Example 4).

EXAMPLE 4   Find the inverse of the function $y = x^2, x \ge 0$, expressed as a function of x.

Solution   We first solve for x in terms of y:

$$y = x^2$$
$$\sqrt{y} = \sqrt{x^2} = |x| = x \qquad |x| = x \text{ because } x \ge 0$$

We then interchange x and y, obtaining

$$y = \sqrt{x}.$$

The inverse of the function $y = x^2, x \ge 0$, is the function $y = \sqrt{x}$ (Figure 1.63).

Notice that the function $y = x^2, x \ge 0$, with domain *restricted* to the nonnegative real numbers, *is* one-to-one (Figure 1.63) and has an inverse. On the other hand, the function $y = x^2$, with no domain restrictions, *is not* one-to-one (Figure 1.60b) and therefore has no inverse. $\qquad \blacksquare$

Logarithmic Functions

If a is any positive real number other than 1, the base a exponential function $f(x) = a^x$ is one-to-one. It therefore has an inverse. Its inverse is called the *logarithm function with base a*.

DEFINITION   The **logarithm function with base** *a*, $y = \log_a x$, is the inverse of the base a exponential function $y = a^x \, (a > 0, a \ne 1)$.

(a)

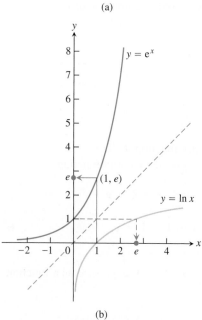

(b)

FIGURE 1.64 (a) The graph of 2^x and its inverse, $\log_2 x$. (b) The graph of e^x and its inverse, $\ln x$.

HISTORICAL BIOGRAPHY*

John Napier
(1550–1617)

The domain of $\log_a x$ is $(0, \infty)$, the range of a^x. The range of $\log_a x$ is $(-\infty, \infty)$, the domain of a^x.

Figure 1.23 in Section 1.1 shows the graphs of four logarithmic functions with $a > 1$. Figure 1.64a shows the graph of $y = \log_2 x$. The graph of $y = a^x$, $a > 1$, increases rapidly for $x > 0$, so its inverse, $y = \log_a x$, increases slowly for $x > 1$.

Because we have no technique yet for solving the equation $y = a^x$ for x in terms of y, we do not have an explicit formula for computing the logarithm at a given value of x. Nevertheless, we can obtain the graph of $y = \log_a x$ by reflecting the graph of the exponential $y = a^x$ across the line $y = x$. Figure 1.64 shows the graphs for $a = 2$ and $a = e$.

Logarithms with base 2 are commonly used in computer science. Logarithms with base e and base 10 are so important in applications that calculators have special keys for them. They also have their own special notation and names:

$$\log_e x \quad \text{is written as} \quad \ln x.$$
$$\log_{10} x \quad \text{is written as} \quad \log x.$$

The function $y = \ln x$ is called the **natural logarithm function**, and $y = \log x$ is often called the **common logarithm function**. For the natural logarithm,

$$\ln x = y \iff e^y = x.$$

In particular, if we set $x = e$, we obtain

$$\ln e = 1$$

because $e^1 = e$.

Properties of Logarithms

Logarithms, invented by John Napier, were the single most important improvement in arithmetic calculation before the modern electronic computer. What made them so useful is that the properties of logarithms reduce multiplication of positive numbers to addition of their logarithms, division of positive numbers to subtraction of their logarithms, and exponentiation of a number to multiplying its logarithm by the exponent.

We summarize these properties for the natural logarithm as a series of rules that we prove in Chapter 3. Although here we state the Power Rule for all real powers r, the case when r is an irrational number cannot be dealt with properly until Chapter 4. We also establish the validity of the rules for logarithmic functions with any base a in Chapter 7.

THEOREM 1—Algebraic Properties of the Natural Logarithm   For any numbers $b > 0$ and $x > 0$, the natural logarithm satisfies the following rules:

1. *Product Rule:* $\ln bx = \ln b + \ln x$

2. *Quotient Rule:* $\ln \dfrac{b}{x} = \ln b - \ln x$

3. *Reciprocal Rule:* $\ln \dfrac{1}{x} = -\ln x$ Rule 2 with $b = 1$

4. *Power Rule:* $\ln x^r = r \ln x$

*To learn more about the historical figures mentioned in the text and the development of many major elements and topics of calculus, visit **www.aw.com/thomas**.

EXAMPLE 5   Here are examples of the properties in Theorem 1.

(a) $\ln 4 + \ln \sin x = \ln (4 \sin x)$ Product Rule

(b) $\ln \dfrac{x + 1}{2x - 3} = \ln (x + 1) - \ln (2x - 3)$ Quotient Rule

(c) $\ln \dfrac{1}{8} = -\ln 8$ Reciprocal Rule

$\quad\quad\; = -\ln 2^3 = -3 \ln 2$ Power Rule   ∎

Because a^x and $\log_a x$ are inverses, composing them in either order gives the identity function.

Inverse Properties for a^x and $\log_a x$

1. Base a: $a^{\log_a x} = x$, $\log_a a^x = x$, $a > 0, a \neq 1, x > 0$

2. Base e: $e^{\ln x} = x$, $\ln e^x = x$, $x > 0$

Substituting a^x for x in the equation $x = e^{\ln x}$ enables us to rewrite a^x as a power of e:

$$a^x = e^{\ln (a^x)}\quad\quad \text{Substitute } a^x \text{ for } x \text{ in } x = e^{\ln x}.$$
$$ = e^{x \ln a}\quad\quad \text{Power Rule for logs}$$
$$ = e^{(\ln a)x}.\quad\quad \text{Exponent rearranged}$$

Thus, the exponential function a^x is the same as e^{kx} for $k = \ln a$.

Every exponential function is a power of the natural exponential function.

$$a^x = e^{x \ln a}$$

That is, a^x is the same as e^x raised to the power $\ln a$: $a^x = e^{kx}$ for $k = \ln a$.

For example,

$$2^x = e^{(\ln 2)x} = e^{x \ln 2}, \quad\quad \text{and} \quad\quad 5^{-3x} = e^{(\ln 5)(-3x)} = e^{-3x \ln 5}.$$

Returning once more to the properties of a^x and $\log_a x$, we have

$$\ln x = \ln (a^{\log_a x})\quad\quad \text{Inverse Property for } a^x \text{ and } \log_a x$$
$$ = (\log_a x)(\ln a).\quad\quad \text{Power Rule for logarithms, with } r = \log_a x$$

Rewriting this equation as $\log_a x = (\ln x)/(\ln a)$ shows that every logarithmic function is a constant multiple of the natural logarithm $\ln x$. This allows us to extend the algebraic properties for $\ln x$ to $\log_a x$. For instance, $\log_a bx = \log_a b + \log_a x$.

Change of Base Formula

Every logarithmic function is a constant multiple of the natural logarithm.

$$\log_a x = \frac{\ln x}{\ln a} \quad\quad (a > 0, a \neq 1)$$

Applications

In Section 1.5 we looked at examples of exponential growth and decay problems. Here we use properties of logarithms to answer more questions concerning such problems.

EXAMPLE 6   If \$1000 is invested in an account that earns 5.25% interest compounded annually, how long will it take the account to reach \$2500?

Solution From Example 1, Section 1.5 with $P = 1000$ and $r = 0.0525$, the amount in the account at any time t in years is $1000(1.0525)^t$, so we need to solve the equation

$$1000(1.0525)^t = 2500.$$

Thus we have

$$(1.0525)^t = 2.5 \qquad \text{Divide by 1000.}$$

$$\ln (1.0525)^t = \ln 2.5 \qquad \text{Take logarithms of both sides.}$$

$$t \ln 1.0525 = \ln 2.5 \qquad \text{Power Rule}$$

$$t = \frac{\ln 2.5}{\ln 1.0525} \approx 17.9 \qquad \text{Values obtained by calculator}$$

The amount in the account will reach $2500 in 18 years, when the annual interest payment is deposited for that year. ∎

EXAMPLE 7 The **half-life** of a radioactive element is the time required for half of the radioactive nuclei present in a sample to decay. It is a remarkable fact that the half-life is a constant that does not depend on the number of radioactive nuclei initially present in the sample, but only on the radioactive substance.

To see why, let y_0 be the number of radioactive nuclei initially present in the sample. Then the number y present at any later time t will be $y = y_0 e^{-kt}$. We seek the value of t at which the number of radioactive nuclei present equals half the original number:

$$y_0 e^{-kt} = \frac{1}{2} y_0$$

$$e^{-kt} = \frac{1}{2}$$

$$-kt = \ln \frac{1}{2} = -\ln 2 \qquad \text{Reciprocal Rule for logarithms}$$

$$t = \frac{\ln 2}{k}. \tag{1}$$

This value of t is the half-life of the element. It depends only on the value of k; the number y_0 does not have any effect.

The effective radioactive lifetime of polonium-210 is so short that we measure it in days rather than years. The number of radioactive atoms remaining after t days in a sample that starts with y_0 radioactive atoms is

$$y = y_0 e^{-5 \times 10^{-3} t}.$$

The element's half-life is

$$\text{Half-life} = \frac{\ln 2}{k} \qquad \text{Eq. (1)}$$

$$= \frac{\ln 2}{5 \times 10^{-3}} \qquad \text{The } k \text{ from polonium's decay equation}$$

$$\approx 139 \text{ days.} \qquad \blacksquare$$

Inverse Trigonometric Functions

The six basic trigonometric functions of a general radian angle x were reviewed in Section 1.3. These functions are not one-to-one (their values repeat periodically). However, we can restrict their domains to intervals on which they are one-to-one. The sine function increases from -1 at $x = -\pi/2$ to $+1$ at $x = \pi/2$. By restricting its domain to the interval $[-\pi/2, \pi/2]$ we make it one-to-one, so that it has an inverse $\sin^{-1} x$ (Figure 1.65). Similar domain restrictions can be applied to all six trigonometric functions.

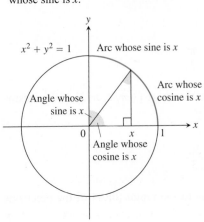

FIGURE 1.65 The graph of $y = \sin^{-1} x$.

Domain restrictions that make the trigonometric functions one-to-one

$y = \sin x$
Domain: $[-\pi/2, \pi/2]$
Range: $[-1, 1]$

$y = \cos x$
Domain: $[0, \pi]$
Range: $[-1, 1]$

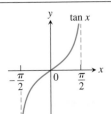

$y = \tan x$
Domain: $(-\pi/2, \pi/2)$
Range: $(-\infty, \infty)$

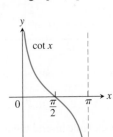

$y = \cot x$
Domain: $(0, \pi)$
Range: $(-\infty, \infty)$

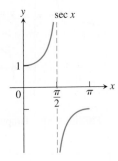

$y = \sec x$
Domain: $[0, \pi/2) \cup (\pi/2, \pi]$
Range: $(-\infty, -1] \cup [1, \infty)$

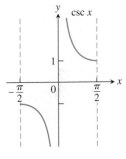

$y = \csc x$
Domain: $[-\pi/2, 0) \cup (0, \pi/2]$
Range: $(-\infty, -1] \cup [1, \infty)$

The "Arc" in Arcsine and Arccosine

The accompanying figure gives a geometric interpretation of $y = \sin^{-1} x$ and $y = \cos^{-1} x$ for radian angles in the first quadrant. For a unit circle, the equation $s = r\theta$ becomes $s = \theta$, so central angles and the arcs they subtend have the same measure. If $x = \sin y$, then, in addition to being the angle whose sine is x, y is also the length of arc on the unit circle that subtends an angle whose sine is x. So we call y "the arc whose sine is x."

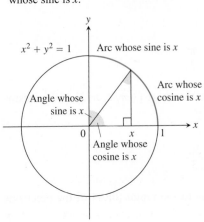

Since these restricted functions are now one-to-one, they have inverses, which we denote by

$$y = \sin^{-1} x \quad \text{or} \quad y = \arcsin x$$
$$y = \cos^{-1} x \quad \text{or} \quad y = \arccos x$$
$$y = \tan^{-1} x \quad \text{or} \quad y = \arctan x$$
$$y = \cot^{-1} x \quad \text{or} \quad y = \text{arccot } x$$
$$y = \sec^{-1} x \quad \text{or} \quad y = \text{arcsec } x$$
$$y = \csc^{-1} x \quad \text{or} \quad y = \text{arccsc } x$$

These equations are read "y equals the arcsine of x" or "y equals arcsin x" and so on.

Caution The -1 in the expressions for the inverse means "inverse." It does *not* mean reciprocal. For example, the *reciprocal* of $\sin x$ is $(\sin x)^{-1} = 1/\sin x = \csc x$.

The graphs of the six inverse trigonometric functions are shown in Figure 1.66. We can obtain these graphs by reflecting the graphs of the restricted trigonometric functions through the line $y = x$. We now take a closer look at two of these functions.

The Arcsine and Arccosine Functions

We define the arcsine and arccosine as functions whose values are angles (measured in radians) that belong to restricted domains of the sine and cosine functions.

Domain: $-1 \le x \le 1$
Range: $-\dfrac{\pi}{2} \le y \le \dfrac{\pi}{2}$

(a)

Domain: $-1 \le x \le 1$
Range: $0 \le y \le \pi$

(b)

Domain: $-\infty < x < \infty$
Range: $-\dfrac{\pi}{2} < y < \dfrac{\pi}{2}$

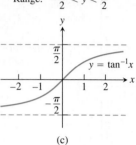

(c)

Domain: $x \le -1$ or $x \ge 1$
Range: $0 \le y \le \pi,\, y \ne \dfrac{\pi}{2}$

(d)

Domain: $x \le -1$ or $x \ge 1$
Range: $-\dfrac{\pi}{2} \le y \le \dfrac{\pi}{2},\, y \ne 0$

(e)

Domain: $-\infty < x < \infty$
Range: $0 < y < \pi$

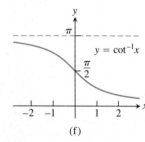

(f)

FIGURE 1.66 Graphs of the six basic inverse trigonometric functions.

FIGURE 1.67 The graphs of
(a) $y = \sin x$, $-\pi/2 \le x \le \pi/2$, and
(b) its inverse, $y = \sin^{-1} x$. The graph of
$\sin^{-1} x$, obtained by reflection across the
line $y = x$, is a portion of the curve
$x = \sin y$.

DEFINITION

$y = \sin^{-1} x$ is the number in $[-\pi/2, \pi/2]$ for which $\sin y = x$.
$y = \cos^{-1} x$ is the number in $[0, \pi]$ for which $\cos y = x$.

The graph of $y = \sin^{-1} x$ (Figure 1.67b) is symmetric about the origin (it lies along the graph of $x = \sin y$). The arcsine is therefore an odd function:

$$\sin^{-1}(-x) = -\sin^{-1} x. \tag{2}$$

The graph of $y = \cos^{-1} x$ (Figure 1.68b) has no such symmetry.

EXAMPLE 8 Evaluate **(a)** $\sin^{-1}\left(\dfrac{\sqrt{3}}{2}\right)$ and **(b)** $\cos^{-1}\left(-\dfrac{1}{2}\right)$.

Solution

(a) We see that

$$\sin^{-1}\left(\frac{\sqrt{3}}{2}\right) = \frac{\pi}{3}$$

because $\sin(\pi/3) = \sqrt{3}/2$ and $\pi/3$ belongs to the range $[-\pi/2, \pi/2]$ of the arcsine function. See Figure 1.69a.

(b) We have

$$\cos^{-1}\left(-\frac{1}{2}\right) = \frac{2\pi}{3}$$

because $\cos(2\pi/3) = -1/2$ and $2\pi/3$ belongs to the range $[0, \pi]$ of the arccosine function. See Figure 1.69b. ∎

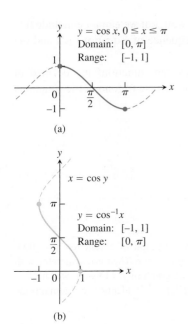

FIGURE 1.68 The graphs of
(a) $y = \cos x$, $0 \le x \le \pi$, and
(b) its inverse, $y = \cos^{-1} x$. The graph of
$\cos^{-1} x$, obtained by reflection across the
line $y = x$, is a portion of the curve
$x = \cos y$.

FIGURE 1.70 Diagram for drift
correction (Example 9), with distances
rounded to the nearest mile (drawing not
to scale).

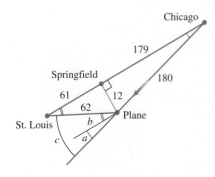

FIGURE 1.71 $\cos^{-1} x$ and $\cos^{-1}(-x)$ are
supplementary angles (so their sum is π).

Using the same procedure illustrated in Example 8, we can create the following table of
common values for the arcsine and arccosine functions.

x	$\sin^{-1} x$	$\cos^{-1} x$
$\sqrt{3}/2$	$\pi/3$	$\pi/6$
$\sqrt{2}/2$	$\pi/4$	$\pi/4$
$1/2$	$\pi/6$	$\pi/3$
$-1/2$	$-\pi/6$	$2\pi/3$
$-\sqrt{2}/2$	$-\pi/4$	$3\pi/4$
$-\sqrt{3}/2$	$-\pi/3$	$5\pi/6$

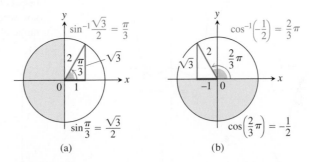

FIGURE 1.69 Values of the arcsine and arccosine functions
(Example 8).

EXAMPLE 9 During an airplane flight from Chicago to St. Louis, the navigator determines that the plane is 12 mi off course, as shown in Figure 1.70. Find the angle a for a
course parallel to the original correct course, the angle b, and the drift correction angle
$c = a + b$.

Solution From Figure 1.70 and elementary geometry, we see that $180 \sin a = 12$ and
$62 \sin b = 12$, so

$$a = \sin^{-1} \frac{12}{180} \approx 0.067 \text{ radian} \approx 3.8°$$

$$b = \sin^{-1} \frac{12}{62} \approx 0.195 \text{ radian} \approx 11.2°$$

$$c = a + b \approx 15°. \qquad \blacksquare$$

Identities Involving Arcsine and Arccosine

As we can see from Figure 1.71, the arccosine of x satisfies the identity

$$\cos^{-1} x + \cos^{-1}(-x) = \pi, \qquad (3)$$

or

$$\cos^{-1}(-x) = \pi - \cos^{-1} x. \qquad (4)$$

Also, we can see from the triangle in Figure 1.72 that for $x > 0$,

$$\sin^{-1} x + \cos^{-1} x = \pi/2. \qquad (5)$$

FIGURE 1.72 $\sin^{-1} x$ and $\cos^{-1} x$ are complementary angles (so their sum is $\pi/2$).

Equation (5) holds for the other values of x in $[-1, 1]$ as well, but we cannot conclude this from the triangle in Figure 1.72. It is, however, a consequence of Equations (2) and (4) (Exercise 74).

The arctangent, arccotangent, arcsecant, and arccosecant functions are defined in Section 3.9. There we develop additional properties of the inverse trigonometric functions in a calculus setting using the identities discussed here.

Exercises 1.6

Identifying One-to-One Functions Graphically

Which of the functions graphed in Exercises 1–6 are one-to-one, and which are not?

1.

$y = -3x^3$

2.

$y = x^4 - x^2$

3.

$y = 2|x|$

4.

$y = \text{int } x$

5.

$y = \dfrac{1}{x}$

6.

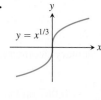

$y = x^{1/3}$

In Exercises 7–10, determine from its graph if the function is one-to-one.

7. $f(x) = \begin{cases} 3 - x, & x < 0 \\ 3, & x \geq 0 \end{cases}$

8. $f(x) = \begin{cases} 2x + 6, & x \leq -3 \\ x + 4, & x > -3 \end{cases}$

9. $f(x) = \begin{cases} 1 - \dfrac{x}{2}, & x \leq 0 \\ \dfrac{x}{x + 2}, & x > 0 \end{cases}$

10. $f(x) = \begin{cases} 2 - x^2, & x \leq 1 \\ x^2, & x > 1 \end{cases}$

Graphing Inverse Functions

Each of Exercises 11–16 shows the graph of a function $y = f(x)$. Copy the graph and draw in the line $y = x$. Then use symmetry with respect to the line $y = x$ to add the graph of f^{-1} to your sketch. (It is not necessary to find a formula for f^{-1}.) Identify the domain and range of f^{-1}.

11.

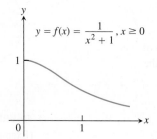

$y = f(x) = \dfrac{1}{x^2 + 1}, x \geq 0$

12.

$y = f(x) = 1 - \dfrac{1}{x}, x > 0$

13.

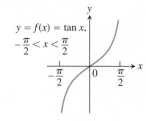

$y = f(x) = \sin x,$ $-\dfrac{\pi}{2} \leq x \leq \dfrac{\pi}{2}$

14.

$y = f(x) = \tan x,$ $-\dfrac{\pi}{2} < x < \dfrac{\pi}{2}$

15.

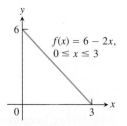

$f(x) = 6 - 2x,$ $0 \leq x \leq 3$

16.

$f(x) = \begin{cases} x + 1, & -1 \leq x \leq 0 \\ -2 + \dfrac{2}{3}x, & 0 < x < 3 \end{cases}$

17. a. Graph the function $f(x) = \sqrt{1 - x^2}, 0 \leq x \leq 1$. What symmetry does the graph have?

b. Show that f is its own inverse. (Remember that $\sqrt{x^2} = x$ if $x \geq 0$.)

18. a. Graph the function $f(x) = 1/x$. What symmetry does the graph have?

b. Show that f is its own inverse.

Formulas for Inverse Functions

Each of Exercises 19–24 gives a formula for a function $y = f(x)$ and shows the graphs of f and f^{-1}. Find a formula for f^{-1} in each case.

19. $f(x) = x^2 + 1, \quad x \geq 0$ **20.** $f(x) = x^2, \quad x \leq 0$

     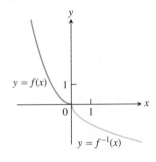

21. $f(x) = x^3 - 1$ **22.** $f(x) = x^2 - 2x + 1, \quad x \geq 1$

     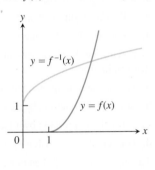

23. $f(x) = (x + 1)^2, \quad x \geq -1$ **24.** $f(x) = x^{2/3}, \quad x \geq 0$

     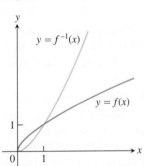

Each of Exercises 25–34 gives a formula for a function $y = f(x)$. In each case, find $f^{-1}(x)$ and identify the domain and range of f^{-1}. As a check, show that $f(f^{-1}(x)) = f^{-1}(f(x)) = x$.

25. $f(x) = x^5$ **26.** $f(x) = x^4, \quad x \geq 0$

27. $f(x) = x^3 + 1$ **28.** $f(x) = (1/2)x - 7/2$

29. $f(x) = 1/x^2, \quad x > 0$ **30.** $f(x) = 1/x^3, \quad x \neq 0$

31. $f(x) = \dfrac{x + 3}{x - 2}$ **32.** $f(x) = \dfrac{\sqrt{x}}{\sqrt{x} - 3}$

33. $f(x) = x^2 - 2x, \quad x \leq 1$

(*Hint:* Complete the square.)

34. $f(x) = (2x^3 + 1)^{1/5}$

Inverses of Lines

35. a. Find the inverse of the function $f(x) = mx$, where m is a constant different from zero.

   b. What can you conclude about the inverse of a function $y = f(x)$ whose graph is a line through the origin with a nonzero slope m?

36. Show that the graph of the inverse of $f(x) = mx + b$, where m and b are constants and $m \neq 0$, is a line with slope $1/m$ and y-intercept $-b/m$.

37. a. Find the inverse of $f(x) = x + 1$. Graph f and its inverse together. Add the line $y = x$ to your sketch, drawing it with dashes or dots for contrast.

   b. Find the inverse of $f(x) = x + b$ (b constant). How is the graph of f^{-1} related to the graph of f?

   c. What can you conclude about the inverses of functions whose graphs are lines parallel to the line $y = x$?

38. a. Find the inverse of $f(x) = -x + 1$. Graph the line $y = -x + 1$ together with the line $y = x$. At what angle do the lines intersect?

   b. Find the inverse of $f(x) = -x + b$ (b constant). What angle does the line $y = -x + b$ make with the line $y = x$?

   c. What can you conclude about the inverses of functions whose graphs are lines perpendicular to the line $y = x$?

Logarithms and Exponentials

39. Express the following logarithms in terms of $\ln 2$ and $\ln 3$.

   a. $\ln 0.75$ **b.** $\ln (4/9)$

   c. $\ln (1/2)$ **d.** $\ln \sqrt[3]{9}$

   e. $\ln 3\sqrt{2}$ **f.** $\ln \sqrt{13.5}$

40. Express the following logarithms in terms of $\ln 5$ and $\ln 7$.

   a. $\ln (1/125)$ **b.** $\ln 9.8$

   c. $\ln 7\sqrt{7}$ **d.** $\ln 1225$

   e. $\ln 0.056$ **f.** $(\ln 35 + \ln (1/7))/(\ln 25)$

Use the properties of logarithms to simplify the expressions in Exercises 41 and 42.

41. a. $\ln \sin \theta - \ln \left(\dfrac{\sin \theta}{5}\right)$ **b.** $\ln (3x^2 - 9x) + \ln \left(\dfrac{1}{3x}\right)$

   c. $\dfrac{1}{2} \ln (4t^4) - \ln 2$

42. a. $\ln \sec \theta + \ln \cos \theta$ **b.** $\ln (8x + 4) - 2 \ln 2$

   c. $3 \ln \sqrt[3]{t^2 - 1} - \ln (t + 1)$

Find simpler expressions for the quantities in Exercises 43–46.

43. a. $e^{\ln 7.2}$ **b.** $e^{-\ln x^2}$ **c.** $e^{\ln x - \ln y}$

44. a. $e^{\ln (x^2 + y^2)}$ **b.** $e^{-\ln 0.3}$ **c.** $e^{\ln \pi x - \ln 2}$

45. a. $2 \ln \sqrt{e}$ **b.** $\ln (\ln e^e)$ **c.** $\ln (e^{-x^2 - y^2})$

46. a. $\ln (e^{\sec \theta})$ **b.** $\ln (e^{(e^x)})$ **c.** $\ln (e^{2 \ln x})$

In Exercises 47–52, solve for y in terms of t or x, as appropriate.

47. $\ln y = 2t + 4$ **48.** $\ln y = -t + 5$

49. $\ln (y - 40) = 5t$ **50.** $\ln (1 - 2y) = t$

51. $\ln (y - 1) - \ln 2 = x + \ln x$

52. $\ln (y^2 - 1) - \ln (y + 1) = \ln (\sin x)$

In Exercises 53 and 54, solve for k.

53. a. $e^{2k} = 4$ **b.** $100e^{10k} = 200$ **c.** $e^{k/1000} = a$

54. a. $e^{5k} = \dfrac{1}{4}$ **b.** $80e^{k} = 1$ **c.** $e^{(\ln 0.8)k} = 0.8$

In Exercises 55–58, solve for t.

55. a. $e^{-0.3t} = 27$ **b.** $e^{kt} = \dfrac{1}{2}$ **c.** $e^{(\ln 0.2)t} = 0.4$

56. a. $e^{-0.01t} = 1000$ **b.** $e^{kt} = \dfrac{1}{10}$ **c.** $e^{(\ln 2)t} = \dfrac{1}{2}$

57. $e^{\sqrt{t}} = x^2$ **58.** $e^{(x^2)}e^{(2x+1)} = e^t$

Simplify the expressions in Exercises 59–62.

59. a. $5^{\log_5 7}$ **b.** $8^{\log_8 \sqrt{2}}$ **c.** $1.3^{\log_{1.3} 75}$

     d. $\log_4 16$ **e.** $\log_3 \sqrt{3}$ **f.** $\log_4 \left(\dfrac{1}{4}\right)$

60. a. $2^{\log_2 3}$ **b.** $10^{\log_{10}(1/2)}$ **c.** $\pi^{\log_\pi 7}$

     d. $\log_{11} 121$ **e.** $\log_{121} 11$ **f.** $\log_3 \left(\dfrac{1}{9}\right)$

61. a. $2^{\log_4 x}$ **b.** $9^{\log_3 x}$ **c.** $\log_2 \left(e^{(\ln 2)(\sin x)}\right)$

62. a. $25^{\log_5(3x^2)}$ **b.** $\log_e (e^x)$ **c.** $\log_4 (2^{e^x \sin x})$

Express the ratios in Exercises 63 and 64 as ratios of natural logarithms and simplify.

63. a. $\dfrac{\log_2 x}{\log_3 x}$ **b.** $\dfrac{\log_2 x}{\log_8 x}$ **c.** $\dfrac{\log_x a}{\log_{x^2} a}$

64. a. $\dfrac{\log_9 x}{\log_3 x}$ **b.** $\dfrac{\log_{\sqrt{10}} x}{\log_{\sqrt{2}} x}$ **c.** $\dfrac{\log_a b}{\log_b a}$

Arcsine and Arccosine

In Exercises 65–68, find the exact value of each expression.

65. a. $\sin^{-1}\left(\dfrac{-1}{2}\right)$ **b.** $\sin^{-1}\left(\dfrac{1}{\sqrt{2}}\right)$ **c.** $\sin^{-1}\left(\dfrac{-\sqrt{3}}{2}\right)$

66. a. $\cos^{-1}\left(\dfrac{1}{2}\right)$ **b.** $\cos^{-1}\left(\dfrac{-1}{\sqrt{2}}\right)$ **c.** $\cos^{-1}\left(\dfrac{\sqrt{3}}{2}\right)$

67. a. $\arccos(-1)$ **b.** $\arccos(0)$

68. a. $\arcsin(-1)$ **b.** $\arcsin\left(-\dfrac{1}{\sqrt{2}}\right)$

Theory and Examples

69. If $f(x)$ is one-to-one, can anything be said about $g(x) = -f(x)$? Is it also one-to-one? Give reasons for your answer.

70. If $f(x)$ is one-to-one and $f(x)$ is never zero, can anything be said about $h(x) = 1/f(x)$? Is it also one-to-one? Give reasons for your answer.

71. Suppose that the range of g lies in the domain of f so that the composite $f \circ g$ is defined. If f and g are one-to-one, can anything be said about $f \circ g$? Give reasons for your answer.

72. If a composite $f \circ g$ is one-to-one, must g be one-to-one? Give reasons for your answer.

73. Find a formula for the inverse function f^{-1} and verify that $(f \circ f^{-1})(x) = (f^{-1} \circ f)(x) = x$.

     a. $f(x) = \dfrac{100}{1 + 2^{-x}}$ **b.** $f(x) = \dfrac{50}{1 + 1.1^{-x}}$

74. The identity $\sin^{-1} x + \cos^{-1} x = \pi/2$ Figure 1.72 establishes the identity for $0 < x < 1$. To establish it for the rest of $[-1, 1]$, verify by direct calculation that it holds for $x = 1$, 0, and -1. Then, for values of x in $(-1, 0)$, let $x = -a$, $a > 0$, and apply Eqs. (3) and (5) to the sum $\sin^{-1}(-a) + \cos^{-1}(-a)$.

75. Start with the graph of $y = \ln x$. Find an equation of the graph that results from

     a. shifting down 3 units.

     b. shifting right 1 unit.

     c. shifting left 1, up 3 units.

     d. shifting down 4, right 2 units.

     e. reflecting about the y-axis.

     f. reflecting about the line $y = x$.

76. Start with the graph of $y = \ln x$. Find an equation of the graph that results from

     a. vertical stretching by a factor of 2.

     b. horizontal stretching by a factor of 3.

     c. vertical compression by a factor of 4.

     d. horizontal compression by a factor of 2.

T 77. The equation $x^2 = 2^x$ has three solutions: $x = 2$, $x = 4$, and one other. Estimate the third solution as accurately as you can by graphing.

T 78. Could $x^{\ln 2}$ possibly be the same as $2^{\ln x}$ for $x > 0$? Graph the two functions and explain what you see.

79. Radioactive decay   The half-life of a certain radioactive substance is 12 hours. There are 8 grams present initially.

     a. Express the amount of substance remaining as a function of time t.

     b. When will there be 1 gram remaining?

80. Doubling your money   Determine how much time is required for a \$500 investment to double in value if interest is earned at the rate of 4.75% compounded annually.

81. Population growth   The population of Glenbrook is 375,000 and is increasing at the rate of 2.25% per year. Predict when the population will be 1 million.

82. Radon-222   The decay equation for radon-222 gas is known to be $y = y_0 e^{-0.18t}$, with t in days. About how long will it take the radon in a sealed sample of air to fall to 90% of its original value?

Chapter 1   Questions to Guide Your Review

1. What is a function? What is its domain? Its range? What is an arrow diagram for a function? Give examples.

2. What is the graph of a real-valued function of a real variable? What is the vertical line test?

3. What is a piecewise-defined function? Give examples.

4. What are the important types of functions frequently encountered in calculus? Give an example of each type.

5. What is meant by an increasing function? A decreasing function? Give an example of each.

6. What is an even function? An odd function? What symmetry properties do the graphs of such functions have? What advantage can we take of this? Give an example of a function that is neither even nor odd.

7. If f and g are real-valued functions, how are the domains of $f + g, f - g, fg,$ and f/g related to the domains of f and g? Give examples.

8. When is it possible to compose one function with another? Give examples of composites and their values at various points. Does the order in which functions are composed ever matter?

9. How do you change the equation $y = f(x)$ to shift its graph vertically up or down by $|k|$ units? Horizontally to the left or right? Give examples.

10. How do you change the equation $y = f(x)$ to compress or stretch the graph by a factor $c > 1$? Reflect the graph across a coordinate axis? Give examples.

11. What is the standard equation of an ellipse with center (h, k)? What is its major axis? Its minor axis? Give examples.

12. What is radian measure? How do you convert from radians to degrees? Degrees to radians?

13. Graph the six basic trigonometric functions. What symmetries do the graphs have?

14. What is a periodic function? Give examples. What are the periods of the six basic trigonometric functions?

15. Starting with the identity $\sin^2\theta + \cos^2\theta = 1$ and the formulas for $\cos(A + B)$ and $\sin(A + B)$, show how a variety of other trigonometric identities may be derived.

16. How does the formula for the general sine function $f(x) = A \sin((2\pi/B)(x - C)) + D$ relate to the shifting, stretching, compressing, and reflection of its graph? Give examples. Graph the general sine curve and identify the constants $A, B, C,$ and D.

17. Name three issues that arise when functions are graphed using a calculator or computer with graphing software. Give examples.

18. What is an exponential function? Give examples. What laws of exponents does it obey? How does it differ from a simple power function like $f(x) = x^n$? What kind of real-world phenomena are modeled by exponential functions?

19. What is the number e, and how is it defined? What are the domain and range of $f(x) = e^x$? What does its graph look like? How do the values of e^x relate to $x^2, x^3,$ and so on?

20. What functions have inverses? How do you know if two functions f and g are inverses of one another? Give examples of functions that are (are not) inverses of one another.

21. How are the domains, ranges, and graphs of functions and their inverses related? Give an example.

22. What procedure can you sometimes use to express the inverse of a function of x as a function of x?

23. What is a logarithmic function? What properties does it satisfy? What is the natural logarithm function? What are the domain and range of $y = \ln x$? What does its graph look like?

24. How is the graph of $\log_a x$ related to the graph of $\ln x$? What truth is there in the statement that there is really only one exponential function and one logarithmic function?

25. How are the inverse trigonometric functions defined? How can you sometimes use right triangles to find values of these functions? Give examples.

Chapter 1 Practice Exercises

Functions and Graphs

1. Express the area and circumference of a circle as functions of the circle's radius. Then express the area as a function of the circumference.

2. Express the radius of a sphere as a function of the sphere's surface area. Then express the surface area as a function of the volume.

3. A point P in the first quadrant lies on the parabola $y = x^2$. Express the coordinates of P as functions of the angle of inclination of the line joining P to the origin.

4. A hot-air balloon rising straight up from a level field is tracked by a range finder located 500 ft from the point of liftoff. Express the balloon's height as a function of the angle the line from the range finder to the balloon makes with the ground.

In Exercises 5–8, determine whether the graph of the function is symmetric about the y-axis, the origin, or neither.

5. $y = x^{1/5}$ 6. $y = x^{2/5}$

7. $y = x^2 - 2x - 1$ 8. $y = e^{-x^2}$

In Exercises 9–16, determine whether the function is even, odd, or neither.

9. $y = x^2 + 1$ 10. $y = x^5 - x^3 - x$

11. $y = 1 - \cos x$ 12. $y = \sec x \tan x$

13. $y = \dfrac{x^4 + 1}{x^3 - 2x}$ 14. $y = x - \sin x$

15. $y = x + \cos x$ 16. $y = x \cos x$

17. Suppose that f and g are both odd functions defined on the entire real line. Which of the following (where defined) are even? odd?

    a. fg **b.** f^3 **c.** $f(\sin x)$ **d.** $g(\sec x)$ **e.** $|g|$

18. If $f(a - x) = f(a + x)$, show that $g(x) = f(x + a)$ is an even function.

In Exercises 19–28, find the **(a)** domain and **(b)** range.

19. $y = |x| - 2$ 20. $y = -2 + \sqrt{1 - x}$

21. $y = \sqrt{16 - x^2}$ 22. $y = 3^{2-x} + 1$

23. $y = 2e^{-x} - 3$ 24. $y = \tan(2x - \pi)$

25. $y = 2 \sin(3x + \pi) - 1$ 26. $y = x^{2/5}$

27. $y = \ln(x - 3) + 1$ **28.** $y = -1 + \sqrt[3]{2 - x}$

29. State whether each function is increasing, decreasing, or neither.

   a. Volume of a sphere as a function of its radius

   b. Greatest integer function

   c. Height above Earth's sea level as a function of atmospheric pressure (assumed nonzero)

   d. Kinetic energy as a function of a particle's velocity

30. Find the largest interval on which the given function is increasing.

   a. $f(x) = |x - 2| + 1$ **b.** $f(x) = (x + 1)^4$

   c. $g(x) = (3x - 1)^{1/3}$ **d.** $R(x) = \sqrt{2x - 1}$

Piecewise-Defined Functions

In Exercises 31 and 32, find the **(a)** domain and **(b)** range.

31. $y = \begin{cases} \sqrt{-x}, & -4 \le x \le 0 \\ \sqrt{x}, & 0 < x \le 4 \end{cases}$

32. $y = \begin{cases} -x - 2, & -2 \le x \le -1 \\ x, & -1 < x \le 1 \\ -x + 2, & 1 < x \le 2 \end{cases}$

In Exercises 33 and 34, write a piecewise formula for the function.

33.

34.

Composition of Functions

In Exercises 35 and 36, find

   a. $(f \circ g)(-1)$.   **b.** $(g \circ f)(2)$.

   c. $(f \circ f)(x)$.   **d.** $(g \circ g)(x)$.

35. $f(x) = \dfrac{1}{x}, \quad g(x) = \dfrac{1}{\sqrt{x + 2}}$

36. $f(x) = 2 - x, \quad g(x) = \sqrt[3]{x + 1}$

In Exercises 37 and 38, **(a)** write formulas for $f \circ g$ and $g \circ f$ and find the **(b)** domain and **(c)** range of each.

37. $f(x) = 2 - x^2, \quad g(x) = \sqrt{x + 2}$

38. $f(x) = \sqrt{x}, \quad g(x) = \sqrt{1 - x}$

For Exercises 39 and 40, sketch the graphs of f and $f \circ f$.

39. $f(x) = \begin{cases} -x - 2, & -4 \le x \le -1 \\ -1, & -1 < x \le 1 \\ x - 2, & 1 < x \le 2 \end{cases}$

40. $f(x) = \begin{cases} x + 1, & -2 \le x < 0 \\ x - 1, & 0 \le x \le 2 \end{cases}$

Composition with absolute values In Exercises 41–48, graph f_1 and f_2 together. Then describe how applying the absolute value function in f_2 affects the graph of f_1.

	$f_1(x)$	$f_2(x)$
41.	x	$|x|$
42.	x^2	$|x|^2$
43.	x^3	$|x^3|$
44.	$x^2 + x$	$|x^2 + x|$
45.	$4 - x^2$	$|4 - x^2|$
46.	$\dfrac{1}{x}$	$\dfrac{1}{|x|}$
47.	$\sqrt{x}$	$\sqrt{|x|}$
48.	$\sin x$	$\sin|x|$

Shifting and Scaling Graphs

49. Suppose the graph of g is given. Write equations for the graphs that are obtained from the graph of g by shifting, scaling, or reflecting, as indicated.

   a. Up $\dfrac{1}{2}$ unit, right 3

   b. Down 2 units, left $\dfrac{2}{3}$

   c. Reflect about the y-axis

   d. Reflect about the x-axis

   e. Stretch vertically by a factor of 5

   f. Compress horizontally by a factor of 5

50. Describe how each graph is obtained from the graph of $y = f(x)$.

   a. $y = f(x - 5)$ **b.** $y = f(4x)$

   c. $y = f(-3x)$ **d.** $y = f(2x + 1)$

   e. $y = f\left(\dfrac{x}{3}\right) - 4$ **f.** $y = -3f(x) + \dfrac{1}{4}$

In Exercises 51–54, graph each function, not by plotting points, but by starting with the graph of one of the standard functions presented in Figures 1.15–1.17, and applying an appropriate transformation.

51. $y = -\sqrt{1 + \dfrac{x}{2}}$ **52.** $y = 1 - \dfrac{x}{3}$

53. $y = \dfrac{1}{2x^2} + 1$ **54.** $y = (-5x)^{1/3}$

Trigonometry

In Exercises 55–58, sketch the graph of the given function. What is the period of the function?

55. $y = \cos 2x$ **56.** $y = \sin\dfrac{x}{2}$

57. $y = \sin \pi x$ **58.** $y = \cos\dfrac{\pi x}{2}$

59. Sketch the graph $y = 2\cos\left(x - \dfrac{\pi}{3}\right)$.

60. Sketch the graph $y = 1 + \sin\left(x + \dfrac{\pi}{4}\right)$.

In Exercises 61–64, ABC is a right triangle with the right angle at C. The sides opposite angles A, B, and C are a, b, and c, respectively.

61. a. Find a and b if $c = 2$, $B = \pi/3$.

b. Find a and c if $b = 2$, $B = \pi/3$.

62. a. Express a in terms of A and c.

b. Express a in terms of A and b.

63. a. Express a in terms of B and b.

b. Express c in terms of A and a.

64. a. Express $\sin A$ in terms of a and c.

b. Express $\sin A$ in terms of b and c.

65. Height of a pole Two wires stretch from the top T of a vertical pole to points B and C on the ground, where C is 10 m closer to the base of the pole than is B. If wire BT makes an angle of $35°$ with the horizontal and wire CT makes an angle of $50°$ with the horizontal, how high is the pole?

66. Height of a weather balloon Observers at positions A and B 2 km apart simultaneously measure the angle of elevation of a weather balloon to be $40°$ and $70°$, respectively. If the balloon is directly above a point on the line segment between A and B, find the height of the balloon.

T 67. a. Graph the function $f(x) = \sin x + \cos(x/2)$.

b. What appears to be the period of this function?

c. Confirm your finding in part (b) algebraically.

T 68. a. Graph $f(x) = \sin(1/x)$.

b. What are the domain and range of f?

c. Is f periodic? Give reasons for your answer.

Transcendental Functions

In Exercises 69–72, find the domain of each function.

69. a. $f(x) = 1 + e^{-\sin x}$ **b.** $g(x) = e^x + \ln \sqrt{x}$

70. a. $f(x) = e^{1/x^2}$ **b.** $g(x) = \ln|4 - x^2|$

71. a. $h(x) = \sin^{-1}\left(\dfrac{x}{3}\right)$ **b.** $f(x) = \cos^{-1}(\sqrt{x} - 1)$

72. a. $h(x) = \ln(\cos^{-1} x)$ **b.** $f(x) = \sqrt{\pi - \sin^{-1} x}$

73. If $f(x) = \ln x$ and $g(x) = 4 - x^2$, find the functions $f \circ g$, $g \circ f$, $f \circ f$, $g \circ g$, and their domains.

74. Determine whether f is even, odd, or neither.

a. $f(x) = e^{-x^2}$ **b.** $f(x) = 1 + \sin^{-1}(-x)$

c. $f(x) = |e^x|$ **d.** $f(x) = e^{\ln|x|+1}$

T 75. Graph $\ln x$, $\ln 2x$, $\ln 4x$, $\ln 8x$, and $\ln 16x$ (as many as you can) together for $0 < x \le 10$. What is going on? Explain.

T 76. Graph $y = \ln(x^2 + c)$ for $c = -4, -2, 0, 3$, and 5. How does the graph change when c changes?

T 77. Graph $y = \ln|\sin x|$ in the window $0 \le x \le 22$, $-2 \le y \le 0$. Explain what you see. How could you change the formula to turn the arches upside down?

T 78. Graph the three functions $y = x^a$, $y = a^x$, and $y = \log_a x$ together on the same screen for $a = 2, 10$, and 20. For large values of x, which of these functions has the largest values and which has the smallest values?

Theory and Examples

In Exercises 79 and 80, find the domain and range of each composite function. Then graph the composites on separate screens. Do the graphs make sense in each case? Give reasons for your answers and comment on any differences you see.

79. a. $y = \sin^{-1}(\sin x)$ **b.** $y = \sin(\sin^{-1} x)$

80. a. $y = \cos^{-1}(\cos x)$ **b.** $y = \cos(\cos^{-1} x)$

81. Use a graph to decide whether f is one-to-one.

a. $f(x) = x^3 - \dfrac{x}{2}$ **b.** $f(x) = x^3 + \dfrac{x}{2}$

T 82. Use a graph to find to 3 decimal places the values of x for which $e^x > 10{,}000{,}000$.

83. a. Show that $f(x) = x^3$ and $g(x) = \sqrt[3]{x}$ are inverses of one another.

T b. Graph f and g over an x-interval large enough to show the graphs intersecting at $(1, 1)$ and $(-1, -1)$. Be sure the picture shows the required symmetry in the line $y = x$.

84. a. Show that $h(x) = x^3/4$ and $k(x) = (4x)^{1/3}$ are inverses of one another.

T b. Graph h and k over an x-interval large enough to show the graphs intersecting at $(2, 2)$ and $(-2, -2)$. Be sure the picture shows the required symmetry in the line $y = x$.

Chapter 1 Additional and Advanced Exercises

Functions and Graphs

1. Are there two functions f and g such that $f \circ g = g \circ f$? Give reasons for your answer.

2. Are there two functions f and g with the following property? The graphs of f and g are not straight lines but the graph of $f \circ g$ is a straight line. Give reasons for your answer.

3. If $f(x)$ is odd, can anything be said of $g(x) = f(x) - 2$? What if f is even instead? Give reasons for your answer.

4. If $g(x)$ is an odd function defined for all values of x, can anything be said about $g(0)$? Give reasons for your answer.

5. Graph the equation $|x| + |y| = 1 + x$.

6. Graph the equation $y + |y| = x + |x|$.

Derivations and Proofs

7. Prove the following identities.

a. $\dfrac{1 - \cos x}{\sin x} = \dfrac{\sin x}{1 + \cos x}$ **b.** $\dfrac{1 - \cos x}{1 + \cos x} = \tan^2\dfrac{x}{2}$

8. Explain the following "proof without words" of the law of cosines. (Source: "Proof without Words: The Law of Cosines," Sidney H. Kung, *Mathematics Magazine*, Vol. 63, No. 5, Dec. 1990, p. 342.)

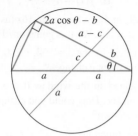

9. Show that the area of triangle *ABC* is given by $(1/2)ab \sin C = (1/2)bc \sin A = (1/2)ca \sin B$.

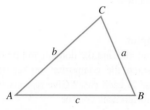

10. Show that the area of triangle *ABC* is given by $\sqrt{s(s-a)(s-b)(s-c)}$ where $s = (a+b+c)/2$ is the semiperimeter of the triangle.

11. Show that if f is both even and odd, then $f(x) = 0$ for every x in the domain of f.

12. a. **Even-odd decompositions** Let f be a function whose domain is symmetric about the origin, that is, $-x$ belongs to the domain whenever x does. Show that f is the sum of an even function and an odd function:

$$f(x) = E(x) + O(x),$$

where E is an even function and O is an odd function. (*Hint:* Let $E(x) = (f(x) + f(-x))/2$. Show that $E(-x) = E(x)$, so that E is even. Then show that $O(x) = f(x) - E(x)$ is odd.)

   b. **Uniqueness** Show that there is only one way to write f as the sum of an even and an odd function. (*Hint:* One way is given in part (a). If also $f(x) = E_1(x) + O_1(x)$ where E_1 is even and O_1 is odd, show that $E - E_1 = O_1 - O$. Then use Exercise 11 to show that $E = E_1$ and $O = O_1$.)

Grapher Explorations—Effects of Parameters

13. What happens to the graph of $y = ax^2 + bx + c$ as
   a. a changes while b and c remain fixed?
   b. b changes (a and c fixed, $a \neq 0$)?
   c. c changes (a and b fixed, $a \neq 0$)?

14. What happens to the graph of $y = a(x+b)^3 + c$ as
   a. a changes while b and c remain fixed?
   b. b changes (a and c fixed, $a \neq 0$)?
   c. c changes (a and b fixed, $a \neq 0$)?

Geometry

15. An object's center of mass moves at a constant velocity v along a straight line past the origin. The accompanying figure shows the coordinate system and the line of motion. The dots show positions that are 1 sec apart. Why are the areas $A_1, A_2, \ldots, A_5$ in the figure all equal? As in Kepler's equal area law (see Section 13.6), the line that joins the object's center of mass to the origin sweeps out equal areas in equal times.

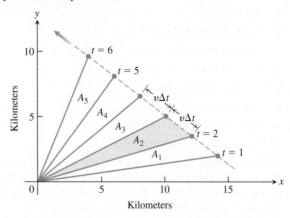

16. a. Find the slope of the line from the origin to the midpoint P of side AB in the triangle in the accompanying figure ($a, b > 0$).

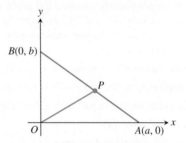

   b. When is OP perpendicular to AB?

17. Consider the quarter-circle of radius 1 and right triangles ABE and ACD given in the accompanying figure. Use standard area formulas to conclude that

$$\frac{1}{2} \sin \theta \cos \theta < \frac{\theta}{2} < \frac{1}{2} \frac{\sin \theta}{\cos \theta}.$$

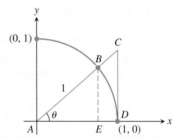

18. Let $f(x) = ax + b$ and $g(x) = cx + d$. What condition must be satisfied by the constants a, b, c, d in order that $(f \circ g)(x) = (g \circ f)(x)$ for every value of x?

Theory and Examples

19. **Domain and range** Suppose that $a \neq 0, b \neq 1$, and $b > 0$. Determine the domain and range of the function.
   a. $y = a(b^{c-x}) + d$ b. $y = a \log_b(x - c) + d$

20. **Inverse functions** Let

$$f(x) = \frac{ax + b}{cx + d}, \qquad c \neq 0, \qquad ad - bc \neq 0.$$

   a. Give a convincing argument that f is one-to-one.
   b. Find a formula for the inverse of f.

21. Depreciation Smith Hauling purchased an 18-wheel truck for $100,000. The truck depreciates at the constant rate of $10,000 per year for 10 years.

  a. Write an expression that gives the value y after x years.

  b. When is the value of the truck $55,000?

22. Drug absorption A drug is administered intravenously for pain. The function

$$f(t) = 90 - 52 \ln (1 + t), \qquad 0 \le t \le 4$$

gives the number of units of the drug remaining in the body after t hours.

  a. What was the initial number of units of the drug administered?

  b. How much is present after 2 hours?

  c. Draw the graph of f.

23. Finding investment time If Juanita invests $1500 in a retirement account that earns 8% compounded annually, how long will it take this single payment to grow to $5000?

24. The rule of 70 If you use the approximation $\ln 2 \approx 0.70$ (in place of 0.69314 . . .), you can derive a rule of thumb that says,

"To estimate how many years it will take an amount of money to double when invested at r percent compounded continuously, divide r into 70." For instance, an amount of money invested at 5% will double in about $70/5 = 14$ years. If you want it to double in 10 years instead, you have to invest it at $70/10 = 7\%$. Show how the rule of 70 is derived. (A similar "rule of 72" uses 72 instead of 70, because 72 has more integer factors.)

25. For what $x > 0$ does $x^{(x^x)} = (x^x)^x$? Give reasons for your answer.

T 26. a. If $(\ln x)/x = (\ln 2)/2$, must $x = 2$?

  b. If $(\ln x)/x = -2 \ln 2$, must $x = 1/2$?

    Give reasons for your answers.

27. The quotient $(\log_4 x)/(\log_2 x)$ has a constant value. What value? Give reasons for your answer.

T 28. $\log_x (2)$ vs. $\log_2 (x)$ How does $f(x) = \log_x (2)$ compare with $g(x) = \log_2 (x)$? Here is one way to find out.

  a. Use the equation $\log_a b = (\ln b)/(\ln a)$ to express $f(x)$ and $g(x)$ in terms of natural logarithms.

  b. Graph f and g together. Comment on the behavior of f in relation to the signs and values of g.

Chapter 1   Technology Application Projects

An Overview of Mathematica
An overview of *Mathematica* sufficient to complete the *Mathematica* modules appearing on the Web site.

Mathematica/Maple Module:

Modeling Change: Springs, Driving Safety, Radioactivity, Trees, Fish, and Mammals
Construct and interpret mathematical models, analyze and improve them, and make predictions using them.

2
LIMITS AND CONTINUITY

OVERVIEW Mathematicians of the seventeenth century were keenly interested in the study of motion for objects on or near the earth and the motion of planets and stars. This study involved both the speed of the object and its direction of motion at any instant, and they knew the direction was tangent to the path of motion. The concept of a limit is fundamental to finding the velocity of a moving object and the tangent to a curve. In this chapter we develop the limit, first intuitively and then formally. We use limits to describe the way a function varies. Some functions vary *continuously*; small changes in *x* produce only small changes in $f(x)$. Other functions can have values that jump, vary erratically, or tend to increase or decrease without bound. The notion of limit gives a precise way to distinguish between these behaviors.

2.1 | Rates of Change and Tangents to Curves

Calculus is a tool to help us understand how functional relationships change, such as the position or speed of a moving object as a function of time, or the changing slope of a curve being traversed by a point moving along it. In this section we introduce the ideas of average and instantaneous rates of change, and show that they are closely related to the slope of a curve at a point *P* on the curve. We give precise developments of these important concepts in the next chapter, but for now we use an informal approach so you will see how they lead naturally to the main idea of the chapter, the *limit*. You will see that limits play a major role in calculus and the study of change.

Average and Instantaneous Speed

HISTORICAL BIOGRAPHY

Galileo Galilei
(1564–1642)

In the late sixteenth century, Galileo discovered that a solid object dropped from rest (not moving) near the surface of the earth and allowed to fall freely will fall a distance proportional to the square of the time it has been falling. This type of motion is called **free fall**. It assumes negligible air resistance to slow the object down, and that gravity is the only force acting on the falling body. If *y* denotes the distance fallen in feet after *t* seconds, then Galileo's law is

$$y = 16t^2,$$

where 16 is the (approximate) constant of proportionality. (If *y* is measured in meters, the constant is 4.9.)

A moving body's **average speed** during an interval of time is found by dividing the distance covered by the time elapsed. The unit of measure is length per unit time: kilometers per hour, feet (or meters) per second, or whatever is appropriate to the problem at hand.

EXAMPLE 1   A rock breaks loose from the top of a tall cliff. What is its average speed

(a) during the first 2 sec of fall?

(b) during the 1-sec interval between second 1 and second 2?

Solution   The average speed of the rock during a given time interval is the change in distance, Δy, divided by the length of the time interval, Δt. (Increments like Δy and Δt are reviewed in Appendix 3.) Measuring distance in feet and time in seconds, we have the following calculations:

(a) For the first 2 sec: $\dfrac{\Delta y}{\Delta t} = \dfrac{16(2)^2 - 16(0)^2}{2 - 0} = 32 \dfrac{\text{ft}}{\text{sec}}$

(b) From sec 1 to sec 2: $\dfrac{\Delta y}{\Delta t} = \dfrac{16(2)^2 - 16(1)^2}{2 - 1} = 48 \dfrac{\text{ft}}{\text{sec}}$ ∎

We want a way to determine the speed of a falling object at a single instant t_0, instead of using its average speed over an interval of time. To do this, we examine what happens when we calculate the average speed over shorter and shorter time intervals starting at t_0. The next example illustrates this process. Our discussion is informal here, but it will be made precise in Chapter 3.

EXAMPLE 2   Find the speed of the falling rock in Example 1 at $t = 1$ and $t = 2$ sec.

Solution   We can calculate the average speed of the rock over a time interval $[t_0, t_0 + h]$, having length $\Delta t = h$, as

$$\frac{\Delta y}{\Delta t} = \frac{16(t_0 + h)^2 - 16{t_0}^2}{h}. \tag{1}$$

We cannot use this formula to calculate the "instantaneous" speed at the exact moment t_0 by simply substituting $h = 0$, because we cannot divide by zero. But we *can* use it to calculate average speeds over increasingly short time intervals starting at $t_0 = 1$ and $t_0 = 2$. When we do so, we see a pattern (Table 2.1).

TABLE 2.1   Average speeds over short time intervals $[t_0, t_0 + h]$

$$\text{Average speed:} \frac{\Delta y}{\Delta t} = \frac{16(t_0 + h)^2 - 16{t_0}^2}{h}$$

Length of time interval h	Average speed over interval of length h starting at $t_0 = 1$	Average speed over interval of length h starting at $t_0 = 2$
1	48	80
0.1	33.6	65.6
0.01	32.16	64.16
0.001	32.016	64.016
0.0001	32.0016	64.0016

The average speed on intervals starting at $t_0 = 1$ seems to approach a limiting value of 32 as the length of the interval decreases. This suggests that the rock is falling at a speed of 32 ft/sec at $t_0 = 1$ sec. Let's confirm this algebraically.

If we set $t_0 = 1$ and then expand the numerator in Equation (1) and simplify, we find that

$$\frac{\Delta y}{\Delta t} = \frac{16(1 + h)^2 - 16(1)^2}{h} = \frac{16(1 + 2h + h^2) - 16}{h}$$

$$= \frac{32h + 16h^2}{h} = 32 + 16h.$$

For values of h different from 0, the expressions on the right and left are equivalent and the average speed is $32 + 16h$ ft/sec. We can now see why the average speed has the limiting value $32 + 16(0) = 32$ ft/sec as h approaches 0.

Similarly, setting $t_0 = 2$ in Equation (1), the procedure yields

$$\frac{\Delta y}{\Delta t} = 64 + 16h$$

for values of h different from 0. As h gets closer and closer to 0, the average speed has the limiting value 64 ft/sec when $t_0 = 2$ sec, as suggested by Table 2.1. ∎

The average speed of a falling object is an example of a more general idea which we discuss next.

Average Rates of Change and Secant Lines

Given an arbitrary function $y = f(x)$, we calculate the average rate of change of y with respect to x over the interval $[x_1, x_2]$ by dividing the change in the value of y, $\Delta y = f(x_2) - f(x_1)$, by the length $\Delta x = x_2 - x_1 = h$ of the interval over which the change occurs. (We use the symbol h for Δx to simplify the notation here and later on.)

DEFINITION   The **average rate of change** of $y = f(x)$ with respect to x over the interval $[x_1, x_2]$ is

$$\frac{\Delta y}{\Delta x} = \frac{f(x_2) - f(x_1)}{x_2 - x_1} = \frac{f(x_1 + h) - f(x_1)}{h}, \qquad h \neq 0.$$

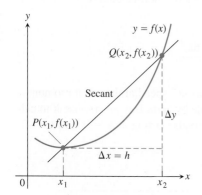

FIGURE 2.1   A secant to the graph $y = f(x)$. Its slope is $\Delta y / \Delta x$, the average rate of change of f over the interval $[x_1, x_2]$.

Geometrically, the rate of change of f over $[x_1, x_2]$ is the slope of the line through the points $P(x_1, f(x_1))$ and $Q(x_2, f(x_2))$ (Figure 2.1). In geometry, a line joining two points of a curve is a **secant** to the curve. Thus, the average rate of change of f from x_1 to x_2 is identical with the slope of secant PQ. Let's consider what happens as the point Q approaches the point P along the curve, so the length h of the interval over which the change occurs approaches zero.

Defining the Slope of a Curve

We know what is meant by the slope of a straight line, which tells us the rate at which it rises or falls—its rate of change as the graph of a linear function. But what is meant by the *slope of a curve* at a point P on the curve? If there is a *tangent* line to the curve at P—a line that just touches the curve like the tangent to a circle—it would be reasonable to identify *the slope of the tangent* as the slope of the curve at P. So we need a precise meaning for the tangent at a point on a curve.

For circles, tangency is straightforward. A line L is tangent to a circle at a point P if L passes through P perpendicular to the radius at P (Figure 2.2). Such a line just *touches* the circle. But what does it mean to say that a line L is tangent to some other curve C at a point P?

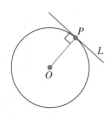

FIGURE 2.2 L is tangent to the circle at P if it passes through P perpendicular to radius OP.

To define tangency for general curves, we need an approach that takes into account the behavior of the secants through P and nearby points Q as Q moves toward P along the curve (Figure 2.3). Here is the idea:

1. Start with what we *can* calculate, namely the slope of the secant PQ.

2. Investigate the limiting value of the secant slope as Q approaches P along the curve. (We clarify the *limit* idea in the next section.)

3. If the *limit* exists, take it to be the slope of the curve at P and *define* the tangent to the curve at P to be the line through P with this slope.

This procedure is what we were doing in the falling-rock problem discussed in Example 2. The next example illustrates the geometric idea for the tangent to a curve.

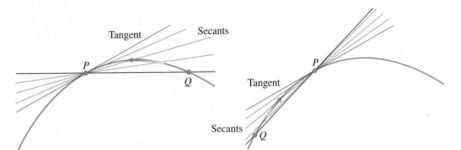

FIGURE 2.3 The tangent to the curve at P is the line through P whose slope is the limit of the secant slopes as $Q \rightarrow P$ from either side.

EXAMPLE 3   Find the slope of the parabola $y = x^2$ at the point $P(2, 4)$. Write an equation for the tangent to the parabola at this point.

Solution   We begin with a secant line through $P(2, 4)$ and $Q(2 + h, (2 + h)^2)$ nearby. We then write an expression for the slope of the secant PQ and investigate what happens to the slope as Q approaches P along the curve:

$$\text{Secant slope} = \frac{\Delta y}{\Delta x} = \frac{(2 + h)^2 - 2^2}{h} = \frac{h^2 + 4h + 4 - 4}{h}$$

$$= \frac{h^2 + 4h}{h} = h + 4.$$

If $h > 0$, then Q lies above and to the right of P, as in Figure 2.4. If $h < 0$, then Q lies to the left of P (not shown). In either case, as Q approaches P along the curve, h approaches zero and the secant slope $h + 4$ approaches 4. We take 4 to be the parabola's slope at P.

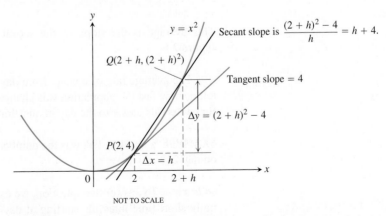

NOT TO SCALE

FIGURE 2.4 Finding the slope of the parabola $y = x^2$ at the point $P(2, 4)$ as the limit of secant slopes (Example 3).

The tangent to the parabola at P is the line through P with slope 4:

$$y = 4 + 4(x - 2) \qquad \text{Point-slope equation}$$
$$y = 4x - 4.$$

Instantaneous Rates of Change and Tangent Lines

The rates at which the rock in Example 2 was falling at the instants $t = 1$ and $t = 2$ are called *instantaneous rates of change*. Instantaneous rates and slopes of tangent lines are intimately connected, as we will now see in the following examples.

EXAMPLE 4   Figure 2.5 shows how a population p of fruit flies (*Drosophila*) grew in a 50-day experiment. The number of flies was counted at regular intervals, the counted values plotted with respect to time t, and the points joined by a smooth curve (colored blue in Figure 2.5). Find the average growth rate from day 23 to day 45.

Solution   There were 150 flies on day 23 and 340 flies on day 45. Thus the number of flies increased by $340 - 150 = 190$ in $45 - 23 = 22$ days. The average rate of change of the population from day 23 to day 45 was

$$\text{Average rate of change: } \frac{\Delta p}{\Delta t} = \frac{340 - 150}{45 - 23} = \frac{190}{22} \approx 8.6 \text{ flies/day}.$$

FIGURE 2.5   Growth of a fruit fly population in a controlled experiment. The average rate of change over 22 days is the slope $\Delta p / \Delta t$ of the secant line (Example 4).

This average is the slope of the secant through the points P and Q on the graph in Figure 2.5.

The average rate of change from day 23 to day 45 calculated in Example 4 does not tell us how fast the population was changing on day 23 itself. For that we need to examine time intervals closer to the day in question.

EXAMPLE 5   How fast was the number of flies in the population of Example 4 growing on day 23?

Solution   To answer this question, we examine the average rates of change over increasingly short time intervals starting at day 23. In geometric terms, we find these rates by calculating the slopes of secants from P to Q, for a sequence of points Q approaching P along the curve (Figure 2.6).

Q	Slope of $PQ = \Delta p / \Delta t$ (flies/day)
(45, 340)	$\dfrac{340 - 150}{45 - 23} \approx 8.6$
(40, 330)	$\dfrac{330 - 150}{40 - 23} \approx 10.6$
(35, 310)	$\dfrac{310 - 150}{35 - 23} \approx 13.3$
(30, 265)	$\dfrac{265 - 150}{30 - 23} \approx 16.4$

FIGURE 2.6 The positions and slopes of four secants through the point P on the fruit fly graph (Example 5).

The values in the table show that the secant slopes rise from 8.6 to 16.4 as the t-coordinate of Q decreases from 45 to 30, and we would expect the slopes to rise slightly higher as t continued on toward 23. Geometrically, the secants rotate about P and seem to approach the red tangent line in the figure. Since the line appears to pass through the points (14, 0) and (35, 350), it has slope

$$\frac{350 - 0}{35 - 14} = 16.7 \text{ flies/day (approximately)}.$$

On day 23 the population was increasing at a rate of about 16.7 flies/day. ∎

The instantaneous rates in Example 2 were found to be the values of the average speeds, or average rates of change, as the time interval of length h approached 0. That is, the instantaneous rate is the value the average rate approaches as the length h of the interval over which the change occurs approaches zero. The average rate of change corresponds to the slope of a secant line; the instantaneous rate corresponds to the slope of the tangent line as the independent variable approaches a fixed value. In Example 2, the independent variable t approached the values $t = 1$ and $t = 2$. In Example 3, the independent variable x approached the value $x = 2$. So we see that instantaneous rates and slopes of tangent lines are closely connected. We investigate this connection thoroughly in the next chapter, but to do so we need the concept of a *limit*.

Exercises 2.1

Average Rates of Change

In Exercises 1–6, find the average rate of change of the function over the given interval or intervals.

1. $f(x) = x^3 + 1$
   a. $[2, 3]$　　　　b. $[-1, 1]$
2. $g(x) = x^2$
   a. $[-1, 1]$　　　b. $[-2, 0]$
3. $h(t) = \cot t$
   a. $[\pi/4, 3\pi/4]$　b. $[\pi/6, \pi/2]$
4. $g(t) = 2 + \cos t$
   a. $[0, \pi]$　　　　b. $[-\pi, \pi]$

5. $R(\theta) = \sqrt{4\theta + 1}$;　$[0, 2]$
6. $P(\theta) = \theta^3 - 4\theta^2 + 5\theta$;　$[1, 2]$

Slope of a Curve at a Point

In Exercises 7–14, use the method in Example 3 to find **(a)** the slope of the curve at the given point P, and **(b)** an equation of the tangent line at P.

7. $y = x^2 - 3$,　$P(2, 1)$
8. $y = 5 - x^2$,　$P(1, 4)$
9. $y = x^2 - 2x - 3$,　$P(2, -3)$
10. $y = x^2 - 4x$,　$P(1, -3)$
11. $y = x^3$,　$P(2, 8)$

12. $y = 2 - x^3$, $P(1, 1)$

13. $y = x^3 - 12x$, $P(1, -11)$

14. $y = x^3 - 3x^2 + 4$, $P(2, 0)$

Instantaneous Rates of Change

15. Speed of a car The accompanying figure shows the time-to-distance graph for a sports car accelerating from a standstill.

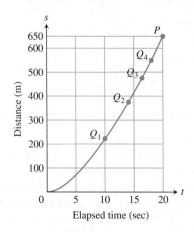

Elapsed time (sec)

a. Estimate the slopes of secants PQ_1, PQ_2, PQ_3, and PQ_4, arranging them in order in a table like the one in Figure 2.6. What are the appropriate units for these slopes?

b. Then estimate the car's speed at time $t = 20$ sec.

16. The accompanying figure shows the plot of distance fallen versus time for an object that fell from the lunar landing module a distance 80 m to the surface of the moon.

a. Estimate the slopes of the secants PQ_1, PQ_2, PQ_3, and PQ_4, arranging them in a table like the one in Figure 2.6.

b. About how fast was the object going when it hit the surface?

Elapsed time (sec)

T 17. The profits of a small company for each of the first five years of its operation are given in the following table:

Year	Profit in $1000s
2000	6
2001	27
2002	62
2003	111
2004	174

a. Plot points representing the profit as a function of year, and join them by as smooth a curve as you can.

b. What is the average rate of increase of the profits between 2002 and 2004?

c. Use your graph to estimate the rate at which the profits were changing in 2002.

T 18. Make a table of values for the function $F(x) = (x + 2)/(x - 2)$ at the points $x = 1.2$, $x = 11/10$, $x = 101/100$, $x = 1001/1000$, $x = 10001/10000$, and $x = 1$.

a. Find the average rate of change of $F(x)$ over the intervals $[1, x]$ for each $x \neq 1$ in your table.

b. Extending the table if necessary, try to determine the rate of change of $F(x)$ at $x = 1$.

T 19. Let $g(x) = \sqrt{x}$ for $x \geq 0$.

a. Find the average rate of change of $g(x)$ with respect to x over the intervals $[1, 2]$, $[1, 1.5]$ and $[1, 1 + h]$.

b. Make a table of values of the average rate of change of g with respect to x over the interval $[1, 1 + h]$ for some values of h approaching zero, say $h = 0.1, 0.01, 0.001, 0.0001, 0.00001$, and 0.000001.

c. What does your table indicate is the rate of change of $g(x)$ with respect to x at $x = 1$?

d. Calculate the limit as h approaches zero of the average rate of change of $g(x)$ with respect to x over the interval $[1, 1 + h]$.

T 20. Let $f(t) = 1/t$ for $t \neq 0$.

a. Find the average rate of change of f with respect to t over the intervals (i) from $t = 2$ to $t = 3$, and (ii) from $t = 2$ to $t = T$.

b. Make a table of values of the average rate of change of f with respect to t over the interval $[2, T]$, for some values of T approaching 2, say $T = 2.1, 2.01, 2.001, 2.0001, 2.00001$, and 2.000001.

c. What does your table indicate is the rate of change of f with respect to t at $t = 2$?

d. Calculate the limit as T approaches 2 of the average rate of change of f with respect to t over the interval from 2 to T. You will have to do some algebra before you can substitute $T = 2$.

21. The accompanying graph shows the total distance s traveled by a bicyclist after t hours.

Elapsed time (hr)

a. Estimate the bicyclist's average speed over the time intervals $[0, 1]$, $[1, 2.5]$, and $[2.5, 3.5]$.

b. Estimate the bicyclist's instantaneous speed at the times $t = \frac{1}{2}$, $t = 2$, and $t = 3$.

c. Estimate the bicyclist's maximum speed and the specific time at which it occurs.

22. The accompanying graph shows the total amount of gasoline A in the gas tank of an automobile after being driven for t days.

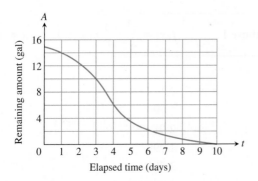

Elapsed time (days)

a. Estimate the average rate of gasoline consumption over the time intervals $[0, 3]$, $[0, 5]$, and $[7, 10]$.

b. Estimate the instantaneous rate of gasoline consumption at the times $t = 1$, $t = 4$, and $t = 8$.

c. Estimate the maximum rate of gasoline consumption and the specific time at which it occurs.

2.2 Limit of a Function and Limit Laws

In Section 2.1 we saw that limits arise when finding the instantaneous rate of change of a function or the tangent to a curve. Here we begin with an informal definition of *limit* and show how we can calculate the values of limits. A precise definition is presented in the next section.

Limits of Function Values

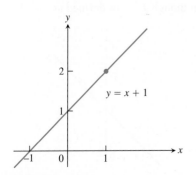

FIGURE 2.7 The graph of f is identical with the line $y = x + 1$ except at $x = 1$, where f is not defined (Example 1).

Frequently when studying a function $y = f(x)$, we find ourselves interested in the function's behavior *near* a particular point x_0, but not *at* x_0. This might be the case, for instance, if x_0 is an irrational number, like π or $\sqrt{2}$, whose values can only be approximated by "close" rational numbers at which we actually evaluate the function instead. Another situation occurs when trying to evaluate a function at x_0 leads to division by zero, which is undefined. We encountered this last circumstance when seeking the instantaneous rate of change in y by considering the quotient function $\Delta y/h$ for h closer and closer to zero. Here's a specific example where we explore numerically how a function behaves near a particular point at which we cannot directly evaluate the function.

EXAMPLE 1   How does the function

$$f(x) = \frac{x^2 - 1}{x - 1}$$

behave near $x = 1$?

Solution   The given formula defines f for all real numbers x except $x = 1$ (we cannot divide by zero). For any $x \neq 1$, we can simplify the formula by factoring the numerator and canceling common factors:

$$f(x) = \frac{(x - 1)(x + 1)}{x - 1} = x + 1 \quad \text{for} \quad x \neq 1.$$

The graph of f is the line $y = x + 1$ with the point $(1, 2)$ *removed*. This removed point is shown as a "hole" in Figure 2.7. Even though $f(1)$ is not defined, it is clear that we can make the value of $f(x)$ *as close as we want* to 2 by choosing x *close enough* to 1 (Table 2.2). ∎

TABLE 2.2 The closer x gets to 1, the closer $f(x) = (x^2 - 1)/(x - 1)$ seems to get to 2

Values of x below and above 1	$f(x) = \dfrac{x^2 - 1}{x - 1} = x + 1, \quad x \neq 1$
0.9	1.9
1.1	2.1
0.99	1.99
1.01	2.01
0.999	1.999
1.001	2.001
0.999999	1.999999
1.000001	2.000001

Let's generalize the idea illustrated in Example 1.

Suppose $f(x)$ is defined on an open interval about x_0, *except possibly at x_0 itself*. If $f(x)$ is arbitrarily close to L (as close to L as we like) for all x sufficiently close to x_0, we say that f approaches the **limit** L as x approaches x_0, and write

$$\lim_{x \to x_0} f(x) = L,$$

which is read "the limit of $f(x)$ as x approaches x_0 is L." For instance, in Example 1 we would say that $f(x)$ approaches the *limit* 2 as x approaches 1, and write

$$\lim_{x \to 1} f(x) = 2, \quad \text{or} \quad \lim_{x \to 1} \frac{x^2 - 1}{x - 1} = 2.$$

Essentially, the definition says that the values of $f(x)$ are close to the number L whenever x is close to x_0 (on either side of x_0). This definition is "informal" because phrases like *arbitrarily close* and *sufficiently close* are imprecise; their meaning depends on the context. (To a machinist manufacturing a piston, *close* may mean *within a few thousandths of an inch*. To an astronomer studying distant galaxies, *close* may mean *within a few thousand light-years*.) Nevertheless, the definition is clear enough to enable us to recognize and evaluate limits of specific functions. We will need the precise definition of Section 2.3, however, when we set out to prove theorems about limits. Here are several more examples exploring the idea of limits.

EXAMPLE 2 This example illustrates that the limit value of a function does not depend on how the function is defined at the point being approached. Consider the three functions in Figure 2.8. The function f has limit 2 as $x \to 1$ even though f is not defined at $x = 1$.

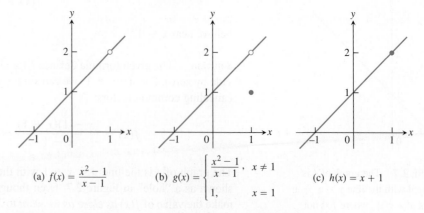

(a) $f(x) = \dfrac{x^2 - 1}{x - 1}$

(b) $g(x) = \begin{cases} \dfrac{x^2 - 1}{x - 1}, & x \neq 1 \\ 1, & x = 1 \end{cases}$

(c) $h(x) = x + 1$

FIGURE 2.8 The limits of $f(x)$, $g(x)$, and $h(x)$ all equal 2 as x approaches 1. However, only $h(x)$ has the same function value as its limit at $x = 1$ (Example 2).

The function g has limit 2 as $x \to 1$ even though $2 \neq g(1)$. The function h is the only one of the three functions in Figure 2.8 whose limit as $x \to 1$ equals its value at $x = 1$. For h, we have $\lim_{x \to 1} h(x) = h(1)$. This equality of limit and function value is significant, and we return to it in Section 2.5.   ■

EXAMPLE 3

(a) If f is the **identity function** $f(x) = x$, then for any value of x_0 (Figure 2.9a),

$$\lim_{x \to x_0} f(x) = \lim_{x \to x_0} x = x_0.$$

(b) If f is the **constant function** $f(x) = k$ (function with the constant value k), then for any value of x_0 (Figure 2.9b),

$$\lim_{x \to x_0} f(x) = \lim_{x \to x_0} k = k.$$

For instances of each of these rules we have

$$\lim_{x \to 3} x = 3 \qquad \text{and} \qquad \lim_{x \to -7}(4) = \lim_{x \to 2}(4) = 4.$$

We prove these rules in Example 3 in Section 2.3.   ■

Some ways that limits can fail to exist are illustrated in Figure 2.10 and described in the next example.

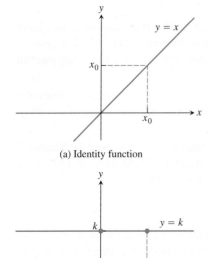

(a) Identity function

(b) Constant function

FIGURE 2.9 The functions in Example 3 have limits at all points x_0.

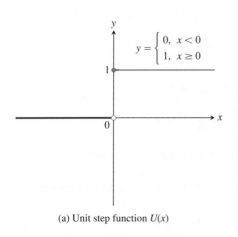

(a) Unit step function $U(x)$

(b) $g(x)$

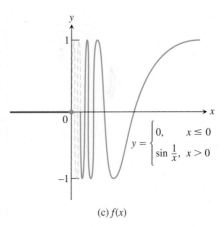

(c) $f(x)$

FIGURE 2.10 None of these functions has a limit as x approaches 0 (Example 4).

EXAMPLE 4 Discuss the behavior of the following functions as $x \to 0$.

(a) $U(x) = \begin{cases} 0, & x < 0 \\ 1, & x \geq 0 \end{cases}$

(b) $g(x) = \begin{cases} \dfrac{1}{x}, & x \neq 0 \\ 0, & x = 0 \end{cases}$

(c) $f(x) = \begin{cases} 0, & x \leq 0 \\ \sin \dfrac{1}{x}, & x > 0 \end{cases}$

Solution

(a) It *jumps*: The **unit step function** $U(x)$ has no limit as $x \to 0$ because its values jump at $x = 0$. For negative values of x arbitrarily close to zero, $U(x) = 0$. For positive values of x arbitrarily close to zero, $U(x) = 1$. There is no *single* value L approached by $U(x)$ as $x \to 0$ (Figure 2.10a).

(b) It *grows too "large" to have a limit*: $g(x)$ has no limit as $x \to 0$ because the values of g grow arbitrarily large in absolute value as $x \to 0$ and do not stay close to *any* fixed real number (Figure 2.10b).

(c) It *oscillates too much to have a limit*: $f(x)$ has no limit as $x \to 0$ because the function's values oscillate between $+1$ and -1 in every open interval containing 0. The values do not stay close to any one number as $x \to 0$ (Figure 2.10c). ∎

The Limit Laws

When discussing limits, sometimes we use the notation $x \to x_0$ if we want to emphasize the point x_0 that is being approached in the limit process (usually to enhance the clarity of a particular discussion or example). Other times, such as in the statements of the following theorem, we use the simpler notation $x \to c$ or $x \to a$ which avoids the subscript in x_0. In every case, the symbols x_0, c, and a refer to a single point on the x-axis that may or may not belong to the domain of the function involved. To calculate limits of functions that are arithmetic combinations of functions having known limits, we can use several easy rules.

THEOREM 1—Limit Laws   If L, M, c, and k are real numbers and

$$\lim_{x \to c} f(x) = L \quad \text{and} \quad \lim_{x \to c} g(x) = M, \quad \text{then}$$

1. *Sum Rule:* $\qquad\qquad\qquad$ $\lim\limits_{x \to c}(f(x) + g(x)) = L + M$

2. *Difference Rule:* $\qquad\qquad$ $\lim\limits_{x \to c}(f(x) - g(x)) = L - M$

3. *Constant Multiple Rule:* $\qquad$ $\lim\limits_{x \to c}(k \cdot f(x)) = k \cdot L$

4. *Product Rule:* $\qquad\qquad\quad$ $\lim\limits_{x \to c}(f(x) \cdot g(x)) = L \cdot M$

5. *Quotient Rule:* $\qquad\qquad\quad$ $\lim\limits_{x \to c} \dfrac{f(x)}{g(x)} = \dfrac{L}{M}, \quad M \neq 0$

6. *Power Rule:* $\qquad\qquad\quad$ $\lim\limits_{x \to c}[f(x)]^n = L^n$, n a positive integer

7. *Root Rule:* $\qquad\qquad\qquad$ $\lim\limits_{x \to c} \sqrt[n]{f(x)} = \sqrt[n]{L} = L^{1/n}$, n a positive integer

(If n is even, we assume that $\lim\limits_{x \to c} f(x) = L > 0$.)

In words, the Sum Rule says that the limit of a sum is the sum of the limits. Similarly, the next rules say that the limit of a difference is the difference of the limits; the limit of a constant times a function is the constant times the limit of the function; the limit of a product is the product of the limits; the limit of a quotient is the quotient of the limits (provided that the limit of the denominator is not 0); the limit of a positive integer power (or root) of a function is the integer power (or root) of the limit (provided that the root of the limit is a real number).

It is reasonable that the properties in Theorem 1 are true (although these intuitive arguments do not constitute proofs). If x is sufficiently close to c, then $f(x)$ is close to L and $g(x)$ is close to M, from our informal definition of a limit. It is then reasonable that $f(x) + g(x)$ is close to $L + M$; $f(x) - g(x)$ is close to $L - M$; $kf(x)$ is close to kL; $f(x)g(x)$ is close to LM; and $f(x)/g(x)$ is close to L/M if M is not zero. We prove the Sum Rule in Section 2.3, based on a precise definition of limit. Rules 2–5 are proved in

Appendix 4. Rule 6 is obtained by applying Rule 4 repeatedly. Rule 7 is proved in more advanced texts. The sum, difference, and product rules can be extended to any number of functions, not just two.

EXAMPLE 5   Use the observations $\lim_{x\to c} k = k$ and $\lim_{x\to c} x = c$ (Example 3) and the properties of limits to find the following limits.

(a) $\displaystyle\lim_{x\to c}(x^3 + 4x^2 - 3)$ **(b)** $\displaystyle\lim_{x\to c}\frac{x^4 + x^2 - 1}{x^2 + 5}$ **(c)** $\displaystyle\lim_{x\to -2}\sqrt{4x^2 - 3}$

Solution

(a) $\displaystyle\lim_{x\to c}(x^3 + 4x^2 - 3) = \lim_{x\to c} x^3 + \lim_{x\to c} 4x^2 - \lim_{x\to c} 3$ Sum and Difference Rules

$$= c^3 + 4c^2 - 3$$ Power and Multiple Rules

(b) $\displaystyle\lim_{x\to c}\frac{x^4 + x^2 - 1}{x^2 + 5} = \frac{\lim_{x\to c}(x^4 + x^2 - 1)}{\lim_{x\to c}(x^2 + 5)}$ Quotient Rule

$$= \frac{\lim_{x\to c} x^4 + \lim_{x\to c} x^2 - \lim_{x\to c} 1}{\lim_{x\to c} x^2 + \lim_{x\to c} 5}$$ Sum and Difference Rules

$$= \frac{c^4 + c^2 - 1}{c^2 + 5}$$ Power or Product Rule

(c) $\displaystyle\lim_{x\to -2}\sqrt{4x^2 - 3} = \sqrt{\lim_{x\to -2}(4x^2 - 3)}$ Root Rule with $n = 2$

$$= \sqrt{\lim_{x\to -2} 4x^2 - \lim_{x\to -2} 3}$$ Difference Rule

$$= \sqrt{4(-2)^2 - 3}$$ Product and Multiple Rules

$$= \sqrt{16 - 3}$$

$$= \sqrt{13}$$ ∎

Two consequences of Theorem 1 further simplify the task of calculating limits of polynomials and rational functions. To evaluate the limit of a polynomial function as x approaches c, merely substitute c for x in the formula for the function. To evaluate the limit of a rational function as x approaches a point c *at which the denominator is not zero*, substitute c for x in the formula for the function. (See Examples 5a and 5b.) We state these results formally as theorems.

THEOREM 2—Limits of Polynomials

If $P(x) = a_n x^n + a_{n-1} x^{n-1} + \cdots + a_0$, then

$$\lim_{x\to c} P(x) = P(c) = a_n c^n + a_{n-1} c^{n-1} + \cdots + a_0.$$

THEOREM 3—Limits of Rational Functions

If $P(x)$ and $Q(x)$ are polynomials and $Q(c) \neq 0$, then

$$\lim_{x\to c}\frac{P(x)}{Q(x)} = \frac{P(c)}{Q(c)}.$$

EXAMPLE 6   The following calculation illustrates Theorems 2 and 3:

$$\lim_{x \to -1} \frac{x^3 + 4x^2 - 3}{x^2 + 5} = \frac{(-1)^3 + 4(-1)^2 - 3}{(-1)^2 + 5} = \frac{0}{6} = 0 \qquad \blacksquare$$

> **Identifying Common Factors**
> It can be shown that if $Q(x)$ is a polynomial and $Q(c) = 0$, then $(x - c)$ is a factor of $Q(x)$. Thus, if the numerator and denominator of a rational function of x are both zero at $x = c$, they have $(x - c)$ as a common factor.

Eliminating Zero Denominators Algebraically

Theorem 3 applies only if the denominator of the rational function is not zero at the limit point c. If the denominator is zero, canceling common factors in the numerator and denominator may reduce the fraction to one whose denominator is no longer zero at c. If this happens, we can find the limit by substitution in the simplified fraction.

EXAMPLE 7   Evaluate

$$\lim_{x \to 1} \frac{x^2 + x - 2}{x^2 - x}.$$

Solution   We cannot substitute $x = 1$ because it makes the denominator zero. We test the numerator to see if it, too, is zero at $x = 1$. It is, so it has a factor of $(x - 1)$ in common with the denominator. Canceling the $(x - 1)$'s gives a simpler fraction with the same values as the original for $x \neq 1$:

$$\frac{x^2 + x - 2}{x^2 - x} = \frac{(x - 1)(x + 2)}{x(x - 1)} = \frac{x + 2}{x}, \qquad \text{if } x \neq 1.$$

Using the simpler fraction, we find the limit of these values as $x \to 1$ by substitution:

$$\lim_{x \to 1} \frac{x^2 + x - 2}{x^2 - x} = \lim_{x \to 1} \frac{x + 2}{x} = \frac{1 + 2}{1} = 3.$$

See Figure 2.11. $\blacksquare$

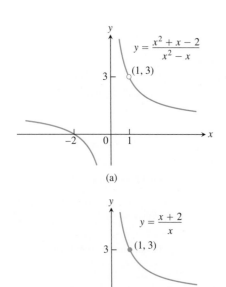

FIGURE 2.11   The graph of $f(x) = (x^2 + x - 2)/(x^2 - x)$ in part (a) is the same as the graph of $g(x) = (x + 2)/x$ in part (b) except at $x = 1$, where f is undefined. The functions have the same limit as $x \to 1$ (Example 7).

Using Calculators and Computers to Estimate Limits

When we cannot use the Quotient Rule in Theorem 1 because the limit of the denominator is zero, we can try using a calculator or computer to guess the limit numerically as x gets closer and closer to c. We used this approach in Example 1, but calculators and computers can sometimes give false values and misleading impressions for functions that are undefined at a point or fail to have a limit there, as we now illustrate.

EXAMPLE 8   Estimate the value of $\displaystyle\lim_{x \to 0} \frac{\sqrt{x^2 + 100} - 10}{x^2}$.

Solution   Table 2.3 lists values of the function for several values near $x = 0$. As x approaches 0 through the values ± 1, ± 0.5, ± 0.10, and ± 0.01, the function seems to approach the number 0.05.

As we take even smaller values of x, ± 0.0005, ± 0.0001, ± 0.00001, and ± 0.000001, the function appears to approach the value 0.

Is the answer 0.05 or 0, or some other value? We resolve this question in the next example. $\blacksquare$

TABLE 2.3 Computer values of $f(x) = \dfrac{\sqrt{x^2 + 100} - 10}{x^2}$ near $x = 0$

x	$f(x)$	
± 1	0.049876	
± 0.5	0.049969	approaches 0.05?
± 0.1	0.049999	
± 0.01	0.050000	
± 0.0005	0.050000	
± 0.0001	0.000000	approaches 0?
± 0.00001	0.000000	
± 0.000001	0.000000	

Using a computer or calculator may give ambiguous results, as in the last example. We cannot substitute $x = 0$ in the problem, and the numerator and denominator have no obvious common factors (as they did in Example 7). Sometimes, however, we can create a common factor algebraically.

EXAMPLE 9   Evaluate

$$\lim_{x \to 0} \frac{\sqrt{x^2 + 100} - 10}{x^2}.$$

Solution   This is the limit we considered in Example 8. We can create a common factor by multiplying both numerator and denominator by the conjugate radical expression $\sqrt{x^2 + 100} + 10$ (obtained by changing the sign after the square root). The preliminary algebra rationalizes the numerator:

$$\frac{\sqrt{x^2 + 100} - 10}{x^2} = \frac{\sqrt{x^2 + 100} - 10}{x^2} \cdot \frac{\sqrt{x^2 + 100} + 10}{\sqrt{x^2 + 100} + 10}$$

$$= \frac{x^2 + 100 - 100}{x^2 \left(\sqrt{x^2 + 100} + 10 \right)}$$

$$= \frac{x^2}{x^2 \left(\sqrt{x^2 + 100} + 10 \right)} \qquad \text{Common factor } x^2$$

$$= \frac{1}{\sqrt{x^2 + 100} + 10}. \qquad \text{Cancel } x^2 \text{ for } x \neq 0$$

Therefore,

$$\lim_{x \to 0} \frac{\sqrt{x^2 + 100} - 10}{x^2} = \lim_{x \to 0} \frac{1}{\sqrt{x^2 + 100} + 10}$$

$$= \frac{1}{\sqrt{0^2 + 100} + 10} \qquad \begin{array}{l} \text{Denominator not 0 at} \\ x = 0; \text{ substitute} \end{array}$$

$$= \frac{1}{20} = 0.05.$$

This calculation provides the correct answer, in contrast to the ambiguous computer results in Example 8.   ∎

We cannot always algebraically resolve the problem of finding the limit of a quotient where the denominator becomes zero. In some cases the limit might then be found with the

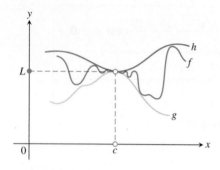

FIGURE 2.12 The graph of f is sandwiched between the graphs of g and h.

aid of some geometry applied to the problem (see the proof of Theorem 7 in Section 2.4), or through methods of calculus (illustrated in Section 7.5). The next theorem is also useful.

The Sandwich Theorem

The following theorem enables us to calculate a variety of limits. It is called the Sandwich Theorem because it refers to a function f whose values are sandwiched between the values of two other functions g and h that have the same limit L at a point c. Being trapped between the values of two functions that approach L, the values of f must also approach L (Figure 2.12). You will find a proof in Appendix 4.

THEOREM 4—The Sandwich Theorem   Suppose that $g(x) \leq f(x) \leq h(x)$ for all x in some open interval containing c, except possibly at $x = c$ itself. Suppose also that

$$\lim_{x \to c} g(x) = \lim_{x \to c} h(x) = L.$$

Then $\lim_{x \to c} f(x) = L$.

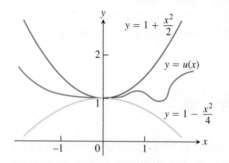

FIGURE 2.13 Any function $u(x)$ whose graph lies in the region between $y = 1 + (x^2/2)$ and $y = 1 - (x^2/4)$ has limit 1 as $x \to 0$ (Example 10).

The Sandwich Theorem is also called the Squeeze Theorem or the Pinching Theorem.

EXAMPLE 10   Given that

$$1 - \frac{x^2}{4} \leq u(x) \leq 1 + \frac{x^2}{2} \qquad \text{for all } x \neq 0,$$

find $\lim_{x \to 0} u(x)$, no matter how complicated u is.

Solution   Since

$$\lim_{x \to 0}(1 - (x^2/4)) = 1 \qquad \text{and} \qquad \lim_{x \to 0}(1 + (x^2/2)) = 1,$$

the Sandwich Theorem implies that $\lim_{x \to 0} u(x) = 1$ (Figure 2.13). ∎

EXAMPLE 11   The Sandwich Theorem helps us establish several important limit rules:

(a) $\displaystyle \lim_{\theta \to 0} \sin \theta = 0$

(b) $\displaystyle \lim_{\theta \to 0} \cos \theta = 1$

(c) For any function f, $\displaystyle \lim_{x \to c} |f(x)| = 0$ implies $\displaystyle \lim_{x \to c} f(x) = 0$.

Solution

(a) In Section 1.3 we established that $-|\theta| \leq \sin \theta \leq |\theta|$ for all θ (see Figure 2.14a). Since $\lim_{\theta \to 0}(-|\theta|) = \lim_{\theta \to 0}|\theta| = 0$, we have

$$\lim_{\theta \to 0} \sin \theta = 0.$$

(a)

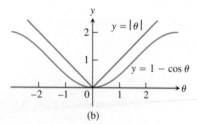

(b)

FIGURE 2.14 The Sandwich Theorem confirms the limits in Example 11.

(b) From Section 1.3, $0 \leq 1 - \cos \theta \leq |\theta|$ for all θ (see Figure 2.14b), and we have $\lim_{\theta \to 0}(1 - \cos \theta) = 0$ or

$$\lim_{\theta \to 0} \cos \theta = 1.$$

(c) Since $-|f(x)| \leq f(x) \leq |f(x)|$ and $-|f(x)|$ and $|f(x)|$ have limit 0 as $x \to c$, it follows that $\lim_{x \to c} f(x) = 0$. ∎

Another important property of limits is given by the next theorem. A proof is given in the next section.

THEOREM 5   If $f(x) \le g(x)$ for all x in some open interval containing c, except possibly at $x = c$ itself, and the limits of f and g both exist as x approaches c, then

$$\lim_{x \to c} f(x) \le \lim_{x \to c} g(x).$$

The assertion resulting from replacing the less than or equal to ($\le$) inequality by the strict less than ($<$) inequality in Theorem 5 is false. Figure 2.14a shows that for $\theta \ne 0$, $-|\theta| < \sin \theta < |\theta|$, but in the limit as $\theta \to 0$, equality holds.

Exercises 2.2

Limits from Graphs

1. For the function $g(x)$ graphed here, find the following limits or explain why they do not exist.

   a. $\lim_{x \to 1} g(x)$ **b.** $\lim_{x \to 2} g(x)$ **c.** $\lim_{x \to 3} g(x)$ **d.** $\lim_{x \to 2.5} g(x)$

   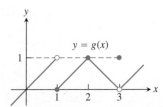

2. For the function $f(t)$ graphed here, find the following limits or explain why they do not exist.

   a. $\lim_{t \to -2} f(t)$ **b.** $\lim_{t \to -1} f(t)$ **c.** $\lim_{t \to 0} f(t)$ **d.** $\lim_{t \to -0.5} f(t)$

   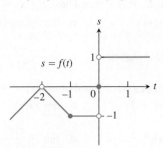

3. Which of the following statements about the function $y = f(x)$ graphed here are true, and which are false?

   a. $\lim_{x \to 0} f(x)$ exists.

   b. $\lim_{x \to 0} f(x) = 0$

   c. $\lim_{x \to 0} f(x) = 1$

   d. $\lim_{x \to 1} f(x) = 1$

   e. $\lim_{x \to 1} f(x) = 0$

   f. $\lim_{x \to x_0} f(x)$ exists at every point x_0 in $(-1, 1)$.

   g. $\lim_{x \to 1} f(x)$ does not exist.

   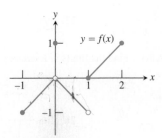

4. Which of the following statements about the function $y = f(x)$ graphed here are true, and which are false?

   a. $\lim_{x \to 2} f(x)$ does not exist.

   b. $\lim_{x \to 2} f(x) = 2$

   c. $\lim_{x \to 1} f(x)$ does not exist.

   d. $\lim_{x \to x_0} f(x)$ exists at every point x_0 in $(-1, 1)$.

   e. $\lim_{x \to x_0} f(x)$ exists at every point x_0 in $(1, 3)$.

Existence of Limits

In Exercises 5 and 6, explain why the limits do not exist.

5. $\lim_{x \to 0} \dfrac{x}{|x|}$ 6. $\lim_{x \to 1} \dfrac{1}{x - 1}$

7. Suppose that a function $f(x)$ is defined for all real values of x except $x = x_0$. Can anything be said about the existence of $\lim_{x \to x_0} f(x)$? Give reasons for your answer.

8. Suppose that a function $f(x)$ is defined for all x in $[-1, 1]$. Can anything be said about the existence of $\lim_{x \to 0} f(x)$? Give reasons for your answer.

9. If $\lim_{x \to 1} f(x) = 5$, must f be defined at $x = 1$? If it is, must $f(1) = 5$? Can we conclude *anything* about the values of f at $x = 1$? Explain.

10. If $f(1) = 5$, must $\lim_{x \to 1} f(x)$ exist? If it does, then must $\lim_{x \to 1} f(x) = 5$? Can we conclude *anything* about $\lim_{x \to 1} f(x)$? Explain.

Calculating Limits

Find the limits in Exercises 11–22.

11. $\lim_{x \to -7} (2x + 5)$

12. $\lim_{x \to 2} (-x^2 + 5x - 2)$

13. $\lim_{t \to 6} 8(t - 5)(t - 7)$

14. $\lim_{x \to -2} (x^3 - 2x^2 + 4x + 8)$

15. $\lim_{x \to 2} \dfrac{x + 3}{x + 6}$

16. $\lim_{s \to 2/3} 3s(2s - 1)$

17. $\lim_{x \to -1} 3(2x - 1)^2$

18. $\lim_{y \to 2} \dfrac{y + 2}{y^2 + 5y + 6}$

19. $\lim_{y \to -3} (5 - y)^{4/3}$

20. $\lim_{z \to 0} (2z - 8)^{1/3}$

21. $\lim_{h \to 0} \dfrac{3}{\sqrt{3h + 1} + 1}$

22. $\lim_{h \to 0} \dfrac{\sqrt{5h + 4} - 2}{h}$

Limits of quotients   Find the limits in Exercises 23–42.

23. $\lim_{x \to 5} \dfrac{x - 5}{x^2 - 25}$

24. $\lim_{x \to -3} \dfrac{x + 3}{x^2 + 4x + 3}$

25. $\lim_{x \to -5} \dfrac{x^2 + 3x - 10}{x + 5}$

26. $\lim_{x \to 2} \dfrac{x^2 - 7x + 10}{x - 2}$

27. $\lim_{t \to 1} \dfrac{t^2 + t - 2}{t^2 - 1}$

28. $\lim_{t \to -1} \dfrac{t^2 + 3t + 2}{t^2 - t - 2}$

29. $\lim_{x \to -2} \dfrac{-2x - 4}{x^3 + 2x^2}$

30. $\lim_{y \to 0} \dfrac{5y^3 + 8y^2}{3y^4 - 16y^2}$

31. $\lim_{x \to 1} \dfrac{\frac{1}{x} - 1}{x - 1}$

32. $\lim_{x \to 0} \dfrac{\frac{1}{x - 1} + \frac{1}{x + 1}}{x}$

33. $\lim_{u \to 1} \dfrac{u^4 - 1}{u^3 - 1}$

34. $\lim_{v \to 2} \dfrac{v^3 - 8}{v^4 - 16}$

35. $\lim_{x \to 9} \dfrac{\sqrt{x} - 3}{x - 9}$

36. $\lim_{x \to 4} \dfrac{4x - x^2}{2 - \sqrt{x}}$

37. $\lim_{x \to 1} \dfrac{x - 1}{\sqrt{x + 3} - 2}$

38. $\lim_{x \to -1} \dfrac{\sqrt{x^2 + 8} - 3}{x + 1}$

39. $\lim_{x \to 2} \dfrac{\sqrt{x^2 + 12} - 4}{x - 2}$

40. $\lim_{x \to -2} \dfrac{x + 2}{\sqrt{x^2 + 5} - 3}$

41. $\lim_{x \to -3} \dfrac{2 - \sqrt{x^2 - 5}}{x + 3}$

42. $\lim_{x \to 4} \dfrac{4 - x}{5 - \sqrt{x^2 + 9}}$

Limits with trigonometric functions   Find the limits in Exercises 43–50.

43. $\lim_{x \to 0} (2 \sin x - 1)$

44. $\lim_{x \to 0} \sin^2 x$

45. $\lim_{x \to 0} \sec x$

46. $\lim_{x \to 0} \tan x$

47. $\lim_{x \to 0} \dfrac{1 + x + \sin x}{3 \cos x}$

48. $\lim_{x \to 0} (x^2 - 1)(2 - \cos x)$

49. $\lim_{x \to -\pi} \sqrt{x + 4} \cos (x + \pi)$ **50.** $\lim_{x \to 0} \sqrt{7 + \sec^2 x}$

Using Limit Rules

51. Suppose $\lim_{x \to 0} f(x) = 1$ and $\lim_{x \to 0} g(x) = -5$. Name the rules in Theorem 1 that are used to accomplish steps (a), (b), and (c) of the following calculation.

$$\lim_{x \to 0} \frac{2f(x) - g(x)}{(f(x) + 7)^{2/3}} = \frac{\lim_{x \to 0} (2f(x) - g(x))}{\lim_{x \to 0} (f(x) + 7)^{2/3}} \quad \text{(a)}$$

$$= \frac{\lim_{x \to 0} 2f(x) - \lim_{x \to 0} g(x)}{\left(\lim_{x \to 0} (f(x) + 7)\right)^{2/3}} \quad \text{(b)}$$

$$= \frac{2 \lim_{x \to 0} f(x) - \lim_{x \to 0} g(x)}{\left(\lim_{x \to 0} f(x) + \lim_{x \to 0} 7\right)^{2/3}} \quad \text{(c)}$$

$$= \frac{(2)(1) - (-5)}{(1 + 7)^{2/3}} = \frac{7}{4}$$

52. Let $\lim_{x \to 1} h(x) = 5$, $\lim_{x \to 1} p(x) = 1$, and $\lim_{x \to 1} r(x) = 2$. Name the rules in Theorem 1 that are used to accomplish steps (a), (b), and (c) of the following calculation.

$$\lim_{x \to 1} \frac{\sqrt{5h(x)}}{p(x)(4 - r(x))} = \frac{\lim_{x \to 1} \sqrt{5h(x)}}{\lim_{x \to 1} (p(x)(4 - r(x)))} \quad \text{(a)}$$

$$= \frac{\sqrt{\lim_{x \to 1} 5h(x)}}{\left(\lim_{x \to 1} p(x)\right)\left(\lim_{x \to 1} (4 - r(x))\right)} \quad \text{(b)}$$

$$= \frac{\sqrt{5 \lim_{x \to 1} h(x)}}{\left(\lim_{x \to 1} p(x)\right)\left(\lim_{x \to 1} 4 - \lim_{x \to 1} r(x)\right)} \quad \text{(c)}$$

$$= \frac{\sqrt{(5)(5)}}{(1)(4 - 2)} = \frac{5}{2}$$

53. Suppose $\lim_{x \to c} f(x) = 5$ and $\lim_{x \to c} g(x) = -2$. Find

   a. $\lim_{x \to c} f(x)g(x)$ **b.** $\lim_{x \to c} 2f(x)g(x)$

   c. $\lim_{x \to c} (f(x) + 3g(x))$ **d.** $\lim_{x \to c} \dfrac{f(x)}{f(x) - g(x)}$

54. Suppose $\lim_{x \to 4} f(x) = 0$ and $\lim_{x \to 4} g(x) = -3$. Find

   a. $\lim_{x \to 4} (g(x) + 3)$ **b.** $\lim_{x \to 4} xf(x)$

   c. $\lim_{x \to 4} (g(x))^2$ **d.** $\lim_{x \to 4} \dfrac{g(x)}{f(x) - 1}$

55. Suppose $\lim_{x \to b} f(x) = 7$ and $\lim_{x \to b} g(x) = -3$. Find

   a. $\lim_{x \to b} (f(x) + g(x))$ **b.** $\lim_{x \to b} f(x) \cdot g(x)$

   c. $\lim_{x \to b} 4g(x)$ **d.** $\lim_{x \to b} f(x)/g(x)$

56. Suppose that $\lim_{x \to -2} p(x) = 4$, $\lim_{x \to -2} r(x) = 0$, and $\lim_{x \to -2} s(x) = -3$. Find

   a. $\lim_{x \to -2} (p(x) + r(x) + s(x))$

   b. $\lim_{x \to -2} p(x) \cdot r(x) \cdot s(x)$

   c. $\lim_{x \to -2} (-4p(x) + 5r(x))/s(x)$

Limits of Average Rates of Change

Because of their connection with secant lines, tangents, and instantaneous rates, limits of the form

$$\lim_{h \to 0} \frac{f(x + h) - f(x)}{h}$$

occur frequently in calculus. In Exercises 57–62, evaluate this limit for the given value of x and function f.

57. $f(x) = x^2$, $x = 1$

58. $f(x) = x^2$, $x = -2$

59. $f(x) = 3x - 4$, $x = 2$

60. $f(x) = 1/x$, $x = -2$

61. $f(x) = \sqrt{x}$, $x = 7$

62. $f(x) = \sqrt{3x + 1}$, $x = 0$

Using the Sandwich Theorem

63. If $\sqrt{5 - 2x^2} \le f(x) \le \sqrt{5 - x^2}$ for $-1 \le x \le 1$, find $\lim_{x \to 0} f(x)$.

64. If $2 - x^2 \le g(x) \le 2 \cos x$ for all x, find $\lim_{x \to 0} g(x)$.

65. a. It can be shown that the inequalities

$$1 - \frac{x^2}{6} < \frac{x \sin x}{2 - 2 \cos x} < 1$$

hold for all values of x close to zero. What, if anything, does this tell you about

$$\lim_{x \to 0} \frac{x \sin x}{2 - 2 \cos x}?$$

Give reasons for your answer.

T **b.** Graph $y = 1 - (x^2/6)$, $y = (x \sin x)/(2 - 2 \cos x)$, and $y = 1$ together for $-2 \le x \le 2$. Comment on the behavior of the graphs as $x \to 0$.

66. a. Suppose that the inequalities

$$\frac{1}{2} - \frac{x^2}{24} < \frac{1 - \cos x}{x^2} < \frac{1}{2}$$

hold for values of x close to zero. (They do, as you will see in Section 10.9.) What, if anything, does this tell you about

$$\lim_{x \to 0} \frac{1 - \cos x}{x^2}?$$

Give reasons for your answer.

T **b.** Graph the equations $y = (1/2) - (x^2/24)$, $y = (1 - \cos x)/x^2$, and $y = 1/2$ together for $-2 \le x \le 2$. Comment on the behavior of the graphs as $x \to 0$.

Estimating Limits

T You will find a graphing calculator useful for Exercises 67–76.

67. Let $f(x) = (x^2 - 9)/(x + 3)$.

a. Make a table of the values of f at the points $x = -3.1$, -3.01, -3.001, and so on as far as your calculator can go. Then estimate $\lim_{x \to -3} f(x)$. What estimate do you arrive at if you evaluate f at $x = -2.9, -2.99, -2.999, \ldots$ instead?

b. Support your conclusions in part (a) by graphing f near $x_0 = -3$ and using Zoom and Trace to estimate y-values on the graph as $x \to -3$.

c. Find $\lim_{x \to -3} f(x)$ algebraically, as in Example 7.

68. Let $g(x) = (x^2 - 2)/(x - \sqrt{2})$.

a. Make a table of the values of g at the points $x = 1.4, 1.41$, 1.414, and so on through successive decimal approximations of $\sqrt{2}$. Estimate $\lim_{x \to \sqrt{2}} g(x)$.

b. Support your conclusion in part (a) by graphing g near $x_0 = \sqrt{2}$ and using Zoom and Trace to estimate y-values on the graph as $x \to \sqrt{2}$.

c. Find $\lim_{x \to \sqrt{2}} g(x)$ algebraically.

69. Let $G(x) = (x + 6)/(x^2 + 4x - 12)$.

a. Make a table of the values of G at $x = -5.9, -5.99, -5.999$, and so on. Then estimate $\lim_{x \to -6} G(x)$. What estimate do you arrive at if you evaluate G at $x = -6.1, -6.01$, $-6.001, \ldots$ instead?

b. Support your conclusions in part (a) by graphing G and using Zoom and Trace to estimate y-values on the graph as $x \to -6$.

c. Find $\lim_{x \to -6} G(x)$ algebraically.

70. Let $h(x) = (x^2 - 2x - 3)/(x^2 - 4x + 3)$.

a. Make a table of the values of h at $x = 2.9, 2.99, 2.999$, and so on. Then estimate $\lim_{x \to 3} h(x)$. What estimate do you arrive at if you evaluate h at $x = 3.1, 3.01, 3.001, \ldots$ instead?

b. Support your conclusions in part (a) by graphing h near $x_0 = 3$ and using Zoom and Trace to estimate y-values on the graph as $x \to 3$.

c. Find $\lim_{x \to 3} h(x)$ algebraically.

71. Let $f(x) = (x^2 - 1)/(|x| - 1)$.

a. Make tables of the values of f at values of x that approach $x_0 = -1$ from above and below. Then estimate $\lim_{x \to -1} f(x)$.

b. Support your conclusion in part (a) by graphing f near $x_0 = -1$ and using Zoom and Trace to estimate y-values on the graph as $x \to -1$.

c. Find $\lim_{x \to -1} f(x)$ algebraically.

72. Let $F(x) = (x^2 + 3x + 2)/(2 - |x|)$.

a. Make tables of values of F at values of x that approach $x_0 = -2$ from above and below. Then estimate $\lim_{x \to -2} F(x)$.

b. Support your conclusion in part (a) by graphing F near $x_0 = -2$ and using Zoom and Trace to estimate y-values on the graph as $x \to -2$.

c. Find $\lim_{x \to -2} F(x)$ algebraically.

73. Let $g(\theta) = (\sin \theta)/\theta$.

a. Make a table of the values of g at values of θ that approach $\theta_0 = 0$ from above and below. Then estimate $\lim_{\theta \to 0} g(\theta)$.

b. Support your conclusion in part (a) by graphing g near $\theta_0 = 0$.

74. Let $G(t) = (1 - \cos t)/t^2$.

a. Make tables of values of G at values of t that approach $t_0 = 0$ from above and below. Then estimate $\lim_{t \to 0} G(t)$.

b. Support your conclusion in part (a) by graphing G near $t_0 = 0$.

75. Let $f(x) = x^{1/(1-x)}$.

a. Make tables of values of f at values of x that approach $x_0 = 1$ from above and below. Does f appear to have a limit as $x \to 1$? If so, what is it? If not, why not?

b. Support your conclusions in part (a) by graphing f near $x_0 = 1$.

76. Let $f(x) = (3^x - 1)/x$.

  a. Make tables of values of f at values of x that approach $x_0 = 0$ from above and below. Does f appear to have a limit as $x \to 0$? If so, what is it? If not, why not?

  b. Support your conclusions in part (a) by graphing f near $x_0 = 0$.

Theory and Examples

77. If $x^4 \le f(x) \le x^2$ for x in $[-1, 1]$ and $x^2 \le f(x) \le x^4$ for $x < -1$ and $x > 1$, at what points c do you automatically know $\lim_{x \to c} f(x)$? What can you say about the value of the limit at these points?

78. Suppose that $g(x) \le f(x) \le h(x)$ for all $x \ne 2$ and suppose that

$$\lim_{x \to 2} g(x) = \lim_{x \to 2} h(x) = -5.$$

Can we conclude anything about the values of f, g, and h at $x = 2$? Could $f(2) = 0$? Could $\lim_{x \to 2} f(x) = 0$? Give reasons for your answers.

79. If $\lim_{x \to 4} \dfrac{f(x) - 5}{x - 2} = 1$, find $\lim_{x \to 4} f(x)$.

80. If $\lim_{x \to -2} \dfrac{f(x)}{x^2} = 1$, find

  a. $\lim_{x \to -2} f(x)$ **b.** $\lim_{x \to -2} \dfrac{f(x)}{x}$

81. a. If $\lim_{x \to 2} \dfrac{f(x) - 5}{x - 2} = 3$, find $\lim_{x \to 2} f(x)$.

  b. If $\lim_{x \to 2} \dfrac{f(x) - 5}{x - 2} = 4$, find $\lim_{x \to 2} f(x)$.

82. If $\lim_{x \to 0} \dfrac{f(x)}{x^2} = 1$, find

  a. $\lim_{x \to 0} f(x)$ **b.** $\lim_{x \to 0} \dfrac{f(x)}{x}$

T 83. a. Graph $g(x) = x \sin (1/x)$ to estimate $\lim_{x \to 0} g(x)$, zooming in on the origin as necessary.

  b. Confirm your estimate in part (a) with a proof.

T 84. a. Graph $h(x) = x^2 \cos (1/x^3)$ to estimate $\lim_{x \to 0} h(x)$, zooming in on the origin as necessary.

  b. Confirm your estimate in part (a) with a proof.

COMPUTER EXPLORATIONS

Graphical Estimates of Limits

In Exercises 85–90, use a CAS to perform the following steps:

  a. Plot the function near the point x_0 being approached.

  b. From your plot guess the value of the limit.

85. $\lim_{x \to 2} \dfrac{x^4 - 16}{x - 2}$ **86.** $\lim_{x \to -1} \dfrac{x^3 - x^2 - 5x - 3}{(x + 1)^2}$

87. $\lim_{x \to 0} \dfrac{\sqrt[3]{1 + x} - 1}{x}$ **88.** $\lim_{x \to 3} \dfrac{x^2 - 9}{\sqrt{x^2 + 7} - 4}$

89. $\lim_{x \to 0} \dfrac{1 - \cos x}{x \sin x}$ **90.** $\lim_{x \to 0} \dfrac{2x^2}{3 - 3 \cos x}$

2.3

The Precise Definition of a Limit

We now turn our attention to the precise definition of a limit. We replace vague phrases like "gets arbitrarily close to" in the informal definition with specific conditions that can be applied to any particular example. With a precise definition, we can prove the limit properties given in the preceding section and establish many important limits.

To show that the limit of $f(x)$ as $x \to x_0$ equals the number L, we need to show that the gap between $f(x)$ and L can be made "as small as we choose" if x is kept "close enough" to x_0. Let us see what this would require if we specified the size of the gap between $f(x)$ and L.

EXAMPLE 1   Consider the function $y = 2x - 1$ near $x_0 = 4$. Intuitively it appears that y is close to 7 when x is close to 4, so $\lim_{x \to 4}(2x - 1) = 7$. However, how close to $x_0 = 4$ does x have to be so that $y = 2x - 1$ differs from 7 by, say, less than 2 units?

Solution   We are asked: For what values of x is $|y - 7| < 2$? To find the answer we first express $|y - 7|$ in terms of x:

$$|y - 7| = |(2x - 1) - 7| = |2x - 8|.$$

The question then becomes: what values of x satisfy the inequality $|2x - 8| < 2$? To find out, we solve the inequality:

$$|2x - 8| < 2$$
$$-2 < 2x - 8 < 2$$
$$6 < 2x < 10$$
$$3 < x < 5$$
$$-1 < x - 4 < 1.$$

Keeping x within 1 unit of $x_0 = 4$ will keep y within 2 units of $y_0 = 7$ (Figure 2.15). ∎

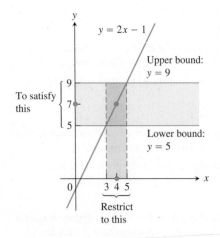

FIGURE 2.15   Keeping x within 1 unit of $x_0 = 4$ will keep y within 2 units of $y_0 = 7$ (Example 1).

In the previous example we determined how close x must be to a particular value x_0 to ensure that the outputs $f(x)$ of some function lie within a prescribed interval about a limit value L. To show that the limit of $f(x)$ as $x \to x_0$ actually equals L, we must be able to show that the gap between $f(x)$ and L can be made less than *any prescribed error*, no matter how small, by holding x close enough to x_0.

Definition of Limit

Suppose we are watching the values of a function $f(x)$ as x approaches x_0 (without taking on the value of x_0 itself). Certainly we want to be able to say that $f(x)$ stays within one-tenth of a unit from L as soon as x stays within some distance δ of x_0 (Figure 2.16). But that in itself is not enough, because as x continues on its course toward x_0, what is to prevent $f(x)$ from jittering about within the interval from $L - (1/10)$ to $L + (1/10)$ without tending toward L?

We can be told that the error can be no more than $1/100$ or $1/1000$ or $1/100,000$. Each time, we find a new δ-interval about x_0 so that keeping x within that interval satisfies the new error tolerance. And each time the possibility exists that $f(x)$ jitters away from L at some stage.

The figures on the next page illustrate the problem. You can think of this as a quarrel between a skeptic and a scholar. The skeptic presents ϵ-challenges to prove that the limit does not exist or, more precisely, that there is room for doubt. The scholar answers every challenge with a δ-interval around x_0 that keeps the function values within ϵ of L.

How do we stop this seemingly endless series of challenges and responses? By proving that for every error tolerance ϵ that the challenger can produce, we can find, calculate, or conjure a matching distance δ that keeps x "close enough" to x_0 to keep $f(x)$ within that tolerance of L (Figure 2.17). This leads us to the precise definition of a limit.

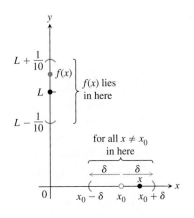

FIGURE 2.16 How should we define $\delta > 0$ so that keeping x within the interval $(x_0 - \delta, x_0 + \delta)$ will keep $f(x)$ within the interval $\left(L - \dfrac{1}{10}, L + \dfrac{1}{10}\right)$?

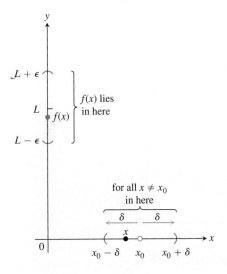

FIGURE 2.17 The relation of δ and ϵ in the definition of limit.

DEFINITION   Let $f(x)$ be defined on an open interval about x_0, except possibly at x_0 itself. We say that the **limit of $f(x)$ as x approaches x_0 is the number L**, and write
$$\lim_{x \to x_0} f(x) = L,$$
if, for every number $\epsilon > 0$, there exists a corresponding number $\delta > 0$ such that for all x,
$$0 < |x - x_0| < \delta \quad \Rightarrow \quad |f(x) - L| < \epsilon.$$

One way to think about the definition is to suppose we are machining a generator shaft to a close tolerance. We may try for diameter L, but since nothing is perfect, we must be satisfied with a diameter $f(x)$ somewhere between $L - \epsilon$ and $L + \epsilon$. The δ is the measure of how accurate our control setting for x must be to guarantee this degree of accuracy in the diameter of the shaft. Notice that as the tolerance for error becomes stricter, we may have to adjust δ. That is, the value of δ, how tight our control setting must be, depends on the value of ϵ, the error tolerance.

Examples: Testing the Definition

The formal definition of limit does not tell how to find the limit of a function, but it enables us to verify that a suspected limit is correct. The following examples show how the definition can be used to verify limit statements for specific functions. However, the real purpose of the definition is not to do calculations like this, but rather to prove general theorems so that the calculation of specific limits can be simplified.

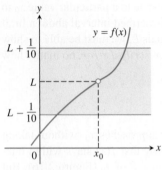

The challenge:
Make $\left| f(x) - L \right| < \epsilon = \dfrac{1}{10}$

Response:
$\left| x - x_0 \right| < \delta_{1/10}$ (a number)

New challenge:
Make $\left| f(x) - L \right| < \epsilon = \dfrac{1}{100}$

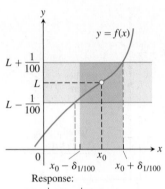

Response:
$\left| x - x_0 \right| < \delta_{1/100}$

New challenge:
$\epsilon = \dfrac{1}{1000}$

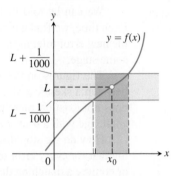

Response:
$\left| x - x_0 \right| < \delta_{1/1000}$

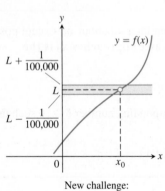

New challenge:
$\epsilon = \dfrac{1}{100,000}$

Response:
$\left| x - x_0 \right| < \delta_{1/100,000}$

New challenge:
$\epsilon = \cdots$

EXAMPLE 2 Show that

$$\lim_{x \to 1} (5x - 3) = 2.$$

Solution Set $x_0 = 1$, $f(x) = 5x - 3$, and $L = 2$ in the definition of limit. For any given $\epsilon > 0$, we have to find a suitable $\delta > 0$ so that if $x \neq 1$ and x is within distance δ of $x_0 = 1$, that is, whenever

$$0 < |x - 1| < \delta,$$

it is true that $f(x)$ is within distance ϵ of $L = 2$, so

$$|f(x) - 2| < \epsilon.$$

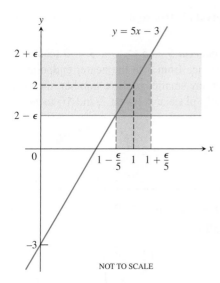

FIGURE 2.18 If $f(x) = 5x - 3$, then $0 < |x - 1| < \epsilon/5$ guarantees that $|f(x) - 2| < \epsilon$ (Example 2).

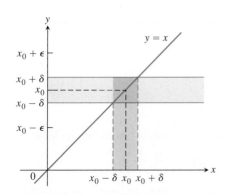

FIGURE 2.19 For the function $f(x) = x$, we find that $0 < |x - x_0| < \delta$ will guarantee $|f(x) - x_0| < \epsilon$ whenever $\delta \le \epsilon$ (Example 3a).

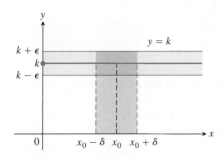

FIGURE 2.20 For the function $f(x) = k$, we find that $|f(x) - k| < \epsilon$ for any positive δ (Example 3b).

We find δ by working backward from the ϵ-inequality:

$$|(5x - 3) - 2| = |5x - 5| < \epsilon$$
$$5|x - 1| < \epsilon$$
$$|x - 1| < \epsilon/5.$$

Thus, we can take $\delta = \epsilon/5$ (Figure 2.18). If $0 < |x - 1| < \delta = \epsilon/5$, then

$$|(5x - 3) - 2| = |5x - 5| = 5|x - 1| < 5(\epsilon/5) = \epsilon,$$

which proves that $\lim_{x \to 1}(5x - 3) = 2$.

The value of $\delta = \epsilon/5$ is not the only value that will make $0 < |x - 1| < \delta$ imply $|5x - 5| < \epsilon$. Any smaller positive δ will do as well. The definition does not ask for a "best" positive δ, just one that will work. ∎

EXAMPLE 3 Prove the following results presented graphically in Section 2.2.

(a) $\lim\limits_{x \to x_0} x = x_0$ **(b)** $\lim\limits_{x \to x_0} k = k$ (k constant)

Solution

(a) Let $\epsilon > 0$ be given. We must find $\delta > 0$ such that for all x

$$0 < |x - x_0| < \delta \qquad \text{implies} \qquad |x - x_0| < \epsilon.$$

The implication will hold if δ equals ϵ or any smaller positive number (Figure 2.19). This proves that $\lim_{x \to x_0} x = x_0$.

(b) Let $\epsilon > 0$ be given. We must find $\delta > 0$ such that for all x

$$0 < |x - x_0| < \delta \qquad \text{implies} \qquad |k - k| < \epsilon.$$

Since $k - k = 0$, we can use any positive number for δ and the implication will hold (Figure 2.20). This proves that $\lim_{x \to x_0} k = k$. ∎

Finding Deltas Algebraically for Given Epsilons

In Examples 2 and 3, the interval of values about x_0 for which $|f(x) - L|$ was less than ϵ was symmetric about x_0 and we could take δ to be half the length of that interval. When such symmetry is absent, as it usually is, we can take δ to be the distance from x_0 to the interval's *nearer* endpoint.

EXAMPLE 4 For the limit $\lim_{x \to 5} \sqrt{x - 1} = 2$, find a $\delta > 0$ that works for $\epsilon = 1$. That is, find a $\delta > 0$ such that for all x

$$0 < |x - 5| < \delta \qquad \Rightarrow \qquad |\sqrt{x - 1} - 2| < 1.$$

Solution We organize the search into two steps, as discussed below.

1. *Solve the inequality* $|\sqrt{x - 1} - 2| < 1$ *to find an interval containing* $x_0 = 5$ *on which the inequality holds for all* $x \neq x_0$.

$$|\sqrt{x - 1} - 2| < 1$$
$$-1 < \sqrt{x - 1} - 2 < 1$$
$$1 < \sqrt{x - 1} < 3$$
$$1 < x - 1 < 9$$
$$2 < x < 10$$

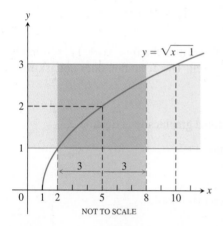

FIGURE 2.21 An open interval of radius 3 about $x_0 = 5$ will lie inside the open interval $(2, 10)$.

FIGURE 2.22 The function and intervals in Example 4.

The inequality holds for all x in the open interval $(2, 10)$, so it holds for all $x \neq 5$ in this interval as well.

2. *Find a value of $\delta > 0$ to place the centered interval $5 - \delta < x < 5 + \delta$ (centered at $x_0 = 5$) inside the interval $(2, 10)$.* The distance from 5 to the nearer endpoint of $(2, 10)$ is 3 (Figure 2.21). If we take $\delta = 3$ or any smaller positive number, then the inequality $0 < |x - 5| < \delta$ will automatically place x between 2 and 10 to make $|\sqrt{x - 1} - 2| < 1$ (Figure 2.22):

$$0 < |x - 5| < 3 \quad \Rightarrow \quad |\sqrt{x - 1} - 2| < 1. \quad \blacksquare$$

How to Find Algebraically a δ for a Given f, L, x_0, and $\epsilon > 0$

The process of finding a $\delta > 0$ such that for all x

$$0 < |x - x_0| < \delta \quad \Rightarrow \quad |f(x) - L| < \epsilon$$

can be accomplished in two steps.

1. *Solve the inequality $|f(x) - L| < \epsilon$ to find an open interval (a, b) containing x_0 on which the inequality holds for all $x \neq x_0$.*

2. *Find a value of $\delta > 0$ that places the open interval $(x_0 - \delta, x_0 + \delta)$ centered at x_0 inside the interval (a, b). The inequality $|f(x) - L| < \epsilon$ will hold for all $x \neq x_0$ in this δ-interval.*

EXAMPLE 5 Prove that $\lim_{x \to 2} f(x) = 4$ if

$$f(x) = \begin{cases} x^2, & x \neq 2 \\ 1, & x = 2. \end{cases}$$

Solution Our task is to show that given $\epsilon > 0$ there exists a $\delta > 0$ such that for all x

$$0 < |x - 2| < \delta \quad \Rightarrow \quad |f(x) - 4| < \epsilon.$$

1. *Solve the inequality $|f(x) - 4| < \epsilon$ to find an open interval containing $x_0 = 2$ on which the inequality holds for all $x \neq x_0$.*

   For $x \neq x_0 = 2$, we have $f(x) = x^2$, and the inequality to solve is $|x^2 - 4| < \epsilon$:

   $$|x^2 - 4| < \epsilon$$
   $$-\epsilon < x^2 - 4 < \epsilon$$
   $$4 - \epsilon < x^2 < 4 + \epsilon$$
   $$\sqrt{4 - \epsilon} < |x| < \sqrt{4 + \epsilon} \qquad \text{Assumes } \epsilon < 4; \text{ see below.}$$
   $$\sqrt{4 - \epsilon} < x < \sqrt{4 + \epsilon}. \qquad \text{An open interval about } x_0 = 2 \text{ that solves the inequality}$$

   The inequality $|f(x) - 4| < \epsilon$ holds for all $x \neq 2$ in the open interval $\left(\sqrt{4 - \epsilon}, \sqrt{4 + \epsilon}\right)$ (Figure 2.23).

2. *Find a value of $\delta > 0$ that places the centered interval $(2 - \delta, 2 + \delta)$ inside the interval $\left(\sqrt{4 - \epsilon}, \sqrt{4 + \epsilon}\right)$.*

FIGURE 2.23 An interval containing $x = 2$ so that the function in Example 5 satisfies $|f(x) - 4| < \epsilon$.

Take δ to be the distance from $x_0 = 2$ to the nearer endpoint of $\left(\sqrt{4 - \epsilon}, \sqrt{4 + \epsilon}\right)$. In other words, take $\delta = \min\left\{2 - \sqrt{4 - \epsilon}, \sqrt{4 + \epsilon} - 2\right\}$, the *minimum* (the smaller) of the two numbers $2 - \sqrt{4 - \epsilon}$ and $\sqrt{4 + \epsilon} - 2$. If δ has this or any smaller positive value, the inequality $0 < |x - 2| < \delta$ will automatically place x between $\sqrt{4 - \epsilon}$ and $\sqrt{4 + \epsilon}$ to make $|f(x) - 4| < \epsilon$. For all x,

$$0 < |x - 2| < \delta \quad \Rightarrow \quad |f(x) - 4| < \epsilon.$$

This completes the proof for $\epsilon < 4$.

If $\epsilon \geq 4$, then we take δ to be the distance from $x_0 = 2$ to the nearer endpoint of the interval $\left(0, \sqrt{4 + \epsilon}\right)$. In other words, take $\delta = \min\left\{2, \sqrt{4 + \epsilon} - 2\right\}$. (See Figure 2.23.) ∎

Using the Definition to Prove Theorems

We do not usually rely on the formal definition of limit to verify specific limits such as those in the preceding examples. Rather we appeal to general theorems about limits, in particular the theorems of Section 2.2. The definition is used to prove these theorems (Appendix 4). As an example, we prove part 1 of Theorem 1, the Sum Rule.

EXAMPLE 6 Given that $\lim_{x \to c} f(x) = L$ and $\lim_{x \to c} g(x) = M$, prove that

$$\lim_{x \to c} (f(x) + g(x)) = L + M.$$

Solution Let $\epsilon > 0$ be given. We want to find a positive number δ such that for all x

$$0 < |x - c| < \delta \quad \Rightarrow \quad |f(x) + g(x) - (L + M)| < \epsilon.$$

Regrouping terms, we get

$$|f(x) + g(x) - (L + M)| = |(f(x) - L) + (g(x) - M)|$$
$$\leq |f(x) - L| + |g(x) - M|.$$

Triangle Inequality:
$|a + b| \leq |a| + |b|$

Since $\lim_{x \to c} f(x) = L$, there exists a number $\delta_1 > 0$ such that for all x

$$0 < |x - c| < \delta_1 \quad \Rightarrow \quad |f(x) - L| < \epsilon/2.$$

Similarly, since $\lim_{x \to c} g(x) = M$, there exists a number $\delta_2 > 0$ such that for all x

$$0 < |x - c| < \delta_2 \quad \Rightarrow \quad |g(x) - M| < \epsilon/2.$$

Let $\delta = \min\{\delta_1, \delta_2\}$, the smaller of δ_1 and δ_2. If $0 < |x - c| < \delta$ then $|x - c| < \delta_1$, so $|f(x) - L| < \epsilon/2$, and $|x - c| < \delta_2$, so $|g(x) - M| < \epsilon/2$. Therefore

$$|f(x) + g(x) - (L + M)| < \frac{\epsilon}{2} + \frac{\epsilon}{2} = \epsilon.$$

This shows that $\lim_{x \to c} (f(x) + g(x)) = L + M$. ∎

Next we prove Theorem 5 of Section 2.2.

EXAMPLE 7 Given that $\lim_{x \to c} f(x) = L$ and $\lim_{x \to c} g(x) = M$, and that $f(x) \leq g(x)$ for all x in an open interval containing c (except possibly c itself), prove that $L \leq M$.

Solution We use the method of proof by contradiction. Suppose, on the contrary, that $L > M$. Then by the limit of a difference property in Theorem 1,

$$\lim_{x \to c} (g(x) - f(x)) = M - L.$$

Therefore, for any $\epsilon > 0$, there exists $\delta > 0$ such that

$$|(g(x) - f(x)) - (M - L)| < \epsilon \qquad \text{whenever} \quad 0 < |x - c| < \delta.$$

Since $L - M > 0$ by hypothesis, we take $\epsilon = L - M$ in particular and we have a number $\delta > 0$ such that

$$|(g(x) - f(x)) - (M - L)| < L - M \qquad \text{whenever} \quad 0 < |x - c| < \delta.$$

Since $a \le |a|$ for any number a, we have

$$(g(x) - f(x)) - (M - L) < L - M \qquad \text{whenever} \quad 0 < |x - c| < \delta$$

which simplifies to

$$g(x) < f(x) \qquad \text{whenever} \quad 0 < |x - c| < \delta.$$

But this contradicts $f(x) \le g(x)$. Thus the inequality $L > M$ must be false. Therefore $L \le M$. ∎

Exercises 2.3

Centering Intervals About a Point

In Exercises 1–6, sketch the interval (a, b) on the x-axis with the point x_0 inside. Then find a value of $\delta > 0$ such that for all $x, 0 < |x - x_0| < \delta \;\Rightarrow\; a < x < b$.

1. $a = 1, \quad b = 7, \quad x_0 = 5$

2. $a = 1, \quad b = 7, \quad x_0 = 2$

3. $a = -7/2, \quad b = -1/2, \quad x_0 = -3$

4. $a = -7/2, \quad b = -1/2, \quad x_0 = -3/2$

5. $a = 4/9, \quad b = 4/7, \quad x_0 = 1/2$

6. $a = 2.7591, \quad b = 3.2391, \quad x_0 = 3$

Finding Deltas Graphically

In Exercises 7–14, use the graphs to find a $\delta > 0$ such that for all x
$$0 < |x - x_0| < \delta \;\Rightarrow\; |f(x) - L| < \epsilon.$$

7.

NOT TO SCALE

8.

NOT TO SCALE

9.

10.

NOT TO SCALE

11.

NOT TO SCALE

12.

NOT TO SCALE

58. Let $h(x) = \begin{cases} x^2, & x < 2 \\ 3, & x = 2 \\ 2, & x > 2. \end{cases}$

Show that

a. $\lim_{x \to 2} h(x) \neq 4$

b. $\lim_{x \to 2} h(x) \neq 3$

c. $\lim_{x \to 2} h(x) \neq 2$

59. For the function graphed here, explain why

a. $\lim_{x \to 3} f(x) \neq 4$

b. $\lim_{x \to 3} f(x) \neq 4.8$

c. $\lim_{x \to 3} f(x) \neq 3$

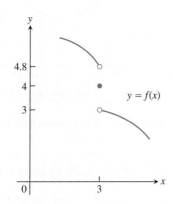

60. a. For the function graphed here, show that $\lim_{x \to -1} g(x) \neq 2$.

b. Does $\lim_{x \to -1} g(x)$ appear to exist? If so, what is the value of the limit? If not, why not?

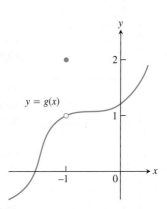

COMPUTER EXPLORATIONS

In Exercises 61–66, you will further explore finding deltas graphically. Use a CAS to perform the following steps:

a. Plot the function $y = f(x)$ near the point x_0 being approached.

b. Guess the value of the limit L and then evaluate the limit symbolically to see if you guessed correctly.

c. Using the value $\epsilon = 0.2$, graph the banding lines $y_1 = L - \epsilon$ and $y_2 = L + \epsilon$ together with the function f near x_0.

d. From your graph in part (c), estimate a $\delta > 0$ such that for all x

$$0 < |x - x_0| < \delta \quad \Rightarrow \quad |f(x) - L| < \epsilon.$$

Test your estimate by plotting f, y_1, and y_2 over the interval $0 < |x - x_0| < \delta$. For your viewing window use $x_0 - 2\delta \leq x \leq x_0 + 2\delta$ and $L - 2\epsilon \leq y \leq L + 2\epsilon$. If any function values lie outside the interval $[L - \epsilon, L + \epsilon]$, your choice of δ was too large. Try again with a smaller estimate.

e. Repeat parts (c) and (d) successively for $\epsilon = 0.1, 0.05$, and 0.001.

61. $f(x) = \dfrac{x^4 - 81}{x - 3}, \quad x_0 = 3$

62. $f(x) = \dfrac{5x^3 + 9x^2}{2x^5 + 3x^2}, \quad x_0 = 0$

63. $f(x) = \dfrac{\sin 2x}{3x}, \quad x_0 = 0$

64. $f(x) = \dfrac{x(1 - \cos x)}{x - \sin x}, \quad x_0 = 0$

65. $f(x) = \dfrac{\sqrt[3]{x} - 1}{x - 1}, \quad x_0 = 1$

66. $f(x) = \dfrac{3x^2 - (7x + 1)\sqrt{x} + 5}{x - 1}, \quad x_0 = 1$

2.4 | One-Sided Limits

In this section we extend the limit concept to *one-sided limits*, which are limits as x approaches the number c from the left-hand side (where $x < c$) or the right-hand side ($x > c$) only.

One-Sided Limits

To have a limit L as x approaches c, a function f must be defined on *both sides* of c and its values $f(x)$ must approach L as x approaches c from either side. Because of this, ordinary limits are called **two-sided**.

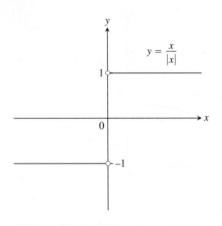

FIGURE 2.24 Different right-hand and left-hand limits at the origin.

If f fails to have a two-sided limit at c, it may still have a one-sided limit, that is, a limit if the approach is only from one side. If the approach is from the right, the limit is a **right-hand limit**. From the left, it is a **left-hand limit**.

The function $f(x) = x/|x|$ (Figure 2.24) has limit 1 as x approaches 0 from the right, and limit -1 as x approaches 0 from the left. Since these one-sided limit values are not the same, there is no single number that $f(x)$ approaches as x approaches 0. So $f(x)$ does not have a (two-sided) limit at 0.

Intuitively, if $f(x)$ is defined on an interval (c, b), where $c < b$, and approaches arbitrarily close to L as x approaches c from within that interval, then f has **right-hand limit** L at c. We write

$$\lim_{x \to c^+} f(x) = L.$$

The symbol "$x \to c^+$" means that we consider only values of x greater than c.

Similarly, if $f(x)$ is defined on an interval (a, c), where $a < c$ and approaches arbitrarily close to M as x approaches c from within that interval, then f has **left-hand limit** M at c. We write

$$\lim_{x \to c^-} f(x) = M.$$

The symbol "$x \to c^-$" means that we consider only x values less than c.

These informal definitions of one-sided limits are illustrated in Figure 2.25. For the function $f(x) = x/|x|$ in Figure 2.24 we have

$$\lim_{x \to 0^+} f(x) = 1 \qquad \text{and} \qquad \lim_{x \to 0^-} f(x) = -1.$$

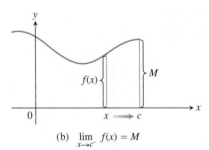

(a) $\lim_{x \to c^+} f(x) = L$ (b) $\lim_{x \to c^-} f(x) = M$

FIGURE 2.25 (a) Right-hand limit as x approaches c. (b) Left-hand limit as x approaches c.

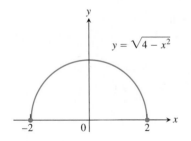

FIGURE 2.26 $\lim_{x \to 2^-} \sqrt{4 - x^2} = 0$ and $\lim_{x \to -2^+} \sqrt{4 - x^2} = 0$ (Example 1).

EXAMPLE 1 The domain of $f(x) = \sqrt{4 - x^2}$ is $[-2, 2]$; its graph is the semicircle in Figure 2.26. We have

$$\lim_{x \to -2^+} \sqrt{4 - x^2} = 0 \qquad \text{and} \qquad \lim_{x \to 2^-} \sqrt{4 - x^2} = 0.$$

The function does not have a left-hand limit at $x = -2$ or a right-hand limit at $x = 2$. It does not have ordinary two-sided limits at either -2 or 2. ∎

One-sided limits have all the properties listed in Theorem 1 in Section 2.2. The right-hand limit of the sum of two functions is the sum of their right-hand limits, and so on. The theorems for limits of polynomials and rational functions hold with one-sided limits, as do the Sandwich Theorem and Theorem 5. One-sided limits are related to limits in the following way.

> **THEOREM 6** A function $f(x)$ has a limit as x approaches c if and only if it has left-hand and right-hand limits there and these one-sided limits are equal:
>
> $$\lim_{x \to c} f(x) = L \quad \Leftrightarrow \quad \lim_{x \to c^-} f(x) = L \quad \text{and} \quad \lim_{x \to c^+} f(x) = L.$$

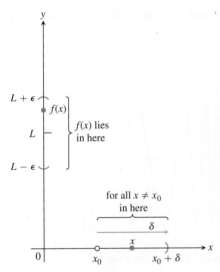

FIGURE 2.28 Intervals associated with the definition of right-hand limit.

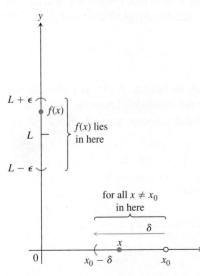

FIGURE 2.29 Intervals associated with the definition of left-hand limit.

EXAMPLE 2 For the function graphed in Figure 2.27,

At $x = 0$: $\lim_{x \to 0^+} f(x) = 1$,

$\lim_{x \to 0^-} f(x)$ and $\lim_{x \to 0} f(x)$ do not exist. The function is not defined to the left of $x = 0$.

At $x = 1$: $\lim_{x \to 1^-} f(x) = 0$ even though $f(1) = 1$,

$\lim_{x \to 1^+} f(x) = 1$,

$\lim_{x \to 1} f(x)$ does not exist. The right- and left-hand limits are not equal.

At $x = 2$: $\lim_{x \to 2^-} f(x) = 1$,

$\lim_{x \to 2^+} f(x) = 1$,

$\lim_{x \to 2} f(x) = 1$ even though $f(2) = 2$.

At $x = 3$: $\lim_{x \to 3^-} f(x) = \lim_{x \to 3^+} f(x) = \lim_{x \to 3} f(x) = f(3) = 2$.

At $x = 4$: $\lim_{x \to 4^-} f(x) = 1$ even though $f(4) \neq 1$,

$\lim_{x \to 4^+} f(x)$ and $\lim_{x \to 4} f(x)$ do not exist. The function is not defined to the right of $x = 4$.

At every other point c in $[0, 4]$, $f(x)$ has limit $f(c)$. ∎

Precise Definitions of One-Sided Limits

The formal definition of the limit in Section 2.3 is readily modified for one-sided limits.

DEFINITIONS We say that $f(x)$ has **right-hand limit L at x_0**, and write

$$\lim_{x \to x_0^+} f(x) = L \qquad \text{(see Figure 2.28)}$$

if for every number $\epsilon > 0$ there exists a corresponding number $\delta > 0$ such that for all x

$$x_0 < x < x_0 + \delta \quad \Rightarrow \quad |f(x) - L| < \epsilon.$$

We say that f has **left-hand limit L at x_0**, and write

$$\lim_{x \to x_0^-} f(x) = L \qquad \text{(see Figure 2.29)}$$

if for every number $\epsilon > 0$ there exists a corresponding number $\delta > 0$ such that for all x

$$x_0 - \delta < x < x_0 \quad \Rightarrow \quad |f(x) - L| < \epsilon.$$

EXAMPLE 3 Prove that

$$\lim_{x \to 0^+} \sqrt{x} = 0.$$

Solution Let $\epsilon > 0$ be given. Here $x_0 = 0$ and $L = 0$, so we want to find a $\delta > 0$ such that for all x

$$0 < x < \delta \quad \Rightarrow \quad |\sqrt{x} - 0| < \epsilon,$$

or

$$0 < x < \delta \quad \Rightarrow \quad \sqrt{x} < \epsilon.$$

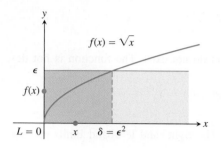

FIGURE 2.30 $\lim\limits_{x \to 0^+} \sqrt{x} = 0$ in Example 3.

Squaring both sides of this last inequality gives

$$x < \epsilon^2 \quad \text{if} \quad 0 < x < \delta.$$

If we choose $\delta = \epsilon^2$ we have

$$0 < x < \delta = \epsilon^2 \quad \Longrightarrow \quad \sqrt{x} < \epsilon,$$

or

$$0 < x < \epsilon^2 \quad \Longrightarrow \quad |\sqrt{x} - 0| < \epsilon.$$

According to the definition, this shows that $\lim_{x \to 0^+}\sqrt{x} = 0$ (Figure 2.30). ∎

The functions examined so far have had some kind of limit at each point of interest. In general, that need not be the case.

EXAMPLE 4 Show that $y = \sin(1/x)$ has no limit as x approaches zero from either side (Figure 2.31).

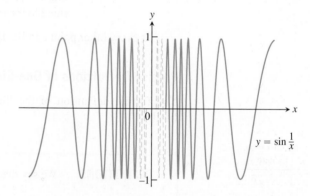

FIGURE 2.31 The function $y = \sin(1/x)$ has neither a right-hand nor a left-hand limit as x approaches zero (Example 4). The graph here omits values very near the y-axis.

Solution As x approaches zero, its reciprocal, $1/x$, grows without bound and the values of $\sin(1/x)$ cycle repeatedly from -1 to 1. There is no single number L that the function's values stay increasingly close to as x approaches zero. This is true even if we restrict x to positive values or to negative values. The function has neither a right-hand limit nor a left-hand limit at $x = 0$. ∎

Limits Involving $(\sin\theta)/\theta$

A central fact about $(\sin\theta)/\theta$ is that in radian measure its limit as $\theta \to 0$ is 1. We can see this in Figure 2.32 and confirm it algebraically using the Sandwich Theorem. You will see the importance of this limit in Section 3.5, where instantaneous rates of change of the trigonometric functions are studied.

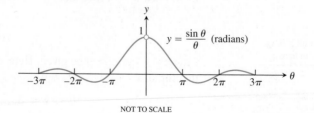

NOT TO SCALE

FIGURE 2.32 The graph of $f(\theta) = (\sin\theta)/\theta$ suggests that the right- and left-hand limits as θ approaches 0 are both 1.

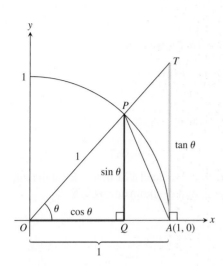

FIGURE 2.33 The figure for the proof of Theorem 7. By definition, $TA/OA = \tan\theta$, but $OA = 1$, so $TA = \tan\theta$.

Equation (2) is where radian measure comes in: The area of sector OAP is $\theta/2$ only if θ is measured in radians.

> **THEOREM 7**
>
> $$\lim_{\theta \to 0} \frac{\sin\theta}{\theta} = 1 \qquad (\theta \text{ in radians}) \qquad (1)$$

Proof The plan is to show that the right-hand and left-hand limits are both 1. Then we will know that the two-sided limit is 1 as well.

To show that the right-hand limit is 1, we begin with positive values of θ less than $\pi/2$ (Figure 2.33). Notice that

$$\text{Area } \Delta OAP < \text{ area sector } OAP < \text{ area } \Delta OAT.$$

We can express these areas in terms of θ as follows:

$$\text{Area } \Delta OAP = \frac{1}{2} \text{ base} \times \text{height} = \frac{1}{2}(1)(\sin\theta) = \frac{1}{2}\sin\theta$$

$$\text{Area sector } OAP = \frac{1}{2}r^2\theta = \frac{1}{2}(1)^2\theta = \frac{\theta}{2} \qquad (2)$$

$$\text{Area } \Delta OAT = \frac{1}{2} \text{ base} \times \text{height} = \frac{1}{2}(1)(\tan\theta) = \frac{1}{2}\tan\theta.$$

Thus,

$$\frac{1}{2}\sin\theta < \frac{1}{2}\theta < \frac{1}{2}\tan\theta.$$

This last inequality goes the same way if we divide all three terms by the number $(1/2)\sin\theta$, which is positive since $0 < \theta < \pi/2$:

$$1 < \frac{\theta}{\sin\theta} < \frac{1}{\cos\theta}.$$

Taking reciprocals reverses the inequalities:

$$1 > \frac{\sin\theta}{\theta} > \cos\theta.$$

Since $\lim_{\theta \to 0^+} \cos\theta = 1$ (Example 11b, Section 2.2), the Sandwich Theorem gives

$$\lim_{\theta \to 0^+} \frac{\sin\theta}{\theta} = 1.$$

Recall that $\sin\theta$ and θ are both *odd functions* (Section 1.1). Therefore, $f(\theta) = (\sin\theta)/\theta$ is an *even function*, with a graph symmetric about the y-axis (see Figure 2.32). This symmetry implies that the left-hand limit at 0 exists and has the same value as the right-hand limit:

$$\lim_{\theta \to 0^-} \frac{\sin\theta}{\theta} = 1 = \lim_{\theta \to 0^+} \frac{\sin\theta}{\theta},$$

so $\lim_{\theta \to 0} (\sin\theta)/\theta = 1$ by Theorem 6. ∎

EXAMPLE 5 Show that **(a)** $\displaystyle\lim_{h \to 0} \frac{\cos h - 1}{h} = 0$ and **(b)** $\displaystyle\lim_{x \to 0} \frac{\sin 2x}{5x} = \frac{2}{5}$.

Solution

(a) Using the half-angle formula $\cos h = 1 - 2\sin^2(h/2)$, we calculate

$$\lim_{h\to 0}\frac{\cos h - 1}{h} = \lim_{h\to 0} -\frac{2\sin^2(h/2)}{h}$$

$$= -\lim_{\theta\to 0}\frac{\sin\theta}{\theta}\sin\theta \qquad \text{Let } \theta = h/2.$$

$$= -(1)(0) = 0. \qquad \begin{array}{l}\text{Eq. (1) and Example 11a}\\\text{in Section 2.2}\end{array}$$

(b) Equation (1) does not apply to the original fraction. We need a $2x$ in the denominator, not a $5x$. We produce it by multiplying numerator and denominator by $2/5$:

$$\lim_{x\to 0}\frac{\sin 2x}{5x} = \lim_{x\to 0}\frac{(2/5)\cdot\sin 2x}{(2/5)\cdot 5x}$$

$$= \frac{2}{5}\lim_{x\to 0}\frac{\sin 2x}{2x} \qquad \begin{array}{l}\text{Now, Eq. (1) applies with}\\\theta = 2x.\end{array}$$

$$= \frac{2}{5}(1) = \frac{2}{5}$$

EXAMPLE 6 Find $\lim\limits_{t\to 0}\dfrac{\tan t \sec 2t}{3t}$.

Solution From the definition of $\tan t$ and $\sec 2t$, we have

$$\lim_{t\to 0}\frac{\tan t \sec 2t}{3t} = \frac{1}{3}\lim_{t\to 0}\frac{\sin t}{t}\cdot\frac{1}{\cos t}\cdot\frac{1}{\cos 2t}$$

$$= \frac{1}{3}(1)(1)(1) = \frac{1}{3}. \qquad \begin{array}{l}\text{Eq. (1) and Example 11b}\\\text{in Section 2.2}\end{array}$$

Exercises 2.4

Finding Limits Graphically

1. Which of the following statements about the function $y = f(x)$ graphed here are true, and which are false?

a. $\lim\limits_{x\to -1^+} f(x) = 1$ **b.** $\lim\limits_{x\to 0^-} f(x) = 0$

c. $\lim\limits_{x\to 0^-} f(x) = 1$ **d.** $\lim\limits_{x\to 0^-} f(x) = \lim\limits_{x\to 0^+} f(x)$

e. $\lim\limits_{x\to 0} f(x)$ exists.   **f.** $\lim\limits_{x\to 0} f(x) = 0$

g. $\lim\limits_{x\to 0} f(x) = 1$ **h.** $\lim\limits_{x\to 1} f(x) = 1$

i. $\lim\limits_{x\to 1} f(x) = 0$ **j.** $\lim\limits_{x\to 2^-} f(x) = 2$

k. $\lim\limits_{x\to -1^-} f(x)$ does not exist.   **l.** $\lim\limits_{x\to 2^+} f(x) = 0$

2. Which of the following statements about the function $y = f(x)$ graphed here are true, and which are false?

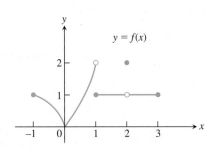

a. $\lim\limits_{x\to -1^+} f(x) = 1$ **b.** $\lim\limits_{x\to 2} f(x)$ does not exist.

c. $\lim\limits_{x\to 2} f(x) = 2$ **d.** $\lim\limits_{x\to 1^-} f(x) = 2$

e. $\lim\limits_{x\to 1^+} f(x) = 1$ **f.** $\lim\limits_{x\to 1} f(x)$ does not exist.

g. $\lim\limits_{x\to 0^+} f(x) = \lim\limits_{x\to 0^-} f(x)$

h. $\lim\limits_{x\to c} f(x)$ exists at every c in the open interval $(-1, 1)$.

i. $\lim\limits_{x\to c} f(x)$ exists at every c in the open interval $(1, 3)$.

j. $\lim\limits_{x\to -1^-} f(x) = 0$ **k.** $\lim\limits_{x\to 3^+} f(x)$ does not exist.

3. Let $f(x) = \begin{cases} 3 - x, & x < 2 \\ \dfrac{x}{2} + 1, & x > 2. \end{cases}$

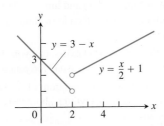

a. Find $\lim_{x \to 2^+} f(x)$ and $\lim_{x \to 2^-} f(x)$.

b. Does $\lim_{x \to 2} f(x)$ exist? If so, what is it? If not, why not?

c. Find $\lim_{x \to 4^-} f(x)$ and $\lim_{x \to 4^+} f(x)$.

d. Does $\lim_{x \to 4} f(x)$ exist? If so, what is it? If not, why not?

4. Let $f(x) = \begin{cases} 3 - x, & x < 2 \\ 2, & x = 2 \\ \dfrac{x}{2}, & x > 2. \end{cases}$

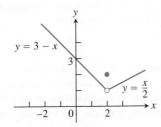

a. Find $\lim_{x \to 2^+} f(x)$, $\lim_{x \to 2^-} f(x)$, and $f(2)$.

b. Does $\lim_{x \to 2} f(x)$ exist? If so, what is it? If not, why not?

c. Find $\lim_{x \to -1^-} f(x)$ and $\lim_{x \to -1^+} f(x)$.

d. Does $\lim_{x \to -1} f(x)$ exist? If so, what is it? If not, why not?

5. Let $f(x) = \begin{cases} 0, & x \leq 0 \\ \sin \dfrac{1}{x}, & x > 0. \end{cases}$

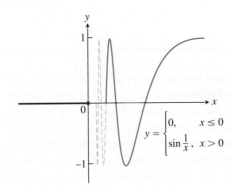

$y = \begin{cases} 0, & x \leq 0 \\ \sin \dfrac{1}{x}, & x > 0 \end{cases}$

a. Does $\lim_{x \to 0^+} f(x)$ exist? If so, what is it? If not, why not?

b. Does $\lim_{x \to 0^-} f(x)$ exist? If so, what is it? If not, why not?

c. Does $\lim_{x \to 0} f(x)$ exist? If so, what is it? If not, why not?

6. Let $g(x) = \sqrt{x} \sin(1/x)$.

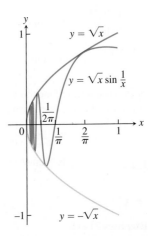

a. Does $\lim_{x \to 0^+} g(x)$ exist? If so, what is it? If not, why not?

b. Does $\lim_{x \to 0^-} g(x)$ exist? If so, what is it? If not, why not?

c. Does $\lim_{x \to 0} g(x)$ exist? If so, what is it? If not, why not?

7. a. Graph $f(x) = \begin{cases} x^3, & x \neq 1 \\ 0, & x = 1. \end{cases}$

b. Find $\lim_{x \to 1^-} f(x)$ and $\lim_{x \to 1^+} f(x)$.

c. Does $\lim_{x \to 1} f(x)$ exist? If so, what is it? If not, why not?

8. a. Graph $f(x) = \begin{cases} 1 - x^2, & x \neq 1 \\ 2, & x = 1. \end{cases}$

b. Find $\lim_{x \to 1^+} f(x)$ and $\lim_{x \to 1^-} f(x)$.

c. Does $\lim_{x \to 1} f(x)$ exist? If so, what is it? If not, why not?

Graph the functions in Exercises 9 and 10. Then answer these questions.

a. What are the domain and range of f?

b. At what points c, if any, does $\lim_{x \to c} f(x)$ exist?

c. At what points does only the left-hand limit exist?

d. At what points does only the right-hand limit exist?

9. $f(x) = \begin{cases} \sqrt{1 - x^2}, & 0 \leq x < 1 \\ 1, & 1 \leq x < 2 \\ 2, & x = 2 \end{cases}$

10. $f(x) = \begin{cases} x, & -1 \leq x < 0, \text{ or } 0 < x \leq 1 \\ 1, & x = 0 \\ 0, & x < -1 \text{ or } x > 1 \end{cases}$

Finding One-Sided Limits Algebraically
Find the limits in Exercises 11–18.

11. $\displaystyle\lim_{x \to -0.5^-} \sqrt{\dfrac{x + 2}{x + 1}}$

12. $\displaystyle\lim_{x \to 1^+} \sqrt{\dfrac{x - 1}{x + 2}}$

13. $\displaystyle\lim_{x \to -2^+} \left(\dfrac{x}{x + 1} \right) \left(\dfrac{2x + 5}{x^2 + x} \right)$

14. $\displaystyle\lim_{x \to 1^-} \left(\dfrac{1}{x + 1} \right) \left(\dfrac{x + 6}{x} \right) \left(\dfrac{3 - x}{7} \right)$

15. $\displaystyle\lim_{h \to 0^+} \dfrac{\sqrt{h^2 + 4h + 5} - \sqrt{5}}{h}$

16. $\lim\limits_{h \to 0^-} \dfrac{\sqrt{6} - \sqrt{5h^2 + 11h + 6}}{h}$

17. a. $\lim\limits_{x \to -2^+} (x + 3) \dfrac{|x + 2|}{x + 2}$ **b.** $\lim\limits_{x \to -2^-} (x + 3) \dfrac{|x + 2|}{x + 2}$

18. a. $\lim\limits_{x \to 1^+} \dfrac{\sqrt{2x}\,(x - 1)}{|x - 1|}$ **b.** $\lim\limits_{x \to 1^-} \dfrac{\sqrt{2x}\,(x - 1)}{|x - 1|}$

Use the graph of the greatest integer function $y = \lfloor x \rfloor$, Figure 1.10 in Section 1.1, to help you find the limits in Exercises 19 and 20.

19. a. $\lim\limits_{\theta \to 3^+} \dfrac{\lfloor \theta \rfloor}{\theta}$ **b.** $\lim\limits_{\theta \to 3^-} \dfrac{\lfloor \theta \rfloor}{\theta}$

20. a. $\lim\limits_{t \to 4^+} (t - \lfloor t \rfloor)$ **b.** $\lim\limits_{t \to 4^-} (t - \lfloor t \rfloor)$

Using $\lim\limits_{\theta \to 0} \dfrac{\sin \theta}{\theta} = 1$

Find the limits in Exercises 21–42.

21. $\lim\limits_{\theta \to 0} \dfrac{\sin \sqrt{2}\theta}{\sqrt{2}\theta}$

22. $\lim\limits_{t \to 0} \dfrac{\sin kt}{t}$ (k constant)

23. $\lim\limits_{y \to 0} \dfrac{\sin 3y}{4y}$

24. $\lim\limits_{h \to 0^-} \dfrac{h}{\sin 3h}$

25. $\lim\limits_{x \to 0} \dfrac{\tan 2x}{x}$

26. $\lim\limits_{t \to 0} \dfrac{2t}{\tan t}$

27. $\lim\limits_{x \to 0} \dfrac{x \csc 2x}{\cos 5x}$

28. $\lim\limits_{x \to 0} 6x^2 (\cot x)(\csc 2x)$

29. $\lim\limits_{x \to 0} \dfrac{x + x \cos x}{\sin x \cos x}$

30. $\lim\limits_{x \to 0} \dfrac{x^2 - x + \sin x}{2x}$

31. $\lim\limits_{\theta \to 0} \dfrac{1 - \cos \theta}{\sin 2\theta}$

32. $\lim\limits_{x \to 0} \dfrac{x - x \cos x}{\sin^2 3x}$

33. $\lim\limits_{t \to 0} \dfrac{\sin (1 - \cos t)}{1 - \cos t}$

34. $\lim\limits_{h \to 0} \dfrac{\sin (\sin h)}{\sin h}$

35. $\lim\limits_{\theta \to 0} \dfrac{\sin \theta}{\sin 2\theta}$

36. $\lim\limits_{x \to 0} \dfrac{\sin 5x}{\sin 4x}$

37. $\lim\limits_{\theta \to 0} \theta \cos \theta$

38. $\lim\limits_{\theta \to 0} \sin \theta \cot 2\theta$

39. $\lim\limits_{x \to 0} \dfrac{\tan 3x}{\sin 8x}$

40. $\lim\limits_{y \to 0} \dfrac{\sin 3y \cot 5y}{y \cot 4y}$

41. $\lim\limits_{\theta \to 0} \dfrac{\tan \theta}{\theta^2 \cot 3\theta}$

42. $\lim\limits_{\theta \to 0} \dfrac{\theta \cot 4\theta}{\sin^2 \theta \cot^2 2\theta}$

Theory and Examples

43. Once you know $\lim_{x \to a^+} f(x)$ and $\lim_{x \to a^-} f(x)$ at an interior point of the domain of f, do you then know $\lim_{x \to a} f(x)$? Give reasons for your answer.

44. If you know that $\lim_{x \to c} f(x)$ exists, can you find its value by calculating $\lim_{x \to c^+} f(x)$? Give reasons for your answer.

45. Suppose that f is an odd function of x. Does knowing that $\lim_{x \to 0^+} f(x) = 3$ tell you anything about $\lim_{x \to 0^-} f(x)$? Give reasons for your answer.

46. Suppose that f is an even function of x. Does knowing that $\lim_{x \to 2^-} f(x) = 7$ tell you anything about either $\lim_{x \to -2^-} f(x)$ or $\lim_{x \to -2^+} f(x)$? Give reasons for your answer.

Formal Definitions of One-Sided Limits

47. Given $\epsilon > 0$, find an interval $I = (5, 5 + \delta)$, $\delta > 0$, such that if x lies in I, then $\sqrt{x - 5} < \epsilon$. What limit is being verified and what is its value?

48. Given $\epsilon > 0$, find an interval $I = (4 - \delta, 4)$, $\delta > 0$, such that if x lies in I, then $\sqrt{4 - x} < \epsilon$. What limit is being verified and what is its value?

Use the definitions of right-hand and left-hand limits to prove the limit statements in Exercises 49 and 50.

49. $\lim\limits_{x \to 0^-} \dfrac{x}{|x|} = -1$ **50.** $\lim\limits_{x \to 2^+} \dfrac{x - 2}{|x - 2|} = 1$

51. Greatest integer function Find **(a)** $\lim_{x \to 400^+} \lfloor x \rfloor$ and **(b)** $\lim_{x \to 400^-} \lfloor x \rfloor$; then use limit definitions to verify your findings. **(c)** Based on your conclusions in parts (a) and (b), can you say anything about $\lim_{x \to 400} \lfloor x \rfloor$? Give reasons for your answer.

52. One-sided limits   Let $f(x) = \begin{cases} x^2 \sin (1/x), & x < 0 \\ \sqrt{x}, & x > 0. \end{cases}$

Find **(a)** $\lim_{x \to 0^+} f(x)$ and **(b)** $\lim_{x \to 0^-} f(x)$; then use limit definitions to verify your findings. **(c)** Based on your conclusions in parts (a) and (b), can you say anything about $\lim_{x \to 0} f(x)$? Give reasons for your answer.

2.5 | Continuity

When we plot function values generated in a laboratory or collected in the field, we often connect the plotted points with an unbroken curve to show what the function's values are likely to have been at the times we did not measure (Figure 2.34). In doing so, we are assuming that we are working with a *continuous function*, so its outputs vary continuously with the inputs and do not jump from one value to another without taking on the values in between. The limit of a continuous function as x approaches c can be found simply by calculating the value of the function at c. (We found this to be true for polynomials in Theorem 2.)

Intuitively, any function $y = f(x)$ whose graph can be sketched over its domain in one continuous motion without lifting the pencil is an example of a continuous function. In this section we investigate more precisely what it means for a function to be continuous.

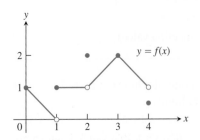

FIGURE 2.34 Connecting plotted points by an unbroken curve from experimental data $Q_1, Q_2, Q_3, \ldots$ for a falling object.

We also study the properties of continuous functions, and see that many of the function types presented in Section 1.1 are continuous.

Continuity at a Point

To understand continuity, it helps to consider a function like that in Figure 2.35, whose limits we investigated in Example 2 in the last section.

EXAMPLE 1   Find the points at which the function f in Figure 2.35 is continuous and the points at which f is not continuous.

Solution   The function f is continuous at every point in its domain $[0, 4]$ except at $x = 1, x = 2$, and $x = 4$. At these points, there are breaks in the graph. Note the relationship between the limit of f and the value of f at each point of the function's domain.

Points at which f is continuous:

At $x = 0$, $\lim\limits_{x \to 0^+} f(x) = f(0)$.

At $x = 3$, $\lim\limits_{x \to 3} f(x) = f(3)$.

At $0 < c < 4, c \neq 1, 2$, $\lim\limits_{x \to c} f(x) = f(c)$.

Points at which f is not continuous:

At $x = 1$, $\lim\limits_{x \to 1} f(x)$ does not exist.

At $x = 2$, $\lim\limits_{x \to 2} f(x) = 1$, but $1 \neq f(2)$.

At $x = 4$, $\lim\limits_{x \to 4^-} f(x) = 1$, but $1 \neq f(4)$.

At $c < 0, c > 4$,   these points are not in the domain of f. ∎

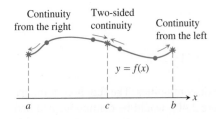

FIGURE 2.35 The function is continuous on $[0, 4]$ except at $x = 1, x = 2$, and $x = 4$ (Example 1).

To define continuity at a point in a function's domain, we need to define continuity at an interior point (which involves a two-sided limit) and continuity at an endpoint (which involves a one-sided limit) (Figure 2.36).

DEFINITION

Interior point: A function $y = f(x)$ is **continuous at an interior point c** of its domain if

$$\lim_{x \to c} f(x) = f(c).$$

Endpoint: A function $y = f(x)$ is **continuous at a left endpoint a** or is **continuous at a right endpoint b** of its domain if

$$\lim_{x \to a^+} f(x) = f(a) \quad \text{or} \quad \lim_{x \to b^-} f(x) = f(b), \quad \text{respectively.}$$

FIGURE 2.36 Continuity at points a, b, and c.

If a function f is not continuous at a point c, we say that f is **discontinuous** at c and that c is a **point of discontinuity** of f. Note that c need not be in the domain of f.

A function f is **right-continuous (continuous from the right)** at a point $x = c$ in its domain if $\lim_{x \to c^+} f(x) = f(c)$. It is **left-continuous (continuous from the left)** at c if $\lim_{x \to c^-} f(x) = f(c)$. Thus, a function is continuous at a left endpoint a of its domain if it

FIGURE 2.37 A function that is continuous at every domain point (Example 2).

FIGURE 2.38 A function that has a jump discontinuity at the origin (Example 3).

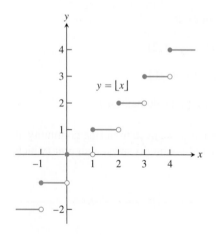

FIGURE 2.39 The greatest integer function is continuous at every noninteger point. It is right-continuous, but not left-continuous, at every integer point (Example 4).

is right-continuous at a and continuous at a right endpoint b of its domain if it is left-continuous at b. A function is continuous at an interior point c of its domain if and only if it is both right-continuous and left-continuous at c (Figure 2.36).

EXAMPLE 2 The function $f(x) = \sqrt{4 - x^2}$ is continuous at every point of its domain $[-2, 2]$ (Figure 2.37), including $x = -2$, where f is right-continuous, and $x = 2$, where f is left-continuous. ∎

EXAMPLE 3 The unit step function $U(x)$, graphed in Figure 2.38, is right-continuous at $x = 0$, but is neither left-continuous nor continuous there. It has a jump discontinuity at $x = 0$. ∎

We summarize continuity at a point in the form of a test.

Continuity Test

A function $f(x)$ is continuous at an interior point $x = c$ of its domain if and only if it meets the following three conditions.

1. $f(c)$ exists   (c lies in the domain of f).
2. $\lim_{x \to c} f(x)$ exists   (f has a limit as $x \to c$).
3. $\lim_{x \to c} f(x) = f(c)$ (the limit equals the function value).

For one-sided continuity and continuity at an endpoint, the limits in parts 2 and 3 of the test should be replaced by the appropriate one-sided limits.

EXAMPLE 4 The function $y = \lfloor x \rfloor$ introduced in Section 1.1 is graphed in Figure 2.39. It is discontinuous at every integer because the left-hand and right-hand limits are not equal as $x \to n$:

$$\lim_{x \to n^-} \lfloor x \rfloor = n - 1 \quad \text{and} \quad \lim_{x \to n^+} \lfloor x \rfloor = n.$$

Since $\lfloor n \rfloor = n$, the greatest integer function is right-continuous at every integer n (but not left-continuous).

The greatest integer function is continuous at every real number other than the integers. For example,

$$\lim_{x \to 1.5} \lfloor x \rfloor = 1 = \lfloor 1.5 \rfloor.$$

In general, if $n - 1 < c < n$, n an integer, then

$$\lim_{x \to c} \lfloor x \rfloor = n - 1 = \lfloor c \rfloor. \qquad ∎$$

Figure 2.40 displays several common types of discontinuities. The function in Figure 2.40a is continuous at $x = 0$. The function in Figure 2.40b would be continuous if it had $f(0) = 1$. The function in Figure 2.40c would be continuous if $f(0)$ were 1 instead of 2. The discontinuities in Figure 2.40b and c are **removable**. Each function has a limit as $x \to 0$, and we can remove the discontinuity by setting $f(0)$ equal to this limit.

The discontinuities in Figure 2.40d through f are more serious: $\lim_{x \to 0} f(x)$ does not exist, and there is no way to improve the situation by changing f at 0. The step function in Figure 2.40d has a **jump discontinuity**: The one-sided limits exist but have different values. The function $f(x) = 1/x^2$ in Figure 2.40e has an **infinite discontinuity**. The function in Figure 2.40f has an **oscillating discontinuity**: It oscillates too much to have a limit as $x \to 0$.

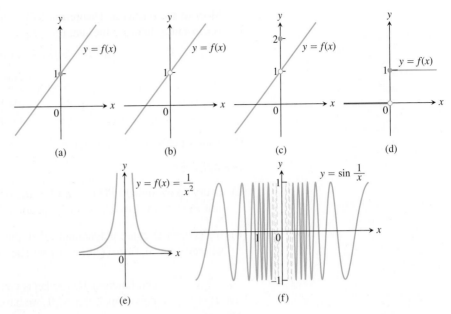

FIGURE 2.40   The function in (a) is continuous at $x = 0$; the functions in (b) through (f) are not.

Continuous Functions

A function is **continuous on an interval** if and only if it is continuous at every point of the interval. For example, the semicircle function graphed in Figure 2.37 is continuous on the interval $[-2, 2]$, which is its domain. A **continuous function** is one that is continuous at every point of its domain. A continuous function need not be continuous on every interval.

EXAMPLE 5

(a) The function $y = 1/x$ (Figure 2.41) is a continuous function because it is continuous at every point of its domain. It has a point of discontinuity at $x = 0$, however, because it is not defined there; that is, it is discontinuous on any interval containing $x = 0$.

(b) The identity function $f(x) = x$ and constant functions are continuous everywhere by Example 3, Section 2.3.   ∎

Algebraic combinations of continuous functions are continuous wherever they are defined.

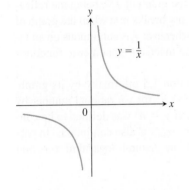

FIGURE 2.41   The function $y = 1/x$ is continuous at every value of x except $x = 0$. It has a point of discontinuity at $x = 0$ (Example 5).

> **THEOREM 8—Properties of Continuous Functions**   If the functions f and g are continuous at $x = c$, then the following combinations are continuous at $x = c$.
>
> 1. *Sums:* $f + g$
> 2. *Differences:* $f - g$
> 3. *Constant multiples:* $k \cdot f$, for any number k
> 4. *Products:* $f \cdot g$
> 5. *Quotients:* f/g, provided $g(c) \neq 0$
> 6. *Powers:* f^n, n a positive integer
> 7. *Roots:* $\sqrt[n]{f}$, provided it is defined on an open interval containing c, where n is a positive integer

Most of the results in Theorem 8 follow from the limit rules in Theorem 1, Section 2.2. For instance, to prove the sum property we have

$$\lim_{x \to c}(f + g)(x) = \lim_{x \to c}(f(x) + g(x))$$
$$= \lim_{x \to c} f(x) + \lim_{x \to c} g(x), \qquad \text{Sum Rule, Theorem 1}$$
$$= f(c) + g(c) \qquad \text{Continuity of } f, g \text{ at } c$$
$$= (f + g)(c).$$

This shows that $f + g$ is continuous.

EXAMPLE 6

(a) Every polynomial $P(x) = a_n x^n + a_{n-1} x^{n-1} + \cdots + a_0$ is continuous because $\lim_{x \to c} P(x) = P(c)$ by Theorem 2, Section 2.2.

(b) If $P(x)$ and $Q(x)$ are polynomials, then the rational function $P(x)/Q(x)$ is continuous wherever it is defined $(Q(c) \neq 0)$ by Theorem 3, Section 2.2. ∎

EXAMPLE 7   The function $f(x) = |x|$ is continuous at every value of x. If $x > 0$, we have $f(x) = x$, a polynomial. If $x < 0$, we have $f(x) = -x$, another polynomial. Finally, at the origin, $\lim_{x \to 0} |x| = 0 = |0|$. ∎

The functions $y = \sin x$ and $y = \cos x$ are continuous at $x = 0$ by Example 11 of Section 2.2. Both functions are, in fact, continuous everywhere (see Exercise 70). It follows from Theorem 8 that all six trigonometric functions are then continuous wherever they are defined. For example, $y = \tan x$ is continuous on $\cdots \cup (-\pi/2, \pi/2) \cup (\pi/2, 3\pi/2) \cup \cdots$.

Inverse Functions and Continuity

The inverse function of any function continuous on an interval is continuous over its domain. This result is suggested from the observation that the graph of f^{-1}, being the reflection of the graph of f across the line $y = x$, cannot have any breaks in it when the graph of f has no breaks. A rigorous proof that f^{-1} is continuous whenever f is continuous on an interval is given in more advanced texts. It follows that the inverse trigonometric functions are all continuous over their domains.

We defined the exponential function $y = a^x$ in Section 1.5 informally by its graph. Recall that the graph was obtained from the graph of $y = a^x$ for x a rational number by filling in the holes at the irrational points x, so the function $y = a^x$ was defined to be continuous over the entire real line. The inverse function $y = \log_a x$ is also continuous. In particular, the natural exponential function $y = e^x$ and the natural logarithm function $y = \ln x$ are both continuous over their domains.

Composites

All composites of continuous functions are continuous. The idea is that if $f(x)$ is continuous at $x = c$ and $g(x)$ is continuous at $x = f(c)$, then $g \circ f$ is continuous at $x = c$ (Figure 2.42). In this case, the limit as $x \to c$ is $g(f(c))$.

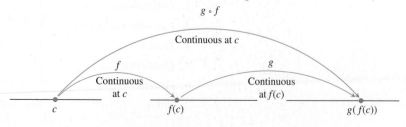

FIGURE 2.42   Composites of continuous functions are continuous.

> **THEOREM 9—Composite of Continuous Functions**   If f is continuous at c and g is continuous at $f(c)$, then the composite $g \circ f$ is continuous at c.

Intuitively, Theorem 9 is reasonable because if x is close to c, then $f(x)$ is close to $f(c)$, and since g is continuous at $f(c)$, it follows that $g(f(x))$ is close to $g(f(c))$.

The continuity of composites holds for any finite number of functions. The only requirement is that each function be continuous where it is applied. For an outline of the proof of Theorem 9, see Exercise 6 in Appendix 4.

EXAMPLE 8   Show that the following functions are continuous everywhere on their respective domains.

(a) $y = \sqrt{x^2 - 2x - 5}$

(b) $y = \dfrac{x^{2/3}}{1 + x^4}$

(c) $y = \left| \dfrac{x - 2}{x^2 - 2} \right|$

(d) $y = \left| \dfrac{x \sin x}{x^2 + 2} \right|$

Solution

(a) The square root function is continuous on $[0, \infty)$ because it is a root of the continuous identity function $f(x) = x$ (Part 7, Theorem 8). The given function is then the composite of the polynomial $f(x) = x^2 - 2x - 5$ with the square root function $g(t) = \sqrt{t}$, and is continuous on its domain.

(b) The numerator is the cube root of the identity function squared; the denominator is an everywhere-positive polynomial. Therefore, the quotient is continuous.

(c) The quotient $(x - 2)/(x^2 - 2)$ is continuous for all $x \neq \pm\sqrt{2}$, and the function is the composition of this quotient with the continuous absolute value function (Example 7).

(d) Because the sine function is everywhere-continuous (Exercise 70), the numerator term $x \sin x$ is the product of continuous functions, and the denominator term $x^2 + 2$ is an everywhere-positive polynomial. The given function is the composite of a quotient of continuous functions with the continuous absolute value function (Figure 2.43). ∎

FIGURE 2.43   The graph suggests that $y = |(x \sin x)/(x^2 + 2)|$ is continuous (Example 8d).

Theorem 9 is actually a consequence of a more general result which we now state and prove.

> **THEOREM 10—Limits of Continuous Functions**   If g is continuous at the point b and $\lim_{x \to c} f(x) = b$, then
> $$\lim_{x \to c} g(f(x)) = g(b) = g(\lim_{x \to c} f(x)).$$

Proof   Let $\epsilon > 0$ be given. Since g is continuous at b, there exists a number $\delta_1 > 0$ such that
$$|g(y) - g(b)| < \epsilon \quad \text{whenever} \quad 0 < |y - b| < \delta_1.$$
Since $\lim_{x \to c} f(x) = b$, there exists a $\delta > 0$ such that
$$|f(x) - b| < \delta_1 \quad \text{whenever} \quad 0 < |x - c| < \delta.$$
If we let $y = f(x)$, we then have that
$$|y - b| < \delta_1 \quad \text{whenever} \quad 0 < |x - c| < \delta,$$
which implies from the first statement that $|g(y) - g(b)| = |g(f(x)) - g(b)| < \epsilon$ whenever $0 < |x - c| < \delta$. From the definition of limit, this proves that $\lim_{x \to c} g(f(x)) = g(b)$. ∎

EXAMPLE 9   As an application of Theorem 10, we have the following calculations.

(a) $\displaystyle\lim_{x\to\pi/2}\cos\left(2x + \sin\left(\frac{3\pi}{2} + x\right)\right) = \cos\left(\lim_{x\to\pi/2}2x + \lim_{x\to\pi/2}\sin\left(\frac{3\pi}{2} + x\right)\right)$

$$= \cos(\pi + \sin 2\pi) = \cos\pi = -1.$$

(b) $\displaystyle\lim_{x\to1}\sin^{-1}\left(\frac{1-x}{1-x^2}\right) = \sin^{-1}\left(\lim_{x\to1}\frac{1-x}{1-x^2}\right)$ Arcsine is continuous.

$$= \sin^{-1}\left(\lim_{x\to1}\frac{1}{1+x}\right)$$ Cancel common factor $(1-x)$.

$$= \sin^{-1}\frac{1}{2} = \frac{\pi}{6}$$

We sometimes denote e^u by exp u when u is a complicated mathematical expression.

(c) $\displaystyle\lim_{x\to0}\sqrt{x+1}\,e^{\tan x} = \lim_{x\to0}\sqrt{x+1}\cdot\exp\left(\lim_{x\to0}\tan x\right)$ Exponential is continuous.

$$= 1\cdot e^0 = 1$$ ∎

Continuous Extension to a Point

The function $y = f(x) = (\sin x)/x$ is continuous at every point except $x = 0$. In this it is like the function $y = 1/x$. But $y = (\sin x)/x$ is different from $y = 1/x$ in that it has a finite limit as $x \to 0$ (Theorem 7). It is therefore possible to extend the function's domain to include the point $x = 0$ in such a way that the extended function is continuous at $x = 0$. We define a new function

$$F(x) = \begin{cases} \dfrac{\sin x}{x}, & x \neq 0 \\[2mm] 1, & x = 0. \end{cases}$$

The function $F(x)$ is continuous at $x = 0$ because

$$\lim_{x\to0}\frac{\sin x}{x} = F(0)$$

(Figure 2.44).

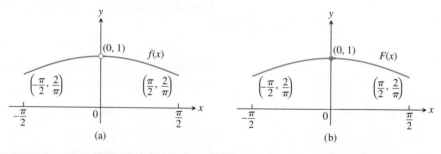

FIGURE 2.44   The graph (a) of $f(x) = (\sin x)/x$ for $-\pi/2 \le x \le \pi/2$ does not include the point $(0, 1)$ because the function is not defined at $x = 0$. (b) We can remove the discontinuity from the graph by defining the new function $F(x)$ with $F(0) = 1$ and $F(x) = f(x)$ everywhere else. Note that $F(0) = \lim_{x\to0}f(x)$.

More generally, a function (such as a rational function) may have a limit even at a point where it is not defined. If $f(c)$ is not defined, but $\lim_{x\to c}f(x) = L$ exists, we can define a new function $F(x)$ by the rule

$$F(x) = \begin{cases} f(x), & \text{if } x \text{ is in the domain of } f \\ L, & \text{if } x = c. \end{cases}$$

EXAMPLE 2   The properties in Theorem 12 are used to calculate limits in the same way as when x approaches a finite number c.

(a) $\displaystyle\lim_{x\to\infty}\left(5+\frac{1}{x}\right)=\lim_{x\to\infty}5+\lim_{x\to\infty}\frac{1}{x}$ Sum Rule

$\qquad\qquad\qquad\quad =5+0=5$ Known limits

(b) $\displaystyle\lim_{x\to-\infty}\frac{\pi\sqrt{3}}{x^2}=\lim_{x\to-\infty}\pi\sqrt{3}\cdot\frac{1}{x}\cdot\frac{1}{x}$

$\qquad\qquad\quad =\lim_{x\to-\infty}\pi\sqrt{3}\cdot\lim_{x\to-\infty}\frac{1}{x}\cdot\lim_{x\to-\infty}\frac{1}{x}$ Product Rule

$\qquad\qquad\quad =\pi\sqrt{3}\cdot 0\cdot 0=0$ Known limits   ∎

Limits at Infinity of Rational Functions

To determine the limit of a rational function as $x\to\pm\infty$, we first divide the numerator and denominator by the highest power of x in the denominator. The result then depends on the degrees of the polynomials involved.

EXAMPLE 3   These examples illustrate what happens when the degree of the numerator is less than or equal to the degree of the denominator.

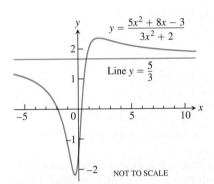

$y=\dfrac{5x^2+8x-3}{3x^2+2}$

Line $y=\dfrac{5}{3}$

NOT TO SCALE

FIGURE 2.51   The graph of the function in Example 3a. The graph approaches the line $y=5/3$ as $|x|$ increases.

(a) $\displaystyle\lim_{x\to\infty}\frac{5x^2+8x-3}{3x^2+2}=\lim_{x\to\infty}\frac{5+(8/x)-(3/x^2)}{3+(2/x^2)}$ Divide numerator and denominator by x^2.

$\qquad\qquad\qquad\qquad\quad =\dfrac{5+0-0}{3+0}=\dfrac{5}{3}$ See Fig. 2.51.

(b) $\displaystyle\lim_{x\to-\infty}\frac{11x+2}{2x^3-1}=\lim_{x\to-\infty}\frac{(11/x^2)+(2/x^3)}{2-(1/x^3)}$ Divide numerator and denominator by x^3.

$\qquad\qquad\qquad\qquad\quad =\dfrac{0+0}{2-0}=0$ See Fig. 2.52.   ∎

A case for which the degree of the numerator is greater than the degree of the denominator is illustrated in Example 10.

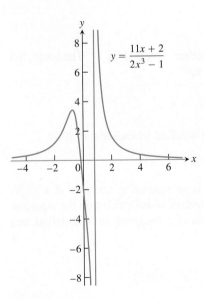

$y=\dfrac{11x+2}{2x^3-1}$

FIGURE 2.52   The graph of the function in Example 3b. The graph approaches the x-axis as $|x|$ increases.

Horizontal Asymptotes

If the distance between the graph of a function and some fixed line approaches zero as a point on the graph moves increasingly far from the origin, we say that the graph approaches the line asymptotically and that the line is an *asymptote* of the graph.

Looking at $f(x)=1/x$ (see Figure 2.49), we observe that the x-axis is an asymptote of the curve on the right because

$$\lim_{x\to\infty}\frac{1}{x}=0$$

and on the left because

$$\lim_{x\to-\infty}\frac{1}{x}=0.$$

We say that the x-axis is a *horizontal asymptote* of the graph of $f(x)=1/x$.

DEFINITION   A line $y=b$ is a **horizontal asymptote** of the graph of a function $y=f(x)$ if either

$$\lim_{x\to\infty}f(x)=b\qquad\text{or}\qquad\lim_{x\to-\infty}f(x)=b.$$

The graph of the function

$$f(x) = \frac{5x^2 + 8x - 3}{3x^2 + 2}$$

sketched in Figure 2.51 (Example 3a) has the line $y = 5/3$ as a horizontal asymptote on both the right and the left because

$$\lim_{x\to\infty} f(x) = \frac{5}{3} \quad \text{and} \quad \lim_{x\to-\infty} f(x) = \frac{5}{3}.$$

EXAMPLE 4   Find the horizontal asymptotes of the graph of

$$f(x) = \frac{x^3 - 2}{|x|^3 + 1}.$$

Solution   We calculate the limits as $x \to \pm\infty$.

For $x \geq 0$: $\displaystyle\lim_{x\to\infty} \frac{x^3 - 2}{|x|^3 + 1} = \lim_{x\to\infty} \frac{x^3 - 2}{x^3 + 1} = \lim_{x\to\infty} \frac{1 - (2/x^3)}{1 + (1/x^3)} = 1.$

For $x < 0$: $\displaystyle\lim_{x\to-\infty} \frac{x^3 - 2}{|x|^3 + 1} = \lim_{x\to-\infty} \frac{x^3 - 2}{(-x)^3 + 1} = \lim_{x\to-\infty} \frac{1 - (2/x^3)}{-1 + (1/x^3)} = -1.$

The horizontal asymptotes are $y = -1$ and $y = 1$. The graph is displayed in Figure 2.53. Notice that the graph crosses the horizontal asymptote $y = -1$ for a positive value of x.   ∎

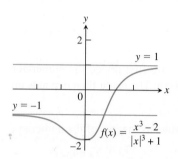

FIGURE 2.53   The graph of the function in Example 4 has two horizontal asymptotes.

EXAMPLE 5   The x-axis (the line $y = 0$) is a horizontal asymptote of the graph of $y = e^x$ because

$$\lim_{x\to-\infty} e^x = 0.$$

To see this, we use the definition of a limit as x approaches $-\infty$. So let $\epsilon > 0$ be given, but arbitrary. We must find a constant N such that for all x,

$$x < N \quad \Rightarrow \quad |e^x - 0| < \epsilon.$$

Now $|e^x - 0| = e^x$, so the condition that needs to be satisfied whenever $x < N$ is

$$e^x < \epsilon.$$

Let $x = N$ be the number where $e^x = \epsilon$. Since e^x is an increasing function, if $x < N$, then $e^x < \epsilon$. We find N by taking the natural logarithm of both sides of the equation $e^N = \epsilon$, so $N = \ln \epsilon$ (see Figure 2.54). With this value of N the condition is satisfied, and we conclude that $\lim_{x\to-\infty} e^x = 0$.   ∎

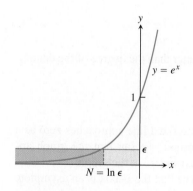

FIGURE 2.54   The graph of $y = e^x$ approaches the x-axis as $x \to -\infty$ (Example 5).

EXAMPLE 6   Find **(a)** $\displaystyle\lim_{x\to\infty} \sin(1/x)$ and   **(b)** $\displaystyle\lim_{x\to\pm\infty} x \sin(1/x)$.

Solution

(a)   We introduce the new variable $t = 1/x$. From Example 1, we know that $t \to 0^+$ as $x \to \infty$ (see Figure 2.49). Therefore,

$$\lim_{x\to\infty} \sin\frac{1}{x} = \lim_{t\to0^+} \sin t = 0.$$

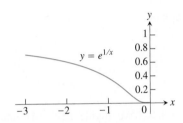

FIGURE 2.55 The line $y = 1$ is a horizontal asymptote of the function graphed here (Example 6b).

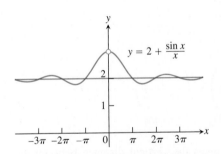

FIGURE 2.56 The graph of $y = e^{1/x}$ for $x < 0$ shows $\lim_{x\to 0^-} e^{1/x} = 0$ (Example 7).

FIGURE 2.57 A curve may cross one of its asymptotes infinitely often (Example 8).

(b) We calculate the limits as $x \to \infty$ and $x \to -\infty$:

$$\lim_{x\to\infty} x \sin\frac{1}{x} = \lim_{t\to 0^+} \frac{\sin t}{t} = 1 \quad\text{and}\quad \lim_{x\to-\infty} x \sin\frac{1}{x} = \lim_{t\to 0^-} \frac{\sin t}{t} = 1.$$

The graph is shown in Figure 2.55, and we see that the line $y = 1$ is a horizontal asymptote. ∎

Likewise, we can investigate the behavior of $y = f(1/x)$ as $x \to 0$ by investigating $y = f(t)$ as $t \to \pm\infty$, where $t = 1/x$.

EXAMPLE 7 Find $\lim_{x\to 0^-} e^{1/x}$.

Solution We let $t = 1/x$. From Figure 2.49, we can see that $t \to -\infty$ as $x \to 0^-$. (We make this idea more precise further on.) Therefore,

$$\lim_{x\to 0^-} e^{1/x} = \lim_{t\to-\infty} e^t = 0 \qquad \text{Example 5}$$

(Figure 2.56). ∎

The Sandwich Theorem also holds for limits as $x \to \pm\infty$. You must be sure, though, that the function whose limit you are trying to find stays between the bounding functions at very large values of x in magnitude consistent with whether $x \to \infty$ or $x \to -\infty$.

EXAMPLE 8 Using the Sandwich Theorem, find the horizontal asymptote of the curve

$$y = 2 + \frac{\sin x}{x}.$$

Solution We are interested in the behavior as $x \to \pm\infty$. Since

$$0 \le \left|\frac{\sin x}{x}\right| \le \left|\frac{1}{x}\right|$$

and $\lim_{x\to\pm\infty} |1/x| = 0$, we have $\lim_{x\to\pm\infty} (\sin x)/x = 0$ by the Sandwich Theorem. Hence,

$$\lim_{x\to\pm\infty}\left(2 + \frac{\sin x}{x}\right) = 2 + 0 = 2,$$

and the line $y = 2$ is a horizontal asymptote of the curve on both left and right (Figure 2.57). This example illustrates that a curve may cross one of its horizontal asymptotes many times. ∎

EXAMPLE 9 Find $\lim_{x\to\infty}\left(x - \sqrt{x^2 + 16}\right)$.

Solution Both of the terms x and $\sqrt{x^2 + 16}$ approach infinity as $x \to \infty$, so what happens to the difference in the limit is unclear (we cannot subtract ∞ from ∞ because the symbol does not represent a real number). In this situation we can multiply the numerator and the denominator by the conjugate radical expression to obtain an equivalent algebraic result:

$$\lim_{x\to\infty}\left(x - \sqrt{x^2 + 16}\right) = \lim_{x\to\infty}\left(x - \sqrt{x^2 + 16}\right)\frac{x + \sqrt{x^2 + 16}}{x + \sqrt{x^2 + 16}}$$

$$= \lim_{x\to\infty}\frac{x^2 - (x^2 + 16)}{x + \sqrt{x^2 + 16}} = \lim_{x\to\infty}\frac{-16}{x + \sqrt{x^2 + 16}}.$$

As $x \to \infty$, the denominator in this last expression becomes arbitrarily large, so we see that the limit is 0. We can also obtain this result by a direct calculation using the Limit Laws:

$$\lim_{x \to \infty} \frac{-16}{x + \sqrt{x^2 + 16}} = \lim_{x \to \infty} \frac{-\dfrac{16}{x}}{1 + \sqrt{\dfrac{x^2}{x^2} + \dfrac{16}{x^2}}} = \frac{0}{1 + \sqrt{1 + 0}} = 0.$$

∎

Oblique Asymptotes

If the degree of the numerator of a rational function is 1 greater than the degree of the denominator, the graph has an **oblique** or **slant line asymptote**. We find an equation for the asymptote by dividing numerator by denominator to express f as a linear function plus a remainder that goes to zero as $x \to \pm\infty$.

EXAMPLE 10   Find the oblique asymptote of the graph of

$$f(x) = \frac{x^2 - 3}{2x - 4}$$

in Figure 2.58.

Solution   We are interested in the behavior as $x \to \pm\infty$. We divide $(2x - 4)$ into $(x^2 - 3)$:

$$
\begin{array}{r}
\frac{x}{2} + 1 \\
2x - 4 \overline{\big)\, x^2 - 3 } \\
\underline{x^2 - 2x } \\
2x - 3 \\
\underline{2x - 4} \\
1
\end{array}
$$

This tells us that

$$f(x) = \frac{x^2 - 3}{2x - 4} = \underbrace{\left(\frac{x}{2} + 1\right)}_{\text{linear } g(x)} + \underbrace{\left(\frac{1}{2x - 4}\right)}_{\text{remainder}}.$$

As $x \to \pm\infty$, the remainder, whose magnitude gives the vertical distance between the graphs of f and g, goes to zero, making the slanted line

$$g(x) = \frac{x}{2} + 1$$

an asymptote of the graph of f (Figure 2.58). The line $y = g(x)$ is an asymptote both to the right and to the left. The next subsection will confirm that the function $f(x)$ grows arbitrarily large in absolute value as $x \to 2$ (where the denominator is zero), as shown in the graph. ∎

Notice in Example 10 that if the degree of the numerator in a rational function is greater than the degree of the denominator, then the limit as $|x|$ becomes large is $+\infty$ or $-\infty$, depending on the signs assumed by the numerator and denominator.

Infinite Limits

Let us look again at the function $f(x) = 1/x$. As $x \to 0^+$, the values of f grow without bound, eventually reaching and surpassing every positive real number. That is, given any positive real number B, however large, the values of f become larger still (Figure 2.59).

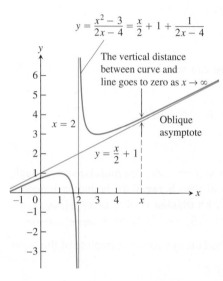

$$y = \frac{x^2 - 3}{2x - 4} = \frac{x}{2} + 1 + \frac{1}{2x - 4}$$

The vertical distance between curve and line goes to zero as $x \to \infty$

$x = 2$

Oblique asymptote

$y = \frac{x}{2} + 1$

FIGURE 2.58   The graph of the function in Example 10 has an oblique asymptote.

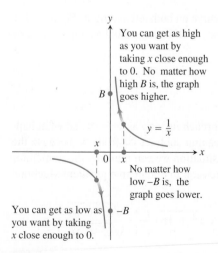

You can get as high as you want by taking x close enough to 0. No matter how high B is, the graph goes higher.

$y = \frac{1}{x}$

No matter how low $-B$ is, the graph goes lower.

You can get as low as you want by taking x close enough to 0.

FIGURE 2.59   One-sided infinite limits:
$$\lim_{x \to 0^+} \frac{1}{x} = \infty \quad \text{and} \quad \lim_{x \to 0^-} \frac{1}{x} = -\infty.$$

Thus, f has no limit as $x \to 0^+$. It is nevertheless convenient to describe the behavior of f by saying that $f(x)$ approaches ∞ as $x \to 0^+$. We write

$$\lim_{x \to 0^+} f(x) = \lim_{x \to 0^+} \frac{1}{x} = \infty.$$

In writing this equation, we are *not* saying that the limit exists. Nor are we saying that there is a real number ∞, for there is no such number. Rather, we are saying that $\lim_{x \to 0^+} (1/x)$ *does not exist because $1/x$ becomes arbitrarily large and positive as $x \to 0^+$.*

As $x \to 0^-$, the values of $f(x) = 1/x$ become arbitrarily large and negative. Given any negative real number $-B$, the values of f eventually lie below $-B$. (See Figure 2.59.) We write

$$\lim_{x \to 0^-} f(x) = \lim_{x \to 0^-} \frac{1}{x} = -\infty.$$

Again, we are not saying that the limit exists and equals the number $-\infty$. There *is* no real number $-\infty$. We are describing the behavior of a function whose limit as $x \to 0^-$ *does not exist because its values become arbitrarily large and negative.*

EXAMPLE 11   Find $\displaystyle\lim_{x \to 1^+} \frac{1}{x-1}$ and $\displaystyle\lim_{x \to 1^-} \frac{1}{x-1}$.

Geometric Solution   The graph of $y = 1/(x-1)$ is the graph of $y = 1/x$ shifted 1 unit to the right (Figure 2.60). Therefore, $y = 1/(x-1)$ behaves near 1 exactly the way $y = 1/x$ behaves near 0:

$$\lim_{x \to 1^+} \frac{1}{x-1} = \infty \qquad \text{and} \qquad \lim_{x \to 1^-} \frac{1}{x-1} = -\infty.$$

Analytic Solution   Think about the number $x - 1$ and its reciprocal. As $x \to 1^+$, we have $(x-1) \to 0^+$ and $1/(x-1) \to \infty$. As $x \to 1^-$, we have $(x-1) \to 0^-$ and $1/(x-1) \to -\infty$.   ∎

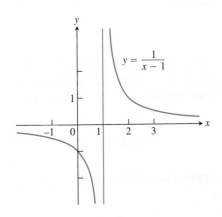

FIGURE 2.60 Near $x = 1$, the function $y = 1/(x-1)$ behaves the way the function $y = 1/x$ behaves near $x = 0$. Its graph is the graph of $y = 1/x$ shifted 1 unit to the right (Example 11).

EXAMPLE 12   Discuss the behavior of

$$f(x) = \frac{1}{x^2} \qquad \text{as} \qquad x \to 0.$$

Solution   As x approaches zero from either side, the values of $1/x^2$ are positive and become arbitrarily large (Figure 2.61). This means that

$$\lim_{x \to 0} f(x) = \lim_{x \to 0} \frac{1}{x^2} = \infty.$$ ∎

The function $y = 1/x$ shows no consistent behavior as $x \to 0$. We have $1/x \to \infty$ if $x \to 0^+$, but $1/x \to -\infty$ if $x \to 0^-$. All we can say about $\lim_{x \to 0} (1/x)$ is that it does not exist. The function $y = 1/x^2$ is different. Its values approach infinity as x approaches zero from either side, so we can say that $\lim_{x \to 0} (1/x^2) = \infty$.

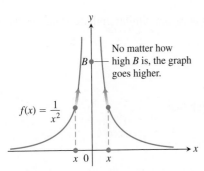

FIGURE 2.61 The graph of $f(x)$ in Example 12 approaches infinity as $x \to 0$.

EXAMPLE 13   These examples illustrate that rational functions can behave in various ways near zeros of the denominator.

(a) $\displaystyle\lim_{x \to 2} \frac{(x-2)^2}{x^2-4} = \lim_{x \to 2} \frac{(x-2)^2}{(x-2)(x+2)} = \lim_{x \to 2} \frac{x-2}{x+2} = 0$

(b) $\displaystyle\lim_{x \to 2} \frac{x-2}{x^2-4} = \lim_{x \to 2} \frac{x-2}{(x-2)(x+2)} = \lim_{x \to 2} \frac{1}{x+2} = \frac{1}{4}$

(c) $\lim\limits_{x\to 2^+} \dfrac{x-3}{x^2-4} = \lim\limits_{x\to 2^+} \dfrac{x-3}{(x-2)(x+2)} = -\infty$ The values are negative for $x > 2$, x near 2.

(d) $\lim\limits_{x\to 2^-} \dfrac{x-3}{x^2-4} = \lim\limits_{x\to 2^-} \dfrac{x-3}{(x-2)(x+2)} = \infty$ The values are positive for $x < 2$, x near 2.

(e) $\lim\limits_{x\to 2} \dfrac{x-3}{x^2-4} = \lim\limits_{x\to 2} \dfrac{x-3}{(x-2)(x+2)}$ does not exist.   See parts (c) and (d).

(f) $\lim\limits_{x\to 2} \dfrac{2-x}{(x-2)^3} = \lim\limits_{x\to 2} \dfrac{-(x-2)}{(x-2)^3} = \lim\limits_{x\to 2} \dfrac{-1}{(x-2)^2} = -\infty$

In parts (a) and (b) the effect of the zero in the denominator at $x = 2$ is canceled because the numerator is zero there also. Thus a finite limit exists. This is not true in part (f), where cancellation still leaves a zero factor in the denominator. ∎

Precise Definitions of Infinite Limits

Instead of requiring $f(x)$ to lie arbitrarily close to a finite number L for all x sufficiently close to x_0, the definitions of infinite limits require $f(x)$ to lie arbitrarily far from zero. Except for this change, the language is very similar to what we have seen before. Figures 2.62 and 2.63 accompany these definitions.

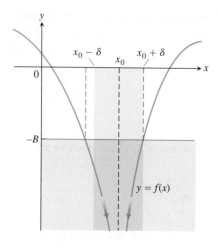

FIGURE 2.62 For $x_0 - \delta < x < x_0 + \delta$, the graph of $f(x)$ lies above the line $y = B$.

FIGURE 2.63 For $x_0 - \delta < x < x_0 + \delta$, the graph of $f(x)$ lies below the line $y = -B$.

DEFINITIONS

1. We say that **$f(x)$ approaches infinity as x approaches x_0**, and write
$$\lim_{x\to x_0} f(x) = \infty,$$
if for every positive real number B there exists a corresponding $\delta > 0$ such that for all x
$$0 < |x - x_0| < \delta \quad \Rightarrow \quad f(x) > B.$$

2. We say that **$f(x)$ approaches minus infinity as x approaches x_0**, and write
$$\lim_{x\to x_0} f(x) = -\infty,$$
if for every negative real number $-B$ there exists a corresponding $\delta > 0$ such that for all x
$$0 < |x - x_0| < \delta \quad \Rightarrow \quad f(x) < -B.$$

The precise definitions of one-sided infinite limits at x_0 are similar and are stated in the exercises.

EXAMPLE 14   Prove that $\lim\limits_{x\to 0} \dfrac{1}{x^2} = \infty$.

Solution   Given $B > 0$, we want to find $\delta > 0$ such that
$$0 < |x - 0| < \delta \quad \text{implies} \quad \frac{1}{x^2} > B.$$

Now,
$$\frac{1}{x^2} > B \qquad \text{if and only if} \quad x^2 < \frac{1}{B}$$

or, equivalently,
$$|x| < \frac{1}{\sqrt{B}}.$$

Thus, choosing $\delta = 1/\sqrt{B}$ (or any smaller positive number), we see that

$$|x| < \delta \quad \text{implies} \quad \frac{1}{x^2} > \frac{1}{\delta^2} \geq B.$$

Therefore, by definition,

$$\lim_{x \to 0} \frac{1}{x^2} = \infty. \qquad \blacksquare$$

Vertical Asymptotes

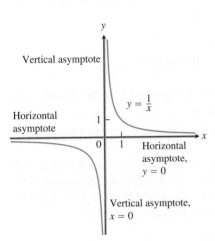

Notice that the distance between a point on the graph of $f(x) = 1/x$ and the y-axis approaches zero as the point moves vertically along the graph and away from the origin (Figure 2.64). The function $f(x) = 1/x$ is unbounded as x approaches 0 because

$$\lim_{x \to 0^+} \frac{1}{x} = \infty \quad \text{and} \quad \lim_{x \to 0^-} \frac{1}{x} = -\infty.$$

We say that the line $x = 0$ (the y-axis) is a *vertical asymptote* of the graph of $f(x) = 1/x$. Observe that the denominator is zero at $x = 0$ and the function is undefined there.

FIGURE 2.64 The coordinate axes are asymptotes of both branches of the hyperbola $y = 1/x$.

DEFINITION A line $x = a$ is a **vertical asymptote** of the graph of a function $y = f(x)$ if either

$$\lim_{x \to a^+} f(x) = \pm\infty \quad \text{or} \quad \lim_{x \to a^-} f(x) = \pm\infty.$$

EXAMPLE 15 Find the horizontal and vertical asymptotes of the curve

$$y = \frac{x + 3}{x + 2}.$$

Solution We are interested in the behavior as $x \to \pm\infty$ and the behavior as $x \to -2$, where the denominator is zero.

The asymptotes are quickly revealed if we recast the rational function as a polynomial with a remainder, by dividing $(x + 2)$ into $(x + 3)$:

$$\begin{array}{r} 1 \\ x + 2 \overline{)x + 3} \\ \underline{x + 2} \\ 1 \end{array}$$

This result enables us to rewrite y as:

$$y = 1 + \frac{1}{x + 2}.$$

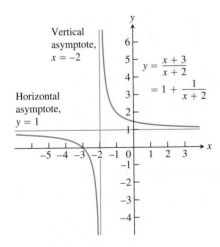

FIGURE 2.65 The lines $y = 1$ and $x = -2$ are asymptotes of the curve in Example 15.

As $x \to \pm\infty$, the curve approaches the horizontal asymptote $y = 1$; as $x \to -2$, the curve approaches the vertical asymptote $x = -2$. We see that the curve in question is the graph of $f(x) = 1/x$ shifted 1 unit up and 2 units left (Figure 2.65). The asymptotes, instead of being the coordinate axes, are now the lines $y = 1$ and $x = -2$. $\qquad \blacksquare$

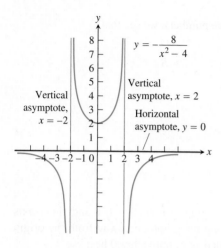

FIGURE 2.66 Graph of the function in Example 16. Notice that the curve approaches the x-axis from only one side. Asymptotes do not have to be two-sided.

EXAMPLE 16 Find the horizontal and vertical asymptotes of the graph of

$$f(x) = -\frac{8}{x^2 - 4}.$$

Solution We are interested in the behavior as $x \to \pm\infty$ and as $x \to \pm 2$, where the denominator is zero. Notice that f is an even function of x, so its graph is symmetric with respect to the y-axis.

(a) *The behavior as $x \to \pm\infty$.* Since $\lim_{x\to\infty} f(x) = 0$, the line $y = 0$ is a horizontal asymptote of the graph to the right. By symmetry it is an asymptote to the left as well (Figure 2.66). Notice that the curve approaches the x-axis from only the negative side (or from below). Also, $f(0) = 2$.

(b) *The behavior as $x \to \pm 2$.* Since

$$\lim_{x\to 2^+} f(x) = -\infty \qquad \text{and} \qquad \lim_{x\to 2^-} f(x) = \infty,$$

the line $x = 2$ is a vertical asymptote both from the right and from the left. By symmetry, the line $x = -2$ is also a vertical asymptote.

There are no other asymptotes because f has a finite limit at every other point. ∎

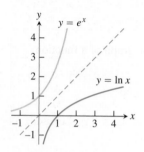

FIGURE 2.67 The line $x = 0$ is a vertical asymptote of the natural logarithm function (Example 17).

EXAMPLE 17 The graph of the natural logarithm function has the y-axis (the line $x = 0$) as a vertical asymptote. We see this from the graph sketched in Figure 2.67 (which is the reflection of the graph of the natural exponential function across the line $y = x$) and the fact that the x-axis is a horizontal asymptote of $y = e^x$ (Example 5). Thus,

$$\lim_{x\to 0^+} \ln x = -\infty.$$

The same result is true for $y = \log_a x$ whenever $a > 1$. ∎

EXAMPLE 18 The curves

$$y = \sec x = \frac{1}{\cos x} \qquad \text{and} \qquad y = \tan x = \frac{\sin x}{\cos x}$$

both have vertical asymptotes at odd-integer multiples of $\pi/2$, where $\cos x = 0$ (Figure 2.68).

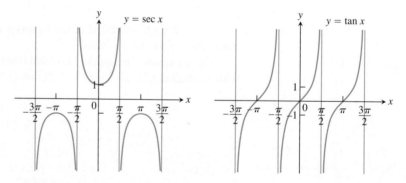

FIGURE 2.68 The graphs of $\sec x$ and $\tan x$ have infinitely many vertical asymptotes (Example 18). ∎

Dominant Terms

In Example 10 we saw that by long division we could rewrite the function

$$f(x) = \frac{x^2 - 3}{2x - 4}$$

as a linear function plus a remainder term:

$$f(x) = \left(\frac{x}{2} + 1\right) + \left(\frac{1}{2x - 4}\right).$$

This tells us immediately that

$$f(x) \approx \frac{x}{2} + 1 \qquad \text{For } x \text{ numerically large, } \frac{1}{2x - 4} \text{ is near 0.}$$

$$f(x) \approx \frac{1}{2x - 4} \qquad \text{For } x \text{ near 2, this term is very large.}$$

If we want to know how f behaves, this is the way to find out. It behaves like $y = (x/2) + 1$ when x is numerically large and the contribution of $1/(2x - 4)$ to the total value of f is insignificant. It behaves like $1/(2x - 4)$ when x is so close to 2 that $1/(2x - 4)$ makes the dominant contribution.

We say that $(x/2) + 1$ **dominates** when x is numerically large, and we say that $1/(2x - 4)$ dominates when x is near 2. **Dominant terms** like these help us predict a function's behavior.

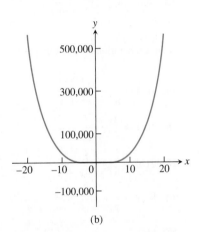

FIGURE 2.69 The graphs of f and g are (a) distinct for $|x|$ small, and (b) nearly identical for $|x|$ large (Example 19).

EXAMPLE 19 Let $f(x) = 3x^4 - 2x^3 + 3x^2 - 5x + 6$ and $g(x) = 3x^4$. Show that although f and g are quite different for numerically small values of x, they are virtually identical for $|x|$ very large, in the sense that their ratios approach 1 as $x \to \infty$ or $x \to -\infty$.

Solution The graphs of f and g behave quite differently near the origin (Figure 2.69a), but appear as virtually identical on a larger scale (Figure 2.69b).

We can test that the term $3x^4$ in f, represented graphically by g, dominates the polynomial f for numerically large values of x by examining the ratio of the two functions as $x \to \pm\infty$. We find that

$$\lim_{x \to \pm\infty} \frac{f(x)}{g(x)} = \lim_{x \to \pm\infty} \frac{3x^4 - 2x^3 + 3x^2 - 5x + 6}{3x^4}$$

$$= \lim_{x \to \pm\infty} \left(1 - \frac{2}{3x} + \frac{1}{x^2} - \frac{5}{3x^3} + \frac{2}{x^4}\right)$$

$$= 1,$$

which means that f and g appear nearly identical for $|x|$ large. ∎

Summary

In this chapter we presented several important calculus ideas that are made meaningful and precise by the concept of the limit. These include the three ideas of the exact rate of change of a function, the slope of the graph of a function at a point, and the continuity of a function. The primary methods used for calculating limits of many functions are captured in the algebraic limit laws of Theorem 1 and in the Sandwich Theorem, all of which are proved from the precise definition of the limit. We saw that these computational rules also apply to one-sided limits and to limits at infinity. Moreover, we can sometimes apply these rules to calculating limits of simple transcendental functions, as illustrated by our examples or in cases like the following:

$$\lim_{x \to 0} \frac{e^x - 1}{e^{2x} - 1} = \lim_{x \to 0} \frac{e^x - 1}{(e^x - 1)(e^x + 1)} = \lim_{x \to 0} \frac{1}{e^x + 1} = \frac{1}{1 + 1} = \frac{1}{2}.$$

However, calculating more complicated limits involving transcendental functions such as

$$\lim_{x \to 0} \frac{x}{e^{2x} - 1}, \qquad \lim_{x \to 0} \frac{\ln x}{x}, \qquad \text{and} \qquad \lim_{x \to 0} \left(1 + \frac{1}{x}\right)^{x}$$

requires more than simple algebraic techniques. The *derivative* is exactly the tool we need to calculate limits in these kinds of cases (see Section 4.5), and this notion is the main subject of our next chapter.

Exercises 2.6

Finding Limits

1. For the function f whose graph is given, determine the following limits.

  a. $\lim\limits_{x \to 2} f(x)$ **b.** $\lim\limits_{x \to -3^{+}} f(x)$ **c.** $\lim\limits_{x \to -3^{-}} f(x)$

  d. $\lim\limits_{x \to -3} f(x)$ **e.** $\lim\limits_{x \to 0^{+}} f(x)$ **f.** $\lim\limits_{x \to 0^{-}} f(x)$

  g. $\lim\limits_{x \to 0} f(x)$ **h.** $\lim\limits_{x \to \infty} f(x)$ **i.** $\lim\limits_{x \to -\infty} f(x)$

2. For the function f whose graph is given, determine the following limits.

  a. $\lim\limits_{x \to 4} f(x)$ **b.** $\lim\limits_{x \to 2^{+}} f(x)$ **c.** $\lim\limits_{x \to 2^{-}} f(x)$

  d. $\lim\limits_{x \to 2} f(x)$ **e.** $\lim\limits_{x \to -3^{+}} f(x)$ **f.** $\lim\limits_{x \to -3^{-}} f(x)$

  g. $\lim\limits_{x \to -3} f(x)$ **h.** $\lim\limits_{x \to 0^{+}} f(x)$ **i.** $\lim\limits_{x \to 0^{-}} f(x)$

  j. $\lim\limits_{x \to 0} f(x)$ **k.** $\lim\limits_{x \to \infty} f(x)$ **l.** $\lim\limits_{x \to -\infty} f(x)$

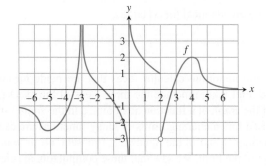

In Exercises 3–8, find the limit of each function **(a)** as $x \to \infty$ and **(b)** as $x \to -\infty$. (You may wish to visualize your answer with a graphing calculator or computer.)

3. $f(x) = \dfrac{2}{x} - 3$

4. $f(x) = \pi - \dfrac{2}{x^2}$

5. $g(x) = \dfrac{1}{2 + (1/x)}$

6. $g(x) = \dfrac{1}{8 - (5/x^2)}$

7. $h(x) = \dfrac{-5 + (7/x)}{3 - (1/x^2)}$

8. $h(x) = \dfrac{3 - (2/x)}{4 + (\sqrt{2}/x^2)}$

Find the limits in Exercises 9–12.

9. $\lim\limits_{x \to \infty} \dfrac{\sin 2x}{x}$

10. $\lim\limits_{\theta \to -\infty} \dfrac{\cos \theta}{3\theta}$

11. $\lim\limits_{t \to -\infty} \dfrac{2 - t + \sin t}{t + \cos t}$

12. $\lim\limits_{r \to \infty} \dfrac{r + \sin r}{2r + 7 - 5 \sin r}$

Limits of Rational Functions

In Exercises 13–22, find the limit of each rational function **(a)** as $x \to \infty$ and **(b)** as $x \to -\infty$.

13. $f(x) = \dfrac{2x + 3}{5x + 7}$

14. $f(x) = \dfrac{2x^3 + 7}{x^3 - x^2 + x + 7}$

15. $f(x) = \dfrac{x + 1}{x^2 + 3}$

16. $f(x) = \dfrac{3x + 7}{x^2 - 2}$

17. $h(x) = \dfrac{7x^3}{x^3 - 3x^2 + 6x}$

18. $g(x) = \dfrac{1}{x^3 - 4x + 1}$

19. $g(x) = \dfrac{10x^5 + x^4 + 31}{x^6}$

20. $h(x) = \dfrac{9x^4 + x}{2x^4 + 5x^2 - x + 6}$

21. $h(x) = \dfrac{-2x^3 - 2x + 3}{3x^3 + 3x^2 - 5x}$

22. $h(x) = \dfrac{-x^4}{x^4 - 7x^3 + 7x^2 + 9}$

Limits as $x \to \infty$ or $x \to -\infty$

The process by which we determine limits of rational functions applies equally well to ratios containing noninteger or negative powers of x: divide numerator and denominator by the highest power of x in the denominator and proceed from there. Find the limits in Exercises 23–36.

23. $\lim\limits_{x \to \infty} \sqrt{\dfrac{8x^2 - 3}{2x^2 + x}}$

24. $\lim\limits_{x \to -\infty} \left(\dfrac{x^2 + x - 1}{8x^2 - 3}\right)^{1/3}$

25. $\lim\limits_{x \to -\infty} \left(\dfrac{1 - x^3}{x^2 + 7x}\right)^5$

26. $\lim\limits_{x \to \infty} \sqrt{\dfrac{x^2 - 5x}{x^3 + x - 2}}$

27. $\lim\limits_{x \to \infty} \dfrac{2\sqrt{x} + x^{-1}}{3x - 7}$

28. $\lim\limits_{x \to \infty} \dfrac{2 + \sqrt{x}}{2 - \sqrt{x}}$

29. $\lim\limits_{x \to -\infty} \dfrac{\sqrt[3]{x} - \sqrt[5]{x}}{\sqrt[3]{x} + \sqrt[5]{x}}$

30. $\lim\limits_{x \to \infty} \dfrac{x^{-1} + x^{-4}}{x^{-2} - x^{-3}}$

31. $\lim\limits_{x \to \infty} \dfrac{2x^{5/3} - x^{1/3} + 7}{x^{8/5} + 3x + \sqrt{x}}$

32. $\lim\limits_{x \to -\infty} \dfrac{\sqrt[3]{x} - 5x + 3}{2x + x^{2/3} - 4}$

33. $\lim\limits_{x\to\infty} \dfrac{\sqrt{x^2+1}}{x+1}$

34. $\lim\limits_{x\to-\infty} \dfrac{\sqrt{x^2+1}}{x+1}$

35. $\lim\limits_{x\to\infty} \dfrac{x-3}{\sqrt{4x^2+25}}$

36. $\lim\limits_{x\to-\infty} \dfrac{4-3x^3}{\sqrt{x^6+9}}$

Infinite Limits

Find the limits in Exercises 37–48.

37. $\lim\limits_{x\to0^+} \dfrac{1}{3x}$

38. $\lim\limits_{x\to0^-} \dfrac{5}{2x}$

39. $\lim\limits_{x\to2^-} \dfrac{3}{x-2}$

40. $\lim\limits_{x\to3^+} \dfrac{1}{x-3}$

41. $\lim\limits_{x\to-8^+} \dfrac{2x}{x+8}$

42. $\lim\limits_{x\to-5^-} \dfrac{3x}{2x+10}$

43. $\lim\limits_{x\to7} \dfrac{4}{(x-7)^2}$

44. $\lim\limits_{x\to0} \dfrac{-1}{x^2(x+1)}$

45. a. $\lim\limits_{x\to0^+} \dfrac{2}{3x^{1/3}}$ **b.** $\lim\limits_{x\to0^-} \dfrac{2}{3x^{1/3}}$

46. a. $\lim\limits_{x\to0^+} \dfrac{2}{x^{1/5}}$ **b.** $\lim\limits_{x\to0^-} \dfrac{2}{x^{1/5}}$

47. $\lim\limits_{x\to0} \dfrac{4}{x^{2/5}}$

48. $\lim\limits_{x\to0} \dfrac{1}{x^{2/3}}$

Find the limits in Exercises 49–52.

49. $\lim\limits_{x\to(\pi/2)^-} \tan x$

50. $\lim\limits_{x\to(-\pi/2)^+} \sec x$

51. $\lim\limits_{\theta\to0^-} (1+\csc\theta)$

52. $\lim\limits_{\theta\to0} (2-\cot\theta)$

Find the limits in Exercises 53–58.

53. $\lim \dfrac{1}{x^2-4}$ as

  a. $x\to2^+$ **b.** $x\to2^-$
  c. $x\to-2^+$ **d.** $x\to-2^-$

54. $\lim \dfrac{x}{x^2-1}$ as

  a. $x\to1^+$ **b.** $x\to1^-$
  c. $x\to-1^+$ **d.** $x\to-1^-$

55. $\lim \left(\dfrac{x^2}{2}-\dfrac{1}{x}\right)$ as

  a. $x\to0^+$ **b.** $x\to0^-$
  c. $x\to\sqrt[3]{2}$ **d.** $x\to-1$

56. $\lim \dfrac{x^2-1}{2x+4}$ as

  a. $x\to-2^+$ **b.** $x\to-2^-$
  c. $x\to1^+$ **d.** $x\to0^-$

57. $\lim \dfrac{x^2-3x+2}{x^3-2x^2}$ as

  a. $x\to0^+$ **b.** $x\to2^+$
  c. $x\to2^-$ **d.** $x\to2$
  e. What, if anything, can be said about the limit as $x\to0$?

58. $\lim \dfrac{x^2-3x+2}{x^3-4x}$ as

  a. $x\to2^+$ **b.** $x\to-2^+$
  c. $x\to0^-$ **d.** $x\to1^+$
  e. What, if anything, can be said about the limit as $x\to0$?

Find the limits in Exercises 59–62.

59. $\lim \left(2-\dfrac{3}{t^{1/3}}\right)$ as

  a. $t\to0^+$ **b.** $t\to0^-$

60. $\lim \left(\dfrac{1}{t^{3/5}}+7\right)$ as

  a. $t\to0^+$ **b.** $t\to0^-$

61. $\lim \left(\dfrac{1}{x^{2/3}}+\dfrac{2}{(x-1)^{2/3}}\right)$ as

  a. $x\to0^+$ **b.** $x\to0^-$
  c. $x\to1^+$ **d.** $x\to1^-$

62. $\lim \left(\dfrac{1}{x^{1/3}}-\dfrac{1}{(x-1)^{4/3}}\right)$ as

  a. $x\to0^+$ **b.** $x\to0^-$
  c. $x\to1^+$ **d.** $x\to1^-$

Graphing Simple Rational Functions

Graph the rational functions in Exercises 63–68. Include the graphs and equations of the asymptotes and dominant terms.

63. $y=\dfrac{1}{x-1}$

64. $y=\dfrac{1}{x+1}$

65. $y=\dfrac{1}{2x+4}$

66. $y=\dfrac{-3}{x-3}$

67. $y=\dfrac{x+3}{x+2}$

68. $y=\dfrac{2x}{x+1}$

Inventing Graphs and Functions

In Exercises 69–72, sketch the graph of a function $y=f(x)$ that satisfies the given conditions. No formulas are required—just label the coordinate axes and sketch an appropriate graph. (The answers are not unique, so your graphs may not be exactly like those in the answer section.)

69. $f(0)=0$, $f(1)=2$, $f(-1)=-2$, $\lim\limits_{x\to-\infty} f(x)=-1$, and $\lim\limits_{x\to\infty} f(x)=1$

70. $f(0)=0$, $\lim\limits_{x\to\pm\infty} f(x)=0$, $\lim\limits_{x\to0^+} f(x)=2$, and $\lim\limits_{x\to0^-} f(x)=-2$

71. $f(0)=0$, $\lim\limits_{x\to\pm\infty} f(x)=0$, $\lim\limits_{x\to1^-} f(x)=\lim\limits_{x\to-1^+} f(x)=\infty$, $\lim\limits_{x\to1^+} f(x)=-\infty$, and $\lim\limits_{x\to-1^-} f(x)=-\infty$

72. $f(2)=1$, $f(-1)=0$, $\lim\limits_{x\to\infty} f(x)=0$, $\lim\limits_{x\to0^+} f(x)=\infty$, $\lim\limits_{x\to0^-} f(x)=-\infty$, and $\lim\limits_{x\to-\infty} f(x)=1$

In Exercises 73–76, find a function that satisfies the given conditions and sketch its graph. (The answers here are not unique. Any function that satisfies the conditions is acceptable. Feel free to use formulas defined in pieces if that will help.)

73. $\lim\limits_{x\to\pm\infty} f(x)=0$, $\lim\limits_{x\to2^-} f(x)=\infty$, and $\lim\limits_{x\to2^+} f(x)=\infty$

74. $\lim\limits_{x\to\pm\infty} g(x)=0$, $\lim\limits_{x\to3^-} g(x)=-\infty$, and $\lim\limits_{x\to3^+} g(x)=\infty$

75. $\lim\limits_{x\to-\infty} h(x)=-1$, $\lim\limits_{x\to\infty} h(x)=1$, $\lim\limits_{x\to0^-} h(x)=-1$, and $\lim\limits_{x\to0^+} h(x)=1$

76. $\lim\limits_{x\to\pm\infty} k(x)=1$, $\lim\limits_{x\to1^-} k(x)=\infty$, and $\lim\limits_{x\to1^+} k(x)=-\infty$

77. Suppose that $f(x)$ and $g(x)$ are polynomials in x and that $\lim_{x \to \infty} (f(x)/g(x)) = 2$. Can you conclude anything about $\lim_{x \to -\infty} (f(x)/g(x))$? Give reasons for your answer.

78. Suppose that $f(x)$ and $g(x)$ are polynomials in x. Can the graph of $f(x)/g(x)$ have an asymptote if $g(x)$ is never zero? Give reasons for your answer.

79. How many horizontal asymptotes can the graph of a given rational function have? Give reasons for your answer.

Finding Limits of Differences when $x \to \pm\infty$

Find the limits in Exercises 80–86.

80. $\lim_{x \to \infty} \left(\sqrt{x + 9} - \sqrt{x + 4} \right)$

81. $\lim_{x \to \infty} \left(\sqrt{x^2 + 25} - \sqrt{x^2 - 1} \right)$

82. $\lim_{x \to -\infty} \left(\sqrt{x^2 + 3} + x \right)$

83. $\lim_{x \to -\infty} \left(2x + \sqrt{4x^2 + 3x - 2} \right)$

84. $\lim_{x \to \infty} \left(\sqrt{9x^2 - x} - 3x \right)$

85. $\lim_{x \to \infty} \left(\sqrt{x^2 + 3x} - \sqrt{x^2 - 2x} \right)$

86. $\lim_{x \to \infty} \left(\sqrt{x^2 + x} - \sqrt{x^2 - x} \right)$

Using the Formal Definitions

Use the formal definitions of limits as $x \to \pm\infty$ to establish the limits in Exercises 87 and 88.

87. If f has the constant value $f(x) = k$, then $\lim_{x \to \infty} f(x) = k$.

88. If f has the constant value $f(x) = k$, then $\lim_{x \to -\infty} f(x) = k$.

Use formal definitions to prove the limit statements in Exercises 89–92.

89. $\lim_{x \to 0} \dfrac{-1}{x^2} = -\infty$

90. $\lim_{x \to 0} \dfrac{1}{|x|} = \infty$

91. $\lim_{x \to 3} \dfrac{-2}{(x - 3)^2} = -\infty$

92. $\lim_{x \to -5} \dfrac{1}{(x + 5)^2} = \infty$

93. Here is the definition of **infinite right-hand limit**.

> We say that $f(x)$ approaches infinity as x approaches x_0 from the right, and write
> $$\lim_{x \to x_0^+} f(x) = \infty,$$
> if, for every positive real number B, there exists a corresponding number $\delta > 0$ such that for all x
> $$x_0 < x < x_0 + \delta \quad \Rightarrow \quad f(x) > B.$$

Modify the definition to cover the following cases.

a. $\lim_{x \to x_0^-} f(x) = \infty$

b. $\lim_{x \to x_0^+} f(x) = -\infty$

c. $\lim_{x \to x_0^-} f(x) = -\infty$

Use the formal definitions from Exercise 93 to prove the limit statements in Exercises 94–98.

94. $\lim_{x \to 0^+} \dfrac{1}{x} = \infty$

95. $\lim_{x \to 0^-} \dfrac{1}{x} = -\infty$

96. $\lim_{x \to 2^-} \dfrac{1}{x - 2} = -\infty$

97. $\lim_{x \to 2^+} \dfrac{1}{x - 2} = \infty$

98. $\lim_{x \to 1^-} \dfrac{1}{1 - x^2} = \infty$

Oblique Asymptotes

Graph the rational functions in Exercises 99–104. Include the graphs and equations of the asymptotes.

99. $y = \dfrac{x^2}{x - 1}$

100. $y = \dfrac{x^2 + 1}{x - 1}$

101. $y = \dfrac{x^2 - 4}{x - 1}$

102. $y = \dfrac{x^2 - 1}{2x + 4}$

103. $y = \dfrac{x^2 - 1}{x}$

104. $y = \dfrac{x^3 + 1}{x^2}$

Additional Graphing Exercises

T Graph the curves in Exercises 105–108. Explain the relationship between the curve's formula and what you see.

105. $y = \dfrac{x}{\sqrt{4 - x^2}}$

106. $y = \dfrac{-1}{\sqrt{4 - x^2}}$

107. $y = x^{2/3} + \dfrac{1}{x^{1/3}}$

108. $y = \sin\left(\dfrac{\pi}{x^2 + 1} \right)$

T Graph the functions in Exercises 109 and 110. Then answer the following questions.

a. How does the graph behave as $x \to 0^+$?

b. How does the graph behave as $x \to \pm\infty$?

c. How does the graph behave near $x = 1$ and $x = -1$?

Give reasons for your answers.

109. $y = \dfrac{3}{2} \left(x - \dfrac{1}{x} \right)^{2/3}$

110. $y = \dfrac{3}{2} \left(\dfrac{x}{x - 1} \right)^{2/3}$

Chapter 2 Questions to Guide Your Review

1. What is the average rate of change of the function $g(t)$ over the interval from $t = a$ to $t = b$? How is it related to a secant line?

2. What limit must be calculated to find the rate of change of a function $g(t)$ at $t = t_0$?

3. Give an informal or intuitive definition of the limit
$$\lim_{x \to x_0} f(x) = L.$$
Why is the definition "informal"? Give examples.

4. Does the existence and value of the limit of a function $f(x)$ as x approaches x_0 ever depend on what happens at $x = x_0$? Explain and give examples.

5. What function behaviors might occur for which the limit may fail to exist? Give examples.

6. What theorems are available for calculating limits? Give examples of how the theorems are used.

7. How are one-sided limits related to limits? How can this relationship sometimes be used to calculate a limit or prove it does not exist? Give examples.

8. What is the value of $\lim_{\theta \to 0} ((\sin \theta)/\theta)$? Does it matter whether θ is measured in degrees or radians? Explain.

9. What exactly does $\lim_{x \to x_0} f(x) = L$ mean? Give an example in which you find a $\delta > 0$ for a given f, L, x_0, and $\epsilon > 0$ in the precise definition of limit.

10. Give precise definitions of the following statements.

   a. $\lim_{x \to 2^-} f(x) = 5$ **b.** $\lim_{x \to 2^+} f(x) = 5$

   c. $\lim_{x \to 2} f(x) = \infty$ **d.** $\lim_{x \to 2} f(x) = -\infty$

11. What conditions must be satisfied by a function if it is to be continuous at an interior point of its domain? At an endpoint?

12. How can looking at the graph of a function help you tell where the function is continuous?

13. What does it mean for a function to be right-continuous at a point? Left-continuous? How are continuity and one-sided continuity related?

14. What does it mean for a function to be continuous on an interval? Give examples to illustrate the fact that a function that is not continuous on its entire domain may still be continuous on selected intervals within the domain.

15. What are the basic types of discontinuity? Give an example of each. What is a removable discontinuity? Give an example.

16. What does it mean for a function to have the Intermediate Value Property? What conditions guarantee that a function has this property over an interval? What are the consequences for graphing and solving the equation $f(x) = 0$?

17. Under what circumstances can you extend a function $f(x)$ to be continuous at a point $x = c$? Give an example.

18. What exactly do $\lim_{x \to \infty} f(x) = L$ and $\lim_{x \to -\infty} f(x) = L$ mean? Give examples.

19. What are $\lim_{x \to \pm\infty} k$ (k a constant) and $\lim_{x \to \pm\infty} (1/x)$? How do you extend these results to other functions? Give examples.

20. How do you find the limit of a rational function as $x \to \pm\infty$? Give examples.

21. What are horizontal and vertical asymptotes? Give examples.

Chapter 2 Practice Exercises

Limits and Continuity

1. Graph the function

$$f(x) = \begin{cases} 1, & x \le -1 \\ -x, & -1 < x < 0 \\ 1, & x = 0 \\ -x, & 0 < x < 1 \\ 1, & x \ge 1. \end{cases}$$

Then discuss, in detail, limits, one-sided limits, continuity, and one-sided continuity of f at $x = -1, 0,$ and 1. Are any of the discontinuities removable? Explain.

2. Repeat the instructions of Exercise 1 for

$$f(x) = \begin{cases} 0, & x \le -1 \\ 1/x, & 0 < |x| < 1 \\ 0, & x = 1 \\ 1, & x > 1. \end{cases}$$

3. Suppose that $f(t)$ and $g(t)$ are defined for all t and that $\lim_{t \to t_0} f(t) = -7$ and $\lim_{t \to t_0} g(t) = 0$. Find the limit as $t \to t_0$ of the following functions.

   a. $3f(t)$ **b.** $(f(t))^2$

   c. $f(t) \cdot g(t)$ **d.** $\dfrac{f(t)}{g(t) - 7}$

   e. $\cos (g(t))$ **f.** $|f(t)|$

   g. $f(t) + g(t)$ **h.** $1/f(t)$

4. Suppose the functions $f(x)$ and $g(x)$ are defined for all x and that $\lim_{x \to 0} f(x) = 1/2$ and $\lim_{x \to 0} g(x) = \sqrt{2}$. Find the limits as $x \to 0$ of the following functions.

   a. $-g(x)$ **b.** $g(x) \cdot f(x)$

   c. $f(x) + g(x)$ **d.** $1/f(x)$

   e. $x + f(x)$ **f.** $\dfrac{f(x) \cdot \cos x}{x - 1}$

In Exercises 5 and 6, find the value that $\lim_{x \to 0} g(x)$ must have if the given limit statements hold.

5. $\lim_{x \to 0} \left(\dfrac{4 - g(x)}{x} \right) = 1$

6. $\lim_{x \to -4} \left(x \lim_{x \to 0} g(x) \right) = 2$

7. On what intervals are the following functions continuous?

   a. $f(x) = x^{1/3}$ **b.** $g(x) = x^{3/4}$

   c. $h(x) = x^{-2/3}$ **d.** $k(x) = x^{-1/6}$

8. On what intervals are the following functions continuous?

   a. $f(x) = \tan x$ **b.** $g(x) = \csc x$

   c. $h(x) = \dfrac{\cos x}{x - \pi}$ **d.** $k(x) = \dfrac{\sin x}{x}$

Finding Limits

In Exercises 9–28, find the limit or explain why it does not exist.

9. $\lim \dfrac{x^2 - 4x + 4}{x^3 + 5x^2 - 14x}$

    a. as $x \to 0$ **b.** as $x \to 2$

10. $\lim \dfrac{x^2 + x}{x^5 + 2x^4 + x^3}$

    a. as $x \to 0$ **b.** as $x \to -1$

11. $\lim\limits_{x \to 1} \dfrac{1 - \sqrt{x}}{1 - x}$

12. $\lim\limits_{x \to a} \dfrac{x^2 - a^2}{x^4 - a^4}$

13. $\lim\limits_{h \to 0} \dfrac{(x + h)^2 - x^2}{h}$

14. $\lim\limits_{x \to 0} \dfrac{(x + h)^2 - x^2}{h}$

15. $\lim\limits_{x \to 0} \dfrac{\dfrac{1}{2 + x} - \dfrac{1}{2}}{x}$

16. $\lim\limits_{x \to 0} \dfrac{(2 + x)^3 - 8}{x}$

17. $\lim\limits_{x \to 1} \dfrac{x^{1/3} - 1}{\sqrt{x} - 1}$

18. $\lim\limits_{x \to 64} \dfrac{x^{2/3} - 16}{\sqrt{x} - 8}$

19. $\lim\limits_{x \to 0} \dfrac{\tan (2x)}{\tan (\pi x)}$

20. $\lim\limits_{x \to \pi^-} \csc x$

21. $\lim\limits_{x \to \pi} \sin \left(\dfrac{x}{2} + \sin x \right)$

22. $\lim\limits_{x \to \pi} \cos^2 (x - \tan x)$

23. $\lim\limits_{x \to 0} \dfrac{8x}{3 \sin x - x}$

24. $\lim\limits_{x \to 0} \dfrac{\cos 2x - 1}{\sin x}$

25. $\lim\limits_{t \to 3^+} \ln (t - 3)$

26. $\lim\limits_{t \to 1} t^2 \ln \left(2 - \sqrt{t}\right)$

27. $\lim\limits_{\theta \to 0^+} \sqrt{\theta} \, e^{\cos (\pi/\theta)}$

28. $\lim\limits_{z \to 0^+} \dfrac{2e^{1/z}}{e^{1/z} + 1}$

In Exercises 29–32, find the limit of $g(x)$ as x approaches the indicated value.

29. $\lim\limits_{x \to 0^+} (4g(x))^{1/3} = 2$

30. $\lim\limits_{x \to \sqrt{5}} \dfrac{1}{x + g(x)} = 2$

31. $\lim\limits_{x \to 1} \dfrac{3x^2 + 1}{g(x)} = \infty$

32. $\lim\limits_{x \to -2} \dfrac{5 - x^2}{\sqrt{g(x)}} = 0$

Continuous Extension

33. Can $f(x) = x(x^2 - 1)/|x^2 - 1|$ be extended to be continuous at $x = 1$ or -1? Give reasons for your answers. (Graph the function—you will find the graph interesting.)

34. Explain why the function $f(x) = \sin(1/x)$ has no continuous extension to $x = 0$.

T In Exercises 35–38, graph the function to see whether it appears to have a continuous extension to the given point a. If it does, use Trace and Zoom to find a good candidate for the extended function's value at a. If the function does not appear to have a continuous extension, can it be extended to be continuous from the right or left? If so, what do you think the extended function's value should be?

35. $f(x) = \dfrac{x - 1}{x - \sqrt[4]{x}}, \quad a = 1$

36. $g(\theta) = \dfrac{5 \cos \theta}{4\theta - 2\pi}, \quad a = \pi/2$

37. $h(t) = (1 + |t|)^{1/t}, \quad a = 0$

38. $k(x) = \dfrac{x}{1 - 2^{|x|}}, \quad a = 0$

Roots

T 39. Let $f(x) = x^3 - x - 1$.

    a. Use the Intermediate Value Theorem to show that f has a zero between -1 and 2.

    b. Solve the equation $f(x) = 0$ graphically with an error of magnitude at most 10^{-8}.

    c. It can be shown that the exact value of the solution in part (b) is

$$\left(\dfrac{1}{2} + \dfrac{\sqrt{69}}{18} \right)^{1/3} + \left(\dfrac{1}{2} - \dfrac{\sqrt{69}}{18} \right)^{1/3}.$$

    Evaluate this exact answer and compare it with the value you found in part (b).

T 40. Let $f(\theta) = \theta^3 - 2\theta + 2$.

    a. Use the Intermediate Value Theorem to show that f has a zero between -2 and 0.

    b. Solve the equation $f(\theta) = 0$ graphically with an error of magnitude at most 10^{-4}.

    c. It can be shown that the exact value of the solution in part (b) is

$$\left(\sqrt{\dfrac{19}{27}} - 1 \right)^{1/3} - \left(\sqrt{\dfrac{19}{27}} + 1 \right)^{1/3}.$$

    Evaluate this exact answer and compare it with the value you found in part (b).

Limits at Infinity

Find the limits in Exercises 41–54.

41. $\lim\limits_{x \to \infty} \dfrac{2x + 3}{5x + 7}$

42. $\lim\limits_{x \to -\infty} \dfrac{2x^2 + 3}{5x^2 + 7}$

43. $\lim\limits_{x \to -\infty} \dfrac{x^2 - 4x + 8}{3x^3}$

44. $\lim\limits_{x \to \infty} \dfrac{1}{x^2 - 7x + 1}$

45. $\lim\limits_{x \to -\infty} \dfrac{x^2 - 7x}{x + 1}$

46. $\lim\limits_{x \to \infty} \dfrac{x^4 + x^3}{12x^3 + 128}$

47. $\lim\limits_{x \to \infty} \dfrac{\sin x}{\lfloor x \rfloor}$ (If you have a grapher, try graphing the function for $-5 \le x \le 5$.)

48. $\lim\limits_{\theta \to \infty} \dfrac{\cos \theta - 1}{\theta}$

    (If you have a grapher, try graphing $f(x) = x(\cos (1/x) - 1)$ near the origin to "see" the limit at infinity.)

49. $\lim\limits_{x \to \infty} \dfrac{x + \sin x + 2\sqrt{x}}{x + \sin x}$

50. $\lim\limits_{x \to \infty} \dfrac{x^{2/3} + x^{-1}}{x^{2/3} + \cos^2 x}$

51. $\lim\limits_{x \to \infty} e^{1/x} \cos \dfrac{1}{x}$

52. $\lim\limits_{t \to \infty} \ln \left(1 + \dfrac{1}{t} \right)$

53. $\lim\limits_{x \to -\infty} \tan^{-1} x$

54. $\lim\limits_{t \to -\infty} e^{3t} \sin^{-1} \dfrac{1}{t}$

Horizontal and Vertical Asymptotes

55. Use limits to determine the equations for all vertical asymptotes.

a. $y = \dfrac{x^2 + 4}{x - 3}$ **b.** $f(x) = \dfrac{x^2 - x - 2}{x^2 - 2x + 1}$

c. $y = \dfrac{x^2 + x - 6}{x^2 + 2x - 8}$

56. Use limits to determine the equations for all horizontal asymptotes.

a. $y = \dfrac{1 - x^2}{x^2 + 1}$ **b.** $f(x) = \dfrac{\sqrt{x} + 4}{\sqrt{x} + 4}$

c. $g(x) = \dfrac{\sqrt{x^2 + 4}}{x}$ **d.** $y = \sqrt{\dfrac{x^2 + 9}{9x^2 + 1}}$

Chapter 2 Additional and Advanced Exercises

T **1. Assigning a value to 0^0** The rules of exponents tell us that $a^0 = 1$ if a is any number different from zero. They also tell us that $0^n = 0$ if n is any positive number.

If we tried to extend these rules to include the case 0^0, we would get conflicting results. The first rule would say $0^0 = 1$, whereas the second would say $0^0 = 0$.

We are not dealing with a question of right or wrong here. Neither rule applies as it stands, so there is no contradiction. We could, in fact, define 0^0 to have any value we wanted as long as we could persuade others to agree.

What value would you like 0^0 to have? Here is an example that might help you to decide. (See Exercise 2 below for another example.)

a. Calculate x^x for $x = 0.1, 0.01, 0.001$, and so on as far as your calculator can go. Record the values you get. What pattern do you see?

b. Graph the function $y = x^x$ for $0 < x \le 1$. Even though the function is not defined for $x \le 0$, the graph will approach the y-axis from the right. Toward what y-value does it seem to be headed? Zoom in to further support your idea.

T **2. A reason you might want 0^0 to be something other than 0 or 1** As the number x increases through positive values, the numbers $1/x$ and $1/(\ln x)$ both approach zero. What happens to the number

$$f(x) = \left(\frac{1}{x}\right)^{1/(\ln x)}$$

as x increases? Here are two ways to find out.

a. Evaluate f for $x = 10, 100, 1000$, and so on as far as your calculator can reasonably go. What pattern do you see?

b. Graph f in a variety of graphing windows, including windows that contain the origin. What do you see? Trace the y-values along the graph. What do you find?

3. Lorentz contraction In relativity theory, the length of an object, say a rocket, appears to an observer to depend on the speed at which the object is traveling with respect to the observer. If the observer measures the rocket's length as L_0 at rest, then at speed v the length will appear to be

$$L = L_0\sqrt{1 - \frac{v^2}{c^2}}.$$

This equation is the Lorentz contraction formula. Here, c is the speed of light in a vacuum, about 3×10^8 m/sec. What happens to L as v increases? Find $\lim_{v \to c^-} L$. Why was the left-hand limit needed?

4. Controlling the flow from a draining tank Torricelli's law says that if you drain a tank like the one in the figure shown, the rate y at which water runs out is a constant times the square root of the water's depth x. The constant depends on the size and shape of the exit valve.

Exit rate y ft^3/min

Suppose that $y = \sqrt{x}/2$ for a certain tank. You are trying to maintain a fairly constant exit rate by adding water to the tank with a hose from time to time. How deep must you keep the water if you want to maintain the exit rate

a. within 0.2 ft^3/min of the rate $y_0 = 1$ ft^3/min?

b. within 0.1 ft^3/min of the rate $y_0 = 1$ ft^3/min?

5. Thermal expansion in precise equipment As you may know, most metals expand when heated and contract when cooled. The dimensions of a piece of laboratory equipment are sometimes so critical that the shop where the equipment is made must be held at the same temperature as the laboratory where the equipment is to be used. A typical aluminum bar that is 10 cm wide at 70°F will be

$$y = 10 + (t - 70) \times 10^{-4}$$

centimeters wide at a nearby temperature t. Suppose that you are using a bar like this in a gravity wave detector, where its width must stay within 0.0005 cm of the ideal 10 cm. How close to $t_0 = 70°F$ must you maintain the temperature to ensure that this tolerance is not exceeded?

6. Stripes on a measuring cup The interior of a typical 1-L measuring cup is a right circular cylinder of radius 6 cm (see accompanying figure). The volume of water we put in the cup is therefore a function of the level h to which the cup is filled, the formula being

$$V = \pi 6^2 h = 36\pi h.$$

How closely must we measure h to measure out 1 L of water (1000 cm^3) with an error of no more than 1% (10 cm^3)?

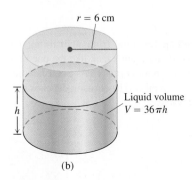

(a)

(b)

A 1-L measuring cup (a), modeled as a right circular cylinder (b) of radius $r = 6$ cm

Precise Definition of Limit

In Exercises 7–10, use the formal definition of limit to prove that the function is continuous at x_0.

7. $f(x) = x^2 - 7$, $x_0 = 1$ **8.** $g(x) = 1/(2x)$, $x_0 = 1/4$

9. $h(x) = \sqrt{2x - 3}$, $x_0 = 2$ **10.** $F(x) = \sqrt{9 - x}$, $x_0 = 5$

11. Uniqueness of limits Show that a function cannot have two different limits at the same point. That is, if $\lim_{x \to x_0} f(x) = L_1$ and $\lim_{x \to x_0} f(x) = L_2$, then $L_1 = L_2$.

12. Prove the limit Constant Multiple Rule:

$$\lim_{x \to c} k f(x) = k \lim_{x \to c} f(x) \quad \text{for any constant } k.$$

13. One-sided limits If $\lim_{x \to 0^+} f(x) = A$ and $\lim_{x \to 0^-} f(x) = B$, find

a. $\lim_{x \to 0^+} f(x^3 - x)$

b. $\lim_{x \to 0^-} f(x^3 - x)$

c. $\lim_{x \to 0^+} f(x^2 - x^4)$

d. $\lim_{x \to 0^-} f(x^2 - x^4)$

14. Limits and continuity Which of the following statements are true, and which are false? If true, say why; if false, give a counterexample (that is, an example confirming the falsehood).

a. If $\lim_{x \to a} f(x)$ exists but $\lim_{x \to a} g(x)$ does not exist, then $\lim_{x \to a} (f(x) + g(x))$ does not exist.

b. If neither $\lim_{x \to a} f(x)$ nor $\lim_{x \to a} g(x)$ exists, then $\lim_{x \to a} (f(x) + g(x))$ does not exist.

c. If f is continuous at x, then so is $|f|$.

d. If $|f|$ is continuous at a, then so is f.

In Exercises 15 and 16, use the formal definition of limit to prove that the function has a continuous extension to the given value of x.

15. $f(x) = \dfrac{x^2 - 1}{x + 1}$, $x = -1$ **16.** $g(x) = \dfrac{x^2 - 2x - 3}{2x - 6}$, $x = 3$

17. A function continuous at only one point Let

$$f(x) = \begin{cases} x, & \text{if } x \text{ is rational} \\ 0, & \text{if } x \text{ is irrational.} \end{cases}$$

a. Show that f is continuous at $x = 0$.

b. Use the fact that every nonempty open interval of real numbers contains both rational and irrational numbers to show that f is not continuous at any nonzero value of x.

18. The Dirichlet ruler function If x is a rational number, then x can be written in a unique way as a quotient of integers m/n where $n > 0$ and m and n have no common factors greater than 1. (We say that such a fraction is in *lowest terms*. For example, $6/4$ written in lowest terms is $3/2$.) Let $f(x)$ be defined for all x in the interval $[0, 1]$ by

$$f(x) = \begin{cases} 1/n, & \text{if } x = m/n \text{ is a rational number in lowest terms} \\ 0, & \text{if } x \text{ is irrational.} \end{cases}$$

For instance, $f(0) = f(1) = 1, f(1/2) = 1/2, f(1/3) = f(2/3) = 1/3, f(1/4) = f(3/4) = 1/4$, and so on.

a. Show that f is discontinuous at every rational number in $[0, 1]$.

b. Show that f is continuous at every irrational number in $[0, 1]$. (*Hint*: If ϵ is a given positive number, show that there are only finitely many rational numbers r in $[0, 1]$ such that $f(r) \geq \epsilon$.)

c. Sketch the graph of f. Why do you think f is called the "ruler function"?

19. Antipodal points Is there any reason to believe that there is always a pair of antipodal (diametrically opposite) points on Earth's equator where the temperatures are the same? Explain.

20. If $\lim_{x \to c} (f(x) + g(x)) = 3$ and $\lim_{x \to c} (f(x) - g(x)) = -1$, find $\lim_{x \to c} f(x)g(x)$.

21. Roots of a quadratic equation that is almost linear The equation $ax^2 + 2x - 1 = 0$, where a is a constant, has two roots if $a > -1$ and $a \neq 0$, one positive and one negative:

$$r_+(a) = \frac{-1 + \sqrt{1 + a}}{a}, \qquad r_-(a) = \frac{-1 - \sqrt{1 + a}}{a}.$$

a. What happens to $r_+(a)$ as $a \to 0$? As $a \to -1^+$?

b. What happens to $r_-(a)$ as $a \to 0$? As $a \to -1^+$?

c. Support your conclusions by graphing $r_+(a)$ and $r_-(a)$ as functions of a. Describe what you see.

d. For added support, graph $f(x) = ax^2 + 2x - 1$ simultaneously for $a = 1, 0.5, 0.2, 0.1$, and 0.05.

22. Root of an equation Show that the equation $x + 2 \cos x = 0$ has at least one solution.

23. Bounded functions A real-valued function f is **bounded from above** on a set D if there exists a number N such that $f(x) \leq N$ for all x in D. We call N, when it exists, an **upper bound** for f on D and say that f is bounded from above by N. In a similar manner, we say that f is **bounded from below** on D if there exists a number M such that $f(x) \geq M$ for all x in D. We call M, when it exists, a **lower bound** for f on D and say that f is bounded from below by M. We say that f is **bounded** on D if it is bounded from both above and below.

a. Show that f is bounded on D if and only if there exists a number B such that $|f(x)| \leq B$ for all x in D.

b. Suppose that f is bounded from above by N. Show that if $\lim_{x \to x_0} f(x) = L$, then $L \leq N$.

c. Suppose that f is bounded from below by M. Show that if $\lim_{x \to x_0} f(x) = L$, then $L \geq M$.

24. Max $\{a, b\}$ and min $\{a, b\}$

a. Show that the expression

$$\max \{a, b\} = \frac{a + b}{2} + \frac{|a - b|}{2}$$

equals a if $a \geq b$ and equals b if $b \geq a$. In other words, $\max \{a, b\}$ gives the larger of the two numbers a and b.

b. Find a similar expression for $\min \{a, b\}$, the smaller of a and b.

Generalized Limits Involving $\dfrac{\sin \theta}{\theta}$

The formula $\lim_{\theta \to 0}(\sin \theta)/\theta = 1$ can be generalized. If $\lim_{x \to c} f(x) = 0$ and $f(x)$ is never zero in an open interval containing the point $x = c$, except possibly c itself, then

$$\lim_{x \to c} \frac{\sin f(x)}{f(x)} = 1.$$

Here are several examples.

a. $\displaystyle\lim_{x \to 0} \frac{\sin x^2}{x^2} = 1$

b. $\displaystyle\lim_{x \to 0} \frac{\sin x^2}{x} = \lim_{x \to 0} \frac{\sin x^2}{x^2} \lim_{x \to 0} \frac{x^2}{x} = 1 \cdot 0 = 0$

c. $\displaystyle\lim_{x \to -1} \frac{\sin (x^2 - x - 2)}{x + 1} = \lim_{x \to -1} \frac{\sin (x^2 - x - 2)}{(x^2 - x - 2)} \cdot$

$\displaystyle\lim_{x \to -1} \frac{(x^2 - x - 2)}{x + 1} = 1 \cdot \lim_{x \to -1} \frac{(x + 1)(x - 2)}{x + 1} = -3$

d. $\displaystyle\lim_{x \to 1} \frac{\sin \left(1 - \sqrt{x}\right)}{x - 1} = \lim_{x \to 1} \frac{\sin \left(1 - \sqrt{x}\right)}{1 - \sqrt{x}} \frac{1 - \sqrt{x}}{x - 1} =$

$\displaystyle 1 \cdot \lim_{x \to 1} \frac{\left(1 - \sqrt{x}\right)\left(1 + \sqrt{x}\right)}{(x - 1)\left(1 + \sqrt{x}\right)} = \lim_{x \to 1} \frac{1 - x}{(x - 1)\left(1 + \sqrt{x}\right)} = -\frac{1}{2}$

Find the limits in Exercises 25–30.

25. $\displaystyle\lim_{x \to 0} \frac{\sin (1 - \cos x)}{x}$

26. $\displaystyle\lim_{x \to 0^+} \frac{\sin x}{\sin \sqrt{x}}$

27. $\displaystyle\lim_{x \to 0} \frac{\sin (\sin x)}{x}$

28. $\displaystyle\lim_{x \to 0} \frac{\sin (x^2 + x)}{x}$

29. $\displaystyle\lim_{x \to 2} \frac{\sin (x^2 - 4)}{x - 2}$

30. $\displaystyle\lim_{x \to 9} \frac{\sin \left(\sqrt{x} - 3\right)}{x - 9}$

Oblique Asymptotes

Find all possible oblique asymptotes in Exercises 31–34.

31. $y = \dfrac{2x^{3/2} + 2x - 3}{\sqrt{x} + 1}$

32. $y = x + x \sin (1/x)$

33. $y = \sqrt{x^2 + 1}$

34. $y = \sqrt{x^2 + 2x}$

Chapter 2 Technology Application Projects

Mathematica/Maple Modules:

Take It to the Limit

Part I

Part II (Zero Raised to the Power Zero: What Does it Mean?)

Part III (One-Sided Limits)

Visualize and interpret the limit concept through graphical and numerical explorations.

Part IV (What a Difference a Power Makes)

See how sensitive limits can be with various powers of x.

Going to Infinity

Part I (Exploring Function Behavior as $x \to \infty$ or $x \to -\infty$)

This module provides four examples to explore the behavior of a function as $x \to \infty$ or $x \to -\infty$.

Part II (Rates of Growth)

Observe graphs that *appear* to be continuous, yet the function is not continuous. Several issues of continuity are explored to obtain results that you may find surprising.

3

DIFFERENTIATION

OVERVIEW   In the beginning of Chapter 2 we discussed how to determine the slope of a curve at a point and how to measure the rate at which a function changes. Now that we have studied limits, we can define these ideas precisely and see that both are interpretations of the *derivative* of a function at a point. We then extend this concept from a single point to the *derivative function*, and we develop rules for finding this derivative function easily, without having to calculate any limits directly. These rules are used to find derivatives of most of the common functions reviewed in Chapter 1, as well as various combinations of them. The derivative is one of the key ideas in calculus, and we use it to solve a wide range of problems involving tangents and rates of change.

3.1 Tangents and the Derivative at a Point

In this section we define the slope and tangent to a curve at a point, and the derivative of a function at a point. Later in the chapter we interpret the derivative as the instantaneous rate of change of a function, and apply this interpretation to the study of certain types of motion.

Finding a Tangent to the Graph of a Function

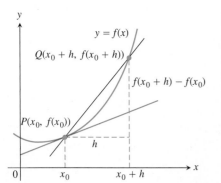

FIGURE 3.1   The slope of the tangent line at P is $\lim\limits_{h \to 0} \dfrac{f(x_0 + h) - f(x_0)}{h}$.

To find a tangent to an arbitrary curve $y = f(x)$ at a point $P(x_0, f(x_0))$, we use the procedure introduced in Section 2.1. We calculate the slope of the secant through P and a nearby point $Q(x_0 + h, f(x_0 + h))$. We then investigate the limit of the slope as $h \to 0$ (Figure 3.1). If the limit exists, we call it the slope of the curve at P and define the tangent at P to be the line through P having this slope.

DEFINITIONS   The **slope of the curve** $y = f(x)$ at the point $P(x_0, f(x_0))$ is the number

$$m = \lim_{h \to 0} \frac{f(x_0 + h) - f(x_0)}{h} \qquad \text{(provided the limit exists)}.$$

The **tangent line** to the curve at P is the line through P with this slope.

In Section 2.1, Example 3, we applied these definitions to find the slope of the parabola $f(x) = x^2$ at the point $P(2, 4)$ and the tangent line to the parabola at P. Let's look at another example.

EXAMPLE 1

(a) Find the slope of the curve $y = 1/x$ at any point $x = a \neq 0$. What is the slope at the point $x = -1$?

(b) Where does the slope equal $-1/4$?

(c) What happens to the tangent to the curve at the point $(a, 1/a)$ as a changes?

Solution

(a) Here $f(x) = 1/x$. The slope at $(a, 1/a)$ is

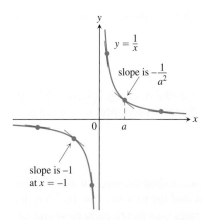

$$\lim_{h \to 0} \frac{f(a + h) - f(a)}{h} = \lim_{h \to 0} \frac{\frac{1}{a + h} - \frac{1}{a}}{h} = \lim_{h \to 0} \frac{1}{h} \frac{a - (a + h)}{a(a + h)}$$

$$= \lim_{h \to 0} \frac{-h}{ha(a + h)} = \lim_{h \to 0} \frac{-1}{a(a + h)} = -\frac{1}{a^2}.$$

FIGURE 3.2 The tangent slopes, steep near the origin, become more gradual as the point of tangency moves away (Example 1).

Notice how we had to keep writing "$\lim_{h \to 0}$" before each fraction until the stage where we could evaluate the limit by substituting $h = 0$. The number a may be positive or negative, but not 0. When $a = -1$, the slope is $-1/(-1)^2 = -1$ (Figure 3.2).

(b) The slope of $y = 1/x$ at the point where $x = a$ is $-1/a^2$. It will be $-1/4$ provided that

$$-\frac{1}{a^2} = -\frac{1}{4}.$$

This equation is equivalent to $a^2 = 4$, so $a = 2$ or $a = -2$. The curve has slope $-1/4$ at the two points $(2, 1/2)$ and $(-2, -1/2)$ (Figure 3.3).

(c) The slope $-1/a^2$ is always negative if $a \neq 0$. As $a \to 0^+$, the slope approaches $-\infty$ and the tangent becomes increasingly steep (Figure 3.2). We see this situation again as $a \to 0^-$. As a moves away from the origin in either direction, the slope approaches 0 and the tangent levels off to become horizontal. ∎

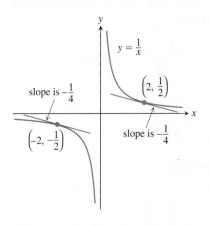

FIGURE 3.3 The two tangent lines to $y = 1/x$ having slope $-1/4$ (Example 1).

Rates of Change: Derivative at a Point

The expression

$$\frac{f(x_0 + h) - f(x_0)}{h}, \quad h \neq 0$$

is called the **difference quotient of f at x_0 with increment h**. If the difference quotient has a limit as h approaches zero, that limit is given a special name and notation.

DEFINITION The **derivative of a function f at a point x_0**, denoted $f'(x_0)$, is

$$f'(x_0) = \lim_{h \to 0} \frac{f(x_0 + h) - f(x_0)}{h}$$

provided this limit exists.

If we interpret the difference quotient as the slope of a secant line, then the derivative gives the slope of the curve $y = f(x)$ at the point $P(x_0, f(x_0))$. Exercise 31 shows that the derivative of the linear function $f(x) = mx + b$ at any point x_0 is simply the slope of the line, so

$$f'(x_0) = m,$$

which is consistent with our definition of slope.

If we interpret the difference quotient as an average rate of change (Section 2.1), the derivative gives the function's instantaneous rate of change with respect to x at the point $x = x_0$. We study this interpretation in Section 3.4.

EXAMPLE 2   In Examples 1 and 2 in Section 2.1, we studied the speed of a rock falling freely from rest near the surface of the earth. We knew that the rock fell $y = 16t^2$ feet during the first t sec, and we used a sequence of average rates over increasingly short intervals to estimate the rock's speed at the instant $t = 1$. What was the rock's *exact* speed at this time?

Solution   We let $f(t) = 16t^2$. The average speed of the rock over the interval between $t = 1$ and $t = 1 + h$ seconds, for $h > 0$, was found to be

$$\frac{f(1 + h) - f(1)}{h} = \frac{16(1 + h)^2 - 16(1)^2}{h} = \frac{16(h^2 + 2h)}{h} = 16(h + 2).$$

The rock's speed at the instant $t = 1$ is then

$$\lim_{h \to 0} 16(h + 2) = 16(0 + 2) = 32 \text{ ft/sec.}$$

Our original estimate of 32 ft/ sec in Section 2.1 was right.   ∎

Summary

We have been discussing slopes of curves, lines tangent to a curve, the rate of change of a function, and the derivative of a function at a point. All of these ideas refer to the same limit.

The following are all interpretations for the limit of the difference quotient,

$$\lim_{h \to 0} \frac{f(x_0 + h) - f(x_0)}{h}.$$

1. The slope of the graph of $y = f(x)$ at $x = x_0$
2. The slope of the tangent to the curve $y = f(x)$ at $x = x_0$
3. The rate of change of $f(x)$ with respect to x at $x = x_0$
4. The derivative $f'(x_0)$ at a point

In the next sections, we allow the point x_0 to vary across the domain of the function f.

Exercises 3.1

Slopes and Tangent Lines

In Exercises 1–4, use the grid and a straight edge to make a rough estimate of the slope of the curve (in y-units per x-unit) at the points P_1 and P_2.

1.

2.

3.

4.

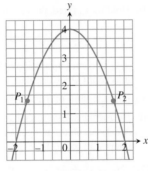

In Exercises 5–10, find an equation for the tangent to the curve at the given point. Then sketch the curve and tangent together.

5. $y = 4 - x^2$, $(-1, 3)$
6. $y = (x - 1)^2 + 1$, $(1, 1)$

7. $y = 2\sqrt{x}$, $(1, 2)$
8. $y = \dfrac{1}{x^2}$, $(-1, 1)$

9. $y = x^3$, $(-2, -8)$
10. $y = \dfrac{1}{x^3}$, $\left(-2, -\dfrac{1}{8}\right)$

In Exercises 11–18, find the slope of the function's graph at the given point. Then find an equation for the line tangent to the graph there.

11. $f(x) = x^2 + 1$, $(2, 5)$
12. $f(x) = x - 2x^2$, $(1, -1)$

13. $g(x) = \dfrac{x}{x - 2}$, $(3, 3)$
14. $g(x) = \dfrac{8}{x^2}$, $(2, 2)$

15. $h(t) = t^3$, $(2, 8)$
16. $h(t) = t^3 + 3t$, $(1, 4)$

17. $f(x) = \sqrt{x}$, $(4, 2)$
18. $f(x) = \sqrt{x + 1}$, $(8, 3)$

In Exercises 19–22, find the slope of the curve at the point indicated.

19. $y = 5x^2$, $x = -1$
20. $y = 1 - x^2$, $x = 2$

21. $y = \dfrac{1}{x - 1}$, $x = 3$
22. $y = \dfrac{x - 1}{x + 1}$, $x = 0$

Tangent Lines with Specified Slopes

At what points do the graphs of the functions in Exercises 23 and 24 have horizontal tangents?

23. $f(x) = x^2 + 4x - 1$
24. $g(x) = x^3 - 3x$

25. Find equations of all lines having slope -1 that are tangent to the curve $y = 1/(x - 1)$.

26. Find an equation of the straight line having slope $1/4$ that is tangent to the curve $y = \sqrt{x}$.

Rates of Change

27. Object dropped from a tower An object is dropped from the top of a 100-m-high tower. Its height above ground after t sec is $100 - 4.9t^2$ m. How fast is it falling 2 sec after it is dropped?

28. Speed of a rocket At t sec after liftoff, the height of a rocket is $3t^2$ ft. How fast is the rocket climbing 10 sec after liftoff?

29. Circle's changing area What is the rate of change of the area of a circle ($A = \pi r^2$) with respect to the radius when the radius is $r = 3$?

30. Ball's changing volume What is the rate of change of the volume of a ball ($V = (4/3)\pi r^3$) with respect to the radius when the radius is $r = 2$?

31. Show that the line $y = mx + b$ is its own tangent line at any point $(x_0, mx_0 + b)$.

32. Find the slope of the tangent to the curve $y = 1/\sqrt{x}$ at the point where $x = 4$.

Testing for Tangents

33. Does the graph of

$$f(x) = \begin{cases} x^2 \sin(1/x), & x \neq 0 \\ 0, & x = 0 \end{cases}$$

have a tangent at the origin? Give reasons for your answer.

34. Does the graph of

$$g(x) = \begin{cases} x \sin(1/x), & x \neq 0 \\ 0, & x = 0 \end{cases}$$

have a tangent at the origin? Give reasons for your answer.

Vertical Tangents

We say that a continuous curve $y = f(x)$ has a **vertical tangent** at the point where $x = x_0$ if $\lim_{h \to 0} (f(x_0 + h) - f(x_0))/h = \infty$ or $-\infty$. For example, $y = x^{1/3}$ has a vertical tangent at $x = 0$ (see accompanying figure):

$$\lim_{h \to 0} \frac{f(0 + h) - f(0)}{h} = \lim_{h \to 0} \frac{h^{1/3} - 0}{h}$$

$$= \lim_{h \to 0} \frac{1}{h^{2/3}} = \infty.$$

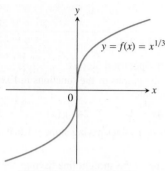

$$y = f(x) = x^{1/3}$$

VERTICAL TANGENT AT ORIGIN

However, $y = x^{2/3}$ has *no* vertical tangent at $x = 0$ (see next figure):

$$\lim_{h \to 0} \frac{g(0 + h) - g(0)}{h} = \lim_{h \to 0} \frac{h^{2/3} - 0}{h}$$

$$= \lim_{h \to 0} \frac{1}{h^{1/3}}$$

does not exist, because the limit is ∞ from the right and $-\infty$ from the left.

$$y = g(x) = x^{2/3}$$

NO VERTICAL TANGENT AT ORIGIN

35. Does the graph of

$$f(x) = \begin{cases} -1, & x < 0 \\ 0, & x = 0 \\ 1, & x > 0 \end{cases}$$

have a vertical tangent at the origin? Give reasons for your answer.

36. Does the graph of

$$U(x) = \begin{cases} 0, & x < 0 \\ 1, & x \ge 0 \end{cases}$$

have a vertical tangent at the point $(0, 1)$? Give reasons for your answer.

T Graph the curves in Exercises 37–46.

    a. Where do the graphs appear to have vertical tangents?

    b. Confirm your findings in part (a) with limit calculations. But before you do, read the introduction to Exercises 35 and 36.

37. $y = x^{2/5}$ **38.** $y = x^{4/5}$

39. $y = x^{1/5}$ **40.** $y = x^{3/5}$

41. $y = 4x^{2/5} - 2x$ **42.** $y = x^{5/3} - 5x^{2/3}$

43. $y = x^{2/3} - (x - 1)^{1/3}$ **44.** $y = x^{1/3} + (x - 1)^{1/3}$

45. $y = \begin{cases} -\sqrt{|x|}, & x \le 0 \\ \sqrt{x}, & x > 0 \end{cases}$ **46.** $y = \sqrt{|4 - x|}$

COMPUTER EXPLORATIONS

Use a CAS to perform the following steps for the functions in Exercises 47–50:

    a. Plot $y = f(x)$ over the interval $(x_0 - 1/2) \le x \le (x_0 + 3)$.

    b. Holding x_0 fixed, the difference quotient

$$q(h) = \frac{f(x_0 + h) - f(x_0)}{h}$$

    at x_0 becomes a function of the step size h. Enter this function into your CAS workspace.

    c. Find the limit of q as $h \to 0$.

    d. Define the secant lines $y = f(x_0) + q \cdot (x - x_0)$ for $h = 3, 2$, and 1. Graph them together with f and the tangent line over the interval in part (a).

47. $f(x) = x^3 + 2x, \quad x_0 = 0$ **48.** $f(x) = x + \dfrac{5}{x}, \quad x_0 = 1$

49. $f(x) = x + \sin(2x), \quad x_0 = \pi/2$

50. $f(x) = \cos x + 4\sin(2x), \quad x_0 = \pi$

3.2   The Derivative as a Function

HISTORICAL ESSAY

The Derivative

In the last section we defined the derivative of $y = f(x)$ at the point $x = x_0$ to be the limit

$$f'(x_0) = \lim_{h \to 0} \frac{f(x_0 + h) - f(x_0)}{h}.$$

We now investigate the derivative as a *function* derived from f by considering the limit at each point x in the domain of f.

> **DEFINITION**   The **derivative** of the function $f(x)$ with respect to the variable x is the function f' whose value at x is
>
> $$f'(x) = \lim_{h \to 0} \frac{f(x + h) - f(x)}{h},$$
>
> provided the limit exists.

We use the notation $f(x)$ in the definition to emphasize the independent variable x with respect to which the derivative function $f'(x)$ is being defined. The domain of f' is the set of points in the domain of f for which the limit exists, which means that the domain may be the same as or smaller than the domain of f. If f' exists at a particular x, we say that f is **differentiable (has a derivative) at x**. If f' exists at every point in the domain of f, we call f **differentiable**.

If we write $z = x + h$, then $h = z - x$ and h approaches 0 if and only if z approaches x. Therefore, an equivalent definition of the derivative is as follows (see Figure 3.4). This formula is sometimes more convenient to use when finding a derivative function.

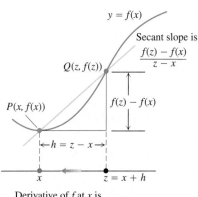

Derivative of f at x is
$$f'(x) = \lim_{h \to 0} \frac{f(x + h) - f(x)}{h}$$
$$= \lim_{z \to x} \frac{f(z) - f(x)}{z - x}$$

FIGURE 3.4 Two forms for the difference quotient.

Alternative Formula for the Derivative

$$f'(x) = \lim_{z \to x} \frac{f(z) - f(x)}{z - x}.$$

Calculating Derivatives from the Definition

The process of calculating a derivative is called **differentiation**. To emphasize the idea that differentiation is an operation performed on a function $y = f(x)$, we use the notation

$$\frac{d}{dx} f(x)$$

as another way to denote the derivative $f'(x)$. Example 1 of Section 3.1 illustrated the differentiation process for the function $y = 1/x$ when $x = a$. For x representing any point in the domain, we get the formula

$$\frac{d}{dx}\left(\frac{1}{x}\right) = -\frac{1}{x^2}.$$

Derivative of the Reciprocal Function

$$\frac{d}{dx}\left(\frac{1}{x}\right) = -\frac{1}{x^2}, \quad x \neq 0$$

Here are two more examples in which we allow x to be any point in the domain of f.

EXAMPLE 1 Differentiate $f(x) = \dfrac{x}{x - 1}$.

Solution We use the definition of derivative, which requires us to calculate $f(x + h)$ and then subtract $f(x)$ to obtain the numerator in the difference quotient. We have

$$f(x) = \frac{x}{x - 1} \quad \text{and} \quad f(x + h) = \frac{(x + h)}{(x + h) - 1}, \text{ so}$$

$$f'(x) = \lim_{h \to 0} \frac{f(x + h) - f(x)}{h} \qquad \text{Definition}$$

$$= \lim_{h \to 0} \frac{\dfrac{x + h}{x + h - 1} - \dfrac{x}{x - 1}}{h}$$

$$= \lim_{h \to 0} \frac{1}{h} \cdot \frac{(x + h)(x - 1) - x(x + h - 1)}{(x + h - 1)(x - 1)} \qquad \frac{a}{b} - \frac{c}{d} = \frac{ad - cb}{bd}$$

$$= \lim_{h \to 0} \frac{1}{h} \cdot \frac{-h}{(x + h - 1)(x - 1)} \qquad \text{Simplify.}$$

$$= \lim_{h \to 0} \frac{-1}{(x + h - 1)(x - 1)} = \frac{-1}{(x - 1)^2}. \qquad \text{Cancel } h \neq 0.$$

EXAMPLE 2

(a) Find the derivative of $f(x) = \sqrt{x}$ for $x > 0$.

(b) Find the tangent line to the curve $y = \sqrt{x}$ at $x = 4$.

Solution

(a) We use the alternative formula to calculate f':

$$f'(x) = \lim_{z \to x} \frac{f(z) - f(x)}{z - x}$$

$$= \lim_{z \to x} \frac{\sqrt{z} - \sqrt{x}}{z - x}$$

$$= \lim_{z \to x} \frac{\sqrt{z} - \sqrt{x}}{\left(\sqrt{z} - \sqrt{x}\right)\left(\sqrt{z} + \sqrt{x}\right)}$$

$$= \lim_{z \to x} \frac{1}{\sqrt{z} + \sqrt{x}} = \frac{1}{2\sqrt{x}}.$$

(b) The slope of the curve at $x = 4$ is

$$f'(4) = \frac{1}{2\sqrt{4}} = \frac{1}{4}.$$

The tangent is the line through the point $(4, 2)$ with slope $1/4$ (Figure 3.5):

$$y = 2 + \frac{1}{4}(x - 4)$$

$$y = \frac{1}{4}x + 1.$$ ∎

Derivative of the Square Root Function

$$\frac{d}{dx}\sqrt{x} = \frac{1}{2\sqrt{x}}, \quad x > 0$$

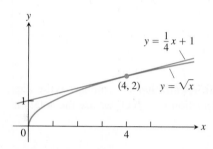

$y = \frac{1}{4}x + 1$

$(4, 2)$ $y = \sqrt{x}$

FIGURE 3.5 The curve $y = \sqrt{x}$ and its tangent at $(4, 2)$. The tangent's slope is found by evaluating the derivative at $x = 4$ (Example 2).

Notations

There are many ways to denote the derivative of a function $y = f(x)$, where the independent variable is x and the dependent variable is y. Some common alternative notations for the derivative are

$$f'(x) = y' = \frac{dy}{dx} = \frac{df}{dx} = \frac{d}{dx}f(x) = D(f)(x) = D_x f(x).$$

The symbols d/dx and D indicate the operation of differentiation. We read dy/dx as "the derivative of y with respect to x," and df/dx and $(d/dx)f(x)$ as "the derivative of f with respect to x." The "prime" notations y' and f' come from notations that Newton used for derivatives. The d/dx notations are similar to those used by Leibniz. The symbol dy/dx should not be regarded as a ratio (until we introduce the idea of "differentials" in Section 3.11).

To indicate the value of a derivative at a specified number $x = a$, we use the notation

$$f'(a) = \frac{dy}{dx}\bigg|_{x=a} = \frac{df}{dx}\bigg|_{x=a} = \frac{d}{dx}f(x)\bigg|_{x=a}.$$

For instance, in Example 2

$$f'(4) = \frac{d}{dx}\sqrt{x}\bigg|_{x=4} = \frac{1}{2\sqrt{x}}\bigg|_{x=4} = \frac{1}{2\sqrt{4}} = \frac{1}{4}.$$

Graphing the Derivative

We can often make a reasonable plot of the derivative of $y = f(x)$ by estimating the slopes on the graph of f. That is, we plot the points $(x, f'(x))$ in the xy-plane and connect them with a smooth curve, which represents $y = f'(x)$.

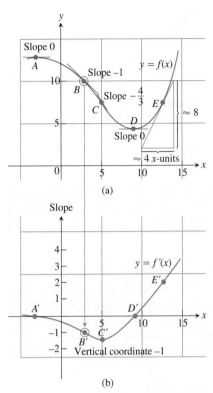

(a)

(b)

FIGURE 3.6 We made the graph of $y = f'(x)$ in (b) by plotting slopes from the graph of $y = f(x)$ in (a). The vertical coordinate of B' is the slope at B and so on. The slope at E is approximately $8/4 = 2$. In (b) we see that the rate of change of f is negative for x between A' and D'; the rate of change is positive for x to the right of D'.

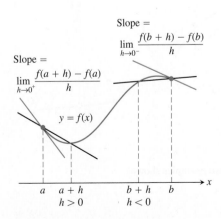

FIGURE 3.7 Derivatives at endpoints are one-sided limits.

EXAMPLE 3   Graph the derivative of the function $y = f(x)$ in Figure 3.6a.

Solution   We sketch the tangents to the graph of f at frequent intervals and use their slopes to estimate the values of $f'(x)$ at these points. We plot the corresponding $(x, f'(x))$ pairs and connect them with a smooth curve as sketched in Figure 3.6b.   ∎

What can we learn from the graph of $y = f'(x)$? At a glance we can see

1.   where the rate of change of f is positive, negative, or zero;
2.   the rough size of the growth rate at any x and its size in relation to the size of $f(x)$;
3.   where the rate of change itself is increasing or decreasing.

Differentiable on an Interval; One-Sided Derivatives

A function $y = f(x)$ is **differentiable on an open interval** (finite or infinite) if it has a derivative at each point of the interval. It is **differentiable on a closed interval** $[a, b]$ if it is differentiable on the interior (a, b) and if the limits

$$\lim_{h \to 0^+} \frac{f(a + h) - f(a)}{h} \qquad \textbf{Right-hand derivative at } a$$

$$\lim_{h \to 0^-} \frac{f(b + h) - f(b)}{h} \qquad \textbf{Left-hand derivative at } b$$

exist at the endpoints (Figure 3.7).

Right-hand and left-hand derivatives may be defined at any point of a function's domain. Because of Theorem 6, Section 2.4, a function has a derivative at a point if and only if it has left-hand and right-hand derivatives there, and these one-sided derivatives are equal.

EXAMPLE 4   Show that the function $y = |x|$ is differentiable on $(-\infty, 0)$ and $(0, \infty)$ but has no derivative at $x = 0$.

Solution   From Section 3.1, the derivative of $y = mx + b$ is the slope m. Thus, to the right of the origin,

$$\frac{d}{dx}(|x|) = \frac{d}{dx}(x) = \frac{d}{dx}(1 \cdot x) = 1. \qquad \tfrac{d}{dx}(mx + b) = m, \; |x| = x$$

To the left,

$$\frac{d}{dx}(|x|) = \frac{d}{dx}(-x) = \frac{d}{dx}(-1 \cdot x) = -1 \qquad |x| = -x$$

(Figure 3.8). There is no derivative at the origin because the one-sided derivatives differ there:

$$\text{Right-hand derivative of } |x| \text{ at zero} = \lim_{h \to 0^+} \frac{|0 + h| - |0|}{h} = \lim_{h \to 0^+} \frac{|h|}{h}$$

$$= \lim_{h \to 0^+} \frac{h}{h} \qquad |h| = h \text{ when } h > 0$$

$$= \lim_{h \to 0^+} 1 = 1$$

$$\text{Left-hand derivative of } |x| \text{ at zero} = \lim_{h \to 0^-} \frac{|0 + h| - |0|}{h} = \lim_{h \to 0^-} \frac{|h|}{h}$$

$$= \lim_{h \to 0^-} \frac{-h}{h} \qquad |h| = -h \text{ when } h < 0$$

$$= \lim_{h \to 0^-} -1 = -1. \qquad ∎$$

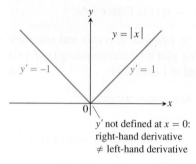

FIGURE 3.8 The function $y = |x|$ is not differentiable at the origin where the graph has a "corner" (Example 4).

EXAMPLE 5 In Example 2 we found that for $x > 0$,

$$\frac{d}{dx}\sqrt{x} = \frac{1}{2\sqrt{x}}.$$

We apply the definition to examine if the derivative exists at $x = 0$:

$$\lim_{h \to 0^+} \frac{\sqrt{0 + h} - \sqrt{0}}{h} = \lim_{h \to 0^+} \frac{1}{\sqrt{h}} = \infty.$$

Since the (right-hand) limit is not finite, there is no derivative at $x = 0$. Since the slopes of the secant lines joining the origin to the points $(h, \sqrt{h})$ on a graph of $y = \sqrt{x}$ approach ∞, the graph has a *vertical tangent* at the origin. (See Figure 1.17 on page 9.) ∎

When Does a Function *Not* Have a Derivative at a Point?

A function has a derivative at a point x_0 if the slopes of the secant lines through $P(x_0, f(x_0))$ and a nearby point Q on the graph approach a finite limit as Q approaches P. Whenever the secants fail to take up a limiting position or become vertical as Q approaches P, the derivative does not exist. Thus differentiability is a "smoothness" condition on the graph of f. A function can fail to have a derivative at a point for many reasons, including the existence of points where the graph has

1. a *corner*, where the one-sided derivatives differ.

2. a *cusp*, where the slope of PQ approaches ∞ from one side and $-\infty$ from the other.

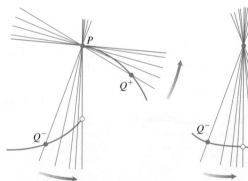

3. a *vertical tangent*, where the slope of PQ approaches ∞ from both sides or approaches $-\infty$ from both sides (here, $-\infty$).

4. a *discontinuity* (two examples shown).

Another case in which the derivative may fail to exist occurs when the function's slope is oscillating rapidly near P, as with $f(x) = \sin(1/x)$ near the origin, where it is discontinuous (see Figure 2.31).

Differentiable Functions Are Continuous

A function is continuous at every point where it has a derivative.

THEOREM 1—Differentiability Implies Continuity   If f has a derivative at $x = c$, then f is continuous at $x = c$.

Proof   Given that $f'(c)$ exists, we must show that $\lim_{x \to c} f(x) = f(c)$, or equivalently, that $\lim_{h \to 0} f(c + h) = f(c)$. If $h \neq 0$, then

$$f(c + h) = f(c) + (f(c + h) - f(c))$$

$$= f(c) + \frac{f(c + h) - f(c)}{h} \cdot h.$$

Now take limits as $h \to 0$. By Theorem 1 of Section 2.2,

$$\lim_{h \to 0} f(c + h) = \lim_{h \to 0} f(c) + \lim_{h \to 0} \frac{f(c + h) - f(c)}{h} \cdot \lim_{h \to 0} h$$

$$= f(c) + f'(c) \cdot 0$$

$$= f(c) + 0$$

$$= f(c). \qquad \blacksquare$$

Similar arguments with one-sided limits show that if f has a derivative from one side (right or left) at $x = c$ then f is continuous from that side at $x = c$.

Theorem 1 says that if a function has a discontinuity at a point (for instance, a jump discontinuity), then it cannot be differentiable there. The greatest integer function $y = \lfloor x \rfloor$ fails to be differentiable at every integer $x = n$ (Example 4, Section 2.5).

Caution   The converse of Theorem 1 is false. A function need not have a derivative at a point where it is continuous, as we saw in Example 4.

Exercises 3.2

Finding Derivative Functions and Values

Using the definition, calculate the derivatives of the functions in Exercises 1–6. Then find the values of the derivatives as specified.

1. $f(x) = 4 - x^2$; $f'(-3), f'(0), f'(1)$

2. $F(x) = (x - 1)^2 + 1$; $F'(-1), F'(0), F'(2)$

3. $g(t) = \dfrac{1}{t^2}$; $g'(-1), g'(2), g'(\sqrt{3})$

4. $k(z) = \dfrac{1 - z}{2z}$; $k'(-1), k'(1), k'(\sqrt{2})$

5. $p(\theta) = \sqrt{3\theta}$; $p'(1), p'(3), p'(2/3)$

6. $r(s) = \sqrt{2s + 1}$; $r'(0), r'(1), r'(1/2)$

In Exercises 7–12, find the indicated derivatives.

7. $\dfrac{dy}{dx}$ if $y = 2x^3$

8. $\dfrac{dr}{ds}$ if $r = s^3 - 2s^2 + 3$

9. $\dfrac{ds}{dt}$ if $s = \dfrac{t}{2t + 1}$

10. $\dfrac{dv}{dt}$ if $v = t - \dfrac{1}{t}$

11. $\dfrac{dp}{dq}$ if $p = \dfrac{1}{\sqrt{q + 1}}$

12. $\dfrac{dz}{dw}$ if $z = \dfrac{1}{\sqrt{3w - 2}}$

Slopes and Tangent Lines

In Exercises 13–16, differentiate the functions and find the slope of the tangent line at the given value of the independent variable.

13. $f(x) = x + \dfrac{9}{x}, \quad x = -3$ **14.** $k(x) = \dfrac{1}{2 + x}, \quad x = 2$

15. $s = t^3 - t^2, \quad t = -1$ **16.** $y = \dfrac{x + 3}{1 - x}, \quad x = -2$

In Exercises 17–18, differentiate the functions. Then find an equation of the tangent line at the indicated point on the graph of the function.

17. $y = f(x) = \dfrac{8}{\sqrt{x - 2}}, \quad (x, y) = (6, 4)$

18. $w = g(z) = 1 + \sqrt{4 - z}, \quad (z, w) = (3, 2)$

In Exercises 19–22, find the values of the derivatives.

19. $\left. \dfrac{ds}{dt} \right|_{t=-1} \quad$ if $\quad s = 1 - 3t^2$

20. $\left. \dfrac{dy}{dx} \right|_{x=\sqrt{3}} \quad$ if $\quad y = 1 - \dfrac{1}{x}$

21. $\left. \dfrac{dr}{d\theta} \right|_{\theta=0} \quad$ if $\quad r = \dfrac{2}{\sqrt{4 - \theta}}$

22. $\left. \dfrac{dw}{dz} \right|_{z=4} \quad$ if $\quad w = z + \sqrt{z}$

Using the Alternative Formula for Derivatives

Use the formula

$$f'(x) = \lim_{z \to x} \frac{f(z) - f(x)}{z - x}$$

to find the derivative of the functions in Exercises 23–26.

23. $f(x) = \dfrac{1}{x + 2}$ **24.** $f(x) = x^2 - 3x + 4$

25. $g(x) = \dfrac{x}{x - 1}$ **26.** $g(x) = 1 + \sqrt{x}$

Graphs

Match the functions graphed in Exercises 27–30 with the derivatives graphed in the accompanying figures (a)–(d).

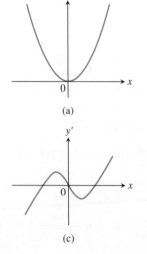

(a)   (b)

(c)   (d)

27.   **28.**

29.   **30.**

31. a. The graph in the accompanying figure is made of line segments joined end to end. At which points of the interval $[-4, 6]$ is f' not defined? Give reasons for your answer.

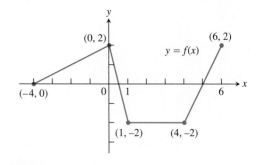

b. Graph the derivative of f.
The graph should show a step function.

32. Recovering a function from its derivative

a. Use the following information to graph the function f over the closed interval $[-2, 5]$.

i) The graph of f is made of closed line segments joined end to end.

ii) The graph starts at the point $(-2, 3)$.

iii) The derivative of f is the step function in the figure shown here.

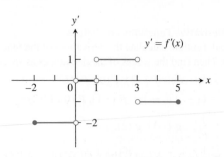

b. Repeat part (a) assuming that the graph starts at $(-2, 0)$ instead of $(-2, 3)$.

33. Growth in the economy   The graph in the accompanying figure shows the average annual percentage change $y = f(t)$ in the U.S. gross national product (GNP) for the years 1983–1988. Graph dy/dt (where defined).

34. Fruit flies   (*Continuation of Example 4, Section 2.1.*) Populations starting out in closed environments grow slowly at first, when there are relatively few members, then more rapidly as the number of reproducing individuals increases and resources are still abundant, then slowly again as the population reaches the carrying capacity of the environment.

a. Use the graphical technique of Example 3 to graph the derivative of the fruit fly population. The graph of the population is reproduced here.

b. During what days does the population seem to be increasing fastest? Slowest?

35. Temperature   The given graph shows the temperature T in °F at Davis, CA, on April 18, 2008, between 6 A.M. and 6 P.M.

a. Estimate the rate of temperature change at the times

i) 7 A.M.   **ii)** 9 A.M.   **iii)** 2 P.M.   **iv)** 4 P.M.

b. At what time does the temperature increase most rapidly? Decrease most rapidly? What is the rate for each of those times?

c. Use the graphical technique of Example 3 to graph the derivative of temperature T versus time t.

36. Weight loss   Jared Fogle, also known as the "Subway Sandwich Guy," weighed 425 lb in 1997 before losing more than 240 lb in 12 months (http://en.wikipedia.org/wiki/Jared_Fogle). A chart showing his possible dramatic weight loss is given in the accompanying figure.

a. Estimate Jared's rate of weight loss when

i) $t = 1$ **ii)** $t = 4$ **iii)** $t = 11$

b. When does Jared lose weight most rapidly and what is this rate of weight loss?

c. Use the graphical technique of Example 3 to graph the derivative of weight W.

One-Sided Derivatives

Compute the right-hand and left-hand derivatives as limits to show that the functions in Exercises 37–40 are not differentiable at the point P.

37.

38.

39.

40.

In Exercises 41 and 42, determine if the piecewise defined function is differentiable at the origin.

41. $f(x) = \begin{cases} 2x - 1, & x \geq 0 \\ x^2 + 2x + 7, & x < 0 \end{cases}$

42. $g(x) = \begin{cases} x^{2/3}, & x \geq 0 \\ x^{1/3}, & x < 0 \end{cases}$

Differentiability and Continuity on an Interval

Each figure in Exercises 43–48 shows the graph of a function over a closed interval D. At what domain points does the function appear to be

a. differentiable?

b. continuous but not differentiable?

c. neither continuous nor differentiable?

Give reasons for your answers.

43.

44.

45.

46.

47.

48.

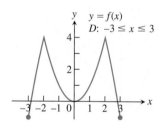

Theory and Examples

In Exercises 49–52,

a. Find the derivative $f'(x)$ of the given function $y = f(x)$.

b. Graph $y = f(x)$ and $y = f'(x)$ side by side using separate sets of coordinate axes, and answer the following questions.

c. For what values of x, if any, is f' positive? Zero? Negative?

d. Over what intervals of x-values, if any, does the function $y = f(x)$ increase as x increases? Decrease as x increases? How is this related to what you found in part (c)? (We will say more about this relationship in Section 4.3.)

49. $y = -x^2$

50. $y = -1/x$

51. $y = x^3/3$

52. $y = x^4/4$

53. Tangent to a parabola Does the parabola $y = 2x^2 - 13x + 5$ have a tangent whose slope is -1? If so, find an equation for the line and the point of tangency. If not, why not?

54. Tangent to $y = \sqrt{x}$ Does any tangent to the curve $y = \sqrt{x}$ cross the x-axis at $x = -1$? If so, find an equation for the line and the point of tangency. If not, why not?

55. Derivative of $-f$ Does knowing that a function $f(x)$ is differentiable at $x = x_0$ tell you anything about the differentiability of the function $-f$ at $x = x_0$? Give reasons for your answer.

56. Derivative of multiples Does knowing that a function $g(t)$ is differentiable at $t = 7$ tell you anything about the differentiability of the function $3g$ at $t = 7$? Give reasons for your answer.

57. Limit of a quotient Suppose that functions $g(t)$ and $h(t)$ are defined for all values of t and $g(0) = h(0) = 0$. Can $\lim_{t \to 0} (g(t))/(h(t))$ exist? If it does exist, must it equal zero? Give reasons for your answers.

58. a. Let $f(x)$ be a function satisfying $|f(x)| \le x^2$ for $-1 \le x \le 1$. Show that f is differentiable at $x = 0$ and find $f'(0)$.

b. Show that

$$f(x) = \begin{cases} x^2 \sin \dfrac{1}{x}, & x \ne 0 \\ 0, & x = 0 \end{cases}$$

is differentiable at $x = 0$ and find $f'(0)$.

T 59. Graph $y = 1/(2\sqrt{x})$ in a window that has $0 \le x \le 2$. Then, on the same screen, graph

$$y = \frac{\sqrt{x + h} - \sqrt{x}}{h}$$

for $h = 1, 0.5, 0.1$. Then try $h = -1, -0.5, -0.1$. Explain what is going on.

T 60. Graph $y = 3x^2$ in a window that has $-2 \le x \le 2, 0 \le y \le 3$. Then, on the same screen, graph

$$y = \frac{(x + h)^3 - x^3}{h}$$

for $h = 2, 1, 0.2$. Then try $h = -2, -1, -0.2$. Explain what is going on.

61. Derivative of $y = |x|$ Graph the derivative of $f(x) = |x|$. Then graph $y = (|x| - 0)/(x - 0) = |x|/x$. What can you conclude?

T 62. Weierstrass's nowhere differentiable continuous function The sum of the first eight terms of the Weierstrass function $f(x) = \sum_{n=0}^{\infty} (2/3)^n \cos(9^n \pi x)$ is

$$g(x) = \cos(\pi x) + (2/3)^1 \cos(9\pi x) + (2/3)^2 \cos(9^2 \pi x)$$
$$+ (2/3)^3 \cos(9^3 \pi x) + \cdots + (2/3)^7 \cos(9^7 \pi x).$$

Graph this sum. Zoom in several times. How wiggly and bumpy is this graph? Specify a viewing window in which the displayed portion of the graph is smooth.

COMPUTER EXPLORATIONS

Use a CAS to perform the following steps for the functions in Exercises 63–68.

a. Plot $y = f(x)$ to see that function's global behavior.

b. Define the difference quotient q at a general point x, with general step size h.

c. Take the limit as $h \to 0$. What formula does this give?

d. Substitute the value $x = x_0$ and plot the function $y = f(x)$ together with its tangent line at that point.

e. Substitute various values for x larger and smaller than x_0 into the formula obtained in part (c). Do the numbers make sense with your picture?

f. Graph the formula obtained in part (c). What does it mean when its values are negative? Zero? Positive? Does this make sense with your plot from part (a)? Give reasons for your answer.

63. $f(x) = x^3 + x^2 - x, \quad x_0 = 1$

64. $f(x) = x^{1/3} + x^{2/3}, \quad x_0 = 1$

65. $f(x) = \dfrac{4x}{x^2 + 1}, \quad x_0 = 2$

66. $f(x) = \dfrac{x - 1}{3x^2 + 1}, \quad x_0 = -1$

67. $f(x) = \sin 2x, \quad x_0 = \pi/2$

68. $f(x) = x^2 \cos x, \quad x_0 = \pi/4$

3.3 Differentiation Rules

This section introduces several rules that allow us to differentiate constant functions, power functions, polynomials, exponential functions, rational functions, and certain combinations of them, simply and directly, without having to take limits each time.

Powers, Multiples, Sums, and Differences

A simple rule of differentiation is that the derivative of every constant function is zero.

FIGURE 3.9 The rule $(d/dx)(c) = 0$ is another way to say that the values of constant functions never change and that the slope of a horizontal line is zero at every point.

Derivative of a Constant Function

If f has the constant value $f(x) = c$, then

$$\frac{df}{dx} = \frac{d}{dx}(c) = 0.$$

Proof We apply the definition of the derivative to $f(x) = c$, the function whose outputs have the constant value c (Figure 3.9). At every value of x, we find that

$$f'(x) = \lim_{h \to 0} \frac{f(x + h) - f(x)}{h} = \lim_{h \to 0} \frac{c - c}{h} = \lim_{h \to 0} 0 = 0. \qquad \blacksquare$$

From Section 3.1, we know that

$$\frac{d}{dx}\left(\frac{1}{x}\right) = -\frac{1}{x^2}, \quad \text{or} \quad \frac{d}{dx}\left(x^{-1}\right) = -x^{-2}.$$

From Example 2 of the last section we also know that

$$\frac{d}{dx}\left(\sqrt{x}\right) = \frac{1}{2\sqrt{x}}, \quad \text{or} \quad \frac{d}{dx}\left(x^{1/2}\right) = \frac{1}{2}x^{-1/2}.$$

These two examples illustrate a general rule for differentiating a power x^n. We first prove the rule when n is a positive integer.

Power Rule for Positive Integers:

If n is a positive integer, then

$$\frac{d}{dx}x^n = nx^{n-1}.$$

Proof of the Positive Integer Power Rule   The formula

$$z^n - x^n = (z - x)(z^{n-1} + z^{n-2}x + \cdots + zx^{n-2} + x^{n-1})$$

can be verified by multiplying out the right-hand side. Then from the alternative formula for the definition of the derivative,

$$f'(x) = \lim_{z \to x} \frac{f(z) - f(x)}{z - x} = \lim_{z \to x} \frac{z^n - x^n}{z - x}$$

$$= \lim_{z \to x}(z^{n-1} + z^{n-2}x + \cdots + zx^{n-2} + x^{n-1}) \qquad n \text{ terms}$$

$$= nx^{n-1}. \qquad \blacksquare$$

The Power Rule is actually valid for all real numbers n. We have seen examples for a negative integer and fractional power, but n could be an irrational number as well. To apply the Power Rule, we subtract 1 from the original exponent n and multiply the result by n. Here we state the general version of the rule, but postpone its proof until Section 3.8.

Power Rule (General Version)

If n is any real number, then

$$\frac{d}{dx}x^n = nx^{n-1},$$

for all x where the powers x^n and x^{n-1} are defined.

EXAMPLE 1   Differentiate the following powers of x.

(a) x^3 **(b)** $x^{2/3}$ **(c)** $x^{\sqrt{2}}$ **(d)** $\dfrac{1}{x^4}$ **(e)** $x^{-4/3}$ **(f)** $\sqrt{x^{2+\pi}}$

Solution

(a) $\dfrac{d}{dx}(x^3) = 3x^{3-1} = 3x^2$ **(b)** $\dfrac{d}{dx}(x^{2/3}) = \dfrac{2}{3}x^{(2/3)-1} = \dfrac{2}{3}x^{-1/3}$

(c) $\dfrac{d}{dx}\left(x^{\sqrt{2}}\right) = \sqrt{2}x^{\sqrt{2}-1}$ **(d)** $\dfrac{d}{dx}\left(\dfrac{1}{x^4}\right) = \dfrac{d}{dx}(x^{-4}) = -4x^{-4-1} = -4x^{-5} = -\dfrac{4}{x^5}$

(e) $\dfrac{d}{dx}(x^{-4/3}) = -\dfrac{4}{3}x^{-(4/3)-1} = -\dfrac{4}{3}x^{-7/3}$

(f) $\dfrac{d}{dx}\left(\sqrt{x^{2+\pi}}\right) = \dfrac{d}{dx}\left(x^{1+(\pi/2)}\right) = \left(1 + \dfrac{\pi}{2}\right)x^{1+(\pi/2)-1} = \dfrac{1}{2}(2 + \pi)\sqrt{x^\pi}$ $\blacksquare$

The next rule says that when a differentiable function is multiplied by a constant, its derivative is multiplied by the same constant.

Derivative Constant Multiple Rule

If u is a differentiable function of x, and c is a constant, then

$$\frac{d}{dx}(cu) = c\frac{du}{dx}.$$

In particular, if n is any real number, then

$$\frac{d}{dx}(cx^n) = cnx^{n-1}.$$

Proof

$$\frac{d}{dx}\,cu = \lim_{h \to 0} \frac{cu(x + h) - cu(x)}{h} \qquad \text{Derivative definition with } f(x) = cu(x)$$

$$= c \lim_{h \to 0} \frac{u(x + h) - u(x)}{h} \qquad \text{Constant Multiple Limit Property}$$

$$= c\,\frac{du}{dx} \qquad u \text{ is differentiable.} \qquad \blacksquare$$

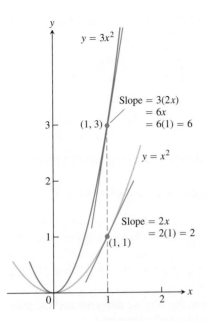

FIGURE 3.10 The graphs of $y = x^2$ and $y = 3x^2$. Tripling the y-coordinate triples the slope (Example 2).

Denoting Functions by u and v
The functions we are working with when we need a differentiation formula are likely to be denoted by letters like f and g. We do not want to use these same letters when stating general differentiation rules, so we use letters like u and v instead that are not likely to be already in use.

EXAMPLE 2

(a) The derivative formula

$$\frac{d}{dx}(3x^2) = 3 \cdot 2x = 6x$$

says that if we rescale the graph of $y = x^2$ by multiplying each y-coordinate by 3, then we multiply the slope at each point by 3 (Figure 3.10).

(b) Negative of a function

The derivative of the negative of a differentiable function u is the negative of the function's derivative. The Constant Multiple Rule with $c = -1$ gives

$$\frac{d}{dx}(-u) = \frac{d}{dx}(-1 \cdot u) = -1 \cdot \frac{d}{dx}(u) = -\frac{du}{dx}. \qquad \blacksquare$$

The next rule says that the derivative of the sum of two differentiable functions is the sum of their derivatives.

Derivative Sum Rule

If u and v are differentiable functions of x, then their sum $u + v$ is differentiable at every point where u and v are both differentiable. At such points,

$$\frac{d}{dx}(u + v) = \frac{du}{dx} + \frac{dv}{dx}.$$

For example, if $y = x^4 + 12x$, then y is the sum of $u(x) = x^4$ and $v(x) = 12x$. We then have

$$\frac{dy}{dx} = \frac{d}{dx}(x^4) + \frac{d}{dx}(12x) = 4x^3 + 12.$$

Proof We apply the definition of the derivative to $f(x) = u(x) + v(x)$:

$$\frac{d}{dx}[u(x) + v(x)] = \lim_{h \to 0} \frac{[u(x + h) + v(x + h)] - [u(x) + v(x)]}{h}$$

$$= \lim_{h \to 0} \left[\frac{u(x + h) - u(x)}{h} + \frac{v(x + h) - v(x)}{h} \right]$$

$$= \lim_{h \to 0} \frac{u(x + h) - u(x)}{h} + \lim_{h \to 0} \frac{v(x + h) - v(x)}{h} = \frac{du}{dx} + \frac{dv}{dx}. \qquad \blacksquare$$

Combining the Sum Rule with the Constant Multiple Rule gives the **Difference Rule**, which says that the derivative of a *difference* of differentiable functions is the difference of their derivatives:

$$\frac{d}{dx}(u - v) = \frac{d}{dx}[u + (-1)v] = \frac{du}{dx} + (-1)\frac{dv}{dx} = \frac{du}{dx} - \frac{dv}{dx}.$$

The Sum Rule also extends to finite sums of more than two functions. If $u_1, u_2, \ldots, u_n$ are differentiable at x, then so is $u_1 + u_2 + \cdots + u_n$, and

$$\frac{d}{dx}(u_1 + u_2 + \cdots + u_n) = \frac{du_1}{dx} + \frac{du_2}{dx} + \cdots + \frac{du_n}{dx}.$$

For instance, to see that the rule holds for three functions we compute

$$\frac{d}{dx}(u_1 + u_2 + u_3) = \frac{d}{dx}((u_1 + u_2) + u_3) = \frac{d}{dx}(u_1 + u_2) + \frac{du_3}{dx} = \frac{du_1}{dx} + \frac{du_2}{dx} + \frac{du_3}{dx}.$$

A proof by mathematical induction for any finite number of terms is given in Appendix 2.

EXAMPLE 3   Find the derivative of the polynomial $y = x^3 + \frac{4}{3}x^2 - 5x + 1$.

Solution $\dfrac{dy}{dx} = \dfrac{d}{dx}x^3 + \dfrac{d}{dx}\left(\dfrac{4}{3}x^2\right) - \dfrac{d}{dx}(5x) + \dfrac{d}{dx}(1)$ Sum and Difference Rules

$$= 3x^2 + \frac{4}{3} \cdot 2x - 5 + 0 = 3x^2 + \frac{8}{3}x - 5 \qquad\blacksquare$$

We can differentiate any polynomial term by term, the way we differentiated the polynomial in Example 3. All polynomials are differentiable at all values of x.

EXAMPLE 4   Does the curve $y = x^4 - 2x^2 + 2$ have any horizontal tangents? If so, where?

Solution   The horizontal tangents, if any, occur where the slope dy/dx is zero. We have

$$\frac{dy}{dx} = \frac{d}{dx}(x^4 - 2x^2 + 2) = 4x^3 - 4x.$$

Now solve the equation $\dfrac{dy}{dx} = 0$ for x:

$$4x^3 - 4x = 0$$
$$4x(x^2 - 1) = 0$$
$$x = 0, 1, -1.$$

The curve $y = x^4 - 2x^2 + 2$ has horizontal tangents at $x = 0, 1$, and -1. The corresponding points on the curve are $(0, 2)$, $(1, 1)$ and $(-1, 1)$. See Figure 3.11. We will see in Chapter 4 that finding the values of x where the derivative of a function is equal to zero is an important and useful procedure. $\qquad\blacksquare$

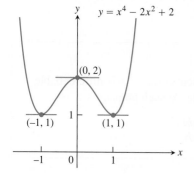

FIGURE 3.11   The curve in Example 4 and its horizontal tangents.

Derivatives of Exponential Functions

We briefly reviewed exponential functions in Section 1.5. When we apply the definition of the derivative to $f(x) = a^x$, we get

$$\frac{d}{dx}(a^x) = \lim_{h \to 0} \frac{a^{x+h} - a^x}{h} \qquad \text{Derivative definition}$$

$$= \lim_{h \to 0} \frac{a^x \cdot a^h - a^x}{h} \qquad a^{x+h} = a^x \cdot a^h$$

$$= \lim_{h \to 0} a^x \cdot \frac{a^h - 1}{h} \qquad \text{Factoring out } a^x$$

$$= a^x \cdot \lim_{h \to 0} \frac{a^h - 1}{h} \qquad a^x \text{ is constant as } h \to 0.$$

$$= \underbrace{\left(\lim_{h \to 0} \frac{a^h - 1}{h}\right)}_{\text{a fixed number } L} \cdot a^x. \tag{1}$$

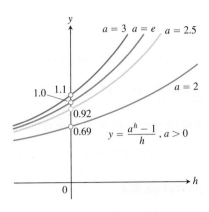

FIGURE 3.12 The position of the curve $y = (a^h - 1)/h$, $a > 0$, varies continuously with a.

Thus we see that the derivative of a^x is a constant multiple L of a^x. The constant L is a limit unlike any we have encountered before. Note, however, that it equals the derivative of $f(x) = a^x$ at $x = 0$:

$$f'(0) = \lim_{h \to 0} \frac{a^h - a^0}{h} = \lim_{h \to 0} \frac{a^h - 1}{h} = L.$$

The limit L is therefore the slope of the graph of $f(x) = a^x$ where it crosses the y-axis. In Chapter 7, where we carefully develop the logarithmic and exponential functions, we prove that the limit L exists and has the value $\ln a$. For now we investigate values of L by graphing the function $y = (a^h - 1)/h$ and studying its behavior as h approaches 0.

Figure 3.12 shows the graphs of $y = (a^h - 1)/h$ for four different values of a. The limit L is approximately 0.69 if $a = 2$, about 0.92 if $a = 2.5$, and about 1.1 if $a = 3$. It appears that the value of L is 1 at some number a chosen between 2.5 and 3. That number is given by $a = e \approx 2.718281828$. With this choice of base we obtain the natural exponential function $f(x) = e^x$ as in Section 1.5, and see that it satisfies the property

$$f'(0) = \lim_{h \to 0} \frac{e^h - 1}{h} = 1. \tag{2}$$

That the limit is 1 implies an important relationship between the natural exponential function e^x and its derivative:

$$\frac{d}{dx}(e^x) = \lim_{h \to 0} \left(\frac{e^h - 1}{h} \right) \cdot e^x \qquad \text{Eq. (1) with } a = e$$

$$= 1 \cdot e^x = e^x. \qquad \text{Eq. (2)}$$

Therefore the natural exponential function is its own derivative.

Derivative of the Natural Exponential Function

$$\frac{d}{dx}(e^x) = e^x$$

EXAMPLE 5   Find an equation for a line that is tangent to the graph of $y = e^x$ and goes through the origin.

Solution   Since the line passes through the origin, its equation is of the form $y = mx$, where m is the slope. If it is tangent to the graph at the point (a, e^a), the slope is $m = (e^a - 0)/(a - 0)$. The slope of the natural exponential at $x = a$ is e^a. Because these slopes are the same, we then have that $e^a = e^a/a$. It follows that $a = 1$ and $m = e$, so the equation of the tangent line is $y = ex$. See Figure 3.13.   ∎

We might ask if there are functions *other* than the natural exponential function that are their own derivatives. The answer is that the only functions that satisfy the property that $f'(x) = f(x)$ are functions that are constant multiples of the natural exponential function, $f(x) = c \cdot e^x$, c any constant. We prove this fact in Section 7.2. Note from the Constant Multiple Rule that indeed

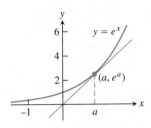

FIGURE 3.13 The line through the origin is tangent to the graph of $y = e^x$ when $a = 1$ (Example 5).

$$\frac{d}{dx}(c \cdot e^x) = c \cdot \frac{d}{dx}(e^x) = c \cdot e^x.$$

Products and Quotients

While the derivative of the sum of two functions is the sum of their derivatives, the derivative of the product of two functions is *not* the product of their derivatives. For instance,

$$\frac{d}{dx}(x \cdot x) = \frac{d}{dx}(x^2) = 2x, \quad \text{while} \quad \frac{d}{dx}(x) \cdot \frac{d}{dx}(x) = 1 \cdot 1 = 1.$$

The derivative of a product of two functions is the sum of *two* products, as we now explain.

Derivative Product Rule

If u and v are differentiable at x, then so is their product uv, and

$$\frac{d}{dx}(uv) = u\frac{dv}{dx} + v\frac{du}{dx}.$$

The derivative of the product uv is u times the derivative of v plus v times the derivative of u. In *prime notation*, $(uv)' = uv' + vu'$. In function notation,

$$\frac{d}{dx}[f(x)g(x)] = f(x)g'(x) + g(x)f'(x).$$

EXAMPLE 6 Find the derivative of **(a)** $y = \frac{1}{x}\left(x^2 + e^x\right)$, **(b)** $y = e^{2x}$.

Solution

(a) We apply the Product Rule with $u = 1/x$ and $v = x^2 + e^x$:

$$\frac{d}{dx}\left[\frac{1}{x}\left(x^2 + e^x\right)\right] = \frac{1}{x}\left(2x + e^x\right) + \left(x^2 + e^x\right)\left(-\frac{1}{x^2}\right) \qquad \frac{d}{dx}(uv) = u\frac{dv}{dx} + v\frac{du}{dx}, \text{ and}$$

$$= 2 + \frac{e^x}{x} - 1 - \frac{e^x}{x^2} \qquad\qquad \frac{d}{dx}\left(\frac{1}{x}\right) = -\frac{1}{x^2}$$

$$= 1 + (x - 1)\frac{e^x}{x^2}.$$

(b) $\dfrac{d}{dx}(e^{2x}) = \dfrac{d}{dx}(e^x \cdot e^x) = e^x \cdot \dfrac{d}{dx}(e^x) + e^x \cdot \dfrac{d}{dx}(e^x) = 2e^x \cdot e^x = 2e^{2x}$ ∎

Proof of the Derivative Product Rule

$$\frac{d}{dx}(uv) = \lim_{h \to 0}\frac{u(x + h)v(x + h) - u(x)v(x)}{h}$$

To change this fraction into an equivalent one that contains difference quotients for the derivatives of u and v, we subtract and add $u(x + h)v(x)$ in the numerator:

$$\frac{d}{dx}(uv) = \lim_{h \to 0}\frac{u(x + h)v(x + h) - u(x + h)v(x) + u(x + h)v(x) - u(x)v(x)}{h}$$

$$= \lim_{h \to 0}\left[u(x + h)\frac{v(x + h) - v(x)}{h} + v(x)\frac{u(x + h) - u(x)}{h}\right]$$

$$= \lim_{h \to 0}u(x + h) \cdot \lim_{h \to 0}\frac{v(x + h) - v(x)}{h} + v(x) \cdot \lim_{h \to 0}\frac{u(x + h) - u(x)}{h}.$$

As h approaches zero, $u(x + h)$ approaches $u(x)$ because u, being differentiable at x, is continuous at x. The two fractions approach the values of dv/dx at x and du/dx at x. In short,

$$\frac{d}{dx}(uv) = u\frac{dv}{dx} + v\frac{du}{dx}.$$

∎

Picturing the Product Rule

Suppose $u(x)$ and $v(x)$ are positive and increase when x increases, and $h > 0$.

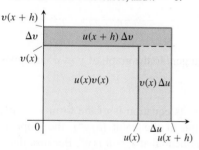

Then the change in the product uv is the difference in areas of the larger and smaller "squares," which is the sum of the upper and right-hand reddish-shaded rectangles. That is,

$$\Delta(uv) = u(x + h)v(x + h) - u(x)v(x)$$
$$= u(x + h)\Delta v + v(x)\Delta u.$$

Division by h gives

$$\frac{\Delta(uv)}{h} = u(x + h)\frac{\Delta v}{h} + v(x)\frac{\Delta u}{h}.$$

The limit as $h \to 0^+$ gives the Product Rule.

EXAMPLE 7 Find the derivative of $y = (x^2 + 1)(x^3 + 3)$.

Solution

(a) From the Product Rule with $u = x^2 + 1$ and $v = x^3 + 3$, we find

$$\frac{d}{dx}\left[(x^2 + 1)(x^3 + 3)\right] = (x^2 + 1)(3x^2) + (x^3 + 3)(2x) \qquad \frac{d}{dx}(uv) = u\frac{dv}{dx} + v\frac{du}{dx}$$

$$= 3x^4 + 3x^2 + 2x^4 + 6x$$

$$= 5x^4 + 3x^2 + 6x.$$

(b) This particular product can be differentiated as well (perhaps better) by multiplying out the original expression for y and differentiating the resulting polynomial:

$$y = (x^2 + 1)(x^3 + 3) = x^5 + x^3 + 3x^2 + 3$$

$$\frac{dy}{dx} = 5x^4 + 3x^2 + 6x.$$

This is in agreement with our first calculation. ∎

The derivative of the quotient of two functions is given by the Quotient Rule.

Derivative Quotient Rule

If u and v are differentiable at x and if $v(x) \neq 0$, then the quotient u/v is differentiable at x, and

$$\frac{d}{dx}\left(\frac{u}{v}\right) = \frac{v\dfrac{du}{dx} - u\dfrac{dv}{dx}}{v^2}.$$

In function notation,

$$\frac{d}{dx}\left[\frac{f(x)}{g(x)}\right] = \frac{g(x)f'(x) - f(x)g'(x)}{g^2(x)}.$$

EXAMPLE 8 Find the derivative of **(a)** $y = \dfrac{t^2 - 1}{t^3 + 1}$, **(b)** $y = e^{-x}$.

Solution

(a) We apply the Quotient Rule with $u = t^2 - 1$ and $v = t^3 + 1$:

$$\frac{dy}{dt} = \frac{(t^3 + 1)\cdot 2t - (t^2 - 1)\cdot 3t^2}{(t^3 + 1)^2} \qquad \frac{d}{dt}\left(\frac{u}{v}\right) = \frac{v(du/dt) - u(dv/dt)}{v^2}$$

$$= \frac{2t^4 + 2t - 3t^4 + 3t^2}{(t^3 + 1)^2}$$

$$= \frac{-t^4 + 3t^2 + 2t}{(t^3 + 1)^2}.$$

(b) $\dfrac{d}{dx}(e^{-x}) = \dfrac{d}{dx}\left(\dfrac{1}{e^x}\right) = \dfrac{e^x \cdot 0 - 1 \cdot e^x}{(e^x)^2} = \dfrac{-1}{e^x} = -e^{-x}$ ∎

Proof of the Derivative Quotient Rule

$$\frac{d}{dx}\left(\frac{u}{v}\right) = \lim_{h \to 0} \frac{\dfrac{u(x+h)}{v(x+h)} - \dfrac{u(x)}{v(x)}}{h}$$

$$= \lim_{h \to 0} \frac{v(x)u(x+h) - u(x)v(x+h)}{hv(x+h)v(x)}$$

To change the last fraction into an equivalent one that contains the difference quotients for the derivatives of u and v, we subtract and add $v(x)u(x)$ in the numerator. We then get

$$\frac{d}{dx}\left(\frac{u}{v}\right) = \lim_{h \to 0} \frac{v(x)u(x+h) - v(x)u(x) + v(x)u(x) - u(x)v(x+h)}{hv(x+h)v(x)}$$

$$= \lim_{h \to 0} \frac{v(x)\dfrac{u(x+h) - u(x)}{h} - u(x)\dfrac{v(x+h) - v(x)}{h}}{v(x+h)v(x)}.$$

Taking the limits in the numerator and denominator now gives the Quotient Rule. ∎

The choice of which rules to use in solving a differentiation problem can make a difference in how much work you have to do. Here is an example.

EXAMPLE 9   Rather than using the Quotient Rule to find the derivative of

$$y = \frac{(x-1)(x^2 - 2x)}{x^4},$$

expand the numerator and divide by x^4:

$$y = \frac{(x-1)(x^2 - 2x)}{x^4} = \frac{x^3 - 3x^2 + 2x}{x^4} = x^{-1} - 3x^{-2} + 2x^{-3}.$$

Then use the Sum and Power Rules:

$$\frac{dy}{dx} = -x^{-2} - 3(-2)x^{-3} + 2(-3)x^{-4}$$

$$= -\frac{1}{x^2} + \frac{6}{x^3} - \frac{6}{x^4}. \qquad ∎$$

Second- and Higher-Order Derivatives

If $y = f(x)$ is a differentiable function, then its derivative $f'(x)$ is also a function. If f' is also differentiable, then we can differentiate f' to get a new function of x denoted by f''. So $f'' = (f')'$. The function f'' is called the **second derivative** of f because it is the derivative of the first derivative. It is written in several ways:

$$f''(x) = \frac{d^2y}{dx^2} = \frac{d}{dx}\left(\frac{dy}{dx}\right) = \frac{dy'}{dx} = y'' = D^2(f)(x) = D_x^2 f(x).$$

The symbol D^2 means the operation of differentiation is performed twice.

If $y = x^6$, then $y' = 6x^5$ and we have

$$y'' = \frac{dy'}{dx} = \frac{d}{dx}(6x^5) = 30x^4.$$

Thus $D^2(x^6) = 30x^4$.

If y'' is differentiable, its derivative, $y''' = dy''/dx = d^3y/dx^3$, is the **third derivative** of y with respect to x. The names continue as you imagine, with

$$y^{(n)} = \frac{d}{dx} y^{(n-1)} = \frac{d^n y}{dx^n} = D^n y$$

denoting the **nth derivative** of y with respect to x for any positive integer n.

We can interpret the second derivative as the rate of change of the slope of the tangent to the graph of $y = f(x)$ at each point. You will see in the next chapter that the second derivative reveals whether the graph bends upward or downward from the tangent line as we move off the point of tangency. In the next section, we interpret both the second and third derivatives in terms of motion along a straight line.

How to Read the Symbols for Derivatives	
y'	"y prime"
y''	"y double prime"
$\dfrac{d^2y}{dx^2}$	"d squared y dx squared"
y'''	"y triple prime"
$y^{(n)}$	"y super n"
$\dfrac{d^n y}{dx^n}$	"d to the n of y by dx to the n"
D^n	"D to the n"

EXAMPLE 10   The first four derivatives of $y = x^3 - 3x^2 + 2$ are

First derivative: $y' = 3x^2 - 6x$

Second derivative: $y'' = 6x - 6$

Third derivative: $y''' = 6$

Fourth derivative: $y^{(4)} = 0$.

The function has derivatives of all orders, the fifth and later derivatives all being zero.   ∎

Exercises 3.3

Derivative Calculations

In Exercises 1–12, find the first and second derivatives.

1. $y = -x^2 + 3$

2. $y = x^2 + x + 8$

3. $s = 5t^3 - 3t^5$

4. $w = 3z^7 - 7z^3 + 21z^2$

5. $y = \dfrac{4x^3}{3} - x + 2e^x$

6. $y = \dfrac{x^3}{3} + \dfrac{x^2}{2} + \dfrac{x}{4}$

7. $w = 3z^{-2} - \dfrac{1}{z}$

8. $s = -2t^{-1} + \dfrac{4}{t^2}$

9. $y = 6x^2 - 10x - 5x^{-2}$

10. $y = 4 - 2x - x^{-3}$

11. $r = \dfrac{1}{3s^2} - \dfrac{5}{2s}$

12. $r = \dfrac{12}{\theta} - \dfrac{4}{\theta^3} + \dfrac{1}{\theta^4}$

In Exercises 13–16, find y' **(a)** by applying the Product Rule and **(b)** by multiplying the factors to produce a sum of simpler terms to differentiate.

13. $y = (3 - x^2)(x^3 - x + 1)$ **14.** $y = (2x + 3)(5x^2 - 4x)$

15. $y = (x^2 + 1)\left(x + 5 + \dfrac{1}{x}\right)$ **16.** $y = (1 + x^2)(x^{3/4} - x^{-3})$

Find the derivatives of the functions in Exercises 17–40.

17. $y = \dfrac{2x + 5}{3x - 2}$

18. $z = \dfrac{4 - 3x}{3x^2 + x}$

19. $g(x) = \dfrac{x^2 - 4}{x + 0.5}$

20. $f(t) = \dfrac{t^2 - 1}{t^2 + t - 2}$

21. $v = (1 - t)(1 + t^2)^{-1}$

22. $w = (2x - 7)^{-1}(x + 5)$

23. $f(s) = \dfrac{\sqrt{s} - 1}{\sqrt{s} + 1}$

24. $u = \dfrac{5x + 1}{2\sqrt{x}}$

25. $v = \dfrac{1 + x - 4\sqrt{x}}{x}$

26. $r = 2\left(\dfrac{1}{\sqrt{\theta}} + \sqrt{\theta}\right)$

27. $y = \dfrac{1}{(x^2 - 1)(x^2 + x + 1)}$

28. $y = \dfrac{(x + 1)(x + 2)}{(x - 1)(x - 2)}$

29. $y = 2e^{-x} + e^{3x}$

30. $y = \dfrac{x^2 + 3e^x}{2e^x - x}$

31. $y = x^3 e^x$

32. $w = re^{-r}$

33. $y = x^{9/4} + e^{-2x}$

34. $y = x^{-3/5} + \pi^{3/2}$

35. $s = 2t^{3/2} + 3e^2$

36. $w = \dfrac{1}{z^{1.4}} + \dfrac{\pi}{\sqrt{z}}$

37. $y = \sqrt[3]{x^2} - x^e$

38. $y = \sqrt[3]{x^{9.6}} + 2e^{1.3}$

39. $r = \dfrac{e^s}{s}$

40. $r = e^\theta\left(\dfrac{1}{\theta^2} + \theta^{-\pi/2}\right)$

Find the derivatives of all orders of the functions in Exercises 41–44.

41. $y = \dfrac{x^4}{2} - \dfrac{3}{2}x^2 - x$

42. $y = \dfrac{x^5}{120}$

43. $y = (x - 1)(x^2 + 3x - 5)$ **44.** $y = (4x^3 + 3x)(2 - x)$

Find the first and second derivatives of the functions in Exercises 45–52.

45. $y = \dfrac{x^3 + 7}{x}$

46. $s = \dfrac{t^2 + 5t - 1}{t^2}$

47. $r = \dfrac{(\theta - 1)(\theta^2 + \theta + 1)}{\theta^3}$

48. $u = \dfrac{(x^2 + x)(x^2 - x + 1)}{x^4}$

49. $w = \left(\dfrac{1 + 3z}{3z}\right)(3 - z)$

50. $p = \dfrac{q^2 + 3}{(q - 1)^3 + (q + 1)^3}$

51. $w = 3z^2 e^{2z}$

52. $w = e^z(z - 1)(z^2 + 1)$

53. Suppose u and v are functions of x that are differentiable at $x = 0$ and that

$$u(0) = 5, \quad u'(0) = -3, \quad v(0) = -1, \quad v'(0) = 2.$$

Find the values of the following derivatives at $x = 0$.

a. $\dfrac{d}{dx}(uv)$ **b.** $\dfrac{d}{dx}\left(\dfrac{u}{v}\right)$ **c.** $\dfrac{d}{dx}\left(\dfrac{v}{u}\right)$ **d.** $\dfrac{d}{dx}(7v - 2u)$

54. Suppose u and v are differentiable functions of x and that

$$u(1) = 2, \quad u'(1) = 0, \quad v(1) = 5, \quad v'(1) = -1.$$

Find the values of the following derivatives at $x = 1$.

a. $\dfrac{d}{dx}(uv)$ **b.** $\dfrac{d}{dx}\left(\dfrac{u}{v}\right)$ **c.** $\dfrac{d}{dx}\left(\dfrac{v}{u}\right)$ **d.** $\dfrac{d}{dx}(7v - 2u)$

Slopes and Tangents

55. a. Normal to a curve   Find an equation for the line perpendicular to the tangent to the curve $y = x^3 - 4x + 1$ at the point $(2, 1)$.

b. Smallest slope   What is the smallest slope on the curve? At what point on the curve does the curve have this slope?

c. Tangents having specified slope   Find equations for the tangents to the curve at the points where the slope of the curve is 8.

56. a. Horizontal tangents   Find equations for the horizontal tangents to the curve $y = x^3 - 3x - 2$. Also find equations for the lines that are perpendicular to these tangents at the points of tangency.

b. Smallest slope   What is the smallest slope on the curve? At what point on the curve does the curve have this slope? Find an equation for the line that is perpendicular to the curve's tangent at this point.

57. Find the tangents to *Newton's serpentine* (graphed here) at the origin and the point $(1, 2)$.

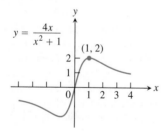

$$y = \frac{4x}{x^2 + 1}$$

58. Find the tangent to the *Witch of Agnesi* (graphed here) at the point $(2, 1)$.

$$y = \frac{8}{x^2 + 4}$$

59. Quadratic tangent to identity function   The curve $y = ax^2 + bx + c$ passes through the point $(1, 2)$ and is tangent to the line $y = x$ at the origin. Find a, b, and c.

60. Quadratics having a common tangent   The curves $y = x^2 + ax + b$ and $y = cx - x^2$ have a common tangent line at the point $(1, 0)$. Find a, b, and c.

61. Find all points (x, y) on the graph of $f(x) = 3x^2 - 4x$ with tangent lines parallel to the line $y = 8x + 5$.

62. Find all points (x, y) on the graph of $g(x) = \frac{1}{3}x^3 - \frac{3}{2}x^2 + 1$ with tangent lines parallel to the line $8x - 2y = 1$.

63. Find all points (x, y) on the graph of $y = x/(x - 2)$ with tangent lines perpendicular to the line $y = 2x + 3$.

64. Find all points (x, y) on the graph of $f(x) = x^2$ with tangent lines passing through the point $(3, 8)$.

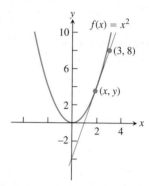

65. a. Find an equation for the line that is tangent to the curve $y = x^3 - x$ at the point $(-1, 0)$.

T b. Graph the curve and tangent line together. The tangent intersects the curve at another point. Use Zoom and Trace to estimate the point's coordinates.

T c. Confirm your estimates of the coordinates of the second intersection point by solving the equations for the curve and tangent simultaneously (Solver key).

66. a. Find an equation for the line that is tangent to the curve $y = x^3 - 6x^2 + 5x$ at the origin.

T b. Graph the curve and tangent together. The tangent intersects the curve at another point. Use Zoom and Trace to estimate the point's coordinates.

T c. Confirm your estimates of the coordinates of the second intersection point by solving the equations for the curve and tangent simultaneously (Solver key).

Theory and Examples

For Exercises 67 and 68 evaluate each limit by first converting each to a derivative at a particular x-value.

67. $\displaystyle\lim_{x \to 1} \frac{x^{50} - 1}{x - 1}$ **68.** $\displaystyle\lim_{x \to -1} \frac{x^{2/9} - 1}{x + 1}$

69. Find the value of a that makes the following function differentiable for all x-values.

$$g(x) = \begin{cases} ax, & \text{if } x < 0 \\ x^2 - 3x, & \text{if } x \geq 0 \end{cases}$$

70. Find the values of a and b that make the following function differentiable for all x-values.

$$f(x) = \begin{cases} ax + b, & x > -1 \\ bx^2 - 3, & x \leq -1 \end{cases}$$

71. The general polynomial of degree n has the form

$$P(x) = a_n x^n + a_{n-1} x^{n-1} + \cdots + a_2 x^2 + a_1 x + a_0$$

where $a_n \neq 0$. Find $P'(x)$.

72. **The body's reaction to medicine** The reaction of the body to a dose of medicine can sometimes be represented by an equation of the form

$$R = M^2 \left(\frac{C}{2} - \frac{M}{3} \right),$$

where C is a positive constant and M is the amount of medicine absorbed in the blood. If the reaction is a change in blood pressure, R is measured in millimeters of mercury. If the reaction is a change in temperature, R is measured in degrees, and so on.

Find dR/dM. This derivative, as a function of M, is called the sensitivity of the body to the medicine. In Section 4.5, we will see how to find the amount of medicine to which the body is most sensitive.

73. Suppose that the function v in the Derivative Product Rule has a constant value c. What does the Derivative Product Rule then say? What does this say about the Derivative Constant Multiple Rule?

74. **The Reciprocal Rule**

a. The *Reciprocal Rule* says that at any point where the function $v(x)$ is differentiable and different from zero,

$$\frac{d}{dx} \left(\frac{1}{v} \right) = - \frac{1}{v^2} \frac{dv}{dx}.$$

Show that the Reciprocal Rule is a special case of the Derivative Quotient Rule.

b. Show that the Reciprocal Rule and the Derivative Product Rule together imply the Derivative Quotient Rule.

75. **Generalizing the Product Rule** The Derivative Product Rule gives the formula

$$\frac{d}{dx} (uv) = u \frac{dv}{dx} + v \frac{du}{dx}$$

for the derivative of the product uv of two differentiable functions of x.

a. What is the analogous formula for the derivative of the product uvw of *three* differentiable functions of x?

b. What is the formula for the derivative of the product $u_1 u_2 u_3 u_4$ of *four* differentiable functions of x?

c. What is the formula for the derivative of a product $u_1 u_2 u_3 \cdots u_n$ of a finite number n of differentiable functions of x?

76. **Power Rule for negative integers** Use the Derivative Quotient Rule to prove the Power Rule for negative integers, that is,

$$\frac{d}{dx} (x^{-m}) = -mx^{-m-1}$$

where m is a positive integer.

77. **Cylinder pressure** If gas in a cylinder is maintained at a constant temperature T, the pressure P is related to the volume V by a formula of the form

$$P = \frac{nRT}{V - nb} - \frac{an^2}{V^2},$$

in which a, b, n, and R are constants. Find dP/dV. (See accompanying figure.)

78. **The best quantity to order** One of the formulas for inventory management says that the average weekly cost of ordering, paying for, and holding merchandise is

$$A(q) = \frac{km}{q} + cm + \frac{hq}{2},$$

where q is the quantity you order when things run low (shoes, radios, brooms, or whatever the item might be); k is the cost of placing an order (the same, no matter how often you order); c is the cost of one item (a constant); m is the number of items sold each week (a constant); and h is the weekly holding cost per item (a constant that takes into account things such as space, utilities, insurance, and security). Find dA/dq and d^2A/dq^2.

3.4 The Derivative as a Rate of Change

In Section 2.1 we introduced average and instantaneous rates of change. In this section we study further applications in which derivatives model the rates at which things change. It is natural to think of a quantity changing with respect to time, but other variables can be treated in the same way. For example, an economist may want to study how the cost of producing steel varies with the number of tons produced, or an engineer may want to know how the power output of a generator varies with its temperature.

Instantaneous Rates of Change

If we interpret the difference quotient $(f(x + h) - f(x))/h$ as the average rate of change in f over the interval from x to $x + h$, we can interpret its limit as $h \to 0$ as the rate at which f is changing at the point x.

DEFINITION The **instantaneous rate of change** of f with respect to x at x_0 is the derivative

$$f'(x_0) = \lim_{h \to 0} \frac{f(x_0 + h) - f(x_0)}{h},$$

provided the limit exists.

Thus, instantaneous rates are limits of average rates.

It is conventional to use the word *instantaneous* even when x does not represent time. The word is, however, frequently omitted. When we say *rate of change*, we mean *instantaneous rate of change*.

EXAMPLE 1 The area A of a circle is related to its diameter by the equation

$$A = \frac{\pi}{4} D^2.$$

How fast does the area change with respect to the diameter when the diameter is 10 m?

Solution The rate of change of the area with respect to the diameter is

$$\frac{dA}{dD} = \frac{\pi}{4} \cdot 2D = \frac{\pi D}{2}.$$

When $D = 10$ m, the area is changing with respect to the diameter at the rate of $(\pi/2)10 = 5\pi$ m^2/m ≈ 15.71 m^2/m. ∎

Motion Along a Line: Displacement, Velocity, Speed, Acceleration, and Jerk

Suppose that an object is moving along a coordinate line (an s-axis), usually horizontal or vertical, so that we know its position s on that line as a function of time t:

$$s = f(t).$$

The **displacement** of the object over the time interval from t to $t + \Delta t$ (Figure 3.14) is

$$\Delta s = f(t + \Delta t) - f(t),$$

and the **average velocity** of the object over that time interval is

$$v_{av} = \frac{\text{displacement}}{\text{travel time}} = \frac{\Delta s}{\Delta t} = \frac{f(t + \Delta t) - f(t)}{\Delta t}.$$

To find the body's velocity at the exact instant t, we take the limit of the average velocity over the interval from t to $t + \Delta t$ as Δt shrinks to zero. This limit is the derivative of f with respect to t.

Position at time t ... and at time $t + \Delta t$

Δs

$s = f(t)$ $s + \Delta s = f(t + \Delta t)$ s

FIGURE 3.14 The positions of a body moving along a coordinate line at time t and shortly later at time $t + \Delta t$. Here the coordinate line is horizontal.

DEFINITION **Velocity** (**instantaneous velocity**) is the derivative of position with respect to time. If a body's position at time t is $s = f(t)$, then the body's velocity at time t is

$$v(t) = \frac{ds}{dt} = \lim_{\Delta t \to 0} \frac{f(t + \Delta t) - f(t)}{\Delta t}.$$

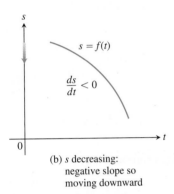

FIGURE 3.15 For motion $s = f(t)$ along a straight line (the vertical axis), $v = ds/dt$ is (a) positive when s increases and (b) negative when s decreases.

HISTORICAL BIOGRAPHY

Bernard Bolzano
(1781–1848)

Besides telling how fast an object is moving along the horizontal line in Figure 3.14, its velocity tells the direction of motion. When the object is moving forward (s increasing), the velocity is positive; when the object is moving backward (s decreasing), the velocity is negative. If the coordinate line is vertical, the object moves upward for positive velocity and downward for negative velocity. The blue curves in Figure 3.15 represent position along the line over time; they do not portray the path of motion, which lies along the s-axis.

If we drive to a friend's house and back at 30 mph, say, the speedometer will show 30 on the way over but it will not show -30 on the way back, even though our distance from home is decreasing. The speedometer always shows *speed*, which is the absolute value of velocity. Speed measures the rate of progress regardless of direction.

DEFINITION   **Speed** is the absolute value of velocity.

$$\text{Speed} = |v(t)| = \left|\frac{ds}{dt}\right|$$

EXAMPLE 2   Figure 3.16 shows the graph of the velocity $v = f'(t)$ of a particle moving along a horizontal line (as opposed to showing a position function $s = f(t)$ such as in Figure 3.15). In the graph of the velocity function, it's not the slope of the curve that tells us if the particle is moving forward or backward along the line (which is not shown in the figure), but rather the sign of the velocity. Looking at Figure 3.16, we see that the particle moves forward for the first 3 sec (when the velocity is positive), moves backward for the next 2 sec (the velocity is negative), stands motionless for a full second, and then moves forward again. The particle is speeding up when its positive velocity increases during the first second, moves at a steady speed during the next second, and then slows down as the velocity decreases to zero during the third second. It stops for an instant at $t = 3$ sec (when the velocity is zero) and reverses direction as the velocity starts to become negative. The particle is now moving backward and gaining in speed until $t = 4$ sec, at which time it achieves its greatest speed during its backward motion. Continuing its backward motion at time $t = 4$, the particle starts to slow down again until it finally stops at time $t = 5$ (when the velocity is once again zero). The particle now remains motionless for one full second, and then moves forward again at $t = 6$ sec, speeding up during the final second of the forward motion indicated in the velocity graph.   ■

The rate at which a body's velocity changes is the body's *acceleration*. The acceleration measures how quickly the body picks up or loses speed.

A sudden change in acceleration is called a *jerk*. When a ride in a car or a bus is jerky, it is not that the accelerations involved are necessarily large but that the changes in acceleration are abrupt.

DEFINITIONS   **Acceleration** is the derivative of velocity with respect to time. If a body's position at time t is $s = f(t)$, then the body's acceleration at time t is

$$a(t) = \frac{dv}{dt} = \frac{d^2s}{dt^2}.$$

Jerk is the derivative of acceleration with respect to time:

$$j(t) = \frac{da}{dt} = \frac{d^3s}{dt^3}.$$

Near the surface of the Earth all bodies fall with the same constant acceleration. Galileo's experiments with free fall (see Section 2.1) lead to the equation

$$s = \frac{1}{2}gt^2,$$

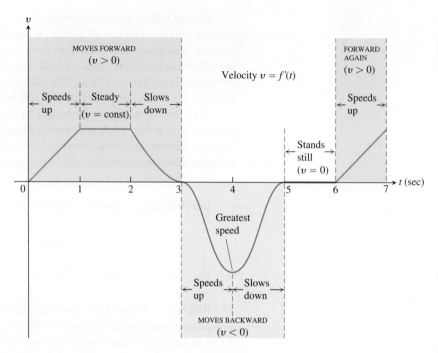

FIGURE 3.16 The velocity graph of a particle moving along a horizontal line, discussed in Example 2.

where s is the distance fallen and g is the acceleration due to Earth's gravity. This equation holds in a vacuum, where there is no air resistance, and closely models the fall of dense, heavy objects, such as rocks or steel tools, for the first few seconds of their fall, before the effects of air resistance are significant.

The value of g in the equation $s = (1/2)gt^2$ depends on the units used to measure t and s. With t in seconds (the usual unit), the value of g determined by measurement at sea level is approximately 32 ft/sec^2 (feet per second squared) in English units, and $g = 9.8$ m/sec^2 (meters per second squared) in metric units. (These gravitational constants depend on the distance from Earth's center of mass, and are slightly lower on top of Mt. Everest, for example.)

The jerk associated with the constant acceleration of gravity ($g = 32$ ft/sec^2) is zero:

$$j = \frac{d}{dt}(g) = 0.$$

An object does not exhibit jerkiness during free fall.

EXAMPLE 3 Figure 3.17 shows the free fall of a heavy ball bearing released from rest at time $t = 0$ sec.

(a) How many meters does the ball fall in the first 2 sec?

(b) What is its velocity, speed, and acceleration when $t = 2$?

Solution

(a) The metric free-fall equation is $s = 4.9t^2$. During the first 2 sec, the ball falls

$$s(2) = 4.9(2)^2 = 19.6 \text{ m}.$$

(b) At any time t, *velocity* is the derivative of position:

$$v(t) = s'(t) = \frac{d}{dt}(4.9t^2) = 9.8t.$$

FIGURE 3.17 A ball bearing falling from rest (Example 3).

At $t = 2$, the velocity is

$$v(2) = 19.6 \text{ m/sec}$$

in the downward (increasing s) direction. The *speed* at $t = 2$ is

$$\text{speed} = |v(2)| = 19.6 \text{ m/sec}.$$

The *acceleration* at any time t is

$$a(t) = v'(t) = s''(t) = 9.8 \text{ m/sec}^2.$$

At $t = 2$, the acceleration is 9.8 m/sec^2. ∎

EXAMPLE 4   A dynamite blast blows a heavy rock straight up with a launch velocity of 160 ft/sec (about 109 mph) (Figure 3.18a). It reaches a height of $s = 160t - 16t^2$ ft after t sec.

(a) How high does the rock go?

(b) What are the velocity and speed of the rock when it is 256 ft above the ground on the way up? On the way down?

(c) What is the acceleration of the rock at any time t during its flight (after the blast)?

(d) When does the rock hit the ground again?

Solution

(a) In the coordinate system we have chosen, s measures height from the ground up, so the velocity is positive on the way up and negative on the way down. The instant the rock is at its highest point is the one instant during the flight when the velocity is 0. To find the maximum height, all we need to do is to find when $v = 0$ and evaluate s at this time.

At any time t during the rock's motion, its velocity is

$$v = \frac{ds}{dt} = \frac{d}{dt}(160t - 16t^2) = 160 - 32t \text{ ft/sec}.$$

The velocity is zero when

$$160 - 32t = 0 \quad \text{or} \quad t = 5 \text{ sec}.$$

The rock's height at $t = 5$ sec is

$$s_{\max} = s(5) = 160(5) - 16(5)^2 = 800 - 400 = 400 \text{ ft}.$$

See Figure 3.18b.

(b) To find the rock's velocity at 256 ft on the way up and again on the way down, we first find the two values of t for which

$$s(t) = 160t - 16t^2 = 256.$$

To solve this equation, we write

$$16t^2 - 160t + 256 = 0$$
$$16(t^2 - 10t + 16) = 0$$
$$(t - 2)(t - 8) = 0$$
$$t = 2 \text{ sec}, t = 8 \text{ sec}.$$

The rock is 256 ft above the ground 2 sec after the explosion and again 8 sec after the explosion. The rock's velocities at these times are

$$v(2) = 160 - 32(2) = 160 - 64 = 96 \text{ ft/sec}.$$
$$v(8) = 160 - 32(8) = 160 - 256 = -96 \text{ ft/sec}.$$

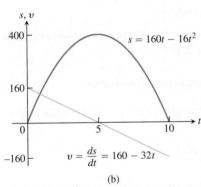

FIGURE 3.18 (a) The rock in Example 4. (b) The graphs of s and v as functions of time; s is largest when $v = ds/dt = 0$. The graph of s is *not* the path of the rock: It is a plot of height versus time. The slope of the plot is the rock's velocity, graphed here as a straight line.

At both instants, the rock's speed is 96 ft/sec. Since $v(2) > 0$, the rock is moving upward (s is increasing) at $t = 2$ sec; it is moving downward (s is decreasing) at $t = 8$ because $v(8) < 0$.

(c) At any time during its flight following the explosion, the rock's acceleration is a constant

$$a = \frac{dv}{dt} = \frac{d}{dt}(160 - 32t) = -32 \text{ ft/sec}^2.$$

The acceleration is always downward. As the rock rises, it slows down; as it falls, it speeds up.

(d) The rock hits the ground at the positive time t for which $s = 0$. The equation $160t - 16t^2 = 0$ factors to give $16t(10 - t) = 0$, so it has solutions $t = 0$ and $t = 10$. At $t = 0$, the blast occurred and the rock was thrown upward. It returned to the ground 10 sec later. ∎

Derivatives in Economics

Engineers use the terms *velocity* and *acceleration* to refer to the derivatives of functions describing motion. Economists, too, have a specialized vocabulary for rates of change and derivatives. They call them *marginals*.

In a manufacturing operation, the *cost of production* $c(x)$ is a function of x, the number of units produced. The **marginal cost of production** is the rate of change of cost with respect to level of production, so it is dc/dx.

Suppose that $c(x)$ represents the dollars needed to produce x tons of steel in one week. It costs more to produce $x + h$ tons per week, and the cost difference, divided by h, is the average cost of producing each additional ton:

$$\frac{c(x + h) - c(x)}{h} = \begin{array}{l}\text{average cost of each of the additional}\\ h \text{ tons of steel produced.}\end{array}$$

The limit of this ratio as $h \to 0$ is the *marginal cost* of producing more steel per week when the current weekly production is x tons (Figure 3.19):

$$\frac{dc}{dx} = \lim_{h \to 0} \frac{c(x + h) - c(x)}{h} = \text{marginal cost of production.}$$

Sometimes the marginal cost of production is loosely defined to be the extra cost of producing one additional unit:

$$\frac{\Delta c}{\Delta x} = \frac{c(x + 1) - c(x)}{1},$$

which is approximated by the value of dc/dx at x. This approximation is acceptable if the slope of the graph of c does not change quickly near x. Then the difference quotient will be close to its limit dc/dx, which is the rise in the tangent line if $\Delta x = 1$ (Figure 3.20). The approximation works best for large values of x.

Economists often represent a total cost function by a cubic polynomial

$$c(x) = \alpha x^3 + \beta x^2 + \gamma x + \delta$$

where δ represents *fixed costs* such as rent, heat, equipment capitalization, and management costs. The other terms represent *variable costs* such as the costs of raw materials, taxes, and labor. Fixed costs are independent of the number of units produced, whereas variable costs depend on the quantity produced. A cubic polynomial is usually adequate to capture the cost behavior on a realistic quantity interval.

EXAMPLE 5   Suppose that it costs

$$c(x) = x^3 - 6x^2 + 15x$$

Cost y (dollars)

FIGURE 3.19   Weekly steel production: $c(x)$ is the cost of producing x tons per week. The cost of producing an additional h tons is $c(x + h) - c(x)$.

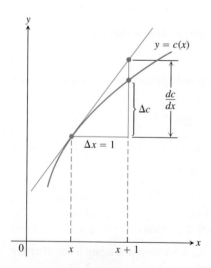

FIGURE 3.20   The marginal cost dc/dx is approximately the extra cost Δc of producing $\Delta x = 1$ more unit.

dollars to produce x radiators when 8 to 30 radiators are produced and that

$$r(x) = x^3 - 3x^2 + 12x$$

gives the dollar revenue from selling x radiators. Your shop currently produces 10 radiators a day. About how much extra will it cost to produce one more radiator a day, and what is your estimated increase in revenue for selling 11 radiators a day?

Solution   The cost of producing one more radiator a day when 10 are produced is about $c'(10)$:

$$c'(x) = \frac{d}{dx}\left(x^3 - 6x^2 + 15x\right) = 3x^2 - 12x + 15$$

$$c'(10) = 3(100) - 12(10) + 15 = 195.$$

The additional cost will be about \$195. The marginal revenue is

$$r'(x) = \frac{d}{dx}(x^3 - 3x^2 + 12x) = 3x^2 - 6x + 12.$$

The marginal revenue function estimates the increase in revenue that will result from selling one additional unit. If you currently sell 10 radiators a day, you can expect your revenue to increase by about

$$r'(10) = 3(100) - 6(10) + 12 = \$252$$

if you increase sales to 11 radiators a day.   ■

EXAMPLE 6   To get some feel for the language of marginal rates, consider marginal tax rates. If your marginal income tax rate is 28% and your income increases by \$1000, you can expect to pay an extra \$280 in taxes. This does not mean that you pay 28% of your entire income in taxes. It just means that at your current income level I, the rate of increase of taxes T with respect to income is $dT/dI = 0.28$. You will pay \$0.28 in taxes out of every extra dollar you earn. Of course, if you earn a lot more, you may land in a higher tax bracket and your marginal rate will increase.   ■

Sensitivity to Change

When a small change in x produces a large change in the value of a function $f(x)$, we say that the function is relatively **sensitive** to changes in x. The derivative $f'(x)$ is a measure of this sensitivity.

EXAMPLE 7   Genetic Data and Sensitivity to Change

The Austrian monk Gregor Johann Mendel (1822–1884), working with garden peas and other plants, provided the first scientific explanation of hybridization.

His careful records showed that if p (a number between 0 and 1) is the frequency of the gene for smooth skin in peas (dominant) and $(1 - p)$ is the frequency of the gene for wrinkled skin in peas, then the proportion of smooth-skinned peas in the next generation will be

$$y = 2p(1 - p) + p^2 = 2p - p^2.$$

The graph of y versus p in Figure 3.21a suggests that the value of y is more sensitive to a change in p when p is small than when p is large. Indeed, this fact is borne out by the derivative graph in Figure 3.21b, which shows that dy/dp is close to 2 when p is near 0 and close to 0 when p is near 1.

The implication for genetics is that introducing a few more smooth skin genes into a population where the frequency of wrinkled skin peas is large will have a more dramatic effect on later generations than will a similar increase when the population has a large proportion of smooth skin peas.   ■

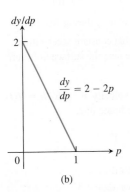

FIGURE 3.21   (a) The graph of $y = 2p - p^2$, describing the proportion of smooth-skinned peas in the next generation. (b) The graph of dy/dp (Example 7).

Exercises 3.4

Motion Along a Coordinate Line

Exercises 1–6 give the positions $s = f(t)$ of a body moving on a coordinate line, with s in meters and t in seconds.

a. Find the body's displacement and average velocity for the given time interval.

b. Find the body's speed and acceleration at the endpoints of the interval.

c. When, if ever, during the interval does the body change direction?

1. $s = t^2 - 3t + 2, \quad 0 \le t \le 2$

2. $s = 6t - t^2, \quad 0 \le t \le 6$

3. $s = -t^3 + 3t^2 - 3t, \quad 0 \le t \le 3$

4. $s = (t^4/4) - t^3 + t^2, \quad 0 \le t \le 3$

5. $s = \dfrac{25}{t^2} - \dfrac{5}{t}, \quad 1 \le t \le 5$

6. $s = \dfrac{25}{t + 5}, \quad -4 \le t \le 0$

7. Particle motion At time t, the position of a body moving along the s-axis is $s = t^3 - 6t^2 + 9t$ m.

a. Find the body's acceleration each time the velocity is zero.

b. Find the body's speed each time the acceleration is zero.

c. Find the total distance traveled by the body from $t = 0$ to $t = 2$.

8. Particle motion At time $t \ge 0$, the velocity of a body moving along the horizontal s-axis is $v = t^2 - 4t + 3$.

a. Find the body's acceleration each time the velocity is zero.

b. When is the body moving forward? Backward?

c. When is the body's velocity increasing? Decreasing?

Free-Fall Applications

9. Free fall on Mars and Jupiter The equations for free fall at the surfaces of Mars and Jupiter (s in meters, t in seconds) are $s = 1.86t^2$ on Mars and $s = 11.44t^2$ on Jupiter. How long does it take a rock falling from rest to reach a velocity of 27.8 m/sec (about 100 km/h) on each planet?

10. Lunar projectile motion A rock thrown vertically upward from the surface of the moon at a velocity of 24 m/sec (about 86 km/h) reaches a height of $s = 24t - 0.8t^2$ m in t sec.

a. Find the rock's velocity and acceleration at time t. (The acceleration in this case is the acceleration of gravity on the moon.)

b. How long does it take the rock to reach its highest point?

c. How high does the rock go?

d. How long does it take the rock to reach half its maximum height?

e. How long is the rock aloft?

11. Finding g on a small airless planet Explorers on a small airless planet used a spring gun to launch a ball bearing vertically upward from the surface at a launch velocity of 15 m/sec. Because the acceleration of gravity at the planet's surface was g_s m/sec², the explorers expected the ball bearing to reach a height of $s = 15t - (1/2)g_s t^2$ m t sec later. The ball bearing reached its maximum height 20 sec after being launched. What was the value of g_s?

12. Speeding bullet A 45-caliber bullet shot straight up from the surface of the moon would reach a height of $s = 832t - 2.6t^2$ ft after t sec. On Earth, in the absence of air, its height would be $s = 832t - 16t^2$ ft after t sec. How long will the bullet be aloft in each case? How high will the bullet go?

13. Free fall from the Tower of Pisa Had Galileo dropped a cannonball from the Tower of Pisa, 179 ft above the ground, the ball's height above the ground t sec into the fall would have been $s = 179 - 16t^2$.

a. What would have been the ball's velocity, speed, and acceleration at time t?

b. About how long would it have taken the ball to hit the ground?

c. What would have been the ball's velocity at the moment of impact?

14. Galileo's free-fall formula Galileo developed a formula for a body's velocity during free fall by rolling balls from rest down increasingly steep inclined planks and looking for a limiting formula that would predict a ball's behavior when the plank was vertical and the ball fell freely; see part (a) of the accompanying figure. He found that, for any given angle of the plank, the ball's velocity t sec into motion was a constant multiple of t. That is, the velocity was given by a formula of the form $v = kt$. The value of the constant k depended on the inclination of the plank.

In modern notation—part (b) of the figure—with distance in meters and time in seconds, what Galileo determined by experiment was that, for any given angle θ, the ball's velocity t sec into the roll was

$$v = 9.8(\sin \theta)t \text{ m/sec}.$$

Free-fall
position

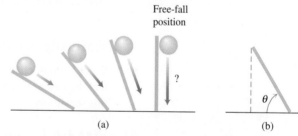

(a)   (b)

a. What is the equation for the ball's velocity during free fall?

b. Building on your work in part (a), what constant acceleration does a freely falling body experience near the surface of Earth?

Understanding Motion from Graphs

15. The accompanying figure shows the velocity $v = ds/dt = f(t)$ (m/sec) of a body moving along a coordinate line.

a. When does the body reverse direction?

b. When (approximately) is the body moving at a constant speed?

c. Graph the body's speed for $0 \le t \le 10$.

d. Graph the acceleration, where defined.

16. A particle P moves on the number line shown in part (a) of the accompanying figure. Part (b) shows the position of P as a function of time t.

(a)

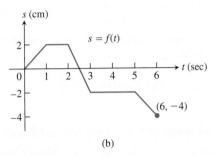

(b)

a. When is P moving to the left? Moving to the right? Standing still?

b. Graph the particle's velocity and speed (where defined).

17. Launching a rocket When a model rocket is launched, the propellant burns for a few seconds, accelerating the rocket upward. After burnout, the rocket coasts upward for a while and then begins to fall. A small explosive charge pops out a parachute shortly after the rocket starts down. The parachute slows the rocket to keep it from breaking when it lands.

The figure here shows velocity data from the flight of the model rocket. Use the data to answer the following.

a. How fast was the rocket climbing when the engine stopped?

b. For how many seconds did the engine burn?

c. When did the rocket reach its highest point? What was its velocity then?

d. When did the parachute pop out? How fast was the rocket falling then?

e. How long did the rocket fall before the parachute opened?

f. When was the rocket's acceleration greatest?

g. When was the acceleration constant? What was its value then (to the nearest integer)?

18. The accompanying figure shows the velocity $v = f(t)$ of a particle moving on a horizontal coordinate line.

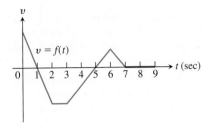

a. When does the particle move forward? Move backward? Speed up? Slow down?

b. When is the particle's acceleration positive? Negative? Zero?

c. When does the particle move at its greatest speed?

d. When does the particle stand still for more than an instant?

19. Two falling balls The multiflash photograph in the accompanying figure shows two balls falling from rest. The vertical rulers are marked in centimeters. Use the equation $s = 490t^2$ (the free-fall equation for s in centimeters and t in seconds) to answer the following questions.

a. How long did it take the balls to fall the first 160 cm? What was their average velocity for the period?

b. How fast were the balls falling when they reached the 160-cm mark? What was their acceleration then?

c. About how fast was the light flashing (flashes per second)?

20. A traveling truck The accompanying graph shows the position s of a truck traveling on a highway. The truck starts at $t = 0$ and returns 15 h later at $t = 15$.

   a. Use the technique described in Section 3.2, Example 3, to graph the truck's velocity $v = ds/dt$ for $0 \le t \le 15$. Then repeat the process, with the velocity curve, to graph the truck's acceleration dv/dt.

   b. Suppose that $s = 15t^2 - t^3$. Graph ds/dt and d^2s/dt^2 and compare your graphs with those in part (a).

21. The graphs in the accompanying figure show the position s, velocity $v = ds/dt$, and acceleration $a = d^2s/dt^2$ of a body moving along a coordinate line as functions of time t. Which graph is which? Give reasons for your answers.

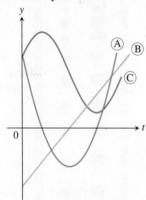

22. The graphs in the accompanying figure show the position s, the velocity $v = ds/dt$, and the acceleration $a = d^2s/dt^2$ of a body moving along the coordinate line as functions of time t. Which graph is which? Give reasons for your answers.

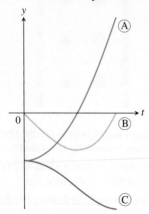

Economics

23. Marginal cost Suppose that the dollar cost of producing x washing machines is $c(x) = 2000 + 100x - 0.1x^2$.

   a. Find the average cost per machine of producing the first 100 washing machines.

   b. Find the marginal cost when 100 washing machines are produced.

   c. Show that the marginal cost when 100 washing machines are produced is approximately the cost of producing one more washing machine after the first 100 have been made, by calculating the latter cost directly.

24. Marginal revenue Suppose that the revenue from selling x washing machines is

$$r(x) = 20{,}000\left(1 - \frac{1}{x}\right)$$

dollars.

   a. Find the marginal revenue when 100 machines are produced.

   b. Use the function $r'(x)$ to estimate the increase in revenue that will result from increasing production from 100 machines a week to 101 machines a week.

   c. Find the limit of $r'(x)$ as $x \to \infty$. How would you interpret this number?

Additional Applications

25. Bacterium population When a bactericide was added to a nutrient broth in which bacteria were growing, the bacterium population continued to grow for a while, but then stopped growing and began to decline. The size of the population at time t (hours) was $b = 10^6 + 10^4t - 10^3t^2$. Find the growth rates at

   a. $t = 0$ hours.

   b. $t = 5$ hours.

   c. $t = 10$ hours.

26. Draining a tank The number of gallons of water in a tank t minutes after the tank has started to drain is $Q(t) = 200(30 - t)^2$. How fast is the water running out at the end of 10 min? What is the average rate at which the water flows out during the first 10 min?

T 27. Draining a tank It takes 12 hours to drain a storage tank by opening the valve at the bottom. The depth y of fluid in the tank t hours after the valve is opened is given by the formula

$$y = 6\left(1 - \frac{t}{12}\right)^2 \text{ m.}$$

   a. Find the rate dy/dt (m/h) at which the tank is draining at time t.

   b. When is the fluid level in the tank falling fastest? Slowest? What are the values of dy/dt at these times?

   c. Graph y and dy/dt together and discuss the behavior of y in relation to the signs and values of dy/dt.

28. Inflating a balloon The volume $V = (4/3)\pi r^3$ of a spherical balloon changes with the radius.

   a. At what rate (ft³/ft) does the volume change with respect to the radius when $r = 2$ ft?

   b. By approximately how much does the volume increase when the radius changes from 2 to 2.2 ft?

29. Airplane takeoff Suppose that the distance an aircraft travels along a runway before takeoff is given by $D = (10/9)t^2$, where D is measured in meters from the starting point and t is measured in seconds from the time the brakes are released. The aircraft will become airborne when its speed reaches 200 km/h. How long will it take to become airborne, and what distance will it travel in that time?

30. Volcanic lava fountains Although the November 1959 Kilauea Iki eruption on the island of Hawaii began with a line of fountains along the wall of the crater, activity was later confined to a single vent in the crater's floor, which at one point shot lava 1900 ft straight into the air (a Hawaiian record). What was the lava's exit velocity in feet per second? In miles per hour? (*Hint:* If v_0 is the exit velocity of a particle of lava, its height t sec later will be $s = v_0 t - 16t^2$ ft. Begin by finding the time at which $ds/dt = 0$. Neglect air resistance.)

Analyzing Motion Using Graphs

T Exercises 31–34 give the position function $s = f(t)$ of an object moving along the s-axis as a function of time t. Graph f together with the velocity function $v(t) = ds/dt = f'(t)$ and the acceleration function $a(t) = d^2s/dt^2 = f''(t)$. Comment on the object's behavior in relation to the signs and values of v and a. Include in your commentary such topics as the following:

a. When is the object momentarily at rest?

b. When does it move to the left (down) or to the right (up)?

c. When does it change direction?

d. When does it speed up and slow down?

e. When is it moving fastest (highest speed)? Slowest?

f. When is it farthest from the axis origin?

31. $s = 200t - 16t^2$, $0 \le t \le 12.5$ (a heavy object fired straight up from Earth's surface at 200 ft/sec)

32. $s = t^2 - 3t + 2$, $0 \le t \le 5$

33. $s = t^3 - 6t^2 + 7t$, $0 \le t \le 4$

34. $s = 4 - 7t + 6t^2 - t^3$, $0 \le t \le 4$

3.5 Derivatives of Trigonometric Functions

Many phenomena of nature are approximately periodic (electromagnetic fields, heart rhythms, tides, weather). The derivatives of sines and cosines play a key role in describing periodic changes. This section shows how to differentiate the six basic trigonometric functions.

Derivative of the Sine Function

To calculate the derivative of $f(x) = \sin x$, for x measured in radians, we combine the limits in Example 5a and Theorem 7 in Section 2.4 with the angle sum identity for the sine function:

$$\sin(x + h) = \sin x \cos h + \cos x \sin h.$$

If $f(x) = \sin x$, then

$$f'(x) = \lim_{h \to 0} \frac{f(x + h) - f(x)}{h} = \lim_{h \to 0} \frac{\sin(x + h) - \sin x}{h} \qquad \text{Derivative definition}$$

$$= \lim_{h \to 0} \frac{(\sin x \cos h + \cos x \sin h) - \sin x}{h} = \lim_{h \to 0} \frac{\sin x (\cos h - 1) + \cos x \sin h}{h}$$

$$= \lim_{h \to 0} \left(\sin x \cdot \frac{\cos h - 1}{h} \right) + \lim_{h \to 0} \left(\cos x \cdot \frac{\sin h}{h} \right)$$

$$= \sin x \cdot \underbrace{\lim_{h \to 0} \frac{\cos h - 1}{h}}_{\text{limit } 0} + \cos x \cdot \underbrace{\lim_{h \to 0} \frac{\sin h}{h}}_{\text{limit } 1} = \sin x \cdot 0 + \cos x \cdot 1 = \cos x.$$

Example 5a and Theorem 7, Section 2.4

> **The derivative of the sine function is the cosine function:**
>
> $$\frac{d}{dx}(\sin x) = \cos x.$$

EXAMPLE 1 We find derivatives of the sine function involving differences, products, and quotients.

(a) $y = x^2 - \sin x$:

$$\frac{dy}{dx} = 2x - \frac{d}{dx}(\sin x) \qquad \text{Difference Rule}$$

$$= 2x - \cos x$$

(b) $y = e^x \sin x$:

$$\frac{dy}{dx} = e^x \frac{d}{dx}(\sin x) + \frac{d}{dx}(e^x)\sin x \qquad \text{Product Rule}$$

$$= e^x \cos x + e^x \sin x$$

$$= e^x (\cos x + \sin x)$$

(c) $y = \dfrac{\sin x}{x}$:

$$\frac{dy}{dx} = \frac{x \cdot \frac{d}{dx}(\sin x) - \sin x \cdot 1}{x^2} \qquad \text{Quotient Rule}$$

$$= \frac{x \cos x - \sin x}{x^2}$$

∎

Derivative of the Cosine Function

With the help of the angle sum formula for the cosine function,

$$\cos(x + h) = \cos x \cos h - \sin x \sin h,$$

we can compute the limit of the difference quotient:

$$\frac{d}{dx}(\cos x) = \lim_{h \to 0} \frac{\cos(x + h) - \cos x}{h} \qquad \text{Derivative definition}$$

$$= \lim_{h \to 0} \frac{(\cos x \cos h - \sin x \sin h) - \cos x}{h} \qquad \begin{array}{l}\text{Cosine angle sum} \\ \text{identity}\end{array}$$

$$= \lim_{h \to 0} \frac{\cos x(\cos h - 1) - \sin x \sin h}{h}$$

$$= \lim_{h \to 0} \cos x \cdot \frac{\cos h - 1}{h} - \lim_{h \to 0} \sin x \cdot \frac{\sin h}{h}$$

$$= \cos x \cdot \lim_{h \to 0} \frac{\cos h - 1}{h} - \sin x \cdot \lim_{h \to 0} \frac{\sin h}{h}$$

$$= \cos x \cdot 0 - \sin x \cdot 1 \qquad \begin{array}{l}\text{Example 5a and} \\ \text{Theorem 7, Section 2.4}\end{array}$$

$$= -\sin x.$$

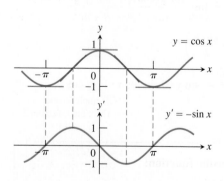

FIGURE 3.22 The curve $y' = -\sin x$ as the graph of the slopes of the tangents to the curve $y = \cos x$.

The derivative of the cosine function is the negative of the sine function:

$$\frac{d}{dx}(\cos x) = -\sin x.$$

Figure 3.22 shows a way to visualize this result in the same way we did for graphing derivatives in Section 3.2, Figure 3.6.

EXAMPLE 2 We find derivatives of the cosine function in combinations with other functions.

(a) $y = 5e^x + \cos x$:

$$\frac{dy}{dx} = \frac{d}{dx}(5e^x) + \frac{d}{dx}(\cos x) \qquad \text{Sum Rule}$$

$$= 5e^x - \sin x$$

(b) $y = \sin x \cos x$:

$$\frac{dy}{dx} = \sin x \frac{d}{dx}(\cos x) + \cos x \frac{d}{dx}(\sin x) \qquad \text{Product Rule}$$

$$= \sin x(-\sin x) + \cos x(\cos x)$$

$$= \cos^2 x - \sin^2 x$$

(c) $y = \dfrac{\cos x}{1 - \sin x}$:

$$\frac{dy}{dx} = \frac{(1 - \sin x)\dfrac{d}{dx}(\cos x) - \cos x \dfrac{d}{dx}(1 - \sin x)}{(1 - \sin x)^2} \qquad \text{Quotient Rule}$$

$$= \frac{(1 - \sin x)(-\sin x) - \cos x(0 - \cos x)}{(1 - \sin x)^2}$$

$$= \frac{1 - \sin x}{(1 - \sin x)^2} \qquad \sin^2 x + \cos^2 x = 1$$

$$= \frac{1}{1 - \sin x} \qquad\qquad\blacksquare$$

Simple Harmonic Motion

The motion of an object or weight bobbing freely up and down with no resistance on the end of a spring is an example of *simple harmonic motion*. The motion is periodic and repeats indefinitely, so we represent it using trigonometric functions. The next example describes a case in which there are no opposing forces such as friction or buoyancy to slow the motion.

EXAMPLE 3 A weight hanging from a spring (Figure 3.23) is stretched down 5 units beyond its rest position and released at time $t = 0$ to bob up and down. Its position at any later time t is

$$s = 5 \cos t.$$

What are its velocity and acceleration at time t?

Solution We have

Position: $s = 5 \cos t$

Velocity: $v = \dfrac{ds}{dt} = \dfrac{d}{dt}(5 \cos t) = -5 \sin t$

Acceleration: $a = \dfrac{dv}{dt} = \dfrac{d}{dt}(-5 \sin t) = -5 \cos t.$

FIGURE 3.23 A weight hanging from a vertical spring and then displaced oscillates above and below its rest position (Example 3).

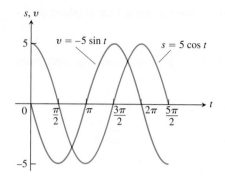

FIGURE 3.24 The graphs of the position and velocity of the weight in Example 3.

Notice how much we can learn from these equations:

1. As time passes, the weight moves down and up between $s = -5$ and $s = 5$ on the s-axis. The amplitude of the motion is 5. The period of the motion is 2π, the period of the cosine function.

2. The velocity $v = -5 \sin t$ attains its greatest magnitude, 5, when $\cos t = 0$, as the graphs show in Figure 3.24. Hence, the speed of the weight, $|v| = 5|\sin t|$, is greatest when $\cos t = 0$, that is, when $s = 0$ (the rest position). The speed of the weight is zero when $\sin t = 0$. This occurs when $s = 5 \cos t = \pm 5$, at the endpoints of the interval of motion.

3. The acceleration value is always the exact opposite of the position value. When the weight is above the rest position, gravity is pulling it back down; when the weight is below the rest position, the spring is pulling it back up.

4. The acceleration, $a = -5 \cos t$, is zero only at the rest position, where $\cos t = 0$ and the force of gravity and the force from the spring balance each other. When the weight is anywhere else, the two forces are unequal and acceleration is nonzero. The acceleration is greatest in magnitude at the points farthest from the rest position, where $\cos t = \pm 1$. ∎

EXAMPLE 4   The jerk associated with the simple harmonic motion in Example 3 is

$$j = \frac{da}{dt} = \frac{d}{dt}(-5 \cos t) = 5 \sin t.$$

It has its greatest magnitude when $\sin t = \pm 1$, not at the extremes of the displacement but at the rest position, where the acceleration changes direction and sign. ∎

Derivatives of the Other Basic Trigonometric Functions

Because $\sin x$ and $\cos x$ are differentiable functions of x, the related functions

$$\tan x = \frac{\sin x}{\cos x}, \qquad \cot x = \frac{\cos x}{\sin x}, \qquad \sec x = \frac{1}{\cos x}, \qquad \text{and} \qquad \csc x = \frac{1}{\sin x}$$

are differentiable at every value of x at which they are defined. Their derivatives, calculated from the Quotient Rule, are given by the following formulas. Notice the negative signs in the derivative formulas for the cofunctions.

The derivatives of the other trigonometric functions:

$$\frac{d}{dx}(\tan x) = \sec^2 x \qquad\qquad \frac{d}{dx}(\cot x) = -\csc^2 x$$

$$\frac{d}{dx}(\sec x) = \sec x \tan x \qquad\qquad \frac{d}{dx}(\csc x) = -\csc x \cot x$$

To show a typical calculation, we find the derivative of the tangent function. The other derivations are left to Exercise 60.

EXAMPLE 5   Find $d(\tan x)/dx$.

Solution   We use the Derivative Quotient Rule to calculate the derivative:

$$\frac{d}{dx}(\tan x) = \frac{d}{dx}\left(\frac{\sin x}{\cos x}\right) = \frac{\cos x \dfrac{d}{dx}(\sin x) - \sin x \dfrac{d}{dx}(\cos x)}{\cos^2 x} \qquad \text{Quotient Rule}$$

$$= \frac{\cos x \cos x - \sin x(-\sin x)}{\cos^2 x}$$

$$= \frac{\cos^2 x + \sin^2 x}{\cos^2 x}$$

$$= \frac{1}{\cos^2 x} = \sec^2 x. \qquad \blacksquare$$

EXAMPLE 6   Find y'' if $y = \sec x$.

Solution   Finding the second derivative involves a combination of trigonometric derivatives.

$$y = \sec x$$

$$y' = \sec x \tan x \qquad \text{Derivative rule for secant function}$$

$$y'' = \frac{d}{dx}(\sec x \tan x)$$

$$= \sec x \frac{d}{dx}(\tan x) + \tan x \frac{d}{dx}(\sec x) \qquad \text{Derivative Product Rule}$$

$$= \sec x(\sec^2 x) + \tan x(\sec x \tan x) \qquad \text{Derivative rules}$$

$$= \sec^3 x + \sec x \tan^2 x \qquad \blacksquare$$

The differentiability of the trigonometric functions throughout their domains gives another proof of their continuity at every point in their domains (Theorem 1, Section 3.2). So we can calculate limits of algebraic combinations and composites of trigonometric functions by direct substitution.

EXAMPLE 7   We can use direct substitution in computing limits provided there is no division by zero, which is algebraically undefined.

$$\lim_{x \to 0} \frac{\sqrt{2 + \sec x}}{\cos(\pi - \tan x)} = \frac{\sqrt{2 + \sec 0}}{\cos(\pi - \tan 0)} = \frac{\sqrt{2 + 1}}{\cos(\pi - 0)} = \frac{\sqrt{3}}{-1} = -\sqrt{3} \qquad \blacksquare$$

Exercises 3.5

Derivatives

In Exercises 1–18, find dy/dx.

1. $y = -10x + 3\cos x$

2. $y = \dfrac{3}{x} + 5\sin x$

3. $y = x^2 \cos x$

4. $y = \sqrt{x}\,\sec x + 3$

5. $y = \csc x - 4\sqrt{x} + 7$

6. $y = x^2 \cot x - \dfrac{1}{x^2}$

7. $f(x) = \sin x \tan x$

8. $g(x) = \csc x \cot x$

9. $y = (\sec x + \tan x)(\sec x - \tan x)$

10. $y = (\sin x + \cos x)\sec x$

11. $y = \dfrac{\cot x}{1 + \cot x}$

12. $y = \dfrac{\cos x}{1 + \sin x}$

13. $y = \dfrac{4}{\cos x} + \dfrac{1}{\tan x}$

14. $y = \dfrac{\cos x}{x} + \dfrac{x}{\cos x}$

15. $y = x^2 \sin x + 2x \cos x - 2 \sin x$

16. $y = x^2 \cos x - 2x \sin x - 2 \cos x$

17. $f(x) = x^3 \sin x \cos x$

18. $g(x) = (2 - x) \tan^2 x$

In Exercises 19–22, find ds/dt.

19. $s = \tan t - e^{-t}$

20. $s = t^2 - \sec t + 5e^t$

21. $s = \dfrac{1 + \csc t}{1 - \csc t}$

22. $s = \dfrac{\sin t}{1 - \cos t}$

In Exercises 23–26, find $dr/d\theta$.

23. $r = 4 - \theta^2 \sin \theta$

24. $r = \theta \sin \theta + \cos \theta$

25. $r = \sec \theta \csc \theta$

26. $r = (1 + \sec \theta) \sin \theta$

In Exercises 27–32, find dp/dq.

27. $p = 5 + \dfrac{1}{\cot q}$

28. $p = (1 + \csc q) \cos q$

29. $p = \dfrac{\sin q + \cos q}{\cos q}$

30. $p = \dfrac{\tan q}{1 + \tan q}$

31. $p = \dfrac{q \sin q}{q^2 - 1}$

32. $p = \dfrac{3q + \tan q}{q \sec q}$

33. Find y'' if

    a. $y = \csc x$.

    b. $y = \sec x$.

34. Find $y^{(4)} = d^4 y/dx^4$ if

    a. $y = -2 \sin x$.

    b. $y = 9 \cos x$.

Tangent Lines

In Exercises 35–38, graph the curves over the given intervals, together with their tangents at the given values of x. Label each curve and tangent with its equation.

35. $y = \sin x, \quad -3\pi/2 \le x \le 2\pi$

    $x = -\pi, 0, 3\pi/2$

36. $y = \tan x, \quad -\pi/2 < x < \pi/2$

    $x = -\pi/3, 0, \pi/3$

37. $y = \sec x, \quad -\pi/2 < x < \pi/2$

    $x = -\pi/3, \pi/4$

38. $y = 1 + \cos x, \quad -3\pi/2 \le x \le 2\pi$

    $x = -\pi/3, 3\pi/2$

T Do the graphs of the functions in Exercises 39–42 have any horizontal tangents in the interval $0 \le x \le 2\pi$? If so, where? If not, why not? Visualize your findings by graphing the functions with a grapher.

39. $y = x + \sin x$

40. $y = 2x + \sin x$

41. $y = x - \cot x$

42. $y = x + 2 \cos x$

43. Find all points on the curve $y = \tan x, -\pi/2 < x < \pi/2$, where the tangent line is parallel to the line $y = 2x$. Sketch the curve and tangent(s) together, labeling each with its equation.

44. Find all points on the curve $y = \cot x, 0 < x < \pi$, where the tangent line is parallel to the line $y = -x$. Sketch the curve and tangent(s) together, labeling each with its equation.

In Exercises 45 and 46, find an equation for **(a)** the tangent to the curve at P and **(b)** the horizontal tangent to the curve at Q.

45.

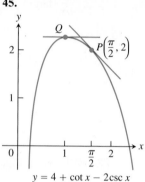

$y = 4 + \cot x - 2\csc x$

46.

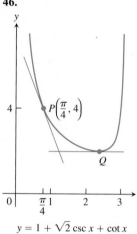

$y = 1 + \sqrt{2} \csc x + \cot x$

Trigonometric Limits

Find the limits in Exercises 47–54.

47. $\lim\limits_{x \to 2} \sin \left(\dfrac{1}{x} - \dfrac{1}{2} \right)$

48. $\lim\limits_{x \to -\pi/6} \sqrt{1 + \cos (\pi \csc x)}$

49. $\lim\limits_{\theta \to \pi/6} \dfrac{\sin \theta - \frac{1}{2}}{\theta - \frac{\pi}{6}}$

50. $\lim\limits_{\theta \to \pi/4} \dfrac{\tan \theta - 1}{\theta - \frac{\pi}{4}}$

51. $\lim\limits_{x \to 0} \sec \left[e^x + \pi \tan \left(\dfrac{\pi}{4 \sec x} \right) - 1 \right]$

52. $\lim\limits_{x \to 0} \sin \left(\dfrac{\pi + \tan x}{\tan x - 2 \sec x} \right)$

53. $\lim\limits_{t \to 0} \tan \left(1 - \dfrac{\sin t}{t} \right)$

54. $\lim\limits_{\theta \to 0} \cos \left(\dfrac{\pi \theta}{\sin \theta} \right)$

Theory and Examples

The equations in Exercises 55 and 56 give the position $s = f(t)$ of a body moving on a coordinate line (s in meters, t in seconds). Find the body's velocity, speed, acceleration, and jerk at time $t = \pi/4$ sec.

55. $s = 2 - 2 \sin t$

56. $s = \sin t + \cos t$

57. Is there a value of c that will make

$$f(x) = \begin{cases} \dfrac{\sin^2 3x}{x^2}, & x \ne 0 \\ c, & x = 0 \end{cases}$$

continuous at $x = 0$? Give reasons for your answer.

58. Is there a value of b that will make

$$g(x) = \begin{cases} x + b, & x < 0 \\ \cos x, & x \ge 0 \end{cases}$$

continuous at $x = 0$? Differentiable at $x = 0$? Give reasons for your answers.

THEOREM 3—The Derivative Rule for Inverses If f has an interval I as domain and $f'(x)$ exists and is never zero on I, then f^{-1} is differentiable at every point in its domain (the range of f). The value of $(f^{-1})'$ at a point b in the domain of f^{-1} is the reciprocal of the value of f' at the point $a = f^{-1}(b)$:

$$(f^{-1})'(b) = \frac{1}{f'(f^{-1}(b))} \qquad (1)$$

or

$$\left. \frac{df^{-1}}{dx} \right|_{x=b} = \frac{1}{\left. \dfrac{df}{dx} \right|_{x=f^{-1}(b)}}$$

Theorem 3 makes two assertions. The first of these has to do with the conditions under which f^{-1} is differentiable; the second assertion is a formula for the derivative of f^{-1} when it exists. While we omit the proof of the first assertion, the second one is proved in the following way:

$$f(f^{-1}(x)) = x \qquad \text{Inverse function relationship}$$

$$\frac{d}{dx} f(f^{-1}(x)) = 1 \qquad \text{Differentiating both sides}$$

$$f'(f^{-1}(x)) \cdot \frac{d}{dx} f^{-1}(x) = 1 \qquad \text{Chain Rule}$$

$$\frac{d}{dx} f^{-1}(x) = \frac{1}{f'(f^{-1}(x))}. \qquad \text{Solving for the derivative}$$

EXAMPLE 1 The function $f(x) = x^2, x \geq 0$ and its inverse $f^{-1}(x) = \sqrt{x}$ have derivatives $f'(x) = 2x$ and $(f^{-1})'(x) = 1/(2\sqrt{x})$.

Let's verify that Theorem 3 gives the same formula for the derivative of $f^{-1}(x)$:

$$(f^{-1})'(x) = \frac{1}{f'(f^{-1}(x))}$$

$$= \frac{1}{2(f^{-1}(x))} \qquad \begin{array}{l} f'(x) = 2x \text{ with } x \text{ replaced} \\ \text{by } f^{-1}(x) \end{array}$$

$$= \frac{1}{2(\sqrt{x})}.$$

Theorem 3 gives a derivative that agrees with the known derivative of the square root function.

Let's examine Theorem 3 at a specific point. We pick $x = 2$ (the number a) and $f(2) = 4$ (the value b). Theorem 3 says that the derivative of f at 2, $f'(2) = 4$, and the derivative of f^{-1} at $f(2)$, $(f^{-1})'(4)$, are reciprocals. It states that

$$(f^{-1})'(4) = \frac{1}{f'(f^{-1}(4))} = \frac{1}{f'(2)} = \left. \frac{1}{2x} \right|_{x=2} = \frac{1}{4}.$$

See Figure 3.36. ∎

FIGURE 3.36 The derivative of $f^{-1}(x) = \sqrt{x}$ at the point $(4, 2)$ is the reciprocal of the derivative of $f(x) = x^2$ at $(2, 4)$ (Example 1).

We will use the procedure illustrated in Example 1 to calculate formulas for the derivatives of many inverse functions throughout this chapter. Equation (1) sometimes enables us to find specific values of df^{-1}/dx without knowing a formula for f^{-1}.

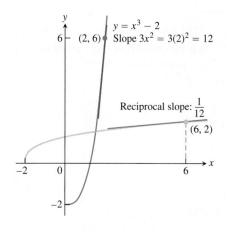

FIGURE 3.37 The derivative of $f(x) = x^3 - 2$ at $x = 2$ tells us the derivative of f^{-1} at $x = 6$ (Example 2).

EXAMPLE 2 Let $f(x) = x^3 - 2$. Find the value of df^{-1}/dx at $x = 6 = f(2)$ without finding a formula for $f^{-1}(x)$.

Solution We apply Theorem 3 to obtain the value of the derivative of f^{-1} at $x = 6$:

$$\frac{df}{dx}\bigg|_{x=2} = 3x^2\bigg|_{x=2} = 12$$

$$\frac{df^{-1}}{dx}\bigg|_{x=f(2)} = \frac{1}{\dfrac{df}{dx}\bigg|_{x=2}} = \frac{1}{12}. \qquad \text{Eq. (1)}$$

See Figure 3.37. ∎

Derivative of the Natural Logarithm Function

Since we know the exponential function $f(x) = e^x$ is differentiable everywhere, we can apply Theorem 3 to find the derivative of its inverse $f^{-1}(x) = \ln x$:

$$(f^{-1})'(x) = \frac{1}{f'(f^{-1}(x))} \qquad \text{Theorem 3}$$

$$= \frac{1}{e^{f^{-1}(x)}} \qquad f'(u) = e^u$$

$$= \frac{1}{e^{\ln x}}$$

$$= \frac{1}{x}. \qquad \text{Inverse function relationship}$$

Alternate Derivation Instead of applying Theorem 3 directly, we can find the derivative of $y = \ln x$ using implicit differentiation, as follows:

$$y = \ln x$$

$$e^y = x \qquad \text{Inverse function relationship}$$

$$\frac{d}{dx}(e^y) = \frac{d}{dx}(x) \qquad \text{Differentiate implicitly}$$

$$e^y \frac{dy}{dx} = 1 \qquad \text{Chain Rule}$$

$$\frac{dy}{dx} = \frac{1}{e^y} = \frac{1}{x}. \qquad e^y = x$$

No matter which derivation we use, the derivative of $y = \ln x$ with respect to x is

$$\frac{d}{dx}(\ln x) = \frac{1}{x}, \quad x > 0.$$

The Chain Rule extends this formula for positive functions $u(x)$:

$$\frac{d}{dx}\ln u = \frac{d}{du}\ln u \cdot \frac{du}{dx}$$

$$\boxed{\frac{d}{dx}\ln u = \frac{1}{u}\frac{du}{dx}, \quad u > 0.} \qquad (2)$$

EXAMPLE 3   We use Equation (2) to find derivatives.

(a) $\dfrac{d}{dx} \ln 2x = \dfrac{1}{2x} \dfrac{d}{dx} (2x) = \dfrac{1}{2x} (2) = \dfrac{1}{x}, \quad x > 0$

(b) Equation (2) with $u = x^2 + 3$ gives

$$\frac{d}{dx} \ln (x^2 + 3) = \frac{1}{x^2 + 3} \cdot \frac{d}{dx} (x^2 + 3) = \frac{1}{x^2 + 3} \cdot 2x = \frac{2x}{x^2 + 3}. \qquad \blacksquare$$

Notice the remarkable occurrence in Example 3a. The function $y = \ln 2x$ has the same derivative as the function $y = \ln x$. This is true of $y = \ln bx$ for any constant b, provided that $bx > 0$:

$$\frac{d}{dx} \ln bx = \frac{1}{bx} \cdot \frac{d}{dx} (bx) = \frac{1}{bx} (b) = \frac{1}{x}. \tag{3}$$

If $x < 0$ and $b < 0$, then $bx > 0$ and Equation (3) still applies. In particular, if $x < 0$ and $b = -1$ we get

$$\frac{d}{dx} \ln (-x) = \frac{1}{x} \qquad \text{for } x < 0.$$

Since $|x| = x$ when $x > 0$ and $|x| = -x$ when $x < 0$, we have the following important result.

$$\frac{d}{dx} \ln |x| = \frac{1}{x}, \qquad x \neq 0 \tag{4}$$

EXAMPLE 4   A line with slope m passes through the origin and is tangent to the graph of $y = \ln x$. What is the value of m?

Solution   Suppose the point of tangency occurs at the unknown point $x = a > 0$. Then we know that the point $(a, \ln a)$ lies on the graph and that the tangent line at that point has slope $m = 1/a$ (Figure 3.38). Since the tangent line passes through the origin, its slope is

$$m = \frac{\ln a - 0}{a - 0} = \frac{\ln a}{a}.$$

Setting these two formulas for m equal to each other, we have

$$\frac{\ln a}{a} = \frac{1}{a}$$

$$\ln a = 1$$

$$e^{\ln a} = e^1$$

$$a = e$$

$$m = \frac{1}{e}. \qquad \blacksquare$$

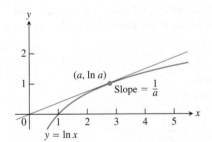

FIGURE 3.38 The tangent line intersects the curve at some point $(a, \ln a)$, where the slope of the curve is $1/a$ (Example 4).

The Derivatives of a^u and $\log_a u$

We start with the equation $a^x = e^{\ln (a^x)} = e^{x \ln a}$, which was established in Section 1.6:

$$\frac{d}{dx} a^x = \frac{d}{dx} e^{x \ln a} = e^{x \ln a} \cdot \frac{d}{dx} (x \ln a) \qquad \frac{d}{dx} e^u = e^u \frac{du}{dx}$$

$$= a^x \ln a.$$

If $a > 0$, then

$$\frac{d}{dx}a^x = a^x \ln a.$$

This equation shows why e^x is the exponential function preferred in calculus. If $a = e$, then $\ln a = 1$ and the derivative of a^x simplifies to

$$\frac{d}{dx}e^x = e^x \ln e = e^x.$$

With the Chain Rule, we get a more general form for the derivative of a general exponential function.

If $a > 0$ and u is a differentiable function of x, then a^u is a differentiable function of x and

$$\frac{d}{dx}a^u = a^u \ln a \frac{du}{dx}. \tag{5}$$

EXAMPLE 5   We illustrate using Equation (5).

(a) $\dfrac{d}{dx}3^x = 3^x \ln 3$ Eq. (5) with $a = 3$, $u = x$

(b) $\dfrac{d}{dx}3^{-x} = 3^{-x}(\ln 3)\dfrac{d}{dx}(-x) = -3^{-x}\ln 3$ Eq. (5) with $a = 3$, $u = -x$

(c) $\dfrac{d}{dx}3^{\sin x} = 3^{\sin x}(\ln 3)\dfrac{d}{dx}(\sin x) = 3^{\sin x}(\ln 3)\cos x$ $\ldots$, $u = \sin x$ ∎

In Section 3.3 we looked at the derivative $f'(0)$ for the exponential functions $f(x) = a^x$ at various values of the base a. The number $f'(0)$ is the limit, $\lim_{h \to 0}(a^h - 1)/h$, and gives the slope of the graph of a^x when it crosses the y-axis at the point $(0, 1)$. We now see that the value of this slope is

$$\lim_{h \to 0}\frac{a^h - 1}{h} = \ln a. \tag{6}$$

In particular, when $a = e$ we obtain

$$\lim_{h \to 0}\frac{e^h - 1}{h} = \ln e = 1.$$

However, we have not fully justified that these limits actually exist. While all of the arguments given in deriving the derivatives of the exponential and logarithmic functions are correct, they do assume the existence of these limits. In Chapter 7 we will give another development of the theory of logarithmic and exponential functions which fully justifies that both limits do in fact exist and have the values derived above.

To find the derivative of $\log_a u$ for an arbitrary base ($a > 0$, $a \neq 1$), we start with the change-of-base formula for logarithms (reviewed in Section 1.6) and express $\log_a u$ in terms of natural logarithms,

$$\log_a x = \frac{\ln x}{\ln a}.$$

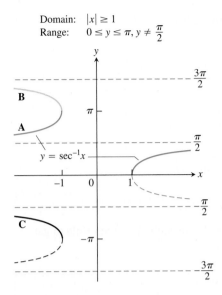

Domain: $|x| \geq 1$
Range: $0 \leq y \leq \pi, y \neq \dfrac{\pi}{2}$

$y = \sec^{-1}x$

FIGURE 3.40 There are several logical choices for the left-hand branch of $y = \sec^{-1} x$. With choice **A**, $\sec^{-1} x = \cos^{-1}(1/x)$, a useful identity employed by many calculators.

We use open intervals to avoid values where the tangent and cotangent are undefined.

The graph of $y = \tan^{-1} x$ is symmetric about the origin because it is a branch of the graph $x = \tan y$ that is symmetric about the origin (Figure 3.39c). Algebraically this means that

$$\tan^{-1}(-x) = -\tan^{-1} x;$$

the arctangent is an odd function. The graph of $y = \cot^{-1} x$ has no such symmetry (Figure 3.39f). Notice from Figure 3.39c that the graph of the arctangent function has two horizontal asymptotes; one at $y = \pi/2$ and the other at $y = -\pi/2$.

The inverses of the restricted forms of $\sec x$ and $\csc x$ are chosen to be the functions graphed in Figures 3.39d and 3.39e.

Caution There is no general agreement about how to define $\sec^{-1} x$ for negative values of x. We chose angles in the second quadrant between $\pi/2$ and π. This choice makes $\sec^{-1} x = \cos^{-1}(1/x)$. It also makes $\sec^{-1} x$ an increasing function on each interval of its domain. Some tables choose $\sec^{-1} x$ to lie in $[-\pi, -\pi/2)$ for $x < 0$ and some texts choose it to lie in $[\pi, 3\pi/2)$ (Figure 3.40). These choices simplify the formula for the derivative (our formula needs absolute value signs) but fail to satisfy the computational equation $\sec^{-1} x = \cos^{-1}(1/x)$. From this, we can derive the identity

$$\sec^{-1} x = \cos^{-1}\left(\frac{1}{x}\right) = \frac{\pi}{2} - \sin^{-1}\left(\frac{1}{x}\right) \tag{1}$$

by applying Equation (5) in Section 1.6.

EXAMPLE 1 The accompanying figures show two values of $\tan^{-1} x$.

x	$\tan^{-1} x$
$\sqrt{3}$	$\pi/3$
1	$\pi/4$
$\sqrt{3}/3$	$\pi/6$
$-\sqrt{3}/3$	$-\pi/6$
-1	$-\pi/4$
$-\sqrt{3}$	$-\pi/3$

$$\tan^{-1}\frac{1}{\sqrt{3}} = \tan^{-1}\frac{\sqrt{3}}{3} = \frac{\pi}{6}$$

$$\tan\frac{\pi}{6} = \frac{1}{\sqrt{3}}$$

$$\tan^{-1}\left(-\sqrt{3}\right) = -\frac{\pi}{3}$$

$$\tan\left(-\frac{\pi}{3}\right) = -\sqrt{3}$$

The angles come from the first and fourth quadrants because the range of $\tan^{-1} x$ is $(-\pi/2, \pi/2)$. ∎

The Derivative of $y = \sin^{-1} u$

We know that the function $x = \sin y$ is differentiable in the interval $-\pi/2 < y < \pi/2$ and that its derivative, the cosine, is positive there. Theorem 3 in Section 3.8 therefore assures us that the inverse function $y = \sin^{-1} x$ is differentiable throughout the interval $-1 < x < 1$. We cannot expect it to be differentiable at $x = 1$ or $x = -1$ because the tangents to the graph are vertical at these points (see Figure 3.41).

FIGURE 3.41 The graph of $y = \sin^{-1} x$ has vertical tangents at $x = -1$ and $x = 1$.

We find the derivative of $y = \sin^{-1} x$ by applying Theorem 3 with $f(x) = \sin x$ and $f^{-1}(x) = \sin^{-1} x$:

$$(f^{-1})'(x) = \frac{1}{f'(f^{-1}(x))} \qquad \text{Theorem 3}$$

$$= \frac{1}{\cos(\sin^{-1} x)} \qquad f'(u) = \cos u$$

$$= \frac{1}{\sqrt{1 - \sin^2(\sin^{-1} x)}} \qquad \cos u = \sqrt{1 - \sin^2 u}$$

$$= \frac{1}{\sqrt{1 - x^2}}. \qquad \sin(\sin^{-1} x) = x$$

If u is a differentiable function of x with $|u| < 1$, we apply the Chain Rule to get

$$\frac{d}{dx}(\sin^{-1} u) = \frac{1}{\sqrt{1 - u^2}} \frac{du}{dx}, \qquad |u| < 1.$$

EXAMPLE 2 Using the Chain Rule, we calculate the derivative

$$\frac{d}{dx}(\sin^{-1} x^2) = \frac{1}{\sqrt{1 - (x^2)^2}} \cdot \frac{d}{dx}(x^2) = \frac{2x}{\sqrt{1 - x^4}}. \qquad \blacksquare$$

The Derivative of $y = \tan^{-1} u$

We find the derivative of $y = \tan^{-1} x$ by applying Theorem 3 with $f(x) = \tan x$ and $f^{-1}(x) = \tan^{-1} x$. Theorem 3 can be applied because the derivative of $\tan x$ is positive for $-\pi/2 < x < \pi/2$:

$$(f^{-1})'(x) = \frac{1}{f'(f^{-1}(x))} \qquad \text{Theorem 3}$$

$$= \frac{1}{\sec^2(\tan^{-1} x)} \qquad f'(u) = \sec^2 u$$

$$= \frac{1}{1 + \tan^2(\tan^{-1} x)} \qquad \sec^2 u = 1 + \tan^2 u$$

$$= \frac{1}{1 + x^2}. \qquad \tan(\tan^{-1} x) = x$$

The derivative is defined for all real numbers. If u is a differentiable function of x, we get the Chain Rule form:

$$\frac{d}{dx}(\tan^{-1} u) = \frac{1}{1 + u^2} \frac{du}{dx}.$$

The Derivative of $y = \sec^{-1} u$

Since the derivative of $\sec x$ is positive for $0 < x < \pi/2$ and $\pi/2 < x < \pi$, Theorem 3 says that the inverse function $y = \sec^{-1} x$ is differentiable. Instead of applying the formula in Theorem 3 directly, we find the derivative of $y = \sec^{-1} x$, $|x| > 1$, using implicit differentiation and the Chain Rule as follows:

$$y = \sec^{-1} x$$

$$\sec y = x \qquad \text{Inverse function relationship}$$

$$\frac{d}{dx}(\sec y) = \frac{d}{dx}x \qquad \text{Differentiate both sides.}$$

$$\sec y \tan y \frac{dy}{dx} = 1 \qquad \text{Chain Rule}$$

$$\frac{dy}{dx} = \frac{1}{\sec y \tan y}. \qquad \begin{array}{l}\text{Since } |x| > 1, y \text{ lies in} \\ (0, \pi/2) \cup (\pi/2, \pi) \text{ and} \\ \sec y \tan y \neq 0.\end{array}$$

To express the result in terms of x, we use the relationships

$$\sec y = x \qquad \text{and} \qquad \tan y = \pm\sqrt{\sec^2 y - 1} = \pm\sqrt{x^2 - 1}$$

to get

$$\frac{dy}{dx} = \pm\frac{1}{x\sqrt{x^2 - 1}}.$$

Can we do anything about the $\pm$ sign? A glance at Figure 3.42 shows that the slope of the graph $y = \sec^{-1} x$ is always positive. Thus,

$$\frac{d}{dx}\sec^{-1} x = \begin{cases} +\dfrac{1}{x\sqrt{x^2 - 1}} & \text{if } x > 1 \\ -\dfrac{1}{x\sqrt{x^2 - 1}} & \text{if } x < -1. \end{cases}$$

FIGURE 3.42 The slope of the curve $y = \sec^{-1} x$ is positive for both $x < -1$ and $x > 1$.

With the absolute value symbol, we can write a single expression that eliminates the "$\pm$" ambiguity:

$$\frac{d}{dx}\sec^{-1} x = \frac{1}{|x|\sqrt{x^2 - 1}}.$$

If u is a differentiable function of x with $|u| > 1$, we have the formula

$$\frac{d}{dx}(\sec^{-1} u) = \frac{1}{|u|\sqrt{u^2 - 1}}\frac{du}{dx}, \qquad |u| > 1.$$

EXAMPLE 3   Using the Chain Rule and derivative of the arcsecant function, we find

$$\frac{d}{dx}\sec^{-1}(5x^4) = \frac{1}{|5x^4|\sqrt{(5x^4)^2 - 1}}\frac{d}{dx}(5x^4)$$

$$= \frac{1}{5x^4\sqrt{25x^8 - 1}}(20x^3) \qquad 5x^4 > 1 > 0$$

$$= \frac{4}{x\sqrt{25x^8 - 1}}. \qquad \blacksquare$$

Derivatives of the Other Three Inverse Trigonometric Functions

We could use the same techniques to find the derivatives of the other three inverse trigonometric functions—arccosine, arccotangent, and arccosecant—but there is an easier way, thanks to the following identities.

Inverse Function–Inverse Cofunction Identities

$$\cos^{-1} x = \pi/2 - \sin^{-1} x$$

$$\cot^{-1} x = \pi/2 - \tan^{-1} x$$

$$\csc^{-1} x = \pi/2 - \sec^{-1} x$$

We saw the first of these identities in Equation (5) of Section 1.6. The others are derived in a similar way. It follows easily that the derivatives of the inverse cofunctions are the negatives of the derivatives of the corresponding inverse functions. For example, the derivative of $\cos^{-1} x$ is calculated as follows:

$$\frac{d}{dx}(\cos^{-1} x) = \frac{d}{dx}\left(\frac{\pi}{2} - \sin^{-1} x\right) \qquad \text{Identity}$$

$$= -\frac{d}{dx}(\sin^{-1} x)$$

$$= -\frac{1}{\sqrt{1 - x^2}}. \qquad \text{Derivative of arcsine}$$

The derivatives of the inverse trigonometric functions are summarized in Table 3.1.

TABLE 3.1 Derivatives of the inverse trigonometric functions

1. $\dfrac{d(\sin^{-1} u)}{dx} = \dfrac{1}{\sqrt{1 - u^2}}\dfrac{du}{dx}, \quad |u| < 1$

2. $\dfrac{d(\cos^{-1} u)}{dx} = -\dfrac{1}{\sqrt{1 - u^2}}\dfrac{du}{dx}, \quad |u| < 1$

3. $\dfrac{d(\tan^{-1} u)}{dx} = \dfrac{1}{1 + u^2}\dfrac{du}{dx}$

4. $\dfrac{d(\cot^{-1} u)}{dx} = -\dfrac{1}{1 + u^2}\dfrac{du}{dx}$

5. $\dfrac{d(\sec^{-1} u)}{dx} = \dfrac{1}{|u|\sqrt{u^2 - 1}}\dfrac{du}{dx}, \quad |u| > 1$

6. $\dfrac{d(\csc^{-1} u)}{dx} = -\dfrac{1}{|u|\sqrt{u^2 - 1}}\dfrac{du}{dx}, \quad |u| > 1$

Exercises 3.9

Common Values

Use reference triangles like those in Example 1 to find the angles in Exercises 1–8.

1. a. $\tan^{-1} 1$ **b.** $\tan^{-1}(-\sqrt{3})$ **c.** $\tan^{-1}\left(\dfrac{1}{\sqrt{3}}\right)$

2. a. $\tan^{-1}(-1)$ **b.** $\tan^{-1}\sqrt{3}$ **c.** $\tan^{-1}\left(\dfrac{-1}{\sqrt{3}}\right)$

3. a. $\sin^{-1}\left(\dfrac{-1}{2}\right)$ **b.** $\sin^{-1}\left(\dfrac{1}{\sqrt{2}}\right)$ **c.** $\sin^{-1}\left(\dfrac{-\sqrt{3}}{2}\right)$

4. a. $\sin^{-1}\left(\dfrac{1}{2}\right)$ **b.** $\sin^{-1}\left(\dfrac{-1}{\sqrt{2}}\right)$ **c.** $\sin^{-1}\left(\dfrac{\sqrt{3}}{2}\right)$

5. a. $\cos^{-1}\left(\dfrac{1}{2}\right)$ **b.** $\cos^{-1}\left(\dfrac{-1}{\sqrt{2}}\right)$ **c.** $\cos^{-1}\left(\dfrac{\sqrt{3}}{2}\right)$

6. a. $\csc^{-1}\sqrt{2}$ **b.** $\csc^{-1}\left(\dfrac{-2}{\sqrt{3}}\right)$ **c.** $\csc^{-1} 2$

7. a. $\sec^{-1}(-\sqrt{2})$ **b.** $\sec^{-1}\left(\dfrac{2}{\sqrt{3}}\right)$ **c.** $\sec^{-1}(-2)$

8. a. $\cot^{-1}(-1)$ **b.** $\cot^{-1}(\sqrt{3})$ **c.** $\cot^{-1}\left(\dfrac{-1}{\sqrt{3}}\right)$

Evaluations

Find the values in Exercises 9–12.

9. $\sin\left(\cos^{-1}\left(\dfrac{\sqrt{2}}{2}\right)\right)$ **10.** $\sec\left(\cos^{-1}\dfrac{1}{2}\right)$

11. $\tan\left(\sin^{-1}\left(-\dfrac{1}{2}\right)\right)$ **12.** $\cot\left(\sin^{-1}\left(-\dfrac{\sqrt{3}}{2}\right)\right)$

Limits

Find the limits in Exercises 13–20. (If in doubt, look at the function's graph.)

13. $\lim\limits_{x \to 1^-} \sin^{-1} x$ **14.** $\lim\limits_{x \to -1^+} \cos^{-1} x$

15. $\lim\limits_{x \to \infty} \tan^{-1} x$ **16.** $\lim\limits_{x \to -\infty} \tan^{-1} x$

17. $\lim\limits_{x \to \infty} \sec^{-1} x$ **18.** $\lim\limits_{x \to -\infty} \sec^{-1} x$

19. $\lim\limits_{x \to \infty} \csc^{-1} x$ **20.** $\lim\limits_{x \to -\infty} \csc^{-1} x$

Finding Derivatives

In Exercises 21–42, find the derivative of y with respect to the appropriate variable.

21. $y = \cos^{-1}(x^2)$ **22.** $y = \cos^{-1}(1/x)$

23. $y = \sin^{-1}\sqrt{2}\,t$ **24.** $y = \sin^{-1}(1 - t)$

25. $y = \sec^{-1}(2s + 1)$ **26.** $y = \sec^{-1} 5s$

27. $y = \csc^{-1}(x^2 + 1), \quad x > 0$

28. $y = \csc^{-1}\dfrac{x}{2}$

29. $y = \sec^{-1}\dfrac{1}{t}, \quad 0 < t < 1$ **30.** $y = \sin^{-1}\dfrac{3}{t^2}$

31. $y = \cot^{-1}\sqrt{t}$ **32.** $y = \cot^{-1}\sqrt{t - 1}$

33. $y = \ln(\tan^{-1} x)$ **34.** $y = \tan^{-1}(\ln x)$

35. $y = \csc^{-1}(e^t)$ **36.** $y = \cos^{-1}(e^{-t})$

37. $y = s\sqrt{1 - s^2} + \cos^{-1} s$ **38.** $y = \sqrt{s^2 - 1} - \sec^{-1} s$

39. $y = \tan^{-1}\sqrt{x^2 - 1} + \csc^{-1} x, \quad x > 1$

40. $y = \cot^{-1}\dfrac{1}{x} - \tan^{-1} x$ **41.** $y = x\sin^{-1} x + \sqrt{1 - x^2}$

42. $y = \ln(x^2 + 4) - x\tan^{-1}\left(\dfrac{x}{2}\right)$

Theory and Examples

43. You are sitting in a classroom next to the wall looking at the blackboard at the front of the room. The blackboard is 12 ft long and starts 3 ft from the wall you are sitting next to. Show that your viewing angle is

$$\alpha = \cot^{-1}\dfrac{x}{15} - \cot^{-1}\dfrac{x}{3}$$

if you are x ft from the front wall.

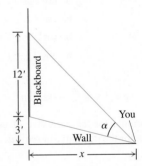

44. Find the angle α.

45. Here is an informal proof that $\tan^{-1} 1 + \tan^{-1} 2 + \tan^{-1} 3 = \pi$. Explain what is going on.

46. Two derivations of the identity $\sec^{-1}(-x) = \pi - \sec^{-1}x$

  a. (*Geometric*) Here is a pictorial proof that $\sec^{-1}(-x) = \pi - \sec^{-1}x$. See if you can tell what is going on.

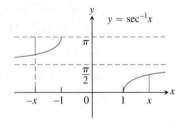

  b. (*Algebraic*) Derive the identity $\sec^{-1}(-x) = \pi - \sec^{-1}x$ by combining the following two equations from the text:

$$\cos^{-1}(-x) = \pi - \cos^{-1}x \qquad \text{Eq. (4), Section 1.6}$$

$$\sec^{-1}x = \cos^{-1}(1/x) \qquad \text{Eq. (1)}$$

Which of the expressions in Exercises 47–50 are defined, and which are not? Give reasons for your answers.

47. a. $\tan^{-1}2$ **b.** $\cos^{-1}2$

48. a. $\csc^{-1}(1/2)$ **b.** $\csc^{-1}2$

49. a. $\sec^{-1}0$ **b.** $\sin^{-1}\sqrt{2}$

50. a. $\cot^{-1}(-1/2)$ **b.** $\cos^{-1}(-5)$

51. Use the identity

$$\csc^{-1}u = \frac{\pi}{2} - \sec^{-1}u$$

to derive the formula for the derivative of $\csc^{-1}u$ in Table 3.1 from the formula for the derivative of $\sec^{-1}u$.

52. Derive the formula

$$\frac{dy}{dx} = \frac{1}{1 + x^2}$$

for the derivative of $y = \tan^{-1}x$ by differentiating both sides of the equivalent equation $\tan y = x$.

53. Use the Derivative Rule in Section 3.8, Theorem 3, to derive

$$\frac{d}{dx}\sec^{-1}x = \frac{1}{|x|\sqrt{x^2 - 1}}, \qquad |x| > 1.$$

54. Use the identity

$$\cot^{-1}u = \frac{\pi}{2} - \tan^{-1}u$$

to derive the formula for the derivative of $\cot^{-1}u$ in Table 3.1 from the formula for the derivative of $\tan^{-1}u$.

55. What is special about the functions

$$f(x) = \sin^{-1}\frac{x - 1}{x + 1}, \quad x \geq 0, \quad \text{and} \quad g(x) = 2\tan^{-1}\sqrt{x}?$$

Explain.

56. What is special about the functions

$$f(x) = \sin^{-1}\frac{1}{\sqrt{x^2 + 1}} \quad \text{and} \quad g(x) = \tan^{-1}\frac{1}{x}?$$

Explain.

T 57. Find the values of

  a. $\sec^{-1}1.5$ **b.** $\csc^{-1}(-1.5)$ **c.** $\cot^{-1}2$

T 58. Find the values of

  a. $\sec^{-1}(-3)$ **b.** $\csc^{-1}1.7$ **c.** $\cot^{-1}(-2)$

T In Exercises 59–61, find the domain and range of each composite function. Then graph the composites on separate screens. Do the graphs make sense in each case? Give reasons for your answers. Comment on any differences you see.

59. a. $y = \tan^{-1}(\tan x)$ **b.** $y = \tan(\tan^{-1}x)$

60. a. $y = \sin^{-1}(\sin x)$ **b.** $y = \sin(\sin^{-1}x)$

61. a. $y = \cos^{-1}(\cos x)$ **b.** $y = \cos(\cos^{-1}x)$

T Use your graphing utility for Exercises 62–66.

62. Graph $y = \sec(\sec^{-1}x) = \sec(\cos^{-1}(1/x))$. Explain what you see.

63. Newton's serpentine Graph Newton's serpentine, $y = 4x/(x^2 + 1)$. Then graph $y = 2\sin(2\tan^{-1}x)$ in the same graphing window. What do you see? Explain.

64. Graph the rational function $y = (2 - x^2)/x^2$. Then graph $y = \cos(2\sec^{-1}x)$ in the same graphing window. What do you see? Explain.

65. Graph $f(x) = \sin^{-1}x$ together with its first two derivatives. Comment on the behavior of f and the shape of its graph in relation to the signs and values of f' and f''.

66. Graph $f(x) = \tan^{-1}x$ together with its first two derivatives. Comment on the behavior of f and the shape of its graph in relation to the signs and values of f' and f''.

3.10   Related Rates

In this section we look at problems that ask for the rate at which some variable changes when it is known how the rate of some other related variable (or perhaps several variables) changes. The problem of finding a rate of change from other known rates of change is called a *related rates problem*.

Related Rates Equations

Suppose we are pumping air into a spherical balloon. Both the volume and radius of the balloon are increasing over time. If V is the volume and r is the radius of the balloon at an instant of time, then

$$V = \frac{4}{3}\pi r^3.$$

Using the Chain Rule, we differentiate both sides with respect to t to find an equation relating the rates of change of V and r,

$$\frac{dV}{dt} = \frac{dV}{dr}\frac{dr}{dt} = 4\pi r^2 \frac{dr}{dt}.$$

So if we know the radius r of the balloon and the rate dV/dt at which the volume is increasing at a given instant of time, then we can solve this last equation for dr/dt to find how fast the radius is increasing at that instant. Note that it is easier to directly measure the rate of increase of the volume (the rate at which air is being pumped into the balloon) than it is to measure the increase in the radius. The related rates equation allows us to calculate dr/dt from dV/dt.

Very often the key to relating the variables in a related rates problem is drawing a picture that shows the geometric relations between them, as illustrated in the following example.

EXAMPLE 1   Water runs into a conical tank at the rate of 9 ft³/min. The tank stands point down and has a height of 10 ft and a base radius of 5 ft. How fast is the water level rising when the water is 6 ft deep?

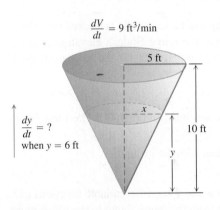

$\frac{dV}{dt} = 9$ ft³/min

5 ft

x

$\frac{dy}{dt} = ?$
when $y = 6$ ft

10 ft

y

FIGURE 3.43   The geometry of the conical tank and the rate at which water fills the tank determine how fast the water level rises (Example 1).

Solution   Figure 3.43 shows a partially filled conical tank. The variables in the problem are

V = volume (ft³) of the water in the tank at time t (min)

x = radius (ft) of the surface of the water at time t

y = depth (ft) of the water in the tank at time t.

We assume that V, x, and y are differentiable functions of t. The constants are the dimensions of the tank. We are asked for dy/dt when

$$y = 6 \text{ ft} \qquad \text{and} \qquad \frac{dV}{dt} = 9 \text{ ft}^3/\text{min}.$$

The water forms a cone with volume

$$V = \frac{1}{3}\pi x^2 y.$$

This equation involves x as well as V and y. Because no information is given about x and dx/dt at the time in question, we need to eliminate x. The similar triangles in Figure 3.43 give us a way to express x in terms of y:

$$\frac{x}{y} = \frac{5}{10} \qquad \text{or} \qquad x = \frac{y}{2}.$$

Therefore, find

$$V = \frac{1}{3}\pi\left(\frac{y}{2}\right)^2 y = \frac{\pi}{12}y^3$$

to give the derivative

$$\frac{dV}{dt} = \frac{\pi}{12}\cdot 3y^2\frac{dy}{dt} = \frac{\pi}{4}y^2\frac{dy}{dt}.$$

Finally, use $y = 6$ and $dV/dt = 9$ to solve for dy/dt.

$$9 = \frac{\pi}{4}(6)^2 \frac{dy}{dt}$$

$$\frac{dy}{dt} = \frac{1}{\pi} \approx 0.32$$

At the moment in question, the water level is rising at about 0.32 ft/min. ∎

Related Rates Problem Strategy

1. *Draw a picture and name the variables and constants.* Use t for time. Assume that all variables are differentiable functions of t.
2. *Write down the numerical information* (in terms of the symbols you have chosen).
3. *Write down what you are asked to find* (usually a rate, expressed as a derivative).
4. *Write an equation that relates the variables.* You may have to combine two or more equations to get a single equation that relates the variable whose rate you want to the variables whose rates you know.
5. *Differentiate with respect to t.* Then express the rate you want in terms of the rates and variables whose values you know.
6. *Evaluate.* Use known values to find the unknown rate.

$\frac{d\theta}{dt} = 0.14$ rad/min
when $\theta = \pi/4$

$\frac{dy}{dt} = ?$
when $\theta = \pi/4$

Range finder

θ

500 ft

FIGURE 3.44 The rate of change of the balloon's height is related to the rate of change of the angle the range finder makes with the ground (Example 2).

EXAMPLE 2   A hot air balloon rising straight up from a level field is tracked by a range finder 500 ft from the liftoff point. At the moment the range finder's elevation angle is $\pi/4$, the angle is increasing at the rate of 0.14 rad/min. How fast is the balloon rising at that moment?

Solution   We answer the question in six steps.

1. *Draw a picture and name the variables and constants* (Figure 3.44). The variables in the picture are

   θ = the angle in radians the range finder makes with the ground.

   y = the height in feet of the balloon.

   We let t represent time in minutes and assume that θ and y are differentiable functions of t.
       The one constant in the picture is the distance from the range finder to the liftoff point (500 ft). There is no need to give it a special symbol.

2. *Write down the additional numerical information.*

   $$\frac{d\theta}{dt} = 0.14 \text{ rad/min} \qquad \text{when} \qquad \theta = \frac{\pi}{4}$$

3. *Write down what we are to find.* We want dy/dt when $\theta = \pi/4$.

4. *Write an equation that relates the variables y and θ.*

   $$\frac{y}{500} = \tan \theta \qquad \text{or} \qquad y = 500 \tan \theta$$

5. *Differentiate with respect to t using the Chain Rule.* The result tells how dy/dt (which we want) is related to $d\theta/dt$ (which we know).

   $$\frac{dy}{dt} = 500 \, (\sec^2 \theta) \frac{d\theta}{dt}$$

6. *Evaluate with $\theta = \pi/4$ and $d\theta/dt = 0.14$ to find dy/dt.*

   $$\frac{dy}{dt} = 500 \left(\sqrt{2} \right)^2 (0.14) = 140 \qquad \sec \frac{\pi}{4} = \sqrt{2}$$

At the moment in question, the balloon is rising at the rate of 140 ft/min. ∎

FIGURE 3.45 The speed of the car is related to the speed of the police cruiser and the rate of change of the distance between them (Example 3).

EXAMPLE 3 A police cruiser, approaching a right-angled intersection from the north, is chasing a speeding car that has turned the corner and is now moving straight east. When the cruiser is 0.6 mi north of the intersection and the car is 0.8 mi to the east, the police determine with radar that the distance between them and the car is increasing at 20 mph. If the cruiser is moving at 60 mph at the instant of measurement, what is the speed of the car?

Solution We picture the car and cruiser in the coordinate plane, using the positive x-axis as the eastbound highway and the positive y-axis as the southbound highway (Figure 3.45). We let t represent time and set

$$x = \text{position of car at time } t$$
$$y = \text{position of cruiser at time } t$$
$$s = \text{distance between car and cruiser at time } t.$$

We assume that x, y, and s are differentiable functions of t.

We want to find dx/dt when

$$x = 0.8 \text{ mi}, \quad y = 0.6 \text{ mi}, \quad \frac{dy}{dt} = -60 \text{ mph}, \quad \frac{ds}{dt} = 20 \text{ mph}.$$

Note that dy/dt is negative because y is decreasing.

We differentiate the distance equation

$$s^2 = x^2 + y^2$$

(we could also use $s = \sqrt{x^2 + y^2}$), and obtain

$$2s\frac{ds}{dt} = 2x\frac{dx}{dt} + 2y\frac{dy}{dt}$$

$$\frac{ds}{dt} = \frac{1}{s}\left(x\frac{dx}{dt} + y\frac{dy}{dt}\right)$$

$$= \frac{1}{\sqrt{x^2 + y^2}}\left(x\frac{dx}{dt} + y\frac{dy}{dt}\right).$$

Finally, we use $x = 0.8$, $y = 0.6$, $dy/dt = -60$, $ds/dt = 20$, and solve for dx/dt.

$$20 = \frac{1}{\sqrt{(0.8)^2 + (0.6)^2}}\left(0.8\frac{dx}{dt} + (0.6)(-60)\right)$$

$$\frac{dx}{dt} = \frac{20\sqrt{(0.8)^2 + (0.6)^2} + (0.6)(60)}{0.8} = 70$$

At the moment in question, the car's speed is 70 mph. ∎

EXAMPLE 4 A particle P moves clockwise at a constant rate along a circle of radius 10 ft centered at the origin. The particle's initial position is $(0, 10)$ on the y-axis and its final destination is the point $(10, 0)$ on the x-axis. Once the particle is in motion, the tangent line at P intersects the x-axis at a point Q (which moves over time). If it takes the particle 30 sec to travel from start to finish, how fast is the point Q moving along the x-axis when it is 20 ft from the center of the circle?

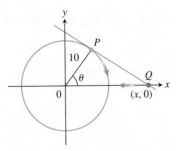

FIGURE 3.46 The particle P travels clockwise along the circle (Example 4).

Solution We picture the situation in the coordinate plane with the circle centered at the origin (see Figure 3.46). We let t represent time and let θ denote the angle from the x-axis to the radial line joining the origin to P. Since the particle travels from start to finish in 30 sec, it is traveling along the circle at a constant rate of $\pi/2$ radians in 1/2 min, or π rad/min. In other words, $d\theta/dt = -\pi$, with t being measured in minutes. The negative sign appears because θ is decreasing over time.

Setting $x(t)$ to be the distance at time t from the point Q to the origin, we want to find dx/dt when

$$x = 20 \text{ ft} \qquad \text{and} \qquad \frac{d\theta}{dt} = -\pi \text{ rad/min}.$$

To relate the variables x and θ, we see from Figure 3.46 that $x \cos \theta = 10$, or $x = 10 \sec \theta$. Differentiation of this last equation gives

$$\frac{dx}{dt} = 10 \sec \theta \tan \theta \frac{d\theta}{dt} = -10\pi \sec \theta \tan \theta.$$

Note that dx/dt is negative because x is decreasing (Q is moving towards the origin).

When $x = 20$, $\cos \theta = 1/2$ and $\sec \theta = 2$. Also, $\tan \theta = \sqrt{\sec^2 \theta - 1} = \sqrt{3}$. It follows that

$$\frac{dx}{dt} = (-10\pi)(2)\left(\sqrt{3}\right) = -20\sqrt{3}\pi.$$

At the moment in question, the point Q is moving towards the origin at the speed of $20\sqrt{3}\pi \approx 108.8 \text{ ft/min}.$ ∎

EXAMPLE 5   A jet airliner is flying at a constant altitude of 12,000 ft above sea level as it approaches a Pacific island. The aircraft comes within the direct line of sight of a radar station located on the island, and the radar indicates the initial angle between sea level and its line of sight to the aircraft is 30°. How fast (in miles per hour) is the aircraft approaching the island when first detected by the radar instrument if it is turning upward (counterclockwise) at the rate of 2/3 deg/sec in order to keep the aircraft within its direct line of sight?

Solution   The aircraft A and radar station R are pictured in the coordinate plane, using the positive x-axis as the horizontal distance at sea level from R to A, and the positive y-axis as the vertical altitude above sea level. We let t represent time and observe that $y = 12,000$ is a constant. The general situation and line-of-sight angle θ are depicted in Figure 3.47. We want to find dx/dt when $\theta = \pi/6$ rad and $d\theta/dt = 2/3$ deg/sec.

From Figure 3.47, we see that

$$\frac{12,000}{x} = \tan \theta \qquad \text{or} \qquad x = 12,000 \cot \theta.$$

Using miles instead of feet for our distance units, the last equation translates to

$$x = \frac{12,000}{5280} \cot \theta.$$

Differentiation with respect to t gives

$$\frac{dx}{dt} = -\frac{1200}{528} \csc^2 \theta \frac{d\theta}{dt}.$$

When $\theta = \pi/6$, $\sin^2 \theta = 1/4$, so $\csc^2 \theta = 4$. Converting $d\theta/dt = 2/3$ deg/sec to radians per hour, we find

$$\frac{d\theta}{dt} = \frac{2}{3}\left(\frac{\pi}{180}\right)(3600) \text{ rad/hr.} \qquad \text{\small 1 hr = 3600 sec, 1 deg = } \pi/180 \text{ \small rad}$$

Substitution into the equation for dx/dt then gives

$$\frac{dx}{dt} = \left(-\frac{1200}{528}\right)(4)\left(\frac{2}{3}\right)\left(\frac{\pi}{180}\right)(3600) \approx -380.$$

The negative sign appears because the distance x is decreasing, so the aircraft is approaching the island at a speed of approximately 380 mi/hr when first detected by the radar. ∎

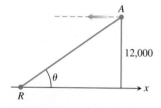

FIGURE 3.47   Jet airliner A traveling at constant altitude toward radar station R (Example 5).

(a)

(b)

FIGURE 3.48 A worker at M walks to the right pulling the weight W upwards as the rope moves through the pulley P (Example 6).

EXAMPLE 6   Figure 3.48(a) shows a rope running through a pulley at P and bearing a weight W at one end. The other end is held 5 ft above the ground in the hand M of a worker. Suppose the pulley is 25 ft above ground, the rope is 45 ft long, and the worker is walking rapidly away from the vertical line PW at the rate of 6 ft/sec. How fast is the weight being raised when the worker's hand is 21 ft away from PW?

Solution   We let OM be the horizontal line of length x ft from a point O directly below the pulley to the worker's hand M at any instant of time (Figure 3.48). Let h be the height of the weight W above O, and let z denote the length of rope from the pulley P to the worker's hand. We want to know dh/dt when $x = 21$ given that $dx/dt = 6$. Note that the height of P above O is 20 ft because O is 5 ft above the ground. We assume the angle at O is a right angle.

At any instant of time t we have the following relationships (see Figure 3.48b):

$$20 - h + z = 45 \qquad \text{Total length of rope is 45 ft.}$$
$$20^2 + x^2 = z^2. \qquad \text{Angle at } O \text{ is a right angle.}$$

If we solve for $z = 25 + h$ in the first equation, and substitute into the second equation, we have

$$20^2 + x^2 = (25 + h)^2. \qquad (1)$$

Differentiating both sides with respect to t gives

$$2x \frac{dx}{dt} = 2(25 + h) \frac{dh}{dt},$$

and solving this last equation for dh/dt we find

$$\frac{dh}{dt} = \frac{x}{25 + h} \frac{dx}{dt}. \qquad (2)$$

Since we know dx/dt, it remains only to find $25 + h$ at the instant when $x = 21$. From Equation (1),

$$20^2 + 21^2 = (25 + h)^2$$

so that

$$(25 + h)^2 = 841, \qquad \text{or} \qquad 25 + h = 29.$$

Equation (2) now gives

$$\frac{dh}{dt} = \frac{21}{29} \cdot 6 = \frac{126}{29} \approx 4.3 \text{ ft/sec}$$

as the rate at which the weight is being raised when $x = 21$ ft.   ∎

Exercises 3.10

1. **Area**   Suppose that the radius r and area $A = \pi r^2$ of a circle are differentiable functions of t. Write an equation that relates dA/dt to dr/dt.

2. **Surface area**   Suppose that the radius r and surface area $S = 4\pi r^2$ of a sphere are differentiable functions of t. Write an equation that relates dS/dt to dr/dt.

3. Assume that $y = 5x$ and $dx/dt = 2$. Find dy/dt.

4. Assume that $2x + 3y = 12$ and $dy/dt = -2$. Find dx/dt.

5. If $y = x^2$ and $dx/dt = 3$, then what is dy/dt when $x = -1$?

6. If $x = y^3 - y$ and $dy/dt = 5$, then what is dx/dt when $y = 2$?

7. If $x^2 + y^2 = 25$ and $dx/dt = -2$, then what is dy/dt when $x = 3$ and $y = -4$?

8. If $x^2 y^3 = 4/27$ and $dy/dt = 1/2$, then what is dx/dt when $x = 2$?

9. If $L = \sqrt{x^2 + y^2}$, $dx/dt = -1$, and $dy/dt = 3$, find dL/dt when $x = 5$ and $y = 12$.

10. If $r + s^2 + v^3 = 12$, $dr/dt = 4$, and $ds/dt = -3$, find dv/dt when $r = 3$ and $s = 1$.

11. If the original 24 m edge length x of a cube decreases at the rate of 5 m/min, when $x = 3$ m at what rate does the cube's

   a. surface area change?

   b. volume change?

12. A cube's surface area increases at the rate of 72 in²/sec. At what rate is the cube's volume changing when the edge length is $x = 3$ in?

13. **Volume**   The radius r and height h of a right circular cylinder are related to the cylinder's volume V by the formula $V = \pi r^2 h$.

   a. How is dV/dt related to dh/dt if r is constant?

   b. How is dV/dt related to dr/dt if h is constant?

   c. How is dV/dt related to dr/dt and dh/dt if neither r nor h is constant?

14. **Volume**   The radius r and height h of a right circular cone are related to the cone's volume V by the equation $V = (1/3)\pi r^2 h$.

   a. How is dV/dt related to dh/dt if r is constant?

   b. How is dV/dt related to dr/dt if h is constant?

   c. How is dV/dt related to dr/dt and dh/dt if neither r nor h is constant?

15. **Changing voltage**   The voltage V (volts), current I (amperes), and resistance R (ohms) of an electric circuit like the one shown here are related by the equation $V = IR$. Suppose that V is increasing at the rate of 1 volt/sec while I is decreasing at the rate of 1/3 amp/sec. Let t denote time in seconds.

   a. What is the value of dV/dt?

   b. What is the value of dI/dt?

   c. What equation relates dR/dt to dV/dt and dI/dt?

   d. Find the rate at which R is changing when $V = 12$ volts and $I = 2$ amp. Is R increasing, or decreasing?

16. **Electrical power**   The power P (watts) of an electric circuit is related to the circuit's resistance R (ohms) and current I (amperes) by the equation $P = RI^2$.

   a. How are dP/dt, dR/dt, and dI/dt related if none of P, R, and I are constant?

   b. How is dR/dt related to dI/dt if P is constant?

17. **Distance**   Let x and y be differentiable functions of t and let $s = \sqrt{x^2 + y^2}$ be the distance between the points $(x, 0)$ and $(0, y)$ in the xy-plane.

   a. How is ds/dt related to dx/dt if y is constant?

   b. How is ds/dt related to dx/dt and dy/dt if neither x nor y is constant?

   c. How is dx/dt related to dy/dt if s is constant?

18. **Diagonals**   If x, y, and z are lengths of the edges of a rectangular box, the common length of the box's diagonals is $s = \sqrt{x^2 + y^2 + z^2}$.

   a. Assuming that x, y, and z are differentiable functions of t, how is ds/dt related to dx/dt, dy/dt, and dz/dt?

   b. How is ds/dt related to dy/dt and dz/dt if x is constant?

   c. How are dx/dt, dy/dt, and dz/dt related if s is constant?

19. **Area**   The area A of a triangle with sides of lengths a and b enclosing an angle of measure θ is

$$A = \frac{1}{2} ab \sin \theta.$$

   a. How is dA/dt related to $d\theta/dt$ if a and b are constant?

   b. How is dA/dt related to $d\theta/dt$ and da/dt if only b is constant?

   c. How is dA/dt related to $d\theta/dt$, da/dt, and db/dt if none of a, b, and θ are constant?

20. **Heating a plate**   When a circular plate of metal is heated in an oven, its radius increases at the rate of 0.01 cm/min. At what rate is the plate's area increasing when the radius is 50 cm?

21. **Changing dimensions in a rectangle**   The length l of a rectangle is decreasing at the rate of 2 cm/sec while the width w is increasing at the rate of 2 cm/sec. When $l = 12$ cm and $w = 5$ cm, find the rates of change of (a) the area, (b) the perimeter, and (c) the lengths of the diagonals of the rectangle. Which of these quantities are decreasing, and which are increasing?

22. **Changing dimensions in a rectangular box**   Suppose that the edge lengths x, y, and z of a closed rectangular box are changing at the following rates:

$$\frac{dx}{dt} = 1 \text{ m/sec}, \quad \frac{dy}{dt} = -2 \text{ m/sec}, \quad \frac{dz}{dt} = 1 \text{ m/sec}.$$

Find the rates at which the box's (a) volume, (b) surface area, and (c) diagonal length $s = \sqrt{x^2 + y^2 + z^2}$ are changing at the instant when $x = 4$, $y = 3$, and $z = 2$.

23. **A sliding ladder**   A 13-ft ladder is leaning against a house when its base starts to slide away (see accompanying figure). By the time the base is 12 ft from the house, the base is moving at the rate of 5 ft/sec.

   a. How fast is the top of the ladder sliding down the wall then?

   b. At what rate is the area of the triangle formed by the ladder, wall, and ground changing then?

   c. At what rate is the angle θ between the ladder and the ground changing then?

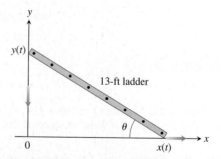

24. **Commercial air traffic**   Two commercial airplanes are flying at an altitude of 40,000 ft along straight-line courses that intersect at right angles. Plane A is approaching the intersection point at a speed of 442 knots (nautical miles per hour; a nautical mile is 2000 yd). Plane B is approaching the intersection at 481 knots. At what rate is the distance between the planes changing when A is 5

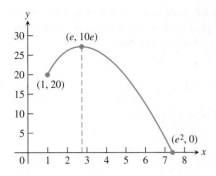

FIGURE 4.8 The extreme values of $f(x) = 10x(2 - \ln x)$ on $[1, e^2]$ occur at $x = e$ and $x = e^2$ (Example 3).

Solution   Figure 4.8 suggests that f has its absolute maximum value near $x = 3$ and its absolute minimum value of 0 at $x = e^2$. Let's verify this observation.

We evaluate the function at the critical points and endpoints and take the largest and smallest of the resulting values.

The first derivative is

$$f'(x) = 10(2 - \ln x) - 10x\left(\frac{1}{x}\right) = 10(1 - \ln x).$$

The only critical point in the domain $[1, e^2]$ is the point $x = e$, where $\ln x = 1$. The values of f at this one critical point and at the endpoints are

Critical point value: $f(e) = 10e$

Endpoint values: $f(1) = 10(2 - \ln 1) = 20$

$f(e^2) = 10e^2(2 - 2 \ln e) = 0.$

We can see from this list that the function's absolute maximum value is $10e \approx 27.2$; it occurs at the critical interior point $x = e$. The absolute minimum value is 0 and occurs at the right endpoint $x = e^2$. ∎

EXAMPLE 4   Find the absolute maximum and minimum values of $f(x) = x^{2/3}$ on the interval $[-2, 3]$.

Solution   We evaluate the function at the critical points and endpoints and take the largest and smallest of the resulting values.

The first derivative

$$f'(x) = \frac{2}{3}x^{-1/3} = \frac{2}{3\sqrt[3]{x}}$$

has no zeros but is undefined at the interior point $x = 0$. The values of f at this one critical point and at the endpoints are

Critical point value: $f(0) = 0$

Endpoint values: $f(-2) = (-2)^{2/3} = \sqrt[3]{4}$

$f(3) = (3)^{2/3} = \sqrt[3]{9}.$

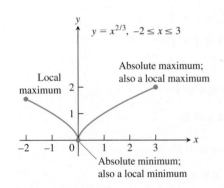

FIGURE 4.9   The extreme values of $f(x) = x^{2/3}$ on $[-2, 3]$ occur at $x = 0$ and $x = 3$ (Example 4).

We can see from this list that the function's absolute maximum value is $\sqrt[3]{9} \approx 2.08$, and it occurs at the right endpoint $x = 3$. The absolute minimum value is 0, and it occurs at the interior point $x = 0$ where the graph has a cusp (Figure 4.9). ∎

Exercises 4.1

Finding Extrema from Graphs
In Exercises 1–6, determine from the graph whether the function has any absolute extreme values on $[a, b]$. Then explain how your answer is consistent with Theorem 1.

1.

2.

3.
4.

5.

6.

In Exercises 7–10, find the absolute extreme values and where they occur.

7.

8.

9.

10.

In Exercises 11–14, match the table with a graph.

11.

x	$f'(x)$
a	0
b	0
c	5

12.

x	$f'(x)$
a	0
b	0
c	-5

13.

x	$f'(x)$
a	does not exist
b	0
c	-2

14.

x	$f'(x)$
a	does not exist
b	does not exist
c	-1.7

(a)

(b)

(c)

(d)

In Exercises 15–20, sketch the graph of each function and determine whether the function has any absolute extreme values on its domain. Explain how your answer is consistent with Theorem 1.

15. $f(x) = |x|, \quad -1 < x < 2$

16. $y = \dfrac{6}{x^2 + 2}, \quad -1 < x < 1$

17. $g(x) = \begin{cases} -x, & 0 \le x < 1 \\ x - 1, & 1 \le x \le 2 \end{cases}$

18. $h(x) = \begin{cases} \dfrac{1}{x}, & -1 \le x < 0 \\ \sqrt{x}, & 0 \le x \le 4 \end{cases}$

19. $y = 3 \sin x, \quad 0 < x < 2\pi$

20. $f(x) = \begin{cases} x + 1, & -1 \le x < 0 \\ \cos x, & 0 \le x \le \dfrac{\pi}{2} \end{cases}$

Absolute Extrema on Finite Closed Intervals

In Exercises 21–40, find the absolute maximum and minimum values of each function on the given interval. Then graph the function. Identify the points on the graph where the absolute extrema occur, and include their coordinates.

21. $f(x) = \dfrac{2}{3}x - 5, \quad -2 \le x \le 3$

22. $f(x) = -x - 4, \quad -4 \le x \le 1$

23. $f(x) = x^2 - 1, \quad -1 \le x \le 2$

24. $f(x) = 4 - x^2, \quad -3 \le x \le 1$

25. $F(x) = -\dfrac{1}{x^2}, \quad 0.5 \le x \le 2$

26. $F(x) = -\dfrac{1}{x}, \quad -2 \le x \le -1$

27. $h(x) = \sqrt[3]{x}, \quad -1 \le x \le 8$

28. $h(x) = -3x^{2/3}, \quad -1 \le x \le 1$

29. $g(x) = \sqrt{4 - x^2}, \quad -2 \le x \le 1$

30. $g(x) = -\sqrt{5 - x^2}, \quad -\sqrt{5} \le x \le 0$

31. $f(\theta) = \sin \theta, \quad -\dfrac{\pi}{2} \le \theta \le \dfrac{5\pi}{6}$

32. $f(\theta) = \tan \theta, \quad -\dfrac{\pi}{3} \le \theta \le \dfrac{\pi}{4}$

33. $g(x) = \csc x, \quad \dfrac{\pi}{3} \le x \le \dfrac{2\pi}{3}$

34. $g(x) = \sec x, \quad -\dfrac{\pi}{3} \le x \le \dfrac{\pi}{6}$

35. $f(t) = 2 - |t|, \quad -1 \le t \le 3$

36. $f(t) = |t - 5|, \quad 4 \le t \le 7$

37. $g(x) = xe^{-x}, \quad -1 \le x \le 1$

38. $h(x) = \ln (x + 1), \quad 0 \le x \le 3$

39. $f(x) = \dfrac{1}{x} + \ln x, \quad 0.5 \le x \le 4$

40. $g(x) = e^{-x^2}, \quad -2 \le x \le 1$

In Exercises 41–44, find the function's absolute maximum and minimum values and say where they are assumed.

41. $f(x) = x^{4/3}, \quad -1 \le x \le 8$

42. $f(x) = x^{5/3}, \quad -1 \le x \le 8$

43. $g(\theta) = \theta^{3/5}, \quad -32 \le \theta \le 1$

44. $h(\theta) = 3\theta^{2/3}, \quad -27 \le \theta \le 8$

Finding Critical Points

In Exercises 45–52, determine all critical points for each function.

45. $y = x^2 - 6x + 7$

46. $f(x) = 6x^2 - x^3$

47. $f(x) = x(4 - x)^3$

48. $g(x) = (x - 1)^2(x - 3)^2$

49. $y = x^2 + \dfrac{2}{x}$

50. $f(x) = \dfrac{x^2}{x - 2}$

51. $y = x^2 - 32\sqrt{x}$

52. $g(x) = \sqrt{2x - x^2}$

Finding Extreme Values

In Exercises 53–68, find the extreme values (absolute and local) of the function and where they occur.

53. $y = 2x^2 - 8x + 9$

54. $y = x^3 - 2x + 4$

55. $y = x^3 + x^2 - 8x + 5$

56. $y = x^3(x - 5)^2$

57. $y = \sqrt{x^2 - 1}$

58. $y = x - 4\sqrt{x}$

59. $y = \dfrac{1}{\sqrt[3]{1 - x^2}}$

60. $y = \sqrt{3 + 2x - x^2}$

61. $y = \dfrac{x}{x^2 + 1}$

62. $y = \dfrac{x + 1}{x^2 + 2x + 2}$

63. $y = e^x + e^{-x}$

64. $y = e^x - e^{-x}$

65. $y = x \ln x$

66. $y = x^2 \ln x$

67. $y = \cos^{-1}(x^2)$

68. $y = \sin^{-1}(e^x)$

Local Extrema and Critical Points

In Exercises 69–76, find the critical points, domain endpoints, and extreme values (absolute and local) for each function.

69. $y = x^{2/3}(x + 2)$

70. $y = x^{2/3}(x^2 - 4)$

71. $y = x\sqrt{4 - x^2}$

72. $y = x^2\sqrt{3 - x}$

73. $y = \begin{cases} 4 - 2x, & x \le 1 \\ x + 1, & x > 1 \end{cases}$

74. $y = \begin{cases} 3 - x, & x < 0 \\ 3 + 2x - x^2, & x \ge 0 \end{cases}$

75. $y = \begin{cases} -x^2 - 2x + 4, & x \le 1 \\ -x^2 + 6x - 4, & x > 1 \end{cases}$

76. $y = \begin{cases} -\dfrac{1}{4}x^2 - \dfrac{1}{2}x + \dfrac{15}{4}, & x \le 1 \\ x^3 - 6x^2 + 8x, & x > 1 \end{cases}$

In Exercises 77 and 78, give reasons for your answers.

77. Let $f(x) = (x - 2)^{2/3}$.

a. Does $f'(2)$ exist?

b. Show that the only local extreme value of f occurs at $x = 2$.

c. Does the result in part (b) contradict the Extreme Value Theorem?

d. Repeat parts (a) and (b) for $f(x) = (x - a)^{2/3}$, replacing 2 by a.

78. Let $f(x) = |x^3 - 9x|$.

a. Does $f'(0)$ exist?   **b.** Does $f'(3)$ exist?

c. Does $f'(-3)$ exist?   **d.** Determine all extrema of f.

Theory and Examples

79. A minimum with no derivative The function $f(x) = |x|$ has an absolute minimum value at $x = 0$ even though f is not differentiable at $x = 0$. Is this consistent with Theorem 2? Give reasons for your answer.

80. Even functions If an even function $f(x)$ has a local maximum value at $x = c$, can anything be said about the value of f at $x = -c$? Give reasons for your answer.

81. Odd functions If an odd function $g(x)$ has a local minimum value at $x = c$, can anything be said about the value of g at $x = -c$? Give reasons for your answer.

82. We know how to find the extreme values of a continuous function $f(x)$ by investigating its values at critical points and endpoints. But what if there *are* no critical points or endpoints? What happens then? Do such functions really exist? Give reasons for your answers.

83. The function

$$V(x) = x(10 - 2x)(16 - 2x), \qquad 0 < x < 5,$$

models the volume of a box.

a. Find the extreme values of V.

b. Interpret any values found in part (a) in terms of the volume of the box.

84. Cubic functions Consider the cubic function

$$f(x) = ax^3 + bx^2 + cx + d.$$

a. Show that f can have 0, 1, or 2 critical points. Give examples and graphs to support your argument.

b. How many local extreme values can f have?

85. Maximum height of a vertically moving body The height of a body moving vertically is given by

$$s = -\dfrac{1}{2}gt^2 + v_0 t + s_0, \qquad g > 0,$$

with s in meters and t in seconds. Find the body's maximum height.

86. Peak alternating current Suppose that at any given time t (in seconds) the current i (in amperes) in an alternating current circuit is $i = 2 \cos t + 2 \sin t$. What is the peak current for this circuit (largest magnitude)?

T Graph the functions in Exercises 87–90. Then find the extreme values of the function on the interval and say where they occur.

87. $f(x) = |x - 2| + |x + 3|, \quad -5 \le x \le 5$

88. $g(x) = |x - 1| - |x - 5|, \quad -2 \le x \le 7$

89. $h(x) = |x + 2| - |x - 3|, \quad -\infty < x < \infty$

90. $k(x) = |x + 1| + |x - 3|, \quad -\infty < x < \infty$

COMPUTER EXPLORATIONS

In Exercises 91–98, you will use a CAS to help find the absolute extrema of the given function over the specified closed interval. Perform the following steps.

a. Plot the function over the interval to see its general behavior there.

b. Find the interior points where $f' = 0$. (In some exercises, you may have to use the numerical equation solver to approximate a solution.) You may want to plot f' as well.

c. Find the interior points where f' does not exist.

d. Evaluate the function at all points found in parts (b) and (c) and at the endpoints of the interval.

e. Find the function's absolute extreme values on the interval and identify where they occur.

91. $f(x) = x^4 - 8x^2 + 4x + 2$, $[-20/25, 64/25]$

92. $f(x) = -x^4 + 4x^3 - 4x + 1$, $[-3/4, 3]$

93. $f(x) = x^{2/3}(3 - x)$, $[-2, 2]$

94. $f(x) = 2 + 2x - 3x^{2/3}$, $[-1, 10/3]$

95. $f(x) = \sqrt{x} + \cos x$, $[0, 2\pi]$

96. $f(x) = x^{3/4} - \sin x + \dfrac{1}{2}$, $[0, 2\pi]$

97. $f(x) = \pi x^2 e^{-3x/2}$, $[0, 5]$

98. $f(x) = \ln(2x + x \sin x)$, $[1, 15]$

4.2 The Mean Value Theorem

We know that constant functions have zero derivatives, but could there be a more complicated function whose derivative is always zero? If two functions have identical derivatives over an interval, how are the functions related? We answer these and other questions in this chapter by applying the Mean Value Theorem. First we introduce a special case, known as Rolle's Theorem, which is used to prove the Mean Value Theorem.

Rolle's Theorem

As suggested by its graph, if a differentiable function crosses a horizontal line at two different points, there is at least one point between them where the tangent to the graph is horizontal and the derivative is zero (Figure 4.10). We now state and prove this result.

> **THEOREM 3—Rolle's Theorem**   Suppose that $y = f(x)$ is continuous at every point of the closed interval $[a, b]$ and differentiable at every point of its interior (a, b). If $f(a) = f(b)$, then there is at least one number c in (a, b) at which $f'(c) = 0$.

FIGURE 4.10   Rolle's Theorem says that a differentiable curve has at least one horizontal tangent between any two points where it crosses a horizontal line. It may have just one (a), or it may have more (b).

Proof   Being continuous, f assumes absolute maximum and minimum values on $[a, b]$ by Theorem 1. These can occur only

1.   at interior points where f' is zero,

2.   at interior points where f' does not exist,

3.   at the endpoints of the function's domain, in this case a and b.

By hypothesis, f has a derivative at every interior point. That rules out possibility (2), leaving us with interior points where $f' = 0$ and with the two endpoints a and b.

If either the maximum or the minimum occurs at a point c between a and b, then $f'(c) = 0$ by Theorem 2 in Section 4.1, and we have found a point for Rolle's Theorem.

If both the absolute maximum and the absolute minimum occur at the endpoints, then because $f(a) = f(b)$ it must be the case that f is a constant function with $f(x) = f(a) = f(b)$ for every $x \in [a, b]$. Therefore $f'(x) = 0$ and the point c can be taken anywhere in the interior (a, b).   ∎

The hypotheses of Theorem 3 are essential. If they fail at even one point, the graph may not have a horizontal tangent (Figure 4.11).

Rolle's Theorem may be combined with the Intermediate Value Theorem to show when there is only one real solution of an equation $f(x) = 0$, as we illustrate in the next example.

HISTORICAL BIOGRAPHY

Michel Rolle
(1652–1719)

EXAMPLE 1   Show that the equation

$$x^3 + 3x + 1 = 0$$

has exactly one real solution.

(a) Discontinuous at an endpoint of [a, b]

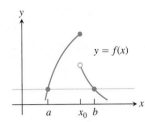

(b) Discontinuous at an interior point of [a, b]

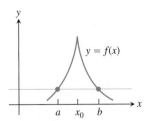

(c) Continuous on [a, b] but not differentiable at an interior point

FIGURE 4.11   There may be no horizontal tangent if the hypotheses of Rolle's Theorem do not hold.

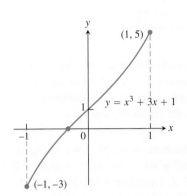

FIGURE 4.12   The only real zero of the polynomial $y = x^3 + 3x + 1$ is the one shown here where the curve crosses the x-axis between -1 and 0 (Example 1).

Solution   We define the continuous function

$$f(x) = x^3 + 3x + 1.$$

Since $f(-1) = -3$ and $f(0) = 1$, the Intermediate Value Theorem tells us that the graph of f crosses the x-axis somewhere in the open interval $(-1, 0)$. (See Figure 4.12.) The derivative

$$f'(x) = 3x^2 + 3$$

is never zero (because it is always positive). Now, if there were even two points $x = a$ and $x = b$ where $f(x)$ was zero, Rolle's Theorem would guarantee the existence of a point $x = c$ in between them where f' was zero. Therefore, f has no more than one zero.   ■

Our main use of Rolle's Theorem is in proving the Mean Value Theorem.

The Mean Value Theorem

The Mean Value Theorem, which was first stated by Joseph-Louis Lagrange, is a slanted version of Rolle's Theorem (Figure 4.13). The Mean Value Theorem guarantees that there is a point where the tangent line is parallel to the chord AB.

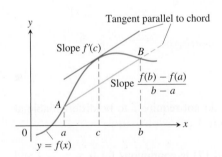

FIGURE 4.13   Geometrically, the Mean Value Theorem says that somewhere between a and b the curve has at least one tangent parallel to chord AB.

THEOREM 4—The Mean Value Theorem   Suppose $y = f(x)$ is continuous on a closed interval $[a, b]$ and differentiable on the interval's interior (a, b). Then there is at least one point c in (a, b) at which

$$\frac{f(b) - f(a)}{b - a} = f'(c). \qquad (1)$$

Proof   We picture the graph of f and draw a line through the points $A(a, f(a))$ and $B(b, f(b))$. (See Figure 4.14.) The line is the graph of the function

$$g(x) = f(a) + \frac{f(b) - f(a)}{b - a}(x - a) \qquad (2)$$

(point-slope equation). The vertical difference between the graphs of f and g at x is

$$h(x) = f(x) - g(x)$$

$$= f(x) - f(a) - \frac{f(b) - f(a)}{b - a}(x - a). \qquad (3)$$

Figure 4.15 shows the graphs of f, g, and h together.

HISTORICAL BIOGRAPHY

Joseph-Louis Lagrange
(1736–1813)

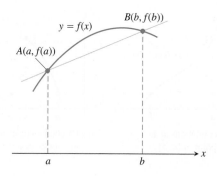

FIGURE 4.14 The graph of f and the chord AB over the interval $[a, b]$.

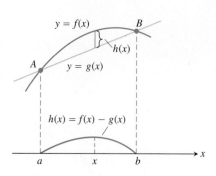

FIGURE 4.15 The chord AB is the graph of the function $g(x)$. The function $h(x) = f(x) - g(x)$ gives the vertical distance between the graphs of f and g at x.

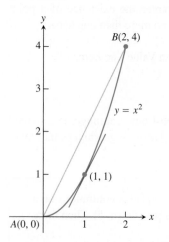

FIGURE 4.16 The function $f(x) = \sqrt{1 - x^2}$ satisfies the hypotheses (and conclusion) of the Mean Value Theorem on $[-1, 1]$ even though f is not differentiable at -1 and 1.

The function h satisfies the hypotheses of Rolle's Theorem on $[a, b]$. It is continuous on $[a, b]$ and differentiable on (a, b) because both f and g are. Also, $h(a) = h(b) = 0$ because the graphs of f and g both pass through A and B. Therefore $h'(c) = 0$ at some point $c \in (a, b)$. This is the point we want for Equation (1).

To verify Equation (1), we differentiate both sides of Equation (3) with respect to x and then set $x = c$:

$$h'(x) = f'(x) - \frac{f(b) - f(a)}{b - a} \qquad \text{Derivative of Eq. (3)} \ldots$$

$$h'(c) = f'(c) - \frac{f(b) - f(a)}{b - a} \qquad \ldots \text{with } x = c$$

$$0 = f'(c) - \frac{f(b) - f(a)}{b - a} \qquad h'(c) = 0$$

$$f'(c) = \frac{f(b) - f(a)}{b - a}, \qquad \text{Rearranged}$$

which is what we set out to prove. ∎

The hypotheses of the Mean Value Theorem do not require f to be differentiable at either a or b. Continuity at a and b is enough (Figure 4.16).

EXAMPLE 2 The function $f(x) = x^2$ (Figure 4.17) is continuous for $0 \le x \le 2$ and differentiable for $0 < x < 2$. Since $f(0) = 0$ and $f(2) = 4$, the Mean Value Theorem says that at some point c in the interval, the derivative $f'(x) = 2x$ must have the value $(4 - 0)/(2 - 0) = 2$. In this case we can identify c by solving the equation $2c = 2$ to get $c = 1$. However, it is not always easy to find c algebraically, even though we know it always exists. ∎

FIGURE 4.17 As we find in Example 2, $c = 1$ is where the tangent is parallel to the chord.

A Physical Interpretation

We can think of the number $(f(b) - f(a))/(b - a)$ as the average change in f over $[a, b]$ and $f'(c)$ as an instantaneous change. Then the Mean Value Theorem says that at some interior point the instantaneous change must equal the average change over the entire interval.

EXAMPLE 3 If a car accelerating from zero takes 8 sec to go 352 ft, its average velocity for the 8-sec interval is $352/8 = 44$ ft/sec. The Mean Value Theorem says that at some point during the acceleration the speedometer must read exactly 30 mph (44 ft/sec) (Figure 4.18). ∎

FIGURE 4.18 Distance versus elapsed time for the car in Example 3.

Mathematical Consequences

At the beginning of the section, we asked what kind of function has a zero derivative over an interval. The first corollary of the Mean Value Theorem provides the answer that only constant functions have zero derivatives.

COROLLARY 1   If $f'(x) = 0$ at each point x of an open interval (a, b), then $f(x) = C$ for all $x \in (a, b)$, where C is a constant.

Proof   We want to show that f has a constant value on the interval (a, b). We do so by showing that if x_1 and x_2 are any two points in (a, b) with $x_1 < x_2$, then $f(x_1) = f(x_2)$. Now f satisfies the hypotheses of the Mean Value Theorem on $[x_1, x_2]$: It is differentiable at every point of $[x_1, x_2]$ and hence continuous at every point as well. Therefore,

$$\frac{f(x_2) - f(x_1)}{x_2 - x_1} = f'(c)$$

at some point c between x_1 and x_2. Since $f' = 0$ throughout (a, b), this equation implies successively that

$$\frac{f(x_2) - f(x_1)}{x_2 - x_1} = 0, \qquad f(x_2) - f(x_1) = 0, \qquad \text{and} \qquad f(x_1) = f(x_2). \qquad ■$$

At the beginning of this section, we also asked about the relationship between two functions that have identical derivatives over an interval. The next corollary tells us that their values on the interval have a constant difference.

COROLLARY 2   If $f'(x) = g'(x)$ at each point x in an open interval (a, b), then there exists a constant C such that $f(x) = g(x) + C$ for all $x \in (a, b)$. That is, $f - g$ is a constant function on (a, b).

Proof   At each point $x \in (a, b)$ the derivative of the difference function $h = f - g$ is

$$h'(x) = f'(x) - g'(x) = 0.$$

Thus, $h(x) = C$ on (a, b) by Corollary 1. That is, $f(x) - g(x) = C$ on (a, b), so $f(x) = g(x) + C$. ■

Corollaries 1 and 2 are also true if the open interval (a, b) fails to be finite. That is, they remain true if the interval is (a, ∞), $(-\infty, b)$, or $(-\infty, \infty)$.

Corollary 2 plays an important role when we discuss antiderivatives in Section 4.8. It tells us, for instance, that since the derivative of $f(x) = x^2$ on $(-\infty, \infty)$ is $2x$, any other function with derivative $2x$ on $(-\infty, \infty)$ must have the formula $x^2 + C$ for some value of C (Figure 4.19).

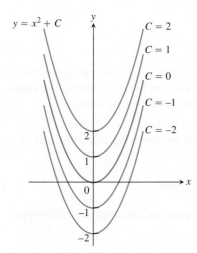

$y = x^2 + C$

$C = 2$
$C = 1$
$C = 0$
$C = -1$
$C = -2$

FIGURE 4.19   From a geometric point of view, Corollary 2 of the Mean Value Theorem says that the graphs of functions with identical derivatives on an interval can differ only by a vertical shift there. The graphs of the functions with derivative $2x$ are the parabolas $y = x^2 + C$, shown here for selected values of C.

EXAMPLE 4   Find the function $f(x)$ whose derivative is $\sin x$ and whose graph passes through the point $(0, 2)$.

Solution   Since the derivative of $g(x) = -\cos x$ is $g'(x) = \sin x$, we see that f and g have the same derivative. Corollary 2 then says that $f(x) = -\cos x + C$ for some constant C. Since the graph of f passes through the point $(0, 2)$, the value of C is determined from the condition that $f(0) = 2$:

$$f(0) = -\cos(0) + C = 2, \quad \text{so} \quad C = 3.$$

The function is $f(x) = -\cos x + 3$. ∎

Finding Velocity and Position from Acceleration

We can use Corollary 2 to find the velocity and position functions of an object moving along a vertical line. Assume the object or body is falling freely from rest with acceleration 9.8 m/sec². We assume the position $s(t)$ of the body is measured positive downward from the rest position (so the vertical coordinate line points *downward*, in the direction of the motion, with the rest position at 0).

We know that the velocity $v(t)$ is some function whose derivative is 9.8. We also know that the derivative of $g(t) = 9.8t$ is 9.8. By Corollary 2,

$$v(t) = 9.8t + C$$

for some constant C. Since the body falls from rest, $v(0) = 0$. Thus

$$9.8(0) + C = 0, \quad \text{and} \quad C = 0.$$

The velocity function must be $v(t) = 9.8t$. What about the position function $s(t)$?

We know that $s(t)$ is some function whose derivative is $9.8t$. We also know that the derivative of $f(t) = 4.9t^2$ is $9.8t$. By Corollary 2,

$$s(t) = 4.9t^2 + C$$

for some constant C. Since $s(0) = 0$,

$$4.9(0)^2 + C = 0, \quad \text{and} \quad C = 0.$$

The position function is $s(t) = 4.9t^2$ until the body hits the ground.

The ability to find functions from their rates of change is one of the very powerful tools of calculus. As we will see, it lies at the heart of the mathematical developments in Chapter 5.

Proofs of the Laws of Logarithms

The algebraic properties of logarithms were stated in Section 1.6. We can prove those properties by applying Corollary 2 of the Mean Value Theorem to each of them. The steps in the proofs are similar to those used in solving problems involving logarithms.

Proof that $\ln bx = \ln b + \ln x$ The argument starts by observing that $\ln bx$ and $\ln x$ have the same derivative:

$$\frac{d}{dx} \ln (bx) = \frac{b}{bx} = \frac{1}{x} = \frac{d}{dx} \ln x.$$

According to Corollary 2 of the Mean Value Theorem, then, the functions must differ by a constant, which means that

$$\ln bx = \ln x + C$$

for some C.

Since this last equation holds for all positive values of x, it must hold for $x = 1$. Hence,

$$\ln (b \cdot 1) = \ln 1 + C$$
$$\ln b = 0 + C \qquad \ln 1 = 0$$
$$C = \ln b.$$

By substituting we conclude,

$$\ln bx = \ln b + \ln x.$$ ∎

Proof that $\ln x^r = r \ln x$ We use the same-derivative argument again. For all positive values of x,

$$\frac{d}{dx} \ln x^r = \frac{1}{x^r} \frac{d}{dx} (x^r) \qquad \text{Chain Rule}$$

$$= \frac{1}{x^r} rx^{r-1} \qquad \text{Derivative Power Rule}$$

$$= r \cdot \frac{1}{x} = \frac{d}{dx} (r \ln x).$$

Since $\ln x^r$ and $r \ln x$ have the same derivative,

$$\ln x^r = r \ln x + C$$

for some constant C. Taking x to be 1 identifies C as zero, and we're done. ∎

You are asked to prove the Quotient Rule for logarithms,

$$\ln \left(\frac{b}{x} \right) = \ln b - \ln x,$$

in Exercise 75. The Reciprocal Rule, $\ln (1/x) = -\ln x$, is a special case of the Quotient Rule, obtained by taking $b = 1$ and noting that $\ln 1 = 0$.

Laws of Exponents

The laws of exponents for the natural exponential e^x are consequences of the algebraic properties of $\ln x$. They follow from the inverse relationship between these functions.

Laws of Exponents for e^x

For all numbers x, x_1, and x_2, the natural exponential e^x obeys the following laws:

1. $e^{x_1} \cdot e^{x_2} = e^{x_1 + x_2}$ **2.** $e^{-x} = \dfrac{1}{e^x}$

3. $\dfrac{e^{x_1}}{e^{x_2}} = e^{x_1 - x_2}$ **4.** $(e^{x_1})^{x_2} = e^{x_1 x_2} = (e^{x_2})^{x_1}$

Proof of Law 1 Let

$$y_1 = e^{x_1} \qquad \text{and} \qquad y_2 = e^{x_2}. \tag{4}$$

Then

$$x_1 = \ln y_1 \quad \text{and} \quad x_2 = \ln y_2 \qquad \text{Take logs of both sides of Eqs. (4).}$$

$$x_1 + x_2 = \ln y_1 + \ln y_2$$

$$= \ln y_1 y_2 \qquad \text{Product Rule for logarithms}$$

$$e^{x_1 + x_2} = e^{\ln y_1 y_2} \qquad \text{Exponentiate.}$$

$$= y_1 y_2 \qquad e^{\ln u} = u$$

$$= e^{x_1} e^{x_2}.$$ ∎

The proof of Law 4 is similar. Laws 2 and 3 follow from Law 1 (Exercises 77 and 78).

Exercises 4.2

Checking the Mean Value Theorem

Find the value or values of c that satisfy the equation

$$\frac{f(b) - f(a)}{b - a} = f'(c)$$

in the conclusion of the Mean Value Theorem for the functions and intervals in Exercises 1–8.

1. $f(x) = x^2 + 2x - 1, \quad [0, 1]$

2. $f(x) = x^{2/3}, \quad [0, 1]$

3. $f(x) = x + \dfrac{1}{x}, \quad \left[\dfrac{1}{2}, 2\right]$

4. $f(x) = \sqrt{x - 1}, \quad [1, 3]$

5. $f(x) = \sin^{-1} x, \quad [-1, 1]$

6. $f(x) = \ln(x - 1), \quad [2, 4]$

7. $f(x) = x^3 - x^2, \quad [-1, 2]$

8. $g(x) = \begin{cases} x^3, & -2 \le x \le 0 \\ x^2, & 0 < x \le 2 \end{cases}$

Which of the functions in Exercises 9–14 satisfy the hypotheses of the Mean Value Theorem on the given interval, and which do not? Give reasons for your answers.

9. $f(x) = x^{2/3}, \quad [-1, 8]$

10. $f(x) = x^{4/5}, \quad [0, 1]$

11. $f(x) = \sqrt{x(1 - x)}, \quad [0, 1]$

12. $f(x) = \begin{cases} \dfrac{\sin x}{x}, & -\pi \le x < 0 \\ 0, & x = 0 \end{cases}$

13. $f(x) = \begin{cases} x^2 - x, & -2 \le x \le -1 \\ 2x^2 - 3x - 3, & -1 < x \le 0 \end{cases}$

14. $f(x) = \begin{cases} 2x - 3, & 0 \le x \le 2 \\ 6x - x^2 - 7, & 2 < x \le 3 \end{cases}$

15. The function

$$f(x) = \begin{cases} x, & 0 \le x < 1 \\ 0, & x = 1 \end{cases}$$

is zero at $x = 0$ and $x = 1$ and differentiable on $(0, 1)$, but its derivative on $(0, 1)$ is never zero. How can this be? Doesn't Rolle's Theorem say the derivative has to be zero somewhere in $(0, 1)$? Give reasons for your answer.

16. For what values of a, m, and b does the function

$$f(x) = \begin{cases} 3, & x = 0 \\ -x^2 + 3x + a, & 0 < x < 1 \\ mx + b, & 1 \le x \le 2 \end{cases}$$

satisfy the hypotheses of the Mean Value Theorem on the interval $[0, 2]$?

Roots (Zeros)

17. **a.** Plot the zeros of each polynomial on a line together with the zeros of its first derivative.

  i) $y = x^2 - 4$

  ii) $y = x^2 + 8x + 15$

  iii) $y = x^3 - 3x^2 + 4 = (x + 1)(x - 2)^2$

  iv) $y = x^3 - 33x^2 + 216x = x(x - 9)(x - 24)$

  b. Use Rolle's Theorem to prove that between every two zeros of $x^n + a_{n-1}x^{n-1} + \cdots + a_1x + a_0$ there lies a zero of

  $$nx^{n-1} + (n - 1)a_{n-1}x^{n-2} + \cdots + a_1.$$

18. Suppose that f'' is continuous on $[a, b]$ and that f has three zeros in the interval. Show that f'' has at least one zero in (a, b). Generalize this result.

19. Show that if $f'' > 0$ throughout an interval $[a, b]$, then f' has at most one zero in $[a, b]$. What if $f'' < 0$ throughout $[a, b]$ instead?

20. Show that a cubic polynomial can have at most three real zeros.

Show that the functions in Exercises 21–28 have exactly one zero in the given interval.

21. $f(x) = x^4 + 3x + 1, \quad [-2, -1]$

22. $f(x) = x^3 + \dfrac{4}{x^2} + 7, \quad (-\infty, 0)$

23. $g(t) = \sqrt{t} + \sqrt{1 + t} - 4, \quad (0, \infty)$

24. $g(t) = \dfrac{1}{1 - t} + \sqrt{1 + t} - 3.1, \quad (-1, 1)$

25. $r(\theta) = \theta + \sin^2\left(\dfrac{\theta}{3}\right) - 8, \quad (-\infty, \infty)$

26. $r(\theta) = 2\theta - \cos^2 \theta + \sqrt{2}, \quad (-\infty, \infty)$

27. $r(\theta) = \sec \theta - \dfrac{1}{\theta^3} + 5, \quad (0, \pi/2)$

28. $r(\theta) = \tan \theta - \cot \theta - \theta, \quad (0, \pi/2)$

Finding Functions from Derivatives

29. Suppose that $f(-1) = 3$ and that $f'(x) = 0$ for all x. Must $f(x) = 3$ for all x? Give reasons for your answer.

30. Suppose that $f(0) = 5$ and that $f'(x) = 2$ for all x. Must $f(x) = 2x + 5$ for all x? Give reasons for your answer.

31. Suppose that $f'(x) = 2x$ for all x. Find $f(2)$ if

  a. $f(0) = 0$ **b.** $f(1) = 0$ **c.** $f(-2) = 3$.

32. What can be said about functions whose derivatives are constant? Give reasons for your answer.

In Exercises 33–38, find all possible functions with the given derivative.

33. **a.** $y' = x$ **b.** $y' = x^2$ **c.** $y' = x^3$

34. **a.** $y' = 2x$ **b.** $y' = 2x - 1$ **c.** $y' = 3x^2 + 2x - 1$

35. **a.** $y' = -\dfrac{1}{x^2}$ **b.** $y' = 1 - \dfrac{1}{x^2}$ **c.** $y' = 5 + \dfrac{1}{x^2}$

36. a. $y' = \dfrac{1}{2\sqrt{x}}$ **b.** $y' = \dfrac{1}{\sqrt{x}}$ **c.** $y' = 4x - \dfrac{1}{\sqrt{x}}$

37. a. $y' = \sin 2t$ **b.** $y' = \cos \dfrac{t}{2}$ **c.** $y' = \sin 2t + \cos \dfrac{t}{2}$

38. a. $y' = \sec^2 \theta$ **b.** $y' = \sqrt{\theta}$ **c.** $y' = \sqrt{\theta} - \sec^2 \theta$

In Exercises 39–42, find the function with the given derivative whose graph passes through the point P.

39. $f'(x) = 2x - 1$, $P(0, 0)$

40. $g'(x) = \dfrac{1}{x^2} + 2x$, $P(-1, 1)$

41. $f'(x) = e^{2x}$, $P\left(0, \dfrac{3}{2}\right)$

42. $r'(t) = \sec t \tan t - 1$, $P(0, 0)$

Finding Position from Velocity or Acceleration

Exercises 43–46 give the velocity $v = ds/dt$ and initial position of a body moving along a coordinate line. Find the body's position at time t.

43. $v = 9.8t + 5$, $s(0) = 10$

44. $v = 32t - 2$, $s(0.5) = 4$

45. $v = \sin \pi t$, $s(0) = 0$

46. $v = \dfrac{2}{\pi} \cos \dfrac{2t}{\pi}$, $s(\pi^2) = 1$

Exercises 47–50 give the acceleration $a = d^2s/dt^2$, initial velocity, and initial position of a body moving on a coordinate line. Find the body's position at time t.

47. $a = e^t$, $v(0) = 20$, $s(0) = 5$

48. $a = 9.8$, $v(0) = -3$, $s(0) = 0$

49. $a = -4 \sin 2t$, $v(0) = 2$, $s(0) = -3$

50. $a = \dfrac{9}{\pi^2} \cos \dfrac{3t}{\pi}$, $v(0) = 0$, $s(0) = -1$

Applications

51. Temperature change It took 14 sec for a mercury thermometer to rise from $-19°C$ to $100°C$ when it was taken from a freezer and placed in boiling water. Show that somewhere along the way the mercury was rising at the rate of $8.5°C/\sec$.

52. A trucker handed in a ticket at a toll booth showing that in 2 hours she had covered 159 mi on a toll road with speed limit 65 mph. The trucker was cited for speeding. Why?

53. Classical accounts tell us that a 170-oar trireme (ancient Greek or Roman warship) once covered 184 sea miles in 24 hours. Explain why at some point during this feat the trireme's speed exceeded 7.5 knots (sea miles per hour).

54. A marathoner ran the 26.2-mi New York City Marathon in 2.2 hours. Show that at least twice the marathoner was running at exactly 11 mph, assuming the initial and final speeds are zero.

55. Show that at some instant during a 2-hour automobile trip the car's speedometer reading will equal the average speed for the trip.

56. Free fall on the moon On our moon, the acceleration of gravity is 1.6 m/sec^2. If a rock is dropped into a crevasse, how fast will it be going just before it hits bottom 30 sec later?

Theory and Examples

57. The geometric mean of a and b The *geometric mean* of two positive numbers a and b is the number $\sqrt{ab}$. Show that the value of c in the conclusion of the Mean Value Theorem for $f(x) = 1/x$ on an interval of positive numbers $[a, b]$ is $c = \sqrt{ab}$.

58. The arithmetic mean of a and b The *arithmetic mean* of two numbers a and b is the number $(a + b)/2$. Show that the value of c in the conclusion of the Mean Value Theorem for $f(x) = x^2$ on any interval $[a, b]$ is $c = (a + b)/2$.

T **59.** Graph the function

$$f(x) = \sin x \sin (x + 2) - \sin^2 (x + 1).$$

What does the graph do? Why does the function behave this way? Give reasons for your answers.

60. Rolle's Theorem

   a. Construct a polynomial $f(x)$ that has zeros at $x = -2, -1, 0, 1,$ and 2.

   b. Graph f and its derivative f' together. How is what you see related to Rolle's Theorem?

   c. Do $g(x) = \sin x$ and its derivative g' illustrate the same phenomenon as f and f'?

61. Unique solution Assume that f is continuous on $[a, b]$ and differentiable on (a, b). Also assume that $f(a)$ and $f(b)$ have opposite signs and that $f' \neq 0$ between a and b. Show that $f(x) = 0$ exactly once between a and b.

62. Parallel tangents Assume that f and g are differentiable on $[a, b]$ and that $f(a) = g(a)$ and $f(b) = g(b)$. Show that there is at least one point between a and b where the tangents to the graphs of f and g are parallel or the same line. Illustrate with a sketch.

63. Suppose that $f'(x) \leq 1$ for $1 \leq x \leq 4$. Show that $f(4) - f(1) \leq 3$.

64. Suppose that $0 < f'(x) < 1/2$ for all x-values. Show that $f(-1) < f(1) < 2 + f(-1)$.

65. Show that $|\cos x - 1| \leq |x|$ for all x-values. (*Hint:* Consider $f(t) = \cos t$ on $[0, x]$.)

66. Show that for any numbers a and b, the sine inequality $|\sin b - \sin a| \leq |b - a|$ is true.

67. If the graphs of two differentiable functions $f(x)$ and $g(x)$ start at the same point in the plane and the functions have the same rate of change at every point, do the graphs have to be identical? Give reasons for your answer.

68. If $|f(w) - f(x)| \leq |w - x|$ for all values w and x and f is a differentiable function, show that $-1 \leq f'(x) \leq 1$ for all x-values.

69. Assume that f is differentiable on $a \leq x \leq b$ and that $f(b) < f(a)$. Show that f' is negative at some point between a and b.

70. Let f be a function defined on an interval $[a, b]$. What conditions could you place on f to guarantee that

$$\min f' \leq \frac{f(b) - f(a)}{b - a} \leq \max f',$$

where $\min f'$ and $\max f'$ refer to the minimum and maximum values of f' on $[a, b]$? Give reasons for your answers.

T **71.** Use the inequalities in Exercise 70 to estimate $f(0.1)$ if $f'(x) = 1/(1 + x^4 \cos x)$ for $0 \le x \le 0.1$ and $f(0) = 1$.

T **72.** Use the inequalities in Exercise 70 to estimate $f(0.1)$ if $f'(x) = 1/(1 - x^4)$ for $0 \le x \le 0.1$ and $f(0) = 2$.

73. Let f be differentiable at every value of x and suppose that $f(1) = 1$, that $f' < 0$ on $(-\infty, 1)$, and that $f' > 0$ on $(1, \infty)$.

a. Show that $f(x) \ge 1$ for all x.

b. Must $f'(1) = 0$? Explain.

74. Let $f(x) = px^2 + qx + r$ be a quadratic function defined on a closed interval $[a, b]$. Show that there is exactly one point c in (a, b) at which f satisfies the conclusion of the Mean Value Theorem.

75. Use the same-derivative argument, as was done to prove the Product and Power Rules for logarithms, to prove the Quotient Rule property.

76. Use the same-derivative argument to prove the identities

a. $\tan^{-1} x + \cot^{-1} x = \dfrac{\pi}{2}$ **b.** $\sec^{-1} x + \csc^{-1} x = \dfrac{\pi}{2}$

77. Starting with the equation $e^{x_1} e^{x_2} = e^{x_1 + x_2}$, derived in the text, show that $e^{-x} = 1/e^x$ for any real number x. Then show that $e^{x_1}/e^{x_2} = e^{x_1 - x_2}$ for any numbers x_1 and x_2.

78. Show that $(e^{x_1})^{x_2} = e^{x_1 x_2} = (e^{x_2})^{x_1}$ for any numbers x_1 and x_2.

4.3 Monotonic Functions and the First Derivative Test

In sketching the graph of a differentiable function it is useful to know where it increases (rises from left to right) and where it decreases (falls from left to right) over an interval. This section gives a test to determine where it increases and where it decreases. We also show how to test the critical points of a function to identify whether local extreme values are present.

Increasing Functions and Decreasing Functions

As another corollary to the Mean Value Theorem, we show that functions with positive derivatives are increasing functions and functions with negative derivatives are decreasing functions. A function that is increasing or decreasing on an interval is said to be **monotonic** on the interval.

COROLLARY 3   Suppose that f is continuous on $[a, b]$ and differentiable on (a, b).

If $f'(x) > 0$ at each point $x \in (a, b)$, then f is increasing on $[a, b]$.

If $f'(x) < 0$ at each point $x \in (a, b)$, then f is decreasing on $[a, b]$.

Proof   Let x_1 and x_2 be any two points in $[a, b]$ with $x_1 < x_2$. The Mean Value Theorem applied to f on $[x_1, x_2]$ says that

$$f(x_2) - f(x_1) = f'(c)(x_2 - x_1)$$

for some c between x_1 and x_2. The sign of the right-hand side of this equation is the same as the sign of $f'(c)$ because $x_2 - x_1$ is positive. Therefore, $f(x_2) > f(x_1)$ if f' is positive on (a, b) and $f(x_2) < f(x_1)$ if f' is negative on (a, b).   ∎

Corollary 3 is valid for infinite as well as finite intervals. To find the intervals where a function f is increasing or decreasing, we first find all of the critical points of f. If $a < b$ are two critical points for f, and if the derivative f' is continuous but never zero on the interval (a, b), then by the Intermediate Value Theorem applied to f', the derivative must be everywhere positive on (a, b), or everywhere negative there. One way we can determine the sign of f' on (a, b) is simply by evaluating the derivative at a single point c in (a, b). If $f'(c) > 0$, then $f'(x) > 0$ for all x in (a, b) so f is increasing on $[a, b]$ by Corollary 3; if $f'(c) < 0$, then f is decreasing on $[a, b]$. The next example illustrates how we use this procedure.

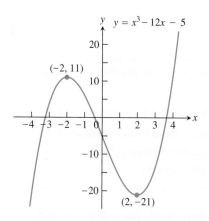

FIGURE 4.20 The function $f(x) = x^3 - 12x - 5$ is monotonic on three separate intervals (Example 1).

EXAMPLE 1 Find the critical points of $f(x) = x^3 - 12x - 5$ and identify the intervals on which f is increasing and on which f is decreasing.

Solution The function f is everywhere continuous and differentiable. The first derivative

$$f'(x) = 3x^2 - 12 = 3(x^2 - 4)$$
$$= 3(x + 2)(x - 2)$$

is zero at $x = -2$ and $x = 2$. These critical points subdivide the domain of f to create nonoverlapping open intervals $(-\infty, -2)$, $(-2, 2)$, and $(2, \infty)$ on which f' is either positive or negative. We determine the sign of f' by evaluating f' at a convenient point in each subinterval. The behavior of f is determined by then applying Corollary 3 to each subinterval. The results are summarized in the following table, and the graph of f is given in Figure 4.20.

Interval	$-\infty < x < -2$	$-2 < x < 2$	$2 < x < \infty$
f' evaluated	$f'(-3) = 15$	$f'(0) = -12$	$f'(3) = 15$
Sign of f'	$+$	$-$	$+$
Behavior of f	increasing	decreasing	increasing

We used "strict" less-than inequalities to specify the intervals in the summary table for Example 1. Corollary 3 says that we could use $\leq$ inequalities as well. That is, the function f in the example is increasing on $-\infty < x \leq -2$, decreasing on $-2 \leq x \leq 2$, and increasing on $2 \leq x < \infty$. We do not talk about whether a function is increasing or decreasing at a single point.

HISTORICAL BIOGRAPHY

Edmund Halley
(1656–1742)

First Derivative Test for Local Extrema

In Figure 4.21, at the points where f has a minimum value, $f' < 0$ immediately to the left and $f' > 0$ immediately to the right. (If the point is an endpoint, there is only one side to consider.) Thus, the function is decreasing on the left of the minimum value and it is increasing on its right. Similarly, at the points where f has a maximum value, $f' > 0$ immediately to the left and $f' < 0$ immediately to the right. Thus, the function is increasing on the left of the maximum value and decreasing on its right. In summary, at a local extreme point, the sign of $f'(x)$ changes.

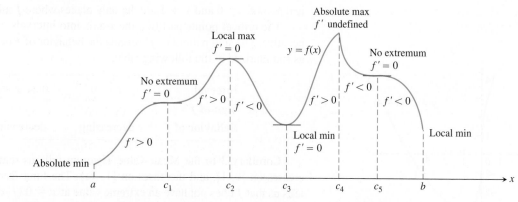

FIGURE 4.21 The critical points of a function locate where it is increasing and where it is decreasing. The first derivative changes sign at a critical point where a local extremum occurs.

These observations lead to a test for the presence and nature of local extreme values of differentiable functions.

First Derivative Test for Local Extrema

Suppose that c is a critical point of a continuous function f, and that f is differentiable at every point in some interval containing c except possibly at c itself. Moving across this interval from left to right,

1. if f' changes from negative to positive at c, then f has a local minimum at c;
2. if f' changes from positive to negative at c, then f has a local maximum at c;
3. if f' does not change sign at c (that is, f' is positive on both sides of c or negative on both sides), then f has no local extremum at c.

The test for local extrema at endpoints is similar, but there is only one side to consider.

Proof of the First Derivative Test   Part (1). Since the sign of f' changes from negative to positive at c, there are numbers a and b such that $a < c < b$, $f' < 0$ on (a, c), and $f' > 0$ on (c, b). If $x \in (a, c)$, then $f(c) < f(x)$ because $f' < 0$ implies that f is decreasing on $[a, c]$. If $x \in (c, b)$, then $f(c) < f(x)$ because $f' > 0$ implies that f is increasing on $[c, b]$. Therefore, $f(x) \geq f(c)$ for every $x \in (a, b)$. By definition, f has a local minimum at c.

Parts (2) and (3) are proved similarly.   ∎

EXAMPLE 2   Find the critical points of

$$f(x) = x^{1/3}(x - 4) = x^{4/3} - 4x^{1/3}.$$

Identify the intervals on which f is increasing and decreasing. Find the function's local and absolute extreme values.

Solution   The function f is continuous at all x since it is the product of two continuous functions, $x^{1/3}$ and $(x - 4)$. The first derivative

$$f'(x) = \frac{d}{dx}\left(x^{4/3} - 4x^{1/3}\right) = \frac{4}{3}x^{1/3} - \frac{4}{3}x^{-2/3}$$

$$= \frac{4}{3}x^{-2/3}\left(x - 1\right) = \frac{4(x - 1)}{3x^{2/3}}$$

is zero at $x = 1$ and undefined at $x = 0$. There are no endpoints in the domain, so the critical points $x = 0$ and $x = 1$ are the only places where f might have an extreme value.

The critical points partition the x-axis into intervals on which f' is either positive or negative. The sign pattern of f' reveals the behavior of f between and at the critical points, as summarized in the following table.

Interval	$x < 0$	$0 < x < 1$	$x > 1$
Sign of f'	−	−	+
Behavior of f	decreasing	decreasing	increasing

Corollary 3 to the Mean Value Theorem tells us that f decreases on $(-\infty, 0]$, decreases on $[0, 1]$, and increases on $[1, \infty)$. The First Derivative Test for Local Extrema tells us that f does not have an extreme value at $x = 0$ (f' does not change sign) and that f has a local minimum at $x = 1$ (f' changes from negative to positive).

The value of the local minimum is $f(1) = 1^{1/3}(1 - 4) = -3$. This is also an absolute minimum since f is decreasing on $(-\infty, 1]$ and increasing on $[1, \infty)$. Figure 4.22 shows this value in relation to the function's graph.

Note that $\lim_{x \to 0} f'(x) = -\infty$, so the graph of f has a vertical tangent at the origin.   ∎

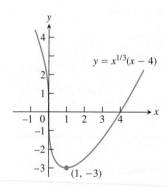

FIGURE 4.22   The function $f(x) = x^{1/3}(x - 4)$ decreases when $x < 1$ and increases when $x > 1$ (Example 2).

EXAMPLE 3 Find the critical points of

$$f(x) = (x^2 - 3)e^x.$$

Identify the intervals on which f is increasing and decreasing. Find the function's local and absolute extreme values.

Solution The function f is continuous and differentiable for all real numbers, so the critical points occur only at the zeros of f'.

Using the Derivative Product Rule, we find the derivative

$$f'(x) = (x^2 - 3) \cdot \frac{d}{dx} e^x + \frac{d}{dx}(x^2 - 3) \cdot e^x$$

$$= (x^2 - 3) \cdot e^x + (2x) \cdot e^x$$

$$= (x^2 + 2x - 3)e^x.$$

Since e^x is never zero, the first derivative is zero if and only if

$$x^2 + 2x - 3 = 0$$

$$(x + 3)(x - 1) = 0.$$

The zeros $x = -3$ and $x = 1$ partition the x-axis into intervals as follows.

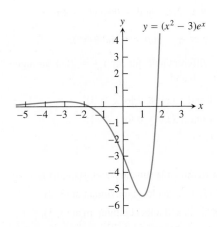

FIGURE 4.23 The graph of $f(x) = (x^2 - 3)e^x$ (Example 3).

Interval	$x < -3$	$-3 < x < 1$	$1 < x$
Sign of f'	$+$	$-$	$+$
Behavior of f	increasing	decreasing	increasing

We can see from the table that there is a local maximum (about 0.299) at $x = -3$ and a local minimum (about -5.437) at $x = 1$. The local minimum value is also an absolute minimum because $f(x) > 0$ for $|x| > \sqrt{3}$. There is no absolute maximum. The function increases on $(-\infty, -3)$ and $(1, \infty)$ and decreases on $(-3, 1)$. Figure 4.23 shows the graph. ∎

Exercises 4.3

Analyzing Functions from Derivatives

Answer the following questions about the functions whose derivatives are given in Exercises 1–14:

  a. What are the critical points of f?

  b. On what intervals is f increasing or decreasing?

  c. At what points, if any, does f assume local maximum and minimum values?

1. $f'(x) = x(x - 1)$ **2.** $f'(x) = (x - 1)(x + 2)$

3. $f'(x) = (x - 1)^2(x + 2)$ **4.** $f'(x) = (x - 1)^2(x + 2)^2$

5. $f'(x) = (x - 1)e^{-x}$

6. $f'(x) = (x - 7)(x + 1)(x + 5)$

7. $f'(x) = \dfrac{x^2(x - 1)}{x + 2}, \quad x \neq -2$

8. $f'(x) = \dfrac{(x - 2)(x + 4)}{(x + 1)(x - 3)}, \quad x \neq -1, 3$

9. $f'(x) = 1 - \dfrac{4}{x^2}, \quad x \neq 0$ **10.** $f'(x) = 3 - \dfrac{6}{\sqrt{x}}, \quad x \neq 0$

11. $f'(x) = x^{-1/3}(x + 2)$ **12.** $f'(x) = x^{-1/2}(x - 3)$

13. $f'(x) = (\sin x - 1)(2 \cos x + 1), 0 \leq x \leq 2\pi$

14. $f'(x) = (\sin x + \cos x)(\sin x - \cos x), 0 \leq x \leq 2\pi$

Identifying Extrema

In Exercises 15–44:

  a. Find the open intervals on which the function is increasing and decreasing.

  b. Identify the function's local and absolute extreme values, if any, saying where they occur.

15.

16.

17.

18.

19. $g(t) = -t^2 - 3t + 3$

20. $g(t) = -3t^2 + 9t + 5$

21. $h(x) = -x^3 + 2x^2$

22. $h(x) = 2x^3 - 18x$

23. $f(\theta) = 3\theta^2 - 4\theta^3$

24. $f(\theta) = 6\theta - \theta^3$

25. $f(r) = 3r^3 + 16r$

26. $h(r) = (r + 7)^3$

27. $f(x) = x^4 - 8x^2 + 16$

28. $g(x) = x^4 - 4x^3 + 4x^2$

29. $H(t) = \frac{3}{2}t^4 - t^6$

30. $K(t) = 15t^3 - t^5$

31. $f(x) = x - 6\sqrt{x - 1}$

32. $g(x) = 4\sqrt{x} - x^2 + 3$

33. $g(x) = x\sqrt{8 - x^2}$

34. $g(x) = x^2\sqrt{5 - x}$

35. $f(x) = \frac{x^2 - 3}{x - 2}, \quad x \neq 2$

36. $f(x) = \frac{x^3}{3x^2 + 1}$

37. $f(x) = x^{1/3}(x + 8)$

38. $g(x) = x^{2/3}(x + 5)$

39. $h(x) = x^{1/3}(x^2 - 4)$

40. $k(x) = x^{2/3}(x^2 - 4)$

41. $f(x) = e^{2x} + e^{-x}$

42. $f(x) = e^{\sqrt{x}}$

43. $f(x) = x \ln x$

44. $f(x) = x^2 \ln x$

In Exercises 45–56:

   a. Identify the function's local extreme values in the given domain, and say where they occur.

   b. Which of the extreme values, if any, are absolute?

   T **c.** Support your findings with a graphing calculator or computer grapher.

45. $f(x) = 2x - x^2, \quad -\infty < x \leq 2$

46. $f(x) = (x + 1)^2, \quad -\infty < x \leq 0$

47. $g(x) = x^2 - 4x + 4, \quad 1 \leq x < \infty$

48. $g(x) = -x^2 - 6x - 9, \quad -4 \leq x < \infty$

49. $f(t) = 12t - t^3, \quad -3 \leq t < \infty$

50. $f(t) = t^3 - 3t^2, \quad -\infty < t \leq 3$

51. $h(x) = \frac{x^3}{3} - 2x^2 + 4x, \quad 0 \leq x < \infty$

52. $k(x) = x^3 + 3x^2 + 3x + 1, \quad -\infty < x \leq 0$

53. $f(x) = \sqrt{25 - x^2}, \quad -5 \leq x \leq 5$

54. $f(x) = \sqrt{x^2 - 2x - 3}, \quad 3 \leq x < \infty$

55. $g(x) = \frac{x - 2}{x^2 - 1}, \quad 0 \leq x < 1$

56. $g(x) = \frac{x^2}{4 - x^2}, \quad -2 < x \leq 1$

In Exercises 57–64:

   a. Find the local extrema of each function on the given interval, and say where they occur.

   T **b.** Graph the function and its derivative together. Comment on the behavior of f in relation to the signs and values of f'.

57. $f(x) = \sin 2x, \quad 0 \leq x \leq \pi$

58. $f(x) = \sin x - \cos x, \quad 0 \leq x \leq 2\pi$

59. $f(x) = \sqrt{3} \cos x + \sin x, \quad 0 \leq x \leq 2\pi$

60. $f(x) = -2x + \tan x, \quad \frac{-\pi}{2} < x < \frac{\pi}{2}$

61. $f(x) = \frac{x}{2} - 2\sin\frac{x}{2}, \quad 0 \leq x \leq 2\pi$

62. $f(x) = -2\cos x - \cos^2 x, \quad -\pi \leq x \leq \pi$

63. $f(x) = \csc^2 x - 2\cot x, \quad 0 < x < \pi$

64. $f(x) = \sec^2 x - 2\tan x, \quad \frac{-\pi}{2} < x < \frac{\pi}{2}$

Theory and Examples

Show that the functions in Exercises 65 and 66 have local extreme values at the given values of θ, and say which kind of local extreme the function has.

65. $h(\theta) = 3\cos\frac{\theta}{2}, \quad 0 \leq \theta \leq 2\pi, \quad$ at $\theta = 0$ and $\theta = 2\pi$

66. $h(\theta) = 5\sin\frac{\theta}{2}, \quad 0 \leq \theta \leq \pi, \quad$ at $\theta = 0$ and $\theta = \pi$

67. Sketch the graph of a differentiable function $y = f(x)$ through the point $(1, 1)$ if $f'(1) = 0$ and

   a. $f'(x) > 0$ for $x < 1$ and $f'(x) < 0$ for $x > 1$;

   b. $f'(x) < 0$ for $x < 1$ and $f'(x) > 0$ for $x > 1$;

   c. $f'(x) > 0$ for $x \neq 1$;

   d. $f'(x) < 0$ for $x \neq 1$.

68. Sketch the graph of a differentiable function $y = f(x)$ that has

   a. a local minimum at $(1, 1)$ and a local maximum at $(3, 3)$;

   b. a local maximum at $(1, 1)$ and a local minimum at $(3, 3)$;

   c. local maxima at $(1, 1)$ and $(3, 3)$;

   d. local minima at $(1, 1)$ and $(3, 3)$.

69. Sketch the graph of a continuous function $y = g(x)$ such that

   a. $g(2) = 2, 0 < g' < 1$ for $x < 2, g'(x) \to 1^-$ as $x \to 2^-$, $-1 < g' < 0$ for $x > 2$, and $g'(x) \to -1^+$ as $x \to 2^+$;

   b. $g(2) = 2, g' < 0$ for $x < 2, g'(x) \to -\infty$ as $x \to 2^-$, $g' > 0$ for $x > 2$, and $g'(x) \to \infty$ as $x \to 2^+$.

70. Sketch the graph of a continuous function $y = h(x)$ such that

   a. $h(0) = 0, -2 \leq h(x) \leq 2$ for all $x, h'(x) \to \infty$ as $x \to 0^-$, and $h'(x) \to \infty$ as $x \to 0^+$;

   b. $h(0) = 0, -2 \leq h(x) \leq 0$ for all $x, h'(x) \to \infty$ as $x \to 0^-$, and $h'(x) \to -\infty$ as $x \to 0^+$.

71. Discuss the extreme-value behavior of the function $f(x) = x \sin(1/x), x \neq 0$. How many critical points does this function have? Where are they located on the x-axis? Does f have an absolute minimum? An absolute maximum? (See Exercise 49 in Section 2.3.)

72. Find the intervals on which the function $f(x) = ax^2 + bx + c$, $a \neq 0$, is increasing and decreasing. Describe the reasoning behind your answer.

73. Determine the values of constants a and b so that $f(x) = ax^2 + bx$ has an absolute maximum at the point $(1, 2)$.

74. Determine the values of constants a, b, c, and d so that $f(x) = ax^3 + bx^2 + cx + d$ has a local maximum at the point $(0, 0)$ and a local minimum at the point $(1, -1)$.

75. Locate and identify the absolute extreme values of

    a. $\ln(\cos x)$ on $[-\pi/4, \pi/3]$,

    b. $\cos(\ln x)$ on $[1/2, 2]$.

76. a. Prove that $f(x) = x - \ln x$ is increasing for $x > 1$.

    b. Using part (a), show that $\ln x < x$ if $x > 1$.

77. Find the absolute maximum and minimum values of $f(x) = e^x - 2x$ on $[0, 1]$.

78. Where does the periodic function $f(x) = 2e^{\sin(x/2)}$ take on its extreme values and what are these values?

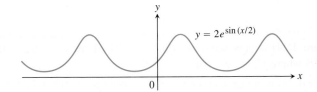

$y = 2e^{\sin(x/2)}$

79. Find the absolute maximum value of $f(x) = x^2 \ln(1/x)$ and say where it is assumed.

80. a. Prove that $e^x \geq 1 + x$ if $x \geq 0$.

    b. Use the result in part (a) to show that

$$e^x \geq 1 + x + \frac{1}{2}x^2.$$

81. Show that increasing functions and decreasing functions are one-to-one. That is, show that for any x_1 and x_2 in I, $x_2 \neq x_1$ implies $f(x_2) \neq f(x_1)$.

Use the results of Exercise 81 to show that the functions in Exercises 82–86 have inverses over their domains. Find a formula for df^{-1}/dx using Theorem 3, Section 3.8.

82. $f(x) = (1/3)x + (5/6)$ **83.** $f(x) = 27x^3$

84. $f(x) = 1 - 8x^3$ **85.** $f(x) = (1 - x)^3$

86. $f(x) = x^{5/3}$

4.4   Concavity and Curve Sketching

We have seen how the first derivative tells us where a function is increasing, where it is decreasing, and whether a local maximum or local minimum occurs at a critical point. In this section we see that the second derivative gives us information about how the graph of a differentiable function bends or turns. With this knowledge about the first and second derivatives, coupled with our previous understanding of asymptotic behavior and symmetry studied in Sections 2.6 and 1.1, we can now draw an accurate graph of a function. By organizing all of these ideas into a coherent procedure, we give a method for sketching graphs and revealing visually the key features of functions. Identifying and knowing the locations of these features is of major importance in mathematics and its applications to science and engineering, especially in the graphical analysis and interpretation of data.

Concavity

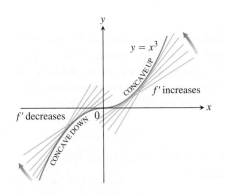

FIGURE 4.24 The graph of $f(x) = x^3$ is concave down on $(-\infty, 0)$ and concave up on $(0, \infty)$ (Example 1a).

As you can see in Figure 4.24, the curve $y = x^3$ rises as x increases, but the portions defined on the intervals $(-\infty, 0)$ and $(0, \infty)$ turn in different ways. As we approach the origin from the left along the curve, the curve turns to our right and falls below its tangents. The slopes of the tangents are decreasing on the interval $(-\infty, 0)$. As we move away from the origin along the curve to the right, the curve turns to our left and rises above its tangents. The slopes of the tangents are increasing on the interval $(0, \infty)$. This turning or bending behavior defines the *concavity* of the curve.

DEFINITION   The graph of a differentiable function $y = f(x)$ is

(a) concave up on an open interval I if f' is increasing on I;

(b) concave down on an open interval I if f' is decreasing on I.

If $y = f(x)$ has a second derivative, we can apply Corollary 3 of the Mean Value Theorem to the first derivative function. We conclude that f' increases if $f'' > 0$ on I, and decreases if $f'' < 0$.

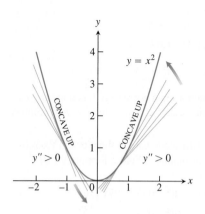

FIGURE 4.25 The graph of $f(x) = x^2$ is concave up on every interval (Example 1b).

If $y = f(x)$ is twice-differentiable, we will use the notations f'' and y'' interchangeably when denoting the second derivative.

EXAMPLE 1

(a) The curve $y = x^3$ (Figure 4.24) is concave down on $(-\infty, 0)$ where $y'' = 6x < 0$ and concave up on $(0, \infty)$ where $y'' = 6x > 0$.

(b) The curve $y = x^2$ (Figure 4.25) is concave up on $(-\infty, \infty)$ because its second derivative $y'' = 2$ is always positive. ∎

EXAMPLE 2

Determine the concavity of $y = 3 + \sin x$ on $[0, 2\pi]$.

Solution The first derivative of $y = 3 + \sin x$ is $y' = \cos x$, and the second derivative is $y'' = -\sin x$. The graph of $y = 3 + \sin x$ is concave down on $(0, \pi)$, where $y'' = -\sin x$ is negative. It is concave up on $(\pi, 2\pi)$, where $y'' = -\sin x$ is positive (Figure 4.26). ∎

Points of Inflection

The curve $y = 3 + \sin x$ in Example 2 changes concavity at the point $(\pi, 3)$. Since the first derivative $y' = \cos x$ exists for all x, we see that the curve has a tangent line of slope -1 at the point $(\pi, 3)$. This point is called a *point of inflection* of the curve. Notice from Figure 4.26 that the graph crosses its tangent line at this point and that the second derivative $y'' = -\sin x$ has value 0 when $x = \pi$. In general, we have the following definition.

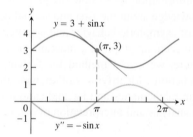

FIGURE 4.26 Using the sign of y'' to determine the concavity of y (Example 2).

We observed that the second derivative of $f(x) = 3 + \sin x$ is equal to zero at the inflection point $(\pi, 3)$. Generally, if the second derivative exists at a point of inflection $(c, f(c))$, then $f''(c) = 0$. This follows immediately from the Intermediate Value Theorem whenever f'' is continuous over an interval containing $x = c$ because the second derivative changes sign moving across this interval. Even if the continuity assumption is dropped, it is still true that $f''(c) = 0$, provided the second derivative exists (although a more advanced agrument is required in this noncontinuous case). Since a tangent line must exist at the point of inflection, either the first derivative $f'(c)$ exists (is finite) or a vertical tangent exists at the point. At a vertical tangent neither the first nor second derivative exists. In summary, we conclude the following result.

The next example illustrates a function having a point of inflection where the first derivative exists, but the second derivative fails to exist.

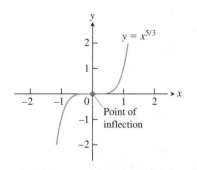

FIGURE 4.27 The graph of $f(x) = x^{5/3}$ has a horizontal tangent at the origin where the concavity changes, although f'' does not exist at $x = 0$ (Example 3).

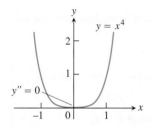

FIGURE 4.28 The graph of $y = x^4$ has no inflection point at the origin, even though $y'' = 0$ there (Example 4).

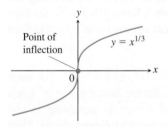

FIGURE 4.29 A point of inflection where y' and y'' fail to exist (Example 5).

EXAMPLE 3   The graph of $f(x) = x^{5/3}$ has a horizontal tangent at the origin because $f'(x) = (5/3)x^{2/3} = 0$ when $x = 0$. However, the second derivative

$$f''(x) = \frac{d}{dx}\left(\frac{5}{3}x^{2/3}\right) = \frac{10}{9}x^{-1/3}$$

fails to exist at $x = 0$. Nevertheless, $f''(x) < 0$ for $x < 0$ and $f''(x) > 0$ for $x > 0$, so the second derivative changes sign at $x = 0$ and there is a point of inflection at the origin. The graph is shown in Figure 4.27.   ∎

Here is an example showing that an inflection point need not occur even though both derivatives exist and $f'' = 0$.

EXAMPLE 4   The curve $y = x^4$ has no inflection point at $x = 0$ (Figure 4.28). Even though the second derivative $y'' = 12x^2$ is zero there, it does not change sign.   ∎

As our final illustration, we show a situation in which a point of inflection occurs at a vertical tangent to the curve where neither the first nor the second derivative exists.

EXAMPLE 5   The graph of $y = x^{1/3}$ has a point of inflection at the origin because the second derivative is positive for $x < 0$ and negative for $x > 0$:

$$y'' = \frac{d^2}{dx^2}\left(x^{1/3}\right) = \frac{d}{dx}\left(\frac{1}{3}x^{-2/3}\right) = -\frac{2}{9}x^{-5/3}.$$

However, both $y' = x^{-2/3}/3$ and y'' fail to exist at $x = 0$, and there is a vertical tangent there. See Figure 4.29.   ∎

To study the motion of an object moving along a line as a function of time, we often are interested in knowing when the object's acceleration, given by the second derivative, is positive or negative. The points of inflection on the graph of the object's position function reveal where the acceleration changes sign.

EXAMPLE 6   A particle is moving along a horizontal coordinate line (positive to the right) with position function

$$s(t) = 2t^3 - 14t^2 + 22t - 5, \qquad t \geq 0.$$

Find the velocity and acceleration, and describe the motion of the particle.

Solution   The velocity is

$$v(t) = s'(t) = 6t^2 - 28t + 22 = 2(t - 1)(3t - 11),$$

and the acceleration is

$$a(t) = v'(t) = s''(t) = 12t - 28 = 4(3t - 7).$$

When the function $s(t)$ is increasing, the particle is moving to the right; when $s(t)$ is decreasing, the particle is moving to the left.

Notice that the first derivative ($v = s'$) is zero at the critical points $t = 1$ and $t = 11/3$.

Interval	$0 < t < 1$	$1 < t < 11/3$	$11/3 < t$
Sign of $v = s'$	+	−	+
Behavior of s	increasing	decreasing	increasing
Particle motion	right	left	right

The particle is moving to the right in the time intervals $[0, 1)$ and $(11/3, \infty)$, and moving to the left in $(1, 11/3)$. It is momentarily stationary (at rest) at $t = 1$ and $t = 11/3$.

The acceleration $a(t) = s''(t) = 4(3t - 7)$ is zero when $t = 7/3$.

Interval	$0 < t < 7/3$	$7/3 < t$
Sign of $a = s''$	$-$	$+$
Graph of s	concave down	concave up

The particle starts out moving to the right while slowing down, and then reverses and begins moving to the left at $t = 1$ under the influence of the leftward acceleration over the time interval $[0, 7/3]$. The acceleration then changes direction at $t = 7/3$ but the particle continues moving leftward, while slowing down under the rightward acceleration. At $t = 11/3$ the particle reverses direction again: moving to the right in the same direction as the acceleration. ∎

Second Derivative Test for Local Extrema

Instead of looking for sign changes in f' at critical points, we can sometimes use the following test to determine the presence and nature of local extrema.

> **THEOREM 5—Second Derivative Test for Local Extrema**   Suppose f'' is continuous on an open interval that contains $x = c$.
>
> 1. If $f'(c) = 0$ and $f''(c) < 0$, then f has a local maximum at $x = c$.
> 2. If $f'(c) = 0$ and $f''(c) > 0$, then f has a local minimum at $x = c$.
> 3. If $f'(c) = 0$ and $f''(c) = 0$, then the test fails. The function f may have a local maximum, a local minimum, or neither.

Proof   Part (1). If $f''(c) < 0$, then $f''(x) < 0$ on some open interval I containing the point c, since f'' is continuous. Therefore, f' is decreasing on I. Since $f'(c) = 0$, the sign of f' changes from positive to negative at c so f has a local maximum at c by the First Derivative Test.

The proof of Part (2) is similar.

For Part (3), consider the three functions $y = x^4$, $y = -x^4$, and $y = x^3$. For each function, the first and second derivatives are zero at $x = 0$. Yet the function $y = x^4$ has a local minimum there, $y = -x^4$ has a local maximum, and $y = x^3$ is increasing in any open interval containing $x = 0$ (having neither a maximum nor a minimum there). Thus the test fails. ∎

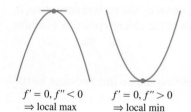

$f' = 0, f'' < 0$
$\Rightarrow$ local max

$f' = 0, f'' > 0$
$\Rightarrow$ local min

This test requires us to know f'' *only at c itself* and not in an interval about c. This makes the test easy to apply. That's the good news. The bad news is that the test is inconclusive if $f'' = 0$ or if f'' does not exist at $x = c$. When this happens, use the First Derivative Test for local extreme values.

Together f' and f'' tell us the shape of the function's graph—that is, where the critical points are located and what happens at a critical point, where the function is increasing and where it is decreasing, and how the curve is turning or bending as defined by its concavity. We use this information to sketch a graph of the function that captures its key features.

EXAMPLE 7   Sketch a graph of the function

$$f(x) = x^4 - 4x^3 + 10$$

using the following steps.

(a) Identify where the extrema of f occur.

(b) Find the intervals on which f is increasing and the intervals on which f is decreasing.

In more advanced treatments of calculus it is proved that l'Hôpital's Rule applies to the indeterminate form ∞/∞ as well as to $0/0$. If $f(x) \to \pm\infty$ and $g(x) \to \pm\infty$ as $x \to a$, then

$$\lim_{x \to a} \frac{f(x)}{g(x)} = \lim_{x \to a} \frac{f'(x)}{g'(x)}$$

provided the limit on the right exists. In the notation $x \to a$, a may be either finite or infinite. Moreover, $x \to a$ may be replaced by the one-sided limits $x \to a^+$ or $x \to a^-$.

EXAMPLE 4   Find the limits of these ∞/∞ forms:

(a) $\displaystyle\lim_{x \to \pi/2} \frac{\sec x}{1 + \tan x}$ **(b)** $\displaystyle\lim_{x \to \infty} \frac{\ln x}{2\sqrt{x}}$ **(c)** $\displaystyle\lim_{x \to \infty} \frac{e^x}{x^2}$.

Solution

(a) The numerator and denominator are discontinuous at $x = \pi/2$, so we investigate the one-sided limits there. To apply l'Hôpital's Rule, we can choose I to be any open interval with $x = \pi/2$ as an endpoint.

$$\lim_{x \to (\pi/2)^-} \frac{\sec x}{1 + \tan x} \qquad \frac{\infty}{\infty} \text{ from the left}$$

$$= \lim_{x \to (\pi/2)^-} \frac{\sec x \tan x}{\sec^2 x} = \lim_{x \to (\pi/2)^-} \sin x = 1$$

The right-hand limit is 1 also, with $(-\infty)/(-\infty)$ as the indeterminate form. Therefore, the two-sided limit is equal to 1.

(b) $\displaystyle\lim_{x \to \infty} \frac{\ln x}{2\sqrt{x}} = \lim_{x \to \infty} \frac{1/x}{1/\sqrt{x}} = \lim_{x \to \infty} \frac{1}{\sqrt{x}} = 0 \qquad \frac{1/x}{1/\sqrt{x}} = \frac{\sqrt{x}}{x} = \frac{1}{\sqrt{x}}$

(c) $\displaystyle\lim_{x \to \infty} \frac{e^x}{x^2} = \lim_{x \to \infty} \frac{e^x}{2x} = \lim_{x \to \infty} \frac{e^x}{2} = \infty$ ∎

Next we turn our attention to the indeterminate forms $\infty \cdot 0$ and $\infty - \infty$. Sometimes these forms can be handled by using algebra to convert them to a $0/0$ or ∞/∞ form. Here again we do not mean to suggest that $\infty \cdot 0$ or $\infty - \infty$ is a number. They are only notations for functional behaviors when considering limits. Here are examples of how we might work with these indeterminate forms.

EXAMPLE 5   Find the limits of these $\infty \cdot 0$ forms:

(a) $\displaystyle\lim_{x \to \infty} \left(x \sin \frac{1}{x} \right)$ **(b)** $\displaystyle\lim_{x \to 0^+} \sqrt{x} \ln x$

Solution

(a) $\displaystyle\lim_{x \to \infty} \left(x \sin \frac{1}{x} \right) = \lim_{h \to 0^+} \left(\frac{1}{h} \sin h \right) = \lim_{h \to 0^+} \frac{\sin h}{h} = 1$ $\infty \cdot 0$; Let $h = 1/x$.

(b) $\displaystyle\lim_{x \to 0^+} \sqrt{x} \ln x = \lim_{x \to 0^+} \frac{\ln x}{1/\sqrt{x}}$ $\infty \cdot 0$ converted to ∞/∞

$$= \lim_{x \to 0^+} \frac{1/x}{-1/2x^{3/2}} \qquad \text{l'Hôpital's Rule}$$

$$= \lim_{x \to 0^+} \left(-2\sqrt{x} \right) = 0 \qquad\qquad ∎$$

EXAMPLE 6 Find the limit of this $\infty - \infty$ form:

$$\lim_{x \to 0}\left(\frac{1}{\sin x} - \frac{1}{x}\right).$$

Solution If $x \to 0^{+}$, then $\sin x \to 0^{+}$ and

$$\frac{1}{\sin x} - \frac{1}{x} \to \infty - \infty.$$

Similarly, if $x \to 0^{-}$, then $\sin x \to 0^{-}$ and

$$\frac{1}{\sin x} - \frac{1}{x} \to -\infty - (-\infty) = -\infty + \infty.$$

Neither form reveals what happens in the limit. To find out, we first combine the fractions:

$$\frac{1}{\sin x} - \frac{1}{x} = \frac{x - \sin x}{x \sin x} \qquad \text{Common denominator is } x \sin x.$$

Then we apply l'Hôpital's Rule to the result:

$$\lim_{x \to 0}\left(\frac{1}{\sin x} - \frac{1}{x}\right) = \lim_{x \to 0} \frac{x - \sin x}{x \sin x} \qquad \frac{0}{0}$$

$$= \lim_{x \to 0} \frac{1 - \cos x}{\sin x + x \cos x} \qquad \text{Still } \frac{0}{0}$$

$$= \lim_{x \to 0} \frac{\sin x}{2 \cos x - x \sin x} = \frac{0}{2} = 0. \qquad \blacksquare$$

Indeterminate Powers

Limits that lead to the indeterminate forms 1^{∞}, 0^{0}, and ∞^{0} can sometimes be handled by first taking the logarithm of the function. We use l'Hôpital's Rule to find the limit of the logarithm expression and then exponentiate the result to find the original function limit. This procedure is justified by the continuity of the exponential function and Theorem 10 in Section 2.5, and it is formulated as follows. (The formula is also valid for one-sided limits.)

If $\lim_{x \to a} \ln f(x) = L$, then

$$\lim_{x \to a} f(x) = \lim_{x \to a} e^{\ln f(x)} = e^{L}.$$

Here a may be either finite or infinite.

EXAMPLE 7 Apply l'Hôpital's Rule to show that $\lim_{x \to 0^{+}} (1 + x)^{1/x} = e$.

Solution The limit leads to the indeterminate form 1^{∞}. We let $f(x) = (1 + x)^{1/x}$ and find $\lim_{x \to 0^{+}} \ln f(x)$. Since

$$\ln f(x) = \ln (1 + x)^{1/x} = \frac{1}{x} \ln (1 + x),$$

l'Hôpital's Rule now applies to give

$$\lim_{x \to 0^+} \ln f(x) = \lim_{x \to 0^+} \frac{\ln(1+x)}{x} \qquad \frac{0}{0}$$

$$= \lim_{x \to 0^+} \frac{\dfrac{1}{1+x}}{1}$$

$$= \frac{1}{1} = 1.$$

Therefore, $\displaystyle\lim_{x \to 0^+} (1+x)^{1/x} = \lim_{x \to 0^+} f(x) = \lim_{x \to 0^+} e^{\ln f(x)} = e^1 = e.$ ∎

EXAMPLE 8   Find $\lim_{x \to \infty} x^{1/x}$.

Solution   The limit leads to the indeterminate form ∞^0. We let $f(x) = x^{1/x}$ and find $\lim_{x \to \infty} \ln f(x)$. Since

$$\ln f(x) = \ln x^{1/x} = \frac{\ln x}{x},$$

l'Hôpital's Rule gives

$$\lim_{x \to \infty} \ln f(x) = \lim_{x \to \infty} \frac{\ln x}{x} \qquad \frac{\infty}{\infty}$$

$$= \lim_{x \to \infty} \frac{1/x}{1}$$

$$= \frac{0}{1} = 0.$$

Therefore $\displaystyle\lim_{x \to \infty} x^{1/x} = \lim_{x \to \infty} f(x) = \lim_{x \to \infty} e^{\ln f(x)} = e^0 = 1.$ ∎

Proof of L'Hôpital's Rule

The proof of l'Hôpital's Rule is based on Cauchy's Mean Value Theorem, an extension of the Mean Value Theorem that involves two functions instead of one. We prove Cauchy's Theorem first and then show how it leads to l'Hôpital's Rule.

HISTORICAL BIOGRAPHY

Augustin-Louis Cauchy
(1789–1857)

> **THEOREM 7—Cauchy's Mean Value Theorem**   Suppose functions f and g are continuous on $[a, b]$ and differentiable throughout (a, b) and also suppose $g'(x) \neq 0$ throughout (a, b). Then there exists a number c in (a, b) at which
>
> $$\frac{f'(c)}{g'(c)} = \frac{f(b) - f(a)}{g(b) - g(a)}.$$

Proof   We apply the Mean Value Theorem of Section 4.2 twice. First we use it to show that $g(a) \neq g(b)$. For if $g(b)$ did equal $g(a)$, then the Mean Value Theorem would give

$$g'(c) = \frac{g(b) - g(a)}{b - a} = 0$$

for some c between a and b, which cannot happen because $g'(x) \neq 0$ in (a, b).

We next apply the Mean Value Theorem to the function

$$F(x) = f(x) - f(a) - \frac{f(b) - f(a)}{g(b) - g(a)}[g(x) - g(a)].$$

This function is continuous and differentiable where f and g are, and $F(b) = F(a) = 0$. Therefore, there is a number c between a and b for which $F'(c) = 0$. When expressed in terms of f and g, this equation becomes

$$F'(c) = f'(c) - \frac{f(b) - f(a)}{g(b) - g(a)}[g'(c)] = 0$$

so that

$$\frac{f'(c)}{g'(c)} = \frac{f(b) - f(a)}{g(b) - g(a)}.$$

∎

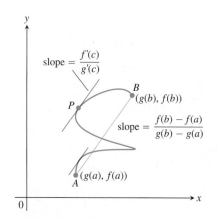

FIGURE 4.34 There is at least one point P on the curve C for which the slope of the tangent to the curve at P is the same as the slope of the secant line joining the points $A(g(a), f(a))$ and $B(g(b), f(b))$.

Notice that the Mean Value Theorem in Section 4.2 is Theorem 7 with $g(x) = x$.

Cauchy's Mean Value Theorem has a geometric interpretation for a general winding curve C in the plane joining the two points $A = (g(a), f(a))$ and $B = (g(b), f(b))$. In Chapter 11 you will learn how the curve C can be formulated so that there is at least one point P on the curve for which the tangent to the curve at P is parallel to the secant line joining the points A and B. The slope of that tangent line turns out to be the quotient f'/g' evaluated at the number c in the interval (a, b), which is the left-hand side of the equation in Theorem 7. Because the slope of the secant line joining A and B is

$$\frac{f(b) - f(a)}{g(b) - g(a)},$$

the equation in Cauchy's Mean Value Theorem says that the slope of the tangent line equals the slope of the secant line. This geometric interpretation is shown in Figure 4.34. Notice from the figure that it is possible for more than one point on the curve C to have a tangent line that is parallel to the secant line joining A and B.

Proof of l'Hôpital's Rule We first establish the limit equation for the case $x \to a^+$. The method needs almost no change to apply to $x \to a^-$, and the combination of these two cases establishes the result.

Suppose that x lies to the right of a. Then $g'(x) \neq 0$, and we can apply Cauchy's Mean Value Theorem to the closed interval from a to x. This step produces a number c between a and x such that

$$\frac{f'(c)}{g'(c)} = \frac{f(x) - f(a)}{g(x) - g(a)}.$$

But $f(a) = g(a) = 0$, so

$$\frac{f'(c)}{g'(c)} = \frac{f(x)}{g(x)}.$$

As x approaches a, c approaches a because it always lies between a and x. Therefore,

$$\lim_{x \to a^+} \frac{f(x)}{g(x)} = \lim_{c \to a^+} \frac{f'(c)}{g'(c)} = \lim_{x \to a^+} \frac{f'(x)}{g'(x)},$$

which establishes l'Hôpital's Rule for the case where x approaches a from above. The case where x approaches a from below is proved by applying Cauchy's Mean Value Theorem to the closed interval $[x, a]$, $x < a$.

∎

Exercises 4.5

Finding Limits in Two Ways

In Exercises 1–6, use l'Hôpital's Rule to evaluate the limit. Then evaluate the limit using a method studied in Chapter 2.

1. $\lim\limits_{x \to -2} \dfrac{x + 2}{x^2 - 4}$

2. $\lim\limits_{x \to 0} \dfrac{\sin 5x}{x}$

3. $\lim\limits_{x \to \infty} \dfrac{5x^2 - 3x}{7x^2 + 1}$

4. $\lim\limits_{x \to 1} \dfrac{x^3 - 1}{4x^3 - x - 3}$

5. $\lim\limits_{x \to 0} \dfrac{1 - \cos x}{x^2}$

6. $\lim\limits_{x \to \infty} \dfrac{2x^2 + 3x}{x^3 + x + 1}$

Applying l'Hôpital's Rule

Use l'Hôpital's rule to find the limits in Exercises 7–50.

7. $\lim\limits_{x \to 2} \dfrac{x - 2}{x^2 - 4}$

8. $\lim\limits_{x \to -5} \dfrac{x^2 - 25}{x + 5}$

9. $\lim\limits_{t \to -3} \dfrac{t^3 - 4t + 15}{t^2 - t - 12}$

10. $\lim\limits_{t \to 1} \dfrac{3t^3 - 3}{4t^3 - t - 3}$

11. $\lim\limits_{x \to \infty} \dfrac{5x^3 - 2x}{7x^3 + 3}$

12. $\lim\limits_{x \to \infty} \dfrac{x - 8x^2}{12x^2 + 5x}$

13. $\lim\limits_{t \to 0} \dfrac{\sin t^2}{t}$

14. $\lim\limits_{t \to 0} \dfrac{\sin 5t}{2t}$

15. $\lim\limits_{x \to 0} \dfrac{8x^2}{\cos x - 1}$

16. $\lim\limits_{x \to 0} \dfrac{\sin x - x}{x^3}$

17. $\lim\limits_{\theta \to \pi/2} \dfrac{2\theta - \pi}{\cos(2\pi - \theta)}$

18. $\lim\limits_{\theta \to -\pi/3} \dfrac{3\theta + \pi}{\sin(\theta + (\pi/3))}$

19. $\lim\limits_{\theta \to \pi/2} \dfrac{1 - \sin \theta}{1 + \cos 2\theta}$

20. $\lim\limits_{x \to 1} \dfrac{x - 1}{\ln x - \sin \pi x}$

21. $\lim\limits_{x \to 0} \dfrac{x^2}{\ln(\sec x)}$

22. $\lim\limits_{x \to \pi/2} \dfrac{\ln(\csc x)}{(x - (\pi/2))^2}$

23. $\lim\limits_{t \to 0} \dfrac{t(1 - \cos t)}{t - \sin t}$

24. $\lim\limits_{t \to 0} \dfrac{t \sin t}{1 - \cos t}$

25. $\lim\limits_{x \to (\pi/2)^-} \left(x - \dfrac{\pi}{2}\right) \sec x$

26. $\lim\limits_{x \to (\pi/2)^-} \left(\dfrac{\pi}{2} - x\right) \tan x$

27. $\lim\limits_{\theta \to 0} \dfrac{3^{\sin \theta} - 1}{\theta}$

28. $\lim\limits_{\theta \to 0} \dfrac{(1/2)^\theta - 1}{\theta}$

29. $\lim\limits_{x \to 0} \dfrac{x2^x}{2^x - 1}$

30. $\lim\limits_{x \to 0} \dfrac{3^x - 1}{2^x - 1}$

31. $\lim\limits_{x \to \infty} \dfrac{\ln(x + 1)}{\log_2 x}$

32. $\lim\limits_{x \to \infty} \dfrac{\log_2 x}{\log_3(x + 3)}$

33. $\lim\limits_{x \to 0^+} \dfrac{\ln(x^2 + 2x)}{\ln x}$

34. $\lim\limits_{x \to 0^+} \dfrac{\ln(e^x - 1)}{\ln x}$

35. $\lim\limits_{y \to 0} \dfrac{\sqrt{5y + 25} - 5}{y}$

36. $\lim\limits_{y \to 0} \dfrac{\sqrt{ay + a^2} - a}{y}, \quad a > 0$

37. $\lim\limits_{x \to \infty} (\ln 2x - \ln(x + 1))$

38. $\lim\limits_{x \to 0^+} (\ln x - \ln \sin x)$

39. $\lim\limits_{x \to 0^+} \dfrac{(\ln x)^2}{\ln(\sin x)}$

40. $\lim\limits_{x \to 0^+} \left(\dfrac{3x + 1}{x} - \dfrac{1}{\sin x}\right)$

41. $\lim\limits_{x \to 1^+} \left(\dfrac{1}{x - 1} - \dfrac{1}{\ln x}\right)$

42. $\lim\limits_{x \to 0^+} (\csc x - \cot x + \cos x)$

43. $\lim\limits_{\theta \to 0} \dfrac{\cos \theta - 1}{e^\theta - \theta - 1}$

44. $\lim\limits_{h \to 0} \dfrac{e^h - (1 + h)}{h^2}$

45. $\lim\limits_{t \to \infty} \dfrac{e^t + t^2}{e^t - t}$

46. $\lim\limits_{x \to \infty} x^2 e^{-x}$

47. $\lim\limits_{x \to 0} \dfrac{x - \sin x}{x \tan x}$

48. $\lim\limits_{x \to 0} \dfrac{(e^x - 1)^2}{x \sin x}$

49. $\lim\limits_{\theta \to 0} \dfrac{\theta - \sin \theta \cos \theta}{\tan \theta - \theta}$

50. $\lim\limits_{x \to 0} \dfrac{\sin 3x - 3x + x^2}{\sin x \sin 2x}$

Indeterminate Powers and Products

Find the limits in Exercise 51–66.

51. $\lim\limits_{x \to 1^+} x^{1/(1-x)}$

52. $\lim\limits_{x \to 1^+} x^{1/(x-1)}$

53. $\lim\limits_{x \to \infty} (\ln x)^{1/x}$

54. $\lim\limits_{x \to e^+} (\ln x)^{1/(x-e)}$

55. $\lim\limits_{x \to 0^+} x^{-1/\ln x}$

56. $\lim\limits_{x \to \infty} x^{1/\ln x}$

57. $\lim\limits_{x \to \infty} (1 + 2x)^{1/(2 \ln x)}$

58. $\lim\limits_{x \to 0} (e^x + x)^{1/x}$

59. $\lim\limits_{x \to 0^+} x^x$

60. $\lim\limits_{x \to 0^+} \left(1 + \dfrac{1}{x}\right)^x$

61. $\lim\limits_{x \to \infty} \left(\dfrac{x + 2}{x - 1}\right)^x$

62. $\lim\limits_{x \to \infty} \left(\dfrac{x^2 + 1}{x + 2}\right)^{1/x}$

63. $\lim\limits_{x \to 0^+} x^2 \ln x$

64. $\lim\limits_{x \to 0^+} x (\ln x)^2$

65. $\lim\limits_{x \to 0^+} x \tan\left(\dfrac{\pi}{2} - x\right)$

66. $\lim\limits_{x \to 0^+} \sin x \cdot \ln x$

Theory and Applications

L'Hôpital's Rule does not help with the limits in Exercises 67–74. Try it—you just keep on cycling. Find the limits some other way.

67. $\lim\limits_{x \to \infty} \dfrac{\sqrt{9x + 1}}{\sqrt{x + 1}}$

68. $\lim\limits_{x \to 0^+} \dfrac{\sqrt{x}}{\sqrt{\sin x}}$

69. $\lim\limits_{x \to (\pi/2)^-} \dfrac{\sec x}{\tan x}$

70. $\lim\limits_{x \to 0^+} \dfrac{\cot x}{\csc x}$

71. $\lim\limits_{x \to \infty} \dfrac{2^x - 3^x}{3^x + 4^x}$

72. $\lim\limits_{x \to -\infty} \dfrac{2^x + 4^x}{5^x - 2^x}$

73. $\lim\limits_{x \to \infty} \dfrac{e^{x^2}}{xe^x}$

74. $\lim\limits_{x \to 0^+} \dfrac{x}{e^{-1/x}}$

75. Which one is correct, and which one is wrong? Give reasons for your answers.

a. $\lim\limits_{x \to 3} \dfrac{x - 3}{x^2 - 3} = \lim\limits_{x \to 3} \dfrac{1}{2x} = \dfrac{1}{6}$ **b.** $\lim\limits_{x \to 3} \dfrac{x - 3}{x^2 - 3} = \dfrac{0}{6} = 0$

76. Which one is correct, and which one is wrong? Give reasons for your answers.

a. $\lim\limits_{x \to 0} \dfrac{x^2 - 2x}{x^2 - \sin x} = \lim\limits_{x \to 0} \dfrac{2x - 2}{2x - \cos x}$

$$= \lim\limits_{x \to 0} \dfrac{2}{2 + \sin x} = \dfrac{2}{2 + 0} = 1$$

b. $\lim\limits_{x \to 0} \dfrac{x^2 - 2x}{x^2 - \sin x} = \lim\limits_{x \to 0} \dfrac{2x - 2}{2x - \cos x} = \dfrac{-2}{0 - 1} = 2$

77. Only one of these calculations is correct. Which one? Why are the others wrong? Give reasons for your answers.

a. $\lim_{x \to 0^+} x \ln x = 0 \cdot (-\infty) = 0$

b. $\lim_{x \to 0^+} x \ln x = 0 \cdot (-\infty) = -\infty$

c. $\lim_{x \to 0^+} x \ln x = \lim_{x \to 0^+} \dfrac{\ln x}{(1/x)} = \dfrac{-\infty}{\infty} = -1$

d. $\lim_{x \to 0^+} x \ln x = \lim_{x \to 0^+} \dfrac{\ln x}{(1/x)}$

$= \lim_{x \to 0^+} \dfrac{(1/x)}{(-1/x^2)} = \lim_{x \to 0^+} (-x) = 0$

78. Find all values of c that satisfy the conclusion of Cauchy's Mean Value Theorem for the given functions and interval.

a. $f(x) = x,$ $g(x) = x^2,$ $(a, b) = (-2, 0)$

b. $f(x) = x,$ $g(x) = x^2,$ (a, b) arbitrary

c. $f(x) = x^3/3 - 4x,$ $g(x) = x^2,$ $(a, b) = (0, 3)$

79. Continuous extension Find a value of c that makes the function

$$f(x) = \begin{cases} \dfrac{9x - 3\sin 3x}{5x^3}, & x \neq 0 \\ c, & x = 0 \end{cases}$$

continuous at $x = 0$. Explain why your value of c works.

80. For what values of a and b is

$$\lim_{x \to 0} \left(\dfrac{\tan 2x}{x^3} + \dfrac{a}{x^2} + \dfrac{\sin bx}{x} \right) = 0?$$

T 81. $\infty - \infty$ **Form**

a. Estimate the value of

$$\lim_{x \to \infty} \left(x - \sqrt{x^2 + x} \right)$$

by graphing $f(x) = x - \sqrt{x^2 + x}$ over a suitably large interval of x-values.

b. Now confirm your estimate by finding the limit with l'Hôpital's Rule. As the first step, multiply $f(x)$ by the fraction $(x + \sqrt{x^2 + x})/(x + \sqrt{x^2 + x})$ and simplify the new numerator.

82. Find $\lim_{x \to \infty} \left(\sqrt{x^2 + 1} - \sqrt{x} \right)$.

T 83. **0/0 Form** Estimate the value of

$$\lim_{x \to 1} \dfrac{2x^2 - (3x + 1)\sqrt{x} + 2}{x - 1}$$

by graphing. Then confirm your estimate with l'Hôpital's Rule.

84. This exercise explores the difference between the limit

$$\lim_{x \to \infty} \left(1 + \dfrac{1}{x^2} \right)^x$$

and the limit

$$\lim_{x \to \infty} \left(1 + \dfrac{1}{x} \right)^x = e.$$

a. Use l'Hôpital's Rule to show that

$$\lim_{x \to \infty} \left(1 + \dfrac{1}{x} \right)^x = e.$$

T b. Graph

$$f(x) = \left(1 + \dfrac{1}{x^2} \right)^x \quad \text{and} \quad g(x) = \left(1 + \dfrac{1}{x} \right)^x$$

together for $x \geq 0$. How does the behavior of f compare with that of g? Estimate the value of $\lim_{x \to \infty} f(x)$.

c. Confirm your estimate of $\lim_{x \to \infty} f(x)$ by calculating it with l'Hôpital's Rule.

85. Show that

$$\lim_{k \to \infty} \left(1 + \dfrac{r}{k} \right)^k = e^r.$$

86. Given that $x > 0$, find the maximum value, if any, of

a. $x^{1/x}$

b. x^{1/x^2}

c. x^{1/x^n} (n a positive integer)

d. Show that $\lim_{x \to \infty} x^{1/x^n} = 1$ for every positive integer n.

87. Use limits to find horizontal asymptotes for each function.

a. $y = x \tan \left(\dfrac{1}{x} \right)$

b. $y = \dfrac{3x + e^{2x}}{2x + e^{3x}}$

88. Find $f'(0)$ for $f(x) = \begin{cases} e^{-1/x^2}, & x \neq 0 \\ 0, & x = 0. \end{cases}$

T 89. **The continuous extension of $(\sin x)^x$ to $[0, \pi]$**

a. Graph $f(x) = (\sin x)^x$ on the interval $0 \leq x \leq \pi$. What value would you assign to f to make it continuous at $x = 0$?

b. Verify your conclusion in part (a) by finding $\lim_{x \to 0^+} f(x)$ with l'Hôpital's Rule.

c. Returning to the graph, estimate the maximum value of f on $[0, \pi]$. About where is max f taken on?

d. Sharpen your estimate in part (c) by graphing f' in the same window to see where its graph crosses the x-axis. To simplify your work, you might want to delete the exponential factor from the expression for f' and graph just the factor that has a zero.

T 90. **The function $(\sin x)^{\tan x}$** (*Continuation of Exercise 89.*)

a. Graph $f(x) = (\sin x)^{\tan x}$ on the interval $-7 \leq x \leq 7$. How do you account for the gaps in the graph? How wide are the gaps?

b. Now graph f on the interval $0 \leq x \leq \pi$. The function is not defined at $x = \pi/2$, but the graph has no break at this point. What is going on? What value does the graph appear to give for f at $x = \pi/2$? (*Hint:* Use l'Hôpital's Rule to find lim f as $x \to (\pi/2)^-$ and $x \to (\pi/2)^+$.)

c. Continuing with the graphs in part (b), find max f and min f as accurately as you can and estimate the values of x at which they are taken on.

4.6 Applied Optimization

What are the dimensions of a rectangle with fixed perimeter having *maximum area*? What are the dimensions for the *least expensive* cylindrical can of a given volume? How many items should be produced for the *most profitable* production run? Each of these questions asks for the best, or optimal, value of a given function. In this section we use derivatives to solve a variety of optimization problems in business, mathematics, physics, and economics.

(a)

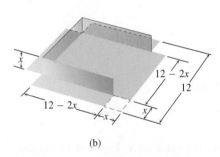

(b)

FIGURE 4.35 An open box made by cutting the corners from a square sheet of tin. What size corners maximize the box's volume (Example 1)?

Solving Applied Optimization Problems

1. *Read the problem.* Read the problem until you understand it. What is given? What is the unknown quantity to be optimized?
2. *Draw a picture.* Label any part that may be important to the problem.
3. *Introduce variables.* List every relation in the picture and in the problem as an equation or algebraic expression, and identify the unknown variable.
4. *Write an equation for the unknown quantity.* If you can, express the unknown as a function of a single variable or in two equations in two unknowns. This may require considerable manipulation.
5. *Test the critical points and endpoints in the domain of the unknown.* Use what you know about the shape of the function's graph. Use the first and second derivatives to identify and classify the function's critical points.

EXAMPLE 1   An open-top box is to be made by cutting small congruent squares from the corners of a 12-in.-by-12-in. sheet of tin and bending up the sides. How large should the squares cut from the corners be to make the box hold as much as possible?

Solution   We start with a picture (Figure 4.35). In the figure, the corner squares are x in. on a side. The volume of the box is a function of this variable:

$$V(x) = x(12 - 2x)^2 = 144x - 48x^2 + 4x^3. \qquad V = hlw$$

Since the sides of the sheet of tin are only 12 in. long, $x \le 6$ and the domain of V is the interval $0 \le x \le 6$.

A graph of V (Figure 4.36) suggests a minimum value of 0 at $x = 0$ and $x = 6$ and a maximum near $x = 2$. To learn more, we examine the first derivative of V with respect to x:

$$\frac{dV}{dx} = 144 - 96x + 12x^2 = 12(12 - 8x + x^2) = 12(2 - x)(6 - x).$$

Of the two zeros, $x = 2$ and $x = 6$, only $x = 2$ lies in the interior of the function's domain and makes the critical-point list. The values of V at this one critical point and two endpoints are

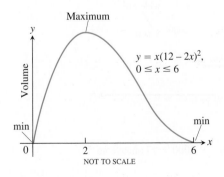

FIGURE 4.36 The volume of the box in Figure 4.35 graphed as a function of x.

$$\text{Critical-point value:} \quad V(2) = 128$$
$$\text{Endpoint values:} \quad V(0) = 0, \qquad V(6) = 0.$$

The maximum volume is 128 in^3. The cutout squares should be 2 in. on a side. ∎

FIGURE 4.37 This one-liter can uses the least material when $h = 2r$ (Example 2).

EXAMPLE 2 You have been asked to design a one-liter can shaped like a right circular cylinder (Figure 4.37). What dimensions will use the least material?

Solution *Volume of can:* If r and h are measured in centimeters, then the volume of the can in cubic centimeters is

$$\pi r^2 h = 1000. \qquad \text{1 liter} = 1000 \text{ cm}^3$$

Surface area of can: $\quad A = \underbrace{2\pi r^2}_{\substack{\text{circular} \\ \text{ends}}} + \underbrace{2\pi rh}_{\substack{\text{cylindrical} \\ \text{wall}}}$

How can we interpret the phrase "least material"? For a first approximation we can ignore the thickness of the material and the waste in manufacturing. Then we ask for dimensions r and h that make the total surface area as small as possible while satisfying the constraint $\pi r^2 h = 1000$.

To express the surface area as a function of one variable, we solve for one of the variables in $\pi r^2 h = 1000$ and substitute that expression into the surface area formula. Solving for h is easier:

$$h = \frac{1000}{\pi r^2}.$$

Thus,

$$A = 2\pi r^2 + 2\pi rh$$

$$= 2\pi r^2 + 2\pi r\left(\frac{1000}{\pi r^2}\right)$$

$$= 2\pi r^2 + \frac{2000}{r}.$$

Our goal is to find a value of $r > 0$ that minimizes the value of A. Figure 4.38 suggests that such a value exists.

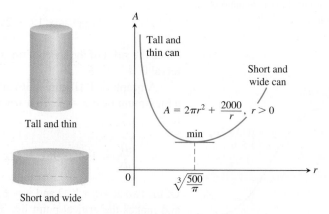

FIGURE 4.38 The graph of $A = 2\pi r^2 + 2000/r$ is concave up.

Notice from the graph that for small r (a tall, thin cylindrical container), the term $2000/r$ dominates (see Section 2.6) and A is large. For large r (a short, wide cylindrical container), the term $2\pi r^2$ dominates and A again is large.

Since A is differentiable on $r > 0$, an interval with no endpoints, it can have a minimum value only where its first derivative is zero.

$$\frac{dA}{dr} = 4\pi r - \frac{2000}{r^2}$$

$$0 = 4\pi r - \frac{2000}{r^2} \qquad \text{Set } dA/dr = 0.$$

$$4\pi r^3 = 2000 \qquad \text{Multiply by } r^2.$$

$$r = \sqrt[3]{\frac{500}{\pi}} \approx 5.42 \qquad \text{Solve for } r.$$

What happens at $r = \sqrt[3]{500/\pi}$?

The second derivative

$$\frac{d^2A}{dr^2} = 4\pi + \frac{4000}{r^3}$$

is positive throughout the domain of A. The graph is therefore everywhere concave up and the value of A at $r = \sqrt[3]{500/\pi}$ is an absolute minimum.

The corresponding value of h (after a little algebra) is

$$h = \frac{1000}{\pi r^2} = 2\sqrt[3]{\frac{500}{\pi}} = 2r.$$

The one-liter can that uses the least material has height equal to twice the radius, here with $r \approx 5.42$ cm and $h \approx 10.84$ cm. ∎

Examples from Mathematics and Physics

EXAMPLE 3 A rectangle is to be inscribed in a semicircle of radius 2. What is the largest area the rectangle can have, and what are its dimensions?

Solution Let $(x, \sqrt{4 - x^2})$ be the coordinates of the corner of the rectangle obtained by placing the circle and rectangle in the coordinate plane (Figure 4.39). The length, height, and area of the rectangle can then be expressed in terms of the position x of the lower right-hand corner:

$$\text{Length: } 2x, \qquad \text{Height: } \sqrt{4 - x^2}, \qquad \text{Area: } 2x\sqrt{4 - x^2}.$$

Notice that the values of x are to be found in the interval $0 \le x \le 2$, where the selected corner of the rectangle lies.

Our goal is to find the absolute maximum value of the function

$$A(x) = 2x\sqrt{4 - x^2}$$

on the domain $[0, 2]$.

The derivative

$$\frac{dA}{dx} = \frac{-2x^2}{\sqrt{4 - x^2}} + 2\sqrt{4 - x^2}$$

is not defined when $x = 2$ and is equal to zero when

$$\frac{-2x^2}{\sqrt{4 - x^2}} + 2\sqrt{4 - x^2} = 0$$

$$-2x^2 + 2(4 - x^2) = 0$$

$$8 - 4x^2 = 0$$

$$x^2 = 2 \text{ or } x = \pm\sqrt{2}.$$

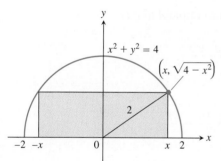

FIGURE 4.39 The rectangle inscribed in the semicircle in Example 3.

Of the two zeros, $x = \sqrt{2}$ and $x = -\sqrt{2}$, only $x = \sqrt{2}$ lies in the interior of A's domain and makes the critical-point list. The values of A at the endpoints and at this one critical point are

Critical-point value: $A(\sqrt{2}) = 2\sqrt{2}\sqrt{4-2} = 4$

Endpoint values: $A(0) = 0, \qquad A(2) = 0.$

The area has a maximum value of 4 when the rectangle is $\sqrt{4 - x^2} = \sqrt{2}$ units high and $2x = 2\sqrt{2}$ units long.   ■

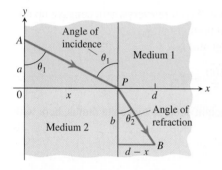

FIGURE 4.40 A light ray refracted (deflected from its path) as it passes from one medium to a denser medium (Example 4).

EXAMPLE 4   The speed of light depends on the medium through which it travels, and is generally slower in denser media.

   Fermat's principle in optics states that light travels from one point to another along a path for which the time of travel is a minimum. Describe the path that a ray of light will follow in going from a point A in a medium where the speed of light is c_1 to a point B in a second medium where its speed is c_2.

Solution   Since light traveling from A to B follows the quickest route, we look for a path that will minimize the travel time. We assume that A and B lie in the xy-plane and that the line separating the two media is the x-axis (Figure 4.40).

   In a uniform medium, where the speed of light remains constant, "shortest time" means "shortest path," and the ray of light will follow a straight line. Thus the path from A to B will consist of a line segment from A to a boundary point P, followed by another line segment from P to B. Distance traveled equals rate times time, so

$$\text{Time} = \frac{\text{distance}}{\text{rate}}.$$

From Figure 4.40, the time required for light to travel from A to P is

$$t_1 = \frac{AP}{c_1} = \frac{\sqrt{a^2 + x^2}}{c_1}.$$

From P to B, the time is

$$t_2 = \frac{PB}{c_2} = \frac{\sqrt{b^2 + (d-x)^2}}{c_2}.$$

The time from A to B is the sum of these:

$$t = t_1 + t_2 = \frac{\sqrt{a^2 + x^2}}{c_1} + \frac{\sqrt{b^2 + (d-x)^2}}{c_2}.$$

This equation expresses t as a differentiable function of x whose domain is $[0, d]$. We want to find the absolute minimum value of t on this closed interval. We find the derivative

$$\frac{dt}{dx} = \frac{x}{c_1\sqrt{a^2 + x^2}} - \frac{d-x}{c_2\sqrt{b^2 + (d-x)^2}}$$

and observe that it is continuous. In terms of the angles θ_1 and θ_2 in Figure 4.40,

$$\frac{dt}{dx} = \frac{\sin \theta_1}{c_1} - \frac{\sin \theta_2}{c_2}.$$

FIGURE 4.41   The sign pattern of dt/dx in Example 4.

The function t has a negative derivative at $x = 0$ and a positive derivative at $x = d$. Since dt/dx is continuous over the interval $[0, d]$, by the Intermediate Value Theorem for continuous functions (Section 2.5), there is a point $x_0 \in [0, d]$ where $dt/dx = 0$ (Figure 4.41).

There is only one such point because dt/dx is an increasing function of x (Exercise 62). At this unique point we then have

$$\frac{\sin \theta_1}{c_1} = \frac{\sin \theta_2}{c_2}.$$

This equation is **Snell's Law** or the **Law of Refraction**, and is an important principle in the theory of optics. It describes the path the ray of light follows. ∎

Examples from Economics

Suppose that

$r(x) = $ the revenue from selling x items

$c(x) = $ the cost of producing the x items

$p(x) = r(x) - c(x) = $ the profit from producing and selling x items.

Although x is usually an integer in many applications, we can learn about the behavior of these functions by defining them for all nonzero real numbers and by assuming they are differentiable functions. Economists use the terms **marginal revenue**, **marginal cost**, and **marginal profit** to name the derivatives $r'(x)$, $c'(x)$, and $p'(x)$ of the revenue, cost, and profit functions. Let's consider the relationship of the profit p to these derivatives.

If $r(x)$ and $c(x)$ are differentiable for x in some interval of production possibilities, and if $p(x) = r(x) - c(x)$ has a maximum value there, it occurs at a critical point of $p(x)$ or at an endpoint of the interval. If it occurs at a critical point, then $p'(x) = r'(x) - c'(x) = 0$ and we see that $r'(x) = c'(x)$. In economic terms, this last equation means that

> At a production level yielding maximum profit, marginal revenue equals marginal cost (Figure 4.42).

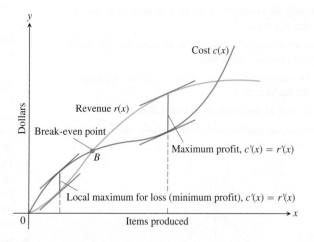

FIGURE 4.42 The graph of a typical cost function starts concave down and later turns concave up. It crosses the revenue curve at the break-even point B. To the left of B, the company operates at a loss. To the right, the company operates at a profit, with the maximum profit occurring where $c'(x) = r'(x)$. Farther to the right, cost exceeds revenue (perhaps because of a combination of rising labor and material costs and market saturation) and production levels become unprofitable again.

EXAMPLE 5   Suppose that $r(x) = 9x$ and $c(x) = x^3 - 6x^2 + 15x$, where x represents millions of MP3 players produced. Is there a production level that maximizes profit? If so, what is it?

Solution   Notice that $r'(x) = 9$ and $c'(x) = 3x^2 - 12x + 15$.

$$3x^2 - 12x + 15 = 9 \qquad \text{Set } c'(x) = r'(x).$$
$$3x^2 - 12x + 6 = 0$$

The two solutions of the quadratic equation are

$$x_1 = \frac{12 - \sqrt{72}}{6} = 2 - \sqrt{2} \approx 0.586 \qquad \text{and}$$

$$x_2 = \frac{12 + \sqrt{72}}{6} = 2 + \sqrt{2} \approx 3.414.$$

The possible production levels for maximum profit are $x \approx 0.586$ million MP3 players or $x \approx 3.414$ million. The second derivative of $p(x) = r(x) - c(x)$ is $p''(x) = -c''(x)$ since $r''(x)$ is everywhere zero. Thus, $p''(x) = 6(2 - x)$, which is negative at $x = 2 + \sqrt{2}$ and positive at $x = 2 - \sqrt{2}$. By the Second Derivative Test, a maximum profit occurs at about $x = 3.414$ (where revenue exceeds costs) and maximum loss occurs at about $x = 0.586$. The graphs of $r(x)$ and $c(x)$ are shown in Figure 4.43.   ■

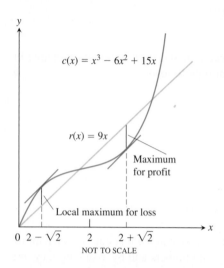

FIGURE 4.43   The cost and revenue curves for Example 5.

Exercises 4.6

Mathematical Applications

Whenever you are maximizing or minimizing a function of a single variable, we urge you to graph it over the domain that is appropriate to the problem you are solving. The graph will provide insight before you calculate and will furnish a visual context for understanding your answer.

1. **Minimizing perimeter**   What is the smallest perimeter possible for a rectangle whose area is 16 in^2, and what are its dimensions?

2. Show that among all rectangles with an 8-m perimeter, the one with largest area is a square.

3. The figure shows a rectangle inscribed in an isosceles right triangle whose hypotenuse is 2 units long.

   a. Express the y-coordinate of P in terms of x. (*Hint:* Write an equation for the line AB.)

   b. Express the area of the rectangle in terms of x.

   c. What is the largest area the rectangle can have, and what are its dimensions?

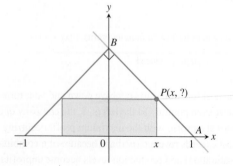

4. A rectangle has its base on the x-axis and its upper two vertices on the parabola $y = 12 - x^2$. What is the largest area the rectangle can have, and what are its dimensions?

5. You are planning to make an open rectangular box from an 8-in.-by-15-in. piece of cardboard by cutting congruent squares from the corners and folding up the sides. What are the dimensions of the box of largest volume you can make this way, and what is its volume?

6. You are planning to close off a corner of the first quadrant with a line segment 20 units long running from $(a, 0)$ to $(0, b)$. Show that the area of the triangle enclosed by the segment is largest when $a = b$.

7. **The best fencing plan**   A rectangular plot of farmland will be bounded on one side by a river and on the other three sides by a single-strand electric fence. With 800 m of wire at your disposal, what is the largest area you can enclose, and what are its dimensions?

8. **The shortest fence**   A 216 m^2 rectangular pea patch is to be enclosed by a fence and divided into two equal parts by another fence parallel to one of the sides. What dimensions for the outer rectangle will require the smallest total length of fence? How much fence will be needed?

9. **Designing a tank**   Your iron works has contracted to design and build a 500 ft^3, square-based, open-top, rectangular steel holding tank for a paper company. The tank is to be made by welding thin stainless steel plates together along their edges. As the production engineer, your job is to find dimensions for the base and height that will make the tank weigh as little as possible.

a. What dimensions do you tell the shop to use?

b. Briefly describe how you took weight into account.

10. **Catching rainwater** A 1125 ft³ open-top rectangular tank with a square base x ft on a side and y ft deep is to be built with its top flush with the ground to catch runoff water. The costs associated with the tank involve not only the material from which the tank is made but also an excavation charge proportional to the product xy.

a. If the total cost is

$$c = 5(x^2 + 4xy) + 10xy,$$

what values of x and y will minimize it?

b. Give a possible scenario for the cost function in part (a).

11. **Designing a poster** You are designing a rectangular poster to contain 50 in² of printing with a 4-in. margin at the top and bottom and a 2-in. margin at each side. What overall dimensions will minimize the amount of paper used?

12. Find the volume of the largest right circular cone that can be inscribed in a sphere of radius 3.

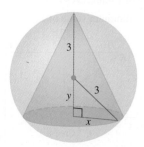

13. Two sides of a triangle have lengths a and b, and the angle between them is θ. What value of θ will maximize the triangle's area? (*Hint:* $A = (1/2)ab \sin \theta$.)

14. **Designing a can** What are the dimensions of the lightest open-top right circular cylindrical can that will hold a volume of 1000 cm³? Compare the result here with the result in Example 2.

15. **Designing a can** You are designing a 1000 cm³ right circular cylindrical can whose manufacture will take waste into account. There is no waste in cutting the aluminum for the side, but the top and bottom of radius r will be cut from squares that measure $2r$ units on a side. The total amount of aluminum used up by the can will therefore be

$$A = 8r^2 + 2\pi rh$$

rather than the $A = 2\pi r^2 + 2\pi rh$ in Example 2. In Example 2, the ratio of h to r for the most economical can was 2 to 1. What is the ratio now?

T 16. **Designing a box with a lid** A piece of cardboard measures 10 in. by 15 in. Two equal squares are removed from the corners of a 10-in. side as shown in the figure. Two equal rectangles are removed from the other corners so that the tabs can be folded to form a rectangular box with lid.

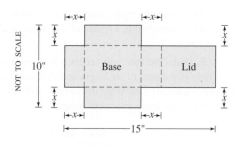

a. Write a formula $V(x)$ for the volume of the box.

b. Find the domain of V for the problem situation and graph V over this domain.

c. Use a graphical method to find the maximum volume and the value of x that gives it.

d. Confirm your result in part (c) analytically.

T 17. **Designing a suitcase** A 24-in.-by-36-in. sheet of cardboard is folded in half to form a 24-in.-by-18-in. rectangle as shown in the accompanying figure. Then four congruent squares of side length x are cut from the corners of the folded rectangle. The sheet is unfolded, and the six tabs are folded up to form a box with sides and a lid.

a. Write a formula $V(x)$ for the volume of the box.

b. Find the domain of V for the problem situation and graph V over this domain.

c. Use a graphical method to find the maximum volume and the value of x that gives it.

d. Confirm your result in part (c) analytically.

e. Find a value of x that yields a volume of 1120 in³.

f. Write a paragraph describing the issues that arise in part (b).

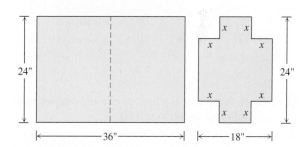

The sheet is then unfolded.

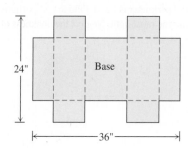

18. A rectangle is to be inscribed under the arch of the curve $y = 4 \cos (0.5x)$ from $x = -\pi$ to $x = \pi$. What are the dimensions of the rectangle with largest area, and what is the largest area?

19. Find the dimensions of a right circular cylinder of maximum volume that can be inscribed in a sphere of radius 10 cm. What is the maximum volume?

20. a. The U.S. Postal Service will accept a box for domestic shipment only if the sum of its length and girth (distance around) does not exceed 108 in. What dimensions will give a box with a square end the largest possible volume?

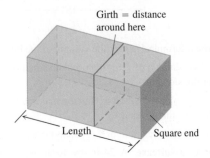

Girth = distance
around here

Length

Square end

T **b.** Graph the volume of a 108-in. box (length plus girth equals 108 in.) as a function of its length and compare what you see with your answer in part (a).

21. (*Continuation of Exercise 20.*)

a. Suppose that instead of having a box with square ends you have a box with square sides so that its dimensions are *h* by *h* by *w* and the girth is $2h + 2w$. What dimensions will give the box its largest volume now?

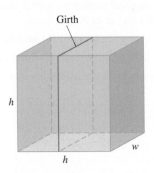

Girth

h

h

w

T **b.** Graph the volume as a function of *h* and compare what you see with your answer in part (a).

22. A window is in the form of a rectangle surmounted by a semicircle. The rectangle is of clear glass, whereas the semicircle is of tinted glass that transmits only half as much light per unit area as clear glass does. The total perimeter is fixed. Find the proportions of the window that will admit the most light. Neglect the thickness of the frame.

23. A silo (base not included) is to be constructed in the form of a cylinder surmounted by a hemisphere. The cost of construction per square unit of surface area is twice as great for the hemisphere as it is for the

cylindrical sidewall. Determine the dimensions to be used if the volume is fixed and the cost of construction is to be kept to a minimum. Neglect the thickness of the silo and waste in construction.

24. The trough in the figure is to be made to the dimensions shown. Only the angle θ can be varied. What value of θ will maximize the trough's volume?

25. Paper folding A rectangular sheet of 8.5-in.-by-11-in. paper is placed on a flat surface. One of the corners is placed on the opposite longer edge, as shown in the figure, and held there as the paper is smoothed flat. The problem is to make the length of the crease as small as possible. Call the length *L*. Try it with paper.

a. Show that $L^2 = 2x^3/(2x - 8.5)$.

b. What value of *x* minimizes L^2?

c. What is the minimum value of *L*?

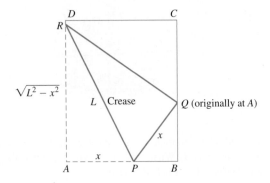

D   *C*

R

$\sqrt{L^2 - x^2}$

L \ Crease

Q (originally at *A*)

x

x

A   *P*   *B*

26. Constructing cylinders Compare the answers to the following two construction problems.

a. A rectangular sheet of perimeter 36 cm and dimensions *x* cm by *y* cm is to be rolled into a cylinder as shown in part (a) of the figure. What values of *x* and *y* give the largest volume?

b. The same sheet is to be revolved about one of the sides of length *y* to sweep out the cylinder as shown in part (b) of the figure. What values of *x* and *y* give the largest volume?

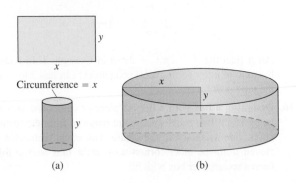

y

x

Circumference = *x*

x

y

y

(a)   (b)

27. Constructing cones A right triangle whose hypotenuse is $\sqrt{3}$ m long is revolved about one of its legs to generate a right circular cone. Find the radius, height, and volume of the cone of greatest volume that can be made this way.

28. Find the point on the line $\dfrac{x}{a} + \dfrac{y}{b} = 1$ that is closest to the origin.

29. Find a positive number for which the sum of it and its reciprocal is the smallest (least) possible.

30. Find a positive number for which the sum of its reciprocal and four times its square is the smallest possible.

31. A wire b m long is cut into two pieces. One piece is bent into an equilateral triangle and the other is bent into a circle. If the sum of the areas enclosed by each part is a minimum, what is the length of each part?

32. Answer Exercise 31 if one piece is bent into a square and the other into a circle.

33. Determine the dimensions of the rectangle of largest area that can be inscribed in the right triangle shown in the accompanying figure.

34. Determine the dimensions of the rectangle of largest area that can be inscribed in a semicircle of radius 3. (See accompanying figure.)

35. What value of a makes $f(x) = x^2 + (a/x)$ have

a. a local minimum at $x = 2$?

b. a point of inflection at $x = 1$?

36. What values of a and b make $f(x) = x^3 + ax^2 + bx$ have

a. a local maximum at $x = -1$ and a local minimum at $x = 3$?

b. a local minimum at $x = 4$ and a point of inflection at $x = 1$?

Physical Applications

37. Vertical motion The height above ground of an object moving vertically is given by

$$s = -16t^2 + 96t + 112,$$

with s in feet and t in seconds. Find

a. the object's velocity when $t = 0$;

b. its maximum height and when it occurs;

c. its velocity when $s = 0$.

38. Quickest route Jane is 2 mi offshore in a boat and wishes to reach a coastal village 6 mi down a straight shoreline from the point nearest the boat. She can row 2 mph and can walk 5 mph. Where should she land her boat to reach the village in the least amount of time?

39. Shortest beam The 8-ft wall shown here stands 27 ft from the building. Find the length of the shortest straight beam that will reach to the side of the building from the ground outside the wall.

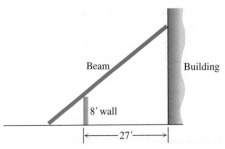

40. Motion on a line The positions of two particles on the s-axis are $s_1 = \sin t$ and $s_2 = \sin (t + \pi/3)$, with s_1 and s_2 in meters and t in seconds.

a. At what time(s) in the interval $0 \le t \le 2\pi$ do the particles meet?

b. What is the farthest apart that the particles ever get?

c. When in the interval $0 \le t \le 2\pi$ is the distance between the particles changing the fastest?

41. The intensity of illumination at any point from a light source is proportional to the square of the reciprocal of the distance between the point and the light source. Two lights, one having an intensity eight times that of the other, are 6 m apart. How far from the stronger light is the total illumination least?

42. Projectile motion The *range R* of a projectile fired from the origin over horizontal ground is the distance from the origin to the point of impact. If the projectile is fired with an initial velocity v_0 at an angle α with the horizontal, then in Chapter 13 we find that

$$R = \frac{v_0^2}{g} \sin 2\alpha,$$

where g is the downward acceleration due to gravity. Find the angle α for which the range R is the largest possible.

T **43. Strength of a beam** The strength S of a rectangular wooden beam is proportional to its width times the square of its depth. (See the accompanying figure.)

a. Find the dimensions of the strongest beam that can be cut from a 12-in.-diameter cylindrical log.

b. Graph S as a function of the beam's width w, assuming the proportionality constant to be $k = 1$. Reconcile what you see with your answer in part (a).

c. On the same screen, graph S as a function of the beam's depth d, again taking $k = 1$. Compare the graphs with one another and with your answer in part (a). What would be the effect of changing to some other value of k? Try it.

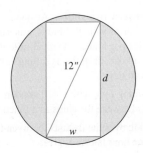

T **44. Stiffness of a beam** The stiffness S of a rectangular beam is proportional to its width times the cube of its depth.

   a. Find the dimensions of the stiffest beam that can be cut from a 12-in.-diameter cylindrical log.

   b. Graph S as a function of the beam's width w, assuming the proportionality constant to be $k = 1$. Reconcile what you see with your answer in part (a).

   c. On the same screen, graph S as a function of the beam's depth d, again taking $k = 1$. Compare the graphs with one another and with your answer in part (a). What would be the effect of changing to some other value of k? Try it.

45. Frictionless cart A small frictionless cart, attached to the wall by a spring, is pulled 10 cm from its rest position and released at time $t = 0$ to roll back and forth for 4 sec. Its position at time t is $s = 10 \cos \pi t$.

   a. What is the cart's maximum speed? When is the cart moving that fast? Where is it then? What is the magnitude of the acceleration then?

   b. Where is the cart when the magnitude of the acceleration is greatest? What is the cart's speed then?

46. Two masses hanging side by side from springs have positions $s_1 = 2 \sin t$ and $s_2 = \sin 2t$, respectively.

   a. At what times in the interval $0 < t$ do the masses pass each other? (*Hint:* $\sin 2t = 2 \sin t \cos t$.)

   b. When in the interval $0 \le t \le 2\pi$ is the vertical distance between the masses the greatest? What is this distance? (*Hint:* $\cos 2t = 2 \cos^2 t - 1$.)

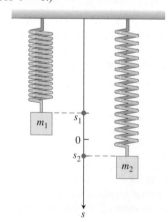

47. Distance between two ships At noon, ship A was 12 nautical miles due north of ship B. Ship A was sailing south at 12 knots (nautical miles per hour; a nautical mile is 2000 yd) and continued to do so all day. Ship B was sailing east at 8 knots and continued to do so all day.

   a. Start counting time with $t = 0$ at noon and express the distance s between the ships as a function of t.

   b. How rapidly was the distance between the ships changing at noon? One hour later?

   c. The visibility that day was 5 nautical miles. Did the ships ever sight each other?

T **d.** Graph s and ds/dt together as functions of t for $-1 \le t \le 3$, using different colors if possible. Compare the graphs and reconcile what you see with your answers in parts (b) and (c).

   e. The graph of ds/dt looks as if it might have a horizontal asymptote in the first quadrant. This in turn suggests that ds/dt approaches a limiting value as $t \to \infty$. What is this value? What is its relation to the ships' individual speeds?

48. Fermat's principle in optics Light from a source A is reflected by a plane mirror to a receiver at point B, as shown in the accompanying figure. Show that for the light to obey Fermat's principle, the angle of incidence must equal the angle of reflection, both measured from the line normal to the reflecting surface. (This result can also be derived without calculus. There is a purely geometric argument, which you may prefer.)

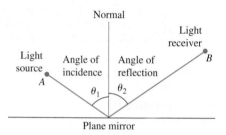

49. Tin pest When metallic tin is kept below 13.2°C, it slowly becomes brittle and crumbles to a gray powder. Tin objects eventually crumble to this gray powder spontaneously if kept in a cold climate for years. The Europeans who saw tin organ pipes in their churches crumble away years ago called the change *tin pest* because it seemed to be contagious, and indeed it was, for the gray powder is a catalyst for its own formation.

A *catalyst* for a chemical reaction is a substance that controls the rate of reaction without undergoing any permanent change in itself. An *autocatalytic reaction* is one whose product is a catalyst for its own formation. Such a reaction may proceed slowly at first if the amount of catalyst present is small and slowly again at the end, when most of the original substance is used up. But in between, when both the substance and its catalyst product are abundant, the reaction proceeds at a faster pace.

In some cases, it is reasonable to assume that the rate $v = dx/dt$ of the reaction is proportional both to the amount of the original substance present and to the amount of product. That is, v may be considered to be a function of x alone, and

$$v = kx(a - x) = kax - kx^2,$$

where

$x =$ the amount of product

$a =$ the amount of substance at the beginning

$k =$ a positive constant.

At what value of x does the rate v have a maximum? What is the maximum value of v?

50. Airplane landing path An airplane is flying at altitude H when it begins its descent to an airport runway that is at horizontal ground distance L from the airplane, as shown in the figure. Assume that the landing path of the airplane is the graph of a cubic polynomial function $y = ax^3 + bx^2 + cx + d$, where $y(-L) = H$ and $y(0) = 0$.

a. What is dy/dx at $x = 0$?

b. What is dy/dx at $x = -L$?

c. Use the values for dy/dx at $x = 0$ and $x = -L$ together with $y(0) = 0$ and $y(-L) = H$ to show that

$$y(x) = H\left[2\left(\frac{x}{L}\right)^3 + 3\left(\frac{x}{L}\right)^2\right].$$

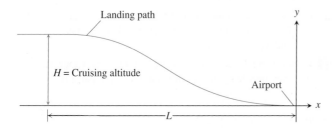

Landing path

H = Cruising altitude

Airport

L

Business and Economics

51. It costs you c dollars each to manufacture and distribute backpacks. If the backpacks sell at x dollars each, the number sold is given by

$$n = \frac{a}{x - c} + b(100 - x),$$

where a and b are positive constants. What selling price will bring a maximum profit?

52. You operate a tour service that offers the following rates:

$200 per person if 50 people (the minimum number to book the tour) go on the tour.

For each additional person, up to a maximum of 80 people total, the rate per person is reduced by $2.

It costs $6000 (a fixed cost) plus $32 per person to conduct the tour. How many people does it take to maximize your profit?

53. Wilson lot size formula One of the formulas for inventory management says that the average weekly cost of ordering, paying for, and holding merchandise is

$$A(q) = \frac{km}{q} + cm + \frac{hq}{2},$$

where q is the quantity you order when things run low (shoes, radios, brooms, or whatever the item might be), k is the cost of placing an order (the same, no matter how often you order), c is the cost of one item (a constant), m is the number of items sold each week (a constant), and h is the weekly holding cost per item (a constant that takes into account things such as space, utilities, insurance, and security).

a. Your job, as the inventory manager for your store, is to find the quantity that will minimize $A(q)$. What is it? (The formula you get for the answer is called the *Wilson lot size formula*.)

b. Shipping costs sometimes depend on order size. When they do, it is more realistic to replace k by $k + bq$, the sum of k and a constant multiple of q. What is the most economical quantity to order now?

54. Production level Prove that the production level (if any) at which average cost is smallest is a level at which the average cost equals marginal cost.

55. Show that if $r(x) = 6x$ and $c(x) = x^3 - 6x^2 + 15x$ are your revenue and cost functions, then the best you can do is break even (have revenue equal cost).

56. Production level Suppose that $c(x) = x^3 - 20x^2 + 20{,}000x$ is the cost of manufacturing x items. Find a production level that will minimize the average cost of making x items.

57. You are to construct an open rectangular box with a square base and a volume of 48 ft³. If material for the bottom costs $6/ft² and material for the sides costs $4/ft², what dimensions will result in the least expensive box? What is the minimum cost?

58. The 800-room Mega Motel chain is filled to capacity when the room charge is $50 per night. For each $10 increase in room charge, 40 fewer rooms are filled each night. What charge per room will result in the maximum revenue per night?

Biology

59. Sensitivity to medicine (*Continuation of Exercise 72, Section 3.3.*) Find the amount of medicine to which the body is most sensitive by finding the value of M that maximizes the derivative dR/dM, where

$$R = M^2\left(\frac{C}{2} - \frac{M}{3}\right)$$

and C is a constant.

60. How we cough

a. When we cough, the trachea (windpipe) contracts to increase the velocity of the air going out. This raises the questions of how much it should contract to maximize the velocity and whether it really contracts that much when we cough.

Under reasonable assumptions about the elasticity of the tracheal wall and about how the air near the wall is slowed by friction, the average flow velocity v can be modeled by the equation

$$v = c(r_0 - r)r^2 \text{ cm/sec}, \qquad \frac{r_0}{2} \le r \le r_0,$$

where r_0 is the rest radius of the trachea in centimeters and c is a positive constant whose value depends in part on the length of the trachea.

Show that v is greatest when $r = (2/3)r_0$; that is, when the trachea is about 33% contracted. The remarkable fact is that X-ray photographs confirm that the trachea contracts about this much during a cough.

T **b.** Take r_0 to be 0.5 and c to be 1 and graph v over the interval $0 \le r \le 0.5$. Compare what you see with the claim that v is at a maximum when $r = (2/3)r_0$.

Theory and Examples

61. An inequality for positive integers Show that if a, b, c, and d are positive integers, then

$$\frac{(a^2 + 1)(b^2 + 1)(c^2 + 1)(d^2 + 1)}{abcd} \ge 16.$$

62. The derivative dt/dx in Example 4

a. Show that

$$f(x) = \frac{x}{\sqrt{a^2 + x^2}}$$

is an increasing function of x.

b. Show that

$$g(x) = \frac{d - x}{\sqrt{b^2 + (d - x)^2}}$$

is a decreasing function of x.

c. Show that

$$\frac{dt}{dx} = \frac{x}{c_1\sqrt{a^2 + x^2}} - \frac{d - x}{c_2\sqrt{b^2 + (d - x)^2}}$$

is an increasing function of x.

63. Let $f(x)$ and $g(x)$ be the differentiable functions graphed here. Point c is the point where the vertical distance between the curves is the greatest. Is there anything special about the tangents to the two curves at c? Give reasons for your answer.

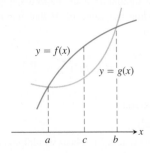

64. You have been asked to determine whether the function $f(x) = 3 + 4\cos x + \cos 2x$ is ever negative.

a. Explain why you need to consider values of x only in the interval $[0, 2\pi]$.

b. Is f ever negative? Explain.

65. a. The function $y = \cot x - \sqrt{2}\csc x$ has an absolute maximum value on the interval $0 < x < \pi$. Find it.

T **b.** Graph the function and compare what you see with your answer in part (a).

66. a. The function $y = \tan x + 3\cot x$ has an absolute minimum value on the interval $0 < x < \pi/2$. Find it.

T **b.** Graph the function and compare what you see with your answer in part (a).

67. a. How close does the curve $y = \sqrt{x}$ come to the point $(3/2, 0)$? (*Hint:* If you minimize the *square* of the distance, you can avoid square roots.)

T **b.** Graph the distance function $D(x)$ and $y = \sqrt{x}$ together and reconcile what you see with your answer in part (a).

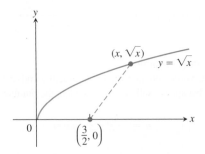

68. a. How close does the semicircle $y = \sqrt{16 - x^2}$ come to the point $\left(1, \sqrt{3}\right)$?

T **b.** Graph the distance function and $y = \sqrt{16 - x^2}$ together and reconcile what you see with your answer in part (a).

4.7 Newton's Method

In this section we study a numerical method, called *Newton's method* or the *Newton–Raphson method*, which is a technique to approximate the solution to an equation $f(x) = 0$. Essentially it uses tangent lines in place of the graph of $y = f(x)$ near the points where f is zero. (A value of x where f is zero is a *root* of the function f and a *solution* of the equation $f(x) = 0$.)

Procedure for Newton's Method

The goal of Newton's method for estimating a solution of an equation $f(x) = 0$ is to produce a sequence of approximations that approach the solution. We pick the first number x_0 of the sequence. Then, under favorable circumstances, the method does the rest by moving step by step toward a point where the graph of f crosses the x-axis (Figure 4.44). At each step the method approximates a zero of f with a zero of one of its linearizations. Here is how it works.

The initial estimate, x_0, may be found by graphing or just plain guessing. The method then uses the tangent to the curve $y = f(x)$ at $(x_0, f(x_0))$ to approximate the curve, calling

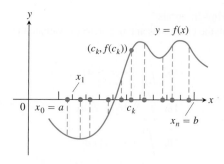

FIGURE 5.14 A sample of values of a function on an interval $[a, b]$.

We divide $[a, b]$ into n subintervals of equal width $\Delta x = (b - a)/n$ and evaluate f at a point c_k in each (Figure 5.14). The average of the n sampled values is

$$\frac{f(c_1) + f(c_2) + \cdots + f(c_n)}{n} = \frac{1}{n} \sum_{k=1}^{n} f(c_k)$$

$$= \frac{\Delta x}{b - a} \sum_{k=1}^{n} f(c_k) \qquad \Delta x = \frac{b - a}{n}, \text{ so } \frac{1}{n} = \frac{\Delta x}{b - a}$$

$$= \frac{1}{b - a} \sum_{k=1}^{n} f(c_k) \, \Delta x \qquad \text{Constant Multiple Rule}$$

The average is obtained by dividing a Riemann sum for f on $[a, b]$ by $(b - a)$. As we increase the size of the sample and let the norm of the partition approach zero, the average approaches $(1/(b - a))\int_a^b f(x) \, dx$. Both points of view lead us to the following definition.

DEFINITION If f is integrable on $[a, b]$, then its **average value on $[a, b]$**, also called its **mean**, is

$$\text{av}(f) = \frac{1}{b - a} \int_a^b f(x) \, dx.$$

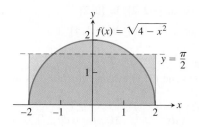

FIGURE 5.15 The average value of $f(x) = \sqrt{4 - x^2}$ on $[-2, 2]$ is $\pi/2$ (Example 5).

EXAMPLE 5 Find the average value of $f(x) = \sqrt{4 - x^2}$ on $[-2, 2]$.

Solution We recognize $f(x) = \sqrt{4 - x^2}$ as a function whose graph is the upper semicircle of radius 2 centered at the origin (Figure 5.15).

The area between the semicircle and the x-axis from -2 to 2 can be computed using the geometry formula

$$\text{Area} = \frac{1}{2} \cdot \pi r^2 = \frac{1}{2} \cdot \pi (2)^2 = 2\pi.$$

Because f is nonnegative, the area is also the value of the integral of f from -2 to 2,

$$\int_{-2}^{2} \sqrt{4 - x^2} \, dx = 2\pi.$$

Therefore, the average value of f is

$$\text{av}(f) = \frac{1}{2 - (-2)} \int_{-2}^{2} \sqrt{4 - x^2} \, dx = \frac{1}{4}(2\pi) = \frac{\pi}{2}.$$

Theorem 3 in the next section asserts that the area of the upper semicircle over $[-2, 2]$ is the same as the area of the rectangle whose height is the average value of f over $[-2, 2]$ (see Figure 5.15). ∎

Exercises 5.3

Interpreting Limits as Integrals
Express the limits in Exercises 1–8 as definite integrals.

1. $\lim\limits_{\|P\| \to 0} \sum\limits_{k=1}^{n} c_k^2 \, \Delta x_k$, where P is a partition of $[0, 2]$

2. $\lim\limits_{\|P\| \to 0} \sum\limits_{k=1}^{n} 2c_k^3 \, \Delta x_k$, where P is a partition of $[-1, 0]$

3. $\lim\limits_{\|P\| \to 0} \sum\limits_{k=1}^{n} (c_k^2 - 3c_k) \, \Delta x_k$, where P is a partition of $[-7, 5]$

4. $\lim\limits_{\|P\| \to 0} \sum\limits_{k=1}^{n} \left(\frac{1}{c_k}\right) \, \Delta x_k$, where P is a partition of $[1, 4]$

5. $\lim\limits_{\|P\| \to 0} \sum\limits_{k=1}^{n} \frac{1}{1 - c_k} \, \Delta x_k$, where P is a partition of $[2, 3]$

6. $\lim\limits_{\|P\|\to 0} \sum\limits_{k=1}^{n} \sqrt{4 - c_k^2}\, \Delta x_k$, where P is a partition of $[0, 1]$

7. $\lim\limits_{\|P\|\to 0} \sum\limits_{k=1}^{n} (\sec c_k)\, \Delta x_k$, where P is a partition of $[-\pi/4, 0]$

8. $\lim\limits_{\|P\|\to 0} \sum\limits_{k=1}^{n} (\tan c_k)\, \Delta x_k$, where P is a partition of $[0, \pi/4]$

Using the Definite Integral Rules

9. Suppose that f and g are integrable and that

$$\int_1^2 f(x)\, dx = -4, \quad \int_1^5 f(x)\, dx = 6, \quad \int_1^5 g(x)\, dx = 8.$$

Use the rules in Table 5.4 to find

a. $\displaystyle\int_2^2 g(x)\, dx$ **b.** $\displaystyle\int_5^1 g(x)\, dx$

c. $\displaystyle\int_1^2 3f(x)\, dx$ **d.** $\displaystyle\int_2^5 f(x)\, dx$

e. $\displaystyle\int_1^5 [f(x) - g(x)]\, dx$ **f.** $\displaystyle\int_1^5 [4f(x) - g(x)]\, dx$

10. Suppose that f and h are integrable and that

$$\int_1^9 f(x)\, dx = -1, \quad \int_7^9 f(x)\, dx = 5, \quad \int_7^9 h(x)\, dx = 4.$$

Use the rules in Table 5.4 to find

a. $\displaystyle\int_1^9 -2f(x)\, dx$ **b.** $\displaystyle\int_7^9 [f(x) + h(x)]\, dx$

c. $\displaystyle\int_7^9 [2f(x) - 3h(x)]\, dx$ **d.** $\displaystyle\int_9^1 f(x)\, dx$

e. $\displaystyle\int_1^7 f(x)\, dx$ **f.** $\displaystyle\int_9^7 [h(x) - f(x)]\, dx$

11. Suppose that $\int_1^2 f(x)\, dx = 5$. Find

a. $\displaystyle\int_1^2 f(u)\, du$ **b.** $\displaystyle\int_1^2 \sqrt{3}f(z)\, dz$

c. $\displaystyle\int_2^1 f(t)\, dt$ **d.** $\displaystyle\int_1^2 [-f(x)]\, dx$

12. Suppose that $\int_{-3}^0 g(t)\, dt = \sqrt{2}$. Find

a. $\displaystyle\int_0^{-3} g(t)\, dt$ **b.** $\displaystyle\int_{-3}^0 g(u)\, du$

c. $\displaystyle\int_{-3}^0 [-g(x)]\, dx$ **d.** $\displaystyle\int_{-3}^0 \frac{g(r)}{\sqrt{2}}\, dr$

13. Suppose that f is integrable and that $\int_0^3 f(z)\, dz = 3$ and $\int_0^4 f(z)\, dz = 7$. Find

a. $\displaystyle\int_3^4 f(z)\, dz$ **b.** $\displaystyle\int_4^3 f(t)\, dt$

14. Suppose that h is integrable and that $\int_{-1}^1 h(r)\, dr = 0$ and $\int_{-1}^3 h(r)\, dr = 6$. Find

a. $\displaystyle\int_1^3 h(r)\, dr$ **b.** $\displaystyle -\int_3^1 h(u)\, du$

Using Known Areas to Find Integrals

In Exercises 15–22, graph the integrands and use areas to evaluate the integrals.

15. $\displaystyle\int_{-2}^4 \left(\frac{x}{2} + 3\right) dx$ **16.** $\displaystyle\int_{1/2}^{3/2} (-2x + 4)\, dx$

17. $\displaystyle\int_{-3}^3 \sqrt{9 - x^2}\, dx$ **18.** $\displaystyle\int_{-4}^0 \sqrt{16 - x^2}\, dx$

19. $\displaystyle\int_{-2}^1 |x|\, dx$ **20.** $\displaystyle\int_{-1}^1 (1 - |x|)\, dx$

21. $\displaystyle\int_{-1}^1 (2 - |x|)\, dx$ **22.** $\displaystyle\int_{-1}^1 \left(1 + \sqrt{1 - x^2}\right) dx$

Use areas to evaluate the integrals in Exercises 23–28.

23. $\displaystyle\int_0^b \frac{x}{2}\, dx, \quad b > 0$ **24.** $\displaystyle\int_0^b 4x\, dx, \quad b > 0$

25. $\displaystyle\int_a^b 2s\, ds, \quad 0 < a < b$ **26.** $\displaystyle\int_a^b 3t\, dt, \quad 0 < a < b$

27. $f(x) = \sqrt{4 - x^2}$ on **a.** $[-2, 2]$, **b.** $[0, 2]$

28. $f(x) = 3x + \sqrt{1 - x^2}$ on **a.** $[-1, 0]$, **b.** $[-1, 1]$

Evaluating Definite Integrals

Use the results of Equations (1) and (3) to evaluate the integrals in Exercises 29–40.

29. $\displaystyle\int_1^{\sqrt{2}} x\, dx$ **30.** $\displaystyle\int_{0.5}^{2.5} x\, dx$ **31.** $\displaystyle\int_{\pi}^{2\pi} \theta\, d\theta$

32. $\displaystyle\int_{\sqrt{2}}^{5\sqrt{2}} r\, dr$ **33.** $\displaystyle\int_0^{\sqrt[3]{7}} x^2\, dx$ **34.** $\displaystyle\int_0^{0.3} s^2\, ds$

35. $\displaystyle\int_0^{1/2} t^2\, dt$ **36.** $\displaystyle\int_0^{\pi/2} \theta^2\, d\theta$ **37.** $\displaystyle\int_a^{2a} x\, dx$

38. $\displaystyle\int_a^{\sqrt{3a}} x\, dx$ **39.** $\displaystyle\int_0^{\sqrt[3]{b}} x^2\, dx$ **40.** $\displaystyle\int_0^{3b} x^2\, dx$

Use the rules in Table 5.4 and Equations (1)–(3) to evaluate the integrals in Exercises 41–50.

41. $\displaystyle\int_3^1 7\, dx$ **42.** $\displaystyle\int_0^2 5x\, dx$

43. $\displaystyle\int_0^2 (2t - 3)\, dt$ **44.** $\displaystyle\int_0^{\sqrt{2}} \left(t - \sqrt{2}\right) dt$

45. $\displaystyle\int_2^1 \left(1 + \frac{z}{2}\right) dz$ **46.** $\displaystyle\int_3^0 (2z - 3)\, dz$

47. $\displaystyle\int_1^2 3u^2\, du$ **48.** $\displaystyle\int_{1/2}^1 24u^2\, du$

49. $\displaystyle\int_0^2 (3x^2 + x - 5)\, dx$ **50.** $\displaystyle\int_1^0 (3x^2 + x - 5)\, dx$

Finding Area by Definite Integrals

In Exercises 51–54, use a definite integral to find the area of the region between the given curve and the x-axis on the interval $[0, b]$.

51. $y = 3x^2$ **52.** $y = \pi x^2$

53. $y = 2x$ **54.** $y = \dfrac{x}{2} + 1$

> **THEOREM 4—The Fundamental Theorem of Calculus, Part 1**   If f is continuous on $[a, b]$, then $F(x) = \int_a^x f(t)\, dt$ is continuous on $[a, b]$ and differentiable on (a, b) and its derivative is $f(x)$:
>
> $$F'(x) = \frac{d}{dx}\int_a^x f(t)\, dt = f(x). \tag{2}$$

Before proving Theorem 4, we look at several examples to gain a better understanding of what it says. In each example, notice that the independent variable appears in a limit of integration, possibly in a formula.

EXAMPLE 2   Use the Fundamental Theorem to find dy/dx if

(a) $y = \displaystyle\int_a^x (t^3 + 1)\, dt$ **(b)** $y = \displaystyle\int_x^5 3t \sin t\, dt$

(c) $y = \displaystyle\int_1^{x^2} \cos t\, dt$ **(d)** $y = \displaystyle\int_{1+3x^2}^4 \frac{1}{2 + e^t}\, dt$

Solution   We calculate the derivatives with respect to the independent variable x.

(a) $\dfrac{dy}{dx} = \dfrac{d}{dx}\displaystyle\int_a^x (t^3 + 1)\, dt = x^3 + 1$ ⟶ Eq. (2) with $f(t) = t^3 + 1$

(b) $\dfrac{dy}{dx} = \dfrac{d}{dx}\displaystyle\int_x^5 3t \sin t\, dt = \dfrac{d}{dx}\left(-\int_5^x 3t \sin t\, dt\right)$ ⟶ Table 5.4, Rule 1

$\qquad = -\dfrac{d}{dx}\displaystyle\int_5^x 3t \sin t\, dt$

$\qquad = -3x \sin x$ ⟶ Eq. (2) with $f(t) = 3t \sin t$

(c) The upper limit of integration is not x but x^2. This makes y a composite of the two functions,

$$y = \int_1^u \cos t\, dt \quad \text{and} \quad u = x^2.$$

We must therefore apply the Chain Rule when finding dy/dx.

$$\frac{dy}{dx} = \frac{dy}{du} \cdot \frac{du}{dx}$$

$$= \left(\frac{d}{du}\int_1^u \cos t\, dt\right) \cdot \frac{du}{dx}$$

$$= \cos u \cdot \frac{du}{dx}$$

$$= \cos(x^2) \cdot 2x$$

$$= 2x \cos x^2$$

(d) $\dfrac{d}{dx}\displaystyle\int_{1+3x^2}^4 \frac{1}{2 + e^t}\, dt = \dfrac{d}{dx}\left(-\int_4^{1+3x^2} \frac{1}{2 + e^t}\, dt\right)$ ⟶ Rule 1

$\qquad = -\dfrac{d}{dx}\displaystyle\int_4^{1+3x^2} \frac{1}{2 + e^t}\, dt$

$\qquad = -\dfrac{1}{2 + e^{(1+3x^2)}} \dfrac{d}{dx}(1 + 3x^2)$ ⟶ Eq. (2) and the Chain Rule

$\qquad = -\dfrac{6x}{2 + e^{(1+3x^2)}}$

∎

Proof of Theorem 4 We prove the Fundamental Theorem, Part 1, by applying the definition of the derivative directly to the function $F(x)$, when x and $x + h$ are in (a, b). This means writing out the difference quotient

$$\frac{F(x + h) - F(x)}{h} \tag{3}$$

and showing that its limit as $h \to 0$ is the number $f(x)$ for each x in (a, b). Thus,

$$F'(x) = \lim_{h \to 0} \frac{F(x + h) - F(x)}{h}$$

$$= \lim_{h \to 0} \frac{1}{h} \left[\int_a^{x+h} f(t)\, dt - \int_a^x f(t)\, dt \right]$$

$$= \lim_{h \to 0} \frac{1}{h} \int_x^{x+h} f(t)\, dt \qquad \text{Table 5.4, Rule 5}$$

According to the Mean Value Theorem for Definite Integrals, the value before taking the limit in the last expression is one of the values taken on by f in the interval between x and $x + h$. That is, for some number c in this interval,

$$\frac{1}{h} \int_x^{x+h} f(t)\, dt = f(c). \tag{4}$$

As $h \to 0$, $x + h$ approaches x, forcing c to approach x also (because c is trapped between x and $x + h$). Since f is continuous at x, $f(c)$ approaches $f(x)$:

$$\lim_{h \to 0} f(c) = f(x). \tag{5}$$

In conclusion, we have

$$F'(x) = \lim_{h \to 0} \frac{1}{h} \int_x^{x+h} f(t)\, dt$$

$$= \lim_{h \to 0} f(c) \qquad \text{Eq. (4)}$$

$$= f(x). \qquad \text{Eq. (5)}$$

If $x = a$ or b, then the limit of Equation (3) is interpreted as a one-sided limit with $h \to 0^+$ or $h \to 0^-$, respectively. Then Theorem 1 in Section 3.2 shows that F is continuous for every point in $[a, b]$. This concludes the proof. ∎

Fundamental Theorem, Part 2 (The Evaluation Theorem)

We now come to the second part of the Fundamental Theorem of Calculus. This part describes how to evaluate definite integrals without having to calculate limits of Riemann sums. Instead we find and evaluate an antiderivative at the upper and lower limits of integration.

THEOREM 4 (Continued)—The Fundamental Theorem of Calculus, Part 2 If f is continuous at every point in $[a, b]$ and F is any antiderivative of f on $[a, b]$, then

$$\int_a^b f(x)\, dx = F(b) - F(a).$$

Proof Part 1 of the Fundamental Theorem tells us that an antiderivative of f exists, namely

$$G(x) = \int_a^x f(t)\, dt.$$

Thus, if F is *any* antiderivative of f, then $F(x) = G(x) + C$ for some constant C for $a < x < b$ (by Corollary 2 of the Mean Value Theorem for Derivatives, Section 4.2).

Since both F and G are continuous on $[a, b]$, we see that $F(x) = G(x) + C$ also holds when $x = a$ and $x = b$ by taking one-sided limits (as $x \to a^+$ and $x \to b^-$).

Evaluating $F(b) - F(a)$, we have

$$F(b) - F(a) = [G(b) + C] - [G(a) + C]$$
$$= G(b) - G(a)$$
$$= \int_a^b f(t)\, dt - \int_a^a f(t)\, dt$$
$$= \int_a^b f(t)\, dt - 0$$
$$= \int_a^b f(t)\, dt. \qquad \blacksquare$$

The Evaluation Theorem is important because it says that to calculate the definite integral of f over an interval $[a, b]$ we need do only two things:

1. Find an antiderivative F of f, and

2. Calculate the number $F(b) - F(a)$, which is equal to $\int_a^b f(x)\, dx$.

This process is much easier than using a Riemann sum computation. The power of the theorem follows from the realization that the definite integral, which is defined by a complicated process involving all of the values of the function f over $[a, b]$, can be found by knowing the values of *any* antiderivative F at only the two endpoints a and b. The usual notation for the difference $F(b) - F(a)$ is

$$F(x)\Big]_a^b \qquad \text{or} \qquad \Big[F(x)\Big]_a^b,$$

depending on whether F has one or more terms.

EXAMPLE 3   We calculate several definite integrals using the Evaluation Theorem, rather than by taking limits of Riemann sums.

(a) $\displaystyle\int_0^\pi \cos x\, dx = \sin x\Big]_0^\pi$ $\qquad\qquad\qquad\qquad \dfrac{d}{dx}\sin x = \cos x$

$\qquad\qquad = \sin \pi - \sin 0 = 0 - 0 = 0$

(b) $\displaystyle\int_{-\pi/4}^0 \sec x \tan x\, dx = \sec x\Big]_{-\pi/4}^0$ $\qquad\qquad \dfrac{d}{dx}\sec x = \sec x \tan x$

$\qquad\qquad = \sec 0 - \sec\left(-\dfrac{\pi}{4}\right) = 1 - \sqrt{2}$

(c) $\displaystyle\int_1^4 \left(\dfrac{3}{2}\sqrt{x} - \dfrac{4}{x^2}\right) dx = \left[x^{3/2} + \dfrac{4}{x}\right]_1^4$ $\qquad \dfrac{d}{dx}\left(x^{3/2} + \dfrac{4}{x}\right) = \dfrac{3}{2}x^{1/2} - \dfrac{4}{x^2}$

$\qquad\qquad = \left[(4)^{3/2} + \dfrac{4}{4}\right] - \left[(1)^{3/2} + \dfrac{4}{1}\right]$

$\qquad\qquad = [8 + 1] - [5] = 4$

(d) $\displaystyle\int_0^1 \dfrac{dx}{x+1} = \ln|x + 1|\Big]_0^1$ $\qquad\qquad\qquad \dfrac{d}{dx}\ln|x+1| = \dfrac{1}{x+1}$

$\qquad\qquad = \ln 2 - \ln 1 = \ln 2$

(e) $\displaystyle\int_0^1 \dfrac{dx}{x^2 + 1} = \tan^{-1} x\Big]_0^1$ $\qquad\qquad\qquad \dfrac{d}{dx}\tan^{-1} x = \dfrac{1}{x^2+1}$

$\qquad\qquad = \tan^{-1} 1 - \tan^{-1} 0 = \dfrac{\pi}{4} - 0 = \dfrac{\pi}{4}.$ $\qquad\qquad\qquad\qquad\qquad\qquad\blacksquare$

Exercise 82 offers another proof of the Evaluation Theorem, bringing together the ideas of Riemann sums, the Mean Value Theorem, and the definition of the definite integral.

The Integral of a Rate

We can interpret Part 2 of the Fundamental Theorem in another way. If F is any antiderivative of f, then $F' = f$. The equation in the theorem can then be rewritten as

$$\int_a^b F'(x)\,dx = F(b) - F(a).$$

Now $F'(x)$ represents the rate of change of the function $F(x)$ with respect to x, so the integral of F' is just the *net change* in F as x changes from a to b. Formally, we have the following result.

THEOREM 5—The Net Change Theorem   The net change in a function $F(x)$ over an interval $a \leq x \leq b$ is the integral of its rate of change:

$$F(b) - F(a) = \int_a^b F'(x)\,dx. \tag{6}$$

EXAMPLE 4   Here are several interpretations of the Net Change Theorem.

(a) If $c(x)$ is the cost of producing x units of a certain commodity, then $c'(x)$ is the marginal cost (Section 3.4). From Theorem 5,

$$\int_{x_1}^{x_2} c'(x)\,dx = c(x_2) - c(x_1),$$

which is the cost of increasing production from x_1 units to x_2 units.

(b) If an object with position function $s(t)$ moves along a coordinate line, its velocity is $v(t) = s'(t)$. Theorem 5 says that

$$\int_{t_1}^{t_2} v(t)\,dt = s(t_2) - s(t_1),$$

so the integral of velocity is the **displacement** over the time interval $t_1 \leq t \leq t_2$. On the other hand, the integral of the speed $|v(t)|$ is the **total distance traveled** over the time interval. This is consistent with our discussion in Section 5.1.   ∎

If we rearrange Equation (6) as

$$F(b) = F(a) + \int_a^b F'(x)\,dx,$$

we see that the Net Change Theorem also says that the final value of a function $F(x)$ over an interval $[a, b]$ equals its initial value $F(a)$ plus its net change over the interval. So if $v(t)$ represents the velocity function of an object moving along a coordinate line, this means that the object's final position $s(t_2)$ over a time interval $t_1 \leq t \leq t_2$ is its initial position $s(t_1)$ plus its net change in position along the line (see Example 4b).

EXAMPLE 5   Consider again our analysis of a heavy rock blown straight up from the ground by a dynamite blast (Example 3, Section 5.1). The velocity of the rock at any time t during its motion was given as $v(t) = 160 - 32t$ ft/sec.

(a) Find the displacement of the rock during the time period $0 \leq t \leq 8$.

(b) Find the total distance traveled during this time period.

Solution

(a) From Example 4b, the displacement is the integral

$$\int_0^8 v(t)\, dt = \int_0^8 (160 - 32t)\, dt = \left[160t - 16t^2\right]_0^8$$
$$= (160)(8) - (16)(64) = 256.$$

This means that the height of the rock is 256 ft above the ground 8 sec after the explosion, which agrees with our conclusion in Example 3, Section 5.1.

(b) As we noted in Table 5.3, the velocity function $v(t)$ is positive over the time interval [0, 5] and negative over the interval [5, 8]. Therefore, from Example 4b, the total distance traveled is the integral

$$\int_0^8 |v(t)|\, dt = \int_0^5 |v(t)|\, dt + \int_5^8 |v(t)|\, dt$$
$$= \int_0^5 (160 - 32t)\, dt - \int_5^8 (160 - 32t)\, dt$$
$$= \left[160t - 16t^2\right]_0^5 - \left[160t - 16t^2\right]_5^8$$
$$= [(160)(5) - (16)(25)] - [(160)(8) - (16)(64) - ((160)(5) - (16)(25))]$$
$$= 400 - (-144) = 544.$$

Again, this calculation agrees with our conclusion in Example 3, Section 5.1. That is, the total distance of 544 ft traveled by the rock during the time period $0 \le t \le 8$ is (i) the maximum height of 400 ft it reached over the time interval [0, 5] plus (ii) the additional distance of 144 ft the rock fell over the time interval [5, 8]. ∎

The Relationship between Integration and Differentiation

The conclusions of the Fundamental Theorem tell us several things. Equation (2) can be rewritten as

$$\frac{d}{dx}\int_a^x f(t)\, dt = f(x),$$

which says that if you first integrate the function f and then differentiate the result, you get the function f back again. Likewise, replacing b by x and x by t in Equation (6) gives

$$\int_a^x F'(t)\, dt = F(x) - F(a),$$

so that if you first differentiate the function F and then integrate the result, you get the function F back (adjusted by an integration constant). In a sense, the processes of integration and differentiation are "inverses" of each other. The Fundamental Theorem also says that every continuous function f has an antiderivative F. It shows the importance of finding antiderivatives in order to evaluate definite integrals easily. Furthermore, it says that the differential equation $dy/dx = f(x)$ has a solution (namely, any of the functions $y = F(x) + C$) for every continuous function f.

Total Area

The Riemann sum contains terms such as $f(c_k)\, \Delta x_k$ that give the area of a rectangle when $f(c_k)$ is positive. When $f(c_k)$ is negative, then the product $f(c_k)\, \Delta x_k$ is the negative of the rectangle's area. When we add up such terms for a negative function we get the negative of the area between the curve and the x-axis. If we then take the absolute value, we obtain the correct positive area.

EXAMPLE 6 Figure 5.20 shows the graph of $f(x) = x^2 - 4$ and its mirror image $g(x) = 4 - x^2$ reflected across the x-axis. For each function, compute

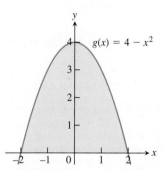

FIGURE 5.20 These graphs enclose the same amount of area with the x-axis, but the definite integrals of the two functions over [−2, 2] differ in sign (Example 6).

(a) the definite integral over the interval $[-2, 2]$, and

(b) the area between the graph and the x-axis over $[-2, 2]$.

Solution

(a) $\displaystyle\int_{-2}^{2} f(x)\, dx = \left[\frac{x^3}{3} - 4x\right]_{-2}^{2} = \left(\frac{8}{3} - 8\right) - \left(-\frac{8}{3} + 8\right) = -\frac{32}{3},$

and

$$\int_{-2}^{2} g(x)\, dx = \left[4x - \frac{x^3}{3}\right]_{-2}^{2} = \frac{32}{3}.$$

(b) In both cases, the area between the curve and the x-axis over $[-2, 2]$ is 32/3 units. Although the definite integral of $f(x)$ is negative, the area is still positive. ∎

To compute the area of the region bounded by the graph of a function $y = f(x)$ and the x-axis when the function takes on both positive and negative values, we must be careful to break up the interval $[a, b]$ into subintervals on which the function doesn't change sign. Otherwise we might get cancellation between positive and negative signed areas, leading to an incorrect total. The correct total area is obtained by adding the absolute value of the definite integral over each subinterval where $f(x)$ does not change sign. The term "area" will be taken to mean this *total area*.

EXAMPLE 7 Figure 5.21 shows the graph of the function $f(x) = \sin x$ between $x = 0$ and $x = 2\pi$. Compute

(a) the definite integral of $f(x)$ over $[0, 2\pi]$.

(b) the area between the graph of $f(x)$ and the x-axis over $[0, 2\pi]$.

Solution The definite integral for $f(x) = \sin x$ is given by

$$\int_{0}^{2\pi} \sin x\, dx = -\cos x\Big]_{0}^{2\pi} = -[\cos 2\pi - \cos 0] = -[1 - 1] = 0.$$

The definite integral is zero because the portions of the graph above and below the x-axis make canceling contributions.

The area between the graph of $f(x)$ and the x-axis over $[0, 2\pi]$ is calculated by breaking up the domain of $\sin x$ into two pieces: the interval $[0, \pi]$ over which it is nonnegative and the interval $[\pi, 2\pi]$ over which it is nonpositive.

$$\int_{0}^{\pi} \sin x\, dx = -\cos x\Big]_{0}^{\pi} = -[\cos \pi - \cos 0] = -[-1 - 1] = 2$$

$$\int_{\pi}^{2\pi} \sin x\, dx = -\cos x\Big]_{\pi}^{2\pi} = -[\cos 2\pi - \cos \pi] = -[1 - (-1)] = -2$$

The second integral gives a negative value. The area between the graph and the axis is obtained by adding the absolute values

$$\text{Area} = |2| + |-2| = 4. \qquad ∎$$

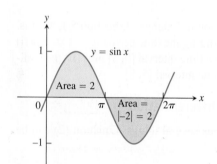

FIGURE 5.21 The total area between $y = \sin x$ and the x-axis for $0 \le x \le 2\pi$ is the sum of the absolute values of two integrals (Example 7).

Summary:

To find the area between the graph of $y = f(x)$ and the x-axis over the interval $[a, b]$:

1. Subdivide $[a, b]$ at the zeros of f.

2. Integrate f over each subinterval.

3. Add the absolute values of the integrals.

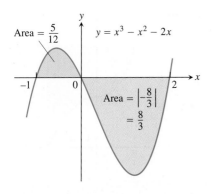

FIGURE 5.22 The region between the curve $y = x^3 - x^2 - 2x$ and the x-axis (Example 8).

EXAMPLE 8 Find the area of the region between the x-axis and the graph of $f(x) = x^3 - x^2 - 2x, -1 \le x \le 2$.

Solution First find the zeros of f. Since

$$f(x) = x^3 - x^2 - 2x = x(x^2 - x - 2) = x(x + 1)(x - 2),$$

the zeros are $x = 0, -1$, and 2 (Figure 5.22). The zeros subdivide $[-1, 2]$ into two subintervals: $[-1, 0]$, on which $f \ge 0$, and $[0, 2]$, on which $f \le 0$. We integrate f over each subinterval and add the absolute values of the calculated integrals.

$$\int_{-1}^{0} (x^3 - x^2 - 2x) \, dx = \left[\frac{x^4}{4} - \frac{x^3}{3} - x^2 \right]_{-1}^{0} = 0 - \left[\frac{1}{4} + \frac{1}{3} - 1 \right] = \frac{5}{12}$$

$$\int_{0}^{2} (x^3 - x^2 - 2x) \, dx = \left[\frac{x^4}{4} - \frac{x^3}{3} - x^2 \right]_{0}^{2} = \left[4 - \frac{8}{3} - 4 \right] - 0 = -\frac{8}{3}$$

The total enclosed area is obtained by adding the absolute values of the calculated integrals.

$$\text{Total enclosed area} = \frac{5}{12} + \left| -\frac{8}{3} \right| = \frac{37}{12} \qquad \blacksquare$$

Exercises 5.4

Evaluating Integrals

Evaluate the integrals in Exercises 1–34.

1. $\int_{-2}^{0} (2x + 5) \, dx$

2. $\int_{-3}^{4} \left(5 - \frac{x}{2} \right) dx$

3. $\int_{0}^{2} x(x - 3) \, dx$

4. $\int_{-1}^{1} (x^2 - 2x + 3) \, dx$

5. $\int_{0}^{4} \left(3x - \frac{x^3}{4} \right) dx$

6. $\int_{-2}^{2} (x^3 - 2x + 3) \, dx$

7. $\int_{0}^{1} \left(x^2 + \sqrt{x} \right) dx$

8. $\int_{1}^{32} x^{-6/5} \, dx$

9. $\int_{0}^{\pi/3} 2 \sec^2 x \, dx$

10. $\int_{0}^{\pi} (1 + \cos x) \, dx$

11. $\int_{\pi/4}^{3\pi/4} \csc \theta \cot \theta \, d\theta$

12. $\int_{0}^{\pi/3} 4 \sec u \tan u \, du$

13. $\int_{\pi/2}^{0} \frac{1 + \cos 2t}{2} \, dt$

14. $\int_{-\pi/3}^{\pi/3} \frac{1 - \cos 2t}{2} \, dt$

15. $\int_{0}^{\pi/4} \tan^2 x \, dx$

16. $\int_{0}^{\pi/6} (\sec x + \tan x)^2 \, dx$

17. $\int_{0}^{\pi/8} \sin 2x \, dx$

18. $\int_{-\pi/3}^{-\pi/4} \left(4 \sec^2 t + \frac{\pi}{t^2} \right) dt$

19. $\int_{1}^{-1} (r + 1)^2 \, dr$

20. $\int_{-\sqrt{3}}^{\sqrt{3}} (t + 1)(t^2 + 4) \, dt$

21. $\int_{\sqrt{2}}^{1} \left(\frac{u^7}{2} - \frac{1}{u^5} \right) du$

22. $\int_{-3}^{-1} \frac{y^5 - 2y}{y^3} \, dy$

23. $\int_{1}^{\sqrt{2}} \frac{s^2 + \sqrt{s}}{s^2} \, ds$

24. $\int_{1}^{8} \frac{(x^{1/3} + 1)(2 - x^{2/3})}{x^{1/3}} \, dx$

25. $\int_{\pi/2}^{\pi} \frac{\sin 2x}{2 \sin x} \, dx$

26. $\int_{0}^{\pi/3} (\cos x + \sec x)^2 \, dx$

27. $\int_{-4}^{4} |x| \, dx$

28. $\int_{0}^{\pi} \frac{1}{2} (\cos x + |\cos x|) \, dx$

29. $\int_{0}^{\ln 2} e^{3x} \, dx$

30. $\int_{1}^{2} \left(\frac{1}{x} - e^{-x} \right) dx$

31. $\int_{0}^{1/2} \frac{4}{\sqrt{1 - x^2}} \, dx$

32. $\int_{0}^{1/\sqrt{3}} \frac{dx}{1 + 4x^2}$

33. $\int_{2}^{4} x^{\pi - 1} \, dx$

34. $\int_{-1}^{0} \pi^{x-1} \, dx$

In Exercises 35–38, guess an antiderivative for the integrand function. Validate your guess by differentiation and then evaluate the given definite integral. (*Hint:* Keep in mind the Chain Rule in guessing an antiderivative. You will learn how to find such antiderivatives in the next section.)

35. $\int_{0}^{1} xe^{x^2} \, dx$

36. $\int_{1}^{2} \frac{\ln x}{x} \, dx$

37. $\int_{2}^{5} \frac{x \, dx}{\sqrt{1 + x^2}}$

38. $\int_{0}^{\pi/3} \sin^2 x \cos x \, dx$

Derivatives of Integrals

Find the derivatives in Exercises 39–44.

a. by evaluating the integral and differentiating the result.

b. by differentiating the integral directly.

39. $\dfrac{d}{dx}\displaystyle\int_0^{\sqrt{x}} \cos t \, dt$

40. $\dfrac{d}{dx}\displaystyle\int_1^{\sin x} 3t^2 \, dt$

41. $\dfrac{d}{dt}\displaystyle\int_0^{t^4} \sqrt{u} \, du$

42. $\dfrac{d}{d\theta}\displaystyle\int_0^{\tan \theta} \sec^2 y \, dy$

43. $\dfrac{d}{dx}\displaystyle\int_0^{x^3} e^{-t} \, dt$

44. $\dfrac{d}{dt}\displaystyle\int_0^{\sqrt{t}} \left(x^4 + \dfrac{3}{\sqrt{1-x^2}}\right) dx$

Find dy/dx in Exercises 45–56.

45. $y = \displaystyle\int_0^x \sqrt{1 + t^2} \, dt$

46. $y = \displaystyle\int_1^x \dfrac{1}{t} \, dt, \quad x > 0$

47. $y = \displaystyle\int_{\sqrt{x}}^0 \sin (t^2) \, dt$

48. $y = x\displaystyle\int_2^{x^2} \sin (t^3) \, dt$

49. $y = \displaystyle\int_{-1}^x \dfrac{t^2}{t^2 + 4} \, dt - \int_3^x \dfrac{t^2}{t^2 + 4} \, dt$

50. $y = \left(\displaystyle\int_0^x (t^3 + 1)^{10} \, dt\right)^3$

51. $y = \displaystyle\int_0^{\sin x} \dfrac{dt}{\sqrt{1 - t^2}}, \quad |x| < \dfrac{\pi}{2}$

52. $y = \displaystyle\int_{\tan x}^0 \dfrac{dt}{1 + t^2}$

53. $y = \displaystyle\int_0^{e^{x^2}} \dfrac{1}{\sqrt{t}} \, dt$

54. $y = \displaystyle\int_{2^x}^1 \sqrt[3]{t} \, dt$

55. $y = \displaystyle\int_0^{\sin^{-1} x} \cos t \, dt$

56. $y = \displaystyle\int_{-1}^{x^{1/\pi}} \sin^{-1} t \, dt$

Area

In Exercises 57–60, find the total area between the region and the x-axis.

57. $y = -x^2 - 2x, \quad -3 \le x \le 2$

58. $y = 3x^2 - 3, \quad -2 \le x \le 2$

59. $y = x^3 - 3x^2 + 2x, \quad 0 \le x \le 2$

60. $y = x^{1/3} - x, \quad -1 \le x \le 8$

Find the areas of the shaded regions in Exercises 61–64.

61.

62.

63.

64.

Initial Value Problems

Each of the following functions solves one of the initial value problems in Exercises 65–68. Which function solves which problem? Give brief reasons for your answers.

a. $y = \displaystyle\int_1^x \dfrac{1}{t} \, dt - 3$

b. $y = \displaystyle\int_0^x \sec t \, dt + 4$

c. $y = \displaystyle\int_{-1}^x \sec t \, dt + 4$

d. $y = \displaystyle\int_{\pi}^x \dfrac{1}{t} \, dt - 3$

65. $\dfrac{dy}{dx} = \dfrac{1}{x}, \quad y(\pi) = -3$

66. $y' = \sec x, \quad y(-1) = 4$

67. $y' = \sec x, \quad y(0) = 4$

68. $y' = \dfrac{1}{x}, \quad y(1) = -3$

Express the solutions of the initial value problems in Exercises 69 and 70 in terms of integrals.

69. $\dfrac{dy}{dx} = \sec x, \quad y(2) = 3$

70. $\dfrac{dy}{dx} = \sqrt{1 + x^2}, \quad y(1) = -2$

Theory and Examples

71. Archimedes' area formula for parabolic arches Archimedes (287–212 B.C.), inventor, military engineer, physicist, and the greatest mathematician of classical times in the Western world, discovered that the area under a parabolic arch is two-thirds the base times the height. Sketch the parabolic arch $y = h - (4h/b^2)x^2$, $-b/2 \le x \le b/2$, assuming that h and b are positive. Then use calculus to find the area of the region enclosed between the arch and the x-axis.

72. Show that if k is a positive constant, then the area between the x-axis and one arch of the curve $y = \sin kx$ is $2/k$.

73. Cost from marginal cost The marginal cost of printing a poster when x posters have been printed is

$$\dfrac{dc}{dx} = \dfrac{1}{2\sqrt{x}}$$

dollars. Find $c(100) - c(1)$, the cost of printing posters 2–100.

74. Revenue from marginal revenue Suppose that a company's marginal revenue from the manufacture and sale of eggbeaters is

$$\dfrac{dr}{dx} = 2 - 2/(x + 1)^2,$$

where r is measured in thousands of dollars and x in thousands of units. How much money should the company expect from a production run of $x = 3$ thousand eggbeaters? To find out, integrate the marginal revenue from $x = 0$ to $x = 3$.

Only the value $x = 4$ satisfies the equation $\sqrt{x} = x - 2$. The value $x = 1$ is an extraneous root introduced by squaring. The right-hand limit is $b = 4$.

$$\text{For } 0 \le x \le 2: \quad f(x) - g(x) = \sqrt{x} - 0 = \sqrt{x}$$
$$\text{For } 2 \le x \le 4: \quad f(x) - g(x) = \sqrt{x} - (x - 2) = \sqrt{x} - x + 2$$

We add the areas of subregions A and B to find the total area:

$$\text{Total area} = \underbrace{\int_0^2 \sqrt{x}\, dx}_{\text{area of } A} + \underbrace{\int_2^4 (\sqrt{x} - x + 2)\, dx}_{\text{area of } B}$$

$$= \left[\frac{2}{3} x^{3/2}\right]_0^2 + \left[\frac{2}{3} x^{3/2} - \frac{x^2}{2} + 2x\right]_2^4$$

$$= \frac{2}{3}(2)^{3/2} - 0 + \left(\frac{2}{3}(4)^{3/2} - 8 + 8\right) - \left(\frac{2}{3}(2)^{3/2} - 2 + 4\right)$$

$$= \frac{2}{3}(8) - 2 = \frac{10}{3}.$$

Integration with Respect to y

If a region's bounding curves are described by functions of y, the approximating rectangles are horizontal instead of vertical and the basic formula has y in place of x.

For regions like these:

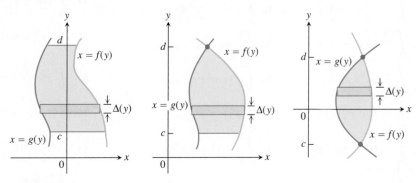

use the formula

$$A = \int_c^d [f(y) - g(y)]\, dy.$$

In this equation f always denotes the right-hand curve and g the left-hand curve, so $f(y) - g(y)$ is nonnegative.

EXAMPLE 7 Find the area of the region in Example 6 by integrating with respect to y.

Solution We first sketch the region and a typical *horizontal* rectangle based on a partition of an interval of y-values (Figure 5.31). The region's right-hand boundary is the line $x = y + 2$, so $f(y) = y + 2$. The left-hand boundary is the curve $x = y^2$, so $g(y) = y^2$. The lower limit of integration is $y = 0$. We find the upper limit by solving $x = y + 2$ and $x = y^2$ simultaneously for y:

$$y + 2 = y^2 \qquad \text{Equate } f(y) = y + 2 \text{ and } g(y) = y^2.$$
$$y^2 - y - 2 = 0 \qquad \text{Rewrite.}$$
$$(y + 1)(y - 2) = 0 \qquad \text{Factor.}$$
$$y = -1, \quad y = 2 \qquad \text{Solve.}$$

The upper limit of integration is $b = 2$. (The value $y = -1$ gives a point of intersection *below* the x-axis.)

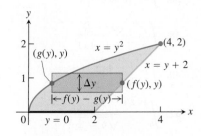

FIGURE 5.31 It takes two integrations to find the area of this region if we integrate with respect to x. It takes only one if we integrate with respect to y (Example 7).

The area of the region is

$$A = \int_c^d [f(y) - g(y)] \, dy = \int_0^2 [y + 2 - y^2] \, dy$$

$$= \int_0^2 [2 + y - y^2] \, dy$$

$$= \left[2y + \frac{y^2}{2} - \frac{y^3}{3} \right]_0^2$$

$$= 4 + \frac{4}{2} - \frac{8}{3} = \frac{10}{3}.$$

This is the result of Example 6, found with less work. ∎

Exercises 5.6

Evaluating Definite Integrals

Use the Substitution Formula in Theorem 7 to evaluate the integrals in Exercises 1–46.

1. a. $\displaystyle\int_0^3 \sqrt{y+1} \, dy$ **b.** $\displaystyle\int_{-1}^0 \sqrt{y+1} \, dy$

2. a. $\displaystyle\int_0^1 r\sqrt{1-r^2} \, dr$ **b.** $\displaystyle\int_{-1}^1 r\sqrt{1-r^2} \, dr$

3. a. $\displaystyle\int_0^{\pi/4} \tan x \sec^2 x \, dx$ **b.** $\displaystyle\int_{-\pi/4}^0 \tan x \sec^2 x \, dx$

4. a. $\displaystyle\int_0^\pi 3\cos^2 x \sin x \, dx$ **b.** $\displaystyle\int_{2\pi}^{3\pi} 3\cos^2 x \sin x \, dx$

5. a. $\displaystyle\int_0^1 t^3(1+t^4)^3 \, dt$ **b.** $\displaystyle\int_{-1}^1 t^3(1+t^4)^3 \, dt$

6. a. $\displaystyle\int_0^{\sqrt{7}} t(t^2+1)^{1/3} \, dt$ **b.** $\displaystyle\int_{-\sqrt{7}}^0 t(t^2+1)^{1/3} \, dt$

7. a. $\displaystyle\int_{-1}^1 \frac{5r}{(4+r^2)^2} \, dr$ **b.** $\displaystyle\int_0^1 \frac{5r}{(4+r^2)^2} \, dr$

8. a. $\displaystyle\int_0^1 \frac{10\sqrt{v}}{(1+v^{3/2})^2} \, dv$ **b.** $\displaystyle\int_1^4 \frac{10\sqrt{v}}{(1+v^{3/2})^2} \, dv$

9. a. $\displaystyle\int_0^{\sqrt{3}} \frac{4x}{\sqrt{x^2+1}} \, dx$ **b.** $\displaystyle\int_{-\sqrt{3}}^{\sqrt{3}} \frac{4x}{\sqrt{x^2+1}} \, dx$

10. a. $\displaystyle\int_0^1 \frac{x^3}{\sqrt{x^4+9}} \, dx$ **b.** $\displaystyle\int_{-1}^0 \frac{x^3}{\sqrt{x^4+9}} \, dx$

11. a. $\displaystyle\int_0^{\pi/6} (1-\cos 3t) \sin 3t \, dt$ **b.** $\displaystyle\int_{\pi/6}^{\pi/3} (1-\cos 3t) \sin 3t \, dt$

12. a. $\displaystyle\int_{-\pi/2}^0 \left(2+\tan \frac{t}{2}\right) \sec^2 \frac{t}{2} \, dt$ **b.** $\displaystyle\int_{-\pi/2}^{\pi/2} \left(2+\tan \frac{t}{2}\right) \sec^2 \frac{t}{2} \, dt$

13. a. $\displaystyle\int_0^{2\pi} \frac{\cos z}{\sqrt{4+3\sin z}} \, dz$ **b.** $\displaystyle\int_{-\pi}^\pi \frac{\cos z}{\sqrt{4+3\sin z}} \, dz$

14. a. $\displaystyle\int_{-\pi/2}^0 \frac{\sin w}{(3+2\cos w)^2} \, dw$ **b.** $\displaystyle\int_0^{\pi/2} \frac{\sin w}{(3+2\cos w)^2} \, dw$

15. $\displaystyle\int_0^1 \sqrt{t^5+2t} \, (5t^4+2) \, dt$ **16.** $\displaystyle\int_1^4 \frac{dy}{2\sqrt{y}\,(1+\sqrt{y})^2}$

17. $\displaystyle\int_0^{\pi/6} \cos^{-3} 2\theta \sin 2\theta \, d\theta$ **18.** $\displaystyle\int_\pi^{3\pi/2} \cot^5 \left(\frac{\theta}{6}\right) \sec^2 \left(\frac{\theta}{6}\right) d\theta$

19. $\displaystyle\int_0^\pi 5(5-4\cos t)^{1/4} \sin t \, dt$ **20.** $\displaystyle\int_0^{\pi/4} (1-\sin 2t)^{3/2} \cos 2t \, dt$

21. $\displaystyle\int_0^1 (4y - y^2 + 4y^3 + 1)^{-2/3} (12y^2 - 2y + 4) \, dy$

22. $\displaystyle\int_0^1 (y^3 + 6y^2 - 12y + 9)^{-1/2} (y^2 + 4y - 4) \, dy$

23. $\displaystyle\int_0^{\sqrt[3]{\pi^2}} \sqrt{\theta} \cos^2 (\theta^{3/2}) \, d\theta$ **24.** $\displaystyle\int_{-1}^{-1/2} t^{-2} \sin^2 \left(1+\frac{1}{t}\right) dt$

25. $\displaystyle\int_0^{\pi/4} (1+e^{\tan\theta}) \sec^2 \theta \, d\theta$ **26.** $\displaystyle\int_{\pi/4}^{\pi/2} (1+e^{\cot\theta}) \csc^2 \theta \, d\theta$

27. $\displaystyle\int_0^\pi \frac{\sin t}{2-\cos t} \, dt$ **28.** $\displaystyle\int_0^{\pi/3} \frac{4\sin\theta}{1-4\cos\theta} \, d\theta$

29. $\displaystyle\int_1^2 \frac{2\ln x}{x} \, dx$ **30.** $\displaystyle\int_2^4 \frac{dx}{x\ln x}$

31. $\displaystyle\int_2^4 \frac{dx}{x(\ln x)^2}$ **32.** $\displaystyle\int_2^{16} \frac{dx}{2x\sqrt{\ln x}}$

33. $\displaystyle\int_0^{\pi/2} \tan \frac{x}{2} \, dx$ **34.** $\displaystyle\int_{\pi/4}^{\pi/2} \cot t \, dt$

35. $\displaystyle\int_{\pi/2}^\pi 2\cot \frac{\theta}{3} \, d\theta$ **36.** $\displaystyle\int_0^{\pi/12} 6\tan 3x \, dx$

37. $\displaystyle\int_{-\pi/2}^{\pi/2} \frac{2\cos\theta \, d\theta}{1+(\sin\theta)^2}$ **38.** $\displaystyle\int_{\pi/6}^{\pi/4} \frac{\csc^2 x \, dx}{1+(\cot x)^2}$

39. $\displaystyle\int_0^{\ln\sqrt{3}} \frac{e^x \, dx}{1+e^{2x}}$ **40.** $\displaystyle\int_1^{e^{\pi/4}} \frac{4 \, dt}{t(1+\ln^2 t)}$

41. $\displaystyle\int_0^1 \frac{4 \, ds}{\sqrt{4-s^2}}$ **42.** $\displaystyle\int_0^{3\sqrt{2}/4} \frac{ds}{\sqrt{9-4s^2}}$

43. $\displaystyle\int_{\sqrt{2}}^2 \frac{\sec^2 (\sec^{-1} x) \, dx}{x\sqrt{x^2-1}}$ **44.** $\displaystyle\int_{2/\sqrt{3}}^2 \frac{\cos (\sec^{-1} x) \, dx}{x\sqrt{x^2-1}}$

45. $\displaystyle\int_{-1}^{-\sqrt{2}/2} \frac{dy}{y\sqrt{4y^2-1}}$ **46.** $\displaystyle\int_{-2/3}^{-\sqrt{2}/3} \frac{dy}{y\sqrt{9y^2-1}}$

Area

Find the total areas of the shaded regions in Exercises 47–62.

47.

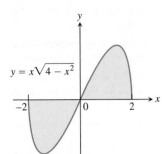

$y = x\sqrt{4 - x^2}$

48.

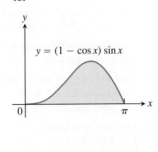

$y = (1 - \cos x)\sin x$

49.

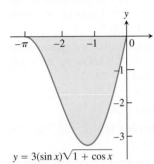

$y = 3(\sin x)\sqrt{1 + \cos x}$

50.

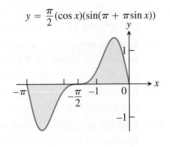

$y = \frac{\pi}{2}(\cos x)(\sin(\pi + \pi \sin x))$

51.

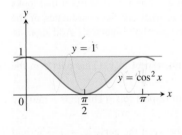

$y = 1$

$y = \cos^2 x$

52.

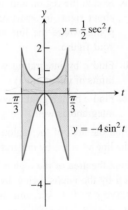

$y = \frac{1}{2}\sec^2 t$

$y = -4\sin^2 t$

53.

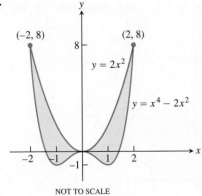

$(-2, 8)$ $\qquad$ $(2, 8)$

$y = 2x^2$

$y = x^4 - 2x^2$

NOT TO SCALE

54.

$x = y^3$

$(1, 1)$

$x = y^2$

55.

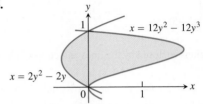

$x = 12y^2 - 12y^3$

$x = 2y^2 - 2y$

56.

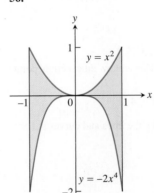

$y = x^2$

$y = -2x^4$

57.

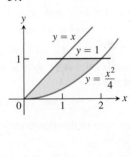

$y = x$

$y = 1$

$y = \frac{x^2}{4}$

58.

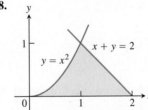

$x + y = 2$

$y = x^2$

59.

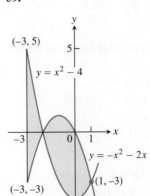

$(-3, 5)$

$y = x^2 - 4$

$y = -x^2 - 2x$

$(-3, -3)$ $\qquad$ $(1, -3)$

60.

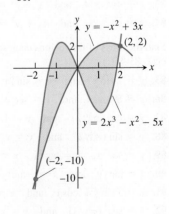

$y = -x^2 + 3x$

$(2, 2)$

$y = 2x^3 - x^2 - 5x$

$(-2, -10)$

61. **62.**

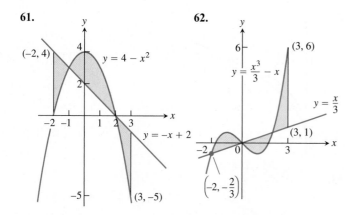

Find the areas of the regions enclosed by the lines and curves in Exercises 63–72.

63. $y = x^2 - 2$ and $y = 2$ **64.** $y = 2x - x^2$ and $y = -3$

65. $y = x^4$ and $y = 8x$ **66.** $y = x^2 - 2x$ and $y = x$

67. $y = x^2$ and $y = -x^2 + 4x$

68. $y = 7 - 2x^2$ and $y = x^2 + 4$

69. $y = x^4 - 4x^2 + 4$ and $y = x^2$

70. $y = x\sqrt{a^2 - x^2}$, $a > 0$, and $y = 0$

71. $y = \sqrt{|x|}$ and $5y = x + 6$ (How many intersection points are there?)

72. $y = |x^2 - 4|$ and $y = (x^2/2) + 4$

Find the areas of the regions enclosed by the lines and curves in Exercises 73–80.

73. $x = 2y^2$, $x = 0$, and $y = 3$

74. $x = y^2$ and $x = y + 2$

75. $y^2 - 4x = 4$ and $4x - y = 16$

76. $x - y^2 = 0$ and $x + 2y^2 = 3$

77. $x + y^2 = 0$ and $x + 3y^2 = 2$

78. $x - y^{2/3} = 0$ and $x + y^4 = 2$

79. $x = y^2 - 1$ and $x = |y|\sqrt{1 - y^2}$

80. $x = y^3 - y^2$ and $x = 2y$

Find the areas of the regions enclosed by the curves in Exercises 81–84.

81. $4x^2 + y = 4$ and $x^4 - y = 1$

82. $x^3 - y = 0$ and $3x^2 - y = 4$

83. $x + 4y^2 = 4$ and $x + y^4 = 1$, for $x \geq 0$

84. $x + y^2 = 3$ and $4x + y^2 = 0$

Find the areas of the regions enclosed by the lines and curves in Exercises 85–92.

85. $y = 2 \sin x$ and $y = \sin 2x$, $0 \leq x \leq \pi$

86. $y = 8 \cos x$ and $y = \sec^2 x$, $-\pi/3 \leq x \leq \pi/3$

87. $y = \cos(\pi x/2)$ and $y = 1 - x^2$

88. $y = \sin(\pi x/2)$ and $y = x$

89. $y = \sec^2 x$, $y = \tan^2 x$, $x = -\pi/4$, and $x = \pi/4$

90. $x = \tan^2 y$ and $x = -\tan^2 y$, $-\pi/4 \leq y \leq \pi/4$

91. $x = 3 \sin y \sqrt{\cos y}$ and $x = 0$, $0 \leq y \leq \pi/2$

92. $y = \sec^2(\pi x/3)$ and $y = x^{1/3}$, $-1 \leq x \leq 1$

Area Between Curves

93. Find the area of the propeller-shaped region enclosed by the curve $x - y^3 = 0$ and the line $x - y = 0$.

94. Find the area of the propeller-shaped region enclosed by the curves $x - y^{1/3} = 0$ and $x - y^{1/5} = 0$.

95. Find the area of the region in the first quadrant bounded by the line $y = x$, the line $x = 2$, the curve $y = 1/x^2$, and the x-axis.

96. Find the area of the "triangular" region in the first quadrant bounded on the left by the y-axis and on the right by the curves $y = \sin x$ and $y = \cos x$.

97. Find the area between the curves $y = \ln x$ and $y = \ln 2x$ from $x = 1$ to $x = 5$.

98. Find the area between the curve $y = \tan x$ and the x-axis from $x = -\pi/4$ to $x = \pi/3$.

99. Find the area of the "triangular" region in the first quadrant that is bounded above by the curve $y = e^{2x}$, below by the curve $y = e^x$, and on the right by the line $x = \ln 3$.

100. Find the area of the "triangular" region in the first quadrant that is bounded above by the curve $y = e^{x/2}$, below by the curve $y = e^{-x/2}$, and on the right by the line $x = 2 \ln 2$.

101. Find the area of the region between the curve $y = 2x/(1 + x^2)$ and the interval $-2 \leq x \leq 2$ of the x-axis.

102. Find the area of the region between the curve $y = 2^{1-x}$ and the interval $-1 \leq x \leq 1$ of the x-axis.

103. The region bounded below by the parabola $y = x^2$ and above by the line $y = 4$ is to be partitioned into two subsections of equal area by cutting across it with the horizontal line $y = c$.

  a. Sketch the region and draw a line $y = c$ across it that looks about right. In terms of c, what are the coordinates of the points where the line and parabola intersect? Add them to your figure.

  b. Find c by integrating with respect to y. (This puts c in the limits of integration.)

  c. Find c by integrating with respect to x. (This puts c into the integrand as well.)

104. Find the area of the region between the curve $y = 3 - x^2$ and the line $y = -1$ by integrating with respect to **a.** x, **b.** y.

105. Find the area of the region in the first quadrant bounded on the left by the y-axis, below by the line $y = x/4$, above left by the curve $y = 1 + \sqrt{x}$, and above right by the curve $y = 2/\sqrt{x}$.

106. Find the area of the region in the first quadrant bounded on the left by the y-axis, below by the curve $x = 2\sqrt{y}$, above left by the curve $x = (y - 1)^2$, and above right by the line $x = 3 - y$.

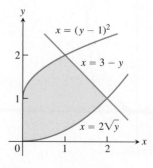

107. The figure here shows triangle AOC inscribed in the region cut from the parabola $y = x^2$ by the line $y = a^2$. Find the limit of the ratio of the area of the triangle to the area of the parabolic region as a approaches zero.

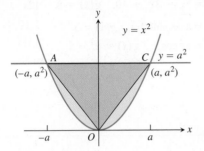

108. Suppose the area of the region between the graph of a positive continuous function f and the x-axis from $x = a$ to $x = b$ is 4 square units. Find the area between the curves $y = f(x)$ and $y = 2f(x)$ from $x = a$ to $x = b$.

109. Which of the following integrals, if either, calculates the area of the shaded region shown here? Give reasons for your answer.

a. $\int_{-1}^{1} (x - (-x)) \, dx = \int_{-1}^{1} 2x \, dx$

b. $\int_{-1}^{1} (-x - (x)) \, dx = \int_{-1}^{1} -2x \, dx$

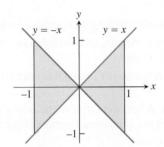

110. True, sometimes true, or never true? The area of the region between the graphs of the continuous functions $y = f(x)$ and $y = g(x)$ and the vertical lines $x = a$ and $x = b$ $(a < b)$ is

$$\int_{a}^{b} [f(x) - g(x)] \, dx.$$

Give reasons for your answer.

Theory and Examples

111. Suppose that $F(x)$ is an antiderivative of $f(x) = (\sin x)/x$, $x > 0$. Express

$$\int_{1}^{3} \frac{\sin 2x}{x} \, dx$$

in terms of F.

112. Show that if f is continuous, then

$$\int_{0}^{1} f(x) \, dx = \int_{0}^{1} f(1 - x) \, dx.$$

113. Suppose that

$$\int_{0}^{1} f(x) \, dx = 3.$$

Find

$$\int_{-1}^{0} f(x) \, dx$$

if **a.** f is odd, **b.** f is even.

114. a. Show that if f is odd on $[-a, a]$, then

$$\int_{-a}^{a} f(x) \, dx = 0.$$

b. Test the result in part (a) with $f(x) = \sin x$ and $a = \pi/2$.

115. If f is a continuous function, find the value of the integral

$$I = \int_{0}^{a} \frac{f(x) \, dx}{f(x) + f(a - x)}$$

by making the substitution $u = a - x$ and adding the resulting integral to I.

116. By using a substitution, prove that for all positive numbers x and y,

$$\int_{x}^{xy} \frac{1}{t} \, dt = \int_{1}^{y} \frac{1}{t} \, dt.$$

The Shift Property for Definite Integrals   A basic property of definite integrals is their invariance under translation, as expressed by the equation.

$$\int_{a}^{b} f(x) \, dx = \int_{a-c}^{b-c} f(x + c) \, dx. \tag{1}$$

The equation holds whenever f is integrable and defined for the necessary values of x. For example in the accompanying figure, show that

$$\int_{-2}^{-1} (x + 2)^3 \, dx = \int_{0}^{1} x^3 \, dx$$

because the areas of the shaded regions are congruent.

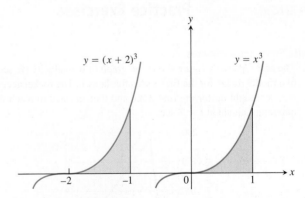

117. Use a substitution to verify Equation (1).

118. For each of the following functions, graph $f(x)$ over $[a, b]$ and $f(x + c)$ over $[a - c, b - c]$ to convince yourself that Equation (1) is reasonable.

a. $f(x) = x^2$, $a = 0$, $b = 1$, $c = 1$

b. $f(x) = \sin x$, $a = 0$, $b = \pi$, $c = \pi/2$

c. $f(x) = \sqrt{x - 4}$, $a = 4$, $b = 8$, $c = 5$

COMPUTER EXPLORATIONS

In Exercises 119–122, you will find the area between curves in the plane when you cannot find their points of intersection using simple algebra. Use a CAS to perform the following steps:

a. Plot the curves together to see what they look like and how many points of intersection they have.

b. Use the numerical equation solver in your CAS to find all the points of intersection.

c. Integrate $|f(x) - g(x)|$ over consecutive pairs of intersection values.

d. Sum together the integrals found in part (c).

119. $f(x) = \dfrac{x^3}{3} - \dfrac{x^2}{2} - 2x + \dfrac{1}{3}, \quad g(x) = x - 1$

120. $f(x) = \dfrac{x^4}{2} - 3x^3 + 10, \quad g(x) = 8 - 12x$

121. $f(x) = x + \sin(2x), \quad g(x) = x^3$

122. $f(x) = x^2 \cos x, \quad g(x) = x^3 - x$

Chapter 5   Questions to Guide Your Review

1. How can you sometimes estimate quantities like distance traveled, area, and average value with finite sums? Why might you want to do so?

2. What is sigma notation? What advantage does it offer? Give examples.

3. What is a Riemann sum? Why might you want to consider such a sum?

4. What is the norm of a partition of a closed interval?

5. What is the definite integral of a function f over a closed interval $[a, b]$? When can you be sure it exists?

6. What is the relation between definite integrals and area? Describe some other interpretations of definite integrals.

7. What is the average value of an integrable function over a closed interval? Must the function assume its average value? Explain.

8. Describe the rules for working with definite integrals (Table 5.4). Give examples.

9. What is the Fundamental Theorem of Calculus? Why is it so important? Illustrate each part of the theorem with an example.

10. What is the Net Change Theorem? What does it say about the integral of velocity? The integral of marginal cost?

11. Discuss how the processes of integration and differentiation can be considered as "inverses" of each other.

12. How does the Fundamental Theorem provide a solution to the initial value problem $dy/dx = f(x)$, $y(x_0) = y_0$, when f is continuous?

13. How is integration by substitution related to the Chain Rule?

14. How can you sometimes evaluate indefinite integrals by substitution? Give examples.

15. How does the method of substitution work for definite integrals? Give examples.

16. How do you define and calculate the area of the region between the graphs of two continuous functions? Give an example.

Chapter 5   Practice Exercises

Finite Sums and Estimates

1. The accompanying figure shows the graph of the velocity (ft/sec) of a model rocket for the first 8 sec after launch. The rocket accelerated straight up for the first 2 sec and then coasted to reach its maximum height at $t = 8$ sec.

Time after launch (sec)

a. Assuming that the rocket was launched from ground level, about how high did it go? (This is the rocket in Section 3.3, Exercise 17, but you do not need to do Exercise 17 to do the exercise here.)

b. Sketch a graph of the rocket's height aboveground as a function of time for $0 \le t \le 8$.

2. a. The accompanying figure shows the velocity (m/sec) of a body moving along the s-axis during the time interval from $t = 0$ to $t = 10$ sec. About how far did the body travel during those 10 sec?

b. Sketch a graph of s as a function of t for $0 \le t \le 10$ assuming $s(0) = 0$.

Time (sec)

3. Suppose that $\sum_{k=1}^{10} a_k = -2$ and $\sum_{k=1}^{10} b_k = 25$. Find the value of

a. $\sum_{k=1}^{10} \dfrac{a_k}{4}$

b. $\sum_{k=1}^{10} (b_k - 3a_k)$

c. $\sum_{k=1}^{10} (a_k + b_k - 1)$

d. $\sum_{k=1}^{10} \left(\dfrac{5}{2} - b_k \right)$

4. Suppose that $\sum_{k=1}^{20} a_k = 0$ and $\sum_{k=1}^{20} b_k = 7$. Find the values of

a. $\sum_{k=1}^{20} 3a_k$

b. $\sum_{k=1}^{20} (a_k + b_k)$

c. $\sum_{k=1}^{20} \left(\dfrac{1}{2} - \dfrac{2b_k}{7} \right)$

d. $\sum_{k=1}^{20} (a_k - 2)$

Definite Integrals

In Exercises 5–8, express each limit as a definite integral. Then evaluate the integral to find the value of the limit. In each case, P is a partition of the given interval and the numbers c_k are chosen from the subintervals of P.

5. $\displaystyle\lim_{\|P\|\to 0} \sum_{k=1}^{n} (2c_k - 1)^{-1/2} \, \Delta x_k$, where P is a partition of $[1, 5]$

6. $\displaystyle\lim_{\|P\|\to 0} \sum_{k=1}^{n} c_k(c_k^{2} - 1)^{1/3} \, \Delta x_k$, where P is a partition of $[1, 3]$

7. $\displaystyle\lim_{\|P\|\to 0} \sum_{k=1}^{n} \left(\cos\left(\dfrac{c_k}{2} \right) \right) \Delta x_k$, where P is a partition of $[-\pi, 0]$

8. $\displaystyle\lim_{\|P\|\to 0} \sum_{k=1}^{n} (\sin c_k)(\cos c_k) \, \Delta x_k$, where P is a partition of $[0, \pi/2]$

9. If $\int_{-2}^{2} 3f(x) \, dx = 12$, $\int_{-2}^{5} f(x) \, dx = 6$, and $\int_{-2}^{5} g(x) \, dx = 2$, find the values of the following.

a. $\displaystyle\int_{-2}^{2} f(x) \, dx$

b. $\displaystyle\int_{2}^{5} f(x) \, dx$

c. $\displaystyle\int_{5}^{-2} g(x) \, dx$

d. $\displaystyle\int_{-2}^{5} (-\pi g(x)) \, dx$

e. $\displaystyle\int_{-2}^{5} \left(\dfrac{f(x) + g(x)}{5} \right) dx$

10. If $\int_{0}^{2} f(x) \, dx = \pi$, $\int_{0}^{2} 7g(x) \, dx = 7$, and $\int_{0}^{1} g(x) \, dx = 2$, find the values of the following.

a. $\displaystyle\int_{0}^{2} g(x) \, dx$

b. $\displaystyle\int_{1}^{2} g(x) \, dx$

c. $\displaystyle\int_{2}^{0} f(x) \, dx$

d. $\displaystyle\int_{0}^{2} \sqrt{2} \, f(x) \, dx$

e. $\displaystyle\int_{0}^{2} (g(x) - 3f(x)) \, dx$

Area

In Exercises 11–14, find the total area of the region between the graph of f and the x-axis.

11. $f(x) = x^2 - 4x + 3, \quad 0 \le x \le 3$

12. $f(x) = 1 - (x^2/4), \quad -2 \le x \le 3$

13. $f(x) = 5 - 5x^{2/3}, \quad -1 \le x \le 8$

14. $f(x) = 1 - \sqrt{x}, \quad 0 \le x \le 4$

Find the areas of the regions enclosed by the curves and lines in Exercises 15–26.

15. $y = x, \quad y = 1/x^2, \quad x = 2$

16. $y = x, \quad y = 1/\sqrt{x}, \quad x = 2$

17. $\sqrt{x} + \sqrt{y} = 1, \quad x = 0, \quad y = 0$

18. $x^3 + \sqrt{y} = 1, \quad x = 0, \quad y = 0, \quad \text{for} \quad 0 \le x \le 1$

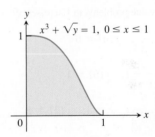

19. $x = 2y^2, \quad x = 0, \quad y = 3$ **20.** $x = 4 - y^2, \quad x = 0$

21. $y^2 = 4x, \quad y = 4x - 2$

22. $y^2 = 4x + 4, \quad y = 4x - 16$

23. $y = \sin x, \quad y = x, \quad 0 \le x \le \pi/4$

24. $y = |\sin x|, \quad y = 1, \quad -\pi/2 \le x \le \pi/2$

25. $y = 2 \sin x, \quad y = \sin 2x, \quad 0 \le x \le \pi$

26. $y = 8 \cos x, \quad y = \sec^2 x, \quad -\pi/3 \le x \le \pi/3$

27. Find the area of the "triangular" region bounded on the left by $x + y = 2$, on the right by $y = x^2$, and above by $y = 2$.

28. Find the area of the "triangular" region bounded on the left by $y = \sqrt{x}$, on the right by $y = 6 - x$, and below by $y = 1$.

29. Find the extreme values of $f(x) = x^3 - 3x^2$ and find the area of the region enclosed by the graph of f and the x-axis.

30. Find the area of the region cut from the first quadrant by the curve $x^{1/2} + y^{1/2} = a^{1/2}$.

31. Find the total area of the region enclosed by the curve $x = y^{2/3}$ and the lines $x = y$ and $y = -1$.

32. Find the total area of the region between the curves $y = \sin x$ and $y = \cos x$ for $0 \le x \le 3\pi/2$.

33. **Area** Find the area between the curve $y = 2(\ln x)/x$ and the x-axis from $x = 1$ to $x = e$.

34. a. Show that the area between the curve $y = 1/x$ and the x-axis from $x = 10$ to $x = 20$ is the same as the area between the curve and the x-axis from $x = 1$ to $x = 2$.

b. Show that the area between the curve $y = 1/x$ and the x-axis from ka to kb is the same as the area between the curve and the x-axis from $x = a$ to $x = b$ $(0 < a < b, k > 0)$.

Initial Value Problems

35. Show that $y = x^2 + \int_1^x \frac{1}{t}\, dt$ solves the initial value problem

$$\frac{d^2 y}{dx^2} = 2 - \frac{1}{x^2}; \quad y'(1) = 3, \quad y(1) = 1.$$

36. Show that $y = \int_0^x \left(1 + 2\sqrt{\sec t}\,\right) dt$ solves the initial value problem

$$\frac{d^2 y}{dx^2} = \sqrt{\sec x}\,\tan x; \quad y'(0) = 3, \quad y(0) = 0.$$

Express the solutions of the initial value problems in Exercises 37 and 38 in terms of integrals.

37. $\dfrac{dy}{dx} = \dfrac{\sin x}{x}, \quad y(5) = -3$

38. $\dfrac{dy}{dx} = \sqrt{2 - \sin^2 x}, \quad y(-1) = 2$

Solve the initial value problems in Exercises 39–42.

39. $\dfrac{dy}{dx} = \dfrac{1}{\sqrt{1 - x^2}}, \quad y(0) = 0$

40. $\dfrac{dy}{dx} = \dfrac{1}{x^2 + 1} - 1, \quad y(0) = 1$

41. $\dfrac{dy}{dx} = \dfrac{1}{x\sqrt{x^2 - 1}}, \quad x > 1; \quad y(2) = \pi$

42. $\dfrac{dy}{dx} = \dfrac{1}{1 + x^2} - \dfrac{2}{\sqrt{1 - x^2}}, \quad y(0) = 2$

Evaluating Indefinite Integrals

Evaluate the integrals in Exercises 43–72.

43. $\displaystyle\int 2(\cos x)^{-1/2} \sin x \, dx$ **44.** $\displaystyle\int (\tan x)^{-3/2} \sec^2 x \, dx$

45. $\displaystyle\int (2\theta + 1 + 2\cos(2\theta + 1)) \, d\theta$

46. $\displaystyle\int \left(\frac{1}{\sqrt{2\theta - \pi}} + 2\sec^2(2\theta - \pi)\right) d\theta$

47. $\displaystyle\int \left(t - \frac{2}{t}\right)\left(t + \frac{2}{t}\right) dt$ **48.** $\displaystyle\int \frac{(t + 1)^2 - 1}{t^4} \, dt$

49. $\displaystyle\int \sqrt{t} \sin(2t^{3/2}) \, dt$ **50.** $\displaystyle\int (\sec\theta \tan\theta)\sqrt{1 + \sec\theta}\, d\theta$

51. $\displaystyle\int e^x \sec^2(e^x - 7) \, dx$

52. $\displaystyle\int e^y \csc(e^y + 1)\cot(e^y + 1) \, dy$

53. $\displaystyle\int (\sec^2 x) e^{\tan x} \, dx$ **54.** $\displaystyle\int (\csc^2 x) e^{\cot x} \, dx$

55. $\displaystyle\int_{-1}^1 \frac{dx}{3x - 4}$ **56.** $\displaystyle\int_1^e \frac{\sqrt{\ln x}}{x} \, dx$

57. $\displaystyle\int_0^4 \frac{2t}{t^2 - 25} \, dt$ **58.** $\displaystyle\int \frac{\tan(\ln v)}{v} \, dv$

59. $\displaystyle\int \frac{(\ln x)^{-3}}{x} \, dx$ **60.** $\displaystyle\int \frac{1}{r}\csc^2(1 + \ln r) \, dr$

61. $\displaystyle\int x 3^{x^2} \, dx$ **62.** $\displaystyle\int 2^{\tan x}\sec^2 x \, dx$

63. $\displaystyle\int \frac{3\, dr}{\sqrt{1 - 4(r - 1)^2}}$ **64.** $\displaystyle\int \frac{6\, dr}{\sqrt{4 - (r + 1)^2}}$

65. $\displaystyle\int \frac{dx}{2 + (x - 1)^2}$ **66.** $\displaystyle\int \frac{dx}{1 + (3x + 1)^2}$

67. $\displaystyle\int \frac{dx}{(2x - 1)\sqrt{(2x - 1)^2 - 4}}$

68. $\displaystyle\int \frac{dx}{(x + 3)\sqrt{(x + 3)^2 - 25}}$

69. $\displaystyle\int \frac{e^{\sin^{-1}\sqrt{x}}\, dx}{2\sqrt{x - x^2}}$ **70.** $\displaystyle\int \frac{\sqrt{\sin^{-1} x}\, dx}{\sqrt{1 - x^2}}$

71. $\displaystyle\int \frac{dy}{\sqrt{\tan^{-1} y}\,(1 + y^2)}$ **72.** $\displaystyle\int \frac{(\tan^{-1} x)^2 \, dx}{1 + x^2}$

Evaluating Definite Integrals

Evaluate the integrals in Exercises 73–112.

73. $\displaystyle\int_{-1}^1 (3x^2 - 4x + 7) \, dx$ **74.** $\displaystyle\int_0^1 (8s^3 - 12s^2 + 5) \, ds$

75. $\displaystyle\int_1^2 \frac{4}{v^2} \, dv$ **76.** $\displaystyle\int_1^{27} x^{-4/3} \, dx$

77. $\displaystyle\int_1^4 \frac{dt}{t\sqrt{t}}$ **78.** $\displaystyle\int_1^4 \frac{(1 + \sqrt{u})^{1/2}}{\sqrt{u}} \, du$

79. $\displaystyle\int_0^1 \frac{36\, dx}{(2x + 1)^3}$ **80.** $\displaystyle\int_0^1 \frac{dr}{\sqrt[3]{(7 - 5r)^2}}$

81. $\displaystyle\int_{1/8}^1 x^{-1/3}(1 - x^{2/3})^{3/2} \, dx$ **82.** $\displaystyle\int_0^{1/2} x^3(1 + 9x^4)^{-3/2} \, dx$

83. $\displaystyle\int_0^{\pi} \sin^2 5r \, dr$ **84.** $\displaystyle\int_0^{\pi/4} \cos^2\left(4t - \frac{\pi}{4}\right) dt$

85. $\displaystyle\int_0^{\pi/3} \sec^2\theta \, d\theta$ **86.** $\displaystyle\int_{\pi/4}^{3\pi/4} \csc^2 x \, dx$

87. $\displaystyle\int_{\pi}^{3\pi} \cot^2\frac{x}{6} \, dx$ **88.** $\displaystyle\int_0^{\pi} \tan^2\frac{\theta}{3} \, d\theta$

89. $\displaystyle\int_{-\pi/3}^0 \sec x \tan x \, dx$ **90.** $\displaystyle\int_{\pi/4}^{3\pi/4} \csc z \cot z \, dz$

91. $\displaystyle\int_0^{\pi/2} 5(\sin x)^{3/2}\cos x \, dx$ **92.** $\displaystyle\int_{-\pi/2}^{\pi/2} 15\sin^4 3x \cos 3x \, dx$

93. $\displaystyle\int_0^{\pi/2} \frac{3\sin x \cos x}{\sqrt{1 + 3\sin^2 x}} \, dx$ **94.** $\displaystyle\int_0^{\pi/4} \frac{\sec^2 x}{(1 + 7\tan x)^{2/3}} \, dx$

95. $\displaystyle\int_1^4 \left(\frac{x}{8} + \frac{1}{2x}\right) dx$ **96.** $\displaystyle\int_1^8 \left(\frac{2}{3x} - \frac{8}{x^2}\right) dx$

97. $\displaystyle\int_{-2}^{-1} e^{-(x+1)} \, dx$ **98.** $\displaystyle\int_{-\ln 2}^0 e^{2w} \, dw$

99. $\displaystyle\int_0^{\ln 5} e^r (3e^r + 1)^{-3/2} \, dr$ **100.** $\displaystyle\int_0^{\ln 9} e^{\theta}(e^{\theta} - 1)^{1/2} \, d\theta$

101. $\displaystyle\int_1^e \frac{1}{x}(1 + 7\ln x)^{-1/3} \, dx$ **102.** $\displaystyle\int_1^3 \frac{(\ln(v + 1))^2}{v + 1} \, dv$

103. $\displaystyle\int_1^8 \frac{\log_4 \theta}{\theta} \, d\theta$ **104.** $\displaystyle\int_1^e \frac{8 \ln 3 \log_3 \theta}{\theta} \, d\theta$

At what rate is the covered area changing? At the instant x, $A(x)$ is increasing by the width $f(v(x))$ of the unrolling carpet times the rate dv/dx at which the carpet is being unrolled. That is, $A(x)$ is being increased at the rate

$$f(v(x)) \frac{dv}{dx}.$$

At the same time, A is being decreased at the rate

$$f(u(x)) \frac{du}{dx},$$

the width at the end that is being rolled up times the rate du/dx. The net rate of change in A is

$$\frac{dA}{dx} = f(v(x)) \frac{dv}{dx} - f(u(x)) \frac{du}{dx},$$

which is precisely Leibniz's Rule.

To prove the rule, let F be an antiderivative of f on $[a, b]$. Then

$$\int_{u(x)}^{v(x)} f(t)\, dt = F(v(x)) - F(u(x)).$$

Differentiating both sides of this equation with respect to x gives the equation we want:

$$\frac{d}{dx} \int_{u(x)}^{v(x)} f(t)\, dt = \frac{d}{dx} \left[F(v(x)) - F(u(x)) \right]$$

$$= F'(v(x)) \frac{dv}{dx} - F'(u(x)) \frac{du}{dx} \qquad \text{Chain Rule}$$

$$= f(v(x)) \frac{dv}{dx} - f(u(x)) \frac{du}{dx}.$$

Use Leibniz's Rule to find the derivatives of the functions in Exercises 31–38.

31. $f(x) = \displaystyle\int_{1/x}^{x} \frac{1}{t}\, dt$

32. $f(x) = \displaystyle\int_{\cos x}^{\sin x} \frac{1}{1 - t^2}\, dt$

33. $g(y) = \displaystyle\int_{\sqrt{y}}^{2\sqrt{y}} \sin t^2\, dt$

34. $g(y) = \displaystyle\int_{\sqrt{y}}^{y^2} \frac{e^t}{t}\, dt$

35. $y = \displaystyle\int_{x^2/2}^{x^2} \ln \sqrt{t}\, dt$

36. $y = \displaystyle\int_{\sqrt{x}}^{\sqrt[3]{x}} \ln t\, dt$

37. $y = \displaystyle\int_{0}^{\ln x} \sin e^t\, dt$

38. $y = \displaystyle\int_{e^{4\sqrt{x}}}^{e^{2x}} \ln t\, dt$

Theory and Examples

39. Use Leibniz's Rule to find the value of x that maximizes the value of the integral

$$\int_{x}^{x+3} t(5 - t)\, dt.$$

40. For what $x > 0$ does $x^{(x^x)} = (x^x)^x$? Give reasons for your answer.

41. Find the areas between the curves $y = 2(\log_2 x)/x$ and $y = 2(\log_4 x)/x$ and the x-axis from $x = 1$ to $x = e$. What is the ratio of the larger area to the smaller?

42. a. Find df/dx if

$$f(x) = \int_{1}^{e^x} \frac{2 \ln t}{t}\, dt.$$

b. Find $f(0)$.

c. What can you conclude about the graph of f? Give reasons for your answer.

43. Find $f'(2)$ if $f(x) = e^{g(x)}$ and $g(x) = \displaystyle\int_{2}^{x} \frac{t}{1 + t^4}\, dt$.

44. Use the accompanying figure to show that

$$\int_{0}^{\pi/2} \sin x\, dx = \frac{\pi}{2} - \int_{0}^{1} \sin^{-1} x\, dx.$$

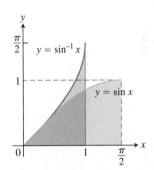

45. Napier's inequality Here are two pictorial proofs that

$$b > a > 0 \quad \Rightarrow \quad \frac{1}{b} < \frac{\ln b - \ln a}{b - a} < \frac{1}{a}.$$

Explain what is going on in each case.

a.

b.

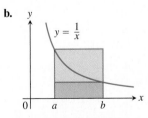

(*Source:* Roger B. Nelson, *College Mathematics Journal*, Vol. 24, No. 2, March 1993, p. 165.)

Technology Application Projects

Mathematica/Maple Modules:

Using Riemann Sums to Estimate Areas, Volumes, and Lengths of Curves
Visualize and approximate areas and volumes in Part I.

Riemann Sums, Definite Integrals, and the Fundamental Theorem of Calculus
Parts I, II, and III develop Riemann sums and definite integrals. Part IV continues the development of the Riemann sum and definite integral using the Fundamental Theorem to solve problems previously investigated.

Rain Catchers, Elevators, and Rockets
Part I illustrates that the area under a curve is the same as the area of an appropriate rectangle for examples taken from the chapter. You will compute the amount of water accumulating in basins of different shapes as the basin is filled and drained.

Motion Along a Straight Line, Part II
You will observe the shape of a graph through dramatic animated visualizations of the derivative relations among position, velocity, and acceleration. Figures in the text can be animated using this software.

Bending of Beams
Study bent shapes of beams, determine their maximum deflections, concavity, and inflection points, and interpret the results in terms of a beam's compression and tension.

6

APPLICATIONS OF
DEFINITE INTEGRALS

OVERVIEW In Chapter 5 we saw that a continuous function over a closed interval has a definite integral, which is the limit of any Riemann sum for the function. We proved that we could evaluate definite integrals using the Fundamental Theorem of Calculus. We also found that the area under a curve and the area between two curves could be computed as definite integrals.

In this chapter we extend the applications of definite integrals to finding volumes, lengths of plane curves, and areas of surfaces of revolution. We also use integrals to solve physical problems involving the work done by a force, the fluid force against a planar wall, and the location of an object's center of mass.

6.1 | Volumes Using Cross-Sections

In this section we define volumes of solids using the areas of their cross-sections. A **cross-section** of a solid S is the plane region formed by intersecting S with a plane (Figure 6.1). We present three different methods for obtaining the cross-sections appropriate to finding the volume of a particular solid: the method of slicing, the disk method, and the washer method.

Suppose we want to find the volume of a solid S like the one in Figure 6.1. We begin by extending the definition of a cylinder from classical geometry to cylindrical solids with arbitrary bases (Figure 6.2). If the cylindrical solid has a known base area A and height h, then the volume of the cylindrical solid is

$$\text{Volume} = \text{area} \times \text{height} = A \cdot h.$$

This equation forms the basis for defining the volumes of many solids that are not cylinders, like the one in Figure 6.1. If the cross-section of the solid S at each point x in the interval $[a, b]$ is a region $S(x)$ of area $A(x)$, and A is a continuous function of x, we can define and calculate the volume of the solid S as the definite integral of $A(x)$. We now show how this integral is obtained by the **method of slicing**.

FIGURE 6.1 A cross-section $S(x)$ of the solid S formed by intersecting S with a plane P_x perpendicular to the x-axis through the point x in the interval $[a, b]$.

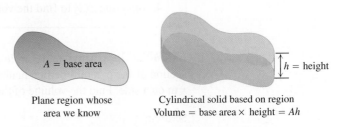

Plane region whose
area we know

Cylindrical solid based on region
Volume = base area × height = Ah

FIGURE 6.2 The volume of a cylindrical solid is always defined to be its base area times its height.

363

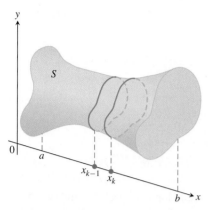

FIGURE 6.3 A typical thin slab in the solid S.

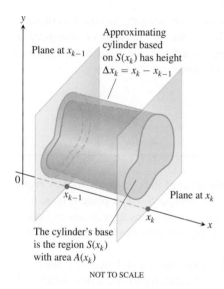

Plane at x_{k-1}

Approximating cylinder based on $S(x_k)$ has height $\Delta x_k = x_k - x_{k-1}$

Plane at x_k

The cylinder's base is the region $S(x_k)$ with area $A(x_k)$

NOT TO SCALE

FIGURE 6.4 The solid thin slab in Figure 6.3 is shown enlarged here. It is approximated by the cylindrical solid with base $S(x_k)$ having area $A(x_k)$ and height $\Delta x_k = x_k - x_{k-1}$.

Slicing by Parallel Planes

We partition $[a, b]$ into subintervals of width (length) Δx_k and slice the solid, as we would a loaf of bread, by planes perpendicular to the x-axis at the partition points $a = x_0 < x_1 < \cdots < x_n = b$. The planes P_{x_k}, perpendicular to the x-axis at the partition points, slice S into thin "slabs" (like thin slices of a loaf of bread). A typical slab is shown in Figure 6.3. We approximate the slab between the plane at x_{k-1} and the plane at x_k by a cylindrical solid with base area $A(x_k)$ and height $\Delta x_k = x_k - x_{k-1}$ (Figure 6.4). The volume V_k of this cylindrical solid is $A(x_k) \cdot \Delta x_k$, which is approximately the same volume as that of the slab:

$$\text{Volume of the } k\text{th slab} \approx V_k = A(x_k) \, \Delta x_k.$$

The volume V of the entire solid S is therefore approximated by the sum of these cylindrical volumes,

$$V \approx \sum_{k=1}^{n} V_k = \sum_{k=1}^{n} A(x_k) \, \Delta x_k.$$

This is a Riemann sum for the function $A(x)$ on $[a, b]$. We expect the approximations from these sums to improve as the norm of the partition of $[a, b]$ goes to zero. Taking a partition of $[a, b]$ into n subintervals with $\|P\| \to 0$ gives

$$\lim_{n \to \infty} \sum_{k=1}^{n} A(x_k) \, \Delta x_k = \int_a^b A(x) \, dx.$$

So we define the limiting definite integral of the Riemann sum to be the volume of the solid S.

DEFINITION The **volume** of a solid of integrable cross-sectional area $A(x)$ from $x = a$ to $x = b$ is the integral of A from a to b,

$$V = \int_a^b A(x) \, dx.$$

This definition applies whenever $A(x)$ is integrable, and in particular when it is continuous. To apply the definition to calculate the volume of a solid, take the following steps:

Calculating the Volume of a Solid

1. *Sketch the solid and a typical cross-section.*
2. *Find a formula for $A(x)$, the area of a typical cross-section.*
3. *Find the limits of integration.*
4. *Integrate $A(x)$ to find the volume.*

EXAMPLE 1 A pyramid 3 m high has a square base that is 3 m on a side. The cross-section of the pyramid perpendicular to the altitude x m down from the vertex is a square x m on a side. Find the volume of the pyramid.

Solution

1. *A sketch.* We draw the pyramid with its altitude along the x-axis and its vertex at the origin and include a typical cross-section (Figure 6.5).

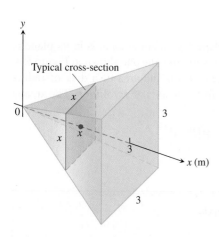

FIGURE 6.5 The cross-sections of the pyramid in Example 1 are squares.

2. *A formula for $A(x)$.* The cross-section at x is a square x meters on a side, so its area is

$$A(x) = x^2.$$

3. *The limits of integration.* The squares lie on the planes from $x = 0$ to $x = 3$.

4. *Integrate to find the volume*:

$$V = \int_0^3 A(x)\,dx = \int_0^3 x^2\,dx = \frac{x^3}{3}\Big]_0^3 = 9\ \text{m}^3.$$ ∎

EXAMPLE 2   A curved wedge is cut from a circular cylinder of radius 3 by two planes. One plane is perpendicular to the axis of the cylinder. The second plane crosses the first plane at a 45° angle at the center of the cylinder. Find the volume of the wedge.

Solution   We draw the wedge and sketch a typical cross-section perpendicular to the x-axis (Figure 6.6). The base of the wedge in the figure is the semicircle with $x \geq 0$ that is cut from the circle $x^2 + y^2 = 9$ by the 45° plane when it intersects the y-axis. For any x in the interval $[0, 3]$, the y-values in this semicircular base vary from $y = -\sqrt{9 - x^2}$ to $y = \sqrt{9 - x^2}$. When we slice through the wedge by a plane perpendicular to the x-axis, we obtain a cross-section at x which is a rectangle of height x whose width extends across the semicircular base. The area of this cross-section is

$$A(x) = (\text{height})(\text{width}) = (x)\left(2\sqrt{9 - x^2}\right)$$
$$= 2x\sqrt{9 - x^2}.$$

The rectangles run from $x = 0$ to $x = 3$, so we have

$$V = \int_a^b A(x)\,dx = \int_0^3 2x\sqrt{9 - x^2}\,dx$$

$$= -\frac{2}{3}(9 - x^2)^{3/2}\Big]_0^3$$

Let $u = 9 - x^2$, $du = -2x\,dx$, integrate, and substitute back.

$$= 0 + \frac{2}{3}(9)^{3/2}$$

$$= 18.$$ ∎

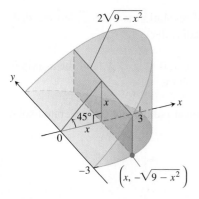

FIGURE 6.6 The wedge of Example 2, sliced perpendicular to the x-axis. The cross-sections are rectangles.

EXAMPLE 3   Cavalieri's principle says that solids with equal altitudes and identical cross-sectional areas at each height have the same volume (Figure 6.7). This follows immediately from the definition of volume, because the cross-sectional area function $A(x)$ and the interval $[a, b]$ are the same for both solids.

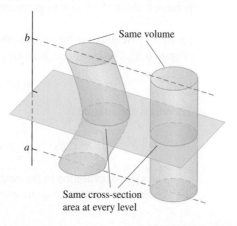

FIGURE 6.7 *Cavalieri's principle:* These solids have the same volume, which can be illustrated with stacks of coins. ∎

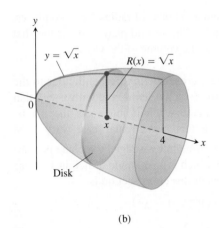

FIGURE 6.8 The region (a) and solid of revolution (b) in Example 4.

Solids of Revolution: The Disk Method

The solid generated by rotating (or revolving) a plane region about an axis in its plane is called a **solid of revolution**. To find the volume of a solid like the one shown in Figure 6.8, we need only observe that the cross-sectional area $A(x)$ is the area of a disk of radius $R(x)$, the distance of the planar region's boundary from the axis of revolution. The area is then

$$A(x) = \pi(\text{radius})^2 = \pi[R(x)]^2.$$

So the definition of volume in this case gives

Volume by Disks for Rotation About the x-axis

$$V = \int_a^b A(x)\, dx = \int_a^b \pi[R(x)]^2\, dx.$$

This method for calculating the volume of a solid of revolution is often called the **disk method** because a cross-section is a circular disk of radius $R(x)$.

EXAMPLE 4   The region between the curve $y = \sqrt{x}$, $0 \le x \le 4$, and the x-axis is revolved about the x-axis to generate a solid. Find its volume.

Solution   We draw figures showing the region, a typical radius, and the generated solid (Figure 6.8). The volume is

$$V = \int_a^b \pi[R(x)]^2\, dx$$

$$= \int_0^4 \pi\left[\sqrt{x}\right]^2 dx \qquad \text{Radius } R(x) = \sqrt{x} \text{ for rotation around } x\text{-axis}$$

$$= \pi \int_0^4 x\, dx = \pi \frac{x^2}{2}\Big]_0^4 = \pi \frac{(4)^2}{2} = 8\pi. \qquad \blacksquare$$

EXAMPLE 5   The circle

$$x^2 + y^2 = a^2$$

is rotated about the x-axis to generate a sphere. Find its volume.

Solution   We imagine the sphere cut into thin slices by planes perpendicular to the x-axis (Figure 6.9). The cross-sectional area at a typical point x between $-a$ and a is

$$A(x) = \pi y^2 = \pi(a^2 - x^2). \qquad R(x) = \sqrt{a^2 - x^2} \text{ for rotation around } x\text{-axis}$$

Therefore, the volume is

$$V = \int_{-a}^a A(x)\, dx = \int_{-a}^a \pi(a^2 - x^2)\, dx = \pi\left[a^2 x - \frac{x^3}{3}\right]_{-a}^a = \frac{4}{3}\pi a^3. \qquad \blacksquare$$

The axis of revolution in the next example is not the x-axis, but the rule for calculating the volume is the same: Integrate $\pi(\text{radius})^2$ between appropriate limits.

EXAMPLE 6   Find the volume of the solid generated by revolving the region bounded by $y = \sqrt{x}$ and the lines $y = 1$, $x = 4$ about the line $y = 1$.

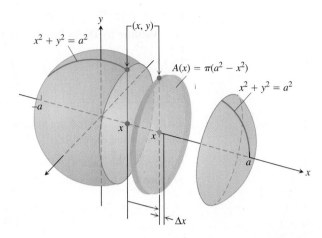

FIGURE 6.9 The sphere generated by rotating the circle $x^2 + y^2 = a^2$ about the x-axis. The radius is $R(x) = y = \sqrt{a^2 - x^2}$ (Example 5).

Solution   We draw figures showing the region, a typical radius, and the generated solid (Figure 6.10). The volume is

$$V = \int_1^4 \pi [R(x)]^2 \, dx$$

$$= \int_1^4 \pi \left[\sqrt{x} - 1 \right]^2 dx \qquad \text{Radius } R(x) = \sqrt{x} - 1 \text{ for rotation around } y = 1$$

$$= \pi \int_1^4 \left[x - 2\sqrt{x} + 1 \right] dx \qquad \text{Expand integrand.}$$

$$= \pi \left[\frac{x^2}{2} - 2 \cdot \frac{2}{3} x^{3/2} + x \right]_1^4 = \frac{7\pi}{6}. \qquad \text{Integrate.}$$

(a)

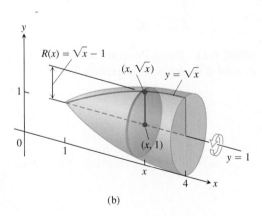

(b)

FIGURE 6.10   The region (a) and solid of revolution (b) in Example 6. ∎

To find the volume of a solid generated by revolving a region between the y-axis and a curve $x = R(y)$, $c \le y \le d$, about the y-axis, we use the same method with x replaced by y. In this case, the circular cross-section is

$$A(y) = \pi [\text{radius}]^2 = \pi [R(y)]^2,$$

and the definition of volume gives

Volume by Disks for Rotation About the y-axis

$$V = \int_c^d A(y)\, dy = \int_c^d \pi [R(y)]^2\, dy.$$

EXAMPLE 7 Find the volume of the solid generated by revolving the region between the y-axis and the curve $x = 2/y$, $1 \le y \le 4$, about the y-axis.

Solution We draw figures showing the region, a typical radius, and the generated solid (Figure 6.11). The volume is

$$V = \int_1^4 \pi [R(y)]^2\, dy$$

$$= \int_1^4 \pi \left(\frac{2}{y}\right)^2 dy \qquad \text{Radius } R(y) = \frac{2}{y} \text{ for rotation around } y\text{-axis}$$

$$= \pi \int_1^4 \frac{4}{y^2}\, dy = 4\pi \left[-\frac{1}{y}\right]_1^4 = 4\pi \left[\frac{3}{4}\right] = 3\pi. \qquad \blacksquare$$

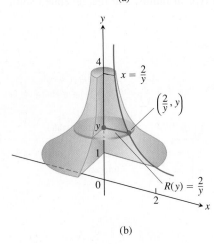

FIGURE 6.11 The region (a) and part of the solid of revolution (b) in Example 7.

EXAMPLE 8 Find the volume of the solid generated by revolving the region between the parabola $x = y^2 + 1$ and the line $x = 3$ about the line $x = 3$.

Solution We draw figures showing the region, a typical radius, and the generated solid (Figure 6.12). Note that the cross-sections are perpendicular to the line $x = 3$ and have y-coordinates from $y = -\sqrt{2}$ to $y = \sqrt{2}$. The volume is

$$V = \int_{-\sqrt{2}}^{\sqrt{2}} \pi [R(y)]^2\, dy \qquad y = \pm\sqrt{2} \text{ when } x = 3$$

$$= \int_{-\sqrt{2}}^{\sqrt{2}} \pi [2 - y^2]^2\, dy \qquad \begin{array}{l}\text{Radius } R(y) = 3 - (y^2 + 1) \\ \text{for rotation around axis } x = 3\end{array}$$

$$= \pi \int_{-\sqrt{2}}^{\sqrt{2}} [4 - 4y^2 + y^4]\, dy \qquad \text{Expand integrand.}$$

$$= \pi \left[4y - \frac{4}{3}y^3 + \frac{y^5}{5}\right]_{-\sqrt{2}}^{\sqrt{2}} \qquad \text{Integrate.}$$

$$= \frac{64\pi\sqrt{2}}{15}.$$

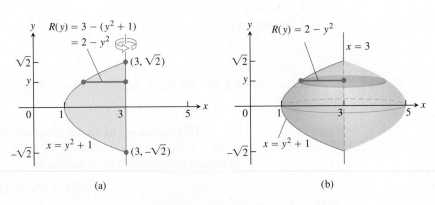

FIGURE 6.12 The region (a) and solid of revolution (b) in Example 8.

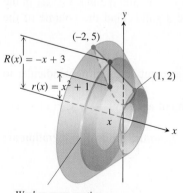

FIGURE 6.13 The cross-sections of the solid of revolution generated here are washers, not disks, so the integral $\int_a^b A(x)\,dx$ leads to a slightly different formula.

Solids of Revolution: The Washer Method

If the region we revolve to generate a solid does not border on or cross the axis of revolution, the solid has a hole in it (Figure 6.13). The cross-sections perpendicular to the axis of revolution are *washers* (the purplish circular surface in Figure 6.13) instead of disks. The dimensions of a typical washer are

$$\text{Outer radius:} \quad R(x)$$

$$\text{Inner radius:} \quad r(x)$$

The washer's area is

$$A(x) = \pi[R(x)]^2 - \pi[r(x)]^2 = \pi([R(x)]^2 - [r(x)]^2).$$

Consequently, the definition of volume in this case gives

Volume by Washers for Rotation About the x-axis

$$V = \int_a^b A(x)\,dx = \int_a^b \pi([R(x)]^2 - [r(x)]^2)\,dx.$$

This method for calculating the volume of a solid of revolution is called the **washer method** because a thin slab of the solid resembles a circular washer of outer radius $R(x)$ and inner radius $r(x)$.

EXAMPLE 9 The region bounded by the curve $y = x^2 + 1$ and the line $y = -x + 3$ is revolved about the x-axis to generate a solid. Find the volume of the solid.

Solution We use the four steps for calculating the volume of a solid as discussed early in this section.

1. Draw the region and sketch a line segment across it perpendicular to the axis of revolution (the red segment in Figure 6.14a).

2. Find the outer and inner radii of the washer that would be swept out by the line segment if it were revolved about the x-axis along with the region.

   These radii are the distances of the ends of the line segment from the axis of revolution (Figure 6.14).

$$\text{Outer radius:} \quad R(x) = -x + 3$$

$$\text{Inner radius:} \quad r(x) = x^2 + 1$$

FIGURE 6.14 (a) The region in Example 9 spanned by a line segment perpendicular to the axis of revolution. (b) When the region is revolved about the x-axis, the line segment generates a washer.

Washer cross-section
Outer radius: $R(x) = -x + 3$
Inner radius: $r(x) = x^2 + 1$
(b)

3. Find the limits of integration by finding the x-coordinates of the intersection points of the curve and line in Figure 6.14a.

$$x^2 + 1 = -x + 3$$
$$x^2 + x - 2 = 0$$
$$(x + 2)(x - 1) = 0$$
$$x = -2, \quad x = 1 \qquad \text{Limits of integration}$$

4. Evaluate the volume integral.

$$V = \int_a^b \pi([R(x)]^2 - [r(x)]^2)\, dx \qquad \text{Rotation around } x\text{-axis}$$

$$= \int_{-2}^1 \pi((-x + 3)^2 - (x^2 + 1)^2)\, dx \qquad \substack{\text{Values from Steps 2}\\ \text{and 3}}$$

$$= \pi \int_{-2}^1 (8 - 6x - x^2 - x^4)\, dx \qquad \text{Simplify algebraically.}$$

$$= \pi \left[8x - 3x^2 - \frac{x^3}{3} - \frac{x^5}{5}\right]_{-2}^1 = \frac{117\pi}{5} \qquad \blacksquare$$

To find the volume of a solid formed by revolving a region about the y-axis, we use the same procedure as in Example 9, but integrate with respect to y instead of x. In this situation the line segment sweeping out a typical washer is perpendicular to the y-axis (the axis of revolution), and the outer and inner radii of the washer are functions of y.

EXAMPLE 10   The region bounded by the parabola $y = x^2$ and the line $y = 2x$ in the first quadrant is revolved about the y-axis to generate a solid. Find the volume of the solid.

Solution   First we sketch the region and draw a line segment across it perpendicular to the axis of revolution (the y-axis). See Figure 6.15a.

The radii of the washer swept out by the line segment are $R(y) = \sqrt{y}$, $r(y) = y/2$ (Figure 6.15).

The line and parabola intersect at $y = 0$ and $y = 4$, so the limits of integration are $c = 0$ and $d = 4$. We integrate to find the volume:

$$V = \int_c^d \pi([R(y)]^2 - [r(y)]^2)\, dy \qquad \text{Rotation around } y\text{-axis}$$

$$= \int_0^4 \pi\left(\left[\sqrt{y}\right]^2 - \left[\frac{y}{2}\right]^2\right) dy \qquad \substack{\text{Substitute for radii and}\\ \text{limits of integration.}}$$

$$= \pi \int_0^4 \left(y - \frac{y^2}{4}\right) dy = \pi \left[\frac{y^2}{2} - \frac{y^3}{12}\right]_0^4 = \frac{8}{3}\pi. \qquad \blacksquare$$

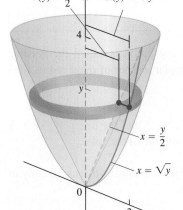

FIGURE 6.15   (a) The region being rotated about the y-axis, the washer radii, and limits of integration in Example 10. (b) The washer swept out by the line segment in part (a).

Exercises 6.1

Volumes by Slicing

Find the volumes of the solids in Exercises 1–10.

1. The solid lies between planes perpendicular to the x-axis at $x = 0$ and $x = 4$. The cross-sections perpendicular to the axis on the interval $0 \leq x \leq 4$ are squares whose diagonals run from the parabola $y = -\sqrt{x}$ to the parabola $y = \sqrt{x}$.

2. The solid lies between planes perpendicular to the x-axis at $x = -1$ and $x = 1$. The cross-sections perpendicular to the x-axis are circular disks whose diameters run from the parabola $y = x^2$ to the parabola $y = 2 - x^2$.

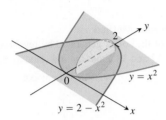

3. The solid lies between planes perpendicular to the x-axis at $x = -1$ and $x = 1$. The cross-sections perpendicular to the x-axis between these planes are squares whose bases run from the semicircle $y = -\sqrt{1 - x^2}$ to the semicircle $y = \sqrt{1 - x^2}$.

4. The solid lies between planes perpendicular to the x-axis at $x = -1$ and $x = 1$. The cross-sections perpendicular to the x-axis between these planes are squares whose diagonals run from the semicircle $y = -\sqrt{1 - x^2}$ to the semicircle $y = \sqrt{1 - x^2}$.

5. The base of a solid is the region between the curve $y = 2\sqrt{\sin x}$ and the interval $[0, \pi]$ on the x-axis. The cross-sections perpendicular to the x-axis are

  a. equilateral triangles with bases running from the x-axis to the curve as shown in the accompanying figure.

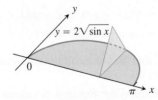

  b. squares with bases running from the x-axis to the curve.

6. The solid lies between planes perpendicular to the x-axis at $x = -\pi/3$ and $x = \pi/3$. The cross-sections perpendicular to the x-axis are

  a. circular disks with diameters running from the curve $y = \tan x$ to the curve $y = \sec x$.

  b. squares whose bases run from the curve $y = \tan x$ to the curve $y = \sec x$.

7. The base of a solid is the region bounded by the graphs of $y = 3x$, $y = 6$, and $x = 0$. The cross-sections perpendicular to the x-axis are

  a. rectangles of height 10.

  b. rectangles of perimeter 20.

8. The base of a solid is the region bounded by the graphs of $y = \sqrt{x}$ and $y = x/2$. The cross-sections perpendicular to the x-axis are

  a. isosceles triangles of height 6.

  b. semi-circles with diameters running across the base of the solid.

9. The solid lies between planes perpendicular to the y-axis at $y = 0$ and $y = 2$. The cross-sections perpendicular to the y-axis are circular disks with diameters running from the y-axis to the parabola $x = \sqrt{5}y^2$.

10. The base of the solid is the disk $x^2 + y^2 \leq 1$. The cross-sections by planes perpendicular to the y-axis between $y = -1$ and $y = 1$ are isosceles right triangles with one leg in the disk.

11. Find the volume of the given tetrahedron. (Hint: Consider slices perpendicular to one of the labeled edges.)

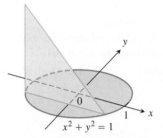

12. Find the volume of the given pyramid, which has a square base of area 9 and height 5.

13. A twisted solid A square of side length s lies in a plane perpendicular to a line L. One vertex of the square lies on L. As this square moves a distance h along L, the square turns one revolution about L to generate a corkscrew-like column with square cross-sections.

  a. Find the volume of the column.

  b. What will the volume be if the square turns twice instead of once? Give reasons for your answer.

14. Cavalieri's principle   A solid lies between planes perpendicular to the x-axis at $x = 0$ and $x = 12$. The cross-sections by planes perpendicular to the x-axis are circular disks whose diameters run from the line $y = x/2$ to the line $y = x$ as shown in the accompanying figure. Explain why the solid has the same volume as a right circular cone with base radius 3 and height 12.

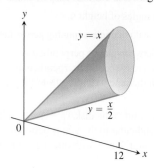

Volumes by the Disk Method

In Exercises 15–18, find the volume of the solid generated by revolving the shaded region about the given axis.

15. About the x-axis

16. About the y-axis

17. About the y-axis

18. About the x-axis

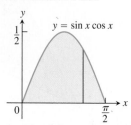

Find the volumes of the solids generated by revolving the regions bounded by the lines and curves in Exercises 19–28 about the x-axis.

19. $y = x^2$, $y = 0$, $x = 2$

20. $y = x^3$, $y = 0$, $x = 2$

21. $y = \sqrt{9 - x^2}$, $y = 0$

22. $y = x - x^2$, $y = 0$

23. $y = \sqrt{\cos x}$, $0 \le x \le \pi/2$, $y = 0$, $x = 0$

24. $y = \sec x$, $y = 0$, $x = -\pi/4$, $x = \pi/4$

25. $y = e^{-x}$, $y = 0$, $x = 0$, $x = 1$

26. The region between the curve $y = \sqrt{\cot x}$ and the x-axis from $x = \pi/6$ to $x = \pi/2$.

27. The region between the curve $y = 1/(2\sqrt{x})$ and the x-axis from $x = 1/4$ to $x = 4$.

28. $y = e^{x-1}$, $y = 0$, $x = 1$, $x = 3$

In Exercises 29 and 30, find the volume of the solid generated by revolving the region about the given line.

29. The region in the first quadrant bounded above by the line $y = \sqrt{2}$, below by the curve $y = \sec x \tan x$, and on the left by the y-axis, about the line $y = \sqrt{2}$

30. The region in the first quadrant bounded above by the line $y = 2$, below by the curve $y = 2 \sin x$, $0 \le x \le \pi/2$, and on the left by the y-axis, about the line $y = 2$

Find the volumes of the solids generated by revolving the regions bounded by the lines and curves in Exercises 31–36 about the y-axis.

31. The region enclosed by $x = \sqrt{5}\, y^2$, $x = 0$, $y = -1$, $y = 1$

32. The region enclosed by $x = y^{3/2}$, $x = 0$, $y = 2$

33. The region enclosed by $x = \sqrt{2 \sin 2y}$, $0 \le y \le \pi/2$, $x = 0$

34. The region enclosed by $x = \sqrt{\cos (\pi y/4)}$, $-2 \le y \le 0$, $x = 0$

35. $x = 2/\sqrt{y + 1}$, $x = 0$, $y = 0$, $y = 3$

36. $x = \sqrt{2y}/(y^2 + 1)$, $x = 0$, $y = 1$

Volumes by the Washer Method

Find the volumes of the solids generated by revolving the shaded regions in Exercises 37 and 38 about the indicated axes.

37. The x-axis

38. The y-axis

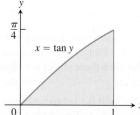

Find the volumes of the solids generated by revolving the regions bounded by the lines and curves in Exercises 39–44 about the x-axis.

39. $y = x$, $y = 1$, $x = 0$

40. $y = 2\sqrt{x}$, $y = 2$, $x = 0$

41. $y = x^2 + 1$, $y = x + 3$

42. $y = 4 - x^2$, $y = 2 - x$

43. $y = \sec x$, $y = \sqrt{2}$, $-\pi/4 \le x \le \pi/4$

44. $y = \sec x$, $y = \tan x$, $x = 0$, $x = 1$

In Exercises 45–48, find the volume of the solid generated by revolving each region about the y-axis.

45. The region enclosed by the triangle with vertices $(1, 0)$, $(2, 1)$, and $(1, 1)$

46. The region enclosed by the triangle with vertices $(0, 1)$, $(1, 0)$, and $(1, 1)$

47. The region in the first quadrant bounded above by the parabola $y = x^2$, below by the x-axis, and on the right by the line $x = 2$

48. The region in the first quadrant bounded on the left by the circle $x^2 + y^2 = 3$, on the right by the line $x = \sqrt{3}$, and above by the line $y = \sqrt{3}$

In Exercises 49 and 50, find the volume of the solid generated by revolving each region about the given axis.

49. The region in the first quadrant bounded above by the curve $y = x^2$, below by the x-axis, and on the right by the line $x = 1$, about the line $x = -1$

50. The region in the second quadrant bounded above by the curve $y = -x^3$, below by the x-axis, and on the left by the line $x = -1$, about the line $x = -2$

Volumes of Solids of Revolution

51. Find the volume of the solid generated by revolving the region bounded by $y = \sqrt{x}$ and the lines $y = 2$ and $x = 0$ about

    a. the x-axis.   **b.** the y-axis.

    c. the line $y = 2$.   **d.** the line $x = 4$.

52. Find the volume of the solid generated by revolving the triangular region bounded by the lines $y = 2x$, $y = 0$, and $x = 1$ about

    a. the line $x = 1$.   **b.** the line $x = 2$.

53. Find the volume of the solid generated by revolving the region bounded by the parabola $y = x^2$ and the line $y = 1$ about

    a. the line $y = 1$.   **b.** the line $y = 2$.

    c. the line $y = -1$.

54. By integration, find the volume of the solid generated by revolving the triangular region with vertices $(0, 0)$, $(b, 0)$, $(0, h)$ about

    a. the x-axis.   **b.** the y-axis.

Theory and Applications

55. The volume of a torus The disk $x^2 + y^2 \le a^2$ is revolved about the line $x = b$ ($b > a$) to generate a solid shaped like a doughnut and called a *torus*. Find its volume. (*Hint:* $\int_{-a}^{a} \sqrt{a^2 - y^2}\, dy = \pi a^2/2$, since it is the area of a semicircle of radius a.)

56. Volume of a bowl A bowl has a shape that can be generated by revolving the graph of $y = x^2/2$ between $y = 0$ and $y = 5$ about the y-axis.

    a. Find the volume of the bowl.

    b. Related rates If we fill the bowl with water at a constant rate of 3 cubic units per second, how fast will the water level in the bowl be rising when the water is 4 units deep?

57. Volume of a bowl

    a. A hemispherical bowl of radius a contains water to a depth h. Find the volume of water in the bowl.

    b. Related rates Water runs into a sunken concrete hemispherical bowl of radius 5 m at the rate of $0.2 \text{ m}^3/\text{sec}$. How fast is the water level in the bowl rising when the water is 4 m deep?

58. Explain how you could estimate the volume of a solid of revolution by measuring the shadow cast on a table parallel to its axis of revolution by a light shining directly above it.

59. Volume of a hemisphere Derive the formula $V = (2/3)\pi R^3$ for the volume of a hemisphere of radius R by comparing its cross-sections with the cross-sections of a solid right circular cylinder of radius R and height R from which a solid right circular cone of base radius R and height R has been removed, as suggested by the accompanying figure.

60. Designing a plumb bob Having been asked to design a brass plumb bob that will weigh in the neighborhood of 190 g, you decide to shape it like the solid of revolution shown here. Find the plumb bob's volume. If you specify a brass that weighs 8.5 g/cm^3, how much will the plumb bob weigh (to the nearest gram)?

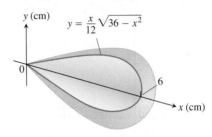

61. Designing a wok You are designing a wok frying pan that will be shaped like a spherical bowl with handles. A bit of experimentation at home persuades you that you can get one that holds about 3 L if you make it 9 cm deep and give the sphere a radius of 16 cm. To be sure, you picture the wok as a solid of revolution, as shown here, and calculate its volume with an integral. To the nearest cubic centimeter, what volume do you really get? ($1 \text{ L} = 1000 \text{ cm}^3$.)

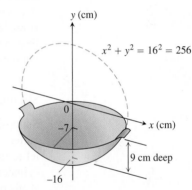

62. Max-min The arch $y = \sin x$, $0 \le x \le \pi$, is revolved about the line $y = c$, $0 \le c \le 1$, to generate the solid in the accompanying figure.

    a. Find the value of c that minimizes the volume of the solid. What is the minimum volume?

    b. What value of c in $[0, 1]$ maximizes the volume of the solid?

    T **c.** Graph the solid's volume as a function of c, first for $0 \le c \le 1$ and then on a larger domain. What happens to the volume of the solid as c moves away from $[0, 1]$? Does this make sense physically? Give reasons for your answers.

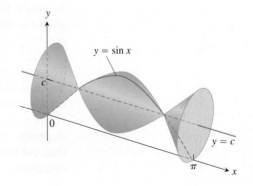

63. Consider the region R bounded by the graphs of $y = f(x) > 0$, $x = a > 0$, $x = b > a$, and $y = 0$ (see accomanying figure). If the volume of the solid formed by revolving R about the x-axis is 4π, and the volume of the solid formed by revolving R about the line $y = -1$ is 8π, find the area of R.

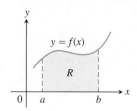

64. Consider the region R given in Exercise 63. If the volume of the solid formed by revolving R around the x-axis is 6π, and the volume of the solid formed by revolving R around the line $y = -2$ is 10π, find the area of R.

6.2 | Volumes Using Cylindrical Shells

In Section 6.1 we defined the volume of a solid as the definite integral $V = \int_a^b A(x)\, dx$, where $A(x)$ is an integrable cross-sectional area of the solid from $x = a$ to $x = b$. The area $A(x)$ was obtained by slicing through the solid with a plane perpendicular to the x-axis. However, this method of slicing is sometimes awkward to apply, as we will illustrate in our first example. To overcome this difficulty, we use the same integral definition for volume, but obtain the area by slicing through the solid in a different way.

Slicing with Cylinders

Suppose we slice through the solid using circular cylinders of increasing radii, like cookie cutters. We slice straight down through the solid so that the axis of each cylinder is parallel to the y-axis. The vertical axis of each cylinder is the same line, but the radii of the cylinders increase with each slice. In this way the solid is sliced up into thin cylindrical shells of constant thickness that grow outward from their common axis, like circular tree rings. Unrolling a cylindrical shell shows that its volume is approximately that of a rectangular slab with area $A(x)$ and thickness Δx. This slab interpretation allows us to apply the same integral definition for volume as before. The following example provides some insight before we derive the general method.

EXAMPLE 1   The region enclosed by the x-axis and the parabola $y = f(x) = 3x - x^2$ is revolved about the vertical line $x = -1$ to generate a solid (Figure 6.16). Find the volume of the solid.

Solution   Using the washer method from Section 6.1 would be awkward here because we would need to express the x-values of the left and right sides of the parabola in Figure 6.16a in terms of y. (These x-values are the inner and outer radii for a typical washer, requiring us to solve $y = 3x - x^2$ for x, which leads to complicated formulas.) Instead of rotating a horizontal strip of thickness Δy, we rotate a *vertical strip* of thickness Δx. This rotation produces a *cylindrical shell* of height y_k above a point x_k within the base of the vertical strip and of thickness Δx. An example of a cylindrical shell is shown as the orange-shaded region in Figure 6.17. We can think of the cylindrical shell shown in the figure as approximating a slice of the solid obtained by cutting straight down through it, parallel to the axis of revolution, all the way around close to the inside hole. We then cut another cylindrical slice around the enlarged hole, then another, and so on, obtaining n cylinders. The radii of the cylinders gradually increase, and the heights of

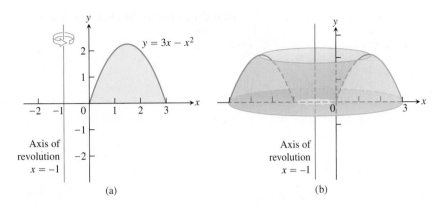

FIGURE 6.16 (a) The graph of the region in Example 1, before revolution. (b) The solid formed when the region in part (a) is revolved about the axis of revolution $x = -1$.

FIGURE 6.17 A cylindrical shell of height y_k obtained by rotating a vertical strip of thickness Δx_k about the line $x = -1$. The outer radius of the cylinder occurs at x_k, where the height of the parabola is $y_k = 3x_k - x_k^2$ (Example 1).

the cylinders follow the contour of the parabola: shorter to taller, then back to shorter (Figure 6.16a).

Each slice is sitting over a subinterval of the x-axis of length (width) Δx_k. Its radius is approximately $(1 + x_k)$, and its height is approximately $3x_k - x_k^2$. If we unroll the cylinder at x_k and flatten it out, it becomes (approximately) a rectangular slab with thickness Δx_k (Figure 6.18). The outer circumference of the kth cylinder is $2\pi \cdot \text{radius} = 2\pi(1 + x_k)$, and this is the length of the rolled-out rectangular slab. Its volume is approximated by that of a rectangular solid,

$$\Delta V_k = \text{circumference} \times \text{height} \times \text{thickness}$$

$$= 2\pi(1 + x_k) \cdot \left(3x_k - x_k^2\right) \cdot \Delta x_k.$$

Summing together the volumes ΔV_k of the individual cylindrical shells over the interval $[0, 3]$ gives the Riemann sum

$$\sum_{k=1}^{n} \Delta V_k = \sum_{k=1}^{n} 2\pi(x_k + 1)\left(3x_k - x_k^2\right) \Delta x_k.$$

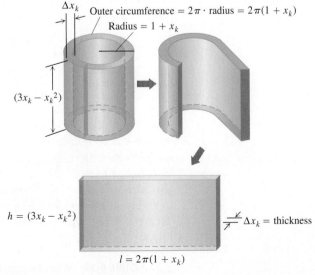

FIGURE 6.18 Cutting and unrolling a cylindrical shell gives a nearly rectangular solid (Example 1).

Taking the limit as the thickness $\Delta x_k \to 0$ and $n \to \infty$ gives the volume integral

$$V = \lim_{n \to \infty} \sum_{k=1}^{n} 2\pi(x_k + 1)\left(3x_k - x_k^2\right) \Delta x_k$$

$$= \int_0^3 2\pi(x + 1)(3x - x^2)\, dx$$

$$= \int_0^3 2\pi(3x^2 + 3x - x^3 - x^2)\, dx$$

$$= 2\pi \int_0^3 (2x^2 + 3x - x^3)\, dx$$

$$= 2\pi\left[\frac{2}{3}x^3 + \frac{3}{2}x^2 - \frac{1}{4}x^4\right]_0^3 = \frac{45\pi}{2}. \qquad \blacksquare$$

We now generalize the procedure used in Example 1.

The Shell Method

Suppose the region bounded by the graph of a nonnegative continuous function $y = f(x)$ and the x-axis over the finite closed interval $[a, b]$ lies to the right of the vertical line $x = L$ (Figure 6.19a). We assume $a \geq L$, so the vertical line may touch the region, but not pass through it. We generate a solid S by rotating this region about the vertical line L.

Let P be a partition of the interval $[a, b]$ by the points $a = x_0 < x_1 < \cdots < x_n = b$, and let c_k be the midpoint of the kth subinterval $[x_{k-1}, x_k]$. We approximate the region in Figure 6.19a with rectangles based on this partition of $[a, b]$. A typical approximating rectangle has height $f(c_k)$ and width $\Delta x_k = x_k - x_{k-1}$. If this rectangle is rotated about the vertical line $x = L$, then a shell is swept out, as in Figure 6.19b. A formula from geometry tells us that the volume of the shell swept out by the rectangle is

> The volume of a cylindrical shell of height h with inner radius r and outer radius R is
>
> $$\pi R^2 h - \pi r^2 h = 2\pi\left(\frac{R + r}{2}\right)(h)(R - r)$$

$$\Delta V_k = 2\pi \times \text{average shell radius} \times \text{shell height} \times \text{thickness}$$

$$= 2\pi \cdot (c_k - L) \cdot f(c_k) \cdot \Delta x_k.$$

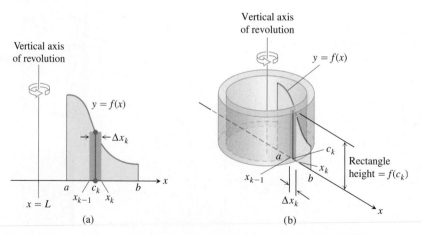

FIGURE 6.19 When the region shown in (a) is revolved about the vertical line $x = L$, a solid is produced which can be sliced into cylindrical shells. A typical shell is shown in (b).

We approximate the volume of the solid S by summing the volumes of the shells swept out by the n rectangles based on P:

$$V \approx \sum_{k=1}^{n} \Delta V_k.$$

The limit of this Riemann sum as each $\Delta x_k \to 0$ and $n \to \infty$ gives the volume of the solid as a definite integral:

$$V = \lim_{n \to \infty} \sum_{k=1}^{n} \Delta V_k = \int_{a}^{b} 2\pi(\text{shell radius})(\text{shell height}) \, dx.$$

$$= \int_{a}^{b} 2\pi(x - L)f(x) \, dx.$$

We refer to the variable of integration, here x, as the **thickness variable**. We use the first integral, rather than the second containing a formula for the integrand, to emphasize the *process* of the shell method. This will allow for rotations about a horizontal line L as well.

Shell Formula for Revolution About a Vertical Line
The volume of the solid generated by revolving the region between the x-axis and the graph of a continuous function $y = f(x) \geq 0, L \leq a \leq x \leq b$, about a vertical line $x = L$ is

$$V = \int_{a}^{b} 2\pi \binom{\text{shell}}{\text{radius}} \binom{\text{shell}}{\text{height}} \, dx.$$

EXAMPLE 2 The region bounded by the curve $y = \sqrt{x}$, the x-axis, and the line $x = 4$ is revolved about the y-axis to generate a solid. Find the volume of the solid.

Solution Sketch the region and draw a line segment across it *parallel* to the axis of revolution (Figure 6.20a). Label the segment's height (shell height) and distance from the axis of revolution (shell radius). (We drew the shell in Figure 6.20b, but you need not do that.)

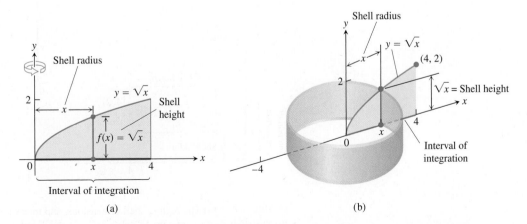

FIGURE 6.20 (a) The region, shell dimensions, and interval of integration in Example 2. (b) The shell swept out by the vertical segment in part (a) with a width Δx.

The shell thickness variable is x, so the limits of integration for the shell formula are $a = 0$ and $b = 4$ (Figure 6.20). The volume is then

$$V = \int_a^b 2\pi \binom{\text{shell}}{\text{radius}}\binom{\text{shell}}{\text{height}} \, dx$$

$$= \int_0^4 2\pi(x)\left(\sqrt{x}\right) dx$$

$$= 2\pi \int_0^4 x^{3/2} \, dx = 2\pi \left[\frac{2}{5} x^{5/2}\right]_0^4 = \frac{128\pi}{5}. \qquad \blacksquare$$

So far, we have used vertical axes of revolution. For horizontal axes, we replace the x's with y's.

EXAMPLE 3   The region bounded by the curve $y = \sqrt{x}$, the x-axis, and the line $x = 4$ is revolved about the x-axis to generate a solid. Find the volume of the solid by the shell method.

Solution   This is the solid whose volume was found by the disk method in Example 4 of Section 6.1. Now we find its volume by the shell method. First, sketch the region and draw a line segment across it *parallel* to the axis of revolution (Figure 6.21a). Label the segment's length (shell height) and distance from the axis of revolution (shell radius). (We drew the shell in Figure 6.21b, but you need not do that.)

In this case, the shell thickness variable is y, so the limits of integration for the shell formula method are $a = 0$ and $b = 2$ (along the y-axis in Figure 6.21). The volume of the solid is

$$V = \int_a^b 2\pi \binom{\text{shell}}{\text{radius}}\binom{\text{shell}}{\text{height}} \, dy$$

$$= \int_0^2 2\pi(y)(4 - y^2) \, dy$$

$$= 2\pi \int_0^2 (4y - y^3) \, dy$$

$$= 2\pi \left[2y^2 - \frac{y^4}{4}\right]_0^2 = 8\pi.$$

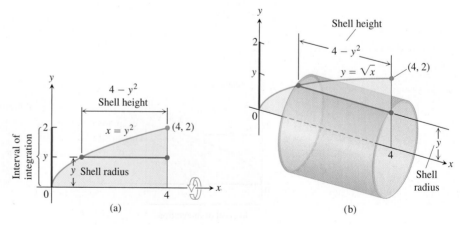

FIGURE 6.21   (a) The region, shell dimensions, and interval of integration in Example 3. (b) The shell swept out by the horizontal segment in part (a) with a width Δy.

Summary of the Shell Method

Regardless of the position of the axis of revolution (horizontal or vertical), the steps for implementing the shell method are these.

1. *Draw the region and sketch a line segment* across it *parallel* to the axis of revolution. *Label* the segment's height or length (shell height) and distance from the axis of revolution (shell radius).

2. *Find the limits of integration* for the thickness variable.

3. *Integrate* the product 2π (shell radius) (shell height) with respect to the thickness variable (x or y) to find the volume.

The shell method gives the same answer as the washer method when both are used to calculate the volume of a region. We do not prove that result here, but it is illustrated in Exercises 37 and 38. (Exercise 45 outlines a proof.) Both volume formulas are actually special cases of a general volume formula we will look at when studying double and triple integrals in Chapter 15. That general formula also allows for computing volumes of solids other than those swept out by regions of revolution.

Exercises 6.2

Revolution About the Axes

In Exercises 1–6, use the shell method to find the volumes of the solids generated by revolving the shaded region about the indicated axis.

1.

$y = 1 + \dfrac{x^2}{4}$

2.

$y = 2 - \dfrac{x^2}{4}$

3.

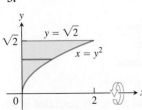

$y = \sqrt{2}$

$x = y^2$

4.

$y = \sqrt{3}$

$x = 3 - y^2$

5. The y-axis

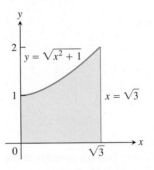

$y = \sqrt{x^2 + 1}$

$x = \sqrt{3}$

6. The y-axis

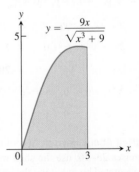

$y = \dfrac{9x}{\sqrt{x^3 + 9}}$

Revolution About the y-Axis

Use the shell method to find the volumes of the solids generated by revolving the regions bounded by the curves and lines in Exercises 7–12 about the y-axis.

7. $y = x$, $y = -x/2$, $x = 2$

8. $y = 2x$, $y = x/2$, $x = 1$

9. $y = x^2$, $y = 2 - x$, $x = 0$, for $x \geq 0$

10. $y = 2 - x^2$, $y = x^2$, $x = 0$

11. $y = 2x - 1$, $y = \sqrt{x}$, $x = 0$

12. $y = 3/(2\sqrt{x})$, $y = 0$, $x = 1$, $x = 4$

13. Let $f(x) = \begin{cases} (\sin x)/x, & 0 < x \le \pi \\ 1, & x = 0 \end{cases}$

a. Show that $x f(x) = \sin x, 0 \le x \le \pi$.

b. Find the volume of the solid generated by revolving the shaded region about the y-axis in the accompanying figure.

14. Let $g(x) = \begin{cases} (\tan x)^2/x, & 0 < x \le \pi/4 \\ 0, & x = 0 \end{cases}$

a. Show that $x g(x) = (\tan x)^2, 0 \le x \le \pi/4$.

b. Find the volume of the solid generated by revolving the shaded region about the y-axis in the accompanying figure.

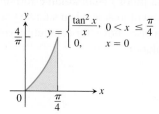

Revolution About the x-Axis

Use the shell method to find the volumes of the solids generated by revolving the regions bounded by the curves and lines in Exercises 15–22 about the x-axis.

15. $x = \sqrt{y}, \quad x = -y, \quad y = 2$

16. $x = y^2, \quad x = -y, \quad y = 2, \quad y \ge 0$

17. $x = 2y - y^2, \quad x = 0$ **18.** $x = 2y - y^2, \quad x = y$

19. $y = |x|, \quad y = 1$ **20.** $y = x, \quad y = 2x, \quad y = 2$

21. $y = \sqrt{x}, \quad y = 0, \quad y = x - 2$

22. $y = \sqrt{x}, \quad y = 0, \quad y = 2 - x$

Revolution About Horizontal and Vertical Lines

In Exercises 23–26, use the shell method to find the volumes of the solids generated by revolving the regions bounded by the given curves about the given lines.

23. $y = 3x, \quad y = 0, \quad x = 2$

   a. The y-axis   **b.** The line $x = 4$

   c. The line $x = -1$ **d.** The x-axis

   e. The line $y = 7$ **f.** The line $y = -2$

24. $y = x^3, \quad y = 8, \quad x = 0$

   a. The y-axis   **b.** The line $x = 3$

   c. The line $x = -2$ **d.** The x-axis

   e. The line $y = 8$ **f.** The line $y = -1$

25. $y = x + 2, \quad y = x^2$

   a. The line $x = 2$ **b.** The line $x = -1$

   c. The x-axis   **d.** The line $y = 4$

26. $y = x^4, \quad y = 4 - 3x^2$

   a. The line $x = 1$ **c.** The x-axis

In Exercises 27 and 28, use the shell method to find the volumes of the solids generated by revolving the shaded regions about the indicated axes.

27. a. The x-axis   **b.** The line $y = 1$

     c. The line $y = 8/5$ **d.** The line $y = -2/5$

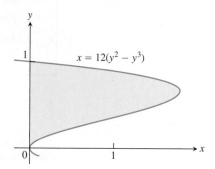

28. a. The x-axis   **b.** The line $y = 2$

     c. The line $y = 5$ **d.** The line $y = -5/8$

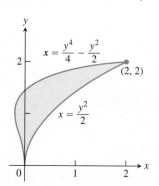

Choosing the Washer Method or Shell Method

For some regions, both the washer and shell methods work well for the solid generated by revolving the region about the coordinate axes, but this is not always the case. When a region is revolved about the y-axis, for example, and washers are used, we must integrate with respect to y. It may not be possible, however, to express the integrand in terms of y. In such a case, the shell method allows us to integrate with respect to x instead. Exercises 29 and 30 provide some insight.

29. Compute the volume of the solid generated by revolving the region bounded by $y = x$ and $y = x^2$ about each coordinate axis using

   a. the shell method.   **b.** the washer method.

30. Compute the volume of the solid generated by revolving the triangular region bounded by the lines $2y = x + 4, y = x$, and $x = 0$ about

   a. the x-axis using the washer method.

   b. the y-axis using the shell method.

   c. the line $x = 4$ using the shell method.

   d. the line $y = 8$ using the washer method.

In Exercises 31–36, find the volumes of the solids generated by revolving the regions about the given axes. If you think it would be better to use washers in any given instance, feel free to do so.

31. The triangle with vertices (1, 1), (1, 2), and (2, 2) about
   a. the x-axis   **b.** the y-axis
   c. the line $x = 10/3$ **d.** the line $y = 1$

32. The region bounded by $y = \sqrt{x}, y = 2, x = 0$ about
   a. the x-axis   **b.** the y-axis
   c. the line $x = 4$ **d.** the line $y = 2$

33. The region in the first quadrant bounded by the curve $x = y - y^3$ and the y-axis about
   a. the x-axis   **b.** the line $y = 1$

34. The region in the first quadrant bounded by $x = y - y^3, x = 1$, and $y = 1$ about
   a. the x-axis   **b.** the y-axis
   c. the line $x = 1$ **d.** the line $y = 1$

35. The region bounded by $y = \sqrt{x}$ and $y = x^2/8$ about
   a. the x-axis   **b.** the y-axis

36. The region bounded by $y = 2x - x^2$ and $y = x$ about
   a. the y-axis   **b.** the line $x = 1$

37. The region in the first quadrant that is bounded above by the curve $y = 1/x^{1/4}$, on the left by the line $x = 1/16$, and below by the line $y = 1$ is revolved about the x-axis to generate a solid. Find the volume of the solid by
   a. the washer method.   **b.** the shell method.

38. The region in the first quadrant that is bounded above by the curve $y = 1/\sqrt{x}$, on the left by the line $x = 1/4$, and below by the line $y = 1$ is revolved about the y-axis to generate a solid. Find the volume of the solid by
   a. the washer method.   **b.** the shell method.

Theory and Examples

39. The region shown here is to be revolved about the x-axis to generate a solid. Which of the methods (disk, washer, shell) could you use to find the volume of the solid? How many integrals would be required in each case? Explain.

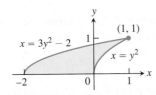

40. The region shown here is to be revolved about the y-axis to generate a solid. Which of the methods (disk, washer, shell) could you use to find the volume of the solid? How many integrals would be required in each case? Give reasons for your answers.

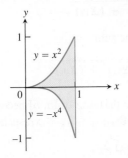

41. A bead is formed from a sphere of radius 5 by drilling through a diameter of the sphere with a drill bit of radius 3.
   a. Find the volume of the bead.
   b. Find the volume of the removed portion of the sphere.

42. A Bundt cake, well known for having a ringed shape, is formed by revolving around the y-axis the region bounded by the graph of $y = \sin(x^2 - 1)$ and the x-axis over the interval $1 \le x \le \sqrt{1 + \pi}$. Find the volume of the cake.

43. Derive the formula for the volume of a right circular cone of height h and radius r using an appropriate solid of revolution.

44. Derive the equation for the volume of a sphere of radius r using the shell method.

45. Equivalence of the washer and shell methods for finding volume Let f be differentiable and increasing on the interval $a \le x \le b$, with $a > 0$, and suppose that f has a differentiable inverse, f^{-1}. Revolve about the y-axis the region bounded by the graph of f and the lines $x = a$ and $y = f(b)$ to generate a solid. Then the values of the integrals given by the washer and shell methods for the volume have identical values:

$$\int_{f(a)}^{f(b)} \pi((f^{-1}(y))^2 - a^2)\, dy = \int_{a}^{b} 2\pi x(f(b) - f(x))\, dx.$$

To prove this equality, define

$$W(t) = \int_{f(a)}^{f(t)} \pi((f^{-1}(y))^2 - a^2)\, dy$$

$$S(t) = \int_{a}^{t} 2\pi x(f(t) - f(x))\, dx.$$

Then show that the functions W and S agree at a point of $[a, b]$ and have identical derivatives on $[a, b]$. As you saw in Section 4.8, Exercise 128, this will guarantee $W(t) = S(t)$ for all t in $[a, b]$. In particular, $W(b) = S(b)$. (*Source:* "Disks and Shells Revisited," by Walter Carlip, *American Mathematical Monthly*, Vol. 98, No. 2, Feb. 1991, pp. 154–156.)

46. The region between the curve $y = \sec^{-1} x$ and the x-axis from $x = 1$ to $x = 2$ (shown here) is revolved about the y-axis to generate a solid. Find the volume of the solid.

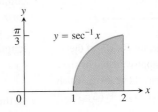

47. Find the volume of the solid generated by revolving the region enclosed by the graphs of $y = e^{-x^2}, y = 0, x = 0$, and $x = 1$ about the y-axis.

48. Find the volume of the solid generated by revolving the region enclosed by the graphs of $y = e^{x/2}, y = 1$, and $x = \ln 3$ about the x-axis.

6.3 | Arc Length

We know what is meant by the length of a straight line segment, but without calculus, we have no precise definition of the length of a general winding curve. If the curve is the graph of a continuous function defined over an interval, then we can find the length of the curve using a procedure similar to that we used for defining the area between the curve and the x-axis. This procedure results in a division of the curve from point A to point B into many pieces and joining successive points of division by straight line segments. We then sum the lengths of all these line segments and define the length of the curve to be the limiting value of this sum as the number of segments goes to infinity.

Length of a Curve $y = f(x)$

Suppose the curve whose length we want to find is the graph of the function $y = f(x)$ from $x = a$ to $x = b$. In order to derive an integral formula for the length of the curve, we assume that f has a continuous derivative at every point of $[a, b]$. Such a function is called **smooth**, and its graph is a **smooth curve** because it does not have any breaks, corners, or cusps.

FIGURE 6.22 The length of the polygonal path $P_0 P_1 P_2 \cdots P_n$ approximates the length of the curve $y = f(x)$ from point A to point B.

We partition the interval $[a, b]$ into n subintervals with $a = x_0 < x_1 < x_2 < \cdots < x_n = b$. If $y_k = f(x_k)$, then the corresponding point $P_k(x_k, y_k)$ lies on the curve. Next we connect successive points P_{k-1} and P_k with straight line segments that, taken together, form a polygonal path whose length approximates the length of the curve (Figure 6.22). If $\Delta x_k = x_k - x_{k-1}$ and $\Delta y_k = y_k - y_{k-1}$, then a representative line segment in the path has length (see Figure 6.23)

$$L_k = \sqrt{(\Delta x_k)^2 + (\Delta y_k)^2},$$

so the length of the curve is approximated by the sum

$$\sum_{k=1}^{n} L_k = \sum_{k=1}^{n} \sqrt{(\Delta x_k)^2 + (\Delta y_k)^2}. \tag{1}$$

We expect the approximation to improve as the partition of $[a, b]$ becomes finer. Now, by the Mean Value Theorem, there is a point c_k, with $x_{k-1} < c_k < x_k$, such that

$$\Delta y_k = f'(c_k) \, \Delta x_k.$$

FIGURE 6.23 The arc $P_{k-1}P_k$ of the curve $y = f(x)$ is approximated by the straight line segment shown here, which has length $L_k = \sqrt{(\Delta x_k)^2 + (\Delta y_k)^2}$.

With this substitution for Δy_k, the sums in Equation (1) take the form

$$\sum_{k=1}^{n} L_k = \sum_{k=1}^{n} \sqrt{(\Delta x_k)^2 + (f'(c_k)\Delta x_k)^2} = \sum_{k=1}^{n} \sqrt{1 + [f'(c_k)]^2}\, \Delta x_k. \qquad (2)$$

Because $\sqrt{1 + [f'(x)]^2}$ is continuous on $[a, b]$, the limit of the Riemann sum on the right-hand side of Equation (2) exists as the norm of the partition goes to zero, giving

$$\lim_{n\to\infty} \sum_{k=1}^{n} L_k = \lim_{n\to\infty} \sum_{k=1}^{n} \sqrt{1 + [f'(c_k)]^2}\, \Delta x_k = \int_a^b \sqrt{1 + [f'(x)]^2}\, dx.$$

We define the value of this limiting integral to be the length of the curve.

DEFINITION   If f' is continuous on $[a, b]$, then the **length (arc length)** of the curve $y = f(x)$ from the point $A = (a, f(a))$ to the point $B = (b, f(b))$ is the value of the integral

$$L = \int_a^b \sqrt{1 + [f'(x)]^2}\, dx = \int_a^b \sqrt{1 + \left(\frac{dy}{dx}\right)^2}\, dx. \qquad (3)$$

EXAMPLE 1   Find the length of the curve (Figure 6.24)

$$y = \frac{4\sqrt{2}}{3} x^{3/2} - 1, \qquad 0 \le x \le 1.$$

Solution   We use Equation (3) with $a = 0$, $b = 1$, and

$$y = \frac{4\sqrt{2}}{3} x^{3/2} - 1 \qquad \scriptstyle x = 1,\ y \approx 0.89$$

$$\frac{dy}{dx} = \frac{4\sqrt{2}}{3} \cdot \frac{3}{2} x^{1/2} = 2\sqrt{2} x^{1/2}$$

$$\left(\frac{dy}{dx}\right)^2 = \left(2\sqrt{2} x^{1/2}\right)^2 = 8x.$$

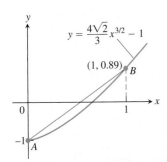

FIGURE 6.24 The length of the curve is slightly larger than the length of the line segment joining points A and B (Example 1).

The length of the curve over $x = 0$ to $x = 1$ is

$$L = \int_0^1 \sqrt{1 + \left(\frac{dy}{dx}\right)^2}\, dx = \int_0^1 \sqrt{1 + 8x}\, dx \qquad \begin{array}{l}\scriptstyle \text{Eq. (3) with} \\ \scriptstyle a = 0,\ b = 1 \\ \scriptstyle \text{Let } u = 1 + 8x, \\ \scriptstyle \text{integrate, and} \\ \scriptstyle \text{replace } u \text{ by} \\ \scriptstyle 1 + 8x.\end{array}$$

$$= \frac{2}{3} \cdot \frac{1}{8} (1 + 8x)^{3/2} \Big]_0^1 = \frac{13}{6} \approx 2.17.$$

Notice that the length of the curve is slightly larger than the length of the straight-line segment joining the points $A = (0, -1)$ and $B = \left(1, 4\sqrt{2}/3 - 1\right)$ on the curve (see Figure 6.24):

$$2.17 > \sqrt{1^2 + (1.89)^2} \approx 2.14 \qquad \scriptstyle \text{Decimal approximations} \qquad \blacksquare$$

EXAMPLE 2   Find the length of the graph of

$$f(x) = \frac{x^3}{12} + \frac{1}{x}, \qquad 1 \le x \le 4.$$

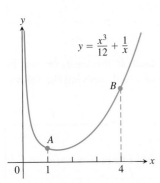

FIGURE 6.25 The curve in Example 2, where $A = (1, 13/12)$ and $B = (4, 67/12)$.

Solution   A graph of the function is shown in Figure 6.25. To use Equation (3), we find

$$f'(x) = \frac{x^2}{4} - \frac{1}{x^2}$$

so

$$1 + [f'(x)]^2 = 1 + \left(\frac{x^2}{4} - \frac{1}{x^2}\right)^2 = 1 + \left(\frac{x^4}{16} - \frac{1}{2} + \frac{1}{x^4}\right)$$

$$= \frac{x^4}{16} + \frac{1}{2} + \frac{1}{x^4} = \left(\frac{x^2}{4} + \frac{1}{x^2}\right)^2.$$

The length of the graph over [1, 4] is

$$L = \int_1^4 \sqrt{1 + [f'(x)]^2}\, dx = \int_1^4 \left(\frac{x^2}{4} + \frac{1}{x^2}\right) dx$$

$$= \left[\frac{x^3}{12} - \frac{1}{x}\right]_1^4 = \left(\frac{64}{12} - \frac{1}{4}\right) - \left(\frac{1}{12} - 1\right) = \frac{72}{12} = 6.$$ ∎

EXAMPLE 3   Find the length of the curve

$$y = \frac{1}{2}(e^x + e^{-x}), \qquad 0 \le x \le 2.$$

Solution   We use Equation (3) with $a = 0$, $b = 2$, and

$$y = \frac{1}{2}(e^x + e^{-x})$$

$$\frac{dy}{dx} = \frac{1}{2}(e^x - e^{-x})$$

$$\left(\frac{dy}{dx}\right)^2 = \frac{1}{4}(e^{2x} - 2 + e^{-2x})$$

$$1 + \left(\frac{dy}{dx}\right)^2 = \frac{1}{4}(e^{2x} + 2 + e^{-2x}) = \left[\frac{1}{2}(e^x + e^{-x})\right]^2.$$

The length of the curve from $x = 0$ to $x = 2$ is

$$L = \int_0^2 \sqrt{1 + \left(\frac{dy}{dx}\right)^2}\, dx = \int_0^2 \frac{1}{2}(e^x + e^{-x})\, dx \qquad \text{Eq. (3) with } a = 0, b = 2$$

$$= \frac{1}{2}\left[e^x - e^{-x}\right]_0^2 = \frac{1}{2}(e^2 - e^{-2}) \approx 3.63.$$ ∎

Dealing with Discontinuities in dy/dx

At a point on a curve where dy/dx fails to exist, dx/dy may exist. In this case, we may be able to find the curve's length by expressing x as a function of y and applying the following analogue of Equation (3):

Formula for the Length of $x = g(y)$, $c \le y \le d$

If g' is continuous on $[c, d]$, the length of the curve $x = g(y)$ from $A = (g(c), c)$ to $B = (g(d), d)$ is

$$L = \int_c^d \sqrt{1 + \left(\frac{dx}{dy}\right)^2}\, dy = \int_c^d \sqrt{1 + [g'(y)]^2}\, dy. \qquad (4)$$

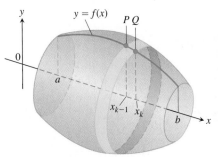

FIGURE 6.30 The surface generated by revolving the graph of a nonnegative function $y = f(x)$, $a \leq x \leq b$, about the x-axis. The surface is a union of bands like the one swept out by the arc PQ.

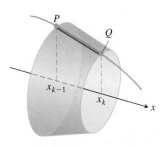

FIGURE 6.31 The line segment joining P and Q sweeps out a frustum of a cone.

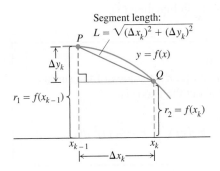

FIGURE 6.32 Dimensions associated with the arc and line segment PQ.

As the arc PQ revolves about the x-axis, the line segment joining P and Q sweeps out a frustum of a cone whose axis lies along the x-axis (Figure 6.31). The surface area of this frustum approximates the surface area of the band swept out by the arc PQ. The surface area of the frustum of the cone shown in Figure 6.31 is $2\pi y^* L$, where y^* is the average height of the line segment joining P and Q, and L is its length (just as before). Since $f \geq 0$, from Figure 6.32 we see that the average height of the line segment is $y^* = (f(x_{k-1}) + f(x_k))/2$, and the slant length is $L = \sqrt{(\Delta x_k)^2 + (\Delta y_k)^2}$. Therefore,

$$\text{Frustum surface area} = 2\pi \cdot \frac{f(x_{k-1}) + f(x_k)}{2} \cdot \sqrt{(\Delta x_k)^2 + (\Delta y_k)^2}$$
$$= \pi(f(x_{k-1}) + f(x_k))\sqrt{(\Delta x_k)^2 + (\Delta y_k)^2}.$$

The area of the original surface, being the sum of the areas of the bands swept out by arcs like arc PQ, is approximated by the frustum area sum

$$\sum_{k=1}^{n} \pi(f(x_{k-1}) + f(x_k))\sqrt{(\Delta x_k)^2 + (\Delta y_k)^2}. \tag{1}$$

We expect the approximation to improve as the partition of $[a, b]$ becomes finer. Moreover, if the function f is differentiable, then by the Mean Value Theorem, there is a point $(c_k, f(c_k))$ on the curve between P and Q where the tangent is parallel to the segment PQ (Figure 6.33). At this point,

$$f'(c_k) = \frac{\Delta y_k}{\Delta x_k},$$
$$\Delta y_k = f'(c_k)\,\Delta x_k.$$

With this substitution for Δy_k, the sums in Equation (1) take the form

$$\sum_{k=1}^{n} \pi(f(x_{k-1}) + f(x_k))\sqrt{(\Delta x_k)^2 + (f'(c_k)\,\Delta x_k)^2}$$
$$= \sum_{k=1}^{n} \pi(f(x_{k-1}) + f(x_k))\sqrt{1 + (f'(c_k))^2}\,\Delta x_k. \tag{2}$$

These sums are not the Riemann sums of any function because the points x_{k-1}, x_k, and c_k are not the same. However, it can be proved that as the norm of the partition of $[a, b]$ goes to zero, the sums in Equation (2) converge to the integral

$$\int_a^b 2\pi f(x)\sqrt{1 + (f'(x))^2}\,dx.$$

We therefore define this integral to be the area of the surface swept out by the graph of f from a to b.

DEFINITION   If the function $f(x) \geq 0$ is continuously differentiable on $[a, b]$, the **area of the surface** generated by revolving the graph of $y = f(x)$ about the x-axis is

$$S = \int_a^b 2\pi y\sqrt{1 + \left(\frac{dy}{dx}\right)^2}\,dx = \int_a^b 2\pi f(x)\sqrt{1 + (f'(x))^2}\,dx. \tag{3}$$

The square root in Equation (3) is the same one that appears in the formula for the arc length differential of the generating curve in Equation (6) of Section 6.3.

EXAMPLE 1   Find the area of the surface generated by revolving the curve $y = 2\sqrt{x}$, $1 \leq x \leq 2$, about the x-axis (Figure 6.34).

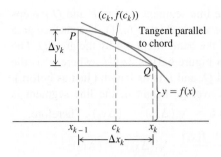

FIGURE 6.33 If f is smooth, the Mean Value Theorem guarantees the existence of a point c_k where the tangent is parallel to segment PQ.

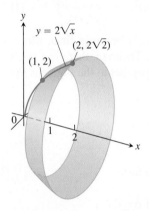

FIGURE 6.34 In Example 1 we calculate the area of this surface.

Solution We evaluate the formula

$$S = \int_a^b 2\pi y \sqrt{1 + \left(\frac{dy}{dx}\right)^2}\, dx \qquad \text{Eq. (3)}$$

with

$$a = 1, \qquad b = 2, \qquad y = 2\sqrt{x}, \qquad \frac{dy}{dx} = \frac{1}{\sqrt{x}}.$$

First, we perform some algebraic manipulation on the radical in the integrand to transform it into an expression that is easier to integrate.

$$\sqrt{1 + \left(\frac{dy}{dx}\right)^2} = \sqrt{1 + \left(\frac{1}{\sqrt{x}}\right)^2}$$

$$= \sqrt{1 + \frac{1}{x}} = \sqrt{\frac{x+1}{x}} = \frac{\sqrt{x+1}}{\sqrt{x}}.$$

With these substitutions, we have

$$S = \int_1^2 2\pi \cdot 2\sqrt{x}\, \frac{\sqrt{x+1}}{\sqrt{x}}\, dx = 4\pi \int_1^2 \sqrt{x+1}\, dx$$

$$= 4\pi \cdot \frac{2}{3}(x+1)^{3/2}\Big]_1^2 = \frac{8\pi}{3}\left(3\sqrt{3} - 2\sqrt{2}\right). \qquad \blacksquare$$

Revolution About the y-Axis

For revolution about the y-axis, we interchange x and y in Equation (3).

Surface Area for Revolution About the y-Axis

If $x = g(y) \geq 0$ is continuously differentiable on $[c, d]$, the area of the surface generated by revolving the graph of $x = g(y)$ about the y-axis is

$$S = \int_c^d 2\pi x \sqrt{1 + \left(\frac{dx}{dy}\right)^2}\, dy = \int_c^d 2\pi g(y) \sqrt{1 + (g'(y))^2}\, dy. \qquad (4)$$

EXAMPLE 2 The line segment $x = 1 - y$, $0 \leq y \leq 1$, is revolved about the y-axis to generate the cone in Figure 6.35. Find its lateral surface area (which excludes the base area).

Solution Here we have a calculation we can check with a formula from geometry:

$$\text{Lateral surface area} = \frac{\text{base circumference}}{2} \times \text{slant height} = \pi\sqrt{2}.$$

To see how Equation (4) gives the same result, we take

$$c = 0, \qquad d = 1, \qquad x = 1 - y, \qquad \frac{dx}{dy} = -1,$$

$$\sqrt{1 + \left(\frac{dx}{dy}\right)^2} = \sqrt{1 + (-1)^2} = \sqrt{2}$$

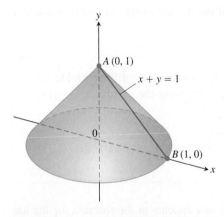

FIGURE 6.35 Revolving line segment AB about the y-axis generates a cone whose lateral surface area we can now calculate in two different ways (Example 2).

and calculate

$$S = \int_c^d 2\pi x \sqrt{1 + \left(\frac{dx}{dy}\right)^2}\, dy = \int_0^1 2\pi(1 - y)\sqrt{2}\, dy$$

$$= 2\pi\sqrt{2}\left[y - \frac{y^2}{2}\right]_0^1 = 2\pi\sqrt{2}\left(1 - \frac{1}{2}\right)$$

$$= \pi\sqrt{2}.$$

The results agree, as they should. ∎

Exercises 6.4

Finding Integrals for Surface Area

In Exercises 1–8:

a. Set up an integral for the area of the surface generated by revolving the given curve about the indicated axis.

T **b.** Graph the curve to see what it looks like. If you can, graph the surface too.

T **c.** Use your grapher's or computer's integral evaluator to find the surface's area numerically.

1. $y = \tan x$, $0 \le x \le \pi/4$; x-axis

2. $y = x^2$, $0 \le x \le 2$; x-axis

3. $xy = 1$, $1 \le y \le 2$; y-axis

4. $x = \sin y$, $0 \le y \le \pi$; y-axis

5. $x^{1/2} + y^{1/2} = 3$ from $(4, 1)$ to $(1, 4)$; x-axis

6. $y + 2\sqrt{y} = x$, $1 \le y \le 2$; y-axis

7. $x = \int_0^y \tan t\, dt$, $0 \le y \le \pi/3$; y-axis

8. $y = \int_1^x \sqrt{t^2 - 1}\, dt$, $1 \le x \le \sqrt{5}$; x-axis

Finding Surface Area

9. Find the lateral (side) surface area of the cone generated by revolving the line segment $y = x/2, 0 \le x \le 4$, about the x-axis. Check your answer with the geometry formula

$$\text{Lateral surface area} = \frac{1}{2} \times \text{base circumference} \times \text{slant height}.$$

10. Find the lateral surface area of the cone generated by revolving the line segment $y = x/2, 0 \le x \le 4$, about the y-axis. Check your answer with the geometry formula

$$\text{Lateral surface area} = \frac{1}{2} \times \text{base circumference} \times \text{slant height}.$$

11. Find the surface area of the cone frustum generated by revolving the line segment $y = (x/2) + (1/2), 1 \le x \le 3$, about the x-axis. Check your result with the geometry formula

$$\text{Frustum surface area} = \pi(r_1 + r_2) \times \text{slant height}.$$

12. Find the surface area of the cone frustum generated by revolving the line segment $y = (x/2) + (1/2), 1 \le x \le 3$, about the y-axis. Check your result with the geometry formula

$$\text{Frustum surface area} = \pi(r_1 + r_2) \times \text{slant height}.$$

Find the areas of the surfaces generated by revolving the curves in Exercises 13–23 about the indicated axes. If you have a grapher, you may want to graph these curves to see what they look like.

13. $y = x^3/9$, $0 \le x \le 2$; x-axis

14. $y = \sqrt{x}$, $3/4 \le x \le 15/4$; x-axis

15. $y = \sqrt{2x - x^2}$, $0.5 \le x \le 1.5$; x-axis

16. $y = \sqrt{x + 1}$, $1 \le x \le 5$; x-axis

17. $x = y^3/3$, $0 \le y \le 1$; y-axis

18. $x = (1/3)y^{3/2} - y^{1/2}$, $1 \le y \le 3$; y-axis

19. $x = 2\sqrt{4 - y}$, $0 \le y \le 15/4$; y-axis

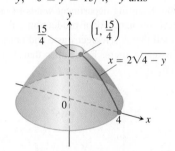

20. $x = \sqrt{2y - 1}$, $5/8 \le y \le 1$; y-axis

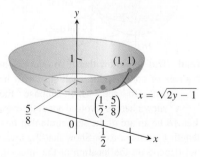

21. $x = (e^y + e^{-y})/2$, $0 \le y \le \ln 2$; y-axis

22. $y = (1/3)(x^2 + 2)^{3/2}$, $0 \le x \le \sqrt{2}$; y-axis (*Hint:* Express $ds = \sqrt{dx^2 + dy^2}$ in terms of dx, and evaluate the integral $S = \int 2\pi x \, ds$ with appropriate limits.)

23. $x = (y^4/4) + 1/(8y^2)$, $1 \le y \le 2$; x-axis (*Hint:* Express $ds = \sqrt{dx^2 + dy^2}$ in terms of dy, and evaluate the integral $S = \int 2\pi y \, ds$ with appropriate limits.)

24. Write an integral for the area of the surface generated by revolving the curve $y = \cos x$, $-\pi/2 \le x \le \pi/2$, about the x-axis. In Section 8.4 we will see how to evaluate such integrals.

25. **Testing the new definition** Show that the surface area of a sphere of radius a is still $4\pi a^2$ by using Equation (3) to find the area of the surface generated by revolving the curve $y = \sqrt{a^2 - x^2}$, $-a \le x \le a$, about the x-axis.

26. **Testing the new definition** The lateral (side) surface area of a cone of height h and base radius r should be $\pi r \sqrt{r^2 + h^2}$, the semiperimeter of the base times the slant height. Show that this is still the case by finding the area of the surface generated by revolving the line segment $y = (r/h)x$, $0 \le x \le h$, about the x-axis.

T 27. **Enameling woks** Your company decided to put out a deluxe version of a wok you designed. The plan is to coat it inside with white enamel and outside with blue enamel. Each enamel will be sprayed on 0.5 mm thick before baking. (See accompanying figure.) Your manufacturing department wants to know how much enamel to have on hand for a production run of 5000 woks. What do you tell them? (Neglect waste and unused material and give your answer in liters. Remember that $1 \text{ cm}^3 = 1 \text{ mL}$, so $1 \text{ L} = 1000 \text{ cm}^3$.)

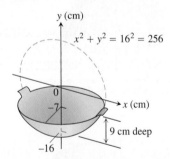

28. **Slicing bread** Did you know that if you cut a spherical loaf of bread into slices of equal width, each slice will have the same amount of crust? To see why, suppose the semicircle $y = \sqrt{r^2 - x^2}$ shown here is revolved about the x-axis to generate a sphere. Let AB be an arc of the semicircle that lies above an interval of length h on the x-axis. Show that the area swept out by AB does not depend on the location of the interval. (It does depend on the length of the interval.)

29. The shaded band shown here is cut from a sphere of radius R by parallel planes h units apart. Show that the surface area of the band is $2\pi Rh$.

30. Here is a schematic drawing of the 90-ft dome used by the U.S. National Weather Service to house radar in Bozeman, Montana.

   a. How much outside surface is there to paint (not counting the bottom)?

   T b. Express the answer to the nearest square foot.

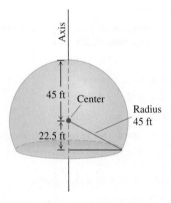

31. **An alternative derivation of the surface area formula** Assume f is smooth on $[a, b]$ and partition $[a, b]$ in the usual way. In the kth subinterval $[x_{k-1}, x_k]$, construct the tangent line to the curve at the midpoint $m_k = (x_{k-1} + x_k)/2$, as in the accompanying figure.

   a. Show that

   $$r_1 = f(m_k) - f'(m_k)\frac{\Delta x_k}{2} \quad \text{and} \quad r_2 = f(m_k) + f'(m_k)\frac{\Delta x_k}{2}.$$

   b. Show that the length L_k of the tangent line segment in the kth subinterval is $L_k = \sqrt{(\Delta x_k)^2 + (f'(m_k)\,\Delta x_k)^2}$.

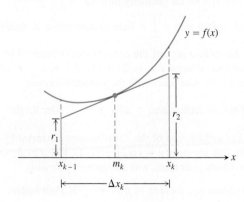

c. Show that the lateral surface area of the frustum of the cone swept out by the tangent line segment as it revolves about the x-axis is $2\pi f(m_k)\sqrt{1 + (f'(m_k))^2}\,\Delta x_k$.

d. Show that the area of the surface generated by revolving $y = f(x)$ about the x-axis over $[a, b]$ is

$$\lim_{n\to\infty}\sum_{k=1}^{n}\binom{\text{lateral surface area}}{\text{of }k\text{th frustum}} = \int_{a}^{b} 2\pi f(x)\sqrt{1 + (f'(x))^2}\,dx.$$

32. The surface of an astroid Find the area of the surface generated by revolving about the x-axis the portion of the astroid $x^{2/3} + y^{2/3} = 1$ shown in the accompanying figure.

(*Hint:* Revolve the first-quadrant portion $y = (1 - x^{2/3})^{3/2}$, $0 \le x \le 1$, about the x-axis and double your result.)

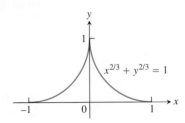

6.5 | Work and Fluid Forces

In everyday life, *work* means an activity that requires muscular or mental effort. In science, the term refers specifically to a force acting on a body (or object) and the body's subsequent displacement. This section shows how to calculate work. The applications run from compressing railroad car springs and emptying subterranean tanks to forcing electrons together and lifting satellites into orbit.

Work Done by a Constant Force

When a body moves a distance d along a straight line as a result of being acted on by a force of constant magnitude F in the direction of motion, we define the **work** W done by the force on the body with the formula

$$W = Fd \qquad \text{(Constant-force formula for work).} \tag{1}$$

From Equation (1) we see that the unit of work in any system is the unit of force multiplied by the unit of distance. In SI units (SI stands for *Système International*, or International System), the unit of force is a newton, the unit of distance is a meter, and the unit of work is a newton-meter ($N \cdot m$). This combination appears so often it has a special name, the **joule**. In the British system, the unit of work is the foot-pound, a unit frequently used by engineers.

Joules

The joule, abbreviated J and pronounced "jewel," is named after the English physicist James Prescott Joule (1818–1889). The defining equation is

1 joule = (1 newton)(1 meter).

In symbols, $1\,J = 1\,N \cdot m$.

EXAMPLE 1   Suppose you jack up the side of a 2000-lb car 1.25 ft to change a tire. The jack applies a constant vertical force of about 1000 lb in lifting the side of the car (but because of the mechanical advantage of the jack, the force you apply to the jack itself is only about 30 lb). The total work performed by the jack on the car is $1000 \times 1.25 = 1250$ ft-lb. In SI units, the jack has applied a force of 4448 N through a distance of 0.381 m to do $4448 \times 0.381 \approx 1695$ J of work. ∎

Work Done by a Variable Force Along a Line

If the force you apply varies along the way, as it will if you are compressing a spring, the formula $W = Fd$ has to be replaced by an integral formula that takes the variation in F into account.

Suppose that the force performing the work acts on an object moving along a straight line, which we take to be the x-axis. We assume that the magnitude of the force is a continuous function F of the object's position x. We want to find the work done over the interval from $x = a$ to $x = b$. We partition $[a, b]$ in the usual way and choose an arbitrary point c_k in each subinterval $[x_{k-1}, x_k]$. If the subinterval is short enough, the continuous function F

will not vary much from x_{k-1} to x_k. The amount of work done across the interval will be about $F(c_k)$ times the distance Δx_k, the same as it would be if F were constant and we could apply Equation (1). The total work done from a to b is therefore approximated by the Riemann sum

$$\text{Work} \approx \sum_{k=1}^{n} F(c_k)\,\Delta x_k.$$

We expect the approximation to improve as the norm of the partition goes to zero, so we define the work done by the force from a to b to be the integral of F from a to b:

$$\lim_{n \to \infty} \sum_{k=1}^{n} F(c_k)\,\Delta x_k = \int_a^b F(x)\,dx.$$

> **DEFINITION** The **work** done by a variable force $F(x)$ in the direction of motion along the x-axis from $x = a$ to $x = b$ is
>
> $$W = \int_a^b F(x)\,dx. \tag{2}$$

The units of the integral are joules if F is in newtons and x is in meters, and foot-pounds if F is in pounds and x is in feet. So the work done by a force of $F(x) = 1/x^2$ newtons in moving an object along the x-axis from $x = 1$ m to $x = 10$ m is

$$W = \int_1^{10} \frac{1}{x^2}\,dx = -\frac{1}{x}\bigg]_1^{10} = -\frac{1}{10} + 1 = 0.9 \text{ J}.$$

Hooke's Law for Springs: $F = kx$

Hooke's Law says that the force required to hold a stretched or compressed spring x units from its natural (unstressed) length is proportional to x. In symbols,

$$F = kx. \tag{3}$$

The constant k, measured in force units per unit length, is a characteristic of the spring, called the **force constant** (or **spring constant**) of the spring. Hooke's Law, Equation (3), gives good results as long as the force doesn't distort the metal in the spring. We assume that the forces in this section are too small to do that.

EXAMPLE 2 Find the work required to compress a spring from its natural length of 1 ft to a length of 0.75 ft if the force constant is $k = 16$ lb/ft.

Solution We picture the uncompressed spring laid out along the x-axis with its movable end at the origin and its fixed end at $x = 1$ ft (Figure 6.36). This enables us to describe the force required to compress the spring from 0 to x with the formula $F = 16x$. To compress the spring from 0 to 0.25 ft, the force must increase from

$$F(0) = 16 \cdot 0 = 0 \text{ lb} \qquad \text{to} \qquad F(0.25) = 16 \cdot 0.25 = 4 \text{ lb}.$$

The work done by F over this interval is

$$W = \int_0^{0.25} 16x\,dx = 8x^2\bigg]_0^{0.25} = 0.5 \text{ ft-lb.}$$

Eq. (2) with $a = 0, b = 0.25,$ $F(x) = 16x$

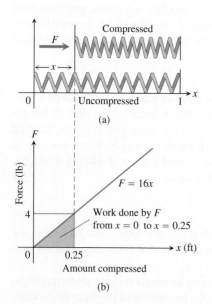

FIGURE 6.36 The force F needed to hold a spring under compression increases linearly as the spring is compressed (Example 2).

EXAMPLE 3   A spring has a natural length of 1 m. A force of 24 N holds the spring stretched to a total length of 1.8 m.

(a) Find the force constant k.

(b) How much work will it take to stretch the spring 2 m beyond its natural length?

(c) How far will a 45-N force stretch the spring?

Solution

(a) *The force constant.* We find the force constant from Equation (3). A force of 24 N maintains the spring at a position where it is stretched 0.8 m from its natural length, so

$$24 = k(0.8) \qquad \text{Eq. (3) with}$$
$$k = 24/0.8 = 30 \text{ N/m.} \qquad F = 24, x = 0.8$$

(b) *The work to stretch the spring* 2 m. We imagine the unstressed spring hanging along the x-axis with its free end at $x = 0$ (Figure 6.37). The force required to stretch the spring x m beyond its natural length is the force required to hold the free end of the spring x units from the origin. Hooke's Law with $k = 30$ says that this force is

$$F(x) = 30x.$$

The work done by F on the spring from $x = 0$ m to $x = 2$ m is

$$W = \int_0^2 30x \, dx = 15x^2 \Big]_0^2 = 60 \text{ J.}$$

(c) *How far will a* 45-N *force stretch the spring?* We substitute $F = 45$ in the equation $F = 30x$ to find

$$45 = 30x, \qquad \text{or} \qquad x = 1.5 \text{ m.}$$

A 45-N force will keep the spring stretched 1.5 m beyond its natural length. ∎

The work integral is useful to calculate the work done in lifting objects whose weights vary with their elevation.

EXAMPLE 4   A 5-lb bucket is lifted from the ground into the air by pulling in 20 ft of rope at a constant speed (Figure 6.38). The rope weighs 0.08 lb/ft. How much work was spent lifting the bucket and rope?

Solution   The bucket has constant weight, so the work done lifting it alone is weight × distance = $5 \cdot 20 = 100$ ft-lb.

The weight of the rope varies with the bucket's elevation, because less of it is freely hanging. When the bucket is x ft off the ground, the remaining proportion of the rope still being lifted weighs $(0.08) \cdot (20 - x)$ lb. So the work in lifting the rope is

$$\text{Work on rope} = \int_0^{20} (0.08)(20 - x) \, dx = \int_0^{20} (1.6 - 0.08x) \, dx$$
$$= \left[1.6x - 0.04x^2 \right]_0^{20} = 32 - 16 = 16 \text{ ft-lb.}$$

The total work for the bucket and rope combined is

$$100 + 16 = 116 \text{ ft-lb.} \qquad ∎$$

Pumping Liquids from Containers

How much work does it take to pump all or part of the liquid from a container? Engineers often need to know the answer in order to design or choose the right pump to transport water or some other liquid from one place to another. To find out how much work is required to pump the liquid, we imagine lifting the liquid out one thin horizontal slab at a time and applying the equation $W = Fd$ to each slab. We then evaluate the integral this leads to as the slabs become thinner and more numerous. The integral we get each time depends on the weight of the liquid and the dimensions of the container, but the way we find the integral is always the same. The next example shows what to do.

FIGURE 6.37   A 24-N weight stretches this spring 0.8 m beyond its unstressed length (Example 3).

FIGURE 6.38   Lifting the bucket in Example 4.

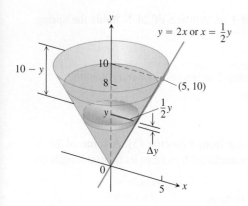

FIGURE 6.39   The olive oil and tank in Example 5.

EXAMPLE 5   The conical tank in Figure 6.39 is filled to within 2 ft of the top with olive oil weighing 57 lb/ft^3. How much work does it take to pump the oil to the rim of the tank?

Solution   We imagine the oil divided into thin slabs by planes perpendicular to the y-axis at the points of a partition of the interval $[0, 8]$.

The typical slab between the planes at y and $y + \Delta y$ has a volume of about

$$\Delta V = \pi(\text{radius})^2(\text{thickness}) = \pi\left(\frac{1}{2}y\right)^2 \Delta y = \frac{\pi}{4}y^2 \Delta y \text{ ft}^3.$$

The force $F(y)$ required to lift this slab is equal to its weight,

$$F(y) = 57 \, \Delta V = \frac{57\pi}{4}y^2 \Delta y \text{ lb.} \qquad \begin{array}{l}\text{Weight} = (\text{weight per unit}\\\text{volume}) \times \text{volume}\end{array}$$

The distance through which $F(y)$ must act to lift this slab to the level of the rim of the cone is about $(10 - y)$ ft, so the work done lifting the slab is about

$$\Delta W = \frac{57\pi}{4}(10 - y)y^2 \Delta y \text{ ft-lb.}$$

Assuming there are n slabs associated with the partition of $[0, 8]$, and that $y = y_k$ denotes the plane associated with the kth slab of thickness Δy_k, we can approximate the work done lifting all of the slabs with the Riemann sum

$$W \approx \sum_{k=1}^{n} \frac{57\pi}{4}(10 - y_k)y_k^2 \Delta y_k \text{ ft-lb.}$$

The work of pumping the oil to the rim is the limit of these sums as the norm of the partition goes to zero and the number of slabs tends to infinity:

$$W = \lim_{n\to\infty} \sum_{k=1}^{n} \frac{57\pi}{4}(10 - y_k)y_k^2 \Delta y_k = \int_0^8 \frac{57\pi}{4}(10 - y)y^2 \, dy$$

$$= \frac{57\pi}{4}\int_0^8 (10y^2 - y^3) \, dy$$

$$= \frac{57\pi}{4}\left[\frac{10y^3}{3} - \frac{y^4}{4}\right]_0^8 \approx 30{,}561 \text{ ft-lb.} \quad \blacksquare$$

Fluid Pressures and Forces

FIGURE 6.40   To withstand the increasing pressure, dams are built thicker as they go down.

Dams are built thicker at the bottom than at the top (Figure 6.40) because the pressure against them increases with depth. The pressure at any point on a dam depends only on how far below the surface the point is and not on how much the surface of the dam happens to be tilted at that point. The pressure, in pounds per square foot at a point h feet below the surface, is always $62.4h$. The number 62.4 is the weight-density of freshwater in pounds per cubic foot. The pressure h feet below the surface of any fluid is the fluid's *weight-density* times h.

Weight-density

A fluid's weight-density w is its weight per unit volume. Typical values (lb/ft^3) are listed below.

Gasoline	42
Mercury	849
Milk	64.5
Molasses	100
Olive oil	57
Seawater	64
Freshwater	62.4

The Pressure-Depth Equation

In a fluid that is standing still, the pressure p at depth h is the fluid's weight-density w times h:

$$p = wh. \tag{4}$$

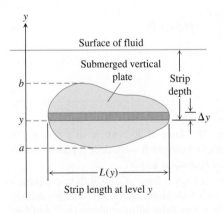

FIGURE 6.41 These containers are filled with water to the same depth and have the same base area. The total force is therefore the same on the bottom of each container. The containers' shapes do not matter here.

In a container of fluid with a flat horizontal base, the total force exerted by the fluid against the base can be calculated by multiplying the area of the base by the pressure at the base. We can do this because total force equals force per unit area (pressure) times area. (See Figure 6.41.) If F, p, and A are the total force, pressure, and area, then

$$F = \text{total force} = \text{force per unit area} \times \text{area}$$
$$= \text{pressure} \times \text{area} = pA$$
$$= whA. \qquad \qquad p = wh \text{ from Eq. (4)}$$

Fluid Force on a Constant-Depth Surface

$$F = pA = whA \qquad \qquad (5)$$

For example, the weight-density of freshwater is 62.4 lb/ft^3, so the fluid force at the bottom of a 10 ft $\times$ 20 ft rectangular swimming pool 3 ft deep is

$$F = whA = (62.4 \text{ lb/ft}^3)(3 \text{ ft})(10 \cdot 20 \text{ ft}^2)$$
$$= 37{,}440 \text{ lb}.$$

For a flat plate submerged *horizontally*, like the bottom of the swimming pool just discussed, the downward force acting on its upper face due to liquid pressure is given by Equation (5). If the plate is submerged *vertically*, however, then the pressure against it will be different at different depths and Equation (5) no longer is usable in that form (because h varies).

Suppose we want to know the force exerted by a fluid against one side of a vertical plate submerged in a fluid of weight-density w. To find it, we model the plate as a region extending from $y = a$ to $y = b$ in the xy-plane (Figure 6.42). We partition $[a, b]$ in the usual way and imagine the region to be cut into thin horizontal strips by planes perpendicular to the y-axis at the partition points. The typical strip from y to $y + \Delta y$ is Δy units wide by $L(y)$ units long. We assume $L(y)$ to be a continuous function of y.

The pressure varies across the strip from top to bottom. If the strip is narrow enough, however, the pressure will remain close to its bottom-edge value of $w \times$ (strip depth). The force exerted by the fluid against one side of the strip will be about

$$\Delta F = (\text{pressure along bottom edge}) \times (\text{area})$$
$$= w \cdot (\text{strip depth}) \cdot L(y)\,\Delta y.$$

Assume there are n strips associated with the partition of $a \le y \le b$ and that y_k is the bottom edge of the kth strip having length $L(y_k)$ and width Δy_k. The force against the entire plate is approximated by summing the forces against each strip, giving the Riemann sum

$$F \approx \sum_{k=1}^{n} (w \cdot (\text{strip depth})_k \cdot L(y_k))\,\Delta y_k. \qquad \qquad (6)$$

The sum in Equation (6) is a Riemann sum for a continuous function on $[a, b]$, and we expect the approximations to improve as the norm of the partition goes to zero. The force against the plate is the limit of these sums:

$$\lim_{n \to \infty} \sum_{k=1}^{n} (w \cdot (\text{strip depth})_k \cdot L(y_k))\,\Delta y_k = \int_a^b w \cdot (\text{strip depth}) \cdot L(y)\,dy.$$

FIGURE 6.42 The force exerted by a fluid against one side of a thin, flat horizontal strip is about $\Delta F = \text{pressure} \times \text{area} = w \times (\text{strip depth}) \times L(y)\,\Delta y$.

The Integral for Fluid Force Against a Vertical Flat Plate

Suppose that a plate submerged vertically in fluid of weight-density w runs from $y = a$ to $y = b$ on the y-axis. Let $L(y)$ be the length of the horizontal strip measured from left to right along the surface of the plate at level y. Then the force exerted by the fluid against one side of the plate is

$$F = \int_a^b w \cdot (\text{strip depth}) \cdot L(y)\, dy. \qquad (7)$$

EXAMPLE 6 A flat isosceles right-triangular plate with base 6 ft and height 3 ft is submerged vertically, base up, 2 ft below the surface of a swimming pool. Find the force exerted by the water against one side of the plate.

Solution We establish a coordinate system to work in by placing the origin at the plate's bottom vertex and running the y-axis upward along the plate's axis of symmetry (Figure 6.43). The surface of the pool lies along the line $y = 5$ and the plate's top edge along the line $y = 3$. The plate's right-hand edge lies along the line $y = x$, with the upper-right vertex at $(3, 3)$. The length of a thin strip at level y is

$$L(y) = 2x = 2y.$$

The depth of the strip beneath the surface is $(5 - y)$. The force exerted by the water against one side of the plate is therefore

$$
\begin{aligned}
F &= \int_a^b w \cdot \left(\begin{array}{c}\text{strip} \\ \text{depth}\end{array}\right) \cdot L(y)\, dy \qquad \text{Eq. (7)} \\
&= \int_0^3 62.4(5 - y)2y\, dy \\
&= 124.8 \int_0^3 (5y - y^2)\, dy \\
&= 124.8 \left[\frac{5}{2}y^2 - \frac{y^3}{3}\right]_0^3 = 1684.8 \text{ lb}. \qquad \blacksquare
\end{aligned}
$$

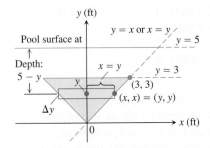

FIGURE 6.43 To find the force on one side of the submerged plate in Example 6, we can use a coordinate system like the one here.

Exercises 6.5

Springs

1. Spring constant It took 1800 J of work to stretch a spring from its natural length of 2 m to a length of 5 m. Find the spring's force constant.

2. Stretching a spring A spring has a natural length of 10 in. An 800-lb force stretches the spring to 14 in.

   a. Find the force constant.

   b. How much work is done in stretching the spring from 10 in. to 12 in.?

   c. How far beyond its natural length will a 1600-lb force stretch the spring?

3. Stretching a rubber band A force of 2 N will stretch a rubber band 2 cm (0.02 m). Assuming that Hooke's Law applies, how far will a 4-N force stretch the rubber band? How much work does it take to stretch the rubber band this far?

4. Stretching a spring If a force of 90 N stretches a spring 1 m beyond its natural length, how much work does it take to stretch the spring 5 m beyond its natural length?

5. Subway car springs It takes a force of 21,714 lb to compress a coil spring assembly on a New York City Transit Authority subway car from its free height of 8 in. to its fully compressed height of 5 in.

   a. What is the assembly's force constant?

   b. How much work does it take to compress the assembly the first half inch? the second half inch? Answer to the nearest in.-lb.

6. Bathroom scale A bathroom scale is compressed 1/16 in. when a 150-lb person stands on it. Assuming that the scale behaves like a spring that obeys Hooke's Law, how much does someone who compresses the scale 1/8 in. weigh? How much work is done compressing the scale 1/8 in.?

Work Done by a Variable Force

7. Lifting a rope A mountain climber is about to haul up a 50 m length of hanging rope. How much work will it take if the rope weighs 0.624 N/m?

8. Leaky sandbag A bag of sand originally weighing 144 lb was lifted at a constant rate. As it rose, sand also leaked out at a constant rate. The sand was half gone by the time the bag had been lifted to 18 ft. How much work was done lifting the sand this far? (Neglect the weight of the bag and lifting equipment.)

9. Lifting an elevator cable An electric elevator with a motor at the top has a multistrand cable weighing 4.5 lb/ft. When the car is at the first floor, 180 ft of cable are paid out, and effectively 0 ft are out when the car is at the top floor. How much work does the motor do just lifting the cable when it takes the car from the first floor to the top?

10. Force of attraction When a particle of mass m is at $(x, 0)$, it is attracted toward the origin with a force whose magnitude is k/x^2. If the particle starts from rest at $x = b$ and is acted on by no other forces, find the work done on it by the time it reaches $x = a$, $0 < a < b$.

11. Leaky bucket Assume the bucket in Example 4 is leaking. It starts with 2 gal of water (16 lb) and leaks at a constant rate. It finishes draining just as it reaches the top. How much work was spent lifting the water alone? (*Hint:* Do not include the rope and bucket, and find the proportion of water left at elevation x ft.)

12. (*Continuation of Exercise 11.*) The workers in Example 4 and Exercise 11 changed to a larger bucket that held 5 gal (40 lb) of water, but the new bucket had an even larger leak so that it, too, was empty by the time it reached the top. Assuming that the water leaked out at a steady rate, how much work was done lifting the water alone? (Do not include the rope and bucket.)

Pumping Liquids from Containers

13. Pumping water The rectangular tank shown here, with its top at ground level, is used to catch runoff water. Assume that the water weighs 62.4 lb/ft³.

a. How much work does it take to empty the tank by pumping the water back to ground level once the tank is full?

b. If the water is pumped to ground level with a (5/11)-horsepower (hp) motor (work output 250 ft-lb/sec), how long will it take to empty the full tank (to the nearest minute)?

c. Show that the pump in part (b) will lower the water level 10 ft (halfway) during the first 25 min of pumping.

d. The weight of water What are the answers to parts (a) and (b) in a location where water weighs 62.26 lb/ft³? 62.59 lb/ft³?

14. Emptying a cistern The rectangular cistern (storage tank for rainwater) shown has its top 10 ft below ground level. The cistern,

currently full, is to be emptied for inspection by pumping its contents to ground level.

a. How much work will it take to empty the cistern?

b. How long will it take a 1/2-hp pump, rated at 275 ft-lb/sec, to pump the tank dry?

c. How long will it take the pump in part (b) to empty the tank halfway? (It will be less than half the time required to empty the tank completely.)

d. The weight of water What are the answers to parts (a) through (c) in a location where water weighs 62.26 lb/ft³? 62.59 lb/ft³?

15. Pumping oil How much work would it take to pump oil from the tank in Example 5 to the level of the top of the tank if the tank were completely full?

16. Pumping a half-full tank Suppose that, instead of being full, the tank in Example 5 is only half full. How much work does it take to pump the remaining oil to a level 4 ft above the top of the tank?

17. Emptying a tank A vertical right-circular cylindrical tank measures 30 ft high and 20 ft in diameter. It is full of kerosene weighing 51.2 lb/ft³. How much work does it take to pump the kerosene to the level of the top of the tank?

18. a. Pumping milk Suppose that the conical container in Example 5 contains milk (weighing 64.5 lb/ft³) instead of olive oil. How much work will it take to pump the contents to the rim?

b. Pumping oil How much work will it take to pump the oil in Example 5 to a level 3 ft above the cone's rim?

19. The graph of $y = x^2$ on $0 \le x \le 2$ is revolved about the y-axis to form a tank that is then filled with salt water from the Dead Sea (weighing approximately 73 lbs/ft³). How much work does it take to pump all of the water to the top of the tank?

20. A right-circular cylindrical tank of height 10 ft and radius 5 ft is lying horizontally and is full of diesel fuel weighing 53 lbs/ft³. How much work is required to pump all of the fuel to a point 15 ft above the top of the tank?

21. Emptying a water reservoir We model pumping from spherical containers the way we do from other containers, with the axis of integration along the vertical axis of the sphere. Use the figure here to find how much work it takes to empty a full hemispherical water reservoir of radius 5 m by pumping the water to a height of 4 m above the top of the reservoir. Water weighs 9800 N/m³.

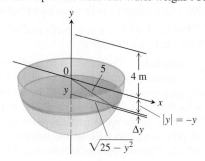

22. You are in charge of the evacuation and repair of the storage tank shown here. The tank is a hemisphere of radius 10 ft and is full of benzene weighing 56 lb/ft³. A firm you contacted says it can empty the tank for 1/2¢ per foot-pound of work. Find the work required to empty the tank by pumping the benzene to an outlet 2 ft above the top of the tank. If you have $5000 budgeted for the job, can you afford to hire the firm?

Work and Kinetic Energy

23. Kinetic energy If a variable force of magnitude $F(x)$ moves a body of mass m along the x-axis from x_1 to x_2, the body's velocity v can be written as dx/dt (where t represents time). Use Newton's second law of motion $F = m(dv/dt)$ and the Chain Rule

$$\frac{dv}{dt} = \frac{dv}{dx}\frac{dx}{dt} = v\frac{dv}{dx}$$

to show that the net work done by the force in moving the body from x_1 to x_2 is

$$W = \int_{x_1}^{x_2} F(x)\, dx = \frac{1}{2}mv_2{}^2 - \frac{1}{2}mv_1{}^2,$$

where v_1 and v_2 are the body's velocities at x_1 and x_2. In physics, the expression $(1/2)mv^2$ is called the *kinetic energy* of a body of mass m moving with velocity v. Therefore, *the work done by the force equals the change in the body's kinetic energy*, and we can find the work by calculating this change.

In Exercises 24–28, use the result of Exercise 23.

24. Tennis A 2-oz tennis ball was served at 160 ft/sec (about 109 mph). How much work was done on the ball to make it go this fast? (To find the ball's mass from its weight, express the weight in pounds and divide by 32 ft/sec², the acceleration of gravity.)

25. Baseball How many foot-pounds of work does it take to throw a baseball 90 mph? A baseball weighs 5 oz, or 0.3125 lb.

26. Golf A 1.6-oz golf ball is driven off the tee at a speed of 280 ft/sec (about 191 mph). How many foot-pounds of work are done on the ball getting it into the air?

27. On June 11, 2004, in a tennis match between Andy Roddick and Paradorn Srichaphan at the Stella Artois tournament in London, England, Roddick hit a serve measured at 153 mi/h. How much work was required by Andy to serve a 2-oz tennis ball at that speed?

28. Softball How much work has to be performed on a 6.5-oz softball to pitch it 132 ft/sec (90 mph)?

29. Drinking a milkshake The truncated conical container shown here is full of strawberry milkshake that weighs 4/9 oz/in³. As you can see, the container is 7 in. deep, 2.5 in. across at the base, and 3.5 in. across at the top (a standard size at Brigham's in Boston). The straw sticks up an inch above the top. About how much work does it take to suck up the milkshake through the straw (neglecting friction)? Answer in inch-ounces.

Dimensions in inches

30. Water tower Your town has decided to drill a well to increase its water supply. As the town engineer, you have determined that a water tower will be necessary to provide the pressure needed for distribution, and you have designed the system shown here. The water is to be pumped from a 300 ft well through a vertical 4 in. pipe into the base of a cylindrical tank 20 ft in diameter and 25 ft high. The base of the tank will be 60 ft above ground. The pump is a 3 hp pump, rated at 1650 ft·lb/sec. To the nearest hour, how long will it take to fill the tank the first time? (Include the time it takes to fill the pipe.) Assume that water weighs 62.4 lb/ft³.

NOT TO SCALE

31. Putting a satellite in orbit The strength of Earth's gravitational field varies with the distance r from Earth's center, and the magnitude of the gravitational force experienced by a satellite of mass m during and after launch is

$$F(r) = \frac{mMG}{r^2}.$$

Here, $M = 5.975 \times 10^{24}$ kg is Earth's mass, $G = 6.6720 \times 10^{-11}$ N·m² kg⁻² is the universal gravitational constant, and r is measured in meters. The work it takes to lift a 1000-kg satellite from Earth's surface to a circular orbit 35,780 km above Earth's center is therefore given by the integral

$$\text{Work} = \int_{6,370,000}^{35,780,000} \frac{1000MG}{r^2}\, dr \text{ joules.}$$

Evaluate the integral. The lower limit of integration is Earth's radius in meters at the launch site. (This calculation does not take into account energy spent lifting the launch vehicle or energy spent bringing the satellite to orbit velocity.)

32. Forcing electrons together Two electrons r meters apart repel each other with a force of

$$F = \frac{23 \times 10^{-29}}{r^2} \text{ newtons.}$$

a. Suppose one electron is held fixed at the point $(1, 0)$ on the x-axis (units in meters). How much work does it take to move a second electron along the x-axis from the point $(-1, 0)$ to the origin?

b. Suppose an electron is held fixed at each of the points $(-1, 0)$ and $(1, 0)$. How much work does it take to move a third electron along the x-axis from $(5, 0)$ to $(3, 0)$?

Finding Fluid Forces

33. Triangular plate Calculate the fluid force on one side of the plate in Example 6 using the coordinate system shown here.

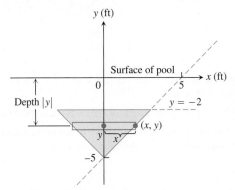

34. Triangular plate Calculate the fluid force on one side of the plate in Example 6 using the coordinate system shown here.

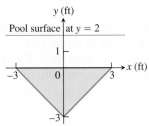

35. Rectangular plate In a pool filled with water to a depth of 10 ft, calculate the fluid force on one side of a 3 ft by 4 ft rectangular plate if the plate rests vertically at the bottom of the pool

a. on its 4-ft edge.　　**b.** on its 3-ft edge.

36. Semicircular plate Calculate the fluid force on one side of a semicircular plate of radius 5 ft that rests vertically on its diameter at the bottom of a pool filled with water to a depth of 6 ft.

37. Triangular plate The isosceles triangular plate shown here is submerged vertically 1 ft below the surface of a freshwater lake.

a. Find the fluid force against one face of the plate.

b. What would be the fluid force on one side of the plate if the water were seawater instead of freshwater?

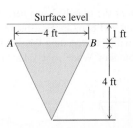

38. Rotated triangular plate The plate in Exercise 37 is revolved 180° about line AB so that part of the plate sticks out of the lake, as shown here. What force does the water exert on one face of the plate now?

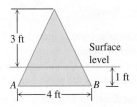

39. New England Aquarium The viewing portion of the rectangular glass window in a typical fish tank at the New England Aquarium in Boston is 63 in. wide and runs from 0.5 in. below the water's surface to 33.5 in. below the surface. Find the fluid force against this portion of the window. The weight-density of seawater is 64 lb/ft³. (In case you were wondering, the glass is 3/4 in. thick and the tank walls extend 4 in. above the water to keep the fish from jumping out.)

40. Semicircular plate A semicircular plate 2 ft in diameter sticks straight down into freshwater with the diameter along the surface. Find the force exerted by the water on one side of the plate.

41. Tilted plate Calculate the fluid force on one side of a 5 ft by 5 ft square plate if the plate is at the bottom of a pool filled with water to a depth of 8 ft and

a. lying flat on its 5 ft by 5 ft face.

b. resting vertically on a 5-ft edge.

c. resting on a 5-ft edge and tilted at 45° to the bottom of the pool.

42. Tilted plate Calculate the fluid force on one side of a right-triangular plate with edges 3 ft, 4 ft, and 5 ft if the plate sits at the bottom of a pool filled with water to a depth of 6 ft on its 3-ft edge and tilted at 60° to the bottom of the pool.

43. The cubical metal tank shown here has a parabolic gate held in place by bolts and designed to withstand a fluid force of 160 lb without rupturing. The liquid you plan to store has a weight-density of 50 lb/ft³.

a. What is the fluid force on the gate when the liquid is 2 ft deep?

b. What is the maximum height to which the container can be filled without exceeding the gate's design limitation?

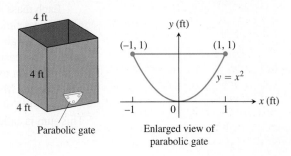

Parabolic gate　　Enlarged view of parabolic gate

44. The end plates of the trough shown here were designed to withstand a fluid force of 6667 lb. How many cubic feet of water can the tank hold without exceeding this limitation? Round down to the nearest cubic foot. What is the value of h?

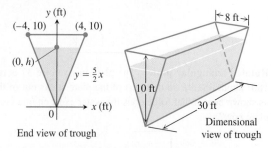

End view of trough

Dimensional view of trough

45. A vertical rectangular plate a units long by b units wide is submerged in a fluid of weight-density w with its long edges parallel to the fluid's surface. Find the average value of the pressure along the vertical dimension of the plate. Explain your answer.

46. (*Continuation of Exercise 45.*) Show that the force exerted by the fluid on one side of the plate is the average value of the pressure (found in Exercise 45) times the area of the plate.

47. Water pours into the tank shown here at the rate of 4 ft³/min. The tank's cross-sections are 4-ft-diameter semicircles. One end of the tank is movable, but moving it to increase the volume compresses a spring. The spring constant is $k = 100$ lb/ft. If the end of the tank moves 5 ft against the spring, the water will drain out of a safety hole in the bottom at the rate of 5 ft³/min. Will the movable end reach the hole before the tank overflows?

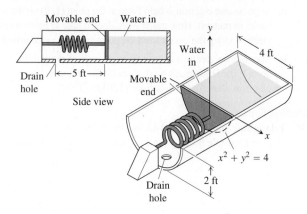

48. Watering trough The vertical ends of a watering trough are squares 3 ft on a side.

a. Find the fluid force against the ends when the trough is full.

b. How many inches do you have to lower the water level in the trough to reduce the fluid force by 25%?

6.6 Moments and Centers of Mass

Many structures and mechanical systems behave as if their masses were concentrated at a single point, called the *center of mass* (Figure 6.44). It is important to know how to locate this point, and doing so is basically a mathematical enterprise. For the moment, we deal with one- and two-dimensional objects. Three-dimensional objects are best done with the multiple integrals of Chapter 15.

Masses Along a Line

We develop our mathematical model in stages. The first stage is to imagine masses m_1, m_2, and m_3 on a rigid x-axis supported by a fulcrum at the origin.

The resulting system might balance, or it might not, depending on how large the masses are and how they are arranged along the x-axis.

Each mass m_k exerts a downward force $m_k g$ (the weight of m_k) equal to the magnitude of the mass times the acceleration due to gravity. Each of these forces has a tendency to turn the axis about the origin, the way a child turns a seesaw. This turning effect, called a **torque**, is measured by multiplying the force $m_k g$ by the signed distance x_k from the point of application to the origin. Masses to the left of the origin exert negative (counterclockwise) torque. Masses to the right of the origin exert positive (clockwise) torque.

The sum of the torques measures the tendency of a system to rotate about the origin. This sum is called the **system torque**.

$$\text{System torque} = m_1 g x_1 + m_2 g x_2 + m_3 g x_3 \tag{1}$$

The system will balance if and only if its torque is zero.

If we factor out the g in Equation (1), we see that the system torque is

$$\underbrace{g}_{\substack{\text{a feature of the} \\ \text{environment}}} \cdot \underbrace{(m_1x_1 + m_2x_2 + m_3x_3)}_{\substack{\text{a feature of} \\ \text{the system}}}.$$

Thus, the torque is the product of the gravitational acceleration g, which is a feature of the environment in which the system happens to reside, and the number $(m_1x_1 + m_2x_2 + m_3x_3)$, which is a feature of the system itself, a constant that stays the same no matter where the system is placed.

The number $(m_1x_1 + m_2x_2 + m_3x_3)$ is called the **moment of the system about the origin**. It is the sum of the **moments** m_1x_1, m_2x_2, m_3x_3 of the individual masses.

$$M_0 = \text{Moment of system about origin} = \sum m_k x_k$$

(We shift to sigma notation here to allow for sums with more terms.)

We usually want to know where to place the fulcrum to make the system balance, that is, at what point $\bar{x}$ to place it to make the torques add to zero.

FIGURE 6.44 A wrench gliding on ice turning about its center of mass as the center glides in a vertical line.

The torque of each mass about the fulcrum in this special location is

$$\text{Torque of } m_k \text{ about } \bar{x} = \begin{pmatrix} \text{signed distance} \\ \text{of } m_k \text{ from } \bar{x} \end{pmatrix} \begin{pmatrix} \text{downward} \\ \text{force} \end{pmatrix}$$
$$= (x_k - \bar{x})m_k g.$$

When we write the equation that says that the sum of these torques is zero, we get an equation we can solve for $\bar{x}$:

$$\sum (x_k - \bar{x})m_k g = 0 \qquad \text{Sum of the torques equals zero.}$$

$$\bar{x} = \frac{\sum m_k x_k}{\sum m_k}. \qquad \text{Solved for } \bar{x}$$

This last equation tells us to find $\bar{x}$ by dividing the system's moment about the origin by the system's total mass:

$$\bar{x} = \frac{\sum m_k x_k}{\sum m_k} = \frac{\text{system moment about origin}}{\text{system mass}}. \tag{2}$$

The point $\bar{x}$ is called the system's **center of mass**.

Masses Distributed over a Plane Region

Suppose that we have a finite collection of masses located in the plane, with mass m_k at the point (x_k, y_k) (see Figure 6.45). The mass of the system is

$$\text{System mass:} \quad M = \sum m_k.$$

Each mass m_k has a moment about each axis. Its moment about the x-axis is $m_k y_k$, and its moment about the y-axis is $m_k x_k$. The moments of the entire system about the two axes are

$$\text{Moment about } x\text{-axis:} \quad M_x = \sum m_k y_k,$$
$$\text{Moment about } y\text{-axis:} \quad M_y = \sum m_k x_k.$$

FIGURE 6.45 Each mass m_k has a moment about each axis.

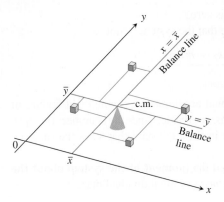

FIGURE 6.46 A two-dimensional array of masses balances on its center of mass.

The x-coordinate of the system's center of mass is defined to be

$$\bar{x} = \frac{M_y}{M} = \frac{\sum m_k x_k}{\sum m_k}. \tag{3}$$

With this choice of $\bar{x}$, as in the one-dimensional case, the system balances about the line $x = \bar{x}$ (Figure 6.46).

The y-coordinate of the system's center of mass is defined to be

$$\bar{y} = \frac{M_x}{M} = \frac{\sum m_k y_k}{\sum m_k}. \tag{4}$$

With this choice of $\bar{y}$, the system balances about the line $y = \bar{y}$ as well. The torques exerted by the masses about the line $y = \bar{y}$ cancel out. Thus, as far as balance is concerned, the system behaves as if all its mass were at the single point $(\bar{x}, \bar{y})$. We call this point the system's **center of mass**.

Thin, Flat Plates

In many applications, we need to find the center of mass of a thin, flat plate: a disk of aluminum, say, or a triangular sheet of steel. In such cases, we assume the distribution of mass to be continuous, and the formulas we use to calculate $\bar{x}$ and $\bar{y}$ contain integrals instead of finite sums. The integrals arise in the following way.

Imagine that the plate occupying a region in the xy-plane is cut into thin strips parallel to one of the axes (in Figure 6.47, the y-axis). The center of mass of a typical strip is $(\tilde{x}, \tilde{y})$. We treat the strip's mass Δm as if it were concentrated at $(\tilde{x}, \tilde{y})$. The moment of the strip about the y-axis is then $\tilde{x}\,\Delta m$. The moment of the strip about the x-axis is $\tilde{y}\,\Delta m$. Equations (3) and (4) then become

$$\bar{x} = \frac{M_y}{M} = \frac{\sum \tilde{x}\,\Delta m}{\sum \Delta m}, \qquad \bar{y} = \frac{M_x}{M} = \frac{\sum \tilde{y}\,\Delta m}{\sum \Delta m}.$$

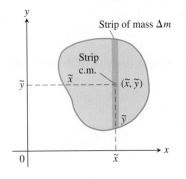

FIGURE 6.47 A plate cut into thin strips parallel to the y-axis. The moment exerted by a typical strip about each axis is the moment its mass Δm would exert if concentrated at the strip's center of mass $(\tilde{x}, \tilde{y})$.

The sums are Riemann sums for integrals and approach these integrals as limiting values as the strips into which the plate is cut become narrower and narrower. We write these integrals symbolically as

$$\bar{x} = \frac{\int \tilde{x}\,dm}{\int dm} \qquad \text{and} \qquad \bar{y} = \frac{\int \tilde{y}\,dm}{\int dm}.$$

Moments, Mass, and Center of Mass of a Thin Plate Covering a Region in the xy-Plane

$$\text{Moment about the } x\text{-axis:} \qquad M_x = \int \tilde{y}\,dm$$

$$\text{Moment about the } y\text{-axis:} \qquad M_y = \int \tilde{x}\,dm \tag{5}$$

$$\text{Mass:} \qquad M = \int dm$$

$$\text{Center of mass:} \qquad \bar{x} = \frac{M_y}{M}, \quad \bar{y} = \frac{M_x}{M}$$

Density

A material's density is its mass per unit area. For wires, rods, and narrow strips, we use mass per unit length.

The differential dm is the mass of the strip. Assuming the density δ of the plate to be a continuous function, the mass differential dm equals the product $\delta\, dA$ (mass per unit area times area). Here dA represents the area of the strip.

To evaluate the integrals in Equations (5), we picture the plate in the coordinate plane and sketch a strip of mass parallel to one of the coordinate axes. We then express the strip's mass dm and the coordinates $(\widetilde{x}, \widetilde{y})$ of the strip's center of mass in terms of x or y. Finally, we integrate $\widetilde{y}\, dm$, $\widetilde{x}\, dm$, and dm between limits of integration determined by the plate's location in the plane.

EXAMPLE 1 The triangular plate shown in Figure 6.48 has a constant density of $\delta = 3\ \text{g/cm}^2$. Find

(a) the plate's moment M_y about the y-axis. **(b)** the plate's mass M.

(c) the x-coordinate of the plate's center of mass (c.m.).

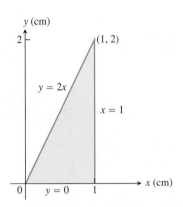

FIGURE 6.48 The plate in Example 1.

Solution Method 1: Vertical Strips (Figure 6.49)

(a) The moment M_y: The typical vertical strip has the following relevant data.

$$\text{center of mass (c.m.):}\quad (\widetilde{x}, \widetilde{y}) = (x, x)$$
$$\text{length:}\quad 2x$$
$$\text{width:}\quad dx$$
$$\text{area:}\quad dA = 2x\, dx$$
$$\text{mass:}\quad dm = \delta\, dA = 3 \cdot 2x\, dx = 6x\, dx$$
$$\text{distance of c.m. from } y\text{-axis:}\quad \widetilde{x} = x$$

The moment of the strip about the y-axis is

$$\widetilde{x}\, dm = x \cdot 6x\, dx = 6x^2\, dx.$$

The moment of the plate about the y-axis is therefore

$$M_y = \int \widetilde{x}\, dm = \int_0^1 6x^2\, dx = 2x^3 \Big]_0^1 = 2\ \text{g} \cdot \text{cm}.$$

(b) The plate's mass:

$$M = \int dm = \int_0^1 6x\, dx = 3x^2 \Big]_0^1 = 3\ \text{g}.$$

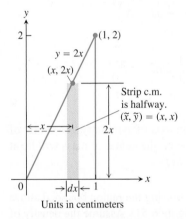

FIGURE 6.49 Modeling the plate in Example 1 with vertical strips.

(c) The x-coordinate of the plate's center of mass:

$$\overline{x} = \frac{M_y}{M} = \frac{2\ \text{g} \cdot \text{cm}}{3\ \text{g}} = \frac{2}{3}\ \text{cm}.$$

By a similar computation, we could find M_x and $\overline{y} = M_x/M$.

Method 2: Horizontal Strips (Figure 6.50)

(a) The moment M_y: The y-coordinate of the center of mass of a typical horizontal strip is y (see the figure), so

$$\widetilde{y} = y.$$

The x-coordinate is the x-coordinate of the point halfway across the triangle. This makes it the average of $y/2$ (the strip's left-hand x-value) and 1 (the strip's right-hand x-value):

$$\widetilde{x} = \frac{(y/2) + 1}{2} = \frac{y}{4} + \frac{1}{2} = \frac{y + 2}{4}.$$

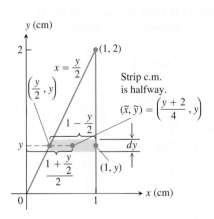

FIGURE 6.50 Modeling the plate in Example 1 with horizontal strips.

We also have

$$\text{length:} \quad 1 - \frac{y}{2} = \frac{2-y}{2}$$

$$\text{width:} \quad dy$$

$$\text{area:} \quad dA = \frac{2-y}{2}\,dy$$

$$\text{mass:} \quad dm = \delta\,dA = 3 \cdot \frac{2-y}{2}\,dy$$

$$\text{distance of c.m. to } y\text{-axis:} \quad \widetilde{x} = \frac{y+2}{4}.$$

The moment of the strip about the y-axis is

$$\widetilde{x}\,dm = \frac{y+2}{4} \cdot 3 \cdot \frac{2-y}{2}\,dy = \frac{3}{8}(4-y^2)\,dy.$$

The moment of the plate about the y-axis is

$$M_y = \int \widetilde{x}\,dm = \int_0^2 \frac{3}{8}(4-y^2)\,dy = \frac{3}{8}\left[4y - \frac{y^3}{3}\right]_0^2 = \frac{3}{8}\left(\frac{16}{3}\right) = 2 \text{ g} \cdot \text{cm}.$$

(b) The plate's mass:

$$M = \int dm = \int_0^2 \frac{3}{2}(2-y)\,dy = \frac{3}{2}\left[2y - \frac{y^2}{2}\right]_0^2 = \frac{3}{2}(4-2) = 3 \text{ g}.$$

(c) The x-coordinate of the plate's center of mass:

$$\bar{x} = \frac{M_y}{M} = \frac{2 \text{ g} \cdot \text{cm}}{3 \text{ g}} = \frac{2}{3} \text{ cm}.$$

By a similar computation, we could find M_x and $\bar{y}$. $\blacksquare$

If the distribution of mass in a thin, flat plate has an axis of symmetry, the center of mass will lie on this axis. If there are two axes of symmetry, the center of mass will lie at their intersection. These facts often help to simplify our work.

EXAMPLE 2   Find the center of mass of a thin plate covering the region bounded above by the parabola $y = 4 - x^2$ and below by the x-axis (Figure 6.51). Assume the density of the plate at the point (x, y) is $\delta = 2x^2$, which is twice the square of the distance from the point to the y-axis.

Solution   The mass distribution is symmetric about the y-axis, so $\bar{x} = 0$. We model the distribution of mass with vertical strips since the density is given as a function of the variable x. The typical vertical strip (see Figure 6.51) has the following relevant data.

$$\text{center of mass (c.m.):} \quad (\widetilde{x}, \widetilde{y}) = \left(x, \frac{4-x^2}{2}\right)$$

$$\text{length:} \quad 4 - x^2$$

$$\text{width:} \quad dx$$

$$\text{area:} \quad dA = (4-x^2)\,dx$$

$$\text{mass:} \quad dm = \delta\,dA = \delta(4-x^2)\,dx$$

$$\text{distance from c.m. to } x\text{-axis:} \quad \widetilde{y} = \frac{4-x^2}{2}$$

The moment of the strip about the x-axis is

$$\widetilde{y}\,dm = \frac{4-x^2}{2} \cdot \delta(4-x^2)\,dx = \frac{\delta}{2}(4-x^2)^2\,dx.$$

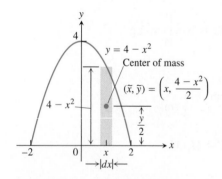

FIGURE 6.51 Modeling the plate in Example 2 with vertical strips.

The moment of the plate about the x-axis is

$$M_x = \int \widetilde{y}\, dm = \int_{-2}^{2} \frac{\delta}{2}(4 - x^2)^2\, dx = \int_{-2}^{2} x^2(4 - x^2)^2\, dx$$

$$= \int_{-2}^{2}(16x^2 - 8x^4 + x^6)\, dx = \frac{2048}{105}$$

$$M = \int dm = \int_{-2}^{2} \delta(4 - x^2)\, dx = \int_{-2}^{2} 2x^2(4 - x^2)\, dx$$

$$= \int_{-2}^{2}(8x^2 - 2x^4)\, dx = \frac{256}{15}.$$

Therefore,

$$\overline{y} = \frac{M_x}{M} = \frac{2048}{105} \cdot \frac{15}{256} = \frac{8}{7}.$$

The plate's center of mass is

$$(\overline{x}, \overline{y}) = \left(0, \frac{8}{7}\right).$$

∎

Plates Bounded by Two Curves

Suppose a plate covers a region that lies between two curves $y = g(x)$ and $y = f(x)$, where $f(x) \geq g(x)$ and $a \leq x \leq b$. The typical vertical strip (see Figure 6.52) has

center of mass (c.m.): $(\widetilde{x}, \widetilde{y}) = (x, \frac{1}{2}[f(x) + g(x)])$
length: $f(x) - g(x)$
width: dx
area: $dA = [f(x) - g(x)]\, dx$
mass: $dm = \delta\, dA = \delta[f(x) - g(x)]\, dx.$

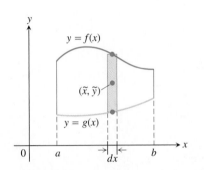

FIGURE 6.52 Modeling the plate bounded by two curves with vertical strips. The strip c.m. is halfway, so $\widetilde{y} = \frac{1}{2}[f(x) + g(x)]$.

The moment of the plate about the y-axis is

$$M_y = \int x\, dm = \int_{a}^{b} x\delta[f(x) - g(x)]\, dx,$$

and the moment about the x-axis is

$$M_x = \int y\, dm = \int_{a}^{b} \frac{1}{2}[f(x) + g(x)] \cdot \delta[f(x) - g(x)]\, dx$$

$$= \int_{a}^{b} \frac{\delta}{2}[f^2(x) - g^2(x)]\, dx.$$

These moments give the formulas

$$\overline{x} = \frac{1}{M} \int_{a}^{b} \delta x\, [f(x) - g(x)]\, dx \tag{6}$$

$$\overline{y} = \frac{1}{M} \int_{a}^{b} \frac{\delta}{2}\, [f^2(x) - g^2(x)]\, dx \tag{7}$$

EXAMPLE 3 Find the center of mass for the thin plate bounded by the curves $g(x) = x/2$ and $f(x) = \sqrt{x}$, $0 \le x \le 1$, (Figure 6.53) using Equations (6) and (7) with the density function $\delta(x) = x^2$.

Solution We first compute the mass of the plate, where $dm = \delta[f(x) - g(x)]\,dx$:

$$M = \int_0^1 x^2\left(\sqrt{x} - \frac{x}{2}\right)dx = \int_0^1 \left(x^{5/2} - \frac{x^3}{2}\right)dx = \left[\frac{2}{7}x^{7/2} - \frac{1}{8}x^4\right]_0^1 = \frac{9}{56}.$$

Then from Equations (6) and (7) we get

$$\bar{x} = \frac{56}{9}\int_0^1 x^2 \cdot x\left(\sqrt{x} - \frac{x}{2}\right)dx$$

$$= \frac{56}{9}\int_0^1 \left(x^{7/2} - \frac{x^4}{2}\right)dx$$

$$= \frac{56}{9}\left[\frac{2}{9}x^{9/2} - \frac{1}{10}x^5\right]_0^1 = \frac{308}{405},$$

and

$$\bar{y} = \frac{56}{9}\int_0^1 \frac{x^2}{2}\left(x - \frac{x^2}{4}\right)dx$$

$$= \frac{28}{9}\int_0^1 \left(x^3 - \frac{x^4}{4}\right)dx$$

$$= \frac{28}{9}\left[\frac{1}{4}x^4 - \frac{1}{20}x^5\right]_0^1 = \frac{252}{405}.$$

The center of mass is shown in Figure 6.53. ■

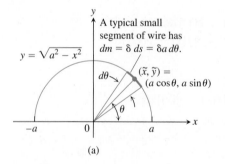

$y = \sqrt{a^2 - x^2}$

A typical small segment of wire has $dm = \delta\,ds = \delta a\,d\theta$.

$(\tilde{x}, \tilde{y}) = (a\cos\theta, a\sin\theta)$

(a)

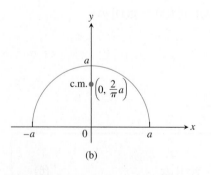

$\left(0, \frac{2}{\pi}a\right)$

(b)

FIGURE 6.54 The semicircular wire in Example 4. (a) The dimensions and variables used in finding the center of mass. (b) The center of mass does not lie on the wire.

FIGURE 6.53 The region in Example 3.

$f(x) = \sqrt{x}$

c.m.

$g(x) = \frac{x}{2}$

Centroids

When the density function is constant, it cancels out of the numerator and denominator of the formulas for $\bar{x}$ and $\bar{y}$. Thus, when the density is constant, the location of the center of mass is a feature of the geometry of the object and not of the material from which it is made. In such cases, engineers may call the center of mass the **centroid** of the shape, as in "Find the centroid of a triangle or a solid cone." To do so, just set δ equal to 1 and proceed to find $\bar{x}$ and $\bar{y}$ as before, by dividing moments by masses.

EXAMPLE 4 Find the center of mass (centroid) of a thin wire of constant density δ shaped like a semicircle of radius a.

Solution We model the wire with the semicircle $y = \sqrt{a^2 - x^2}$ (Figure 6.54). The distribution of mass is symmetric about the y-axis, so $\bar{x} = 0$. To find $\bar{y}$, we imagine the wire divided into short subarc segments. If $(\tilde{x}, \tilde{y})$ is the center of mass of a subarc and θ is the angle between the x-axis and the radial line joining the origin to $(\tilde{x}, \tilde{y})$, then $\tilde{y} = a\sin\theta$ is a function of the angle θ measured in radians (see Figure 6.54a). The length ds of the subarc containing $(\tilde{x}, \tilde{y})$ subtends an angle of $d\theta$ radians, so $ds = a\,d\theta$. Thus a typical subarc segment has these relevant data for calculating $\bar{y}$:

$$\text{length:}\quad ds = a\,d\theta$$

$$\text{mass:}\quad dm = \delta\,ds = \delta a\,d\theta \qquad \text{\small Mass per unit length times length}$$

$$\text{distance of c.m. to } x\text{-axis:}\quad \tilde{y} = a\sin\theta.$$

Hence,

$$\bar{y} = \frac{\int \tilde{y}\, dm}{\int dm} = \frac{\int_0^\pi a \sin\theta \cdot \delta a\, d\theta}{\int_0^\pi \delta a\, d\theta} = \frac{\delta a^2 \big[-\cos\theta\big]_0^\pi}{\delta a \pi} = \frac{2}{\pi}\, a.$$

The center of mass lies on the axis of symmetry at the point $(0, 2a/\pi)$, about two-thirds of the way up from the origin (Figure 6.54b). Notice how δ cancels in the equation for $\bar{y}$, so we could have set $\delta = 1$ everywhere and obtained the same value for $\bar{y}$. ∎

In Example 4 we found the center of mass of a thin wire lying along the graph of a differentiable function in the xy-plane. In Chapter 16 we will learn how to find the center of mass of wires lying along more general smooth curves in the plane (or in space).

Fluid Forces and Centroids

If we know the location of the centroid of a submerged flat vertical plate (Figure 6.55), we can take a shortcut to find the force against one side of the plate. From Equation (7) in Section 6.5,

Surface level of fluid

$\bar{h}$ = centroid depth

Plate centroid

FIGURE 6.55 The force against one side of the plate is $w \cdot \bar{h} \cdot$ plate area.

$$F = \int_a^b w \times (\text{strip depth}) \times L(y)\, dy$$

$$= w \int_a^b (\text{strip depth}) \times L(y)\, dy$$

$$= w \times (\text{moment about surface level line of region occupied by plate})$$

$$= w \times (\text{depth of plate's centroid}) \times (\text{area of plate}).$$

> **Fluid Forces and Centroids**
> The force of a fluid of weight-density w against one side of a submerged flat vertical plate is the product of w, the distance $\bar{h}$ from the plate's centroid to the fluid surface, and the plate's area:
>
> $$F = w\bar{h}A. \tag{8}$$

EXAMPLE 5   A flat isosceles triangular plate with base 6 ft and height 3 ft is submerged vertically, base up with its vertex at the origin, so that the base is 2 ft below the surface of a swimming pool. (This is Example 6, Section 6.5.) Use Equation (8) to find the force exerted by the water against one side of the plate.

Solution   The centroid of the triangle (Figure 6.43) lies on the y-axis, one-third of the way from the base to the vertex, so $\bar{h} = 3$ (where $y = 2$) since the pool's surface is $y = 5$. The triangle's area is

$$A = \frac{1}{2}(\text{base})(\text{height}) = \frac{1}{2}(6)(3) = 9.$$

Hence,

$$F = w\bar{h}A = (62.4)(3)(9) = 1684.8\ \text{lb}. \qquad ∎$$

The Theorems of Pappus

In the fourth century, an Alexandrian Greek named Pappus discovered two formulas that relate centroids to surfaces and solids of revolution. The formulas provide shortcuts to a number of otherwise lengthy calculations.

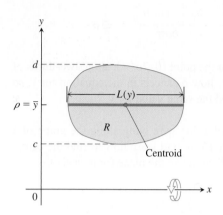

FIGURE 6.56 The region R is to be revolved (once) about the x-axis to generate a solid. A 1700-year-old theorem says that the solid's volume can be calculated by multiplying the region's area by the distance traveled by its centroid during the revolution.

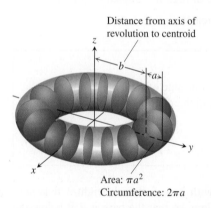

FIGURE 6.57 With Pappus's first theorem, we can find the volume of a torus without having to integrate (Example 6).

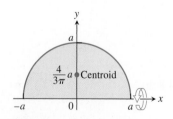

FIGURE 6.58 With Pappus's first theorem, we can locate the centroid of a semicircular region without having to integrate (Example 7).

THEOREM 1   Pappus's Theorem for Volumes

If a plane region is revolved once about a line in the plane that does not cut through the region's interior, then the volume of the solid it generates is equal to the region's area times the distance traveled by the region's centroid during the revolution. If ρ is the distance from the axis of revolution to the centroid, then

$$V = 2\pi\rho A. \qquad (9)$$

Proof   We draw the axis of revolution as the x-axis with the region R in the first quadrant (Figure 6.56). We let $L(y)$ denote the length of the cross-section of R perpendicular to the y-axis at y. We assume $L(y)$ to be continuous.

By the method of cylindrical shells, the volume of the solid generated by revolving the region about the x-axis is

$$V = \int_c^d 2\pi(\text{shell radius})(\text{shell height})\,dy = 2\pi\int_c^d y\,L(y)\,dy. \qquad (10)$$

The y-coordinate of R's centroid is

$$\bar{y} = \frac{\int_c^d \tilde{y}\,dA}{A} = \frac{\int_c^d y\,L(y)\,dy}{A}, \qquad \tilde{y} = y, \; dA = L(y)\,dy$$

so that

$$\int_c^d y\,L(y)\,dy = A\bar{y}.$$

Substituting $A\bar{y}$ for the last integral in Equation (10) gives $V = 2\pi\bar{y}A$. With ρ equal to $\bar{y}$, we have $V = 2\pi\rho A$.   ∎

EXAMPLE 6   Find the volume of the torus (doughnut) generated by revolving a circular disk of radius a about an axis in its plane at a distance $b \geq a$ from its center (Figure 6.57).

Solution   We apply Pappus's Theorem for volumes. The centroid of a disk is located at its center, the area is $A = \pi a^2$, and $\rho = b$ is the distance from the centroid to the axis of revolution (see Figure 6.57). Substituting these values into Equation (9), we find the volume of the torus to be

$$V = 2\pi(b)(\pi a^2) = 2\pi^2 ba^2.$$ ∎

The next example shows how we can use Equation (9) in Pappus's Theorem to find one of the coordinates of the centroid of a plane region of known area A when we also know the volume V of the solid generated by revolving the region about the other coordinate axis. That is, if $\bar{y}$ is the coordinate we want to find, we revolve the region around the x-axis so that $\bar{y} = \rho$ is the distance from the centroid to the axis of revolution. The idea is that the rotation generates a solid of revolution whose volume V is an already known quantity. Then we can solve Equation (9) for ρ, which is the value of the centroid's coordinate $\bar{y}$.

EXAMPLE 7   Locate the centroid of a semicircular region of radius a.

Solution   We consider the region between the semicircle $y = \sqrt{a^2 - x^2}$ (Figure 6.58) and the x-axis and imagine revolving the region about the x-axis to generate a solid sphere. By symmetry, the x-coordinate of the centroid is $\bar{x} = 0$. With $\bar{y} = \rho$ in Equation (9), we have

$$\bar{y} = \frac{V}{2\pi A} = \frac{(4/3)\pi a^3}{2\pi(1/2)\pi a^2} = \frac{4}{3\pi}a.$$ ∎

> **THEOREM 2   Pappus's Theorem for Surface Areas**
>
> If an arc of a smooth plane curve is revolved once about a line in the plane that does not cut through the arc's interior, then the area of the surface generated by the arc equals the length L of the arc times the distance traveled by the arc's centroid during the revolution. If ρ is the distance from the axis of revolution to the centroid, then
>
> $$S = 2\pi\rho L. \tag{11}$$

The proof we give assumes that we can model the axis of revolution as the x-axis and the arc as the graph of a continuously differentiable function of x.

Proof   We draw the axis of revolution as the x-axis with the arc extending from $x = a$ to $x = b$ in the first quadrant (Figure 6.59). The area of the surface generated by the arc is

$$S = \int_{x=a}^{x=b} 2\pi y \, ds = 2\pi \int_{x=a}^{x=b} y \, ds. \tag{12}$$

The y-coordinate of the arc's centroid is

$$\bar{y} = \frac{\int_{x=a}^{x=b} \tilde{y} \, ds}{\int_{x=a}^{x=b} ds} = \frac{\int_{x=a}^{x=b} y \, ds}{L}. \qquad \begin{array}{l} L = \int ds \text{ is the arc's} \\ \text{length and } \tilde{y} = y. \end{array}$$

Hence

$$\int_{x=a}^{x=b} y \, ds = \bar{y}L.$$

Substituting $\bar{y}L$ for the last integral in Equation (12) gives $S = 2\pi\bar{y}L$. With ρ equal to $\bar{y}$, we have $S = 2\pi\rho L$. ∎

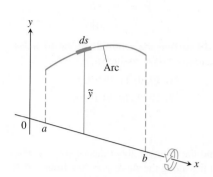

FIGURE 6.59 Figure for proving Pappus's Theorem for surface area. The arc length differential ds is given by Equation (6) in Section 6.3.

EXAMPLE 8   Use Pappus's area theorem to find the surface area of the torus in Example 6.

Solution   From Figure 6.57, the surface of the torus is generated by revolving a circle of radius a about the z-axis, and $b \geq a$ is the distance from the centroid to the axis of revolution. The arc length of the smooth curve generating this surface of revolution is the circumference of the circle, so $L = 2\pi a$. Substituting these values into Equation (11), we find the surface area of the torus to be

$$S = 2\pi(b)(2\pi a) = 4\pi^2 ba. \qquad ■$$

Exercises 6.6

Thin Plates with Constant Density

In Exercises 1–14, find the center of mass of a thin plate of constant density δ covering the given region.

1. The region bounded by the parabola $y = x^2$ and the line $y = 4$
2. The region bounded by the parabola $y = 25 - x^2$ and the x-axis
3. The region bounded by the parabola $y = x - x^2$ and the line $y = -x$
4. The region enclosed by the parabolas $y = x^2 - 3$ and $y = -2x^2$

5. The region bounded by the y-axis and the curve $x = y - y^3$, $0 \leq y \leq 1$
6. The region bounded by the parabola $x = y^2 - y$ and the line $y = x$
7. The region bounded by the x-axis and the curve $y = \cos x$, $-\pi/2 \leq x \leq \pi/2$
8. The region between the curve $y = \sec^2 x$, $-\pi/4 \leq x \leq \pi/4$ and the x-axis

T 9. The region between the curve $y = 1/x$ and the x-axis from $x = 1$ to $x = 2$. Give the coordinates to two decimal places.

10. a. The region cut from the first quadrant by the circle $x^2 + y^2 = 9$

   b. The region bounded by the x-axis and the semicircle $y = \sqrt{9 - x^2}$

   Compare your answer in part (b) with the answer in part (a).

11. The region in the first and fourth quadrants enclosed by the curves $y = 1/(1 + x^2)$ and $y = -1/(1 + x^2)$ and by the lines $x = 0$ and $x = 1$

12. The region bounded by the parabolas $y = 2x^2 - 4x$ and $y = 2x - x^2$

13. The region between the curve $y = 1/\sqrt{x}$ and the x-axis from $x = 1$ to $x = 16$

14. The region bounded above by the curve $y = 1/x^3$, below by the curve $y = -1/x^3$, and on the left and right by the lines $x = 1$ and $x = a > 1$. Also, find $\lim_{a \to \infty} \bar{x}$.

Thin Plates with Varying Density

15. Find the center of mass of a thin plate covering the region between the x-axis and the curve $y = 2/x^2$, $1 \le x \le 2$, if the plate's density at the point (x, y) is $\delta(x) = x^2$.

16. Find the center of mass of a thin plate covering the region bounded below by the parabola $y = x^2$ and above by the line $y = x$ if the plate's density at the point (x, y) is $\delta(x) = 12x$.

17. The region bounded by the curves $y = \pm 4/\sqrt{x}$ and the lines $x = 1$ and $x = 4$ is revolved about the y-axis to generate a solid.

   a. Find the volume of the solid.

   b. Find the center of mass of a thin plate covering the region if the plate's density at the point (x, y) is $\delta(x) = 1/x$.

   c. Sketch the plate and show the center of mass in your sketch.

18. The region between the curve $y = 2/x$ and the x-axis from $x = 1$ to $x = 4$ is revolved about the x-axis to generate a solid.

   a. Find the volume of the solid.

   b. Find the center of mass of a thin plate covering the region if the plate's density at the point (x, y) is $\delta(x) = \sqrt{x}$.

   c. Sketch the plate and show the center of mass in your sketch.

Centroids of Triangles

19. The centroid of a triangle lies at the intersection of the triangle's medians You may recall that the point inside a triangle that lies one-third of the way from each side toward the opposite vertex is the point where the triangle's three medians intersect. Show that the centroid lies at the intersection of the medians by showing that it too lies one-third of the way from each side toward the opposite vertex. To do so, take the following steps.

   i) Stand one side of the triangle on the x-axis as in part (b) of the accompanying figure. Express dm in terms of L and dy.

   ii) Use similar triangles to show that $L = (b/h)(h - y)$. Substitute this expression for L in your formula for dm.

   iii) Show that $\bar{y} = h/3$.

   iv) Extend the argument to the other sides.

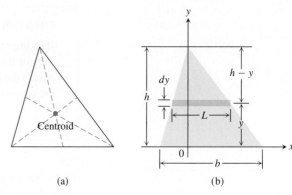

(a)　　　　　(b)

Use the result in Exercise 19 to find the centroids of the triangles whose vertices appear in Exercises 20–24. Assume $a, b > 0$.

20. $(-1, 0), (1, 0), (0, 3)$ 　　**21.** $(0, 0), (1, 0), (0, 1)$

22. $(0, 0), (a, 0), (0, a)$ 　　**23.** $(0, 0), (a, 0), (0, b)$

24. $(0, 0), (a, 0), (a/2, b)$

Thin Wires

25. Constant density Find the moment about the x-axis of a wire of constant density that lies along the curve $y = \sqrt{x}$ from $x = 0$ to $x = 2$.

26. Constant density Find the moment about the x-axis of a wire of constant density that lies along the curve $y = x^3$ from $x = 0$ to $x = 1$.

27. Variable density Suppose that the density of the wire in Example 4 is $\delta = k \sin \theta$ (k constant). Find the center of mass.

28. Variable density Suppose that the density of the wire in Example 4 is $\delta = 1 + k|\cos \theta|$ (k constant). Find the center of mass.

Plates Bounded by Two Curves

In Exercises 29–32, find the centroid of the thin plate bounded by the graphs of the given functions. Use Equations (6) and (7) with $\delta = 1$ and $M = $ area of the region covered by the plate.

29. $g(x) = x^2$ 　and　 $f(x) = x + 6$

30. $g(x) = x^2(x + 1)$, 　$f(x) = 2$, 　and　 $x = 0$

31. $g(x) = x^2(x - 1)$ 　and　 $f(x) = x^2$

32. $g(x) = 0$, 　$f(x) = 2 + \sin x$, 　$x = 0$, 　and 　$x = 2\pi$

   (*Hint:* $\int x \sin x \, dx = \sin x - x \cos x + C$.)

Theory and Examples

Verify the statements and formulas in Exercises 33 and 34.

33. The coordinates of the centroid of a differentiable plane curve are

$$\bar{x} = \frac{\int x \, ds}{\text{length}}, \qquad \bar{y} = \frac{\int y \, ds}{\text{length}}.$$

34. Whatever the value of $p > 0$ in the equation $y = x^2/(4p)$, the y-coordinate of the centroid of the parabolic segment shown here is $\bar{y} = (3/5)a$.

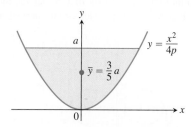

The Theorems of Pappus

35. The square region with vertices $(0, 2)$, $(2, 0)$, $(4, 2)$, and $(2, 4)$ is revolved about the x-axis to generate a solid. Find the volume and surface area of the solid.

36. Use a theorem of Pappus to find the volume generated by revolving about the line $x = 5$ the triangular region bounded by the coordinate axes and the line $2x + y = 6$ (see Exercise 19).

37. Find the volume of the torus generated by revolving the circle $(x - 2)^2 + y^2 = 1$ about the y-axis.

38. Use the theorems of Pappus to find the lateral surface area and the volume of a right-circular cone.

39. Use Pappus's Theorem for surface area and the fact that the surface area of a sphere of radius a is $4\pi a^2$ to find the centroid of the semicircle $y = \sqrt{a^2 - x^2}$.

40. As found in Exercise 39, the centroid of the semicircle $y = \sqrt{a^2 - x^2}$ lies at the point $(0, 2a/\pi)$. Find the area of the surface swept out by revolving the semicircle about the line $y = a$.

41. The area of the region R enclosed by the semiellipse $y = (b/a)\sqrt{a^2 - x^2}$ and the x-axis is $(1/2)\pi ab$, and the volume of the ellipsoid generated by revolving R about the x-axis is $(4/3)\pi ab^2$. Find the centroid of R. Notice that the location is independent of a.

42. As found in Example 7, the centroid of the region enclosed by the x-axis and the semicircle $y = \sqrt{a^2 - x^2}$ lies at the point $(0, 4a/3\pi)$. Find the volume of the solid generated by revolving this region about the line $y = -a$.

43. The region of Exercise 42 is revolved about the line $y = x - a$ to generate a solid. Find the volume of the solid.

44. As found in Exercise 39, the centroid of the semicircle $y = \sqrt{a^2 - x^2}$ lies at the point $(0, 2a/\pi)$. Find the area of the surface generated by revolving the semicircle about the line $y = x - a$.

In Exercises 45 and 46, use a theorem of Pappus to find the centroid of the given triangle. Use the fact that the volume of a cone of radius r and height h is $V = \frac{1}{3}\pi r^2 h$.

45. **46.**

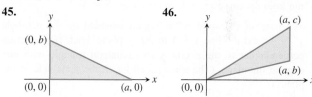

Chapter 6   Questions to Guide Your Review

1. How do you define and calculate the volumes of solids by the method of slicing? Give an example.

2. How are the disk and washer methods for calculating volumes derived from the method of slicing? Give examples of volume calculations by these methods.

3. Describe the method of cylindrical shells. Give an example.

4. How do you find the length of the graph of a smooth function over a closed interval? Give an example. What about functions that do not have continuous first derivatives?

5. How do you define and calculate the area of the surface swept out by revolving the graph of a smooth function $y = f(x)$, $a \le x \le b$, about the x-axis? Give an example.

6. How do you define and calculate the work done by a variable force directed along a portion of the x-axis? How do you calculate the work it takes to pump a liquid from a tank? Give examples.

7. How do you calculate the force exerted by a liquid against a portion of a flat vertical wall? Give an example.

8. What is a center of mass? a centroid?

9. How do you locate the center of mass of a thin flat plate of material? Give an example.

10. How do you locate the center of mass of a thin plate bounded by two curves $y = f(x)$ and $y = g(x)$ over $a \le x \le b$?

Chapter 6   Practice Exercises

Volumes

Find the volumes of the solids in Exercises 1–16.

1. The solid lies between planes perpendicular to the x-axis at $x = 0$ and $x = 1$. The cross-sections perpendicular to the x-axis between these planes are circular disks whose diameters run from the parabola $y = x^2$ to the parabola $y = \sqrt{x}$.

2. The base of the solid is the region in the first quadrant between the line $y = x$ and the parabola $y = 2\sqrt{x}$. The cross-sections of the solid perpendicular to the x-axis are equilateral triangles whose bases stretch from the line to the curve.

3. The solid lies between planes perpendicular to the x-axis at $x = \pi/4$ and $x = 5\pi/4$. The cross-sections between these planes are circular disks whose diameters run from the curve $y = 2 \cos x$ to the curve $y = 2 \sin x$.

4. The solid lies between planes perpendicular to the x-axis at $x = 0$ and $x = 6$. The cross-sections between these planes are squares whose bases run from the x-axis up to the curve $x^{1/2} + y^{1/2} = \sqrt{6}$.

5. The solid lies between planes perpendicular to the x-axis at $x = 0$ and $x = 4$. The cross-sections of the solid perpendicular to the x-axis between these planes are circular disks whose diameters run from the curve $x^2 = 4y$ to the curve $y^2 = 4x$.

6. The base of the solid is the region bounded by the parabola $y^2 = 4x$ and the line $x = 1$ in the xy-plane. Each cross-section perpendicular to the x-axis is an equilateral triangle with one edge in the plane. (The triangles all lie on the same side of the plane.)

7. Find the volume of the solid generated by revolving the region bounded by the x-axis, the curve $y = 3x^4$, and the lines $x = 1$ and $x = -1$ about **(a)** the x-axis; **(b)** the y-axis; **(c)** the line $x = 1$; **(d)** the line $y = 3$.

8. Find the volume of the solid generated by revolving the "triangular" region bounded by the curve $y = 4/x^3$ and the lines $x = 1$ and $y = 1/2$ about **(a)** the x-axis; **(b)** the y-axis; **(c)** the line $x = 2$; **(d)** the line $y = 4$.

9. Find the volume of the solid generated by revolving the region bounded on the left by the parabola $x = y^2 + 1$ and on the right by the line $x = 5$ about **(a)** the x-axis; **(b)** the y-axis; **(c)** the line $x = 5$.

10. Find the volume of the solid generated by revolving the region bounded by the parabola $y^2 = 4x$ and the line $y = x$ about **(a)** the x-axis; **(b)** the y-axis; **(c)** the line $x = 4$; **(d)** the line $y = 4$.

11. Find the volume of the solid generated by revolving the "triangular" region bounded by the x-axis, the line $x = \pi/3$, and the curve $y = \tan x$ in the first quadrant about the x-axis.

12. Find the volume of the solid generated by revolving the region bounded by the curve $y = \sin x$ and the lines $x = 0, x = \pi$, and $y = 2$ about the line $y = 2$.

13. Find the volume of the solid generated by revolving the region bounded by the curve $x = e^{y^2}$ and the lines $y = 0, x = 0$, and $y = 1$ about the x-axis.

14. Find the volume of the solid generated by revolving about the x-axis the region bounded by $y = 2 \tan x, y = 0, x = -\pi/4$, and $x = \pi/4$. (The region lies in the first and third quadrants and resembles a skewed bowtie.)

15. **Volume of a solid sphere hole** A round hole of radius $\sqrt{3}$ ft is bored through the center of a solid sphere of a radius 2 ft. Find the volume of material removed from the sphere.

16. **Volume of a football** The profile of a football resembles the ellipse shown here. Find the football's volume to the nearest cubic inch.

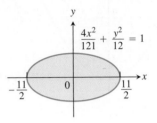

Lengths of Curves
Find the lengths of the curves in Exercises 17–20.

17. $y = x^{1/2} - (1/3)x^{3/2}, \quad 1 \le x \le 4$

18. $x = y^{2/3}, \quad 1 \le y \le 8$

19. $y = x^2 - (\ln x)/8, \quad 1 \le x \le 2$

20. $x = (y^3/12) + (1/y), \quad 1 \le y \le 2$

Areas of Surfaces of Revolution
In Exercises 21–24, find the areas of the surfaces generated by revolving the curves about the given axes.

21. $y = \sqrt{2x + 1}, \quad 0 \le x \le 3; \quad x$-axis

22. $y = x^3/3, \quad 0 \le x \le 1; \quad x$-axis

23. $x = \sqrt{4y - y^2}, \quad 1 \le y \le 2; \quad y$-axis

24. $x = \sqrt{y}, \quad 2 \le y \le 6; \quad y$-axis

Work

25. **Lifting equipment** A rock climber is about to haul up 100 N (about 22.5 lb) of equipment that has been hanging beneath her on 40 m of rope that weighs 0.8 newton per meter. How much work will it take? (*Hint:* Solve for the rope and equipment separately, then add.)

26. **Leaky tank truck** You drove an 800-gal tank truck of water from the base of Mt. Washington to the summit and discovered on arrival that the tank was only half full. You started with a full tank, climbed at a steady rate, and accomplished the 4750-ft elevation change in 50 min. Assuming that the water leaked out at a steady rate, how much work was spent in carrying water to the top? Do not count the work done in getting yourself and the truck there. Water weighs 8 lb/U.S. gal.

27. **Stretching a spring** If a force of 20 lb is required to hold a spring 1 ft beyond its unstressed length, how much work does it take to stretch the spring this far? An additional foot?

28. **Garage door spring** A force of 200 N will stretch a garage door spring 0.8 m beyond its unstressed length. How far will a 300-N force stretch the spring? How much work does it take to stretch the spring this far from its unstressed length?

29. **Pumping a reservoir** A reservoir shaped like a right-circular cone, point down, 20 ft across the top and 8 ft deep, is full of water. How much work does it take to pump the water to a level 6 ft above the top?

30. Pumping a reservoir (*Continuation of Exercise 29.*)   The reservoir is filled to a depth of 5 ft, and the water is to be pumped to the same level as the top. How much work does it take?

31. Pumping a conical tank   A right-circular conical tank, point down, with top radius 5 ft and height 10 ft is filled with a liquid whose weight-density is 60 lb/ft^3. How much work does it take to pump the liquid to a point 2 ft above the tank? If the pump is driven by a motor rated at 275 ft-lb/sec (1/2 hp), how long will it take to empty the tank?

32. Pumping a cylindrical tank   A storage tank is a right-circular cylinder 20 ft long and 8 ft in diameter with its axis horizontal. If the tank is half full of olive oil weighing 57 lb/ft^3, find the work done in emptying it through a pipe that runs from the bottom of the tank to an outlet that is 6 ft above the top of the tank.

Centers of Mass and Centroids

33. Find the centroid of a thin, flat plate covering the region enclosed by the parabolas $y = 2x^2$ and $y = 3 - x^2$.

34. Find the centroid of a thin, flat plate covering the region enclosed by the x-axis, the lines $x = 2$ and $x = -2$, and the parabola $y = x^2$.

35. Find the centroid of a thin, flat plate covering the "triangular" region in the first quadrant bounded by the y-axis, the parabola $y = x^2/4$, and the line $y = 4$.

36. Find the centroid of a thin, flat plate covering the region enclosed by the parabola $y^2 = x$ and the line $x = 2y$.

37. Find the center of mass of a thin, flat plate covering the region enclosed by the parabola $y^2 = x$ and the line $x = 2y$ if the density function is $\delta(y) = 1 + y$. (Use horizontal strips.)

38. a. Find the center of mass of a thin plate of constant density covering the region between the curve $y = 3/x^{3/2}$ and the x-axis from $x = 1$ to $x = 9$.

    b. Find the plate's center of mass if, instead of being constant, the density is $\delta(x) = x$. (Use vertical strips.)

Fluid Force

39. Trough of water   The vertical triangular plate shown here is the end plate of a trough full of water ($w = 62.4$). What is the fluid force against the plate?

UNITS IN FEET

40. Trough of maple syrup   The vertical trapezoidal plate shown here is the end plate of a trough full of maple syrup weighing 75 lb/ft^3. What is the force exerted by the syrup against the end plate of the trough when the syrup is 10 in. deep?

UNITS IN FEET

41. Force on a parabolic gate   A flat vertical gate in the face of a dam is shaped like the parabolic region between the curve $y = 4x^2$ and the line $y = 4$, with measurements in feet. The top of the gate lies 5 ft below the surface of the water. Find the force exerted by the water against the gate ($w = 62.4$).

T 42. You plan to store mercury ($w = 849$ lb/ft^3) in a vertical rectangular tank with a 1 ft square base side whose interior side wall can withstand a total fluid force of 40,000 lb. About how many cubic feet of mercury can you store in the tank at any one time?

Chapter 6   Additional and Advanced Exercises

Volume and Length

1. A solid is generated by revolving about the x-axis the region bounded by the graph of the positive continuous function $y = f(x)$, the x-axis, and the fixed line $x = a$ and the variable line $x = b$, $b > a$. Its volume, for all b, is $b^2 - ab$. Find $f(x)$.

2. A solid is generated by revolving about the x-axis the region bounded by the graph of the positive continuous function $y = f(x)$, the x-axis, and the lines $x = 0$ and $x = a$. Its volume, for all $a > 0$, is $a^2 + a$. Find $f(x)$.

3. Suppose that the increasing function $f(x)$ is smooth for $x \geq 0$ and that $f(0) = a$. Let $s(x)$ denote the length of the graph of f from $(0, a)$ to $(x, f(x))$, $x > 0$. Find $f(x)$ if $s(x) = Cx$ for some constant C. What are the allowable values for C?

4. a. Show that for $0 < \alpha \leq \pi/2$,

$$\int_0^\alpha \sqrt{1 + \cos^2 \theta}\, d\theta > \sqrt{\alpha^2 + \sin^2 \alpha}.$$

    b. Generalize the result in part (a).

5. Find the volume of the solid formed by revolving the region bounded by the graphs of $y = x$ and $y = x^2$ about the line $y = x$.

6. Consider a right-circular cylinder of diameter 1. Form a wedge by making one slice parallel to the base of the cylinder completely through the cylinder, and another slice at an angle of $45°$ to the first slice and intersecting the first slice at the opposite edge of the cylinder (see accompanying diagram). Find the volume of the wedge.

Surface Area

7. At points on the curve $y = 2\sqrt{x}$, line segments of length $h = y$ are drawn perpendicular to the xy-plane. (See accompanying figure.) Find the area of the surface formed by these perpendiculars from $(0, 0)$ to $(3, 2\sqrt{3})$.

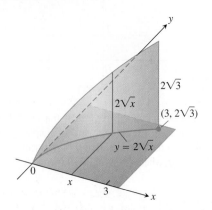

8. At points on a circle of radius a, line segments are drawn perpendicular to the plane of the circle, the perpendicular at each point P being of length ks, where s is the length of the arc of the circle measured counterclockwise from $(a, 0)$ to P and k is a positive constant, as shown here. Find the area of the surface formed by the perpendiculars along the arc beginning at $(a, 0)$ and extending once around the circle.

Work

9. A particle of mass m starts from rest at time $t = 0$ and is moved along the x-axis with constant acceleration a from $x = 0$ to $x = h$ against a variable force of magnitude $F(t) = t^2$. Find the work done.

10. Work and kinetic energy Suppose a 1.6-oz golf ball is placed on a vertical spring with force constant $k = 2$ lb/in. The spring is compressed 6 in. and released. About how high does the ball go (measured from the spring's rest position)?

Centers of Mass

11. Find the centroid of the region bounded below by the x-axis and above by the curve $y = 1 - x^n$, n an even positive integer. What is the limiting position of the centroid as $n \to \infty$?

12. If you haul a telephone pole on a two-wheeled carriage behind a truck, you want the wheels to be 3 ft or so behind the pole's center of mass to provide an adequate "tongue" weight. The 40-ft wooden telephone poles used by Verizon have a 27-in. circumference at the top and a 43.5-in. circumference at the base. About how far from the top is the center of mass?

13. Suppose that a thin metal plate of area A and constant density δ occupies a region R in the xy-plane, and let M_y be the plate's moment about the y-axis. Show that the plate's moment about the line $x = b$ is

   a. $M_y - b\delta A$ if the plate lies to the right of the line, and

   b. $b\delta A - M_y$ if the plate lies to the left of the line.

14. Find the center of mass of a thin plate covering the region bounded by the curve $y^2 = 4ax$ and the line $x = a$, $a =$ positive constant, if the density at (x, y) is directly proportional to **(a)** x, **(b)** $|y|$.

15. a. Find the centroid of the region in the first quadrant bounded by two concentric circles and the coordinate axes, if the circles have radii a and b, $0 < a < b$, and their centers are at the origin.

   b. Find the limits of the coordinates of the centroid as a approaches b and discuss the meaning of the result.

16. A triangular corner is cut from a square 1 ft on a side. The area of the triangle removed is 36 in^2. If the centroid of the remaining region is 7 in. from one side of the original square, how far is it from the remaining sides?

Fluid Force

17. A triangular plate ABC is submerged in water with its plane vertical. The side AB, 4 ft long, is 6 ft below the surface of the water, while the vertex C is 2 ft below the surface. Find the force exerted by the water on one side of the plate.

18. A vertical rectangular plate is submerged in a fluid with its top edge parallel to the fluid's surface. Show that the force exerted by the fluid on one side of the plate equals the average value of the pressure up and down the plate times the area of the plate.

Chapter 6 Technology Application Projects

Mathematica/Maple Modules:

Using Riemann Sums to Estimate Areas, Volumes, and Lengths of Curves
Visualize and approximate areas and volumes in **Part I** and **Part II**: Volumes of Revolution; and **Part III**: Lengths of Curves.

Modeling a Bungee Cord Jump
Collect data (or use data previously collected) to build and refine a model for the force exerted by a jumper's bungee cord. Use the work-energy theorem to compute the distance fallen for a given jumper and a given length of bungee cord.

7

INTEGRALS AND TRANSCENDENTAL FUNCTIONS

OVERVIEW Our treatment of the logarithmic and exponential functions has been rather informal until now, appealing to intuition and graphs to describe what they mean and to explain some of their characteristics. In this chapter, we give a rigorous approach to the definitions and properties of these functions, and we study a wide range of applied problems in which they play a role. We also introduce the hyperbolic functions and their inverses, with their applications to integration and hanging cables.

7.1 The Logarithm Defined as an Integral

In Chapter 1, we introduced the natural logarithm function $\ln x$ as the inverse of the exponential function e^x. The function e^x was chosen as that function in the family of general exponential functions $a^x, a > 0$, whose graph has slope 1 as it crosses the y-axis. The function a^x was presented intuitively, however, based on its graph at rational values of x.

In this section we recreate the theory of logarithmic and exponential functions from an entirely different point of view. Here we define these functions analytically and recover their behaviors. To begin, we use the Fundamental Theorem of Calculus to define the natural logarithm function $\ln x$ as an integral. We quickly develop its properties, including the algebraic, geometric, and analytic properties as seen before. Next we introduce the function e^x as the inverse function of $\ln x$, and establish its previously seen properties. Defining $\ln x$ as an integral and e^x as its inverse is an indirect approach. While it may at first seem strange, it gives an elegant and powerful way to obtain the key properties of logarithmic and exponential functions.

Definition of the Natural Logarithm Function

The natural logarithm of a positive number x, written as $\ln x$, is the value of an integral.

> **DEFINITION** The **natural logarithm** is the function given by
> $$\ln x = \int_1^x \frac{1}{t}\,dt, \qquad x > 0$$

From the Fundamental Theorem of Calculus, $\ln x$ is a continuous function. Geometrically, if $x > 1$, then $\ln x$ is the area under the curve $y = 1/t$ from $t = 1$ to $t = x$ (Figure 7.1). For $0 < x < 1$, $\ln x$ gives the negative of the area under the curve from x to 1.

The function is not defined for $x \leq 0$. From the Zero Width Interval Rule for definite integrals, we also have

$$\ln 1 = \int_1^1 \frac{1}{t} \, dt = 0.$$

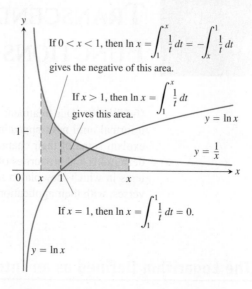

FIGURE 7.1 The graph of $y = \ln x$ and its relation to the function $y = 1/x$, $x > 0$. The graph of the logarithm rises above the x-axis as x moves from 1 to the right, and it falls below the axis as x moves from 1 to the left.

Notice that we show the graph of $y = 1/x$ in Figure 7.1 but use $y = 1/t$ in the integral. Using x for everything would have us writing

$$\ln x = \int_1^x \frac{1}{x} \, dx,$$

with x meaning two different things. So we change the variable of integration to t.

By using rectangles to obtain finite approximations of the area under the graph of $y = 1/t$ and over the interval between $t = 1$ and $t = x$, as in Section 5.1, we can approximate the values of the function $\ln x$. Several values are given in Table 7.1. There is an important number between $x = 2$ and $x = 3$ whose natural logarithm equals 1. This number, which we now define, exists because $\ln x$ is a continuous function and therefore satisfies the Intermediate Value Theorem on $[2, 3]$.

DEFINITION   The **number e** is that number in the domain of the natural logarithm satisfying

$$\ln (e) = \int_1^e \frac{1}{t} \, dt = 1$$

Interpreted geometrically, the number e corresponds to the point on the x-axis for which the area under the graph of $y = 1/t$ and above the interval $[1, e]$ equals the area of the unit square. That is, the area of the region shaded blue in Figure 7.1 is 1 sq unit when $x = e$. We will see further on that this is the same number $e \approx 2.718281828$ we have encountered before.

TABLE 7.1 Typical 2-place values of $\ln x$

x	$\ln x$
0	undefined
0.05	-3.00
0.5	-0.69
1	0
2	0.69
3	1.10
4	1.39
10	2.30

It then follows easily that the arithmetic rules satisfied by $\log_a x$ are the same as the ones for $\ln x$. These rules, given in Table 7.2, can be proved by dividing the corresponding rules for the natural logarithm function by $\ln a$. For example,

$$\ln xy = \ln x + \ln y \qquad \text{Rule 1 for natural logarithms} \ldots$$

$$\frac{\ln xy}{\ln a} = \frac{\ln x}{\ln a} + \frac{\ln y}{\ln a} \qquad \ldots \text{divided by } \ln a \ldots$$

$$\log_a xy = \log_a x + \log_a y. \qquad \ldots \text{gives Rule 1 for base } a \text{ logarithms.}$$

TABLE 7.2 Rules for base *a* logarithms

For any numbers $x > 0$ and $y > 0$,

1. *Product Rule:*
   $$\log_a xy = \log_a x + \log_a y$$

2. *Quotient Rule:*
   $$\log_a \frac{x}{y} = \log_a x - \log_a y$$

3. *Reciprocal Rule:*
   $$\log_a \frac{1}{y} = -\log_a y$$

4. *Power Rule:*
   $$\log_a x^y = y \log_a x$$

Derivatives and Integrals Involving $\log_a x$

To find derivatives or integrals involving base a logarithms, we convert them to natural logarithms. If u is a positive differentiable function of x, then

$$\frac{d}{dx}(\log_a u) = \frac{d}{dx}\left(\frac{\ln u}{\ln a}\right) = \frac{1}{\ln a}\frac{d}{dx}(\ln u) = \frac{1}{\ln a}\cdot\frac{1}{u}\frac{du}{dx}.$$

$$\frac{d}{dx}(\log_a u) = \frac{1}{\ln a}\cdot\frac{1}{u}\frac{du}{dx}$$

EXAMPLE 2 We illustrate the derivative and integral results.

(a) $\dfrac{d}{dx}\log_{10}(3x + 1) = \dfrac{1}{\ln 10}\cdot\dfrac{1}{3x + 1}\dfrac{d}{dx}(3x + 1) = \dfrac{3}{(\ln 10)(3x + 1)}$

(b) $\displaystyle\int \frac{\log_2 x}{x}\,dx = \frac{1}{\ln 2}\int\frac{\ln x}{x}\,dx \qquad \log_2 x = \frac{\ln x}{\ln 2}$

$$= \frac{1}{\ln 2}\int u\,du \qquad u = \ln x, \quad du = \frac{1}{x}dx$$

$$= \frac{1}{\ln 2}\frac{u^2}{2} + C = \frac{1}{\ln 2}\frac{(\ln x)^2}{2} + C = \frac{(\ln x)^2}{2\ln 2} + C \qquad \blacksquare$$

Summary

In this section we used the calculus to give precise definitions of the logarithmic and exponential functions. This approach is somewhat different from our earlier treatments of the polynomial, rational, and trigonometric functions. There we first defined the function and then we studied its derivatives and integrals. Here we started with an integral from which the functions of interest were obtained. The motivation behind this approach was to avoid mathematical difficulties that arise when we attempt to define functions such as a^x for any real number x, rational or irrational. By defining $\ln x$ as the integral of the function $1/t$ from $t = 1$ to $t = x$, we could go on to define all of the exponential and logarithmic functions, and then derive their key algebraic and analytic properties.

Exercises 7.1

Integration

Evaluate the integrals in Exercises 1–46.

1. $\displaystyle\int_{-3}^{-2}\frac{dx}{x}$

2. $\displaystyle\int_{-1}^{0}\frac{3\,dx}{3x - 2}$

3. $\displaystyle\int\frac{2y\,dy}{y^2 - 25}$

4. $\displaystyle\int\frac{8r\,dr}{4r^2 - 5}$

5. $\displaystyle\int\frac{3\sec^2 t}{6 + 3\tan t}\,dt$

6. $\displaystyle\int\frac{\sec y\tan y}{2 + \sec y}\,dy$

7. $\displaystyle\int \frac{dx}{2\sqrt{x} + 2x}$

8. $\displaystyle\int \frac{\sec x \, dx}{\sqrt{\ln(\sec x + \tan x)}}$

9. $\displaystyle\int_{\ln 2}^{\ln 3} e^x \, dx$

10. $\displaystyle\int 8e^{(x+1)} \, dx$

11. $\displaystyle\int_1^4 \frac{(\ln x)^3}{2x} \, dx$

12. $\displaystyle\int \frac{\ln(\ln x)}{x \ln x} \, dx$

13. $\displaystyle\int_{\ln 4}^{\ln 9} e^{x/2} \, dx$

14. $\displaystyle\int \tan x \ln(\cos x) \, dx$

15. $\displaystyle\int \frac{e^{\sqrt{r}}}{\sqrt{r}} \, dr$

16. $\displaystyle\int \frac{e^{-\sqrt{r}}}{\sqrt{r}} \, dr$

17. $\displaystyle\int 2t \, e^{-t^2} \, dt$

18. $\displaystyle\int \frac{\ln x \, dx}{x\sqrt{\ln^2 x + 1}}$

19. $\displaystyle\int \frac{e^{1/x}}{x^2} \, dx$

20. $\displaystyle\int \frac{e^{-1/x^2}}{x^3} \, dx$

21. $\displaystyle\int e^{\sec \pi t} \sec \pi t \tan \pi t \, dt$

22. $\displaystyle\int e^{\csc(\pi + t)} \csc(\pi + t) \cot(\pi + t) \, dt$

23. $\displaystyle\int_{\ln(\pi/6)}^{\ln(\pi/2)} 2e^v \cos e^v \, dv$

24. $\displaystyle\int_0^{\sqrt{\ln \pi}} 2xe^{x^2} \cos(e^{x^2}) \, dx$

25. $\displaystyle\int \frac{e^r}{1 + e^r} \, dr$

26. $\displaystyle\int \frac{dx}{1 + e^x}$

27. $\displaystyle\int_0^1 2^{-\theta} \, d\theta$

28. $\displaystyle\int_{-2}^0 5^{-\theta} \, d\theta$

29. $\displaystyle\int_1^{\sqrt{2}} x2^{(x^2)} \, dx$

30. $\displaystyle\int_1^4 \frac{2^{\sqrt{x}}}{\sqrt{x}} \, dx$

31. $\displaystyle\int_0^{\pi/2} 7^{\cos t} \sin t \, dt$

32. $\displaystyle\int_0^{\pi/4} \left(\frac{1}{3}\right)^{\tan t} \sec^2 t \, dt$

33. $\displaystyle\int_2^4 x^{2x}(1 + \ln x) \, dx$

34. $\displaystyle\int_1^2 \frac{2^{\ln x}}{x} \, dx$

35. $\displaystyle\int_0^3 (\sqrt{2} + 1)x^{\sqrt{2}} \, dx$

36. $\displaystyle\int_1^e x^{(\ln 2) - 1} \, dx$

37. $\displaystyle\int \frac{\log_{10} x}{x} \, dx$

38. $\displaystyle\int_1^4 \frac{\log_2 x}{x} \, dx$

39. $\displaystyle\int_1^4 \frac{\ln 2 \log_2 x}{x} \, dx$

40. $\displaystyle\int_1^e \frac{2 \ln 10 \log_{10} x}{x} \, dx$

41. $\displaystyle\int_0^2 \frac{\log_2(x + 2)}{x + 2} \, dx$

42. $\displaystyle\int_{1/10}^{10} \frac{\log_{10}(10x)}{x} \, dx$

43. $\displaystyle\int_0^9 \frac{2 \log_{10}(x + 1)}{x + 1} \, dx$

44. $\displaystyle\int_2^3 \frac{2 \log_2(x - 1)}{x - 1} \, dx$

45. $\displaystyle\int \frac{dx}{x \log_{10} x}$

46. $\displaystyle\int \frac{dx}{x(\log_8 x)^2}$

Initial Value Problems

Solve the initial value problems in Exercises 47–52.

47. $\dfrac{dy}{dt} = e^t \sin(e^t - 2), \quad y(\ln 2) = 0$

48. $\dfrac{dy}{dt} = e^{-t} \sec^2(\pi e^{-t}), \quad y(\ln 4) = 2/\pi$

49. $\dfrac{d^2y}{dx^2} = 2e^{-x}, \quad y(0) = 1 \quad \text{and} \quad y'(0) = 0$

50. $\dfrac{d^2y}{dt^2} = 1 - e^{2t}, \quad y(1) = -1 \quad \text{and} \quad y'(1) = 0$

51. $\dfrac{dy}{dx} = 1 + \dfrac{1}{x}, \quad y(1) = 3$

52. $\dfrac{d^2y}{dx^2} = \sec^2 x, \quad y(0) = 0 \quad \text{and} \quad y'(0) = 1$

Theory and Applications

53. The region between the curve $y = 1/x^2$ and the x-axis from $x = 1/2$ to $x = 2$ is revolved about the y-axis to generate a solid. Find the volume of the solid.

54. In Section 6.2, Exercise 6, we revolved about the y-axis the region between the curve $y = 9x/\sqrt{x^3 + 9}$ and the x-axis from $x = 0$ to $x = 3$ to generate a solid of volume 36π. What volume do you get if you revolve the region about the x-axis instead? (See Section 6.2, Exercise 6, for a graph.)

Find the lengths of the curves in Exercises 55 and 56.

55. $y = (x^2/8) - \ln x, \quad 4 \le x \le 8$

56. $x = (y/4)^2 - 2\ln(y/4), \quad 4 \le y \le 12$

T 57. The linearization of $\ln(1 + x)$ at $x = 0$ Instead of approximating $\ln x$ near $x = 1$, we approximate $\ln(1 + x)$ near $x = 0$. We get a simpler formula this way.

    a. Derive the linearization $\ln(1 + x) \approx x$ at $x = 0$.

    b. Estimate to five decimal places the error involved in replacing $\ln(1 + x)$ by x on the interval $[0, 0.1]$.

    c. Graph $\ln(1 + x)$ and x together for $0 \le x \le 0.5$. Use different colors, if available. At what points does the approximation of $\ln(1 + x)$ seem best? Least good? By reading coordinates from the graphs, find as good an upper bound for the error as your grapher will allow.

58. The linearization of e^x at $x = 0$

    a. Derive the linear approximation $e^x \approx 1 + x$ at $x = 0$.

    T b. Estimate to five decimal places the magnitude of the error involved in replacing e^x by $1 + x$ on the interval $[0, 0.2]$.

    T c. Graph e^x and $1 + x$ together for $-2 \le x \le 2$. Use different colors, if available. On what intervals does the approximation appear to overestimate e^x? Underestimate e^x?

59. Show that for any number $a > 1$

$$\int_1^a \ln x \, dx + \int_0^{\ln a} e^y \, dy = a \ln a.$$

(See accompanying figure.)

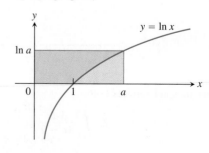

60. The geometric, logarithmic, and arithmetic mean inequality

   a. Show that the graph of e^x is concave up over every interval of x-values.

   b. Show, by reference to the accompanying figure, that if $0 < a < b$ then

$$e^{(\ln a + \ln b)/2} \cdot (\ln b - \ln a) < \int_{\ln a}^{\ln b} e^x \, dx < \frac{e^{\ln a} + e^{\ln b}}{2} \cdot (\ln b - \ln a).$$

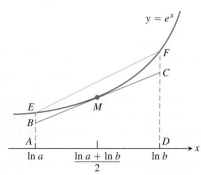

NOT TO SCALE

   c. Use the inequality in part (b) to conclude that

$$\sqrt{ab} < \frac{b - a}{\ln b - \ln a} < \frac{a + b}{2}.$$

This inequality says that the geometric mean of two positive numbers is less than their logarithmic mean, which in turn is less than their arithmetic mean.

   (For more about this inequality, see "The Geometric, Logarithmic, and Arithmetic Mean Inequality" by Frank Burk, *American Mathematical Monthly*, Vol. 94, No. 6, June–July 1987, pp. 527–528.)

Grapher Explorations

61. Graph $\ln x$, $\ln 2x$, $\ln 4x$, $\ln 8x$, and $\ln 16x$ (as many as you can) together for $0 < x \le 10$. What is going on? Explain.

62. Graph $y = \ln |\sin x|$ in the window $0 \le x \le 22$, $-2 \le y \le 0$. Explain what you see. How could you change the formula to turn the arches upside down?

63. a. Graph $y = \sin x$ and the curves $y = \ln(a + \sin x)$ for $a = 2$, 4, 8, 20, and 50 together for $0 \le x \le 23$.

   b. Why do the curves flatten as a increases? (*Hint:* Find an a-dependent upper bound for $|y'|$.)

64. Does the graph of $y = \sqrt{x} - \ln x$, $x > 0$, have an inflection point? Try to answer the question **(a)** by graphing, **(b)** by using calculus.

T 65. The equation $x^2 = 2^x$ has three solutions: $x = 2$, $x = 4$, and one other. Estimate the third solution as accurately as you can by graphing.

T 66. Could $x^{\ln 2}$ possibly be the same as $2^{\ln x}$ for $x > 0$? Graph the two functions and explain what you see.

T 67. Which is bigger, π^e or e^π? Calculators have taken some of the mystery out of this once-challenging question. (Go ahead and check; you will see that it is a surprisingly close call.) You can answer the question without a calculator, though.

   a. Find an equation for the line through the origin tangent to the graph of $y = \ln x$.

[–3, 6] by [–3, 3]

   b. Give an argument based on the graphs of $y = \ln x$ and the tangent line to explain why $\ln x < x/e$ for all positive $x \ne e$.

   c. Show that $\ln(x^e) < x$ for all positive $x \ne e$.

   d. Conclude that $x^e < e^x$ for all positive $x \ne e$.

   e. So which is bigger, π^e or e^π?

T 68. A decimal representation of e Find e to as many decimal places as your calculator allows by solving the equation $\ln x = 1$ using Newton's method in Section 4.7.

Calculations with Other Bases

T 69. Most scientific calculators have keys for $\log_{10} x$ and $\ln x$. To find logarithms to other bases, we use the equation $\log_a x = (\ln x)/(\ln a)$.

Find the following logarithms to five decimal places.

   a. $\log_3 8$ **b.** $\log_7 0.5$ **c.** $\log_{20} 17$ **d.** $\log_{0.5} 7$

   e. $\ln x$, given that $\log_{10} x = 2.3$

   f. $\ln x$, given that $\log_2 x = 1.4$

   g. $\ln x$, given that $\log_2 x = -1.5$

   h. $\ln x$, given that $\log_{10} x = -0.7$

70. Conversion factors

   a. Show that the equation for converting base 10 logarithms to base 2 logarithms is

$$\log_2 x = \frac{\ln 10}{\ln 2} \log_{10} x.$$

   b. Show that the equation for converting base a logarithms to base b logarithms is

$$\log_b x = \frac{\ln a}{\ln b} \log_a x.$$

7.2   Exponential Change and Separable Differential Equations

Exponential functions increase or decrease very rapidly with changes in the independent variable. They describe growth or decay in many natural and industrial situations. The variety of models based on these functions partly accounts for their importance. We now investigate the basic proportionality assumption that leads to such *exponential change*.

Exponential Change

In modeling many real-world situations, a quantity y increases or decreases at a rate proportional to its size at a given time t. Examples of such quantities include the amount of a decaying radioactive material, the size of a population, and the temperature difference between a hot object and its surrounding medium. Such quantities are said to undergo **exponential change**.

If the amount present at time $t = 0$ is called y_0, then we can find y as a function of t by solving the following initial value problem:

$$\text{Differential equation:} \quad \frac{dy}{dt} = ky \qquad\qquad (1a)$$

$$\text{Initial condition:} \quad y = y_0 \quad \text{when} \quad t = 0. \qquad (1b)$$

If y is positive and increasing, then k is positive, and we use Equation (1a) to say that the rate of growth is proportional to what has already been accumulated. If y is positive and decreasing, then k is negative, and we use Equation (1a) to say that the rate of decay is proportional to the amount still left.

We see right away that the constant function $y = 0$ is a solution of Equation (1a) if $y_0 = 0$. To find the nonzero solutions, we divide Equation (1a) by y:

$$\frac{1}{y} \cdot \frac{dy}{dt} = k \qquad\qquad y \neq 0$$

$$\int \frac{1}{y}\frac{dy}{dt}\,dt = \int k\,dt \qquad \text{Integrate with respect to } t;$$

$$\ln |y| = kt + C \qquad \int (1/u)\,du = \ln |u| + C.$$

$$|y| = e^{kt+C} \qquad \text{Exponentiate.}$$

$$|y| = e^C \cdot e^{kt} \qquad e^{a+b} = e^a \cdot e^b$$

$$y = \pm e^C e^{kt} \qquad \text{If } |y| = r, \text{ then } y = \pm r.$$

$$y = Ae^{kt}. \qquad A \text{ is a shorter name for } \pm e^C.$$

By allowing A to take on the value 0 in addition to all possible values $\pm e^C$, we can include the solution $y = 0$ in the formula.

We find the value of A for the initial value problem by solving for A when $y = y_0$ and $t = 0$:

$$y_0 = Ae^{k \cdot 0} = A.$$

The solution of the initial value problem is therefore

$$y = y_0 e^{kt}. \qquad\qquad (2)$$

Quantities changing in this way are said to undergo **exponential growth** if $k > 0$, and **exponential decay** if $k < 0$. The number k is called the **rate constant** of the change.

The derivation of Equation (2) shows also that the only functions that are their own derivatives are constant multiples of the exponential function.

Before presenting several examples of exponential change, let's consider the process we used to derive it.

Separable Differential Equations

Exponential change is modeled by a differential equation of the form $dy/dx = ky$ for some nonzero constant k. More generally, suppose we have a differential equation of the form

$$\frac{dy}{dx} = f(x, y), \qquad\qquad (3)$$

where f is a function of *both* the independent and dependent variables. A **solution** of the equation is a differentiable function $y = y(x)$ defined on an interval of x-values (perhaps infinite) such that

$$\frac{d}{dx} y(x) = f(x, y(x))$$

on that interval. That is, when $y(x)$ and its derivative $y'(x)$ are substituted into the differential equation, the resulting equation is true for all x in the solution interval. The **general solution** is a solution $y(x)$ that contains all possible solutions and it always contains an arbitrary constant.

Equation (3) is **separable** if f can be expressed as a product of a function of x and a function of y. The differential equation then has the form

$$\frac{dy}{dx} = g(x)H(y). \qquad \begin{array}{l} g \text{ is a function of } x; \\ H \text{ is a function of } y. \end{array}$$

When we rewrite this equation in the form

$$\frac{dy}{dx} = \frac{g(x)}{h(y)}, \qquad H(y) = \frac{1}{h(y)}$$

its differential form allows us to collect all y terms with dy and all x terms with dx:

$$h(y)\, dy = g(x)\, dx.$$

Now we simply integrate both sides of this equation:

$$\int h(y)\, dy = \int g(x)\, dx. \qquad (4)$$

After completing the integrations we obtain the solution y defined implicitly as a function of x.

The justification that we can simply integrate both sides in Equation (4) is based on the Substitution Rule (Section 5.5):

$$\int h(y)\, dy = \int h(y(x)) \frac{dy}{dx}\, dx$$

$$= \int h(y(x)) \frac{g(x)}{h(y(x))}\, dx \qquad \frac{dy}{dx} = \frac{g(x)}{h(y)}$$

$$= \int g(x)\, dx.$$

EXAMPLE 1   Solve the differential equation

$$\frac{dy}{dx} = (1 + y)e^x, \quad y > -1.$$

Solution   Since $1 + y$ is never zero for $y > -1$, we can solve the equation by separating the variables.

$$\frac{dy}{dx} = (1 + y)e^x$$

$$dy = (1 + y)e^x\, dx \qquad \begin{array}{l}\text{Treat } dy/dx \text{ as a quotient of} \\ \text{differentials and multiply} \\ \text{both sides by } dx.\end{array}$$

$$\frac{dy}{1 + y} = e^x\, dx \qquad \text{Divide by } (1 + y).$$

$$\int \frac{dy}{1 + y} = \int e^x\, dx \qquad \text{Integrate both sides.}$$

$$\ln (1 + y) = e^x + C \qquad \begin{array}{l} C \text{ represents the combined} \\ \text{constants of integration.}\end{array}$$

The last equation gives y as an implicit function of x. ∎

EXAMPLE 2 Solve the equation $y(x + 1)\dfrac{dy}{dx} = x(y^2 + 1)$.

Solution We change to differential form, separate the variables, and integrate:

$$y(x + 1)\, dy = x(y^2 + 1)\, dx$$

$$\frac{y\, dy}{y^2 + 1} = \frac{x\, dx}{x + 1} \qquad\qquad x \neq -1$$

$$\int \frac{y\, dy}{1 + y^2} = \int \left(1 - \frac{1}{x + 1}\right) dx \qquad \text{Divide } x \text{ by } x + 1.$$

$$\frac{1}{2}\ln(1 + y^2) = x - \ln|x + 1| + C.$$

The last equation gives the solution y as an implicit function of x. ∎

The initial value problem

$$\frac{dy}{dt} = ky, \qquad y(0) = y_0$$

involves a separable differential equation, and the solution $y = y_0 e^{kt}$ expresses exponential change. We now present several examples of such change.

Unlimited Population Growth

Strictly speaking, the number of individuals in a population (of people, plants, animals, or bacteria, for example) is a discontinuous function of time because it takes on discrete values. However, when the number of individuals becomes large enough, the population can be approximated by a continuous function. Differentiability of the approximating function is another reasonable hypothesis in many settings, allowing for the use of calculus to model and predict population sizes.

If we assume that the proportion of reproducing individuals remains constant and assume a constant fertility, then at any instant t the birth rate is proportional to the number $y(t)$ of individuals present. Let's assume, too, that the death rate of the population is stable and proportional to $y(t)$. If, further, we neglect departures and arrivals, the growth rate dy/dt is the birth rate minus the death rate, which is the difference of the two proportionalities under our assumptions. In other words, $dy/dt = ky$ so that $y = y_0 e^{kt}$, where y_0 is the size of the population at time $t = 0$. As with all kinds of growth, there may be limitations imposed by the surrounding environment, but we will not go into these here. The proportionality $dy/dt = ky$ models *unlimited population growth*.

In the following example we assume this population model to look at how the number of individuals infected by a disease within a given population decreases as the disease is appropriately treated.

EXAMPLE 3 One model for the way diseases die out when properly treated assumes that the rate dy/dt at which the number of infected people changes is proportional to the number y. The number of people cured is proportional to the number y that are infected with the disease. Suppose that in the course of any given year the number of cases of a disease is reduced by 20%. If there are 10,000 cases today, how many years will it take to reduce the number to 1000?

Solution We use the equation $y = y_0 e^{kt}$. There are three things to find: the value of y_0, the value of k, and the time t when $y = 1000$.

The value of y_0. We are free to count time beginning anywhere we want. If we count from today, then $y = 10{,}000$ when $t = 0$, so $y_0 = 10{,}000$. Our equation is now

$$y = 10{,}000\, e^{kt}. \tag{5}$$

The value of k. When $t = 1$ year, the number of cases will be 80% of its present value, or 8000. Hence,

$$8000 = 10{,}000\, e^{k(1)} \qquad \text{Eq. (5) with } t = 1 \text{ and } y = 8000$$
$$e^k = 0.8$$
$$\ln\left(e^k\right) = \ln 0.8 \qquad \text{Logs of both sides}$$
$$k = \ln 0.8 < 0.$$

At any given time t,

$$y = 10{,}000\, e^{(\ln 0.8)t}. \tag{6}$$

The value of t that makes $y = 1000$. We set y equal to 1000 in Equation (6) and solve for t:

$$1000 = 10{,}000 e^{(\ln 0.8)t}$$
$$e^{(\ln 0.8)t} = 0.1$$
$$(\ln 0.8)t = \ln 0.1 \qquad \text{Logs of both sides}$$
$$t = \frac{\ln 0.1}{\ln 0.8} \approx 10.32 \text{ years.}$$

It will take a little more than 10 years to reduce the number of cases to 1000. ∎

Radioactivity

Some atoms are unstable and can spontaneously emit mass or radiation. This process is called **radioactive decay**, and an element whose atoms go spontaneously through this process is called **radioactive**. Sometimes when an atom emits some of its mass through this process of radioactivity, the remainder of the atom re-forms to make an atom of some new element. For example, radioactive carbon-14 decays into nitrogen; radium, through a number of intermediate radioactive steps, decays into lead.

Experiments have shown that at any given time the rate at which a radioactive element decays (as measured by the number of nuclei that change per unit time) is approximately proportional to the number of radioactive nuclei present. Thus, the decay of a radioactive element is described by the equation $dy/dt = -ky, k > 0$. It is conventional to use $-k$, with $k > 0$, to emphasize that y is decreasing. If y_0 is the number of radioactive nuclei present at time zero, the number still present at any later time t will be

$$y = y_0 e^{-kt}, \qquad k > 0.$$

For radon-222 gas, t is measured in days and $k = 0.18$. For radium-226, which used to be painted on watch dials to make them glow at night (a dangerous practice), t is measured in years and $k = 4.3 \times 10^{-4}$.

In Section 1.6, we defined the **half-life** of a radioactive element to be the time required for half of the radioactive nuclei present in a sample to decay. It is an interesting fact that the half-life is a constant that does not depend on the number of radioactive nuclei initially present in the sample, but only on the radioactive substance. We found the half-life is given by

$$\text{Half-life} = \frac{\ln 2}{k} \tag{7}$$

For example, the half-life for radon-222 is

$$\text{half-life} = \frac{\ln 2}{0.18} \approx 3.9 \text{ days.}$$

EXAMPLE 4 The decay of radioactive elements can sometimes be used to date events from the Earth's past. In a living organism, the ratio of radioactive carbon, carbon-14, to ordinary carbon stays fairly constant during the lifetime of the organism, being approximately equal to the ratio in the organism's atmosphere at the time. After the organism's death, however, no new carbon is ingested, and the proportion of carbon-14 in the organism's remains decreases as the carbon-14 decays.

Scientists who do carbon-14 dating use a figure of 5700 years for its half-life. Find the age of a sample in which 10% of the radioactive nuclei originally present have decayed.

Solution We use the decay equation $y = y_0 e^{-kt}$. There are two things to find: the value of k and the value of t when y is $0.9y_0$ (90% of the radioactive nuclei are still present). That is, find t when $y_0 e^{-kt} = 0.9y_0$, or $e^{-kt} = 0.9$.

The value of k. We use the half-life Equation (7):

$$k = \frac{\ln 2}{\text{half-life}} = \frac{\ln 2}{5700} \qquad \text{(about } 1.2 \times 10^{-4}\text{)}$$

The value of t that makes $e^{-kt} = 0.9$.

$$e^{-kt} = 0.9$$

$$e^{-(\ln 2/5700)t} = 0.9$$

$$-\frac{\ln 2}{5700} t = \ln 0.9 \qquad \qquad \text{Logs of both sides}$$

$$t = -\frac{5700 \ln 0.9}{\ln 2} \approx 866 \text{ years}$$

The sample is about 866 years old. ∎

Heat Transfer: Newton's Law of Cooling

Hot soup left in a tin cup cools to the temperature of the surrounding air. A hot silver bar immersed in a large tub of water cools to the temperature of the surrounding water. In situations like these, the rate at which an object's temperature is changing at any given time is roughly proportional to the difference between its temperature and the temperature of the surrounding medium. This observation is called *Newton's Law of Cooling*, although it applies to warming as well.

If H is the temperature of the object at time t and H_S is the constant surrounding temperature, then the differential equation is

$$\frac{dH}{dt} = -k(H - H_S). \tag{8}$$

If we substitute y for $(H - H_S)$, then

$$\frac{dy}{dt} = \frac{d}{dt}(H - H_S) = \frac{dH}{dt} - \frac{d}{dt}(H_S)$$

$$= \frac{dH}{dt} - 0 \qquad \qquad H_S \text{ is a constant.}$$

$$= \frac{dH}{dt}$$

$$= -k(H - H_S) \qquad \qquad \text{Eq. (8)}$$

$$= -ky. \qquad \qquad H - H_S = y$$

Now we know that the solution of $dy/dt = -ky$ is $y = y_0e^{-kt}$, where $y(0) = y_0$. Substituting $(H - H_S)$ for y, this says that

$$H - H_S = (H_0 - H_S)e^{-kt}, \qquad (9)$$

where H_0 is the temperature at $t = 0$. This equation is the solution to Newton's Law of Cooling.

EXAMPLE 5   A hard-boiled egg at 98°C is put in a sink of 18°C water. After 5 min, the egg's temperature is 38°C. Assuming that the water has not warmed appreciably, how much longer will it take the egg to reach 20°C?

Solution   We find how long it would take the egg to cool from 98°C to 20°C and subtract the 5 min that have already elapsed. Using Equation (9) with $H_S = 18$ and $H_0 = 98$, the egg's temperature t min after it is put in the sink is

$$H = 18 + (98 - 18)e^{-kt} = 18 + 80e^{-kt}.$$

To find k, we use the information that $H = 38$ when $t = 5$:

$$38 = 18 + 80e^{-5k}$$

$$e^{-5k} = \frac{1}{4}$$

$$-5k = \ln\frac{1}{4} = -\ln 4$$

$$k = \frac{1}{5}\ln 4 = 0.2\ln 4 \qquad \text{(about 0.28)}.$$

The egg's temperature at time t is $H = 18 + 80e^{-(0.2\ln 4)t}$. Now find the time t when $H = 20$:

$$20 = 18 + 80e^{-(0.2\ln 4)t}$$

$$80e^{-(0.2\ln 4)t} = 2$$

$$e^{-(0.2\ln 4)t} = \frac{1}{40}$$

$$-(0.2\ln 4)t = \ln\frac{1}{40} = -\ln 40$$

$$t = \frac{\ln 40}{0.2\ln 4} \approx 13 \text{ min}.$$

The egg's temperature will reach 20°C about 13 min after it is put in the water to cool. Since it took 5 min to reach 38°C, it will take about 8 min more to reach 20°C.   ∎

Exercises 7.2

Verifying Solutions

In Exercises 1–4, show that each function $y = f(x)$ is a solution of the accompanying differential equation.

1. $2y' + 3y = e^{-x}$

  a. $y = e^{-x}$ **b.** $y = e^{-x} + e^{-(3/2)x}$

  c. $y = e^{-x} + Ce^{-(3/2)x}$

2. $y' = y^2$

  a. $y = -\dfrac{1}{x}$ **b.** $y = -\dfrac{1}{x + 3}$ **c.** $y = -\dfrac{1}{x + C}$

3. $y = \dfrac{1}{x}\displaystyle\int_1^x \dfrac{e^t}{t}\,dt, \quad x^2y' + xy = e^x$

4. $y = \dfrac{1}{\sqrt{1 + x^4}}\displaystyle\int_1^x \sqrt{1 + t^4}\,dt, \quad y' + \dfrac{2x^3}{1 + x^4}y = 1$

Initial Value Problems

In Exercises 5–8, show that each function is a solution of the given initial value problem.

Differential equation	Initial condition	Solution candidate
5. $y' + y = \dfrac{2}{1 + 4e^{2x}}$	$y(-\ln 2) = \dfrac{\pi}{2}$	$y = e^{-x}\tan^{-1}(2e^x)$
6. $y' = e^{-x^2} - 2xy$	$y(2) = 0$	$y = (x - 2)e^{-x^2}$
7. $xy' + y = -\sin x,$ $x > 0$	$y\left(\dfrac{\pi}{2}\right) = 0$	$y = \dfrac{\cos x}{x}$
8. $x^2y' = xy - y^2,$ $x > 1$	$y(e) = e$	$y = \dfrac{x}{\ln x}$

Separable Differential Equations

Solve the differential equation in Exercises 9–22.

9. $2\sqrt{xy}\,\dfrac{dy}{dx} = 1,$ $x, y > 0$ **10.** $\dfrac{dy}{dx} = x^2\sqrt{y},$ $y > 0$

11. $\dfrac{dy}{dx} = e^{x-y}$ **12.** $\dfrac{dy}{dx} = 3x^2 e^{-y}$

13. $\dfrac{dy}{dx} = \sqrt{y}\cos^2\sqrt{y}$ **14.** $\sqrt{2xy}\,\dfrac{dy}{dx} = 1$

15. $\sqrt{x}\,\dfrac{dy}{dx} = e^{y+\sqrt{x}},$ $x > 0$ **16.** $(\sec x)\dfrac{dy}{dx} = e^{y+\sin x}$

17. $\dfrac{dy}{dx} = 2x\sqrt{1 - y^2},$ $-1 < y < 1$

18. $\dfrac{dy}{dx} = \dfrac{e^{2x-y}}{e^{x+y}}$

19. $y^2\dfrac{dy}{dx} = 3x^2y^3 - 6x^2$ **20.** $\dfrac{dy}{dx} = xy + 3x - 2y - 6$

21. $\dfrac{1}{x}\dfrac{dy}{dx} = ye^{x^2} + 2\sqrt{y}\,e^{x^2}$ **22.** $\dfrac{dy}{dx} = e^{x-y} + e^x + e^{-y} + 1$

Applications and Examples

The answers to most of the following exercises are in terms of logarithms and exponentials. A calculator can be helpful, enabling you to express the answers in decimal form.

23. Human evolution continues The analysis of tooth shrinkage by C. Loring Brace and colleagues at the University of Michigan's Museum of Anthropology indicates that human tooth size is continuing to decrease and that the evolutionary process did not come to a halt some 30,000 years ago as many scientists contend. In northern Europeans, for example, tooth size reduction now has a rate of 1% per 1000 years.

    a. If t represents time in years and y represents tooth size, use the condition that $y = 0.99y_0$ when $t = 1000$ to find the value of k in the equation $y = y_0 e^{kt}$. Then use this value of k to answer the following questions.

    b. In about how many years will human teeth be 90% of their present size?

    c. What will be our descendants' tooth size 20,000 years from now (as a percentage of our present tooth size)?

24. Atmospheric pressure The earth's atmospheric pressure p is often modeled by assuming that the rate dp/dh at which p changes with the altitude h above sea level is proportional to p. Suppose that the pressure at sea level is 1013 millibars (about 14.7 pounds per square inch) and that the pressure at an altitude of 20 km is 90 millibars.

    a. Solve the initial value problem

      Differential equation: $dp/dh = kp$ (k a constant)

      Initial condition: $p = p_0$ when $h = 0$

    to express p in terms of h. Determine the values of p_0 and k from the given altitude-pressure data.

    b. What is the atmospheric pressure at $h = 50$ km?

    c. At what altitude does the pressure equal 900 millibars?

25. First-order chemical reactions In some chemical reactions, the rate at which the amount of a substance changes with time is proportional to the amount present. For the change of δ-glucono lactone into gluconic acid, for example,

$$\frac{dy}{dt} = -0.6y$$

when t is measured in hours. If there are 100 grams of δ-glucono lactone present when $t = 0$, how many grams will be left after the first hour?

26. The inversion of sugar The processing of raw sugar has a step called "inversion" that changes the sugar's molecular structure. Once the process has begun, the rate of change of the amount of raw sugar is proportional to the amount of raw sugar remaining. If 1000 kg of raw sugar reduces to 800 kg of raw sugar during the first 10 hours, how much raw sugar will remain after another 14 hours?

27. Working underwater The intensity $L(x)$ of light x feet beneath the surface of the ocean satisfies the differential equation

$$\frac{dL}{dx} = -kL.$$

As a diver, you know from experience that diving to 18 ft in the Caribbean Sea cuts the intensity in half. You cannot work without artificial light when the intensity falls below one-tenth of the surface value. About how deep can you expect to work without artificial light?

28. Voltage in a discharging capacitor Suppose that electricity is draining from a capacitor at a rate that is proportional to the voltage V across its terminals and that, if t is measured in seconds,

$$\frac{dV}{dt} = -\frac{1}{40}V.$$

Solve this equation for V, using V_0 to denote the value of V when $t = 0$. How long will it take the voltage to drop to 10% of its original value?

29. Cholera bacteria Suppose that the bacteria in a colony can grow unchecked, by the law of exponential change. The colony starts with 1 bacterium and doubles every half-hour. How many bacteria will the colony contain at the end of 24 hours? (Under favorable laboratory conditions, the number of cholera bacteria can double every 30 min. In an infected person, many bacteria are destroyed, but this example helps explain why a person who feels well in the morning may be dangerously ill by evening.)

30. Growth of bacteria A colony of bacteria is grown under ideal conditions in a laboratory so that the population increases exponentially with time. At the end of 3 hours there are 10,000 bacteria. At the end of 5 hours there are 40,000. How many bacteria were present initially?

31. The incidence of a disease (*Continuation of Example 3.*) Suppose that in any given year the number of cases can be reduced by 25% instead of 20%.

   a. How long will it take to reduce the number of cases to 1000?

   b. How long will it take to eradicate the disease, that is, reduce the number of cases to less than 1?

32. The U.S. population The U.S. Census Bureau keeps a running clock totaling the U.S. population. On March 26, 2008, the total was increasing at the rate of 1 person every 13 sec. The population figure for 2:31 P.M. EST on that day was 303,714,725.

   a. Assuming exponential growth at a constant rate, find the rate constant for the population's growth (people per 365-day year).

   b. At this rate, what will the U.S. population be at 2:31 P.M. EST on March 26, 2015?

33. Oil depletion Suppose the amount of oil pumped from one of the canyon wells in Whittier, California, decreases at the continuous rate of 10% per year. When will the well's output fall to one-fifth of its present value?

34. Continuous price discounting To encourage buyers to place 100-unit orders, your firm's sales department applies a continuous discount that makes the unit price a function $p(x)$ of the number of units x ordered. The discount decreases the price at the rate of $0.01 per unit ordered. The price per unit for a 100-unit order is $p(100) = \$20.09$.

   a. Find $p(x)$ by solving the following initial value problem:

   Differential equation: $\quad \dfrac{dp}{dx} = -\dfrac{1}{100}p$

   Initial condition: $\qquad p(100) = 20.09$.

   b. Find the unit price $p(10)$ for a 10-unit order and the unit price $p(90)$ for a 90-unit order.

   c. The sales department has asked you to find out if it is discounting so much that the firm's revenue, $r(x) = x \cdot p(x)$, will actually be less for a 100-unit order than, say, for a 90-unit order. Reassure them by showing that r has its maximum value at $x = 100$.

   d. Graph the revenue function $r(x) = xp(x)$ for $0 \le x \le 200$.

35. Plutonium-239 The half-life of the plutonium isotope is 24,360 years. If 10 g of plutonium is released into the atmosphere by a nuclear accident, how many years will it take for 80% of the isotope to decay?

36. Polonium-210 The half-life of polonium is 139 days, but your sample will not be useful to you after 95% of the radioactive nuclei present on the day the sample arrives has disintegrated. For about how many days after the sample arrives will you be able to use the polonium?

37. The mean life of a radioactive nucleus Physicists using the radioactivity equation $y = y_0 e^{-kt}$ call the number $1/k$ the *mean life* of a radioactive nucleus. The mean life of a radon nucleus is about $1/0.18 = 5.6$ days. The mean life of a carbon-14 nucleus is more than 8000 years. Show that 95% of the radioactive nuclei originally present in a sample will disintegrate within three mean lifetimes, i.e., by time $t = 3/k$. Thus, the mean life of a nucleus gives a quick way to estimate how long the radioactivity of a sample will last.

38. Californium-252 What costs $27 million per gram and can be used to treat brain cancer, analyze coal for its sulfur content, and detect explosives in luggage? The answer is californium-252, a radioactive isotope so rare that only 8 g of it have been made in the western world since its discovery by Glenn Seaborg in 1950. The half-life of the isotope is 2.645 years—long enough for a useful service life and short enough to have a high radioactivity per unit mass. One microgram of the isotope releases 170 million neutrons per minute.

   a. What is the value of k in the decay equation for this isotope?

   b. What is the isotope's mean life? (See Exercise 37.)

   c. How long will it take 95% of a sample's radioactive nuclei to disintegrate?

39. Cooling soup Suppose that a cup of soup cooled from 90°C to 60°C after 10 min in a room whose temperature was 20°C. Use Newton's law of cooling to answer the following questions.

   a. How much longer would it take the soup to cool to 35°C?

   b. Instead of being left to stand in the room, the cup of 90°C soup is put in a freezer whose temperature is −15°C. How long will it take the soup to cool from 90°C to 35°C?

40. A beam of unknown temperature An aluminum beam was brought from the outside cold into a machine shop where the temperature was held at 65°F. After 10 min, the beam warmed to 35°F and after another 10 min it was 50°F. Use Newton's law of cooling to estimate the beam's initial temperature.

41. Surrounding medium of unknown temperature A pan of warm water (46°C) was put in a refrigerator. Ten minutes later, the water's temperature was 39°C; 10 min after that, it was 33°C. Use Newton's law of cooling to estimate how cold the refrigerator was.

42. Silver cooling in air The temperature of an ingot of silver is 60°C above room temperature right now. Twenty minutes ago, it was 70°C above room temperature. How far above room temperature will the silver be

   a. 15 min from now?   **b.** 2 hours from now?

   c. When will the silver be 10°C above room temperature?

43. The age of Crater Lake The charcoal from a tree killed in the volcanic eruption that formed Crater Lake in Oregon contained 44.5% of the carbon-14 found in living matter. About how old is Crater Lake?

44. The sensitivity of carbon-14 dating to measurement To see the effect of a relatively small error in the estimate of the amount of carbon-14 in a sample being dated, consider this hypothetical situation:

   a. A fossilized bone found in central Illinois in the year A.D. 2000 contains 17% of its original carbon-14 content. Estimate the year the animal died.

   b. Repeat part (a) assuming 18% instead of 17%.

   c. Repeat part (a) assuming 16% instead of 17%.

45. Carbon-14 The oldest known frozen human mummy, discovered in the Schnalstal glacier of the Italian Alps in 1991 and called *Otzi*, was found wearing straw shoes and a leather coat with goat fur, and holding a copper ax and stone dagger. It was estimated that Otzi died 5000 years before he was discovered in the melting glacier. How much of the original carbon-14 remained in Otzi at the time of his discovery?

46. Art forgery A painting attributed to Vermeer (1632–1675), which should contain no more than 96.2% of its original carbon-14, contains 99.5% instead. About how old is the forgery?

7.3 | Hyperbolic Functions

The hyperbolic functions are formed by taking combinations of the two exponential functions e^x and e^{-x}. The hyperbolic functions simplify many mathematical expressions and occur frequently in mathematical applications. In this section we give a brief introduction to these functions, their graphs, and their derivatives.

Definitions and Identities

The hyperbolic sine and hyperbolic cosine functions are defined by the equations

$$\sinh x = \frac{e^x - e^{-x}}{2} \quad \text{and} \quad \cosh x = \frac{e^x + e^{-x}}{2}.$$

We pronounce $\sinh x$ as "cinch x," rhyming with "pinch x," and $\cosh x$ as "kosh x," rhyming with "gosh x." From this basic pair, we define the hyperbolic tangent, cotangent, secant, and cosecant functions. The defining equations and graphs of these functions are shown in Table 7.3. We will see that the hyperbolic functions bear many similarities to the trigonometric functions after which they are named.

TABLE 7.3 The six basic hyperbolic functions

(a)

Hyperbolic sine:

$$\sinh x = \frac{e^x - e^{-x}}{2}$$

(b)

Hyperbolic cosine:

$$\cosh x = \frac{e^x + e^{-x}}{2}$$

(c)

Hyperbolic tangent:

$$\tanh x = \frac{\sinh x}{\cosh x} = \frac{e^x - e^{-x}}{e^x + e^{-x}}$$

Hyperbolic cotangent:

$$\coth x = \frac{\cosh x}{\sinh x} = \frac{e^x + e^{-x}}{e^x - e^{-x}}$$

(d)

Hyperbolic secant:

$$\operatorname{sech} x = \frac{1}{\cosh x} = \frac{2}{e^x + e^{-x}}$$

(e)

Hyperbolic cosecant:

$$\operatorname{csch} x = \frac{1}{\sinh x} = \frac{2}{e^x - e^{-x}}$$

TABLE 7.4 Identities for hyperbolic functions

$$\cosh^2 x - \sinh^2 x = 1$$
$$\sinh 2x = 2 \sinh x \cosh x$$
$$\cosh 2x = \cosh^2 x + \sinh^2 x$$
$$\cosh^2 x = \frac{\cosh 2x + 1}{2}$$
$$\sinh^2 x = \frac{\cosh 2x - 1}{2}$$
$$\tanh^2 x = 1 - \operatorname{sech}^2 x$$
$$\coth^2 x = 1 + \operatorname{csch}^2 x$$

TABLE 7.5 Derivatives of hyperbolic functions

$$\frac{d}{dx}(\sinh u) = \cosh u \frac{du}{dx}$$

$$\frac{d}{dx}(\cosh u) = \sinh u \frac{du}{dx}$$

$$\frac{d}{dx}(\tanh u) = \operatorname{sech}^2 u \frac{du}{dx}$$

$$\frac{d}{dx}(\coth u) = -\operatorname{csch}^2 u \frac{du}{dx}$$

$$\frac{d}{dx}(\operatorname{sech} u) = -\operatorname{sech} u \tanh u \frac{du}{dx}$$

$$\frac{d}{dx}(\operatorname{csch} u) = -\operatorname{csch} u \coth u \frac{du}{dx}$$

TABLE 7.6 Integral formulas for hyperbolic functions

$$\int \sinh u \, du = \cosh u + C$$

$$\int \cosh u \, du = \sinh u + C$$

$$\int \operatorname{sech}^2 u \, du = \tanh u + C$$

$$\int \operatorname{csch}^2 u \, du = -\coth u + C$$

$$\int \operatorname{sech} u \tanh u \, du = -\operatorname{sech} u + C$$

$$\int \operatorname{csch} u \coth u \, du = -\operatorname{csch} u + C$$

Hyperbolic functions satisfy the identities in Table 7.4. Except for differences in sign, these resemble identities we know for the trigonometric functions. The identities are proved directly from the definitions, as we show here for the second one:

$$2 \sinh x \cosh x = 2\left(\frac{e^x - e^{-x}}{2}\right)\left(\frac{e^x + e^{-x}}{2}\right)$$
$$= \frac{e^{2x} - e^{-2x}}{2}$$
$$= \sinh 2x.$$

The other identities are obtained similarly, by substituting in the definitions of the hyperbolic functions and using algebra. Like many standard functions, hyperbolic functions and their inverses are easily evaluated with calculators, which often have special keys for that purpose.

For any real number u, we know the point with coordinates $(\cos u, \sin u)$ lies on the unit circle $x^2 + y^2 = 1$. So the trigonometric functions are sometimes called the *circular* functions. Because of the first identity

$$\cosh^2 u - \sinh^2 u = 1,$$

with u substituted for x in Table 7.4, the point having coordinates $(\cosh u, \sinh u)$ lies on the right-hand branch of the hyperbola $x^2 - y^2 = 1$. This is where the *hyperbolic* functions get their names (see Exercise 86).

Derivatives and Integrals of Hyperbolic Functions

The six hyperbolic functions, being rational combinations of the differentiable functions e^x and e^{-x}, have derivatives at every point at which they are defined (Table 7.5). Again, there are similarities with trigonometric functions.

The derivative formulas are derived from the derivative of e^u:

$$\frac{d}{dx}(\sinh u) = \frac{d}{dx}\left(\frac{e^u - e^{-u}}{2}\right) \qquad \text{Definition of } \sinh u$$

$$= \frac{e^u \, du/dx + e^{-u} \, du/dx}{2} \qquad \text{Derivative of } e^u$$

$$= \cosh u \frac{du}{dx}. \qquad \text{Definition of } \cosh u$$

This gives the first derivative formula. From the definition, we can calculate the derivative of the hyperbolic cosecant function, as follows:

$$\frac{d}{dx}(\operatorname{csch} u) = \frac{d}{dx}\left(\frac{1}{\sinh u}\right) \qquad \text{Definition of } \operatorname{csch} u$$

$$= -\frac{\cosh u}{\sinh^2 u}\frac{du}{dx} \qquad \text{Quotient Rule}$$

$$= -\frac{1}{\sinh u}\frac{\cosh u}{\sinh u}\frac{du}{dx} \qquad \text{Rearrange terms.}$$

$$= -\operatorname{csch} u \coth u \frac{du}{dx} \qquad \text{Definitions of } \operatorname{csch} u \text{ and } \coth u$$

The other formulas in Table 7.5 are obtained similarly.

The derivative formulas lead to the integral formulas in Table 7.6.

EXAMPLE 1

(a) $\dfrac{d}{dt}\left(\tanh\sqrt{1+t^2}\right) = \text{sech}^2\sqrt{1+t^2}\cdot\dfrac{d}{dt}\left(\sqrt{1+t^2}\right)$

$$= \dfrac{t}{\sqrt{1+t^2}}\,\text{sech}^2\sqrt{1+t^2}$$

(b) $\displaystyle\int\coth 5x\,dx = \int\dfrac{\cosh 5x}{\sinh 5x}\,dx = \dfrac{1}{5}\int\dfrac{du}{u}$ $\qquad$ $u = \sinh 5x,$
$\qquad\qquad\qquad\qquad\qquad\qquad\qquad\qquad\qquad\qquad\qquad\quad du = 5\cosh 5x\,dx$

$$= \dfrac{1}{5}\ln|u| + C = \dfrac{1}{5}\ln|\sinh 5x| + C$$

(c) $\displaystyle\int_0^1\sinh^2 x\,dx = \int_0^1\dfrac{\cosh 2x - 1}{2}\,dx$ $\qquad$ Table 7.4

$$= \dfrac{1}{2}\int_0^1(\cosh 2x - 1)\,dx = \dfrac{1}{2}\left[\dfrac{\sinh 2x}{2} - x\right]_0^1$$

$$= \dfrac{\sinh 2}{4} - \dfrac{1}{2} \approx 0.40672$$ $\qquad$ Evaluate with a calculator.

(d) $\displaystyle\int_0^{\ln 2}4e^x\sinh x\,dx = \int_0^{\ln 2}4e^x\dfrac{e^x - e^{-x}}{2}\,dx = \int_0^{\ln 2}(2e^{2x} - 2)\,dx$

$$= \left[e^{2x} - 2x\right]_0^{\ln 2} = (e^{2\ln 2} - 2\ln 2) - (1 - 0)$$

$$= 4 - 2\ln 2 - 1 \approx 1.6137$$ ■

Inverse Hyperbolic Functions

The inverses of the six basic hyperbolic functions are very useful in integration (see Chapter 8). Since $d(\sinh x)/dx = \cosh x > 0$, the hyperbolic sine is an increasing function of x. We denote its inverse by

$$y = \sinh^{-1}x.$$

For every value of x in the interval $-\infty < x < \infty$, the value of $y = \sinh^{-1}x$ is the number whose hyperbolic sine is x. The graphs of $y = \sinh x$ and $y = \sinh^{-1}x$ are shown in Figure 7.5a.

The function $y = \cosh x$ is not one-to-one because its graph in Table 7.3 does not pass the horizontal line test. The restricted function $y = \cosh x, x \geq 0$, however, is one-to-one and therefore has an inverse, denoted by

$$y = \cosh^{-1}x.$$

For every value of $x \geq 1$, $y = \cosh^{-1}x$ is the number in the interval $0 \leq y < \infty$ whose hyperbolic cosine is x. The graphs of $y = \cosh x, x \geq 0$, and $y = \cosh^{-1}x$ are shown in Figure 7.5b.

Like $y = \cosh x$, the function $y = \text{sech}\,x = 1/\cosh x$ fails to be one-to-one, but its restriction to nonnegative values of x does have an inverse, denoted by

$$y = \text{sech}^{-1}x.$$

For every value of x in the interval $(0, 1]$, $y = \text{sech}^{-1}x$ is the nonnegative number whose hyperbolic secant is x. The graphs of $y = \text{sech}\,x, x \geq 0$, and $y = \text{sech}^{-1}x$ are shown in Figure 7.5c.

The hyperbolic tangent, cotangent, and cosecant are one-to-one on their domains and therefore have inverses, denoted by

$$y = \tanh^{-1}x, \qquad y = \coth^{-1}x, \qquad y = \text{csch}^{-1}x.$$

These functions are graphed in Figure 7.6.

FIGURE 7.8 Scale drawings of the graphs of e^x and $\ln x$.

By contrast, with axes scaled in centimeters, you have to go nearly 5 light-years out on the x-axis to find a point where the graph of $y = \ln x$ is even $y = 43$ cm high. See Figure 7.8.

These important comparisons of exponential, polynomial, and logarithmic functions can be made precise by defining what it means for a function $f(x)$ to grow faster than another function $g(x)$ as $x \to \infty$.

DEFINITION Rates of Growth as $x \to \infty$

Let $f(x)$ and $g(x)$ be positive for x sufficiently large.

1. f **grows faster than** g as $x \to \infty$ if

$$\lim_{x \to \infty} \frac{f(x)}{g(x)} = \infty$$

or, equivalently, if

$$\lim_{x \to \infty} \frac{g(x)}{f(x)} = 0.$$

We also say that g **grows slower than** f as $x \to \infty$.

2. f and g **grow at the same rate** as $x \to \infty$ if

$$\lim_{x \to \infty} \frac{f(x)}{g(x)} = L$$

where L is finite and positive.

According to these definitions, $y = 2x$ does not grow faster than $y = x$. The two functions grow at the same rate because

$$\lim_{x \to \infty} \frac{2x}{x} = \lim_{x \to \infty} 2 = 2,$$

which is a finite, positive limit. The reason for this departure from more common usage is that we want "f grows faster than g" to mean that for large x-values g is negligible when compared with f.

EXAMPLE 1 Let's compare the growth rates of several common functions.

(a) e^x grows faster than x^2 as $x \to \infty$ because

$$\underbrace{\lim_{x \to \infty} \frac{e^x}{x^2}}_{\infty \, / \, \infty} = \underbrace{\lim_{x \to \infty} \frac{e^x}{2x}}_{\infty \, / \, \infty} = \lim_{x \to \infty} \frac{e^x}{2} = \infty. \qquad \text{Using l'Hôpital's Rule twice}$$

(b) 3^x grows faster than 2^x as $x \to \infty$ because

$$\lim_{x \to \infty} \frac{3^x}{2^x} = \lim_{x \to \infty} \left(\frac{3}{2}\right)^x = \infty.$$

(c) x^2 grows faster than $\ln x$ as $x \to \infty$ because

$$\lim_{x \to \infty} \frac{x^2}{\ln x} = \lim_{x \to \infty} \frac{2x}{1/x} = \lim_{x \to \infty} 2x^2 = \infty. \qquad \text{l'Hôpital's Rule}$$

(d) $\ln x$ grows slower than $x^{1/n}$ as $x \to \infty$ for any positive integer n because

$$\lim_{x \to \infty} \frac{\ln x}{x^{1/n}} = \lim_{x \to \infty} \frac{1/x}{(1/n)\,x^{(1/n)-1}} \qquad \text{l'Hôpital's Rule}$$

$$= \lim_{x \to \infty} \frac{n}{x^{1/n}} = 0. \qquad \text{n is constant.}$$

(e) As Part (b) suggests, exponential functions with different bases never grow at the same rate as $x \to \infty$. If $a > b > 0$, then a^x grows faster than b^x. Since $(a/b) > 1$,

$$\lim_{x \to \infty} \frac{a^x}{b^x} = \lim_{x \to \infty} \left(\frac{a}{b}\right)^x = \infty.$$

(f) In contrast to exponential functions, logarithmic functions with different bases $a > 1$ and $b > 1$ always grow at the same rate as $x \to \infty$:

$$\lim_{x \to \infty} \frac{\log_a x}{\log_b x} = \lim_{x \to \infty} \frac{\ln x/\ln a}{\ln x/\ln b} = \frac{\ln b}{\ln a}.$$

The limiting ratio is always finite and never zero. ∎

If f grows at the same rate as g as $x \to \infty$, and g grows at the same rate as h as $x \to \infty$, then f grows at the same rate as h as $x \to \infty$. The reason is that

$$\lim_{x \to \infty} \frac{f}{g} = L_1 \qquad \text{and} \qquad \lim_{x \to \infty} \frac{g}{h} = L_2$$

together imply

$$\lim_{x \to \infty} \frac{f}{h} = \lim_{x \to \infty} \frac{f}{g} \cdot \frac{g}{h} = L_1 L_2.$$

If L_1 and L_2 are finite and nonzero, then so is $L_1 L_2$.

EXAMPLE 2 Show that $\sqrt{x^2 + 5}$ and $(2\sqrt{x} - 1)^2$ grow at the same rate as $x \to \infty$.

Solution We show that the functions grow at the same rate by showing that they both grow at the same rate as the function $g(x) = x$:

$$\lim_{x \to \infty} \frac{\sqrt{x^2 + 5}}{x} = \lim_{x \to \infty} \sqrt{1 + \frac{5}{x^2}} = 1,$$

$$\lim_{x \to \infty} \frac{(2\sqrt{x} - 1)^2}{x} = \lim_{x \to \infty} \left(\frac{2\sqrt{x} - 1}{\sqrt{x}}\right)^2 = \lim_{x \to \infty} \left(2 - \frac{1}{\sqrt{x}}\right)^2 = 4. \qquad ∎$$

Order and Oh-Notation

The "little-oh" and "big-oh" notation was invented by number theorists a hundred years ago and is now commonplace in mathematical analysis and computer science.

DEFINITION   A function f is **of smaller order than** g as $x \to \infty$ if $\lim_{x \to \infty} \dfrac{f(x)}{g(x)} = 0$. We indicate this by writing $f = o(g)$ ("f is little-oh of g").

Notice that saying $f = o(g)$ as $x \to \infty$ is another way to say that f grows slower than g as $x \to \infty$.

EXAMPLE 3   Here we use little-oh notation.

(a) $\ln x = o(x)$ as $x \to \infty$ because $\displaystyle \lim_{x \to \infty} \frac{\ln x}{x} = 0$

(b) $x^2 = o(x^3 + 1)$ as $x \to \infty$ because $\displaystyle \lim_{x \to \infty} \frac{x^2}{x^3 + 1} = 0$ ∎

> **DEFINITION**   Let $f(x)$ and $g(x)$ be positive for x sufficiently large. Then f is **of at most the order of g as $x \to \infty$** if there is a positive integer M for which
>
> $$\frac{f(x)}{g(x)} \le M,$$
>
> for x sufficiently large. We indicate this by writing $\boldsymbol{f = O(g)}$ ("f is big-oh of g").

EXAMPLE 4   Here we use big-oh notation.

(a) $x + \sin x = O(x)$ as $x \to \infty$ because $\dfrac{x + \sin x}{x} \le 2$ for x sufficiently large.

(b) $e^x + x^2 = O(e^x)$ as $x \to \infty$ because $\dfrac{e^x + x^2}{e^x} \to 1$ as $x \to \infty$.

(c) $x = O(e^x)$ as $x \to \infty$ because $\dfrac{x}{e^x} \to 0$ as $x \to \infty$.   ∎

If you look at the definitions again, you will see that $f = o(g)$ implies $f = O(g)$ for functions that are positive for x sufficiently large. Also, if f and g grow at the same rate, then $f = O(g)$ and $g = O(f)$ (Exercise 11).

Sequential vs. Binary Search

Computer scientists often measure the efficiency of an algorithm by counting the number of steps a computer must take to execute the algorithm. There can be significant differences in how efficiently algorithms perform, even if they are designed to accomplish the same task. These differences are often described in big-oh notation. Here is an example.

Webster's International Dictionary lists about 26,000 words that begin with the letter *a*. One way to look up a word, or to learn if it is not there, is to read through the list one word at a time until you either find the word or determine that it is not there. This method, called **sequential search**, makes no particular use of the words' alphabetical arrangement. You are sure to get an answer, but it might take 26,000 steps.

Another way to find the word or to learn it is not there is to go straight to the middle of the list (give or take a few words). If you do not find the word, then go to the middle of the half that contains it and forget about the half that does not. (You know which half contains it because you know the list is ordered alphabetically.) This method, called a **binary search**, eliminates roughly 13,000 words in a single step. If you do not find the word on the second try, then jump to the middle of the half that contains it. Continue this way until you have either found the word or divided the list in half so many times there are no words left. How many times do you have to divide the list to find the word or learn that it is not there? At most 15, because

$$(26{,}000/2^{15}) < 1.$$

That certainly beats a possible 26,000 steps.

For a list of length n, a sequential search algorithm takes on the order of n steps to find a word or determine that it is not in the list. A binary search, as the second algorithm is called, takes on the order of $\log_2 n$ steps. The reason is that if $2^{m-1} < n \leq 2^m$, then $m - 1 < \log_2 n \leq m$, and the number of bisections required to narrow the list to one word will be at most $m = \lceil \log_2 n \rceil$, the integer ceiling for $\log_2 n$.

Big-oh notation provides a compact way to say all this. The number of steps in a sequential search of an ordered list is $O(n)$; the number of steps in a binary search is $O(\log_2 n)$. In our example, there is a big difference between the two (26,000 vs. 15), and the difference can only increase with n because n grows faster than $\log_2 n$ as $n \to \infty$.

Summary

The integral definition of the natural logarithm function $\ln x$ in Section 7.1 is the key to obtaining precisely the exponential and logarithmic functions a^x and $\log_a x$ for any base $a > 0$. The differentiability and increasing behavior of $\ln x$ allows us to define its differentiable inverse, the natural exponential function e^x, through Theorem 3 in Chapter 3. Then e^x provides for the definition of the differentiable function $a^x = e^{x \ln a}$, giving a simple and precise meaning of irrational exponents, and from which we see that *every* exponential function is just e^x raised to an appropriate power, $\ln a$. The increasing (or decreasing) behavior of a^x gives its differentiable inverse $\log_a x$, using Theorem 3 again. Moreover, we saw that $\log_a x = (\ln x)/(\ln a)$ is just a multiple of the natural logarithm function. So e^x and $\ln x$ give the entire array of exponential and logarithmic functions using the algebraic operations of taking constant powers and constant multiples. Furthermore, the differentiability of e^x and a^x establish the existence of the limits

$$\lim_{h \to 0} \frac{e^h - 1}{h} = 1 \quad \text{and} \quad \lim_{h \to 0} \frac{a^h - 1}{h} = \ln a$$

(claimed in Section 3.3) as the slopes of those functions where they cross the y-axis. These limits were foundational to defining informally the natural exponential function e^x in Section 3.3, which then gave rise to $\ln x$ as its inverse in Section 3.8.

In this chapter we have seen the important roles the exponential and logarithmic functions play in analyzing problems associated with growth and decay, in comparing the growth rates of various functions, and in measuring the efficiency of a computer algorithm. In Chapters 9 and 17 we will see that exponential functions play a major role in the solutions to differential equations.

Exercises 7.4

Comparisons with the Exponential e^x

1. Which of the following functions grow faster than e^x as $x \to \infty$? Which grow at the same rate as e^x? Which grow slower?

a. $x - 3$

b. $x^3 + \sin^2 x$

c. $\sqrt{x}$

d. 4^x

e. $(3/2)^x$

f. $e^{x/2}$

g. $e^x/2$

h. $\log_{10} x$

2. Which of the following functions grow faster than e^x as $x \to \infty$? Which grow at the same rate as e^x? Which grow slower?

a. $10x^4 + 30x + 1$

b. $x \ln x - x$

c. $\sqrt{1 + x^4}$

d. $(5/2)^x$

e. e^{-x}

f. xe^x

g. $e^{\cos x}$

h. e^{x-1}

Comparisons with the Power x^2

3. Which of the following functions grow faster than x^2 as $x \to \infty$? Which grow at the same rate as x^2? Which grow slower?

a. $x^2 + 4x$

b. $x^5 - x^2$

c. $\sqrt{x^4 + x^3}$

d. $(x + 3)^2$

e. $x \ln x$

f. 2^x

g. $x^3 e^{-x}$

h. $8x^2$

4. Which of the following functions grow faster than x^2 as $x \to \infty$? Which grow at the same rate as x^2? Which grow slower?

a. $x^2 + \sqrt{x}$

b. $10x^2$

c. $x^2 e^{-x}$

d. $\log_{10}(x^2)$

e. $x^3 - x^2$

f. $(1/10)^x$

g. $(1.1)^x$

h. $x^2 + 100x$

Comparisons with the Logarithm ln x

5. Which of the following functions grow faster than $\ln x$ as $x \to \infty$? Which grow at the same rate as $\ln x$? Which grow slower?

a. $\log_3 x$ **b.** $\ln 2x$

c. $\ln \sqrt{x}$ **d.** $\sqrt{x}$

e. x **f.** $5 \ln x$

g. $1/x$ **h.** e^x

6. Which of the following functions grow faster than $\ln x$ as $x \to \infty$? Which grow at the same rate as $\ln x$? Which grow slower?

a. $\log_2 (x^2)$ **b.** $\log_{10} 10x$

c. $1/\sqrt{x}$ **d.** $1/x^2$

e. $x - 2 \ln x$ **f.** e^{-x}

g. $\ln (\ln x)$ **h.** $\ln (2x + 5)$

Ordering Functions by Growth Rates

7. Order the following functions from slowest growing to fastest growing as $x \to \infty$.

a. e^x **b.** x^x

c. $(\ln x)^x$ **d.** $e^{x/2}$

8. Order the following functions from slowest growing to fastest growing as $x \to \infty$.

a. 2^x **b.** x^2

c. $(\ln 2)^x$ **d.** e^x

Big-oh and Little-oh; Order

9. True, or false? As $x \to \infty$,

a. $x = o(x)$ **b.** $x = o(x + 5)$

c. $x = O(x + 5)$ **d.** $x = O(2x)$

e. $e^x = o(e^{2x})$ **f.** $x + \ln x = O(x)$

g. $\ln x = o(\ln 2x)$ **h.** $\sqrt{x^2 + 5} = O(x)$

10. True, or false? As $x \to \infty$,

a. $\dfrac{1}{x + 3} = O\left(\dfrac{1}{x}\right)$ **b.** $\dfrac{1}{x} + \dfrac{1}{x^2} = O\left(\dfrac{1}{x}\right)$

c. $\dfrac{1}{x} - \dfrac{1}{x^2} = o\left(\dfrac{1}{x}\right)$ **d.** $2 + \cos x = O(2)$

e. $e^x + x = O(e^x)$ **f.** $x \ln x = o(x^2)$

g. $\ln (\ln x) = O(\ln x)$ **h.** $\ln (x) = o(\ln (x^2 + 1))$

11. Show that if positive functions $f(x)$ and $g(x)$ grow at the same rate as $x \to \infty$, then $f = O(g)$ and $g = O(f)$.

12. When is a polynomial $f(x)$ of smaller order than a polynomial $g(x)$ as $x \to \infty$? Give reasons for your answer.

13. When is a polynomial $f(x)$ of at most the order of a polynomial $g(x)$ as $x \to \infty$? Give reasons for your answer.

14. What do the conclusions we drew in Section 2.6 about the limits of rational functions tell us about the relative growth of polynomials as $x \to \infty$?

Other Comparisons

T 15. Investigate

$$\lim_{x \to \infty} \frac{\ln (x + 1)}{\ln x} \quad \text{and} \quad \lim_{x \to \infty} \frac{\ln (x + 999)}{\ln x}.$$

Then use l'Hôpital's Rule to explain what you find.

16. (*Continuation of Exercise 15.*) Show that the value of

$$\lim_{x \to \infty} \frac{\ln (x + a)}{\ln x}$$

is the same no matter what value you assign to the constant a. What does this say about the relative rates at which the functions $f(x) = \ln (x + a)$ and $g(x) = \ln x$ grow?

17. Show that $\sqrt{10x + 1}$ and $\sqrt{x + 1}$ grow at the same rate as $x \to \infty$ by showing that they both grow at the same rate as $\sqrt{x}$ as $x \to \infty$.

18. Show that $\sqrt{x^4 + x}$ and $\sqrt{x^4 - x^3}$ grow at the same rate as $x \to \infty$ by showing that they both grow at the same rate as x^2 as $x \to \infty$.

19. Show that e^x grows faster as $x \to \infty$ than x^n for any positive integer n, even $x^{1,000,000}$. (*Hint:* What is the nth derivative of x^n?)

20. The function e^x outgrows any polynomial   Show that e^x grows faster as $x \to \infty$ than any polynomial

$$a_n x^n + a_{n-1} x^{n-1} + \cdots + a_1 x + a_0.$$

21. a. Show that $\ln x$ grows slower as $x \to \infty$ than $x^{1/n}$ for any positive integer n, even $x^{1/1,000,000}$.

T b. Although the values of $x^{1/1,000,000}$ eventually overtake the values of $\ln x$, you have to go way out on the x-axis before this happens. Find a value of x greater than 1 for which $x^{1/1,000,000} > \ln x$. You might start by observing that when $x > 1$ the equation $\ln x = x^{1/1,000,000}$ is equivalent to the equation $\ln (\ln x) = (\ln x)/1,000,000$.

T c. Even $x^{1/10}$ takes a long time to overtake $\ln x$. Experiment with a calculator to find the value of x at which the graphs of $x^{1/10}$ and $\ln x$ cross, or, equivalently, at which $\ln x = 10 \ln (\ln x)$. Bracket the crossing point between powers of 10 and then close in by successive halving.

T d. (*Continuation of part (c).*) The value of x at which $\ln x = 10 \ln (\ln x)$ is too far out for some graphers and root finders to identify. Try it on the equipment available to you and see what happens.

22. The function $\ln x$ grows slower than any polynomial   Show that $\ln x$ grows slower as $x \to \infty$ than any nonconstant polynomial.

Algorithms and Searches

23. a. Suppose you have three different algorithms for solving the same problem and each algorithm takes a number of steps that is of the order of one of the functions listed here:

$$n \log_2 n, \quad n^{3/2}, \quad n(\log_2 n)^2.$$

Which of the algorithms is the most efficient in the long run? Give reasons for your answer.

T b. Graph the functions in part (a) together to get a sense of how rapidly each one grows.

24. Repeat Exercise 23 for the functions

$$n, \quad \sqrt{n} \log_2 n, \quad (\log_2 n)^2.$$

T 25. Suppose you are looking for an item in an ordered list one million items long. How many steps might it take to find that item with a sequential search? A binary search?

T 26. You are looking for an item in an ordered list 450,000 items long (the length of *Webster's Third New International Dictionary*). How many steps might it take to find the item with a sequential search? A binary search?

Questions to Guide Your Review

1. How is the natural logarithm funciton defined as an integral? What are its domain, range, and derivative? What arithmetic properties does it have? Comment on its graph.

2. What integrals lead to logarithms? Give examples.

3. What are the integrals of $\tan x$ and $\cot x$? $\sec x$ and $\csc x$?

4. How is the exponential function e^x defined? What are its domain, range, and derivative? What laws of exponents does it obey? Comment on its graph.

5. How are the functions a^x and $\log_a x$ defined? Are there any restrictions on a? How is the graph of $\log_a x$ related to the graph of $\ln x$? What truth is there in the statement that there is really only one exponential function and one logarithmic function?

6. How do you solve separable first-order differential equations?

7. What is the law of exponential change? How can it be derived from an initial value problem? What are some of the applications of the law?

8. What are the six basic hyperbolic functions? Comment on their domains, ranges, and graphs. What are some of the identities relating them?

9. What are the derivatives of the six basic hyperbolic functions? What are the corresponding integral formulas? What similarities do you see here with the six basic trigonometric functions?

10. How are the inverse hyperbolic functions defined? Comment on their domains, ranges, and graphs. How can you find values of $\text{sech}^{-1} x$, $\text{csch}^{-1} x$, and $\coth^{-1} x$ using a calculator's keys for $\cosh^{-1} x$, $\sinh^{-1} x$, and $\tanh^{-1} x$?

11. What integrals lead naturally to inverse hyperbolic functions?

12. How do you compare the growth rates of positive functions as $x \to \infty$?

13. What roles do the functions e^x and $\ln x$ play in growth comparisons?

14. Describe big-oh and little-oh notation. Give examples.

15. Which is more efficient—a sequential search or a binary search? Explain.

Practice Exercises

Integration

Evaluate the integrals in Exercises 1–12.

1. $\displaystyle\int e^x \sin(e^x)\, dx$

2. $\displaystyle\int e^t \cos(3e^t - 2)\, dt$

3. $\displaystyle\int_0^\pi \tan \frac{x}{3}\, dx$

4. $\displaystyle\int_{1/6}^{1/4} 2 \cot \pi x\, dx$

5. $\displaystyle\int_{-\pi/2}^{\pi/6} \frac{\cos t}{1 - \sin t}\, dt$

6. $\displaystyle\int e^x \sec e^x\, dx$

7. $\displaystyle\int \frac{\ln(x - 5)}{x - 5}\, dx$

8. $\displaystyle\int \frac{\cos(1 - \ln v)}{v}\, dv$

9. $\displaystyle\int_1^7 \frac{3}{x}\, dx$

10. $\displaystyle\int_1^{32} \frac{1}{5x}\, dx$

11. $\displaystyle\int_e^{e^2} \frac{1}{x\sqrt{\ln x}}\, dx$

12. $\displaystyle\int_2^4 (1 + \ln t)t \ln t\, dt$

Solving Equations with Logarithmic or Exponential Terms

In Exercises 13–18, solve for y.

13. $3^y = 2^{y+1}$

14. $4^{-y} = 3^{y+2}$

15. $9e^{2y} = x^2$

16. $3^y = 3 \ln x$

17. $\ln(y - 1) = x + \ln y$

18. $\ln(10 \ln y) = \ln 5x$

Comparing Growth Rates of Functions

19. Does f grow faster, slower, or at the same rate as g as $x \to \infty$? Give reasons for your answers.

   a. $f(x) = \log_2 x$, $\quad g(x) = \log_3 x$

   b. $f(x) = x$, $\quad g(x) = x + \dfrac{1}{x}$

   c. $f(x) = x/100$, $\quad g(x) = xe^{-x}$

   d. $f(x) = x$, $\quad g(x) = \tan^{-1} x$

   e. $f(x) = \csc^{-1} x$, $\quad g(x) = 1/x$

   f. $f(x) = \sinh x$, $\quad g(x) = e^x$

20. Does f grow faster, slower, or at the same rate as g as $x \to \infty$? Give reasons for your answers.

   a. $f(x) = 3^{-x}$, $\quad g(x) = 2^{-x}$

   b. $f(x) = \ln 2x$, $\quad g(x) = \ln x^2$

   c. $f(x) = 10x^3 + 2x^2$, $\quad g(x) = e^x$

   d. $f(x) = \tan^{-1}(1/x)$, $\quad g(x) = 1/x$

   e. $f(x) = \sin^{-1}(1/x)$, $\quad g(x) = 1/x^2$

   f. $f(x) = \text{sech } x$, $\quad g(x) = e^{-x}$

21. True, or false? Give reasons for your answers.

   a. $\dfrac{1}{x^2} + \dfrac{1}{x^4} = O\left(\dfrac{1}{x^2}\right)$

   b. $\dfrac{1}{x^2} + \dfrac{1}{x^4} = O\left(\dfrac{1}{x^4}\right)$

   c. $x = o(x + \ln x)$

   d. $\ln(\ln x) = o(\ln x)$

   e. $\tan^{-1} x = O(1)$

   f. $\cosh x = O(e^x)$

22. True, or false? Give reasons for your answers.

a. $\dfrac{1}{x^4} = O\left(\dfrac{1}{x^2} + \dfrac{1}{x^4}\right)$ **b.** $\dfrac{1}{x^4} = o\left(\dfrac{1}{x^2} + \dfrac{1}{x^4}\right)$

c. $\ln x = o(x + 1)$ **d.** $\ln 2x = O(\ln x)$

e. $\sec^{-1} x = O(1)$ **f.** $\sinh x = O(e^x)$

Theory and Applications

23. The function $f(x) = e^x + x$, being differentiable and one-to-one, has a differentiable inverse $f^{-1}(x)$. Find the value of df^{-1}/dx at the point $f(\ln 2)$.

24. Find the inverse of the function $f(x) = 1 + (1/x), x \neq 0$. Then show that $f^{-1}(f(x)) = f(f^{-1}(x)) = x$ and that

$$\left.\frac{df^{-1}}{dx}\right|_{f(x)} = \frac{1}{f'(x)}.$$

25. A particle is traveling upward and to the right along the curve $y = \ln x$. Its x-coordinate is increasing at the rate $(dx/dt) = \sqrt{x}$ m/sec. At what rate is the y-coordinate changing at the point $(e^2, 2)$?

26. A girl is sliding down a slide shaped like the curve $y = 9e^{-x/3}$. Her y-coordinate is changing at the rate $dy/dt = (-1/4)\sqrt{9 - y}$ ft/sec. At approximately what rate is her x-coordinate changing when she reaches the bottom of the slide at $x = 9$ ft? (Take e^3 to be 20 and round your answer to the nearest ft/sec.)

27. The functions $f(x) = \ln 5x$ and $g(x) = \ln 3x$ differ by a constant. What constant? Give reasons for your answer.

28. a. If $(\ln x)/x = (\ln 2)/2$, must $x = 2$?

b. If $(\ln x)/x = -2 \ln 2$, must $x = 1/2$?

Give reasons for your answers.

29. The quotient $(\log_4 x)/(\log_2 x)$ has a constant value. What value? Give reasons for your answer.

T 30. $\log_x (2)$ **vs.** $\log_2 (x)$ How does $f(x) = \log_x (2)$ compare with $g(x) = \log_2 (x)$? Here is one way to find out.

a. Use the equation $\log_a b = (\ln b)/(\ln a)$ to express $f(x)$ and $g(x)$ in terms of natural logarithms.

b. Graph f and g together. Comment on the behavior of f in relation to the signs and values of g.

In Exercises 31–34, solve the differential equation.

31. $\dfrac{dy}{dx} = \sqrt{y}\cos^2 \sqrt{y}$ **32.** $y' = \dfrac{3y(x + 1)^2}{y - 1}$

33. $yy' = \sec y^2 \sec^2 x$ **34.** $y\cos^2 x \, dy + \sin x \, dx = 0$

In Exercises 35–38, solve the initial value problem.

35. $\dfrac{dy}{dx} = e^{-x-y-2}, \quad y(0) = -2$

36. $\dfrac{dy}{dx} = \dfrac{y \ln y}{1 + x^2}, \quad y(0) = e^2$

37. $x \, dy - \left(y + \sqrt{y}\right) dx = 0, \quad y(1) = 1$

38. $y^{-2}\dfrac{dx}{dy} = \dfrac{e^x}{e^{2x} + 1}, \quad y(0) = 1$

39. What is the age of a sample of charcoal in which 90% of the carbon-14 originally present has decayed?

40. Cooling a pie A deep-dish apple pie, whose internal temperature was 220°F when removed from the oven, was set out on a breezy 40°F porch to cool. Fifteen minutes later, the pie's internal temperature was 180°F. How long did it take the pie to cool from there to 70°F?

Chapter 7   Additional and Advanced Exercises

1. Let $A(t)$ be the area of the region in the first quadrant enclosed by the coordinate axes, the curve $y = e^{-x}$, and the vertical line $x = t, t > 0$. Let $V(t)$ be the volume of the solid generated by revolving the region about the x-axis. Find the following limits.

a. $\lim\limits_{t\to\infty} A(t)$ **b.** $\lim\limits_{t\to\infty} V(t)/A(t)$ **c.** $\lim\limits_{t\to 0^+} V(t)/A(t)$

2. Varying a logarithm's base

a. Find $\lim \log_a 2$ as $a \to 0^+, 1^-, 1^+$, and ∞.

T b. Graph $y = \log_a 2$ as a function of a over the interval $0 < a \leq 4$.

T 3. Graph $f(x) = \tan^{-1} x + \tan^{-1}(1/x)$ for $-5 \leq x \leq 5$. Then use calculus to explain what you see. How would you expect f to behave beyond the interval $[-5, 5]$? Give reasons for your answer.

T 4. Graph $f(x) = (\sin x)^{\sin x}$ over $[0, 3\pi]$. Explain what you see.

5. Even-odd decompositions

a. Suppose that g is an even function of x and h is an odd function of x. Show that if $g(x) + h(x) = 0$ for all x then $g(x) = 0$ for all x and $h(x) = 0$ for all x.

b. Use the result in part (a) to show that if $f(x) = f_E(x) + f_O(x)$ is the sum of an even function $f_E(x)$ and an odd function $f_O(x)$, then

$$f_E(x) = (f(x) + f(-x))/2 \quad \text{and} \quad f_O(x) = (f(x) - f(-x))/2.$$

c. What is the significance of the result in part (b)?

6. Let g be a function that is differentiable throughout an open interval containing the origin. Suppose g has the following properties:

i. $g(x + y) = \dfrac{g(x) + g(y)}{1 - g(x)g(y)}$ for all real numbers x, y, and $x + y$ in the domain of g.

ii. $\lim\limits_{h\to 0} g(h) = 0$

iii. $\lim\limits_{h\to 0} \dfrac{g(h)}{h} = 1$

a. Show that $g(0) = 0$.

b. Show that $g'(x) = 1 + [g(x)]^2$.

c. Find $g(x)$ by solving the differential equation in part (b).

7. Center of mass Find the center of mass of a thin plate of constant density covering the region in the first and fourth quadrants enclosed by the curves $y = 1/(1 + x^2)$ and $y = -1/(1 + x^2)$ and by the lines $x = 0$ and $x = 1$.

8. **Solid of revolution** The region between the curve $y = 1/(2\sqrt{x})$ and the x-axis from $x = 1/4$ to $x = 4$ is revolved about the x-axis to generate a solid.

   a. Find the volume of the solid.

   b. Find the centroid of the region.

9. **The Rule of 70** If you use the approximation $\ln 2 \approx 0.70$ (in place of $0.69314\ldots$), you can derive a rule of thumb that says, "To estimate how many years it will take an amount of money to double when invested at r percent compounded continuously, divide r into 70." For instance, an amount of money invested at 5% will double in about $70/5 = 14$ years. If you want it to double in 10 years instead, you have to invest it at $70/10 = 7\%$. Show how the Rule of 70 is derived. (A similar "Rule of 72" uses 72 instead of 70, because 72 has more integer factors.)

T 10. **Urban gardening** A vegetable garden 50 ft wide is to be grown between two buildings, which are 500 ft apart along an east-west line. If the buildings are 200 ft and 350 ft tall, where should the garden be placed in order to receive the maximum number of hours of sunlight exposure? (*Hint:* Determine the value of x in the accompanying figure that maximizes sunlight exposure for the garden.)

8

TECHNIQUES OF INTEGRATION

OVERVIEW The Fundamental Theorem tells us how to evaluate a definite integral once we have an antiderivative for the integrand function. Table 8.1 summarizes the forms of antiderivatives for many of the functions we have studied so far, and the substitution method helps us use the table to evaluate more complicated functions involving these basic ones. In this chapter we study a number of other important techniques for finding antiderivatives (or indefinite integrals) for many combinations of functions whose antiderivatives cannot be found using the methods presented before.

TABLE 8.1 Basic integration formulas

1. $\int k\,dx = kx + C$ (any number k)

2. $\int x^n\,dx = \dfrac{x^{n+1}}{n+1} + C$ $(n \neq -1)$

3. $\int \dfrac{dx}{x} = \ln|x| + C$

4. $\int e^x\,dx = e^x + C$

5. $\int a^x\,dx = \dfrac{a^x}{\ln a} + C$ $(a > 0, a \neq 1)$

6. $\int \sin x\,dx = -\cos x + C$

7. $\int \cos x\,dx = \sin x + C$

8. $\int \sec^2 x\,dx = \tan x + C$

9. $\int \csc^2 x\,dx = -\cot x + C$

10. $\int \sec x \tan x\,dx = \sec x + C$

11. $\int \csc x \cot x\,dx = -\csc x + C$

12. $\int \tan x\,dx = \ln|\sec x| + C$

13. $\int \cot x\,dx = \ln|\sin x| + C$

14. $\int \sec x\,dx = \ln|\sec x + \tan x| + C$

15. $\int \csc x\,dx = -\ln|\csc x + \cot x| + C$

16. $\int \sinh x\,dx = \cosh x + C$

17. $\int \cosh x\,dx = \sinh x + C$

18. $\int \dfrac{dx}{\sqrt{a^2 - x^2}} = \sin^{-1}\left(\dfrac{x}{a}\right) + C$

19. $\int \dfrac{dx}{a^2 + x^2} = \dfrac{1}{a}\tan^{-1}\left(\dfrac{x}{a}\right) + C$

20. $\int \dfrac{dx}{x\sqrt{x^2 - a^2}} = \dfrac{1}{a}\sec^{-1}\left|\dfrac{x}{a}\right| + C$

21. $\int \dfrac{dx}{\sqrt{a^2 + x^2}} = \sinh^{-1}\left(\dfrac{x}{a}\right) + C$ $(a > 0)$

22. $\int \dfrac{dx}{\sqrt{x^2 - a^2}} = \cosh^{-1}\left(\dfrac{x}{a}\right) + C$ $(x > a > 0)$

8.1 Integration by Parts

Integration by parts is a technique for simplifying integrals of the form

$$\int f(x)g(x)\, dx.$$

It is useful when f can be differentiated repeatedly and g can be integrated repeatedly without difficulty. The integrals

$$\int x \cos x\, dx \qquad \text{and} \qquad \int x^2 e^x\, dx$$

are such integrals because $f(x) = x$ or $f(x) = x^2$ can be differentiated repeatedly to become zero, and $g(x) = \cos x$ or $g(x) = e^x$ can be integrated repeatedly without difficulty. Integration by parts also applies to integrals like

$$\int \ln x\, dx \qquad \text{and} \qquad \int e^x \cos x\, dx.$$

In the first case, $f(x) = \ln x$ is easy to differentiate and $g(x) = 1$ easily integrates to x. In the second case, each part of the integrand appears again after repeated differentiation or integration.

Product Rule in Integral Form

If f and g are differentiable functions of x, the Product Rule says that

$$\frac{d}{dx}[f(x)g(x)] = f'(x)g(x) + f(x)g'(x).$$

In terms of indefinite integrals, this equation becomes

$$\int \frac{d}{dx}[f(x)g(x)]\, dx = \int [f'(x)g(x) + f(x)g'(x)]\, dx$$

or

$$\int \frac{d}{dx}[f(x)g(x)]\, dx = \int f'(x)g(x)\, dx + \int f(x)g'(x)\, dx.$$

Rearranging the terms of this last equation, we get

$$\int f(x)g'(x)\, dx = \int \frac{d}{dx}[f(x)g(x)]\, dx - \int f'(x)g(x)\, dx,$$

leading to the **integration by parts** formula

$$\int f(x)g'(x)\, dx = f(x)g(x) - \int f'(x)g(x)\, dx \qquad (1)$$

Sometimes it is easier to remember the formula if we write it in differential form. Let $u = f(x)$ and $v = g(x)$. Then $du = f'(x)\, dx$ and $dv = g'(x)\, dx$. Using the Substitution Rule, the integration by parts formula becomes

> **Integration by Parts Formula**
>
> $$\int u\,dv = uv - \int v\,du \qquad\qquad (2)$$

This formula expresses one integral, $\int u\,dv$, in terms of a second integral, $\int v\,du$. With a proper choice of u and v, the second integral may be easier to evaluate than the first. In using the formula, various choices may be available for u and dv. The next examples illustrate the technique. To avoid mistakes, we always list our choices for u and dv, then we add to the list our calculated new terms du and v, and finally we apply the formula in Equation (2).

EXAMPLE 1   Find

$$\int x\cos x\,dx.$$

Solution   We use the formula $\int u\,dv = uv - \int v\,du$ with

$$u = x, \qquad dv = \cos x\,dx,$$
$$du = dx, \qquad v = \sin x. \qquad \text{Simplest antiderivative of } \cos x$$

Then

$$\int x\cos x\,dx = x\sin x - \int \sin x\,dx = x\sin x + \cos x + C. \qquad \blacksquare$$

There are four choices available for u and dv in Example 1:

1. Let $u = 1$ and $dv = x\cos x\,dx$.   2. Let $u = x$ and $dv = \cos x\,dx$.
3. Let $u = x\cos x$ and $dv = dx$.   4. Let $u = \cos x$ and $dv = x\,dx$.

Choice 2 was used in Example 1. The other three choices lead to integrals we don't know how to integrate. For instance, Choice 3 leads to the integral

$$\int (x\cos x - x^2\sin x)\,dx.$$

The goal of integration by parts is to go from an integral $\int u\,dv$ that we don't see how to evaluate to an integral $\int v\,du$ that we can evaluate. Generally, you choose dv first to be as much of the integrand, including dx, as you can readily integrate; u is the leftover part. When finding v from dv, any antiderivative will work and we usually pick the simplest one; no arbitrary constant of integration is needed in v because it would simply cancel out of the right-hand side of Equation (2).

EXAMPLE 2   Find

$$\int \ln x\,dx.$$

Solution   Since $\int \ln x\,dx$ can be written as $\int \ln x \cdot 1\,dx$, we use the formula $\int u\,dv = uv - \int v\,du$ with

$$u = \ln x \qquad \text{Simplifies when differentiated} \qquad\qquad dv = dx \qquad \text{Easy to integrate}$$

$$du = \frac{1}{x}\,dx, \qquad\qquad\qquad\qquad\qquad\qquad v = x. \qquad \text{Simplest antiderivative}$$

Then from Equation (2),

$$\int \ln x \, dx = x \ln x - \int x \cdot \frac{1}{x} \, dx = x \ln x - \int dx = x \ln x - x + C. \qquad \blacksquare$$

Sometimes we have to use integration by parts more than once.

EXAMPLE 3   Evaluate

$$\int x^2 e^x \, dx.$$

Solution   With $u = x^2$, $dv = e^x \, dx$, $du = 2x \, dx$, and $v = e^x$, we have

$$\int x^2 e^x \, dx = x^2 e^x - 2\int x e^x \, dx.$$

The new integral is less complicated than the original because the exponent on x is reduced by one. To evaluate the integral on the right, we integrate by parts again with $u = x$, $dv = e^x \, dx$. Then $du = dx$, $v = e^x$, and

$$\int x e^x \, dx = x e^x - \int e^x \, dx = x e^x - e^x + C.$$

Using this last evaluation, we then obtain

$$\int x^2 e^x \, dx = x^2 e^x - 2\int x e^x \, dx$$

$$= x^2 e^x - 2x e^x + 2e^x + C. \qquad \blacksquare$$

The technique of Example 3 works for any integral $\int x^n e^x \, dx$ in which n is a positive integer, because differentiating x^n will eventually lead to zero and integrating e^x is easy.

Integrals like the one in the next example occur in electrical engineering. Their evaluation requires two integrations by parts, followed by solving for the unknown integral.

EXAMPLE 4   Evaluate

$$\int e^x \cos x \, dx.$$

Solution   Let $u = e^x$ and $dv = \cos x \, dx$. Then $du = e^x \, dx$, $v = \sin x$, and

$$\int e^x \cos x \, dx = e^x \sin x - \int e^x \sin x \, dx.$$

The second integral is like the first except that it has $\sin x$ in place of $\cos x$. To evaluate it, we use integration by parts with

$$u = e^x, \qquad dv = \sin x \, dx, \qquad v = -\cos x, \qquad du = e^x \, dx.$$

Then

$$\int e^x \cos x \, dx = e^x \sin x - \left(-e^x \cos x - \int (-\cos x)(e^x \, dx) \right)$$

$$= e^x \sin x + e^x \cos x - \int e^x \cos x \, dx.$$

The unknown integral now appears on both sides of the equation. Adding the integral to both sides and adding the constant of integration give

$$2\int e^x \cos x \, dx = e^x \sin x + e^x \cos x + C_1.$$

Dividing by 2 and renaming the constant of integration give

$$\int e^x \cos x \, dx = \frac{e^x \sin x + e^x \cos x}{2} + C. \qquad \blacksquare$$

EXAMPLE 5   Obtain a formula that expresses the integral

$$\int \cos^n x \, dx$$

in terms of an integral of a lower power of $\cos x$.

Solution   We may think of $\cos^n x$ as $\cos^{n-1} x \cdot \cos x$. Then we let

$$u = \cos^{n-1} x \qquad \text{and} \qquad dv = \cos x \, dx,$$

so that

$$du = (n-1)\cos^{n-2} x \, (-\sin x \, dx) \qquad \text{and} \qquad v = \sin x.$$

Integration by parts then gives

$$\int \cos^n x \, dx = \cos^{n-1} x \sin x + (n-1)\int \sin^2 x \cos^{n-2} x \, dx$$

$$= \cos^{n-1} x \sin x + (n-1)\int (1 - \cos^2 x)\cos^{n-2} x \, dx$$

$$= \cos^{n-1} x \sin x + (n-1)\int \cos^{n-2} x \, dx - (n-1)\int \cos^n x \, dx.$$

If we add

$$(n-1)\int \cos^n x \, dx$$

to both sides of this equation, we obtain

$$n\int \cos^n x \, dx = \cos^{n-1} x \sin x + (n-1)\int \cos^{n-2} x \, dx.$$

We then divide through by n, and the final result is

$$\int \cos^n x \, dx = \frac{\cos^{n-1} x \sin x}{n} + \frac{n-1}{n}\int \cos^{n-2} x \, dx. \qquad \blacksquare$$

The formula found in Example 5 is called a **reduction formula** because it replaces an integral containing some power of a function with an integral of the same form having the power reduced. When n is a positive integer, we may apply the formula repeatedly until the remaining integral is easy to evaluate. For example, the result in Example 5 tells us that

$$\int \cos^3 x \, dx = \frac{\cos^2 x \sin x}{3} + \frac{2}{3}\int \cos x \, dx$$

$$= \frac{1}{3}\cos^2 x \sin x + \frac{2}{3}\sin x + C.$$

Evaluating Definite Integrals by Parts

The integration by parts formula in Equation (1) can be combined with Part 2 of the Fundamental Theorem in order to evaluate definite integrals by parts. Assuming that both f' and g' are continuous over the interval $[a, b]$, Part 2 of the Fundamental Theorem gives

Integration by Parts Formula for Definite Integrals

$$\int_a^b f(x)g'(x)\, dx = f(x)g(x)\Big]_a^b - \int_a^b f'(x)g(x)\, dx \qquad (3)$$

In applying Equation (3), we normally use the u and v notation from Equation (2) because it is easier to remember. Here is an example.

EXAMPLE 6   Find the area of the region bounded by the curve $y = xe^{-x}$ and the x-axis from $x = 0$ to $x = 4$.

Solution   The region is shaded in Figure 8.1. Its area is

$$\int_0^4 xe^{-x}\, dx.$$

Let $u = x$, $dv = e^{-x}\, dx$, $v = -e^{-x}$, and $du = dx$. Then,

$$\int_0^4 xe^{-x}\, dx = -xe^{-x}\Big]_0^4 - \int_0^4 (-e^{-x})\, dx$$

$$= [-4e^{-4} - (0)] + \int_0^4 e^{-x}\, dx$$

$$= -4e^{-4} - e^{-x}\Big]_0^4$$

$$= -4e^{-4} - e^{-4} - (-e^0) = 1 - 5e^{-4} \approx 0.91. \qquad \blacksquare$$

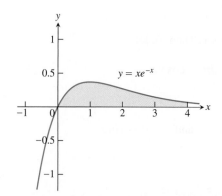

FIGURE 8.1   The region in Example 6.

Tabular Integration

We have seen that integrals of the form $\int f(x)g(x)\, dx$, in which f can be differentiated repeatedly to become zero and g can be integrated repeatedly without difficulty, are natural candidates for integration by parts. However, if many repetitions are required, the calculations can be cumbersome; or, you choose substitutions for a repeated integration by parts that just ends up giving back the original integral you were trying to find. In situations like these, there is a way to organize the calculations that prevents these pitfalls and makes the work much easier. It is called **tabular integration** and is illustrated in the following examples.

EXAMPLE 7   Evaluate

$$\int x^2 e^x\, dx.$$

Solution   With $f(x) = x^2$ and $g(x) = e^x$, we list:

$f(x)$ **and its derivatives**		$g(x)$ **and its integrals**
x^2	$(+)$	e^x
$2x$	$(-)$	e^x
2	$(+)$	e^x
0		e^x

We combine the products of the functions connected by the arrows according to the operation signs above the arrows to obtain

$$\int x^2 e^x \, dx = x^2 e^x - 2x e^x + 2 e^x + C.$$

Compare this with the result in Example 3. ∎

EXAMPLE 8 Evaluate

$$\int x^3 \sin x \, dx.$$

Solution With $f(x) = x^3$ and $g(x) = \sin x$, we list:

$f(x)$ and its derivatives		$g(x)$ and its integrals
x^3	$(+)$	$\sin x$
$3x^2$	$(-)$	$-\cos x$
$6x$	$(+)$	$-\sin x$
6	$(-)$	$\cos x$
0		$\sin x$

Again we combine the products of the functions connected by the arrows according to the operation signs above the arrows to obtain

$$\int x^3 \sin x \, dx = -x^3 \cos x + 3x^2 \sin x + 6x \cos x - 6 \sin x + C.$$ ∎

The Additional Exercises at the end of this chapter show how tabular integration can be used when neither function f nor g can be differentiated repeatedly to become zero.

Exercises 8.1

Integration by Parts
Evaluate the integrals in Exercises 1–24 using integration by parts.

1. $\displaystyle\int x \sin \frac{x}{2} \, dx$

2. $\displaystyle\int \theta \cos \pi\theta \, d\theta$

3. $\displaystyle\int t^2 \cos t \, dt$

4. $\displaystyle\int x^2 \sin x \, dx$

5. $\displaystyle\int_1^2 x \ln x \, dx$

6. $\displaystyle\int_1^e x^3 \ln x \, dx$

7. $\displaystyle\int x e^x \, dx$

8. $\displaystyle\int x e^{3x} \, dx$

9. $\displaystyle\int x^2 e^{-x} \, dx$

10. $\displaystyle\int (x^2 - 2x + 1) e^{2x} \, dx$

11. $\displaystyle\int \tan^{-1} y \, dy$

12. $\displaystyle\int \sin^{-1} y \, dy$

13. $\displaystyle\int x \sec^2 x \, dx$

14. $\displaystyle\int 4x \sec^2 2x \, dx$

15. $\displaystyle\int x^3 e^x \, dx$

16. $\displaystyle\int p^4 e^{-p} \, dp$

17. $\displaystyle\int (x^2 - 5x) e^x \, dx$

18. $\displaystyle\int (r^2 + r + 1) e^r \, dr$

19. $\displaystyle\int x^5 e^x \, dx$

20. $\displaystyle\int t^2 e^{4t} \, dt$

21. $\displaystyle\int e^\theta \sin \theta \, d\theta$

22. $\displaystyle\int e^{-y} \cos y \, dy$

23. $\displaystyle\int e^{2x} \cos 3x \, dx$

24. $\displaystyle\int e^{-2x} \sin 2x \, dx$

Using Substitution
Evaluate the integrals in Exercises 25–30 by using a substitution prior to integration by parts.

25. $\displaystyle\int e^{\sqrt{3s+9}} \, ds$

26. $\displaystyle\int_0^1 x\sqrt{1 - x} \, dx$

27. $\displaystyle\int_0^{\pi/3} x \tan^2 x \, dx$

28. $\displaystyle\int \ln(x + x^2) \, dx$

29. $\displaystyle\int \sin(\ln x) \, dx$

30. $\displaystyle\int z(\ln z)^2 \, dz$

Evaluating Integrals

Evaluate the integrals in Exercises 31–50. Some integrals do not require integration by parts.

31. $\displaystyle\int x \sec x^2 \, dx$

32. $\displaystyle\int \frac{\cos \sqrt{x}}{\sqrt{x}} \, dx$

33. $\displaystyle\int x(\ln x)^2 \, dx$

34. $\displaystyle\int \frac{1}{x(\ln x)^2} \, dx$

35. $\displaystyle\int \frac{\ln x}{x^2} \, dx$

36. $\displaystyle\int \frac{(\ln x)^3}{x} \, dx$

37. $\displaystyle\int x^3 e^{x^4} \, dx$

38. $\displaystyle\int x^5 e^{x^3} \, dx$

39. $\displaystyle\int x^3 \sqrt{x^2 + 1} \, dx$

40. $\displaystyle\int x^2 \sin x^3 \, dx$

41. $\displaystyle\int \sin 3x \cos 2x \, dx$

42. $\displaystyle\int \sin 2x \cos 4x \, dx$

43. $\displaystyle\int e^x \sin e^x \, dx$

44. $\displaystyle\int \frac{e^{\sqrt{x}}}{\sqrt{x}} \, dx$

45. $\displaystyle\int \cos \sqrt{x} \, dx$

46. $\displaystyle\int \sqrt{x} \, e^{\sqrt{x}} \, dx$

47. $\displaystyle\int_0^{\pi/2} \theta^2 \sin 2\theta \, d\theta$

48. $\displaystyle\int_0^{\pi/2} x^3 \cos 2x \, dx$

49. $\displaystyle\int_{2/\sqrt{3}}^{2} t \sec^{-1} t \, dt$

50. $\displaystyle\int_0^{1/\sqrt{2}} 2x \sin^{-1}(x^2) \, dx$

Theory and Examples

51. Finding area   Find the area of the region enclosed by the curve $y = x \sin x$ and the x-axis (see the accompanying figure) for

a. $0 \le x \le \pi$.

b. $\pi \le x \le 2\pi$.

c. $2\pi \le x \le 3\pi$.

d. What pattern do you see here? What is the area between the curve and the x-axis for $n\pi \le x \le (n+1)\pi$, n an arbitrary nonnegative integer? Give reasons for your answer.

52. Finding area   Find the area of the region enclosed by the curve $y = x \cos x$ and the x-axis (see the accompanying figure) for

a. $\pi/2 \le x \le 3\pi/2$.

b. $3\pi/2 \le x \le 5\pi/2$.

c. $5\pi/2 \le x \le 7\pi/2$.

d. What pattern do you see? What is the area between the curve and the x-axis for

$$\left(\frac{2n-1}{2}\right)\pi \le x \le \left(\frac{2n+1}{2}\right)\pi,$$

n an arbitrary positive integer? Give reasons for your answer.

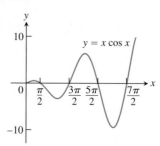

53. Finding volume   Find the volume of the solid generated by revolving the region in the first quadrant bounded by the coordinate axes, the curve $y = e^x$, and the line $x = \ln 2$ about the line $x = \ln 2$.

54. Finding volume   Find the volume of the solid generated by revolving the region in the first quadrant bounded by the coordinate axes, the curve $y = e^{-x}$, and the line $x = 1$

a. about the y-axis.

b. about the line $x = 1$.

55. Finding volume   Find the volume of the solid generated by revolving the region in the first quadrant bounded by the coordinate axes and the curve $y = \cos x$, $0 \le x \le \pi/2$, about

a. the y-axis.

b. the line $x = \pi/2$.

56. Finding volume   Find the volume of the solid generated by revolving the region bounded by the x-axis and the curve $y = x \sin x$, $0 \le x \le \pi$, about

a. the y-axis.

b. the line $x = \pi$.

(See Exercise 51 for a graph.)

57. Consider the region bounded by the graphs of $y = \ln x$, $y = 0$, and $x = e$.

a. Find the area of the region.

b. Find the volume of the solid formed by revolving this region about the x-axis.

c. Find the volume of the solid formed by revolving this region about the line $x = -2$.

d. Find the centroid of the region.

58. Consider the region bounded by the graphs of $y = \tan^{-1} x$, $y = 0$, and $x = 1$.

a. Find the area of the region.

b. Find the volume of the solid formed by revolving this region about the y-axis.

59. Average value   A retarding force, symbolized by the dashpot in the accompanying figure, slows the motion of the weighted spring so that the mass's position at time t is

$$y = 2e^{-t} \cos t, \qquad t \ge 0.$$

Find the average value of y over the interval $0 \le t \le 2\pi$.

Mass

Dashpot

60. **Average value** In a mass-spring-dashpot system like the one in Exercise 59, the mass's position at time t is

$$y = 4e^{-t}(\sin t - \cos t), \qquad t \ge 0.$$

Find the average value of y over the interval $0 \le t \le 2\pi$.

Reduction Formulas
In Exercises 61–64, use integration by parts to establish the reduction formula.

61. $\displaystyle\int x^n \cos x \, dx = x^n \sin x - n \int x^{n-1} \sin x \, dx$

62. $\displaystyle\int x^n \sin x \, dx = -x^n \cos x + n \int x^{n-1} \cos x \, dx$

63. $\displaystyle\int x^n e^{ax} \, dx = \frac{x^n e^{ax}}{a} - \frac{n}{a} \int x^{n-1} e^{ax} \, dx, \quad a \ne 0$

64. $\displaystyle\int (\ln x)^n \, dx = x(\ln x)^n - n \int (\ln x)^{n-1} \, dx$

65. Show that
$$\int_a^b \left(\int_x^b f(t) \, dt \right) dx = \int_a^b (x - a)f(x) \, dx.$$

66. Use integration by parts to obtain the formula
$$\int \sqrt{1 - x^2} \, dx = \frac{1}{2} x \sqrt{1 - x^2} + \frac{1}{2} \int \frac{1}{\sqrt{1 - x^2}} \, dx.$$

Integrating Inverses of Functions
Integration by parts leads to a rule for integrating inverses that usually gives good results:

$$\int f^{-1}(x) \, dx = \int y f'(y) \, dy \qquad \begin{array}{l} y = f^{-1}(x), \quad x = f(y) \\ dx = f'(y) \, dy \end{array}$$

$$= yf(y) - \int f(y) \, dy \qquad \begin{array}{l} \text{Integration by parts with} \\ u = y, \, dv = f'(y) \, dy \end{array}$$

$$= xf^{-1}(x) - \int f(y) \, dy$$

The idea is to take the most complicated part of the integral, in this case $f^{-1}(x)$, and simplify it first. For the integral of $\ln x$, we get

$$\int \ln x \, dx = \int y e^y \, dy \qquad \begin{array}{l} y = \ln x, \quad x = e^y \\ dx = e^y \, dy \end{array}$$

$$= ye^y - e^y + C$$

$$= x \ln x - x + C.$$

For the integral of $\cos^{-1} x$ we get

$$\int \cos^{-1} x \, dx = x \cos^{-1} x - \int \cos y \, dy \qquad y = \cos^{-1} x$$

$$= x \cos^{-1} x - \sin y + C$$

$$= x \cos^{-1} x - \sin(\cos^{-1} x) + C.$$

Use the formula

$$\int f^{-1}(x) \, dx = xf^{-1}(x) - \int f(y) \, dy \qquad y = f^{-1}(x) \qquad (4)$$

to evaluate the integrals in Exercises 67–70. Express your answers in terms of x.

67. $\displaystyle\int \sin^{-1} x \, dx$

68. $\displaystyle\int \tan^{-1} x \, dx$

69. $\displaystyle\int \sec^{-1} x \, dx$

70. $\displaystyle\int \log_2 x \, dx$

Another way to integrate $f^{-1}(x)$ (when f^{-1} is integrable, of course) is to use integration by parts with $u = f^{-1}(x)$ and $dv = dx$ to rewrite the integral of f^{-1} as

$$\int f^{-1}(x) \, dx = xf^{-1}(x) - \int x \left(\frac{d}{dx} f^{-1}(x) \right) dx. \qquad (5)$$

Exercises 71 and 72 compare the results of using Equations (4) and (5).

71. Equations (4) and (5) give different formulas for the integral of $\cos^{-1} x$:

a. $\displaystyle\int \cos^{-1} x \, dx = x \cos^{-1} x - \sin(\cos^{-1} x) + C$ Eq. (4)

b. $\displaystyle\int \cos^{-1} x \, dx = x \cos^{-1} x - \sqrt{1 - x^2} + C$ Eq. (5)

Can both integrations be correct? Explain.

72. Equations (4) and (5) lead to different formulas for the integral of $\tan^{-1} x$:

a. $\displaystyle\int \tan^{-1} x \, dx = x \tan^{-1} x - \ln \sec(\tan^{-1} x) + C$ Eq. (4)

b. $\displaystyle\int \tan^{-1} x \, dx = x \tan^{-1} x - \ln \sqrt{1 + x^2} + C$ Eq. (5)

Can both integrations be correct? Explain.

Evaluate the integrals in Exercises 73 and 74 with (**a**) Eq. (4) and (**b**) Eq. (5). In each case, check your work by differentiating your answer with respect to x.

73. $\displaystyle\int \sinh^{-1} x \, dx$

74. $\displaystyle\int \tanh^{-1} x \, dx$

8.2 | Trigonometric Integrals

Trigonometric integrals involve algebraic combinations of the six basic trigonometric functions. In principle, we can always express such integrals in terms of sines and cosines, but it is often simpler to work with other functions, as in the integral

$$\int \sec^2 x \, dx = \tan x + C.$$

The general idea is to use identities to transform the integrals we have to find into integrals that are easier to work with.

Products of Powers of Sines and Cosines

We begin with integrals of the form:

$$\int \sin^m x \cos^n x \, dx,$$

where m and n are nonnegative integers (positive or zero). We can divide the appropriate substitution into three cases according to m and n being odd or even.

Case 1 If **m is odd**, we write m as $2k + 1$ and use the identity $\sin^2 x = 1 - \cos^2 x$ to obtain

$$\sin^m x = \sin^{2k+1} x = (\sin^2 x)^k \sin x = (1 - \cos^2 x)^k \sin x. \qquad (1)$$

Then we combine the single $\sin x$ with dx in the integral and set $\sin x \, dx$ equal to $-d(\cos x)$.

Case 2 If **m is even and n is odd** in $\int \sin^m x \cos^n x \, dx$, we write n as $2k + 1$ and use the identity $\cos^2 x = 1 - \sin^2 x$ to obtain

$$\cos^n x = \cos^{2k+1} x = (\cos^2 x)^k \cos x = (1 - \sin^2 x)^k \cos x.$$

We then combine the single $\cos x$ with dx and set $\cos x \, dx$ equal to $d(\sin x)$.

Case 3 If **both m and n are even** in $\int \sin^m x \cos^n x \, dx$, we substitute

$$\sin^2 x = \frac{1 - \cos 2x}{2}, \qquad \cos^2 x = \frac{1 + \cos 2x}{2} \qquad (2)$$

to reduce the integrand to one in lower powers of $\cos 2x$.

Here are some examples illustrating each case.

EXAMPLE 1 Evaluate

$$\int \sin^3 x \cos^2 x \, dx.$$

Solution This is an example of Case 1.

$$\int \sin^3 x \cos^2 x \, dx = \int \sin^2 x \cos^2 x \sin x \, dx \qquad m \text{ is odd.}$$

$$= \int (1 - \cos^2 x) \cos^2 x \, (-d(\cos x)) \qquad \sin x \, dx = -d(\cos x)$$

$$= \int (1 - u^2)(u^2)(-du) \qquad u = \cos x$$

$$= \int (u^4 - u^2) \, du \qquad \text{Multiply terms.}$$

$$= \frac{u^5}{5} - \frac{u^3}{3} + C = \frac{\cos^5 x}{5} - \frac{\cos^3 x}{3} + C. \qquad \blacksquare$$

EXAMPLE 2 Evaluate

$$\int \cos^5 x \, dx.$$

Solution This is an example of Case 2, where $m = 0$ is even and $n = 5$ is odd.

$$\int \cos^5 x \, dx = \int \cos^4 x \cos x \, dx = \int (1 - \sin^2 x)^2 \, d(\sin x) \qquad \cos x \, dx = d(\sin x)$$

$$= \int (1 - u^2)^2 \, du \qquad u = \sin x$$

$$= \int (1 - 2u^2 + u^4) \, du \qquad \text{Square } 1 - u^2.$$

$$= u - \frac{2}{3} u^3 + \frac{1}{5} u^5 + C = \sin x - \frac{2}{3} \sin^3 x + \frac{1}{5} \sin^5 x + C. \qquad \blacksquare$$

EXAMPLE 3 Evaluate

$$\int \sin^2 x \cos^4 x \, dx.$$

Solution This is an example of Case 3.

$$\int \sin^2 x \cos^4 x \, dx = \int \left(\frac{1 - \cos 2x}{2} \right) \left(\frac{1 + \cos 2x}{2} \right)^2 dx \qquad m \text{ and } n \text{ both even}$$

$$= \frac{1}{8} \int (1 - \cos 2x)(1 + 2 \cos 2x + \cos^2 2x) \, dx$$

$$= \frac{1}{8} \int (1 + \cos 2x - \cos^2 2x - \cos^3 2x) \, dx$$

$$= \frac{1}{8} \left[x + \frac{1}{2} \sin 2x - \int (\cos^2 2x + \cos^3 2x) \, dx \right].$$

For the term involving $\cos^2 2x$, we use

$$\int \cos^2 2x \, dx = \frac{1}{2} \int (1 + \cos 4x) \, dx$$

$$= \frac{1}{2} \left(x + \frac{1}{4} \sin 4x \right). \qquad \text{Omitting the constant of integration until the final result}$$

For the $\cos^3 2x$ term, we have

$$\int \cos^3 2x \, dx = \int (1 - \sin^2 2x) \cos 2x \, dx$$

u = sin 2x,
du = 2 cos 2x dx

$$= \frac{1}{2} \int (1 - u^2) \, du = \frac{1}{2} \left(\sin 2x - \frac{1}{3} \sin^3 2x \right).$$

Again
omitting C

Combining everything and simplifying, we get

$$\int \sin^2 x \cos^4 x \, dx = \frac{1}{16} \left(x - \frac{1}{4} \sin 4x + \frac{1}{3} \sin^3 2x \right) + C. \qquad \blacksquare$$

Eliminating Square Roots

In the next example, we use the identity $\cos^2 \theta = (1 + \cos 2\theta)/2$ to eliminate a square root.

EXAMPLE 4 Evaluate

$$\int_0^{\pi/4} \sqrt{1 + \cos 4x} \, dx.$$

Solution To eliminate the square root, we use the identity

$$\cos^2 \theta = \frac{1 + \cos 2\theta}{2} \qquad \text{or} \qquad 1 + \cos 2\theta = 2 \cos^2 \theta.$$

With $\theta = 2x$, this becomes

$$1 + \cos 4x = 2 \cos^2 2x.$$

Therefore,

$$\int_0^{\pi/4} \sqrt{1 + \cos 4x} \, dx = \int_0^{\pi/4} \sqrt{2 \cos^2 2x} \, dx = \int_0^{\pi/4} \sqrt{2} \sqrt{\cos^2 2x} \, dx$$

$$= \sqrt{2} \int_0^{\pi/4} |\cos 2x| \, dx = \sqrt{2} \int_0^{\pi/4} \cos 2x \, dx \qquad \begin{array}{l} \cos 2x \geq 0 \\ \text{on } [0, \pi/4] \end{array}$$

$$= \sqrt{2} \left[\frac{\sin 2x}{2} \right]_0^{\pi/4} = \frac{\sqrt{2}}{2} [1 - 0] = \frac{\sqrt{2}}{2}. \qquad \blacksquare$$

Integrals of Powers of tan x and sec x

We know how to integrate the tangent and secant and their squares. To integrate higher powers, we use the identities $\tan^2 x = \sec^2 x - 1$ and $\sec^2 x = \tan^2 x + 1$, and integrate by parts when necessary to reduce the higher powers to lower powers.

EXAMPLE 5 Evaluate

$$\int \tan^4 x \, dx.$$

Solution

$$\int \tan^4 x \, dx = \int \tan^2 x \cdot \tan^2 x \, dx = \int \tan^2 x \cdot (\sec^2 x - 1) \, dx$$

$$= \int \tan^2 x \sec^2 x \, dx - \int \tan^2 x \, dx$$

$$= \int \tan^2 x \sec^2 x \, dx - \int (\sec^2 x - 1) \, dx$$

$$= \int \tan^2 x \sec^2 x \, dx - \int \sec^2 x \, dx + \int dx.$$

In the first integral, we let

$$u = \tan x, \qquad du = \sec^2 x\,dx$$

and have

$$\int u^2\,du = \frac{1}{3}u^3 + C_1.$$

The remaining integrals are standard forms, so

$$\int \tan^4 x\,dx = \frac{1}{3}\tan^3 x - \tan x + x + C. \qquad\blacksquare$$

EXAMPLE 6   Evaluate

$$\int \sec^3 x\,dx.$$

Solution   We integrate by parts using

$$u = \sec x, \qquad dv = \sec^2 x\,dx, \qquad v = \tan x, \qquad du = \sec x \tan x\,dx.$$

Then

$$\int \sec^3 x\,dx = \sec x \tan x - \int (\tan x)(\sec x \tan x\,dx)$$

$$= \sec x \tan x - \int (\sec^2 x - 1)\sec x\,dx \qquad \tan^2 x = \sec^2 x - 1$$

$$= \sec x \tan x + \int \sec x\,dx - \int \sec^3 x\,dx.$$

Combining the two secant-cubed integrals gives

$$2\int \sec^3 x\,dx = \sec x \tan x + \int \sec x\,dx$$

and

$$\int \sec^3 x\,dx = \frac{1}{2}\sec x \tan x + \frac{1}{2}\ln|\sec x + \tan x| + C. \qquad\blacksquare$$

Products of Sines and Cosines

The integrals

$$\int \sin mx \sin nx\,dx, \qquad \int \sin mx \cos nx\,dx, \qquad \text{and} \qquad \int \cos mx \cos nx\,dx$$

arise in many applications involving periodic functions. We can evaluate these integrals through integration by parts, but two such integrations are required in each case. It is simpler to use the identities

$$\sin mx \sin nx = \frac{1}{2}[\cos(m-n)x - \cos(m+n)x], \qquad (3)$$

$$\sin mx \cos nx = \frac{1}{2}[\sin(m-n)x + \sin(m+n)x], \qquad (4)$$

$$\cos mx \cos nx = \frac{1}{2}[\cos(m-n)x + \cos(m+n)x]. \qquad (5)$$

These identities come from the angle sum formulas for the sine and cosine functions (Section 1.3). They give functions whose antiderivatives are easily found.

EXAMPLE 7   Evaluate

$$\int \sin 3x \cos 5x \, dx.$$

Solution   From Equation (4) with $m = 3$ and $n = 5$, we get

$$\int \sin 3x \cos 5x \, dx = \frac{1}{2}\int [\sin(-2x) + \sin 8x] \, dx$$

$$= \frac{1}{2}\int (\sin 8x - \sin 2x) \, dx$$

$$= -\frac{\cos 8x}{16} + \frac{\cos 2x}{4} + C. \qquad\blacksquare$$

Exercises 8.2

Powers of Sines and Cosines
Evaluate the integrals in Exercises 1–22.

1. $\displaystyle\int \cos 2x \, dx$

2. $\displaystyle\int_0^\pi 3 \sin \frac{x}{3} \, dx$

3. $\displaystyle\int \cos^3 x \sin x \, dx$

4. $\displaystyle\int \sin^4 2x \cos 2x \, dx$

5. $\displaystyle\int \sin^3 x \, dx$

6. $\displaystyle\int \cos^3 4x \, dx$

7. $\displaystyle\int \sin^5 x \, dx$

8. $\displaystyle\int_0^\pi \sin^5 \frac{x}{2} \, dx$

9. $\displaystyle\int \cos^3 x \, dx$

10. $\displaystyle\int_0^{\pi/6} 3 \cos^5 3x \, dx$

11. $\displaystyle\int \sin^3 x \cos^3 x \, dx$

12. $\displaystyle\int \cos^3 2x \sin^5 2x \, dx$

13. $\displaystyle\int \cos^2 x \, dx$

14. $\displaystyle\int_0^{\pi/2} \sin^2 x \, dx$

15. $\displaystyle\int_0^{\pi/2} \sin^7 y \, dy$

16. $\displaystyle\int 7 \cos^7 t \, dt$

17. $\displaystyle\int_0^\pi 8 \sin^4 x \, dx$

18. $\displaystyle\int 8 \cos^4 2\pi x \, dx$

19. $\displaystyle\int 16 \sin^2 x \cos^2 x \, dx$

20. $\displaystyle\int_0^\pi 8 \sin^4 y \cos^2 y \, dy$

21. $\displaystyle\int 8 \cos^3 2\theta \sin 2\theta \, d\theta$

22. $\displaystyle\int_0^{\pi/2} \sin^2 2\theta \cos^3 2\theta \, d\theta$

Integrating Square Roots
Evaluate the integrals in Exercises 23–32.

23. $\displaystyle\int_0^{2\pi} \sqrt{\frac{1 - \cos x}{2}} \, dx$

24. $\displaystyle\int_0^\pi \sqrt{1 - \cos 2x} \, dx$

25. $\displaystyle\int_0^\pi \sqrt{1 - \sin^2 t} \, dt$

26. $\displaystyle\int_0^\pi \sqrt{1 - \cos^2 \theta} \, d\theta$

27. $\displaystyle\int_{\pi/3}^{\pi/2} \frac{\sin^2 x}{\sqrt{1 - \cos x}} \, dx$

28. $\displaystyle\int_0^{\pi/6} \sqrt{1 + \sin x} \, dx$

$\left(\textit{Hint: Multiply by } \sqrt{\dfrac{1 - \sin x}{1 - \sin x}}.\right)$

29. $\displaystyle\int_{5\pi/6}^\pi \frac{\cos^4 x}{\sqrt{1 - \sin x}} \, dx$

30. $\displaystyle\int_{\pi/2}^{3\pi/4} \sqrt{1 - \sin 2x} \, dx$

31. $\displaystyle\int_0^{\pi/2} \theta \sqrt{1 - \cos 2\theta} \, d\theta$

32. $\displaystyle\int_{-\pi}^\pi (1 - \cos^2 t)^{3/2} \, dt$

Powers of Tangents and Secants
Evaluate the integrals in Exercises 33–50.

33. $\displaystyle\int \sec^2 x \tan x \, dx$

34. $\displaystyle\int \sec x \tan^2 x \, dx$

35. $\displaystyle\int \sec^3 x \tan x \, dx$

36. $\displaystyle\int \sec^3 x \tan^3 x \, dx$

37. $\displaystyle\int \sec^2 x \tan^2 x \, dx$

38. $\displaystyle\int \sec^4 x \tan^2 x \, dx$

39. $\displaystyle\int_{-\pi/3}^0 2 \sec^3 x \, dx$

40. $\displaystyle\int e^x \sec^3 e^x \, dx$

41. $\displaystyle\int \sec^4 \theta \, d\theta$

42. $\displaystyle\int 3 \sec^4 3x \, dx$

43. $\displaystyle\int_{\pi/4}^{\pi/2} \csc^4 \theta \, d\theta$

44. $\displaystyle\int \sec^6 x \, dx$

45. $\displaystyle\int 4 \tan^3 x \, dx$

46. $\displaystyle\int_{-\pi/4}^{\pi/4} 6 \tan^4 x \, dx$

47. $\displaystyle\int \tan^5 x \, dx$

48. $\displaystyle\int \cot^6 2x \, dx$

49. $\displaystyle\int_{\pi/6}^{\pi/3} \cot^3 x \, dx$

50. $\displaystyle\int 8 \cot^4 t \, dt$

2. Let $x^2 + px + q$ be an irreducible quadratic factor of $g(x)$ so that $x^2 + px + q$ has no real roots. Suppose that $(x^2 + px + q)^n$ is the highest power of this factor that divides $g(x)$. Then, to this factor, assign the sum of the n partial fractions:

$$\frac{B_1 x + C_1}{(x^2 + px + q)} + \frac{B_2 x + C_2}{(x^2 + px + q)^2} + \cdots + \frac{B_n x + C_n}{(x^2 + px + q)^n}.$$

Do this for each distinct quadratic factor of $g(x)$.

3. Set the original fraction $f(x)/g(x)$ equal to the sum of all these partial fractions. Clear the resulting equation of fractions and arrange the terms in decreasing powers of x.

4. Equate the coefficients of corresponding powers of x and solve the resulting equations for the undetermined coefficients.

EXAMPLE 1 Use partial fractions to evaluate

$$\int \frac{x^2 + 4x + 1}{(x - 1)(x + 1)(x + 3)}\, dx.$$

Solution The partial fraction decomposition has the form

$$\frac{x^2 + 4x + 1}{(x - 1)(x + 1)(x + 3)} = \frac{A}{x - 1} + \frac{B}{x + 1} + \frac{C}{x + 3}.$$

To find the values of the undetermined coefficients A, B, and C, we clear fractions and get

$$\begin{aligned}
x^2 + 4x + 1 &= A(x + 1)(x + 3) + B(x - 1)(x + 3) + C(x - 1)(x + 1) \\
&= A(x^2 + 4x + 3) + B(x^2 + 2x - 3) + C(x^2 - 1) \\
&= (A + B + C)x^2 + (4A + 2B)x + (3A - 3B - C).
\end{aligned}$$

The polynomials on both sides of the above equation are identical, so we equate coefficients of like powers of x, obtaining

$$\begin{array}{lll}
\text{Coefficient of } x^2: & A + B + C = 1 \\
\text{Coefficient of } x^1: & 4A + 2B = 4 \\
\text{Coefficient of } x^0: & 3A - 3B - C = 1
\end{array}$$

There are several ways of solving such a system of linear equations for the unknowns A, B, and C, including elimination of variables or the use of a calculator or computer. Whatever method is used, the solution is $A = 3/4$, $B = 1/2$, and $C = -1/4$. Hence we have

$$\int \frac{x^2 + 4x + 1}{(x - 1)(x + 1)(x + 3)}\, dx = \int \left[\frac{3}{4} \frac{1}{x - 1} + \frac{1}{2} \frac{1}{x + 1} - \frac{1}{4} \frac{1}{x + 3} \right] dx$$

$$= \frac{3}{4} \ln |x - 1| + \frac{1}{2} \ln |x + 1| - \frac{1}{4} \ln |x + 3| + K,$$

where K is the arbitrary constant of integration (to avoid confusion with the undetermined coefficient we labeled as C). ∎

EXAMPLE 2 Use partial fractions to evaluate

$$\int \frac{6x + 7}{(x + 2)^2}\, dx.$$

Solution   First we express the integrand as a sum of partial fractions with undetermined coefficients.

$$\frac{6x + 7}{(x + 2)^2} = \frac{A}{x + 2} + \frac{B}{(x + 2)^2}$$

$$6x + 7 = A(x + 2) + B \qquad \text{Multiply both sides by } (x + 2)^2.$$

$$= Ax + (2A + B)$$

Equating coefficients of corresponding powers of x gives

$$A = 6 \quad \text{and} \quad 2A + B = 12 + B = 7, \quad \text{or} \quad A = 6 \quad \text{and} \quad B = -5.$$

Therefore,

$$\int \frac{6x + 7}{(x + 2)^2} \, dx = \int \left(\frac{6}{x + 2} - \frac{5}{(x + 2)^2} \right) dx$$

$$= 6 \int \frac{dx}{x + 2} - 5 \int (x + 2)^{-2} \, dx$$

$$= 6 \ln |x + 2| + 5(x + 2)^{-1} + C. \qquad \blacksquare$$

EXAMPLE 3   Use partial fractions to evaluate

$$\int \frac{2x^3 - 4x^2 - x - 3}{x^2 - 2x - 3} \, dx.$$

Solution   First we divide the denominator into the numerator to get a polynomial plus a proper fraction.

$$\begin{array}{r} 2x \\ x^2 - 2x - 3 \overline{) 2x^3 - 4x^2 - x - 3} \\ \underline{2x^3 - 4x^2 - 6x } \\ 5x - 3 \end{array}$$

Then we write the improper fraction as a polynomial plus a proper fraction.

$$\frac{2x^3 - 4x^2 - x - 3}{x^2 - 2x - 3} = 2x + \frac{5x - 3}{x^2 - 2x - 3}$$

We found the partial fraction decomposition of the fraction on the right in the opening example, so

$$\int \frac{2x^3 - 4x^2 - x - 3}{x^2 - 2x - 3} \, dx = \int 2x \, dx + \int \frac{5x - 3}{x^2 - 2x - 3} \, dx$$

$$= \int 2x \, dx + \int \frac{2}{x + 1} \, dx + \int \frac{3}{x - 3} \, dx$$

$$= x^2 + 2 \ln |x + 1| + 3 \ln |x - 3| + C. \qquad \blacksquare$$

EXAMPLE 4   Use partial fractions to evaluate

$$\int \frac{-2x + 4}{(x^2 + 1)(x - 1)^2} \, dx.$$

Solution   The denominator has an irreducible quadratic factor as well as a repeated linear factor, so we write

$$\frac{-2x + 4}{(x^2 + 1)(x - 1)^2} = \frac{Ax + B}{x^2 + 1} + \frac{C}{x - 1} + \frac{D}{(x - 1)^2}. \qquad (2)$$

Clearing the equation of fractions gives

$$-2x + 4 = (Ax + B)(x - 1)^2 + C(x - 1)(x^2 + 1) + D(x^2 + 1)$$

$$= (A + C)x^3 + (-2A + B - C + D)x^2$$

$$+ (A - 2B + C)x + (B - C + D).$$

Equating coefficients of like terms gives

Coefficients of x^3: $0 = A + C$

Coefficients of x^2: $0 = -2A + B - C + D$

Coefficients of x^1: $-2 = A - 2B + C$

Coefficients of x^0: $4 = B - C + D$

We solve these equations simultaneously to find the values of A, B, C, and D:

$$-4 = -2A, \quad A = 2 \qquad \text{Subtract fourth equation from second.}$$

$$C = -A = -2 \qquad \text{From the first equation}$$

$$B = (A + C + 2)/2 = 1 \qquad \text{From the third equation and } C = -A$$

$$D = 4 - B + C = 1. \qquad \text{From the fourth equation}$$

We substitute these values into Equation (2), obtaining

$$\frac{-2x + 4}{(x^2 + 1)(x - 1)^2} = \frac{2x + 1}{x^2 + 1} - \frac{2}{x - 1} + \frac{1}{(x - 1)^2}.$$

Finally, using the expansion above we can integrate:

$$\int \frac{-2x + 4}{(x^2 + 1)(x - 1)^2} \, dx = \int \left(\frac{2x + 1}{x^2 + 1} - \frac{2}{x - 1} + \frac{1}{(x - 1)^2} \right) dx$$

$$= \int \left(\frac{2x}{x^2 + 1} + \frac{1}{x^2 + 1} - \frac{2}{x - 1} + \frac{1}{(x - 1)^2} \right) dx$$

$$= \ln (x^2 + 1) + \tan^{-1} x - 2 \ln |x - 1| - \frac{1}{x - 1} + C. \quad \blacksquare$$

EXAMPLE 5   Use partial fractions to evaluate

$$\int \frac{dx}{x(x^2 + 1)^2}.$$

Solution   The form of the partial fraction decomposition is

$$\frac{1}{x(x^2 + 1)^2} = \frac{A}{x} + \frac{Bx + C}{x^2 + 1} + \frac{Dx + E}{(x^2 + 1)^2}.$$

Multiplying by $x(x^2 + 1)^2$, we have

$$1 = A(x^2 + 1)^2 + (Bx + C)x(x^2 + 1) + (Dx + E)x$$

$$= A(x^4 + 2x^2 + 1) + B(x^4 + x^2) + C(x^3 + x) + Dx^2 + Ex$$

$$= (A + B)x^4 + Cx^3 + (2A + B + D)x^2 + (C + E)x + A$$

If we equate coefficients, we get the system

$$A + B = 0, \quad C = 0, \quad 2A + B + D = 0, \quad C + E = 0, \quad A = 1.$$

Solving this system gives $A = 1, B = -1, C = 0, D = -1$, and $E = 0$. Thus,

$$\int \frac{dx}{x(x^2 + 1)^2} = \int \left[\frac{1}{x} + \frac{-x}{x^2 + 1} + \frac{-x}{(x^2 + 1)^2} \right] dx$$

$$= \int \frac{dx}{x} - \int \frac{x \, dx}{x^2 + 1} - \int \frac{x \, dx}{(x^2 + 1)^2}$$

$$= \int \frac{dx}{x} - \frac{1}{2} \int \frac{du}{u} - \frac{1}{2} \int \frac{du}{u^2} \qquad \begin{array}{l} u = x^2 + 1, \\ du = 2x \, dx \end{array}$$

$$= \ln |x| - \frac{1}{2} \ln |u| + \frac{1}{2u} + K$$

$$= \ln |x| - \frac{1}{2} \ln (x^2 + 1) + \frac{1}{2(x^2 + 1)} + K$$

$$= \ln \frac{|x|}{\sqrt{x^2 + 1}} + \frac{1}{2(x^2 + 1)} + K. \qquad ■$$

HISTORICAL BIOGRAPHY

Oliver Heaviside
(1850–1925)

The Heaviside "Cover-up" Method for Linear Factors

When the degree of the polynomial $f(x)$ is less than the degree of $g(x)$ and

$$g(x) = (x - r_1)(x - r_2) \cdots (x - r_n)$$

is a product of n distinct linear factors, each raised to the first power, there is a quick way to expand $f(x)/g(x)$ by partial fractions.

EXAMPLE 6   Find A, B, and C in the partial fraction expansion

$$\frac{x^2 + 1}{(x - 1)(x - 2)(x - 3)} = \frac{A}{x - 1} + \frac{B}{x - 2} + \frac{C}{x - 3}. \qquad (3)$$

Solution   If we multiply both sides of Equation (3) by $(x - 1)$ to get

$$\frac{x^2 + 1}{(x - 2)(x - 3)} = A + \frac{B(x - 1)}{x - 2} + \frac{C(x - 1)}{x - 3}$$

and set $x = 1$, the resulting equation gives the value of A:

$$\frac{(1)^2 + 1}{(1 - 2)(1 - 3)} = A + 0 + 0,$$

$$A = 1.$$

Thus, the value of A is the number we would have obtained if we had covered the factor $(x - 1)$ in the denominator of the original fraction

$$\frac{x^2 + 1}{(x - 1)(x - 2)(x - 3)} \qquad (4)$$

and evaluated the rest at $x = 1$:

$$A = \frac{(1)^2 + 1}{\boxed{(x - 1)} \, (1 - 2)(1 - 3)} = \frac{2}{(-1)(-2)} = 1.$$

⇑
Cover

Similarly, we find the value of B in Equation (3) by covering the factor $(x - 2)$ in Expression (4) and evaluating the rest at $x = 2$:

$$B = \frac{(2)^2 + 1}{(2 - 1)\,\boxed{(x - 2)}\,(2 - 3)} = \frac{5}{(1)(-1)} = -5.$$

$$\uparrow$$
$$\text{Cover}$$

Finally, C is found by covering the $(x - 3)$ in Expression (4) and evaluating the rest at $x = 3$:

$$C = \frac{(3)^2 + 1}{(3 - 1)(3 - 2)\,\boxed{(x - 3)}} = \frac{10}{(2)(1)} = 5. \qquad \blacksquare$$

$$\uparrow$$
$$\text{Cover}$$

Heaviside Method

1. *Write the quotient with $g(x)$ factored:*

$$\frac{f(x)}{g(x)} = \frac{f(x)}{(x - r_1)(x - r_2)\cdots(x - r_n)}.$$

2. *Cover the factors $(x - r_i)$ of $g(x)$ one at a time*, each time replacing all the uncovered x's by the number r_i. This gives a number A_i for each root r_i:

$$A_1 = \frac{f(r_1)}{(r_1 - r_2)\cdots(r_1 - r_n)}$$

$$A_2 = \frac{f(r_2)}{(r_2 - r_1)(r_2 - r_3)\cdots(r_2 - r_n)}$$

$$\vdots$$

$$A_n = \frac{f(r_n)}{(r_n - r_1)(r_n - r_2)\cdots(r_n - r_{n-1})}.$$

3. *Write the partial fraction expansion of $f(x)/g(x)$ as*

$$\frac{f(x)}{g(x)} = \frac{A_1}{(x - r_1)} + \frac{A_2}{(x - r_2)} + \cdots + \frac{A_n}{(x - r_n)}.$$

EXAMPLE 7   Use the Heaviside Method to evaluate

$$\int \frac{x + 4}{x^3 + 3x^2 - 10x}\,dx.$$

Solution   The degree of $f(x) = x + 4$ is less than the degree of the cubic polynomial $g(x) = x^3 + 3x^2 - 10x$, and, with $g(x)$ factored,

$$\frac{x + 4}{x^3 + 3x^2 - 10x} = \frac{x + 4}{x(x - 2)(x + 5)}.$$

The roots of $g(x)$ are $r_1 = 0$, $r_2 = 2$, and $r_3 = -5$. We find

$$A_1 = \frac{0 + 4}{\boxed{x}\,(0 - 2)(0 + 5)} = \frac{4}{(-2)(5)} = -\frac{2}{5}$$

$\Uparrow$
Cover

$$A_2 = \frac{2 + 4}{2\,\boxed{(x - 2)}\,(2 + 5)} = \frac{6}{(2)(7)} = \frac{3}{7}$$

$\Uparrow$
Cover

$$A_3 = \frac{-5 + 4}{(-5)(-5 - 2)\,\boxed{(x + 5)}} = \frac{-1}{(-5)(-7)} = -\frac{1}{35}.$$

$\Uparrow$
Cover

Therefore,

$$\frac{x + 4}{x(x - 2)(x + 5)} = -\frac{2}{5x} + \frac{3}{7(x - 2)} - \frac{1}{35(x + 5)},$$

and

$$\int \frac{x + 4}{x(x - 2)(x + 5)}\, dx = -\frac{2}{5} \ln |x| + \frac{3}{7} \ln |x - 2| - \frac{1}{35} \ln |x + 5| + C. \qquad \blacksquare$$

Other Ways to Determine the Coefficients

Another way to determine the constants that appear in partial fractions is to differentiate, as in the next example. Still another is to assign selected numerical values to x.

EXAMPLE 8   Find A, B, and C in the equation

$$\frac{x - 1}{(x + 1)^3} = \frac{A}{x + 1} + \frac{B}{(x + 1)^2} + \frac{C}{(x + 1)^3}$$

by clearing fractions, differentiating the result, and substituting $x = -1$.

Solution   We first clear fractions:

$$x - 1 = A(x + 1)^2 + B(x + 1) + C.$$

Substituting $x = -1$ shows $C = -2$. We then differentiate both sides with respect to x, obtaining

$$1 = 2A(x + 1) + B.$$

Substituting $x = -1$ shows $B = 1$. We differentiate again to get $0 = 2A$, which shows $A = 0$. Hence,

$$\frac{x - 1}{(x + 1)^3} = \frac{1}{(x + 1)^2} - \frac{2}{(x + 1)^3}. \qquad \blacksquare$$

In some problems, assigning small values to x, such as $x = 0, \pm 1, \pm 2$, to get equations in A, B, and C provides a fast alternative to other methods.

EXAMPLE 9   Find A, B, and C in the expression

$$\frac{x^2 + 1}{(x - 1)(x - 2)(x - 3)} = \frac{A}{x - 1} + \frac{B}{x - 2} + \frac{C}{x - 3}$$

by assigning numerical values to x.

Solution Clear fractions to get

$$x^2 + 1 = A(x - 2)(x - 3) + B(x - 1)(x - 3) + C(x - 1)(x - 2).$$

Then let $x = 1, 2, 3$ successively to find A, B, and C:

$$
\begin{aligned}
x = 1: \quad & (1)^2 + 1 = A(-1)(-2) + B(0) + C(0) \\
& 2 = 2A \\
& A = 1 \\
x = 2: \quad & (2)^2 + 1 = A(0) + B(1)(-1) + C(0) \\
& 5 = -B \\
& B = -5 \\
x = 3: \quad & (3)^2 + 1 = A(0) + B(0) + C(2)(1) \\
& 10 = 2C \\
& C = 5.
\end{aligned}
$$

Conclusion:

$$\frac{x^2 + 1}{(x - 1)(x - 2)(x - 3)} = \frac{1}{x - 1} - \frac{5}{x - 2} + \frac{5}{x - 3}. \quad \blacksquare$$

Exercises 8.4

Expanding Quotients into Partial Fractions
Expand the quotients in Exercises 1–8 by partial fractions.

1. $\dfrac{5x - 13}{(x - 3)(x - 2)}$

2. $\dfrac{5x - 7}{x^2 - 3x + 2}$

3. $\dfrac{x + 4}{(x + 1)^2}$

4. $\dfrac{2x + 2}{x^2 - 2x + 1}$

5. $\dfrac{z + 1}{z^2(z - 1)}$

6. $\dfrac{z}{z^3 - z^2 - 6z}$

7. $\dfrac{t^2 + 8}{t^2 - 5t + 6}$

8. $\dfrac{t^4 + 9}{t^4 + 9t^2}$

Nonrepeated Linear Factors
In Exercises 9–16, express the integrand as a sum of partial fractions and evaluate the integrals.

9. $\displaystyle\int \frac{dx}{1 - x^2}$

10. $\displaystyle\int \frac{dx}{x^2 + 2x}$

11. $\displaystyle\int \frac{x + 4}{x^2 + 5x - 6} \, dx$

12. $\displaystyle\int \frac{2x + 1}{x^2 - 7x + 12} \, dx$

13. $\displaystyle\int_4^8 \frac{y \, dy}{y^2 - 2y - 3}$

14. $\displaystyle\int_{1/2}^1 \frac{y + 4}{y^2 + y} \, dy$

15. $\displaystyle\int \frac{dt}{t^3 + t^2 - 2t}$

16. $\displaystyle\int \frac{x + 3}{2x^3 - 8x} \, dx$

Repeated Linear Factors
In Exercises 17–20, express the integrand as a sum of partial fractions and evaluate the integrals.

17. $\displaystyle\int_0^1 \frac{x^3 \, dx}{x^2 + 2x + 1}$

18. $\displaystyle\int_{-1}^0 \frac{x^3 \, dx}{x^2 - 2x + 1}$

19. $\displaystyle\int \frac{dx}{(x^2 - 1)^2}$

20. $\displaystyle\int \frac{x^2 \, dx}{(x - 1)(x^2 + 2x + 1)}$

Irreducible Quadratic Factors
In Exercises 21–32, express the integrand as a sum of partial fractions and evaluate the integrals.

21. $\displaystyle\int_0^1 \frac{dx}{(x + 1)(x^2 + 1)}$

22. $\displaystyle\int_1^{\sqrt{3}} \frac{3t^2 + t + 4}{t^3 + t} \, dt$

23. $\displaystyle\int \frac{y^2 + 2y + 1}{(y^2 + 1)^2} \, dy$

24. $\displaystyle\int \frac{8x^2 + 8x + 2}{(4x^2 + 1)^2} \, dx$

25. $\displaystyle\int \frac{2s + 2}{(s^2 + 1)(s - 1)^3} \, ds$

26. $\displaystyle\int \frac{s^4 + 81}{s(s^2 + 9)^2} \, ds$

27. $\displaystyle\int \frac{x^2 - x + 2}{x^3 - 1} \, dx$

28. $\displaystyle\int \frac{1}{x^4 + x} \, dx$

29. $\displaystyle\int \frac{x^2}{x^4 - 1} \, dx$

30. $\displaystyle\int \frac{x^2 + x}{x^4 - 3x^2 - 4} \, dx$

31. $\displaystyle\int \frac{2\theta^3 + 5\theta^2 + 8\theta + 4}{(\theta^2 + 2\theta + 2)^2} \, d\theta$

32. $\displaystyle\int \frac{\theta^4 - 4\theta^3 + 2\theta^2 - 3\theta + 1}{(\theta^2 + 1)^3} \, d\theta$

Improper Fractions
In Exercises 33–38, perform long division on the integrand, write the proper fraction as a sum of partial fractions, and then evaluate the integral.

33. $\displaystyle\int \frac{2x^3 - 2x^2 + 1}{x^2 - x} \, dx$

34. $\displaystyle\int \frac{x^4}{x^2 - 1} \, dx$

35. $\displaystyle\int \frac{9x^3 - 3x + 1}{x^3 - x^2}\, dx$ **36.** $\displaystyle\int \frac{16x^3}{4x^2 - 4x + 1}\, dx$

37. $\displaystyle\int \frac{y^4 + y^2 - 1}{y^3 + y}\, dy$ **38.** $\displaystyle\int \frac{2y^4}{y^3 - y^2 + y - 1}\, dy$

Evaluating Integrals

Evaluate the integrals in Exercises 39–50.

39. $\displaystyle\int \frac{e^t\, dt}{e^{2t} + 3e^t + 2}$ **40.** $\displaystyle\int \frac{e^{4t} + 2e^{2t} - e^t}{e^{2t} + 1}\, dt$

41. $\displaystyle\int \frac{\cos y\, dy}{\sin^2 y + \sin y - 6}$ **42.** $\displaystyle\int \frac{\sin\theta\, d\theta}{\cos^2\theta + \cos\theta - 2}$

43. $\displaystyle\int \frac{(x - 2)^2 \tan^{-1}(2x) - 12x^3 - 3x}{(4x^2 + 1)(x - 2)^2}\, dx$

44. $\displaystyle\int \frac{(x + 1)^2 \tan^{-1}(3x) + 9x^3 + x}{(9x^2 + 1)(x + 1)^2}\, dx$

45. $\displaystyle\int \frac{1}{x^{3/2} - \sqrt{x}}\, dx$ **46.** $\displaystyle\int \frac{1}{(x^{1/3} - 1)\sqrt{x}}\, dx$

(*Hint*: Let $x = u^6$.)

47. $\displaystyle\int \frac{\sqrt{x + 1}}{x}\, dx$ **48.** $\displaystyle\int \frac{1}{x\sqrt{x + 9}}\, dx$

(*Hint*: Let $x + 1 = u^2$.)

49. $\displaystyle\int \frac{1}{x(x^4 + 1)}\, dx$ **50.** $\displaystyle\int \frac{1}{x^6(x^5 + 4)}\, dx$

$\left(\text{\textit{Hint}: Multiply by }\dfrac{x^3}{x^3}.\right)$

Initial Value Problems

Solve the initial value problems in Exercises 51–54 for x as a function of t.

51. $(t^2 - 3t + 2)\dfrac{dx}{dt} = 1 \quad (t > 2), \quad x(3) = 0$

52. $(3t^4 + 4t^2 + 1)\dfrac{dx}{dt} = 2\sqrt{3}, \quad x(1) = -\pi\sqrt{3}/4$

53. $(t^2 + 2t)\dfrac{dx}{dt} = 2x + 2 \quad (t, x > 0), \quad x(1) = 1$

54. $(t + 1)\dfrac{dx}{dt} = x^2 + 1 \quad (t > -1), \quad x(0) = 0$

Applications and Examples

In Exercises 55 and 56, find the volume of the solid generated by revolving the shaded region about the indicated axis.

55. The x-axis

56. The y-axis

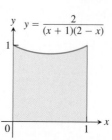

T 57. Find, to two decimal places, the x-coordinate of the centroid of the region in the first quadrant bounded by the x-axis, the curve $y = \tan^{-1} x$, and the line $x = \sqrt{3}$.

T 58. Find the x-coordinate of the centroid of this region to two decimal places.

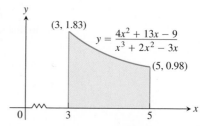

T 59. **Social diffusion** Sociologists sometimes use the phrase "social diffusion" to describe the way information spreads through a population. The information might be a rumor, a cultural fad, or news about a technical innovation. In a sufficiently large population, the number of people x who have the information is treated as a differentiable function of time t, and the rate of diffusion, dx/dt, is assumed to be proportional to the number of people who have the information times the number of people who do not. This leads to the equation

$$\frac{dx}{dt} = kx(N - x),$$

where N is the number of people in the population.

Suppose t is in days, $k = 1/250$, and two people start a rumor at time $t = 0$ in a population of $N = 1000$ people.

a. Find x as a function of t.

b. When will half the population have heard the rumor? (This is when the rumor will be spreading the fastest.)

T 60. **Second-order chemical reactions** Many chemical reactions are the result of the interaction of two molecules that undergo a change to produce a new product. The rate of the reaction typically depends on the concentrations of the two kinds of molecules. If a is the amount of substance A and b is the amount of substance B at time $t = 0$, and if x is the amount of product at time t, then the rate of formation of x may be given by the differential equation

$$\frac{dx}{dt} = k(a - x)(b - x),$$

or

$$\frac{1}{(a - x)(b - x)}\frac{dx}{dt} = k,$$

where k is a constant for the reaction. Integrate both sides of this equation to obtain a relation between x and t **(a)** if $a = b$, and **(b)** if $a \neq b$. Assume in each case that $x = 0$ when $t = 0$.

for some number c between a and b. Thus, as Δx approaches zero, the error defined by

$$E_T = -\frac{b-a}{12} \cdot f''(c)(\Delta x)^2$$

approaches zero as the *square* of Δx.

The inequality

$$|E_T| \le \frac{b-a}{12} \max|f''(x)|(\Delta x)^2$$

where max refers to the interval $[a, b]$, gives an upper bound for the magnitude of the error. In practice, we usually cannot find the exact value of $\max|f''(x)|$ and have to estimate an upper bound or "worst case" value for it instead. If M is any upper bound for the values of $|f''(x)|$ on $[a, b]$, so that $|f''(x)| \le M$ on $[a, b]$, then

$$|E_T| \le \frac{b-a}{12} M(\Delta x)^2.$$

If we substitute $(b-a)/n$ for Δx, we get

$$|E_T| \le \frac{M(b-a)^3}{12n^2}.$$

To estimate the error in Simpson's rule, we start with a result from advanced calculus that says that if the fourth derivative $f^{(4)}$ is continuous, then

$$\int_a^b f(x)\,dx = S - \frac{b-a}{180} \cdot f^{(4)}(c)(\Delta x)^4$$

for some point c between a and b. Thus, as Δx approaches zero, the error,

$$E_S = -\frac{b-a}{180} \cdot f^{(4)}(c)(\Delta x)^4,$$

approaches zero as the *fourth power* of Δx. (This helps to explain why Simpson's Rule is likely to give better results than the Trapezoidal Rule.)

The inequality

$$|E_S| \le \frac{b-a}{180} \max|f^{(4)}(x)|(\Delta x)^4,$$

where max refers to the interval $[a, b]$, gives an upper bound for the magnitude of the error. As with $\max|f''|$ in the error formula for the Trapezoidal Rule, we usually cannot find the exact value of $\max|f^{(4)}(x)|$ and have to replace it with an upper bound. If M is any upper bound for the values of $|f^{(4)}|$ on $[a, b]$, then

$$|E_S| \le \frac{b-a}{180} M(\Delta x)^4.$$

Substituting $(b-a)/n$ for Δx in this last expression gives

$$|E_S| \le \frac{M(b-a)^5}{180n^4}.$$

EXAMPLE 3   Find an upper bound for the error in estimating $\int_0^2 5x^4\,dx$ using Simpson's Rule with $n = 4$ (Example 2).

Solution   To estimate the error, we first find an upper bound M for the magnitude of the fourth derivative of $f(x) = 5x^4$ on the interval $0 \le x \le 2$. Since the fourth derivative has the constant value $f^{(4)}(x) = 120$, we take $M = 120$. With $b - a = 2$ and $n = 4$, the error estimate for Simpson's Rule gives

$$|E_S| \leq \frac{M(b - a)^5}{180n^4} = \frac{120(2)^5}{180 \cdot 4^4} = \frac{1}{12}.$$

This estimate is consistent with the result of Example 2. ■

Theorem 1 can also be used to estimate the number of subintervals required when using the Trapezoidal or Simpson's Rules if we specify a certain tolerance for the error.

EXAMPLE 4 Estimate the minimum number of subintervals needed to approximate the integral in Example 3 using Simpson's Rule with an error of magnitude less than 10^{-4}.

Solution Using the inequality in Theorem 1, if we choose the number of subintervals n to satisfy

$$\frac{M(b - a)^5}{180n^4} < 10^{-4},$$

then the error E_S in Simpson's Rule satisfies $|E_S| < 10^{-4}$ as required.

From the solution in Example 3, we have $M = 120$ and $b - a = 2$, so we want n to satisfy

$$\frac{120(2)^5}{180n^4} < \frac{1}{10^4}$$

or, equivalently,

$$n^4 > \frac{64 \cdot 10^4}{3}.$$

It follows that

$$n > 10\left(\frac{64}{3}\right)^{1/4} \approx 21.5.$$

Since n must be even in Simpson's Rule, we estimate the minimum number of subintervals required for the error tolerance to be $n = 22$. ■

EXAMPLE 5 As we saw in Chapter 7, the value of ln 2 can be calculated from the integral

$$\ln 2 = \int_1^2 \frac{1}{x} \, dx.$$

Table 8.4 shows T and S values for approximations of $\int_1^2 (1/x) \, dx$ using various values of n. Notice how Simpson's Rule dramatically improves over the Trapezoidal Rule.

TABLE 8.4 Trapezoidal Rule approximations (T_n) and Simpson's Rule approximations (S_n) of $\ln 2 = \int_1^2 (1/x) \, dx$

n	T_n	|Error| less than...	S_n	|Error| less than...
10	0.6937714032	0.0006242227	0.6931502307	0.0000030502
20	0.6933033818	0.0001562013	0.6931473747	0.0000001942
30	0.6932166154	0.0000694349	0.6931472190	0.0000000385
40	0.6931862400	0.0000390595	0.6931471927	0.0000000122
50	0.6931721793	0.0000249988	0.6931471856	0.0000000050
100	0.6931534305	0.0000062500	0.6931471809	0.0000000004

In particular, notice that when we double the value of n (thereby halving the value of $h = \Delta x$), the T error is divided by 2 *squared*, whereas the S error is divided by 2 *to the fourth*.

This has a dramatic effect as $\Delta x = (2 - 1)/n$ gets very small. The Simpson approximation for $n = 50$ rounds accurately to seven places and for $n = 100$ agrees to nine decimal places (billionths)! ∎

If $f(x)$ is a polynomial of degree less than four, then its fourth derivative is zero, and

$$E_S = -\frac{b - a}{180} f^{(4)}(c)(\Delta x)^4 = -\frac{b - a}{180}(0)(\Delta x)^4 = 0.$$

Thus, there will be no error in the Simpson approximation of any integral of f. In other words, if f is a constant, a linear function, or a quadratic or cubic polynomial, Simpson's Rule will give the value of any integral of f exactly, whatever the number of subdivisions. Similarly, if f is a constant or a linear function, then its second derivative is zero, and

$$E_T = -\frac{b - a}{12} f''(c)(\Delta x)^2 = -\frac{b - a}{12}(0)(\Delta x)^2 = 0.$$

The Trapezoidal Rule will therefore give the exact value of any integral of f. This is no surprise, for the trapezoids fit the graph perfectly.

Although decreasing the step size Δx reduces the error in the Simpson and Trapezoidal approximations in theory, it may fail to do so in practice. When Δx is very small, say $\Delta x = 10^{-5}$, computer or calculator round-off errors in the arithmetic required to evaluate S and T may accumulate to such an extent that the error formulas no longer describe what is going on. Shrinking Δx below a certain size can actually make things worse. Although this is not an issue in this book, you should consult a text on numerical analysis for alternative methods if you are having problems with round-off.

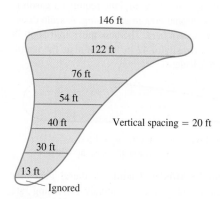

FIGURE 8.11 The dimensions of the swamp in Example 6.

EXAMPLE 6 A town wants to drain and fill a small polluted swamp (Figure 8.11). The swamp averages 5 ft deep. About how many cubic yards of dirt will it take to fill the area after the swamp is drained?

Solution To calculate the volume of the swamp, we estimate the surface area and multiply by 5. To estimate the area, we use Simpson's Rule with $\Delta x = 20$ ft and the y's equal to the distances measured across the swamp, as shown in Figure 8.11.

$$S = \frac{\Delta x}{3}(y_0 + 4y_1 + 2y_2 + 4y_3 + 2y_4 + 4y_5 + y_6)$$

$$= \frac{20}{3}(146 + 488 + 152 + 216 + 80 + 120 + 13) = 8100$$

The volume is about $(8100)(5) = 40{,}500 \text{ ft}^3$ or 1500 yd^3. ∎

Exercises 8.6

Estimating Integrals

The instructions for the integrals in Exercises 1–10 have two parts, one for the Trapezoidal Rule and one for Simpson's Rule.

I. Using the Trapezoidal Rule

a. Estimate the integral with $n = 4$ steps and find an upper bound for $|E_T|$.

b. Evaluate the integral directly and find $|E_T|$.

c. Use the formula $(|E_T|/(\text{true value})) \times 100$ to express $|E_T|$ as a percentage of the integral's true value.

II. Using Simpson's Rule

a. Estimate the integral with $n = 4$ steps and find an upper bound for $|E_S|$.

b. Evaluate the integral directly and find $|E_S|$.

c. Use the formula $(|E_S|/(\text{true value})) \times 100$ to express $|E_S|$ as a percentage of the integral's true value.

1. $\displaystyle\int_1^2 x \, dx$

2. $\displaystyle\int_1^3 (2x - 1) \, dx$

3. $\int_{-1}^{1} (x^2 + 1)\, dx$

4. $\int_{-2}^{0} (x^2 - 1)\, dx$

5. $\int_{0}^{2} (t^3 + t)\, dt$

6. $\int_{-1}^{1} (t^3 + 1)\, dt$

7. $\int_{1}^{2} \frac{1}{s^2}\, ds$

8. $\int_{2}^{4} \frac{1}{(s-1)^2}\, ds$

9. $\int_{0}^{\pi} \sin t\, dt$

10. $\int_{0}^{1} \sin \pi t\, dt$

Estimating the Number of Subintervals

In Exercises 11–22, estimate the minimum number of subintervals needed to approximate the integrals with an error of magnitude less than 10^{-4} by **(a)** the Trapezoidal Rule and **(b)** Simpson's Rule. (The integrals in Exercises 11–18 are the integrals from Exercises 1–8.)

11. $\int_{1}^{2} x\, dx$

12. $\int_{1}^{3} (2x - 1)\, dx$

13. $\int_{-1}^{1} (x^2 + 1)\, dx$

14. $\int_{-2}^{0} (x^2 - 1)\, dx$

15. $\int_{0}^{2} (t^3 + t)\, dt$

16. $\int_{-1}^{1} (t^3 + 1)\, dt$

17. $\int_{1}^{2} \frac{1}{s^2}\, ds$

18. $\int_{2}^{4} \frac{1}{(s-1)^2}\, ds$

19. $\int_{0}^{3} \sqrt{x+1}\, dx$

20. $\int_{0}^{3} \frac{1}{\sqrt{x+1}}\, dx$

21. $\int_{0}^{2} \sin (x + 1)\, dx$

22. $\int_{-1}^{1} \cos (x + \pi)\, dx$

Estimates with Numerical Data

23. Volume of water in a swimming pool A rectangular swimming pool is 30 ft wide and 50 ft long. The accompanying table shows the depth $h(x)$ of the water at 5-ft intervals from one end of the pool to the other. Estimate the volume of water in the pool using the Trapezoidal Rule with $n = 10$ applied to the integral

$$V = \int_{0}^{50} 30 \cdot h(x)\, dx.$$

Position (ft) x	Depth (ft) $h(x)$	Position (ft) x	Depth (ft) $h(x)$
0	6.0	30	11.5
5	8.2	35	11.9
10	9.1	40	12.3
15	9.9	45	12.7
20	10.5	50	13.0
25	11.0		

24. Distance traveled The accompanying table shows time-to-speed data for a sports car accelerating from rest to 130 mph. How far had the car traveled by the time it reached this speed? (Use trapezoids to estimate the area under the velocity curve, but be careful: The time intervals vary in length.)

Speed change	Time (sec)
Zero to 30 mph	2.2
40 mph	3.2
50 mph	4.5
60 mph	5.9
70 mph	7.8
80 mph	10.2
90 mph	12.7
100 mph	16.0
110 mph	20.6
120 mph	26.2
130 mph	37.1

25. Wing design The design of a new airplane requires a gasoline tank of constant cross-sectional area in each wing. A scale drawing of a cross-section is shown here. The tank must hold 5000 lb of gasoline, which has a density of 42 lb/ft³. Estimate the length of the tank by Simpson's Rule.

$y_0 = 1.5$ ft, $y_1 = 1.6$ ft, $y_2 = 1.8$ ft, $y_3 = 1.9$ ft,
$y_4 = 2.0$ ft, $y_5 = y_6 = 2.1$ ft   Horizontal spacing = 1 ft

26. Oil consumption on Pathfinder Island A diesel generator runs continuously, consuming oil at a gradually increasing rate until it must be temporarily shut down to have the filters replaced. Use the Trapezoidal Rule to estimate the amount of oil consumed by the generator during that week.

Day	Oil consumption rate (liters/h)
Sun	0.019
Mon	0.020
Tue	0.021
Wed	0.023
Thu	0.025
Fri	0.028
Sat	0.031
Sun	0.035

Theory and Examples

27. Usable values of the sine-integral function *The sine-integral function,*

$$\text{Si}(x) = \int_{0}^{x} \frac{\sin t}{t}\, dt, \qquad \text{"Sine integral of } x\text{"}$$

is one of the many functions in engineering whose formulas cannot be simplified. There is no elementary formula for the antiderivative of $(\sin t)/t$. The values of $\text{Si}(x)$, however, are readily estimated by numerical integration.

Although the notation does not show it explicitly, the function being integrated is

$$f(t) = \begin{cases} \dfrac{\sin t}{t}, & t \neq 0 \\[2mm] 1, & t = 0, \end{cases}$$

the continuous extension of $(\sin t)/t$ to the interval $[0, x]$. The function has derivatives of all orders at every point of its domain. Its graph is smooth, and you can expect good results from Simpson's Rule.

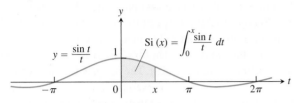

a. Use the fact that $|f^{(4)}| \leq 1$ on $[0, \pi/2]$ to give an upper bound for the error that will occur if

$$\text{Si}\left(\frac{\pi}{2}\right) = \int_0^{\pi/2} \frac{\sin t}{t}\, dt$$

is estimated by Simpson's Rule with $n = 4$.

b. Estimate $\text{Si}(\pi/2)$ by Simpson's Rule with $n = 4$.

c. Express the error bound you found in part (a) as a percentage of the value you found in part (b).

28. **The error function**   *The error function*,

$$\text{erf}(x) = \frac{2}{\sqrt{\pi}} \int_0^x e^{-t^2}\, dt,$$

important in probability and in the theories of heat flow and signal transmission, must be evaluated numerically because there is no elementary expression for the antiderivative of e^{-t^2}.

a. Use Simpson's Rule with $n = 10$ to estimate $\text{erf}(1)$.

b. In $[0, 1]$,

$$\left|\frac{d^4}{dt^4}\left(e^{-t^2}\right)\right| \leq 12.$$

Give an upper bound for the magnitude of the error of the estimate in part (a).

29. Prove that the sum T in the Trapezoidal Rule for $\int_a^b f(x)\, dx$ is a Riemann sum for f continuous on $[a, b]$. (*Hint:* Use the Intermediate Value Theorem to show the existence of c_k in the subinterval $[x_{k-1}, x_k]$ satisfying $f(c_k) = (f(x_{k-1}) + f(x_k))/2$.)

30. Prove that the sum S in Simpson's Rule for $\int_a^b f(x)\, dx$ is a Riemann sum for f continuous on $[a, b]$. (See Exercise 29.)

31. **Elliptic integrals**   The length of the ellipse

$$\frac{x^2}{a^2} + \frac{y^2}{b^2} = 1$$

turns out to be

$$\text{Length} = 4a \int_0^{\pi/2} \sqrt{1 - e^2 \cos^2 t}\, dt,$$

where $e = \sqrt{a^2 - b^2}/a$ is the ellipse's eccentricity. The integral in this formula, called an *elliptic integral*, is nonelementary except when $e = 0$ or 1.

a. Use the Trapezoidal Rule with $n = 10$ to estimate the length of the ellipse when $a = 1$ and $e = 1/2$.

b. Use the fact that the absolute value of the second derivative of $f(t) = \sqrt{1 - e^2 \cos^2 t}$ is less than 1 to find an upper bound for the error in the estimate you obtained in part (a).

Applications

32. The length of one arch of the curve $y = \sin x$ is given by

$$L = \int_0^{\pi} \sqrt{1 + \cos^2 x}\, dx.$$

Estimate L by Simpson's Rule with $n = 8$.

33. Your metal fabrication company is bidding for a contract to make sheets of corrugated iron roofing like the one shown here. The cross-sections of the corrugated sheets are to conform to the curve

$$y = \sin\frac{3\pi}{20}x, \quad 0 \leq x \leq 20 \text{ in.}$$

If the roofing is to be stamped from flat sheets by a process that does not stretch the material, how wide should the original material be? To find out, use numerical integration to approximate the length of the sine curve to two decimal places.

34. Your engineering firm is bidding for the contract to construct the tunnel shown here. The tunnel is 300 ft long and 50 ft wide at the base. The cross-section is shaped like one arch of the curve $y = 25 \cos(\pi x/50)$. Upon completion, the tunnel's inside surface (excluding the roadway) will be treated with a waterproof sealer that costs \$1.75 per square foot to apply. How much will it cost to apply the sealer? (*Hint:* Use numerical integration to find the length of the cosine curve.)

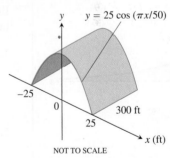

NOT TO SCALE

Find, to two decimal places, the areas of the surfaces generated by revolving the curves in Exercises 35 and 36 about the x-axis.

35. $y = \sin x, \quad 0 \leq x \leq \pi$

36. $y = x^2/4, \quad 0 \leq x \leq 2$

37. Use numerical integration to estimate the value of

$$\sin^{-1} 0.6 = \int_0^{0.6} \frac{dx}{\sqrt{1 - x^2}}.$$

For reference, $\sin^{-1} 0.6 = 0.64350$ to five decimal places.

38. Use numerical integration to estimate the value of

$$\pi = 4 \int_0^1 \frac{1}{1 + x^2}\, dx.$$

8.7 | Improper Integrals

FIGURE 8.12 Are the areas under these infinite curves finite? We will see that the answer is yes for both curves.

Up to now, we have required definite integrals to have two properties. First, that the domain of integration $[a, b]$ be finite. Second, that the range of the integrand be finite on this domain. In practice, we may encounter problems that fail to meet one or both of these conditions. The integral for the area under the curve $y = (\ln x)/x^2$ from $x = 1$ to $x = \infty$ is an example for which the domain is infinite (Figure 8.12a). The integral for the area under the curve of $y = 1/\sqrt{x}$ between $x = 0$ and $x = 1$ is an example for which the range of the integrand is infinite (Figure 8.12b). In either case, the integrals are said to be *improper* and are calculated as limits. We will see in Chapter 10 that improper integrals play an important role when investigating the convergence of certain infinite series.

Infinite Limits of Integration

Consider the infinite region that lies under the curve $y = e^{-x/2}$ in the first quadrant (Figure 8.13a). You might think this region has infinite area, but we will see that the value is finite. We assign a value to the area in the following way. First find the area $A(b)$ of the portion of the region that is bounded on the right by $x = b$ (Figure 8.13b).

$$A(b) = \int_0^b e^{-x/2} \, dx = -2e^{-x/2} \Big]_0^b = -2e^{-b/2} + 2$$

Then find the limit of $A(b)$ as $b \to \infty$

$$\lim_{b \to \infty} A(b) = \lim_{b \to \infty} (-2e^{-b/2} + 2) = 2.$$

The value we assign to the area under the curve from 0 to ∞ is

$$\int_0^{\infty} e^{-x/2} \, dx = \lim_{b \to \infty} \int_0^b e^{-x/2} \, dx = 2.$$

FIGURE 8.13 (a) The area in the first quadrant under the curve $y = e^{-x/2}$ (b) The area is an improper integral of the first type.

DEFINITION   Integrals with infinite limits of integration are **improper integrals of Type I**.

1. If $f(x)$ is continuous on $[a, \infty)$, then

$$\int_a^{\infty} f(x) \, dx = \lim_{b \to \infty} \int_a^b f(x) \, dx.$$

2. If $f(x)$ is continuous on $(-\infty, b]$, then

$$\int_{-\infty}^b f(x) \, dx = \lim_{a \to -\infty} \int_a^b f(x) \, dx.$$

3. If $f(x)$ is continuous on $(-\infty, \infty)$, then

$$\int_{-\infty}^{\infty} f(x) \, dx = \int_{-\infty}^c f(x) \, dx + \int_c^{\infty} f(x) \, dx,$$

where c is any real number.

In each case, if the limit is finite we say that the improper integral **converges** and that the limit is the **value** of the improper integral. If the limit fails to exist, the improper integral **diverges**.

We omit the more advanced proof of Theorem 3.

Although the improper integrals of two functions from a to ∞ may both converge, this does not mean that their integrals necessarily have the same value, as the next example shows.

EXAMPLE 8   Show that

$$\int_1^\infty \frac{dx}{1 + x^2}$$

converges by comparison with $\int_1^\infty (1/x^2)\,dx$. Find and compare the two integral values.

Solution   The functions $f(x) = 1/x^2$ and $g(x) = 1/(1 + x^2)$ are positive and continuous on $[1, \infty)$. Also,

$$\lim_{x \to \infty} \frac{f(x)}{g(x)} = \lim_{x \to \infty} \frac{1/x^2}{1/(1 + x^2)} = \lim_{x \to \infty} \frac{1 + x^2}{x^2}$$

$$= \lim_{x \to \infty} \left(\frac{1}{x^2} + 1 \right) = 0 + 1 = 1,$$

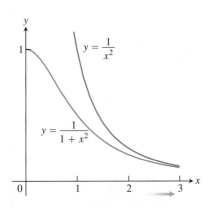

FIGURE 8.20   The functions in Example 8.

a positive finite limit (Figure 8.20). Therefore, $\displaystyle\int_1^\infty \frac{dx}{1 + x^2}$ converges because $\displaystyle\int_1^\infty \frac{dx}{x^2}$ converges.

The integrals converge to different values, however:

$$\int_1^\infty \frac{dx}{x^2} = \frac{1}{2 - 1} = 1 \qquad \text{Example 3}$$

and

$$\int_1^\infty \frac{dx}{1 + x^2} = \lim_{b \to \infty} \int_1^b \frac{dx}{1 + x^2}$$

$$= \lim_{b \to \infty} [\tan^{-1} b - \tan^{-1} 1] = \frac{\pi}{2} - \frac{\pi}{4} = \frac{\pi}{4}. \qquad \blacksquare$$

EXAMPLE 9   Investigate the convergence of $\displaystyle\int_1^\infty \frac{1 - e^{-x}}{x}\,dx$.

Solution   The integrand suggests a comparison of $f(x) = (1 - e^{-x})/x$ with $g(x) = 1/x$. However, we cannot use the Direct Comparison Test because $f(x) \leq g(x)$ and the integral of $g(x)$ *diverges*. On the other hand, using the Limit Comparison Test we find that

$$\lim_{x \to \infty} \frac{f(x)}{g(x)} = \lim_{x \to \infty} \left(\frac{1 - e^{-x}}{x} \right) \left(\frac{x}{1} \right) = \lim_{x \to \infty} (1 - e^{-x}) = 1,$$

which is a positive finite limit. Therefore, $\displaystyle\int_1^\infty \frac{1 - e^{-x}}{x}\,dx$ diverges because $\displaystyle\int_1^\infty \frac{dx}{x}$ diverges. Approximations to the improper integral are given in Table 8.5. Note that the values do not appear to approach any fixed limiting value as $b \to \infty$. $\blacksquare$

TABLE 8.5

b	$\displaystyle\int_1^b \frac{1 - e^{-x}}{x}\,dx$
2	0.5226637569
5	1.3912002736
10	2.0832053156
100	4.3857862516
1000	6.6883713446
10000	8.9909564376
100000	11.2935415306

Types of Improper Integrals Discussed in This Section

INFINITE LIMITS OF INTEGRATION: TYPE I

1. Upper limit

$$\int_1^\infty \frac{\ln x}{x^2}\, dx = \lim_{b\to\infty}\int_1^b \frac{\ln x}{x^2}\, dx$$

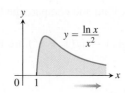

2. Lower limit

$$\int_{-\infty}^0 \frac{dx}{1 + x^2} = \lim_{a\to-\infty}\int_a^0 \frac{dx}{1 + x^2}$$

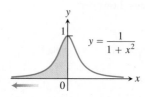

3. Both limits

$$\int_{-\infty}^\infty \frac{dx}{1 + x^2} = \lim_{b\to-\infty}\int_b^0 \frac{dx}{1 + x^2} + \lim_{c\to\infty}\int_0^c \frac{dx}{1 + x^2}$$

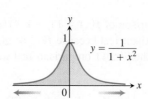

INTEGRAND BECOMES INFINITE: TYPE II

4. Upper endpoint

$$\int_0^1 \frac{dx}{(x - 1)^{2/3}} = \lim_{b\to1^-}\int_0^b \frac{dx}{(x - 1)^{2/3}}$$

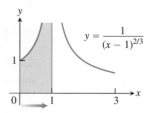

5. Lower endpoint

$$\int_1^3 \frac{dx}{(x - 1)^{2/3}} = \lim_{d\to1^+}\int_d^3 \frac{dx}{(x - 1)^{2/3}}$$

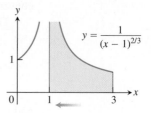

6. Interior point

$$\int_0^3 \frac{dx}{(x - 1)^{2/3}} = \int_0^1 \frac{dx}{(x - 1)^{2/3}} + \int_1^3 \frac{dx}{(x - 1)^{2/3}}$$

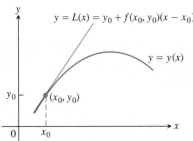

FIGURE 9.4 The linearization $L(x)$ of $y = y(x)$ at $x = x_0$.

FIGURE 9.5 The first Euler step approximates $y(x_1)$ with $y_1 = L(x_1)$.

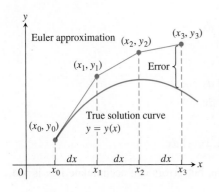

FIGURE 9.6 Three steps in the Euler approximation to the solution of the initial value problem $y' = f(x, y), y(x_0) = y_0$. As we take more steps, the errors involved usually accumulate, but not in the exaggerated way shown here.

FIGURE 9.3 Slope fields (top row) and selected solution curves (bottom row). In computer renditions, slope segments are sometimes portrayed with arrows, as they are here. This is not to be taken as an indication that slopes have directions, however, for they do not.

Constructing a slope field with pencil and paper can be quite tedious. All our examples were generated by a computer.

Euler's Method

If we do not require or cannot immediately find an *exact* solution giving an explicit formula for an initial value problem $y' = f(x, y), y(x_0) = y_0$, we can often use a computer to generate a table of approximate numerical values of y for values of x in an appropriate interval. Such a table is called a **numerical solution** of the problem, and the method by which we generate the table is called a **numerical method**.

Given a differential equation $dy/dx = f(x, y)$ and an initial condition $y(x_0) = y_0$, we can approximate the solution $y = y(x)$ by its linearization

$$L(x) = y(x_0) + y'(x_0)(x - x_0) \qquad \text{or} \qquad L(x) = y_0 + f(x_0, y_0)(x - x_0).$$

The function $L(x)$ gives a good approximation to the solution $y(x)$ in a short interval about x_0 (Figure 9.4). The basis of Euler's method is to patch together a string of linearizations to approximate the curve over a longer stretch. Here is how the method works.

We know the point (x_0, y_0) lies on the solution curve. Suppose that we specify a new value for the independent variable to be $x_1 = x_0 + dx$. (Recall that $dx = \Delta x$ in the definition of differentials.) If the increment dx is small, then

$$y_1 = L(x_1) = y_0 + f(x_0, y_0)\, dx$$

is a good approximation to the exact solution value $y = y(x_1)$. So from the point (x_0, y_0), which lies *exactly* on the solution curve, we have obtained the point (x_1, y_1), which lies very close to the point $(x_1, y(x_1))$ on the solution curve (Figure 9.5).

Using the point (x_1, y_1) and the slope $f(x_1, y_1)$ of the solution curve through (x_1, y_1), we take a second step. Setting $x_2 = x_1 + dx$, we use the linearization of the solution curve through (x_1, y_1) to calculate

$$y_2 = y_1 + f(x_1, y_1)\, dx.$$

This gives the next approximation (x_2, y_2) to values along the solution curve $y = y(x)$ (Figure 9.6). Continuing in this fashion, we take a third step from the point (x_2, y_2) with slope $f(x_2, y_2)$ to obtain the third approximation

$$y_3 = y_2 + f(x_2, y_2)\, dx,$$

and so on. We are literally building an approximation to one of the solutions by following the direction of the slope field of the differential equation.

The steps in Figure 9.6 are drawn large to illustrate the construction process, so the approximation looks crude. In practice, dx would be small enough to make the red curve hug the blue one and give a good approximation throughout.

EXAMPLE 3   Find the first three approximations y_1, y_2, y_3 using Euler's method for the initial value problem

$$y' = 1 + y, \qquad y(0) = 1,$$

starting at $x_0 = 0$ with $dx = 0.1$.

Solution   We have the starting values $x_0 = 0$ and $y_0 = 1$. Next we determine the values of x at which the Euler approximations will take place: $x_1 = x_0 + dx = 0.1$, $x_2 = x_0 + 2\,dx = 0.2$, and $x_3 = x_0 + 3\,dx = 0.3$. Then we find

$$
\begin{aligned}
\textit{First:} \quad y_1 &= y_0 + f(x_0, y_0)\,dx \\
&= y_0 + (1 + y_0)\,dx \\
&= 1 + (1 + 1)(0.1) = 1.2 \\
\textit{Second:} \quad y_2 &= y_1 + f(x_1, y_1)\,dx \\
&= y_1 + (1 + y_1)\,dx \\
&= 1.2 + (1 + 1.2)(0.1) = 1.42 \\
\textit{Third:} \quad y_3 &= y_2 + f(x_2, y_2)\,dx \\
&= y_2 + (1 + y_2)\,dx \\
&= 1.42 + (1 + 1.42)(0.1) = 1.662 \qquad \blacksquare
\end{aligned}
$$

The step-by-step process used in Example 3 can be continued easily. Using equally spaced values for the independent variable in the table for the numerical solution, and generating n of them, set

$$
\begin{aligned}
x_1 &= x_0 + dx \\
x_2 &= x_1 + dx \\
&\;\vdots \\
x_n &= x_{n-1} + dx.
\end{aligned}
$$

Then calculate the approximations to the solution,

$$
\begin{aligned}
y_1 &= y_0 + f(x_0, y_0)\,dx \\
y_2 &= y_1 + f(x_1, y_1)\,dx \\
&\;\vdots \\
y_n &= y_{n-1} + f(x_{n-1}, y_{n-1})\,dx.
\end{aligned}
$$

The number of steps n can be as large as we like, but errors can accumulate if n is too large.

HISTORICAL BIOGRAPHY

Leonhard Euler
(1703–1783)

Euler's method is easy to implement on a computer or calculator. A computer program generates a table of numerical solutions to an initial value problem, allowing us to input x_0 and y_0, the number of steps n, and the step size dx. It then calculates the approximate solution values $y_1, y_2, \ldots, y_n$ in iterative fashion, as just described.

Solving the separable equation in Example 3, we find that the exact solution to the initial value problem is $y = 2e^x - 1$. We use this information in Example 4.

EXAMPLE 4   Use Euler's method to solve

$$y' = 1 + y, \qquad y(0) = 1,$$

on the interval $0 \le x \le 1$, starting at $x_0 = 0$ and taking **(a)** $dx = 0.1$ and **(b)** $dx = 0.05$. Compare the approximations with the values of the exact solution $y = 2e^x - 1$.

Solution

(a) We used a computer to generate the approximate values in Table 9.1. The "error" column is obtained by subtracting the unrounded Euler values from the unrounded values found using the exact solution. All entries are then rounded to four decimal places.

TABLE 9.1	Euler solution of $y' = 1 + y$, $y(0) = 1$, step size $dx = 0.1$		
x	y **(Euler)**	y **(exact)**	**Error**
0	1	1	0
0.1	1.2	1.2103	0.0103
0.2	1.42	1.4428	0.0228
0.3	1.662	1.6997	0.0377
0.4	1.9282	1.9836	0.0554
0.5	2.2210	2.2974	0.0764
0.6	2.5431	2.6442	0.1011
0.7	2.8974	3.0275	0.1301
0.8	3.2872	3.4511	0.1639
0.9	3.7159	3.9192	0.2033
1.0	4.1875	4.4366	0.2491

FIGURE 9.7 The graph of $y = 2e^x - 1$ superimposed on a scatterplot of the Euler approximations shown in Table 9.1 (Example 4).

By the time we reach $x = 1$ (after 10 steps), the error is about 5.6% of the exact solution. A plot of the exact solution curve with the scatterplot of Euler solution points from Table 9.1 is shown in Figure 9.7.

(b) One way to try to reduce the error is to decrease the step size. Table 9.2 shows the results and their comparisons with the exact solutions when we decrease the step size to 0.05, doubling the number of steps to 20. As in Table 9.1, all computations are performed before rounding. This time when we reach $x = 1$, the relative error is only about 2.9%. ∎

It might be tempting to reduce the step size even further in Example 4 to obtain greater accuracy. Each additional calculation, however, not only requires additional computer time but more importantly adds to the buildup of round-off errors due to the approximate representations of numbers inside the computer.

The analysis of error and the investigation of methods to reduce it when making numerical calculations are important but are appropriate for a more advanced course. There are numerical methods more accurate than Euler's method, usually presented in a further study of differential equations.

TABLE 9.2 Euler solution of $y' = 1 + y$, $y(0) = 1$, step size $dx = 0.05$

x	y (Euler)	y (exact)	Error
0	1	1	0
0.05	1.1	1.1025	0.0025
0.10	1.205	1.2103	0.0053
0.15	1.3153	1.3237	0.0084
0.20	1.4310	1.4428	0.0118
0.25	1.5526	1.5681	0.0155
0.30	1.6802	1.6997	0.0195
0.35	1.8142	1.8381	0.0239
0.40	1.9549	1.9836	0.0287
0.45	2.1027	2.1366	0.0340
0.50	2.2578	2.2974	0.0397
0.55	2.4207	2.4665	0.0458
0.60	2.5917	2.6442	0.0525
0.65	2.7713	2.8311	0.0598
0.70	2.9599	3.0275	0.0676
0.75	3.1579	3.2340	0.0761
0.80	3.3657	3.4511	0.0853
0.85	3.5840	3.6793	0.0953
0.90	3.8132	3.9192	0.1060
0.95	4.0539	4.1714	0.1175
1.00	4.3066	4.4366	0.1300

Exercises 9.1

Slope Fields

In Exercises 1–4, match the differential equations with their slope fields, graphed here.

(a)

(b)

(c)

(d)

1. $y' = x + y$

2. $y' = y + 1$

3. $y' = -\dfrac{x}{y}$

4. $y' = y^2 - x^2$

In Exercises 5 and 6, copy the slope fields and sketch in some of the solution curves.

5. $y' = (y + 2)(y - 2)$

6. $y' = y(y + 1)(y - 1)$

Integral Equations

In Exercises 7–10, write an equivalent first-order differential equation and initial condition for y.

7. $y = -1 + \int_1^x (t - y(t)) \, dt$

8. $y = \int_1^x \frac{1}{t} \, dt$

9. $y = 2 - \int_0^x (1 + y(t)) \sin t \, dt$

10. $y = 1 + \int_0^x y(t) \, dt$

Using Euler's Method

In Exercises 11–16, use Euler's method to calculate the first three approximations to the given initial value problem for the specified increment size. Calculate the exact solution and investigate the accuracy of your approximations. Round your results to four decimal places.

11. $y' = 1 - \dfrac{y}{x}, \quad y(2) = -1, \quad dx = 0.5$

12. $y' = x(1 - y), \quad y(1) = 0, \quad dx = 0.2$

13. $y' = 2xy + 2y, \quad y(0) = 3, \quad dx = 0.2$

14. $y' = y^2(1 + 2x), \quad y(-1) = 1, \quad dx = 0.5$

T 15. $y' = 2xe^{x^2}, \quad y(0) = 2, \quad dx = 0.1$

T 16. $y' = ye^x, \quad y(0) = 2, \quad dx = 0.5$

17. Use the Euler method with $dx = 0.2$ to estimate $y(1)$ if $y' = y$ and $y(0) = 1$. What is the exact value of $y(1)$?

18. Use the Euler method with $dx = 0.2$ to estimate $y(2)$ if $y' = y/x$ and $y(1) = 2$. What is the exact value of $y(2)$?

19. Use the Euler method with $dx = 0.5$ to estimate $y(5)$ if $y' = y^2/\sqrt{x}$ and $y(1) = -1$. What is the exact value of $y(5)$?

20. Use the Euler method with $dx = 1/3$ to estimate $y(2)$ if $y' = x \sin y$ and $y(0) = 1$. What is the exact value of $y(2)$?

21. Show that the solution of the initial value problem

$$y' = x + y, \quad y(x_0) = y_0$$

is

$$y = -1 - x + (1 + x_0 + y_0) e^{x - x_0}.$$

22. What integral equation is equivalent to the initial value problem $y' = f(x), \ y(x_0) = y_0$?

COMPUTER EXPLORATIONS

In Exercises 23–28, obtain a slope field and add to it graphs of the solution curves passing through the given points.

23. $y' = y$ with
   a. $(0, 1)$ **b.** $(0, 2)$ **c.** $(0, -1)$

24. $y' = 2(y - 4)$ with
   a. $(0, 1)$ **b.** $(0, 4)$ **c.** $(0, 5)$

25. $y' = y(x + y)$ with
   a. $(0, 1)$ **b.** $(0, -2)$ **c.** $(0, 1/4)$ **d.** $(-1, -1)$

26. $y' = y^2$ with
   a. $(0, 1)$ **b.** $(0, 2)$ **c.** $(0, -1)$ **d.** $(0, 0)$

27. $y' = (y - 1)(x + 2)$ with
   a. $(0, -1)$ **b.** $(0, 1)$ **c.** $(0, 3)$ **d.** $(1, -1)$

28. $y' = \dfrac{xy}{x^2 + 4}$ with
   a. $(0, 2)$ **b.** $(0, -6)$ **c.** $\left(-2\sqrt{3}, -4\right)$

In Exercises 29 and 30, obtain a slope field and graph the particular solution over the specified interval. Use your CAS DE solver to find the general solution of the differential equation.

29. A logistic equation $y' = y(2 - y), \quad y(0) = 1/2$;
   $0 \le x \le 4, \quad 0 \le y \le 3$

30. $y' = (\sin x)(\sin y), \quad y(0) = 2; \quad -6 \le x \le 6, \quad -6 \le y \le 6$

Exercises 31 and 32 have no explicit solution in terms of elementary functions. Use a CAS to explore graphically each of the differential equations.

31. $y' = \cos(2x - y), \quad y(0) = 2; \quad 0 \le x \le 5, \quad 0 \le y \le 5$

32. A Gompertz equation $y' = y(1/2 - \ln y), \quad y(0) = 1/3$;
   $0 \le x \le 4, \quad 0 \le y \le 3$

33. Use a CAS to find the solutions of $y' + y = f(x)$ subject to the initial condition $y(0) = 0$, if $f(x)$ is
   a. $2x$ **b.** $\sin 2x$ **c.** $3e^{x/2}$ **d.** $2e^{-x/2} \cos 2x$.

Graph all four solutions over the interval $-2 \le x \le 6$ to compare the results.

34. a. Use a CAS to plot the slope field of the differential equation

$$y' = \frac{3x^2 + 4x + 2}{2(y - 1)}$$

over the region $-3 \le x \le 3$ and $-3 \le y \le 3$.

b. Separate the variables and use a CAS integrator to find the general solution in implicit form.

c. Using a CAS implicit function grapher, plot solution curves for the arbitrary constant values $C = -6, -4, -2, 0, 2, 4, 6$.

d. Find and graph the solution that satisfies the initial condition $y(0) = -1$.

In Exercises 35–38, use Euler's method with the specified step size to estimate the value of the solution at the given point x^*. Find the value of the exact solution at x^*.

35. $y' = 2xe^{x^2}$, $y(0) = 2$, $dx = 0.1$, $x^* = 1$

36. $y' = 2y^2(x - 1)$, $y(2) = -1/2$, $dx = 0.1$, $x^* = 3$

37. $y' = \sqrt{x}/y$, $y > 0$, $y(0) = 1$, $dx = 0.1$, $x^* = 1$

38. $y' = 1 + y^2$, $y(0) = 0$, $dx = 0.1$, $x^* = 1$

Use a CAS to explore graphically each of the differential equations in Exercises 39–42. Perform the following steps to help with your explorations.

a. Plot a slope field for the differential equation in the given xy-window.

b. Find the general solution of the differential equation using your CAS DE solver.

c. Graph the solutions for the values of the arbitrary constant $C = -2, -1, 0, 1, 2$ superimposed on your slope field plot.

d. Find and graph the solution that satisfies the specified initial condition over the interval $[0, b]$.

e. Find the Euler numerical approximation to the solution of the initial value problem with 4 subintervals of the x-interval and plot the Euler approximation superimposed on the graph produced in part (d).

f. Repeat part (e) for 8, 16, and 32 subintervals. Plot these three Euler approximations superimposed on the graph from part (e).

g. Find the error $(y(\text{exact}) - y(\text{Euler}))$ at the specified point $x = b$ for each of your four Euler approximations. Discuss the improvement in the percentage error.

39. $y' = x + y$, $y(0) = -7/10$; $-4 \leq x \leq 4$, $-4 \leq y \leq 4$; $b = 1$

40. $y' = -x/y$, $y(0) = 2$; $-3 \leq x \leq 3$, $-3 \leq y \leq 3$; $b = 2$

41. $y' = y(2 - y)$, $y(0) = 1/2$; $0 \leq x \leq 4$, $0 \leq y \leq 3$; $b = 3$

42. $y' = (\sin x)(\sin y)$, $y(0) = 2$; $-6 \leq x \leq 6$, $-6 \leq y \leq 6$; $b = 3\pi/2$

9.2 First-Order Linear Equations

A first-order **linear** differential equation is one that can be written in the form

$$\frac{dy}{dx} + P(x)y = Q(x), \tag{1}$$

where P and Q are continuous functions of x. Equation (1) is the linear equation's **standard form**. Since the exponential growth/decay equation $dy/dx = ky$ (Section 7.2) can be put in the standard form

$$\frac{dy}{dx} - ky = 0,$$

we see it is a linear equation with $P(x) = -k$ and $Q(x) = 0$. Equation (1) is *linear* (in y) because y and its derivative dy/dx occur only to the first power, they are not multiplied together, nor do they appear as the argument of a function $\left(\text{such as } \sin y, e^y, \text{ or } \sqrt{dy/dx}\right)$.

EXAMPLE 1   Put the following equation in standard form:

$$x\frac{dy}{dx} = x^2 + 3y, \qquad x > 0.$$

Solution

$$x\frac{dy}{dx} = x^2 + 3y$$

$$\frac{dy}{dx} = x + \frac{3}{x}y \qquad \text{Divide by } x.$$

$$\frac{dy}{dx} - \frac{3}{x}y = x \qquad \text{Standard form with } P(x) = -3/x$$
$$\text{and } Q(x) = x$$

Multiplying both sides of the standard equation by $v(t)$ and integrating both sides gives

$$(2000 - 5t)^{-9} \cdot \left(\frac{dy}{dt} + \frac{45}{2000 - 5t} \, y \right) = 80(2000 - 5t)^{-9}$$

$$(2000 - 5t)^{-9} \frac{dy}{dt} + 45(2000 - 5t)^{-10} \, y = 80(2000 - 5t)^{-9}$$

$$\frac{d}{dt} \left[(2000 - 5t)^{-9} y \right] = 80(2000 - 5t)^{-9}$$

$$(2000 - 5t)^{-9} y = \int 80(2000 - 5t)^{-9} \, dt$$

$$(2000 - 5t)^{-9} y = 80 \cdot \frac{(2000 - 5t)^{-8}}{(-8)(-5)} + C.$$

The general solution is

$$y = 2(2000 - 5t) + C(2000 - 5t)^9.$$

Because $y = 100$ when $t = 0$, we can determine the value of C:

$$100 = 2(2000 - 0) + C(2000 - 0)^9$$

$$C = -\frac{3900}{(2000)^9}.$$

The particular solution of the initial value problem is

$$y = 2(2000 - 5t) - \frac{3900}{(2000)^9} (2000 - 5t)^9.$$

The amount of additive 20 min after the pumping begins is

$$y(20) = 2[2000 - 5(20)] - \frac{3900}{(2000)^9} [2000 - 5(20)]^9 \approx 1342 \text{ lb}. \qquad \blacksquare$$

Exercises 9.3

Motion Along a Line

1. **Coasting bicycle**   A 66-kg cyclist on a 7-kg bicycle starts coasting on level ground at 9 m/sec. The k in Equation (1) is about 3.9 kg/sec.

   a. About how far will the cyclist coast before reaching a complete stop?

   b. How long will it take the cyclist's speed to drop to 1 m/sec?

2. **Coasting battleship**   Suppose that an Iowa class battleship has mass around 51,000 metric tons (51,000,000 kg) and a k value in

Equation (1) of about 59,000 kg/sec. Assume that the ship loses power when it is moving at a speed of 9 m/sec.

   a. About how far will the ship coast before it is dead in the water?

   b. About how long will it take the ship's speed to drop to 1 m/sec?

3. The data in Table 9.4 were collected with a motion detector and a CBL™ by Valerie Sharritts, a mathematics teacher at St. Francis DeSales High School in Columbus, Ohio. The table shows the distance s (meters) coasted on in-line skates in t sec by her daughter Ashley when she was 10 years old. Find a model for Ashley's

position given by the data in Table 9.4 in the form of Equation (2). Her initial velocity was $v_0 = 2.75$ m/sec, her mass $m = 39.92$ kg (she weighed 88 lb), and her total coasting distance was 4.91 m.

TABLE 9.4 Ashley Sharritts skating data

t (sec)	s (m)	t (sec)	s (m)	t (sec)	s (m)
0	0	2.24	3.05	4.48	4.77
0.16	0.31	2.40	3.22	4.64	4.82
0.32	0.57	2.56	3.38	4.80	4.84
0.48	0.80	2.72	3.52	4.96	4.86
0.64	1.05	2.88	3.67	5.12	4.88
0.80	1.28	3.04	3.82	5.28	4.89
0.96	1.50	3.20	3.96	5.44	4.90
1.12	1.72	3.36	4.08	5.60	4.90
1.28	1.93	3.52	4.18	5.76	4.91
1.44	2.09	3.68	4.31	5.92	4.90
1.60	2.30	3.84	4.41	6.08	4.91
1.76	2.53	4.00	4.52	6.24	4.90
1.92	2.73	4.16	4.63	6.40	4.91
2.08	2.89	4.32	4.69	6.56	4.91

4. Coasting to a stop Table 9.5 shows the distance s (meters) coasted on in-line skates in terms of time t (seconds) by Kelly Schmitzer. Find a model for her position in the form of Equation (2). Her initial velocity was $v_0 = 0.80$ m/sec, her mass $m = 49.90$ kg (110 lb), and her total coasting distance was 1.32 m.

TABLE 9.5 Kelly Schmitzer skating data

t (sec)	s (m)	t (sec)	s (m)	t (sec)	s (m)
0	0	1.5	0.89	3.1	1.30
0.1	0.07	1.7	0.97	3.3	1.31
0.3	0.22	1.9	1.05	3.5	1.32
0.5	0.36	2.1	1.11	3.7	1.32
0.7	0.49	2.3	1.17	3.9	1.32
0.9	0.60	2.5	1.22	4.1	1.32
1.1	0.71	2.7	1.25	4.3	1.32
1.3	0.81	2.9	1.28	4.5	1.32

Orthogonal Trajectories

In Exercises 5–10, find the orthogonal trajectories of the family of curves. Sketch several members of each family.

5. $y = mx$

6. $y = cx^2$

7. $kx^2 + y^2 = 1$

8. $2x^2 + y^2 = c^2$

9. $y = ce^{-x}$

10. $y = e^{kx}$

11. Show that the curves $2x^2 + 3y^2 = 5$ and $y^2 = x^3$ are orthogonal.

12. Find the family of solutions of the given differential equation and the family of orthogonal trajectories. Sketch both families.

　　a. $x\,dx + y\,dy = 0$ 　　**b.** $x\,dy - 2y\,dx = 0$

Mixture Problems

13. Salt mixture A tank initially contains 100 gal of brine in which 50 lb of salt are dissolved. A brine containing 2 lb/gal of salt runs into the tank at the rate of 5 gal/min. The mixture is kept uniform by stirring and flows out of the tank at the rate of 4 gal/min.

　　a. At what rate (pounds per minute) does salt enter the tank at time t?

　　b. What is the volume of brine in the tank at time t?

　　c. At what rate (pounds per minute) does salt leave the tank at time t?

　　d. Write down and solve the initial value problem describing the mixing process.

　　e. Find the concentration of salt in the tank 25 min after the process starts.

14. Mixture problem A 200-gal tank is half full of distilled water. At time $t = 0$, a solution containing 0.5 lb/gal of concentrate enters the tank at the rate of 5 gal/min, and the well-stirred mixture is withdrawn at the rate of 3 gal/min.

　　a. At what time will the tank be full?

　　b. At the time the tank is full, how many pounds of concentrate will it contain?

15. Fertilizer mixture A tank contains 100 gal of fresh water. A solution containing 1 lb/gal of soluble lawn fertilizer runs into the tank at the rate of 1 gal/min, and the mixture is pumped out of the tank at the rate of 3 gal/min. Find the maximum amount of fertilizer in the tank and the time required to reach the maximum.

16. Carbon monoxide pollution An executive conference room of a corporation contains 4500 ft³ of air initially free of carbon monoxide. Starting at time $t = 0$, cigarette smoke containing 4% carbon monoxide is blown into the room at the rate of 0.3 ft³/min. A ceiling fan keeps the air in the room well circulated and the air leaves the room at the same rate of 0.3 ft³/min. Find the time when the concentration of carbon monoxide in the room reaches 0.01%.

9.4　Graphical Solutions of Autonomous Equations

In Chapter 4 we learned that the sign of the first derivative tells where the graph of a function is increasing and where it is decreasing. The sign of the second derivative tells the concavity of the graph. We can build on our knowledge of how derivatives determine the shape of a graph to solve differential equations graphically. We will see that the ability to

FIGURE 9.21 The completed phase line for the falling body.

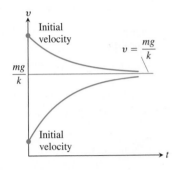

FIGURE 9.22 Typical velocity curves for a falling body encountering resistance. The value $v = mg/k$ is the terminal velocity.

FIGURE 9.23 The initial phase line for logistic growth (Equation 6).

We determine the concavity of the solution curves by differentiating both sides of Equation (4) with respect to t:

$$\frac{d^2v}{dt^2} = \frac{d}{dt}\left(g - \frac{k}{m}v\right) = -\frac{k}{m}\frac{dv}{dt}.$$

We see that $d^2v/dt^2 < 0$ when $v < mg/k$ and $d^2v/dt^2 > 0$ when $v > mg/k$. Figure 9.21 adds this information to the phase line. Notice the similarity to the phase line for Newton's law of cooling (Figure 9.17). The solution curves are similar as well (Figure 9.22).

Figure 9.22 shows two typical solution curves. Regardless of the initial velocity, we see the body's velocity tending toward the limiting value $v = mg/k$. This value, a stable equilibrium point, is called the body's **terminal velocity**. Skydivers can vary their terminal velocity from 95 mph to 180 mph by changing the amount of body area opposing the fall, which affects the value of k.

Logistic Population Growth

In Section 9.3 we examined population growth using the model of exponential change. That is, if P represents the number of individuals and we neglect departures and arrivals, then

$$\frac{dP}{dt} = kP, \tag{5}$$

where $k > 0$ is the birth rate minus the death rate per individual per unit time.

Because the natural environment has only a limited number of resources to sustain life, it is reasonable to assume that only a maximum population M can be accommodated. As the population approaches this **limiting population** or **carrying capacity**, resources become less abundant and the growth rate k decreases. A simple relationship exhibiting this behavior is

$$k = r(M - P),$$

where $r > 0$ is a constant. Notice that k decreases as P increases toward M and that k is negative if P is greater than M. Substituting $r(M - P)$ for k in Equation (5) gives the differential equation

$$\frac{dP}{dt} = r(M - P)P = rMP - rP^2. \tag{6}$$

The model given by Equation (6) is referred to as **logistic growth**.

We can forecast the behavior of the population over time by analyzing the phase line for Equation (6). The equilibrium values are $P = M$ and $P = 0$, and we can see that $dP/dt > 0$ if $0 < P < M$ and $dP/dt < 0$ if $P > M$. These observations are recorded on the phase line in Figure 9.23.

We determine the concavity of the population curves by differentiating both sides of Equation (6) with respect to t:

$$\frac{d^2P}{dt^2} = \frac{d}{dt}(rMP - rP^2)$$

$$= rM\frac{dP}{dt} - 2rP\frac{dP}{dt}$$

$$= r(M - 2P)\frac{dP}{dt}. \tag{7}$$

If $P = M/2$, then $d^2P/dt^2 = 0$. If $P < M/2$, then $(M - 2P)$ and dP/dt are positive and $d^2P/dt^2 > 0$. If $M/2 < P < M$, then $(M - 2P) < 0$, $dP/dt > 0$, and $d^2P/dt^2 < 0$.

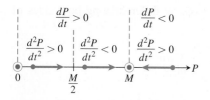

FIGURE 9.24 The completed phase line for logistic growth (Equation 6).

If $P > M$, then $(M - 2P)$ and dP/dt are both negative and $d^2P/dt^2 > 0$. We add this information to the phase line (Figure 9.24).

The lines $P = M/2$ and $P = M$ divide the first quadrant of the tP-plane into horizontal bands in which we know the signs of both dP/dt and d^2P/dt^2. In each band, we know how the solution curves rise and fall, and how they bend as time passes. The equilibrium lines $P = 0$ and $P = M$ are both population curves. Population curves crossing the line $P = M/2$ have an inflection point there, giving them a **sigmoid** shape (curved in two directions like a letter S). Figure 9.25 displays typical population curves. Notice that each population curve approaches the limiting population M as $t \to \infty$.

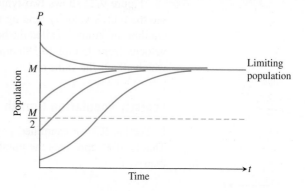

FIGURE 9.25 Population curves for logistic growth.

Exercises 9.4

Phase Lines and Solution Curves
In Exercises 1–8,

    a. Identify the equilibrium values. Which are stable and which are unstable?

    b. Construct a phase line. Identify the signs of y' and y''.

    c. Sketch several solution curves.

1. $\dfrac{dy}{dx} = (y + 2)(y - 3)$ **2.** $\dfrac{dy}{dx} = y^2 - 4$

3. $\dfrac{dy}{dx} = y^3 - y$ **4.** $\dfrac{dy}{dx} = y^2 - 2y$

5. $y' = \sqrt{y}, \quad y > 0$ **6.** $y' = y - \sqrt{y}, \quad y > 0$

7. $y' = (y - 1)(y - 2)(y - 3)$ **8.** $y' = y^3 - y^2$

Models of Population Growth
The autonomous differential equations in Exercises 9–12 represent models for population growth. For each exercise, use a phase line analysis to sketch solution curves for $P(t)$, selecting different starting values $P(0)$. Which equilibria are stable, and which are unstable?

9. $\dfrac{dP}{dt} = 1 - 2P$ **10.** $\dfrac{dP}{dt} = P(1 - 2P)$

11. $\dfrac{dP}{dt} = 2P(P - 3)$ **12.** $\dfrac{dP}{dt} = 3P(1 - P)\left(P - \dfrac{1}{2}\right)$

13. Catastrophic change in logistic growth Suppose that a healthy population of some species is growing in a limited environment and that the current population P_0 is fairly close to the carrying capacity M_0. You might imagine a population of fish living in a freshwater lake in a wilderness area. Suddenly a catastrophe such as the Mount St. Helens volcanic eruption contaminates the lake and destroys a significant part of the food and oxygen on which the fish depend. The result is a new environment with a carrying capacity M_1 considerably less than M_0 and, in fact, less than the current population P_0. Starting at some time before the catastrophe, sketch a "before-and-after" curve that shows how the fish population responds to the change in environment.

14. Controlling a population The fish and game department in a certain state is planning to issue hunting permits to control the deer population (one deer per permit). It is known that if the deer population falls below a certain level m, the deer will become extinct. It is also known that if the deer population rises above the carrying capacity M, the population will decrease back to M through disease and malnutrition.

    a. Discuss the reasonableness of the following model for the growth rate of the deer population as a function of time:

$$\frac{dP}{dt} = rP(M - P)(P - m),$$

    where P is the population of the deer and r is a positive constant of proportionality. Include a phase line.

    b. Explain how this model differs from the logistic model $dP/dt = rP(M - P)$. Is it better or worse than the logistic model?

c. Show that if $P > M$ for all t, then $\lim_{t \to \infty} P(t) = M$.

d. What happens if $P < m$ for all t?

e. Discuss the solutions to the differential equation. What are the equilibrium points of the model? Explain the dependence of the steady-state value of P on the initial values of P. About how many permits should be issued?

Applications and Examples

15. **Skydiving** If a body of mass m falling from rest under the action of gravity encounters an air resistance proportional to the square of velocity, then the body's velocity t seconds into the fall satisfies the equation

$$m \frac{dv}{dt} = mg - kv^2, \qquad k > 0$$

where k is a constant that depends on the body's aerodynamic properties and the density of the air. (We assume that the fall is too short to be affected by changes in the air's density.)

a. Draw a phase line for the equation.

b. Sketch a typical velocity curve.

c. For a 110-lb skydiver ($mg = 110$) and with time in seconds and distance in feet, a typical value of k is 0.005. What is the diver's terminal velocity? Repeat for a 200-lb skydiver.

16. **Resistance proportional to $\sqrt{v}$** A body of mass m is projected vertically downward with initial velocity v_0. Assume that the resisting force is proportional to the square root of the velocity and find the terminal velocity from a graphical analysis.

17. **Sailing** A sailboat is running along a straight course with the wind providing a constant forward force of 50 lb. The only other force acting on the boat is resistance as the boat moves through the water. The resisting force is numerically equal to five times the boat's speed, and the initial velocity is 1 ft/sec. What is the maximum velocity in feet per second of the boat under this wind?

18. **The spread of information** Sociologists recognize a phenomenon called *social diffusion*, which is the spreading of a piece of information, technological innovation, or cultural fad among a population. The members of the population can be divided into two classes: those who have the information and those who do not. In a fixed population whose size is known, it is reasonable to assume that the rate of diffusion is proportional to the number who have the information times the number yet to receive it. If X denotes the number of individuals who have the information in a population of N people, then a mathematical model for social diffusion is given by

$$\frac{dX}{dt} = kX(N - X),$$

where t represents time in days and k is a positive constant.

a. Discuss the reasonableness of the model.

b. Construct a phase line identifying the signs of X' and X''.

c. Sketch representative solution curves.

d. Predict the value of X for which the information is spreading most rapidly. How many people eventually receive the information?

19. **Current in an *RL*-circuit** The accompanying diagram represents an electrical circuit whose total resistance is a constant R ohms and whose self-inductance, shown as a coil, is L henries, also a constant. There is a switch whose terminals at a and b can be closed to connect a constant electrical source of V volts. From Section 9.2, we have

$$L \frac{di}{dt} + Ri = V,$$

where i is the current in amperes and t is the time in seconds.

Use a phase line analysis to sketch the solution curve assuming that the switch in the *RL*-circuit is closed at time $t = 0$. What happens to the current as $t \to \infty$? This value is called the *steady-state solution*.

20. **A pearl in shampoo** Suppose that a pearl is sinking in a thick fluid, like shampoo, subject to a frictional force opposing its fall and proportional to its velocity. Suppose that there is also a resistive buoyant force exerted by the shampoo. According to *Archimedes' principle*, the buoyant force equals the weight of the fluid displaced by the pearl. Using m for the mass of the pearl and P for the mass of the shampoo displaced by the pearl as it descends, complete the following steps.

a. Draw a schematic diagram showing the forces acting on the pearl as it sinks, as in Figure 9.19.

b. Using $v(t)$ for the pearl's velocity as a function of time t, write a differential equation modeling the velocity of the pearl as a falling body.

c. Construct a phase line displaying the signs of v' and v''.

d. Sketch typical solution curves.

e. What is the terminal velocity of the pearl?

9.5 | Systems of Equations and Phase Planes

In some situations we are led to consider not one, but several first-order differential equations. Such a collection is called a **system** of differential equations. In this section we present an approach to understanding systems through a graphical procedure known as a *phase-plane analysis*. We present this analysis in the context of modeling the populations of trout and bass living in a common pond.

Phase Planes

A general system of two first-order differential equations may take the form

$$\frac{dx}{dt} = F(x, y),$$

$$\frac{dy}{dt} = G(x, y).$$

Such a system of equations is called **autonomous** because dx/dt and dy/dt do not depend on the independent variable time t, but only on the dependent variables x and y. A **solution** of such a system consists of a pair of functions $x(t)$ and $y(t)$ that satisfies both of the differential equations simultaneously for every t over some time interval (finite or infinite).

We cannot look at just one of these equations in isolation to find solutions $x(t)$ or $y(t)$ since each derivative depends on both x and y. To gain insight into the solutions, we look at both dependent variables together by plotting the points $(x(t), y(t))$ in the xy-plane starting at some specified point. Therefore the solution functions define a solution curve through the specified point, called a **trajectory** of the system. The xy-plane itself, in which these trajectories reside, is referred to as the **phase plane**. Thus we consider both solutions together and study the behavior of all the solution trajectories in the phase plane. It can be proved that two trajectories can never cross or touch each other. (Solution trajectories are examples of *parametric curves*, which are studied in detail in Chapter 11.)

A Competitive-Hunter Model

Imagine two species of fish, say trout and bass, competing for the same limited resources (such as food and oxygen) in a certain pond. We let $x(t)$ represent the number of trout and $y(t)$ the number of bass living in the pond at time t. In reality $x(t)$ and $y(t)$ are always integer valued, but we will approximate them with real-valued differentiable functions. This allows us to apply the methods of differential equations.

Several factors affect the rates of change of these populations. As time passes, each species breeds, so we assume its population increases proportionally to its size. Taken by itself, this would lead to exponential growth in each of the two populations. However, there is a countervailing effect from the fact that the two species are in competition. A large number of bass tends to cause a decrease in the number of trout, and vice-versa. Our model takes the size of this effect to be proportional to the frequency with which the two species interact, which in turn is proportional to xy, the product of the two populations. These considerations lead to the following model for the growth of the trout and bass in the pond:

$$\frac{dx}{dt} = (a - by)x, \tag{1a}$$

$$\frac{dy}{dt} = (m - nx)y. \tag{1b}$$

Here $x(t)$ represents the trout population, $y(t)$ the bass population, and a, b, m, n are positive constants. A solution of this system then consists of a pair of functions $x(t)$ and $y(t)$ that gives the population of each fish species at time t. Each equation in (1) contains both of the unknown functions x and y, so we are unable to solve them individually. Instead, we will use a graphical analysis to study the solution trajectories of this **competitive-hunter model**.

We now examine the nature of the phase plane in the trout-bass population model. We will be interested in the 1st quadrant of the xy-plane, where $x \geq 0$ and $y \geq 0$, since populations cannot be negative. First, we determine where the bass and trout populations are both constant. Noting that the $(x(t), y(t))$ values remain unchanged when $dx/dt = 0$ and $dy/dt = 0$, Equations (1a and 1b) then become

$$(a - by)x = 0,$$

$$(m - nx)y = 0.$$

This pair of simultaneous equations has two solutions: $(x, y) = (0, 0)$ and $(x, y) = (m/n, a/b)$. At these (x, y) values, called **equilibrium** or **rest points**, the two populations remain at constant values over all time. The point $(0, 0)$ represents a pond containing no members of either fish species; the point $(m/n, a/b)$ corresponds to a pond with an unchanging number of each fish species.

Next, we note that if $y = a/b$, then Equation (1a) implies $dx/dt = 0$, so the trout population $x(t)$ is constant. Similarly, if $x = m/n$, then Equation (1b) implies $dy/dt = 0$, and the bass population $y(t)$ is constant. This information is recorded in Figure 9.26.

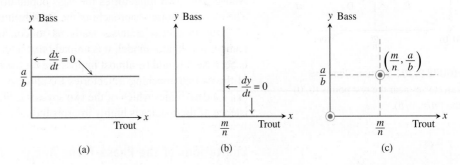

(a)   (b)   (c)

FIGURE 9.26 Rest points in the competitive-hunter model given by Equations (1a) and (1b).

In setting up our competitive-hunter model, precise values of the constants a, b, m, n will not generally be known. Nonetheless, we can analyze the system of Equations (1) to learn the nature of its solution trajectories. We begin by determining the signs of dx/dt and dy/dt throughout the phase plane. Although $x(t)$ represents the number of trout and $y(t)$ the number of bass at time t, we are thinking of the pair of values $(x(t), y(t))$ as a point tracing out a trajectory curve in the phase plane. When dx/dt is positive, $x(t)$ is increasing and the point is moving to the right in the phase plane. If dx/dt is negative, the point is moving to the left. Likewise, the point is moving upward where dy/dt is positive and downward where dy/dt is negative.

We saw that $dy/dt = 0$ along the vertical line $x = m/n$. To the left of this line, dy/dt is positive since $dy/dt = (m - nx)y$ and $x < m/n$. So the trajectories on this side of the line are directed upward. To the right of this line, dy/dt is negative and the trajectories point downward. The directions of the associated trajectories are indicated in Figure 9.27. Similarly, above the horizontal line $y = a/b$, we have $dx/dt < 0$ and the trajectories head leftward; below this line they head rightward, as shown in Figure 9.28. Combining this information gives four distinct regions in the plane A, B, C, D, with their respective trajectory directions shown in Figure 9.29.

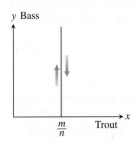

FIGURE 9.27 To the left of the line $x = m/n$ the trajectories move upward, and to the right they move downward.

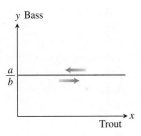

FIGURE 9.28 Above the line $y = a/b$ the trajectories move to the left, and below it they move to the right.

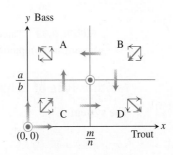

FIGURE 9.29 Composite graphical analysis of the trajectory directions in the four regions determined by $x = m/n$ and $y = a/b$.

Next, we examine what happens near the two equilibrium points. The trajectories near $(0, 0)$ point away from it, upward and to the right. The behavior near the equilibrium point $(m/n, a/b)$ depends on the region in which a trajectory begins. If it starts in region B, for instance, then it will move downward and leftward towards the equilibrium point. Depending on where the trajectory begins, it may move downward into region D, leftward into region A,

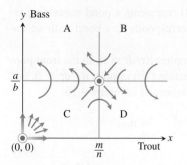

FIGURE 9.30 Motion along the trajectories near the rest points $(0, 0)$ and $(m/n, a/b)$.

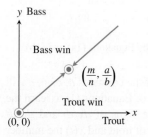

FIGURE 9.31 Qualitative results of analyzing the competitive-hunter model. There are exactly two trajectories approaching the point $(m/n, a/b)$.

FIGURE 9.32 Trajectory direction near the rest point $(0, 0)$.

or perhaps straight into the equilibrium point. If it enters into regions A or D, then it will continue to move away from the rest point. We say that both rest points are **unstable**, meaning (in this setting) there are trajectories near each point that head away from them. These features are indicated in Figure 9.30.

It turns out that in each of the half-planes above and below the line $y = a/b$, there is exactly one trajectory approaching the equilibrium point $(m/n, a/b)$ (see Exercise 7). Above these two trajectories the bass population increases and below them it decreases. The two trajectories approaching the equilibrium point are suggested in Figure 9.31.

Our graphical analysis leads us to conclude that, under the assumptions of the competitive-hunter model, it is unlikely that both species will reach equilibrium levels. This is because it would be almost impossible for the fish populations to move exactly along one of the two approaching trajectories for all time. Furthermore, the initial populations point (x_0, y_0) determines which of the two species is likely to survive over time, and mutual coexistence of the species is highly improbable.

Limitations of the Phase-Plane Analysis Method

Unlike the situation for the competitive-hunter model, it is not always possible to determine the behavior of trajectories near a rest point. For example, suppose we know that the trajectories near a rest point, chosen here to be the origin $(0, 0)$, behave as in Figure 9.32. The information provided by Figure 9.32 is not sufficient to distinguish between the three possible trajectories shown in Figure 9.33. Even if we could determine that a trajectory near an equilibrium point resembles that of Figure 9.33c, we would still not know how the other trajectories behave. It could happen that a trajectory closer to the origin behaves like the motions displayed in Figure 9.33a or 9.33b. The spiraling trajectory in Figure 9.33b can never actually reach the rest point in a finite time period.

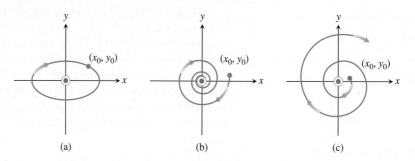

FIGURE 9.33 Three possible trajectory motions: (a) periodic motion, (b) motion toward an asymptotically stable rest point, and (c) motion near an unstable rest point.

Another Type of Behavior

The system

$$\frac{dx}{dt} = y + x - x(x^2 + y^2), \tag{2a}$$

$$\frac{dy}{dt} = -x + y - y(x^2 + y^2) \tag{2b}$$

can be shown to have only one equilibrium point at $(0, 0)$. Yet any trajectory starting on the unit circle traverses it clockwise because, when $x^2 + y^2 = 1$, we have $dy/dx = -x/y$ (see Exercise 2). If a trajectory starts inside the unit circle, it spirals outward, asymptotically approaching the circle as $t \to \infty$. If a trajectory starts outside the unit circle, it spirals inward, again asymptotically approaching the circle as $t \to \infty$. The circle $x^2 + y^2 = 1$ is called a **limit cycle** of the system (Figure 9.34). In this system, the values of x and y eventually become periodic.

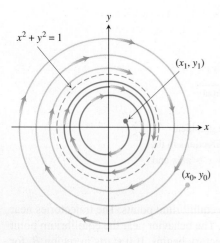

FIGURE 9.34 The solution $x^2 + y^2 = 1$ is a limit cycle.

Exercises 9.5

1. List three important considerations that are ignored in the competitive-hunter model as presented in the text.

2. For the system (2a) and (2b), show that any trajectory starting on the unit circle $x^2 + y^2 = 1$ will traverse the unit circle in a periodic solution. First introduce polar coordinates and rewrite the system as $dr/dt = r(1 - r^2)$ and $-d\theta/dt = -1$.

3. Develop a model for the growth of trout and bass, assuming that in isolation trout demonstrate exponential decay [so that $a < 0$ in Equations (1a) and (1b)] and that the bass population grows logistically with a population limit M. Analyze graphically the motion in the vicinity of the rest points in your model. Is coexistence possible?

4. How might the competitive-hunter model be validated? Include a discussion of how the various constants a, b, m, and n might be estimated. How could state conservation authorities use the model to ensure the survival of both species?

5. Consider another competitive-hunter model defined by

$$\frac{dx}{dt} = a\left(1 - \frac{x}{k_1}\right)x - bxy,$$

$$\frac{dy}{dt} = m\left(1 - \frac{y}{k_2}\right)y - nxy,$$

where x and y represent trout and bass populations, respectively.

a. What assumptions are implicitly being made about the growth of trout and bass in the absence of competition?

b. Interpret the constants a, b, m, n, k_1, and k_2 in terms of the physical problem.

c. Perform a graphical analysis:

  i) Find the possible equilibrium levels.

  ii) Determine whether coexistence is possible.

  iii) Pick several typical starting points and sketch typical trajectories in the phase plane.

  iv) Interpret the outcomes predicted by your graphical analysis in terms of the constants a, b, m, n, k_1, and k_2.

Note: When you get to part (iii), you should realize that five cases exist. You will need to analyze all five cases.

6. **An economic model** Consider the following economic model. Let P be the price of a single item on the market. Let Q be the quantity of the item available on the market. Both P and Q are functions of time. If one considers price and quantity as two interacting species, the following model might be proposed:

$$\frac{dP}{dt} = aP\left(\frac{b}{Q} - P\right),$$

$$\frac{dQ}{dt} = cQ(fP - Q),$$

where a, b, c, and f are positive constants. Justify and discuss the adequacy of the model.

a. If $a = 1$, $b = 20{,}000$, $c = 1$, and $f = 30$, find the equilibrium points of this system. If possible, classify each equilibrium point with respect to its stability. If a point cannot be readily classified, give some explanation.

b. Perform a graphical stability analysis to determine what will happen to the levels of P and Q as time increases.

c. Give an economic interpretation of the curves that determine the equilibrium points.

7. **Two trajectories approach equilibrium** Show that the two trajectories leading to $(m/n, a/b)$ shown in Figure 9.31 are unique by carrying out the following steps.

a. From system (1a) and (1b) apply the Chain Rule to derive the following equation:

$$\frac{dy}{dx} = \frac{(m - nx)y}{(a - by)x}.$$

b. Separate the variables, integrate, and exponentiate to obtain

$$y^a e^{-by} = Kx^m e^{-nx},$$

where K is a constant of integration.

c. Let $f(y) = y^a/e^{by}$ and $g(x) = x^m/e^{nx}$. Show that $f(y)$ has a unique maximum of $M_y = (a/eb)^a$ when $y = a/b$ as shown in Figure 9.35. Similarly, show that $g(x)$ has a unique maximum $M_x = (m/en)^m$ when $x = m/n$, also shown in Figure 9.35.

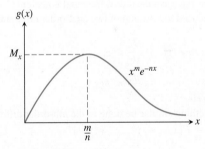

FIGURE 9.35 Graphs of the functions $f(y) = y^a/e^{by}$ and $g(x) = x^m/e^{nx}$.

d. Consider what happens as (x, y) approaches $(m/n, a/b)$. Take limits in part (b) as $x \to m/n$ and $y \to a/b$ to show that either

$$\lim_{\substack{x \to m/n \\ y \to a/b}} \left[\left(\frac{y^a}{e^{by}}\right)\left(\frac{e^{nx}}{x^m}\right)\right] = K$$

or $M_y/M_x = K$. Thus any solution trajectory that approaches $(m/n, a/b)$ must satisfy

$$\frac{y^a}{e^{by}} = \left(\frac{M_y}{M_x}\right)\left(\frac{x^m}{e^{nx}}\right).$$

e. Show that only one trajectory can approach $(m/n, a/b)$ from below the line $y = a/b$. Pick $y_0 < a/b$. From Figure 9.35 you can see that $f(y_0) < M_y$, which implies that

$$\frac{M_y}{M_x}\left(\frac{x^m}{e^{nx}}\right) = y_0{}^a/e^{by_0} < M_y.$$

This in turn implies that

$$\frac{x^m}{e^{nx}} < M_x.$$

Figure 9.35 tells you that for $g(x)$ there is a unique value $x_0 < m/n$ satisfying this last inequality. That is, for each $y < a/b$ there is a unique value of x satisfying the equation in part (d). Thus there can exist only one trajectory solution approaching $(m/n, a/b)$ from below, as shown in Figure 9.36.

f. Use a similar argument to show that the solution trajectory leading to $(m/n, a/b)$ is unique if $y_0 > a/b$.

FIGURE 9.36 For any $y < a/b$ only one solution trajectory leads to the rest point $(m/n, a/b)$.

8. Show that the second-order differential equation $y'' = F(x, y, y')$ can be reduced to a system of two first-order differential equations

$$\frac{dy}{dx} = z,$$

$$\frac{dz}{dx} = F(x, y, z).$$

Can something similar be done to the nth-order differential equation $y^{(n)} = F\big(x, y, y', y'', \dots, y^{(n-1)}\big)$?

Lotka-Volterra Equations for a Predator-Prey Model

In 1925 Lotka and Volterra introduced the *predator-prey* equations, a system of equations that models the populations of two species, one of which preys on the other. Let $x(t)$ represent the number of rabbits living in a region at time t, and $y(t)$ the number of foxes in the same region. As time passes, the number of rabbits increases at a rate proportional to their population, and decreases at a rate proportional to the number of encounters between rabbits and foxes. The foxes, which compete for food, increase in number at a rate proportional to the number of encounters with rabbits but decrease at a rate proportional to the number of foxes. The number of encounters between rabbits and foxes is assumed to be proportional to the product of the two populations. These assumptions lead to the autonomous system

$$\frac{dx}{dt} = (a - by)x$$

$$\frac{dy}{dt} = (-c + dx)y$$

where a, b, c, d are positive constants. The values of these constants vary according to the specific situation being modeled. We can study the nature of the population changes without setting these constants to specific values.

9. What happens to the rabbit population if there are no foxes present?

10. What happens to the fox population if there are no rabbits present?

11. Show that $(0, 0)$ and $(c/d, a/b)$ are equilibrium points. Explain the meaning of each of these points.

12. Show, by differentiating, that the function

$$C(t) = a \ln y(t) - by(t) - dx(t) + c \ln x(t)$$

is constant when $x(t)$ and $y(t)$ are positive and satisfy the predator-prey equations.

While x and y may change over time, $C(t)$ does not. Thus, C is a *conserved quantity* and its existence gives a *conservation law*. A trajectory that begins at a point (x, y) at time $t = 0$ gives a value of C that remains unchanged at future times. Each value of the constant C gives a trajectory for the autonomous system, and these trajectories close up, rather than spiraling inwards or outwards. The rabbit and fox populations oscillate through repeated cycles along a fixed trajectory. Figure 9.37 shows several trajectories for the predator-prey system.

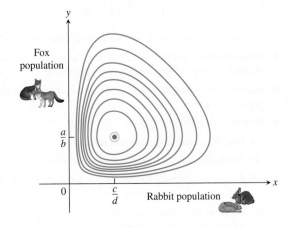

FIGURE 9.37 Some trajectories along which C is conserved.

13. Using a procedure similar to that in the text for the competitive-hunter model, show that each trajectory is traversed in a counterclockwise direction as time t increases.

Along each trajectory, both the rabbit and fox populations fluctuate between their maximum and minimum levels. The maximum and minimum levels for the rabbit population occur where the trajectory intersects the horizontal line $y = a/b$. For the fox population, they occur where the trajectory intersects the vertical line $x = c/d$. When the rabbit population is at its maximum, the fox population is below its maximum value. As the rabbit population declines from this point in time, we move counterclockwise around the trajectory, and the fox population grows until it reaches its maximum value. At this point the rabbit population has declined to $x = c/d$ and is no longer at its peak value. We see that the fox population reaches its maximum value at a later time than the rabbits. The predator population *lags behind* that of the prey in achieving its maximum values. This lag effect is shown in Figure 9.38, which graphs both $x(t)$ and $y(t)$.

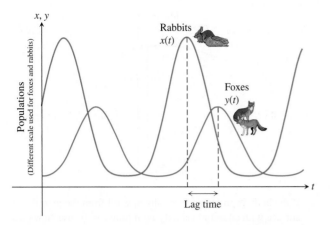

FIGURE 9.38 The fox and rabbit populations oscillate periodically, with the maximum fox population lagging the maximum rabbit population.

14. At some time during a trajectory cycle, a wolf invades the rabbit-fox territory, eats some rabbits, and then leaves. Does this mean that the fox population will from then on have a lower maximum value? Explain your answer.

Chapter 9 Questions to Guide Your Review

1. What is a first-order differential equation? When is a function a solution of such an equation?

2. What is a general solution? A particular solution?

3. What is the slope field of a differential equation $y' = f(x, y)$? What can we learn from such fields?

4. Describe Euler's method for solving the initial value problem $y' = f(x, y), y(x_0) = y_0$ numerically. Give an example. Comment on the method's accuracy. Why might you want to solve an initial value problem numerically?

5. How do you solve linear first-order differential equations?

6. What is an orthogonal trajectory of a family of curves? Describe how one is found for a given family of curves.

7. What is an autonomous differential equation? What are its equilibrium values? How do they differ from critical points? What is a stable equilibrium value? Unstable?

8. How do you construct the phase line for an autonomous differential equation? How does the phase line help you produce a graph which qualitatively depicts a solution to the differential equation?

9. Why is the exponential model unrealistic for predicting long-term population growth? How does the logistic model correct for the deficiency in the exponential model for population growth? What is the logistic differential equation? What is the form of its solution? Describe the graph of the logistic solution.

10. What is an autonomous system of differential equations? What is a solution to such a system? What is a trajectory of the system?

Chapter 9 Practice Exercises

In Exercises 1–16 solve the differential equation.

1. $y' = xe^y \sqrt{x - 2}$

2. $y' = xye^{x^2}$

3. $\sec x \, dy + x \cos^2 y \, dx = 0$

4. $2x^2 \, dx - 3\sqrt{y} \csc x \, dy = 0$

5. $y' = \dfrac{e^y}{xy}$

6. $y' = xe^{x-y} \csc y$

7. $x(x - 1) \, dy - y \, dx = 0$

8. $y' = (y^2 - 1)x^{-1}$

9. $2y' - y = xe^{x/2}$

10. $\dfrac{y'}{2} + y = e^{-x} \sin x$

11. $xy' + 2y = 1 - x^{-1}$

12. $xy' - y = 2x \ln x$

13. $(1 + e^x) \, dy + (ye^x + e^{-x}) \, dx = 0$

14. $e^{-x} \, dy + (e^{-x}y - 4x) \, dx = 0$

15. $(x + 3y^2) \, dy + y \, dx = 0$ (*Hint:* $d(xy) = y \, dx + x \, dy$)

16. $x \, dy + (3y - x^{-2} \cos x) \, dx = 0, \quad x > 0$

Initial Value Problems

In Exercises 17–22 solve the initial value problem.

17. $(x + 1)\dfrac{dy}{dx} + 2y = x, \quad x > -1, \quad y(0) = 1$

18. $x\dfrac{dy}{dx} + 2y = x^2 + 1, \quad x > 0, \quad y(1) = 1$

19. $\dfrac{dy}{dx} + 3x^2y = x^2, \quad y(0) = -1$

20. $x \, dy + (y - \cos x) \, dx = 0, \quad y\left(\dfrac{\pi}{2}\right) = 0$

21. $xy' + (x - 2)y = 3x^3e^{-x}, \quad y(1) = 0$

22. $y \, dx + (3x - xy + 2) \, dy = 0, \quad y(2) = -1, \quad y < 0$

Euler's Method

In Exercises 23 and 24, use Euler's method to solve the initial value problem on the given interval starting at x_0 with $dx = 0.1$.

T **23.** $y' = y + \cos x$, $y(0) = 0$; $0 \le x \le 2$; $x_0 = 0$

T **24.** $y' = (2 - y)(2x + 3)$, $y(-3) = 1$;
$-3 \le x \le -1$; $x_0 = -3$

In Exercises 25 and 26, use Euler's method with $dx = 0.05$ to estimate $y(c)$ where y is the solution to the given initial value problem.

T **25.** $c = 3$; $\dfrac{dy}{dx} = \dfrac{x - 2y}{x + 1}$, $y(0) = 1$

T **26.** $c = 4$; $\dfrac{dy}{dx} = \dfrac{x^2 - 2y + 1}{x}$, $y(1) = 1$

In Exercises 27 and 28, use Euler's method to solve the initial value problem graphically, starting at $x_0 = 0$ with

   a. $dx = 0.1$.   **b.** $dx = -0.1$.

T **27.** $\dfrac{dy}{dx} = \dfrac{1}{e^{x+y+2}}$, $y(0) = -2$

T **28.** $\dfrac{dy}{dx} = -\dfrac{x^2 + y}{e^y + x}$, $y(0) = 0$

Slope Fields

In Exercises 29–32, sketch part of the equation's slope field. Then add to your sketch the solution curve that passes through the point $P(1, -1)$. Use Euler's method with $x_0 = 1$ and $dx = 0.2$ to estimate $y(2)$. Round your answers to four decimal places. Find the exact value of $y(2)$ for comparison.

29. $y' = x$ **30.** $y' = 1/x$

31. $y' = xy$ **32.** $y' = 1/y$

Autonomous Differential Equations and Phase Lines

In Exercises 33 and 34:

   a. Identify the equilibrium values. Which are stable and which are unstable?

   b. Construct a phase line. Identify the signs of y' and y''.

   c. Sketch a representative selection of solution curves.

33. $\dfrac{dy}{dx} = y^2 - 1$ **34.** $\dfrac{dy}{dx} = y - y^2$

Applications

35. Escape velocity The gravitational attraction F exerted by an airless moon on a body of mass m at a distance s from the moon's center is given by the equation $F = -mg R^2 s^{-2}$, where g is the acceleration of gravity at the moon's surface and R is the moon's radius (see accompanying figure). The force F is negative because it acts in the direction of decreasing s.

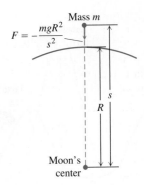

a. If the body is projected vertically upward from the moon's surface with an initial velocity v_0 at time $t = 0$, use Newton's second law, $F = ma$, to show that the body's velocity at position s is given by the equation

$$v^2 = \frac{2gR^2}{s} + v_0{}^2 - 2gR.$$

Thus, the velocity remains positive as long as $v_0 \ge \sqrt{2gR}$. The velocity $v_0 = \sqrt{2gR}$ is the moon's **escape velocity**. A body projected upward with this velocity or a greater one will escape from the moon's gravitational pull.

b. Show that if $v_0 = \sqrt{2gR}$, then

$$s = R\left(1 + \frac{3v_0}{2R}t\right)^{2/3}.$$

36. Coasting to a stop Table 9.6 shows the distance s (meters) coasted on in-line skates in t sec by Johnathon Krueger. Find a model for his position in the form of Equation (2) of Section 9.3. His initial velocity was $v_0 = 0.86$ m/sec, his mass $m = 30.84$ kg (he weighed 68 lb), and his total coasting distance 0.97 m.

TABLE 9.6 Johnathon Krueger skating data

t (sec)	s (m)	t (sec)	s (m)	t (sec)	s (m)
0	0	0.93	0.61	1.86	0.93
0.13	0.08	1.06	0.68	2.00	0.94
0.27	0.19	1.20	0.74	2.13	0.95
0.40	0.28	1.33	0.79	2.26	0.96
0.53	0.36	1.46	0.83	2.39	0.96
0.67	0.45	1.60	0.87	2.53	0.97
0.80	0.53	1.73	0.90	2.66	0.97

Chapter 9   Additional and Advanced Exercises

Theory and Applications

1. Transport through a cell membrane Under some conditions, the result of the movement of a dissolved substance across a cell's membrane is described by the equation

$$\frac{dy}{dt} = k\frac{A}{V}(c - y).$$

In this equation, y is the concentration of the substance inside the cell and dy/dt is the rate at which y changes over time. The letters

k, A, V, and c stand for constants, k being the *permeability coefficient* (a property of the membrane), A the surface area of the membrane, V the cell's volume, and c the concentration of the substance outside the cell. The equation says that the rate at which the concentration changes within the cell is proportional to the difference between it and the outside concentration.

a. Solve the equation for $y(t)$, using y_0 to denote $y(0)$.

b. Find the steady-state concentration, $\lim_{t\to\infty} y(t)$.

2. **Height of a rocket** If an external force F acts upon a system whose mass varies with time, Newton's law of motion is

$$\frac{d(mv)}{dt} = F + (v + u)\frac{dm}{dt}.$$

In this equation, m is the mass of the system at time t, v is its velocity, and $v + u$ is the velocity of the mass that is entering (or leaving) the system at the rate dm/dt. Suppose that a rocket of initial mass m_0 starts from rest, but is driven upward by firing some of its mass directly backward at the constant rate of $dm/dt = -b$ units per second and at constant speed relative to the rocket $u = -c$. The only external force acting on the rocket is $F = -mg$ due to gravity. Under these assumptions, show that the height of the rocket above the ground at the end of t seconds (t small compared to m_0/b) is

$$y = c\left[t + \frac{m_0 - bt}{b}\ln\frac{m_0 - bt}{m_0}\right] - \frac{1}{2}gt^2.$$

3. **a.** Assume that $P(x)$ and $Q(x)$ are continuous over the interval $[a, b]$. Use the Fundamental Theorem of Calculus, Part 1 to show that any function y satisfying the equation

$$v(x)y = \int v(x)Q(x)\,dx + C$$

for $v(x) = e^{\int P(x)\,dx}$ is a solution to the first-order linear equation

$$\frac{dy}{dx} + P(x)y = Q(x).$$

b. If $C = y_0v(x_0) - \int_{x_0}^{x} v(t)Q(t)\,dt$, then show that any solution y in part (a) satisfies the initial condition $y(x_0) = y_0$.

4. (*Continuation of Exercise 3.*) Assume the hypotheses of Exercise 3, and assume that $y_1(x)$ and $y_2(x)$ are both solutions to the first-order linear equation satisfying the initial condition $y(x_0) = y_0$.

a. Verify that $y(x) = y_1(x) - y_2(x)$ satisfies the initial value problem

$$y' + P(x)y = 0, \quad y(x_0) = 0.$$

b. For the integrating factor $v(x) = e^{\int P(x)\,dx}$, show that

$$\frac{d}{dx}\left(v(x)[y_1(x) - y_2(x)]\right) = 0.$$

Conclude that $v(x)[y_1(x) - y_2(x)] \equiv \text{constant}$.

c. From part (a), we have $y_1(x_0) - y_2(x_0) = 0$. Since $v(x) > 0$ for $a < x < b$, use part (b) to establish that $y_1(x) - y_2(x) \equiv 0$ on the interval (a, b). Thus $y_1(x) = y_2(x)$ for all $a < x < b$.

Homogeneous Equations

A first-order differential equation of the form

$$\frac{dy}{dx} = F\left(\frac{y}{x}\right)$$

is called *homogeneous*. It can be transformed into an equation whose variables are separable by defining the new variable $v = y/x$. Then, $y = vx$ and

$$\frac{dy}{dx} = v + x\frac{dv}{dx}.$$

Substitution into the original differential equation and collecting terms with like variables then gives the separable equation

$$\frac{dx}{x} + \frac{dv}{v - F(v)} = 0.$$

After solving this separable equation, the solution of the original equation is obtained when we replace v by y/x.

Solve the homogeneous equations in Exercises 5–10. First put the equation in the form of a homogeneous equation.

5. $(x^2 + y^2)\,dx + xy\,dy = 0$

6. $x^2\,dy + (y^2 - xy)\,dx = 0$

7. $(xe^{y/x} + y)\,dx - x\,dy = 0$

8. $(x + y)\,dy + (x - y)\,dx = 0$

9. $y' = \dfrac{y}{x} + \cos\dfrac{y - x}{x}$

10. $\left(x\sin\dfrac{y}{x} - y\cos\dfrac{y}{x}\right)dx + x\cos\dfrac{y}{x}\,dy = 0$

Chapter 9   Technology Application Projects

Mathematica/Maple Modules:

Drug Dosages: Are They Effective? Are They Safe?

Formulate and solve an initial value model for the absorption of a drug in the bloodstream.

First-Order Differential Equations and Slope Fields

Plot slope fields and solution curves for various initial conditions to selected first-order differential equations.

10

INFINITE SEQUENCES AND SERIES

OVERVIEW Everyone knows how to add two numbers together, or even several. But how do you add infinitely many numbers together? In this chapter we answer this question, which is part of the theory of infinite sequences and series.

An important application of this theory is a method for representing a known differentiable function $f(x)$ as an infinite sum of powers of x, so it looks like a "polynomial with infinitely many terms." Moreover, the method extends our knowledge of how to evaluate, differentiate, and integrate polynomials, so we can work with even more general functions than those encountered so far. These new functions are often solutions to important problems in science and engineering.

10.1 Sequences

HISTORICAL ESSAY

Sequences and Series

Sequences are fundamental to the study of infinite series and many applications of mathematics. We have already seen an example of a sequence when we studied Newton's Method in Section 4.7. There we produced a sequence of approximations x_n that became closer and closer to the root of a differentiable function. Now we will explore general sequences of numbers and the conditions under which they converge.

Representing Sequences

A sequence is a list of numbers

$$a_1, a_2, a_3, \ldots, a_n, \ldots$$

in a given order. Each of a_1, a_2, a_3 and so on represents a number. These are the **terms** of the sequence. For example, the sequence

$$2, 4, 6, 8, 10, 12, \ldots, 2n, \ldots$$

has first term $a_1 = 2$, second term $a_2 = 4$, and nth term $a_n = 2n$. The integer n is called the **index** of a_n, and indicates where a_n occurs in the list. Order is important. The sequence $2, 4, 6, 8 \ldots$ is not the same as the sequence $4, 2, 6, 8 \ldots$.

We can think of the sequence

$$a_1, a_2, a_3, \ldots, a_n, \ldots$$

as a function that sends 1 to a_1, 2 to a_2, 3 to a_3, and in general sends the positive integer n to the nth term a_n. More precisely, an **infinite sequence** of numbers is a function whose domain is the set of positive integers.

The function associated with the sequence

$$2, 4, 6, 8, 10, 12, \ldots, 2n, \ldots$$

sends 1 to $a_1 = 2$, 2 to $a_2 = 4$, and so on. The general behavior of this sequence is described by the formula $a_n = 2n$.

We can equally well make the domain the integers larger than a given number n_0, and we allow sequences of this type also. For example, the sequence

$$12, 14, 16, 18, 20, 22 \ldots$$

is described by the formula $a_n = 10 + 2n$. It can also be described by the simpler formula $b_n = 2n$, where the index n starts at 6 and increases. To allow such simpler formulas, we let the first index of the sequence be any integer. In the sequence above, $\{a_n\}$ starts with a_1 while $\{b_n\}$ starts with b_6.

Sequences can be described by writing rules that specify their terms, such as

$$a_n = \sqrt{n}, \qquad b_n = (-1)^{n+1}\frac{1}{n}, \qquad c_n = \frac{n-1}{n}, \qquad d_n = (-1)^{n+1},$$

or by listing terms:

$$\{a_n\} = \left\{ \sqrt{1}, \sqrt{2}, \sqrt{3}, \ldots, \sqrt{n}, \ldots \right\}$$

$$\{b_n\} = \left\{ 1, -\frac{1}{2}, \frac{1}{3}, -\frac{1}{4}, \ldots, (-1)^{n+1}\frac{1}{n}, \ldots \right\}$$

$$\{c_n\} = \left\{ 0, \frac{1}{2}, \frac{2}{3}, \frac{3}{4}, \frac{4}{5}, \ldots, \frac{n-1}{n}, \ldots \right\}$$

$$\{d_n\} = \left\{ 1, -1, 1, -1, 1, -1, \ldots, (-1)^{n+1}, \ldots \right\}.$$

We also sometimes write

$$\{a_n\} = \left\{ \sqrt{n} \right\}_{n=1}^{\infty}.$$

Figure 10.1 shows two ways to represent sequences graphically. The first marks the first few points from $a_1, a_2, a_3, \ldots, a_n, \ldots$ on the real axis. The second method shows the graph of the function defining the sequence. The function is defined only on integer inputs, and the graph consists of some points in the xy-plane located at $(1, a_1), (2, a_2), \ldots, (n, a_n), \ldots.$

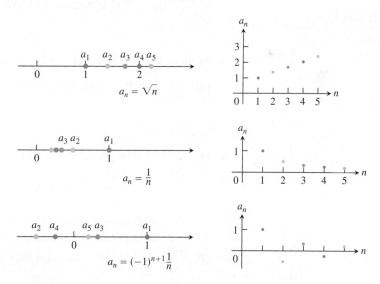

FIGURE 10.1 Sequences can be represented as points on the real line or as points in the plane where the horizontal axis n is the index number of the term and the vertical axis a_n is its value.

Convergence and Divergence

Sometimes the numbers in a sequence approach a single value as the index n increases. This happens in the sequence

$$\left\{ 1, \frac{1}{2}, \frac{1}{3}, \frac{1}{4}, \ldots, \frac{1}{n}, \ldots \right\}$$

whose terms approach 0 as n gets large, and in the sequence

$$\left\{ 0, \frac{1}{2}, \frac{2}{3}, \frac{3}{4}, \frac{4}{5}, \ldots, 1 - \frac{1}{n}, \ldots \right\}$$

whose terms approach 1. On the other hand, sequences like

$$\left\{ \sqrt{1}, \sqrt{2}, \sqrt{3}, \ldots, \sqrt{n}, \ldots \right\}$$

have terms that get larger than any number as n increases, and sequences like

$$\left\{ 1, -1, 1, -1, 1, -1, \ldots, (-1)^{n+1}, \ldots \right\}$$

bounce back and forth between 1 and -1, never converging to a single value. The following definition captures the meaning of having a sequence converge to a limiting value. It says that if we go far enough out in the sequence, by taking the index n to be larger than some value N, the difference between a_n and the limit of the sequence becomes less than any preselected number $\epsilon > 0$.

FIGURE 10.2 In the representation of a sequence as points in the plane, $a_n \rightarrow L$ if $y = L$ is a horizontal asymptote of the sequence of points $\{(n, a_n)\}$. In this figure, all the a_n's after a_N lie within ϵ of L.

> **DEFINITIONS** The sequence $\{a_n\}$ **converges** to the number L if for every positive number ϵ there corresponds an integer N such that for all n,
>
> $$n > N \quad \Rightarrow \quad |a_n - L| < \epsilon.$$
>
> If no such number L exists, we say that $\{a_n\}$ **diverges**.
> If $\{a_n\}$ converges to L, we write $\lim_{n \to \infty} a_n = L$, or simply $a_n \rightarrow L$, and call L the **limit** of the sequence (Figure 10.2).

The definition is very similar to the definition of the limit of a function $f(x)$ as x tends to ∞ ($\lim_{x \to \infty} f(x)$ in Section 2.6). We will exploit this connection to calculate limits of sequences.

EXAMPLE 1 Show that

(a) $\lim\limits_{n \to \infty} \dfrac{1}{n} = 0$ **(b)** $\lim\limits_{n \to \infty} k = k$ (any constant k)

Solution

(a) Let $\epsilon > 0$ be given. We must show that there exists an integer N such that for all n,

$$n > N \quad \Rightarrow \quad \left| \frac{1}{n} - 0 \right| < \epsilon.$$

This implication will hold if $(1/n) < \epsilon$ or $n > 1/\epsilon$. If N is any integer greater than $1/\epsilon$, the implication will hold for all $n > N$. This proves that $\lim_{n \to \infty} (1/n) = 0$.

(b) Let $\epsilon > 0$ be given. We must show that there exists an integer N such that for all n,

$$n > N \quad \Rightarrow \quad |k - k| < \epsilon.$$

Since $k - k = 0$, we can use any positive integer for N and the implication will hold. This proves that $\lim_{n \to \infty} k = k$ for any constant k. ∎

EXAMPLE 2 Show that the sequence $\{1, -1, 1, -1, 1, -1, \ldots, (-1)^{n+1}, \ldots\}$ diverges.

Solution Suppose the sequence converges to some number L. By choosing $\epsilon = 1/2$ in the definition of the limit, all terms a_n of the sequence with index n larger than some N must lie within $\epsilon = 1/2$ of L. Since the number 1 appears repeatedly as every other term of the sequence, we must have that the number 1 lies within the distance $\epsilon = 1/2$ of L.

Factorial Notation

The notation $n!$ ("n factorial") means the product $1 \cdot 2 \cdot 3 \cdots n$ of the integers from 1 to n. Notice that $(n+1)! = (n+1) \cdot n!$. Thus, $4! = 1 \cdot 2 \cdot 3 \cdot 4 = 24$ and $5! = 1 \cdot 2 \cdot 3 \cdot 4 \cdot 5 = 5 \cdot 4! = 120$. We define $0!$ to be 1. Factorials grow even faster than exponentials, as the table suggests. The values in the table are rounded.

n	e^n	$n!$
1	3	1
5	148	120
10	22,026	3,628,800
20	4.9×10^8	2.4×10^{18}

(e) $\left(\dfrac{n-2}{n}\right)^n = \left(1 + \dfrac{-2}{n}\right)^n \to e^{-2}$ Formula 5 with $x = -2$

(f) $\dfrac{100^n}{n!} \to 0$ Formula 6 with $x = 100$ ∎

Recursive Definitions

So far, we have calculated each a_n directly from the value of n. But sequences are often defined **recursively** by giving

1. The value(s) of the initial term or terms, and

2. A rule, called a **recursion formula**, for calculating any later term from terms that precede it.

EXAMPLE 10

(a) The statements $a_1 = 1$ and $a_n = a_{n-1} + 1$ for $n > 1$ define the sequence $1, 2, 3, \ldots, n, \ldots$ of positive integers. With $a_1 = 1$, we have $a_2 = a_1 + 1 = 2$, $a_3 = a_2 + 1 = 3$, and so on.

(b) The statements $a_1 = 1$ and $a_n = n \cdot a_{n-1}$ for $n > 1$ define the sequence $1, 2, 6, 24, \ldots, n!, \ldots$ of factorials. With $a_1 = 1$, we have $a_2 = 2 \cdot a_1 = 2$, $a_3 = 3 \cdot a_2 = 6$, $a_4 = 4 \cdot a_3 = 24$, and so on.

(c) The statements $a_1 = 1$, $a_2 = 1$, and $a_{n+1} = a_n + a_{n-1}$ for $n > 2$ define the sequence $1, 1, 2, 3, 5, \ldots$ of **Fibonacci numbers**. With $a_1 = 1$ and $a_2 = 1$, we have $a_3 = 1 + 1 = 2$, $a_4 = 2 + 1 = 3$, $a_5 = 3 + 2 = 5$, and so on.

(d) As we can see by applying Newton's method (see Exercise 133), the statements $x_0 = 1$ and $x_{n+1} = x_n - [(\sin x_n - x_n^2)/(\cos x_n - 2x_n)]$ for $n > 0$ define a sequence that, when it converges, gives a solution to the equation $\sin x - x^2 = 0$. ∎

Bounded Monotonic Sequences

Two concepts that play a key role in determining the convergence of a sequence are those of a *bounded* sequence and a *monotonic* sequence.

DEFINITIONS A sequence $\{a_n\}$ is **bounded from above** if there exists a number M such that $a_n \leq M$ for all n. The number M is an **upper bound** for $\{a_n\}$. If M is an upper bound for $\{a_n\}$ but no number less than M is an upper bound for $\{a_n\}$, then M is the **least upper bound** for $\{a_n\}$.

A sequence $\{a_n\}$ is **bounded from below** if there exists a number m such that $a_n \geq m$ for all n. The number m is a **lower bound** for $\{a_n\}$. If m is a lower bound for $\{a_n\}$ but no number greater than m is a lower bound for $\{a_n\}$, then m is the **greatest lower bound** for $\{a_n\}$.

If $\{a_n\}$ is bounded from above and below, then $\{a_n\}$ is **bounded**. If $\{a_n\}$ is not bounded, then we say that $\{a_n\}$ is an **unbounded** sequence.

EXAMPLE 11

(a) The sequence $1, 2, 3, \ldots, n, \ldots$ has no upper bound since it eventually surpasses every number M. However, it is bounded below by every real number less than or equal to 1. The number $m = 1$ is the greatest lower bound of the sequence.

(b) The sequence $\dfrac{1}{2}, \dfrac{2}{3}, \dfrac{3}{4}, \ldots, \dfrac{n}{n+1}, \ldots$ is bounded above by every real number greater than or equal to 1. The upper bound $M = 1$ is the least upper bound (Exercise 125). The sequence is also bounded below by every number less than or equal to $\dfrac{1}{2}$, which is its greatest lower bound. ∎

Convergent sequences are bounded

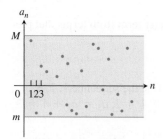

FIGURE 10.6 Some bounded sequences bounce around between their bounds and fail to converge to any limiting value.

If a sequence $\{a_n\}$ converges to the number L, then by definition there is a number N such that $|a_n - L| < 1$ if $n > N$. That is,

$$L - 1 < a_n < L + 1 \quad \text{for } n > N.$$

If M is a number larger than $L + 1$ and all of the finitely many numbers $a_1, a_2, \ldots, a_N$, then for every index n we have $a_n \leq M$ so that $\{a_n\}$ is bounded from above. Similarly, if m is a number smaller than $L - 1$ and all of the numbers $a_1, a_2, \ldots, a_N$, then m is a lower bound of the sequence. Therefore, all convergent sequences are bounded.

Although it is true that every convergent sequence is bounded, there are bounded sequences that fail to converge. One example is the bounded sequence $\{(-1)^{n+1}\}$ discussed in Example 2. The problem here is that some bounded sequences bounce around in the band determined by any lower bound m and any upper bound M (Figure 10.6). An important type of sequence that does not behave that way is one for which each term is at least as large, or at least as small, as its predecessor.

> **DEFINITION** A sequence $\{a_n\}$ is **nondecreasing** if $a_n \leq a_{n+1}$ for all n. That is, $a_1 \leq a_2 \leq a_3 \leq \ldots$. The sequence is **nonincreasing** if $a_n \geq a_{n+1}$ for all n. The sequence $\{a_n\}$ is **monotonic** if it is either nondecreasing or nonincreasing.

EXAMPLE 12

(a) The sequence $1, 2, 3, \ldots, n, \ldots$ is nondecreasing.

(b) The sequence $\dfrac{1}{2}, \dfrac{2}{3}, \dfrac{3}{4}, \ldots, \dfrac{n}{n+1}, \ldots$ is nondecreasing.

(c) The sequence $1, \dfrac{1}{2}, \dfrac{1}{4}, \dfrac{1}{8}, \ldots, \dfrac{1}{2^n}, \ldots$ is nonincreasing.

(d) The constant sequence $3, 3, 3, \ldots, 3, \ldots$ is both nondecreasing and nonincreasing.

(e) The sequence $1, -1, 1, -1, 1, -1, \ldots$ is not monotonic. ■

A nondecreasing sequence that is bounded from above always has a least upper bound. Likewise, a nonincreasing sequence bounded from below always has a greatest lower bound. These results are based on the *completeness property* of the real numbers, discussed in Appendix 6. We now prove that if L is the least upper bound of a nondecreasing sequence then the sequence converges to L, and that if L is the greatest lower bound of a nonincreasing sequence then the sequence converges to L.

> **THEOREM 6—The Monotonic Sequence Theorem** If a sequence $\{a_n\}$ is both bounded and monotonic, then the sequence converges.

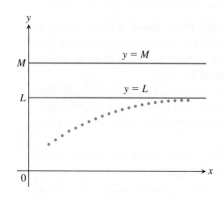

FIGURE 10.7 If the terms of a nondecreasing sequence have an upper bound M, they have a limit $L \leq M$.

Proof Suppose $\{a_n\}$ is nondecreasing, L is its least upper bound, and we plot the points $(1, a_1), (2, a_2), \ldots, (n, a_n), \ldots$ in the xy-plane. If M is an upper bound of the sequence, all these points will lie on or below the line $y = M$ (Figure 10.7). The line $y = L$ is the lowest such line. None of the points (n, a_n) lies above $y = L$, but some do lie above any lower line $y = L - \epsilon$, if ϵ is a positive number. The sequence converges to L because

(a) $a_n \leq L$ for *all* values of n, and

(b) given any $\epsilon > 0$, there exists at least one integer N for which $a_N > L - \epsilon$.

The fact that $\{a_n\}$ is nondecreasing tells us further that

$$a_n \geq a_N > L - \epsilon \quad \text{for all } n \geq N.$$

Thus, *all* the numbers a_n beyond the Nth number lie within ϵ of L. This is precisely the condition for L to be the limit of the sequence $\{a_n\}$.

The proof for nonincreasing sequences bounded from below is similar. ■

Exercises 10.2

Finding nth Partial Sums

In Exercises 1–6, find a formula for the nth partial sum of each series and use it to find the series' sum if the series converges.

1. $2 + \dfrac{2}{3} + \dfrac{2}{9} + \dfrac{2}{27} + \cdots + \dfrac{2}{3^{n-1}} + \cdots$

2. $\dfrac{9}{100} + \dfrac{9}{100^2} + \dfrac{9}{100^3} + \cdots + \dfrac{9}{100^n} + \cdots$

3. $1 - \dfrac{1}{2} + \dfrac{1}{4} - \dfrac{1}{8} + \cdots + (-1)^{n-1}\dfrac{1}{2^{n-1}} + \cdots$

4. $1 - 2 + 4 - 8 + \cdots + (-1)^{n-1}\,2^{n-1} + \cdots$

5. $\dfrac{1}{2 \cdot 3} + \dfrac{1}{3 \cdot 4} + \dfrac{1}{4 \cdot 5} + \cdots + \dfrac{1}{(n+1)(n+2)} + \cdots$

6. $\dfrac{5}{1 \cdot 2} + \dfrac{5}{2 \cdot 3} + \dfrac{5}{3 \cdot 4} + \cdots + \dfrac{5}{n(n+1)} + \cdots$

Series with Geometric Terms

In Exercises 7–14, write out the first few terms of each series to show how the series starts. Then find the sum of the series.

7. $\displaystyle\sum_{n=0}^{\infty} \dfrac{(-1)^n}{4^n}$

8. $\displaystyle\sum_{n=2}^{\infty} \dfrac{1}{4^n}$

9. $\displaystyle\sum_{n=1}^{\infty} \dfrac{7}{4^n}$

10. $\displaystyle\sum_{n=0}^{\infty} (-1)^n \dfrac{5}{4^n}$

11. $\displaystyle\sum_{n=0}^{\infty} \left(\dfrac{5}{2^n} + \dfrac{1}{3^n}\right)$

12. $\displaystyle\sum_{n=0}^{\infty} \left(\dfrac{5}{2^n} - \dfrac{1}{3^n}\right)$

13. $\displaystyle\sum_{n=0}^{\infty} \left(\dfrac{1}{2^n} + \dfrac{(-1)^n}{5^n}\right)$

14. $\displaystyle\sum_{n=0}^{\infty} \left(\dfrac{2^{n+1}}{5^n}\right)$

In Exercises 15–18, determine if the geometric series converges or diverges. If a series converges, find its sum.

15. $1 + \left(\dfrac{2}{5}\right) + \left(\dfrac{2}{5}\right)^2 + \left(\dfrac{2}{5}\right)^3 + \left(\dfrac{2}{5}\right)^4 + \cdots$

16. $1 + (-3) + (-3)^2 + (-3)^3 + (-3)^4 + \cdots$

17. $\left(\dfrac{1}{8}\right) + \left(\dfrac{1}{8}\right)^2 + \left(\dfrac{1}{8}\right)^3 + \left(\dfrac{1}{8}\right)^4 + \left(\dfrac{1}{8}\right)^5 + \cdots$

18. $\left(\dfrac{-2}{3}\right)^2 + \left(\dfrac{-2}{3}\right)^3 + \left(\dfrac{-2}{3}\right)^4 + \left(\dfrac{-2}{3}\right)^5 + \left(\dfrac{-2}{3}\right)^6 + \cdots$

Repeating Decimals

Express each of the numbers in Exercises 19–26 as the ratio of two integers.

19. $0.\overline{23} = 0.23\,23\,23\ldots$

20. $0.\overline{234} = 0.234\,234\,234\ldots$

21. $0.\overline{7} = 0.7777\ldots$

22. $0.\overline{d} = 0.dddd\ldots,$ where d is a digit

23. $0.0\overline{6} = 0.06666\ldots$

24. $1.\overline{414} = 1.414\,414\,414\ldots$

25. $1.24\overline{123} = 1.24\,123\,123\,123\ldots$

26. $3.\overline{142857} = 3.142857\,142857\ldots$

Using the nth-Term Test

In Exercises 27–34, use the nth-Term Test for divergence to show that the series is divergent, or state that the test is inconclusive.

27. $\displaystyle\sum_{n=1}^{\infty} \dfrac{n}{n+10}$

28. $\displaystyle\sum_{n=1}^{\infty} \dfrac{n(n+1)}{(n+2)(n+3)}$

29. $\displaystyle\sum_{n=0}^{\infty} \dfrac{1}{n+4}$

30. $\displaystyle\sum_{n=1}^{\infty} \dfrac{n}{n^2+3}$

31. $\displaystyle\sum_{n=1}^{\infty} \cos\dfrac{1}{n}$

32. $\displaystyle\sum_{n=0}^{\infty} \dfrac{e^n}{e^n+n}$

33. $\displaystyle\sum_{n=1}^{\infty} \ln\dfrac{1}{n}$

34. $\displaystyle\sum_{n=0}^{\infty} \cos n\pi$

Telescoping Series

In Exercises 35–40, find a formula for the nth partial sum of the series and use it to determine if the series converges or diverges. If a series converges, find its sum.

35. $\displaystyle\sum_{n=1}^{\infty} \left(\dfrac{1}{n} - \dfrac{1}{n+1}\right)$

36. $\displaystyle\sum_{n=1}^{\infty} \left(\dfrac{3}{n^2} - \dfrac{3}{(n+1)^2}\right)$

37. $\displaystyle\sum_{n=1}^{\infty} \left(\ln\sqrt{n+1} - \ln\sqrt{n}\right)$

38. $\displaystyle\sum_{n=1}^{\infty} \left(\tan(n) - \tan(n-1)\right)$

39. $\displaystyle\sum_{n=1}^{\infty} \left(\cos^{-1}\left(\dfrac{1}{n+1}\right) - \cos^{-1}\left(\dfrac{1}{n+2}\right)\right)$

40. $\displaystyle\sum_{n=1}^{\infty} \left(\sqrt{n+4} - \sqrt{n+3}\right)$

Find the sum of each series in Exercises 41–48.

41. $\displaystyle\sum_{n=1}^{\infty} \dfrac{4}{(4n-3)(4n+1)}$

42. $\displaystyle\sum_{n=1}^{\infty} \dfrac{6}{(2n-1)(2n+1)}$

43. $\displaystyle\sum_{n=1}^{\infty} \dfrac{40n}{(2n-1)^2(2n+1)^2}$

44. $\displaystyle\sum_{n=1}^{\infty} \dfrac{2n+1}{n^2(n+1)^2}$

45. $\displaystyle\sum_{n=1}^{\infty} \left(\dfrac{1}{\sqrt{n}} - \dfrac{1}{\sqrt{n+1}}\right)$

46. $\displaystyle\sum_{n=1}^{\infty} \left(\dfrac{1}{2^{1/n}} - \dfrac{1}{2^{1/(n+1)}}\right)$

47. $\displaystyle\sum_{n=1}^{\infty} \left(\dfrac{1}{\ln(n+2)} - \dfrac{1}{\ln(n+1)}\right)$

48. $\displaystyle\sum_{n=1}^{\infty} \left(\tan^{-1}(n) - \tan^{-1}(n+1)\right)$

Convergence or Divergence

Which series in Exercises 49–68 converge, and which diverge? Give reasons for your answers. If a series converges, find its sum.

49. $\displaystyle\sum_{n=0}^{\infty} \left(\dfrac{1}{\sqrt{2}}\right)^n$

50. $\displaystyle\sum_{n=0}^{\infty} \left(\sqrt{2}\right)^n$

51. $\displaystyle\sum_{n=1}^{\infty} (-1)^{n+1}\dfrac{3}{2^n}$

52. $\displaystyle\sum_{n=1}^{\infty} (-1)^{n+1}n$

53. $\displaystyle\sum_{n=0}^{\infty} \cos n\pi$

54. $\displaystyle\sum_{n=0}^{\infty} \dfrac{\cos n\pi}{5^n}$

55. $\displaystyle\sum_{n=0}^{\infty} e^{-2n}$

56. $\displaystyle\sum_{n=1}^{\infty} \ln\frac{1}{3^n}$

57. $\displaystyle\sum_{n=1}^{\infty} \frac{2}{10^n}$

58. $\displaystyle\sum_{n=0}^{\infty} \frac{1}{x^n}$, $\quad |x| > 1$

59. $\displaystyle\sum_{n=0}^{\infty} \frac{2^n - 1}{3^n}$

60. $\displaystyle\sum_{n=1}^{\infty} \left(1 - \frac{1}{n}\right)^n$

61. $\displaystyle\sum_{n=0}^{\infty} \frac{n!}{1000^n}$

62. $\displaystyle\sum_{n=1}^{\infty} \frac{n^n}{n!}$

63. $\displaystyle\sum_{n=1}^{\infty} \frac{2^n + 3^n}{4^n}$

64. $\displaystyle\sum_{n=1}^{\infty} \frac{2^n + 4^n}{3^n + 4^n}$

65. $\displaystyle\sum_{n=1}^{\infty} \ln\left(\frac{n}{n + 1}\right)$

66. $\displaystyle\sum_{n=1}^{\infty} \ln\left(\frac{n}{2n + 1}\right)$

67. $\displaystyle\sum_{n=0}^{\infty} \left(\frac{e}{\pi}\right)^n$

68. $\displaystyle\sum_{n=0}^{\infty} \frac{e^{n\pi}}{\pi^{ne}}$

Geometric Series with a Variable x

In each of the geometric series in Exercises 69–72, write out the first few terms of the series to find a and r, and find the sum of the series. Then express the inequality $|r| < 1$ in terms of x and find the values of x for which the inequality holds and the series converges.

69. $\displaystyle\sum_{n=0}^{\infty} (-1)^n x^n$

70. $\displaystyle\sum_{n=0}^{\infty} (-1)^n x^{2n}$

71. $\displaystyle\sum_{n=0}^{\infty} 3\left(\frac{x - 1}{2}\right)^n$

72. $\displaystyle\sum_{n=0}^{\infty} \frac{(-1)^n}{2} \left(\frac{1}{3 + \sin x}\right)^n$

In Exercises 73–78, find the values of x for which the given geometric series converges. Also, find the sum of the series (as a function of x) for those values of x.

73. $\displaystyle\sum_{n=0}^{\infty} 2^n x^n$

74. $\displaystyle\sum_{n=0}^{\infty} (-1)^n x^{-2n}$

75. $\displaystyle\sum_{n=0}^{\infty} (-1)^n (x + 1)^n$

76. $\displaystyle\sum_{n=0}^{\infty} \left(-\frac{1}{2}\right)^n (x - 3)^n$

77. $\displaystyle\sum_{n=0}^{\infty} \sin^n x$

78. $\displaystyle\sum_{n=0}^{\infty} (\ln x)^n$

Theory and Examples

79. The series in Exercise 5 can also be written as

$$\sum_{n=1}^{\infty} \frac{1}{(n + 1)(n + 2)} \quad \text{and} \quad \sum_{n=-1}^{\infty} \frac{1}{(n + 3)(n + 4)}.$$

Write it as a sum beginning with **(a)** $n = -2$, **(b)** $n = 0$, **(c)** $n = 5$.

80. The series in Exercise 6 can also be written as

$$\sum_{n=1}^{\infty} \frac{5}{n(n + 1)} \quad \text{and} \quad \sum_{n=0}^{\infty} \frac{5}{(n + 1)(n + 2)}.$$

Write it as a sum beginning with **(a)** $n = -1$, **(b)** $n = 3$, **(c)** $n = 20$.

81. Make up an infinite series of nonzero terms whose sum is

a. 1   **b.** -3 **c.** 0.

82. (*Continuation of Exercise 81.*) Can you make an infinite series of nonzero terms that converges to any number you want? Explain.

83. Show by example that $\Sigma(a_n/b_n)$ may diverge even though Σa_n and Σb_n converge and no b_n equals 0.

84. Find convergent geometric series $A = \Sigma a_n$ and $B = \Sigma b_n$ that illustrate the fact that $\Sigma a_n b_n$ may converge without being equal to AB.

85. Show by example that $\Sigma(a_n/b_n)$ may converge to something other than A/B even when $A = \Sigma a_n, B = \Sigma b_n \neq 0$, and no b_n equals 0.

86. If Σa_n converges and $a_n > 0$ for all n, can anything be said about $\Sigma(1/a_n)$? Give reasons for your answer.

87. What happens if you add a finite number of terms to a divergent series or delete a finite number of terms from a divergent series? Give reasons for your answer.

88. If Σa_n converges and Σb_n diverges, can anything be said about their term-by-term sum $\Sigma(a_n + b_n)$? Give reasons for your answer.

89. Make up a geometric series Σar^{n-1} that converges to the number 5 if

a. $a = 2$ **b.** $a = 13/2$.

90. Find the value of b for which

$$1 + e^b + e^{2b} + e^{3b} + \cdots = 9.$$

91. For what values of r does the infinite series

$$1 + 2r + r^2 + 2r^3 + r^4 + 2r^5 + r^6 + \cdots$$

converge? Find the sum of the series when it converges.

92. Show that the error $(L - s_n)$ obtained by replacing a convergent geometric series with one of its partial sums s_n is $ar^n/(1 - r)$.

93. The accompanying figure shows the first five of a sequence of squares. The outermost square has an area of 4 m². Each of the other squares is obtained by joining the midpoints of the sides of the squares before it. Find the sum of the areas of all the squares.

94. Helga von Koch's snowflake curve   Helga von Koch's snowflake is a curve of infinite length that encloses a region of finite area. To see why this is so, suppose the curve is generated by starting with an equilateral triangle whose sides have length 1.

a. Find the length L_n of the nth curve C_n and show that $\lim_{n\to\infty} L_n = \infty$.

b. Find the area A_n of the region enclosed by C_n and show that $\lim_{n\to\infty} A_n = (8/5) A_1$.

10.3 | The Integral Test

Given a series, we want to know whether it converges or not. In this section and the next two, we study series with nonnegative terms. Such a series converges if its sequence of partial sums is bounded. If we establish that a given series does converge, we generally do not have a formula available for its sum, so we investigate methods to approximate the sum instead.

Nondecreasing Partial Sums

Suppose that $\sum_{n=1}^{\infty} a_n$ is an infinite series with $a_n \geq 0$ for all n. Then each partial sum is greater than or equal to its predecessor because $s_{n+1} = s_n + a_n$:

$$s_1 \leq s_2 \leq s_3 \leq \cdots \leq s_n \leq s_{n+1} \leq \cdots.$$

Since the partial sums form a nondecreasing sequence, the Monotonic Sequence Theorem (Theorem 6, Section 10.1) gives the following result.

> **Corollary of Theorem 6**   A series $\sum_{n=1}^{\infty} a_n$ of nonnegative terms converges if and only if its partial sums are bounded from above.

EXAMPLE 1   The series

$$\sum_{n=1}^{\infty} \frac{1}{n} = 1 + \frac{1}{2} + \frac{1}{3} + \cdots + \frac{1}{n} + \cdots$$

is called the **harmonic series**. The harmonic series is divergent, but this doesn't follow from the nth-Term Test. The nth term $1/n$ does go to zero, but the series still diverges. The reason it diverges is because there is no upper bound for its partial sums. To see why, group the terms of the series in the following way:

$$1 + \frac{1}{2} + \underbrace{\left(\frac{1}{3} + \frac{1}{4}\right)}_{> \frac{2}{4} = \frac{1}{2}} + \underbrace{\left(\frac{1}{5} + \frac{1}{6} + \frac{1}{7} + \frac{1}{8}\right)}_{> \frac{4}{8} = \frac{1}{2}} + \underbrace{\left(\frac{1}{9} + \frac{1}{10} + \cdots + \frac{1}{16}\right)}_{> \frac{8}{16} = \frac{1}{2}} + \cdots.$$

The sum of the first two terms is 1.5. The sum of the next two terms is $1/3 + 1/4$, which is greater than $1/4 + 1/4 = 1/2$. The sum of the next four terms is $1/5 + 1/6 + 1/7 + 1/8$, which is greater than $1/8 + 1/8 + 1/8 + 1/8 = 1/2$. The sum of the next eight terms is $1/9 + 1/10 + 1/11 + 1/12 + 1/13 + 1/14 + 1/15 + 1/16$, which is greater than $8/16 = 1/2$. The sum of the next 16 terms is greater than $16/32 = 1/2$, and so on. In general, the sum of 2^n terms ending with $1/2^{n+1}$ is greater than $2^n/2^{n+1} = 1/2$. The sequence of partial sums is not bounded from above: If $n = 2^k$, the partial sum s_n is greater than $k/2$. The harmonic series diverges. ∎

The Integral Test

We now introduce the Integral Test with a series that is related to the harmonic series, but whose nth term is $1/n^2$ instead of $1/n$.

EXAMPLE 2   Does the following series converge?

$$\sum_{n=1}^{\infty} \frac{1}{n^2} = 1 + \frac{1}{4} + \frac{1}{9} + \frac{1}{16} + \cdots + \frac{1}{n^2} + \cdots$$

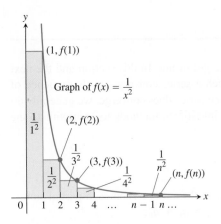

FIGURE 10.10 The sum of the areas of the rectangles under the graph of $f(x) = 1/x^2$ is less than the area under the graph (Example 2).

Caution

The series and integral need not have the same value in the convergent case. As we noted in Example 2, $\sum_{n=1}^{\infty} (1/n^2) = \pi^2/6$ while $\int_1^{\infty} (1/x^2) \, dx = 1$.

(a)

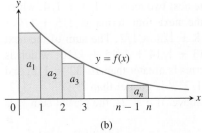

(b)

FIGURE 10.11 Subject to the conditions of the Integral Test, the series $\sum_{n=1}^{\infty} a_n$ and the integral $\int_1^{\infty} (x) \, dx$ both converge or both diverge.

Solution   We determine the convergence of $\sum_{n=1}^{\infty}(1/n^2)$ by comparing it with $\int_1^{\infty}(1/x^2) \, dx$. To carry out the comparison, we think of the terms of the series as values of the function $f(x) = 1/x^2$ and interpret these values as the areas of rectangles under the curve $y = 1/x^2$.

As Figure 10.10 shows,

$$s_n = \frac{1}{1^2} + \frac{1}{2^2} + \frac{1}{3^2} + \cdots + \frac{1}{n^2}$$

$$= f(1) + f(2) + f(3) + \cdots + f(n)$$

$$< f(1) + \int_1^n \frac{1}{x^2} \, dx \qquad \text{Rectangle areas sum to less than area under graph.}$$

$$< 1 + \int_1^{\infty} \frac{1}{x^2} \, dx \qquad \int_1^n (1/x^2) \, dx < \int_1^{\infty} (1/x^2) \, dx$$

$$< 1 + 1 = 2. \qquad \text{As in Section 8.7, Example 3,} \\ \int_1^{\infty} (1/x^2) \, dx = 1.$$

Thus the partial sums of $\sum_{n=1}^{\infty} (1/n^2)$ are bounded from above (by 2) and the series converges. The sum of the series is known to be $\pi^2/6 \approx 1.64493$. ■

> **THEOREM 9—The Integral Test**   Let $\{a_n\}$ be a sequence of positive terms. Suppose that $a_n = f(n)$, where f is a continuous, positive, decreasing function of x for all $x \geq N$ (N a positive integer). Then the series $\sum_{n=N}^{\infty} a_n$ and the integral $\int_N^{\infty} f(x) \, dx$ both converge or both diverge.

Proof   We establish the test for the case $N = 1$. The proof for general N is similar.

We start with the assumption that f is a decreasing function with $f(n) = a_n$ for every n. This leads us to observe that the rectangles in Figure 10.11a, which have areas $a_1, a_2, \ldots, a_n$, collectively enclose more area than that under the curve $y = f(x)$ from $x = 1$ to $x = n + 1$. That is,

$$\int_1^{n+1} f(x) \, dx \leq a_1 + a_2 + \cdots + a_n.$$

In Figure 10.11b the rectangles have been faced to the left instead of to the right. If we momentarily disregard the first rectangle of area a_1, we see that

$$a_2 + a_3 + \cdots + a_n \leq \int_1^n f(x) \, dx.$$

If we include a_1, we have

$$a_1 + a_2 + \cdots + a_n \leq a_1 + \int_1^n f(x) \, dx.$$

Combining these results gives

$$\int_1^{n+1} f(x) \, dx \leq a_1 + a_2 + \cdots + a_n \leq a_1 + \int_1^n f(x) \, dx.$$

These inequalities hold for each n, and continue to hold as $n \to \infty$.

If $\int_1^{\infty} f(x) \, dx$ is finite, the right-hand inequality shows that $\sum a_n$ is finite. If $\int_1^{\infty} f(x) \, dx$ is infinite, the left-hand inequality shows that $\sum a_n$ is infinite. Hence the series and the integral are both finite or both infinite. ■

EXAMPLE 3   Show that the *p*-series

$$\sum_{n=1}^{\infty} \frac{1}{n^p} = \frac{1}{1^p} + \frac{1}{2^p} + \frac{1}{3^p} + \cdots + \frac{1}{n^p} + \cdots$$

(*p* a real constant) converges if $p > 1$, and diverges if $p \leq 1$.

Solution   If $p > 1$, then $f(x) = 1/x^p$ is a positive decreasing function of x. Since

$$\int_1^{\infty} \frac{1}{x^p} dx = \int_1^{\infty} x^{-p} dx = \lim_{b \to \infty} \left[\frac{x^{-p+1}}{-p+1} \right]_1^b$$

$$= \frac{1}{1-p} \lim_{b \to \infty} \left(\frac{1}{b^{p-1}} - 1 \right)$$

$$= \frac{1}{1-p}(0 - 1) = \frac{1}{p-1}, \qquad \begin{array}{l} b^{p-1} \to \infty \text{ as } b \to \infty \\ \text{because } p - 1 > 0. \end{array}$$

> The *p*-series $\sum_{n=1}^{\infty} \dfrac{1}{n^p}$
>
> converges if $p > 1$, diverges if $p \leq 1$.

the series converges by the Integral Test. We emphasize that the sum of the *p*-series is *not* $1/(p-1)$. The series converges, but we don't know the value it converges to.
    If $p < 1$, then $1 - p > 0$ and

$$\int_1^{\infty} \frac{1}{x^p} dx = \frac{1}{1-p} \lim_{b \to \infty} (b^{1-p} - 1) = \infty.$$

The series diverges by the Integral Test.
    If $p = 1$, we have the (divergent) harmonic series

$$1 + \frac{1}{2} + \frac{1}{3} + \cdots + \frac{1}{n} + \cdots.$$

We have convergence for $p > 1$ but divergence for all other values of p. ∎

    The *p*-series with $p = 1$ is the **harmonic series** (Example 1). The *p*-Series Test shows that the harmonic series is just *barely* divergent; if we increase p to 1.000000001, for instance, the series converges!
    The slowness with which the partial sums of the harmonic series approach infinity is impressive. For instance, it takes more than 178 million terms of the harmonic series to move the partial sums beyond 20. (See also Exercise 43b.)

EXAMPLE 4   The series $\sum_{n=1}^{\infty} (1/(n^2 + 1))$ is not a *p*-series, but it converges by the Integral Test. The function $f(x) = 1/(x^2 + 1)$ is positive, continuous, and decreasing for $x \geq 1$, and

$$\int_1^{\infty} \frac{1}{x^2 + 1} dx = \lim_{b \to \infty} \left[\arctan x \right]_1^b$$

$$= \lim_{b \to \infty} [\arctan b - \arctan 1]$$

$$= \frac{\pi}{2} - \frac{\pi}{4} = \frac{\pi}{4}.$$

Again we emphasize that $\pi/4$ is *not* the sum of the series. The series converges, but we do not know the value of its sum. ∎

Error Estimation

If a series Σa_n is shown to be convergent by the Integral Test, we may want to estimate the size of the **remainder** R_n between the total sum S of the series and its *n*th partial sum s_n. That is, we wish to estimate

$$R_n = S - s_n = a_{n+1} + a_{n+2} + a_{n+3} + \cdots.$$

To get a lower bound for the remainder, we compare the sum of the areas of the rectangles with the area under the curve $y = f(x)$ for $x \geq n$ (see Figure 10.11a). We see that

$$R_n = a_{n+1} + a_{n+2} + a_{n+3} + \cdots \geq \int_{n+1}^{\infty} f(x)\, dx.$$

Similarly, from Figure 10.11b, we find an upper bound with

$$R_n = a_{n+1} + a_{n+2} + a_{n+3} + \cdots \leq \int_{n}^{\infty} f(x)\, dx.$$

These comparisons prove the following result giving bounds on the size of the remainder.

Bounds for the Remainder in the Integral Test

Suppose $\{a_k\}$ is a sequence of positive terms with $a_k = f(k)$, where f is a continuous positive decreasing function of x for all $x \geq n$, and that Σa_n converges to S. Then the remainder $R_n = S - s_n$ satisfies the inequalities

$$\int_{n+1}^{\infty} f(x)\, dx \leq R_n \leq \int_{n}^{\infty} f(x)\, dx. \tag{1}$$

If we add the partial sum s_n to each side of the inequalities in (1), we get

$$s_n + \int_{n+1}^{\infty} f(x)\, dx \leq S \leq s_n + \int_{n}^{\infty} f(x)\, dx \tag{2}$$

since $s_n + R_n = S$. The inequalities in (2) are useful for estimating the error in approximating the sum of a convergent series. The error can be no larger than the length of the interval containing S, as given by (2).

EXAMPLE 5 Estimate the sum of the series $\Sigma(1/n^2)$ using the inequalities in (2) and $n = 10$.

Solution We have that

$$\int_{n}^{\infty} \frac{1}{x^2}\, dx = \lim_{b \to \infty} \left[-\frac{1}{x} \right]_{n}^{b} = \lim_{b \to \infty} \left(-\frac{1}{b} + \frac{1}{n} \right) = \frac{1}{n}.$$

Using this result with the inequalities in (2), we get

$$s_{10} + \frac{1}{11} \leq S \leq s_{10} + \frac{1}{10}.$$

Taking $s_{10} = 1 + (1/4) + (1/9) + (1/16) + \cdots + (1/100) \approx 1.54977$, these last inequalities give

$$1.64068 \leq S \leq 1.64997.$$

If we approximate the sum S by the midpoint of this interval, we find that

$$\sum_{n=1}^{\infty} \frac{1}{n^2} \approx 1.6453.$$

The error in this approximation is less than half the length of the interval, so the error is less than 0.005. ∎

Exercises 10.3

Applying the Integral Test

Use the Integral Test to determine if the series in Exercises 1–10 converge or diverge. Be sure to check that the conditions of the Integral Test are satisfied.

1. $\displaystyle\sum_{n=1}^{\infty} \frac{1}{n^2}$ **2.** $\displaystyle\sum_{n=1}^{\infty} \frac{1}{n^{0.2}}$

3. $\displaystyle\sum_{n=1}^{\infty} \frac{1}{n^2 + 4}$ **4.** $\displaystyle\sum_{n=1}^{\infty} \frac{1}{n + 4}$

5. $\displaystyle\sum_{n=1}^{\infty} e^{-2n}$ **6.** $\displaystyle\sum_{n=2}^{\infty} \frac{1}{n(\ln n)^2}$

7. $\displaystyle\sum_{n=1}^{\infty} \frac{n}{n^2 + 4}$ **8.** $\displaystyle\sum_{n=2}^{\infty} \frac{\ln (n^2)}{n}$

9. $\displaystyle\sum_{n=1}^{\infty} \frac{n^2}{e^{n/3}}$ **10.** $\displaystyle\sum_{n=2}^{\infty} \frac{n - 4}{n^2 - 2n + 1}$

Determining Convergence or Divergence

Which of the series in Exercises 11–40 converge, and which diverge? Give reasons for your answers. (When you check an answer, remember that there may be more than one way to determine the series' convergence or divergence.)

11. $\displaystyle\sum_{n=1}^{\infty} \frac{1}{10^n}$ **12.** $\displaystyle\sum_{n=1}^{\infty} e^{-n}$ **13.** $\displaystyle\sum_{n=1}^{\infty} \frac{n}{n + 1}$

14. $\displaystyle\sum_{n=1}^{\infty} \frac{5}{n + 1}$ **15.** $\displaystyle\sum_{n=1}^{\infty} \frac{3}{\sqrt{n}}$ **16.** $\displaystyle\sum_{n=1}^{\infty} \frac{-2}{n\sqrt{n}}$

17. $\displaystyle\sum_{n=1}^{\infty} -\frac{1}{8^n}$ **18.** $\displaystyle\sum_{n=1}^{\infty} \frac{-8}{n}$ **19.** $\displaystyle\sum_{n=2}^{\infty} \frac{\ln n}{n}$

20. $\displaystyle\sum_{n=2}^{\infty} \frac{\ln n}{\sqrt{n}}$ **21.** $\displaystyle\sum_{n=1}^{\infty} \frac{2^n}{3^n}$ **22.** $\displaystyle\sum_{n=1}^{\infty} \frac{5^n}{4^n + 3}$

23. $\displaystyle\sum_{n=0}^{\infty} \frac{-2}{n + 1}$ **24.** $\displaystyle\sum_{n=1}^{\infty} \frac{1}{2n - 1}$ **25.** $\displaystyle\sum_{n=1}^{\infty} \frac{2^n}{n + 1}$

26. $\displaystyle\sum_{n=1}^{\infty} \frac{1}{\sqrt{n}(\sqrt{n} + 1)}$ **27.** $\displaystyle\sum_{n=2}^{\infty} \frac{\sqrt{n}}{\ln n}$ **28.** $\displaystyle\sum_{n=1}^{\infty} \left(1 + \frac{1}{n}\right)^n$

29. $\displaystyle\sum_{n=1}^{\infty} \frac{1}{(\ln 2)^n}$ **30.** $\displaystyle\sum_{n=1}^{\infty} \frac{1}{(\ln 3)^n}$

31. $\displaystyle\sum_{n=3}^{\infty} \frac{(1/n)}{(\ln n)\sqrt{\ln^2 n - 1}}$ **32.** $\displaystyle\sum_{n=1}^{\infty} \frac{1}{n(1 + \ln^2 n)}$

33. $\displaystyle\sum_{n=1}^{\infty} n \sin \frac{1}{n}$ **34.** $\displaystyle\sum_{n=1}^{\infty} n \tan \frac{1}{n}$

35. $\displaystyle\sum_{n=1}^{\infty} \frac{e^n}{1 + e^{2n}}$ **36.** $\displaystyle\sum_{n=1}^{\infty} \frac{2}{1 + e^n}$

37. $\displaystyle\sum_{n=1}^{\infty} \frac{8 \tan^{-1} n}{1 + n^2}$ **38.** $\displaystyle\sum_{n=1}^{\infty} \frac{n}{n^2 + 1}$

39. $\displaystyle\sum_{n=1}^{\infty} \operatorname{sech} n$ **40.** $\displaystyle\sum_{n=1}^{\infty} \operatorname{sech}^2 n$

Theory and Examples

For what values of a, if any, do the series in Exercises 41 and 42 converge?

41. $\displaystyle\sum_{n=1}^{\infty} \left(\frac{a}{n + 2} - \frac{1}{n + 4} \right)$ **42.** $\displaystyle\sum_{n=3}^{\infty} \left(\frac{1}{n - 1} - \frac{2a}{n + 1} \right)$

43. a. Draw illustrations like those in Figures 10.7 and 10.8 to show that the partial sums of the harmonic series satisfy the inequalities

$$\ln (n + 1) = \int_1^{n+1} \frac{1}{x} dx \le 1 + \frac{1}{2} + \cdots + \frac{1}{n}$$

$$\le 1 + \int_1^n \frac{1}{x} dx = 1 + \ln n.$$

T **b.** There is absolutely no empirical evidence for the divergence of the harmonic series even though we know it diverges. The partial sums just grow too slowly. To see what we mean, suppose you had started with $s_1 = 1$ the day the universe was formed, 13 billion years ago, and added a new term every *second*. About how large would the partial sum s_n be today, assuming a 365-day year?

44. Are there any values of x for which $\sum_{n=1}^{\infty}(1/(nx))$ converges? Give reasons for your answer.

45. Is it true that if $\sum_{n=1}^{\infty} a_n$ is a divergent series of positive numbers, then there is also a divergent series $\sum_{n=1}^{\infty} b_n$ of positive numbers with $b_n < a_n$ for every n? Is there a "smallest" divergent series of positive numbers? Give reasons for your answers.

46. (*Continuation of Exercise 45.*) Is there a "largest" convergent series of positive numbers? Explain.

47. $\sum_{n=1}^{\infty} \left(1/(\sqrt{n} + 1)\right)$ **diverges**

   a. Use the accompanying graph to show that the partial sum

   $s_{50} = \sum_{n=1}^{50} \left(1/(\sqrt{n} + 1)\right)$ satisfies

   $$\int_1^{51} \frac{1}{\sqrt{x} + 1} dx < s_{50} < \int_0^{50} \frac{1}{\sqrt{x} + 1} dx.$$

   Conclude that $11.5 < s_{50} < 12.3$.

   b. What should n be in order that the partial sum

   $$s_n = \sum_{i=1}^{n} \left(1/(\sqrt{i} + 1)\right) \text{ satisfy } s_n > 1000?$$

48. $\sum_{n=1}^{\infty} (1/n^4)$ **converges**

a. Use the accompanying graph to determine the error if $s_{30} = \sum_{n=1}^{30} (1/n^4)$ is used to estimate the value of $\sum_{n=1}^{\infty} (1/n^4)$.

b. Find n so that the partial sum $s_n = \sum_{i=1}^{n} (1/i^4)$ estimates the value of $\sum_{n=1}^{\infty} (1/n^4)$ with an error of at most 0.000001.

49. Estimate the value of $\sum_{n=1}^{\infty} (1/n^3)$ to within 0.01 of its exact value.

50. Estimate the value of $\sum_{n=2}^{\infty} (1/(n^2 + 4))$ to within 0.1 of its exact value.

51. How many terms of the convergent series $\sum_{n=1}^{\infty} (1/n^{1.1})$ should be used to estimate its value with error at most 0.00001?

52. How many terms of the convergent series $\sum_{n=4}^{\infty} (1/n(\ln n)^3)$ should be used to estimate its value with error at most 0.01?

53. The Cauchy condensation test The Cauchy condensation test says: Let $\{a_n\}$ be a nonincreasing sequence ($a_n \geq a_{n+1}$ for all n) of positive terms that converges to 0. Then $\sum a_n$ converges if and only if $\sum 2^n a_{2^n}$ converges. For example, $\sum (1/n)$ diverges because $\sum 2^n \cdot (1/2^n) = \sum 1$ diverges. Show why the test works.

54. Use the Cauchy condensation test from Exercise 53 to show that

a. $\sum_{n=2}^{\infty} \dfrac{1}{n \ln n}$ diverges;

b. $\sum_{n=1}^{\infty} \dfrac{1}{n^p}$ converges if $p > 1$ and diverges if $p \leq 1$.

55. Logarithmic p-series

a. Show that the improper integral

$$\int_{2}^{\infty} \frac{dx}{x(\ln x)^p} \quad (p \text{ a positive constant})$$

converges if and only if $p > 1$.

b. What implications does the fact in part (a) have for the convergence of the series

$$\sum_{n=2}^{\infty} \frac{1}{n(\ln n)^p} ?$$

Give reasons for your answer.

56. (*Continuation of Exercise 55.*) Use the result in Exercise 55 to determine which of the following series converge and which diverge. Support your answer in each case.

a. $\sum_{n=2}^{\infty} \dfrac{1}{n(\ln n)}$

b. $\sum_{n=2}^{\infty} \dfrac{1}{n(\ln n)^{1.01}}$

c. $\sum_{n=2}^{\infty} \dfrac{1}{n \ln(n^3)}$

d. $\sum_{n=2}^{\infty} \dfrac{1}{n(\ln n)^3}$

57. Euler's constant Graphs like those in Figure 10.11 suggest that as n increases there is little change in the difference between the sum

$$1 + \frac{1}{2} + \cdots + \frac{1}{n}$$

and the integral

$$\ln n = \int_{1}^{n} \frac{1}{x} \, dx.$$

To explore this idea, carry out the following steps.

a. By taking $f(x) = 1/x$ in the proof of Theorem 9, show that

$$\ln (n + 1) \leq 1 + \frac{1}{2} + \cdots + \frac{1}{n} \leq 1 + \ln n$$

or

$$0 < \ln (n + 1) - \ln n \leq 1 + \frac{1}{2} + \cdots + \frac{1}{n} - \ln n \leq 1.$$

Thus, the sequence

$$a_n = 1 + \frac{1}{2} + \cdots + \frac{1}{n} - \ln n$$

is bounded from below and from above.

b. Show that

$$\frac{1}{n + 1} < \int_{n}^{n+1} \frac{1}{x} \, dx = \ln (n + 1) - \ln n,$$

and use this result to show that the sequence $\{a_n\}$ in part (a) is decreasing.

Since a decreasing sequence that is bounded from below converges, the numbers a_n defined in part (a) converge:

$$1 + \frac{1}{2} + \cdots + \frac{1}{n} - \ln n \rightarrow \gamma.$$

The number γ, whose value is $0.5772 \ldots$, is called *Euler's constant*.

58. Use the Integral Test to show that the series

$$\sum_{n=0}^{\infty} e^{-n^2}$$

converges.

59. a. For the series $\sum (1/n^3)$, use the inequalities in Equation (2) with $n = 10$ to find an interval containing the sum S.

b. As in Example 5, use the midpoint of the interval found in part (a) to approximate the sum of the series. What is the maximum error for your approximation?

60. Repeat Exercise 59 using the series $\sum (1/n^4)$.

10.4 Comparison Tests

We have seen how to determine the convergence of geometric series, p-series, and a few others. We can test the convergence of many more series by comparing their terms to those of a series whose convergence is known.

> **THEOREM 10—The Comparison Test**   Let $\sum a_n$, $\sum c_n$, and $\sum d_n$ be series with nonnegative terms. Suppose that for some integer N
>
> $$d_n \le a_n \le c_n \qquad \text{for all} \qquad n > N.$$
>
> **(a)** If $\sum c_n$ converges, then $\sum a_n$ also converges.
> **(b)** If $\sum d_n$ diverges, then $\sum a_n$ also diverges.

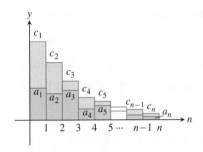

FIGURE 10.12 If the total area $\sum c_n$ of the taller c_n rectangles is finite, then so is the total area $\sum a_n$ of the shorter a_n rectangles.

Proof   In Part (a), the partial sums of $\sum a_n$ are bounded above by

$$M = a_1 + a_2 + \cdots + a_N + \sum_{n=N+1}^{\infty} c_n.$$

They therefore form a nondecreasing sequence with a limit $L \le M$. That is, if $\sum c_n$ converges, then so does $\sum a_n$. Figure 10.12 depicts this result, where each term of each series is interpreted as the area of a rectangle (just like we did for the integral test in Figure 10.11).

In Part (b), the partial sums of $\sum a_n$ are not bounded from above. If they were, the partial sums for $\sum d_n$ would be bounded by

$$M^* = d_1 + d_2 + \cdots + d_N + \sum_{n=N+1}^{\infty} a_n$$

and $\sum d_n$ would have to converge instead of diverge.   ∎

EXAMPLE 1   We apply Theorem 10 to several series.

(a) The series

$$\sum_{n=1}^{\infty} \frac{5}{5n - 1}$$

diverges because its nth term

$$\frac{5}{5n - 1} = \frac{1}{n - \frac{1}{5}} > \frac{1}{n}$$

is greater than the nth term of the divergent harmonic series.

(b) The series

$$\sum_{n=0}^{\infty} \frac{1}{n!} = 1 + \frac{1}{1!} + \frac{1}{2!} + \frac{1}{3!} + \cdots$$

converges because its terms are all positive and less than or equal to the corresponding terms of

$$1 + \sum_{n=0}^{\infty} \frac{1}{2^n} = 1 + 1 + \frac{1}{2} + \frac{1}{2^2} + \cdots.$$

The geometric series on the left converges and we have

$$1 + \sum_{n=0}^{\infty} \frac{1}{2^n} = 1 + \frac{1}{1 - (1/2)} = 3.$$

The fact that 3 is an upper bound for the partial sums of $\sum_{n=0}^{\infty}(1/n!)$ does not mean that the series converges to 3. As we will see in Section 10.9, the series converges to e.

(c) The series

$$5 + \frac{2}{3} + \frac{1}{7} + 1 + \frac{1}{2 + \sqrt{1}} + \frac{1}{4 + \sqrt{2}} + \frac{1}{8 + \sqrt{3}} + \cdots + \frac{1}{2^n + \sqrt{n}} + \cdots$$

converges. To see this, we ignore the first three terms and compare the remaining terms with those of the convergent geometric series $\sum_{n=0}^{\infty}(1/2^n)$. The term $1/\left(2^n + \sqrt{n}\right)$ of the truncated sequence is less than the corresponding term $1/2^n$ of the geometric series. We see that term by term we have the comparison

$$1 + \frac{1}{2 + \sqrt{1}} + \frac{1}{4 + \sqrt{2}} + \frac{1}{8 + \sqrt{3}} + \cdots \leq 1 + \frac{1}{2} + \frac{1}{4} + \frac{1}{8} + \cdots.$$

So the truncated series and the original series converge by an application of the Comparison Test. ∎

The Limit Comparison Test

We now introduce a comparison test that is particularly useful for series in which a_n is a rational function of n.

THEOREM 11—Limit Comparison Test   Suppose that $a_n > 0$ and $b_n > 0$ for all $n \geq N$ (N an integer).

1. If $\displaystyle\lim_{n \to \infty} \frac{a_n}{b_n} = c > 0$, then Σa_n and Σb_n both converge or both diverge.

2. If $\displaystyle\lim_{n \to \infty} \frac{a_n}{b_n} = 0$ and Σb_n converges, then Σa_n converges.

3. If $\displaystyle\lim_{n \to \infty} \frac{a_n}{b_n} = \infty$ and Σb_n diverges, then Σa_n diverges.

Proof   We will prove Part 1. Parts 2 and 3 are left as Exercises 55a and b.
Since $c/2 > 0$, there exists an integer N such that for all n

$$n > N \implies \left| \frac{a_n}{b_n} - c \right| < \frac{c}{2}.$$

Limit definition with $\epsilon = c/2$, $L = c$, and a_n replaced by a_n/b_n

Thus, for $n > N$,

$$-\frac{c}{2} < \frac{a_n}{b_n} - c < \frac{c}{2},$$

$$\frac{c}{2} < \frac{a_n}{b_n} < \frac{3c}{2},$$

$$\left(\frac{c}{2}\right)b_n < a_n < \left(\frac{3c}{2}\right)b_n.$$

If Σb_n converges, then $\Sigma(3c/2)b_n$ converges and Σa_n converges by the Direct Comparison Test. If Σb_n diverges, then $\Sigma(c/2)b_n$ diverges and Σa_n diverges by the Direct Comparison Test. ∎

EXAMPLE 2   Which of the following series converge, and which diverge?

(a) $\dfrac{3}{4} + \dfrac{5}{9} + \dfrac{7}{16} + \dfrac{9}{25} + \cdots = \displaystyle\sum_{n=1}^{\infty} \frac{2n + 1}{(n + 1)^2} = \sum_{n=1}^{\infty} \frac{2n + 1}{n^2 + 2n + 1}$

(b) $\dfrac{1}{1} + \dfrac{1}{3} + \dfrac{1}{7} + \dfrac{1}{15} + \cdots = \displaystyle\sum_{n=1}^{\infty} \frac{1}{2^n - 1}$

(c) $\dfrac{1 + 2 \ln 2}{9} + \dfrac{1 + 3 \ln 3}{14} + \dfrac{1 + 4 \ln 4}{21} + \cdots = \displaystyle\sum_{n=2}^{\infty} \frac{1 + n \ln n}{n^2 + 5}$

Solution We apply the Limit Comparison Test to each series.

(a) Let $a_n = (2n + 1)/(n^2 + 2n + 1)$. For large n, we expect a_n to behave like $2n/n^2 = 2/n$ since the leading terms dominate for large n, so we let $b_n = 1/n$. Since

$$\sum_{n=1}^{\infty} b_n = \sum_{n=1}^{\infty} \frac{1}{n} \text{ diverges}$$

and

$$\lim_{n \to \infty} \frac{a_n}{b_n} = \lim_{n \to \infty} \frac{2n^2 + n}{n^2 + 2n + 1} = 2,$$

$\sum a_n$ diverges by Part 1 of the Limit Comparison Test. We could just as well have taken $b_n = 2/n$, but $1/n$ is simpler.

(b) Let $a_n = 1/(2^n - 1)$. For large n, we expect a_n to behave like $1/2^n$, so we let $b_n = 1/2^n$. Since

$$\sum_{n=1}^{\infty} b_n = \sum_{n=1}^{\infty} \frac{1}{2^n} \text{ converges}$$

and

$$\lim_{n \to \infty} \frac{a_n}{b_n} = \lim_{n \to \infty} \frac{2^n}{2^n - 1}$$

$$= \lim_{n \to \infty} \frac{1}{1 - (1/2^n)}$$

$$= 1,$$

$\sum a_n$ converges by Part 1 of the Limit Comparison Test.

(c) Let $a_n = (1 + n \ln n)/(n^2 + 5)$. For large n, we expect a_n to behave like $(n \ln n)/n^2 = (\ln n)/n$, which is greater than $1/n$ for $n \geq 3$, so we let $b_n = 1/n$. Since

$$\sum_{n=2}^{\infty} b_n = \sum_{n=2}^{\infty} \frac{1}{n} \text{ diverges}$$

and

$$\lim_{n \to \infty} \frac{a_n}{b_n} = \lim_{n \to \infty} \frac{n + n^2 \ln n}{n^2 + 5}$$

$$= \infty,$$

$\sum a_n$ diverges by Part 3 of the Limit Comparison Test. ∎

EXAMPLE 3 Does $\displaystyle\sum_{n=1}^{\infty} \frac{\ln n}{n^{3/2}}$ converge?

Solution Because $\ln n$ grows more slowly than n^c for any positive constant c (Section 10.1, Exercise 105), we can compare the series to a convergent p-series. To get the p-series, we see that

$$\frac{\ln n}{n^{3/2}} < \frac{n^{1/4}}{n^{3/2}} = \frac{1}{n^{5/4}}$$

for n sufficiently large. Then taking $a_n = (\ln n)/n^{3/2}$ and $b_n = 1/n^{5/4}$, we have

$$\lim_{n \to \infty} \frac{a_n}{b_n} = \lim_{n \to \infty} \frac{\ln n}{n^{1/4}}$$

$$= \lim_{n \to \infty} \frac{1/n}{(1/4)n^{-3/4}} \qquad \text{l'Hôpital's Rule}$$

$$= \lim_{n \to \infty} \frac{4}{n^{1/4}} = 0.$$

Since $\sum b_n = \sum(1/n^{5/4})$ is a p-series with $p > 1$, it converges, so $\sum a_n$ converges by Part 2 of the Limit Comparison Test. ∎

Exercises 10.4

Comparison Test

In Exercises 1–8, use the Comparison Test to determine if each series converges or diverges.

1. $\displaystyle\sum_{n=1}^{\infty} \frac{1}{n^2 + 30}$

2. $\displaystyle\sum_{n=1}^{\infty} \frac{n-1}{n^4 + 2}$

3. $\displaystyle\sum_{n=2}^{\infty} \frac{1}{\sqrt{n-1}}$

4. $\displaystyle\sum_{n=2}^{\infty} \frac{n+2}{n^2 - n}$

5. $\displaystyle\sum_{n=1}^{\infty} \frac{\cos^2 n}{n^{3/2}}$

6. $\displaystyle\sum_{n=1}^{\infty} \frac{1}{n3^n}$

7. $\displaystyle\sum_{n=1}^{\infty} \sqrt{\frac{n+4}{n^4 + 4}}$

8. $\displaystyle\sum_{n=1}^{\infty} \frac{\sqrt{n}+1}{\sqrt{n^2 + 3}}$

Limit Comparison Test

In Exercises 9–16, use the Limit Comparison Test to determine if each series converges or diverges.

9. $\displaystyle\sum_{n=1}^{\infty} \frac{n-2}{n^3 - n^2 + 3}$

(*Hint:* Limit Comparison with $\sum_{n=1}^{\infty} (1/n^2)$)

10. $\displaystyle\sum_{n=1}^{\infty} \sqrt{\frac{n+1}{n^2 + 2}}$

$\left(\text{*Hint:* Limit Comparison with } \sum_{n=1}^{\infty} \left(1/\sqrt{n}\right)\right)$

11. $\displaystyle\sum_{n=2}^{\infty} \frac{n(n+1)}{(n^2 + 1)(n-1)}$

12. $\displaystyle\sum_{n=1}^{\infty} \frac{2^n}{3 + 4^n}$

13. $\displaystyle\sum_{n=1}^{\infty} \frac{5^n}{\sqrt{n}\, 4^n}$

14. $\displaystyle\sum_{n=1}^{\infty} \left(\frac{2n+3}{5n+4}\right)^n$

15. $\displaystyle\sum_{n=2}^{\infty} \frac{1}{\ln n}$

(*Hint:* Limit Comparison with $\sum_{n=2}^{\infty} (1/n)$)

16. $\displaystyle\sum_{n=1}^{\infty} \ln\left(1 + \frac{1}{n^2}\right)$

(*Hint:* Limit Comparison with $\sum_{n=1}^{\infty} (1/n^2)$)

Determining Convergence or Divergence

Which of the series in Exercises 17–54 converge, and which diverge? Use any method, and give reasons for your answers.

17. $\displaystyle\sum_{n=1}^{\infty} \frac{1}{2\sqrt{n} + \sqrt[3]{n}}$

18. $\displaystyle\sum_{n=1}^{\infty} \frac{3}{n + \sqrt{n}}$

19. $\displaystyle\sum_{n=1}^{\infty} \frac{\sin^2 n}{2^n}$

20. $\displaystyle\sum_{n=1}^{\infty} \frac{1 + \cos n}{n^2}$

21. $\displaystyle\sum_{n=1}^{\infty} \frac{2n}{3n-1}$

22. $\displaystyle\sum_{n=1}^{\infty} \frac{n+1}{n^2\sqrt{n}}$

23. $\displaystyle\sum_{n=1}^{\infty} \frac{10n+1}{n(n+1)(n+2)}$

24. $\displaystyle\sum_{n=3}^{\infty} \frac{5n^3 - 3n}{n^2(n-2)(n^2 + 5)}$

25. $\displaystyle\sum_{n=1}^{\infty} \left(\frac{n}{3n+1}\right)^n$

26. $\displaystyle\sum_{n=1}^{\infty} \frac{1}{\sqrt{n^3 + 2}}$

27. $\displaystyle\sum_{n=3}^{\infty} \frac{1}{\ln(\ln n)}$

28. $\displaystyle\sum_{n=1}^{\infty} \frac{(\ln n)^2}{n^3}$

29. $\displaystyle\sum_{n=2}^{\infty} \frac{1}{\sqrt{n}\ln n}$

30. $\displaystyle\sum_{n=1}^{\infty} \frac{(\ln n)^2}{n^{3/2}}$

31. $\displaystyle\sum_{n=1}^{\infty} \frac{1}{1 + \ln n}$

32. $\displaystyle\sum_{n=2}^{\infty} \frac{\ln(n+1)}{n+1}$

33. $\displaystyle\sum_{n=2}^{\infty} \frac{1}{n\sqrt{n^2 - 1}}$

34. $\displaystyle\sum_{n=1}^{\infty} \frac{\sqrt{n}}{n^2 + 1}$

35. $\displaystyle\sum_{n=1}^{\infty} \frac{1-n}{n2^n}$

36. $\displaystyle\sum_{n=1}^{\infty} \frac{n+2^n}{n^2 2^n}$

37. $\displaystyle\sum_{n=1}^{\infty} \frac{1}{3^{n-1} + 1}$

38. $\displaystyle\sum_{n=1}^{\infty} \frac{3^{n-1} + 1}{3^n}$

39. $\displaystyle\sum_{n=1}^{\infty} \frac{n+1}{n^2 + 3n} \cdot \frac{1}{5n}$

40. $\displaystyle\sum_{n=1}^{\infty} \frac{2^n + 3^n}{3^n + 4^n}$

41. $\displaystyle\sum_{n=1}^{\infty} \frac{2^n - n}{n2^n}$

42. $\displaystyle\sum_{n=1}^{\infty} \ln\left(\frac{n}{n+1}\right)$

43. $\displaystyle\sum_{n=2}^{\infty} \frac{1}{n!}$

(*Hint:* First show that $(1/n!) \le (1/n(n-1))$ for $n \ge 2$.)

44. $\displaystyle\sum_{n=1}^{\infty} \frac{(n-1)!}{(n+2)!}$

45. $\displaystyle\sum_{n=1}^{\infty} \sin\frac{1}{n}$

46. $\displaystyle\sum_{n=1}^{\infty} \tan\frac{1}{n}$

47. $\displaystyle\sum_{n=1}^{\infty} \frac{\tan^{-1} n}{n^{1.1}}$

48. $\displaystyle\sum_{n=1}^{\infty} \frac{\sec^{-1} n}{n^{1.3}}$

49. $\displaystyle\sum_{n=1}^{\infty} \frac{\coth n}{n^2}$

50. $\displaystyle\sum_{n=1}^{\infty} \frac{\tanh n}{n^2}$

51. $\displaystyle\sum_{n=1}^{\infty} \frac{1}{n\sqrt[n]{n}}$

52. $\displaystyle\sum_{n=1}^{\infty} \frac{\sqrt[n]{n}}{n^2}$

53. $\displaystyle\sum_{n=1}^{\infty} \frac{1}{1 + 2 + 3 + \cdots + n}$

54. $\displaystyle\sum_{n=1}^{\infty} \frac{1}{1 + 2^2 + 3^2 + \cdots + n^2}$

Theory and Examples

55. Prove **(a)** Part 2 and **(b)** Part 3 of the Limit Comparison Test.

56. If $\sum_{n=1}^{\infty} a_n$ is a convergent series of nonnegative numbers, can anything be said about $\sum_{n=1}^{\infty}(a_n/n)$? Explain.

57. Suppose that $a_n > 0$ and $b_n > 0$ for $n \ge N$ (N an integer). If $\lim_{n\to\infty}(a_n/b_n) = \infty$ and $\sum a_n$ converges, can anything be said about $\sum b_n$? Give reasons for your answer.

58. Prove that if $\sum a_n$ is a convergent series of nonnegative terms, then $\sum a_n^2$ converges.

59. Suppose that $a_n > 0$ and $\lim_{n\to\infty} a_n = \infty$. Prove that $\sum a_n$ diverges.

60. Suppose that $a_n > 0$ and $\lim_{n\to\infty} n^2 a_n = 0$. Prove that $\sum a_n$ converges.

61. Show that $\sum_{n=2}^{\infty} ((\ln n)^q/n^p)$ converges for $-\infty < q < \infty$ and $p > 1$.

(*Hint:* Limit Comparison with $\sum_{n=2}^{\infty} 1/n^r$ for $1 < r < p$.)

62. (*Continuation of Exercise 61.*) Show that $\sum_{n=2}^{\infty} ((\ln n)^q/n^p)$ diverges for $-\infty < q < \infty$ and $0 < p \le 1$.

(*Hint:* Limit Comparison with an appropriate p-series.)

In Exercises 63–68, use the results of Exercises 61 and 62 to determine if each series converges or diverges.

63. $\displaystyle\sum_{n=2}^{\infty} \frac{(\ln n)^3}{n^4}$

64. $\displaystyle\sum_{n=2}^{\infty} \sqrt{\frac{\ln n}{n}}$

65. $\displaystyle\sum_{n=2}^{\infty} \frac{(\ln n)^{1000}}{n^{1.001}}$

66. $\displaystyle\sum_{n=2}^{\infty} \frac{(\ln n)^{1/5}}{n^{0.99}}$

67. $\displaystyle\sum_{n=2}^{\infty} \frac{1}{n^{1.1}(\ln n)^3}$

68. $\displaystyle\sum_{n=2}^{\infty} \frac{1}{\sqrt{n} \cdot \ln n}$

COMPUTER EXPLORATIONS

69. It is not yet known whether the series

$$\sum_{n=1}^{\infty} \frac{1}{n^3 \sin^2 n}$$

converges or diverges. Use a CAS to explore the behavior of the series by performing the following steps.

a. Define the sequence of partial sums

$$s_k = \sum_{n=1}^{k} \frac{1}{n^3 \sin^2 n}.$$

What happens when you try to find the limit of s_k as $k \to \infty$? Does your CAS find a closed form answer for this limit?

b. Plot the first 100 points (k, s_k) for the sequence of partial sums. Do they appear to converge? What would you estimate the limit to be?

c. Next plot the first 200 points (k, s_k). Discuss the behavior in your own words.

d. Plot the first 400 points (k, s_k). What happens when $k = 355$? Calculate the number $355/113$. Explain from your calculation what happened at $k = 355$. For what values of k would you guess this behavior might occur again?

70. a. Use Theorem 8 to show that

$$S = \sum_{n=1}^{\infty} \frac{1}{n(n+1)} + \sum_{n=1}^{\infty} \left(\frac{1}{n^2} - \frac{1}{n(n+1)} \right)$$

where $S = \sum_{n=1}^{\infty} (1/n^2)$, the sum of a convergent p-series.

b. From Example 5, Section 10.2, show that

$$S = 1 + \sum_{n=1}^{\infty} \frac{1}{n^2(n+1)}.$$

c. Explain why taking the first M terms in the series in part (b) gives a better approximation to S than taking the first M terms in the original series $\sum_{n=1}^{\infty} (1/n^2)$.

d. The exact value of S is known to be $\pi^2/6$. Which of the sums

$$\sum_{n=1}^{1000000} \frac{1}{n^2} \quad \text{or} \quad 1 + \sum_{n=1}^{1000} \frac{1}{n^2(n+1)}$$

gives a better approximation to S?

10.5 The Ratio and Root Tests

The Ratio Test measures the rate of growth (or decline) of a series by examining the ratio a_{n+1}/a_n. For a geometric series $\sum ar^n$, this rate is a constant $((ar^{n+1})/(ar^n) = r)$, and the series converges if and only if its ratio is less than 1 in absolute value. The Ratio Test is a powerful rule extending that result.

THEOREM 12—The Ratio Test Let $\sum a_n$ be a series with positive terms and suppose that

$$\lim_{n \to \infty} \frac{a_{n+1}}{a_n} = \rho.$$

Then **(a)** the series *converges* if $\rho < 1$, **(b)** the series *diverges* if $\rho > 1$ or ρ is infinite, **(c)** the test is *inconclusive* if $\rho = 1$.

Proof

(a) $\rho < 1$. Let r be a number between ρ and 1. Then the number $\epsilon = r - \rho$ is positive. Since

$$\frac{a_{n+1}}{a_n} \to \rho,$$

a_{n+1}/a_n must lie within ϵ of ρ when n is large enough, say for all $n \geq N$. In particular,

$$\frac{a_{n+1}}{a_n} < \rho + \epsilon = r, \qquad \text{when } n \geq N.$$

That is,

$$a_{N+1} < ra_N,$$
$$a_{N+2} < ra_{N+1} < r^2 a_N,$$
$$a_{N+3} < ra_{N+2} < r^3 a_N,$$
$$\vdots$$
$$a_{N+m} < ra_{N+m-1} < r^m a_N.$$

These inequalities show that the terms of our series, after the Nth term, approach zero more rapidly than the terms in a geometric series with ratio $r < 1$. More precisely, consider the series Σc_n, where $c_n = a_n$ for $n = 1, 2, \ldots, N$ and $c_{N+1} = ra_N$, $c_{N+2} = r^2 a_N, \ldots, c_{N+m} = r^m a_N, \ldots$. Now $a_n \le c_n$ for all n, and

$$\sum_{n=1}^{\infty} c_n = a_1 + a_2 + \cdots + a_{N-1} + a_N + ra_N + r^2 a_N + \cdots$$
$$= a_1 + a_2 + \cdots + a_{N-1} + a_N(1 + r + r^2 + \cdots).$$

The geometric series $1 + r + r^2 + \cdots$ converges because $|r| < 1$, so Σc_n converges. Since $a_n \le c_n$, Σa_n also converges.

(b) $1 < \rho \le \infty$. From some index M on,

$$\frac{a_{n+1}}{a_n} > 1 \qquad \text{and} \qquad a_M < a_{M+1} < a_{M+2} < \cdots.$$

The terms of the series do not approach zero as n becomes infinite, and the series diverges by the nth-Term Test.

(c) $\rho = 1$. The two series

$$\sum_{n=1}^{\infty} \frac{1}{n} \qquad \text{and} \qquad \sum_{n=1}^{\infty} \frac{1}{n^2}$$

show that some other test for convergence must be used when $\rho = 1$.

$$\text{For } \sum_{n=1}^{\infty} \frac{1}{n}: \qquad \frac{a_{n+1}}{a_n} = \frac{1/(n+1)}{1/n} = \frac{n}{n+1} \to 1.$$

$$\text{For } \sum_{n=1}^{\infty} \frac{1}{n^2}: \qquad \frac{a_{n+1}}{a_n} = \frac{1/(n+1)^2}{1/n^2} = \left(\frac{n}{n+1}\right)^2 \to 1^2 = 1.$$

In both cases, $\rho = 1$, yet the first series diverges, whereas the second converges. ∎

The Ratio Test is often effective when the terms of a series contain factorials of expressions involving n or expressions raised to a power involving n.

EXAMPLE 1   Investigate the convergence of the following series.

(a) $\displaystyle\sum_{n=0}^{\infty} \frac{2^n + 5}{3^n}$ **(b)** $\displaystyle\sum_{n=1}^{\infty} \frac{(2n)!}{n!n!}$ **(c)** $\displaystyle\sum_{n=1}^{\infty} \frac{4^n n! n!}{(2n)!}$

Solution   We apply the Ratio Test to each series.

(a) For the series $\sum_{n=0}^{\infty} (2^n + 5)/3^n$,

$$\frac{a_{n+1}}{a_n} = \frac{(2^{n+1} + 5)/3^{n+1}}{(2^n + 5)/3^n} = \frac{1}{3} \cdot \frac{2^{n+1} + 5}{2^n + 5} = \frac{1}{3} \cdot \left(\frac{2 + 5 \cdot 2^{-n}}{1 + 5 \cdot 2^{-n}}\right) \to \frac{1}{3} \cdot \frac{2}{1} = \frac{2}{3}.$$

The series converges because $\rho = 2/3$ is less than 1. This does *not* mean that $2/3$ is the sum of the series. In fact,

$$\sum_{n=0}^{\infty} \frac{2^n + 5}{3^n} = \sum_{n=0}^{\infty} \left(\frac{2}{3}\right)^n + \sum_{n=0}^{\infty} \frac{5}{3^n} = \frac{1}{1 - (2/3)} + \frac{5}{1 - (1/3)} = \frac{21}{2}.$$

The first equality shows that s_{2m} is the sum of m nonnegative terms since each term in parentheses is positive or zero. Hence $s_{2m+2} \geq s_{2m}$, and the sequence $\{s_{2m}\}$ is nondecreasing. The second equality shows that $s_{2m} \leq u_1$. Since $\{s_{2m}\}$ is nondecreasing and bounded from above, it has a limit, say

$$\lim_{m \to \infty} s_{2m} = L. \tag{4}$$

If n is an odd integer, say $n = 2m + 1$, then the sum of the first n terms is $s_{2m+1} = s_{2m} + u_{2m+1}$. Since $u_n \to 0$,

$$\lim_{m \to \infty} u_{2m+1} = 0$$

and, as $m \to \infty$,

$$s_{2m+1} = s_{2m} + u_{2m+1} \to L + 0 = L. \tag{5}$$

Combining the results of Equations (4) and (5) gives $\lim_{n \to \infty} s_n = L$ (Section 10.1, Exercise 131). ■

EXAMPLE 1   The alternating harmonic series

$$\sum_{n=1}^{\infty} (-1)^{n+1} \frac{1}{n} = 1 - \frac{1}{2} + \frac{1}{3} - \frac{1}{4} + \cdots$$

clearly satisfies the three requirements of Theorem 14 with $N = 1$; it therefore converges. ■

Rather than directly verifying the definition $u_n \geq u_{n+1}$, a second way to show that the sequence $\{u_n\}$ is nonincreasing is to define a differentiable function $f(x)$ satisfying $f(n) = u_n$. That is, the values of f match the values of the sequence at every positive integer n. If $f'(x) \leq 0$ for all x greater than or equal to some positive integer N, then $f(x)$ is nonincreasing for $x \geq N$. It follows that $f(n) \geq f(n + 1)$, or $u_n \geq u_{n+1}$, for $n \geq N$.

EXAMPLE 2   Consider the sequence where $u_n = 10n/(n^2 + 16)$. Define $f(x) = 10x/(x^2 + 16)$. Then from the Derivative Quotient Rule,

$$f'(x) = \frac{10(16 - x^2)}{(x^2 + 16)^2} \leq 0 \qquad \text{whenever } x \geq 4.$$

It follows that $u_n \geq u_{n+1}$ for $n \geq 4$. That is, the sequence $\{u_n\}$ is nonincreasing for $n \geq 4$. ■

FIGURE 10.13 The partial sums of an alternating series that satisfies the hypotheses of Theorem 14 for $N = 1$ straddle the limit from the beginning.

A graphical interpretation of the partial sums (Figure 10.13) shows how an alternating series converges to its limit L when the three conditions of Theorem 14 are satisfied with $N = 1$. Starting from the origin of the x-axis, we lay off the positive distance $s_1 = u_1$. To find the point corresponding to $s_2 = u_1 - u_2$, we back up a distance equal to u_2. Since $u_2 \leq u_1$, we do not back up any farther than the origin. We continue in this seesaw fashion, backing up or going forward as the signs in the series demand. But for $n \geq N$, each forward or backward step is shorter than (or at most the same size as) the preceding step because $u_{n+1} \leq u_n$. And since the nth term approaches zero as n increases, the size of step we take forward or backward gets smaller and smaller. We oscillate across the limit L, and the amplitude of oscillation approaches zero. The limit L lies between any two successive sums s_n and s_{n+1} and hence differs from s_n by an amount less than u_{n+1}.
Because

$$|L - s_n| < u_{n+1} \qquad \text{for } n \geq N,$$

we can make useful estimates of the sums of convergent alternating series.

> **THEOREM 15—The Alternating Series Estimation Theorem**   If the alternating series $\sum_{n=1}^{\infty}(-1)^{n+1}u_n$ satisfies the three conditions of Theorem 14, then for $n \geq N$,
>
> $$s_n = u_1 - u_2 + \cdots + (-1)^{n+1}u_n$$
>
> approximates the sum L of the series with an error whose absolute value is less than u_{n+1}, the absolute value of the first unused term. Furthermore, the sum L lies between any two successive partial sums s_n and s_{n+1}, and the remainder, $L - s_n$, has the same sign as the first unused term.

We leave the verification of the sign of the remainder for Exercise 61.

EXAMPLE 3   We try Theorem 15 on a series whose sum we know:

$$\sum_{n=0}^{\infty}(-1)^n\frac{1}{2^n} = 1 - \frac{1}{2} + \frac{1}{4} - \frac{1}{8} + \frac{1}{16} - \frac{1}{32} + \frac{1}{64} - \frac{1}{128} + \frac{1}{256} - \cdots.$$

The theorem says that if we truncate the series after the eighth term, we throw away a total that is positive and less than $1/256$. The sum of the first eight terms is $s_8 = 0.6640625$ and the sum of the first nine terms is $s_9 = 0.66796875$. The sum of the geometric series is

$$\frac{1}{1 - (-1/2)} = \frac{1}{3/2} = \frac{2}{3},$$

and we note that $0.6640625 < (2/3) < 0.66796875$. The difference, $(2/3) - 0.6640625 = 0.0026041666\ldots$, is positive and is less than $(1/256) = 0.00390625$. ∎

Absolute and Conditional Convergence

We can apply the tests for convergence studied before to the series of absolute values of a series with both positive and negative terms.

> **DEFINITION**   A series $\sum a_n$ **converges absolutely** (is **absolutely convergent**) if the corresponding series of absolute values, $\sum|a_n|$, converges.

The geometric series in Example 3 converges absolutely because the corresponding series of absolute values

$$\sum_{n=0}^{\infty}\frac{1}{2^n} = 1 + \frac{1}{2} + \frac{1}{4} + \frac{1}{8} + \cdots$$

converges. The alternating harmonic series does not converge absolutely because the corresponding series of absolute values is the (divergent) harmonic series.

> **DEFINITION**   A series that converges but does not converge absolutely **converges conditionally**.

The alternating harmonic series converges conditionally.

Absolute convergence is important for two reasons. First, we have good tests for convergence of series of positive terms. Second, if a series converges absolutely, then it converges, as we now prove.

> **THEOREM 16—The Absolute Convergence Test**   If $\sum_{n=1}^{\infty} |a_n|$ converges, then $\sum_{n=1}^{\infty} a_n$ converges.

Proof For each n,

$$-|a_n| \le a_n \le |a_n|, \quad \text{so} \quad 0 \le a_n + |a_n| \le 2|a_n|.$$

If $\sum_{n=1}^{\infty} |a_n|$ converges, then $\sum_{n=1}^{\infty} 2|a_n|$ converges and, by the Direct Comparison Test, the nonnegative series $\sum_{n=1}^{\infty} (a_n + |a_n|)$ converges. The equality $a_n = (a_n + |a_n|) - |a_n|$ now lets us express $\sum_{n=1}^{\infty} a_n$ as the difference of two convergent series:

$$\sum_{n=1}^{\infty} a_n = \sum_{n=1}^{\infty} (a_n + |a_n| - |a_n|) = \sum_{n=1}^{\infty} (a_n + |a_n|) - \sum_{n=1}^{\infty} |a_n|.$$

Therefore, $\sum_{n=1}^{\infty} a_n$ converges. ∎

Caution We can rephrase Theorem 16 to say that every absolutely convergent series converges. However, the converse statement is false: Many convergent series do not converge absolutely (such as the alternating harmonic series in Example 1).

EXAMPLE 4   This example gives two series that converge absolutely.

(a) For $\sum_{n=1}^{\infty} (-1)^{n+1} \dfrac{1}{n^2} = 1 - \dfrac{1}{4} + \dfrac{1}{9} - \dfrac{1}{16} + \cdots$, the corresponding series of absolute values is the convergent series

$$\sum_{n=1}^{\infty} \frac{1}{n^2} = 1 + \frac{1}{4} + \frac{1}{9} + \frac{1}{16} + \cdots.$$

The original series converges because it converges absolutely.

(b) For $\sum_{n=1}^{\infty} \dfrac{\sin n}{n^2} = \dfrac{\sin 1}{1} + \dfrac{\sin 2}{4} + \dfrac{\sin 3}{9} + \cdots$, which contains both positive and negative terms, the corresponding series of absolute values is

$$\sum_{n=1}^{\infty} \left| \frac{\sin n}{n^2} \right| = \frac{|\sin 1|}{1} + \frac{|\sin 2|}{4} + \cdots,$$

which converges by comparison with $\sum_{n=1}^{\infty} (1/n^2)$ because $|\sin n| \le 1$ for every n. The original series converges absolutely; therefore it converges. ∎

EXAMPLE 5   If p is a positive constant, the sequence $\{1/n^p\}$ is a decreasing sequence with limit zero. Therefore the alternating p-series

$$\sum_{n=1}^{\infty} \frac{(-1)^{n-1}}{n^p} = 1 - \frac{1}{2^p} + \frac{1}{3^p} - \frac{1}{4^p} + \cdots, \qquad p > 0$$

converges.

If $p > 1$, the series converges absolutely. If $0 < p \le 1$, the series converges conditionally.

$$\text{Conditional convergence:} \quad 1 - \frac{1}{\sqrt{2}} + \frac{1}{\sqrt{3}} - \frac{1}{\sqrt{4}} + \cdots$$

$$\text{Absolute convergence:} \quad 1 - \frac{1}{2^{3/2}} + \frac{1}{3^{3/2}} - \frac{1}{4^{3/2}} + \cdots \qquad ∎$$

Rearranging Series

We can always rearrange the terms of a *finite* sum. The same result is true for an infinite series that is absolutely convergent (see Exercise 68 for an outline of the proof).

THEOREM 17—The Rearrangement Theorem for Absolutely Convergent Series If $\sum_{n=1}^{\infty} a_n$ converges absolutely, and $b_1, b_2, \ldots, b_n, \ldots$ is any arrangement of the sequence $\{a_n\}$, then $\sum b_n$ converges absolutely and

$$\sum_{n=1}^{\infty} b_n = \sum_{n=1}^{\infty} a_n.$$

If we rearrange the terms of a conditionally convergent series, we get different results. In fact, it can be proved that for any real number r, a given conditionally convergent series can be rearranged so its sum is equal to r. (We omit the proof of this fact.) Here's an example of summing the terms of a conditionally convergent series with different orderings, with each ordering giving a different value for the sum.

EXAMPLE 6 We know that the alternating harmonic series $\sum_{n=1}^{\infty} (-1)^{n+1}/n$ converges to some number L. Moreover, by Theorem 15, L lies between the successive partial sums $s_2 = 1/2$ and $s_3 = 5/6$, so $L \neq 0$. If we multiply the series by 2 we obtain

$$2L = 2 \sum_{n=1}^{\infty} \frac{(-1)^{n+1}}{n} = 2\left(1 - \frac{1}{2} + \frac{1}{3} - \frac{1}{4} + \frac{1}{5} - \frac{1}{6} + \frac{1}{7} - \frac{1}{8} + \frac{1}{9} - \frac{1}{10} + \frac{1}{11} - \cdots\right)$$

$$= 2 - 1 + \frac{2}{3} - \frac{1}{2} + \frac{2}{5} - \frac{1}{3} + \frac{2}{7} - \frac{1}{4} + \frac{2}{9} - \frac{1}{5} + \frac{2}{11} - \cdots.$$

Now we change the order of this last sum by grouping each pair of terms with the same odd denominator, but leaving the negative terms with the even denominators as they are placed (so the denominators are the positive integers in their natural order). This rearrangement gives

$$(2 - 1) - \frac{1}{2} + \left(\frac{2}{3} - \frac{1}{3}\right) - \frac{1}{4} + \left(\frac{2}{5} - \frac{1}{5}\right) - \frac{1}{6} + \left(\frac{2}{7} - \frac{1}{7}\right) - \frac{1}{8} + \cdots$$

$$= \left(1 - \frac{1}{2} + \frac{1}{3} - \frac{1}{4} + \frac{1}{5} - \frac{1}{6} + \frac{1}{7} - \frac{1}{8} + \frac{1}{9} - \frac{1}{10} + \frac{1}{11} - \cdots\right)$$

$$= \sum_{n=1}^{\infty} \frac{(-1)^{n+1}}{n} = L.$$

So by rearranging the terms of the conditionally convergent series $\sum_{n=1}^{\infty} 2(-1)^{n+1}/n$, the series becomes $\sum_{n=1}^{\infty} (-1)^{n+1}/n$, which is the alternating harmonic series itself. If the two series are the same, it would imply that $2L = L$, which is clearly false since $L \neq 0$. ■

Example 6 shows that we cannot rearrange the terms of a conditionally convergent series and expect the new series to be the same as the original one. When we are using a conditionally convergent series, the terms must be added together in the order they are given to obtain a correct result. On the other hand, Theorem 17 guarantees that the terms of an absolutely convergent series can be summed in any order without affecting the result.

Summary of Tests

We have developed a variety of tests to determine convergence or divergence for an infinite series of constants. There are other tests we have not presented which are sometimes given in more advanced courses. Here is a summary of the tests we have considered.

1. **The nth-Term Test:** Unless $a_n \to 0$, the series diverges.

2. **Geometric series:** $\sum ar^n$ converges if $|r| < 1$; otherwise it diverges.

3. **p-series:** $\sum 1/n^p$ converges if $p > 1$; otherwise it diverges.

4. **Series with nonnegative terms:** Try the Integral Test, Ratio Test, or Root Test. Try comparing to a known series with the Comparison Test or the Limit Comparison Test.

5. **Series with some negative terms:** Does $\sum |a_n|$ converge? If yes, so does $\sum a_n$ since absolute convergence implies convergence.

6. **Alternating series:** $\sum a_n$ converges if the series satisfies the conditions of the Alternating Series Test.

Exercises 10.6

Determining Convergence or Divergence

In Exercises 1–14, determine if the alternating series converges or diverges. Some of the series do not satisfy the conditions of the Alternating Series Test.

1. $\displaystyle\sum_{n=1}^{\infty} (-1)^{n+1} \frac{1}{\sqrt{n}}$

2. $\displaystyle\sum_{n=1}^{\infty} (-1)^{n+1} \frac{1}{n^{3/2}}$

3. $\displaystyle\sum_{n=1}^{\infty} (-1)^{n+1} \frac{1}{n3^n}$

4. $\displaystyle\sum_{n=2}^{\infty} (-1)^n \frac{4}{(\ln n)^2}$

5. $\displaystyle\sum_{n=1}^{\infty} (-1)^n \frac{n}{n^2 + 1}$

6. $\displaystyle\sum_{n=1}^{\infty} (-1)^{n+1} \frac{n^2 + 5}{n^2 + 4}$

7. $\displaystyle\sum_{n=1}^{\infty} (-1)^{n+1} \frac{2^n}{n^2}$

8. $\displaystyle\sum_{n=1}^{\infty} (-1)^n \frac{10^n}{(n+1)!}$

9. $\displaystyle\sum_{n=1}^{\infty} (-1)^{n+1} \left(\frac{n}{10}\right)^n$

10. $\displaystyle\sum_{n=2}^{\infty} (-1)^{n+1} \frac{1}{\ln n}$

11. $\displaystyle\sum_{n=1}^{\infty} (-1)^{n+1} \frac{\ln n}{n}$

12. $\displaystyle\sum_{n=1}^{\infty} (-1)^n \ln\left(1 + \frac{1}{n}\right)$

13. $\displaystyle\sum_{n=1}^{\infty} (-1)^{n+1} \frac{\sqrt{n} + 1}{n + 1}$

14. $\displaystyle\sum_{n=1}^{\infty} (-1)^{n+1} \frac{3\sqrt{n} + 1}{\sqrt{n} + 1}$

Absolute and Conditional Convergence

Which of the series in Exercises 15–48 converge absolutely, which converge, and which diverge? Give reasons for your answers.

15. $\displaystyle\sum_{n=1}^{\infty} (-1)^{n+1} (0.1)^n$

16. $\displaystyle\sum_{n=1}^{\infty} (-1)^{n+1} \frac{(0.1)^n}{n}$

17. $\displaystyle\sum_{n=1}^{\infty} (-1)^n \frac{1}{\sqrt{n}}$

18. $\displaystyle\sum_{n=1}^{\infty} \frac{(-1)^n}{1 + \sqrt{n}}$

19. $\displaystyle\sum_{n=1}^{\infty} (-1)^{n+1} \frac{n}{n^3 + 1}$

20. $\displaystyle\sum_{n=1}^{\infty} (-1)^{n+1} \frac{n!}{2^n}$

21. $\displaystyle\sum_{n=1}^{\infty} (-1)^n \frac{1}{n + 3}$

22. $\displaystyle\sum_{n=1}^{\infty} (-1)^n \frac{\sin n}{n^2}$

23. $\displaystyle\sum_{n=1}^{\infty} (-1)^{n+1} \frac{3 + n}{5 + n}$

24. $\displaystyle\sum_{n=1}^{\infty} \frac{(-2)^{n+1}}{n + 5^n}$

25. $\displaystyle\sum_{n=1}^{\infty} (-1)^{n+1} \frac{1 + n}{n^2}$

26. $\displaystyle\sum_{n=1}^{\infty} (-1)^{n+1} \left(\sqrt[n]{10}\right)$

27. $\displaystyle\sum_{n=1}^{\infty} (-1)^n n^2 (2/3)^n$

28. $\displaystyle\sum_{n=2}^{\infty} (-1)^{n+1} \frac{1}{n \ln n}$

29. $\displaystyle\sum_{n=1}^{\infty} (-1)^n \frac{\tan^{-1} n}{n^2 + 1}$

30. $\displaystyle\sum_{n=1}^{\infty} (-1)^n \frac{\ln n}{n - \ln n}$

31. $\displaystyle\sum_{n=1}^{\infty} (-1)^n \frac{n}{n + 1}$

32. $\displaystyle\sum_{n=1}^{\infty} (-5)^{-n}$

33. $\displaystyle\sum_{n=1}^{\infty} \frac{(-100)^n}{n!}$

34. $\displaystyle\sum_{n=1}^{\infty} \frac{(-1)^{n-1}}{n^2 + 2n + 1}$

35. $\displaystyle\sum_{n=1}^{\infty} \frac{\cos n\pi}{n\sqrt{n}}$

36. $\displaystyle\sum_{n=1}^{\infty} \frac{\cos n\pi}{n}$

37. $\displaystyle\sum_{n=1}^{\infty} \frac{(-1)^n (n + 1)^n}{(2n)^n}$

38. $\displaystyle\sum_{n=1}^{\infty} \frac{(-1)^{n+1} (n!)^2}{(2n)!}$

39. $\displaystyle\sum_{n=1}^{\infty} (-1)^n \frac{(2n)!}{2^n n! n}$

40. $\displaystyle\sum_{n=1}^{\infty} (-1)^n \frac{(n!)^2 3^n}{(2n + 1)!}$

41. $\displaystyle\sum_{n=1}^{\infty} (-1)^n \left(\sqrt{n + 1} - \sqrt{n}\right)$

42. $\displaystyle\sum_{n=1}^{\infty} (-1)^n \left(\sqrt{n^2 + n} - n\right)$

43. $\displaystyle\sum_{n=1}^{\infty} (-1)^n \left(\sqrt{n + \sqrt{n}} - \sqrt{n}\right)$

44. $\displaystyle\sum_{n=1}^{\infty} \frac{(-1)^n}{\sqrt{n} + \sqrt{n + 1}}$

45. $\displaystyle\sum_{n=1}^{\infty} (-1)^n \operatorname{sech} n$

46. $\displaystyle\sum_{n=1}^{\infty} (-1)^n \operatorname{csch} n$

47. $\dfrac{1}{4} - \dfrac{1}{6} + \dfrac{1}{8} - \dfrac{1}{10} + \dfrac{1}{12} - \dfrac{1}{14} + \cdots$

48. $1 + \dfrac{1}{4} - \dfrac{1}{9} - \dfrac{1}{16} + \dfrac{1}{25} + \dfrac{1}{36} - \dfrac{1}{49} - \dfrac{1}{64} + \cdots$

Error Estimation

In Exercises 49–52, estimate the magnitude of the error involved in using the sum of the first four terms to approximate the sum of the entire series.

49. $\displaystyle\sum_{n=1}^{\infty} (-1)^{n+1} \frac{1}{n}$

50. $\displaystyle\sum_{n=1}^{\infty} (-1)^{n+1} \frac{1}{10^n}$

51. $\displaystyle\sum_{n=1}^{\infty} (-1)^{n+1} \frac{(0.01)^n}{n}$ As you will see in Section 10.7, the sum is $\ln(1.01)$.

52. $\displaystyle\frac{1}{1+t} = \sum_{n=0}^{\infty} (-1)^n t^n, \quad 0 < t < 1$

In Exercises 53–56, determine how many terms should be used to estimate the sum of the entire series with an error of less than 0.001.

53. $\displaystyle\sum_{n=1}^{\infty} (-1)^n \frac{1}{n^2 + 3}$

54. $\displaystyle\sum_{n=1}^{\infty} (-1)^{n+1} \frac{n}{n^2 + 1}$

55. $\displaystyle\sum_{n=1}^{\infty} (-1)^{n+1} \frac{1}{\left(n + 3\sqrt{n}\right)^3}$

56. $\displaystyle\sum_{n=1}^{\infty} (-1)^n \frac{1}{\ln(\ln(n+2))}$

T Approximate the sums in Exercises 57 and 58 with an error of magnitude less than 5×10^{-6}.

57. $\displaystyle\sum_{n=0}^{\infty} (-1)^n \frac{1}{(2n)!}$ As you will see in Section 10.9, the sum is $\cos 1$, the cosine of 1 radian.

58. $\displaystyle\sum_{n=0}^{\infty} (-1)^n \frac{1}{n!}$ As you will see in Section 10.9, the sum is e^{-1}.

Theory and Examples

59. a. The series

$$\frac{1}{3} - \frac{1}{2} + \frac{1}{9} - \frac{1}{4} + \frac{1}{27} - \frac{1}{8} + \cdots + \frac{1}{3^n} - \frac{1}{2^n} + \cdots$$

does not meet one of the conditions of Theorem 14. Which one?

b. Use Theorem 17 to find the sum of the series in part (a).

T 60. The limit L of an alternating series that satisfies the conditions of Theorem 14 lies between the values of any two consecutive partial sums. This suggests using the average

$$\frac{s_n + s_{n+1}}{2} = s_n + \frac{1}{2}(-1)^{n+2} a_{n+1}$$

to estimate L. Compute

$$s_{20} + \frac{1}{2} \cdot \frac{1}{21}$$

as an approximation to the sum of the alternating harmonic series. The exact sum is $\ln 2 = 0.69314718 \ldots$.

61. The sign of the remainder of an alternating series that satisfies the conditions of Theorem 14   Prove the assertion in Theorem 15 that whenever an alternating series satisfying the conditions of Theorem 14 is approximated with one of its partial sums, then the remainder (sum of the unused terms) has the same sign as the first unused term. (*Hint:* Group the remainder's terms in consecutive pairs.)

62. Show that the sum of the first $2n$ terms of the series

$$1 - \frac{1}{2} + \frac{1}{2} - \frac{1}{3} + \frac{1}{3} - \frac{1}{4} + \frac{1}{4} - \frac{1}{5} + \frac{1}{5} - \frac{1}{6} + \cdots$$

is the same as the sum of the first n terms of the series

$$\frac{1}{1 \cdot 2} + \frac{1}{2 \cdot 3} + \frac{1}{3 \cdot 4} + \frac{1}{4 \cdot 5} + \frac{1}{5 \cdot 6} + \cdots.$$

Do these series converge? What is the sum of the first $2n + 1$ terms of the first series? If the series converge, what is their sum?

63. Show that if $\sum_{n=1}^{\infty} a_n$ diverges, then $\sum_{n=1}^{\infty} |a_n|$ diverges.

64. Show that if $\sum_{n=1}^{\infty} a_n$ converges absolutely, then

$$\left| \sum_{n=1}^{\infty} a_n \right| \le \sum_{n=1}^{\infty} |a_n|.$$

65. Show that if $\sum_{n=1}^{\infty} a_n$ and $\sum_{n=1}^{\infty} b_n$ both converge absolutely, then so do the following.

a. $\displaystyle\sum_{n=1}^{\infty} (a_n + b_n)$

b. $\displaystyle\sum_{n=1}^{\infty} (a_n - b_n)$

c. $\displaystyle\sum_{n=1}^{\infty} k a_n$ (k any number)

66. Show by example that $\sum_{n=1}^{\infty} a_n b_n$ may diverge even if $\sum_{n=1}^{\infty} a_n$ and $\sum_{n=1}^{\infty} b_n$ both converge.

T 67. In the alternating harmonic series, suppose the goal is to arrange the terms to get a new series that converges to $-1/2$. Start the new arrangement with the first negative term, which is $-1/2$. Whenever you have a sum that is less than or equal to $-1/2$, start introducing positive terms, taken in order, until the new total is greater than $-1/2$. Then add negative terms until the total is less than or equal to $-1/2$ again. Continue this process until your partial sums have been above the target at least three times and finish at or below it. If s_n is the sum of the first n terms of your new series, plot the points (n, s_n) to illustrate how the sums are behaving.

68. Outline of the proof of the Rearrangement Theorem (Theorem 17)

a. Let ϵ be a positive real number, let $L = \sum_{n=1}^{\infty} a_n$, and let $s_k = \sum_{n=1}^{k} a_n$. Show that for some index N_1 and for some index $N_2 \ge N_1$,

$$\sum_{n=N_1}^{\infty} |a_n| < \frac{\epsilon}{2} \quad \text{and} \quad |s_{N_2} - L| < \frac{\epsilon}{2}.$$

Since all the terms $a_1, a_2, \ldots, a_{N_2}$ appear somewhere in the sequence $\{b_n\}$, there is an index $N_3 \ge N_2$ such that if $n \ge N_3$, then $\left(\sum_{k=1}^{n} b_k \right) - s_{N_2}$ is at most a sum of terms a_m with $m \ge N_1$. Therefore, if $n \ge N_3$,

$$\left| \sum_{k=1}^{n} b_k - L \right| \le \left| \sum_{k=1}^{n} b_k - s_{N_2} \right| + |s_{N_2} - L|$$

$$\le \sum_{k=N_1}^{\infty} |a_k| + |s_{N_2} - L| < \epsilon.$$

b. The argument in part (a) shows that if $\sum_{n=1}^{\infty} a_n$ converges absolutely then $\sum_{n=1}^{\infty} b_n$ converges and $\sum_{n=1}^{\infty} b_n = \sum_{n=1}^{\infty} a_n$. Now show that because $\sum_{n=1}^{\infty} |a_n|$ converges, $\sum_{n=1}^{\infty} |b_n|$ converges to $\sum_{n=1}^{\infty} |a_n|$.

where for some c between 0 and 1,

$$R_n(1) = e^c \frac{1}{(n+1)!} < \frac{3}{(n+1)!}. \qquad e^c < e^1 < 3$$

Estimating the Remainder

It is often possible to estimate $R_n(x)$ as we did in Example 1. This method of estimation is so convenient that we state it as a theorem for future reference.

THEOREM 24—The Remainder Estimation Theorem   If there is a positive constant M such that $|f^{(n+1)}(t)| \leq M$ for all t between x and a, inclusive, then the remainder term $R_n(x)$ in Taylor's Theorem satisfies the inequality

$$|R_n(x)| \leq M \frac{|x-a|^{n+1}}{(n+1)!}.$$

If this inequality holds for every n and the other conditions of Taylor's Theorem are satisfied by f, then the series converges to $f(x)$.

The next two examples use Theorem 24 to show that the Taylor series generated by the sine and cosine functions do in fact converge to the functions themselves.

EXAMPLE 2   Show that the Taylor series for $\sin x$ at $x = 0$ converges for all x.

Solution   The function and its derivatives are

$$\begin{array}{llll}
f(x) = & \sin x, & f'(x) = & \cos x, \\
f''(x) = & -\sin x, & f'''(x) = & -\cos x, \\
\vdots & & \vdots & \\
f^{(2k)}(x) = (-1)^k \sin x, & & f^{(2k+1)}(x) = (-1)^k \cos x,
\end{array}$$

so

$$f^{(2k)}(0) = 0 \qquad \text{and} \qquad f^{(2k+1)}(0) = (-1)^k.$$

The series has only odd-powered terms and, for $n = 2k + 1$, Taylor's Theorem gives

$$\sin x = x - \frac{x^3}{3!} + \frac{x^5}{5!} - \cdots + \frac{(-1)^k x^{2k+1}}{(2k+1)!} + R_{2k+1}(x).$$

All the derivatives of $\sin x$ have absolute values less than or equal to 1, so we can apply the Remainder Estimation Theorem with $M = 1$ to obtain

$$|R_{2k+1}(x)| \leq 1 \cdot \frac{|x|^{2k+2}}{(2k+2)!}.$$

From Theorem 5, Rule 6, we have $(|x|^{2k+2}/(2k+2)!) \to 0$ as $k \to \infty$, whatever the value of x, so $R_{2k+1}(x) \to 0$ and the Maclaurin series for $\sin x$ converges to $\sin x$ for every x. Thus,

$$\sin x = \sum_{k=0}^{\infty} \frac{(-1)^k x^{2k+1}}{(2k+1)!} = x - \frac{x^3}{3!} + \frac{x^5}{5!} - \frac{x^7}{7!} + \cdots. \tag{4}$$

∎

EXAMPLE 3   Show that the Taylor series for $\cos x$ at $x = 0$ converges to $\cos x$ for every value of x.

Solution   We add the remainder term to the Taylor polynomial for $\cos x$ (Section 10.8, Example 3) to obtain Taylor's formula for $\cos x$ with $n = 2k$:

$$\cos x = 1 - \frac{x^2}{2!} + \frac{x^4}{4!} - \cdots + (-1)^k \frac{x^{2k}}{(2k)!} + R_{2k}(x).$$

Because the derivatives of the cosine have absolute value less than or equal to 1, the Remainder Estimation Theorem with $M = 1$ gives

$$|R_{2k}(x)| \leq 1 \cdot \frac{|x|^{2k+1}}{(2k+1)!}.$$

For every value of x, $R_{2k}(x) \to 0$ as $k \to \infty$. Therefore, the series converges to $\cos x$ for every value of x. Thus,

$$\cos x = \sum_{k=0}^{\infty} \frac{(-1)^k x^{2k}}{(2k)!} = 1 - \frac{x^2}{2!} + \frac{x^4}{4!} - \frac{x^6}{6!} + \cdots. \tag{5}$$

■

Using Taylor Series

Since every Taylor series is a power series, the operations of adding, subtracting, and multiplying Taylor series are all valid on the intersection of their intervals of convergence.

EXAMPLE 4   Using known series, find the first few terms of the Taylor series for the given function using power series operations.

(a) $\frac{1}{3}(2x + x \cos x)$ **(b)** $e^x \cos x$

Solution

(a) $\frac{1}{3}(2x + x \cos x) = \frac{2}{3}x + \frac{1}{3}x \left(1 - \frac{x^2}{2!} + \frac{x^4}{4!} - \cdots + (-1)^k \frac{x^{2k}}{(2k)!} + \cdots \right)$

$$= \frac{2}{3}x + \frac{1}{3}x - \frac{x^3}{3!} + \frac{x^5}{3 \cdot 4!} - \cdots = x - \frac{x^3}{6} + \frac{x^5}{72} - \cdots$$

(b) $e^x \cos x = \left(1 + x + \frac{x^2}{2!} + \frac{x^3}{3!} + \frac{x^4}{4!} + \cdots \right) \cdot \left(1 - \frac{x^2}{2!} + \frac{x^4}{4!} - \cdots \right)$ Multiply the first series by each term of the second series.

$$= \left(1 + x + \frac{x^2}{2!} + \frac{x^3}{3!} + \frac{x^4}{4!} + \cdots \right) - \left(\frac{x^2}{2!} + \frac{x^3}{2!} + \frac{x^4}{2!2!} + \frac{x^5}{2!3!} + \cdots \right)$$

$$+ \left(\frac{x^4}{4!} + \frac{x^5}{4!} + \frac{x^6}{2!4!} + \cdots \right) + \cdots$$

$$= 1 + x - \frac{x^3}{3} - \frac{x^4}{6} + \cdots$$

■

By Theorem 20, we can use the Taylor series of the function f to find the Taylor series of $f(u(x))$ where $u(x)$ is any continuous function. The Taylor series resulting from this substitution will converge for all x such that $u(x)$ lies within the interval of convergence of the Taylor series of f. For instance, we can find the Taylor series for $\cos 2x$ by substituting $2x$ for x in the Taylor series for $\cos x$:

$$\cos 2x = \sum_{k=0}^{\infty} \frac{(-1)^k (2x)^{2k}}{(2k)!} = 1 - \frac{(2x)^2}{2!} + \frac{(2x)^4}{4!} - \frac{(2x)^6}{6!} + \cdots \qquad \text{Eq. (5) with } 2x \text{ for } x$$

$$= 1 - \frac{2^2 x^2}{2!} + \frac{2^4 x^4}{4!} - \frac{2^6 x^6}{6!} + \cdots$$

$$= \sum_{k=0}^{\infty} (-1)^k \frac{2^{2k} x^{2k}}{(2k)!}.$$

EXAMPLE 5 For what values of x can we replace $\sin x$ by $x - (x^3/3!)$ with an error of magnitude no greater than 3×10^{-4}?

Solution Here we can take advantage of the fact that the Taylor series for $\sin x$ is an alternating series for every nonzero value of x. According to the Alternating Series Estimation Theorem (Section 10.6), the error in truncating

$$\sin x = x - \frac{x^3}{3!} + \frac{x^5}{5!} - \cdots$$

after $(x^3/3!)$ is no greater than

$$\left| \frac{x^5}{5!} \right| = \frac{|x|^5}{120}.$$

Therefore the error will be less than or equal to 3×10^{-4} if

$$\frac{|x|^5}{120} < 3 \times 10^{-4} \qquad \text{or} \qquad |x| < \sqrt[5]{360 \times 10^{-4}} \approx 0.514. \qquad \begin{array}{l}\text{Rounded down,} \\ \text{to be safe}\end{array}$$

The Alternating Series Estimation Theorem tells us something that the Remainder Estimation Theorem does not: namely, that the estimate $x - (x^3/3!)$ for $\sin x$ is an underestimate when x is positive, because then $x^5/120$ is positive.

Figure 10.20 shows the graph of $\sin x$, along with the graphs of a number of its approximating Taylor polynomials. The graph of $P_3(x) = x - (x^3/3!)$ is almost indistinguishable from the sine curve when $0 \le x \le 1$. ∎

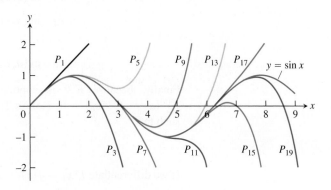

FIGURE 10.20 The polynomials

$$P_{2n+1}(x) = \sum_{k=0}^{n} \frac{(-1)^k x^{2k+1}}{(2k+1)!}$$

converge to $\sin x$ as $n \to \infty$. Notice how closely $P_3(x)$ approximates the sine curve for $x \le 1$ (Example 5).

A Proof of Taylor's Theorem

We prove Taylor's theorem assuming $a < b$. The proof for $a > b$ is nearly the same.

The Taylor polynomial

$$P_n(x) = f(a) + f'(a)(x - a) + \frac{f''(a)}{2!}(x - a)^2 + \cdots + \frac{f^{(n)}(a)}{n!}(x - a)^n$$

and its first n derivatives match the function f and its first n derivatives at $x = a$. We do not disturb that matching if we add another term of the form $K(x - a)^{n+1}$, where K is any constant, because such a term and its first n derivatives are all equal to zero at $x = a$. The new function

$$\phi_n(x) = P_n(x) + K(x - a)^{n+1}$$

and its first n derivatives still agree with f and its first n derivatives at $x = a$.

We now choose the particular value of K that makes the curve $y = \phi_n(x)$ agree with the original curve $y = f(x)$ at $x = b$. In symbols,

$$f(b) = P_n(b) + K(b - a)^{n+1}, \qquad \text{or} \qquad K = \frac{f(b) - P_n(b)}{(b - a)^{n+1}}. \tag{7}$$

With K defined by Equation (7), the function

$$F(x) = f(x) - \phi_n(x)$$

measures the difference between the original function f and the approximating function ϕ_n for each x in $[a, b]$.

We now use Rolle's Theorem (Section 4.2). First, because $F(a) = F(b) = 0$ and both F and F' are continuous on $[a, b]$, we know that

$$F'(c_1) = 0 \qquad \text{for some } c_1 \text{ in } (a, b).$$

Next, because $F'(a) = F'(c_1) = 0$ and both F' and F'' are continuous on $[a, c_1]$, we know that

$$F''(c_2) = 0 \qquad \text{for some } c_2 \text{ in } (a, c_1).$$

Rolle's Theorem, applied successively to $F'', F''', \ldots, F^{(n-1)}$ implies the existence of

$$c_3 \quad \text{in } (a, c_2) \qquad \text{such that } F'''(c_3) = 0,$$
$$c_4 \quad \text{in } (a, c_3) \qquad \text{such that } F^{(4)}(c_4) = 0,$$
$$\vdots$$
$$c_n \quad \text{in } (a, c_{n-1}) \qquad \text{such that } F^{(n)}(c_n) = 0.$$

Finally, because $F^{(n)}$ is continuous on $[a, c_n]$ and differentiable on (a, c_n), and $F^{(n)}(a) = F^{(n)}(c_n) = 0$, Rolle's Theorem implies that there is a number c_{n+1} in (a, c_n) such that

$$F^{(n+1)}(c_{n+1}) = 0. \tag{8}$$

If we differentiate $F(x) = f(x) - P_n(x) - K(x - a)^{n+1}$ a total of $n + 1$ times, we get

$$F^{(n+1)}(x) = f^{(n+1)}(x) - 0 - (n + 1)!K. \tag{9}$$

Equations (8) and (9) together give

$$K = \frac{f^{(n+1)}(c)}{(n + 1)!} \qquad \text{for some number } c = c_{n+1} \text{ in } (a, b). \tag{10}$$

Equations (7) and (10) give

$$f(b) = P_n(b) + \frac{f^{(n+1)}(c)}{(n+1)!}(b-a)^{n+1}.$$

This concludes the proof. ∎

Exercises 10.9

Finding Taylor Series

Use substitution (as in Example 4) to find the Taylor series at $x = 0$ of the functions in Exercises 1–10.

1. e^{-5x} **2.** $e^{-x/2}$ **3.** $5 \sin(-x)$

4. $\sin\left(\dfrac{\pi x}{2}\right)$ **5.** $\cos 5x^2$ **6.** $\cos\left(x^{2/3}/\sqrt{2}\right)$

7. $\ln(1 + x^2)$ **8.** $\tan^{-1}(3x^4)$ **9.** $\dfrac{1}{1 + \frac{3}{4}x^3}$

10. $\dfrac{1}{2 - x}$

Use power series operations to find the Taylor series at $x = 0$ for the functions in Exercises 11–28.

11. xe^x **12.** $x^2 \sin x$ **13.** $\dfrac{x^2}{2} - 1 + \cos x$

14. $\sin x - x + \dfrac{x^3}{3!}$ **15.** $x \cos \pi x$ **16.** $x^2 \cos(x^2)$

17. $\cos^2 x$ (*Hint:* $\cos^2 x = (1 + \cos 2x)/2$.)

18. $\sin^2 x$ **19.** $\dfrac{x^2}{1 - 2x}$ **20.** $x \ln(1 + 2x)$

21. $\dfrac{1}{(1 - x)^2}$ **22.** $\dfrac{2}{(1 - x)^3}$ **23.** $x \tan^{-1} x^2$

24. $\sin x \cdot \cos x$ **25.** $e^x + \dfrac{1}{1 + x}$ **26.** $\cos x - \sin x$

27. $\dfrac{x}{3} \ln(1 + x^2)$ **28.** $\ln(1 + x) - \ln(1 - x)$

Find the first four nonzero terms in the Maclaurin series for the functions in Exercises 29–34.

29. $e^x \sin x$ **30.** $\dfrac{\ln(1 + x)}{1 - x}$ **31.** $(\tan^{-1} x)^2$

32. $\cos^2 x \cdot \sin x$ **33.** $e^{\sin x}$ **34.** $\sin(\tan^{-1} x)$

Error Estimates

35. Estimate the error if $P_3(x) = x - (x^3/6)$ is used to estimate the value of $\sin x$ at $x = 0.1$.

36. Estimate the error if $P_4(x) = 1 + x + (x^2/2) + (x^3/6) + (x^4/24)$ is used to estimate the value of e^x at $x = 1/2$.

37. For approximately what values of x can you replace $\sin x$ by $x - (x^3/6)$ with an error of magnitude no greater than 5×10^{-4}? Give reasons for your answer.

38. If $\cos x$ is replaced by $1 - (x^2/2)$ and $|x| < 0.5$, what estimate can be made of the error? Does $1 - (x^2/2)$ tend to be too large, or too small? Give reasons for your answer.

39. How close is the approximation $\sin x = x$ when $|x| < 10^{-3}$? For which of these values of x is $x < \sin x$?

40. The estimate $\sqrt{1 + x} = 1 + (x/2)$ is used when x is small. Estimate the error when $|x| < 0.01$.

41. The approximation $e^x = 1 + x + (x^2/2)$ is used when x is small. Use the Remainder Estimation Theorem to estimate the error when $|x| < 0.1$.

42. (*Continuation of Exercise 41.*) When $x < 0$, the series for e^x is an alternating series. Use the Alternating Series Estimation Theorem to estimate the error that results from replacing e^x by $1 + x + (x^2/2)$ when $-0.1 < x < 0$. Compare your estimate with the one you obtained in Exercise 41.

Theory and Examples

43. Use the identity $\sin^2 x = (1 - \cos 2x)/2$ to obtain the Maclaurin series for $\sin^2 x$. Then differentiate this series to obtain the Maclaurin series for $2 \sin x \cos x$. Check that this is the series for $\sin 2x$.

44. (*Continuation of Exercise 43.*) Use the identity $\cos^2 x = \cos 2x + \sin^2 x$ to obtain a power series for $\cos^2 x$.

45. Taylor's Theorem and the Mean Value Theorem Explain how the Mean Value Theorem (Section 4.2, Theorem 4) is a special case of Taylor's Theorem.

46. Linearizations at inflection points Show that if the graph of a twice-differentiable function $f(x)$ has an inflection point at $x = a$, then the linearization of f at $x = a$ is also the quadratic approximation of f at $x = a$. This explains why tangent lines fit so well at inflection points.

47. The (second) second derivative test Use the equation

$$f(x) = f(a) + f'(a)(x - a) + \frac{f''(c_2)}{2}(x - a)^2$$

to establish the following test.

Let f have continuous first and second derivatives and suppose that $f'(a) = 0$. Then

a. f has a local maximum at a if $f'' \leq 0$ throughout an interval whose interior contains a;

b. f has a local minimum at a if $f'' \geq 0$ throughout an interval whose interior contains a.

48. A cubic approximation Use Taylor's formula with $a = 0$ and $n = 3$ to find the standard cubic approximation of $f(x) = 1/(1 - x)$ at $x = 0$. Give an upper bound for the magnitude of the error in the approximation when $|x| \leq 0.1$.

49. a. Use Taylor's formula with $n = 2$ to find the quadratic approximation of $f(x) = (1 + x)^k$ at $x = 0$ (k a constant).

b. If $k = 3$, for approximately what values of x in the interval $[0, 1]$ will the error in the quadratic approximation be less than $1/100$?

50. Improving approximations of π

a. Let P be an approximation of π accurate to n decimals. Show that $P + \sin P$ gives an approximation correct to $3n$ decimals. (*Hint:* Let $P = \pi + x$.)

T b. Try it with a calculator.

51. The Taylor series generated by $f(x) = \sum_{n=0}^{\infty} a_n x^n$ **is** $\sum_{n=0}^{\infty} a_n x^n$ A function defined by a power series $\sum_{n=0}^{\infty} a_n x^n$ with a radius of convergence $R > 0$ has a Taylor series that converges to the function at every point of $(-R, R)$. Show this by showing that the Taylor series generated by $f(x) = \sum_{n=0}^{\infty} a_n x^n$ is the series $\sum_{n=0}^{\infty} a_n x^n$ itself.

An immediate consequence of this is that series like

$$x \sin x = x^2 - \frac{x^4}{3!} + \frac{x^6}{5!} - \frac{x^8}{7!} + \cdots$$

and

$$x^2 e^x = x^2 + x^3 + \frac{x^4}{2!} + \frac{x^5}{3!} + \cdots,$$

obtained by multiplying Taylor series by powers of x, as well as series obtained by integration and differentiation of convergent power series, are themselves the Taylor series generated by the functions they represent.

52. Taylor series for even functions and odd functions (*Continuation of Section 10.7, Exercise 55.*) Suppose that $f(x) = \sum_{n=0}^{\infty} a_n x^n$ converges for all x in an open interval $(-R, R)$. Show that

a. If f is even, then $a_1 = a_3 = a_5 = \cdots = 0$, i.e., the Taylor series for f at $x = 0$ contains only even powers of x.

b. If f is odd, then $a_0 = a_2 = a_4 = \cdots = 0$, i.e., the Taylor series for f at $x = 0$ contains only odd powers of x.

COMPUTER EXPLORATIONS

Taylor's formula with $n = 1$ and $a = 0$ gives the linearization of a function at $x = 0$. With $n = 2$ and $n = 3$ we obtain the standard quadratic and cubic approximations. In these exercises we explore the errors associated with these approximations. We seek answers to two questions:

a. For what values of x can the function be replaced by each approximation with an error less than 10^{-2}?

b. What is the maximum error we could expect if we replace the function by each approximation over the specified interval?

Using a CAS, perform the following steps to aid in answering questions (a) and (b) for the functions and intervals in Exercises 53–58.

Step 1: Plot the function over the specified interval.

Step 2: Find the Taylor polynomials $P_1(x)$, $P_2(x)$, and $P_3(x)$ at $x = 0$.

Step 3: Calculate the $(n + 1)$st derivative $f^{(n+1)}(c)$ associated with the remainder term for each Taylor polynomial. Plot the derivative as a function of c over the specified interval and estimate its maximum absolute value, M.

Step 4: Calculate the remainder $R_n(x)$ for each polynomial. Using the estimate M from Step 3 in place of $f^{(n+1)}(c)$, plot $R_n(x)$ over the specified interval. Then estimate the values of x that answer question (a).

Step 5: Compare your estimated error with the actual error $E_n(x) = |f(x) - P_n(x)|$ by plotting $E_n(x)$ over the specified interval. This will help answer question (b).

Step 6: Graph the function and its three Taylor approximations together. Discuss the graphs in relation to the information discovered in Steps 4 and 5.

53. $f(x) = \dfrac{1}{\sqrt{1 + x}}, \quad |x| \leq \dfrac{3}{4}$

54. $f(x) = (1 + x)^{3/2}, \quad -\dfrac{1}{2} \leq x \leq 2$

55. $f(x) = \dfrac{x}{x^2 + 1}, \quad |x| \leq 2$

56. $f(x) = (\cos x)(\sin 2x), \quad |x| \leq 2$

57. $f(x) = e^{-x} \cos 2x, \quad |x| \leq 1$

58. $f(x) = e^{x/3} \sin 2x, \quad |x| \leq 2$

10.10 The Binomial Series and Applications of Taylor Series

In this section we introduce the binomial series for estimating powers and roots of binomial expressions $(1 + x)^m$. We also show how series can be used to evaluate nonelementary integrals and limits that lead to indeterminate forms, and we provide a derivation of the Taylor series for $\tan^{-1} x$. This section concludes with a reference table of frequently used series.

The Binomial Series for Powers and Roots

The Taylor series generated by $f(x) = (1 + x)^m$, when m is constant, is

$$1 + mx + \frac{m(m-1)}{2!}x^2 + \frac{m(m-1)(m-2)}{3!}x^3 + \cdots$$

$$+ \frac{m(m-1)(m-2)\cdots(m-k+1)}{k!}x^k + \cdots. \qquad (1)$$

This series, called the **binomial series**, converges absolutely for $|x| < 1$. To derive the series, we first list the function and its derivatives:

$$f(x) = (1 + x)^m$$

$$f'(x) = m(1 + x)^{m-1}$$

$$f''(x) = m(m - 1)(1 + x)^{m-2}$$

$$f'''(x) = m(m - 1)(m - 2)(1 + x)^{m-3}$$

$$\vdots$$

$$f^{(k)}(x) = m(m - 1)(m - 2)\cdots(m - k + 1)(1 + x)^{m-k}.$$

We then evaluate these at $x = 0$ and substitute into the Taylor series formula to obtain Series (1).

If m is an integer greater than or equal to zero, the series stops after $(m + 1)$ terms because the coefficients from $k = m + 1$ on are zero.

If m is not a positive integer or zero, the series is infinite and converges for $|x| < 1$. To see why, let u_k be the term involving x^k. Then apply the Ratio Test for absolute convergence to see that

$$\left| \frac{u_{k+1}}{u_k} \right| = \left| \frac{m - k}{k + 1}x \right| \rightarrow |x| \qquad \text{as } k \rightarrow \infty.$$

Our derivation of the binomial series shows only that it is generated by $(1 + x)^m$ and converges for $|x| < 1$. The derivation does not show that the series converges to $(1 + x)^m$. It does, but we omit the proof. (See Exercise 64.)

The Binomial Series

For $-1 < x < 1$,

$$(1 + x)^m = 1 + \sum_{k=1}^{\infty} \binom{m}{k} x^k,$$

where we define

$$\binom{m}{1} = m, \qquad \binom{m}{2} = \frac{m(m-1)}{2!},$$

and

$$\binom{m}{k} = \frac{m(m-1)(m-2)\cdots(m-k+1)}{k!} \qquad \text{for } k \geq 3.$$

EXAMPLE 1   If $m = -1$,

$$\binom{-1}{1} = -1, \qquad \binom{-1}{2} = \frac{-1(-2)}{2!} = 1,$$

and

$$\binom{-1}{k} = \frac{-1(-2)(-3)\cdots(-1-k+1)}{k!} = (-1)^k \left(\frac{k!}{k!}\right) = (-1)^k.$$

With these coefficient values and with x replaced by $-x$, the binomial series formula gives the familiar geometric series

$$(1+x)^{-1} = 1 + \sum_{k=1}^{\infty}(-1)^k x^k = 1 - x + x^2 - x^3 + \cdots + (-1)^k x^k + \cdots. \quad \blacksquare$$

EXAMPLE 2   We know from Section 3.11, Example 1, that $\sqrt{1+x} \approx 1 + (x/2)$ for $|x|$ small. With $m = 1/2$, the binomial series gives quadratic and higher-order approximations as well, along with error estimates that come from the Alternating Series Estimation Theorem:

$$(1+x)^{1/2} = 1 + \frac{x}{2} + \frac{\left(\frac{1}{2}\right)\left(-\frac{1}{2}\right)}{2!}x^2 + \frac{\left(\frac{1}{2}\right)\left(-\frac{1}{2}\right)\left(-\frac{3}{2}\right)}{3!}x^3$$

$$+ \frac{\left(\frac{1}{2}\right)\left(-\frac{1}{2}\right)\left(-\frac{3}{2}\right)\left(-\frac{5}{2}\right)}{4!}x^4 + \cdots$$

$$= 1 + \frac{x}{2} - \frac{x^2}{8} + \frac{x^3}{16} - \frac{5x^4}{128} + \cdots.$$

Substitution for x gives still other approximations. For example,

$$\sqrt{1-x^2} \approx 1 - \frac{x^2}{2} - \frac{x^4}{8} \qquad \text{for } |x^2| \text{ small}$$

$$\sqrt{1 - \frac{1}{x}} \approx 1 - \frac{1}{2x} - \frac{1}{8x^2} \qquad \text{for } \left|\frac{1}{x}\right| \text{ small, that is, } |x| \text{ large.} \quad \blacksquare$$

Sometimes we can use the binomial series to find the sum of a given power series in terms of a known function. For example,

$$x^2 - \frac{x^6}{3!} + \frac{x^{10}}{5!} - \frac{x^{14}}{7!} + \cdots = (x^2) - \frac{(x^2)^3}{3!} + \frac{(x^2)^5}{5!} - \frac{(x^2)^7}{7!} + \cdots = \sin x^2.$$

Additional examples are provided in Exercises 59–62.

Evaluating Nonelementary Integrals

Taylor series can be used to express nonelementary integrals in terms of series. Integrals like $\int \sin x^2\, dx$ arise in the study of the diffraction of light.

EXAMPLE 3   Express $\int \sin x^2\, dx$ as a power series.

Solution   From the series for $\sin x$ we substitute x^2 for x to obtain

$$\sin x^2 = x^2 - \frac{x^6}{3!} + \frac{x^{10}}{5!} - \frac{x^{14}}{7!} + \frac{x^{18}}{9!} - \cdots.$$

Therefore,

$$\int \sin x^2\, dx = C + \frac{x^3}{3} - \frac{x^7}{7\cdot 3!} + \frac{x^{11}}{11\cdot 5!} - \frac{x^{15}}{15\cdot 7!} + \frac{x^{10}}{19\cdot 9!} - \cdots \quad \blacksquare$$

EXAMPLE 4 Estimate $\int_0^1 \sin x^2 \, dx$ with an error of less than 0.001.

Solution From the indefinite integral in Example 3,

$$\int_0^1 \sin x^2 \, dx = \frac{1}{3} - \frac{1}{7 \cdot 3!} + \frac{1}{11 \cdot 5!} - \frac{1}{15 \cdot 7!} + \frac{1}{19 \cdot 9!} - \cdots.$$

The series alternates, and we find by experiment that

$$\frac{1}{11 \cdot 5!} \approx 0.00076$$

is the first term to be numerically less than 0.001. The sum of the preceding two terms gives

$$\int_0^1 \sin x^2 \, dx \approx \frac{1}{3} - \frac{1}{42} \approx 0.310.$$

With two more terms we could estimate

$$\int_0^1 \sin x^2 \, dx \approx 0.310268$$

with an error of less than 10^{-6}. With only one term beyond that we have

$$\int_0^1 \sin x^2 \, dx \approx \frac{1}{3} - \frac{1}{42} + \frac{1}{1320} - \frac{1}{75600} + \frac{1}{6894720} \approx 0.310268303,$$

with an error of about 1.08×10^{-9}. To guarantee this accuracy with the error formula for the Trapezoidal Rule would require using about 8000 subintervals. ∎

Arctangents

In Section 10.7, Example 5, we found a series for $\tan^{-1} x$ by differentiating to get

$$\frac{d}{dx} \tan^{-1} x = \frac{1}{1 + x^2} = 1 - x^2 + x^4 - x^6 + \cdots$$

and then integrating to get

$$\tan^{-1} x = x - \frac{x^3}{3} + \frac{x^5}{5} - \frac{x^7}{7} + \cdots.$$

However, we did not prove the term-by-term integration theorem on which this conclusion depended. We now derive the series again by integrating both sides of the finite formula

$$\frac{1}{1 + t^2} = 1 - t^2 + t^4 - t^6 + \cdots + (-1)^n t^{2n} + \frac{(-1)^{n+1} t^{2n+2}}{1 + t^2}, \qquad (2)$$

in which the last term comes from adding the remaining terms as a geometric series with first term $a = (-1)^{n+1} t^{2n+2}$ and ratio $r = -t^2$. Integrating both sides of Equation (2) from $t = 0$ to $t = x$ gives

$$\tan^{-1} x = x - \frac{x^3}{3} + \frac{x^5}{5} - \frac{x^7}{7} + \cdots + (-1)^n \frac{x^{2n+1}}{2n + 1} + R_n(x),$$

where

$$R_n(x) = \int_0^x \frac{(-1)^{n+1} t^{2n+2}}{1 + t^2} \, dt.$$

The denominator of the integrand is greater than or equal to 1; hence

$$|R_n(x)| \le \int_0^{|x|} t^{2n+2} \, dt = \frac{|x|^{2n+3}}{2n + 3}.$$

If $|x| \leq 1$, the right side of this inequality approaches zero as $n \to \infty$. Therefore $\lim_{n \to \infty} R_n(x) = 0$ if $|x| \leq 1$ and

$$\tan^{-1} x = \sum_{n=0}^{\infty} \frac{(-1)^n x^{2n+1}}{2n+1}, \qquad |x| \leq 1.$$

$$\tan^{-1} x = x - \frac{x^3}{3} + \frac{x^5}{5} - \frac{x^7}{7} + \cdots, \qquad |x| \leq 1.$$

(3)

We take this route instead of finding the Taylor series directly because the formulas for the higher-order derivatives of $\tan^{-1} x$ are unmanageable. When we put $x = 1$ in Equation (3), we get **Leibniz's formula**:

$$\frac{\pi}{4} = 1 - \frac{1}{3} + \frac{1}{5} - \frac{1}{7} + \frac{1}{9} - \cdots + \frac{(-1)^n}{2n+1} + \cdots.$$

Because this series converges very slowly, it is not used in approximating π to many decimal places. The series for $\tan^{-1} x$ converges most rapidly when x is near zero. For that reason, people who use the series for $\tan^{-1} x$ to compute π use various trigonometric identities.

For example, if

$$\alpha = \tan^{-1} \frac{1}{2} \qquad \text{and} \qquad \beta = \tan^{-1} \frac{1}{3},$$

then

$$\tan (\alpha + \beta) = \frac{\tan \alpha + \tan \beta}{1 - \tan \alpha \tan \beta} = \frac{\frac{1}{2} + \frac{1}{3}}{1 - \frac{1}{6}} = 1 = \tan \frac{\pi}{4}$$

and

$$\frac{\pi}{4} = \alpha + \beta = \tan^{-1} \frac{1}{2} + \tan^{-1} \frac{1}{3}.$$

Now Equation (3) may be used with $x = 1/2$ to evaluate $\tan^{-1}(1/2)$ and with $x = 1/3$ to give $\tan^{-1}(1/3)$. The sum of these results, multiplied by 4, gives π.

Evaluating Indeterminate Forms

We can sometimes evaluate indeterminate forms by expressing the functions involved as Taylor series.

EXAMPLE 5   Evaluate

$$\lim_{x \to 1} \frac{\ln x}{x - 1}.$$

Solution   We represent $\ln x$ as a Taylor series in powers of $x - 1$. This can be accomplished by calculating the Taylor series generated by $\ln x$ at $x = 1$ directly or by replacing x by $x - 1$ in the series for $\ln (1 + x)$ in Section 10.7, Example 6. Either way, we obtain

$$\ln x = (x - 1) - \frac{1}{2} (x - 1)^2 + \cdots,$$

from which we find that

$$\lim_{x \to 1} \frac{\ln x}{x - 1} = \lim_{x \to 1} \left(1 - \frac{1}{2} (x - 1) + \cdots \right) = 1. \qquad \blacksquare$$

EXAMPLE 6   Evaluate

$$\lim_{x \to 0} \frac{\sin x - \tan x}{x^3}.$$

Solution   The Taylor series for $\sin x$ and $\tan x$, to terms in x^5, are

$$\sin x = x - \frac{x^3}{3!} + \frac{x^5}{5!} - \cdots, \qquad \tan x = x + \frac{x^3}{3} + \frac{2x^5}{15} + \cdots.$$

Hence,

$$\sin x - \tan x = -\frac{x^3}{2} - \frac{x^5}{8} - \cdots = x^3\left(-\frac{1}{2} - \frac{x^2}{8} - \cdots\right)$$

and

$$\lim_{x\to 0} \frac{\sin x - \tan x}{x^3} = \lim_{x\to 0}\left(-\frac{1}{2} - \frac{x^2}{8} - \cdots\right)$$

$$= -\frac{1}{2}.$$ ∎

If we apply series to calculate $\lim_{x\to 0}((1/\sin x) - (1/x))$, we not only find the limit successfully but also discover an approximation formula for $\csc x$.

EXAMPLE 7   Find $\displaystyle\lim_{x\to 0}\left(\frac{1}{\sin x} - \frac{1}{x}\right)$.

Solution

$$\frac{1}{\sin x} - \frac{1}{x} = \frac{x - \sin x}{x\sin x} = \frac{x - \left(x - \dfrac{x^3}{3!} + \dfrac{x^5}{5!} - \cdots\right)}{x\cdot\left(x - \dfrac{x^3}{3!} + \dfrac{x^5}{5!} - \cdots\right)}$$

$$= \frac{x^3\left(\dfrac{1}{3!} - \dfrac{x^2}{5!} + \cdots\right)}{x^2\left(1 - \dfrac{x^2}{3!} + \cdots\right)} = x\,\frac{\dfrac{1}{3!} - \dfrac{x^2}{5!} + \cdots}{1 - \dfrac{x^2}{3!} + \cdots}$$

Therefore,

$$\lim_{x\to 0}\left(\frac{1}{\sin x} - \frac{1}{x}\right) = \lim_{x\to 0}\left(x\,\frac{\dfrac{1}{3!} - \dfrac{x^2}{5!} + \cdots}{1 - \dfrac{x^2}{3!} + \cdots}\right) = 0.$$

From the quotient on the right, we can see that if $|x|$ is small, then

$$\frac{1}{\sin x} - \frac{1}{x} \approx x\cdot\frac{1}{3!} = \frac{x}{6} \qquad \text{or} \qquad \csc x \approx \frac{1}{x} + \frac{x}{6}.$$ ∎

Euler's Identity

As you may recall, a complex number is a number of the form $a + bi$, where a and b are real numbers and $i = \sqrt{-1}$. If we substitute $x = i\theta$ (θ real) in the Taylor series for e^x and use the relations

$$i^2 = -1, \qquad i^3 = i^2 i = -i, \qquad i^4 = i^2 i^2 = 1, \qquad i^5 = i^4 i = i,$$

and so on, to simplify the result, we obtain

$$e^{i\theta} = 1 + \frac{i\theta}{1!} + \frac{i^2\theta^2}{2!} + \frac{i^3\theta^3}{3!} + \frac{i^4\theta^4}{4!} + \frac{i^5\theta^5}{5!} + \frac{i^6\theta^6}{6!} + \cdots$$

$$= \left(1 - \frac{\theta^2}{2!} + \frac{\theta^4}{4!} - \frac{\theta^6}{6!} + \cdots\right) + i\left(\theta - \frac{\theta^3}{3!} + \frac{\theta^5}{5!} - \cdots\right) = \cos\theta + i\sin\theta.$$

This does not *prove* that $e^{i\theta} = \cos\theta + i\sin\theta$ because we have not yet defined what it means to raise e to an imaginary power. Rather, it says how to define $e^{i\theta}$ to be consistent with other things we know.

DEFINITION

For any real number θ, $e^{i\theta} = \cos\theta + i\sin\theta$. $\qquad(4)$

Equation (4), called **Euler's identity**, enables us to define e^{a+bi} to be $e^a \cdot e^{bi}$ for any complex number $a + bi$. One consequence of the identity is the equation

$$e^{i\pi} = -1.$$

When written in the form $e^{i\pi} + 1 = 0$, this equation combines five of the most important constants in mathematics.

TABLE 10.1 Frequently used Taylor series

$$\frac{1}{1-x} = 1 + x + x^2 + \cdots + x^n + \cdots = \sum_{n=0}^{\infty} x^n, \quad |x| < 1$$

$$\frac{1}{1+x} = 1 - x + x^2 - \cdots + (-x)^n + \cdots = \sum_{n=0}^{\infty} (-1)^n x^n, \quad |x| < 1$$

$$e^x = 1 + x + \frac{x^2}{2!} + \cdots + \frac{x^n}{n!} + \cdots = \sum_{n=0}^{\infty} \frac{x^n}{n!}, \quad |x| < \infty$$

$$\sin x = x - \frac{x^3}{3!} + \frac{x^5}{5!} - \cdots + (-1)^n \frac{x^{2n+1}}{(2n+1)!} + \cdots = \sum_{n=0}^{\infty} \frac{(-1)^n x^{2n+1}}{(2n+1)!}, \quad |x| < \infty$$

$$\cos x = 1 - \frac{x^2}{2!} + \frac{x^4}{4!} - \cdots + (-1)^n \frac{x^{2n}}{(2n)!} + \cdots = \sum_{n=0}^{\infty} \frac{(-1)^n x^{2n}}{(2n)!}, \quad |x| < \infty$$

$$\ln(1+x) = x - \frac{x^2}{2} + \frac{x^3}{3} - \cdots + (-1)^{n-1}\frac{x^n}{n} + \cdots = \sum_{n=1}^{\infty} \frac{(-1)^{n-1} x^n}{n}, \quad -1 < x \le 1$$

$$\tan^{-1} x = x - \frac{x^3}{3} + \frac{x^5}{5} - \cdots + (-1)^n \frac{x^{2n+1}}{2n+1} + \cdots = \sum_{n=0}^{\infty} \frac{(-1)^n x^{2n+1}}{2n+1}, \quad |x| \le 1$$

Exercises 10.10

Binomial Series

Find the first four terms of the binomial series for the functions in Exercises 1–10.

1. $(1+x)^{1/2}$
2. $(1+x)^{1/3}$
3. $(1-x)^{-1/2}$
4. $(1-2x)^{1/2}$
5. $\left(1 + \frac{x}{2}\right)^{-2}$
6. $\left(1 - \frac{x}{3}\right)^{4}$
7. $(1+x^3)^{-1/2}$
8. $(1+x^2)^{-1/3}$

9. $\left(1 + \dfrac{1}{x}\right)^{1/2}$

10. $\dfrac{x}{\sqrt[3]{1 + x}}$

Find the binomial series for the functions in Exercises 11–14.

11. $(1 + x)^4$

12. $(1 + x^2)^3$

13. $(1 - 2x)^3$

14. $\left(1 - \dfrac{x}{2}\right)^4$

Approximations and Nonelementary Integrals

T In Exercises 15–18, use series to estimate the integrals' values with an error of magnitude less than 10^{-3}. (The answer section gives the integrals' values rounded to five decimal places.)

15. $\displaystyle\int_0^{0.2} \sin x^2 \, dx$

16. $\displaystyle\int_0^{0.2} \dfrac{e^{-x} - 1}{x} \, dx$

17. $\displaystyle\int_0^{0.1} \dfrac{1}{\sqrt{1 + x^4}} \, dx$

18. $\displaystyle\int_0^{0.25} \sqrt[3]{1 + x^2} \, dx$

T Use series to approximate the values of the integrals in Exercises 19–22 with an error of magnitude less than 10^{-8}.

19. $\displaystyle\int_0^{0.1} \dfrac{\sin x}{x} \, dx$

20. $\displaystyle\int_0^{0.1} e^{-x^2} \, dx$

21. $\displaystyle\int_0^{0.1} \sqrt{1 + x^4} \, dx$

22. $\displaystyle\int_0^{1} \dfrac{1 - \cos x}{x^2} \, dx$

23. Estimate the error if $\cos t^2$ is approximated by $1 - \dfrac{t^4}{2} + \dfrac{t^8}{4!}$ in the integral $\int_0^1 \cos t^2 \, dt$.

24. Estimate the error if $\cos \sqrt{t}$ is approximated by $1 - \dfrac{t}{2} + \dfrac{t^2}{4!} - \dfrac{t^3}{6!}$ in the integral $\int_0^1 \cos \sqrt{t} \, dt$.

In Exercises 25–28, find a polynomial that will approximate $F(x)$ throughout the given interval with an error of magnitude less than 10^{-3}.

25. $F(x) = \displaystyle\int_0^x \sin t^2 \, dt, \quad [0, 1]$

26. $F(x) = \displaystyle\int_0^x t^2 e^{-t^2} \, dt, \quad [0, 1]$

27. $F(x) = \displaystyle\int_0^x \tan^{-1} t \, dt, \quad$ **(a)** $[0, 0.5]$ **(b)** $[0, 1]$

28. $F(x) = \displaystyle\int_0^x \dfrac{\ln(1 + t)}{t} \, dt, \quad$ **(a)** $[0, 0.5]$ **(b)** $[0, 1]$

Indeterminate Forms

Use series to evaluate the limits in Exercises 29–40.

29. $\displaystyle\lim_{x \to 0} \dfrac{e^x - (1 + x)}{x^2}$

30. $\displaystyle\lim_{x \to 0} \dfrac{e^x - e^{-x}}{x}$

31. $\displaystyle\lim_{t \to 0} \dfrac{1 - \cos t - (t^2/2)}{t^4}$

32. $\displaystyle\lim_{\theta \to 0} \dfrac{\sin \theta - \theta + (\theta^3/6)}{\theta^5}$

33. $\displaystyle\lim_{y \to 0} \dfrac{y - \tan^{-1} y}{y^3}$

34. $\displaystyle\lim_{y \to 0} \dfrac{\tan^{-1} y - \sin y}{y^3 \cos y}$

35. $\displaystyle\lim_{x \to \infty} x^2 (e^{-1/x^2} - 1)$

36. $\displaystyle\lim_{x \to \infty} (x + 1) \sin \dfrac{1}{x + 1}$

37. $\displaystyle\lim_{x \to 0} \dfrac{\ln(1 + x^2)}{1 - \cos x}$

38. $\displaystyle\lim_{x \to 2} \dfrac{x^2 - 4}{\ln(x - 1)}$

39. $\displaystyle\lim_{x \to 0} \dfrac{\sin 3x^2}{1 - \cos 2x}$

40. $\displaystyle\lim_{x \to 0} \dfrac{\ln(1 + x^3)}{x \cdot \sin x^2}$

Using Table 10.1

In Exercises 41–52, use Table 10.1 to find the sum of each series.

41. $1 + 1 + \dfrac{1}{2!} + \dfrac{1}{3!} + \dfrac{1}{4!} + \cdots$

42. $\left(\dfrac{1}{4}\right)^3 + \left(\dfrac{1}{4}\right)^4 + \left(\dfrac{1}{4}\right)^5 + \left(\dfrac{1}{4}\right)^6 + \cdots$

43. $1 - \dfrac{3^2}{4^2 \cdot 2!} + \dfrac{3^4}{4^4 \cdot 4!} - \dfrac{3^6}{4^6 \cdot 6!} + \cdots$

44. $\dfrac{1}{2} - \dfrac{1}{2 \cdot 2^2} + \dfrac{1}{3 \cdot 2^3} - \dfrac{1}{4 \cdot 2^4} + \cdots$

45. $\dfrac{\pi}{3} - \dfrac{\pi^3}{3^3 \cdot 3!} + \dfrac{\pi^5}{3^5 \cdot 5!} - \dfrac{\pi^7}{3^7 \cdot 7!} + \cdots$

46. $\dfrac{2}{3} - \dfrac{2^3}{3^3 \cdot 3} + \dfrac{2^5}{3^5 \cdot 5} - \dfrac{2^7}{3^7 \cdot 7} + \cdots$

47. $x^3 + x^4 + x^5 + x^6 + \cdots$

48. $1 - \dfrac{3^2 x^2}{2!} + \dfrac{3^4 x^4}{4!} - \dfrac{3^6 x^6}{6!} + \cdots$

49. $x^3 - x^5 + x^7 - x^9 + x^{11} - \cdots$

50. $x^2 - 2x^3 + \dfrac{2^2 x^4}{2!} - \dfrac{2^3 x^5}{3!} + \dfrac{2^4 x^6}{4!} - \cdots$

51. $-1 + 2x - 3x^2 + 4x^3 - 5x^4 + \cdots$

52. $1 + \dfrac{x}{2} + \dfrac{x^2}{3} + \dfrac{x^3}{4} + \dfrac{x^4}{5} + \cdots$

Theory and Examples

53. Replace x by $-x$ in the Taylor series for $\ln(1 + x)$ to obtain a series for $\ln(1 - x)$. Then subtract this from the Taylor series for $\ln(1 + x)$ to show that for $|x| < 1$,

$$\ln\dfrac{1 + x}{1 - x} = 2\left(x + \dfrac{x^3}{3} + \dfrac{x^5}{5} + \cdots\right).$$

54. How many terms of the Taylor series for $\ln(1 + x)$ should you add to be sure of calculating $\ln(1.1)$ with an error of magnitude less than 10^{-8}? Give reasons for your answer.

55. According to the Alternating Series Estimation Theorem, how many terms of the Taylor series for $\tan^{-1} 1$ would you have to add to be sure of finding $\pi/4$ with an error of magnitude less than 10^{-3}? Give reasons for your answer.

56. Show that the Taylor series for $f(x) = \tan^{-1} x$ diverges for $|x| > 1$.

T **57. Estimating Pi** About how many terms of the Taylor series for $\tan^{-1} x$ would you have to use to evaluate each term on the right-hand side of the equation

$$\pi = 48 \tan^{-1} \dfrac{1}{18} + 32 \tan^{-1} \dfrac{1}{57} - 20 \tan^{-1} \dfrac{1}{239}$$

with an error of magnitude less than 10^{-6}? In contrast, the convergence of $\sum_{n=1}^{\infty}(1/n^2)$ to $\pi^2/6$ is so slow that even 50 terms will not yield two-place accuracy.

58. Integrate the first three nonzero terms of the Taylor series for $\tan t$ from 0 to x to obtain the first three nonzero terms of the Taylor series for $\ln \sec x$.

59. a. Use the binomial series and the fact that

$$\frac{d}{dx}\sin^{-1}x = (1 - x^2)^{-1/2}$$

to generate the first four nonzero terms of the Taylor series for $\sin^{-1}x$. What is the radius of convergence?

   b. Series for $\cos^{-1}x$ Use your result in part (a) to find the first five nonzero terms of the Taylor series for $\cos^{-1}x$.

60. a. Series for $\sinh^{-1}x$ Find the first four nonzero terms of the Taylor series for

$$\sinh^{-1}x = \int_0^x \frac{dt}{\sqrt{1 + t^2}}.$$

   T b. Use the first *three* terms of the series in part (a) to estimate $\sinh^{-1} 0.25$. Give an upper bound for the magnitude of the estimation error.

61. Obtain the Taylor series for $1/(1 + x)^2$ from the series for $-1/(1 + x)$.

62. Use the Taylor series for $1/(1 - x^2)$ to obtain a series for $2x/(1 - x^2)^2$.

T 63. Estimating Pi   The English mathematician Wallis discovered the formula

$$\frac{\pi}{4} = \frac{2 \cdot 4 \cdot 4 \cdot 6 \cdot 6 \cdot 8 \cdot \cdots}{3 \cdot 3 \cdot 5 \cdot 5 \cdot 7 \cdot 7 \cdot \cdots}.$$

Find π to two decimal places with this formula.

64. Use the following steps to prove Equation (1).

   a. Differentiate the series

$$f(x) = 1 + \sum_{k=1}^{\infty} \binom{m}{k}x^k$$

to show that

$$f'(x) = \frac{mf(x)}{1 + x}, \quad -1 < x < 1.$$

   b. Define $g(x) = (1 + x)^{-m}f(x)$ and show that $g'(x) = 0$.

   c. From part (b), show that

$$f(x) = (1 + x)^m.$$

65. Series for $\sin^{-1}x$ Integrate the binomial series for $(1 - x^2)^{-1/2}$ to show that for $|x| < 1$,

$$\sin^{-1}x = x + \sum_{n=1}^{\infty} \frac{1 \cdot 3 \cdot 5 \cdot \cdots \cdot (2n - 1)}{2 \cdot 4 \cdot 6 \cdot \cdots \cdot (2n)} \frac{x^{2n+1}}{2n + 1}.$$

66. Series for $\tan^{-1}x$ for $|x| > 1$ Derive the series

$$\tan^{-1}x = \frac{\pi}{2} - \frac{1}{x} + \frac{1}{3x^3} - \frac{1}{5x^5} + \cdots, \quad x > 1$$

$$\tan^{-1}x = -\frac{\pi}{2} - \frac{1}{x} + \frac{1}{3x^3} - \frac{1}{5x^5} + \cdots, \quad x < -1,$$

by integrating the series

$$\frac{1}{1 + t^2} = \frac{1}{t^2} \cdot \frac{1}{1 + (1/t^2)} = \frac{1}{t^2} - \frac{1}{t^4} + \frac{1}{t^6} - \frac{1}{t^8} + \cdots$$

in the first case from x to ∞ and in the second case from $-\infty$ to x.

Euler's Identity

67. Use Equation (4) to write the following powers of e in the form $a + bi$.

   a. $e^{-i\pi}$ **b.** $e^{i\pi/4}$ **c.** $e^{-i\pi/2}$

68. Use Equation (4) to show that

$$\cos \theta = \frac{e^{i\theta} + e^{-i\theta}}{2} \quad \text{and} \quad \sin \theta = \frac{e^{i\theta} - e^{-i\theta}}{2i}.$$

69. Establish the equations in Exercise 68 by combining the formal Taylor series for $e^{i\theta}$ and $e^{-i\theta}$.

70. Show that

   a. $\cosh i\theta = \cos \theta$,   **b.** $\sinh i\theta = i \sin \theta$.

71. By multiplying the Taylor series for e^x and $\sin x$, find the terms through x^5 of the Taylor series for $e^x \sin x$. This series is the imaginary part of the series for

$$e^x \cdot e^{ix} = e^{(1+i)x}.$$

Use this fact to check your answer. For what values of x should the series for $e^x \sin x$ converge?

72. When a and b are real, we define $e^{(a+ib)x}$ with the equation

$$e^{(a+ib)x} = e^{ax} \cdot e^{ibx} = e^{ax}(\cos bx + i \sin bx).$$

Differentiate the right-hand side of this equation to show that

$$\frac{d}{dx}e^{(a+ib)x} = (a + ib)e^{(a+ib)x}.$$

Thus the familiar rule $(d/dx)e^{kx} = ke^{kx}$ holds for k complex as well as real.

73. Use the definition of $e^{i\theta}$ to show that for any real numbers θ, θ_1, and θ_2,

   a. $e^{i\theta_1}e^{i\theta_2} = e^{i(\theta_1+\theta_2)}$,   **b.** $e^{-i\theta} = 1/e^{i\theta}$.

74. Two complex numbers $a + ib$ and $c + id$ are equal if and only if $a = c$ and $b = d$. Use this fact to evaluate

$$\int e^{ax} \cos bx \, dx \quad \text{and} \quad \int e^{ax} \sin bx \, dx$$

from

$$\int e^{(a+ib)x} \, dx = \frac{a - ib}{a^2 + b^2} e^{(a+ib)x} + C,$$

where $C = C_1 + iC_2$ is a complex constant of integration.

Chapter **10**   Questions to Guide Your Review

1. What is an infinite sequence? What does it mean for such a sequence to converge? To diverge? Give examples.

2. What is a monotonic sequence? Under what circumstances does such a sequence have a limit? Give examples.

3. What theorems are available for calculating limits of sequences? Give examples.

4. What theorem sometimes enables us to use l'Hôpital's Rule to calculate the limit of a sequence? Give an example.

5. What are the six commonly occurring limits in Theorem 5 that arise frequently when you work with sequences and series?

6. What is an infinite series? What does it mean for such a series to converge? To diverge? Give examples.

7. What is a geometric series? When does such a series converge? Diverge? When it does converge, what is its sum? Give examples.

8. Besides geometric series, what other convergent and divergent series do you know?

9. What is the nth-Term Test for Divergence? What is the idea behind the test?

10. What can be said about term-by-term sums and differences of convergent series? About constant multiples of convergent and divergent series?

11. What happens if you add a finite number of terms to a convergent series? A divergent series? What happens if you delete a finite number of terms from a convergent series? A divergent series?

12. How do you reindex a series? Why might you want to do this?

13. Under what circumstances will an infinite series of nonnegative terms converge? Diverge? Why study series of nonnegative terms?

14. What is the Integral Test? What is the reasoning behind it? Give an example of its use.

15. When do p-series converge? Diverge? How do you know? Give examples of convergent and divergent p-series.

16. What are the Direct Comparison Test and the Limit Comparison Test? What is the reasoning behind these tests? Give examples of their use.

17. What are the Ratio and Root Tests? Do they always give you the information you need to determine convergence or divergence? Give examples.

18. What is an alternating series? What theorem is available for determining the convergence of such a series?

19. How can you estimate the error involved in approximating the sum of an alternating series with one of the series' partial sums? What is the reasoning behind the estimate?

20. What is absolute convergence? Conditional convergence? How are the two related?

21. What do you know about rearranging the terms of an absolutely convergent series? Of a conditionally convergent series?

22. What is a power series? How do you test a power series for convergence? What are the possible outcomes?

23. What are the basic facts about

    a. sums, differences, and products of power series?

    b. substitution of a function for x in a power series?

    c. term-by-term differentiation of power series?

    d. term-by-term integration of power series?

    Give examples.

24. What is the Taylor series generated by a function $f(x)$ at a point $x = a$? What information do you need about f to construct the series? Give an example.

25. What is a Maclaurin series?

26. Does a Taylor series always converge to its generating function? Explain.

27. What are Taylor polynomials? Of what use are they?

28. What is Taylor's formula? What does it say about the errors involved in using Taylor polynomials to approximate functions? In particular, what does Taylor's formula say about the error in a linearization? A quadratic approximation?

29. What is the binomial series? On what interval does it converge? How is it used?

30. How can you sometimes use power series to estimate the values of nonelementary definite integrals?

31. What are the Taylor series for $1/(1 - x)$, $1/(1 + x)$, e^x, $\sin x$, $\cos x$, $\ln (1 + x)$, and $\tan^{-1} x$? How do you estimate the errors involved in replacing these series with their partial sums?

Chapter **10**   Practice Exercises

Determining Convergence of Sequences

Which of the sequences whose nth terms appear in Exercises 1–18 converge, and which diverge? Find the limit of each convergent sequence.

1. $a_n = 1 + \dfrac{(-1)^n}{n}$

2. $a_n = \dfrac{1 - (-1)^n}{\sqrt{n}}$

3. $a_n = \dfrac{1 - 2^n}{2^n}$

4. $a_n = 1 + (0.9)^n$

5. $a_n = \sin \dfrac{n\pi}{2}$

6. $a_n = \sin n\pi$

7. $a_n = \dfrac{\ln (n^2)}{n}$

8. $a_n = \dfrac{\ln (2n + 1)}{n}$

9. $a_n = \dfrac{n + \ln n}{n}$

10. $a_n = \dfrac{\ln (2n^3 + 1)}{n}$

11. $a_n = \left(\dfrac{n - 5}{n}\right)^n$

12. $a_n = \left(1 + \dfrac{1}{n}\right)^{-n}$

13. $a_n = \sqrt[n]{\dfrac{3^n}{n}}$

14. $a_n = \left(\dfrac{3}{n}\right)^{1/n}$

15. $a_n = n(2^{1/n} - 1)$

16. $a_n = \sqrt[n]{2n + 1}$

17. $a_n = \dfrac{(n + 1)!}{n!}$

18. $a_n = \dfrac{(-4)^n}{n!}$

Convergent Series

Find the sums of the series in Exercises 19–24.

19. $\displaystyle\sum_{n=3}^{\infty} \dfrac{1}{(2n - 3)(2n - 1)}$

20. $\displaystyle\sum_{n=2}^{\infty} \dfrac{-2}{n(n + 1)}$

21. $\displaystyle\sum_{n=1}^{\infty} \dfrac{9}{(3n - 1)(3n + 2)}$

22. $\displaystyle\sum_{n=3}^{\infty} \dfrac{-8}{(4n - 3)(4n + 1)}$

23. $\displaystyle\sum_{n=0}^{\infty} e^{-n}$

24. $\displaystyle\sum_{n=1}^{\infty} (-1)^n \dfrac{3}{4^n}$

Determining Convergence of Series

Which of the series in Exercises 25–40 converge absolutely, which converge conditionally, and which diverge? Give reasons for your answers.

25. $\displaystyle\sum_{n=1}^{\infty} \dfrac{1}{\sqrt{n}}$

26. $\displaystyle\sum_{n=1}^{\infty} \dfrac{-5}{n}$

27. $\displaystyle\sum_{n=1}^{\infty} \dfrac{(-1)^n}{\sqrt{n}}$

28. $\displaystyle\sum_{n=1}^{\infty} \dfrac{1}{2n^3}$

29. $\displaystyle\sum_{n=1}^{\infty} \dfrac{(-1)^n}{\ln(n + 1)}$

30. $\displaystyle\sum_{n=2}^{\infty} \dfrac{1}{n(\ln n)^2}$

31. $\displaystyle\sum_{n=1}^{\infty} \dfrac{\ln n}{n^3}$

32. $\displaystyle\sum_{n=3}^{\infty} \dfrac{\ln n}{\ln(\ln n)}$

33. $\displaystyle\sum_{n=1}^{\infty} \dfrac{(-1)^n}{n\sqrt{n^2 + 1}}$

34. $\displaystyle\sum_{n=1}^{\infty} \dfrac{(-1)^n 3n^2}{n^3 + 1}$

35. $\displaystyle\sum_{n=1}^{\infty} \dfrac{n + 1}{n!}$

36. $\displaystyle\sum_{n=1}^{\infty} \dfrac{(-1)^n(n^2 + 1)}{2n^2 + n - 1}$

37. $\displaystyle\sum_{n=1}^{\infty} \dfrac{(-3)^n}{n!}$

38. $\displaystyle\sum_{n=1}^{\infty} \dfrac{2^n 3^n}{n^n}$

39. $\displaystyle\sum_{n=1}^{\infty} \dfrac{1}{\sqrt{n(n + 1)(n + 2)}}$

40. $\displaystyle\sum_{n=2}^{\infty} \dfrac{1}{n\sqrt{n^2 - 1}}$

Power Series

In Exercises 41–50, **(a)** find the series' radius and interval of convergence. Then identify the values of x for which the series converges **(b)** absolutely and **(c)** conditionally.

41. $\displaystyle\sum_{n=1}^{\infty} \dfrac{(x + 4)^n}{n3^n}$

42. $\displaystyle\sum_{n=1}^{\infty} \dfrac{(x - 1)^{2n-2}}{(2n - 1)!}$

43. $\displaystyle\sum_{n=1}^{\infty} \dfrac{(-1)^{n-1}(3x - 1)^n}{n^2}$

44. $\displaystyle\sum_{n=0}^{\infty} \dfrac{(n + 1)(2x + 1)^n}{(2n + 1)2^n}$

45. $\displaystyle\sum_{n=1}^{\infty} \dfrac{x^n}{n^n}$

46. $\displaystyle\sum_{n=1}^{\infty} \dfrac{x^n}{\sqrt{n}}$

47. $\displaystyle\sum_{n=0}^{\infty} \dfrac{(n + 1)x^{2n-1}}{3^n}$

48. $\displaystyle\sum_{n=0}^{\infty} \dfrac{(-1)^n(x - 1)^{2n+1}}{2n + 1}$

49. $\displaystyle\sum_{n=1}^{\infty} (\operatorname{csch} n)x^n$

50. $\displaystyle\sum_{n=1}^{\infty} (\coth n)x^n$

Maclaurin Series

Each of the series in Exercises 51–56 is the value of the Taylor series at $x = 0$ of a function $f(x)$ at a particular point. What function and what point? What is the sum of the series?

51. $1 - \dfrac{1}{4} + \dfrac{1}{16} - \cdots + (-1)^n \dfrac{1}{4^n} + \cdots$

52. $\dfrac{2}{3} - \dfrac{4}{18} + \dfrac{8}{81} - \cdots + (-1)^{n-1} \dfrac{2^n}{n3^n} + \cdots$

53. $\pi - \dfrac{\pi^3}{3!} + \dfrac{\pi^5}{5!} - \cdots + (-1)^n \dfrac{\pi^{2n+1}}{(2n + 1)!} + \cdots$

54. $1 - \dfrac{\pi^2}{9 \cdot 2!} + \dfrac{\pi^4}{81 \cdot 4!} - \cdots + (-1)^n \dfrac{\pi^{2n}}{3^{2n}(2n)!} + \cdots$

55. $1 + \ln 2 + \dfrac{(\ln 2)^2}{2!} + \cdots + \dfrac{(\ln 2)^n}{n!} + \cdots$

56. $\dfrac{1}{\sqrt{3}} - \dfrac{1}{9\sqrt{3}} + \dfrac{1}{45\sqrt{3}} - \cdots$
$\qquad + (-1)^{n-1} \dfrac{1}{(2n - 1)(\sqrt{3})^{2n-1}} + \cdots$

Find Taylor series at $x = 0$ for the functions in Exercises 57–64.

57. $\dfrac{1}{1 - 2x}$

58. $\dfrac{1}{1 + x^3}$

59. $\sin \pi x$

60. $\sin \dfrac{2x}{3}$

61. $\cos (x^{5/3})$

62. $\cos \dfrac{x^3}{\sqrt{5}}$

63. $e^{(\pi x/2)}$

64. e^{-x^2}

Taylor Series

In Exercises 65–68, find the first four nonzero terms of the Taylor series generated by f at $x = a$.

65. $f(x) = \sqrt{3 + x^2}$ at $x = -1$

66. $f(x) = 1/(1 - x)$ at $x = 2$

67. $f(x) = 1/(x + 1)$ at $x = 3$

68. $f(x) = 1/x$ at $x = a > 0$

Nonelementary Integrals

Use series to approximate the values of the integrals in Exercises 69–72 with an error of magnitude less than 10^{-8}. (The answer section gives the integrals' values rounded to 10 decimal places.)

69. $\displaystyle\int_0^{1/2} e^{-x^3} \, dx$

70. $\displaystyle\int_0^1 x \sin(x^3) \, dx$

71. $\displaystyle\int_0^{1/2} \dfrac{\tan^{-1} x}{x} \, dx$

72. $\displaystyle\int_0^{1/64} \dfrac{\tan^{-1} x}{\sqrt{x}} \, dx$

Using Series to Find Limits

In Exercises 73–78:

  a. Use power series to evaluate the limit.

T **b.** Then use a grapher to support your calculation.

73. $\displaystyle\lim_{x \to 0} \dfrac{7 \sin x}{e^{2x} - 1}$

74. $\displaystyle\lim_{\theta \to 0} \dfrac{e^{\theta} - e^{-\theta} - 2\theta}{\theta - \sin \theta}$

75. $\displaystyle\lim_{t \to 0} \left(\dfrac{1}{2 - 2\cos t} - \dfrac{1}{t^2}\right)$

76. $\displaystyle\lim_{h \to 0} \dfrac{(\sin h)/h - \cos h}{h^2}$

77. $\lim\limits_{z \to 0} \dfrac{1 - \cos^2 z}{\ln(1 - z) + \sin z}$ **78.** $\lim\limits_{y \to 0} \dfrac{y^2}{\cos y - \cosh y}$

Theory and Examples

79. Use a series representation of $\sin 3x$ to find values of r and s for which

$$\lim_{x \to 0}\left(\frac{\sin 3x}{x^3} + \frac{r}{x^2} + s\right) = 0.$$

T 80. Compare the accuracies of the approximations $\sin x \approx x$ and $\sin x \approx 6x/(6 + x^2)$ by comparing the graphs of $f(x) = \sin x - x$ and $g(x) = \sin x - (6x/(6 + x^2))$. Describe what you find.

81. Find the radius of convergence of the series

$$\sum_{n=1}^{\infty} \frac{2 \cdot 5 \cdot 8 \cdot \cdots \cdot (3n - 1)}{2 \cdot 4 \cdot 6 \cdot \cdots \cdot (2n)} x^n.$$

82. Find the radius of convergence of the series

$$\sum_{n=1}^{\infty} \frac{3 \cdot 5 \cdot 7 \cdot \cdots \cdot (2n + 1)}{4 \cdot 9 \cdot 14 \cdot \cdots \cdot (5n - 1)} (x - 1)^n.$$

83. Find a closed-form formula for the nth partial sum of the series $\sum_{n=2}^{\infty} \ln(1 - (1/n^2))$ and use it to determine the convergence or divergence of the series.

84. Evaluate $\sum_{k=2}^{\infty} (1/(k^2 - 1))$ by finding the limits as $n \to \infty$ of the series' nth partial sum.

85. a. Find the interval of convergence of the series

$$y = 1 + \frac{1}{6}x^3 + \frac{1}{180}x^6 + \cdots$$
$$+ \frac{1 \cdot 4 \cdot 7 \cdot \cdots \cdot (3n - 2)}{(3n)!} x^{3n} + \cdots.$$

b. Show that the function defined by the series satisfies a differential equation of the form

$$\frac{d^2y}{dx^2} = x^a y + b$$

and find the values of the constants a and b.

86. a. Find the Maclaurin series for the function $x^2/(1 + x)$.

b. Does the series converge at $x = 1$? Explain.

87. If $\sum_{n=1}^{\infty} a_n$ and $\sum_{n=1}^{\infty} b_n$ are convergent series of nonnegative numbers, can anything be said about $\sum_{n=1}^{\infty} a_n b_n$? Give reasons for your answer.

88. If $\sum_{n=1}^{\infty} a_n$ and $\sum_{n=1}^{\infty} b_n$ are divergent series of nonnegative numbers, can anything be said about $\sum_{n=1}^{\infty} a_n b_n$? Give reasons for your answer.

89. Prove that the sequence $\{x_n\}$ and the series $\sum_{k=1}^{\infty} (x_{k+1} - x_k)$ both converge or both diverge.

90. Prove that $\sum_{n=1}^{\infty} (a_n/(1 + a_n))$ converges if $a_n > 0$ for all n and $\sum_{n=1}^{\infty} a_n$ converges.

91. Suppose that $a_1, a_2, a_3, \ldots, a_n$ are positive numbers satisfying the following conditions:

i) $a_1 \geq a_2 \geq a_3 \geq \cdots$;

ii) the series $a_2 + a_4 + a_8 + a_{16} + \cdots$ diverges.

Show that the series

$$\frac{a_1}{1} + \frac{a_2}{2} + \frac{a_3}{3} + \cdots$$

diverges.

92. Use the result in Exercise 91 to show that

$$1 + \sum_{n=2}^{\infty} \frac{1}{n \ln n}$$

diverges.

Chapter 10   Additional and Advanced Exercises

Determining Convergence of Series

Which of the series $\sum_{n=1}^{\infty} a_n$ defined by the formulas in Exercises 1–4 converge, and which diverge? Give reasons for your answers.

1. $\sum\limits_{n=1}^{\infty} \dfrac{1}{(3n - 2)^{n+(1/2)}}$ **2.** $\sum\limits_{n=1}^{\infty} \dfrac{(\tan^{-1} n)^2}{n^2 + 1}$

3. $\sum\limits_{n=1}^{\infty} (-1)^n \tanh n$ **4.** $\sum\limits_{n=2}^{\infty} \dfrac{\log_n(n!)}{n^3}$

Which of the series $\sum_{n=1}^{\infty} a_n$ defined by the formulas in Exercises 5–8 converge, and which diverge? Give reasons for your answers.

5. $a_1 = 1, \quad a_{n+1} = \dfrac{n(n + 1)}{(n + 2)(n + 3)} a_n$

(*Hint:* Write out several terms, see which factors cancel, and then generalize.)

6. $a_1 = a_2 = 7, \quad a_{n+1} = \dfrac{n}{(n - 1)(n + 1)} a_n \quad$ if $n \geq 2$

7. $a_1 = a_2 = 1, \quad a_{n+1} = \dfrac{1}{1 + a_n} \quad$ if $n \geq 2$

8. $a_n = 1/3^n$ if n is odd, $a_n = n/3^n$ if n is even

Choosing Centers for Taylor Series

Taylor's formula

$$f(x) = f(a) + f'(a)(x - a) + \frac{f''(a)}{2!}(x - a)^2 + \cdots$$
$$+ \frac{f^{(n)}(a)}{n!}(x - a)^n + \frac{f^{(n+1)}(c)}{(n + 1)!}(x - a)^{n+1}$$

expresses the value of f at x in terms of the values of f and its derivatives at $x = a$. In numerical computations, we therefore need a to be a point where we know the values of f and its derivatives. We also need a to be close enough to the values of f we are interested in to make $(x - a)^{n+1}$ so small we can neglect the remainder.

In Exercises 9–14, what Taylor series would you choose to represent the function near the given value of x? (There may be more than one good answer.) Write out the first four nonzero terms of the series you choose.

9. $\cos x$ near $x = 1$ **10.** $\sin x$ near $x = 6.3$

11. e^x near $x = 0.4$ **12.** $\ln x$ near $x = 1.3$

13. $\cos x$ near $x = 69$ **14.** $\tan^{-1} x$ near $x = 2$

Theory and Examples

15. Let a and b be constants with $0 < a < b$. Does the sequence $\{(a^n + b^n)^{1/n}\}$ converge? If it does converge, what is the limit?

16. Find the sum of the infinite series

$$1 + \frac{2}{10} + \frac{3}{10^2} + \frac{7}{10^3} + \frac{2}{10^4} + \frac{3}{10^5} + \frac{7}{10^6} + \frac{2}{10^7}$$

$$+ \frac{3}{10^8} + \frac{7}{10^9} + \cdots.$$

17. Evaluate

$$\sum_{n=0}^{\infty} \int_{n}^{n+1} \frac{1}{1 + x^2}\, dx.$$

18. Find all values of x for which

$$\sum_{n=1}^{\infty} \frac{nx^n}{(n + 1)(2x + 1)^n}$$

converges absolutely.

T 19. a. Does the value of

$$\lim_{n \to \infty} \left(1 - \frac{\cos (a/n)}{n}\right)^n, \quad a \text{ constant},$$

appear to depend on the value of a? If so, how?

b. Does the value of

$$\lim_{n \to \infty} \left(1 - \frac{\cos (a/n)}{bn}\right)^n, \quad a \text{ and } b \text{ constant}, b \neq 0,$$

appear to depend on the value of b? If so, how?

c. Use calculus to confirm your findings in parts (a) and (b).

20. Show that if $\sum_{n=1}^{\infty} a_n$ converges, then

$$\sum_{n=1}^{\infty} \left(\frac{1 + \sin (a_n)}{2}\right)^n$$

converges.

21. Find a value for the constant b that will make the radius of convergence of the power series

$$\sum_{n=2}^{\infty} \frac{b^n x^n}{\ln n}$$

equal to 5.

22. How do you know that the functions $\sin x$, $\ln x$, and e^x are not polynomials? Give reasons for your answer.

23. Find the value of a for which the limit

$$\lim_{x \to 0} \frac{\sin (ax) - \sin x - x}{x^3}$$

is finite and evaluate the limit.

24. Find values of a and b for which

$$\lim_{x \to 0} \frac{\cos (ax) - b}{2x^2} = -1.$$

25. Raabe's (or Gauss's) Test The following test, which we state without proof, is an extension of the Ratio Test.

Raabe's Test: If $\sum_{n=1}^{\infty} u_n$ is a series of positive constants and there exist constants C, K, and N such that

$$\frac{u_n}{u_{n+1}} = 1 + \frac{C}{n} + \frac{f(n)}{n^2},$$

where $|f(n)| < K$ for $n \geq N$, then $\sum_{n=1}^{\infty} u_n$ converges if $C > 1$ and diverges if $C \leq 1$.

Show that the results of Raabe's Test agree with what you know about the series $\sum_{n=1}^{\infty} (1/n^2)$ and $\sum_{n=1}^{\infty} (1/n)$.

26. (*Continuation of Exercise 25.*) Suppose that the terms of $\sum_{n=1}^{\infty} u_n$ are defined recursively by the formulas

$$u_1 = 1, \quad u_{n+1} = \frac{(2n - 1)^2}{(2n)(2n + 1)} u_n.$$

Apply Raabe's Test to determine whether the series converges.

27. If $\sum_{n=1}^{\infty} a_n$ converges, and if $a_n \neq 1$ and $a_n > 0$ for all n,

a. Show that $\sum_{n=1}^{\infty} a_n^2$ converges.

b. Does $\sum_{n=1}^{\infty} a_n/(1 - a_n)$ converge? Explain.

28. (*Continuation of Exercise 27.*) If $\sum_{n=1}^{\infty} a_n$ converges, and if $1 > a_n > 0$ for all n, show that $\sum_{n=1}^{\infty} \ln (1 - a_n)$ converges.

(*Hint:* First show that $|\ln (1 - a_n)| \leq a_n/(1 - a_n)$.)

29. Nicole Oresme's Theorem Prove Nicole Oresme's Theorem that

$$1 + \frac{1}{2} \cdot 2 + \frac{1}{4} \cdot 3 + \cdots + \frac{n}{2^{n-1}} + \cdots = 4.$$

(*Hint:* Differentiate both sides of the equation $1/(1 - x) = 1 + \sum_{n=1}^{\infty} x^n$.)

30. a. Show that

$$\sum_{n=1}^{\infty} \frac{n(n + 1)}{x^n} = \frac{2x^2}{(x - 1)^3}$$

for $|x| > 1$ by differentiating the identity

$$\sum_{n=1}^{\infty} x^{n+1} = \frac{x^2}{1 - x}$$

twice, multiplying the result by x, and then replacing x by $1/x$.

b. Use part (a) to find the real solution greater than 1 of the equation

$$x = \sum_{n=1}^{\infty} \frac{n(n + 1)}{x^n}.$$

31. Quality control

a. Differentiate the series

$$\frac{1}{1 - x} = 1 + x + x^2 + \cdots + x^n + \cdots$$

to obtain a series for $1/(1 - x)^2$.

b. In one throw of two dice, the probability of getting a roll of 7 is $p = 1/6$. If you throw the dice repeatedly, the probability that a 7 will appear for the first time at the nth throw is $q^{n-1}p$, where $q = 1 - p = 5/6$. The expected number of throws until a 7 first appears is $\sum_{n=1}^{\infty} nq^{n-1}p$. Find the sum of this series.

c. As an engineer applying statistical control to an industrial operation, you inspect items taken at random from the assembly line. You classify each sampled item as either "good" or "bad." If the probability of an item's being good is p and of an item's being bad is $q = 1 - p$, the probability that the first bad item found is the nth one inspected is $p^{n-1}q$. The average number inspected up to and including the first bad item found is $\sum_{n=1}^{\infty} np^{n-1}q$. Evaluate this sum, assuming $0 < p < 1$.

32. **Expected value** Suppose that a random variable X may assume the values 1, 2, 3, $\ldots$, with probabilities $p_1, p_2, p_3, \ldots$, where p_k is the probability that X equals k ($k = 1, 2, 3, \ldots$). Suppose also that $p_k \geq 0$ and that $\sum_{k=1}^{\infty} p_k = 1$. The **expected value** of X, denoted by $E(X)$, is the number $\sum_{k=1}^{\infty} kp_k$, provided the series converges. In each of the following cases, show that $\sum_{k=1}^{\infty} p_k = 1$ and find $E(X)$ if it exists. (*Hint:* See Exercise 31.)

a. $p_k = 2^{-k}$
b. $p_k = \dfrac{5^{k-1}}{6^k}$

c. $p_k = \dfrac{1}{k(k + 1)} = \dfrac{1}{k} - \dfrac{1}{k + 1}$

T 33. **Safe and effective dosage** The concentration in the blood resulting from a single dose of a drug normally decreases with time as the drug is eliminated from the body. Doses may therefore need to be repeated periodically to keep the concentration from dropping below some particular level. One model for the effect of repeated doses gives the residual concentration just before the $(n + 1)$st dose as

$$R_n = C_0 e^{-kt_0} + C_0 e^{-2kt_0} + \cdots + C_0 e^{-nkt_0},$$

where C_0 = the change in concentration achievable by a single dose (mg/mL), k = the *elimination constant* (h^{-1}), and t_0 = time between doses (h). See the accompanying figure.

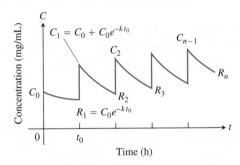

a. Write R_n in closed form as a single fraction, and find $R = \lim_{n \to \infty} R_n$.

b. Calculate R_1 and R_{10} for $C_0 = 1$ mg/mL, $k = 0.1$ h^{-1}, and $t_0 = 10$ h. How good an estimate of R is R_{10}?

c. If $k = 0.01$ h^{-1} and $t_0 = 10$ h, find the smallest n such that $R_n > (1/2)R$.

(*Source: Prescribing Safe and Effective Dosage*, B. Horelick and S. Koont, COMAP, Inc., Lexington, MA.)

34. **Time between drug doses** (*Continuation of Exercise 33.*) If a drug is known to be ineffective below a concentration C_L and harmful above some higher concentration C_H, one needs to find values of C_0 and t_0 that will produce a concentration that is safe (not above C_H) but effective (not below C_L). See the accompanying figure. We therefore want to find values for C_0 and t_0 for which

$$R = C_L \quad \text{and} \quad C_0 + R = C_H.$$

Thus $C_0 = C_H - C_L$. When these values are substituted in the equation for R obtained in part (a) of Exercise 33, the resulting equation simplifies to

$$t_0 = \frac{1}{k} \ln \frac{C_H}{C_L}.$$

To reach an effective level rapidly, one might administer a "loading" dose that would produce a concentration of C_H mg/mL. This could be followed every t_0 hours by a dose that raises the concentration by $C_0 = C_H - C_L$ mg/mL.

a. Verify the preceding equation for t_0.

b. If $k = 0.05$ h^{-1} and the highest safe concentration is e times the lowest effective concentration, find the length of time between doses that will assure safe and effective concentrations.

c. Given $C_H = 2$ mg/mL, $C_L = 0.5$ mg/mL, and $k = 0.02$ h^{-1}, determine a scheme for administering the drug.

d. Suppose that $k = 0.2$ h^{-1} and that the smallest effective concentration is 0.03 mg/mL. A single dose that produces a concentration of 0.1 mg/mL is administered. About how long will the drug remain effective?

Chapter 10 Technology Application Projects

Mathematica/Maple Modules:

Bouncing Ball
The model predicts the height of a bouncing ball, and the time until it stops bouncing.

Taylor Polynomial Approximations of a Function
A graphical animation shows the convergence of the Taylor polynomials to functions having derivatives of all orders over an interval in their domains.

11

PARAMETRIC EQUATIONS
AND POLAR COORDINATES

OVERVIEW In this chapter we study new ways to define curves in the plane. Instead of thinking of a curve as the graph of a function or equation, we consider a more general way of thinking of a curve as the path of a moving particle whose position is changing over time. Then each of the x- and y-coordinates of the particle's position becomes a function of a third variable t. We can also change the way in which points in the plane themselves are described by using *polar coordinates* rather than the rectangular or Cartesian system. Both of these new tools are useful for describing motion, like that of planets and satellites, or projectiles moving in the plane or space. In addition, we review the geometric definitions and standard equations of parabolas, ellipses, and hyperbolas. These curves are called *conic sections,* or *conics,* and model the paths traveled by projectiles, planets, or any other object moving under the sole influence of a gravitational or electromagnetic force.

11.1 | Parametrizations of Plane Curves

In previous chapters, we have studied curves as the graphs of functions or equations involving the two variables x and y. We are now going to introduce another way to describe a curve by expressing both coordinates as functions of a third variable t.

Parametric Equations

Figure 11.1 shows the path of a moving particle in the xy-plane. Notice that the path fails the vertical line test, so it cannot be described as the graph of a function of the variable x. However, we can sometimes describe the path by a pair of equations, $x = f(t)$ and $y = g(t)$, where f and g are continuous functions. When studying motion, t usually denotes time. Equations like these describe more general curves than those like $y = f(x)$ and provide not only the graph of the path traced out but also the location of the particle $(x, y) = (f(t), g(t))$ at any time t.

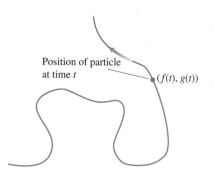

Position of particle at time t — $(f(t), g(t))$

FIGURE 11.1 The curve or path traced by a particle moving in the xy-plane is not always the graph of a function or single equation.

DEFINITION   If x and y are given as functions

$$x = f(t), \qquad y = g(t)$$

over an interval I of t-values, then the set of points $(x, y) = (f(t), g(t))$ defined by these equations is a **parametric curve**. The equations are **parametric equations** for the curve.

The variable t is a **parameter** for the curve, and its domain I is the **parameter interval**. If I is a closed interval, $a \le t \le b$, the point $(f(a), g(a))$ is the **initial point** of the curve and $(f(b), g(b))$ is the **terminal point**. When we give parametric equations and a parameter interval for a curve, we say that we have **parametrized** the curve. The equations and interval together constitute a **parametrization** of the curve. A given curve can be represented by different sets of parametric equations. (See Exercises 19 and 20.)

EXAMPLE 1 Sketch the curve defined by the parametric equations

$$x = t^2, \qquad y = t + 1, \qquad -\infty < t < \infty.$$

Solution We make a brief table of values (Table 11.1), plot the points (x, y), and draw a smooth curve through them (Figure 11.2). Each value of t gives a point (x, y) on the curve, such as $t = 1$ giving the point $(1, 2)$ recorded in Table 11.1. If we think of the curve as the path of a moving particle, then the particle moves along the curve in the direction of the arrows shown in Figure 11.2. Although the time intervals in the table are equal, the consecutive points plotted along the curve are not at equal arc length distances. The reason for this is that the particle slows down at it gets nearer to the y-axis along the lower branch of the curve as t increases, and then speeds up after reaching the y-axis at $(0, 1)$ and moving along the upper branch. Since the interval of values for t is all real numbers, there is no initial point and no terminal point for the curve. ∎

TABLE 11.1 Values of $x = t^2$ and $y = t + 1$ for selected values of t.		
t	x	y
-3	9	-2
-2	4	-1
-1	1	0
0	0	1
1	1	2
2	4	3
3	9	4

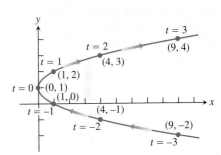

FIGURE 11.2 The curve given by the parametric equations $x = t^2$ and $y = t + 1$ (Example 1).

EXAMPLE 2 Identify geometrically the curve in Example 1 (Figure 11.2) by eliminating the parameter t and obtaining an algebraic equation in x and y.

Solution We solve the equation $y = t + 1$ for the parameter t and substitute the result into the parametric equation for x. This procedure gives $t = y - 1$ and

$$x = t^2 = (y - 1)^2 = y^2 - 2y + 1.$$

The equation $x = y^2 - 2y + 1$ represents a parabola, as displayed in Figure 11.2. It is sometimes quite difficult, or even impossible, to eliminate the parameter from a pair of parametric equations, as we did here. ∎

EXAMPLE 3 Graph the parametric curves

(a) $x = \cos t, \qquad y = \sin t, \qquad 0 \le t \le 2\pi.$

(b) $x = a \cos t, \qquad y = a \sin t, \qquad 0 \le t \le 2\pi.$

Solution

(a) Since $x^2 + y^2 = \cos^2 t + \sin^2 t = 1$, the parametric curve lies along the unit circle $x^2 + y^2 = 1$. As t increases from 0 to 2π, the point $(x, y) = (\cos t, \sin t)$ starts at $(1, 0)$ and traces the entire circle once counterclockwise (Figure 11.3).

FIGURE 11.3 The equations $x = \cos t$ and $y = \sin t$ describe motion on the circle $x^2 + y^2 = 1$. The arrow shows the direction of increasing t (Example 3).

(b) For $x = a \cos t$, $y = a \sin t$, $0 \le t \le 2\pi$, we have $x^2 + y^2 = a^2 \cos^2 t + a^2 \sin^2 t = a^2$. The parametrization describes a motion that begins at the point $(a, 0)$ and traverses the circle $x^2 + y^2 = a^2$ once counterclockwise, returning to $(a, 0)$ at $t = 2\pi$. The graph is a circle centered at the origin with radius $r = a$ and coordinate points $(a \cos t, a \sin t)$. ∎

EXAMPLE 4 The position $P(x, y)$ of a particle moving in the xy-plane is given by the equations and parameter interval

$$x = \sqrt{t}, \qquad y = t, \qquad t \ge 0.$$

Identify the path traced by the particle and describe the motion.

Solution We try to identify the path by eliminating t between the equations $x = \sqrt{t}$ and $y = t$. With any luck, this will produce a recognizable algebraic relation between x and y. We find that

$$y = t = \left(\sqrt{t}\right)^2 = x^2.$$

Thus, the particle's position coordinates satisfy the equation $y = x^2$, so the particle moves along the parabola $y = x^2$.

It would be a mistake, however, to conclude that the particle's path is the entire parabola $y = x^2$; it is only half the parabola. The particle's x-coordinate is never negative. The particle starts at $(0, 0)$ when $t = 0$ and rises into the first quadrant as t increases (Figure 11.4). The parameter interval is $[0, \infty)$ and there is no terminal point. ∎

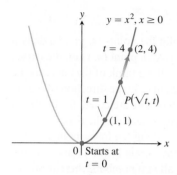

FIGURE 11.4 The equations $x = \sqrt{t}$ and $y = t$ and the interval $t \ge 0$ describe the path of a particle that traces the right-hand half of the parabola $y = x^2$ (Example 4).

The graph of any function $y = f(x)$ can always be given a natural parametrization $x = t$ and $y = f(t)$. The domain of the parameter in this case is the same as the domain of the function f.

EXAMPLE 5 A parametrization of the graph of the function $f(x) = x^2$ is given by

$$x = t, \qquad y = f(t) = t^2, \qquad -\infty < t < \infty.$$

When $t \ge 0$, this parametrization gives the same path in the xy-plane as we had in Example 4. However, since the parameter t here can now also be negative, we obtain the left-hand part of the parabola as well; that is, we have the entire parabolic curve. For this parametrization, there is no starting point and no terminal point (Figure 11.5). ∎

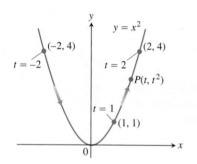

FIGURE 11.5 The path defined by $x = t, y = t^2, -\infty < t < \infty$ is the entire parabola $y = x^2$ (Example 5).

Notice that a parametrization also specifies *when* (the value of the parameter) a particle moving along the curve is *located* at a specific point along the curve. In Example 4, the point $(2, 4)$ is reached when $t = 4$; in Example 5, it is reached "earlier" when $t = 2$. You can see the implications of this aspect of parametrizations when considering the possibility of two objects coming into collision: they have to be at the exact same location point $P(x, y)$ for some (possibly different) values of their respective parameters. We will say more about this aspect of parametrizations when we study motion in Chapter 13.

EXAMPLE 6 Find a parametrization for the line through the point (a, b) having slope m.

Solution A Cartesian equation of the line is $y - b = m(x - a)$. If we set the parameter $t = x - a$, we find that $x = a + t$ and $y - b = mt$. That is,

$$x = a + t, \qquad y = b + mt, \qquad -\infty < t < \infty$$

parametrizes the line. This parametrization differs from the one we would obtain by the technique used in Example 5 when $t = x$. However, both parametrizations give the same line.

∎

TABLE 11.2 Values of $x = t + (1/t)$ and $y = t - (1/t)$ for selected values of t.

t	$1/t$	x	y
0.1	10.0	10.1	−9.9
0.2	5.0	5.2	−4.8
0.4	2.5	2.9	−2.1
1.0	1.0	2.0	0.0
2.0	0.5	2.5	1.5
5.0	0.2	5.2	4.8
10.0	0.1	10.1	9.9

FIGURE 11.6 The curve for $x = t + (1/t)$, $y = t - (1/t)$, $t > 0$ in Example 7. (The part shown is for $0.1 \le t \le 10$.)

EXAMPLE 7 Sketch and identify the path traced by the point $P(x, y)$ if

$$x = t + \frac{1}{t}, \qquad y = t - \frac{1}{t}, \qquad t > 0.$$

Solution We make a brief table of values in Table 11.2, plot the points, and draw a smooth curve through them, as we did in Example 1. Next we eliminate the parameter t from the equations. The procedure is more complicated than in Example 2. Taking the difference between x and y as given by the parametric equations, we find that

$$x - y = \left(t + \frac{1}{t}\right) - \left(t - \frac{1}{t}\right) = \frac{2}{t}.$$

If we add the two parametric equations, we get

$$x + y = \left(t + \frac{1}{t}\right) + \left(t - \frac{1}{t}\right) = 2t.$$

We can then eliminate the parameter t by multiplying these last equations together:

$$(x - y)(x + y) = \left(\frac{2}{t}\right)(2t) = 4,$$

or, multiplying together the terms on the left-hand side, we obtain a standard equation for a hyperbola (reviewed in Section 11.6):

$$x^2 - y^2 = 4. \tag{1}$$

Thus the coordinates of all the points $P(x, y)$ described by the parametric equations satisfy Equation (1). However, Equation (1) does not require that the x-coordinate be positive. So there are points (x, y) on the hyperbola that do not satisfy the parametric equation $x = t + (1/t)$, $t > 0$, for which x is always positive. That is, the parametric equations do not yield any points on the left branch of the hyperbola given by Equation (1), points where the x-coordinate would be negative. For small positive values of t, the path lies in the fourth quadrant and rises into the first quadrant as t increases, crossing the x-axis when $t = 1$ (see Figure 11.6). The parameter domain is $(0, \infty)$ and there is no starting point and no terminal point for the path. ∎

Examples 4, 5, and 6 illustrate that a given curve, or portion of it, can be represented by different parametrizations. In the case of Example 7, we can also represent the right-hand branch of the hyperbola by the parametrization

$$x = \sqrt{4 + t^2}, \qquad y = t, \qquad -\infty < t < \infty,$$

which is obtained by solving Equation (1) for $x \ge 0$ and letting y be the parameter. Still another parametrization for the right-hand branch of the hyperbola given by Equation (1) is

$$x = 2 \sec t, \qquad y = 2 \tan t, \qquad -\frac{\pi}{2} < t < \frac{\pi}{2}.$$

This parametrization follows from the trigonometric identity $\sec^2 t - \tan^2 t = 1$, so

$$x^2 - y^2 = 4 \sec^2 t - 4 \tan^2 t = 4(\sec^2 t - \tan^2 t) = 4.$$

As t runs between $-\pi/2$ and $\pi/2$, $x = \sec t$ remains positive and $y = \tan t$ runs between $-\infty$ and ∞, so P traverses the hyperbola's right-hand branch. It comes in along the branch's lower half as $t \to 0^-$, reaches $(2, 0)$ at $t = 0$, and moves out into the first quadrant as t increases steadily toward $\pi/2$. This is the same hyperbola branch for which a portion is shown in Figure 11.6.

HISTORICAL BIOGRAPHY

Christian Huygens
(1629–1695)

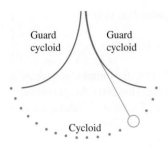

Guard
cycloid

Guard
cycloid

Cycloid

FIGURE 11.7 In Huygens' pendulum clock, the bob swings in a cycloid, so the frequency is independent of the amplitude.

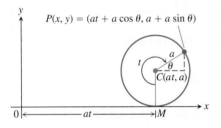

$P(x, y) = (at + a \cos \theta, a + a \sin \theta)$

$C(at, a)$

$0 \longleftarrow at \longrightarrow |M$

FIGURE 11.8 The position of $P(x, y)$ on the rolling wheel at angle t (Example 8).

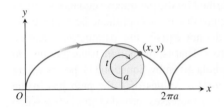

(x, y)

$2\pi a$

FIGURE 11.9 The cycloid curve $x = a(t - \sin t), y = a(1 - \cos t)$, for $t \geq 0$.

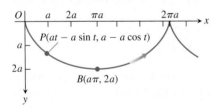

$O \quad a \quad 2a \quad \pi a \qquad 2\pi a$

$P(at - a \sin t, a - a \cos t)$

$2a$

$B(a\pi, 2a)$

FIGURE 11.10 To study motion along an upside-down cycloid under the influence of gravity, we turn Figure 11.9 upside down. This points the y-axis in the direction of the gravitational force and makes the downward y-coordinates positive. The equations and parameter interval for the cycloid are still

$$x = a(t - \sin t),$$
$$y = a(1 - \cos t), \quad t \geq 0.$$

The arrow shows the direction of increasing t.

Cycloids

The problem with a pendulum clock whose bob swings in a circular arc is that the frequency of the swing depends on the amplitude of the swing. The wider the swing, the longer it takes the bob to return to center (its lowest position).

This does not happen if the bob can be made to swing in a *cycloid*. In 1673, Christian Huygens designed a pendulum clock whose bob would swing in a cycloid, a curve we define in Example 8. He hung the bob from a fine wire constrained by guards that caused it to draw up as it swung away from center (Figure 11.7).

EXAMPLE 8 A wheel of radius a rolls along a horizontal straight line. Find parametric equations for the path traced by a point P on the wheel's circumference. The path is called a **cycloid**.

Solution We take the line to be the x-axis, mark a point P on the wheel, start the wheel with P at the origin, and roll the wheel to the right. As parameter, we use the angle t through which the wheel turns, measured in radians. Figure 11.8 shows the wheel a short while later when its base lies at units from the origin. The wheel's center C lies at (at, a) and the coordinates of P are

$$x = at + a \cos \theta, \qquad y = a + a \sin \theta.$$

To express θ in terms of t, we observe that $t + \theta = 3\pi/2$ in the figure, so that

$$\theta = \frac{3\pi}{2} - t.$$

This makes

$$\cos \theta = \cos\left(\frac{3\pi}{2} - t\right) = -\sin t, \qquad \sin \theta = \sin\left(\frac{3\pi}{2} - t\right) = -\cos t.$$

The equations we seek are

$$x = at - a \sin t, \qquad y = a - a \cos t.$$

These are usually written with the a factored out:

$$x = a(t - \sin t), \qquad y = a(1 - \cos t). \tag{2}$$

Figure 11.9 shows the first arch of the cycloid and part of the next. ∎

Brachistochrones and Tautochrones

If we turn Figure 11.9 upside down, Equations (2) still apply and the resulting curve (Figure 11.10) has two interesting physical properties. The first relates to the origin O and the point B at the bottom of the first arch. Among all smooth curves joining these points, the cycloid is the curve along which a frictionless bead, subject only to the force of gravity, will slide from O to B fastest. This makes the cycloid a **brachistochrone** ("brah-*kiss*-toe-krone"), or shortest-time curve for these points. The second property is that even if you start the bead partway down the curve toward B, it will still take the bead the same amount of time to reach B. This makes the cycloid a **tautochrone** ("*taw*-toe-krone"), or same-time curve for O and B.

Are there any other brachistochrones joining O and B, or is the cycloid the only one? We can formulate this as a mathematical question in the following way. At the start, the kinetic energy of the bead is zero, since its velocity is zero. The work done by gravity in moving the bead from $(0, 0)$ to any other point (x, y) in the plane is mgy, and this must equal the change in kinetic energy. That is,

$$mgy = \frac{1}{2}mv^2 - \frac{1}{2}m(0)^2.$$

Thus, the velocity of the bead when it reaches (x, y) has to be

$$v = \sqrt{2gy}.$$

That is,

$$\frac{ds}{dt} = \sqrt{2gy} \qquad \text{ds is the arc length differential along the bead's path.}$$

or

$$dt = \frac{ds}{\sqrt{2gy}} = \frac{\sqrt{1 + (dy/dx)^2}\, dx}{\sqrt{2gy}}.$$

The time T_f it takes the bead to slide along a particular path $y = f(x)$ from O to $B(a\pi, 2a)$ is

$$T_f = \int_{x=0}^{x=a\pi} \sqrt{\frac{1 + (dy/dx)^2}{2gy}}\, dx. \tag{3}$$

What curves $y = f(x)$, if any, minimize the value of this integral?

At first sight, we might guess that the straight line joining O and B would give the shortest time, but perhaps not. There might be some advantage in having the bead fall vertically at first to build up its velocity faster. With a higher velocity, the bead could travel a longer path and still reach B first. Indeed, this is the right idea. The solution, from a branch of mathematics known as the *calculus of variations*, is that the original cycloid from O to B is the one and only brachistochrone for O and B.

While the solution of the brachistochrone problem is beyond our present reach, we can still show why the cycloid is a tautochrone. In the next section we show that the derivative dy/dx is simply the derivative dy/dt divided by the derivative dx/dt. Making the derivative calculations and substituting into Equation (3) (we omit the details of the calculations here) gives

$$T_{\text{cycloid}} = \int_{x=0}^{x=a\pi} \sqrt{\frac{1 + (dy/dx)^2}{2gy}}\, dx$$

$$= \int_{t=0}^{t=\pi} \sqrt{\frac{a^2(2 - 2\cos t)}{2ga(1 - \cos t)}}\, dt \qquad \begin{array}{l} \text{From Equations (2),} \\ dx/dt = a(1 - \cos t), \\ dy/dt = a\sin t, \text{ and} \\ y = a(1 - \cos t) \end{array}$$

$$= \int_0^\pi \sqrt{\frac{a}{g}}\, dt = \pi\sqrt{\frac{a}{g}}.$$

Thus, the amount of time it takes the frictionless bead to slide down the cycloid to B after it is released from rest at O is $\pi\sqrt{a/g}$.

Suppose that instead of starting the bead at O we start it at some lower point on the cycloid, a point (x_0, y_0) corresponding to the parameter value $t_0 > 0$. The bead's velocity at any later point (x, y) on the cycloid is

$$v = \sqrt{2g(y - y_0)} = \sqrt{2ga(\cos t_0 - \cos t)}. \qquad y = a(1 - \cos t)$$

Accordingly, the time required for the bead to slide from (x_0, y_0) down to B is

$$T = \int_{t_0}^{\pi} \sqrt{\frac{a^2(2 - 2\cos t)}{2ga(\cos t_0 - \cos t)}}\, dt = \sqrt{\frac{a}{g}} \int_{t_0}^{\pi} \sqrt{\frac{1 - \cos t}{\cos t_0 - \cos t}}\, dt$$

$$= \sqrt{\frac{a}{g}} \int_{t_0}^{\pi} \sqrt{\frac{2\sin^2(t/2)}{(2\cos^2(t_0/2) - 1) - (2\cos^2(t/2) - 1)}}\, dt$$

$$= \sqrt{\frac{a}{g}} \int_{t_0}^{\pi} \frac{\sin(t/2)\, dt}{\sqrt{\cos^2(t_0/2) - \cos^2(t/2)}}$$

$$= \sqrt{\frac{a}{g}} \int_{t=t_0}^{t=\pi} \frac{-2\, du}{\sqrt{c^2 - u^2}} \qquad \begin{array}{l} u = \cos(t/2) \\ -2\, du = \sin(t/2)\, dt \\ c = \cos(t_0/2) \end{array}$$

$$= 2\sqrt{\frac{a}{g}} \left[-\sin^{-1}\frac{u}{c} \right]_{t=t_0}^{t=\pi}$$

$$= 2\sqrt{\frac{a}{g}} \left[-\sin^{-1}\frac{\cos(t/2)}{\cos(t_0/2)} \right]_{t_0}^{\pi}$$

$$= 2\sqrt{\frac{a}{g}} (-\sin^{-1} 0 + \sin^{-1} 1) = \pi\sqrt{\frac{a}{g}}.$$

FIGURE 11.11 Beads released simultaneously on the upside-down cycloid at O, A, and C will reach B at the same time.

This is precisely the time it takes the bead to slide to B from O. It takes the bead the same amount of time to reach B no matter where it starts. Beads starting simultaneously from O, A, and C in Figure 11.11, for instance, will all reach B at the same time. This is the reason that Huygens' pendulum clock is independent of the amplitude of the swing.

Exercises 11.1

Finding Cartesian from Parametric Equations

Exercises 1–18 give parametric equations and parameter intervals for the motion of a particle in the xy-plane. Identify the particle's path by finding a Cartesian equation for it. Graph the Cartesian equation. (The graphs will vary with the equation used.) Indicate the portion of the graph traced by the particle and the direction of motion.

1. $x = 3t, \quad y = 9t^2, \quad -\infty < t < \infty$
2. $x = -\sqrt{t}, \quad y = t, \quad t \geq 0$
3. $x = 2t - 5, \quad y = 4t - 7, \quad -\infty < t < \infty$
4. $x = 3 - 3t, \quad y = 2t, \quad 0 \leq t \leq 1$
5. $x = \cos 2t, \quad y = \sin 2t, \quad 0 \leq t \leq \pi$
6. $x = \cos(\pi - t), \quad y = \sin(\pi - t), \quad 0 \leq t \leq \pi$
7. $x = 4\cos t, \quad y = 2\sin t, \quad 0 \leq t \leq 2\pi$
8. $x = 4\sin t, \quad y = 5\cos t, \quad 0 \leq t \leq 2\pi$
9. $x = \sin t, \quad y = \cos 2t, \quad -\dfrac{\pi}{2} \leq t \leq \dfrac{\pi}{2}$
10. $x = 1 + \sin t, \quad y = \cos t - 2, \quad 0 \leq t \leq \pi$
11. $x = t^2, \quad y = t^6 - 2t^4, \quad -\infty < t < \infty$
12. $x = \dfrac{t}{t - 1}, \quad y = \dfrac{t - 2}{t + 1}, \quad -1 < t < 1$
13. $x = t, \quad y = \sqrt{1 - t^2}, \quad -1 \leq t \leq 0$
14. $x = \sqrt{t + 1}, \quad y = \sqrt{t}, \quad t \geq 0$

15. $x = \sec^2 t - 1, \quad y = \tan t, \quad -\pi/2 < t < \pi/2$
16. $x = -\sec t, \quad y = \tan t, \quad -\pi/2 < t < \pi/2$
17. $x = -\cosh t, \quad y = \sinh t, \quad -\infty < t < \infty$
18. $x = 2\sinh t, \quad y = 2\cosh t, \quad -\infty < t < \infty$

Finding Parametric Equations

19. Find parametric equations and a parameter interval for the motion of a particle that starts at $(a, 0)$ and traces the circle $x^2 + y^2 = a^2$

    a. once clockwise.   b. once counterclockwise.

    c. twice clockwise.   d. twice counterclockwise.

    (There are many ways to do these, so your answers may not be the same as the ones in the back of the book.)

20. Find parametric equations and a parameter interval for the motion of a particle that starts at $(a, 0)$ and traces the ellipse $(x^2/a^2) + (y^2/b^2) = 1$

    a. once clockwise.   b. once counterclockwise.

    c. twice clockwise.   d. twice counterclockwise.

    (As in Exercise 19, there are many correct answers.)

In Exercises 21–26, find a parametrization for the curve.

21. the line segment with endpoints $(-1, -3)$ and $(4, 1)$
22. the line segment with endpoints $(-1, 3)$ and $(3, -2)$

23. the lower half of the parabola $x - 1 = y^2$
24. the left half of the parabola $y = x^2 + 2x$
25. the ray (half line) with initial point (2, 3) that passes through the point $(-1, -1)$
26. the ray (half line) with initial point $(-1, 2)$ that passes through the point (0, 0)
27. Find parametric equations and a parameter interval for the motion of a particle starting at the point (2, 0) and tracing the top half of the circle $x^2 + y^2 = 4$ four times.
28. Find parametric equations and a parameter interval for the motion of a particle that moves along the graph of $y = x^2$ in the following way: beginning at (0, 0) it moves to (3, 9), and then travels back and forth from (3, 9) to $(-3, 9)$ infinitely many times.
29. Find parametric equations for the semicircle

$$x^2 + y^2 = a^2, \quad y > 0,$$

using as parameter the slope $t = dy/dx$ of the tangent to the curve at (x, y).
30. Find parametric equations for the circle

$$x^2 + y^2 = a^2,$$

using as parameter the arc length s measured counterclockwise from the point $(a, 0)$ to the point (x, y).
31. Find a parametrization for the line segment joining points (0, 2) and (4, 0) using the angle θ in the accompanying figure as the parameter.

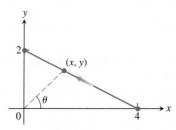

32. Find a parametrization for the curve $y = \sqrt{x}$ with terminal point (0, 0) using the angle θ in the accompanying figure as the parameter.

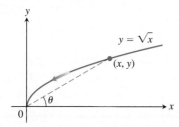

33. Find a parametrization for the circle $(x - 2)^2 + y^2 = 1$ starting at (1, 0) and moving clockwise once around the circle, using the central angle θ in the accompanying figure as the parameter.

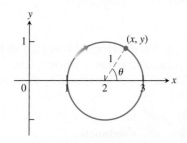

34. Find a parametrization for the circle $x^2 + y^2 = 1$ starting at (1, 0) and moving counterclockwise to the terminal point (0, 1), using the angle θ in the accompanying figure as the parameter.

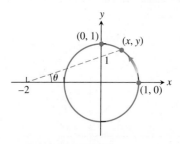

35. **The witch of Maria Agnesi** The bell-shaped witch of Maria Agnesi can be constructed in the following way. Start with a circle of radius 1, centered at the point (0, 1), as shown in the accompanying figure. Choose a point A on the line $y = 2$ and connect it to the origin with a line segment. Call the point where the segment crosses the circle B. Let P be the point where the vertical line through A crosses the horizontal line through B. The witch is the curve traced by P as A moves along the line $y = 2$. Find parametric equations and a parameter interval for the witch by expressing the coordinates of P in terms of t, the radian measure of the angle that segment OA makes with the positive x-axis. The following equalities (which you may assume) will help.

a. $x = AQ$ **b.** $y = 2 - AB \sin t$

c. $AB \cdot OA = (AQ)^2$

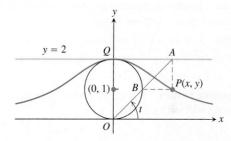

36. **Hypocycloid** When a circle rolls on the inside of a fixed circle, any point P on the circumference of the rolling circle describes a *hypocycloid*. Let the fixed circle be $x^2 + y^2 = a^2$, let the radius of the rolling circle be b, and let the initial position of the tracing point P be A $(a, 0)$. Find parametric equations for the hypocycloid, using as the parameter the angle θ from the positive x-axis to the line joining the circles' centers. In particular, if $b = a/4$, as in the accompanying figure, show that the hypocycloid is the astroid

$$x = a \cos^3 \theta, \quad y = a \sin^3 \theta.$$

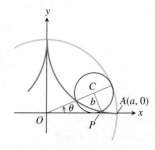

37. As the point N moves along the line $y = a$ in the accompanying figure, P moves in such a way that $OP = MN$. Find parametric equations for the coordinates of P as functions of the angle t that the line ON makes with the positive y-axis.

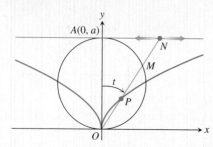

38. **Trochoids** A wheel of radius a rolls along a horizontal straight line without slipping. Find parametric equations for the curve traced out by a point P on a spoke of the wheel b units from its center. As parameter, use the angle θ through which the wheel turns. The curve is called a *trochoid*, which is a cycloid when $b = a$.

Distance Using Parametric Equations

39. Find the point on the parabola $x = t, y = t^2, -\infty < t < \infty$, closest to the point $(2, 1/2)$. (*Hint:* Minimize the square of the distance as a function of t.)

40. Find the point on the ellipse $x = 2\cos t, y = \sin t, 0 \le t \le 2\pi$ closest to the point $(3/4, 0)$. (*Hint:* Minimize the square of the distance as a function of t.)

T GRAPHER EXPLORATIONS

If you have a parametric equation grapher, graph the equations over the given intervals in Exercises 41–48.

41. **Ellipse** $x = 4\cos t, \quad y = 2\sin t, \quad$ over
   a. $0 \le t \le 2\pi$
   b. $0 \le t \le \pi$
   c. $-\pi/2 \le t \le \pi/2$.

42. **Hyperbola branch** $x = \sec t$ (enter as $1/\cos (t)$), $y = \tan t$ (enter as $\sin (t)/\cos (t)$), over
   a. $-1.5 \le t \le 1.5$
   b. $-0.5 \le t \le 0.5$
   c. $-0.1 \le t \le 0.1$.

43. **Parabola** $x = 2t + 3, \quad y = t^2 - 1, \quad -2 \le t \le 2$

44. **Cycloid** $x = t - \sin t, \quad y = 1 - \cos t, \quad$ over
   a. $0 \le t \le 2\pi$
   b. $0 \le t \le 4\pi$
   c. $\pi \le t \le 3\pi$.

45. **Deltoid**

$$x = 2\cos t + \cos 2t, \quad y = 2\sin t - \sin 2t; \quad 0 \le t \le 2\pi$$

What happens if you replace 2 with -2 in the equations for x and y? Graph the new equations and find out.

46. **A nice curve**

$$x = 3\cos t + \cos 3t, \quad y = 3\sin t - \sin 3t; \quad 0 \le t \le 2\pi$$

What happens if you replace 3 with -3 in the equations for x and y? Graph the new equations and find out.

47. a. **Epicycloid**

$$x = 9\cos t - \cos 9t, \quad y = 9\sin t - \sin 9t; \quad 0 \le t \le 2\pi$$

   b. **Hypocycloid**

$$x = 8\cos t + 2\cos 4t, \quad y = 8\sin t - 2\sin 4t; \quad 0 \le t \le 2\pi$$

   c. **Hypotrochoid**

$$x = \cos t + 5\cos 3t, \quad y = 6\cos t - 5\sin 3t; \quad 0 \le t \le 2\pi$$

48. a. $x = 6\cos t + 5\cos 3t, \quad y = 6\sin t - 5\sin 3t;$
      $0 \le t \le 2\pi$

   b. $x = 6\cos 2t + 5\cos 6t, \quad y = 6\sin 2t - 5\sin 6t;$
      $0 \le t \le \pi$

   c. $x = 6\cos t + 5\cos 3t, \quad y = 6\sin 2t - 5\sin 3t;$
      $0 \le t \le 2\pi$

   d. $x = 6\cos 2t + 5\cos 6t, \quad y = 6\sin 4t - 5\sin 6t;$
      $0 \le t \le \pi$

11.2 Calculus with Parametric Curves

In this section we apply calculus to parametric curves. Specifically, we find slopes, lengths, and areas associated with parametrized curves.

Tangents and Areas

A parametrized curve $x = f(t)$ and $y = g(t)$ is **differentiable** at t if f and g are differentiable at t. At a point on a differentiable parametrized curve where y is also a differentiable function of x, the derivatives dy/dt, dx/dt, and dy/dx are related by the Chain Rule:

$$\frac{dy}{dt} = \frac{dy}{dx} \cdot \frac{dx}{dt}.$$

If $dx/dt \neq 0$, we may divide both sides of this equation by dx/dt to solve for dy/dx.

Parametric Formula for dy/dx

If all three derivatives exist and $dx/dt \neq 0$,

$$\frac{dy}{dx} = \frac{dy/dt}{dx/dt}. \tag{1}$$

If parametric equations define y as a twice-differentiable function of x, we can apply Equation (1) to the function $dy/dx = y'$ to calculate d^2y/dx^2 as a function of t:

$$\frac{d^2y}{dx^2} = \frac{d}{dx}(y') = \frac{dy'/dt}{dx/dt}. \qquad \text{Eq. (1) with } y' \text{ in place of } y$$

Parametric Formula for d^2y/dx^2

If the equations $x = f(t)$, $y = g(t)$ define y as a twice-differentiable function of x, then at any point where $dx/dt \neq 0$ and $y' = dy/dx$,

$$\frac{d^2y}{dx^2} = \frac{dy'/dt}{dx/dt}. \tag{2}$$

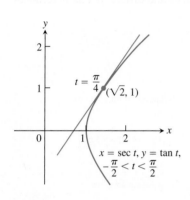

FIGURE 11.12 The curve in Example 1 is the right-hand branch of the hyperbola $x^2 - y^2 = 1$.

EXAMPLE 1 Find the tangent to the curve

$$x = \sec t, \qquad y = \tan t, \qquad -\frac{\pi}{2} < t < \frac{\pi}{2},$$

at the point $(\sqrt{2}, 1)$, where $t = \pi/4$ (Figure 11.12).

Solution The slope of the curve at t is

$$\frac{dy}{dx} = \frac{dy/dt}{dx/dt} = \frac{\sec^2 t}{\sec t \tan t} = \frac{\sec t}{\tan t}. \qquad \text{Eq. (1)}$$

Setting t equal to $\pi/4$ gives

$$\frac{dy}{dx}\Big|_{t=\pi/4} = \frac{\sec(\pi/4)}{\tan(\pi/4)}$$

$$= \frac{\sqrt{2}}{1} = \sqrt{2}.$$

The tangent line is

$$y - 1 = \sqrt{2}(x - \sqrt{2})$$

$$y = \sqrt{2}x - 2 + 1$$

$$y = \sqrt{2}x - 1. \qquad \blacksquare$$

EXAMPLE 2 Find d^2y/dx^2 as a function of t if $x = t - t^2$, $y = t - t^3$.

Solution

1. Express $y' = dy/dx$ in terms of t.

$$y' = \frac{dy}{dx} = \frac{dy/dt}{dx/dt} = \frac{1 - 3t^2}{1 - 2t}$$

Finding d^2y/dx^2 in Terms of t
1. Express $y' = dy/dx$ in terms of t.
2. Find dy'/dt.
3. Divide dy'/dt by dx/dt.

2. Differentiate y' with respect to t.

$$\frac{dy'}{dt} = \frac{d}{dt}\left(\frac{1 - 3t^2}{1 - 2t}\right) = \frac{2 - 6t + 6t^2}{(1 - 2t)^2} \qquad \text{Derivative Quotient Rule}$$

3. Divide dy'/dt by dx/dt.

$$\frac{d^2y}{dx^2} = \frac{dy'/dt}{dx/dt} = \frac{(2 - 6t + 6t^2)/(1 - 2t)^2}{1 - 2t} = \frac{2 - 6t + 6t^2}{(1 - 2t)^3} \qquad \text{Eq. (2)} \qquad ■$$

EXAMPLE 3 Find the area enclosed by the astroid (Figure 11.13)

$$x = \cos^3 t, \qquad y = \sin^3 t, \qquad 0 \le t \le 2\pi.$$

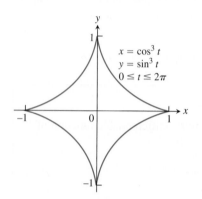

$x = \cos^3 t$
$y = \sin^3 t$
$0 \le t \le 2\pi$

Figure 11.13 The astroid in Example 3.

Solution By symmetry, the enclosed area is 4 times the area beneath the curve in the first quadrant where $0 \le t \le \pi/2$. We can apply the definite integral formula for area studied in Chapter 5, using substitution to express the curve and differential dx in terms of the parameter t. So,

$$A = 4 \int_0^1 y \, dx$$

$$= 4 \int_0^{\pi/2} \sin^3 t \cdot 3 \cos^2 t \sin t \, dt \qquad \text{Substitution for } y \text{ and } dx$$

$$= 12 \int_0^{\pi/2} \left(\frac{1 - \cos 2t}{2}\right)^2 \left(\frac{1 + \cos 2t}{2}\right) dt \qquad \sin^4 t = \left(\frac{1 - \cos 2t}{2}\right)^2$$

$$= \frac{3}{2} \int_0^{\pi/2} (1 - 2\cos 2t + \cos^2 2t)(1 + \cos 2t) \, dt \qquad \text{Expand square term.}$$

$$= \frac{3}{2} \int_0^{\pi/2} (1 - \cos 2t - \cos^2 2t + \cos^3 2t) \, dt \qquad \text{Multiply terms.}$$

$$= \frac{3}{2}\left[\int_0^{\pi/2} (1 - \cos 2t) \, dt - \int_0^{\pi/2} \cos^2 2t \, dt + \int_0^{\pi/2} \cos^3 2t \, dt\right]$$

$$= \frac{3}{2}\left[\left(t - \frac{1}{2}\sin 2t\right) - \frac{1}{2}\left(t + \frac{1}{4}\sin 2t\right) + \frac{1}{2}\left(\sin 2t - \frac{1}{3}\sin^3 2t\right)\right]_0^{\pi/2} \qquad \begin{array}{l}\text{Section 8.2,}\\\text{Example 3}\end{array}$$

$$= \frac{3}{2}\left[\left(\frac{\pi}{2} - 0 - 0 - 0\right) - \frac{1}{2}\left(\frac{\pi}{2} + 0 - 0 - 0\right) + \frac{1}{2}(0 - 0 - 0 + 0)\right] \qquad \text{Evaluate.}$$

$$= \frac{3\pi}{8}. \qquad ■$$

Length of a Parametrically Defined Curve

Let C be a curve given parametrically by the equations

$$x = f(t) \qquad \text{and} \qquad y = g(t), \qquad a \le t \le b.$$

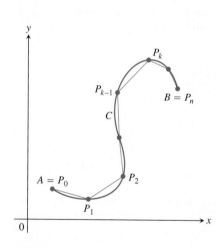

FIGURE 11.14 The smooth curve C defined parametrically by the equations $x = f(t)$ and $y = g(t)$, $a \le t \le b$. The length of the curve from A to B is approximated by the sum of the lengths of the polygonal path (straight line segments) starting at $A = P_0$, then to P_1, and so on, ending at $B = P_n$.

We assume the functions f and g are **continuously differentiable** (meaning they have continuous first derivatives) on the interval $[a, b]$. We also assume that the derivatives $f'(t)$ and $g'(t)$ are not simultaneously zero, which prevents the curve C from having any corners or cusps. Such a curve is called a **smooth curve**. We subdivide the path (or arc) AB into n pieces at points $A = P_0, P_1, P_2, \ldots, P_n = B$ (Figure 11.14). These points correspond to a partition of the interval $[a, b]$ by $a = t_0 < t_1 < t_2 < \cdots < t_n = b$, where $P_k = (f(t_k), g(t_k))$. Join successive points of this subdivision by straight line segments (Figure 11.14). A representative line segment has length

$$L_k = \sqrt{(\Delta x_k)^2 + (\Delta y_k)^2}$$
$$= \sqrt{[f(t_k) - f(t_{k-1})]^2 + [g(t_k) - g(t_{k-1})]^2}$$

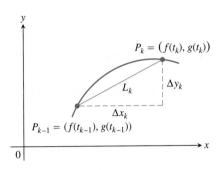

FIGURE 11.15 The arc $P_{k-1}P_k$ is approximated by the straight line segment shown here, which has length $L_k = \sqrt{(\Delta x_k)^2 + (\Delta y_k)^2}$.

(see Figure 11.15). If Δt_k is small, the length L_k is approximately the length of arc $P_{k-1}P_k$. By the Mean Value Theorem there are numbers t_k^* and t_k^{**} in $[t_{k-1}, t_k]$ such that

$$\Delta x_k = f(t_k) - f(t_{k-1}) = f'(t_k^*)\, \Delta t_k,$$
$$\Delta y_k = g(t_k) - g(t_{k-1}) = g'(t_k^{**})\, \Delta t_k.$$

Assuming the path from A to B is traversed exactly once as t increases from $t = a$ to $t = b$, with no doubling back or retracing, an approximation to the (yet to be defined) "length" of the curve AB is the sum of all the lengths L_k:

$$\sum_{k=1}^{n} L_k = \sum_{k=1}^{n} \sqrt{(\Delta x_k)^2 + (\Delta y_k)^2}$$
$$= \sum_{k=1}^{n} \sqrt{[f'(t_k^*)]^2 + [g'(t_k^{**})]^2}\, \Delta t_k.$$

Although this last sum on the right is not exactly a Riemann sum (because f' and g' are evaluated at different points), it can be shown that its limit, as the norm of the partition tends to zero and the number of segments $n \to \infty$, is the definite integral

$$\lim_{\|P\| \to 0} \sum_{k=1}^{n} \sqrt{[f'(t_k^*)]^2 + [g'(t_k^{**})]^2}\, \Delta t_k = \int_a^b \sqrt{[f'(t)]^2 + [g'(t)]^2}\, dt.$$

Therefore, it is reasonable to define the length of the curve from A to B as this integral.

DEFINITION   If a curve C is defined parametrically by $x = f(t)$ and $y = g(t)$, $a \le t \le b$, where f' and g' are continuous and not simultaneously zero on $[a, b]$, and C is traversed exactly once as t increases from $t = a$ to $t = b$, then **the length of C** is the definite integral

$$L = \int_a^b \sqrt{[f'(t)]^2 + [g'(t)]^2}\, dt.$$

A smooth curve C does not double back or reverse the direction of motion over the time interval $[a, b]$ since $(f')^2 + (g')^2 > 0$ throughout the interval. At a point where a curve does start to double back on itself, either the curve fails to be differentiable or both derivatives must simultaneously equal zero. We will examine this phenomenon in Chapter 13, where we study tangent vectors to curves.

If $x = f(t)$ and $y = g(t)$, then using the Leibniz notation we have the following result for arc length:

$$L = \int_a^b \sqrt{\left(\frac{dx}{dt}\right)^2 + \left(\frac{dy}{dt}\right)^2}\, dt. \tag{3}$$

What if there are two different parametrizations for a curve C whose length we want to find; does it matter which one we use? The answer is no, as long as the parametrization we choose meets the conditions stated in the definition of the length of C (see Exercise 41 for an example).

EXAMPLE 4   Using the definition, find the length of the circle of radius r defined parametrically by

$$x = r \cos t \quad \text{and} \quad y = r \sin t, \quad 0 \le t \le 2\pi.$$

Solution   As t varies from 0 to 2π, the circle is traversed exactly once, so the circumference is

$$L = \int_0^{2\pi} \sqrt{\left(\frac{dx}{dt}\right)^2 + \left(\frac{dy}{dt}\right)^2}\, dt.$$

We find

$$\frac{dx}{dt} = -r \sin t, \qquad \frac{dy}{dt} = r \cos t$$

and

$$\left(\frac{dx}{dt}\right)^2 + \left(\frac{dy}{dt}\right)^2 = r^2(\sin^2 t + \cos^2 t) = r^2.$$

So

$$L = \int_0^{2\pi} \sqrt{r^2}\, dt = r\left[t\right]_0^{2\pi} = 2\pi r.$$

EXAMPLE 5   Find the length of the astroid (Figure 11.13)

$$x = \cos^3 t, \quad y = \sin^3 t, \quad 0 \le t \le 2\pi.$$

Solution   Because of the curve's symmetry with respect to the coordinate axes, its length is four times the length of the first-quadrant portion. We have

$$x = \cos^3 t, \quad y = \sin^3 t$$

$$\left(\frac{dx}{dt}\right)^2 = [3 \cos^2 t(-\sin t)]^2 = 9 \cos^4 t \sin^2 t$$

$$\left(\frac{dy}{dt}\right)^2 = [3 \sin^2 t(\cos t)]^2 = 9 \sin^4 t \cos^2 t$$

$$\sqrt{\left(\frac{dx}{dt}\right)^2 + \left(\frac{dy}{dt}\right)^2} = \sqrt{9 \cos^2 t \sin^2 t \underbrace{(\cos^2 t + \sin^2 t)}_{1}}$$

$$= \sqrt{9 \cos^2 t \sin^2 t}$$

$$= 3|\cos t \sin t| \qquad \qquad \cos t \sin t \ge 0 \text{ for }$$

$$= 3 \cos t \sin t. \qquad \qquad 0 \le t \le \pi/2$$

Therefore,

$$\text{Length of first-quadrant portion} = \int_0^{\pi/2} 3 \cos t \sin t\, dt$$

$$= \frac{3}{2} \int_0^{\pi/2} \sin 2t\, dt \qquad \qquad \cos t \sin t = \\ (1/2) \sin 2t$$

$$= -\frac{3}{4} \cos 2t \Big]_0^{\pi/2} = \frac{3}{2}.$$

The length of the astroid is four times this: $4(3/2) = 6$.

Length of a Curve $y = f(x)$

The length formula in Section 6.3 is a special case of Equation (3). Given a continuously differentiable function $y = f(x)$, $a \leq x \leq b$, we can assign $x = t$ as a parameter. The graph of the function f is then the curve C defined parametrically by

$$x = t \quad \text{and} \quad y = f(t), \quad a \leq t \leq b,$$

a special case of what we considered before. Then,

$$\frac{dx}{dt} = 1 \quad \text{and} \quad \frac{dy}{dt} = f'(t).$$

From Equation (1), we have

$$\frac{dy}{dx} = \frac{dy/dt}{dx/dt} = f'(t),$$

giving

$$\left(\frac{dx}{dt}\right)^2 + \left(\frac{dy}{dt}\right)^2 = 1 + [f'(t)]^2$$

$$= 1 + [f'(x)]^2. \qquad t = x$$

Substitution into Equation (3) gives the arc length formula for the graph of $y = f(x)$, consistent with Equation (3) in Section 6.3.

The Arc Length Differential

Consistent with our discussion in Section 6.3, we can define the arc length function for a parametrically defined curve $x = f(t)$ and $y = g(t)$, $a \leq t \leq b$, by

$$s(t) = \int_a^t \sqrt{[f'(z)]^2 + [g'(z)]^2} \, dz.$$

Then, by the Fundamental Theorem of Calculus,

$$\frac{ds}{dt} = \sqrt{[f'(t)]^2 + [g'(t)]^2} = \sqrt{\left(\frac{dx}{dt}\right)^2 + \left(\frac{dy}{dt}\right)^2}.$$

The differential of arc length is

$$ds = \sqrt{\left(\frac{dx}{dt}\right)^2 + \left(\frac{dy}{dt}\right)^2} \, dt. \qquad (4)$$

Equation (4) is often abbreviated to

$$ds = \sqrt{dx^2 + dy^2}.$$

Just as in Section 6.3, we can integrate the differential ds between appropriate limits to find the total length of a curve.

Here's an example where we use the arc length formula to find the centroid of an arc.

EXAMPLE 6   Find the centroid of the first-quadrant arc of the astroid in Example 5.

Solution   We take the curve's density to be $\delta = 1$ and calculate the curve's mass and moments about the coordinate axes as we did in Section 6.6.

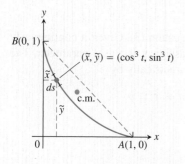

FIGURE 11.16 The centroid (c.m.) of the astroid arc in Example 6.

The distribution of mass is symmetric about the line $y = x$, so $\bar{x} = \bar{y}$. A typical segment of the curve (Figure 11.16) has mass

$$dm = 1 \cdot ds = \sqrt{\left(\frac{dx}{dt}\right)^2 + \left(\frac{dy}{dt}\right)^2} \, dt = 3 \cos t \sin t \, dt. \qquad \text{From Example 5}$$

The curve's mass is

$$M = \int_0^{\pi/2} dm = \int_0^{\pi/2} 3 \cos t \sin t \, dt = \frac{3}{2}. \qquad \text{Again from Example 5}$$

The curve's moment about the x-axis is

$$M_x = \int \tilde{y} \, dm = \int_0^{\pi/2} \sin^3 t \cdot 3 \cos t \sin t \, dt$$

$$= 3 \int_0^{\pi/2} \sin^4 t \cos t \, dt = 3 \cdot \frac{\sin^5 t}{5} \bigg]_0^{\pi/2} = \frac{3}{5}.$$

It follows that

$$\bar{y} = \frac{M_x}{M} = \frac{3/5}{3/2} = \frac{2}{5}.$$

The centroid is the point $(2/5, 2/5)$. ∎

Areas of Surfaces of Revolution

In Section 6.4 we found integral formulas for the area of a surface when a curve is revolved about a coordinate axis. Specifically, we found that the surface area is $S = \int 2\pi y \, ds$ for revolution about the x-axis, and $S = \int 2\pi x \, ds$ for revolution about the y-axis. If the curve is parametrized by the equations $x = f(t)$ and $y = g(t)$, $a \le t \le b$, where f and g are continuously differentiable and $(f')^2 + (g')^2 > 0$ on $[a, b]$, then the arc length differential ds is given by Equation (4). This observation leads to the following formulas for area of surfaces of revolution for smooth parametrized curves.

Area of Surface of Revolution for Parametrized Curves

If a smooth curve $x = f(t), y = g(t), a \le t \le b$, is traversed exactly once as t increases from a to b, then the areas of the surfaces generated by revolving the curve about the coordinate axes are as follows.

1. Revolution about the x-axis ($y \ge 0$):

$$S = \int_a^b 2\pi y \sqrt{\left(\frac{dx}{dt}\right)^2 + \left(\frac{dy}{dt}\right)^2} \, dt \qquad (5)$$

2. Revolution about the y-axis ($x \ge 0$):

$$S = \int_a^b 2\pi x \sqrt{\left(\frac{dx}{dt}\right)^2 + \left(\frac{dy}{dt}\right)^2} \, dt \qquad (6)$$

As with length, we can calculate surface area from any convenient parametrization that meets the stated criteria.

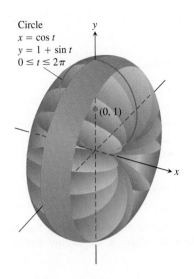

Circle
$x = \cos t$
$y = 1 + \sin t$
$0 \le t \le 2\pi$

(0, 1)

FIGURE 11.17 In Example 7 we calculate the area of the surface of revolution swept out by this parametrized curve.

EXAMPLE 7 The standard parametrization of the circle of radius 1 centered at the point (0, 1) in the xy-plane is

$$x = \cos t, \qquad y = 1 + \sin t, \qquad 0 \le t \le 2\pi.$$

Use this parametrization to find the area of the surface swept out by revolving the circle about the x-axis (Figure 11.17).

Solution We evaluate the formula

$$S = \int_a^b 2\pi y \sqrt{\left(\frac{dx}{dt}\right)^2 + \left(\frac{dy}{dt}\right)^2}\, dt$$

Eq. (5) for revolution about the x-axis; $y = 1 + \sin t \ge 0$

$$= \int_0^{2\pi} 2\pi(1 + \sin t)\sqrt{\underbrace{(-\sin t)^2 + (\cos t)^2}_{1}}\, dt$$

$$= 2\pi \int_0^{2\pi} (1 + \sin t)\, dt$$

$$= 2\pi\Big[t - \cos t\Big]_0^{2\pi} = 4\pi^2. \qquad \blacksquare$$

Exercises 11.2

Tangents to Parametrized Curves

In Exercises 1–14, find an equation for the line tangent to the curve at the point defined by the given value of t. Also, find the value of d^2y/dx^2 at this point.

1. $x = 2\cos t$, $y = 2\sin t$, $t = \pi/4$
2. $x = \sin 2\pi t$, $y = \cos 2\pi t$, $t = -1/6$
3. $x = 4\sin t$, $y = 2\cos t$, $t = \pi/4$
4. $x = \cos t$, $y = \sqrt{3}\cos t$, $t = 2\pi/3$
5. $x = t$, $y = \sqrt{t}$, $t = 1/4$
6. $x = \sec^2 t - 1$, $y = \tan t$, $t = -\pi/4$
7. $x = \sec t$, $y = \tan t$, $t = \pi/6$
8. $x = -\sqrt{t + 1}$, $y = \sqrt{3t}$, $t = 3$
9. $x = 2t^2 + 3$, $y = t^4$, $t = -1$
10. $x = 1/t$, $y = -2 + \ln t$, $t = 1$
11. $x = t - \sin t$, $y = 1 - \cos t$, $t = \pi/3$
12. $x = \cos t$, $y = 1 + \sin t$, $t = \pi/2$
13. $x = \dfrac{1}{t + 1}$, $y = \dfrac{t}{t - 1}$, $t = 2$
14. $x = t + e^t$, $y = 1 - e^t$, $t = 0$

Implicitly Defined Parametrizations

Assuming that the equations in Exercises 15–20 define x and y implicitly as differentiable functions $x = f(t), y = g(t)$, find the slope of the curve $x = f(t), y = g(t)$ at the given value of t.

15. $x^3 + 2t^2 = 9$, $2y^3 - 3t^2 = 4$, $t = 2$
16. $x = \sqrt{5 - \sqrt{t}}$, $y(t - 1) = \sqrt{t}$, $t = 4$
17. $x + 2x^{3/2} = t^2 + t$, $y\sqrt{t + 1} + 2t\sqrt{y} = 4$, $t = 0$
18. $x \sin t + 2x = t$, $t \sin t - 2t = y$, $t = \pi$
19. $x = t^3 + t$, $y + 2t^3 = 2x + t^2$, $t = 1$
20. $t = \ln(x - t)$, $y = te^t$, $t = 0$

Area

21. Find the area under one arch of the cycloid
$$x = a(t - \sin t), \quad y = a(1 - \cos t).$$

22. Find the area enclosed by the y-axis and the curve
$$x = t - t^2, \quad y = 1 + e^{-t}.$$

23. Find the area enclosed by the ellipse
$$x = a\cos t, \quad y = b\sin t, \quad 0 \le t \le 2\pi.$$

24. Find the area under $y = x^3$ over [0, 1] using the following parametrizations.
   a. $x = t^2$, $y = t^6$ **b.** $x = t^3$, $y = t^9$

Lengths of Curves

Find the lengths of the curves in Exercises 25–30.
25. $x = \cos t$, $y = t + \sin t$, $0 \le t \le \pi$
26. $x = t^3$, $y = 3t^2/2$, $0 \le t \le \sqrt{3}$
27. $x = t^2/2$, $y = (2t + 1)^{3/2}/3$, $0 \le t \le 4$
28. $x = (2t + 3)^{3/2}/3$, $y = t + t^2/2$, $0 \le t \le 3$
29. $x = 8\cos t + 8t\sin t$
    $y = 8\sin t - 8t\cos t$,
    $0 \le t \le \pi/2$
30. $x = \ln(\sec t + \tan t) - \sin t$
    $y = \cos t$, $0 \le t \le \pi/3$

Surface Area

Find the areas of the surfaces generated by revolving the curves in Exercises 31–34 about the indicated axes.
31. $x = \cos t$, $y = 2 + \sin t$, $0 \le t \le 2\pi$; x-axis

32. $x = (2/3)t^{3/2}, \quad y = 2\sqrt{t}, \quad 0 \le t \le \sqrt{3}; \quad y\text{-axis}$

33. $x = t + \sqrt{2}, \quad y = (t^2/2) + \sqrt{2}t, \quad -\sqrt{2} \le t \le \sqrt{2}; \quad y\text{-axis}$

34. $x = \ln(\sec t + \tan t) - \sin t, \quad y = \cos t, \quad 0 \le t \le \pi/3; \quad x\text{-axis}$

35. A cone frustum The line segment joining the points (0, 1) and (2, 2) is revolved about the x-axis to generate a frustum of a cone. Find the surface area of the frustum using the parametrization $x = 2t, y = t + 1, 0 \le t \le 1$. Check your result with the geometry formula: Area = $\pi(r_1 + r_2)$(slant height).

36. A cone The line segment joining the origin to the point (h, r) is revolved about the x-axis to generate a cone of height h and base radius r. Find the cone's surface area with the parametric equations $x = ht, y = rt, 0 \le t \le 1$. Check your result with the geometry formula: Area = πr(slant height).

Centroids

37. Find the coordinates of the centroid of the curve

$$x = \cos t + t \sin t, \quad y = \sin t - t \cos t, \quad 0 \le t \le \pi/2.$$

38. Find the coordinates of the centroid of the curve

$$x = e^t \cos t, \quad y = e^t \sin t, \quad 0 \le t \le \pi.$$

39. Find the coordinates of the centroid of the curve

$$x = \cos t, \quad y = t + \sin t, \quad 0 \le t \le \pi.$$

T 40. Most centroid calculations for curves are done with a calculator or computer that has an integral evaluation program. As a case in point, find, to the nearest hundredth, the coordinates of the centroid of the curve

$$x = t^3, \quad y = 3t^2/2, \quad 0 \le t \le \sqrt{3}.$$

Theory and Examples

41. Length is independent of parametrization To illustrate the fact that the numbers we get for length do not depend on the way we parametrize our curves (except for the mild restrictions preventing doubling back mentioned earlier), calculate the length of the semicircle $y = \sqrt{1 - x^2}$ with these two different parametrizations:

a. $x = \cos 2t, \quad y = \sin 2t, \quad 0 \le t \le \pi/2$

b. $x = \sin \pi t, \quad y = \cos \pi t, \quad -1/2 \le t \le 1/2$

42. a. Show that the Cartesian formula

$$L = \int_c^d \sqrt{1 + \left(\frac{dx}{dy}\right)^2}\, dy$$

for the length of the curve $x = g(y), c \le y \le d$ (Section 6.3, Equation 4), is a special case of the parametric length formula

$$L = \int_a^b \sqrt{\left(\frac{dx}{dt}\right)^2 + \left(\frac{dy}{dt}\right)^2}\, dt.$$

Use this result to find the length of each curve.

b. $x = y^{3/2}, \quad 0 \le y \le 4/3$

c. $x = \frac{3}{2}y^{2/3}, \quad 0 \le y \le 1$

43. The curve with parametric equations

$$x = (1 + 2\sin\theta)\cos\theta, \quad y = (1 + 2\sin\theta)\sin\theta$$

is called a *limaçon* and is shown in the accompanying figure. Find the points (x, y) and the slopes of the tangent lines at these points for

a. $\theta = 0$.　**b.** $\theta = \pi/2$.　**c.** $\theta = 4\pi/3$.

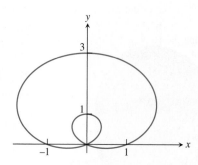

44. The curve with parametric equations

$$x = t, \quad y = 1 - \cos t, \quad 0 \le t \le 2\pi$$

is called a *sinusoid* and is shown in the accompanying figure. Find the point (x, y) where the slope of the tangent line is
a. largest　**b.** smallest.

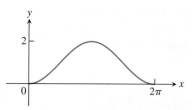

T The curves in Exercises 45 and 46 are called *Bowditch curves* or *Lissajous figures*. In each case, find the point in the interior of the first quadrant where the tangent to the curve is horizontal, and find the equations of the two tangents at the origin.

45.　　　　　　**46.**

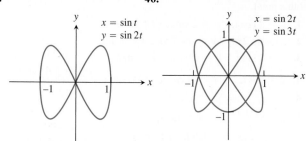

47. Cycloid

a. Find the length of one arch of the cycloid

$$x = a(t - \sin t), \quad y = a(1 - \cos t).$$

b. Find the area of the surface generated by revolving one arch of the cycloid in part (a) about the x-axis for $a = 1$.

48. Volume Find the volume swept out by revolving the region bounded by the x-axis and one arch of the cycloid

$$x = t - \sin t, \quad y = 1 - \cos t$$

about the x-axis.

COMPUTER EXPLORATIONS

In Exercises 49–52, use a CAS to perform the following steps for the given curve over the closed interval.

a. Plot the curve together with the polygonal path approximations for $n = 2, 4, 8$ partition points over the interval. (See Figure 11.14.)

b. Find the corresponding approximation to the length of the curve by summing the lengths of the line segments.

c. Evaluate the length of the curve using an integral. Compare your approximations for $n = 2, 4, 8$ with the actual length given by the integral. How does the actual length compare with the approximations as n increases? Explain your answer.

49. $x = \frac{1}{3}t^3, \quad y = \frac{1}{2}t^2, \quad 0 \le t \le 1$

50. $x = 2t^3 - 16t^2 + 25t + 5, \quad y = t^2 + t - 3,$ $0 \le t \le 6$

51. $x = t - \cos t, \quad y = 1 + \sin t, \quad -\pi \le t \le \pi$

52. $x = e^t \cos t, \quad y = e^t \sin t, \quad 0 \le t \le \pi$

11.3 Polar Coordinates

In this section we study polar coordinates and their relation to Cartesian coordinates. You will see that polar coordinates are very useful for calculating many multiple integrals studied in Chapter 15.

Definition of Polar Coordinates

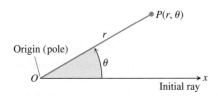

FIGURE 11.18 To define polar coordinates for the plane, we start with an origin, called the pole, and an initial ray.

To define polar coordinates, we first fix an **origin** O (called the **pole**) and an **initial ray** from O (Figure 11.18). Then each point P can be located by assigning to it a **polar coordinate pair** (r, θ) in which r gives the directed distance from O to P and θ gives the directed angle from the initial ray to ray OP.

Polar Coordinates

$$P(r, \theta)$$

Directed distance from O to P — Directed angle from initial ray to OP

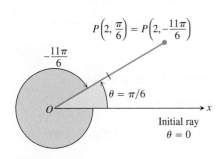

FIGURE 11.19 Polar coordinates are not unique.

As in trigonometry, θ is positive when measured counterclockwise and negative when measured clockwise. The angle associated with a given point is not unique. While a point in the plane has just one pair of Cartesian coordinates, it has infinitely many pairs of polar coordinates. For instance, the point 2 units from the origin along the ray $\theta = \pi/6$ has polar coordinates $r = 2, \theta = \pi/6$. It also has coordinates $r = 2, \theta = -11\pi/6$ (Figure 11.19). In some situations we allow r to be negative. That is why we use directed distance in defining $P(r, \theta)$. The point $P(2, 7\pi/6)$ can be reached by turning $7\pi/6$ radians counterclockwise from the initial ray and going forward 2 units (Figure 11.20). It can also be reached by turning $\pi/6$ radians counterclockwise from the initial ray and going *backward* 2 units. So the point also has polar coordinates $r = -2, \theta = \pi/6$.

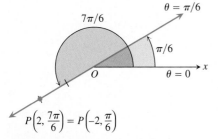

FIGURE 11.20 Polar coordinates can have negative r-values.

EXAMPLE 1 Find all the polar coordinates of the point $P(2, \pi/6)$.

Solution We sketch the initial ray of the coordinate system, draw the ray from the origin that makes an angle of $\pi/6$ radians with the initial ray, and mark the point $(2, \pi/6)$ (Figure 11.21). We then find the angles for the other coordinate pairs of P in which $r = 2$ and $r = -2$.

For $r = 2$, the complete list of angles is

$$\frac{\pi}{6}, \quad \frac{\pi}{6} \pm 2\pi, \quad \frac{\pi}{6} \pm 4\pi, \quad \frac{\pi}{6} \pm 6\pi, \ldots.$$

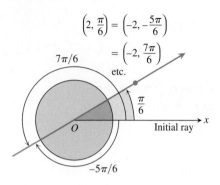

FIGURE 11.21 The point $P(2, \pi/6)$ has infinitely many polar coordinate pairs (Example 1).

For $r = -2$, the angles are

$$-\frac{5\pi}{6}, \quad -\frac{5\pi}{6} \pm 2\pi, \quad -\frac{5\pi}{6} \pm 4\pi, \quad -\frac{5\pi}{6} \pm 6\pi, \ldots.$$

The corresponding coordinate pairs of P are

$$\left(2, \frac{\pi}{6} + 2n\pi\right), \qquad n = 0, \pm 1, \pm 2, \ldots$$

and

$$\left(-2, -\frac{5\pi}{6} + 2n\pi\right), \qquad n = 0, \pm 1, \pm 2, \ldots.$$

When $n = 0$, the formulas give $(2, \pi/6)$ and $(-2, -5\pi/6)$. When $n = 1$, they give $(2, 13\pi/6)$ and $(-2, 7\pi/6)$, and so on. ∎

Polar Equations and Graphs

If we hold r fixed at a constant value $r = a \neq 0$, the point $P(r, \theta)$ will lie $|a|$ units from the origin O. As θ varies over any interval of length 2π, P then traces a circle of radius $|a|$ centered at O (Figure 11.22).

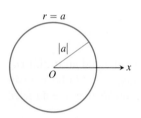

FIGURE 11.22 The polar equation for a circle is $r = a$.

If we hold θ fixed at a constant value $\theta = \theta_0$ and let r vary between $-\infty$ and ∞, the point $P(r, \theta)$ traces the line through O that makes an angle of measure θ_0 with the initial ray.

Equation	Graph
$r = a$	Circle of radius $|a|$ centered at O
$\theta = \theta_0$	Line through O making an angle θ_0 with the initial ray

EXAMPLE 2

(a) $r = 1$ and $r = -1$ are equations for the circle of radius 1 centered at O.

(b) $\theta = \pi/6, \theta = 7\pi/6$, and $\theta = -5\pi/6$ are equations for the line in Figure 11.21. ∎

Equations of the form $r = a$ and $\theta = \theta_0$ can be combined to define regions, segments, and rays.

EXAMPLE 3
Graph the sets of points whose polar coordinates satisfy the following conditions.

(a) $1 \leq r \leq 2$ and $0 \leq \theta \leq \dfrac{\pi}{2}$

(b) $-3 \leq r \leq 2$ and $\theta = \dfrac{\pi}{4}$

(c) $\dfrac{2\pi}{3} \leq \theta \leq \dfrac{5\pi}{6}$ (no restriction on r)

Solution The graphs are shown in Figure 11.23. ∎

(a)

(b)

(c)

FIGURE 11.23 The graphs of typical inequalities in r and θ (Example 3).

Relating Polar and Cartesian Coordinates

When we use both polar and Cartesian coordinates in a plane, we place the two origins together and take the initial polar ray as the positive x-axis. The ray $\theta = \pi/2, r > 0$,

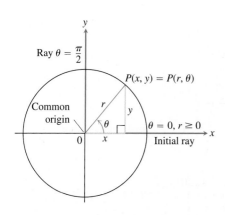

FIGURE 11.24 The usual way to relate polar and Cartesian coordinates.

becomes the positive y-axis (Figure 11.24). The two coordinate systems are then related by the following equations.

Equations Relating Polar and Cartesian Coordinates

$$x = r \cos \theta, \qquad y = r \sin \theta, \qquad r^2 = x^2 + y^2, \qquad \tan \theta = \frac{y}{x}$$

The first two of these equations uniquely determine the Cartesian coordinates x and y given the polar coordinates r and θ. On the other hand, if x and y are given, the third equation gives two possible choices for r (a positive and a negative value). For each $(x, y) \neq (0, 0)$, there is a unique $\theta \in [0, 2\pi)$ satisfying the first two equations, each then giving a polar coordinate representation of the Cartesian point (x, y). The other polar coordinate representations for the point can be determined from these two, as in Example 1.

EXAMPLE 4 Here are some equivalent equations expressed in terms of both polar coordinates and Cartesian coordinates.

Polar equation	Cartesian equivalent
$r \cos \theta = 2$	$x = 2$
$r^2 \cos \theta \sin \theta = 4$	$xy = 4$
$r^2 \cos^2 \theta - r^2 \sin^2 \theta = 1$	$x^2 - y^2 = 1$
$r = 1 + 2r \cos \theta$	$y^2 - 3x^2 - 4x - 1 = 0$
$r = 1 - \cos \theta$	$x^4 + y^4 + 2x^2y^2 + 2x^3 + 2xy^2 - y^2 = 0$

Some curves are more simply expressed with polar coordinates; others are not. ∎

EXAMPLE 5 Find a polar equation for the circle $x^2 + (y - 3)^2 = 9$ (Figure 11.25).

Solution We apply the equations relating polar and Cartesian coordinates:

$$
\begin{aligned}
x^2 + (y - 3)^2 &= 9 \\
x^2 + y^2 - 6y + 9 &= 9 \qquad &\text{Expand } (y - 3)^2. \\
x^2 + y^2 - 6y &= 0 \qquad &\text{Cancellation} \\
r^2 - 6r \sin \theta &= 0 \qquad &x^2 + y^2 = r^2 \\
r = 0 \quad \text{or} \quad r - 6 \sin \theta &= 0 \\
r &= 6 \sin \theta \qquad &\text{Includes both possibilities}
\end{aligned}
$$

∎

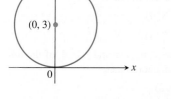

FIGURE 11.25 The circle in Example 5.

EXAMPLE 6 Replace the following polar equations by equivalent Cartesian equations and identify their graphs.

(a) $r \cos \theta = -4$

(b) $r^2 = 4r \cos \theta$

(c) $r = \dfrac{4}{2 \cos \theta - \sin \theta}$

Solution We use the substitutions $r \cos \theta = x$, $r \sin \theta = y$, $r^2 = x^2 + y^2$.

(a) $r \cos \theta = -4$

The Cartesian equation: $r \cos \theta = -4$
$$x = -4$$

The graph: Vertical line through $x = -4$ on the x-axis

(b) $r^2 = 4r \cos \theta$

The Cartesian equation:
$$r^2 = 4r \cos \theta$$
$$x^2 + y^2 = 4x$$
$$x^2 - 4x + y^2 = 0$$
$$x^2 - 4x + 4 + y^2 = 4 \qquad \text{Completing the square}$$
$$(x - 2)^2 + y^2 = 4$$

The graph: Circle, radius 2, center $(h, k) = (2, 0)$

(c) $r = \dfrac{4}{2 \cos \theta - \sin \theta}$

The Cartesian equation:
$$r(2 \cos \theta - \sin \theta) = 4$$
$$2r \cos \theta - r \sin \theta = 4$$
$$2x - y = 4$$
$$y = 2x - 4$$

The graph: Line, slope $m = 2$, y-intercept $b = -4$ ∎

Exercises 11.3

Polar Coordinates

1. Which polar coordinate pairs label the same point?

a. $(3, 0)$ **b.** $(-3, 0)$ **c.** $(2, 2\pi/3)$

d. $(2, 7\pi/3)$ **e.** $(-3, \pi)$ **f.** $(2, \pi/3)$

g. $(-3, 2\pi)$ **h.** $(-2, -\pi/3)$

2. Which polar coordinate pairs label the same point?

a. $(-2, \pi/3)$ **b.** $(2, -\pi/3)$ **c.** (r, θ)

d. $(r, \theta + \pi)$ **e.** $(-r, \theta)$ **f.** $(2, -2\pi/3)$

g. $(-r, \theta + \pi)$ **h.** $(-2, 2\pi/3)$

3. Plot the following points (given in polar coordinates). Then find all the polar coordinates of each point.

a. $(2, \pi/2)$ **b.** $(2, 0)$

c. $(-2, \pi/2)$ **d.** $(-2, 0)$

4. Plot the following points (given in polar coordinates). Then find all the polar coordinates of each point.

a. $(3, \pi/4)$ **b.** $(-3, \pi/4)$

c. $(3, -\pi/4)$ **d.** $(-3, -\pi/4)$

Polar to Cartesian Coordinates

5. Find the Cartesian coordinates of the points in Exercise 1.

6. Find the Cartesian coordinates of the following points (given in polar coordinates).

a. $\left(\sqrt{2}, \pi/4\right)$ **b.** $(1, 0)$

c. $(0, \pi/2)$ **d.** $\left(-\sqrt{2}, \pi/4\right)$

e. $(-3, 5\pi/6)$ **f.** $(5, \tan^{-1}(4/3))$

g. $(-1, 7\pi)$ **h.** $\left(2\sqrt{3}, 2\pi/3\right)$

Cartesian to Polar Coordinates

7. Find the polar coordinates, $0 \leq \theta < 2\pi$ and $r \geq 0$, of the following points given in Cartesian coordinates.

a. $(1, 1)$ **b.** $(-3, 0)$

c. $(\sqrt{3}, -1)$ **d.** $(-3, 4)$

8. Find the polar coordinates, $-\pi \leq \theta < \pi$ and $r \geq 0$, of the following points given in Cartesian coordinates.

a. $(-2, -2)$ **b.** $(0, 3)$

c. $(-\sqrt{3}, 1)$ **d.** $(5, -12)$

9. Find the polar coordinates, $0 \leq \theta < 2\pi$ and $r \leq 0$, of the following points given in Cartesian coordinates.

a. $(3, 3)$ **b.** $(-1, 0)$

c. $(-1, \sqrt{3})$ **d.** $(4, -3)$

10. Find the polar coordinates, $-\pi \leq \theta < 2\pi$ and $r \leq 0$, of the following points given in Cartesian coordinates.

a. $(-2, 0)$ **b.** $(1, 0)$

c. $(0, -3)$ **d.** $\left(\dfrac{\sqrt{3}}{2}, \dfrac{1}{2}\right)$

Graphing in Polar Coordinates

Graph the sets of points whose polar coordinates satisfy the equations and inequalities in Exercises 11–26.

11. $r = 2$ **12.** $0 \leq r \leq 2$

13. $r \geq 1$ **14.** $1 \leq r \leq 2$

15. $0 \le \theta \le \pi/6, \quad r \ge 0$ **16.** $\theta = 2\pi/3, \quad r \le -2$

17. $\theta = \pi/3, \quad -1 \le r \le 3$ **18.** $\theta = 11\pi/4, \quad r \ge -1$

19. $\theta = \pi/2, \quad r \ge 0$ **20.** $\theta = \pi/2, \quad r \le 0$

21. $0 \le \theta \le \pi, \quad r = 1$ **22.** $0 \le \theta \le \pi, \quad r = -1$

23. $\pi/4 \le \theta \le 3\pi/4, \quad 0 \le r \le 1$

24. $-\pi/4 \le \theta \le \pi/4, \quad -1 \le r \le 1$

25. $-\pi/2 \le \theta \le \pi/2, \quad 1 \le r \le 2$

26. $0 \le \theta \le \pi/2, \quad 1 \le |r| \le 2$

Polar to Cartesian Equations

Replace the polar equations in Exercises 27–52 with equivalent Cartesian equations. Then describe or identify the graph.

27. $r \cos \theta = 2$ **28.** $r \sin \theta = -1$

29. $r \sin \theta = 0$ **30.** $r \cos \theta = 0$

31. $r = 4 \csc \theta$ **32.** $r = -3 \sec \theta$

33. $r \cos \theta + r \sin \theta = 1$ **34.** $r \sin \theta = r \cos \theta$

35. $r^2 = 1$ **36.** $r^2 = 4r \sin \theta$

37. $r = \dfrac{5}{\sin \theta - 2 \cos \theta}$ **38.** $r^2 \sin 2\theta = 2$

39. $r = \cot \theta \csc \theta$ **40.** $r = 4 \tan \theta \sec \theta$

41. $r = \csc \theta \, e^{r \cos \theta}$ **42.** $r \sin \theta = \ln r + \ln \cos \theta$

43. $r^2 + 2r^2 \cos \theta \sin \theta = 1$ **44.** $\cos^2 \theta = \sin^2 \theta$

45. $r^2 = -4r \cos \theta$ **46.** $r^2 = -6r \sin \theta$

47. $r = 8 \sin \theta$ **48.** $r = 3 \cos \theta$

49. $r = 2 \cos \theta + 2 \sin \theta$ **50.** $r = 2 \cos \theta - \sin \theta$

51. $r \sin \left(\theta + \dfrac{\pi}{6} \right) = 2$ **52.** $r \sin \left(\dfrac{2\pi}{3} - \theta \right) = 5$

Cartesian to Polar Equations

Replace the Cartesian equations in Exercises 53–66 with equivalent polar equations.

53. $x = 7$ **54.** $y = 1$ **55.** $x = y$

56. $x - y = 3$ **57.** $x^2 + y^2 = 4$ **58.** $x^2 - y^2 = 1$

59. $\dfrac{x^2}{9} + \dfrac{y^2}{4} = 1$ **60.** $xy = 2$

61. $y^2 = 4x$ **62.** $x^2 + xy + y^2 = 1$

63. $x^2 + (y - 2)^2 = 4$ **64.** $(x - 5)^2 + y^2 = 25$

65. $(x - 3)^2 + (y + 1)^2 = 4$ **66.** $(x + 2)^2 + (y - 5)^2 = 16$

67. Find all polar coordinates of the origin.

68. Vertical and horizontal lines

     a. Show that every vertical line in the xy-plane has a polar equation of the form $r = a \sec \theta$.

     b. Find the analogous polar equation for horizontal lines in the xy-plane.

11.4   Graphing in Polar Coordinates

It is often helpful to have the graph of an equation in polar coordinates. This section describes techniques for graphing these equations using symmetries and tangents to the graph.

Symmetry

Figure 11.26 illustrates the standard polar coordinate tests for symmetry. The following summary says how the symmetric points are related.

Symmetry Tests for Polar Graphs

1. *Symmetry about the x-axis:* If the point (r, θ) lies on the graph, then the point $(r, -\theta)$ or $(-r, \pi - \theta)$ lies on the graph (Figure 11.26a).

2. *Symmetry about the y-axis:* If the point (r, θ) lies on the graph, then the point $(r, \pi - \theta)$ or $(-r, -\theta)$ lies on the graph (Figure 11.26b).

3. *Symmetry about the origin:* If the point (r, θ) lies on the graph, then the point $(-r, \theta)$ or $(r, \theta + \pi)$ lies on the graph (Figure 11.26c).

(a) About the *x*-axis

(b) About the *y*-axis

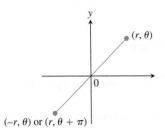

(c) About the origin

FIGURE 11.26 Three tests for symmetry in polar coordinates.

Slope

The slope of a polar curve $r = f(\theta)$ in the *xy*-plane is still given by dy/dx, which is not $r' = df/d\theta$. To see why, think of the graph of f as the graph of the parametric equations

$$x = r\cos\theta = f(\theta)\cos\theta, \qquad y = r\sin\theta = f(\theta)\sin\theta.$$

If f is a differentiable function of θ, then so are x and y and, when $dx/d\theta \neq 0$, we can calculate dy/dx from the parametric formula

$$\frac{dy}{dx} = \frac{dy/d\theta}{dx/d\theta} \qquad \text{Section 11.2, Eq. (1)} \atop \text{with } t = \theta$$

$$= \frac{\dfrac{d}{d\theta}(f(\theta)\cdot\sin\theta)}{\dfrac{d}{d\theta}(f(\theta)\cdot\cos\theta)}$$

$$= \frac{\dfrac{df}{d\theta}\sin\theta + f(\theta)\cos\theta}{\dfrac{df}{d\theta}\cos\theta - f(\theta)\sin\theta} \qquad \text{Product Rule for derivatives}$$

Therefore we see that dy/dx is not the same as $df/d\theta$.

Slope of the Curve $r = f(\theta)$

$$\left.\frac{dy}{dx}\right|_{(r,\theta)} = \frac{f'(\theta)\sin\theta + f(\theta)\cos\theta}{f'(\theta)\cos\theta - f(\theta)\sin\theta},$$

provided $dx/d\theta \neq 0$ at (r,θ).

If the curve $r = f(\theta)$ passes through the origin at $\theta = \theta_0$, then $f(\theta_0) = 0$, and the slope equation gives

$$\left.\frac{dy}{dx}\right|_{(0,\theta_0)} = \frac{f'(\theta_0)\sin\theta_0}{f'(\theta_0)\cos\theta_0} = \tan\theta_0.$$

If the graph of $r = f(\theta)$ passes through the origin at the value $\theta = \theta_0$, the slope of the curve there is $\tan\theta_0$. The reason we say "slope at $(0, \theta_0)$" and not just "slope at the origin" is that a polar curve may pass through the origin (or any point) more than once, with different slopes at different θ-values. This is not the case in our first example, however.

EXAMPLE 1 Graph the curve $r = 1 - \cos\theta$.

Solution The curve is symmetric about the *x*-axis because

$$(r, \theta) \text{ on the graph} \Rightarrow r = 1 - \cos\theta$$

$$\Rightarrow r = 1 - \cos(-\theta) \qquad \cos\theta = \cos(-\theta)$$

$$\Rightarrow (r, -\theta) \text{ on the graph}.$$

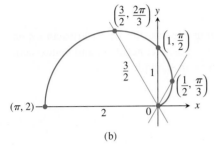

θ	$r = 1 - \cos\theta$
0	0
$\dfrac{\pi}{3}$	$\dfrac{1}{2}$
$\dfrac{\pi}{2}$	1
$\dfrac{2\pi}{3}$	$\dfrac{3}{2}$
π	2

(a)

(b)

(c)

FIGURE 11.27 The steps in graphing the cardioid $r = 1 - \cos\theta$ (Example 1). The arrow shows the direction of increasing θ.

As θ increases from 0 to π, $\cos\theta$ decreases from 1 to -1, and $r = 1 - \cos\theta$ increases from a minimum value of 0 to a maximum value of 2. As θ continues on from π to 2π, $\cos\theta$ increases from -1 back to 1 and r decreases from 2 back to 0. The curve starts to repeat when $\theta = 2\pi$ because the cosine has period 2π.

The curve leaves the origin with slope $\tan(0) = 0$ and returns to the origin with slope $\tan(2\pi) = 0$.

We make a table of values from $\theta = 0$ to $\theta = \pi$, plot the points, draw a smooth curve through them with a horizontal tangent at the origin, and reflect the curve across the x-axis to complete the graph (Figure 11.27). The curve is called a *cardioid* because of its heart shape. ∎

EXAMPLE 2   Graph the curve $r^2 = 4\cos\theta$.

Solution   The equation $r^2 = 4\cos\theta$ requires $\cos\theta \geq 0$, so we get the entire graph by running θ from $-\pi/2$ to $\pi/2$. The curve is symmetric about the x-axis because

$$(r, \theta) \text{ on the graph} \Rightarrow r^2 = 4\cos\theta$$

$$\Rightarrow r^2 = 4\cos(-\theta) \qquad \cos\theta = \cos(-\theta)$$

$$\Rightarrow (r, -\theta) \text{ on the graph.}$$

The curve is also symmetric about the origin because

$$(r, \theta) \text{ on the graph} \Rightarrow r^2 = 4\cos\theta$$

$$\Rightarrow (-r)^2 = 4\cos\theta$$

$$\Rightarrow (-r, \theta) \text{ on the graph.}$$

Together, these two symmetries imply symmetry about the y-axis.

The curve passes through the origin when $\theta = -\pi/2$ and $\theta = \pi/2$. It has a vertical tangent both times because $\tan\theta$ is infinite.

For each value of θ in the interval between $-\pi/2$ and $\pi/2$, the formula $r^2 = 4\cos\theta$ gives two values of r:

$$r = \pm 2\sqrt{\cos\theta}.$$

We make a short table of values, plot the corresponding points, and use information about symmetry and tangents to guide us in connecting the points with a smooth curve (Figure 11.28).

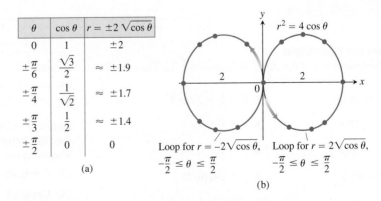

θ	$\cos\theta$	$r = \pm 2\sqrt{\cos\theta}$
0	1	± 2
$\pm\dfrac{\pi}{6}$	$\dfrac{\sqrt{3}}{2}$	$\approx \pm 1.9$
$\pm\dfrac{\pi}{4}$	$\dfrac{1}{\sqrt{2}}$	$\approx \pm 1.7$
$\pm\dfrac{\pi}{3}$	$\dfrac{1}{2}$	$\approx \pm 1.4$
$\pm\dfrac{\pi}{2}$	0	0

(a)

Loop for $r = -2\sqrt{\cos\theta}$, $-\dfrac{\pi}{2} \leq \theta \leq \dfrac{\pi}{2}$ Loop for $r = 2\sqrt{\cos\theta}$, $-\dfrac{\pi}{2} \leq \theta \leq \dfrac{\pi}{2}$

(b)

FIGURE 11.28 The graph of $r^2 = 4\cos\theta$. The arrows show the direction of increasing θ. The values of r in the table are rounded (Example 2). ∎

(a)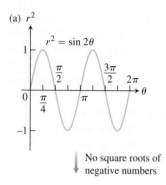

No square roots of negative numbers

(b)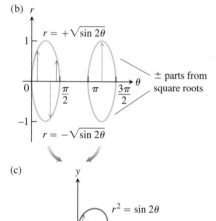

(c)

A Technique for Graphing

One way to graph a polar equation $r = f(\theta)$ is to make a table of (r, θ)-values, plot the corresponding points, and connect them in order of increasing θ. This can work well if enough points have been plotted to reveal all the loops and dimples in the graph. Another method of graphing that is usually quicker and more reliable is to

1. first graph $r = f(\theta)$ in the *Cartesian $r\theta$-plane*,
2. then use the Cartesian graph as a "table" and guide to sketch the *polar* coordinate graph.

This method is better than simple point plotting because the first Cartesian graph, even when hastily drawn, shows at a glance where r is positive, negative, and nonexistent, as well as where r is increasing and decreasing. Here's an example.

EXAMPLE 3   Graph the *lemniscate* curve

$$r^2 = \sin 2\theta.$$

Solution   Here we begin by plotting r^2 (not r) as a function of θ in the Cartesian $r^2\theta$-plane. See Figure 11.29a. We pass from there to the graph of $r = \pm\sqrt{\sin 2\theta}$ in the $r\theta$-plane (Figure 11.29b), and then draw the polar graph (Figure 11.29c). The graph in Figure 11.29b "covers" the final polar graph in Figure 11.29c twice. We could have managed with either loop alone, with the two upper halves, or with the two lower halves. The double covering does no harm, however, and we actually learn a little more about the behavior of the function this way.   ∎

FIGURE 11.29   To plot $r = f(\theta)$ in the Cartesian $r\theta$-plane in (b), we first plot $r^2 = \sin 2\theta$ in the $r^2\theta$-plane in (a) and then ignore the values of θ for which $\sin 2\theta$ is negative. The radii from the sketch in (b) cover the polar graph of the lemniscate in (c) twice (Example 3).

USING TECHNOLOGY   Graphing Polar Curves Parametrically

For complicated polar curves we may need to use a graphing calculator or computer to graph the curve. If the device does not plot polar graphs directly, we can convert $r = f(\theta)$ into parametric form using the equations

$$x = r\cos\theta = f(\theta)\cos\theta, \qquad y = r\sin\theta = f(\theta)\sin\theta.$$

Then we use the device to draw a parametrized curve in the Cartesian xy-plane. It may be necessary to use the parameter t rather than θ for the graphing device.

Exercises 11.4

Symmetries and Polar Graphs

Identify the symmetries of the curves in Exercises 1–12. Then sketch the curves.

1. $r = 1 + \cos\theta$ 2. $r = 2 - 2\cos\theta$

3. $r = 1 - \sin\theta$ 4. $r = 1 + \sin\theta$

5. $r = 2 + \sin\theta$ 6. $r = 1 + 2\sin\theta$

7. $r = \sin(\theta/2)$ 8. $r = \cos(\theta/2)$

9. $r^2 = \cos\theta$ 10. $r^2 = \sin\theta$

11. $r^2 = -\sin\theta$ 12. $r^2 = -\cos\theta$

Graph the lemniscates in Exercises 13–16. What symmetries do these curves have?

13. $r^2 = 4\cos 2\theta$ 14. $r^2 = 4\sin 2\theta$

15. $r^2 = -\sin 2\theta$ 16. $r^2 = -\cos 2\theta$

Slopes of Polar Curves

Find the slopes of the curves in Exercises 17–20 at the given points. Sketch the curves along with their tangents at these points.

17. **Cardioid** $r = -1 + \cos\theta$; $\theta = \pm\pi/2$

18. **Cardioid** $r = -1 + \sin\theta$; $\theta = 0, \pi$

19. **Four-leaved rose** $r = \sin 2\theta$; $\theta = \pm\pi/4, \pm 3\pi/4$

20. **Four-leaved rose** $r = \cos 2\theta$; $\theta = 0, \pm\pi/2, \pi$

Graphing Limaçons

Graph the limaçons in Exercises 21–24. Limaçon ("*lee*-ma-sahn") is Old French for "snail." You will understand the name when you graph the limaçons in Exercise 21. Equations for limaçons have the form $r = a \pm b \cos \theta$ or $r = a \pm b \sin \theta$. There are four basic shapes.

21. Limaçons with an inner loop

   a. $r = \dfrac{1}{2} + \cos \theta$ **b.** $r = \dfrac{1}{2} + \sin \theta$

22. Cardioids

   a. $r = 1 - \cos \theta$ **b.** $r = -1 + \sin \theta$

23. Dimpled limaçons

   a. $r = \dfrac{3}{2} + \cos \theta$ **b.** $r = \dfrac{3}{2} - \sin \theta$

24. Oval limaçons

   a. $r = 2 + \cos \theta$ **b.** $r = -2 + \sin \theta$

Graphing Polar Regions and Curves

25. Sketch the region defined by the inequalities $-1 \le r \le 2$ and $-\pi/2 \le \theta \le \pi/2$.

26. Sketch the region defined by the inequalities $0 \le r \le 2 \sec \theta$ and $-\pi/4 \le \theta \le \pi/4$.

In Exercises 27 and 28, sketch the region defined by the inequality.

27. $0 \le r \le 2 - 2 \cos \theta$ **28.** $0 \le r^2 \le \cos \theta$

T 29. Which of the following has the same graph as $r = 1 - \cos \theta$?

   a. $r = -1 - \cos \theta$

   b. $r = 1 + \cos \theta$

   Confirm your answer with algebra.

T 30. Which of the following has the same graph as $r = \cos 2\theta$?

   a. $r = -\sin (2\theta + \pi/2)$

   b. $r = -\cos (\theta/2)$

   Confirm your answer with algebra.

T 31. A rose within a rose   Graph the equation $r = 1 - 2 \sin 3\theta$.

T 32. The nephroid of Freeth   Graph the nephroid of Freeth:

$$r = 1 + 2 \sin \frac{\theta}{2}.$$

T 33. Roses   Graph the roses $r = \cos m\theta$ for $m = 1/3, 2, 3$, and 7.

T 34. Spirals   Polar coordinates are just the thing for defining spirals. Graph the following spirals.

   a. $r = \theta$

   b. $r = -\theta$

   c. *A logarithmic spiral:* $r = e^{\theta/10}$

   d. *A hyperbolic spiral:* $r = 8/\theta$

   e. *An equilateral hyperbola:* $r = \pm 10/\sqrt{\theta}$

   (Use different colors for the two branches.)

11.5 Areas and Lengths in Polar Coordinates

This section shows how to calculate areas of plane regions and lengths of curves in polar coordinates. The defining ideas are the same as before, but the formulas are different in polar versus Cartesian coordinates.

Area in the Plane

The region *OTS* in Figure 11.30 is bounded by the rays $\theta = \alpha$ and $\theta = \beta$ and the curve $r = f(\theta)$. We approximate the region with n nonoverlapping fan-shaped circular sectors based on a partition P of angle *TOS*. The typical sector has radius $r_k = f(\theta_k)$ and central angle of radian measure $\Delta\theta_k$. Its area is $\Delta\theta_k/2\pi$ times the area of a circle of radius r_k, or

$$A_k = \frac{1}{2} r_k^2 \, \Delta\theta_k = \frac{1}{2} \left(f(\theta_k) \right)^2 \Delta\theta_k.$$

The area of region *OTS* is approximately

$$\sum_{k=1}^{n} A_k = \sum_{k=1}^{n} \frac{1}{2} \left(f(\theta_k) \right)^2 \Delta\theta_k.$$

If f is continuous, we expect the approximations to improve as the norm of the partition P goes to zero, where the norm of P is the largest value of $\Delta\theta_k$. We are then led to the following formula defining the region's area:

FIGURE 11.30 To derive a formula for the area of region *OTS*, we approximate the region with fan-shaped circular sectors.

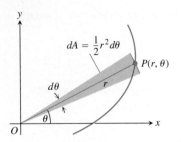

FIGURE 11.31 The area differential dA for the curve $r = f(\theta)$.

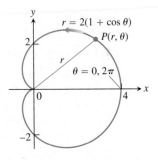

FIGURE 11.32 The cardioid in Example 1.

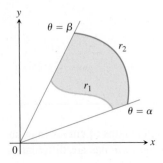

FIGURE 11.33 The area of the shaded region is calculated by subtracting the area of the region between r_1 and the origin from the area of the region between r_2 and the origin.

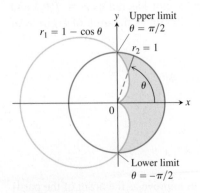

FIGURE 11.34 The region and limits of integration in Example 2.

$$A = \lim_{\|P\| \to 0} \sum_{k=1}^{n} \frac{1}{2} \left(f(\theta_k)\right)^2 \Delta\theta_k$$

$$= \int_{\alpha}^{\beta} \frac{1}{2} \left(f(\theta)\right)^2 d\theta.$$

Area of the Fan-Shaped Region Between the Origin and the Curve
$r = f(\theta), \; \alpha \le \theta \le \beta$

$$A = \int_{\alpha}^{\beta} \frac{1}{2} r^2 \, d\theta.$$

This is the integral of the **area differential** (Figure 11.31)

$$dA = \frac{1}{2} r^2 \, d\theta = \frac{1}{2} \left(f(\theta)\right)^2 d\theta.$$

EXAMPLE 1   Find the area of the region in the plane enclosed by the cardioid $r = 2(1 + \cos\theta)$.

Solution   We graph the cardioid (Figure 11.32) and determine that the radius OP sweeps out the region exactly once as θ runs from 0 to 2π. The area is therefore

$$\int_{\theta=0}^{\theta=2\pi} \frac{1}{2} r^2 \, d\theta = \int_{0}^{2\pi} \frac{1}{2} \cdot 4(1 + \cos\theta)^2 \, d\theta$$

$$= \int_{0}^{2\pi} 2(1 + 2\cos\theta + \cos^2\theta) \, d\theta$$

$$= \int_{0}^{2\pi} \left(2 + 4\cos\theta + 2\frac{1 + \cos 2\theta}{2}\right) d\theta$$

$$= \int_{0}^{2\pi} (3 + 4\cos\theta + \cos 2\theta) \, d\theta$$

$$= \left[3\theta + 4\sin\theta + \frac{\sin 2\theta}{2}\right]_{0}^{2\pi} = 6\pi - 0 = 6\pi. \qquad \blacksquare$$

To find the area of a region like the one in Figure 11.33, which lies between two polar curves $r_1 = r_1(\theta)$ and $r_2 = r_2(\theta)$ from $\theta = \alpha$ to $\theta = \beta$, we subtract the integral of $(1/2)r_1^2 \, d\theta$ from the integral of $(1/2)r_2^2 \, d\theta$. This leads to the following formula.

Area of the Region $0 \le r_1(\theta) \le r \le r_2(\theta), \; \alpha \le \theta \le \beta$

$$A = \int_{\alpha}^{\beta} \frac{1}{2} r_2^2 \, d\theta - \int_{\alpha}^{\beta} \frac{1}{2} r_1^2 \, d\theta = \int_{\alpha}^{\beta} \frac{1}{2} \left(r_2^2 - r_1^2\right) d\theta \qquad (1)$$

EXAMPLE 2   Find the area of the region that lies inside the circle $r = 1$ and outside the cardioid $r = 1 - \cos\theta$.

Solution   We sketch the region to determine its boundaries and find the limits of integration (Figure 11.34). The outer curve is $r_2 = 1$, the inner curve is $r_1 = 1 - \cos\theta$, and θ runs from $-\pi/2$ to $\pi/2$. The area, from Equation (1), is

$$A = \int_{-\pi/2}^{\pi/2} \frac{1}{2}\left(r_2{}^2 - r_1{}^2\right) d\theta$$

$$= 2\int_{0}^{\pi/2} \frac{1}{2}\left(r_2{}^2 - r_1{}^2\right) d\theta \qquad \text{Symmetry}$$

$$= \int_{0}^{\pi/2} (1 - (1 - 2\cos\theta + \cos^2\theta)) \, d\theta \qquad \text{Square } r_1.$$

$$= \int_{0}^{\pi/2} (2\cos\theta - \cos^2\theta) \, d\theta = \int_{0}^{\pi/2} \left(2\cos\theta - \frac{1 + \cos 2\theta}{2}\right) d\theta$$

$$= \left[2\sin\theta - \frac{\theta}{2} - \frac{\sin 2\theta}{4}\right]_{0}^{\pi/2} = 2 - \frac{\pi}{4}.$$ ∎

The fact that we can represent a point in different ways in polar coordinates requires extra care in deciding when a point lies on the graph of a polar equation and in determining the points in which polar graphs intersect. (We needed intersection points in Example 2.) In Cartesian coordinates, we can always find the points where two curves cross by solving their equations simultaneously. In polar coordinates, the story is different. Simultaneous solution may reveal some intersection points without revealing others, so it is sometimes difficult to find all points of intersection of two polar curves. One way to identify all the points of intersection is to graph the equations.

Length of a Polar Curve

We can obtain a polar coordinate formula for the length of a curve $r = f(\theta)$, $\alpha \le \theta \le \beta$, by parametrizing the curve as

$$x = r\cos\theta = f(\theta)\cos\theta, \qquad y = r\sin\theta = f(\theta)\sin\theta, \qquad \alpha \le \theta \le \beta. \qquad (2)$$

The parametric length formula, Equation (3) from Section 11.2, then gives the length as

$$L = \int_{\alpha}^{\beta} \sqrt{\left(\frac{dx}{d\theta}\right)^2 + \left(\frac{dy}{d\theta}\right)^2} \, d\theta.$$

This equation becomes

$$L = \int_{\alpha}^{\beta} \sqrt{r^2 + \left(\frac{dr}{d\theta}\right)^2} \, d\theta$$

when Equations (2) are substituted for x and y (Exercise 29).

Length of a Polar Curve

If $r = f(\theta)$ has a continuous first derivative for $\alpha \le \theta \le \beta$ and if the point $P(r, \theta)$ traces the curve $r = f(\theta)$ exactly once as θ runs from α to β, then the length of the curve is

$$L = \int_{\alpha}^{\beta} \sqrt{r^2 + \left(\frac{dr}{d\theta}\right)^2} \, d\theta. \qquad (3)$$

EXAMPLE 3   Find the length of the cardioid $r = 1 - \cos\theta$.

Solution   We sketch the cardioid to determine the limits of integration (Figure 11.35). The point $P(r, \theta)$ traces the curve once, counterclockwise as θ runs from 0 to 2π, so these are the values we take for α and β.

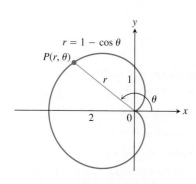

FIGURE 11.35 Calculating the length of a cardioid (Example 3).

With

$$r = 1 - \cos\theta, \qquad \frac{dr}{d\theta} = \sin\theta,$$

we have

$$r^2 + \left(\frac{dr}{d\theta}\right)^2 = (1 - \cos\theta)^2 + (\sin\theta)^2$$

$$= 1 - 2\cos\theta + \underbrace{\cos^2\theta + \sin^2\theta}_{1} = 2 - 2\cos\theta$$

and

$$L = \int_\alpha^\beta \sqrt{r^2 + \left(\frac{dr}{d\theta}\right)^2}\, d\theta = \int_0^{2\pi} \sqrt{2 - 2\cos\theta}\, d\theta$$

$$= \int_0^{2\pi} \sqrt{4\sin^2\frac{\theta}{2}}\, d\theta \qquad 1 - \cos\theta = 2\sin^2(\theta/2)$$

$$= \int_0^{2\pi} 2\left|\sin\frac{\theta}{2}\right| d\theta$$

$$= \int_0^{2\pi} 2\sin\frac{\theta}{2}\, d\theta \qquad \sin(\theta/2) \geq 0 \quad \text{for} \quad 0 \leq \theta \leq 2\pi$$

$$= \left[-4\cos\frac{\theta}{2}\right]_0^{2\pi} = 4 + 4 = 8. \qquad \blacksquare$$

EXERCISES 11.5

Finding Polar Areas

Find the areas of the regions in Exercises 1–8.

1. Bounded by the spiral $r = \theta$ for $0 \leq \theta \leq \pi$

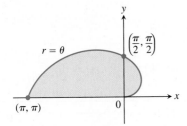

2. Bounded by the circle $r = 2\sin\theta$ for $\pi/4 \leq \theta \leq \pi/2$

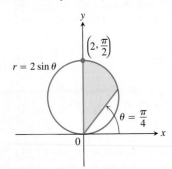

3. Inside the oval limaçon $r = 4 + 2\cos\theta$

4. Inside the cardioid $r = a(1 + \cos\theta), \quad a > 0$

5. Inside one leaf of the four-leaved rose $r = \cos 2\theta$

6. Inside one leaf of the three-leaved rose $r = \cos 3\theta$

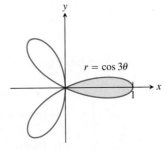

7. Inside one loop of the lemniscate $r^2 = 4\sin 2\theta$

8. Inside the six-leaved rose $r^2 = 2\sin 3\theta$

Find the areas of the regions in Exercises 9–16.

9. Shared by the circles $r = 2\cos\theta$ and $r = 2\sin\theta$

10. Shared by the circles $r = 1$ and $r = 2\sin\theta$

11. Shared by the circle $r = 2$ and the cardioid $r = 2(1 - \cos\theta)$

12. Shared by the cardioids $r = 2(1 + \cos\theta)$ and $r = 2(1 - \cos\theta)$

13. Inside the lemniscate $r^2 = 6\cos 2\theta$ and outside the circle $r = \sqrt{3}$

14. Inside the circle $r = 3a \cos \theta$ and outside the cardioid $r = a(1 + \cos \theta), a > 0$

15. Inside the circle $r = -2 \cos \theta$ and outside the circle $r = 1$

16. Inside the circle $r = 6$ above the line $r = 3 \csc \theta$

17. Inside the circle $r = 4 \cos \theta$ and to the right of the vertical line $r = \sec \theta$

18. Inside the circle $r = 4 \sin \theta$ and below the horizontal line $r = 3 \csc \theta$

19. a. Find the area of the shaded region in the accompanying figure.

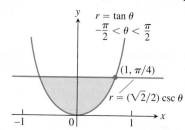

b. It looks as if the graph of $r = \tan \theta, -\pi/2 < \theta < \pi/2$, could be asymptotic to the lines $x = 1$ and $x = -1$. Is it? Give reasons for your answer.

20. The area of the region that lies inside the cardioid curve $r = \cos \theta + 1$ and outside the circle $r = \cos \theta$ is not

$$\frac{1}{2} \int_0^{2\pi} [(\cos \theta + 1)^2 - \cos^2 \theta] \, d\theta = \pi.$$

Why not? What *is* the area? Give reasons for your answers.

Finding Lengths of Polar Curves

Find the lengths of the curves in Exercises 21–28.

21. The spiral $r = \theta^2, \quad 0 \le \theta \le \sqrt{5}$

22. The spiral $r = e^\theta/\sqrt{2}, \quad 0 \le \theta \le \pi$

23. The cardioid $r = 1 + \cos \theta$

24. The curve $r = a \sin^2 (\theta/2), \quad 0 \le \theta \le \pi, \quad a > 0$

25. The parabolic segment $r = 6/(1 + \cos \theta), \quad 0 \le \theta \le \pi/2$

26. The parabolic segment $r = 2/(1 - \cos \theta), \quad \pi/2 \le \theta \le \pi$

27. The curve $r = \cos^3 (\theta/3), \quad 0 \le \theta \le \pi/4$

28. The curve $r = \sqrt{1 + \sin 2\theta}, \quad 0 \le \theta \le \pi\sqrt{2}$

29. The length of the curve $r = f(\theta), \; \alpha \le \theta \le \beta$ Assuming that the necessary derivatives are continuous, show how the substitutions

$$x = f(\theta) \cos \theta, \quad y = f(\theta) \sin \theta$$

(Equations 2 in the text) transform

$$L = \int_\alpha^\beta \sqrt{\left(\frac{dx}{d\theta}\right)^2 + \left(\frac{dy}{d\theta}\right)^2} \, d\theta$$

into

$$L = \int_\alpha^\beta \sqrt{r^2 + \left(\frac{dr}{d\theta}\right)^2} \, d\theta.$$

30. Circumferences of circles As usual, when faced with a new formula, it is a good idea to try it on familiar objects to be sure it gives results consistent with past experience. Use the length formula in Equation (3) to calculate the circumferences of the following circles ($a > 0$).

a. $r = a$ **b.** $r = a \cos \theta$ **c.** $r = a \sin \theta$

Theory and Examples

31. Average value If f is continuous, the average value of the polar coordinate r over the curve $r = f(\theta), \alpha \le \theta \le \beta$, with respect to θ is given by the formula

$$r_{av} = \frac{1}{\beta - \alpha} \int_\alpha^\beta f(\theta) \, d\theta.$$

Use this formula to find the average value of r with respect to θ over the following curves ($a > 0$).

a. The cardioid $r = a(1 - \cos \theta)$

b. The circle $r = a$

c. The circle $r = a \cos \theta, \quad -\pi/2 \le \theta \le \pi/2$

32. $r = f(\theta)$ **vs.** $r = 2f(\theta)$ Can anything be said about the relative lengths of the curves $r = f(\theta), \alpha \le \theta \le \beta$, and $r = 2f(\theta), \alpha \le \theta \le \beta$? Give reasons for your answer.

11.6   Conic Sections

In this section we define and review parabolas, ellipses, and hyperbolas geometrically and derive their standard Cartesian equations. These curves are called *conic sections* or *conics* because they are formed by cutting a double cone with a plane (Figure 11.36). This geometry method was the only way they could be described by Greek mathematicians who did not have our tools of Cartesian or polar coordinates. In the next section we express the conics in polar coordinates.

Parabolas

> **DEFINITIONS**   A set that consists of all the points in a plane equidistant from a given fixed point and a given fixed line in the plane is a **parabola**. The fixed point is the **focus** of the parabola. The fixed line is the **directrix**.

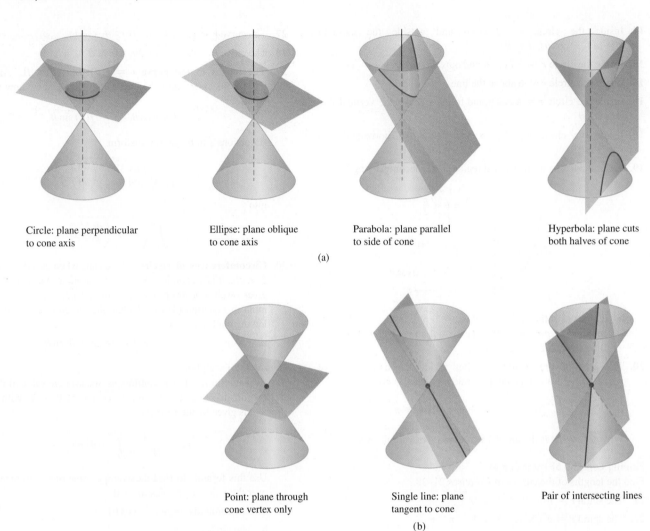

Circle: plane perpendicular
to cone axis

Ellipse: plane oblique
to cone axis

Parabola: plane parallel
to side of cone

Hyperbola: plane cuts
both halves of cone

(a)

Point: plane through
cone vertex only

Single line: plane
tangent to cone

Pair of intersecting lines

(b)

FIGURE 11.36 The standard conic sections (a) are the curves in which a plane cuts a *double* cone. Hyperbolas come in two parts, called *branches*. The point and lines obtained by passing the plane through the cone's vertex (b) are *degenerate* conic sections.

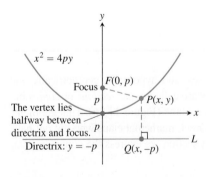

FIGURE 11.37 The standard form of the parabola $x^2 = 4py$, $p > 0$.

If the focus F lies on the directrix L, the parabola is the line through F perpendicular to L. We consider this to be a degenerate case and assume henceforth that F does not lie on L.

A parabola has its simplest equation when its focus and directrix straddle one of the coordinate axes. For example, suppose that the focus lies at the point $F(0, p)$ on the positive y-axis and that the directrix is the line $y = -p$ (Figure 11.37). In the notation of the figure, a point $P(x, y)$ lies on the parabola if and only if $PF = PQ$. From the distance formula,

$$PF = \sqrt{(x - 0)^2 + (y - p)^2} = \sqrt{x^2 + (y - p)^2}$$
$$PQ = \sqrt{(x - x)^2 + (y - (-p))^2} = \sqrt{(y + p)^2}.$$

When we equate these expressions, square, and simplify, we get

$$y = \frac{x^2}{4p} \qquad \text{or} \qquad x^2 = 4py. \qquad \text{Standard form} \qquad (1)$$

These equations reveal the parabola's symmetry about the y-axis. We call the y-axis the **axis** of the parabola (short for "axis of symmetry").

The point where a parabola crosses its axis is the **vertex**. The vertex of the parabola $x^2 = 4py$ lies at the origin (Figure 11.37). The positive number p is the parabola's **focal length**.

If the parabola opens downward, with its focus at $(0, -p)$ and its directrix the line $y = p$, then Equations (1) become

$$y = -\frac{x^2}{4p} \quad \text{and} \quad x^2 = -4py.$$

By interchanging the variables x and y, we obtain similar equations for parabolas opening to the right or to the left (Figure 11.38).

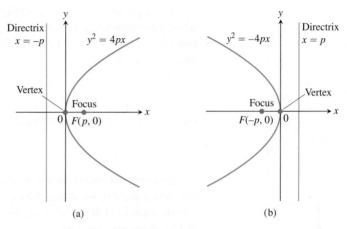

FIGURE 11.38 (a) The parabola $y^2 = 4px$. (b) The parabola $y^2 = -4px$.

EXAMPLE 1 Find the focus and directrix of the parabola $y^2 = 10x$.

Solution We find the value of p in the standard equation $y^2 = 4px$:

$$4p = 10, \quad \text{so} \quad p = \frac{10}{4} = \frac{5}{2}.$$

Then we find the focus and directrix for this value of p:

$$\text{Focus:} \quad (p, 0) = \left(\frac{5}{2}, 0\right)$$

$$\text{Directrix:} \quad x = -p \quad \text{or} \quad x = -\frac{5}{2}. \qquad \blacksquare$$

Ellipses

FIGURE 11.39 Points on the focal axis of an ellipse.

> **DEFINITIONS** An **ellipse** is the set of points in a plane whose distances from two fixed points in the plane have a constant sum. The two fixed points are the **foci** of the ellipse.
> The line through the foci of an ellipse is the ellipse's **focal axis**. The point on the axis halfway between the foci is the **center**. The points where the focal axis and ellipse cross are the ellipse's **vertices** (Figure 11.39).

FIGURE 11.40 The ellipse defined by the equation $PF_1 + PF_2 = 2a$ is the graph of the equation $(x^2/a^2) + (y^2/b^2) = 1$, where $b^2 = a^2 - c^2$.

If the foci are $F_1(-c, 0)$ and $F_2(c, 0)$ (Figure 11.40), and $PF_1 + PF_2$ is denoted by $2a$, then the coordinates of a point P on the ellipse satisfy the equation

$$\sqrt{(x + c)^2 + y^2} + \sqrt{(x - c)^2 + y^2} = 2a.$$

To simplify this equation, we move the second radical to the right-hand side, square, isolate the remaining radical, and square again, obtaining

$$\frac{x^2}{a^2} + \frac{y^2}{a^2 - c^2} = 1. \tag{2}$$

Since $PF_1 + PF_2$ is greater than the length F_1F_2 (by the triangle inequality for triangle PF_1F_2), the number $2a$ is greater than $2c$. Accordingly, $a > c$ and the number $a^2 - c^2$ in Equation (2) is positive.

The algebraic steps leading to Equation (2) can be reversed to show that every point P whose coordinates satisfy an equation of this form with $0 < c < a$ also satisfies the equation $PF_1 + PF_2 = 2a$. A point therefore lies on the ellipse if and only if its coordinates satisfy Equation (2).

If

$$b = \sqrt{a^2 - c^2}, \tag{3}$$

then $a^2 - c^2 = b^2$ and Equation (2) takes the form

$$\frac{x^2}{a^2} + \frac{y^2}{b^2} = 1. \tag{4}$$

Equation (4) reveals that this ellipse is symmetric with respect to the origin and both coordinate axes. It lies inside the rectangle bounded by the lines $x = \pm a$ and $y = \pm b$. It crosses the axes at the points $(\pm a, 0)$ and $(0, \pm b)$. The tangents at these points are perpendicular to the axes because

$$\frac{dy}{dx} = -\frac{b^2 x}{a^2 y} \qquad \text{Obtained from Eq. (4)}$$
$$\text{by implicit differentiation}$$

is zero if $x = 0$ and infinite if $y = 0$.

The **major axis** of the ellipse in Equation (4) is the line segment of length $2a$ joining the points $(\pm a, 0)$. The **minor axis** is the line segment of length $2b$ joining the points $(0, \pm b)$. The number a itself is the **semimajor axis**, the number b the **semiminor axis**. The number c, found from Equation (3) as

$$c = \sqrt{a^2 - b^2},$$

is the **center-to-focus distance** of the ellipse. If $a = b$, the ellipse is a circle.

EXAMPLE 2 The ellipse

$$\frac{x^2}{16} + \frac{y^2}{9} = 1 \tag{5}$$

(Figure 11.41) has

Semimajor axis: $a = \sqrt{16} = 4$, Semiminor axis: $b = \sqrt{9} = 3$

Center-to-focus distance: $c = \sqrt{16 - 9} = \sqrt{7}$

Foci: $(\pm c, 0) = \left(\pm\sqrt{7}, 0\right)$

Vertices: $(\pm a, 0) = (\pm 4, 0)$

Center: $(0, 0)$. ∎

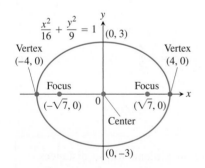

FIGURE 11.41 An ellipse with its major axis horizontal (Example 2).

If we interchange x and y in Equation (5), we have the equation

$$\frac{x^2}{9} + \frac{y^2}{16} = 1. \tag{6}$$

The major axis of this ellipse is now vertical instead of horizontal, with the foci and vertices on the y-axis. There is no confusion in analyzing Equations (5) and (6). If we find the intercepts on the coordinate axes, we will know which way the major axis runs because it is the longer of the two axes.

Standard-Form Equations for Ellipses Centered at the Origin

Foci on the x-axis: $\dfrac{x^2}{a^2} + \dfrac{y^2}{b^2} = 1$ $(a > b)$

Center-to-focus distance: $c = \sqrt{a^2 - b^2}$

Foci: $(\pm c, 0)$

Vertices: $(\pm a, 0)$

Foci on the y-axis: $\dfrac{x^2}{b^2} + \dfrac{y^2}{a^2} = 1$ $(a > b)$

Center-to-focus distance: $c = \sqrt{a^2 - b^2}$

Foci: $(0, \pm c)$

Vertices: $(0, \pm a)$

In each case, a is the semimajor axis and b is the semiminor axis.

Hyperbolas

DEFINITIONS   A **hyperbola** is the set of points in a plane whose distances from two fixed points in the plane have a constant difference. The two fixed points are the **foci** of the hyperbola.

The line through the foci of a hyperbola is the **focal axis**. The point on the axis halfway between the foci is the hyperbola's **center**. The points where the focal axis and hyperbola cross are the **vertices** (Figure 11.42).

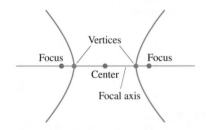

FIGURE 11.42 Points on the focal axis of a hyperbola.

If the foci are $F_1(-c, 0)$ and $F_2(c, 0)$ (Figure 11.43) and the constant difference is $2a$, then a point (x, y) lies on the hyperbola if and only if

$$\sqrt{(x + c)^2 + y^2} - \sqrt{(x - c)^2 + y^2} = \pm 2a. \tag{7}$$

To simplify this equation, we move the second radical to the right-hand side, square, isolate the remaining radical, and square again, obtaining

$$\frac{x^2}{a^2} + \frac{y^2}{a^2 - c^2} = 1. \tag{8}$$

So far, this looks just like the equation for an ellipse. But now $a^2 - c^2$ is negative because $2a$, being the difference of two sides of triangle PF_1F_2, is less than $2c$, the third side.

The algebraic steps leading to Equation (8) can be reversed to show that every point P whose coordinates satisfy an equation of this form with $0 < a < c$ also satisfies Equation (7). A point therefore lies on the hyperbola if and only if its coordinates satisfy Equation (8).

If we let b denote the positive square root of $c^2 - a^2$,

$$b = \sqrt{c^2 - a^2}, \tag{9}$$

then $a^2 - c^2 = -b^2$ and Equation (8) takes the more compact form

$$\frac{x^2}{a^2} - \frac{y^2}{b^2} = 1. \tag{10}$$

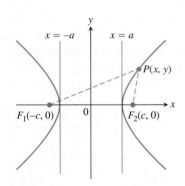

FIGURE 11.43 Hyperbolas have two branches. For points on the right-hand branch of the hyperbola shown here, $PF_1 - PF_2 = 2a$. For points on the left-hand branch, $PF_2 - PF_1 = 2a$. We then let $b = \sqrt{c^2 - a^2}$.

The differences between Equation (10) and the equation for an ellipse (Equation 4) are the minus sign and the new relation

$$c^2 = a^2 + b^2. \qquad \text{From Eq. (9)}$$

Like the ellipse, the hyperbola is symmetric with respect to the origin and coordinate axes. It crosses the x-axis at the points $(\pm a, 0)$. The tangents at these points are vertical because

$$\frac{dy}{dx} = \frac{b^2 x}{a^2 y} \qquad \begin{array}{l}\text{Obtained from Eq. (10)}\\ \text{by implicit differentiation}\end{array}$$

is infinite when $y = 0$. The hyperbola has no y-intercepts; in fact, no part of the curve lies between the lines $x = -a$ and $x = a$.

The lines

$$y = \pm \frac{b}{a} x$$

are the two **asymptotes** of the hyperbola defined by Equation (10). The fastest way to find the equations of the asymptotes is to replace the 1 in Equation (10) by 0 and solve the new equation for y:

$$\underbrace{\frac{x^2}{a^2} - \frac{y^2}{b^2} = 1}_{\text{hyperbola}} \rightarrow \underbrace{\frac{x^2}{a^2} - \frac{y^2}{b^2} = 0}_{\text{0 for 1}} \rightarrow \underbrace{y = \pm \frac{b}{a} x.}_{\text{asymptotes}}$$

EXAMPLE 3 The equation

$$\frac{x^2}{4} - \frac{y^2}{5} = 1 \tag{11}$$

is Equation (10) with $a^2 = 4$ and $b^2 = 5$ (Figure 11.44). We have

Center-to-focus distance: $\quad c = \sqrt{a^2 + b^2} = \sqrt{4 + 5} = 3$

Foci: $\quad (\pm c, 0) = (\pm 3, 0), \qquad$ Vertices: $\quad (\pm a, 0) = (\pm 2, 0)$

Center: $\quad (0, 0)$

Asymptotes: $\quad \dfrac{x^2}{4} - \dfrac{y^2}{5} = 0 \quad \text{or} \quad y = \pm \dfrac{\sqrt{5}}{2} x.$ ■

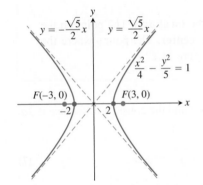

FIGURE 11.44 The hyperbola and its asymptotes in Example 3.

If we interchange x and y in Equation (11), the foci and vertices of the resulting hyperbola will lie along the y-axis. We still find the asymptotes in the same way as before, but now their equations will be $y = \pm 2x/\sqrt{5}$.

Standard-Form Equations for Hyperbolas Centered at the Origin

Foci on the x-axis: $\quad \dfrac{x^2}{a^2} - \dfrac{y^2}{b^2} = 1$

Center-to-focus distance: $\quad c = \sqrt{a^2 + b^2}$

Foci: $\quad (\pm c, 0)$

Vertices: $\quad (\pm a, 0)$

Asymptotes: $\quad \dfrac{x^2}{a^2} - \dfrac{y^2}{b^2} = 0 \quad \text{or} \quad y = \pm \dfrac{b}{a} x$

Foci on the y-axis: $\quad \dfrac{y^2}{a^2} - \dfrac{x^2}{b^2} = 1$

Center-to-focus distance: $\quad c = \sqrt{a^2 + b^2}$

Foci: $\quad (0, \pm c)$

Vertices: $\quad (0, \pm a)$

Asymptotes: $\quad \dfrac{y^2}{a^2} - \dfrac{x^2}{b^2} = 0 \quad \text{or} \quad y = \pm \dfrac{a}{b} x$

Notice the difference in the asymptote equations (b/a in the first, a/b in the second).

We shift conics using the principles reviewed in Section 1.2, replacing x by $x + h$ and y by $y + k$.

EXAMPLE 4   Show that the equation $x^2 - 4y^2 + 2x + 8y - 7 = 0$ represents a hyperbola. Find its center, asymptotes, and foci.

Solution   We reduce the equation to standard form by completing the square in x and y as follows:

$$(x^2 + 2x) - 4(y^2 - 2y) = 7$$

$$(x^2 + 2x + 1) - 4(y^2 - 2y + 1) = 7 + 1 - 4$$

$$\frac{(x + 1)^2}{4} - (y - 1)^2 = 1.$$

This is the standard form Equation (10) of a hyperbola with x replaced by $x + 1$ and y replaced by $y - 1$. The hyperbola is shifted one unit to the left and one unit upward, and it has center $x + 1 = 0$ and $y - 1 = 0$, or $x = -1$ and $y = 1$. Moreover,

$$a^2 = 4, \qquad b^2 = 1, \qquad c^2 = a^2 + b^2 = 5,$$

so the asymptotes are the two lines

$$\frac{x + 1}{2} - (y - 1) = 0 \qquad \text{and} \qquad \frac{x + 1}{2} + (y - 1) = 0.$$

The shifted foci have coordinates $\left(-1 \pm \sqrt{5}, 1\right)$.   ∎

Exercises 11.6

Identifying Graphs

Match the parabolas in Exercises 1–4 with the following equations:

$$x^2 = 2y, \quad x^2 = -6y, \quad y^2 = 8x, \quad y^2 = -4x.$$

Then find each parabola's focus and directrix.

1.

2.

3.

4.

Match each conic section in Exercises 5–8 with one of these equations:

$$\frac{x^2}{4} + \frac{y^2}{9} = 1, \qquad \frac{x^2}{2} + y^2 = 1,$$

$$\frac{y^2}{4} - x^2 = 1, \qquad \frac{x^2}{4} - \frac{y^2}{9} = 1.$$

Then find the conic section's foci and vertices. If the conic section is a hyperbola, find its asymptotes as well.

5.

6.

7.

8.

Parabolas

Exercises 9–16 give equations of parabolas. Find each parabola's focus and directrix. Then sketch the parabola. Include the focus and directrix in your sketch.

9. $y^2 = 12x$　　**10.** $x^2 = 6y$　　**11.** $x^2 = -8y$

12. $y^2 = -2x$　　**13.** $y = 4x^2$　　**14.** $y = -8x^2$

15. $x = -3y^2$　　**16.** $x = 2y^2$

Ellipses

Exercises 17–24 give equations for ellipses. Put each equation in standard form. Then sketch the ellipse. Include the foci in your sketch.

17. $16x^2 + 25y^2 = 400$　　　**18.** $7x^2 + 16y^2 = 112$

19. $2x^2 + y^2 = 2$　　　**20.** $2x^2 + y^2 = 4$

21. $3x^2 + 2y^2 = 6$　　　**22.** $9x^2 + 10y^2 = 90$

23. $6x^2 + 9y^2 = 54$　　　**24.** $169x^2 + 25y^2 = 4225$

Exercises 25 and 26 give information about the foci and vertices of ellipses centered at the origin of the xy-plane. In each case, find the ellipse's standard-form equation from the given information.

25. Foci: $\left(\pm\sqrt{2}, 0\right)$　Vertices: $(\pm 2, 0)$

26. Foci: $(0, \pm 4)$　Vertices: $(0, \pm 5)$

Hyperbolas

Exercises 27–34 give equations for hyperbolas. Put each equation in standard form and find the hyperbola's asymptotes. Then sketch the hyperbola. Include the asymptotes and foci in your sketch.

27. $x^2 - y^2 = 1$　　　**28.** $9x^2 - 16y^2 = 144$

29. $y^2 - x^2 = 8$　　　**30.** $y^2 - x^2 = 4$

31. $8x^2 - 2y^2 = 16$　　　**32.** $y^2 - 3x^2 = 3$

33. $8y^2 - 2x^2 = 16$　　　**34.** $64x^2 - 36y^2 = 2304$

Exercises 35–38 give information about the foci, vertices, and asymptotes of hyperbolas centered at the origin of the xy-plane. In each case, find the hyperbola's standard-form equation from the information given.

35. Foci: $\left(0, \pm\sqrt{2}\right)$　　　**36.** Foci: $(\pm 2, 0)$

　　Asymptotes: $y = \pm x$　　　Asymptotes: $y = \pm\dfrac{1}{\sqrt{3}}x$

37. Vertices: $(\pm 3, 0)$　　　**38.** Vertices: $(0, \pm 2)$

　　Asymptotes: $y = \pm\dfrac{4}{3}x$　　　Asymptotes: $y = \pm\dfrac{1}{2}x$

Shifting Conic Sections

You may wish to review Section 1.2 before solving Exercises 39–56.

39. The parabola $y^2 = 8x$ is shifted down 2 units and right 1 unit to generate the parabola $(y + 2)^2 = 8(x - 1)$.

　　a. Find the new parabola's vertex, focus, and directrix.

　　b. Plot the new vertex, focus, and directrix, and sketch in the parabola.

40. The parabola $x^2 = -4y$ is shifted left 1 unit and up 3 units to generate the parabola $(x + 1)^2 = -4(y - 3)$.

　　a. Find the new parabola's vertex, focus, and directrix.

　　b. Plot the new vertex, focus, and directrix, and sketch in the parabola.

41. The ellipse $(x^2/16) + (y^2/9) = 1$ is shifted 4 units to the right and 3 units up to generate the ellipse

$$\frac{(x - 4)^2}{16} + \frac{(y - 3)^2}{9} = 1.$$

　　a. Find the foci, vertices, and center of the new ellipse.

　　b. Plot the new foci, vertices, and center, and sketch in the new ellipse.

42. The ellipse $(x^2/9) + (y^2/25) = 1$ is shifted 3 units to the left and 2 units down to generate the ellipse

$$\frac{(x + 3)^2}{9} + \frac{(y + 2)^2}{25} = 1.$$

　　a. Find the foci, vertices, and center of the new ellipse.

　　b. Plot the new foci, vertices, and center, and sketch in the new ellipse.

43. The hyperbola $(x^2/16) - (y^2/9) = 1$ is shifted 2 units to the right to generate the hyperbola

$$\frac{(x - 2)^2}{16} - \frac{y^2}{9} = 1.$$

　　a. Find the center, foci, vertices, and asymptotes of the new hyperbola.

　　b. Plot the new center, foci, vertices, and asymptotes, and sketch in the hyperbola.

44. The hyperbola $(y^2/4) - (x^2/5) = 1$ is shifted 2 units down to generate the hyperbola

$$\frac{(y + 2)^2}{4} - \frac{x^2}{5} = 1.$$

　　a. Find the center, foci, vertices, and asymptotes of the new hyperbola.

　　b. Plot the new center, foci, vertices, and asymptotes, and sketch in the hyperbola.

Exercises 45–48 give equations for parabolas and tell how many units up or down and to the right or left each parabola is to be shifted. Find an equation for the new parabola, and find the new vertex, focus, and directrix.

45. $y^2 = 4x$,　left 2, down 3　**46.** $y^2 = -12x$,　right 4, up 3

47. $x^2 = 8y$,　right 1, down 7　**48.** $x^2 = 6y$,　left 3, down 2

Exercises 49–52 give equations for ellipses and tell how many units up or down and to the right or left each ellipse is to be shifted. Find an equation for the new ellipse, and find the new foci, vertices, and center.

49. $\dfrac{x^2}{6} + \dfrac{y^2}{9} = 1$,   left 2, down 1

50. $\dfrac{x^2}{2} + y^2 = 1$,   right 3, up 4

51. $\dfrac{x^2}{3} + \dfrac{y^2}{2} = 1$,   right 2, up 3

52. $\dfrac{x^2}{16} + \dfrac{y^2}{25} = 1$,   left 4, down 5

Exercises 53–56 give equations for hyperbolas and tell how many units up or down and to the right or left each hyperbola is to be shifted. Find an equation for the new hyperbola, and find the new center, foci, vertices, and asymptotes.

53. $\dfrac{x^2}{4} - \dfrac{y^2}{5} = 1$,   right 2, up 2

54. $\dfrac{x^2}{16} - \dfrac{y^2}{9} = 1$,   left 2, down 1

55. $y^2 - x^2 = 1$,   left 1, down 1

56. $\dfrac{y^2}{3} - x^2 = 1$,   right 1, up 3

Find the center, foci, vertices, asymptotes, and radius, as appropriate, of the conic sections in Exercises 57–68.

57. $x^2 + 4x + y^2 = 12$

58. $2x^2 + 2y^2 - 28x + 12y + 114 = 0$

59. $x^2 + 2x + 4y - 3 = 0$ **60.** $y^2 - 4y - 8x - 12 = 0$

61. $x^2 + 5y^2 + 4x = 1$ **62.** $9x^2 + 6y^2 + 36y = 0$

63. $x^2 + 2y^2 - 2x - 4y = -1$

64. $4x^2 + y^2 + 8x - 2y = -1$

65. $x^2 - y^2 - 2x + 4y = 4$ **66.** $x^2 - y^2 + 4x - 6y = 6$

67. $2x^2 - y^2 + 6y = 3$ **68.** $y^2 - 4x^2 + 16x = 24$

Theory and Examples

69. If lines are drawn parallel to the coordinate axes through a point P on the parabola $y^2 = kx$, $k > 0$, the parabola partitions the rectangular region bounded by these lines and the coordinate axes into two smaller regions, A and B.

   a. If the two smaller regions are revolved about the y-axis, show that they generate solids whose volumes have the ratio 4:1.

   b. What is the ratio of the volumes generated by revolving the regions about the x-axis?

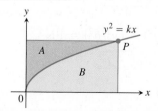

70. Suspension bridge cables hang in parabolas   The suspension bridge cable shown in the accompanying figure supports a uniform load of w pounds per horizontal foot. It can be shown that if H is the horizontal tension of the cable at the origin, then the curve of the cable satisfies the equation

$$\frac{dy}{dx} = \frac{w}{H}x.$$

Show that the cable hangs in a parabola by solving this differential equation subject to the initial condition that $y = 0$ when $x = 0$.

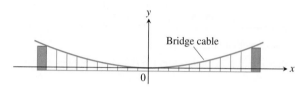

71. The width of a parabola at the focus   Show that the number $4p$ is the *width* of the parabola $x^2 = 4py$ ($p > 0$) at the focus by showing that the line $y = p$ cuts the parabola at points that are $4p$ units apart.

72. The asymptotes of $(x^2/a^2) - (y^2/b^2) = 1$ Show that the vertical distance between the line $y = (b/a)x$ and the upper half of the right-hand branch $y = (b/a)\sqrt{x^2 - a^2}$ of the hyperbola $(x^2/a^2) - (y^2/b^2) = 1$ approaches 0 by showing that

$$\lim_{x \to \infty} \left(\frac{b}{a}x - \frac{b}{a}\sqrt{x^2 - a^2} \right) = \frac{b}{a} \lim_{x \to \infty} \left(x - \sqrt{x^2 - a^2} \right) = 0.$$

Similar results hold for the remaining portions of the hyperbola and the lines $y = \pm(b/a)x$.

73. Area   Find the dimensions of the rectangle of largest area that can be inscribed in the ellipse $x^2 + 4y^2 = 4$ with its sides parallel to the coordinate axes. What is the area of the rectangle?

74. Volume   Find the volume of the solid generated by revolving the region enclosed by the ellipse $9x^2 + 4y^2 = 36$ about the **(a)** x-axis, **(b)** y-axis.

75. Volume   The "triangular" region in the first quadrant bounded by the x-axis, the line $x = 4$, and the hyperbola $9x^2 - 4y^2 = 36$ is revolved about the x-axis to generate a solid. Find the volume of the solid.

76. Tangents   Show that the tangents to the curve $y^2 = 4px$ from any point on the line $x = -p$ are perpendicular.

77. Tangents   Find equations for the tangents to the circle $(x - 2)^2 + (y - 1)^2 = 5$ at the points where the circle crosses the coordinate axes.

78. Volume   The region bounded on the left by the y-axis, on the right by the hyperbola $x^2 - y^2 = 1$, and above and below by the lines $y = \pm 3$ is revolved about the y-axis to generate a solid. Find the volume of the solid.

79. Centroid   Find the centroid of the region that is bounded below by the x-axis and above by the ellipse $(x^2/9) + (y^2/16) = 1$.

80. Surface area   The curve $y = \sqrt{x^2 + 1}$, $0 \le x \le \sqrt{2}$, which is part of the upper branch of the hyperbola $y^2 - x^2 = 1$, is revolved about the x-axis to generate a surface. Find the area of the surface.

81. The reflective property of parabolas The accompanying figure shows a typical point $P(x_0, y_0)$ on the parabola $y^2 = 4px$. The line L is tangent to the parabola at P. The parabola's focus lies at $F(p, 0)$. The ray L' extending from P to the right is parallel to the x-axis. We show that light from F to P will be reflected out along L' by showing that β equals α. Establish this equality by taking the following steps.

a. Show that $\tan \beta = 2p/y_0$.

b. Show that $\tan \phi = y_0/(x_0 - p)$.

c. Use the identity

$$\tan \alpha = \frac{\tan \phi - \tan \beta}{1 + \tan \phi \tan \beta}$$

to show that $\tan \alpha = 2p/y_0$.

Since α and β are both acute, $\tan \beta = \tan \alpha$ implies $\beta = \alpha$.

This reflective property of parabolas is used in applications like car headlights, radio telescopes, and satellite TV dishes.

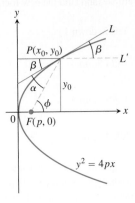

11.7 | Conics in Polar Coordinates

Polar coordinates are especially important in astronomy and astronautical engineering because satellites, moons, planets, and comets all move approximately along ellipses, parabolas, and hyperbolas that can be described with a single relatively simple polar coordinate equation. We develop that equation here after first introducing the idea of a conic section's *eccentricity*. The eccentricity reveals the conic section's type (circle, ellipse, parabola, or hyperbola) and the degree to which it is "squashed" or flattened.

Eccentricity

Although the center-to-focus distance c does not appear in the equation

$$\frac{x^2}{a^2} + \frac{y^2}{b^2} = 1, \quad (a > b)$$

for an ellipse, we can still determine c from the equation $c = \sqrt{a^2 - b^2}$. If we fix a and vary c over the interval $0 \leq c \leq a$, the resulting ellipses will vary in shape. They are circles if $c = 0$ (so that $a = b$) and flatten as c increases. If $c = a$, the foci and vertices overlap and the ellipse degenerates into a line segment. Thus we are led to consider the ratio $e = c/a$. We use this ratio for hyperbolas as well, only in this case c equals $\sqrt{a^2 + b^2}$ instead of $\sqrt{a^2 - b^2}$, and define these ratios with the somewhat familiar term *eccentricity*.

DEFINITION

The **eccentricity** of the ellipse $(x^2/a^2) + (y^2/b^2) = 1 \, (a > b)$ is

$$e = \frac{c}{a} = \frac{\sqrt{a^2 - b^2}}{a}.$$

The **eccentricity** of the hyperbola $(x^2/a^2) - (y^2/b^2) = 1$ is

$$e = \frac{c}{a} = \frac{\sqrt{a^2 + b^2}}{a}.$$

The **eccentricity** of a parabola is $e = 1$.

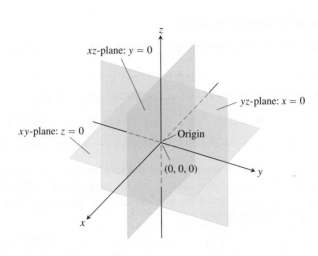

FIGURE 12.2 The planes $x = 0$, $y = 0$, and $z = 0$ divide space into eight octants.

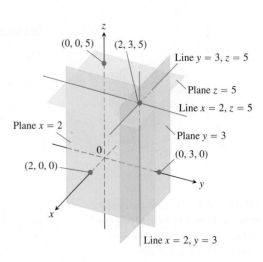

FIGURE 12.3 The planes $x = 2$, $y = 3$, and $z = 5$ determine three lines through the point $(2, 3, 5)$.

at $y = 3$. The plane $z = 5$ is the plane perpendicular to the z-axis at $z = 5$. Figure 12.3 shows the planes $x = 2$, $y = 3$, and $z = 5$, together with their intersection point $(2, 3, 5)$.

The planes $x = 2$ and $y = 3$ in Figure 12.3 intersect in a line parallel to the z-axis. This line is described by the *pair* of equations $x = 2$, $y = 3$. A point (x, y, z) lies on the line if and only if $x = 2$ and $y = 3$. Similarly, the line of intersection of the planes $y = 3$ and $z = 5$ is described by the equation pair $y = 3$, $z = 5$. This line runs parallel to the x-axis. The line of intersection of the planes $x = 2$ and $z = 5$, parallel to the y-axis, is described by the equation pair $x = 2$, $z = 5$.

In the following examples, we match coordinate equations and inequalities with the sets of points they define in space.

EXAMPLE 1 We interpret these equations and inequalities geometrically.

(a) $z \geq 0$ — The half-space consisting of the points on and above the xy-plane.

(b) $x = -3$ — The plane perpendicular to the x-axis at $x = -3$. This plane lies parallel to the yz-plane and 3 units behind it.

(c) $z = 0, x \leq 0, y \geq 0$ — The second quadrant of the xy-plane.

(d) $x \geq 0, y \geq 0, z \geq 0$ — The first octant.

(e) $-1 \leq y \leq 1$ — The slab between the planes $y = -1$ and $y = 1$ (planes included).

(f) $y = -2, z = 2$ — The line in which the planes $y = -2$ and $z = 2$ intersect. Alternatively, the line through the point $(0, -2, 2)$ parallel to the x-axis. ∎

EXAMPLE 2 What points $P(x, y, z)$ satisfy the equations

$$x^2 + y^2 = 4 \quad \text{and} \quad z = 3?$$

Solution The points lie in the horizontal plane $z = 3$ and, in this plane, make up the circle $x^2 + y^2 = 4$. We call this set of points "the circle $x^2 + y^2 = 4$ in the plane $z = 3$" or, more simply, "the circle $x^2 + y^2 = 4$, $z = 3$" (Figure 12.4). ∎

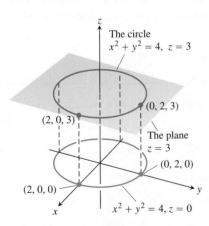

FIGURE 12.4 The circle $x^2 + y^2 = 4$ in the plane $z = 3$ (Example 2).

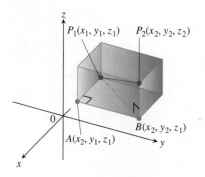

FIGURE 12.5 We find the distance between P_1 and P_2 by applying the Pythagorean theorem to the right triangles P_1AB and P_1BP_2.

Distance and Spheres in Space

The formula for the distance between two points in the xy-plane extends to points in space.

> **The Distance Between $P_1(x_1, y_1, z_1)$ and $P_2(x_2, y_2, z_2)$ is**
> $$|P_1P_2| = \sqrt{(x_2 - x_1)^2 + (y_2 - y_1)^2 + (z_2 - z_1)^2}$$

Proof We construct a rectangular box with faces parallel to the coordinate planes and the points P_1 and P_2 at opposite corners of the box (Figure 12.5). If $A(x_2, y_1, z_1)$ and $B(x_2, y_2, z_1)$ are the vertices of the box indicated in the figure, then the three box edges P_1A, AB, and BP_2 have lengths

$$|P_1A| = |x_2 - x_1|, \qquad |AB| = |y_2 - y_1|, \qquad |BP_2| = |z_2 - z_1|.$$

Because triangles P_1BP_2 and P_1AB are both right-angled, two applications of the Pythagorean theorem give

$$|P_1P_2|^2 = |P_1B|^2 + |BP_2|^2 \qquad \text{and} \qquad |P_1B|^2 = |P_1A|^2 + |AB|^2$$

(see Figure 12.5).
So

$$\begin{aligned}
|P_1P_2|^2 &= |P_1B|^2 + |BP_2|^2 \\
&= |P_1A|^2 + |AB|^2 + |BP_2|^2 && \text{Substitute } |P_1B|^2 = |P_1A|^2 + |AB|^2. \\
&= |x_2 - x_1|^2 + |y_2 - y_1|^2 + |z_2 - z_1|^2 \\
&= (x_2 - x_1)^2 + (y_2 - y_1)^2 + (z_2 - z_1)^2
\end{aligned}$$

Therefore

$$|P_1P_2| = \sqrt{(x_2 - x_1)^2 + (y_2 - y_1)^2 + (z_2 - z_1)^2} \qquad \blacksquare$$

EXAMPLE 3 The distance between $P_1(2, 1, 5)$ and $P_2(-2, 3, 0)$ is

$$\begin{aligned}
|P_1P_2| &= \sqrt{(-2 - 2)^2 + (3 - 1)^2 + (0 - 5)^2} \\
&= \sqrt{16 + 4 + 25} \\
&= \sqrt{45} \approx 6.708.
\end{aligned}$$
$\blacksquare$

We can use the distance formula to write equations for spheres in space (Figure 12.6). A point $P(x, y, z)$ lies on the sphere of radius a centered at $P_0(x_0, y_0, z_0)$ precisely when $|P_0P| = a$ or

$$(x - x_0)^2 + (y - y_0)^2 + (z - z_0)^2 = a^2.$$

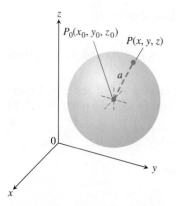

FIGURE 12.6 The sphere of radius a centered at the point (x_0, y_0, z_0).

> **The Standard Equation for the Sphere of Radius a and Center (x_0, y_0, z_0)**
> $$(x - x_0)^2 + (y - y_0)^2 + (z - z_0)^2 = a^2$$

EXAMPLE 4 Find the center and radius of the sphere

$$x^2 + y^2 + z^2 + 3x - 4z + 1 = 0.$$

Solution We find the center and radius of a sphere the way we find the center and radius of a circle: Complete the squares on the x-, y-, and z-terms as necessary and write each quadratic as a squared linear expression. Then, from the equation in standard form, read off the center and radius. For the sphere here, we have

$$x^2 + y^2 + z^2 + 3x - 4z + 1 = 0$$
$$(x^2 + 3x) + y^2 + (z^2 - 4z) = -1$$
$$\left(x^2 + 3x + \left(\frac{3}{2}\right)^2\right) + y^2 + \left(z^2 - 4z + \left(\frac{-4}{2}\right)^2\right) = -1 + \left(\frac{3}{2}\right)^2 + \left(\frac{-4}{2}\right)^2$$
$$\left(x + \frac{3}{2}\right)^2 + y^2 + (z - 2)^2 = -1 + \frac{9}{4} + 4 = \frac{21}{4}.$$

From this standard form, we read that $x_0 = -3/2, y_0 = 0, z_0 = 2$, and $a = \sqrt{21}/2$. The center is $(-3/2, 0, 2)$. The radius is $\sqrt{21}/2$. ∎

EXAMPLE 5   Here are some geometric interpretations of inequalities and equations involving spheres.

(a) $x^2 + y^2 + z^2 < 4$	The interior of the sphere $x^2 + y^2 + z^2 = 4$.
(b) $x^2 + y^2 + z^2 \leq 4$	The solid ball bounded by the sphere $x^2 + y^2 + z^2 = 4$. Alternatively, the sphere $x^2 + y^2 + z^2 = 4$ together with its interior.
(c) $x^2 + y^2 + z^2 > 4$	The exterior of the sphere $x^2 + y^2 + z^2 = 4$.
(d) $x^2 + y^2 + z^2 = 4, z \leq 0$	The lower hemisphere cut from the sphere $x^2 + y^2 + z^2 = 4$ by the xy-plane (the plane $z = 0$).

∎

Just as polar coordinates give another way to locate points in the xy-plane (Section 11.3), alternative coordinate systems, different from the Cartesian coordinate system developed here, exist for three-dimensional space. We examine two of these coordinate systems in Section 15.7.

Exercises 12.1

Geometric Interpretations of Equations

In Exercises 1–16, give a geometric description of the set of points in space whose coordinates satisfy the given pairs of equations.

1. $x = 2, \quad y = 3$
2. $x = -1, \quad z = 0$
3. $y = 0, \quad z = 0$
4. $x = 1, \quad y = 0$
5. $x^2 + y^2 = 4, \quad z = 0$
6. $x^2 + y^2 = 4, \quad z = -2$
7. $x^2 + z^2 = 4, \quad y = 0$
8. $y^2 + z^2 = 1, \quad x = 0$
9. $x^2 + y^2 + z^2 = 1, \quad x = 0$
10. $x^2 + y^2 + z^2 = 25, \quad y = -4$
11. $x^2 + y^2 + (z + 3)^2 = 25, \quad z = 0$
12. $x^2 + (y - 1)^2 + z^2 = 4, \quad y = 0$
13. $x^2 + y^2 = 4, \quad z = y$
14. $x^2 + y^2 + z^2 = 4, \quad y = x$
15. $y = x^2, \quad z = 0$
16. $z = y^2, \quad x = 1$

Geometric Interpretations of Inequalities and Equations

In Exercises 17–24, describe the sets of points in space whose coordinates satisfy the given inequalities or combinations of equations and inequalities.

17. **a.** $x \geq 0, \quad y \geq 0, \quad z = 0$ **b.** $x \geq 0, \quad y \leq 0, \quad z = 0$
18. **a.** $0 \leq x \leq 1$ **b.** $0 \leq x \leq 1, \quad 0 \leq y \leq 1$
    c. $0 \leq x \leq 1, \quad 0 \leq y \leq 1, \quad 0 \leq z \leq 1$
19. **a.** $x^2 + y^2 + z^2 \leq 1$ **b.** $x^2 + y^2 + z^2 > 1$
20. **a.** $x^2 + y^2 \leq 1, \quad z = 0$ **b.** $x^2 + y^2 \leq 1, \quad z = 3$
    c. $x^2 + y^2 \leq 1, \quad$ no restriction on z
21. **a.** $1 \leq x^2 + y^2 + z^2 \leq 4$
    b. $x^2 + y^2 + z^2 \leq 1, \quad z \geq 0$
22. **a.** $x = y, \quad z = 0$ **b.** $x = y, \quad$ no restriction on z
23. **a.** $y \geq x^2, \quad z \geq 0$ **b.** $x \leq y^2, \quad 0 \leq z \leq 2$
24. **a.** $z = 1 - y, \quad$ no restriction on x
    b. $z = y^3, \quad x = 2$

In Exercises 25–34, describe the given set with a single equation or with a pair of equations.

25. The plane perpendicular to the

  a. x-axis at $(3, 0, 0)$ **b.** y-axis at $(0, -1, 0)$

  c. z-axis at $(0, 0, -2)$

26. The plane through the point $(3, -1, 2)$ perpendicular to the

  a. x-axis   **b.** y-axis   **c.** z-axis

27. The plane through the point $(3, -1, 1)$ parallel to the

  a. xy-plane   **b.** yz-plane   **c.** xz-plane

28. The circle of radius 2 centered at $(0, 0, 0)$ and lying in the

  a. xy-plane   **b.** yz-plane   **c.** xz-plane

29. The circle of radius 2 centered at $(0, 2, 0)$ and lying in the

  a. xy-plane   **b.** yz-plane   **c.** plane $y = 2$

30. The circle of radius 1 centered at $(-3, 4, 1)$ and lying in a plane parallel to the

  a. xy-plane   **b.** yz-plane   **c.** xz-plane

31. The line through the point $(1, 3, -1)$ parallel to the

  a. x-axis   **b.** y-axis   **c.** z-axis

32. The set of points in space equidistant from the origin and the point $(0, 2, 0)$

33. The circle in which the plane through the point $(1, 1, 3)$ perpendicular to the z-axis meets the sphere of radius 5 centered at the origin

34. The set of points in space that lie 2 units from the point $(0, 0, 1)$ and, at the same time, 2 units from the point $(0, 0, -1)$

Inequalities to Describe Sets of Points

Write inequalities to describe the sets in Exercises 35–40.

35. The slab bounded by the planes $z = 0$ and $z = 1$ (planes included)

36. The solid cube in the first octant bounded by the coordinate planes and the planes $x = 2$, $y = 2$, and $z = 2$

37. The half-space consisting of the points on and below the xy-plane

38. The upper hemisphere of the sphere of radius 1 centered at the origin

39. The **(a)** interior and **(b)** exterior of the sphere of radius 1 centered at the point $(1, 1, 1)$

40. The closed region bounded by the spheres of radius 1 and radius 2 centered at the origin. (*Closed* means the spheres are to be included. Had we wanted the spheres left out, we would have asked for the *open* region bounded by the spheres. This is analogous to the way we use *closed* and *open* to describe intervals: *closed* means endpoints included, *open* means endpoints left out. Closed sets include boundaries; open sets leave them out.)

Distance

In Exercises 41–46, find the distance between points P_1 and P_2.

41. $P_1(1, 1, 1)$, $P_2(3, 3, 0)$

42. $P_1(-1, 1, 5)$, $P_2(2, 5, 0)$

43. $P_1(1, 4, 5)$, $P_2(4, -2, 7)$

44. $P_1(3, 4, 5)$, $P_2(2, 3, 4)$

45. $P_1(0, 0, 0)$, $P_2(2, -2, -2)$

46. $P_1(5, 3, -2)$, $P_2(0, 0, 0)$

Spheres

Find the centers and radii of the spheres in Exercises 47–50.

47. $(x + 2)^2 + y^2 + (z - 2)^2 = 8$

48. $(x - 1)^2 + \left(y + \dfrac{1}{2}\right)^2 + (z + 3)^2 = 25$

49. $\left(x - \sqrt{2}\right)^2 + \left(y - \sqrt{2}\right)^2 + \left(z + \sqrt{2}\right)^2 = 2$

50. $x^2 + \left(y + \dfrac{1}{3}\right)^2 + \left(z - \dfrac{1}{3}\right)^2 = \dfrac{16}{9}$

Find equations for the spheres whose centers and radii are given in Exercises 51–54.

	Center	Radius
51.	$(1, 2, 3)$	$\sqrt{14}$
52.	$(0, -1, 5)$	2
53.	$\left(-1, \dfrac{1}{2}, -\dfrac{2}{3}\right)$	$\dfrac{4}{9}$
54.	$(0, -7, 0)$	7

Find the centers and radii of the spheres in Exercises 55–58.

55. $x^2 + y^2 + z^2 + 4x - 4z = 0$

56. $x^2 + y^2 + z^2 - 6y + 8z = 0$

57. $2x^2 + 2y^2 + 2z^2 + x + y + z = 9$

58. $3x^2 + 3y^2 + 3z^2 + 2y - 2z = 9$

Theory and Examples

59. Find a formula for the distance from the point $P(x, y, z)$ to the

  a. x-axis   **b.** y-axis   **c.** z-axis

60. Find a formula for the distance from the point $P(x, y, z)$ to the

  a. xy-plane   **b.** yz-plane   **c.** xz-plane

61. Find the perimeter of the triangle with vertices $A(-1, 2, 1)$, $B(1, -1, 3)$, and $C(3, 4, 5)$.

62. Show that the point $P(3, 1, 2)$ is equidistant from the points $A(2, -1, 3)$ and $B(4, 3, 1)$.

63. Find an equation for the set of all points equidistant from the planes $y = 3$ and $y = -1$.

64. Find an equation for the set of all points equidistant from the point $(0, 0, 2)$ and the xy-plane.

65. Find the point on the sphere $x^2 + (y - 3)^2 + (z + 5)^2 = 4$ nearest

  a. the xy-plane.   **b.** the point $(0, 7, -5)$.

66. Find the point equidistant from the points $(0, 0, 0)$, $(0, 4, 0)$, $(3, 0, 0)$, and $(2, 2, -3)$.

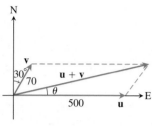

FIGURE 12.17 Vectors representing the velocities of the airplane **u** and tailwind **v** in Example 8.

Applications

An important application of vectors occurs in navigation.

EXAMPLE 8   A jet airliner, flying due east at 500 mph in still air, encounters a 70-mph tailwind blowing in the direction 60° north of east. The airplane holds its compass heading due east but, because of the wind, acquires a new ground speed and direction. What are they?

Solution   If **u** = the velocity of the airplane alone and **v** = the velocity of the tailwind, then $|\mathbf{u}| = 500$ and $|\mathbf{v}| = 70$ (Figure 12.17). The velocity of the airplane with respect to the ground is given by the magnitude and direction of the resultant vector **u** + **v**. If we let the positive *x*-axis represent east and the positive *y*-axis represent north, then the component forms of **u** and **v** are

$$\mathbf{u} = \langle 500, 0 \rangle \quad \text{and} \quad \mathbf{v} = \langle 70 \cos 60°, 70 \sin 60° \rangle = \langle 35, 35\sqrt{3} \rangle.$$

Therefore,

$$\mathbf{u} + \mathbf{v} = \langle 535, 35\sqrt{3} \rangle = 535\mathbf{i} + 35\sqrt{3}\,\mathbf{j}$$

$$|\mathbf{u} + \mathbf{v}| = \sqrt{535^2 + (35\sqrt{3})^2} \approx 538.4$$

and

$$\theta = \tan^{-1} \frac{35\sqrt{3}}{535} \approx 6.5°. \qquad \text{Figure 12.17}$$

The new ground speed of the airplane is about 538.4 mph, and its new direction is about 6.5° north of east. ∎

Another important application occurs in physics and engineering when several forces are acting on a single object.

EXAMPLE 9   A 75-N weight is suspended by two wires, as shown in Figure 12.18a. Find the forces **F**$_1$ and **F**$_2$ acting in both wires.

Solution   The force vectors **F**$_1$ and **F**$_2$ have magnitudes $|\mathbf{F}_1|$ and $|\mathbf{F}_2|$ and components that are measured in Newtons. The resultant force is the sum **F**$_1$ + **F**$_2$ and must be equal in magnitude and acting in the opposite (or upward) direction to the weight vector **w** (see Figure 12.18b). It follows from the figure that

$$\mathbf{F}_1 = \langle -|\mathbf{F}_1|\cos 55°, |\mathbf{F}_1|\sin 55° \rangle \quad \text{and} \quad \mathbf{F}_2 = \langle |\mathbf{F}_2|\cos 40°, |\mathbf{F}_2|\sin 40° \rangle.$$

Since **F**$_1$ + **F**$_2$ = $\langle 0, 75 \rangle$, the resultant vector leads to the system of equations

$$-|\mathbf{F}_1|\cos 55° + |\mathbf{F}_2|\cos 40° = 0$$
$$|\mathbf{F}_1|\sin 55° + |\mathbf{F}_2|\sin 40° = 75.$$

Solving for $|\mathbf{F}_2|$ in the first equation and substituting the result into the second equation, we get

$$|\mathbf{F}_2| = \frac{|\mathbf{F}_1|\cos 55°}{\cos 40°} \quad \text{and} \quad |\mathbf{F}_1|\sin 55° + \frac{|\mathbf{F}_1|\cos 55°}{\cos 40°} \sin 40° = 75.$$

It follows that

$$|\mathbf{F}_1| = \frac{75}{\sin 55° + \cos 55° \tan 40°} \approx 57.67 \text{ N},$$

FIGURE 12.18   The suspended weight in Example 9.

and

$$|\mathbf{F}_2| = \frac{75\cos 55°}{\sin 55° \cos 40° + \cos 55° \sin 40°}$$

$$= \frac{75\cos 55°}{\sin(55° + 40°)} \approx 43.18 \text{ N.}$$

The force vectors are then $\mathbf{F}_1 = \langle -33.08, 47.24 \rangle$ and $\mathbf{F}_2 = \langle 33.08, 27.76 \rangle$.   ∎

Exercises 12.2

Vectors in the Plane

In Exercises 1–8, let $\mathbf{u} = \langle 3, -2 \rangle$ and $\mathbf{v} = \langle -2, 5 \rangle$. Find the **(a)** component form and **(b)** magnitude (length) of the vector.

1. $3\mathbf{u}$
2. $-2\mathbf{v}$
3. $\mathbf{u} + \mathbf{v}$
4. $\mathbf{u} - \mathbf{v}$
5. $2\mathbf{u} - 3\mathbf{v}$
6. $-2\mathbf{u} + 5\mathbf{v}$
7. $\frac{3}{5}\mathbf{u} + \frac{4}{5}\mathbf{v}$
8. $-\frac{5}{13}\mathbf{u} + \frac{12}{13}\mathbf{v}$

In Exercises 9–16, find the component form of the vector.

9. The vector $\overrightarrow{PQ}$, where $P = (1, 3)$ and $Q = (2, -1)$
10. The vector $\overrightarrow{OP}$ where O is the origin and P is the midpoint of segment RS, where $R = (2, -1)$ and $S = (-4, 3)$
11. The vector from the point $A = (2, 3)$ to the origin
12. The sum of $\overrightarrow{AB}$ and $\overrightarrow{CD}$, where $A = (1, -1), B = (2, 0)$, $C = (-1, 3)$, and $D = (-2, 2)$
13. The unit vector that makes an angle $\theta = 2\pi/3$ with the positive x-axis
14. The unit vector that makes an angle $\theta = -3\pi/4$ with the positive x-axis
15. The unit vector obtained by rotating the vector $\langle 0, 1 \rangle$ 120° counterclockwise about the origin
16. The unit vector obtained by rotating the vector $\langle 1, 0 \rangle$ 135° counterclockwise about the origin

Vectors in Space

In Exercises 17–22, express each vector in the form $\mathbf{v} = v_1\mathbf{i} + v_2\mathbf{j} + v_3\mathbf{k}$.

17. $\overrightarrow{P_1P_2}$ if P_1 is the point $(5, 7, -1)$ and P_2 is the point $(2, 9, -2)$
18. $\overrightarrow{P_1P_2}$ if P_1 is the point $(1, 2, 0)$ and P_2 is the point $(-3, 0, 5)$
19. $\overrightarrow{AB}$ if A is the point $(-7, -8, 1)$ and B is the point $(-10, 8, 1)$
20. $\overrightarrow{AB}$ if A is the point $(1, 0, 3)$ and B is the point $(-1, 4, 5)$
21. $5\mathbf{u} - \mathbf{v}$ if $\mathbf{u} = \langle 1, 1, -1 \rangle$ and $\mathbf{v} = \langle 2, 0, 3 \rangle$
22. $-2\mathbf{u} + 3\mathbf{v}$ if $\mathbf{u} = \langle -1, 0, 2 \rangle$ and $\mathbf{v} = \langle 1, 1, 1 \rangle$

Geometric Representations

In Exercises 23 and 24, copy vectors $\mathbf{u}$, $\mathbf{v}$, and $\mathbf{w}$ head to tail as needed to sketch the indicated vector.

23.

a. $\mathbf{u} + \mathbf{v}$ **b.** $\mathbf{u} + \mathbf{v} + \mathbf{w}$

c. $\mathbf{u} - \mathbf{v}$ **d.** $\mathbf{u} - \mathbf{w}$

24.

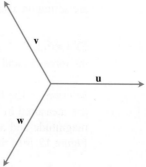

a. $\mathbf{u} - \mathbf{v}$ **b.** $\mathbf{u} - \mathbf{v} + \mathbf{w}$

c. $2\mathbf{u} - \mathbf{v}$ **d.** $\mathbf{u} + \mathbf{v} + \mathbf{w}$

Length and Direction

In Exercises 25–30, express each vector as a product of its length and direction.

25. $2\mathbf{i} + \mathbf{j} - 2\mathbf{k}$
26. $9\mathbf{i} - 2\mathbf{j} + 6\mathbf{k}$
27. $5\mathbf{k}$
28. $\frac{3}{5}\mathbf{i} + \frac{4}{5}\mathbf{k}$
29. $\frac{1}{\sqrt{6}}\mathbf{i} - \frac{1}{\sqrt{6}}\mathbf{j} - \frac{1}{\sqrt{6}}\mathbf{k}$
30. $\frac{\mathbf{i}}{\sqrt{3}} + \frac{\mathbf{j}}{\sqrt{3}} + \frac{\mathbf{k}}{\sqrt{3}}$

31. Find the vectors whose lengths and directions are given. Try to do the calculations without writing.

Length	Direction
a. 2	$\mathbf{i}$
b. $\sqrt{3}$	$-\mathbf{k}$
c. $\dfrac{1}{2}$	$\dfrac{3}{5}\mathbf{j} + \dfrac{4}{5}\mathbf{k}$
d. 7	$\dfrac{6}{7}\mathbf{i} - \dfrac{2}{7}\mathbf{j} + \dfrac{3}{7}\mathbf{k}$

32. Find the vectors whose lengths and directions are given. Try to do the calculations without writing.

Length	Direction
a. 7	$-\mathbf{j}$
b. $\sqrt{2}$	$-\dfrac{3}{5}\mathbf{i} - \dfrac{4}{5}\mathbf{k}$
c. $\dfrac{13}{12}$	$\dfrac{3}{13}\mathbf{i} - \dfrac{4}{13}\mathbf{j} - \dfrac{12}{13}\mathbf{k}$
d. $a > 0$	$\dfrac{1}{\sqrt{2}}\mathbf{i} + \dfrac{1}{\sqrt{3}}\mathbf{j} - \dfrac{1}{\sqrt{6}}\mathbf{k}$

33. Find a vector of magnitude 7 in the direction of $\mathbf{v} = 12\mathbf{i} - 5\mathbf{k}$.

34. Find a vector of magnitude 3 in the direction opposite to the direction of $\mathbf{v} = (1/2)\mathbf{i} - (1/2)\mathbf{j} - (1/2)\mathbf{k}$.

Direction and Midpoints
In Exercises 35–38, find

    a. the direction of $\overrightarrow{P_1P_2}$ and

    b. the midpoint of line segment P_1P_2.

35. $P_1(-1, 1, 5)$ $P_2(2, 5, 0)$

36. $P_1(1, 4, 5)$ $P_2(4, -2, 7)$

37. $P_1(3, 4, 5)$ $P_2(2, 3, 4)$

38. $P_1(0, 0, 0)$ $P_2(2, -2, -2)$

39. If $\overrightarrow{AB} = \mathbf{i} + 4\mathbf{j} - 2\mathbf{k}$ and B is the point $(5, 1, 3)$, find A.

40. If $\overrightarrow{AB} = -7\mathbf{i} + 3\mathbf{j} + 8\mathbf{k}$ and A is the point $(-2, -3, 6)$, find B.

Theory and Applications

41. Linear combination Let $\mathbf{u} = 2\mathbf{i} + \mathbf{j}$, $\mathbf{v} = \mathbf{i} + \mathbf{j}$, and $\mathbf{w} = \mathbf{i} - \mathbf{j}$. Find scalars a and b such that $\mathbf{u} = a\mathbf{v} + b\mathbf{w}$.

42. Linear combination Let $\mathbf{u} = \mathbf{i} - 2\mathbf{j}$, $\mathbf{v} = 2\mathbf{i} + 3\mathbf{j}$, and $\mathbf{w} = \mathbf{i} + \mathbf{j}$. Write $\mathbf{u} = \mathbf{u}_1 + \mathbf{u}_2$, where $\mathbf{u}_1$ is parallel to $\mathbf{v}$ and $\mathbf{u}_2$ is parallel to $\mathbf{w}$. (See Exercise 41.)

43. Velocity An airplane is flying in the direction $25°$ west of north at 800 km/h. Find the component form of the velocity of the airplane, assuming that the positive x-axis represents due east and the positive y-axis represents due north.

44. (*Continuation of Example 8.*) What speed and direction should the jetliner in Example 8 have in order for the resultant vector to be 500 mph due east?

45. Consider a 100-N weight suspended by two wires as shown in the accompanying figure. Find the magnitudes and components of the force vectors $\mathbf{F}_1$ and $\mathbf{F}_2$.

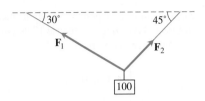

46. Consider a 50-N weight suspended by two wires as shown in the accompanying figure. If the magnitude of vector $\mathbf{F}_1$ is 35 N, find angle α and the magnitude of vector $\mathbf{F}_2$.

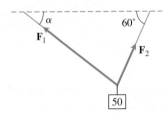

47. Consider a w-N weight suspended by two wires as shown in the accompanying figure. If the magnitude of vector $\mathbf{F}_2$ is 100 N, find w and the magnitude of vector $\mathbf{F}_1$.

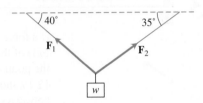

48. Consider a 25-N weight suspended by two wires as shown in the accompanying figure. If the magnitudes of vectors $\mathbf{F}_1$ and $\mathbf{F}_2$ are both 75 N, then angles α and β are equal. Find α.

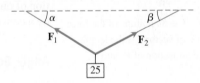

49. Location A bird flies from its nest 5 km in the direction $60°$ north of east, where it stops to rest on a tree. It then flies 10 km in the direction due southeast and lands atop a telephone pole. Place an xy-coordinate system so that the origin is the bird's nest, the x-axis points east, and the y-axis points north.

    a. At what point is the tree located?

    b. At what point is the telephone pole?

50. Use similar triangles to find the coordinates of the point Q that divides the segment from $P_1(x_1, y_1, z_1)$ to $P_2(x_2, y_2, z_2)$ into two lengths whose ratio is $p/q = r$.

51. Medians of a triangle Suppose that A, B, and C are the corner points of the thin triangular plate of constant density shown here.

    a. Find the vector from C to the midpoint M of side AB.

    b. Find the vector from C to the point that lies two-thirds of the way from C to M on the median CM.

c. Find the coordinates of the point in which the medians of ΔABC intersect. According to Exercise 17, Section 6.6, this point is the plate's center of mass.

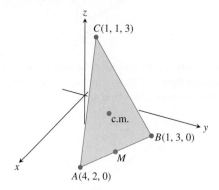

52. Find the vector from the origin to the point of intersection of the medians of the triangle whose vertices are

$$A(1, -1, 2), \quad B(2, 1, 3), \quad \text{and} \quad C(-1, 2, -1).$$

53. Let $ABCD$ be a general, not necessarily planar, quadrilateral in space. Show that the two segments joining the midpoints of opposite sides of $ABCD$ bisect each other. (*Hint:* Show that the segments have the same midpoint.)

54. Vectors are drawn from the center of a regular n-sided polygon in the plane to the vertices of the polygon. Show that the sum of the vectors is zero. (*Hint:* What happens to the sum if you rotate the polygon about its center?)

55. Suppose that A, B, and C are vertices of a triangle and that a, b, and c are, respectively, the midpoints of the opposite sides. Show that $\overrightarrow{Aa} + \overrightarrow{Bb} + \overrightarrow{Cc} = 0$.

56. **Unit vectors in the plane** Show that a unit vector in the plane can be expressed as $\mathbf{u} = (\cos \theta)\mathbf{i} + (\sin \theta)\mathbf{j}$, obtained by rotating $\mathbf{i}$ through an angle θ in the counterclockwise direction. Explain why this form gives *every* unit vector in the plane.

12.3 The Dot Product

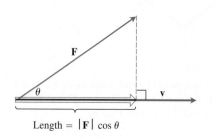

FIGURE 12.19 The magnitude of the force $\mathbf{F}$ in the direction of vector $\mathbf{v}$ is the length $|\mathbf{F}| \cos \theta$ of the projection of $\mathbf{F}$ onto $\mathbf{v}$.

If a force $\mathbf{F}$ is applied to a particle moving along a path, we often need to know the magnitude of the force in the direction of motion. If $\mathbf{v}$ is parallel to the tangent line to the path at the point where $\mathbf{F}$ is applied, then we want the magnitude of $\mathbf{F}$ in the direction of $\mathbf{v}$. Figure 12.19 shows that the scalar quantity we seek is the length $|\mathbf{F}| \cos \theta$, where θ is the angle between the two vectors $\mathbf{F}$ and $\mathbf{v}$.

In this section we show how to calculate easily the angle between two vectors directly from their components. A key part of the calculation is an expression called the *dot product*. Dot products are also called *inner* or *scalar* products because the product results in a scalar, not a vector. After investigating the dot product, we apply it to finding the projection of one vector onto another (as displayed in Figure 12.19) and to finding the work done by a constant force acting through a displacement.

Angle Between Vectors

When two nonzero vectors $\mathbf{u}$ and $\mathbf{v}$ are placed so their initial points coincide, they form an angle θ of measure $0 \le \theta \le \pi$ (Figure 12.20). If the vectors do not lie along the same line, the angle θ is measured in the plane containing both of them. If they do lie along the same line, the angle between them is 0 if they point in the same direction and π if they point in opposite directions. The angle θ is the **angle between u and v**. Theorem 1 gives a formula to determine this angle.

FIGURE 12.20 The angle between $\mathbf{u}$ and $\mathbf{v}$.

> **THEOREM 1—Angle Between Two Vectors** The angle θ between two nonzero vectors $\mathbf{u} = \langle u_1, u_2, u_3 \rangle$ and $\mathbf{v} = \langle v_1, v_2, v_3 \rangle$ is given by
> $$\theta = \cos^{-1}\left(\frac{u_1 v_1 + u_2 v_2 + u_3 v_3}{|\mathbf{u}||\mathbf{v}|}\right).$$

Before proving Theorem 1, we focus attention on the expression $u_1 v_1 + u_2 v_2 + u_3 v_3$ in the calculation for θ. This expression is the sum of the products of the corresponding components for the vectors $\mathbf{u}$ and $\mathbf{v}$.

DEFINITION The **dot product** $\mathbf{u} \cdot \mathbf{v}$ ("**u** dot **v**") of vectors $\mathbf{u} = \langle u_1, u_2, u_3 \rangle$ and $\mathbf{v} = \langle v_1, v_2, v_3 \rangle$ is

$$\mathbf{u} \cdot \mathbf{v} = u_1 v_1 + u_2 v_2 + u_3 v_3.$$

EXAMPLE 1

(a) $\langle 1, -2, -1 \rangle \cdot \langle -6, 2, -3 \rangle = (1)(-6) + (-2)(2) + (-1)(-3)$
$$= -6 - 4 + 3 = -7$$

(b) $\left(\dfrac{1}{2}\mathbf{i} + 3\mathbf{j} + \mathbf{k} \right) \cdot (4\mathbf{i} - \mathbf{j} + 2\mathbf{k}) = \left(\dfrac{1}{2} \right)(4) + (3)(-1) + (1)(2) = 1$ ∎

The dot product of a pair of two-dimensional vectors is defined in a similar fashion:

$$\langle u_1, u_2 \rangle \cdot \langle v_1, v_2 \rangle = u_1 v_1 + u_2 v_2.$$

We will see throughout the remainder of the book that the dot product is a key tool for many important geometric and physical calculations in space (and the plane), not just for finding the angle between two vectors.

Proof of Theorem 1 Applying the law of cosines (Equation (8), Section 1.3) to the triangle in Figure 12.21, we find that

$$|\mathbf{w}|^2 = |\mathbf{u}|^2 + |\mathbf{v}|^2 - 2|\mathbf{u}||\mathbf{v}| \cos \theta \qquad \text{Law of cosines}$$

$$2|\mathbf{u}||\mathbf{v}| \cos \theta = |\mathbf{u}|^2 + |\mathbf{v}|^2 - |\mathbf{w}|^2.$$

Because $\mathbf{w} = \mathbf{u} - \mathbf{v}$, the component form of $\mathbf{w}$ is $\langle u_1 - v_1, u_2 - v_2, u_3 - v_3 \rangle$. So

$$|\mathbf{u}|^2 = \left(\sqrt{u_1^2 + u_2^2 + u_3^2} \right)^2 = u_1^2 + u_2^2 + u_3^2$$

$$|\mathbf{v}|^2 = \left(\sqrt{v_1^2 + v_2^2 + v_3^2} \right)^2 = v_1^2 + v_2^2 + v_3^2$$

$$|\mathbf{w}|^2 = \left(\sqrt{(u_1 - v_1)^2 + (u_2 - v_2)^2 + (u_3 - v_3)^2} \right)^2$$

$$= (u_1 - v_1)^2 + (u_2 - v_2)^2 + (u_3 - v_3)^2$$

$$= u_1^2 - 2u_1 v_1 + v_1^2 + u_2^2 - 2u_2 v_2 + v_2^2 + u_3^2 - 2u_3 v_3 + v_3^2$$

and

$$|\mathbf{u}|^2 + |\mathbf{v}|^2 - |\mathbf{w}|^2 = 2(u_1 v_1 + u_2 v_2 + u_3 v_3).$$

Therefore,

$$2|\mathbf{u}||\mathbf{v}| \cos \theta = |\mathbf{u}|^2 + |\mathbf{v}|^2 - |\mathbf{w}|^2 = 2(u_1 v_1 + u_2 v_2 + u_3 v_3)$$

$$|\mathbf{u}||\mathbf{v}| \cos \theta = u_1 v_1 + u_2 v_2 + u_3 v_3$$

$$\cos \theta = \frac{u_1 v_1 + u_2 v_2 + u_3 v_3}{|\mathbf{u}||\mathbf{v}|}.$$

Since $0 \leq \theta < \pi$, we have

$$\theta = \cos^{-1} \left(\frac{u_1 v_1 + u_2 v_2 + u_3 v_3}{|\mathbf{u}||\mathbf{v}|} \right).$$ ∎

In the notation of the dot product, the angle between two vectors $\mathbf{u}$ and $\mathbf{v}$ is

$$\theta = \cos^{-1} \left(\frac{\mathbf{u} \cdot \mathbf{v}}{|\mathbf{u}||\mathbf{v}|} \right).$$

FIGURE 12.21 The parallelogram law of addition of vectors gives $\mathbf{w} = \mathbf{u} - \mathbf{v}$.

EXAMPLE 2 Find the angle between $\mathbf{u} = \mathbf{i} - 2\mathbf{j} - 2\mathbf{k}$ and $\mathbf{v} = 6\mathbf{i} + 3\mathbf{j} + 2\mathbf{k}$.

Solution We use the formula above:

$$\mathbf{u} \cdot \mathbf{v} = (1)(6) + (-2)(3) + (-2)(2) = 6 - 6 - 4 = -4$$

$$|\mathbf{u}| = \sqrt{(1)^2 + (-2)^2 + (-2)^2} = \sqrt{9} = 3$$

$$|\mathbf{v}| = \sqrt{(6)^2 + (3)^2 + (2)^2} = \sqrt{49} = 7$$

$$\theta = \cos^{-1}\left(\frac{\mathbf{u} \cdot \mathbf{v}}{|\mathbf{u}||\mathbf{v}|}\right) = \cos^{-1}\left(\frac{-4}{(3)(7)}\right) \approx 1.76 \text{ radians}. \qquad \blacksquare$$

The angle formula applies to two-dimensional vectors as well.

EXAMPLE 3 Find the angle θ in the triangle ABC determined by the vertices $A = (0, 0)$, $B = (3, 5)$, and $C = (5, 2)$ (Figure 12.22).

Solution The angle θ is the angle between the vectors $\overrightarrow{CA}$ and $\overrightarrow{CB}$. The component forms of these two vectors are

$$\overrightarrow{CA} = \langle -5, -2 \rangle \qquad \text{and} \qquad \overrightarrow{CB} = \langle -2, 3 \rangle.$$

First we calculate the dot product and magnitudes of these two vectors.

$$\overrightarrow{CA} \cdot \overrightarrow{CB} = (-5)(-2) + (-2)(3) = 4$$

$$|\overrightarrow{CA}| = \sqrt{(-5)^2 + (-2)^2} = \sqrt{29}$$

$$|\overrightarrow{CB}| = \sqrt{(-2)^2 + (3)^2} = \sqrt{13}$$

Then applying the angle formula, we have

$$\theta = \cos^{-1}\left(\frac{\overrightarrow{CA} \cdot \overrightarrow{CB}}{|\overrightarrow{CA}||\overrightarrow{CB}|}\right)$$

$$= \cos^{-1}\left(\frac{4}{(\sqrt{29})(\sqrt{13})}\right)$$

$$\approx 78.1° \qquad \text{or} \qquad 1.36 \text{ radians}. \qquad \blacksquare$$

FIGURE 12.22 The triangle in Example 3.

Perpendicular (Orthogonal) Vectors

Two nonzero vectors $\mathbf{u}$ and $\mathbf{v}$ are perpendicular or **orthogonal** if the angle between them is $\pi/2$. For such vectors, we have $\mathbf{u} \cdot \mathbf{v} = 0$ because $\cos(\pi/2) = 0$. The converse is also true. If $\mathbf{u}$ and $\mathbf{v}$ are nonzero vectors with $\mathbf{u} \cdot \mathbf{v} = |\mathbf{u}||\mathbf{v}| \cos\theta = 0$, then $\cos\theta = 0$ and $\theta = \cos^{-1} 0 = \pi/2$.

DEFINITION Vectors $\mathbf{u}$ and $\mathbf{v}$ are **orthogonal** (or **perpendicular**) if and only if $\mathbf{u} \cdot \mathbf{v} = 0$.

EXAMPLE 4 To determine if two vectors are orthogonal, calculate their dot product.

(a) $\mathbf{u} = \langle 3, -2 \rangle$ and $\mathbf{v} = \langle 4, 6 \rangle$ are orthogonal because $\mathbf{u} \cdot \mathbf{v} = (3)(4) + (-2)(6) = 0$.

(b) $\mathbf{u} = 3\mathbf{i} - 2\mathbf{j} + \mathbf{k}$ and $\mathbf{v} = 2\mathbf{j} + 4\mathbf{k}$ are orthogonal because $\mathbf{u} \cdot \mathbf{v} = (3)(0) + (-2)(2) + (1)(4) = 0$.

(c) 0 is orthogonal to every vector **u** since

$$\mathbf{0} \cdot \mathbf{u} = \langle 0, 0, 0 \rangle \cdot \langle u_1, u_2, u_3 \rangle$$
$$= (0)(u_1) + (0)(u_2) + (0)(u_3)$$
$$= 0.$$

∎

Dot Product Properties and Vector Projections

The dot product obeys many of the laws that hold for ordinary products of real numbers (scalars).

Properties of the Dot Product

If **u**, **v**, and **w** are any vectors and c is a scalar, then

1. $\mathbf{u} \cdot \mathbf{v} = \mathbf{v} \cdot \mathbf{u}$
2. $(c\mathbf{u}) \cdot \mathbf{v} = \mathbf{u} \cdot (c\mathbf{v}) = c(\mathbf{u} \cdot \mathbf{v})$
3. $\mathbf{u} \cdot (\mathbf{v} + \mathbf{w}) = \mathbf{u} \cdot \mathbf{v} + \mathbf{u} \cdot \mathbf{w}$
4. $\mathbf{u} \cdot \mathbf{u} = |\mathbf{u}|^2$
5. $\mathbf{0} \cdot \mathbf{u} = 0.$

HISTORICAL BIOGRAPHY

Carl Friedrich Gauss
(1777–1855)

Proofs of Properties 1 and 3   The properties are easy to prove using the definition. For instance, here are the proofs of Properties 1 and 3.

1. $\mathbf{u} \cdot \mathbf{v} = u_1 v_1 + u_2 v_2 + u_3 v_3 = v_1 u_1 + v_2 u_2 + v_3 u_3 = \mathbf{v} \cdot \mathbf{u}$

3. $\mathbf{u} \cdot (\mathbf{v} + \mathbf{w}) = \langle u_1, u_2, u_3 \rangle \cdot \langle v_1 + w_1, v_2 + w_2, v_3 + w_3 \rangle$

$$= u_1(v_1 + w_1) + u_2(v_2 + w_2) + u_3(v_3 + w_3)$$

$$= u_1 v_1 + u_1 w_1 + u_2 v_2 + u_2 w_2 + u_3 v_3 + u_3 w_3$$

$$= (u_1 v_1 + u_2 v_2 + u_3 v_3) + (u_1 w_1 + u_2 w_2 + u_3 w_3)$$

$$= \mathbf{u} \cdot \mathbf{v} + \mathbf{u} \cdot \mathbf{w}$$

∎

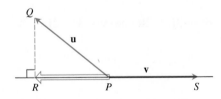

FIGURE 12.23 The vector projection of **u** onto **v**.

We now return to the problem of projecting one vector onto another, posed in the opening to this section. The **vector projection** of $\mathbf{u} = \overrightarrow{PQ}$ onto a nonzero vector $\mathbf{v} = \overrightarrow{PS}$ (Figure 12.23) is the vector $\overrightarrow{PR}$ determined by dropping a perpendicular from Q to the line PS. The notation for this vector is

$$\text{proj}_\mathbf{v}\, \mathbf{u} \qquad (\text{"the vector projection of } \mathbf{u} \text{ onto } \mathbf{v}\text{"}).$$

If **u** represents a force, then $\text{proj}_\mathbf{v}\, \mathbf{u}$ represents the effective force in the direction of **v** (Figure 12.24).

If the angle θ between **u** and **v** is acute, $\text{proj}_\mathbf{v}\, \mathbf{u}$ has length $|\mathbf{u}| \cos \theta$ and direction $\mathbf{v}/|\mathbf{v}|$ (Figure 12.25). If θ is obtuse, $\cos \theta < 0$ and $\text{proj}_\mathbf{v}\, \mathbf{u}$ has length $-|\mathbf{u}| \cos \theta$ and direction $-\mathbf{v}/|\mathbf{v}|$. In both cases,

$$\text{proj}_\mathbf{v}\, \mathbf{u} = (|\mathbf{u}| \cos \theta) \frac{\mathbf{v}}{|\mathbf{v}|}$$

$$= \left(\frac{\mathbf{u} \cdot \mathbf{v}}{|\mathbf{v}|} \right) \frac{\mathbf{v}}{|\mathbf{v}|} \qquad |\mathbf{u}| \cos \theta = \frac{|\mathbf{u}||\mathbf{v}| \cos \theta}{|\mathbf{v}|} = \frac{\mathbf{u} \cdot \mathbf{v}}{|\mathbf{v}|}$$

$$= \left(\frac{\mathbf{u} \cdot \mathbf{v}}{|\mathbf{v}|^2} \right) \mathbf{v}.$$

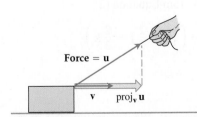

FIGURE 12.24 If we pull on the box with force **u**, the effective force moving the box forward in the direction **v** is the projection of **u** onto **v**.

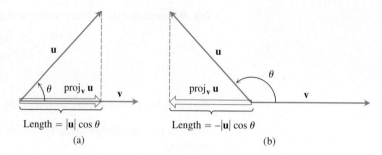

FIGURE 12.25 The length of $\text{proj}_\mathbf{v}\,\mathbf{u}$ is (a) $|\mathbf{u}|\cos\theta$ if $\cos\theta \geq 0$ and (b) $-|\mathbf{u}|\cos\theta$ if $\cos\theta < 0$.

The number $|\mathbf{u}|\cos\theta$ is called the **scalar component of u in the direction of v** (or of **u** **onto v**). To summarize,

---

The vector projection of **u** onto **v** is the vector

$$\text{proj}_\mathbf{v}\,\mathbf{u} = \left(\frac{\mathbf{u}\cdot\mathbf{v}}{|\mathbf{v}|^2}\right)\mathbf{v}. \tag{1}$$

The scalar component of **u** in the direction of **v** is the scalar

$$|\mathbf{u}|\cos\theta = \frac{\mathbf{u}\cdot\mathbf{v}}{|\mathbf{v}|} = \mathbf{u}\cdot\frac{\mathbf{v}}{|\mathbf{v}|}. \tag{2}$$

Note that both the vector projection of **u** onto **v** and the scalar component of **u** onto **v** depend only on the direction of the vector **v** and not its length (because we dot **u** with $\mathbf{v}/|\mathbf{v}|$, which is the direction of **v**).

EXAMPLE 5 Find the vector projection of $\mathbf{u} = 6\mathbf{i} + 3\mathbf{j} + 2\mathbf{k}$ onto $\mathbf{v} = \mathbf{i} - 2\mathbf{j} - 2\mathbf{k}$ and the scalar component of **u** in the direction of **v**.

Solution We find $\text{proj}_\mathbf{v}\,\mathbf{u}$ from Equation (1):

$$\text{proj}_\mathbf{v}\,\mathbf{u} = \frac{\mathbf{u}\cdot\mathbf{v}}{\mathbf{v}\cdot\mathbf{v}}\mathbf{v} = \frac{6 - 6 - 4}{1 + 4 + 4}(\mathbf{i} - 2\mathbf{j} - 2\mathbf{k})$$

$$= -\frac{4}{9}(\mathbf{i} - 2\mathbf{j} - 2\mathbf{k}) = -\frac{4}{9}\mathbf{i} + \frac{8}{9}\mathbf{j} + \frac{8}{9}\mathbf{k}.$$

We find the scalar component of **u** in the direction of **v** from Equation (2):

$$|\mathbf{u}|\cos\theta = \mathbf{u}\cdot\frac{\mathbf{v}}{|\mathbf{v}|} = (6\mathbf{i} + 3\mathbf{j} + 2\mathbf{k})\cdot\left(\frac{1}{3}\mathbf{i} - \frac{2}{3}\mathbf{j} - \frac{2}{3}\mathbf{k}\right)$$

$$= 2 - 2 - \frac{4}{3} = -\frac{4}{3}. \qquad\blacksquare$$

Equations (1) and (2) also apply to two-dimensional vectors. We demonstrate this in the next example.

EXAMPLE 6 Find the vector projection of a force $\mathbf{F} = 5\mathbf{i} + 2\mathbf{j}$ onto $\mathbf{v} = \mathbf{i} - 3\mathbf{j}$ and the scalar component of **F** in the direction of **v**.

FIGURE 12.28 The construction of $\mathbf{v} \times \mathbf{u}$.

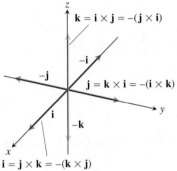

FIGURE 12.29 The pairwise cross products of $\mathbf{i}$, $\mathbf{j}$, and $\mathbf{k}$.

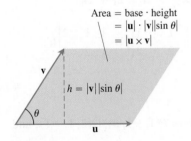

FIGURE 12.30 The parallelogram determined by $\mathbf{u}$ and $\mathbf{v}$.

To visualize Property 3, for example, notice that when the fingers of your right hand curl through the angle θ from $\mathbf{v}$ to $\mathbf{u}$, your thumb points the opposite way; the unit vector we choose in forming $\mathbf{v} \times \mathbf{u}$ is the negative of the one we choose in forming $\mathbf{u} \times \mathbf{v}$ (Figure 12.28).

Property 1 can be verified by applying the definition of cross product to both sides of the equation and comparing the results. Property 2 is proved in Appendix 8. Property 4 follows by multiplying both sides of the equation in Property 2 by -1 and reversing the order of the products using Property 3. Property 5 is a definition. As a rule, cross product multiplication is *not associative* so $(\mathbf{u} \times \mathbf{v}) \times \mathbf{w}$ does not generally equal $\mathbf{u} \times (\mathbf{v} \times \mathbf{w})$. (See Additional Exercise 17.)

When we apply the definition to calculate the pairwise cross products of $\mathbf{i}$, $\mathbf{j}$, and $\mathbf{k}$, we find (Figure 12.29)

$$\mathbf{i} \times \mathbf{j} = -(\mathbf{j} \times \mathbf{i}) = \mathbf{k}$$

$$\mathbf{j} \times \mathbf{k} = -(\mathbf{k} \times \mathbf{j}) = \mathbf{i}$$

$$\mathbf{k} \times \mathbf{i} = -(\mathbf{i} \times \mathbf{k}) = \mathbf{j}$$

Diagram for recalling these products

and

$$\mathbf{i} \times \mathbf{i} = \mathbf{j} \times \mathbf{j} = \mathbf{k} \times \mathbf{k} = \mathbf{0}.$$

$|\mathbf{u} \times \mathbf{v}|$ Is the Area of a Parallelogram

Because $\mathbf{n}$ is a unit vector, the magnitude of $\mathbf{u} \times \mathbf{v}$ is

$$|\mathbf{u} \times \mathbf{v}| = |\mathbf{u}||\mathbf{v}| \,|\sin \theta||\mathbf{n}| = |\mathbf{u}||\mathbf{v}| \sin \theta.$$

This is the area of the parallelogram determined by $\mathbf{u}$ and $\mathbf{v}$ (Figure 12.30), $|\mathbf{u}|$ being the base of the parallelogram and $|\mathbf{v}||\sin \theta|$ the height.

Determinant Formula for $\mathbf{u} \times \mathbf{v}$

Our next objective is to calculate $\mathbf{u} \times \mathbf{v}$ from the components of $\mathbf{u}$ and $\mathbf{v}$ relative to a Cartesian coordinate system.

Suppose that

$$\mathbf{u} = u_1\mathbf{i} + u_2\mathbf{j} + u_3\mathbf{k} \quad \text{and} \quad \mathbf{v} = v_1\mathbf{i} + v_2\mathbf{j} + v_3\mathbf{k}.$$

Then the distributive laws and the rules for multiplying $\mathbf{i}$, $\mathbf{j}$, and $\mathbf{k}$ tell us that

$$\begin{aligned}
\mathbf{u} \times \mathbf{v} &= (u_1\mathbf{i} + u_2\mathbf{j} + u_3\mathbf{k}) \times (v_1\mathbf{i} + v_2\mathbf{j} + v_3\mathbf{k}) \\
&= u_1v_1\mathbf{i} \times \mathbf{i} + u_1v_2\mathbf{i} \times \mathbf{j} + u_1v_3\mathbf{i} \times \mathbf{k} \\
&\quad + u_2v_1\mathbf{j} \times \mathbf{i} + u_2v_2\mathbf{j} \times \mathbf{j} + u_2v_3\mathbf{j} \times \mathbf{k} \\
&\quad + u_3v_1\mathbf{k} \times \mathbf{i} + u_3v_2\mathbf{k} \times \mathbf{j} + u_3v_3\mathbf{k} \times \mathbf{k} \\
&= (u_2v_3 - u_3v_2)\mathbf{i} - (u_1v_3 - u_3v_1)\mathbf{j} + (u_1v_2 - u_2v_1)\mathbf{k}.
\end{aligned}$$

The component terms in the last line are hard to remember, but they are the same as the terms in the expansion of the symbolic determinant

$$\begin{vmatrix} \mathbf{i} & \mathbf{j} & \mathbf{k} \\ u_1 & u_2 & u_3 \\ v_1 & v_2 & v_3 \end{vmatrix}.$$

Determinants

2×2 and 3×3 determinants are evaluated as follows:

$$\begin{vmatrix} a & b \\ c & d \end{vmatrix} = ad - bc$$

EXAMPLE

$$\begin{vmatrix} 2 & 1 \\ -4 & 3 \end{vmatrix} = (2)(3) - (1)(-4)$$

$$= 6 + 4 = 10$$

$$\begin{vmatrix} a_1 & a_2 & a_3 \\ b_1 & b_2 & b_3 \\ c_1 & c_2 & c_3 \end{vmatrix} = a_1 \begin{vmatrix} b_2 & b_3 \\ c_2 & c_3 \end{vmatrix}$$

$$- a_2 \begin{vmatrix} b_1 & b_3 \\ c_1 & c_3 \end{vmatrix} + a_3 \begin{vmatrix} b_1 & b_2 \\ c_1 & c_2 \end{vmatrix}$$

EXAMPLE

$$\begin{vmatrix} -5 & 3 & 1 \\ 2 & 1 & 1 \\ -4 & 3 & 1 \end{vmatrix} = (-5) \begin{vmatrix} 1 & 1 \\ 3 & 1 \end{vmatrix}$$

$$- (3) \begin{vmatrix} 2 & 1 \\ -4 & 1 \end{vmatrix} + (1) \begin{vmatrix} 2 & 1 \\ -4 & 3 \end{vmatrix}$$

$$= -5(1 - 3) - 3(2 + 4)$$

$$+ 1(6 + 4)$$

$$= 10 - 18 + 10 = 2$$

(For more information, see the Web site at **www.aw.com/thomas**.)

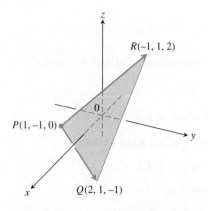

FIGURE 12.31 The vector $\overrightarrow{PQ} \times \overrightarrow{PR}$ is perpendicular to the plane of triangle PQR (Example 2). The area of triangle PQR is half of $|\overrightarrow{PQ} \times \overrightarrow{PR}|$ (Example 3).

So we restate the calculation in this easy-to-remember form.

Calculating the Cross Product as a Determinant

If $\mathbf{u} = u_1\mathbf{i} + u_2\mathbf{j} + u_3\mathbf{k}$ and $\mathbf{v} = v_1\mathbf{i} + v_2\mathbf{j} + v_3\mathbf{k}$, then

$$\mathbf{u} \times \mathbf{v} = \begin{vmatrix} \mathbf{i} & \mathbf{j} & \mathbf{k} \\ u_1 & u_2 & u_3 \\ v_1 & v_2 & v_3 \end{vmatrix}.$$

EXAMPLE 1   Find $\mathbf{u} \times \mathbf{v}$ and $\mathbf{v} \times \mathbf{u}$ if $\mathbf{u} = 2\mathbf{i} + \mathbf{j} + \mathbf{k}$ and $\mathbf{v} = -4\mathbf{i} + 3\mathbf{j} + \mathbf{k}$.

Solution

$$\mathbf{u} \times \mathbf{v} = \begin{vmatrix} \mathbf{i} & \mathbf{j} & \mathbf{k} \\ 2 & 1 & 1 \\ -4 & 3 & 1 \end{vmatrix} = \begin{vmatrix} 1 & 1 \\ 3 & 1 \end{vmatrix}\mathbf{i} - \begin{vmatrix} 2 & 1 \\ -4 & 1 \end{vmatrix}\mathbf{j} + \begin{vmatrix} 2 & 1 \\ -4 & 3 \end{vmatrix}\mathbf{k}$$

$$= -2\mathbf{i} - 6\mathbf{j} + 10\mathbf{k}$$

$$\mathbf{v} \times \mathbf{u} = -(\mathbf{u} \times \mathbf{v}) = 2\mathbf{i} + 6\mathbf{j} - 10\mathbf{k} \qquad \blacksquare$$

EXAMPLE 2   Find a vector perpendicular to the plane of $P(1, -1, 0)$, $Q(2, 1, -1)$, and $R(-1, 1, 2)$ (Figure 12.31).

Solution   The vector $\overrightarrow{PQ} \times \overrightarrow{PR}$ is perpendicular to the plane because it is perpendicular to both vectors. In terms of components,

$$\overrightarrow{PQ} = (2 - 1)\mathbf{i} + (1 + 1)\mathbf{j} + (-1 - 0)\mathbf{k} = \mathbf{i} + 2\mathbf{j} - \mathbf{k}$$

$$\overrightarrow{PR} = (-1 - 1)\mathbf{i} + (1 + 1)\mathbf{j} + (2 - 0)\mathbf{k} = -2\mathbf{i} + 2\mathbf{j} + 2\mathbf{k}$$

$$\overrightarrow{PQ} \times \overrightarrow{PR} = \begin{vmatrix} \mathbf{i} & \mathbf{j} & \mathbf{k} \\ 1 & 2 & -1 \\ -2 & 2 & 2 \end{vmatrix} = \begin{vmatrix} 2 & -1 \\ 2 & 2 \end{vmatrix}\mathbf{i} - \begin{vmatrix} 1 & -1 \\ -2 & 2 \end{vmatrix}\mathbf{j} + \begin{vmatrix} 1 & 2 \\ -2 & 2 \end{vmatrix}\mathbf{k}$$

$$= 6\mathbf{i} + 6\mathbf{k}. \qquad \blacksquare$$

EXAMPLE 3   Find the area of the triangle with vertices $P(1, -1, 0)$, $Q(2, 1, -1)$, and $R(-1, 1, 2)$ (Figure 12.31).

Solution   The area of the parallelogram determined by P, Q, and R is

$$|\overrightarrow{PQ} \times \overrightarrow{PR}| = |6\mathbf{i} + 6\mathbf{k}| \qquad \text{Values from Example 2}$$

$$= \sqrt{(6)^2 + (6)^2} = \sqrt{2 \cdot 36} = 6\sqrt{2}.$$

The triangle's area is half of this, or $3\sqrt{2}$. $\qquad \blacksquare$

EXAMPLE 4   Find a unit vector perpendicular to the plane of $P(1, -1, 0)$, $Q(2, 1, -1)$, and $R(-1, 1, 2)$.

Solution   Since $\overrightarrow{PQ} \times \overrightarrow{PR}$ is perpendicular to the plane, its direction $\mathbf{n}$ is a unit vector perpendicular to the plane. Taking values from Examples 2 and 3, we have

$$\mathbf{n} = \frac{\overrightarrow{PQ} \times \overrightarrow{PR}}{|\overrightarrow{PQ} \times \overrightarrow{PR}|} = \frac{6\mathbf{i} + 6\mathbf{k}}{6\sqrt{2}} = \frac{1}{\sqrt{2}}\mathbf{i} + \frac{1}{\sqrt{2}}\mathbf{k}. \qquad \blacksquare$$

DEFINITIONS If **r** is the position vector of a particle moving along a smooth curve in space, then

$$\mathbf{v}(t) = \frac{d\mathbf{r}}{dt}$$

is the particle's **velocity vector**, tangent to the curve. At any time t, the direction of **v** is the **direction of motion**, the magnitude of **v** is the particle's **speed**, and the derivative $\mathbf{a} = d\mathbf{v}/dt$, when it exists, is the particle's **acceleration vector**. In summary,

1. Velocity is the derivative of position: $\mathbf{v} = \dfrac{d\mathbf{r}}{dt}$.

2. Speed is the magnitude of velocity: Speed $= |\mathbf{v}|$.

3. Acceleration is the derivative of velocity: $\mathbf{a} = \dfrac{d\mathbf{v}}{dt} = \dfrac{d^2\mathbf{r}}{dt^2}$.

4. The unit vector $\mathbf{v}/|\mathbf{v}|$ is the direction of motion at time t.

EXAMPLE 4 Find the velocity, speed, and acceleration of a particle whose motion in space is given by the position vector $\mathbf{r}(t) = 2 \cos t\, \mathbf{i} + 2 \sin t\, \mathbf{j} + 5 \cos^2 t\, \mathbf{k}$. Sketch the velocity vector $\mathbf{v}(7\pi/4)$.

Solution The velocity and acceleration vectors at time t are

$$\mathbf{v}(t) = \mathbf{r}'(t) = -2 \sin t\, \mathbf{i} + 2 \cos t\, \mathbf{j} - 10 \cos t \sin t\, \mathbf{k}$$
$$= -2 \sin t\, \mathbf{i} + 2 \cos t\, \mathbf{j} - 5 \sin 2t\, \mathbf{k},$$
$$\mathbf{a}(t) = \mathbf{r}''(t) = -2 \cos t\, \mathbf{i} - 2 \sin t\, \mathbf{j} - 10 \cos 2t\, \mathbf{k},$$

and the speed is

$$|\mathbf{v}(t)| = \sqrt{(-2 \sin t)^2 + (2 \cos t)^2 + (-5 \sin 2t)^2} = \sqrt{4 + 25 \sin^2 2t}.$$

When $t = 7\pi/4$, we have

$$\mathbf{v}\left(\frac{7\pi}{4}\right) = \sqrt{2}\, \mathbf{i} + \sqrt{2}\, \mathbf{j} + 5\, \mathbf{k}, \qquad \mathbf{a}\left(\frac{7\pi}{4}\right) = -\sqrt{2}\, \mathbf{i} + \sqrt{2}\, \mathbf{j}, \qquad \left|\mathbf{v}\left(\frac{7\pi}{4}\right)\right| = \sqrt{29}.$$

A sketch of the curve of motion, and the velocity vector when $t = 7\pi/4$, can be seen in Figure 13.7. ∎

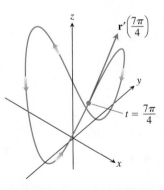

FIGURE 13.7 The curve and the velocity vector when $t = 7\pi/4$ for the motion given in Example 4.

We can express the velocity of a moving particle as the product of its speed and direction:

$$\text{Velocity} = |\mathbf{v}|\left(\frac{\mathbf{v}}{|\mathbf{v}|}\right) = (\text{speed})(\text{direction}).$$

Differentiation Rules

Because the derivatives of vector functions may be computed component by component, the rules for differentiating vector functions have the same form as the rules for differentiating scalar functions.

Differentiation Rules for Vector Functions

Let $\mathbf{u}$ and $\mathbf{v}$ be differentiable vector functions of t, $\mathbf{C}$ a constant vector, c any scalar, and f any differentiable scalar function.

1. *Constant Function Rule:* $\quad \dfrac{d}{dt}\mathbf{C} = \mathbf{0}$

2. *Scalar Multiple Rules:* $\quad \dfrac{d}{dt}[c\mathbf{u}(t)] = c\mathbf{u}'(t)$

   $\qquad\qquad\qquad\qquad\quad \dfrac{d}{dt}[f(t)\mathbf{u}(t)] = f'(t)\mathbf{u}(t) + f(t)\mathbf{u}'(t)$

3. *Sum Rule:* $\quad \dfrac{d}{dt}[\mathbf{u}(t) + \mathbf{v}(t)] = \mathbf{u}'(t) + \mathbf{v}'(t)$

4. *Difference Rule:* $\quad \dfrac{d}{dt}[\mathbf{u}(t) - \mathbf{v}(t)] = \mathbf{u}'(t) - \mathbf{v}'(t)$

5. *Dot Product Rule:* $\quad \dfrac{d}{dt}[\mathbf{u}(t) \cdot \mathbf{v}(t)] = \mathbf{u}'(t) \cdot \mathbf{v}(t) + \mathbf{u}(t) \cdot \mathbf{v}'(t)$

6. *Cross Product Rule:* $\quad \dfrac{d}{dt}[\mathbf{u}(t) \times \mathbf{v}(t)] = \mathbf{u}'(t) \times \mathbf{v}(t) + \mathbf{u}(t) \times \mathbf{v}'(t)$

7. *Chain Rule:* $\quad \dfrac{d}{dt}[\mathbf{u}(f(t))] = f'(t)\mathbf{u}'(f(t))$

When you use the Cross Product Rule, remember to preserve the order of the factors. If $\mathbf{u}$ comes first on the left side of the equation, it must also come first on the right or the signs will be wrong.

We will prove the product rules and Chain Rule but leave the rules for constants, scalar multiples, sums, and differences as exercises.

Proof of the Dot Product Rule   Suppose that

$$\mathbf{u} = u_1(t)\mathbf{i} + u_2(t)\mathbf{j} + u_3(t)\mathbf{k}$$

and

$$\mathbf{v} = v_1(t)\mathbf{i} + v_2(t)\mathbf{j} + v_3(t)\mathbf{k}.$$

Then

$$\frac{d}{dt}(\mathbf{u} \cdot \mathbf{v}) = \frac{d}{dt}(u_1 v_1 + u_2 v_2 + u_3 v_3)$$

$$= \underbrace{u_1' v_1 + u_2' v_2 + u_3' v_3}_{\mathbf{u}' \cdot \mathbf{v}} + \underbrace{u_1 v_1' + u_2 v_2' + u_3 v_3'}_{\mathbf{u} \cdot \mathbf{v}'}. \qquad \blacksquare$$

Proof of the Cross Product Rule   We model the proof after the proof of the Product Rule for scalar functions. According to the definition of derivative,

$$\frac{d}{dt}(\mathbf{u} \times \mathbf{v}) = \lim_{h \to 0} \frac{\mathbf{u}(t + h) \times \mathbf{v}(t + h) - \mathbf{u}(t) \times \mathbf{v}(t)}{h}.$$

To change this fraction into an equivalent one that contains the difference quotients for the derivatives of $\mathbf{u}$ and $\mathbf{v}$, we subtract and add $\mathbf{u}(t) \times \mathbf{v}(t + h)$ in the numerator. Then

$$\frac{d}{dt}(\mathbf{u} \times \mathbf{v})$$

$$= \lim_{h \to 0} \frac{\mathbf{u}(t + h) \times \mathbf{v}(t + h) - \mathbf{u}(t) \times \mathbf{v}(t + h) + \mathbf{u}(t) \times \mathbf{v}(t + h) - \mathbf{u}(t) \times \mathbf{v}(t)}{h}$$

$$= \lim_{h \to 0} \left[\frac{\mathbf{u}(t + h) - \mathbf{u}(t)}{h} \times \mathbf{v}(t + h) + \mathbf{u}(t) \times \frac{\mathbf{v}(t + h) - \mathbf{v}(t)}{h} \right]$$

$$= \lim_{h \to 0} \frac{\mathbf{u}(t + h) - \mathbf{u}(t)}{h} \times \lim_{h \to 0} \mathbf{v}(t + h) + \lim_{h \to 0} \mathbf{u}(t) \times \lim_{h \to 0} \frac{\mathbf{v}(t + h) - \mathbf{v}(t)}{h}.$$

The last of these equalities holds because the limit of the cross product of two vector functions is the cross product of their limits if the latter exist (Exercise 32). As h approaches zero, $\mathbf{v}(t + h)$ approaches $\mathbf{v}(t)$ because $\mathbf{v}$, being differentiable at t, is continuous at t (Exercise 33). The two fractions approach the values of $d\mathbf{u}/dt$ and $d\mathbf{v}/dt$ at t. In short,

$$\frac{d}{dt}(\mathbf{u} \times \mathbf{v}) = \frac{d\mathbf{u}}{dt} \times \mathbf{v} + \mathbf{u} \times \frac{d\mathbf{v}}{dt}. \qquad \blacksquare$$

Proof of the Chain Rule   Suppose that $\mathbf{u}(s) = a(s)\mathbf{i} + b(s)\mathbf{j} + c(s)\mathbf{k}$ is a differentiable vector function of s and that $s = f(t)$ is a differentiable scalar function of t. Then a, b, and c are differentiable functions of t, and the Chain Rule for differentiable real-valued functions gives

As an algebraic convenience, we sometimes write the product of a scalar c and a vector $\mathbf{v}$ as $\mathbf{v}c$ instead of $c\mathbf{v}$. This permits us, for instance, to write the Chain Rule in a familiar form:

$$\frac{d\mathbf{u}}{dt} = \frac{d\mathbf{u}}{ds}\frac{ds}{dt},$$

where $s = f(t)$.

$$\frac{d}{dt}[\mathbf{u}(s)] = \frac{da}{dt}\mathbf{i} + \frac{db}{dt}\mathbf{j} + \frac{dc}{dt}\mathbf{k}$$

$$= \frac{da}{ds}\frac{ds}{dt}\mathbf{i} + \frac{db}{ds}\frac{ds}{dt}\mathbf{j} + \frac{dc}{ds}\frac{ds}{dt}\mathbf{k}$$

$$= \frac{ds}{dt}\left(\frac{da}{ds}\mathbf{i} + \frac{db}{ds}\mathbf{j} + \frac{dc}{ds}\mathbf{k}\right)$$

$$= \frac{ds}{dt}\frac{d\mathbf{u}}{ds}$$

$$= f'(t)\mathbf{u}'(f(t)). \qquad \qquad s = f(t) \qquad \blacksquare$$

Vector Functions of Constant Length

When we track a particle moving on a sphere centered at the origin (Figure 13.8), the position vector has a constant length equal to the radius of the sphere. The velocity vector $d\mathbf{r}/dt$, tangent to the path of motion, is tangent to the sphere and hence perpendicular to $\mathbf{r}$. This is always the case for a differentiable vector function of constant length: The vector and its first derivative are orthogonal. By direct calculation,

$$\mathbf{r}(t) \cdot \mathbf{r}(t) = c^2 \qquad |\mathbf{r}(t)| = c \text{ is constant.}$$

$$\frac{d}{dt}[\mathbf{r}(t) \cdot \mathbf{r}(t)] = 0 \qquad \text{Differentiate both sides.}$$

$$\mathbf{r}'(t) \cdot \mathbf{r}(t) + \mathbf{r}(t) \cdot \mathbf{r}'(t) = 0 \qquad \text{Rule 5 with } \mathbf{r}(t) = \mathbf{u}(t) = \mathbf{v}(t)$$

$$2\mathbf{r}'(t) \cdot \mathbf{r}(t) = 0.$$

The vectors $\mathbf{r}'(t)$ and $\mathbf{r}(t)$ are orthogonal because their dot product is 0. In summary,

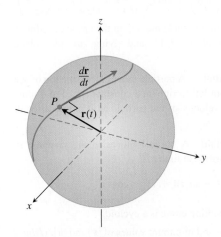

FIGURE 13.8   If a particle moves on a sphere in such a way that its position $\mathbf{r}$ is a differentiable function of time, then $\mathbf{r} \cdot (d\mathbf{r}/dt) = 0$.

> If $\mathbf{r}$ is a differentiable vector function of t of constant length, then
>
> $$\mathbf{r} \cdot \frac{d\mathbf{r}}{dt} = 0. \qquad (4)$$

We will use this observation repeatedly in Section 13.4. The converse is also true (see Exercise 27).

Exercises 13.1

Motion in the Plane

In Exercises 1–4, $\mathbf{r}(t)$ is the position of a particle in the xy-plane at time t. Find an equation in x and y whose graph is the path of the particle. Then find the particle's velocity and acceleration vectors at the given value of t.

1. $\mathbf{r}(t) = (t + 1)\mathbf{i} + (t^2 - 1)\mathbf{j}, \quad t = 1$

2. $\mathbf{r}(t) = \dfrac{t}{t + 1}\mathbf{i} + \dfrac{1}{t}\mathbf{j}, \quad t = -1/2$

3. $\mathbf{r}(t) = e^t\mathbf{i} + \dfrac{2}{9}e^{2t}\mathbf{j}, \quad t = \ln 3$

4. $\mathbf{r}(t) = (\cos 2t)\mathbf{i} + (3 \sin 2t)\mathbf{j}, \quad t = 0$

Exercises 5–8 give the position vectors of particles moving along various curves in the xy-plane. In each case, find the particle's velocity and acceleration vectors at the stated times and sketch them as vectors on the curve.

5. Motion on the circle $x^2 + y^2 = 1$

$$\mathbf{r}(t) = (\sin t)\mathbf{i} + (\cos t)\mathbf{j}; \quad t = \pi/4 \text{ and } \pi/2$$

6. Motion on the circle $x^2 + y^2 = 16$

$$\mathbf{r}(t) = \left(4\cos\frac{t}{2}\right)\mathbf{i} + \left(4\sin\frac{t}{2}\right)\mathbf{j}; \quad t = \pi \text{ and } 3\pi/2$$

7. Motion on the cycloid $x = t - \sin t, \quad y = 1 - \cos t$

$$\mathbf{r}(t) = (t - \sin t)\mathbf{i} + (1 - \cos t)\mathbf{j}; \quad t = \pi \text{ and } 3\pi/2$$

8. Motion on the parabola $y = x^2 + 1$

$$\mathbf{r}(t) = t\mathbf{i} + (t^2 + 1)\mathbf{j}; \quad t = -1, 0, \text{ and } 1$$

Motion in Space

In Exercises 9–14, $\mathbf{r}(t)$ is the position of a particle in space at time t. Find the particle's velocity and acceleration vectors. Then find the particle's speed and direction of motion at the given value of t. Write the particle's velocity at that time as the product of its speed and direction.

9. $\mathbf{r}(t) = (t + 1)\mathbf{i} + (t^2 - 1)\mathbf{j} + 2t\mathbf{k}, \quad t = 1$

10. $\mathbf{r}(t) = (1 + t)\mathbf{i} + \dfrac{t^2}{\sqrt{2}}\mathbf{j} + \dfrac{t^3}{3}\mathbf{k}, \quad t = 1$

11. $\mathbf{r}(t) = (2\cos t)\mathbf{i} + (3\sin t)\mathbf{j} + 4t\mathbf{k}, \quad t = \pi/2$

12. $\mathbf{r}(t) = (\sec t)\mathbf{i} + (\tan t)\mathbf{j} + \dfrac{4}{3}t\mathbf{k}, \quad t = \pi/6$

13. $\mathbf{r}(t) = (2\ln(t + 1))\mathbf{i} + t^2\mathbf{j} + \dfrac{t^2}{2}\mathbf{k}, \quad t = 1$

14. $\mathbf{r}(t) = (e^{-t})\mathbf{i} + (2\cos 3t)\mathbf{j} + (2\sin 3t)\mathbf{k}, \quad t = 0$

In Exercises 15–18, $\mathbf{r}(t)$ is the position of a particle in space at time t. Find the angle between the velocity and acceleration vectors at time $t = 0$.

15. $\mathbf{r}(t) = (3t + 1)\mathbf{i} + \sqrt{3}t\mathbf{j} + t^2\mathbf{k}$

16. $\mathbf{r}(t) = \left(\dfrac{\sqrt{2}}{2}t\right)\mathbf{i} + \left(\dfrac{\sqrt{2}}{2}t - 16t^2\right)\mathbf{j}$

17. $\mathbf{r}(t) = (\ln(t^2 + 1))\mathbf{i} + (\tan^{-1} t)\mathbf{j} + \sqrt{t^2 + 1}\,\mathbf{k}$

18. $\mathbf{r}(t) = \dfrac{4}{9}(1 + t)^{3/2}\mathbf{i} + \dfrac{4}{9}(1 - t)^{3/2}\mathbf{j} + \dfrac{1}{3}t\mathbf{k}$

Tangents to Curves

As mentioned in the text, the **tangent line** to a smooth curve $\mathbf{r}(t) = f(t)\mathbf{i} + g(t)\mathbf{j} + h(t)\mathbf{k}$ at $t = t_0$ is the line that passes through the point $(f(t_0), g(t_0), h(t_0))$ parallel to $\mathbf{v}(t_0)$, the curve's velocity vector at t_0. In Exercises 19–22, find parametric equations for the line that is tangent to the given curve at the given parameter value $t = t_0$.

19. $\mathbf{r}(t) = (\sin t)\mathbf{i} + (t^2 - \cos t)\mathbf{j} + e^t\mathbf{k}, \quad t_0 = 0$

20. $\mathbf{r}(t) = t^2\mathbf{i} + (2t - 1)\mathbf{j} + t^3\mathbf{k}, \quad t_0 = 2$

21. $\mathbf{r}(t) = \ln t\,\mathbf{i} + \dfrac{t - 1}{t + 2}\mathbf{j} + t\ln t\,\mathbf{k}, \quad t_0 = 1$

22. $\mathbf{r}(t) = (\cos t)\mathbf{i} + (\sin t)\mathbf{j} + (\sin 2t)\mathbf{k}, \quad t_0 = \dfrac{\pi}{2}$

Theory and Examples

23. Motion along a circle Each of the following equations in parts (a)–(e) describes the motion of a particle having the same path, namely the unit circle $x^2 + y^2 = 1$. Although the path of each particle in parts (a)–(e) is the same, the behavior, or "dynamics," of each particle is different. For each particle, answer the following questions.

   i) Does the particle have constant speed? If so, what is its constant speed?

   ii) Is the particle's acceleration vector always orthogonal to its velocity vector?

   iii) Does the particle move clockwise or counterclockwise around the circle?

   iv) Does the particle begin at the point $(1, 0)$?

   a. $\mathbf{r}(t) = (\cos t)\mathbf{i} + (\sin t)\mathbf{j}, \quad t \geq 0$

   b. $\mathbf{r}(t) = \cos(2t)\mathbf{i} + \sin(2t)\mathbf{j}, \quad t \geq 0$

   c. $\mathbf{r}(t) = \cos(t - \pi/2)\mathbf{i} + \sin(t - \pi/2)\mathbf{j}, \quad t \geq 0$

   d. $\mathbf{r}(t) = (\cos t)\mathbf{i} - (\sin t)\mathbf{j}, \quad t \geq 0$

   e. $\mathbf{r}(t) = \cos(t^2)\mathbf{i} + \sin(t^2)\mathbf{j}, \quad t \geq 0$

24. Motion along a circle Show that the vector-valued function

$$\mathbf{r}(t) = (2\mathbf{i} + 2\mathbf{j} + \mathbf{k})$$

$$+ \cos t\left(\frac{1}{\sqrt{2}}\mathbf{i} - \frac{1}{\sqrt{2}}\mathbf{j}\right) + \sin t\left(\frac{1}{\sqrt{3}}\mathbf{i} + \frac{1}{\sqrt{3}}\mathbf{j} + \frac{1}{\sqrt{3}}\mathbf{k}\right)$$

describes the motion of a particle moving in the circle of radius 1 centered at the point $(2, 2, 1)$ and lying in the plane $x + y - 2z = 2$.

25. Motion along a parabola A particle moves along the top of the parabola $y^2 = 2x$ from left to right at a constant speed of 5 units per second. Find the velocity of the particle as it moves through the point $(2, 2)$.

26. Motion along a cycloid A particle moves in the xy-plane in such a way that its position at time t is

$$\mathbf{r}(t) = (t - \sin t)\mathbf{i} + (1 - \cos t)\mathbf{j}.$$

   T **a.** Graph $\mathbf{r}(t)$. The resulting curve is a cycloid.

   b. Find the maximum and minimum values of $|\mathbf{v}|$ and $|\mathbf{a}|$. (*Hint:* Find the extreme values of $|\mathbf{v}|^2$ and $|\mathbf{a}|^2$ first and take square roots later.)

27. Let $\mathbf{r}$ be a differentiable vector function of t. Show that if $\mathbf{r} \cdot (d\mathbf{r}/dt) = 0$ for all t, then $|\mathbf{r}|$ is constant.

28. Derivatives of triple scalar products

   a. Show that if $\mathbf{u}$, $\mathbf{v}$, and $\mathbf{w}$ are differentiable vector functions of t, then

$$\frac{d}{dt}(\mathbf{u} \cdot \mathbf{v} \times \mathbf{w}) = \frac{d\mathbf{u}}{dt} \cdot \mathbf{v} \times \mathbf{w} + \mathbf{u} \cdot \frac{d\mathbf{v}}{dt} \times \mathbf{w} + \mathbf{u} \cdot \mathbf{v} \times \frac{d\mathbf{w}}{dt}.$$

   b. Show that

$$\frac{d}{dt}\left(\mathbf{r} \cdot \frac{d\mathbf{r}}{dt} \times \frac{d^2\mathbf{r}}{dt^2}\right) = \mathbf{r} \cdot \left(\frac{d\mathbf{r}}{dt} \times \frac{d^3\mathbf{r}}{dt^3}\right).$$

(*Hint:* Differentiate on the left and look for vectors whose products are zero.)

29. Prove the two Scalar Multiple Rules for vector functions.

30. Prove the Sum and Difference Rules for vector functions.

31. Component Test for Continuity at a Point Show that the vector function $\mathbf{r}$ defined by $\mathbf{r}(t) = f(t)\mathbf{i} + g(t)\mathbf{j} + h(t)\mathbf{k}$ is continuous at $t = t_0$ if and only if f, g, and h are continuous at t_0.

32. Limits of cross products of vector functions Suppose that $\mathbf{r}_1(t) = f_1(t)\mathbf{i} + f_2(t)\mathbf{j} + f_3(t)\mathbf{k}$, $\mathbf{r}_2(t) = g_1(t)\mathbf{i} + g_2(t)\mathbf{j} + g_3(t)\mathbf{k}$, $\lim_{t \to t_0} \mathbf{r}_1(t) = \mathbf{A}$, and $\lim_{t \to t_0} \mathbf{r}_2(t) = \mathbf{B}$. Use the determinant formula for cross products and the Limit Product Rule for scalar functions to show that

$$\lim_{t \to t_0}(\mathbf{r}_1(t) \times \mathbf{r}_2(t)) = \mathbf{A} \times \mathbf{B}.$$

33. Differentiable vector functions are continuous Show that if $\mathbf{r}(t) = f(t)\mathbf{i} + g(t)\mathbf{j} + h(t)\mathbf{k}$ is differentiable at $t = t_0$, then it is continuous at t_0 as well.

34. Constant Function Rule Prove that if $\mathbf{u}$ is the vector function with the constant value $\mathbf{C}$, then $d\mathbf{u}/dt = \mathbf{0}$.

COMPUTER EXPLORATIONS

Use a CAS to perform the following steps in Exercises 35–38.

a. Plot the space curve traced out by the position vector $\mathbf{r}$.

b. Find the components of the velocity vector $d\mathbf{r}/dt$.

c. Evaluate $d\mathbf{r}/dt$ at the given point t_0 and determine the equation of the tangent line to the curve at $\mathbf{r}(t_0)$.

d. Plot the tangent line together with the curve over the given interval.

35. $\mathbf{r}(t) = (\sin t - t \cos t)\mathbf{i} + (\cos t + t \sin t)\mathbf{j} + t^2\mathbf{k}$, $0 \le t \le 6\pi$, $t_0 = 3\pi/2$

36. $\mathbf{r}(t) = \sqrt{2}t\mathbf{i} + e^t\mathbf{j} + e^{-t}\mathbf{k}$, $-2 \le t \le 3$, $t_0 = 1$

37. $\mathbf{r}(t) = (\sin 2t)\mathbf{i} + (\ln(1 + t))\mathbf{j} + t\mathbf{k}$, $0 \le t \le 4\pi$, $t_0 = \pi/4$

38. $\mathbf{r}(t) = (\ln(t^2 + 2))\mathbf{i} + (\tan^{-1} 3t)\mathbf{j} + \sqrt{t^2 + 1}\,\mathbf{k}$, $-3 \le t \le 5$, $t_0 = 3$

In Exercises 39 and 40, you will explore graphically the behavior of the helix

$$\mathbf{r}(t) = (\cos at)\mathbf{i} + (\sin at)\mathbf{j} + bt\mathbf{k}$$

as you change the values of the constants a and b. Use a CAS to perform the steps in each exercise.

39. Set $b = 1$. Plot the helix $\mathbf{r}(t)$ together with the tangent line to the curve at $t = 3\pi/2$ for $a = 1$, 2, 4, and 6 over the interval $0 \le t \le 4\pi$. Describe in your own words what happens to the graph of the helix and the position of the tangent line as a increases through these positive values.

40. Set $a = 1$. Plot the helix $\mathbf{r}(t)$ together with the tangent line to the curve at $t = 3\pi/2$ for $b = 1/4$, 1/2, 2, and 4 over the interval $0 \le t \le 4\pi$. Describe in your own words what happens to the graph of the helix and the position of the tangent line as b increases through these positive values.

13.2 Integrals of Vector Functions; Projectile Motion

In this section we investigate integrals of vector functions and their application to motion along a path in space or in the plane.

Integrals of Vector Functions

A differentiable vector function $\mathbf{R}(t)$ is an **antiderivative** of a vector function $\mathbf{r}(t)$ on an interval I if $d\mathbf{R}/dt = \mathbf{r}$ at each point of I. If $\mathbf{R}$ is an antiderivative of $\mathbf{r}$ on I, it can be shown, working one component at a time, that every antiderivative of $\mathbf{r}$ on I has the form $\mathbf{R} + \mathbf{C}$ for some constant vector $\mathbf{C}$ (Exercise 41). The set of all antiderivatives of $\mathbf{r}$ on I is the **indefinite integral** of $\mathbf{r}$ on I.

> **DEFINITION** The **indefinite integral** of $\mathbf{r}$ with respect to t is the set of all antiderivatives of $\mathbf{r}$, denoted by $\int \mathbf{r}(t)\, dt$. If $\mathbf{R}$ is any antiderivative of $\mathbf{r}$, then
>
> $$\int \mathbf{r}(t)\, dt = \mathbf{R}(t) + \mathbf{C}.$$

The usual arithmetic rules for indefinite integrals apply.

EXAMPLE 1 To integrate a vector function, we integrate each of its components.

$$\int ((\cos t)\mathbf{i} + \mathbf{j} - 2t\mathbf{k})\,dt = \left(\int \cos t\,dt\right)\mathbf{i} + \left(\int dt\right)\mathbf{j} - \left(\int 2t\,dt\right)\mathbf{k} \qquad (1)$$

$$= (\sin t + C_1)\mathbf{i} + (t + C_2)\mathbf{j} - (t^2 + C_3)\mathbf{k} \qquad (2)$$

$$= (\sin t)\mathbf{i} + t\mathbf{j} - t^2\mathbf{k} + \mathbf{C} \qquad \mathbf{C} = C_1\mathbf{i} + C_2\mathbf{j} - C_3\mathbf{k}$$

As in the integration of scalar functions, we recommend that you skip the steps in Equations (1) and (2) and go directly to the final form. Find an antiderivative for each component and add a *constant vector* at the end. ∎

Definite integrals of vector functions are best defined in terms of components. The definition is consistent with how we compute limits and derivatives of vector functions.

> **DEFINITION** If the components of $\mathbf{r}(t) = f(t)\mathbf{i} + g(t)\mathbf{j} + h(t)\mathbf{k}$ are integrable over $[a, b]$, then so is $\mathbf{r}$, and the **definite integral** of $\mathbf{r}$ from a to b is
>
> $$\int_a^b \mathbf{r}(t)\,dt = \left(\int_a^b f(t)\,dt\right)\mathbf{i} + \left(\int_a^b g(t)\,dt\right)\mathbf{j} + \left(\int_a^b h(t)\,dt\right)\mathbf{k}.$$

EXAMPLE 2 As in Example 1, we integrate each component.

$$\int_0^\pi ((\cos t)\mathbf{i} + \mathbf{j} - 2t\mathbf{k})\,dt = \left(\int_0^\pi \cos t\,dt\right)\mathbf{i} + \left(\int_0^\pi dt\right)\mathbf{j} - \left(\int_0^\pi 2t\,dt\right)\mathbf{k}$$

$$= \left[\sin t\right]_0^\pi \mathbf{i} + \left[t\right]_0^\pi \mathbf{j} - \left[t^2\right]_0^\pi \mathbf{k}$$

$$= [0 - 0]\mathbf{i} + [\pi - 0]\mathbf{j} - [\pi^2 - 0^2]\mathbf{k}$$

$$= \pi\mathbf{j} - \pi^2\mathbf{k} \qquad\qquad ∎$$

The Fundamental Theorem of Calculus for continuous vector functions says that

$$\int_a^b \mathbf{r}(t)\,dt = \mathbf{R}(t)\Big]_a^b = \mathbf{R}(b) - \mathbf{R}(a)$$

where $\mathbf{R}$ is any antiderivative of $\mathbf{r}$, so that $\mathbf{R}'(t) = \mathbf{r}(t)$ (Exercise 42).

EXAMPLE 3 Suppose we do not know the path of a hang glider, but only its acceleration vector $\mathbf{a}(t) = -(3\cos t)\mathbf{i} - (3\sin t)\mathbf{j} + 2\mathbf{k}$. We also know that initially (at time $t = 0$) the glider departed from the point $(3, 0, 0)$ with velocity $\mathbf{v}(0) = 3\mathbf{j}$. Find the glider's position as a function of t.

Solution Our goal is to find $\mathbf{r}(t)$ knowing

The differential equation: $\quad \mathbf{a} = \dfrac{d^2\mathbf{r}}{dt^2} = -(3\cos t)\mathbf{i} - (3\sin t)\mathbf{j} + 2\mathbf{k}$

The initial conditions: $\quad \mathbf{v}(0) = 3\mathbf{j} \quad$ and $\quad \mathbf{r}(0) = 3\mathbf{i} + 0\mathbf{j} + 0\mathbf{k}.$

Integrating both sides of the differential equation with respect to t gives

$$\mathbf{v}(t) = -(3\sin t)\mathbf{i} + (3\cos t)\mathbf{j} + 2t\mathbf{k} + \mathbf{C}_1.$$

We use $\mathbf{v}(0) = 3\mathbf{j}$ to find $\mathbf{C}_1$:

$$3\mathbf{j} = -(3\sin 0)\mathbf{i} + (3\cos 0)\mathbf{j} + (0)\mathbf{k} + \mathbf{C}_1$$
$$3\mathbf{j} = 3\mathbf{j} + \mathbf{C}_1$$
$$\mathbf{C}_1 = \mathbf{0}.$$

The glider's velocity as a function of time is

$$\frac{d\mathbf{r}}{dt} = \mathbf{v}(t) = -(3\sin t)\mathbf{i} + (3\cos t)\mathbf{j} + 2t\mathbf{k}.$$

Integrating both sides of this last differential equation gives

$$\mathbf{r}(t) = (3\cos t)\mathbf{i} + (3\sin t)\mathbf{j} + t^2\mathbf{k} + \mathbf{C}_2.$$

We then use the initial condition $\mathbf{r}(0) = 3\mathbf{i}$ to find $\mathbf{C}_2$:

$$3\mathbf{i} = (3\cos 0)\mathbf{i} + (3\sin 0)\mathbf{j} + (0^2)\mathbf{k} + \mathbf{C}_2$$
$$3\mathbf{i} = 3\mathbf{i} + (0)\mathbf{j} + (0)\mathbf{k} + \mathbf{C}_2$$
$$\mathbf{C}_2 = \mathbf{0}.$$

The glider's position as a function of t is

$$\mathbf{r}(t) = (3\cos t)\mathbf{i} + (3\sin t)\mathbf{j} + t^2\mathbf{k}.$$

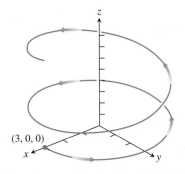

FIGURE 13.9 The path of the hang glider in Example 3. Although the path spirals around the z-axis, it is not a helix.

This is the path of the glider shown in Figure 13.9. Although the path resembles that of a helix due to its spiraling nature around the z-axis, it is not a helix because of the way it is rising. (We say more about this in Section 13.5.)

Note: It turned out in this example that both of the constant vectors of integration, $\mathbf{C}_1$ and $\mathbf{C}_2$, are $\mathbf{0}$. Exercises 15 and 16 give examples for which the constant vectors of integration are not $\mathbf{0}$.   ∎

The Vector and Parametric Equations for Ideal Projectile Motion

A classic example of integrating vector functions is the derivation of the equations for the motion of a projectile. In physics, projectile motion describes how an object fired at some angle from an initial position, and acted upon by only the force of gravity, moves in a vertical coordinate plane. In the classic example, we ignore the effects of any frictional drag on the object, which may vary with its speed and altitude, and also the fact that the force of gravity changes slightly with the projectile's changing height. In addition, we ignore the long-distance effects of the Earth turning beneath the projectile, such as in a rocket launch or the firing of a projectile from a cannon. Ignoring these effects gives us a reasonable approximation of the motion in most cases.

To derive equations for projectile motion, we assume that the projectile behaves like a particle moving in a vertical coordinate plane and that the only force acting on the projectile during its flight is the constant force of gravity, which always points straight down. We assume that the projectile is launched from the origin at time $t = 0$ into the first quadrant with an initial velocity $\mathbf{v}_0$ (Figure 13.10). If $\mathbf{v}_0$ makes an angle α with the horizontal, then

$$\mathbf{v}_0 = (|\mathbf{v}_0|\cos\alpha)\mathbf{i} + (|\mathbf{v}_0|\sin\alpha)\mathbf{j}.$$

If we use the simpler notation v_0 for the initial speed $|\mathbf{v}_0|$, then

$$\mathbf{v}_0 = (v_0\cos\alpha)\mathbf{i} + (v_0\sin\alpha)\mathbf{j}. \tag{3}$$

The projectile's initial position is

$$\mathbf{r}_0 = 0\mathbf{i} + 0\mathbf{j} = \mathbf{0}. \tag{4}$$

Newton's second law of motion says that the force acting on the projectile is equal to the projectile's mass m times its acceleration, or $m(d^2\mathbf{r}/dt^2)$ if $\mathbf{r}$ is the projectile's position vector and t is time. If the force is solely the gravitational force $-mg\mathbf{j}$, then

$$m\frac{d^2\mathbf{r}}{dt^2} = -mg\mathbf{j} \qquad \text{and} \qquad \frac{d^2\mathbf{r}}{dt^2} = -g\mathbf{j}$$

(a)

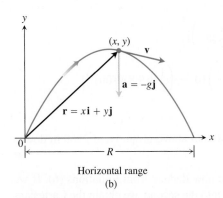

(b)

FIGURE 13.10 (a) Position, velocity, acceleration, and launch angle at $t = 0$. (b) Position, velocity, and acceleration at a later time t.

where g is the acceleration due to gravity. We find $\mathbf{r}$ as a function of t by solving the following initial value problem.

Differential equation: $\quad \dfrac{d^2\mathbf{r}}{dt^2} = -g\mathbf{j}$

Initial conditions: $\quad \mathbf{r} = \mathbf{r}_0 \quad$ and $\quad \dfrac{d\mathbf{r}}{dt} = \mathbf{v}_0 \quad$ when $t = 0$

The first integration gives

$$\frac{d\mathbf{r}}{dt} = -(gt)\mathbf{j} + \mathbf{v}_0.$$

A second integration gives

$$\mathbf{r} = -\frac{1}{2}gt^2\mathbf{j} + \mathbf{v}_0 t + \mathbf{r}_0.$$

Substituting the values of $\mathbf{v}_0$ and $\mathbf{r}_0$ from Equations (3) and (4) gives

$$\mathbf{r} = -\frac{1}{2}gt^2\mathbf{j} + \underbrace{(v_0\cos\alpha)t\mathbf{i} + (v_0\sin\alpha)t\mathbf{j}}_{\mathbf{v}_0 t} + \mathbf{0}.$$

Collecting terms, we have

Ideal Projectile Motion Equation

$$\mathbf{r} = (v_0\cos\alpha)t\mathbf{i} + \left((v_0\sin\alpha)t - \frac{1}{2}gt^2\right)\mathbf{j}. \qquad (5)$$

Equation (5) is the *vector equation* for ideal projectile motion. The angle α is the projectile's **launch angle (firing angle, angle of elevation)**, and v_0, as we said before, is the projectile's **initial speed**. The components of $\mathbf{r}$ give the parametric equations

$$x = (v_0\cos\alpha)t \qquad \text{and} \qquad y = (v_0\sin\alpha)t - \frac{1}{2}gt^2, \qquad (6)$$

where x is the distance downrange and y is the height of the projectile at time $t \geq 0$.

EXAMPLE 4 A projectile is fired from the origin over horizontal ground at an initial speed of 500 m/sec and a launch angle of 60°. Where will the projectile be 10 sec later?

Solution We use Equation (5) with $v_0 = 500$, $\alpha = 60°$, $g = 9.8$, and $t = 10$ to find the projectile's components 10 sec after firing.

$$\mathbf{r} = (v_0\cos\alpha)t\mathbf{i} + \left((v_0\sin\alpha)t - \frac{1}{2}gt^2\right)\mathbf{j}$$

$$= (500)\left(\frac{1}{2}\right)(10)\mathbf{i} + \left((500)\left(\frac{\sqrt{3}}{2}\right)10 - \left(\frac{1}{2}\right)(9.8)(100)\right)\mathbf{j}$$

$$\approx 2500\mathbf{i} + 3840\mathbf{j}$$

Ten seconds after firing, the projectile is about 3840 m above ground and 2500 m downrange from the origin. ∎

Ideal projectiles move along parabolas, as we now deduce from Equations (6). If we substitute $t = x/(v_0\cos\alpha)$ from the first equation into the second, we obtain the Cartesian-coordinate equation

$$y = -\left(\frac{g}{2v_0{}^2\cos^2\alpha}\right)x^2 + (\tan\alpha)x.$$

This equation has the form $y = ax^2 + bx$, so its graph is a parabola.

A projectile reaches its highest point when its vertical velocity component is zero. When fired over horizontal ground, the projectile lands when its vertical component equals zero in Equation (5), and the **range** R is the distance from the origin to the point of impact. We summarize the results here, which you are asked to verify in Exercise 27.

Height, Flight Time, and Range for Ideal Projectile Motion
For ideal projectile motion when an object is launched from the origin over a horizontal surface with initial speed v_0 and launch angle α:

$$\text{Maximum height:} \qquad y_{\max} = \frac{(v_0 \sin \alpha)^2}{2g}$$

$$\text{Flight time:} \qquad t = \frac{2v_0 \sin \alpha}{g}$$

$$\text{Range:} \qquad R = \frac{v_0^2}{g} \sin 2\alpha.$$

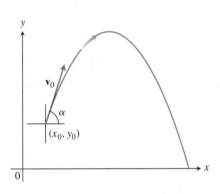

FIGURE 13.11 The path of a projectile fired from (x_0, y_0) with an initial velocity $\mathbf{v}_0$ at an angle of α degrees with the horizontal.

If we fire our ideal projectile from the point (x_0, y_0) instead of the origin (Figure 13.11), the position vector for the path of motion is

$$\mathbf{r} = \left(x_0 + (v_0 \cos \alpha)t\right)\mathbf{i} + \left(y_0 + (v_0 \sin \alpha)t - \frac{1}{2}gt^2\right)\mathbf{j}, \qquad (7)$$

as you are asked to show in Exercise 29.

Projectile Motion with Wind Gusts

The next example shows how to account for another force acting on a projectile, due to a gust of wind. We also assume that the path of the baseball in Example 5 lies in a vertical plane.

EXAMPLE 5 A baseball is hit when it is 3 ft above the ground. It leaves the bat with initial speed of 152 ft/sec, making an angle of 20° with the horizontal. At the instant the ball is hit, an instantaneous gust of wind blows in the horizontal direction directly opposite the direction the ball is taking toward the outfield, adding a component of $-8.8\mathbf{i}$ (ft/sec) to the ball's initial velocity (8.8 ft/sec = 6 mph).

(a) Find a vector equation (position vector) for the path of the baseball.

(b) How high does the baseball go, and when does it reach maximum height?

(c) Assuming that the ball is not caught, find its range and flight time.

Solution

(a) Using Equation (3) and accounting for the gust of wind, the initial velocity of the baseball is

$$\mathbf{v}_0 = (v_0 \cos \alpha)\mathbf{i} + (v_0 \sin \alpha)\mathbf{j} - 8.8\mathbf{i}$$

$$= (152 \cos 20°)\mathbf{i} + (152 \sin 20°)\mathbf{j} - (8.8)\mathbf{i}$$

$$= (152 \cos 20° - 8.8)\mathbf{i} + (152 \sin 20°)\mathbf{j}.$$

The initial position is $\mathbf{r}_0 = 0\mathbf{i} + 3\mathbf{j}$. Integration of $d^2\mathbf{r}/dt^2 = -g\mathbf{j}$ gives

$$\frac{d\mathbf{r}}{dt} = -(gt)\mathbf{j} + \mathbf{v}_0.$$

A second integration gives

$$\mathbf{r} = -\frac{1}{2}gt^2\mathbf{j} + \mathbf{v}_0 t + \mathbf{r}_0.$$

Substituting the values of $\mathbf{v}_0$ and $\mathbf{r}_0$ into the last equation gives the position vector of the baseball.

$$\mathbf{r} = -\frac{1}{2}gt^2\mathbf{j} + \mathbf{v}_0 t + \mathbf{r}_0$$

$$= -16t^2\mathbf{j} + (152\cos 20° - 8.8)t\mathbf{i} + (152\sin 20°)t\mathbf{j} + 3\mathbf{j}$$

$$= (152\cos 20° - 8.8)t\mathbf{i} + \left(3 + (152\sin 20°)t - 16t^2\right)\mathbf{j}.$$

(b) The baseball reaches its highest point when the vertical component of velocity is zero, or

$$\frac{dy}{dt} = 152\sin 20° - 32t = 0.$$

Solving for t we find

$$t = \frac{152\sin 20°}{32} \approx 1.62 \text{ sec}.$$

Substituting this time into the vertical component for $\mathbf{r}$ gives the maximum height

$$y_{max} = 3 + (152\sin 20°)(1.62) - 16(1.62)^2$$

$$\approx 45.2 \text{ ft}.$$

That is, the maximum height of the baseball is about 45.2 ft, reached about 1.6 sec after leaving the bat.

(c) To find when the baseball lands, we set the vertical component for $\mathbf{r}$ equal to 0 and solve for t:

$$3 + (152\sin 20°)t - 16t^2 = 0$$

$$3 + (51.99)t - 16t^2 = 0.$$

The solution values are about $t = 3.3$ sec and $t = -0.06$ sec. Substituting the positive time into the horizontal component for $\mathbf{r}$, we find the range

$$R = (152\cos 20° - 8.8)(3.3)$$

$$\approx 442 \text{ ft}.$$

Thus, the horizontal range is about 442 ft, and the flight time is about 3.3 sec. ∎

In Exercises 37 and 38, we consider projectile motion when there is air resistance slowing down the flight.

Exercises 13.2

Integrating Vector-Valued Functions
Evaluate the integrals in Exercises 1–10.

1. $\int_0^1 [t^3\mathbf{i} + 7\mathbf{j} + (t + 1)\mathbf{k}]\, dt$

2. $\int_1^2 \left[(6 - 6t)\mathbf{i} + 3\sqrt{t}\mathbf{j} + \left(\frac{4}{t^2}\right)\mathbf{k}\right] dt$

3. $\int_{-\pi/4}^{\pi/4} [(\sin t)\mathbf{i} + (1 + \cos t)\mathbf{j} + (\sec^2 t)\mathbf{k}]\, dt$

4. $\int_0^{\pi/3} [(\sec t \tan t)\mathbf{i} + (\tan t)\mathbf{j} + (2\sin t \cos t)\mathbf{k}]\, dt$

5. $\int_1^4 \left[\frac{1}{t}\mathbf{i} + \frac{1}{5 - t}\mathbf{j} + \frac{1}{2t}\mathbf{k}\right] dt$

6. $\int_0^1 \left[\dfrac{2}{\sqrt{1 - t^2}}\,\mathbf{i} + \dfrac{\sqrt{3}}{1 + t^2}\,\mathbf{k} \right] dt$

7. $\int_0^1 [te^{t^2}\,\mathbf{i} + e^{-t}\,\mathbf{j} + \mathbf{k}]\,dt$

8. $\int_1^{\ln 3} [te^t\,\mathbf{i} + e^t\,\mathbf{j} + \ln t\,\mathbf{k}]\,dt$

9. $\int_0^{\pi/2} [\cos t\,\mathbf{i} - \sin 2t\,\mathbf{j} + \sin^2 t\,\mathbf{k}]\,dt$

10. $\int_0^{\pi/4} [\sec t\,\mathbf{i} + \tan^2 t\,\mathbf{j} - t\sin t\,\mathbf{k}]\,dt$

Initial Value Problems

Solve the initial value problems in Exercises 11–16 for $\mathbf{r}$ as a vector function of t.

11. Differential equation: $\dfrac{d\mathbf{r}}{dt} = -t\mathbf{i} - t\mathbf{j} - t\mathbf{k}$

Initial condition: $\mathbf{r}(0) = \mathbf{i} + 2\mathbf{j} + 3\mathbf{k}$

12. Differential equation: $\dfrac{d\mathbf{r}}{dt} = (180t)\mathbf{i} + (180t - 16t^2)\mathbf{j}$

Initial condition: $\mathbf{r}(0) = 100\mathbf{j}$

13. Differential equation: $\dfrac{d\mathbf{r}}{dt} = \dfrac{3}{2}(t + 1)^{1/2}\mathbf{i} + e^{-t}\mathbf{j} + \dfrac{1}{t + 1}\mathbf{k}$

Initial condition: $\mathbf{r}(0) = \mathbf{k}$

14. Differential equation: $\dfrac{d\mathbf{r}}{dt} = (t^3 + 4t)\mathbf{i} + t\mathbf{j} + 2t^2\mathbf{k}$

Initial condition: $\mathbf{r}(0) = \mathbf{i} + \mathbf{j}$

15. Differential equation: $\dfrac{d^2\mathbf{r}}{dt^2} = -32\mathbf{k}$

Initial conditions: $\mathbf{r}(0) = 100\mathbf{k}$ and

$\left. \dfrac{d\mathbf{r}}{dt} \right|_{t=0} = 8\mathbf{i} + 8\mathbf{j}$

16. Differential equation: $\dfrac{d^2\mathbf{r}}{dt^2} = -(\mathbf{i} + \mathbf{j} + \mathbf{k})$

Initial conditions: $\mathbf{r}(0) = 10\mathbf{i} + 10\mathbf{j} + 10\mathbf{k}$ and

$\left. \dfrac{d\mathbf{r}}{dt} \right|_{t=0} = \mathbf{0}$

Motion Along a Straight Line

17. At time $t = 0$, a particle is located at the point $(1, 2, 3)$. It travels in a straight line to the point $(4, 1, 4)$, has speed 2 at $(1, 2, 3)$ and constant acceleration $3\mathbf{i} - \mathbf{j} + \mathbf{k}$. Find an equation for the position vector $\mathbf{r}(t)$ of the particle at time t.

18. A particle traveling in a straight line is located at the point $(1, -1, 2)$ and has speed 2 at time $t = 0$. The particle moves toward the point $(3, 0, 3)$ with constant acceleration $2\mathbf{i} + \mathbf{j} + \mathbf{k}$. Find its position vector $\mathbf{r}(t)$ at time t.

Projectile Motion

Projectile flights in the following exercises are to be treated as ideal unless stated otherwise. All launch angles are assumed to be measured from the horizontal. All projectiles are assumed to be launched from the origin over a horizontal surface unless stated otherwise.

19. Travel time A projectile is fired at a speed of 840 m/sec at an angle of 60°. How long will it take to get 21 km downrange?

20. Finding muzzle speed Find the muzzle speed of a gun whose maximum range is 24.5 km.

21. Flight time and height A projectile is fired with an initial speed of 500 m/sec at an angle of elevation of 45°.

  a. When and how far away will the projectile strike?

  b. How high overhead will the projectile be when it is 5 km downrange?

  c. What is the greatest height reached by the projectile?

22. Throwing a baseball A baseball is thrown from the stands 32 ft above the field at an angle of 30° up from the horizontal. When and how far away will the ball strike the ground if its initial speed is 32 ft/sec?

23. Firing golf balls A spring gun at ground level fires a golf ball at an angle of 45°. The ball lands 10 m away.

  a. What was the ball's initial speed?

  b. For the same initial speed, find the two firing angles that make the range 6 m.

24. Beaming electrons An electron in a TV tube is beamed horizontally at a speed of 5×10^6 m/sec toward the face of the tube 40 cm away. About how far will the electron drop before it hits?

25. Equal-range firing angles What two angles of elevation will enable a projectile to reach a target 16 km downrange on the same level as the gun if the projectile's initial speed is 400 m/sec?

26. Range and height versus speed

  a. Show that doubling a projectile's initial speed at a given launch angle multiplies its range by 4.

  b. By about what percentage should you increase the initial speed to double the height and range?

27. Verify the results given in the text (following Example 4) for the maximum height, flight time, and range for ideal projectile motion.

28. Colliding marbles The accompanying figure shows an experiment with two marbles. Marble A was launched toward marble B with launch angle α and initial speed v_0. At the same instant, marble B was released to fall from rest at $R \tan \alpha$ units directly above a spot R units downrange from A. The marbles were found to collide regardless of the value of v_0. Was this mere coincidence, or must this happen? Give reasons for your answer.

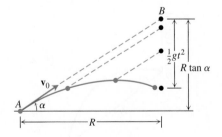

29. Firing from (x_0, y_0) Derive the equations

$$x = x_0 + (v_0 \cos \alpha)t,$$

$$y = y_0 + (v_0 \sin \alpha)t - \frac{1}{2}gt^2$$

(see Equation (7) in the text) by solving the following initial value problem for a vector $\mathbf{r}$ in the plane.

Differential equation: $\dfrac{d^2\mathbf{r}}{dt^2} = -g\mathbf{j}$

Initial conditions: $\mathbf{r}(0) = x_0\mathbf{i} + y_0\mathbf{j}$

$\dfrac{d\mathbf{r}}{dt}(0) = (v_0\cos\alpha)\mathbf{i} + (v_0\sin\alpha)\mathbf{j}$

30. Where trajectories crest   For a projectile fired from the ground at launch angle α with initial speed v_0, consider α as a variable and v_0 as a fixed constant. For each α, $0 < \alpha < \pi/2$, we obtain a parabolic trajectory as shown in the accompanying figure. Show that the points in the plane that give the maximum heights of these parabolic trajectories all lie on the ellipse

$$x^2 + 4\left(y - \frac{v_0^2}{4g}\right)^2 = \frac{v_0^4}{4g^2},$$

where $x \geq 0$.

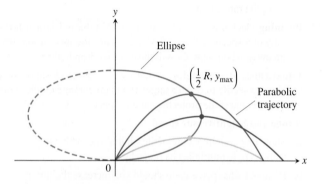

31. Launching downhill   An ideal projectile is launched straight down an inclined plane as shown in the accompanying figure.

a. Show that the greatest downhill range is achieved when the initial velocity vector bisects angle AOR.

b. If the projectile were fired uphill instead of down, what launch angle would maximize its range? Give reasons for your answer.

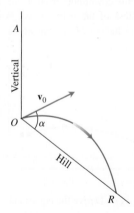

32. Elevated green   A golf ball is hit with an initial speed of 116 ft/sec at an angle of elevation of 45° from the tee to a green that is elevated 45 ft above the tee as shown in the diagram. Assuming that the pin, 369 ft downrange, does not get in the way, where will the ball land in relation to the pin?

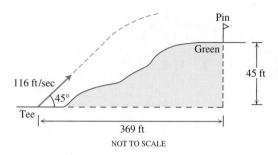

33. Volleyball   A volleyball is hit when it is 4 ft above the ground and 12 ft from a 6-ft-high net. It leaves the point of impact with an initial velocity of 35 ft/sec at an angle of 27° and slips by the opposing team untouched.

a. Find a vector equation for the path of the volleyball.

b. How high does the volleyball go, and when does it reach maximum height?

c. Find its range and flight time.

d. When is the volleyball 7 ft above the ground? How far (ground distance) is the volleyball from where it will land?

e. Suppose that the net is raised to 8 ft. Does this change things? Explain.

34. Shot put   In Moscow in 1987, Natalya Lisouskaya set a women's world record by putting an 8 lb 13 oz shot 73 ft 10 in. Assuming that she launched the shot at a 40° angle to the horizontal from 6.5 ft above the ground, what was the shot's initial speed?

35. Model train   The accompanying multiflash photograph shows a model train engine moving at a constant speed on a straight horizontal track. As the engine moved along, a marble was fired into the air by a spring in the engine's smokestack. The marble, which continued to move with the same forward speed as the engine, rejoined the engine 1 sec after it was fired. Measure the angle the marble's path made with the horizontal and use the information to find how high the marble went and how fast the engine was moving.

36. Hitting a baseball under a wind gust   A baseball is hit when it is 2.5 ft above the ground. It leaves the bat with an initial velocity of 145 ft/sec at a launch angle of 23°. At the instant the ball is hit, an instantaneous gust of wind blows against the ball, adding a component of $-14\mathbf{i}$ (ft/sec) to the ball's initial velocity. A 15-ft-high fence lies 300 ft from home plate in the direction of the flight.

a. Find a vector equation for the path of the baseball.

b. How high does the baseball go, and when does it reach maximum height?

c. Find the range and flight time of the baseball, assuming that the ball is not caught.

d. When is the baseball 20 ft high? How far (ground distance) is the baseball from home plate at that height?

e. Has the batter hit a home run? Explain.

Projectile Motion with Linear Drag

The main force affecting the motion of a projectile, other than gravity, is air resistance. This slowing down force is **drag force**, and it acts in a direction *opposite* to the velocity of the projectile (see accompanying figure). For projectiles moving through the air at relatively low speeds, however, the drag force is (very nearly) proportional to the speed (to the first power) and so is called **linear**.

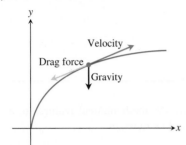

37. Linear drag Derive the equations

$$x = \frac{v_0}{k}(1 - e^{-kt})\cos\alpha$$

$$y = \frac{v_0}{k}(1 - e^{-kt})(\sin\alpha) + \frac{g}{k^2}(1 - kt - e^{-kt})$$

by solving the following initial value problem for a vector **r** in the plane.

Differential equation: $\quad \dfrac{d^2\mathbf{r}}{dt^2} = -g\mathbf{j} - k\mathbf{v} = -g\mathbf{j} - k\dfrac{d\mathbf{r}}{dt}$

Initial conditions: $\quad \mathbf{r}(0) = \mathbf{0}$

$$\dfrac{d\mathbf{r}}{dt}\bigg|_{t=0} = \mathbf{v}_0 = (v_0\cos\alpha)\mathbf{i} + (v_0\sin\alpha)\mathbf{j}$$

The **drag coefficient** k is a positive constant representing resistance due to air density, v_0 and α are the projectile's initial speed and launch angle, and g is the acceleration of gravity.

38. Hitting a baseball with linear drag Consider the baseball problem in Example 5 when there is linear drag (see Exercise 37). Assume a drag coefficient $k = 0.12$, but no gust of wind.

a. From Exercise 37, find a vector form for the path of the baseball.

b. How high does the baseball go, and when does it reach maximum height?

c. Find the range and flight time of the baseball.

d. When is the baseball 30 ft high? How far (ground distance) is the baseball from home plate at that height?

e. A 10-ft-high outfield fence is 340 ft from home plate in the direction of the flight of the baseball. The outfielder can jump and catch any ball up to 11 ft off the ground to stop it from going over the fence. Has the batter hit a home run?

Theory and Examples

39. Establish the following properties of integrable vector functions.

a. The *Constant Scalar Multiple Rule:*

$$\int_a^b k\mathbf{r}(t)\,dt = k\int_a^b \mathbf{r}(t)\,dt \quad \text{(any scalar } k\text{)}$$

The *Rule for Negatives,*

$$\int_a^b (-\mathbf{r}(t))\,dt = -\int_a^b \mathbf{r}(t)\,dt,$$

is obtained by taking $k = -1$.

b. The *Sum and Difference Rules:*

$$\int_a^b (\mathbf{r}_1(t) \pm \mathbf{r}_2(t))\,dt = \int_a^b \mathbf{r}_1(t)\,dt \pm \int_a^b \mathbf{r}_2(t)\,dt$$

c. The *Constant Vector Multiple Rules:*

$$\int_a^b \mathbf{C}\cdot\mathbf{r}(t)\,dt = \mathbf{C}\cdot\int_a^b \mathbf{r}(t)\,dt \quad \text{(any constant vector } \mathbf{C}\text{)}$$

and

$$\int_a^b \mathbf{C}\times\mathbf{r}(t)\,dt = \mathbf{C}\times\int_a^b \mathbf{r}(t)\,dt \quad \text{(any constant vector } \mathbf{C}\text{)}$$

40. Products of scalar and vector functions Suppose that the scalar function $u(t)$ and the vector function $\mathbf{r}(t)$ are both defined for $a \le t \le b$.

a. Show that $u\mathbf{r}$ is continuous on $[a, b]$ if u and $\mathbf{r}$ are continuous on $[a, b]$.

b. If u and $\mathbf{r}$ are both differentiable on $[a, b]$, show that $u\mathbf{r}$ is differentiable on $[a, b]$ and that

$$\frac{d}{dt}(u\mathbf{r}) = u\frac{d\mathbf{r}}{dt} + \mathbf{r}\frac{du}{dt}.$$

41. Antiderivatives of vector functions

a. Use Corollary 2 of the Mean Value Theorem for scalar functions to show that if two vector functions $\mathbf{R}_1(t)$ and $\mathbf{R}_2(t)$ have identical derivatives on an interval I, then the functions differ by a constant vector value throughout I.

b. Use the result in part (a) to show that if $\mathbf{R}(t)$ is any antiderivative of $\mathbf{r}(t)$ on I, then any other antiderivative of $\mathbf{r}$ on I equals $\mathbf{R}(t) + \mathbf{C}$ for some constant vector $\mathbf{C}$.

42. The Fundamental Theorem of Calculus The Fundamental Theorem of Calculus for scalar functions of a real variable holds for vector functions of a real variable as well. Prove this by using the theorem for scalar functions to show first that if a vector function $\mathbf{r}(t)$ is continuous for $a \le t \le b$, then

$$\frac{d}{dt}\int_a^t \mathbf{r}(\tau)\,d\tau = \mathbf{r}(t)$$

at every point t of (a, b). Then use the conclusion in part (b) of Exercise 41 to show that if $\mathbf{R}$ is any antiderivative of $\mathbf{r}$ on $[a, b]$ then

$$\int_a^b \mathbf{r}(t)\,dt = \mathbf{R}(b) - \mathbf{R}(a).$$

43. Hitting a baseball with linear drag under a wind gust Consider again the baseball problem in Example 5. This time assume a drag coefficient of 0.08 *and* an instantaneous gust of wind that adds a component of $-17.6\mathbf{i}$ (ft/sec) to the initial velocity at the instant the baseball is hit.

  a. Find a vector equation for the path of the baseball.

  b. How high does the baseball go, and when does it reach maximum height?

  c. Find the range and flight time of the baseball.

  d. When is the baseball 35 ft high? How far (ground distance) is the baseball from home plate at that height?

  e. A 20-ft-high outfield fence is 380 ft from home plate in the direction of the flight of the baseball. Has the batter hit a home run? If "yes," what change in the horizontal component of the ball's initial velocity would have kept the ball in the park? If "no," what change would have allowed it to be a home run?

44. Height versus time Show that a projectile attains three-quarters of its maximum height in half the time it takes to reach the maximum height.

13.3 Arc Length in Space

In this and the next two sections, we study the mathematical features of a curve's shape that describe the sharpness of its turning and its twisting.

Arc Length Along a Space Curve

One of the features of smooth space and plane curves is that they have a measurable length. This enables us to locate points along these curves by giving their directed distance s along the curve from some base point, the way we locate points on coordinate axes by giving their directed distance from the origin (Figure 13.12). This is what we did for plane curves in Section 11.2.

FIGURE 13.12 Smooth curves can be scaled like number lines, the coordinate of each point being its directed distance along the curve from a preselected base point.

To measure distance along a smooth curve in space, we add a *z*-term to the formula we use for curves in the plane.

DEFINITION The **length** of a smooth curve $\mathbf{r}(t) = x(t)\mathbf{i} + y(t)\mathbf{j} + z(t)\mathbf{k}$, $a \leq t \leq b$, that is traced exactly once as t increases from $t = a$ to $t = b$, is

$$L = \int_a^b \sqrt{\left(\frac{dx}{dt}\right)^2 + \left(\frac{dy}{dt}\right)^2 + \left(\frac{dz}{dt}\right)^2}\, dt. \tag{1}$$

Just as for plane curves, we can calculate the length of a curve in space from any convenient parametrization that meets the stated conditions. We omit the proof.

The square root in Equation (1) is $|\mathbf{v}|$, the length of a velocity vector $d\mathbf{r}/dt$. This enables us to write the formula for length a shorter way.

Arc Length Formula

$$L = \int_a^b |\mathbf{v}|\, dt \tag{2}$$

EXAMPLE 1 A glider is soaring upward along the helix $\mathbf{r}(t) = (\cos t)\mathbf{i} + (\sin t)\mathbf{j} + t\mathbf{k}$. How long is the glider's path from $t = 0$ to $t = 2\pi$?

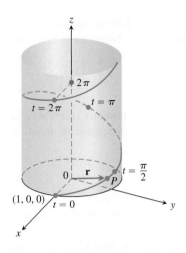

FIGURE 13.13 The helix in Example 1, $\mathbf{r}(t) = (\cos t)\mathbf{i} + (\sin t)\mathbf{j} + t\mathbf{k}$.

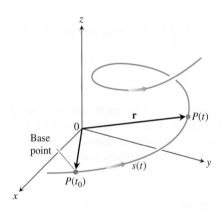

FIGURE 13.14 The directed distance along the curve from $P(t_0)$ to any point $P(t)$ is

$$s(t) = \int_{t_0}^{t} |\mathbf{v}(\tau)| \, d\tau.$$

Solution The path segment during this time corresponds to one full turn of the helix (Figure 13.13). The length of this portion of the curve is

$$L = \int_a^b |\mathbf{v}| \, dt = \int_0^{2\pi} \sqrt{(-\sin t)^2 + (\cos t)^2 + (1)^2} \, dt$$

$$= \int_0^{2\pi} \sqrt{2} \, dt = 2\pi\sqrt{2} \text{ units of length.}$$

This is $\sqrt{2}$ times the circumference of the circle in the xy-plane over which the helix stands. ∎

If we choose a base point $P(t_0)$ on a smooth curve C parametrized by t, each value of t determines a point $P(t) = (x(t), y(t), z(t))$ on C and a "directed distance"

$$s(t) = \int_{t_0}^{t} |\mathbf{v}(\tau)| \, d\tau,$$

measured along C from the base point (Figure 13.14). This is the arc length function we defined in Section 11.2 for plane curves that have no z-component. If $t > t_0$, $s(t)$ is the distance along the curve from $P(t_0)$ to $P(t)$. If $t < t_0$, $s(t)$ is the negative of the distance. Each value of s determines a point on C and this parametrizes C with respect to s. We call s an **arc length parameter** for the curve. The parameter's value increases in the direction of increasing t. We will see that the arc length parameter is particularly effective for investigating the turning and twisting nature of a space curve.

Arc Length Parameter with Base Point $P(t_0)$

$$s(t) = \int_{t_0}^{t} \sqrt{[x'(\tau)]^2 + [y'(\tau)]^2 + [z'(\tau)]^2} \, d\tau = \int_{t_0}^{t} |\mathbf{v}(\tau)| \, d\tau \qquad (3)$$

We use the Greek letter τ ("tau") as the variable of integration in Equation (3) because the letter t is already in use as the upper limit.

If a curve $\mathbf{r}(t)$ is already given in terms of some parameter t and $s(t)$ is the arc length function given by Equation (3), then we may be able to solve for t as a function of s: $t = t(s)$. Then the curve can be reparametrized in terms of s by substituting for t: $\mathbf{r} = \mathbf{r}(t(s))$. The new parametrization identifies a point on the curve with its directed distance along the curve from the base point.

EXAMPLE 2 This is an example for which we can actually find the arc length parametrization of a curve. If $t_0 = 0$, the arc length parameter along the helix

$$\mathbf{r}(t) = (\cos t)\mathbf{i} + (\sin t)\mathbf{j} + t\mathbf{k}$$

from t_0 to t is

$$s(t) = \int_{t_0}^{t} |\mathbf{v}(\tau)| \, d\tau \qquad \text{Eq. (3)}$$

$$= \int_0^{t} \sqrt{2} \, d\tau \qquad \text{Value from Example 1}$$

$$= \sqrt{2}\, t.$$

Solving this equation for t gives $t = s/\sqrt{2}$. Substituting into the position vector $\mathbf{r}$ gives the following arc length parametrization for the helix:

$$\mathbf{r}(t(s)) = \left(\cos \frac{s}{\sqrt{2}} \right)\mathbf{i} + \left(\sin \frac{s}{\sqrt{2}} \right)\mathbf{j} + \frac{s}{\sqrt{2}}\mathbf{k}. \qquad ∎$$

Unlike Example 2, the arc length parametrization is generally difficult to find analytically for a curve already given in terms of some other parameter t. Fortunately, however, we rarely need an exact formula for $s(t)$ or its inverse $t(s)$.

HISTORICAL BIOGRAPHY

Josiah Willard Gibbs
(1839–1903)

Speed on a Smooth Curve

Since the derivatives beneath the radical in Equation (3) are continuous (the curve is smooth), the Fundamental Theorem of Calculus tells us that s is a differentiable function of t with derivative

$$\frac{ds}{dt} = |\mathbf{v}(t)|. \tag{4}$$

Equation (4) says that the speed with which a particle moves along its path is the magnitude of $\mathbf{v}$, consistent with what we know.

Although the base point $P(t_0)$ plays a role in defining s in Equation (3), it plays no role in Equation (4). The rate at which a moving particle covers distance along its path is independent of how far away it is from the base point.

Notice that $ds/dt > 0$ since, by definition, $|\mathbf{v}|$ is never zero for a smooth curve. We see once again that s is an increasing function of t.

Unit Tangent Vector

We already know the velocity vector $\mathbf{v} = d\mathbf{r}/dt$ is tangent to the curve $\mathbf{r}(t)$ and that the vector

$$\mathbf{T} = \frac{\mathbf{v}}{|\mathbf{v}|}$$

is therefore a unit vector tangent to the (smooth) curve, called the **unit tangent vector** (Figure 13.15). The unit tangent vector $\mathbf{T}$ is a differentiable function of t whenever $\mathbf{v}$ is a differentiable function of t. As we will see in Section 13.5, $\mathbf{T}$ is one of three unit vectors in a traveling reference frame that is used to describe the motion of objects traveling in three dimensions.

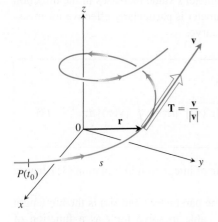

FIGURE 13.15 We find the unit tangent vector $\mathbf{T}$ by dividing $\mathbf{v}$ by $|\mathbf{v}|$.

EXAMPLE 3 Find the unit tangent vector of the curve

$$\mathbf{r}(t) = (3\cos t)\mathbf{i} + (3\sin t)\mathbf{j} + t^2\mathbf{k}$$

representing the path of the glider in Example 3, Section 13.2.

Solution In that example, we found

$$\mathbf{v} = \frac{d\mathbf{r}}{dt} = -(3\sin t)\mathbf{i} + (3\cos t)\mathbf{j} + 2t\mathbf{k}$$

and

$$|\mathbf{v}| = \sqrt{9 + 4t^2}.$$

Thus,

$$\mathbf{T} = \frac{\mathbf{v}}{|\mathbf{v}|} = -\frac{3\sin t}{\sqrt{9 + 4t^2}}\mathbf{i} + \frac{3\cos t}{\sqrt{9 + 4t^2}}\mathbf{j} + \frac{2t}{\sqrt{9 + 4t^2}}\mathbf{k}. \qquad\blacksquare$$

For the counterclockwise motion

$$\mathbf{r}(t) = (\cos t)\mathbf{i} + (\sin t)\mathbf{j}$$

around the unit circle, we see that

$$\mathbf{v} = (-\sin t)\mathbf{i} + (\cos t)\mathbf{j}$$

is already a unit vector, so $\mathbf{T} = \mathbf{v}$ (Figure 13.16).

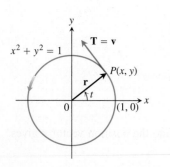

FIGURE 13.16 Counterclockwise motion around the unit circle.

EXAMPLE 6 Find **N** for the helix in Example 5 and describe how the vector is pointing.

Solution We have

$$\frac{d\mathbf{T}}{dt} = -\frac{1}{\sqrt{a^2 + b^2}} [(a \cos t)\mathbf{i} + (a \sin t)\mathbf{j}] \qquad \text{Example 5}$$

$$\left|\frac{d\mathbf{T}}{dt}\right| = \frac{1}{\sqrt{a^2 + b^2}} \sqrt{a^2 \cos^2 t + a^2 \sin^2 t} = \frac{a}{\sqrt{a^2 + b^2}}$$

$$\mathbf{N} = \frac{d\mathbf{T}/dt}{|d\mathbf{T}/dt|} \qquad \text{Eq. (4)}$$

$$= -\frac{\sqrt{a^2 + b^2}}{a} \cdot \frac{1}{\sqrt{a^2 + b^2}} [(a \cos t)\mathbf{i} + (a \sin t)\mathbf{j}]$$

$$= -(\cos t)\mathbf{i} - (\sin t)\mathbf{j}.$$

Thus, **N** is parallel to the *xy*-plane and always points toward the *z*-axis. ∎

Exercises 13.4

Plane Curves

Find **T**, **N**, and κ for the plane curves in Exercises 1–4.

1. $\mathbf{r}(t) = t\mathbf{i} + (\ln \cos t)\mathbf{j}, \quad -\pi/2 < t < \pi/2$

2. $\mathbf{r}(t) = (\ln \sec t)\mathbf{i} + t\mathbf{j}, \quad -\pi/2 < t < \pi/2$

3. $\mathbf{r}(t) = (2t + 3)\mathbf{i} + (5 - t^2)\mathbf{j}$

4. $\mathbf{r}(t) = (\cos t + t \sin t)\mathbf{i} + (\sin t - t \cos t)\mathbf{j}, \quad t > 0$

5. **A formula for the curvature of the graph of a function in the *xy*-plane**

   a. The graph $y = f(x)$ in the *xy*-plane automatically has the parametrization $x = x$, $y = f(x)$, and the vector formula $\mathbf{r}(x) = x\mathbf{i} + f(x)\mathbf{j}$. Use this formula to show that if f is a twice-differentiable function of x, then

   $$\kappa(x) = \frac{|f''(x)|}{\left[1 + (f'(x))^2\right]^{3/2}}.$$

   b. Use the formula for κ in part (a) to find the curvature of $y = \ln(\cos x)$, $-\pi/2 < x < \pi/2$. Compare your answer with the answer in Exercise 1.

   c. Show that the curvature is zero at a point of inflection.

6. **A formula for the curvature of a parametrized plane curve**

   a. Show that the curvature of a smooth curve $\mathbf{r}(t) = f(t)\mathbf{i} + g(t)\mathbf{j}$ defined by twice-differentiable functions $x = f(t)$ and $y = g(t)$ is given by the formula

   $$\kappa = \frac{|\dot{x}\ddot{y} - \dot{y}\ddot{x}|}{(\dot{x}^2 + \dot{y}^2)^{3/2}}.$$

   The dots in the formula denote differentiation with respect to t, one derivative for each dot. Apply the formula to find the curvatures of the following curves.

   b. $\mathbf{r}(t) = t\mathbf{i} + (\ln \sin t)\mathbf{j}, \quad 0 < t < \pi$

   c. $\mathbf{r}(t) = [\tan^{-1}(\sinh t)]\mathbf{i} + (\ln \cosh t)\mathbf{j}$.

7. **Normals to plane curves**

   a. Show that $\mathbf{n}(t) = -g'(t)\mathbf{i} + f'(t)\mathbf{j}$ and $-\mathbf{n}(t) = g'(t)\mathbf{i} - f'(t)\mathbf{j}$ are both normal to the curve $\mathbf{r}(t) = f(t)\mathbf{i} + g(t)\mathbf{j}$ at the point $(f(t), g(t))$.

   To obtain **N** for a particular plane curve, we can choose the one of **n** or $-\mathbf{n}$ from part (a) that points toward the concave side of the curve, and make it into a unit vector. (See Figure 13.19.) Apply this method to find **N** for the following curves.

   b. $\mathbf{r}(t) = t\mathbf{i} + e^{2t}\mathbf{j}$

   c. $\mathbf{r}(t) = \sqrt{4 - t^2}\,\mathbf{i} + t\mathbf{j}, \quad -2 \le t \le 2$

8. (*Continuation of Exercise 7.*)

   a. Use the method of Exercise 7 to find **N** for the curve $\mathbf{r}(t) = t\mathbf{i} + (1/3)t^3\mathbf{j}$ when $t < 0$; when $t > 0$.

   b. Calculate **N** for $t \neq 0$ directly from **T** using Equation (4) for the curve in part (a). Does **N** exist at $t = 0$? Graph the curve and explain what is happening to **N** as t passes from negative to positive values.

Space Curves

Find **T**, **N**, and κ for the space curves in Exercises 9–16.

9. $\mathbf{r}(t) = (3 \sin t)\mathbf{i} + (3 \cos t)\mathbf{j} + 4t\mathbf{k}$

10. $\mathbf{r}(t) = (\cos t + t \sin t)\mathbf{i} + (\sin t - t \cos t)\mathbf{j} + 3\mathbf{k}$

11. $\mathbf{r}(t) = (e^t \cos t)\mathbf{i} + (e^t \sin t)\mathbf{j} + 2\mathbf{k}$

12. $\mathbf{r}(t) = (6 \sin 2t)\mathbf{i} + (6 \cos 2t)\mathbf{j} + 5t\mathbf{k}$

13. $\mathbf{r}(t) = (t^3/3)\mathbf{i} + (t^2/2)\mathbf{j}, \quad t > 0$

14. $\mathbf{r}(t) = (\cos^3 t)\mathbf{i} + (\sin^3 t)\mathbf{j}, \quad 0 < t < \pi/2$

15. $\mathbf{r}(t) = t\mathbf{i} + (a \cosh(t/a))\mathbf{j}, \quad a > 0$

16. $\mathbf{r}(t) = (\cosh t)\mathbf{i} - (\sinh t)\mathbf{j} + t\mathbf{k}$

More on Curvature

17. Show that the parabola $y = ax^2$, $a \neq 0$, has its largest curvature at its vertex and has no minimum curvature. (*Note:* Since the curvature of a curve remains the same if the curve is translated or rotated, this result is true for any parabola.)

18. Show that the ellipse $x = a \cos t$, $y = b \sin t$, $a > b > 0$, has its largest curvature on its major axis and its smallest curvature on its minor axis. (As in Exercise 17, the same is true for any ellipse.)

19. **Maximizing the curvature of a helix** In Example 5, we found the curvature of the helix $\mathbf{r}(t) = (a \cos t)\mathbf{i} + (a \sin t)\mathbf{j} + bt\mathbf{k}$ ($a, b \geq 0$) to be $\kappa = a/(a^2 + b^2)$. What is the largest value κ can have for a given value of b? Give reasons for your answer.

20. **Total curvature** We find the **total curvature** of the portion of a smooth curve that runs from $s = s_0$ to $s = s_1 > s_0$ by integrating κ from s_0 to s_1. If the curve has some other parameter, say t, then the total curvature is

$$K = \int_{s_0}^{s_1} \kappa \, ds = \int_{t_0}^{t_1} \kappa \frac{ds}{dt} \, dt = \int_{t_0}^{t_1} \kappa |\mathbf{v}| \, dt,$$

where t_0 and t_1 correspond to s_0 and s_1. Find the total curvatures of

a. The portion of the helix $\mathbf{r}(t) = (3 \cos t)\mathbf{i} + (3 \sin t)\mathbf{j} + t\mathbf{k}$, $0 \leq t \leq 4\pi$.

b. The parabola $y = x^2$, $-\infty < x < \infty$.

21. Find an equation for the circle of curvature of the curve $\mathbf{r}(t) = t\mathbf{i} + (\sin t)\mathbf{j}$ at the point $(\pi/2, 1)$. (The curve parametrizes the graph of $y = \sin x$ in the xy-plane.)

22. Find an equation for the circle of curvature of the curve $\mathbf{r}(t) = (2 \ln t)\mathbf{i} - [t + (1/t)]\mathbf{j}$, $e^{-2} \leq t \leq e^2$, at the point $(0, -2)$, where $t = 1$.

T The formula

$$\kappa(x) = \frac{|f''(x)|}{\left[1 + (f'(x))^2\right]^{3/2}},$$

derived in Exercise 5, expresses the curvature $\kappa(x)$ of a twice-differentiable plane curve $y = f(x)$ as a function of x. Find the curvature function of each of the curves in Exercises 23–26. Then graph $f(x)$ together with $\kappa(x)$ over the given interval. You will find some surprises.

23. $y = x^2$, $-2 \leq x \leq 2$ 24. $y = x^4/4$, $-2 \leq x \leq 2$

25. $y = \sin x$, $0 \leq x \leq 2\pi$ 26. $y = e^x$, $-1 \leq x \leq 2$

COMPUTER EXPLORATIONS

In Exercises 27–34 you will use a CAS to explore the osculating circle at a point P on a plane curve where $\kappa \neq 0$. Use a CAS to perform the following steps:

a. Plot the plane curve given in parametric or function form over the specified interval to see what it looks like.

b. Calculate the curvature κ of the curve at the given value t_0 using the appropriate formula from Exercise 5 or 6. Use the parametrization $x = t$ and $y = f(t)$ if the curve is given as a function $y = f(x)$.

c. Find the unit normal vector $\mathbf{N}$ at t_0. Notice that the signs of the components of $\mathbf{N}$ depend on whether the unit tangent vector $\mathbf{T}$ is turning clockwise or counterclockwise at $t = t_0$. (See Exercise 7.)

d. If $\mathbf{C} = a\mathbf{i} + b\mathbf{j}$ is the vector from the origin to the center (a, b) of the osculating circle, find the center $\mathbf{C}$ from the vector equation

$$\mathbf{C} = \mathbf{r}(t_0) + \frac{1}{\kappa(t_0)} \mathbf{N}(t_0).$$

The point $P(x_0, y_0)$ on the curve is given by the position vector $\mathbf{r}(t_0)$.

e. Plot implicitly the equation $(x - a)^2 + (y - b)^2 = 1/\kappa^2$ of the osculating circle. Then plot the curve and osculating circle together. You may need to experiment with the size of the viewing window, but be sure it is square.

27. $\mathbf{r}(t) = (3 \cos t)\mathbf{i} + (5 \sin t)\mathbf{j}$, $0 \leq t \leq 2\pi$, $t_0 = \pi/4$

28. $\mathbf{r}(t) = (\cos^3 t)\mathbf{i} + (\sin^3 t)\mathbf{j}$, $0 \leq t \leq 2\pi$, $t_0 = \pi/4$

29. $\mathbf{r}(t) = t^2\mathbf{i} + (t^3 - 3t)\mathbf{j}$, $-4 \leq t \leq 4$, $t_0 = 3/5$

30. $\mathbf{r}(t) = (t^3 - 2t^2 - t)\mathbf{i} + \dfrac{3t}{\sqrt{1 + t^2}}\mathbf{j}$, $-2 \leq t \leq 5$, $t_0 = 1$

31. $\mathbf{r}(t) = (2t - \sin t)\mathbf{i} + (2 - 2 \cos t)\mathbf{j}$, $0 \leq t \leq 3\pi$, $t_0 = 3\pi/2$

32. $\mathbf{r}(t) = (e^{-t} \cos t)\mathbf{i} + (e^{-t} \sin t)\mathbf{j}$, $0 \leq t \leq 6\pi$, $t_0 = \pi/4$

33. $y = x^2 - x$, $-2 \leq x \leq 5$, $x_0 = 1$

34. $y = x(1 - x)^{2/5}$, $-1 \leq x \leq 2$, $x_0 = 1/2$

13.5 Tangential and Normal Components of Acceleration

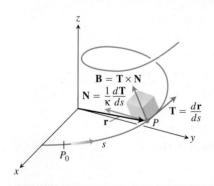

FIGURE 13.23 The **TNB** frame of mutually orthogonal unit vectors traveling along a curve in space.

If you are traveling along a space curve, the Cartesian $\mathbf{i}$, $\mathbf{j}$, and $\mathbf{k}$ coordinate system for representing the vectors describing your motion is not truly relevant to you. What is meaningful instead are the vectors representative of your forward direction (the unit tangent vector $\mathbf{T}$), the direction in which your path is turning (the unit normal vector $\mathbf{N}$), and the tendency of your motion to "twist" out of the plane created by these vectors in the direction perpendicular to this plane (defined by the *unit binormal vector* $\mathbf{B} = \mathbf{T} \times \mathbf{N}$). Expressing the acceleration vector along the curve as a linear combination of this **TNB** frame of mutually orthogonal unit vectors traveling with the motion (Figure 13.23) is particularly revealing of the nature of the path and motion along it.

The TNB Frame

The **binormal vector** of a curve in space is $\mathbf{B} = \mathbf{T} \times \mathbf{N}$, a unit vector orthogonal to both $\mathbf{T}$ and $\mathbf{N}$ (Figure 13.24). Together $\mathbf{T}$, $\mathbf{N}$, and $\mathbf{B}$ define a moving right-handed vector frame that plays a significant role in calculating the paths of particles moving through space. It is called the **Frenet** ("fre-*nay*") **frame** (after Jean-Frédéric Frenet, 1816–1900), or the **TNB frame**.

Tangential and Normal Components of Acceleration

When an object is accelerated by gravity, brakes, or a combination of rocket motors, we usually want to know how much of the acceleration acts in the direction of motion, in the tangential direction **T**. We can calculate this using the Chain Rule to rewrite **v** as

$$\mathbf{v} = \frac{d\mathbf{r}}{dt} = \frac{d\mathbf{r}}{ds}\frac{ds}{dt} = \mathbf{T}\frac{ds}{dt}.$$

Then we differentiate both ends of this string of equalities to get

$$\mathbf{a} = \frac{d\mathbf{v}}{dt} = \frac{d}{dt}\left(\mathbf{T}\frac{ds}{dt}\right) = \frac{d^2s}{dt^2}\mathbf{T} + \frac{ds}{dt}\frac{d\mathbf{T}}{dt}$$

$$= \frac{d^2s}{dt^2}\mathbf{T} + \frac{ds}{dt}\left(\frac{d\mathbf{T}}{ds}\frac{ds}{dt}\right) = \frac{d^2s}{dt^2}\mathbf{T} + \frac{ds}{dt}\left(\kappa\mathbf{N}\frac{ds}{dt}\right) \qquad \frac{d\mathbf{T}}{ds} = \kappa\mathbf{N}$$

$$= \frac{d^2s}{dt^2}\mathbf{T} + \kappa\left(\frac{ds}{dt}\right)^2\mathbf{N}.$$

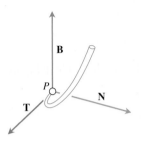

FIGURE 13.24 The vectors **T**, **N**, and **B** (in that order) make a right-handed frame of mutually orthogonal unit vectors in space.

> **DEFINITION** If the acceleration vector is written as
>
> $$\mathbf{a} = a_{\mathrm{T}}\mathbf{T} + a_{\mathrm{N}}\mathbf{N}, \tag{1}$$
>
> then
>
> $$a_{\mathrm{T}} = \frac{d^2s}{dt^2} = \frac{d}{dt}|\mathbf{v}| \qquad \text{and} \qquad a_{\mathrm{N}} = \kappa\left(\frac{ds}{dt}\right)^2 = \kappa|\mathbf{v}|^2 \tag{2}$$
>
> are the **tangential** and **normal** scalar components of acceleration.

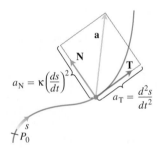

FIGURE 13.25 The tangential and normal components of acceleration. The acceleration **a** always lies in the plane of **T** and **N**, orthogonal to **B**.

Notice that the binormal vector **B** does not appear in Equation (1). No matter how the path of the moving object we are watching may appear to twist and turn in space, the acceleration **a** *always lies in the plane of* **T** and **N** orthogonal to **B**. The equation also tells us exactly how much of the acceleration takes place tangent to the motion (d^2s/dt^2) and how much takes place normal to the motion [$\kappa(ds/dt)^2$] (Figure 13.25).

What information can we discover from Equations (2)? By definition, acceleration **a** is the rate of change of velocity **v**, and in general, both the length and direction of **v** change as an object moves along its path. The tangential component of acceleration a_{T} measures the rate of change of the *length* of **v** (that is, the change in the speed). The normal component of acceleration a_{N} measures the rate of change of the *direction* of **v**.

Notice that the normal scalar component of the acceleration is the curvature times the *square* of the speed. This explains why you have to hold on when your car makes a sharp (large κ), high-speed (large $|\mathbf{v}|$) turn. If you double the speed of your car, you will experience four times the normal component of acceleration for the same curvature.

If an object moves in a circle at a constant speed, d^2s/dt^2 is zero and all the acceleration points along **N** toward the circle's center. If the object is speeding up or slowing down, **a** has a nonzero tangential component (Figure 13.26).

To calculate a_{N}, we usually use the formula $a_{\mathrm{N}} = \sqrt{|\mathbf{a}|^2 - a_{\mathrm{T}}^2}$, which comes from solving the equation $|\mathbf{a}|^2 = \mathbf{a}\cdot\mathbf{a} = a_{\mathrm{T}}^2 + a_{\mathrm{N}}^2$ for a_{N}. With this formula, we can find a_{N} without having to calculate κ first.

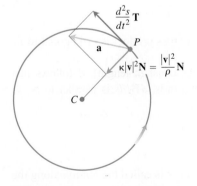

FIGURE 13.26 The tangential and normal components of the acceleration of an object that is speeding up as it moves counterclockwise around a circle of radius ρ.

Formula for Calculating the Normal Component of Acceleration

$$a_{\mathrm{N}} = \sqrt{|\mathbf{a}|^2 - a_{\mathrm{T}}^2} \tag{3}$$

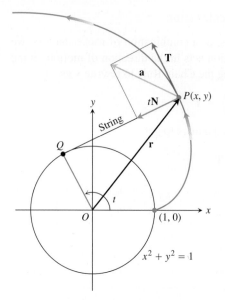

FIGURE 13.27 The tangential and normal components of the acceleration of the motion $\mathbf{r}(t) = (\cos t + t \sin t)\mathbf{i} + (\sin t - t \cos t)\mathbf{j}$, for $t > 0$. If a string wound around a fixed circle is unwound while held taut in the plane of the circle, its end P traces an involute of the circle (Example 1).

EXAMPLE 1 Without finding $\mathbf{T}$ and $\mathbf{N}$, write the acceleration of the motion

$$\mathbf{r}(t) = (\cos t + t \sin t)\mathbf{i} + (\sin t - t \cos t)\mathbf{j}, \qquad t > 0$$

in the form $\mathbf{a} = a_T\mathbf{T} + a_N\mathbf{N}$. (The path of the motion is the involute of the circle in Figure 13.27. See also Section 13.3, Exercise 19.)

Solution We use the first of Equations (2) to find a_T:

$$\mathbf{v} = \frac{d\mathbf{r}}{dt} = (-\sin t + \sin t + t \cos t)\mathbf{i} + (\cos t - \cos t + t \sin t)\mathbf{j}$$

$$= (t \cos t)\mathbf{i} + (t \sin t)\mathbf{j}$$

$$|\mathbf{v}| = \sqrt{t^2 \cos^2 t + t^2 \sin^2 t} = \sqrt{t^2} = |t| = t \qquad\qquad t > 0$$

$$a_T = \frac{d}{dt}|\mathbf{v}| = \frac{d}{dt}(t) = 1. \qquad\qquad \text{Eq. (2)}$$

Knowing a_T, we use Equation (3) to find a_N:

$$\mathbf{a} = (\cos t - t \sin t)\mathbf{i} + (\sin t + t \cos t)\mathbf{j}$$

$$|\mathbf{a}|^2 = t^2 + 1 \qquad\qquad \text{After some algebra}$$

$$a_N = \sqrt{|\mathbf{a}|^2 - a_T{}^2}$$

$$= \sqrt{(t^2 + 1) - (1)} = \sqrt{t^2} = t.$$

We then use Equation (1) to find $\mathbf{a}$:

$$\mathbf{a} = a_T\mathbf{T} + a_N\mathbf{N} = (1)\mathbf{T} + (t)\mathbf{N} = \mathbf{T} + t\mathbf{N}. \qquad \blacksquare$$

Torsion

How does $d\mathbf{B}/ds$ behave in relation to $\mathbf{T}$, $\mathbf{N}$, and $\mathbf{B}$? From the rule for differentiating a cross product, we have

$$\frac{d\mathbf{B}}{ds} = \frac{d(\mathbf{T} \times \mathbf{N})}{ds} = \frac{d\mathbf{T}}{ds} \times \mathbf{N} + \mathbf{T} \times \frac{d\mathbf{N}}{ds}.$$

Since $\mathbf{N}$ is the direction of $d\mathbf{T}/ds$, $(d\mathbf{T}/ds) \times \mathbf{N} = \mathbf{0}$ and

$$\frac{d\mathbf{B}}{ds} = \mathbf{0} + \mathbf{T} \times \frac{d\mathbf{N}}{ds} = \mathbf{T} \times \frac{d\mathbf{N}}{ds}.$$

From this we see that $d\mathbf{B}/ds$ is orthogonal to $\mathbf{T}$ since a cross product is orthogonal to its factors.

Since $d\mathbf{B}/ds$ is also orthogonal to $\mathbf{B}$ (the latter has constant length), it follows that $d\mathbf{B}/ds$ is orthogonal to the plane of $\mathbf{B}$ and $\mathbf{T}$. In other words, $d\mathbf{B}/ds$ is parallel to $\mathbf{N}$, so $d\mathbf{B}/ds$ is a scalar multiple of $\mathbf{N}$. In symbols,

$$\frac{d\mathbf{B}}{ds} = -\tau\mathbf{N}.$$

The negative sign in this equation is traditional. The scalar τ is called the *torsion* along the curve. Notice that

$$\frac{d\mathbf{B}}{ds} \cdot \mathbf{N} = -\tau\mathbf{N} \cdot \mathbf{N} = -\tau(1) = -\tau.$$

We use this equation for our next definition.

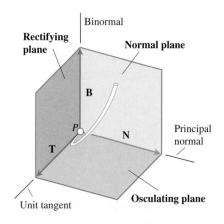

FIGURE 13.28   The names of the three planes determined by **T**, **N**, and **B**.

> **DEFINITION**   Let $\mathbf{B} = \mathbf{T} \times \mathbf{N}$. The **torsion** function of a smooth curve is
>
> $$\tau = -\frac{d\mathbf{B}}{ds} \cdot \mathbf{N}. \tag{4}$$

Unlike the curvature κ, which is never negative, the torsion τ may be positive, negative, or zero.

The three planes determined by **T**, **N**, and **B** are named and shown in Figure 13.28. The curvature $\kappa = |d\mathbf{T}/ds|$ can be thought of as the rate at which the normal plane turns as the point P moves along its path. Similarly, the torsion $\tau = -(d\mathbf{B}/ds) \cdot \mathbf{N}$ is the rate at which the osculating plane turns about **T** as P moves along the curve. Torsion measures how the curve twists.

Look at Figure 13.29. If P is a train climbing up a curved track, the rate at which the headlight turns from side to side per unit distance is the curvature of the track. The rate at which the engine tends to twist out of the plane formed by **T** and **N** is the torsion. In a more advanced course it can be shown that a space curve is a helix if and only if it has constant nonzero curvature and constant nonzero torsion.

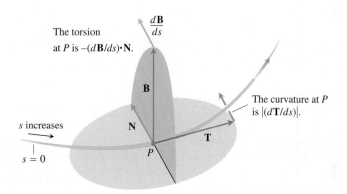

FIGURE 13.29   Every moving body travels with a **TNB** frame that characterizes the geometry of its path of motion.

Computational Formulas

The most widely used formula for torsion, derived in more advanced texts, is

$$\tau = \frac{\begin{vmatrix} \dot{x} & \dot{y} & \dot{z} \\ \ddot{x} & \ddot{y} & \ddot{z} \\ \dddot{x} & \dddot{y} & \dddot{z} \end{vmatrix}}{|\mathbf{v} \times \mathbf{a}|^2} \qquad (\text{if } \mathbf{v} \times \mathbf{a} \neq \mathbf{0}). \tag{5}$$

The dots in Equation (5) denote differentiation with respect to t, one derivative for each dot. Thus, $\dot{x}$ ("x dot") means dx/dt, $\ddot{x}$ ("x double dot") means d^2x/dt^2, and $\dddot{x}$ ("x triple dot") means d^3x/dt^3. Similarly, $\dot{y} = dy/dt$, and so on.

There is also an easy-to-use formula for curvature, as given in the following summary table (see Exercise 21).

Computation Formulas for Curves in Space

Unit tangent vector:
$$\mathbf{T} = \frac{\mathbf{v}}{|\mathbf{v}|}$$

Principal unit normal vector:
$$\mathbf{N} = \frac{d\mathbf{T}/dt}{|d\mathbf{T}/dt|}$$

Binormal vector:
$$\mathbf{B} = \mathbf{T} \times \mathbf{N}$$

Curvature:
$$\kappa = \left|\frac{d\mathbf{T}}{ds}\right| = \frac{|\mathbf{v} \times \mathbf{a}|}{|\mathbf{v}|^3}$$

Torsion:
$$\tau = -\frac{d\mathbf{B}}{ds} \cdot \mathbf{N} = \frac{\begin{vmatrix} \dot{x} & \dot{y} & \dot{z} \\ \ddot{x} & \ddot{y} & \ddot{z} \\ \dddot{x} & \dddot{y} & \dddot{z} \end{vmatrix}}{|\mathbf{v} \times \mathbf{a}|^2}$$

Tangential and normal scalar components of acceleration:
$$\mathbf{a} = a_T\mathbf{T} + a_N\mathbf{N}$$
$$a_T = \frac{d}{dt}|\mathbf{v}|$$
$$a_N = \kappa|\mathbf{v}|^2 = \sqrt{|\mathbf{a}|^2 - a_T^2}$$

Exercises 13.5

Finding Tangential and Normal Components

In Exercises 1 and 2, write $\mathbf{a}$ in the form $\mathbf{a} = a_T\mathbf{T} + a_N\mathbf{N}$ without finding $\mathbf{T}$ and $\mathbf{N}$.

1. $\mathbf{r}(t) = (a\cos t)\mathbf{i} + (a\sin t)\mathbf{j} + bt\mathbf{k}$

2. $\mathbf{r}(t) = (1 + 3t)\mathbf{i} + (t - 2)\mathbf{j} - 3t\mathbf{k}$

In Exercises 3–6, write $\mathbf{a}$ in the form $\mathbf{a} = a_T\mathbf{T} + a_N\mathbf{N}$ at the given value of t without finding $\mathbf{T}$ and $\mathbf{N}$.

3. $\mathbf{r}(t) = (t + 1)\mathbf{i} + 2t\mathbf{j} + t^2\mathbf{k}, \quad t = 1$

4. $\mathbf{r}(t) = (t\cos t)\mathbf{i} + (t\sin t)\mathbf{j} + t^2\mathbf{k}, \quad t = 0$

5. $\mathbf{r}(t) = t^2\mathbf{i} + (t + (1/3)t^3)\mathbf{j} + (t - (1/3)t^3)\mathbf{k}, \quad t = 0$

6. $\mathbf{r}(t) = (e^t\cos t)\mathbf{i} + (e^t\sin t)\mathbf{j} + \sqrt{2}e^t\mathbf{k}, \quad t = 0$

Finding the TNB Frame

In Exercises 7 and 8, find $\mathbf{r}$, $\mathbf{T}$, $\mathbf{N}$, and $\mathbf{B}$ at the given value of t. Then find equations for the osculating, normal, and rectifying planes at that value of t.

7. $\mathbf{r}(t) = (\cos t)\mathbf{i} + (\sin t)\mathbf{j} - \mathbf{k}, \quad t = \pi/4$

8. $\mathbf{r}(t) = (\cos t)\mathbf{i} + (\sin t)\mathbf{j} + t\mathbf{k}, \quad t = 0$

In Exercises 9–16 of Section 13.4, you found $\mathbf{T}$, $\mathbf{N}$, and κ. Now, in the following Exercises 9–16, find $\mathbf{B}$ and τ for these space curves.

9. $\mathbf{r}(t) = (3\sin t)\mathbf{i} + (3\cos t)\mathbf{j} + 4t\mathbf{k}$

10. $\mathbf{r}(t) = (\cos t + t\sin t)\mathbf{i} + (\sin t - t\cos t)\mathbf{j} + 3\mathbf{k}$

11. $\mathbf{r}(t) = (e^t\cos t)\mathbf{i} + (e^t\sin t)\mathbf{j} + 2\mathbf{k}$

12. $\mathbf{r}(t) = (6\sin 2t)\mathbf{i} + (6\cos 2t)\mathbf{j} + 5t\mathbf{k}$

13. $\mathbf{r}(t) = (t^3/3)\mathbf{i} + (t^2/2)\mathbf{j}, \quad t > 0$

14. $\mathbf{r}(t) = (\cos^3 t)\mathbf{i} + (\sin^3 t)\mathbf{j}, \quad 0 < t < \pi/2$

15. $\mathbf{r}(t) = t\mathbf{i} + (a\cosh(t/a))\mathbf{j}, \quad a > 0$

16. $\mathbf{r}(t) = (\cosh t)\mathbf{i} - (\sinh t)\mathbf{j} + t\mathbf{k}$

Physical Applications

17. The speedometer on your car reads a steady 35 mph. Could you be accelerating? Explain.

18. Can anything be said about the acceleration of a particle that is moving at a constant speed? Give reasons for your answer.

19. Can anything be said about the speed of a particle whose acceleration is always orthogonal to its velocity? Give reasons for your answer.

20. An object of mass m travels along the parabola $y = x^2$ with a constant speed of 10 units/sec. What is the force on the object due to its acceleration at $(0, 0)$? at $(2^{1/2}, 2)$? Write your answers in terms of $\mathbf{i}$ and $\mathbf{j}$. (Remember Newton's law, $\mathbf{F} = m\mathbf{a}$.)

Theory and Examples

21. Vector formula for curvature For a smooth curve, use Equation (1) to derive the curvature formula
$$\kappa = \frac{|\mathbf{v} \times \mathbf{a}|}{|\mathbf{v}|^3}.$$

22. Show that a moving particle will move in a straight line if the normal component of its acceleration is zero.

23. A sometime shortcut to curvature If you already know $|a_N|$ and $|\mathbf{v}|$, then the formula $a_N = \kappa|\mathbf{v}|^2$ gives a convenient way to find the curvature. Use it to find the curvature and radius of curvature of the curve

$$\mathbf{r}(t) = (\cos t + t \sin t)\mathbf{i} + (\sin t - t \cos t)\mathbf{j}, \quad t > 0.$$

(Take a_N and $|\mathbf{v}|$ from Example 1.)

24. Show that κ and τ are both zero for the line

$$\mathbf{r}(t) = (x_0 + At)\mathbf{i} + (y_0 + Bt)\mathbf{j} + (z_0 + Ct)\mathbf{k}.$$

25. What can be said about the torsion of a smooth plane curve $\mathbf{r}(t) = f(t)\mathbf{i} + g(t)\mathbf{j}$? Give reasons for your answer.

26. The torsion of a helix Show that the torsion of the helix

$$\mathbf{r}(t) = (a \cos t)\mathbf{i} + (a \sin t)\mathbf{j} + bt\mathbf{k}, \quad a, b \geq 0$$

is $\tau = b/(a^2 + b^2)$. What is the largest value τ can have for a given value of a? Give reasons for your answer.

27. Differentiable curves with zero torsion lie in planes That a sufficiently differentiable curve with zero torsion lies in a plane is a special case of the fact that a particle whose velocity remains perpendicular to a fixed vector $\mathbf{C}$ moves in a plane perpendicular to $\mathbf{C}$. This, in turn, can be viewed as the following result.

Suppose $\mathbf{r}(t) = f(t)\mathbf{i} + g(t)\mathbf{j} + h(t)\mathbf{k}$ is twice differentiable for all t in an interval $[a, b]$, that $\mathbf{r} = 0$ when $t = a$, and that

$\mathbf{v} \cdot \mathbf{k} = 0$ for all t in $[a, b]$. Show that $h(t) = 0$ for all t in $[a, b]$. (*Hint:* Start with $\mathbf{a} = d^2\mathbf{r}/dt^2$ and apply the initial conditions in reverse order.).

28. A formula that calculates τ from B and v If we start with the definition $\tau = -(d\mathbf{B}/ds) \cdot \mathbf{N}$ and apply the Chain Rule to rewrite $d\mathbf{B}/ds$ as

$$\frac{d\mathbf{B}}{ds} = \frac{d\mathbf{B}}{dt}\frac{dt}{ds} = \frac{d\mathbf{B}}{dt}\frac{1}{|\mathbf{v}|},$$

we arrive at the formula

$$\tau = -\frac{1}{|\mathbf{v}|}\left(\frac{d\mathbf{B}}{dt} \cdot \mathbf{N}\right).$$

The advantage of this formula over Equation (5) is that it is easier to derive and state. The disadvantage is that it can take a lot of work to evaluate without a computer. Use the new formula to find the torsion of the helix in Exercise 26.

COMPUTER EXPLORATIONS

Rounding the answers to four decimal places, use a CAS to find $\mathbf{v}$, $\mathbf{a}$, speed, $\mathbf{T}$, $\mathbf{N}$, $\mathbf{B}$, κ, τ, and the tangential and normal components of acceleration for the curves in Exercises 29–32 at the given values of t.

29. $\mathbf{r}(t) = (t \cos t)\mathbf{i} + (t \sin t)\mathbf{j} + t\mathbf{k}, \quad t = \sqrt{3}$

30. $\mathbf{r}(t) = (e^t \cos t)\mathbf{i} + (e^t \sin t)\mathbf{j} + e^t\mathbf{k}, \quad t = \ln 2$

31. $\mathbf{r}(t) = (t - \sin t)\mathbf{i} + (1 - \cos t)\mathbf{j} + \sqrt{-t}\,\mathbf{k}, \quad t = -3\pi$

32. $\mathbf{r}(t) = (3t - t^2)\mathbf{i} + (3t^2)\mathbf{j} + (3t + t^3)\mathbf{k}, \quad t = 1$

13.6 | Velocity and Acceleration in Polar Coordinates

In this section we derive equations for velocity and acceleration in polar coordinates. These equations are useful for calculating the paths of planets and satellites in space, and we use them to examine Kepler's three laws of planetary motion.

Motion in Polar and Cylindrical Coordinates

When a particle at $P(r, \theta)$ moves along a curve in the polar coordinate plane, we express its position, velocity, and acceleration in terms of the moving unit vectors

$$\mathbf{u}_r = (\cos \theta)\mathbf{i} + (\sin \theta)\mathbf{j}, \qquad \mathbf{u}_\theta = -(\sin \theta)\mathbf{i} + (\cos \theta)\mathbf{j}, \tag{1}$$

shown in Figure 13.30. The vector $\mathbf{u}_r$ points along the position vector $\overrightarrow{OP}$, so $\mathbf{r} = r\mathbf{u}_r$. The vector $\mathbf{u}_\theta$, orthogonal to $\mathbf{u}_r$, points in the direction of increasing θ.

We find from Equations (1) that

$$\frac{d\mathbf{u}_r}{d\theta} = -(\sin \theta)\mathbf{i} + (\cos \theta)\mathbf{j} = \mathbf{u}_\theta$$

$$\frac{d\mathbf{u}_\theta}{d\theta} = -(\cos \theta)\mathbf{i} - (\sin \theta)\mathbf{j} = -\mathbf{u}_r.$$

When we differentiate $\mathbf{u}_r$ and $\mathbf{u}_\theta$ with respect to t to find how they change with time, the Chain Rule gives

$$\dot{\mathbf{u}}_r = \frac{d\mathbf{u}_r}{d\theta}\dot{\theta} = \dot{\theta}\mathbf{u}_\theta, \qquad \dot{\mathbf{u}}_\theta = \frac{d\mathbf{u}_\theta}{d\theta}\dot{\theta} = -\dot{\theta}\mathbf{u}_r. \tag{2}$$

FIGURE 13.30 The length of $\mathbf{r}$ is the positive polar coordinate r of the point P. Thus, $\mathbf{u}_r$, which is $\mathbf{r}/|\mathbf{r}|$, is also $\mathbf{r}/r$. Equations (1) express $\mathbf{u}_r$ and $\mathbf{u}_\theta$ in terms of $\mathbf{i}$ and $\mathbf{j}$.

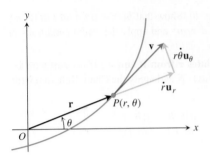

FIGURE 13.31 In polar coordinates, the velocity vector is

$$\mathbf{v} = \dot{r}\mathbf{u}_r + r\dot{\theta}\mathbf{u}_\theta.$$

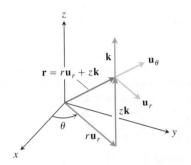

FIGURE 13.32 Position vector and basic unit vectors in cylindrical coordinates. Notice that $|\mathbf{r}| \neq r$ if $z \neq 0$.

Hence, we can express the velocity vector in terms of $\mathbf{u}_r$ and $\mathbf{u}_\theta$ as

$$\mathbf{v} = \dot{\mathbf{r}} = \frac{d}{dt}\left(r\mathbf{u}_r\right) = \dot{r}\mathbf{u}_r + r\dot{\mathbf{u}}_r = \dot{r}\mathbf{u}_r + r\dot{\theta}\mathbf{u}_\theta.$$

See Figure 13.31. As in the previous section, we use Newton's dot notation for time derivatives to keep the formulas as simple as we can: $\dot{\mathbf{u}}_r$ means $d\mathbf{u}_r/dt$, $\dot{\theta}$ means $d\theta/dt$, and so on.

The acceleration is

$$\mathbf{a} = \dot{\mathbf{v}} = (\ddot{r}\mathbf{u}_r + \dot{r}\dot{\mathbf{u}}_r) + (\dot{r}\dot{\theta}\mathbf{u}_\theta + r\ddot{\theta}\mathbf{u}_\theta + r\dot{\theta}\dot{\mathbf{u}}_\theta).$$

When Equations (2) are used to evaluate $\dot{\mathbf{u}}_r$ and $\dot{\mathbf{u}}_\theta$ and the components are separated, the equation for acceleration in terms of $\mathbf{u}_r$ and $\mathbf{u}_\theta$ becomes

$$\mathbf{a} = (\ddot{r} - r\dot{\theta}^2)\mathbf{u}_r + (r\ddot{\theta} + 2\dot{r}\dot{\theta})\mathbf{u}_\theta.$$

To extend these equations of motion to space, we add $z\mathbf{k}$ to the right-hand side of the equation $\mathbf{r} = r\mathbf{u}_r$. Then, in these *cylindrical coordinates*, we have

$$\mathbf{r} = r\mathbf{u}_r + z\mathbf{k}$$
$$\mathbf{v} = \dot{r}\mathbf{u}_r + r\dot{\theta}\mathbf{u}_\theta + \dot{z}\mathbf{k} \tag{3}$$
$$\mathbf{a} = (\ddot{r} - r\dot{\theta}^2)\mathbf{u}_r + (r\ddot{\theta} + 2\dot{r}\dot{\theta})\mathbf{u}_\theta + \ddot{z}\mathbf{k}.$$

The vectors $\mathbf{u}_r$, $\mathbf{u}_\theta$, and $\mathbf{k}$ make a right-handed frame (Figure 13.32) in which

$$\mathbf{u}_r \times \mathbf{u}_\theta = \mathbf{k}, \qquad \mathbf{u}_\theta \times \mathbf{k} = \mathbf{u}_r, \qquad \mathbf{k} \times \mathbf{u}_r = \mathbf{u}_\theta.$$

Planets Move in Planes

Newton's law of gravitation says that if $\mathbf{r}$ is the radius vector from the center of a sun of mass M to the center of a planet of mass m, then the force $\mathbf{F}$ of the gravitational attraction between the planet and sun is

$$\mathbf{F} = -\frac{GmM}{|\mathbf{r}|^2}\frac{\mathbf{r}}{|\mathbf{r}|}$$

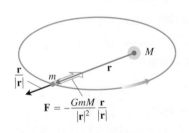

FIGURE 13.33 The force of gravity is directed along the line joining the centers of mass.

(Figure 13.33). The number G is the **universal gravitational constant**. If we measure mass in kilograms, force in newtons, and distance in meters, G is about $6.6726 \times 10^{-11}\ \text{Nm}^2\ \text{kg}^{-2}$.

Combining the gravitation law with Newton's second law, $\mathbf{F} = m\ddot{\mathbf{r}}$, for the force acting on the planet gives

$$m\ddot{\mathbf{r}} = -\frac{GmM}{|\mathbf{r}|^2}\frac{\mathbf{r}}{|\mathbf{r}|},$$
$$\ddot{\mathbf{r}} = -\frac{GM}{|\mathbf{r}|^2}\frac{\mathbf{r}}{|\mathbf{r}|}.$$

The planet is accelerated toward the sun's center of mass at all times.

Since $\ddot{\mathbf{r}}$ is a scalar multiple of $\mathbf{r}$, we have

$$\mathbf{r} \times \ddot{\mathbf{r}} = \mathbf{0}.$$

From this last equation,

$$\frac{d}{dt}(\mathbf{r} \times \dot{\mathbf{r}}) = \underbrace{\dot{\mathbf{r}} \times \dot{\mathbf{r}}}_{0} + \mathbf{r} \times \ddot{\mathbf{r}} = \mathbf{r} \times \ddot{\mathbf{r}} = \mathbf{0}.$$

It follows that

$$\mathbf{r} \times \dot{\mathbf{r}} = \mathbf{C} \tag{4}$$

for some constant vector C.

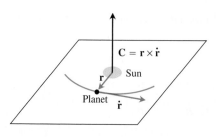

FIGURE 13.34 A planet that obeys Newton's laws of gravitation and motion travels in the plane through the sun's center of mass perpendicular to $\mathbf{C} = \mathbf{r} \times \dot{\mathbf{r}}$.

HISTORICAL BIOGRAPHY

Johannes Kepler
(1571–1630)

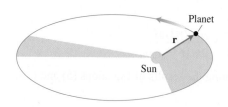

FIGURE 13.35 The line joining a planet to its sun sweeps over equal areas in equal times.

Equation (4) tells us that $\mathbf{r}$ and $\dot{\mathbf{r}}$ always lie in a plane perpendicular to $\mathbf{C}$. Hence, the planet moves in a fixed plane through the center of its sun (Figure 13.34). We next see how Kepler's laws describe the motion in a precise way.

Kepler's First Law (Ellipse Law)

Kepler's first law says that a planet's path is an ellipse with the sun at one focus. The eccentricity of the ellipse is

$$e = \frac{r_0 v_0^2}{GM} - 1 \tag{5}$$

and the polar equation (see Section 11.7, Equation (5)) is

$$r = \frac{(1 + e)r_0}{1 + e \cos \theta}. \tag{6}$$

Here v_0 is the speed when the planet is positioned at its minimum distance r_0 from the sun. We omit the lengthy proof. The sun's mass M is 1.99×10^{30} kg.

Kepler's Second Law (Equal Area Law)

Kepler's second law says that the radius vector from the sun to a planet (the vector $\mathbf{r}$ in our model) sweeps out equal areas in equal times (Figure 13.35). To derive the law, we use Equation (3) to evaluate the cross product $\mathbf{C} = \mathbf{r} \times \dot{\mathbf{r}}$ from Equation (4):

$$
\begin{aligned}
\mathbf{C} = \mathbf{r} \times \dot{\mathbf{r}} &= \mathbf{r} \times \mathbf{v} \\
&= r\mathbf{u}_r \times (\dot{r}\mathbf{u}_r + r\dot{\theta}\mathbf{u}_\theta) && \text{Eq. (3), } \dot{z} = 0 \\
&= r\dot{r}\underbrace{(\mathbf{u}_r \times \mathbf{u}_r)}_{0} + r(r\dot{\theta})\underbrace{(\mathbf{u}_r \times \mathbf{u}_\theta)}_{\mathbf{k}} \\
&= r(r\dot{\theta})\mathbf{k}.
\end{aligned}
\tag{7}
$$

Setting t equal to zero shows that

$$\mathbf{C} = [r(r\dot{\theta})]_{t=0}\,\mathbf{k} = r_0 v_0 \mathbf{k}.$$

Substituting this value for $\mathbf{C}$ in Equation (7) gives

$$r_0 v_0 \mathbf{k} = r^2 \dot{\theta} \mathbf{k}, \qquad \text{or} \qquad r^2 \dot{\theta} = r_0 v_0.$$

This is where the area comes in. The area differential in polar coordinates is

$$dA = \frac{1}{2} r^2 \, d\theta$$

(Section 11.5). Accordingly, dA/dt has the constant value

$$\frac{dA}{dt} = \frac{1}{2} r^2 \dot{\theta} = \frac{1}{2} r_0 v_0. \tag{8}$$

So dA/dt is constant, giving Kepler's second law.

Kepler's Third Law (Time–Distance Law)

The time T it takes a planet to go around its sun once is the planet's **orbital period**. *Kepler's third law* says that T and the orbit's semimajor axis a are related by the equation

$$\frac{T^2}{a^3} = \frac{4\pi^2}{GM}.$$

Since the right-hand side of this equation is constant within a given solar system, the ratio of T^2 to a^3 *is the same for every planet in the system.*

Here is a partial derivation of Kepler's third law. The area enclosed by the planet's elliptical orbit is calculated as follows:

$$\text{Area} = \int_0^T dA$$

$$= \int_0^T \frac{1}{2} r_0 v_0 \, dt \qquad \text{Eq. (8)}$$

$$= \frac{1}{2} T r_0 v_0.$$

If b is the semiminor axis, the area of the ellipse is πab, so

$$T = \frac{2\pi ab}{r_0 v_0} = \frac{2\pi a^2}{r_0 v_0} \sqrt{1 - e^2}. \qquad \begin{array}{c} \text{For any ellipse,} \\ b = a\sqrt{1 - e^2} \end{array} \qquad (9)$$

It remains only to express a and e in terms of r_0, v_0, G, and M. Equation (5) does this for e. For a, we observe that setting θ equal to π in Equation (6) gives

$$r_{\max} = r_0 \frac{1 + e}{1 - e}.$$

Hence, from Figure 13.36,

$$2a = r_0 + r_{\max} = \frac{2r_0}{1 - e} = \frac{2r_0 GM}{2GM - r_0 v_0^2}. \qquad (10)$$

Squaring both sides of Equation (9) and substituting the results of Equations (5) and (10) produces Kepler's third law (Exercise 9).

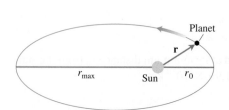

FIGURE 13.36 The length of the major axis of the ellipse is $2a = r_0 + r_{\max}$.

Exercises 13.6

In Exercises 1–5, find the velocity and acceleration vectors in terms of $\mathbf{u}_r$ and $\mathbf{u}_\theta$.

1. $r = a(1 - \cos\theta)$ and $\dfrac{d\theta}{dt} = 3$

2. $r = a\sin 2\theta$ and $\dfrac{d\theta}{dt} = 2t$

3. $r = e^{a\theta}$ and $\dfrac{d\theta}{dt} = 2$

4. $r = a(1 + \sin t)$ and $\theta = 1 - e^{-t}$

5. $r = 2\cos 4t$ and $\theta = 2t$

6. **Type of orbit** For what values of v_0 in Equation (5) is the orbit in Equation (6) a circle? An ellipse? A parabola? A hyperbola?

7. **Circular orbits** Show that a planet in a circular orbit moves with a constant speed. (*Hint:* This is a consequence of one of Kepler's laws.)

8. Suppose that $\mathbf{r}$ is the position vector of a particle moving along a plane curve and dA/dt is the rate at which the vector sweeps out area. Without introducing coordinates, and assuming the necessary derivatives exist, give a geometric argument based on increments and limits for the validity of the equation

$$\frac{dA}{dt} = \frac{1}{2} |\mathbf{r} \times \dot{\mathbf{r}}|.$$

9. **Kepler's third law** Complete the derivation of Kepler's third law (the part following Equation (10)).

10. Find the length of the major axis of Earth's orbit using Kepler's third law and the fact that Earth's orbital period is 365.256 days.

Chapter 13   Questions to Guide Your Review

1. State the rules for differentiating and integrating vector functions. Give examples.

2. How do you define and calculate the velocity, speed, direction of motion, and acceleration of a body moving along a sufficiently differentiable space curve? Give an example.

3. What is special about the derivatives of vector functions of constant length? Give an example.

4. What are the vector and parametric equations for ideal projectile motion? How do you find a projectile's maximum height, flight time, and range? Give examples.

5. How do you define and calculate the length of a segment of a smooth space curve? Give an example. What mathematical assumptions are involved in the definition?

6. How do you measure distance along a smooth curve in space from a preselected base point? Give an example.

7. What is a differentiable curve's unit tangent vector? Give an example.

8. Define curvature, circle of curvature (osculating circle), center of curvature, and radius of curvature for twice-differentiable curves in the plane. Give examples. What curves have zero curvature? Constant curvature?

9. What is a plane curve's principal normal vector? When is it defined? Which way does it point? Give an example.

10. How do you define **N** and κ for curves in space? How are these quantities related? Give examples.

11. What is a curve's binormal vector? Give an example. How is this vector related to the curve's torsion? Give an example.

12. What formulas are available for writing a moving body's acceleration as a sum of its tangential and normal components? Give an example. Why might one want to write the acceleration this way? What if the body moves at a constant speed? At a constant speed around a circle?

13. State Kepler's laws.

Chapter 13 Practice Exercises

Motion in the Plane

In Exercises 1 and 2, graph the curves and sketch their velocity and acceleration vectors at the given values of t. Then write **a** in the form $\mathbf{a} = a_T\mathbf{T} + a_N\mathbf{N}$ without finding **T** and **N**, and find the value of κ at the given values of t.

1. $\mathbf{r}(t) = (4 \cos t)\mathbf{i} + \left(\sqrt{2} \sin t\right)\mathbf{j}, \quad t = 0$ and $\pi/4$

2. $\mathbf{r}(t) = \left(\sqrt{3} \sec t\right)\mathbf{i} + \left(\sqrt{3} \tan t\right)\mathbf{j}, \quad t = 0$

3. The position of a particle in the plane at time t is

$$\mathbf{r} = \frac{1}{\sqrt{1 + t^2}}\mathbf{i} + \frac{t}{\sqrt{1 + t^2}}\mathbf{j}.$$

Find the particle's highest speed.

4. Suppose $\mathbf{r}(t) = (e^t \cos t)\mathbf{i} + (e^t \sin t)\mathbf{j}$. Show that the angle between **r** and **a** never changes. What *is* the angle?

5. Finding curvature At point P, the velocity and acceleration of a particle moving in the plane are $\mathbf{v} = 3\mathbf{i} + 4\mathbf{j}$ and $\mathbf{a} = 5\mathbf{i} + 15\mathbf{j}$. Find the curvature of the particle's path at P.

6. Find the point on the curve $y = e^x$ where the curvature is greatest.

7. A particle moves around the unit circle in the xy-plane. Its position at time t is $\mathbf{r} = x\mathbf{i} + y\mathbf{j}$, where x and y are differentiable functions of t. Find dy/dt if $\mathbf{v} \cdot \mathbf{i} = y$. Is the motion clockwise or counterclockwise?

8. You send a message through a pneumatic tube that follows the curve $9y = x^3$ (distance in meters). At the point $(3, 3)$, $\mathbf{v} \cdot \mathbf{i} = 4$ and $\mathbf{a} \cdot \mathbf{i} = -2$. Find the values of $\mathbf{v} \cdot \mathbf{j}$ and $\mathbf{a} \cdot \mathbf{j}$ at $(3, 3)$.

9. Characterizing circular motion A particle moves in the plane so that its velocity and position vectors are always orthogonal. Show that the particle moves in a circle centered at the origin.

10. Speed along a cycloid A circular wheel with radius 1 ft and center C rolls to the right along the x-axis at a half-turn per second. (See the accompanying figure.) At time t seconds, the position vector of the point P on the wheel's circumference is

$$\mathbf{r} = (\pi t - \sin \pi t)\mathbf{i} + (1 - \cos \pi t)\mathbf{j}.$$

a. Sketch the curve traced by P during the interval $0 \leq t \leq 3$.

b. Find **v** and **a** at $t = 0, 1, 2$, and 3 and add these vectors to your sketch.

c. At any given time, what is the forward speed of the topmost point of the wheel? Of C?

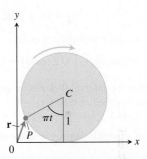

Projectile Motion

11. Shot put A shot leaves the thrower's hand 6.5 ft above the ground at a 45° angle at 44 ft/sec. Where is it 3 sec later?

12. Javelin A javelin leaves the thrower's hand 7 ft above the ground at a 45° angle at 80 ft/sec. How high does it go?

13. A golf ball is hit with an initial speed v_0 at an angle α to the horizontal from a point that lies at the foot of a straight-sided hill that is inclined at an angle ϕ to the horizontal, where

$$0 < \phi < \alpha < \frac{\pi}{2}.$$

Show that the ball lands at a distance

$$\frac{2v_0^2 \cos \alpha}{g \cos^2 \phi} \sin (\alpha - \phi),$$

measured up the face of the hill. Hence, show that the greatest range that can be achieved for a given v_0 occurs when $\alpha = (\phi/2) + (\pi/4)$, i.e., when the initial velocity vector bisects the angle between the vertical and the hill.

T **14. Javelin** In Potsdam in 1988, Petra Felke of (then) East Germany set a women's world record by throwing a javelin 262 ft 5 in.

a. Assuming that Felke launched the javelin at a 40° angle to the horizontal 6.5 ft above the ground, what was the javelin's initial speed?

b. How high did the javelin go?

Motion in Space

Find the lengths of the curves in Exercises 15 and 16.

15. $\mathbf{r}(t) = (2 \cos t)\mathbf{i} + (2 \sin t)\mathbf{j} + t^2\mathbf{k}, \quad 0 \le t \le \pi/4$

16. $\mathbf{r}(t) = (3 \cos t)\mathbf{i} + (3 \sin t)\mathbf{j} + 2t^{3/2}\mathbf{k}, \quad 0 \le t \le 3$

In Exercises 17–20, find $\mathbf{T}, \mathbf{N}, \mathbf{B}, \kappa,$ and τ at the given value of t.

17. $\mathbf{r}(t) = \dfrac{4}{9}(1 + t)^{3/2}\mathbf{i} + \dfrac{4}{9}(1 - t)^{3/2}\mathbf{j} + \dfrac{1}{3}t\mathbf{k}, \quad t = 0$

18. $\mathbf{r}(t) = (e^t \sin 2t)\mathbf{i} + (e^t \cos 2t)\mathbf{j} + 2e^t\mathbf{k}, \quad t = 0$

19. $\mathbf{r}(t) = t\mathbf{i} + \dfrac{1}{2}e^{2t}\mathbf{j}, \quad t = \ln 2$

20. $\mathbf{r}(t) = (3 \cosh 2t)\mathbf{i} + (3 \sinh 2t)\mathbf{j} + 6t\mathbf{k}, \quad t = \ln 2$

In Exercises 21 and 22, write $\mathbf{a}$ in the form $\mathbf{a} = a_{\mathrm{T}}\mathbf{T} + a_{\mathrm{N}}\mathbf{N}$ at $t = 0$ without finding $\mathbf{T}$ and $\mathbf{N}$.

21. $\mathbf{r}(t) = (2 + 3t + 3t^2)\mathbf{i} + (4t + 4t^2)\mathbf{j} - (6 \cos t)\mathbf{k}$

22. $\mathbf{r}(t) = (2 + t)\mathbf{i} + (t + 2t^2)\mathbf{j} + (1 + t^2)\mathbf{k}$

23. Find $\mathbf{T}, \mathbf{N}, \mathbf{B}, \kappa,$ and τ as functions of t if

$$\mathbf{r}(t) = (\sin t)\mathbf{i} + \left(\sqrt{2} \cos t\right)\mathbf{j} + (\sin t)\mathbf{k}.$$

24. At what times in the interval $0 \le t \le \pi$ are the velocity and acceleration vectors of the motion $\mathbf{r}(t) = \mathbf{i} + (5 \cos t)\mathbf{j} + (3 \sin t)\mathbf{k}$ orthogonal?

25. The position of a particle moving in space at time $t \ge 0$ is

$$\mathbf{r}(t) = 2\mathbf{i} + \left(4 \sin \frac{t}{2}\right)\mathbf{j} + \left(3 - \frac{t}{\pi}\right)\mathbf{k}.$$

Find the first time $\mathbf{r}$ is orthogonal to the vector $\mathbf{i} - \mathbf{j}$.

26. Find equations for the osculating, normal, and rectifying planes of the curve $\mathbf{r}(t) = t\mathbf{i} + t^2\mathbf{j} + t^3\mathbf{k}$ at the point $(1, 1, 1)$.

27. Find parametric equations for the line that is tangent to the curve $\mathbf{r}(t) = e^t\mathbf{i} + (\sin t)\mathbf{j} + \ln(1 - t)\mathbf{k}$ at $t = 0$.

28. Find parametric equations for the line tangent to the helix $\mathbf{r}(t) = \left(\sqrt{2} \cos t\right)\mathbf{i} + \left(\sqrt{2} \sin t\right)\mathbf{j} + t\mathbf{k}$ at the point where $t = \pi/4$.

Theory and Examples

29. Synchronous curves By eliminating α from the ideal projectile equations

$$x = (v_0 \cos \alpha)t, \quad y = (v_0 \sin \alpha)t - \frac{1}{2}gt^2,$$

show that $x^2 + (y + gt^2/2)^2 = v_0^2 t^2$. This shows that projectiles launched simultaneously from the origin at the same initial speed will, at any given instant, all lie on the circle of radius $v_0 t$ centered at $(0, -gt^2/2)$, regardless of their launch angle. These circles are the *synchronous curves* of the launching.

30. Radius of curvature Show that the radius of curvature of a twice-differentiable plane curve $\mathbf{r}(t) = f(t)\mathbf{i} + g(t)\mathbf{j}$ is given by the formula

$$\rho = \frac{\dot{x}^2 + \dot{y}^2}{\sqrt{\dot{x}^2 + \dot{y}^2 - \ddot{s}^2}}, \quad \text{where} \quad \ddot{s} = \frac{d}{dt}\sqrt{\dot{x}^2 + \dot{y}^2}.$$

31. An alternative definition of curvature in the plane An alternative definition gives the curvature of a sufficiently differentiable plane curve to be $|d\phi/ds|$, where ϕ is the angle between $\mathbf{T}$ and $\mathbf{i}$ (Figure 13.37a). Figure 13.37b shows the distance s measured counterclockwise around the circle $x^2 + y^2 = a^2$ from the point $(a, 0)$ to a point P, along with the angle ϕ at P. Calculate the circle's curvature using the alternative definition. (*Hint:* $\phi = \theta + \pi/2$.)

(a)

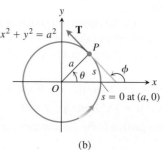

(b)

FIGURE 13.37 Figures for Exercise 31.

32. The view from *Skylab 4* What percentage of Earth's surface area could the astronauts see when *Skylab 4* was at its apogee height, 437 km above the surface? To find out, model the visible surface as the surface generated by revolving the circular arc GT, shown here, about the y-axis. Then carry out these steps:

1. Use similar triangles in the figure to show that $y_0/6380 = 6380/(6380 + 437)$. Solve for y_0.

2. To four significant digits, calculate the visible area as

$$VA = \int_{y_0}^{6380} 2\pi x \sqrt{1 + \left(\frac{dx}{dy}\right)^2}\, dy.$$

3. Express the result as a percentage of Earth's surface area.

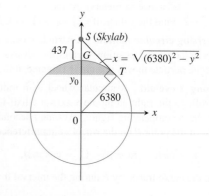

Chapter **13**   Additional and Advanced Exercises

Applications

1. A frictionless particle P, starting from rest at time $t = 0$ at the point $(a, 0, 0)$, slides down the helix

$$\mathbf{r}(\theta) = (a \cos \theta)\mathbf{i} + (a \sin \theta)\mathbf{j} + b\theta\mathbf{k} \quad (a, b > 0)$$

under the influence of gravity, as in the accompanying figure. The θ in this equation is the cylindrical coordinate θ and the helix is the curve $r = a, z = b\theta, \theta \geq 0$, in cylindrical coordinates. We assume θ to be a differentiable function of t for the motion. The law of conservation of energy tells us that the particle's speed after it has fallen straight down a distance z is $\sqrt{2gz}$, where g is the constant acceleration of gravity.

a. Find the angular velocity $d\theta/dt$ when $\theta = 2\pi$.

b. Express the particle's θ- and z-coordinates as functions of t.

c. Express the tangential and normal components of the velocity $d\mathbf{r}/dt$ and acceleration $d^2\mathbf{r}/dt^2$ as functions of t. Does the acceleration have any nonzero component in the direction of the binormal vector $\mathbf{B}$?

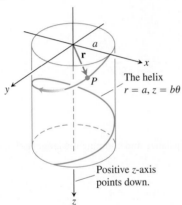

2. Suppose the curve in Exercise 1 is replaced by the conical helix $r = a\theta, z = b\theta$ shown in the accompanying figure.

a. Express the angular velocity $d\theta/dt$ as a function of θ.

b. Express the distance the particle travels along the helix as a function of θ.

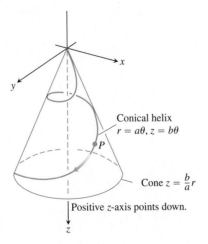

Motion in Polar and Cylindrical Coordinates

3. Deduce from the orbit equation

$$r = \frac{(1 + e)r_0}{1 + e \cos \theta}$$

that a planet is closest to its sun when $\theta = 0$ and show that $r = r_0$ at that time.

T 4. **A Kepler equation**   The problem of locating a planet in its orbit at a given time and date eventually leads to solving "Kepler" equations of the form

$$f(x) = x - 1 - \frac{1}{2} \sin x = 0.$$

a. Show that this particular equation has a solution between $x = 0$ and $x = 2$.

b. With your computer or calculator in radian mode, use Newton's method to find the solution to as many places as you can.

5. In Section 13.6, we found the velocity of a particle moving in the plane to be

$$\mathbf{v} = \dot{x}\mathbf{i} + \dot{y}\mathbf{j} = \dot{r}\mathbf{u}_r + r\dot{\theta}\mathbf{u}_\theta.$$

a. Express $\dot{x}$ and $\dot{y}$ in terms of $\dot{r}$ and $r\dot{\theta}$ by evaluating the dot products $\mathbf{v} \cdot \mathbf{i}$ and $\mathbf{v} \cdot \mathbf{j}$.

b. Express $\dot{r}$ and $r\dot{\theta}$ in terms of $\dot{x}$ and $\dot{y}$ by evaluating the dot products $\mathbf{v} \cdot \mathbf{u}_r$ and $\mathbf{v} \cdot \mathbf{u}_\theta$.

6. Express the curvature of a twice-differentiable curve $r = f(\theta)$ in the polar coordinate plane in terms of f and its derivatives.

7. A slender rod through the origin of the polar coordinate plane rotates (in the plane) about the origin at the rate of 3 rad/min. A beetle starting from the point (2, 0) crawls along the rod toward the origin at the rate of 1 in./min.

a. Find the beetle's acceleration and velocity in polar form when it is halfway to (1 in. from) the origin.

T b. To the nearest tenth of an inch, what will be the length of the path the beetle has traveled by the time it reaches the origin?

8. **Arc length in cylindrical coordinates**

a. Show that when you express $ds^2 = dx^2 + dy^2 + dz^2$ in terms of cylindrical coordinates, you get $ds^2 = dr^2 + r^2 \, d\theta^2 + dz^2$.

b. Interpret this result geometrically in terms of the edges and a diagonal of a box. Sketch the box.

c. Use the result in part (a) to find the length of the curve $r = e^\theta, z = e^\theta, 0 \leq \theta \leq \theta \ln 8$.

9. **Unit vectors for position and motion in cylindrical coordinates**   When the position of a particle moving in space is given in cylindrical coordinates, the unit vectors we use to describe its position and motion are

$$\mathbf{u}_r = (\cos \theta)\mathbf{i} + (\sin \theta)\mathbf{j}, \qquad \mathbf{u}_\theta = -(\sin \theta)\mathbf{i} + (\cos \theta)\mathbf{j},$$

and $\mathbf{k}$ (see accompanying figure). The particle's position vector is then $\mathbf{r} = r\mathbf{u}_r + z\mathbf{k}$, where r is the positive polar distance coordinate of the particle's position.

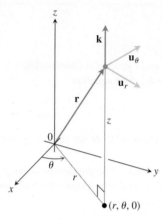

a. Show that $\mathbf{u}_r$, $\mathbf{u}_\theta$, and $\mathbf{k}$, in this order, form a right-handed frame of unit vectors.

b. Show that

$$\frac{d\mathbf{u}_r}{d\theta} = \mathbf{u}_\theta \quad \text{and} \quad \frac{d\mathbf{u}_\theta}{d\theta} = -\mathbf{u}_r.$$

c. Assuming that the necessary derivatives with respect to t exist, express $\mathbf{v} = \dot{\mathbf{r}}$ and $\mathbf{a} = \ddot{\mathbf{r}}$ in terms of $\mathbf{u}_r$, $\mathbf{u}_\theta$, $\mathbf{k}$, $\dot{r}$, and $\dot{\theta}$.

10. Conservation of angular momentum Let $\mathbf{r}(t)$ denote the position in space of a moving object at time t. Suppose the force acting on the object at time t is

$$\mathbf{F}(t) = -\frac{c}{|\mathbf{r}(t)|^3}\mathbf{r}(t),$$

where c is a constant. In physics the **angular momentum** of an object at time t is defined to be $\mathbf{L}(t) = \mathbf{r}(t) \times m\mathbf{v}(t)$, where m is the mass of the object and $\mathbf{v}(t)$ is the velocity. Prove that angular momentum is a conserved quantity; i.e., prove that $\mathbf{L}(t)$ is a constant vector, independent of time. Remember Newton's law $\mathbf{F} = m\mathbf{a}$. (This is a calculus problem, not a physics problem.)

Chapter 13 Technology Application Projects

Mathematica/Maple Module:

Radar Tracking of a Moving Object
Visualize position, velocity, and acceleration vectors to analyze motion.

Parametric and Polar Equations with a Figure Skater
Visualize position, velocity, and acceleration vectors to analyze motion.

Moving in Three Dimensions
Compute distance traveled, speed, curvature, and torsion for motion along a space curve. Visualize and compute the tangential, normal, and binormal vectors associated with motion along a space curve.

14

PARTIAL DERIVATIVES

OVERVIEW Many functions depend on more than one independent variable. For instance, the volume of a right circular cylinder is a function $V = \pi r^2 h$ of its radius and its height, so it is a function $V(r, h)$ of two variables r and h. In this chapter we extend the basic ideas of single variable calculus to functions of several variables. Their derivatives are more varied and interesting because of the different ways the variables can interact. The applications of these derivatives are also more varied than for single-variable calculus, and in the next chapter we will see that the same is true for integrals involving several variables.

14.1 | Functions of Several Variables

In this section we define functions of more than one independent variable and discuss ways to graph them.

Real-valued functions of several independent real variables are defined similarly to functions in the single-variable case. Points in the domain are ordered pairs (triples, quadruples, n-tuples) of real numbers, and values in the range are real numbers as we have worked with all along.

DEFINITIONS Suppose D is a set of n-tuples of real numbers $(x_1, x_2, \ldots, x_n)$. A **real-valued function** f on D is a rule that assigns a unique (single) real number

$$w = f(x_1, x_2, \ldots, x_n)$$

to each element in D. The set D is the function's **domain**. The set of w-values taken on by f is the function's **range**. The symbol w is the **dependent variable** of f, and f is said to be a function of the n **independent variables** x_1 to x_n. We also call the x_j's the function's **input variables** and call w the function's **output variable**.

If f is a function of two independent variables, we usually call the independent variables x and y and the dependent variable z, and we picture the domain of f as a region in the xy-plane (Figure 14.1). If f is a function of three independent variables, we call the independent variables x, y, and z and the dependent variable w, and we picture the domain as a region in space.

In applications, we tend to use letters that remind us of what the variables stand for. To say that the volume of a right circular cylinder is a function of its radius and height, we might write $V = f(r, h)$. To be more specific, we might replace the notation $f(r, h)$ by the formula that calculates the value of V from the values of r and h, and write $V = \pi r^2 h$. In either case, r and h would be the independent variables and V the dependent variable of the function.

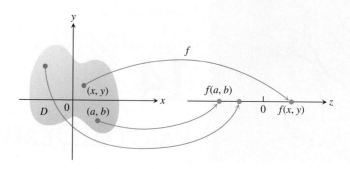

FIGURE 14.1 An arrow diagram for the function $z = f(x, y)$.

As usual, we evaluate functions defined by formulas by substituting the values of the independent variables in the formula and calculating the corresponding value of the dependent variable. For example, the value of $f(x, y, z) = \sqrt{x^2 + y^2 + z^2}$ at the point $(3, 0, 4)$ is

$$f(3, 0, 4) = \sqrt{(3)^2 + (0)^2 + (4)^2} = \sqrt{25} = 5.$$

Domains and Ranges

In defining a function of more than one variable, we follow the usual practice of excluding inputs that lead to complex numbers or division by zero. If $f(x, y) = \sqrt{y - x^2}$, y cannot be less than x^2. If $f(x, y) = 1/(xy)$, xy cannot be zero. The domain of a function is assumed to be the largest set for which the defining rule generates real numbers, unless the domain is otherwise specified explicitly. The range consists of the set of output values for the dependent variable.

EXAMPLE 1 **(a)** These are functions of two variables. Note the restrictions that may apply to their domains in order to obtain a real value for the dependent variable z.

Function	Domain	Range
$z = \sqrt{y - x^2}$	$y \geq x^2$	$[0, \infty)$
$z = \dfrac{1}{xy}$	$xy \neq 0$	$(-\infty, 0) \cup (0, \infty)$
$z = \sin xy$	Entire plane	$[-1, 1]$

(b) These are functions of three variables with restrictions on some of their domains.

Function	Domain	Range
$w = \sqrt{x^2 + y^2 + z^2}$	Entire space	$[0, \infty)$
$w = \dfrac{1}{x^2 + y^2 + z^2}$	$(x, y, z) \neq (0, 0, 0)$	$(0, \infty)$
$w = xy \ln z$	Half-space $z > 0$	$(-\infty, \infty)$

Functions of Two Variables

Regions in the plane can have interior points and boundary points just like intervals on the real line. Closed intervals $[a, b]$ include their boundary points, open intervals (a, b) don't include their boundary points, and intervals such as $[a, b)$ are neither open nor closed.

(a) Interior point

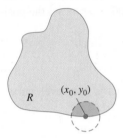

(b) Boundary point

FIGURE 14.2 Interior points and boundary points of a plane region R. An interior point is necessarily a point of R. A boundary point of R need not belong to R.

DEFINITIONS A point (x_0, y_0) in a region (set) R in the xy-plane is an **interior point** of R if it is the center of a disk of positive radius that lies entirely in R (Figure 14.2). A point (x_0, y_0) is a **boundary point** of R if every disk centered at (x_0, y_0) contains points that lie outside of R as well as points that lie in R. (The boundary point itself need not belong to R.)

The interior points of a region, as a set, make up the **interior** of the region. The region's boundary points make up its **boundary**. A region is **open** if it consists entirely of interior points. A region is **closed** if it contains all its boundary points (Figure 14.3).

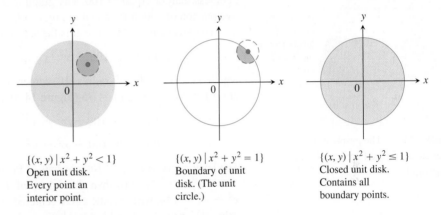

$\{(x, y) \mid x^2 + y^2 < 1\}$
Open unit disk.
Every point an
interior point.

$\{(x, y) \mid x^2 + y^2 = 1\}$
Boundary of unit
disk. (The unit
circle.)

$\{(x, y) \mid x^2 + y^2 \le 1\}$
Closed unit disk.
Contains all
boundary points.

FIGURE 14.3 Interior points and boundary points of the unit disk in the plane.

As with a half-open interval of real numbers $[a, b)$, some regions in the plane are neither open nor closed. If you start with the open disk in Figure 14.3 and add to it some of but not all its boundary points, the resulting set is neither open nor closed. The boundary points that *are* there keep the set from being open. The absence of the remaining boundary points keeps the set from being closed.

DEFINITIONS A region in the plane is **bounded** if it lies inside a disk of fixed radius. A region is **unbounded** if it is not bounded.

Examples of *bounded* sets in the plane include line segments, triangles, interiors of triangles, rectangles, circles, and disks. Examples of *unbounded* sets in the plane include lines, coordinate axes, the graphs of functions defined on infinite intervals, quadrants, half-planes, and the plane itself.

EXAMPLE 2 Describe the domain of the function $f(x, y) = \sqrt{y - x^2}$.

Solution Since f is defined only where $y - x^2 \ge 0$, the domain is the closed, unbounded region shown in Figure 14.4. The parabola $y = x^2$ is the boundary of the domain. The points above the parabola make up the domain's interior. ∎

Graphs, Level Curves, and Contours of Functions of Two Variables

There are two standard ways to picture the values of a function $f(x, y)$. One is to draw and label curves in the domain on which f has a constant value. The other is to sketch the surface $z = f(x, y)$ in space.

FIGURE 14.4 The domain of $f(x, y)$ in Example 2 consists of the shaded region and its bounding parabola.

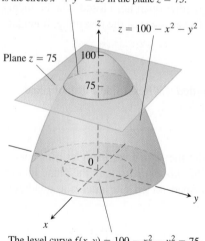

FIGURE 14.5 The graph and selected level curves of the function $f(x, y)$ in Example 3.

DEFINITIONS The set of points in the plane where a function $f(x, y)$ has a constant value $f(x, y) = c$ is called a **level curve** of f. The set of all points $(x, y, f(x, y))$ in space, for (x, y) in the domain of f, is called the **graph** of f. The graph of f is also called the **surface** $z = f(x, y)$.

EXAMPLE 3 Graph $f(x, y) = 100 - x^2 - y^2$ and plot the level curves $f(x, y) = 0$, $f(x, y) = 51$, and $f(x, y) = 75$ in the domain of f in the plane.

Solution The domain of f is the entire xy-plane, and the range of f is the set of real numbers less than or equal to 100. The graph is the paraboloid $z = 100 - x^2 - y^2$, the positive portion of which is shown in Figure 14.5.

The level curve $f(x, y) = 0$ is the set of points in the xy-plane at which

$$f(x, y) = 100 - x^2 - y^2 = 0, \quad \text{or} \quad x^2 + y^2 = 100,$$

which is the circle of radius 10 centered at the origin. Similarly, the level curves $f(x, y) = 51$ and $f(x, y) = 75$ (Figure 14.5) are the circles

$$f(x, y) = 100 - x^2 - y^2 = 51, \quad \text{or} \quad x^2 + y^2 = 49$$

$$f(x, y) = 100 - x^2 - y^2 = 75, \quad \text{or} \quad x^2 + y^2 = 25.$$

The level curve $f(x, y) = 100$ consists of the origin alone. (It is still a level curve.)

If $x^2 + y^2 > 100$, then the values of $f(x, y)$ are negative. For example, the circle $x^2 + y^2 = 144$, which is the circle centered at the origin with radius 12, gives the constant value $f(x, y) = -44$ and is a level curve of f. ∎

The curve in space in which the plane $z = c$ cuts a surface $z = f(x, y)$ is made up of the points that represent the function value $f(x, y) = c$. It is called the **contour curve** $f(x, y) = c$ to distinguish it from the level curve $f(x, y) = c$ in the domain of f. Figure 14.6 shows the contour curve $f(x, y) = 75$ on the surface $z = 100 - x^2 - y^2$ defined by the function $f(x, y) = 100 - x^2 - y^2$. The contour curve lies directly above the circle $x^2 + y^2 = 25$, which is the level curve $f(x, y) = 75$ in the function's domain.

Not everyone makes this distinction, however, and you may wish to call both kinds of curves by a single name and rely on context to convey which one you have in mind. On most maps, for example, the curves that represent constant elevation (height above sea level) are called contours, not level curves (Figure 14.7).

Functions of Three Variables

In the plane, the points where a function of two independent variables has a constant value $f(x, y) = c$ make a curve in the function's domain. In space, the points where a function of three independent variables has a constant value $f(x, y, z) = c$ make a surface in the function's domain.

DEFINITION The set of points (x, y, z) in space where a function of three independent variables has a constant value $f(x, y, z) = c$ is called a **level surface** of f.

Since the graphs of functions of three variables consist of points $(x, y, z, f(x, y, z))$ lying in a four-dimensional space, we cannot sketch them effectively in our three-dimensional frame of reference. We can see how the function behaves, however, by looking at its three-dimensional level surfaces.

EXAMPLE 4 Describe the level surfaces of the function

$$f(x, y, z) = \sqrt{x^2 + y^2 + z^2}.$$

The contour curve $f(x, y) = 100 - x^2 - y^2 = 75$ is the circle $x^2 + y^2 = 25$ in the plane $z = 75$.

The level curve $f(x, y) = 100 - x^2 - y^2 = 75$ is the circle $x^2 + y^2 = 25$ in the xy-plane.

FIGURE 14.6 A plane $z = c$ parallel to the xy-plane intersecting a surface $z = f(x, y)$ produces a contour curve.

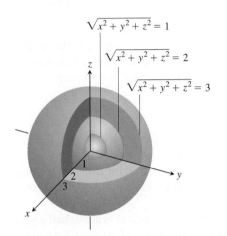

$$\sqrt{x^2 + y^2 + z^2} = 1$$
$$\sqrt{x^2 + y^2 + z^2} = 2$$
$$\sqrt{x^2 + y^2 + z^2} = 3$$

FIGURE 14.8 The level surfaces of $f(x, y, z) = \sqrt{x^2 + y^2 + z^2}$ are concentric spheres (Example 4).

FIGURE 14.7 Contours on Mt. Washington in New Hampshire. (Reproduced by permission from the Appalachian Mountain Club.)

Solution The value of f is the distance from the origin to the point (x, y, z). Each level surface $\sqrt{x^2 + y^2 + z^2} = c$, $c > 0$, is a sphere of radius c centered at the origin. Figure 14.8 shows a cutaway view of three of these spheres. The level surface $\sqrt{x^2 + y^2 + z^2} = 0$ consists of the origin alone.

We are not graphing the function here; we are looking at level surfaces in the function's domain. The level surfaces show how the function's values change as we move through its domain. If we remain on a sphere of radius c centered at the origin, the function maintains a constant value, namely c. If we move from a point on one sphere to a point on another, the function's value changes. It increases if we move away from the origin and decreases if we move toward the origin. The way the values change depends on the direction we take. The dependence of change on direction is important. We return to it in Section 14.5.

The definitions of interior, boundary, open, closed, bounded, and unbounded for regions in space are similar to those for regions in the plane. To accommodate the extra dimension, we use solid balls of positive radius instead of disks.

(a) Interior point

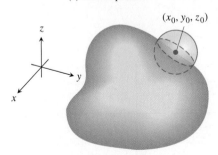

(b) Boundary point

FIGURE 14.9 Interior points and boundary points of a region in space. As with regions in the plane, a boundary point need not belong to the space region R.

DEFINITIONS A point (x_0, y_0, z_0) in a region R in space is an **interior point** of R if it is the center of a solid ball that lies entirely in R (Figure 14.9a). A point (x_0, y_0, z_0) is a **boundary point** of R if every solid ball centered at (x_0, y_0, z_0) contains points that lie outside of R as well as points that lie inside R (Figure 14.9b). The **interior** of R is the set of interior points of R. The **boundary** of R is the set of boundary points of R.

A region is **open** if it consists entirely of interior points. A region is **closed** if it contains its entire boundary.

Examples of *open* sets in space include the interior of a sphere, the open half-space $z > 0$, the first octant (where x, y, and z are all positive), and space itself. Examples of *closed* sets in space include lines, planes, and the closed half-space $z \geq 0$. A solid sphere with part of its boundary removed or a solid cube with a missing face, edge, or corner point is *neither open nor closed*.

Functions of more than three independent variables are also important. For example, the temperature on a surface in space may depend not only on the location of the point $P(x, y, z)$ on the surface but also on the time t when it is visited, so we would write $T = f(x, y, z, t)$.

Computer Graphing

Three-dimensional graphing programs for computers and calculators make it possible to graph functions of two variables with only a few keystrokes. We can often get information more quickly from a graph than from a formula.

EXAMPLE 5 The temperature w beneath the Earth's surface is a function of the depth x beneath the surface and the time t of the year. If we measure x in feet and t as the number of days elapsed from the expected date of the yearly highest surface temperature, we can model the variation in temperature with the function

$$w = \cos(1.7 \times 10^{-2}t - 0.2x)e^{-0.2x}.$$

(The temperature at 0 ft is scaled to vary from $+1$ to -1, so that the variation at x feet can be interpreted as a fraction of the variation at the surface.)

Figure 14.10 shows a graph of the function. At a depth of 15 ft, the variation (change in vertical amplitude in the figure) is about 5% of the surface variation. At 25 ft, there is almost no variation during the year.

The graph also shows that the temperature 15 ft below the surface is about half a year out of phase with the surface temperature. When the temperature is lowest on the surface (late January, say), it is at its highest 15 ft below. Fifteen feet below the ground, the seasons are reversed. ∎

Figure 14.11 shows computer-generated graphs of a number of functions of two variables together with their level curves.

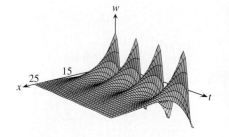

FIGURE 14.10 This graph shows the seasonal variation of the temperature below ground as a fraction of surface temperature (Example 5).

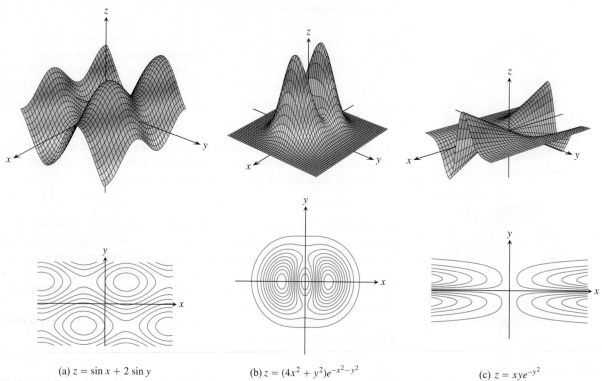

(a) $z = \sin x + 2 \sin y$ (b) $z = (4x^2 + y^2)e^{-x^2-y^2}$ (c) $z = xye^{-y^2}$

FIGURE 14.11 Computer-generated graphs and level curves of typical functions of two variables.

Exercises 14.1

Domain, Range, and Level Curves

In Exercises 1–4, find the specific function values.

1. $f(x, y) = x^2 + xy^3$

   a. $f(0, 0)$ **b.** $f(-1, 1)$

   c. $f(2, 3)$ **d.** $f(-3, -2)$

2. $f(x, y) = \sin(xy)$

   a. $f\left(2, \dfrac{\pi}{6}\right)$ **b.** $f\left(-3, \dfrac{\pi}{12}\right)$

   c. $f\left(\pi, \dfrac{1}{4}\right)$ **d.** $f\left(-\dfrac{\pi}{2}, -7\right)$

3. $f(x, y, z) = \dfrac{x - y}{y^2 + z^2}$

   a. $f(3, -1, 2)$ **b.** $f\left(1, \dfrac{1}{2}, -\dfrac{1}{4}\right)$

   c. $f\left(0, -\dfrac{1}{3}, 0\right)$ **d.** $f(2, 2, 100)$

4. $f(x, y, z) = \sqrt{49 - x^2 - y^2 - z^2}$

   a. $f(0, 0, 0)$ **b.** $f(2, -3, 6)$

   c. $f(-1, 2, 3)$ **d.** $f\left(\dfrac{4}{\sqrt{2}}, \dfrac{5}{\sqrt{2}}, \dfrac{6}{\sqrt{2}}\right)$

In Exercises 5–12, find and sketch the domain for each function.

5. $f(x, y) = \sqrt{y - x - 2}$

6. $f(x, y) = \ln(x^2 + y^2 - 4)$

7. $f(x, y) = \dfrac{(x - 1)(y + 2)}{(y - x)(y - x^3)}$

8. $f(x, y) = \dfrac{\sin(xy)}{x^2 + y^2 - 25}$

9. $f(x, y) = \cos^{-1}(y - x^2)$

10. $f(x, y) = \ln(xy + x - y - 1)$

11. $f(x, y) = \sqrt{(x^2 - 4)(y^2 - 9)}$

12. $f(x, y) = \dfrac{1}{\ln(4 - x^2 - y^2)}$

In Exercises 13–16, find and sketch the level curves $f(x, y) = c$ on the same set of coordinate axes for the given values of c. We refer to these level curves as a contour map.

13. $f(x, y) = x + y - 1$, $c = -3, -2, -1, 0, 1, 2, 3$

14. $f(x, y) = x^2 + y^2$, $c = 0, 1, 4, 9, 16, 25$

15. $f(x, y) = xy$, $c = -9, -4, -1, 0, 1, 4, 9$

16. $f(x, y) = \sqrt{25 - x^2 - y^2}$, $c = 0, 1, 2, 3, 4$

In Exercises 17–30, **(a)** find the function's domain, **(b)** find the function's range, **(c)** describe the function's level curves, **(d)** find the boundary of the function's domain, **(e)** determine if the domain is an open region, a closed region, or neither, and **(f)** decide if the domain is bounded or unbounded.

17. $f(x, y) = y - x$ **18.** $f(x, y) = \sqrt{y - x}$

19. $f(x, y) = 4x^2 + 9y^2$ **20.** $f(x, y) = x^2 - y^2$

21. $f(x, y) = xy$ **22.** $f(x, y) = y/x^2$

23. $f(x, y) = \dfrac{1}{\sqrt{16 - x^2 - y^2}}$ **24.** $f(x, y) = \sqrt{9 - x^2 - y^2}$

25. $f(x, y) = \ln(x^2 + y^2)$ **26.** $f(x, y) = e^{-(x^2+y^2)}$

27. $f(x, y) = \sin^{-1}(y - x)$ **28.** $f(x, y) = \tan^{-1}\left(\dfrac{y}{x}\right)$

29. $f(x, y) = \ln(x^2 + y^2 - 1)$ **30.** $f(x, y) = \ln(9 - x^2 - y^2)$

Matching Surfaces with Level Curves

Exercises 31–36 show level curves for the functions graphed in (a)–(f) on the following page. Match each set of curves with the appropriate function.

31.

32.

33.

34.

35.

36.

a.

$$z = (\cos x)(\cos y) \, e^{-\sqrt{x^2 + y^2}/4}$$

b.

$$z = -\frac{xy^2}{x^2 + y^2}$$

c.

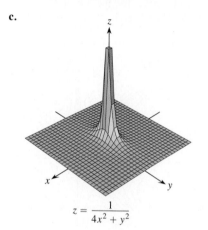

$$z = \frac{1}{4x^2 + y^2}$$

d.

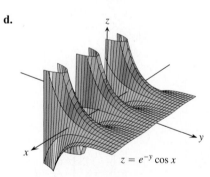

$$z = e^{-y} \cos x$$

e.

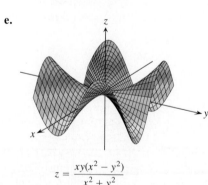

$$z = \frac{xy(x^2 - y^2)}{x^2 + y^2}$$

f.

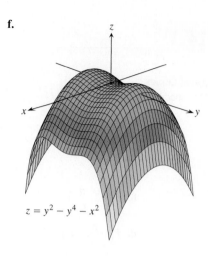

$$z = y^2 - y^4 - x^2$$

Functions of Two Variables

Display the values of the functions in Exercises 37–48 in two ways: **(a)** by sketching the surface $z = f(x, y)$ and **(b)** by drawing an assortment of level curves in the function's domain. Label each level curve with its function value.

37. $f(x, y) = y^2$

38. $f(x, y) = \sqrt{x}$

39. $f(x, y) = x^2 + y^2$

40. $f(x, y) = \sqrt{x^2 + y^2}$

41. $f(x, y) = x^2 - y$

42. $f(x, y) = 4 - x^2 - y^2$

43. $f(x, y) = 4x^2 + y^2$

44. $f(x, y) = 6 - 2x - 3y$

45. $f(x, y) = 1 - |y|$

46. $f(x, y) = 1 - |x| - |y|$

47. $f(x, y) = \sqrt{x^2 + y^2 + 4}$

48. $f(x, y) = \sqrt{x^2 + y^2 - 4}$

Finding Level Curves

In Exercises 49–52, find an equation for and sketch the graph of the level curve of the function $f(x, y)$ that passes through the given point.

49. $f(x, y) = 16 - x^2 - y^2, \quad \left(2\sqrt{2}, \sqrt{2}\right)$

50. $f(x, y) = \sqrt{x^2 - 1}, \quad (1, 0)$

51. $f(x, y) = \sqrt{x + y^2 - 3}, \quad (3, -1)$

52. $f(x, y) = \dfrac{2y - x}{x + y + 1}, \quad (-1, 1)$

Sketching Level Surfaces

In Exercises 53–60, sketch a typical level surface for the function.

53. $f(x, y, z) = x^2 + y^2 + z^2$

54. $f(x, y, z) = \ln(x^2 + y^2 + z^2)$

55. $f(x, y, z) = x + z$

56. $f(x, y, z) = z$

57. $f(x, y, z) = x^2 + y^2$

58. $f(x, y, z) = y^2 + z^2$

59. $f(x, y, z) = z - x^2 - y^2$

60. $f(x, y, z) = (x^2/25) + (y^2/16) + (z^2/9)$

Finding Level Surfaces

In Exercises 61–64, find an equation for the level surface of the function through the given point.

61. $f(x, y, z) = \sqrt{x - y} - \ln z, \quad (3, -1, 1)$

62. $f(x, y, z) = \ln(x^2 + y + z^2), \quad (-1, 2, 1)$

63. $g(x, y, z) = \sqrt{x^2 + y^2 + z^2}, \quad (1, -1, \sqrt{2})$

64. $g(x, y, z) = \dfrac{x - y + z}{2x + y - z}, \quad (1, 0, -2)$

In Exercises 65–68, find and sketch the domain of f. Then find an equation for the level curve or surface of the function passing through the given point.

65. $f(x, y) = \displaystyle\sum_{n=0}^{\infty} \left(\dfrac{x}{y}\right)^n, \quad (1, 2)$

66. $g(x, y, z) = \displaystyle\sum_{n=0}^{\infty} \dfrac{(x + y)^n}{n! z^n}, \quad (\ln 4, \ln 9, 2)$

67. $f(x, y) = \displaystyle\int_{x}^{y} \dfrac{d\theta}{\sqrt{1 - \theta^2}}, \quad (0, 1)$

68. $g(x, y, z) = \displaystyle\int_{x}^{y} \dfrac{dt}{1 + t^2} + \int_{0}^{z} \dfrac{d\theta}{\sqrt{4 - \theta^2}}, \quad (0, 1, \sqrt{3})$

COMPUTER EXPLORATIONS

Use a CAS to perform the following steps for each of the functions in Exercises 69–72.

  a. Plot the surface over the given rectangle.

  b. Plot several level curves in the rectangle.

  c. Plot the level curve of f through the given point.

69. $f(x, y) = x \sin \dfrac{y}{2} + y \sin 2x, \quad 0 \le x \le 5\pi, \quad 0 \le y \le 5\pi,$
  $P(3\pi, 3\pi)$

70. $f(x, y) = (\sin x)(\cos y)e^{\sqrt{x^2 + y^2}/8}, \quad 0 \le x \le 5\pi,$
  $0 \le y \le 5\pi, \quad P(4\pi, 4\pi)$

71. $f(x, y) = \sin(x + 2 \cos y), \quad -2\pi \le x \le 2\pi,$
  $-2\pi \le y \le 2\pi, \quad P(\pi, \pi)$

72. $f(x, y) = e^{(x^{0.1} - y)} \sin(x^2 + y^2), \quad 0 \le x \le 2\pi,$
  $-2\pi \le y \le \pi, \quad P(\pi, -\pi)$

Use a CAS to plot the implicitly defined level surfaces in Exercises 73–76.

73. $4 \ln(x^2 + y^2 + z^2) = 1$ **74.** $x^2 + z^2 = 1$

75. $x + y^2 - 3z^2 = 1$

76. $\sin\left(\dfrac{x}{2}\right) - (\cos y)\sqrt{x^2 + z^2} = 2$

Parametrized Surfaces Just as you describe curves in the plane parametrically with a pair of equations $x = f(t), y = g(t)$ defined on some parameter interval I, you can sometimes describe surfaces in space with a triple of equations $x = f(u, v), y = g(u, v), z = h(u, v)$ defined on some parameter rectangle $a \le u \le b, c \le v \le d$. Many computer algebra systems permit you to plot such surfaces in *parametric mode*. (Parametrized surfaces are discussed in detail in Section 16.5.) Use a CAS to plot the surfaces in Exercises 77–80. Also plot several level curves in the xy-plane.

77. $x = u \cos v, \quad y = u \sin v, \quad z = u, \quad 0 \le u \le 2,$
  $0 \le v \le 2\pi$

78. $x = u \cos v, \quad y = u \sin v, \quad z = v, \quad 0 \le u \le 2,$
  $0 \le v \le 2\pi$

79. $x = (2 + \cos u) \cos v, \quad y = (2 + \cos u) \sin v, \quad z = \sin u,$
  $0 \le u \le 2\pi, \quad 0 \le v \le 2\pi$

80. $x = 2 \cos u \cos v, \quad y = 2 \cos u \sin v, \quad z = 2 \sin u,$
  $0 \le u \le 2\pi, \quad 0 \le v \le \pi$

14.2 Limits and Continuity in Higher Dimensions

This section treats limits and continuity for multivariable functions. These ideas are analogous to limits and continuity for single-variable functions, but including more independent variables leads to additional complexity and important differences requiring some new ideas.

Limits for Functions of Two Variables

If the values of $f(x, y)$ lie arbitrarily close to a fixed real number L for all points (x, y) sufficiently close to a point (x_0, y_0), we say that f approaches the limit L as (x, y) approaches (x_0, y_0). This is similar to the informal definition for the limit of a function of a single variable. Notice, however, that if (x_0, y_0) lies in the interior of f's domain, (x, y) can approach (x_0, y_0) from any direction. For the limit to exist, the same limiting value must be obtained whatever direction of approach is taken. We illustrate this issue in several examples following the definition.

DEFINITION   We say that a function $f(x, y)$ approaches the **limit** L as (x, y) approaches (x_0, y_0), and write

$$\lim_{(x, y) \to (x_0, y_0)} f(x, y) = L$$

if, for every number $\epsilon > 0$, there exists a corresponding number $\delta > 0$ such that for all (x, y) in the domain of f,

$$|f(x, y) - L| < \epsilon \qquad \text{whenever} \qquad 0 < \sqrt{(x - x_0)^2 + (y - y_0)^2} < \delta.$$

The definition of limit says that the distance between $f(x, y)$ and L becomes arbitrarily small whenever the distance from (x, y) to (x_0, y_0) is made sufficiently small (but not 0). The definition applies to interior points (x_0, y_0) as well as boundary points of the domain of f, although a boundary point need not lie within the domain. The points (x, y) that approach (x_0, y_0) are always taken to be in the domain of f. See Figure 14.12.

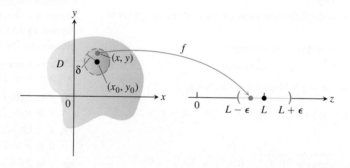

FIGURE 14.12   In the limit definition, δ is the radius of a disk centered at (x_0, y_0). For all points (x, y) within this disk, the function values $f(x, y)$ lie inside the corresponding interval $(L - \epsilon, L + \epsilon)$.

As for functions of a single variable, it can be shown that

$$\lim_{(x, y) \to (x_0, y_0)} x = x_0$$

$$\lim_{(x, y) \to (x_0, y_0)} y = y_0$$

$$\lim_{(x, y) \to (x_0, y_0)} k = k \qquad \text{(any number } k\text{).}$$

For example, in the first limit statement above, $f(x, y) = x$ and $L = x_0$. Using the definition of limit, suppose that $\epsilon > 0$ is chosen. If we let δ equal this ϵ, we see that

$$0 < \sqrt{(x - x_0)^2 + (y - y_0)^2} < \delta = \epsilon$$

implies

$$\sqrt{(x - x_0)^2} < \epsilon \qquad (x - x_0)^2 \leq (x - x_0)^2 + (y - y_0)^2$$

$$|x - x_0| < \epsilon \qquad \sqrt{a^2} = |a|$$

$$|f(x, y) - x_0| < \epsilon \qquad x = f(x, y)$$

That is,

$$|f(x, y) - x_0| < \epsilon \qquad \text{whenever} \qquad 0 < \sqrt{(x - x_0)^2 + (y - y_0)^2} < \delta.$$

So a δ has been found satisfying the requirement of the definition, and

$$\lim_{(x,\,y)\to(x_0,\,y_0)} f(x, y) = \lim_{(x,\,y)\to(x_0,\,y_0)} x = x_0.$$

As with single-variable functions, the limit of the sum of two functions is the sum of their limits (when they both exist), with similar results for the limits of the differences, constant multiples, products, quotients, powers, and roots.

THEOREM 1—Properties of Limits of Functions of Two Variables   The following rules hold if L, M, and k are real numbers and

$$\lim_{(x,\,y)\to(x_0,\,y_0)} f(x, y) = L \qquad \text{and} \qquad \lim_{(x,\,y)\to(x_0,\,y_0)} g(x, y) = M.$$

1. *Sum Rule:* $\displaystyle\lim_{(x,\,y)\to(x_0,\,y_0)} (f(x, y) + g(x, y)) = L + M$

2. *Difference Rule:* $\displaystyle\lim_{(x,\,y)\to(x_0,\,y_0)} (f(x, y) - g(x, y)) = L - M$

3. *Constant Multiple Rule:* $\displaystyle\lim_{(x,\,y)\to(x_0,\,y_0)} kf(x, y) = kL$ (any number k)

4. *Product Rule:* $\displaystyle\lim_{(x,\,y)\to(x_0,\,y_0)} (f(x, y) \cdot g(x, y)) = L \cdot M$

5. *Quotient Rule:* $\displaystyle\lim_{(x,\,y)\to(x_0,\,y_0)} \frac{f(x, y)}{g(x, y)} = \frac{L}{M}, \qquad M \neq 0$

6. *Power Rule:* $\displaystyle\lim_{(x,\,y)\to(x_0,\,y_0)} [f(x, y)]^n = L^n$, n a positive integer

7. *Root Rule:* $\displaystyle\lim_{(x,\,y)\to(x_0,\,y_0)} \sqrt[n]{f(x, y)} = \sqrt[n]{L} = L^{1/n},$

   n a positive integer, and if n is even, we assume that $L > 0$.

While we won't prove Theorem 1 here, we give an informal discussion of why it's true. If (x, y) is sufficiently close to (x_0, y_0), then $f(x, y)$ is close to L and $g(x, y)$ is close to M (from the informal interpretation of limits). It is then reasonable that $f(x, y) + g(x, y)$ is close to $L + M$; $f(x, y) - g(x, y)$ is close to $L - M$; $kf(x, y)$ is close to kL; $f(x, y)g(x, y)$ is close to LM; and $f(x, y)/g(x, y)$ is close to L/M if $M \neq 0$.

When we apply Theorem 1 to polynomials and rational functions, we obtain the useful result that the limits of these functions as $(x, y) \to (x_0, y_0)$ can be calculated by evaluating the functions at (x_0, y_0). The only requirement is that the rational functions be defined at (x_0, y_0).

EXAMPLE 1   In this example, we can combine the three simple results following the limit definition with the results in Theorem 1 to calculate the limits. We simply substitute the x and y values of the point being approached into the functional expression to find the limiting value.

(a) $\displaystyle\lim_{(x,\,y)\to(0,1)} \frac{x - xy + 3}{x^2 y + 5xy - y^3} = \frac{0 - (0)(1) + 3}{(0)^2(1) + 5(0)(1) - (1)^3} = -3$

(b) $\displaystyle\lim_{(x,\,y)\to(3,\,-4)} \sqrt{x^2 + y^2} = \sqrt{(3)^2 + (-4)^2} = \sqrt{25} = 5$ ∎

EXAMPLE 2   Find

$$\lim_{(x,\,y)\to(0,0)} \frac{x^2 - xy}{\sqrt{x} - \sqrt{y}}.$$

Solution Since the denominator $\sqrt{x} - \sqrt{y}$ approaches 0 as $(x, y) \to (0, 0)$, we cannot use the Quotient Rule from Theorem 1. If we multiply numerator and denominator by $\sqrt{x} + \sqrt{y}$, however, we produce an equivalent fraction whose limit we *can* find:

$$\lim_{(x, y) \to (0,0)} \frac{x^2 - xy}{\sqrt{x} - \sqrt{y}} = \lim_{(x, y) \to (0,0)} \frac{\left(x^2 - xy\right)\left(\sqrt{x} + \sqrt{y}\right)}{\left(\sqrt{x} - \sqrt{y}\right)\left(\sqrt{x} + \sqrt{y}\right)}$$

$$= \lim_{(x, y) \to (0,0)} \frac{x\left(x - y\right)\left(\sqrt{x} + \sqrt{y}\right)}{x - y} \qquad \text{Algebra}$$

$$= \lim_{(x, y) \to (0,0)} x\left(\sqrt{x} + \sqrt{y}\right) \qquad \begin{array}{l}\text{Cancel the nonzero}\\\text{factor } (x - y).\end{array}$$

$$= 0\left(\sqrt{0} + \sqrt{0}\right) = 0 \qquad \text{Known limit values}$$

We can cancel the factor $(x - y)$ because the path $y = x$ (along which $x - y = 0$) is *not* in the domain of the function

$$\frac{x^2 - xy}{\sqrt{x} - \sqrt{y}}.$$ ∎

EXAMPLE 3 Find $\displaystyle\lim_{(x, y) \to (0,0)} \frac{4xy^2}{x^2 + y^2}$ if it exists.

Solution We first observe that along the line $x = 0$, the function always has value 0 when $y \neq 0$. Likewise, along the line $y = 0$, the function has value 0 provided $x \neq 0$. So if the limit does exist as (x, y) approaches $(0, 0)$, the value of the limit must be 0. To see if this is true, we apply the definition of limit.

Let $\epsilon > 0$ be given, but arbitrary. We want to find a $\delta > 0$ such that

$$\left|\frac{4xy^2}{x^2 + y^2} - 0\right| < \epsilon \qquad \text{whenever} \qquad 0 < \sqrt{x^2 + y^2} < \delta$$

or

$$\frac{4|x|y^2}{x^2 + y^2} < \epsilon \qquad \text{whenever} \qquad 0 < \sqrt{x^2 + y^2} < \delta.$$

Since $y^2 \leq x^2 + y^2$ we have that

$$\frac{4|x|y^2}{x^2 + y^2} \leq 4|x| = 4\sqrt{x^2} \leq 4\sqrt{x^2 + y^2}. \qquad \frac{y^2}{x^2 + y^2} \leq 1$$

So if we choose $\delta = \epsilon/4$ and let $0 < \sqrt{x^2 + y^2} < \delta$, we get

$$\left|\frac{4xy^2}{x^2 + y^2} - 0\right| \leq 4\sqrt{x^2 + y^2} < 4\delta = 4\left(\frac{\epsilon}{4}\right) = \epsilon.$$

It follows from the definition that

$$\lim_{(x, y) \to (0,0)} \frac{4xy^2}{x^2 + y^2} = 0.$$ ∎

EXAMPLE 4   If $f(x, y) = \dfrac{y}{x}$, does $\displaystyle\lim_{(x, y)\to(0, 0)} f(x, y)$ exist?

Solution   The domain of f does not include the y-axis, so we do not consider any points (x, y) where $x = 0$ in the approach toward the origin $(0, 0)$. Along the x-axis, the value of the function is $f(x, 0) = 0$ for all $x \neq 0$. So if the limit does exist as $(x, y) \to (0, 0)$, the value of the limit must be $L = 0$. On the other hand, along the line $y = x$, the value of the function is $f(x, x) = x/x = 1$ for all $x \neq 0$. That is, the function f approaches the value 1 along the line $y = x$. This means that for every disk of radius δ centered at $(0, 0)$, the disk will contain points $(x, 0)$ on the x-axis where the value of the function is 0, and also points (x, x) along the line $y = x$ where the value of the function is 1. So no matter how small we choose δ as the radius of the disk in Figure 14.12, there will be points within the disk for which the function values differ by 1. Therefore, the limit cannot exist because we can take ϵ to be any number less than 1 in the limit definition and deny that $L = 0$ or 1, or any other real number. The limit does not exist because we have different limiting values along different paths approaching the point $(0, 0)$.   ∎

Continuity

As with functions of a single variable, continuity is defined in terms of limits.

(a)

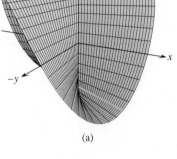

(b)

FIGURE 14.13   (a) The graph of

$$f(x, y) = \begin{cases} \dfrac{2xy}{x^2 + y^2}, & (x, y) \neq (0, 0) \\ 0, & (x, y) = (0, 0). \end{cases}$$

The function is continuous at every point except the origin. (b) The values of f are different constants along each line $y = mx, x \neq 0$ (Example 5).

DEFINITION   A function $f(x, y)$ is **continuous at the point (x_0, y_0)** if

1. f is defined at (x_0, y_0),

2. $\displaystyle\lim_{(x, y)\to(x_0, y_0)} f(x, y)$ exists,

3. $\displaystyle\lim_{(x, y)\to(x_0, y_0)} f(x, y) = f(x_0, y_0)$.

A function is **continuous** if it is continuous at every point of its domain.

As with the definition of limit, the definition of continuity applies at boundary points as well as interior points of the domain of f. The only requirement is that each point (x, y) near (x_0, y_0) be in the domain of f.

A consequence of Theorem 1 is that algebraic combinations of continuous functions are continuous at every point at which all the functions involved are defined. This means that sums, differences, constant multiples, products, quotients, and powers of continuous functions are continuous where defined. In particular, polynomials and rational functions of two variables are continuous at every point at which they are defined.

EXAMPLE 5   Show that

$$f(x, y) = \begin{cases} \dfrac{2xy}{x^2 + y^2}, & (x, y) \neq (0, 0) \\ 0, & (x, y) = (0, 0) \end{cases}$$

is continuous at every point except the origin (Figure 14.13).

Solution   The function f is continuous at any point $(x, y) \neq (0, 0)$ because its values are then given by a rational function of x and y and the limiting value is obtained by substituting the values of x and y into the functional expression.

At $(0, 0)$, the value of f is defined, but f, we claim, has no limit as $(x, y) \to (0, 0)$. The reason is that different paths of approach to the origin can lead to different results, as we now see.

For every value of m, the function f has a constant value on the "punctured" line $y = mx, x \neq 0$, because

$$f(x, y)\Big|_{y=mx} = \frac{2xy}{x^2 + y^2}\Big|_{y=mx} = \frac{2x(mx)}{x^2 + (mx)^2} = \frac{2mx^2}{x^2 + m^2x^2} = \frac{2m}{1 + m^2}.$$

Therefore, f has this number as its limit as (x, y) approaches $(0, 0)$ along the line:

$$\lim_{\substack{(x, y) \to (0,0) \\ \text{along } y=mx}} f(x, y) = \lim_{(x, y) \to (0,0)} \left[f(x, y)\Big|_{y=mx} \right] = \frac{2m}{1 + m^2}.$$

This limit changes with each value of the slope m. There is therefore no single number we may call the limit of f as (x, y) approaches the origin. The limit fails to exist, and the function is not continuous. ∎

Examples 4 and 5 illustrate an important point about limits of functions of two or more variables. For a limit to exist at a point, the limit must be the same along every approach path. This result is analogous to the single-variable case where both the left- and right-sided limits had to have the same value. For functions of two or more variables, if we ever find paths with different limits, we know the function has no limit at the point they approach.

> **Two-Path Test for Nonexistence of a Limit**
> If a function $f(x, y)$ has different limits along two different paths in the domain of f as (x, y) approaches (x_0, y_0), then $\lim_{(x, y) \to (x_0, y_0)} f(x, y)$ does not exist.

EXAMPLE 6 Show that the function

$$f(x, y) = \frac{2x^2y}{x^4 + y^2}$$

(Figure 14.14) has no limit as (x, y) approaches $(0, 0)$.

Solution The limit cannot be found by direct substitution, which gives the indeterminate form $0/0$. We examine the values of f along curves that end at $(0, 0)$. Along the curve $y = kx^2, x \neq 0$, the function has the constant value

$$f(x, y)\Big|_{y=kx^2} = \frac{2x^2y}{x^4 + y^2}\Big|_{y=kx^2} = \frac{2x^2(kx^2)}{x^4 + (kx^2)^2} = \frac{2kx^4}{x^4 + k^2x^4} = \frac{2k}{1 + k^2}.$$

Therefore,

$$\lim_{\substack{(x, y) \to (0,0) \\ \text{along } y=kx^2}} f(x, y) = \lim_{(x, y) \to (0,0)} \left[f(x, y)\Big|_{y=kx^2} \right] = \frac{2k}{1 + k^2}.$$

This limit varies with the path of approach. If (x, y) approaches $(0, 0)$ along the parabola $y = x^2$, for instance, $k = 1$ and the limit is 1. If (x, y) approaches $(0, 0)$ along the x-axis, $k = 0$ and the limit is 0. By the two-path test, f has no limit as (x, y) approaches $(0, 0)$. ∎

It can be shown that the function in Example 6 has limit 0 along every path $y = mx$ (Exercise 53). We conclude that

> Having the same limit along all straight lines approaching (x_0, y_0) does not imply a limit exists at (x_0, y_0).

(a)

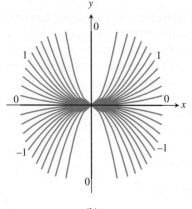

(b)

FIGURE 14.14 (a) The graph of $f(x, y) = 2x^2y/(x^4 + y^2)$. (b) Along each path $y = kx^2$ the value of f is constant, but varies with k (Example 6).

Whenever it is correctly defined, the composite of continuous functions is also continuous. The only requirement is that each function be continuous where it is applied. The proof, omitted here, is similar to that for functions of a single variable (Theorem 9 in Section 2.5).

Continuity of Composites

If f is continuous at (x_0, y_0) and g is a single-variable function continuous at $f(x_0, y_0)$, then the composite function $h = g \circ f$ defined by $h(x, y) = g(f(x, y))$ is continuous at (x_0, y_0).

For example, the composite functions

$$e^{x-y}, \qquad \cos \frac{xy}{x^2 + 1}, \qquad \ln (1 + x^2 y^2)$$

are continuous at every point (x, y).

Functions of More Than Two Variables

The definitions of limit and continuity for functions of two variables and the conclusions about limits and continuity for sums, products, quotients, powers, and composites all extend to functions of three or more variables. Functions like

$$\ln (x + y + z) \qquad \text{and} \qquad \frac{y \sin z}{x - 1}$$

are continuous throughout their domains, and limits like

$$\lim_{P \to (1,0,-1)} \frac{e^{x+z}}{z^2 + \cos \sqrt{xy}} = \frac{e^{1-1}}{(-1)^2 + \cos 0} = \frac{1}{2},$$

where P denotes the point (x, y, z), may be found by direct substitution.

Extreme Values of Continuous Functions on Closed, Bounded Sets

The Extreme Value Theorem (Theorem 1, Section 4.1) states that a function of a single variable that is continuous throughout a closed, bounded interval $[a, b]$ takes on an absolute maximum value and an absolute minimum value at least once in $[a, b]$. The same holds true of a function $z = f(x, y)$ that is continuous on a closed, bounded set R in the plane (like a line segment, a disk, or a filled-in triangle). The function takes on an absolute maximum value at some point in R and an absolute minimum value at some point in R.

Similar results hold for functions of three or more variables. A continuous function $w = f(x, y, z)$, for example, must take on absolute maximum and minimum values on any closed, bounded set (solid ball or cube, spherical shell, rectangular solid) on which it is defined. We will learn how to find these extreme values in Section 14.7.

Exercises 14.2

Limits with Two Variables

Find the limits in Exercises 1–12.

1. $\displaystyle\lim_{(x,y) \to (0,0)} \frac{3x^2 - y^2 + 5}{x^2 + y^2 + 2}$

2. $\displaystyle\lim_{(x,y) \to (0,4)} \frac{x}{\sqrt{y}}$

3. $\displaystyle\lim_{(x,y) \to (3,4)} \sqrt{x^2 + y^2 - 1}$

4. $\displaystyle\lim_{(x,y) \to (2,-3)} \left(\frac{1}{x} + \frac{1}{y} \right)^2$

5. $\displaystyle\lim_{(x,y) \to (0,\pi/4)} \sec x \tan y$

6. $\displaystyle\lim_{(x,y) \to (0,0)} \cos \frac{x^2 + y^3}{x + y + 1}$

7. $\displaystyle\lim_{(x,y)\to(0,\ln 2)} e^{x-y}$

8. $\displaystyle\lim_{(x,y)\to(1,1)} \ln|1 + x^2 y^2|$

9. $\displaystyle\lim_{(x,y)\to(0,0)} \frac{e^y \sin x}{x}$

10. $\displaystyle\lim_{(x,y)\to(1/27,\,\pi^3)} \cos\sqrt[3]{xy}$

11. $\displaystyle\lim_{(x,y)\to(1,\,\pi/6)} \frac{x \sin y}{x^2 + 1}$

12. $\displaystyle\lim_{(x,y)\to(\pi/2,0)} \frac{\cos y + 1}{y - \sin x}$

Limits of Quotients
Find the limits in Exercises 13–24 by rewriting the fractions first.

13. $\displaystyle\lim_{\substack{(x,y)\to(1,1)\\x\neq y}} \frac{x^2 - 2xy + y^2}{x - y}$

14. $\displaystyle\lim_{\substack{(x,y)\to(1,1)\\x\neq y}} \frac{x^2 - y^2}{x - y}$

15. $\displaystyle\lim_{\substack{(x,y)\to(1,1)\\x\neq 1}} \frac{xy - y - 2x + 2}{x - 1}$

16. $\displaystyle\lim_{\substack{(x,y)\to(2,-4)\\y\neq-4,\,x\neq x^2}} \frac{y + 4}{x^2 y - xy + 4x^2 - 4x}$

17. $\displaystyle\lim_{\substack{(x,y)\to(0,0)\\x\neq y}} \frac{x - y + 2\sqrt{x} - 2\sqrt{y}}{\sqrt{x} - \sqrt{y}}$

18. $\displaystyle\lim_{\substack{(x,y)\to(2,2)\\x+y\neq 4}} \frac{x + y - 4}{\sqrt{x + y} - 2}$

19. $\displaystyle\lim_{\substack{(x,y)\to(2,0)\\2x-y\neq 4}} \frac{\sqrt{2x - y} - 2}{2x - y - 4}$

20. $\displaystyle\lim_{\substack{(x,y)\to(4,3)\\x\neq y+1}} \frac{\sqrt{x} - \sqrt{y + 1}}{x - y - 1}$

21. $\displaystyle\lim_{(x,y)\to(0,0)} \frac{\sin(x^2 + y^2)}{x^2 + y^2}$

22. $\displaystyle\lim_{(x,y)\to(0,0)} \frac{1 - \cos(xy)}{xy}$

23. $\displaystyle\lim_{(x,y)\to(1,-1)} \frac{x^3 + y^3}{x + y}$

24. $\displaystyle\lim_{(x,y)\to(2,2)} \frac{x - y}{x^4 - y^4}$

Limits with Three Variables
Find the limits in Exercises 25–30.

25. $\displaystyle\lim_{P\to(1,3,4)} \left(\frac{1}{x} + \frac{1}{y} + \frac{1}{z}\right)$

26. $\displaystyle\lim_{P\to(1,-1,-1)} \frac{2xy + yz}{x^2 + z^2}$

27. $\displaystyle\lim_{P\to(\pi,\pi,0)} (\sin^2 x + \cos^2 y + \sec^2 z)$

28. $\displaystyle\lim_{P\to(-1/4,\pi/2,2)} \tan^{-1} xyz$

29. $\displaystyle\lim_{P\to(\pi,0,3)} z e^{-2y} \cos 2x$

30. $\displaystyle\lim_{P\to(2,-3,6)} \ln\sqrt{x^2 + y^2 + z^2}$

Continuity in the Plane
At what points (x, y) in the plane are the functions in Exercises 31–34 continuous?

31. a. $f(x, y) = \sin(x + y)$ **b.** $f(x, y) = \ln(x^2 + y^2)$

32. a. $f(x, y) = \dfrac{x + y}{x - y}$ **b.** $f(x, y) = \dfrac{y}{x^2 + 1}$

33. a. $g(x, y) = \sin\dfrac{1}{xy}$ **b.** $g(x, y) = \dfrac{x + y}{2 + \cos x}$

34. a. $g(x, y) = \dfrac{x^2 + y^2}{x^2 - 3x + 2}$ **b.** $g(x, y) = \dfrac{1}{x^2 - y}$

Continuity in Space
At what points (x, y, z) in space are the functions in Exercises 35–40 continuous?

35. a. $f(x, y, z) = x^2 + y^2 - 2z^2$

  b. $f(x, y, z) = \sqrt{x^2 + y^2 - 1}$

36. a. $f(x, y, z) = \ln xyz$ **b.** $f(x, y, z) = e^{x+y} \cos z$

37. a. $h(x, y, z) = xy \sin\dfrac{1}{z}$ **b.** $h(x, y, z) = \dfrac{1}{x^2 + z^2 - 1}$

38. a. $h(x, y, z) = \dfrac{1}{|y| + |z|}$ **b.** $h(x, y, z) = \dfrac{1}{|xy| + |z|}$

39. a. $h(x, y, z) = \ln(z - x^2 - y^2 - 1)$

  b. $h(x, y, z) = \dfrac{1}{z - \sqrt{x^2 + y^2}}$

40. a. $h(x, y, z) = \sqrt{4 - x^2 - y^2 - z^2}$

  b. $h(x, y, z) = \dfrac{1}{4 - \sqrt{x^2 + y^2 + z^2 - 9}}$

No Limit at a Point
By considering different paths of approach, show that the functions in Exercises 41–48 have no limit as $(x, y) \to (0, 0)$.

41. $f(x, y) = -\dfrac{x}{\sqrt{x^2 + y^2}}$

42. $f(x, y) = \dfrac{x^4}{x^4 + y^2}$

43. $f(x, y) = \dfrac{x^4 - y^2}{x^4 + y^2}$

44. $f(x, y) = \dfrac{xy}{|xy|}$

45. $g(x, y) = \dfrac{x - y}{x + y}$

46. $g(x, y) = \dfrac{x^2 - y}{x - y}$

47. $h(x, y) = \dfrac{x^2 + y}{y}$

48. $h(x, y) = \dfrac{x^2 y}{x^4 + y^2}$

Theory and Examples
In Exercises 49 and 50, show that the limits do not exist.

49. $\displaystyle\lim_{(x,y)\to(1,1)} \frac{xy^2 - 1}{y - 1}$

50. $\displaystyle\lim_{(x,y)\to(1,-1)} \frac{xy + 1}{x^2 - y^2}$

51. Let $f(x, y) = \begin{cases} 1, & y \geq x^4 \\ 1, & y \leq 0 \\ 0, & \text{otherwise.} \end{cases}$

Find each of the following limits, or explain that the limit does not exist.

a. $\displaystyle\lim_{(x,y)\to(0,1)} f(x, y)$

b. $\displaystyle\lim_{(x,y)\to(2,3)} f(x, y)$

c. $\displaystyle\lim_{(x,y)\to(0,0)} f(x, y)$

To find $\partial f / \partial y$, we treat x as a constant and differentiate with respect to y:

$$\frac{\partial f}{\partial y} = \frac{\partial}{\partial y}(x^2 + 3xy + y - 1) = 0 + 3 \cdot x \cdot 1 + 1 - 0 = 3x + 1.$$

The value of $\partial f / \partial y$ at $(4, -5)$ is $3(4) + 1 = 13$.   ∎

EXAMPLE 2   Find $\partial f / \partial y$ as a function if $f(x, y) = y \sin xy$.

Solution   We treat x as a constant and f as a product of y and $\sin xy$:

$$\frac{\partial f}{\partial y} = \frac{\partial}{\partial y}(y \sin xy) = y\frac{\partial}{\partial y}\sin xy + (\sin xy)\frac{\partial}{\partial y}(y)$$

$$= (y \cos xy)\frac{\partial}{\partial y}(xy) + \sin xy = xy \cos xy + \sin xy.$$ ∎

EXAMPLE 3   Find f_x and f_y as functions if

$$f(x, y) = \frac{2y}{y + \cos x}.$$

Solution   We treat f as a quotient. With y held constant, we get

$$f_x = \frac{\partial}{\partial x}\left(\frac{2y}{y + \cos x}\right) = \frac{(y + \cos x)\frac{\partial}{\partial x}(2y) - 2y\frac{\partial}{\partial x}(y + \cos x)}{(y + \cos x)^2}$$

$$= \frac{(y + \cos x)(0) - 2y(-\sin x)}{(y + \cos x)^2} = \frac{2y \sin x}{(y + \cos x)^2}.$$

With x held constant, we get

$$f_y = \frac{\partial}{\partial y}\left(\frac{2y}{y + \cos x}\right) = \frac{(y + \cos x)\frac{\partial}{\partial y}(2y) - 2y\frac{\partial}{\partial y}(y + \cos x)}{(y + \cos x)^2}$$

$$= \frac{(y + \cos x)(2) - 2y(1)}{(y + \cos x)^2} = \frac{2 \cos x}{(y + \cos x)^2}.$$ ∎

Implicit differentiation works for partial derivatives the way it works for ordinary derivatives, as the next example illustrates.

EXAMPLE 4   Find $\partial z / \partial x$ if the equation

$$yz - \ln z = x + y$$

defines z as a function of the two independent variables x and y and the partial derivative exists.

Solution   We differentiate both sides of the equation with respect to x, holding y constant and treating z as a differentiable function of x:

$$\frac{\partial}{\partial x}(yz) - \frac{\partial}{\partial x}\ln z = \frac{\partial x}{\partial x} + \frac{\partial y}{\partial x}$$

$$y\frac{\partial z}{\partial x} - \frac{1}{z}\frac{\partial z}{\partial x} = 1 + 0 \qquad \text{With } y \text{ constant,}$$
$$\qquad\qquad\qquad\qquad \frac{\partial}{\partial x}(yz) = y\frac{\partial z}{\partial x}.$$

$$\left(y - \frac{1}{z}\right)\frac{\partial z}{\partial x} = 1$$

$$\frac{\partial z}{\partial x} = \frac{z}{yz - 1}.$$ ∎

EXAMPLE 5 The plane $x = 1$ intersects the paraboloid $z = x^2 + y^2$ in a parabola. Find the slope of the tangent to the parabola at $(1, 2, 5)$ (Figure 14.18).

Solution The slope is the value of the partial derivative $\partial z / \partial y$ at $(1, 2)$:

$$\left.\frac{\partial z}{\partial y}\right|_{(1,2)} = \left.\frac{\partial}{\partial y}(x^2 + y^2)\right|_{(1,2)} = \left.2y\right|_{(1,2)} = 2(2) = 4.$$

As a check, we can treat the parabola as the graph of the single-variable function $z = (1)^2 + y^2 = 1 + y^2$ in the plane $x = 1$ and ask for the slope at $y = 2$. The slope, calculated now as an ordinary derivative, is

$$\left.\frac{dz}{dy}\right|_{y=2} = \left.\frac{d}{dy}(1 + y^2)\right|_{y=2} = \left.2y\right|_{y=2} = 4. \qquad \blacksquare$$

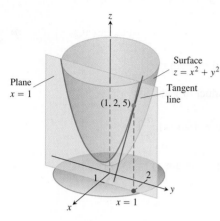

FIGURE 14.18 The tangent to the curve of intersection of the plane $x = 1$ and surface $z = x^2 + y^2$ at the point $(1, 2, 5)$ (Example 5).

Functions of More Than Two Variables

The definitions of the partial derivatives of functions of more than two independent variables are like the definitions for functions of two variables. They are ordinary derivatives with respect to one variable, taken while the other independent variables are held constant.

EXAMPLE 6 If x, y, and z are independent variables and

$$f(x, y, z) = x \sin(y + 3z),$$

then

$$\frac{\partial f}{\partial z} = \frac{\partial}{\partial z}[x \sin(y + 3z)] = x \frac{\partial}{\partial z}\sin(y + 3z)$$

$$= x \cos(y + 3z)\frac{\partial}{\partial z}(y + 3z) = 3x \cos(y + 3z). \qquad \blacksquare$$

EXAMPLE 7 If resistors of R_1, R_2, and R_3 ohms are connected in parallel to make an R-ohm resistor, the value of R can be found from the equation

$$\frac{1}{R} = \frac{1}{R_1} + \frac{1}{R_2} + \frac{1}{R_3}$$

(Figure 14.19). Find the value of $\partial R / \partial R_2$ when $R_1 = 30$, $R_2 = 45$, and $R_3 = 90$ ohms.

Solution To find $\partial R / \partial R_2$, we treat R_1 and R_3 as constants and, using implicit differentiation, differentiate both sides of the equation with respect to R_2:

$$\frac{\partial}{\partial R_2}\left(\frac{1}{R}\right) = \frac{\partial}{\partial R_2}\left(\frac{1}{R_1} + \frac{1}{R_2} + \frac{1}{R_3}\right)$$

$$-\frac{1}{R^2}\frac{\partial R}{\partial R_2} = 0 - \frac{1}{R_2^2} + 0$$

$$\frac{\partial R}{\partial R_2} = \frac{R^2}{R_2^2} = \left(\frac{R}{R_2}\right)^2.$$

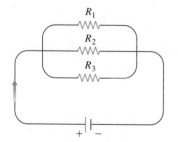

FIGURE 14.19 Resistors arranged this way are said to be connected in parallel (Example 7). Each resistor lets a portion of the current through. Their equivalent resistance R is calculated with the formula

$$\frac{1}{R} = \frac{1}{R_1} + \frac{1}{R_2} + \frac{1}{R_3}.$$

When $R_1 = 30$, $R_2 = 45$, and $R_3 = 90$,

$$\frac{1}{R} = \frac{1}{30} + \frac{1}{45} + \frac{1}{90} = \frac{3 + 2 + 1}{90} = \frac{6}{90} = \frac{1}{15},$$

so $R = 15$ and

$$\frac{\partial R}{\partial R_2} = \left(\frac{15}{45}\right)^2 = \left(\frac{1}{3}\right)^2 = \frac{1}{9}.$$

Thus at the given values, a small change in the resistance R_2 leads to a change in R about 1/9th as large. ∎

Partial Derivatives and Continuity

A function $f(x, y)$ can have partial derivatives with respect to both x and y at a point without the function being continuous there. This is different from functions of a single variable, where the existence of a derivative implies continuity. If the partial derivatives of $f(x, y)$ exist and are continuous throughout a disk centered at (x_0, y_0), however, then f *is* continuous at (x_0, y_0), as we see at the end of this section.

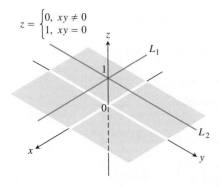

$$z = \begin{cases} 0, & xy \neq 0 \\ 1, & xy = 0 \end{cases}$$

FIGURE 14.20 The graph of

$$f(x, y) = \begin{cases} 0, & xy \neq 0 \\ 1, & xy = 0 \end{cases}$$

consists of the lines L_1 and L_2 and the four open quadrants of the xy-plane. The function has partial derivatives at the origin but is not continuous there (Example 8).

EXAMPLE 8   Let

$$f(x, y) = \begin{cases} 0, & xy \neq 0 \\ 1, & xy = 0 \end{cases}$$

(Figure 14.20).

(a) Find the limit of f as (x, y) approaches $(0, 0)$ along the line $y = x$.

(b) Prove that f is not continuous at the origin.

(c) Show that both partial derivatives $\partial f/\partial x$ and $\partial f/\partial y$ exist at the origin.

Solution

(a) Since $f(x, y)$ is constantly zero along the line $y = x$ (except at the origin), we have

$$\lim_{(x, y) \to (0,0)} f(x, y) \bigg|_{y=x} = \lim_{(x, y) \to (0,0)} 0 = 0.$$

(b) Since $f(0, 0) = 1$, the limit in part (a) proves that f is not continuous at $(0, 0)$.

(c) To find $\partial f/\partial x$ at $(0, 0)$, we hold y fixed at $y = 0$. Then $f(x, y) = 1$ for all x, and the graph of f is the line L_1 in Figure 14.20. The slope of this line at any x is $\partial f/\partial x = 0$. In particular, $\partial f/\partial x = 0$ at $(0, 0)$. Similarly, $\partial f/\partial y$ is the slope of line L_2 at any y, so $\partial f/\partial y = 0$ at $(0, 0)$. ∎

Example 8 notwithstanding, it is still true in higher dimensions that *differentiability* at a point implies continuity. What Example 8 suggests is that we need a stronger requirement for differentiability in higher dimensions than the mere existence of the partial derivatives. We define differentiability for functions of two variables (which is slightly more complicated than for single-variable functions) at the end of this section and then revisit the connection to continuity.

Second-Order Partial Derivatives

When we differentiate a function $f(x, y)$ twice, we produce its second-order derivatives. These derivatives are usually denoted by

$$\frac{\partial^2 f}{\partial x^2} \quad \text{or} \quad f_{xx}, \qquad \frac{\partial^2 f}{\partial y^2} \quad \text{or} \quad f_{yy},$$

$$\frac{\partial^2 f}{\partial x \partial y} \quad \text{or} \quad f_{yx}, \qquad \text{and} \qquad \frac{\partial^2 f}{\partial y \partial x} \quad \text{or} \quad f_{xy}.$$

The defining equations are

$$\frac{\partial^2 f}{\partial x^2} = \frac{\partial}{\partial x}\left(\frac{\partial f}{\partial x}\right), \qquad \frac{\partial^2 f}{\partial x \partial y} = \frac{\partial}{\partial x}\left(\frac{\partial f}{\partial y}\right),$$

and so on. Notice the order in which the mixed partial derivatives are taken:

$$\frac{\partial^2 f}{\partial x \partial y} \qquad \text{Differentiate first with respect to } y, \text{ then with respect to } x.$$

$$f_{yx} = (f_y)_x \qquad \text{Means the same thing.}$$

HISTORICAL BIOGRAPHY

Pierre-Simon Laplace
(1749–1827)

EXAMPLE 9 If $f(x, y) = x \cos y + y e^x$, find the second-order derivatives

$$\frac{\partial^2 f}{\partial x^2}, \qquad \frac{\partial^2 f}{\partial y \partial x}, \qquad \frac{\partial^2 f}{\partial y^2}, \qquad \text{and} \qquad \frac{\partial^2 f}{\partial x \partial y}.$$

Solution The first step is to calculate both first partial derivatives.

$$\frac{\partial f}{\partial x} = \frac{\partial}{\partial x}(x \cos y + y e^x) \qquad\qquad \frac{\partial f}{\partial y} = \frac{\partial}{\partial y}(x \cos y + y e^x)$$

$$= \cos y + y e^x \qquad\qquad\qquad = -x \sin y + e^x$$

Now we find both partial derivatives of each first partial:

$$\frac{\partial^2 f}{\partial y \partial x} = \frac{\partial}{\partial y}\left(\frac{\partial f}{\partial x}\right) = -\sin y + e^x \qquad \frac{\partial^2 f}{\partial x \partial y} = \frac{\partial}{\partial x}\left(\frac{\partial f}{\partial y}\right) = -\sin y + e^x$$

$$\frac{\partial^2 f}{\partial x^2} = \frac{\partial}{\partial x}\left(\frac{\partial f}{\partial x}\right) = y e^x. \qquad \frac{\partial^2 f}{\partial y^2} = \frac{\partial}{\partial y}\left(\frac{\partial f}{\partial y}\right) = -x \cos y.$$

The Mixed Derivative Theorem

You may have noticed that the "mixed" second-order partial derivatives

$$\frac{\partial^2 f}{\partial y \partial x} \qquad \text{and} \qquad \frac{\partial^2 f}{\partial x \partial y}$$

in Example 9 are equal. This is not a coincidence. They must be equal whenever f, f_x, f_y, f_{xy}, and f_{yx} are continuous, as stated in the following theorem.

> **THEOREM 2—The Mixed Derivative Theorem** If $f(x, y)$ and its partial derivatives f_x, f_y, f_{xy}, and f_{yx} are defined throughout an open region containing a point (a, b) and are all continuous at (a, b), then
>
> $$f_{xy}(a, b) = f_{yx}(a, b).$$

HISTORICAL BIOGRAPHY

Alexis Clairaut
(1713–1765)

Theorem 2 is also known as Clairaut's Theorem, named after the French mathematician Alexis Clairaut who discovered it. A proof is given in Appendix 9. Theorem 2 says that to calculate a mixed second-order derivative, we may differentiate in either order, provided the continuity conditions are satisfied. This ability to proceed in different order sometimes simplifies our calculations.

EXAMPLE 10 Find $\partial^2 w / \partial x \partial y$ if

$$w = xy + \frac{e^y}{y^2 + 1}.$$

Solution The symbol $\partial^2 w/\partial x \partial y$ tells us to differentiate first with respect to y and then with respect to x. However, if we interchange the order of differentiation and differentiate first with respect to x we get the answer more quickly. In two steps,

$$\frac{\partial w}{\partial x} = y \quad \text{and} \quad \frac{\partial^2 w}{\partial y \partial x} = 1.$$

If we differentiate first with respect to y, we obtain $\partial^2 w/\partial x \partial y = 1$ as well. We can differentiate in either order because the conditions of Theorem 2 hold for w at all points (x_0, y_0). ∎

Partial Derivatives of Still Higher Order

Although we will deal mostly with first- and second-order partial derivatives, because these appear the most frequently in applications, there is no theoretical limit to how many times we can differentiate a function as long as the derivatives involved exist. Thus, we get third- and fourth-order derivatives denoted by symbols like

$$\frac{\partial^3 f}{\partial x \partial y^2} = f_{yyx},$$

$$\frac{\partial^4 f}{\partial x^2 \partial y^2} = f_{yyxx},$$

and so on. As with second-order derivatives, the order of differentiation is immaterial as long as all the derivatives through the order in question are continuous.

EXAMPLE 11 Find f_{yxyz} if $f(x, y, z) = 1 - 2xy^2z + x^2y$.

Solution We first differentiate with respect to the variable y, then x, then y again, and finally with respect to z:

$$f_y = -4xyz + x^2$$
$$f_{yx} = -4yz + 2x$$
$$f_{yxy} = -4z$$
$$f_{yxyz} = -4$$

∎

Differentiability

The starting point for differentiability is not the difference quotient we saw in studying single-variable functions, but rather the idea of increment. Recall from our work with functions of a single variable in Section 3.11 that if $y = f(x)$ is differentiable at $x = x_0$, then the change in the value of f that results from changing x from x_0 to $x_0 + \Delta x$ is given by an equation of the form

$$\Delta y = f'(x_0)\Delta x + \epsilon \Delta x$$

in which $\epsilon \to 0$ as $\Delta x \to 0$. For functions of two variables, the analogous property becomes the definition of differentiability. The Increment Theorem (proved in Appendix 9) tells us when to expect the property to hold.

THEOREM 3—The Increment Theorem for Functions of Two Variables Suppose that the first partial derivatives of $f(x, y)$ are defined throughout an open region R containing the point (x_0, y_0) and that f_x and f_y are continuous at (x_0, y_0). Then the change

$$\Delta z = f(x_0 + \Delta x, y_0 + \Delta y) - f(x_0, y_0)$$

in the value of f that results from moving from (x_0, y_0) to another point $(x_0 + \Delta x, y_0 + \Delta y)$ in R satisfies an equation of the form

$$\Delta z = f_x(x_0, y_0)\Delta x + f_y(x_0, y_0)\Delta y + \epsilon_1 \Delta x + \epsilon_2 \Delta y$$

in which each of $\epsilon_1, \epsilon_2 \to 0$ as both $\Delta x, \Delta y \to 0$.

You can see where the epsilons come from in the proof given in Appendix 9. Similar results hold for functions of more than two independent variables.

DEFINITION   A function $z = f(x, y)$ is **differentiable at** (x_0, y_0) if $f_x(x_0, y_0)$ and $f_y(x_0, y_0)$ exist and Δz satisfies an equation of the form

$$\Delta z = f_x(x_0, y_0)\Delta x + f_y(x_0, y_0)\Delta y + \epsilon_1 \Delta x + \epsilon_2 \Delta y$$

in which each of $\epsilon_1, \epsilon_2 \to 0$ as both $\Delta x, \Delta y \to 0$. We call f **differentiable** if it is differentiable at every point in its domain, and say that its graph is a **smooth surface**.

Because of this definition, an immediate corollary of Theorem 3 is that a function is differentiable at (x_0, y_0) if its first partial derivatives are *continuous* there.

COROLLARY OF THEOREM 3   If the partial derivatives f_x and f_y of a function $f(x, y)$ are continuous throughout an open region R, then f is differentiable at every point of R.

If $z = f(x, y)$ is differentiable, then the definition of differentiability assures that $\Delta z = f(x_0 + \Delta x, y_0 + \Delta y) - f(x_0, y_0)$ approaches 0 as Δx and Δy approach 0. This tells us that a function of two variables is continuous at every point where it is differentiable.

THEOREM 4—Differentiability Implies Continuity   If a function $f(x, y)$ is differentiable at (x_0, y_0), then f is continuous at (x_0, y_0).

As we can see from Corollary 3 and Theorem 4, a function $f(x, y)$ must be continuous at a point (x_0, y_0) if f_x and f_y are continuous throughout an open region containing (x_0, y_0). Remember, however, that it is still possible for a function of two variables to be discontinuous at a point where its first partial derivatives exist, as we saw in Example 8. Existence alone of the partial derivatives at that point is not enough, but continuity of the partial derivatives guarantees differentiability.

Exercises 14.3

Calculating First-Order Partial Derivatives

In Exercises 1–22, find $\partial f/\partial x$ and $\partial f/\partial y$.

1. $f(x, y) = 2x^2 - 3y - 4$ **2.** $f(x, y) = x^2 - xy + y^2$

3. $f(x, y) = (x^2 - 1)(y + 2)$

4. $f(x, y) = 5xy - 7x^2 - y^2 + 3x - 6y + 2$

5. $f(x, y) = (xy - 1)^2$ **6.** $f(x, y) = (2x - 3y)^3$

7. $f(x, y) = \sqrt{x^2 + y^2}$ **8.** $f(x, y) = (x^3 + (y/2))^{2/3}$

9. $f(x, y) = 1/(x + y)$ **10.** $f(x, y) = x/(x^2 + y^2)$

11. $f(x, y) = (x + y)/(xy - 1)$ **12.** $f(x, y) = \tan^{-1}(y/x)$

13. $f(x, y) = e^{(x+y+1)}$ **14.** $f(x, y) = e^{-x} \sin(x + y)$

15. $f(x, y) = \ln(x + y)$ **16.** $f(x, y) = e^{xy} \ln y$

17. $f(x, y) = \sin^2(x - 3y)$ **18.** $f(x, y) = \cos^2(3x - y^2)$

19. $f(x, y) = x^y$ **20.** $f(x, y) = \log_y x$

21. $f(x, y) = \displaystyle\int_x^y g(t)\, dt$ (g continuous for all t)

22. $f(x, y) = \displaystyle\sum_{n=0}^{\infty} (xy)^n$ ($|xy| < 1$)

In Exercises 23–34, find f_x, f_y, and f_z.

23. $f(x, y, z) = 1 + xy^2 - 2z^2$ **24.** $f(x, y, z) = xy + yz + xz$

25. $f(x, y, z) = x - \sqrt{y^2 + z^2}$

26. $f(x, y, z) = (x^2 + y^2 + z^2)^{-1/2}$

27. $f(x, y, z) = \sin^{-1}(xyz)$ **28.** $f(x, y, z) = \sec^{-1}(x + yz)$

29. $f(x, y, z) = \ln(x + 2y + 3z)$

30. $f(x, y, z) = yz \ln (xy)$ **31.** $f(x, y, z) = e^{-(x^2+y^2+z^2)}$

32. $f(x, y, z) = e^{-xyz}$

33. $f(x, y, z) = \tanh (x + 2y + 3z)$

34. $f(x, y, z) = \sinh (xy - z^2)$

In Exercises 35–40, find the partial derivative of the function with respect to each variable.

35. $f(t, \alpha) = \cos (2\pi t - \alpha)$ **36.** $g(u, v) = v^2 e^{(2u/v)}$

37. $h(\rho, \phi, \theta) = \rho \sin \phi \cos \theta$ **38.** $g(r, \theta, z) = r(1 - \cos \theta) - z$

39. Work done by the heart   (Section 3.11, Exercise 61)

$$W(P, V, \delta, v, g) = PV + \frac{V\delta v^2}{2g}$$

40. Wilson lot size formula   (Section 4.6, Exercise 53)

$$A(c, h, k, m, q) = \frac{km}{q} + cm + \frac{hq}{2}$$

Calculating Second-Order Partial Derivatives

Find all the second-order partial derivatives of the functions in Exercises 41–50.

41. $f(x, y) = x + y + xy$ **42.** $f(x, y) = \sin xy$

43. $g(x, y) = x^2 y + \cos y + y \sin x$

44. $h(x, y) = xe^y + y + 1$ **45.** $r(x, y) = \ln (x + y)$

46. $s(x, y) = \tan^{-1} (y/x)$ **47.** $w = x^2 \tan (xy)$

48. $w = ye^{x^2 - y}$ **49.** $w = x \sin (x^2 y)$

50. $w = \dfrac{x - y}{x^2 + y}$

Mixed Partial Derivatives

In Exercises 51–54, verify that $w_{xy} = w_{yx}$.

51. $w = \ln (2x + 3y)$ **52.** $w = e^x + x \ln y + y \ln x$

53. $w = xy^2 + x^2 y^3 + x^3 y^4$ **54.** $w = x \sin y + y \sin x + xy$

55. Which order of differentiation will calculate f_{xy} faster: x first or y first? Try to answer without writing anything down.

  a. $f(x, y) = x \sin y + e^y$

  b. $f(x, y) = 1/x$

  c. $f(x, y) = y + (x/y)$

  d. $f(x, y) = y + x^2 y + 4y^3 - \ln (y^2 + 1)$

  e. $f(x, y) = x^2 + 5xy + \sin x + 7e^x$

  f. $f(x, y) = x \ln xy$

56. The fifth-order partial derivative $\partial^5 f / \partial x^2 \partial y^3$ is zero for each of the following functions. To show this as quickly as possible, which variable would you differentiate with respect to first: x or y? Try to answer without writing anything down.

  a. $f(x, y) = y^2 x^4 e^x + 2$

  b. $f(x, y) = y^2 + y(\sin x - x^4)$

  c. $f(x, y) = x^2 + 5xy + \sin x + 7e^x$

  d. $f(x, y) = xe^{y^2/2}$

Using the Partial Derivative Definition

In Exercises 57–60, use the limit definition of partial derivative to compute the partial derivatives of the functions at the specified points.

57. $f(x, y) = 1 - x + y - 3x^2 y$, $\dfrac{\partial f}{\partial x}$ and $\dfrac{\partial f}{\partial y}$ at $(1, 2)$

58. $f(x, y) = 4 + 2x - 3y - xy^2$, $\dfrac{\partial f}{\partial x}$ and $\dfrac{\partial f}{\partial y}$ at $(-2, 1)$

59. $f(x, y) = \sqrt{2x + 3y - 1}$, $\dfrac{\partial f}{\partial x}$ and $\dfrac{\partial f}{\partial y}$ at $(-2, 3)$

60. $f(x, y) = \begin{cases} \dfrac{\sin (x^3 + y^4)}{x^2 + y^2}, & (x, y) \neq (0, 0) \\ 0, & (x, y) = (0, 0), \end{cases}$

$\dfrac{\partial f}{\partial x}$ and $\dfrac{\partial f}{\partial y}$ at $(0, 0)$

61. Let $f(x, y) = 2x + 3y - 4$. Find the slope of the line tangent to this surface at the point $(2, -1)$ and lying in the **a.** plane $x = 2$ **b.** plane $y = -1$.

62. Let $f(x, y) = x^2 + y^3$. Find the slope of the line tangent to this surface at the point $(-1, 1)$ and lying in the **a.** plane $x = -1$ **b.** plane $y = 1$.

63. Three variables   Let $w = f(x, y, z)$ be a function of three independent variables and write the formal definition of the partial derivative $\partial f / \partial z$ at (x_0, y_0, z_0). Use this definition to find $\partial f / \partial z$ at $(1, 2, 3)$ for $f(x, y, z) = x^2 yz^2$.

64. Three variables   Let $w = f(x, y, z)$ be a function of three independent variables and write the formal definition of the partial derivative $\partial f / \partial y$ at (x_0, y_0, z_0). Use this definition to find $\partial f / \partial y$ at $(-1, 0, 3)$ for $f(x, y, z) = -2xy^2 + yz^2$.

Differentiating Implicitly

65. Find the value of $\partial z / \partial x$ at the point $(1, 1, 1)$ if the equation

$$xy + z^3 x - 2yz = 0$$

defines z as a function of the two independent variables x and y and the partial derivative exists.

66. Find the value of $\partial x / \partial z$ at the point $(1, -1, -3)$ if the equation

$$xz + y \ln x - x^2 + 4 = 0$$

defines x as a function of the two independent variables y and z and the partial derivative exists.

Exercises 67 and 68 are about the triangle shown here.

67. Express A implicitly as a function of a, b, and c and calculate $\partial A / \partial a$ and $\partial A / \partial b$.

68. Express a implicitly as a function of A, b, and B and calculate $\partial a / \partial A$ and $\partial a / \partial B$.

69. Two dependent variables   Express v_x in terms of u and y if the equations $x = v \ln u$ and $y = u \ln v$ define u and v as functions of the independent variables x and y, and if v_x exists. (*Hint:* Differentiate both equations with respect to x and solve for v_x by eliminating u_x.)

70. Two dependent variables Find $\partial x/\partial u$ and $\partial y/\partial u$ if the equations $u = x^2 - y^2$ and $v = x^2 - y$ define x and y as functions of the independent variables u and v, and the partial derivatives exist. (See the hint in Exercise 69.) Then let $s = x^2 + y^2$ and find $\partial s/\partial u$.

71. Let $f(x, y) = \begin{cases} y^3, & y \geq 0 \\ -y^2, & y < 0. \end{cases}$

Find f_x, f_y, f_{xy}, and f_{yx}, and state the domain for each partial derivative.

72. Let $f(x, y) = \begin{cases} \sqrt{x}, & x \geq 0 \\ x^2, & x < 0. \end{cases}$

Find f_x, f_y, f_{xy}, and f_{yx}, and state the domain for each partial derivative.

Theory and Examples

The **three-dimensional Laplace equation**

$$\frac{\partial^2 f}{\partial x^2} + \frac{\partial^2 f}{\partial y^2} + \frac{\partial^2 f}{\partial z^2} = 0$$

is satisfied by steady-state temperature distributions $T = f(x, y, z)$ in space, by gravitational potentials, and by electrostatic potentials. The **two-dimensional Laplace equation**

$$\frac{\partial^2 f}{\partial x^2} + \frac{\partial^2 f}{\partial y^2} = 0,$$

obtained by dropping the $\partial^2 f/\partial z^2$ term from the previous equation, describes potentials and steady-state temperature distributions in a plane (see the accompanying figure). The plane (a) may be treated as a thin slice of the solid (b) perpendicular to the z-axis.

(a)

$$\frac{\partial^2 f}{\partial x^2} + \frac{\partial^2 f}{\partial y^2} = 0$$

(b)

$$\frac{\partial^2 f}{\partial x^2} + \frac{\partial^2 f}{\partial y^2} + \frac{\partial^2 f}{\partial z^2} = 0$$

Boundary temperatures controlled

Show that each function in Exercises 73–80 satisfies a Laplace equation.

73. $f(x, y, z) = x^2 + y^2 - 2z^2$

74. $f(x, y, z) = 2z^3 - 3(x^2 + y^2)z$

75. $f(x, y) = e^{-2y} \cos 2x$

76. $f(x, y) = \ln\sqrt{x^2 + y^2}$

77. $f(x, y) = 3x + 2y - 4$

78. $f(x, y) = \tan^{-1}\frac{x}{y}$

79. $f(x, y, z) = (x^2 + y^2 + z^2)^{-1/2}$

80. $f(x, y, z) = e^{3x+4y} \cos 5z$

The Wave Equation If we stand on an ocean shore and take a snapshot of the waves, the picture shows a regular pattern of peaks and valleys in an instant of time. We see periodic vertical motion in space, with respect to distance. If we stand in the water, we can feel the rise and fall of the water as the waves go by. We see periodic vertical motion in time. In physics, this beautiful symmetry is expressed by the **one-dimensional wave equation**

$$\frac{\partial^2 w}{\partial t^2} = c^2 \frac{\partial^2 w}{\partial x^2},$$

where w is the wave height, x is the distance variable, t is the time variable, and c is the velocity with which the waves are propagated.

In our example, x is the distance across the ocean's surface, but in other applications, x might be the distance along a vibrating string, distance through air (sound waves), or distance through space (light waves). The number c varies with the medium and type of wave.

Show that the functions in Exercises 81–87 are all solutions of the wave equation.

81. $w = \sin(x + ct)$

82. $w = \cos(2x + 2ct)$

83. $w = \sin(x + ct) + \cos(2x + 2ct)$

84. $w = \ln(2x + 2ct)$ **85.** $w = \tan(2x - 2ct)$

86. $w = 5\cos(3x + 3ct) + e^{x+ct}$

87. $w = f(u)$, where f is a differentiable function of u, and $u = a(x + ct)$, where a is a constant

88. Does a function $f(x, y)$ with continuous first partial derivatives throughout an open region R have to be continuous on R? Give reasons for your answer.

89. If a function $f(x, y)$ has continuous second partial derivatives throughout an open region R, must the first-order partial derivatives of f be continuous on R? Give reasons for your answer.

90. The heat equation An important partial differential equation that describes the distribution of heat in a region at time t can be represented by the *one-dimensional heat equation*

$$\frac{\partial f}{\partial t} = \frac{\partial^2 f}{\partial x^2}.$$

Show that $u(x, t) = \sin(\alpha x) \cdot e^{-\beta t}$ satisfies the heat equation for constants α and β. What is the relationship between α and β for this function to be a solution?

91. Let $f(x, y) = \begin{cases} \dfrac{xy^2}{x^2 + y^4}, & (x, y) \neq (0, 0) \\ 0, & (x, y) = (0, 0). \end{cases}$

Show that $f_x(0, 0)$ and $f_y(0, 0)$ exist, but f is not differentiable at $(0, 0)$. (*Hint:* Use Theorem 4 and show that f is not continuous at $(0, 0)$.)

92. Let $f(x, y) = \begin{cases} 0, & x^2 < y < 2x^2 \\ 1, & \text{otherwise.} \end{cases}$

Show that $f_x(0, 0)$ and $f_y(0, 0)$ exist, but f is not differentiable at $(0, 0)$.

14.4 The Chain Rule

The Chain Rule for functions of a single variable studied in Section 3.6 says that when $w = f(x)$ is a differentiable function of x and $x = g(t)$ is a differentiable function of t, w is a differentiable function of t and dw/dt can be calculated by the formula

$$\frac{dw}{dt} = \frac{dw}{dx}\frac{dx}{dt}.$$

For functions of two or more variables the Chain Rule has several forms. The form depends on how many variables are involved, but once this is taken into account, it works like the Chain Rule in Section 3.6.

Functions of Two Variables

The Chain Rule formula for a differentiable function $w = f(x, y)$ when $x = x(t)$ and $y = y(t)$ are both differentiable functions of t is given in the following theorem.

Each of $\dfrac{\partial f}{\partial x}, \dfrac{\partial w}{\partial x}, f_x$ indicates the partial derivative of f with respect to x.

> **THEOREM 5—Chain Rule for Functions of One Independent Variable and Two Intermediate Variables** If $w = f(x, y)$ is differentiable and if $x = x(t)$, $y = y(t)$ are differentiable functions of t, then the composite $w = f(x(t), y(t))$ is a differentiable function of t and
>
> $$\frac{dw}{dt} = f_x(x(t), y(t)) \cdot x'(t) + f_y(x(t), y(t)) \cdot y'(t),$$
>
> or
>
> $$\frac{dw}{dt} = \frac{\partial f}{\partial x}\frac{dx}{dt} + \frac{\partial f}{\partial y}\frac{dy}{dt}.$$

Proof The proof consists of showing that if x and y are differentiable at $t = t_0$, then w is differentiable at t_0 and

$$\left(\frac{dw}{dt}\right)_{t_0} = \left(\frac{\partial w}{\partial x}\right)_{P_0}\left(\frac{dx}{dt}\right)_{t_0} + \left(\frac{\partial w}{\partial y}\right)_{P_0}\left(\frac{dy}{dt}\right)_{t_0},$$

where $P_0 = (x(t_0), y(t_0))$. The subscripts indicate where each of the derivatives is to be evaluated.

Let Δx, Δy, and Δw be the increments that result from changing t from t_0 to $t_0 + \Delta t$. Since f is differentiable (see the definition in Section 14.3),

$$\Delta w = \left(\frac{\partial w}{\partial x}\right)_{P_0} \Delta x + \left(\frac{\partial w}{\partial y}\right)_{P_0} \Delta y + \epsilon_1 \Delta x + \epsilon_2 \Delta y,$$

where ϵ_1, $\epsilon_2 \to 0$ as Δx, $\Delta y \to 0$. To find dw/dt, we divide this equation through by Δt and let Δt approach zero. The division gives

$$\frac{\Delta w}{\Delta t} = \left(\frac{\partial w}{\partial x}\right)_{P_0} \frac{\Delta x}{\Delta t} + \left(\frac{\partial w}{\partial y}\right)_{P_0} \frac{\Delta y}{\Delta t} + \epsilon_1 \frac{\Delta x}{\Delta t} + \epsilon_2 \frac{\Delta y}{\Delta t}.$$

Letting Δt approach zero gives

$$\left(\frac{dw}{dt}\right)_{t_0} = \lim_{\Delta t \to 0} \frac{\Delta w}{\Delta t}$$

$$= \left(\frac{\partial w}{\partial x}\right)_{P_0} \left(\frac{dx}{dt}\right)_{t_0} + \left(\frac{\partial w}{\partial y}\right)_{P_0} \left(\frac{dy}{dt}\right)_{t_0} + 0 \cdot \left(\frac{dx}{dt}\right)_{t_0} + 0 \cdot \left(\frac{dy}{dt}\right)_{t_0}. \quad \blacksquare$$

Often we write $\partial w/\partial x$ for the partial derivative $\partial f/\partial x$, so we can rewrite the Chain Rule in Theorem 5 in the form

$$\frac{dw}{dt} = \frac{\partial w}{\partial x}\frac{dx}{dt} + \frac{\partial w}{\partial y}\frac{dy}{dt}.$$

To remember the Chain Rule picture the diagram below. To find dw/dt, start at w and read down each route to t, multiplying derivatives along the way. Then add the products.

Chain Rule

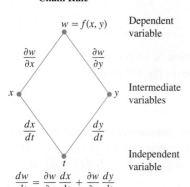

$w = f(x, y)$ — Dependent variable

$\frac{\partial w}{\partial x}$ $\frac{\partial w}{\partial y}$

x y — Intermediate variables

$\frac{dx}{dt}$ $\frac{dy}{dt}$

t — Independent variable

$\frac{dw}{dt} = \frac{\partial w}{\partial x}\frac{dx}{dt} + \frac{\partial w}{\partial y}\frac{dy}{dt}$

However, the meaning of the dependent variable w is different on each side of the preceding equation. On the left-hand side, it refers to the composite function $w = f(x(t), y(t))$ as a function of the single variable t. On the right-hand side, it refers to the function $w = f(x, y)$ as a function of the two variables x and y. Moreover, the single derivatives dw/dt, dx/dt, and dy/dt are being evaluated at a point t_0, whereas the partial derivatives $\partial w/\partial x$ and $\partial w/\partial y$ are being evaluated at the point (x_0, y_0), with $x_0 = x(t_0)$ and $y_0 = y(t_0)$. With that understanding, we will use both of these forms interchangeably throughout the text whenever no confusion will arise.

The **branch diagram** in the margin provides a convenient way to remember the Chain Rule. The "true" independent variable in the composite function is t, whereas x and y are *intermediate variables* (controlled by t) and w is the dependent variable.

A more precise notation for the Chain Rule shows where the various derivatives in Theorem 5 are evaluated:

$$\frac{dw}{dt}(t_0) = \frac{\partial f}{\partial x}(x_0, y_0) \cdot \frac{dx}{dt}(t_0) + \frac{\partial f}{\partial y}(x_0, y_0) \cdot \frac{dy}{dt}(t_0).$$

EXAMPLE 1   Use the Chain Rule to find the derivative of

$$w = xy$$

with respect to t along the path $x = \cos t$, $y = \sin t$. What is the derivative's value at $t = \pi/2$?

Solution   We apply the Chain Rule to find dw/dt as follows:

$$\frac{dw}{dt} = \frac{\partial w}{\partial x}\frac{dx}{dt} + \frac{\partial w}{\partial y}\frac{dy}{dt}$$

$$= \frac{\partial(xy)}{\partial x} \cdot \frac{d}{dt}(\cos t) + \frac{\partial(xy)}{\partial y} \cdot \frac{d}{dt}(\sin t)$$

$$= (y)(-\sin t) + (x)(\cos t)$$

$$= (\sin t)(-\sin t) + (\cos t)(\cos t)$$

$$= -\sin^2 t + \cos^2 t$$

$$= \cos 2t.$$

In this example, we can check the result with a more direct calculation. As a function of t,

$$w = xy = \cos t \sin t = \frac{1}{2} \sin 2t,$$

so

$$\frac{dw}{dt} = \frac{d}{dt}\left(\frac{1}{2} \sin 2t\right) = \frac{1}{2} \cdot 2 \cos 2t = \cos 2t.$$

In either case, at the given value of t,

$$\left(\frac{dw}{dt}\right)_{t=\pi/2} = \cos\left(2 \cdot \frac{\pi}{2}\right) = \cos \pi = -1.$$

Functions of Three Variables

You can probably predict the Chain Rule for functions of three intermediate variables, as it only involves adding the expected third term to the two-variable formula.

> **THEOREM 6—Chain Rule for Functions of One Independent Variable and Three Intermediate Variables** If $w = f(x, y, z)$ is differentiable and x, y, and z are differentiable functions of t, then w is a differentiable function of t and
>
> $$\frac{dw}{dt} = \frac{\partial w}{\partial x}\frac{dx}{dt} + \frac{\partial w}{\partial y}\frac{dy}{dt} + \frac{\partial w}{\partial z}\frac{dz}{dt}.$$

The proof is identical with the proof of Theorem 5 except that there are now three intermediate variables instead of two. The branch diagram we use for remembering the new equation is similar as well, with three routes from w to t.

Here we have three routes from w to t instead of two, but finding dw/dt is still the same. Read down each route, multiplying derivatives along the way; then add.

Chain Rule

$$\frac{dw}{dt} = \frac{\partial w}{\partial x}\frac{dx}{dt} + \frac{\partial w}{\partial y}\frac{dy}{dt} + \frac{\partial w}{\partial z}\frac{dz}{dt}$$

EXAMPLE 2 Find dw/dt if

$$w = xy + z, \qquad x = \cos t, \qquad y = \sin t, \qquad z = t.$$

In this example the values of $w(t)$ are changing along the path of a helix (Section 13.1) as t changes. What is the derivative's value at $t = 0$?

Solution Using the Chain Rule for three intermediate variables, we have

$$\begin{aligned}
\frac{dw}{dt} &= \frac{\partial w}{\partial x}\frac{dx}{dt} + \frac{\partial w}{\partial y}\frac{dy}{dt} + \frac{\partial w}{\partial z}\frac{dz}{dt} \\
&= (y)(-\sin t) + (x)(\cos t) + (1)(1) &&\text{Substitute for the} \\
&= (\sin t)(-\sin t) + (\cos t)(\cos t) + 1 &&\text{intermediate} \\
&= -\sin^2 t + \cos^2 t + 1 = 1 + \cos 2t, &&\text{variables.}
\end{aligned}$$

so

$$\left(\frac{dw}{dt}\right)_{t=0} = 1 + \cos(0) = 2.$$

For a physical interpretation of change along a curve think of an object whose position is changing with time t. If $w = T(x, y, z)$ is the temperature at each point (x, y, z) along a curve C with parametric equations $x = x(t), y = y(t)$, and $z = z(t)$, then the composite function $w = T(x(t), y(t), z(t))$ represents the temperature relative to t along the curve. The derivative dw/dt is then the instantaneous rate of change of temperature due to the motion along the curve, as calculated in Theorem 6.

Functions Defined on Surfaces

If we are interested in the temperature $w = f(x, y, z)$ at points (x, y, z) on the earth's surface, we might prefer to think of x, y, and z as functions of the variables r and s that give the points' longitudes and latitudes. If $x = g(r, s)$, $y = h(r, s)$, and $z = k(r, s)$, we could then express the temperature as a function of r and s with the composite function

$$w = f(g(r, s), h(r, s), k(r, s)).$$

Under the conditions stated below, w has partial derivatives with respect to both r and s that can be calculated in the following way.

THEOREM 7—Chain Rule for Two Independent Variables and Three Intermediate Variables   Suppose that $w = f(x, y, z)$, $x = g(r, s)$, $y = h(r, s)$, and $z = k(r, s)$. If all four functions are differentiable, then w has partial derivatives with respect to r and s, given by the formulas

$$\frac{\partial w}{\partial r} = \frac{\partial w}{\partial x}\frac{\partial x}{\partial r} + \frac{\partial w}{\partial y}\frac{\partial y}{\partial r} + \frac{\partial w}{\partial z}\frac{\partial z}{\partial r}$$

$$\frac{\partial w}{\partial s} = \frac{\partial w}{\partial x}\frac{\partial x}{\partial s} + \frac{\partial w}{\partial y}\frac{\partial y}{\partial s} + \frac{\partial w}{\partial z}\frac{\partial z}{\partial s}.$$

The first of these equations can be derived from the Chain Rule in Theorem 6 by holding s fixed and treating r as t. The second can be derived in the same way, holding r fixed and treating s as t. The branch diagrams for both equations are shown in Figure 14.21.

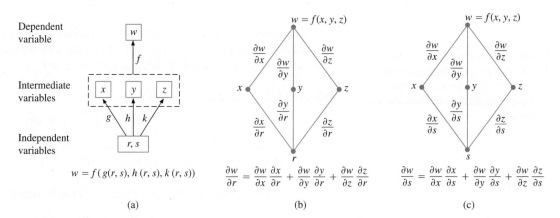

FIGURE 14.21   Composite function and branch diagrams for Theorem 7.

EXAMPLE 3   Express $\partial w/\partial r$ and $\partial w/\partial s$ in terms of r and s if

$$w = x + 2y + z^2, \qquad x = \frac{r}{s}, \qquad y = r^2 + \ln s, \qquad z = 2r.$$

Solution   Using the formulas in Theorem 7, we find

$$\frac{\partial w}{\partial r} = \frac{\partial w}{\partial x}\frac{\partial x}{\partial r} + \frac{\partial w}{\partial y}\frac{\partial y}{\partial r} + \frac{\partial w}{\partial z}\frac{\partial z}{\partial r}$$

$$= (1)\left(\frac{1}{s}\right) + (2)(2r) + (2z)(2)$$

$$= \frac{1}{s} + 4r + (4r)(2) = \frac{1}{s} + 12r \qquad \text{Substitute for intermediate variable } z.$$

$$\frac{\partial w}{\partial s} = \frac{\partial w}{\partial x}\frac{\partial x}{\partial s} + \frac{\partial w}{\partial y}\frac{\partial y}{\partial s} + \frac{\partial w}{\partial z}\frac{\partial z}{\partial s}$$

$$= (1)\left(-\frac{r}{s^2}\right) + (2)\left(\frac{1}{s}\right) + (2z)(0) = \frac{2}{s} - \frac{r}{s^2} \qquad \blacksquare$$

Chain Rule

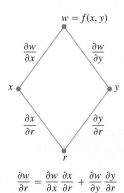

FIGURE 14.22 Branch diagram for the equation

$$\frac{\partial w}{\partial r} = \frac{\partial w}{\partial x}\frac{\partial x}{\partial r} + \frac{\partial w}{\partial y}\frac{\partial y}{\partial r}.$$

If f is a function of two intermediate variables instead of three, each equation in Theorem 7 becomes correspondingly one term shorter.

If $w = f(x, y)$, $x = g(r, s)$, and $y = h(r, s)$, then

$$\frac{\partial w}{\partial r} = \frac{\partial w}{\partial x}\frac{\partial x}{\partial r} + \frac{\partial w}{\partial y}\frac{\partial y}{\partial r} \quad \text{and} \quad \frac{\partial w}{\partial s} = \frac{\partial w}{\partial x}\frac{\partial x}{\partial s} + \frac{\partial w}{\partial y}\frac{\partial y}{\partial s}.$$

Figure 14.22 shows the branch diagram for the first of these equations. The diagram for the second equation is similar; just replace r with s.

EXAMPLE 4　Express $\partial w/\partial r$ and $\partial w/\partial s$ in terms of r and s if

$$w = x^2 + y^2, \quad x = r - s, \quad y = r + s.$$

Solution　The preceding discussion gives the following.

$$\frac{\partial w}{\partial r} = \frac{\partial w}{\partial x}\frac{\partial x}{\partial r} + \frac{\partial w}{\partial y}\frac{\partial y}{\partial r} \qquad\qquad \frac{\partial w}{\partial s} = \frac{\partial w}{\partial x}\frac{\partial x}{\partial s} + \frac{\partial w}{\partial y}\frac{\partial y}{\partial s}$$

$$= (2x)(1) + (2y)(1) \qquad\qquad = (2x)(-1) + (2y)(1) \qquad \text{Substitute for the}$$

$$= 2(r - s) + 2(r + s) \qquad\qquad = -2(r - s) + 2(r + s) \qquad \text{intermediate variables.}$$

$$= 4r \qquad\qquad\qquad\qquad\qquad = 4s$$

If f is a function of a single intermediate variable x, our equations are even simpler.

If $w = f(x)$ and $x = g(r, s)$, then

$$\frac{\partial w}{\partial r} = \frac{dw}{dx}\frac{\partial x}{\partial r} \quad \text{and} \quad \frac{\partial w}{\partial s} = \frac{dw}{dx}\frac{\partial x}{\partial s}.$$

Chain Rule

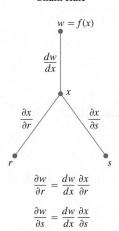

FIGURE 14.23 Branch diagram for differentiating f as a composite function of r and s with one intermediate variable.

In this case, we use the ordinary (single-variable) derivative, dw/dx. The branch diagram is shown in Figure 14.23.

Implicit Differentiation Revisited

The two-variable Chain Rule in Theorem 5 leads to a formula that takes some of the algebra out of implicit differentiation. Suppose that

1. The function $F(x, y)$ is differentiable and
2. The equation $F(x, y) = 0$ defines y implicitly as a differentiable function of x, say $y = h(x)$.

Since $w = F(x, y) = 0$, the derivative dw/dx must be zero. Computing the derivative from the Chain Rule (branch diagram in Figure 14.24), we find

$$0 = \frac{dw}{dx} = F_x\frac{dx}{dx} + F_y\frac{dy}{dx} \qquad \text{Theorem 5 with } t = x \text{ and } f = F$$

$$= F_x \cdot 1 + F_y \cdot \frac{dy}{dx}.$$

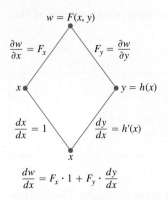

$w = F(x, y)$

$\dfrac{\partial w}{\partial x} = F_x$ $F_y = \dfrac{\partial w}{\partial y}$

x $y = h(x)$

$\dfrac{dx}{dx} = 1$ $\dfrac{dy}{dx} = h'(x)$

x

$\dfrac{dw}{dx} = F_x \cdot 1 + F_y \cdot \dfrac{dy}{dx}$

FIGURE 14.24 Branch diagram for differentiating $w = F(x, y)$ with respect to x. Setting $dw/dx = 0$ leads to a simple computational formula for implicit differentiation (Theorem 8).

If $F_y = \partial w / \partial y \neq 0$, we can solve this equation for dy/dx to get

$$\frac{dy}{dx} = -\frac{F_x}{F_y}.$$

We state this result formally.

THEOREM 8—A Formula for Implicit Differentiation   Suppose that $F(x, y)$ is differentiable and that the equation $F(x, y) = 0$ defines y as a differentiable function of x. Then at any point where $F_y \neq 0$,

$$\frac{dy}{dx} = -\frac{F_x}{F_y}. \tag{1}$$

EXAMPLE 5   Use Theorem 8 to find dy/dx if $y^2 - x^2 - \sin xy = 0$.

Solution   Take $F(x, y) = y^2 - x^2 - \sin xy$. Then

$$\frac{dy}{dx} = -\frac{F_x}{F_y} = -\frac{-2x - y \cos xy}{2y - x \cos xy}$$

$$= \frac{2x + y \cos xy}{2y - x \cos xy}.$$

This calculation is significantly shorter than a single-variable calculation using implicit differentiation.   ∎

The result in Theorem 8 is easily extended to three variables. Suppose that the equation $F(x, y, z) = 0$ defines the variable z implicitly as a function $z = f(x, y)$. Then for all (x, y) in the domain of f, we have $F(x, y, f(x, y)) = 0$. Assuming that F and f are differentiable functions, we can use the Chain Rule to differentiate the equation $F(x, y, z) = 0$ with respect to the independent variable x:

$$0 = \frac{\partial F}{\partial x}\frac{\partial x}{\partial x} + \frac{\partial F}{\partial y}\frac{\partial y}{\partial x} + \frac{\partial F}{\partial z}\frac{\partial z}{\partial x}$$

$$= F_x \cdot 1 + F_y \cdot 0 + F_z \cdot \frac{\partial z}{\partial x}, \qquad \text{\small y is constant when differentiating with respect to x.}$$

so

$$F_x + F_z \frac{\partial z}{\partial x} = 0.$$

A similar calculation for differentiating with respect to the independent variable y gives

$$F_y + F_z \frac{\partial z}{\partial y} = 0.$$

Whenever $F_z \neq 0$, we can solve these last two equations for the partial derivatives of $z = f(x, y)$ to obtain

---

$$\frac{\partial z}{\partial x} = -\frac{F_x}{F_z} \quad \text{and} \quad \frac{\partial z}{\partial y} = -\frac{F_y}{F_z}. \tag{2}$$

An important result from advanced calculus, called the **Implicit Function Theorem**, states the conditions for which our results in Equations (2) are valid. If the partial derivatives F_x, F_y, and F_z are continuous throughout an open region R in space containing the point (x_0, y_0, z_0), and if for some constant c, $F(x_0, y_0, z_0) = c$ and $F_z(x_0, y_0, z_0) \neq 0$, then the equation $F(x, y, z) = c$ defines z implicitly as a differentiable function of x and y near (x_0, y_0, z_0), and the partial derivatives of z are given by Equations (2).

EXAMPLE 6   Find $\dfrac{\partial z}{\partial x}$ and $\dfrac{\partial z}{\partial y}$ at $(0, 0, 0)$ if $x^3 + z^2 + ye^{xz} + z \cos y = 0$.

Solution   Let $F(x, y, z) = x^3 + z^2 + ye^{xz} + z \cos y$. Then

$$F_x = 3x^2 + zye^{xz}, \qquad F_y = e^{xz} - z \sin y, \qquad \text{and} \qquad F_z = 2z + xye^{xz} + \cos y.$$

Since $F(0, 0, 0) = 0$, $F_z(0, 0, 0) = 1 \neq 0$, and all first partial derivatives are continuous, the Implicit Function Theorem says that $F(x, y, z) = 0$ defines z as a differentiable function of x and y near the point $(0, 0, 0)$. From Equations (2),

$$\frac{\partial z}{\partial x} = -\frac{F_x}{F_z} = -\frac{3x^2 + zye^{xz}}{2z + xye^{xz} + \cos y} \qquad \text{and} \qquad \frac{\partial z}{\partial y} = -\frac{F_y}{F_z} = -\frac{e^{xz} - z \sin y}{2z + xye^{xz} + \cos y}.$$

At $(0, 0, 0)$ we find

$$\frac{\partial z}{\partial x} = -\frac{0}{1} = 0 \qquad \text{and} \qquad \frac{\partial z}{\partial y} = -\frac{1}{1} = -1. \qquad \blacksquare$$

Functions of Many Variables

We have seen several different forms of the Chain Rule in this section, but each one is just a special case of one general formula. When solving particular problems, it may help to draw the appropriate branch diagram by placing the dependent variable on top, the intermediate variables in the middle, and the selected independent variable at the bottom. To find the derivative of the dependent variable with respect to the selected independent variable, start at the dependent variable and read down each route of the branch diagram to the independent variable, calculating and multiplying the derivatives along each route. Then add the products found for the different routes.

In general, suppose that $w = f(x, y, \ldots, v)$ is a differentiable function of the variables $x, y, \ldots, v$ (a finite set) and the $x, y, \ldots, v$ are differentiable functions of $p, q, \ldots, t$ (another finite set). Then w is a differentiable function of the variables p through t, and the partial derivatives of w with respect to these variables are given by equations of the form

$$\frac{\partial w}{\partial p} = \frac{\partial w}{\partial x}\frac{\partial x}{\partial p} + \frac{\partial w}{\partial y}\frac{\partial y}{\partial p} + \cdots + \frac{\partial w}{\partial v}\frac{\partial v}{\partial p}.$$

The other equations are obtained by replacing p by $q, \ldots, t$, one at a time.

One way to remember this equation is to think of the right-hand side as the dot product of two vectors with components

$$\underbrace{\left(\frac{\partial w}{\partial x}, \frac{\partial w}{\partial y}, \ldots, \frac{\partial w}{\partial v} \right)}_{\substack{\text{Derivatives of } w \text{ with} \\ \text{respect to the} \\ \text{intermediate variables}}} \qquad \text{and} \qquad \underbrace{\left(\frac{\partial x}{\partial p}, \frac{\partial y}{\partial p}, \ldots, \frac{\partial v}{\partial p} \right)}_{\substack{\text{Derivatives of the intermediate} \\ \text{variables with respect to the} \\ \text{selected independent variable}}}.$$

Exercises 14.4

Chain Rule: One Independent Variable

In Exercises 1–6, (a) express dw/dt as a function of t, both by using the Chain Rule and by expressing w in terms of t and differentiating directly with respect to t. Then (b) evaluate dw/dt at the given value of t.

1. $w = x^2 + y^2$, $x = \cos t$, $y = \sin t$; $t = \pi$

2. $w = x^2 + y^2$, $x = \cos t + \sin t$, $y = \cos t - \sin t$; $t = 0$

3. $w = \dfrac{x}{z} + \dfrac{y}{z}$, $x = \cos^2 t$, $y = \sin^2 t$, $z = 1/t$; $t = 3$

4. $w = \ln(x^2 + y^2 + z^2)$, $x = \cos t$, $y = \sin t$, $z = 4\sqrt{t}$; $t = 3$

5. $w = 2ye^x - \ln z$, $x = \ln(t^2 + 1)$, $y = \tan^{-1} t$, $z = e^t$; $t = 1$

6. $w = z - \sin xy$, $x = t$, $y = \ln t$, $z = e^{t-1}$; $t = 1$

Chain Rule: Two and Three Independent Variables

In Exercises 7 and 8, (a) express $\partial z/\partial u$ and $\partial z/\partial v$ as functions of u and v both by using the Chain Rule and by expressing z directly in terms of u and v before differentiating. Then (b) evaluate $\partial z/\partial u$ and $\partial z/\partial v$ at the given point (u, v).

7. $z = 4e^x \ln y$, $x = \ln(u \cos v)$, $y = u \sin v$;
   $(u, v) = (2, \pi/4)$

8. $z = \tan^{-1}(x/y)$, $x = u \cos v$, $y = u \sin v$;
   $(u, v) = (1.3, \pi/6)$

In Exercises 9 and 10, (a) express $\partial w/\partial u$ and $\partial w/\partial v$ as functions of u and v both by using the Chain Rule and by expressing w directly in terms of u and v before differentiating. Then (b) evaluate $\partial w/\partial u$ and $\partial w/\partial v$ at the given point (u, v).

9. $w = xy + yz + xz$, $x = u + v$, $y = u - v$, $z = uv$;
   $(u, v) = (1/2, 1)$

10. $w = \ln(x^2 + y^2 + z^2)$, $x = ue^v \sin u$, $y = ue^v \cos u$,
    $z = ue^v$; $(u, v) = (-2, 0)$

In Exercises 11 and 12, (a) express $\partial u/\partial x$, $\partial u/\partial y$, and $\partial u/\partial z$ as functions of x, y, and z both by using the Chain Rule and by expressing u directly in terms of x, y, and z before differentiating. Then (b) evaluate $\partial u/\partial x$, $\partial u/\partial y$, and $\partial u/\partial z$ at the given point (x, y, z).

11. $u = \dfrac{p - q}{q - r}$, $p = x + y + z$, $q = x - y + z$,
    $r = x + y - z$; $(x, y, z) = \left(\sqrt{3}, 2, 1\right)$

12. $u = e^{qr} \sin^{-1} p$, $p = \sin x$, $q = z^2 \ln y$, $r = 1/z$;
    $(x, y, z) = (\pi/4, 1/2, -1/2)$

Using a Branch Diagram

In Exercises 13–24, draw a branch diagram and write a Chain Rule formula for each derivative.

13. $\dfrac{dz}{dt}$ for $z = f(x, y)$, $x = g(t)$, $y = h(t)$

14. $\dfrac{dz}{dt}$ for $z = f(u, v, w)$, $u = g(t)$, $v = h(t)$, $w = k(t)$

15. $\dfrac{\partial w}{\partial u}$ and $\dfrac{\partial w}{\partial v}$ for $w = h(x, y, z)$, $x = f(u, v)$, $y = g(u, v)$,
    $z = k(u, v)$

16. $\dfrac{\partial w}{\partial x}$ and $\dfrac{\partial w}{\partial y}$ for $w = f(r, s, t)$, $r = g(x, y)$, $s = h(x, y)$,
    $t = k(x, y)$

17. $\dfrac{\partial w}{\partial u}$ and $\dfrac{\partial w}{\partial v}$ for $w = g(x, y)$, $x = h(u, v)$, $y = k(u, v)$

18. $\dfrac{\partial w}{\partial x}$ and $\dfrac{\partial w}{\partial y}$ for $w = g(u, v)$, $u = h(x, y)$, $v = k(x, y)$

19. $\dfrac{\partial z}{\partial t}$ and $\dfrac{\partial z}{\partial s}$ for $z = f(x, y)$, $x = g(t, s)$, $y = h(t, s)$

20. $\dfrac{\partial y}{\partial r}$ for $y = f(u)$, $u = g(r, s)$

21. $\dfrac{\partial w}{\partial s}$ and $\dfrac{\partial w}{\partial t}$ for $w = g(u)$, $u = h(s, t)$

22. $\dfrac{\partial w}{\partial p}$ for $w = f(x, y, z, v)$, $x = g(p, q)$, $y = h(p, q)$,
    $z = j(p, q)$, $v = k(p, q)$

23. $\dfrac{\partial w}{\partial r}$ and $\dfrac{\partial w}{\partial s}$ for $w = f(x, y)$, $x = g(r)$, $y = h(s)$

24. $\dfrac{\partial w}{\partial s}$ for $w = g(x, y)$, $x = h(r, s, t)$, $y = k(r, s, t)$

Implicit Differentiation

Assuming that the equations in Exercises 25–28 define y as a differentiable function of x, use Theorem 8 to find the value of dy/dx at the given point.

25. $x^3 - 2y^2 + xy = 0$, $(1, 1)$

26. $xy + y^2 - 3x - 3 = 0$, $(-1, 1)$

27. $x^2 + xy + y^2 - 7 = 0$, $(1, 2)$

28. $xe^y + \sin xy + y - \ln 2 = 0$, $(0, \ln 2)$

Find the values of $\partial z/\partial x$ and $\partial z/\partial y$ at the points in Exercises 29–32.

29. $z^3 - xy + yz + y^3 - 2 = 0$, $(1, 1, 1)$

30. $\dfrac{1}{x} + \dfrac{1}{y} + \dfrac{1}{z} - 1 = 0$, $(2, 3, 6)$

31. $\sin(x + y) + \sin(y + z) + \sin(x + z) = 0$, (π, π, π)

32. $xe^y + ye^z + 2\ln x - 2 - 3\ln 2 = 0$, $(1, \ln 2, \ln 3)$

Finding Partial Derivatives at Specified Points

33. Find $\partial w/\partial r$ when $r = 1, s = -1$ if $w = (x + y + z)^2$, $x = r - s$, $y = \cos(r + s)$, $z = \sin(r + s)$.

34. Find $\partial w/\partial v$ when $u = -1, v = 2$ if $w = xy + \ln z$, $x = v^2/u$, $y = u + v$, $z = \cos u$.

35. Find $\partial w/\partial v$ when $u = 0, v = 0$ if $w = x^2 + (y/x)$, $x = u - 2v + 1$, $y = 2u + v - 2$.

36. Find $\partial z/\partial u$ when $u = 0, v = 1$ if $z = \sin xy + x \sin y$, $x = u^2 + v^2$, $y = uv$.

37. Find $\partial z/\partial u$ and $\partial z/\partial v$ when $u = \ln 2, v = 1$ if $z = 5 \tan^{-1} x$ and $x = e^u + \ln v$.

38. Find $\partial z/\partial u$ and $\partial z/\partial v$ when $u = 1, v = -2$ if $z = \ln q$ and $q = \sqrt{v} + 3 \tan^{-1} u$.

Theory and Examples

39. Assume that $w = f(s^3 + t^2)$ and $f'(x) = e^x$. Find $\dfrac{\partial w}{\partial t}$ and $\dfrac{\partial w}{\partial s}$.

40. Assume that $w = f\left(ts^2, \dfrac{s}{t}\right)$, $\dfrac{\partial f}{\partial x}(x, y) = xy$, and $\dfrac{\partial f}{\partial y}(x, y) = \dfrac{x^2}{2}$. Find $\dfrac{\partial w}{\partial t}$ and $\dfrac{\partial w}{\partial s}$.

41. **Changing voltage in a circuit** The voltage V in a circuit that satisfies the law $V = IR$ is slowly dropping as the battery wears out. At the same time, the resistance R is increasing as the resistor heats up. Use the equation

$$\frac{dV}{dt} = \frac{\partial V}{\partial I}\frac{dI}{dt} + \frac{\partial V}{\partial R}\frac{dR}{dt}$$

to find how the current is changing at the instant when $R = 600$ ohms, $I = 0.04$ amp, $dR/dt = 0.5$ ohm/sec, and $dV/dt = -0.01$ volt/sec.

42. **Changing dimensions in a box** The lengths a, b, and c of the edges of a rectangular box are changing with time. At the instant in question, $a = 1$ m, $b = 2$ m, $c = 3$ m, $da/dt = db/dt = 1$ m/sec, and $dc/dt = -3$ m/sec. At what rates are the box's volume V and surface area S changing at that instant? Are the box's interior diagonals increasing in length or decreasing?

43. If $f(u, v, w)$ is differentiable and $u = x - y$, $v = y - z$, and $w = z - x$, show that

$$\frac{\partial f}{\partial x} + \frac{\partial f}{\partial y} + \frac{\partial f}{\partial z} = 0.$$

44. **Polar coordinates** Suppose that we substitute polar coordinates $x = r\cos\theta$ and $y = r\sin\theta$ in a differentiable function $w = f(x, y)$.

   a. Show that

   $$\frac{\partial w}{\partial r} = f_x\cos\theta + f_y\sin\theta$$

   and

   $$\frac{1}{r}\frac{\partial w}{\partial \theta} = -f_x\sin\theta + f_y\cos\theta.$$

   b. Solve the equations in part (a) to express f_x and f_y in terms of $\partial w/\partial r$ and $\partial w/\partial\theta$.

   c. Show that

   $$(f_x)^2 + (f_y)^2 = \left(\frac{\partial w}{\partial r}\right)^2 + \frac{1}{r^2}\left(\frac{\partial w}{\partial\theta}\right)^2.$$

45. **Laplace equations** Show that if $w = f(u, v)$ satisfies the Laplace equation $f_{uu} + f_{vv} = 0$ and if $u = (x^2 - y^2)/2$ and $v = xy$, then w satisfies the Laplace equation $w_{xx} + w_{yy} = 0$.

46. **Laplace equations** Let $w = f(u) + g(v)$, where $u = x + iy$, $v = x - iy$, and $i = \sqrt{-1}$. Show that w satisfies the Laplace equation $w_{xx} + w_{yy} = 0$ if all the necessary functions are differentiable.

47. **Extreme values on a helix** Suppose that the partial derivatives of a function $f(x, y, z)$ at points on the helix $x = \cos t$, $y = \sin t$, $z = t$ are

$$f_x = \cos t, \qquad f_y = \sin t, \qquad f_z = t^2 + t - 2.$$

At what points on the curve, if any, can f take on extreme values?

48. **A space curve** Let $w = x^2 e^{2y} \cos 3z$. Find the value of dw/dt at the point $(1, \ln 2, 0)$ on the curve $x = \cos t$, $y = \ln(t + 2)$, $z = t$.

49. **Temperature on a circle** Let $T = f(x, y)$ be the temperature at the point (x, y) on the circle $x = \cos t$, $y = \sin t$, $0 \le t \le 2\pi$ and suppose that

$$\frac{\partial T}{\partial x} = 8x - 4y, \qquad \frac{\partial T}{\partial y} = 8y - 4x.$$

   a. Find where the maximum and minimum temperatures on the circle occur by examining the derivatives dT/dt and d^2T/dt^2.

   b. Suppose that $T = 4x^2 - 4xy + 4y^2$. Find the maximum and minimum values of T on the circle.

50. **Temperature on an ellipse** Let $T = g(x, y)$ be the temperature at the point (x, y) on the ellipse

$$x = 2\sqrt{2}\cos t, \qquad y = \sqrt{2}\sin t, \qquad 0 \le t \le 2\pi,$$

and suppose that

$$\frac{\partial T}{\partial x} = y, \qquad \frac{\partial T}{\partial y} = x.$$

   a. Locate the maximum and minimum temperatures on the ellipse by examining dT/dt and d^2T/dt^2.

   b. Suppose that $T = xy - 2$. Find the maximum and minimum values of T on the ellipse.

Differentiating Integrals Under mild continuity restrictions, it is true that if

$$F(x) = \int_a^b g(t, x)\, dt,$$

then $F'(x) = \displaystyle\int_a^b g_x(t, x)\, dt$. Using this fact and the Chain Rule, we can find the derivative of

$$F(x) = \int_a^{f(x)} g(t, x)\, dt$$

by letting

$$G(u, x) = \int_a^u g(t, x)\, dt,$$

where $u = f(x)$. Find the derivatives of the functions in Exercises 51 and 52.

51. $F(x) = \displaystyle\int_0^{x^2} \sqrt{t^4 + x^3}\, dt$ 52. $F(x) = \displaystyle\int_{x^2}^1 \sqrt{t^3 + x^2}\, dt$

14.5 Directional Derivatives and Gradient Vectors

If you look at the map (Figure 14.25) showing contours within Yosemite National Park in California, you will notice that the streams flow perpendicular to the contours. The streams are following paths of steepest descent so the waters reach lower elevations as quickly as possible. Therefore, the fastest instantaneous rate of change in a stream's elevation above sea level has a particular direction. In this section, you will see why this direction, called the "downhill" direction, is perpendicular to the contours.

FIGURE 14.25 Contours within Yosemite National Park in California show streams, which follow paths of steepest descent, running perpendicular to the contours. (*Source:* http://www.usgs.gov)

Directional Derivatives in the Plane

We know from Section 14.4 that if $f(x, y)$ is differentiable, then the rate at which f changes with respect to t along a differentiable curve $x = g(t), y = h(t)$ is

$$\frac{df}{dt} = \frac{\partial f}{\partial x}\frac{dx}{dt} + \frac{\partial f}{\partial y}\frac{dy}{dt}.$$

At any point $P_0(x_0, y_0) = P_0(g(t_0), h(t_0))$, this equation gives the rate of change of f with respect to increasing t and therefore depends, among other things, on the direction of motion along the curve. If the curve is a straight line and t is the arc length parameter along the line measured from P_0 in the direction of a given unit vector $\mathbf{u}$, then df/dt is the rate of change of f with respect to distance in its domain in the direction of $\mathbf{u}$. By varying $\mathbf{u}$, we find the rates at which f changes with respect to distance as we move through P_0 in different directions. We now define this idea more precisely.

Suppose that the function $f(x, y)$ is defined throughout a region R in the xy-plane, that $P_0(x_0, y_0)$ is a point in R, and that $\mathbf{u} = u_1\mathbf{i} + u_2\mathbf{j}$ is a unit vector. Then the equations

$$x = x_0 + su_1, \qquad y = y_0 + su_2$$

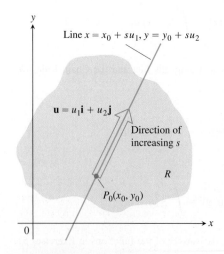

FIGURE 14.26 The rate of change of f in the direction of $\mathbf{u}$ at a point P_0 is the rate at which f changes along this line at P_0.

parametrize the line through P_0 parallel to $\mathbf{u}$. If the parameter s measures arc length from P_0 in the direction of $\mathbf{u}$, we find the rate of change of f at P_0 in the direction of $\mathbf{u}$ by calculating df/ds at P_0 (Figure 14.26).

The tangent is the line

$$(-1)(x + 2) + (2)(y - 1) = 0 \qquad \text{Eq. (6)}$$
$$x - 2y = -4.$$

If we know the gradients of two functions f and g, we automatically know the gradients of their sum, difference, constant multiples, product, and quotient. You are asked to establish the following rules in Exercise 40. Notice that these rules have the same form as the corresponding rules for derivatives of single-variable functions.

Algebra Rules for Gradients

1. *Sum Rule:* $\nabla(f + g) = \nabla f + \nabla g$

2. *Difference Rule:* $\nabla(f - g) = \nabla f - \nabla g$

3. *Constant Multiple Rule:* $\nabla(kf) = k\nabla f \qquad$ (any number k)

4. *Product Rule:* $\nabla(fg) = f\nabla g + g\nabla f$

5. *Quotient Rule:* $\nabla\left(\dfrac{f}{g}\right) = \dfrac{g\nabla f - f\nabla g}{g^2}$

EXAMPLE 5   We illustrate two of the rules with

$$f(x, y) = x - y \qquad g(x, y) = 3y$$
$$\nabla f = \mathbf{i} - \mathbf{j} \qquad \nabla g = 3\mathbf{j}.$$

We have

1. $\nabla(f - g) = \nabla(x - 4y) = \mathbf{i} - 4\mathbf{j} = \nabla f - \nabla g \qquad$ Rule 2

2. $\nabla(fg) = \nabla(3xy - 3y^2) = 3y\mathbf{i} + (3x - 6y)\mathbf{j}$

$\qquad = 3y(\mathbf{i} - \mathbf{j}) + 3y\mathbf{j} + (3x - 6y)\mathbf{j}$

$\qquad = 3y(\mathbf{i} - \mathbf{j}) + (3x - 3y)\mathbf{j}$

$\qquad = 3y(\mathbf{i} - \mathbf{j}) + (x - y)3\mathbf{j} = g\nabla f + f\nabla g \qquad$ Rule 4 ∎

Functions of Three Variables

For a differentiable function $f(x, y, z)$ and a unit vector $\mathbf{u} = u_1\mathbf{i} + u_2\mathbf{j} + u_3\mathbf{k}$ in space, we have

$$\nabla f = \frac{\partial f}{\partial x}\mathbf{i} + \frac{\partial f}{\partial y}\mathbf{j} + \frac{\partial f}{\partial z}\mathbf{k}$$

and

$$D_\mathbf{u}f = \nabla f \cdot \mathbf{u} = \frac{\partial f}{\partial x}u_1 + \frac{\partial f}{\partial y}u_2 + \frac{\partial f}{\partial z}u_3.$$

The directional derivative can once again be written in the form

$$D_\mathbf{u}f = \nabla f \cdot \mathbf{u} = |\nabla f||u|\cos\theta = |\nabla f|\cos\theta,$$

so the properties listed earlier for functions of two variables extend to three variables. At any given point, f increases most rapidly in the direction of ∇f and decreases most rapidly in the direction of $-\nabla f$. In any direction orthogonal to ∇f, the derivative is zero.

EXAMPLE 6

(a) Find the derivative of $f(x, y, z) = x^3 - xy^2 - z$ at $P_0(1, 1, 0)$ in the direction of $\mathbf{v} = 2\mathbf{i} - 3\mathbf{j} + 6\mathbf{k}$.

(b) In what directions does f change most rapidly at P_0, and what are the rates of change in these directions?

Solution

(a) The direction of $\mathbf{v}$ is obtained by dividing $\mathbf{v}$ by its length:

$$|\mathbf{v}| = \sqrt{(2)^2 + (-3)^2 + (6)^2} = \sqrt{49} = 7$$

$$\mathbf{u} = \frac{\mathbf{v}}{|\mathbf{v}|} = \frac{2}{7}\mathbf{i} - \frac{3}{7}\mathbf{j} + \frac{6}{7}\mathbf{k}.$$

The partial derivatives of f at P_0 are

$$f_x = (3x^2 - y^2)_{(1,1,0)} = 2, \qquad f_y = -2xy|_{(1,1,0)} = -2, \qquad f_z = -1|_{(1,1,0)} = -1.$$

The gradient of f at P_0 is

$$\nabla f|_{(1,1,0)} = 2\mathbf{i} - 2\mathbf{j} - \mathbf{k}.$$

The derivative of f at P_0 in the direction of $\mathbf{v}$ is therefore

$$(D_{\mathbf{u}}f)_{(1,1,0)} = \nabla f|_{(1,1,0)} \cdot \mathbf{u} = (2\mathbf{i} - 2\mathbf{j} - \mathbf{k}) \cdot \left(\frac{2}{7}\mathbf{i} - \frac{3}{7}\mathbf{j} + \frac{6}{7}\mathbf{k}\right)$$

$$= \frac{4}{7} + \frac{6}{7} - \frac{6}{7} = \frac{4}{7}.$$

(b) The function increases most rapidly in the direction of $\nabla f = 2\mathbf{i} - 2\mathbf{j} - \mathbf{k}$ and decreases most rapidly in the direction of $-\nabla f$. The rates of change in the directions are, respectively,

$$|\nabla f| = \sqrt{(2)^2 + (-2)^2 + (-1)^2} = \sqrt{9} = 3 \qquad \text{and} \qquad -|\nabla f| = -3. \qquad \blacksquare$$

Exercises 14.5

Calculating Gradients

In Exercises 1–6, find the gradient of the function at the given point. Then sketch the gradient together with the level curve that passes through the point.

1. $f(x, y) = y - x$, $(2, 1)$ 2. $f(x, y) = \ln(x^2 + y^2)$, $(1, 1)$

3. $g(x, y) = xy^2$, $(2, -1)$ 4. $g(x, y) = \frac{x^2}{2} - \frac{y^2}{2}$, $\left(\sqrt{2}, 1\right)$

5. $f(x, y) = \sqrt{2x + 3y}$, $(-1, 2)$

6. $f(x, y) = \tan^{-1}\frac{\sqrt{x}}{y}$, $(4, -2)$

In Exercises 7–10, find ∇f at the given point.

7. $f(x, y, z) = x^2 + y^2 - 2z^2 + z\ln x$, $(1, 1, 1)$

8. $f(x, y, z) = 2z^3 - 3(x^2 + y^2)z + \tan^{-1}xz$, $(1, 1, 1)$

9. $f(x, y, z) = (x^2 + y^2 + z^2)^{-1/2} + \ln(xyz)$, $(-1, 2, -2)$

10. $f(x, y, z) = e^{x+y}\cos z + (y + 1)\sin^{-1}x$, $(0, 0, \pi/6)$

Finding Directional Derivatives

In Exercises 11–18, find the derivative of the function at P_0 in the direction of $\mathbf{u}$.

11. $f(x, y) = 2xy - 3y^2$, $P_0(5, 5)$, $\mathbf{u} = 4\mathbf{i} + 3\mathbf{j}$

12. $f(x, y) = 2x^2 + y^2$, $P_0(-1, 1)$, $\mathbf{u} = 3\mathbf{i} - 4\mathbf{j}$

13. $g(x, y) = \frac{x - y}{xy + 2}$, $P_0(1, -1)$, $\mathbf{u} = 12\mathbf{i} + 5\mathbf{j}$

14. $h(x, y) = \tan^{-1}(y/x) + \sqrt{3}\sin^{-1}(xy/2)$, $P_0(1, 1)$, $\mathbf{u} = 3\mathbf{i} - 2\mathbf{j}$

15. $f(x, y, z) = xy + yz + zx$, $P_0(1, -1, 2)$, $\mathbf{u} = 3\mathbf{i} + 6\mathbf{j} - 2\mathbf{k}$

16. $f(x, y, z) = x^2 + 2y^2 - 3z^2$, $P_0(1, 1, 1)$, $\mathbf{u} = \mathbf{i} + \mathbf{j} + \mathbf{k}$

17. $g(x, y, z) = 3e^x\cos yz$, $P_0(0, 0, 0)$, $\mathbf{u} = 2\mathbf{i} + \mathbf{j} - 2\mathbf{k}$

18. $h(x, y, z) = \cos xy + e^{yz} + \ln zx$, $P_0(1, 0, 1/2)$, $\mathbf{u} = \mathbf{i} + 2\mathbf{j} + 2\mathbf{k}$

Thus,

$$L(x, y, z) = 2 + 3(x - 2) + (-2)(y - 1) + 3(z - 0) = 3x - 2y + 3z - 2.$$

Since

$$f_{xx} = 2, \qquad f_{yy} = 0, \qquad f_{zz} = -3 \sin z, \qquad f_{xy} = -1, \qquad f_{xz} = 0, \qquad f_{yz} = 0,$$

and $|-3 \sin z| \le 3 \sin 0.01 \approx .03$, we may take $M = 2$ as a bound on the second partials. Hence, the error incurred by replacing f by L on R satisfies

$$|E| \le \frac{1}{2}(2)(0.01 + 0.02 + 0.01)^2 = 0.0016. \qquad \blacksquare$$

Exercises 14.6

Tangent Planes and Normal Lines to Surfaces

In Exercises 1–8, find equations for the

(a) tangent plane and

(b) normal line at the point P_0 on the given surface.

1. $x^2 + y^2 + z^2 = 3$, $P_0(1, 1, 1)$

2. $x^2 + y^2 - z^2 = 18$, $P_0(3, 5, -4)$

3. $2z - x^2 = 0$, $P_0(2, 0, 2)$

4. $x^2 + 2xy - y^2 + z^2 = 7$, $P_0(1, -1, 3)$

5. $\cos \pi x - x^2 y + e^{xz} + yz = 4$, $P_0(0, 1, 2)$

6. $x^2 - xy - y^2 - z = 0$, $P_0(1, 1, -1)$

7. $x + y + z = 1$, $P_0(0, 1, 0)$

8. $x^2 + y^2 - 2xy - x + 3y - z = -4$, $P_0(2, -3, 18)$

In Exercises 9–12, find an equation for the plane that is tangent to the given surface at the given point.

9. $z = \ln(x^2 + y^2)$, $(1, 0, 0)$ 10. $z = e^{-(x^2 + y^2)}$, $(0, 0, 1)$

11. $z = \sqrt{y - x}$, $(1, 2, 1)$ 12. $z = 4x^2 + y^2$, $(1, 1, 5)$

Tangent Lines to Space Curves

In Exercises 13–18, find parametric equations for the line tangent to the curve of intersection of the surfaces at the given point.

13. Surfaces: $x + y^2 + 2z = 4$, $x = 1$

    Point: $(1, 1, 1)$

14. Surfaces: $xyz = 1$, $x^2 + 2y^2 + 3z^2 = 6$

    Point: $(1, 1, 1)$

15. Surfaces: $x^2 + 2y + 2z = 4$, $y = 1$

    Point: $(1, 1, 1/2)$

16. Surfaces: $x + y^2 + z = 2$, $y = 1$

    Point: $(1/2, 1, 1/2)$

17. Surfaces: $x^3 + 3x^2 y^2 + y^3 + 4xy - z^2 = 0$,
    $x^2 + y^2 + z^2 = 11$

    Point: $(1, 1, 3)$

18. Surfaces: $x^2 + y^2 = 4$, $x^2 + y^2 - z = 0$

    Point: $\left(\sqrt{2}, \sqrt{2}, 4\right)$

Estimating Change

19. By about how much will

    $$f(x, y, z) = \ln\sqrt{x^2 + y^2 + z^2}$$

    change if the point $P(x, y, z)$ moves from $P_0(3, 4, 12)$ a distance of $ds = 0.1$ unit in the direction of $3\mathbf{i} + 6\mathbf{j} - 2\mathbf{k}$?

20. By about how much will

    $$f(x, y, z) = e^x \cos yz$$

    change as the point $P(x, y, z)$ moves from the origin a distance of $ds = 0.1$ unit in the direction of $2\mathbf{i} + 2\mathbf{j} - 2\mathbf{k}$?

21. By about how much will

    $$g(x, y, z) = x + x \cos z - y \sin z + y$$

    change if the point $P(x, y, z)$ moves from $P_0(2, -1, 0)$ a distance of $ds = 0.2$ unit toward the point $P_1(0, 1, 2)$?

22. By about how much will

    $$h(x, y, z) = \cos(\pi xy) + xz^2$$

    change if the point $P(x, y, z)$ moves from $P_0(-1, -1, -1)$ a distance of $ds = 0.1$ unit toward the origin?

23. **Temperature change along a circle**   Suppose that the Celsius temperature at the point (x, y) in the xy-plane is $T(x, y) = x \sin 2y$ and that distance in the xy-plane is measured in meters. A particle is moving *clockwise* around the circle of radius 1 m centered at the origin at the constant rate of 2 m/sec.

    a. How fast is the temperature experienced by the particle changing in degrees Celsius per meter at the point $P\left(1/2, \sqrt{3}/2\right)$?

    b. How fast is the temperature experienced by the particle changing in degrees Celsius per second at P?

24. **Changing temperature along a space curve**   The Celsius temperature in a region in space is given by $T(x, y, z) = 2x^2 - xyz$. A particle is moving in this region and its position at time t is given by $x = 2t^2$, $y = 3t$, $z = -t^2$, where time is measured in seconds and distance in meters.

    a. How fast is the temperature experienced by the particle changing in degrees Celsius per meter when the particle is at the point $P(8, 6, -4)$?

    b. How fast is the temperature experienced by the particle changing in degrees Celsius per second at P?

Finding Linearizations

In Exercises 25–30, find the linearization $L(x, y)$ of the function at each point.

25. $f(x, y) = x^2 + y^2 + 1$ at   **a.** $(0, 0)$,   **b.** $(1, 1)$

26. $f(x, y) = (x + y + 2)^2$ at   **a.** $(0, 0)$,   **b.** $(1, 2)$

27. $f(x, y) = 3x - 4y + 5$ at   **a.** $(0, 0)$,   **b.** $(1, 1)$

28. $f(x, y) = x^3 y^4$ at   **a.** $(1, 1)$,   **b.** $(0, 0)$

29. $f(x, y) = e^x \cos y$ at   **a.** $(0, 0)$,   **b.** $(0, \pi/2)$

30. $f(x, y) = e^{2y-x}$ at   **a.** $(0, 0)$,   **b.** $(1, 2)$

31. **Wind chill factor** Wind chill, a measure of the apparent temperature felt on exposed skin, is a function of air temperature and wind speed. The precise formula, updated by the National Weather Service in 2001 and based on modern heat transfer theory, a human face model, and skin tissue resistance, is

$$W = W(v, T) = 35.74 + 0.6215\,T - 35.75\,v^{0.16}$$
$$+ 0.4275\,T \cdot v^{0.16},$$

where T is air temperature in °F and v is wind speed in mph. A partial wind chill chart is given.

T(°F)

v (mph)	30	25	20	15	10	5	0	−5	−10
5	25	19	13	7	1	−5	−11	−16	−22
10	21	15	9	3	−4	−10	−16	−22	−28
15	19	13	6	0	−7	−13	−19	−26	−32
20	17	11	4	−2	−9	−15	−22	−29	−35
25	16	9	3	−4	−11	−17	−24	−31	−37
30	15	8	1	−5	−12	−19	−26	−33	−39
35	14	7	0	−7	−14	−21	−27	−34	−41

  a. Use the table to find $W(20, 25)$, $W(30, -10)$, and $W(15, 15)$.

  b. Use the formula to find $W(10, -40)$, $W(50, -40)$, and $W(60, 30)$.

  c. Find the linearization $L(v, T)$ of the function $W(v, T)$ at the point $(25, 5)$.

  d. Use $L(v, T)$ in part (c) to estimate the following wind chill values.

    i) $W(24, 6)$ **ii)** $W(27, 2)$

    iii) $W(5, -10)$ (Explain why this value is much different from the value found in the table.)

32. Find the linearization $L(v, T)$ of the function $W(v, T)$ in Exercise 31 at the point $(50, -20)$. Use it to estimate the following wind chill values.

  a. $W(49, -22)$ **b.** $W(53, -19)$ **c.** $W(60, -30)$

Bounding the Error in Linear Approximations

In Exercises 33–38, find the linearization $L(x, y)$ of the function $f(x, y)$ at P_0. Then find an upper bound for the magnitude $|E|$ of the error in the approximation $f(x, y) \approx L(x, y)$ over the rectangle R.

33. $f(x, y) = x^2 - 3xy + 5$ at $P_0(2, 1)$,

    R: $|x - 2| \le 0.1$, $|y - 1| \le 0.1$

34. $f(x, y) = (1/2)x^2 + xy + (1/4)y^2 + 3x - 3y + 4$ at $P_0(2, 2)$,

    R: $|x - 2| \le 0.1$, $|y - 2| \le 0.1$

35. $f(x, y) = 1 + y + x \cos y$ at $P_0(0, 0)$,

    R: $|x| \le 0.2$, $|y| \le 0.2$

    (Use $|\cos y| \le 1$ and $|\sin y| \le 1$ in estimating E.)

36. $f(x, y) = xy^2 + y \cos(x - 1)$ at $P_0(1, 2)$,

    R: $|x - 1| \le 0.1$, $|y - 2| \le 0.1$

37. $f(x, y) = e^x \cos y$ at $P_0(0, 0)$,

    R: $|x| \le 0.1$, $|y| \le 0.1$

    (Use $e^x \le 1.11$ and $|\cos y| \le 1$ in estimating E.)

38. $f(x, y) = \ln x + \ln y$ at $P_0(1, 1)$,

    R: $|x - 1| \le 0.2$, $|y - 1| \le 0.2$

Linearizations for Three Variables

Find the linearizations $L(x, y, z)$ of the functions in Exercises 39–44 at the given points.

39. $f(x, y, z) = xy + yz + xz$ at

    a. $(1, 1, 1)$ **b.** $(1, 0, 0)$ **c.** $(0, 0, 0)$

40. $f(x, y, z) = x^2 + y^2 + z^2$ at

    a. $(1, 1, 1)$ **b.** $(0, 1, 0)$ **c.** $(1, 0, 0)$

41. $f(x, y, z) = \sqrt{x^2 + y^2 + z^2}$ at

    a. $(1, 0, 0)$ **b.** $(1, 1, 0)$ **c.** $(1, 2, 2)$

42. $f(x, y, z) = (\sin xy)/z$ at

    a. $(\pi/2, 1, 1)$ **b.** $(2, 0, 1)$

43. $f(x, y, z) = e^x + \cos(y + z)$ at

    a. $(0, 0, 0)$ **b.** $\left(0, \dfrac{\pi}{2}, 0\right)$ **c.** $\left(0, \dfrac{\pi}{4}, \dfrac{\pi}{4}\right)$

44. $f(x, y, z) = \tan^{-1}(xyz)$ at

    a. $(1, 0, 0)$ **b.** $(1, 1, 0)$ **c.** $(1, 1, 1)$

In Exercises 45–48, find the linearization $L(x, y, z)$ of the function $f(x, y, z)$ at P_0. Then find an upper bound for the magnitude of the error E in the approximation $f(x, y, z) \approx L(x, y, z)$ over the region R.

45. $f(x, y, z) = xz - 3yz + 2$ at $P_0(1, 1, 2)$,

    R: $|x - 1| \le 0.01$, $|y - 1| \le 0.01$, $|z - 2| \le 0.02$

46. $f(x, y, z) = x^2 + xy + yz + (1/4)z^2$ at $P_0(1, 1, 2)$,

    R: $|x - 1| \le 0.01$, $|y - 1| \le 0.01$, $|z - 2| \le 0.08$

47. $f(x, y, z) = xy + 2yz - 3xz$ at $P_0(1, 1, 0)$,

    R: $|x - 1| \le 0.01$, $|y - 1| \le 0.01$, $|z| \le 0.01$

48. $f(x, y, z) = \sqrt{2} \cos x \sin(y + z)$ at $P_0(0, 0, \pi/4)$,

    R: $|x| \le 0.01$, $|y| \le 0.01$, $|z - \pi/4| \le 0.01$

Estimating Error; Sensitivity to Change

49. **Estimating maximum error** Suppose that T is to be found from the formula $T = x(e^y + e^{-y})$, where x and y are found to be 2 and $\ln 2$ with maximum possible errors of $|dx| = 0.1$ and $|dy| = 0.02$. Estimate the maximum possible error in the computed value of T.

50. **Estimating volume of a cylinder** About how accurately may $V = \pi r^2 h$ be calculated from measurements of r and h that are in error by 1%?

51. Consider a closed rectangular box with a square base as shown in the accompanying figure. If x is measured with error at most 2% and y is measured with error at most 3%, use a differential to estimate the corresponding percentage error in computing the box's

    a. surface area

    b. volume.

52. Consider a closed container in the shape of a cylinder of radius 10 cm and height 15 cm with a hemisphere on each end, as shown in the accompanying figure.

    The container is coated with a layer of ice 1/2 cm thick. Use a differential to estimate the total volume of ice. (*Hint:* Assume r is radius with $dr = 1/2$ and h is height with $dh = 0$.)

53. **Maximum percentage error** If $r = 5.0$ cm and $h = 12.0$ cm to the nearest millimeter, what should we expect the maximum percentage error in calculating $V = \pi r^2 h$ to be?

54. **Variation in electrical resistance** The resistance R produced by wiring resistors of R_1 and R_2 ohms in parallel (see accompanying figure) can be calculated from the formula

$$\frac{1}{R} = \frac{1}{R_1} + \frac{1}{R_2}.$$

    a. Show that

$$dR = \left(\frac{R}{R_1}\right)^2 dR_1 + \left(\frac{R}{R_2}\right)^2 dR_2.$$

    b. You have designed a two-resistor circuit like the one shown to have resistances of $R_1 = 100$ ohms and $R_2 = 400$ ohms, but there is always some variation in manufacturing and the resistors received by your firm will probably not have these exact values. Will the value of R be more sensitive to variation in R_1 or to variation in R_2? Give reasons for your answer.

    c. In another circuit like the one shown you plan to change R_1 from 20 to 20.1 ohms and R_2 from 25 to 24.9 ohms. By about what percentage will this change R?

55. You plan to calculate the area of a long, thin rectangle from measurements of its length and width. Which dimension should you measure more carefully? Give reasons for your answer.

56. **a.** Around the point $(1, 0)$, is $f(x, y) = x^2(y + 1)$ more sensitive to changes in x or to changes in y? Give reasons for your answer.

    b. What ratio of dx to dy will make df equal zero at $(1, 0)$?

57. **Error carryover in coordinate changes**

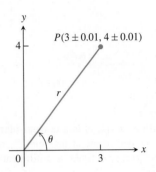

    a. If $x = 3 \pm 0.01$ and $y = 4 \pm 0.01$, as shown here, with approximately what accuracy can you calculate the polar coordinates r and θ of the point $P(x, y)$ from the formulas $r^2 = x^2 + y^2$ and $\theta = \tan^{-1}(y/x)$? Express your estimates as percentage changes of the values that r and θ have at the point $(x_0, y_0) = (3, 4)$.

    b. At the point $(x_0, y_0) = (3, 4)$, are the values of r and θ more sensitive to changes in x or to changes in y? Give reasons for your answer.

58. **Designing a soda can** A standard 12-fl-oz can of soda is essentially a cylinder of radius $r = 1$ in. and height $h = 5$ in.

    a. At these dimensions, how sensitive is the can's volume to a small change in radius versus a small change in height?

    b. Could you design a soda can that *appears* to hold more soda but in fact holds the same 12 fl oz? What might its dimensions be? (There is more than one correct answer.)

59. **Value of a 2 × 2 determinant** If $|a|$ is much greater than $|b|, |c|$, and $|d|$, to which of $a, b, c,$ and d is the value of the determinant

$$f(a, b, c, d) = \begin{vmatrix} a & b \\ c & d \end{vmatrix}$$

    most sensitive? Give reasons for your answer.

60. **Estimating maximum error** Suppose that $u = xe^y + y \sin z$ and that $x, y,$ and z can be measured with maximum possible errors of $\pm 0.2, \pm 0.6,$ and $\pm \pi/180$, respectively. Estimate the maximum possible error in calculating u from the measured values $x = 2, y = \ln 3, z = \pi/2$.

61. **The Wilson lot size formula** The Wilson lot size formula in economics says that the most economical quantity Q of goods (radios, shoes, brooms, whatever) for a store to order is given by the formula $Q = \sqrt{2KM/h}$, where K is the cost of placing the order, M is the number of items sold per week, and h is the weekly holding cost for each item (cost of space, utilities, security, and so on). To which of the variables $K, M,$ and h is Q most sensitive near the point $(K_0, M_0, h_0) = (2, 20, 0.05)$? Give reasons for your answer.

62. Surveying a triangular field The area of a triangle is $(1/2)ab \sin C$, where a and b are the lengths of two sides of the triangle and C is the measure of the included angle. In surveying a triangular plot, you have measured a, b, and C to be 150 ft, 200 ft, and 60°, respectively. By about how much could your area calculation be in error if your values of a and b are off by half a foot each and your measurement of C is off by 2°? See the accompanying figure. Remember to use radians.

Theory and Examples

63. The linearization of $f(x, y)$ is a tangent-plane approximation Show that the tangent plane at the point $P_0(x_0, y_0, f(x_0, y_0))$ on the surface $z = f(x, y)$ defined by a differentiable function f is the plane

$$f_x(x_0, y_0)(x - x_0) + f_y(x_0, y_0)(y - y_0) - (z - f(x_0, y_0)) = 0$$

or

$$z = f(x_0, y_0) + f_x(x_0, y_0)(x - x_0) + f_y(x_0, y_0)(y - y_0).$$

Thus, the tangent plane at P_0 is the graph of the linearization of f at P_0 (see accompanying figure).

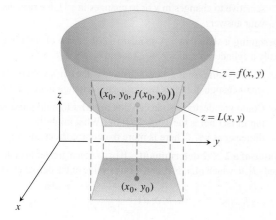

64. Change along the involute of a circle Find the derivative of $f(x, y) = x^2 + y^2$ in the direction of the unit tangent vector of the curve

$$\mathbf{r}(t) = (\cos t + t \sin t)\mathbf{i} + (\sin t - t \cos t)\mathbf{j}, \qquad t > 0.$$

65. Change along a helix Find the derivative of $f(x, y, z) = x^2 + y^2 + z^2$ in the direction of the unit tangent vector of the helix

$$\mathbf{r}(t) = (\cos t)\mathbf{i} + (\sin t)\mathbf{j} + t\mathbf{k}$$

at the points where $t = -\pi/4$, 0, and $\pi/4$. The function f gives the square of the distance from a point $P(x, y, z)$ on the helix to the origin. The derivatives calculated here give the rates at which the square of the distance is changing with respect to t as P moves through the points where $t = -\pi/4$, 0, and $\pi/4$.

66. Normal curves A smooth curve is *normal* to a surface $f(x, y, z) = c$ at a point of intersection if the curve's velocity vector is a nonzero scalar multiple of ∇f at the point.

Show that the curve

$$\mathbf{r}(t) = \sqrt{t}\,\mathbf{i} + \sqrt{t}\,\mathbf{j} - \frac{1}{4}(t + 3)\mathbf{k}$$

is normal to the surface $x^2 + y^2 - z = 3$ when $t = 1$.

67. Tangent curves A smooth curve is *tangent* to the surface at a point of intersection if its velocity vector is orthogonal to ∇f there.

Show that the curve

$$\mathbf{r}(t) = \sqrt{t}\,\mathbf{i} + \sqrt{t}\,\mathbf{j} + (2t - 1)\mathbf{k}$$

is tangent to the surface $x^2 + y^2 - z = 1$ when $t = 1$.

14.7 Extreme Values and Saddle Points

HISTORICAL BIOGRAPHY

Siméon-Denis Poisson
(1781–1840)

Continuous functions of two variables assume extreme values on closed, bounded domains (see Figures 14.38 and 14.39). We see in this section that we can narrow the search for these extreme values by examining the functions' first partial derivatives. A function of two variables can assume extreme values only at domain boundary points or at interior domain points where both first partial derivatives are zero or where one or both of the first partial derivatives fail to exist. However, the vanishing of derivatives at an interior point (a, b) does not always signal the presence of an extreme value. The surface that is the graph of the function might be shaped like a saddle right above (a, b) and cross its tangent plane there.

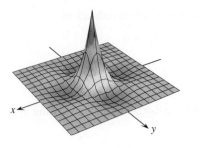

FIGURE 14.38 The function

$$z = (\cos x)(\cos y)e^{-\sqrt{x^2+y^2}}$$

has a maximum value of 1 and a minimum value of about -0.067 on the square region $|x| \le 3\pi/2, |y| \le 3\pi/2$.

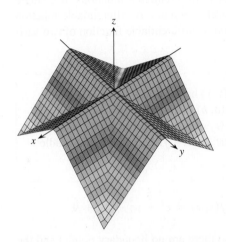

FIGURE 14.39 The "roof surface"

$$z = \frac{1}{2}\left(\big||x| - |y|\big| - |x| - |y|\right)$$

has a maximum value of 0 and a minimum value of $-a$ on the square region $|x| \le a$, $|y| \le a$.

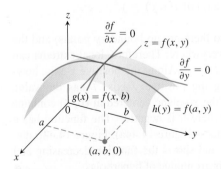

FIGURE 14.41 If a local maximum of f occurs at $x = a$, $y = b$, then the first partial derivatives $f_x(a, b)$ and $f_y(a, b)$ are both zero.

Derivative Tests for Local Extreme Values

To find the local extreme values of a function of a single variable, we look for points where the graph has a horizontal tangent line. At such points, we then look for local maxima, local minima, and points of inflection. For a function $f(x, y)$ of two variables, we look for points where the surface $z = f(x, y)$ has a horizontal tangent *plane*. At such points, we then look for local maxima, local minima, and saddle points. We begin by defining maxima and minima.

> **DEFINITIONS** Let $f(x, y)$ be defined on a region R containing the point (a, b). Then
>
> 1. $f(a, b)$ is a **local maximum** value of f if $f(a, b) \ge f(x, y)$ for all domain points (x, y) in an open disk centered at (a, b).
> 2. $f(a, b)$ is a **local minimum** value of f if $f(a, b) \le f(x, y)$ for all domain points (x, y) in an open disk centered at (a, b).

Local maxima correspond to mountain peaks on the surface $z = f(x, y)$ and local minima correspond to valley bottoms (Figure 14.40). At such points the tangent planes, when they exist, are horizontal. Local extrema are also called **relative extrema**.

As with functions of a single variable, the key to identifying the local extrema is a first derivative test.

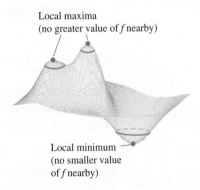

FIGURE 14.40 A local maximum occurs at a mountain peak and a local minimum occurs at a valley low point.

> **THEOREM 10—First Derivative Test for Local Extreme Values** If $f(x, y)$ has a local maximum or minimum value at an interior point (a, b) of its domain and if the first partial derivatives exist there, then $f_x(a, b) = 0$ and $f_y(a, b) = 0$.

Proof If f has a local extremum at (a, b), then the function $g(x) = f(x, b)$ has a local extremum at $x = a$ (Figure 14.41). Therefore, $g'(a) = 0$ (Chapter 4, Theorem 2). Now $g'(a) = f_x(a, b)$, so $f_x(a, b) = 0$. A similar argument with the function $h(y) = f(a, y)$ shows that $f_y(a, b) = 0$. ∎

If we substitute the values $f_x(a, b) = 0$ and $f_y(a, b) = 0$ into the equation

$$f_x(a, b)(x - a) + f_y(a, b)(y - b) - (z - f(a, b)) = 0$$

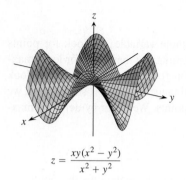

$$z = \frac{xy(x^2 - y^2)}{x^2 + y^2}$$

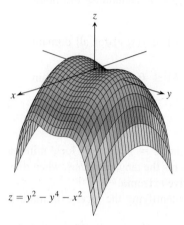

$$z = y^2 - y^4 - x^2$$

FIGURE 14.42 Saddle points at the origin.

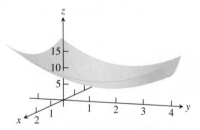

FIGURE 14.43 The graph of the function $f(x, y) = x^2 + y^2 - 4y + 9$ is a paraboloid which has a local minimum value of 5 at the point $(0, 2)$ (Example 1).

for the tangent plane to the surface $z = f(x, y)$ at (a, b), the equation reduces to

$$0 \cdot (x - a) + 0 \cdot (y - b) - z + f(a, b) = 0$$

or

$$z = f(a, b).$$

Thus, Theorem 10 says that the surface does indeed have a horizontal tangent plane at a local extremum, provided there is a tangent plane there.

DEFINITION An interior point of the domain of a function $f(x, y)$ where both f_x and f_y are zero or where one or both of f_x and f_y do not exist is a **critical point** of f.

Theorem 10 says that the only points where a function $f(x, y)$ can assume extreme values are critical points and boundary points. As with differentiable functions of a single variable, not every critical point gives rise to a local extremum. A differentiable function of a single variable might have a point of inflection. A differentiable function of two variables might have a *saddle point*.

DEFINITION A differentiable function $f(x, y)$ has a **saddle point** at a critical point (a, b) if in every open disk centered at (a, b) there are domain points (x, y) where $f(x, y) > f(a, b)$ and domain points (x, y) where $f(x, y) < f(a, b)$. The corresponding point $(a, b, f(a, b))$ on the surface $z = f(x, y)$ is called a saddle point of the surface (Figure 14.42).

EXAMPLE 1 Find the local extreme values of $f(x, y) = x^2 + y^2 - 4y + 9$.

Solution The domain of f is the entire plane (so there are no boundary points) and the partial derivatives $f_x = 2x$ and $f_y = 2y - 4$ exist everywhere. Therefore, local extreme values can occur only where

$$f_x = 2x = 0 \qquad \text{and} \qquad f_y = 2y - 4 = 0.$$

The only possibility is the point $(0, 2)$, where the value of f is 5. Since $f(x, y) = x^2 + (y - 2)^2 + 5$ is never less than 5, we see that the critical point $(0, 2)$ gives a local minimum (Figure 14.43). ∎

EXAMPLE 2 Find the local extreme values (if any) of $f(x, y) = y^2 - x^2$.

Solution The domain of f is the entire plane (so there are no boundary points) and the partial derivatives $f_x = -2x$ and $f_y = 2y$ exist everywhere. Therefore, local extrema can occur only at the origin $(0, 0)$ where $f_x = 0$ and $f_y = 0$. Along the positive x-axis, however, f has the value $f(x, 0) = -x^2 < 0$; along the positive y-axis, f has the value $f(0, y) = y^2 > 0$. Therefore, every open disk in the xy-plane centered at $(0, 0)$ contains points where the function is positive and points where it is negative. The function has a saddle point at the origin and no local extreme values (Figure 14.44a). Figure 14.44b displays the level curves (they are hyperbolas) of f, and shows the function decreasing and increasing in an alternating fashion among the four groupings of hyperbolas. ∎

That $f_x = f_y = 0$ at an interior point (a, b) of R does not guarantee f has a local extreme value there. If f and its first and second partial derivatives are continuous on R, however, we may be able to learn more from the following theorem, proved in Section 14.9.

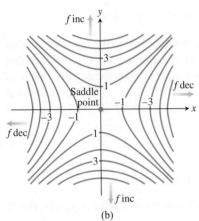

FIGURE 14.44 (a) The origin is a saddle point of the function $f(x, y) = y^2 - x^2$. There are no local extreme values (Example 2). (b) Level curves for the function f in Example 2.

THEOREM 11—Second Derivative Test for Local Extreme Values Suppose that $f(x, y)$ and its first and second partial derivatives are continuous throughout a disk centered at (a, b) and that $f_x(a, b) = f_y(a, b) = 0$. Then

i) f has a **local maximum** at (a, b) if $f_{xx} < 0$ and $f_{xx}f_{yy} - f_{xy}^2 > 0$ at (a, b).

ii) f has a **local minimum** at (a, b) if $f_{xx} > 0$ and $f_{xx}f_{yy} - f_{xy}^2 > 0$ at (a, b).

iii) f has a **saddle point** at (a, b) if $f_{xx}f_{yy} - f_{xy}^2 < 0$ at (a, b).

iv) **the test is inconclusive** at (a, b) if $f_{xx}f_{yy} - f_{xy}^2 = 0$ at (a, b). In this case, we must find some other way to determine the behavior of f at (a, b).

The expression $f_{xx}f_{yy} - f_{xy}^2$ is called the **discriminant** or **Hessian** of f. It is sometimes easier to remember it in determinant form,

$$f_{xx}f_{yy} - f_{xy}^2 = \begin{vmatrix} f_{xx} & f_{xy} \\ f_{xy} & f_{yy} \end{vmatrix}.$$

Theorem 11 says that if the discriminant is positive at the point (a, b), then the surface curves the same way in all directions: downward if $f_{xx} < 0$, giving rise to a local maximum, and upward if $f_{xx} > 0$, giving a local minimum. On the other hand, if the discriminant is negative at (a, b), then the surface curves up in some directions and down in others, so we have a saddle point.

EXAMPLE 3 Find the local extreme values of the function

$$f(x, y) = xy - x^2 - y^2 - 2x - 2y + 4.$$

Solution The function is defined and differentiable for all x and y and its domain has no boundary points. The function therefore has extreme values only at the points where f_x and f_y are simultaneously zero. This leads to

$$f_x = y - 2x - 2 = 0, \qquad f_y = x - 2y - 2 = 0,$$

or

$$x = y = -2.$$

Therefore, the point $(-2, -2)$ is the only point where f may take on an extreme value. To see if it does so, we calculate

$$f_{xx} = -2, \qquad f_{yy} = -2, \qquad f_{xy} = 1.$$

The discriminant of f at $(a, b) = (-2, -2)$ is

$$f_{xx}f_{yy} - f_{xy}^2 = (-2)(-2) - (1)^2 = 4 - 1 = 3.$$

The combination

$$f_{xx} < 0 \qquad \text{and} \qquad f_{xx}f_{yy} - f_{xy}^2 > 0$$

tells us that f has a local maximum at $(-2, -2)$. The value of f at this point is $f(-2, -2) = 8$. ∎

EXAMPLE 4 Find the local extreme values of $f(x, y) = 3y^2 - 2y^3 - 3x^2 + 6xy$.

Solution Since f is differentiable everywhere, it can assume extreme values only where

$$f_x = 6y - 6x = 0 \qquad \text{and} \qquad f_y = 6y - 6y^2 + 6x = 0.$$

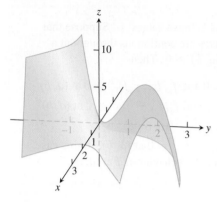

FIGURE 14.45 The surface $z = 3y^2 - 2y^3 - 3x^2 + 6xy$ has a saddle point at the origin and a local maximum at the point $(2, 2)$ (Example 4).

From the first of these equations we find $x = y$, and substitution for y into the second equation then gives

$$6x - 6x^2 + 6x = 0 \qquad \text{or} \qquad 6x(2 - x) = 0.$$

The two critical points are therefore $(0, 0)$ and $(2, 2)$.

To classify the critical points, we calculate the second derivatives:

$$f_{xx} = -6, \qquad f_{yy} = 6 - 12y, \qquad f_{xy} = 6.$$

The discriminant is given by

$$f_{xx}f_{yy} - f_{xy}{}^2 = (-36 + 72y) - 36 = 72(y - 1).$$

At the critical point $(0, 0)$ we see that the value of the discriminant is the negative number -72, so the function has a saddle point at the origin. At the critical point $(2, 2)$ we see that the discriminant has the positive value 72. Combining this result with the negative value of the second partial $f_{xx} = -6$, Theorem 11 says that the critical point $(2, 2)$ gives a local maximum value of $f(2, 2) = 12 - 16 - 12 + 24 = 8$. A graph of the surface is shown in Figure 14.45. ∎

Absolute Maxima and Minima on Closed Bounded Regions

We organize the search for the absolute extrema of a continuous function $f(x, y)$ on a closed and bounded region R into three steps.

1. *List the interior points of R* where f may have local maxima and minima and evaluate f at these points. These are the critical points of f.

2. *List the boundary points of R* where f has local maxima and minima and evaluate f at these points. We show how to do this shortly.

3. *Look through the lists* for the maximum and minimum values of f. These will be the absolute maximum and minimum values of f on R. Since absolute maxima and minima are also local maxima and minima, the absolute maximum and minimum values of f appear somewhere in the lists made in Steps 1 and 2.

EXAMPLE 5   Find the absolute maximum and minimum values of

$$f(x, y) = 2 + 2x + 2y - x^2 - y^2$$

on the triangular region in the first quadrant bounded by the lines $x = 0$, $y = 0$, $y = 9 - x$.

Solution   Since f is differentiable, the only places where f can assume these values are points inside the triangle (Figure 14.46) where $f_x = f_y = 0$ and points on the boundary.

(a) Interior points. For these we have

$$f_x = 2 - 2x = 0, \qquad f_y = 2 - 2y = 0,$$

yielding the single point $(x, y) = (1, 1)$. The value of f there is

$$f(1, 1) = 4.$$

(b) Boundary points. We take the triangle one side at a time:

i) On the segment OA, $y = 0$. The function

$$f(x, y) = f(x, 0) = 2 + 2x - x^2$$

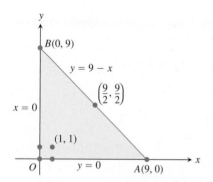

FIGURE 14.46 This triangular region is the domain of the function in Example 5.

may now be regarded as a function of x defined on the closed interval $0 \leq x \leq 9$. Its extreme values (we know from Chapter 4) may occur at the endpoints

$$x = 0 \quad \text{where} \quad f(0, 0) = 2$$

$$x = 9 \quad \text{where} \quad f(9, 0) = 2 + 18 - 81 = -61$$

and at the interior points where $f'(x, 0) = 2 - 2x = 0$. The only interior point where $f'(x, 0) = 0$ is $x = 1$, where

$$f(x, 0) = f(1, 0) = 3.$$

ii) On the segment OB, $x = 0$ and

$$f(x, y) = f(0, y) = 2 + 2y - y^2.$$

We know from the symmetry of f in x and y and from the analysis we just carried out that the candidates on this segment are

$$f(0, 0) = 2, \quad f(0, 9) = -61, \quad f(0, 1) = 3.$$

iii) We have already accounted for the values of f at the endpoints of AB, so we need only look at the interior points of AB. With $y = 9 - x$, we have

$$f(x, y) = 2 + 2x + 2(9 - x) - x^2 - (9 - x)^2 = -61 + 18x - 2x^2.$$

Setting $f'(x, 9 - x) = 18 - 4x = 0$ gives

$$x = \frac{18}{4} = \frac{9}{2}.$$

At this value of x,

$$y = 9 - \frac{9}{2} = \frac{9}{2} \quad \text{and} \quad f(x, y) = f\left(\frac{9}{2}, \frac{9}{2}\right) = -\frac{41}{2}.$$

Summary We list all the candidates: 4, 2, -61, 3, $-(41/2)$. The maximum is 4, which f assumes at (1, 1). The minimum is -61, which f assumes at (0, 9) and (9, 0). ∎

Solving extreme value problems with algebraic constraints on the variables usually requires the method of Lagrange multipliers introduced in the next section. But sometimes we can solve such problems directly, as in the next example.

EXAMPLE 6 A delivery company accepts only rectangular boxes the sum of whose length and girth (perimeter of a cross-section) does not exceed 108 in. Find the dimensions of an acceptable box of largest volume.

Solution Let x, y, and z represent the length, width, and height of the rectangular box, respectively. Then the girth is $2y + 2z$. We want to maximize the volume $V = xyz$ of the box (Figure 14.47) satisfying $x + 2y + 2z = 108$ (the largest box accepted by the delivery company). Thus, we can write the volume of the box as a function of two variables:

$$V(y, z) = (108 - 2y - 2z)yz \qquad \begin{array}{l} V = xyz \text{ and} \\ x = 108 - 2y - 2z \end{array}$$

$$= 108yz - 2y^2z - 2yz^2.$$

Setting the first partial derivatives equal to zero,

$$V_y(y, z) = 108z - 4yz - 2z^2 = (108 - 4y - 2z)z = 0$$

$$V_z(y, z) = 108y - 2y^2 - 4yz = (108 - 2y - 4z)y = 0,$$

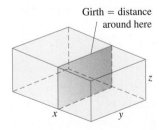

Girth = distance around here

FIGURE 14.47 The box in Example 6.

gives the critical points $(0, 0)$, $(0, 54)$, $(54, 0)$, and $(18, 18)$. The volume is zero at $(0, 0)$, $(0, 54)$, $(54, 0)$, which are not maximum values. At the point $(18, 18)$, we apply the Second Derivative Test (Theorem 11):

$$V_{yy} = -4z, \qquad V_{zz} = -4y, \qquad V_{yz} = 108 - 4y - 4z.$$

Then

$$V_{yy}V_{zz} - V_{yz}{}^2 = 16yz - 16(27 - y - z)^2.$$

Thus,

$$V_{yy}(18, 18) = -4(18) < 0$$

and

$$\left[V_{yy}V_{zz} - V_{yz}{}^2\right]_{(18,18)} = 16(18)(18) - 16(-9)^2 > 0$$

imply that $(18, 18)$ gives a maximum volume. The dimensions of the package are $x = 108 - 2(18) - 2(18) = 36$ in., $y = 18$ in., and $z = 18$ in. The maximum volume is $V = (36)(18)(18) = 11,664$ in^3, or 6.75 ft^3. ∎

Despite the power of Theorem 11, we urge you to remember its limitations. It does not apply to boundary points of a function's domain, where it is possible for a function to have extreme values along with nonzero derivatives. Also, it does not apply to points where either f_x or f_y fails to exist.

Summary of Max-Min Tests

The extreme values of $f(x, y)$ can occur only at

i) **boundary points** of the domain of f

ii) **critical points** (interior points where $f_x = f_y = 0$ or points where f_x or f_y fails to exist).

If the first- and second-order partial derivatives of f are continuous throughout a disk centered at a point (a, b) and $f_x(a, b) = f_y(a, b) = 0$, the nature of $f(a, b)$ can be tested with the **Second Derivative Test**:

i) $f_{xx} < 0$ and $f_{xx}f_{yy} - f_{xy}{}^2 > 0$ at (a, b) ⟹ **local maximum**

ii) $f_{xx} > 0$ and $f_{xx}f_{yy} - f_{xy}{}^2 > 0$ at (a, b) ⟹ **local minimum**

iii) $f_{xx}f_{yy} - f_{xy}{}^2 < 0$ at (a, b) ⟹ **saddle point**

iv) $f_{xx}f_{yy} - f_{xy}{}^2 = 0$ at (a, b) ⟹ **test is inconclusive**

Exercises 14.7

Finding Local Extrema

Find all the local maxima, local minima, and saddle points of the functions in Exercises 1–30.

1. $f(x, y) = x^2 + xy + y^2 + 3x - 3y + 4$

2. $f(x, y) = 2xy - 5x^2 - 2y^2 + 4x + 4y - 4$

3. $f(x, y) = x^2 + xy + 3x + 2y + 5$

4. $f(x, y) = 5xy - 7x^2 + 3x - 6y + 2$

5. $f(x, y) = 2xy - x^2 - 2y^2 + 3x + 4$

6. $f(x, y) = x^2 - 4xy + y^2 + 6y + 2$

7. $f(x, y) = 2x^2 + 3xy + 4y^2 - 5x + 2y$

8. $f(x, y) = x^2 - 2xy + 2y^2 - 2x + 2y + 1$

9. $f(x, y) = x^2 - y^2 - 2x + 4y + 6$

10. $f(x, y) = x^2 + 2xy$

11. $f(x, y) = \sqrt{56x^2 - 8y^2 - 16x - 31} + 1 - 8x$

12. $f(x, y) = 1 - \sqrt[3]{x^2 + y^2}$

13. $f(x, y) = x^3 - y^3 - 2xy + 6$

14. $f(x, y) = x^3 + 3xy + y^3$

15. $f(x, y) = 6x^2 - 2x^3 + 3y^2 + 6xy$

16. $f(x, y) = x^3 + y^3 + 3x^2 - 3y^2 - 8$

17. $f(x, y) = x^3 + 3xy^2 - 15x + y^3 - 15y$

18. $f(x, y) = 2x^3 + 2y^3 - 9x^2 + 3y^2 - 12y$

19. $f(x, y) = 4xy - x^4 - y^4$

20. $f(x, y) = x^4 + y^4 + 4xy$

21. $f(x, y) = \dfrac{1}{x^2 + y^2 - 1}$ 22. $f(x, y) = \dfrac{1}{x} + xy + \dfrac{1}{y}$

23. $f(x, y) = y \sin x$ 24. $f(x, y) = e^{2x} \cos y$

25. $f(x, y) = e^{x^2 + y^2 - 4x}$ 26. $f(x, y) = e^y - ye^x$

27. $f(x, y) = e^{-y}(x^2 + y^2)$ 28. $f(x, y) = e^x(x^2 - y^2)$

29. $f(x, y) = 2 \ln x + \ln y - 4x - y$

30. $f(x, y) = \ln(x + y) + x^2 - y$

Finding Absolute Extrema

In Exercises 31–38, find the absolute maxima and minima of the functions on the given domains.

31. $f(x, y) = 2x^2 - 4x + y^2 - 4y + 1$ on the closed triangular plate bounded by the lines $x = 0$, $y = 2$, $y = 2x$ in the first quadrant

32. $D(x, y) = x^2 - xy + y^2 + 1$ on the closed triangular plate in the first quadrant bounded by the lines $x = 0$, $y = 4$, $y = x$

33. $f(x, y) = x^2 + y^2$ on the closed triangular plate bounded by the lines $x = 0$, $y = 0$, $y + 2x = 2$ in the first quadrant

34. $T(x, y) = x^2 + xy + y^2 - 6x$ on the rectangular plate $0 \le x \le 5$, $-3 \le y \le 3$

35. $T(x, y) = x^2 + xy + y^2 - 6x + 2$ on the rectangular plate $0 \le x \le 5$, $-3 \le y \le 0$

36. $f(x, y) = 48xy - 32x^3 - 24y^2$ on the rectangular plate $0 \le x \le 1$, $0 \le y \le 1$

37. $f(x, y) = (4x - x^2) \cos y$ on the rectangular plate $1 \le x \le 3$, $-\pi/4 \le y \le \pi/4$ (see accompanying figure).

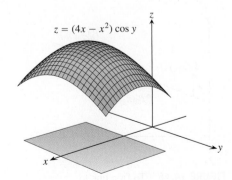

$z = (4x - x^2) \cos y$

38. $f(x, y) = 4x - 8xy + 2y + 1$ on the triangular plate bounded by the lines $x = 0$, $y = 0$, $x + y = 1$ in the first quadrant

39. Find two numbers a and b with $a \le b$ such that

$$\int_a^b (6 - x - x^2)\, dx$$

has its largest value.

40. Find two numbers a and b with $a \le b$ such that

$$\int_a^b (24 - 2x - x^2)^{1/3}\, dx$$

has its largest value.

41. **Temperatures**   A flat circular plate has the shape of the region $x^2 + y^2 \le 1$. The plate, including the boundary where $x^2 + y^2 = 1$, is heated so that the temperature at the point (x, y) is

$$T(x, y) = x^2 + 2y^2 - x.$$

Find the temperatures at the hottest and coldest points on the plate.

42. Find the critical point of

$$f(x, y) = xy + 2x - \ln x^2 y$$

in the open first quadrant ($x > 0$, $y > 0$) and show that f takes on a minimum there.

Theory and Examples

43. Find the maxima, minima, and saddle points of $f(x, y)$, if any, given that

   a. $f_x = 2x - 4y$ and $f_y = 2y - 4x$

   b. $f_x = 2x - 2$ and $f_y = 2y - 4$

   c. $f_x = 9x^2 - 9$ and $f_y = 2y + 4$

   Describe your reasoning in each case.

44. The discriminant $f_{xx}f_{yy} - f_{xy}{}^2$ is zero at the origin for each of the following functions, so the Second Derivative Test fails there. Determine whether the function has a maximum, a minimum, or neither at the origin by imagining what the surface $z = f(x, y)$ looks like. Describe your reasoning in each case.

   a. $f(x, y) = x^2 y^2$ b. $f(x, y) = 1 - x^2 y^2$

   c. $f(x, y) = xy^2$ d. $f(x, y) = x^3 y^2$

   e. $f(x, y) = x^3 y^3$ f. $f(x, y) = x^4 y^4$

45. Show that $(0, 0)$ is a critical point of $f(x, y) = x^2 + kxy + y^2$ no matter what value the constant k has. (*Hint:* Consider two cases: $k = 0$ and $k \ne 0$.)

46. For what values of the constant k does the Second Derivative Test guarantee that $f(x, y) = x^2 + kxy + y^2$ will have a saddle point at $(0, 0)$? A local minimum at $(0, 0)$? For what values of k is the Second Derivative Test inconclusive? Give reasons for your answers.

47. If $f_x(a, b) = f_y(a, b) = 0$, must f have a local maximum or minimum value at (a, b)? Give reasons for your answer.

48. Can you conclude anything about $f(a, b)$ if f and its first and second partial derivatives are continuous throughout a disk centered at the critical point (a, b) and $f_{xx}(a, b)$ and $f_{yy}(a, b)$ differ in sign? Give reasons for your answer.

49. Among all the points on the graph of $z = 10 - x^2 - y^2$ that lie above the plane $x + 2y + 3z = 0$, find the point farthest from the plane.

50. Find the point on the graph of $z = x^2 + y^2 + 10$ nearest the plane $x + 2y - z = 0$.

51. Find the point on the plane $3x + 2y + z = 6$ that is nearest the origin.

52. Find the minimum distance from the point $(2, -1, 1)$ to the plane $x + y - z = 2$.

53. Find three numbers whose sum is 9 and whose sum of squares is a minimum.

54. Find three positive numbers whose sum is 3 and whose product is a maximum.

55. Find the maximum value of $s = xy + yz + xz$ where $x + y + z = 6$.

56. Find the minimum distance from the cone $z = \sqrt{x^2 + y^2}$ to the point $(-6, 4, 0)$.

57. Find the dimensions of the rectangular box of maximum volume that can be inscribed inside the sphere $x^2 + y^2 + z^2 = 4$.

58. Among all closed rectangular boxes of volume 27 cm³, what is the smallest surface area?

59. You are to construct an open rectangular box from 12 ft² of material. What dimensions will result in a box of maximum volume?

60. Consider the function $f(x, y) = x^2 + y^2 + 2xy - x - y + 1$ over the square $0 \le x \le 1$ and $0 \le y \le 1$.

    a. Show that f has an absolute minimum along the line segment $2x + 2y = 1$ in this square. What *is* the absolute minimum value?

    b. Find the absolute maximum value of f over the square.

Extreme Values on Parametrized Curves   To find the extreme values of a function $f(x, y)$ on a curve $x = x(t), y = y(t)$, we treat f as a function of the single variable t and use the Chain Rule to find where df/dt is zero. As in any other single-variable case, the extreme values of f are then found among the values at the

    a. critical points (points where df/dt is zero or fails to exist), and

    b. endpoints of the parameter domain.

Find the absolute maximum and minimum values of the following functions on the given curves.

61. Functions:

    a. $f(x, y) = x + y$ **b.** $g(x, y) = xy$

    c. $h(x, y) = 2x^2 + y^2$

    Curves:

    i) The semicircle $x^2 + y^2 = 4$, $y \ge 0$

    ii) The quarter circle $x^2 + y^2 = 4$, $x \ge 0$, $y \ge 0$

    Use the parametric equations $x = 2 \cos t, y = 2 \sin t$.

62. Functions:

    a. $f(x, y) = 2x + 3y$ **b.** $g(x, y) = xy$

    c. $h(x, y) = x^2 + 3y^2$

    Curves:

    i) The semiellipse $(x^2/9) + (y^2/4) = 1$, $y \ge 0$

    ii) The quarter ellipse $(x^2/9) + (y^2/4) = 1$, $x \ge 0$, $y \ge 0$

    Use the parametric equations $x = 3 \cos t, y = 2 \sin t$.

63. Function: $f(x, y) = xy$

    Curves:

    i) The line $x = 2t$, $y = t + 1$

    ii) The line segment $x = 2t$, $y = t + 1$, $-1 \le t \le 0$

    iii) The line segment $x = 2t$, $y = t + 1$, $0 \le t \le 1$

64. Functions:

    a. $f(x, y) = x^2 + y^2$

    b. $g(x, y) = 1/(x^2 + y^2)$

    Curves:

    i) The line $x = t$, $y = 2 - 2t$

    ii) The line segment $x = t$, $y = 2 - 2t$, $0 \le t \le 1$

65. **Least squares and regression lines**   When we try to fit a line $y = mx + b$ to a set of numerical data points (x_1, y_1), $(x_2, y_2), \ldots, (x_n, y_n)$ (Figure 14.48), we usually choose the line that minimizes the sum of the squares of the vertical distances from the points to the line. In theory, this means finding the values of m and b that minimize the value of the function

$$w = (mx_1 + b - y_1)^2 + \cdots + (mx_n + b - y_n)^2. \quad (1)$$

Show that the values of m and b that do this are

$$m = \frac{\left(\sum x_k\right)\left(\sum y_k\right) - n\sum x_k y_k}{\left(\sum x_k\right)^2 - n\sum x_k^2}, \quad (2)$$

$$b = \frac{1}{n}\left(\sum y_k - m\sum x_k\right), \quad (3)$$

with all sums running from $k = 1$ to $k = n$. Many scientific calculators have these formulas built in, enabling you to find m and b with only a few keystrokes after you have entered the data.

    The line $y = mx + b$ determined by these values of m and b is called the **least squares line**, **regression line**, or **trend line** for the data under study. Finding a least squares line lets you

1. summarize data with a simple expression,

2. predict values of y for other, experimentally untried values of x,

3. handle data analytically.

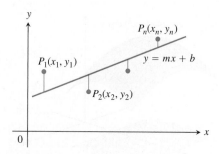

FIGURE 14.48   To fit a line to noncollinear points, we choose the line that minimizes the sum of the squares of the deviations.

In Exercises 66–68, use Equations (2) and (3) to find the least squares line for each set of data points. Then use the linear equation you obtain to predict the value of y that would correspond to $x = 4$.

66. $(-2, 0)$, $(0, 2)$, $(2, 3)$ **67.** $(-1, 2)$, $(0, 1)$, $(3, -4)$

68. $(0, 0)$, $(1, 2)$, $(2, 3)$

COMPUTER EXPLORATIONS

In Exercises 69–74, you will explore functions to identify their local extrema. Use a CAS to perform the following steps:

a. Plot the function over the given rectangle.

b. Plot some level curves in the rectangle.

c. Calculate the function's first partial derivatives and use the CAS equation solver to find the critical points. How do the critical points relate to the level curves plotted in part (b)? Which critical points, if any, appear to give a saddle point? Give reasons for your answer.

d. Calculate the function's second partial derivatives and find the discriminant $f_{xx}f_{yy} - f_{xy}^2$.

e. Using the max-min tests, classify the critical points found in part (c). Are your findings consistent with your discussion in part (c)?

69. $f(x, y) = x^2 + y^3 - 3xy$, $-5 \le x \le 5$, $-5 \le y \le 5$

70. $f(x, y) = x^3 - 3xy^2 + y^2$, $-2 \le x \le 2$, $-2 \le y \le 2$

71. $f(x, y) = x^4 + y^2 - 8x^2 - 6y + 16$, $-3 \le x \le 3$, $-6 \le y \le 6$

72. $f(x, y) = 2x^4 + y^4 - 2x^2 - 2y^2 + 3$, $-3/2 \le x \le 3/2$, $-3/2 \le y \le 3/2$

73. $f(x, y) = 5x^6 + 18x^5 - 30x^4 + 30xy^2 - 120x^3$, $-4 \le x \le 3$, $-2 \le y \le 2$

74. $f(x, y) = \begin{cases} x^5 \ln(x^2 + y^2), & (x, y) \ne (0, 0) \\ 0, & (x, y) = (0, 0) \end{cases}$, $-2 \le x \le 2$, $-2 \le y \le 2$

14.8 Lagrange Multipliers

HISTORICAL BIOGRAPHY

Joseph Louis Lagrange
(1736–1813)

Sometimes we need to find the extreme values of a function whose domain is constrained to lie within some particular subset of the plane—a disk, for example, a closed triangular region, or along a curve. In this section, we explore a powerful method for finding extreme values of constrained functions: the method of *Lagrange multipliers*.

Constrained Maxima and Minima

We first consider a problem where a constrained minimum can be found by eliminating a variable.

EXAMPLE 1 Find the point $P(x, y, z)$ on the plane $2x + y - z - 5 = 0$ that is closest to the origin.

Solution The problem asks us to find the minimum value of the function

$$|\overrightarrow{OP}| = \sqrt{(x - 0)^2 + (y - 0)^2 + (z - 0)^2}$$
$$= \sqrt{x^2 + y^2 + z^2}$$

subject to the constraint that

$$2x + y - z - 5 = 0.$$

Since $|\overrightarrow{OP}|$ has a minimum value wherever the function

$$f(x, y, z) = x^2 + y^2 + z^2$$

has a minimum value, we may solve the problem by finding the minimum value of $f(x, y, z)$ subject to the constraint $2x + y - z - 5 = 0$ (thus avoiding square roots). If we regard x and y as the independent variables in this equation and write z as

$$z = 2x + y - 5,$$

our problem reduces to one of finding the points (x, y) at which the function

$$h(x, y) = f(x, y, 2x + y - 5) = x^2 + y^2 + (2x + y - 5)^2$$

has its minimum value or values. Since the domain of h is the entire xy-plane, the First Derivative Test of Section 14.7 tells us that any minima that h might have must occur at points where

$$h_x = 2x + 2(2x + y - 5)(2) = 0, \qquad h_y = 2y + 2(2x + y - 5) = 0.$$

This leads to

$$10x + 4y = 20, \qquad 4x + 4y = 10,$$

and the solution

$$x = \frac{5}{3}, \qquad y = \frac{5}{6}.$$

We may apply a geometric argument together with the Second Derivative Test to show that these values minimize h. The z-coordinate of the corresponding point on the plane $z = 2x + y - 5$ is

$$z = 2\left(\frac{5}{3}\right) + \frac{5}{6} - 5 = -\frac{5}{6}.$$

Therefore, the point we seek is

$$\text{Closest point:} \qquad P\left(\frac{5}{3}, \frac{5}{6}, -\frac{5}{6}\right).$$

The distance from P to the origin is $5/\sqrt{6} \approx 2.04$. ∎

Attempts to solve a constrained maximum or minimum problem by substitution, as we might call the method of Example 1, do not always go smoothly. This is one of the reasons for learning the new method of this section.

EXAMPLE 2 Find the points on the hyperbolic cylinder $x^2 - z^2 - 1 = 0$ that are closest to the origin.

Solution 1 The cylinder is shown in Figure 14.49. We seek the points on the cylinder closest to the origin. These are the points whose coordinates minimize the value of the function

$$f(x, y, z) = x^2 + y^2 + z^2 \qquad \text{Square of the distance}$$

subject to the constraint that $x^2 - z^2 - 1 = 0$. If we regard x and y as independent variables in the constraint equation, then

$$z^2 = x^2 - 1$$

and the values of $f(x, y, z) = x^2 + y^2 + z^2$ on the cylinder are given by the function

$$h(x, y) = x^2 + y^2 + (x^2 - 1) = 2x^2 + y^2 - 1.$$

To find the points on the cylinder whose coordinates minimize f, we look for the points in the xy-plane whose coordinates minimize h. The only extreme value of h occurs where

$$h_x = 4x = 0 \qquad \text{and} \qquad h_y = 2y = 0,$$

that is, at the point $(0, 0)$. But there are no points on the cylinder where both x and y are zero. What went wrong?

What happened was that the First Derivative Test found (as it should have) the point *in the domain of h* where h has a minimum value. We, on the other hand, want the points *on the cylinder* where h has a minimum value. Although the domain of h is the entire

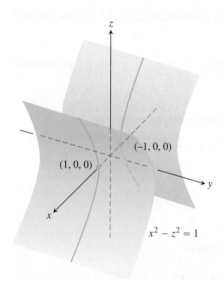

FIGURE 14.49 The hyperbolic cylinder $x^2 - z^2 - 1 = 0$ in Example 2.

$(-1, 0, 0)$

$(1, 0, 0)$

$x^2 - z^2 = 1$

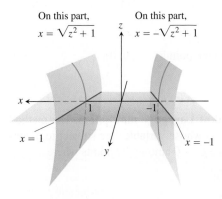

The hyperbolic cylinder $x^2 - z^2 = 1$

On this part,
$x = \sqrt{z^2 + 1}$

On this part,
$x = -\sqrt{z^2 + 1}$

$x = 1$

$x = -1$

FIGURE 14.50 The region in the xy-plane from which the first two coordinates of the points (x, y, z) on the hyperbolic cylinder $x^2 - z^2 = 1$ are selected excludes the band $-1 < x < 1$ in the xy-plane (Example 2).

$x^2 - z^2 - 1 = 0$

$x^2 + y^2 + z^2 - a^2 = 0$

FIGURE 14.51 A sphere expanding like a soap bubble centered at the origin until it just touches the hyperbolic cylinder $x^2 - z^2 - 1 = 0$ (Example 2).

xy-plane, the domain from which we can select the first two coordinates of the points (x, y, z) on the cylinder is restricted to the "shadow" of the cylinder on the xy-plane; it does not include the band between the lines $x = -1$ and $x = 1$ (Figure 14.50).

We can avoid this problem if we treat y and z as independent variables (instead of x and y) and express x in terms of y and z as

$$x^2 = z^2 + 1.$$

With this substitution, $f(x, y, z) = x^2 + y^2 + z^2$ becomes

$$k(y, z) = (z^2 + 1) + y^2 + z^2 = 1 + y^2 + 2z^2$$

and we look for the points where k takes on its smallest value. The domain of k in the yz-plane now matches the domain from which we select the y- and z-coordinates of the points (x, y, z) on the cylinder. Hence, the points that minimize k in the plane will have corresponding points on the cylinder. The smallest values of k occur where

$$k_y = 2y = 0 \qquad \text{and} \qquad k_z = 4z = 0,$$

or where $y = z = 0$. This leads to

$$x^2 = z^2 + 1 = 1, \qquad x = \pm 1.$$

The corresponding points on the cylinder are $(\pm 1, 0, 0)$. We can see from the inequality

$$k(y, z) = 1 + y^2 + 2z^2 \geq 1$$

that the points $(\pm 1, 0, 0)$ give a minimum value for k. We can also see that the minimum distance from the origin to a point on the cylinder is 1 unit.

Solution 2   Another way to find the points on the cylinder closest to the origin is to imagine a small sphere centered at the origin expanding like a soap bubble until it just touches the cylinder (Figure 14.51). At each point of contact, the cylinder and sphere have the same tangent plane and normal line. Therefore, if the sphere and cylinder are represented as the level surfaces obtained by setting

$$f(x, y, z) = x^2 + y^2 + z^2 - a^2 \qquad \text{and} \qquad g(x, y, z) = x^2 - z^2 - 1$$

equal to 0, then the gradients ∇f and ∇g will be parallel where the surfaces touch. At any point of contact, we should therefore be able to find a scalar λ ("lambda") such that

$$\nabla f = \lambda \nabla g,$$

or

$$2x\mathbf{i} + 2y\mathbf{j} + 2z\mathbf{k} = \lambda(2x\mathbf{i} - 2z\mathbf{k}).$$

Thus, the coordinates x, y, and z of any point of tangency will have to satisfy the three scalar equations

$$2x = 2\lambda x, \qquad 2y = 0, \qquad 2z = -2\lambda z.$$

For what values of λ will a point (x, y, z) whose coordinates satisfy these scalar equations also lie on the surface $x^2 - z^2 - 1 = 0$? To answer this question, we use our knowledge that no point on the surface has a zero x-coordinate to conclude that $x \neq 0$. Hence, $2x = 2\lambda x$ only if

$$2 = 2\lambda, \qquad \text{or} \qquad \lambda = 1.$$

For $\lambda = 1$, the equation $2z = -2\lambda z$ becomes $2z = -2z$. If this equation is to be satisfied as well, z must be zero. Since $y = 0$ also (from the equation $2y = 0$), we conclude that the points we seek all have coordinates of the form

$$(x, 0, 0).$$

What points on the surface $x^2 - z^2 = 1$ have coordinates of this form? The answer is the points $(x, 0, 0)$ for which

$$x^2 - (0)^2 = 1, \qquad x^2 = 1, \qquad \text{or} \qquad x = \pm 1.$$

The points on the cylinder closest to the origin are the points $(\pm 1, 0, 0)$. ∎

The Method of Lagrange Multipliers

In Solution 2 of Example 2, we used the **method of Lagrange multipliers**. The method says that the extreme values of a function $f(x, y, z)$ whose variables are subject to a constraint $g(x, y, z) = 0$ are to be found on the surface $g = 0$ among the points where

$$\nabla f = \lambda \nabla g$$

for some scalar λ (called a **Lagrange multiplier**).

To explore the method further and see why it works, we first make the following observation, which we state as a theorem.

THEOREM 12—The Orthogonal Gradient Theorem   Suppose that $f(x, y, z)$ is differentiable in a region whose interior contains a smooth curve

$$C: \quad \mathbf{r}(t) = g(t)\mathbf{i} + h(t)\mathbf{j} + k(t)\mathbf{k}.$$

If P_0 is a point on C where f has a local maximum or minimum relative to its values on C, then ∇f is orthogonal to C at P_0.

Proof   We show that ∇f is orthogonal to the curve's velocity vector at P_0. The values of f on C are given by the composite $f(g(t), h(t), k(t))$, whose derivative with respect to t is

$$\frac{df}{dt} = \frac{\partial f}{\partial x}\frac{dg}{dt} + \frac{\partial f}{\partial y}\frac{dh}{dt} + \frac{\partial f}{\partial z}\frac{dk}{dt} = \nabla f \cdot \mathbf{v}.$$

At any point P_0 where f has a local maximum or minimum relative to its values on the curve, $df/dt = 0$, so

$$\nabla f \cdot \mathbf{v} = 0. \qquad ∎$$

By dropping the z-terms in Theorem 12, we obtain a similar result for functions of two variables.

COROLLARY OF THEOREM 12   At the points on a smooth curve $\mathbf{r}(t) = g(t)\mathbf{i} + h(t)\mathbf{j}$ where a differentiable function $f(x, y)$ takes on its local maxima and minima relative to its values on the curve, $\nabla f \cdot \mathbf{v} = 0$, where $\mathbf{v} = d\mathbf{r}/dt$.

Theorem 12 is the key to the method of Lagrange multipliers. Suppose that $f(x, y, z)$ and $g(x, y, z)$ are differentiable and that P_0 is a point on the surface $g(x, y, z) = 0$ where f has a local maximum or minimum value relative to its other values on the surface. We assume also that $\nabla g \neq \mathbf{0}$ at points on the surface $g(x, y, z) = 0$. Then f takes on a local maximum or minimum at P_0 relative to its values on every differentiable curve through P_0 on the surface $g(x, y, z) = 0$. Therefore, ∇f is orthogonal to the velocity vector of every such differentiable curve through P_0. So is ∇g, moreover (because ∇g is orthogonal to the level surface $g = 0$, as we saw in Section 14.5). Therefore, at P_0, ∇f is some scalar multiple λ of ∇g.

24. Extrema on a sphere Find the points on the sphere $x^2 + y^2 + z^2 = 25$ where $f(x, y, z) = x + 2y + 3z$ has its maximum and minimum values.

25. Minimizing a sum of squares Find three real numbers whose sum is 9 and the sum of whose squares is as small as possible.

26. Maximizing a product Find the largest product the positive numbers x, y, and z can have if $x + y + z^2 = 16$.

27. Rectangular box of largest volume in a sphere Find the dimensions of the closed rectangular box with maximum volume that can be inscribed in the unit sphere.

28. Box with vertex on a plane Find the volume of the largest closed rectangular box in the first octant having three faces in the coordinate planes and a vertex on the plane $x/a + y/b + z/c = 1$, where $a > 0$, $b > 0$, and $c > 0$.

29. Hottest point on a space probe A space probe in the shape of the ellipsoid

$$4x^2 + y^2 + 4z^2 = 16$$

enters Earth's atmosphere and its surface begins to heat. After 1 hour, the temperature at the point (x, y, z) on the probe's surface is

$$T(x, y, z) = 8x^2 + 4yz - 16z + 600 .$$

Find the hottest point on the probe's surface.

30. Extreme temperatures on a sphere Suppose that the Celsius temperature at the point (x, y, z) on the sphere $x^2 + y^2 + z^2 = 1$ is $T = 400xyz^2$. Locate the highest and lowest temperatures on the sphere.

31. Maximizing a utility function: an example from economics In economics, the usefulness or *utility* of amounts x and y of two capital goods G_1 and G_2 is sometimes measured by a function $U(x, y)$. For example, G_1 and G_2 might be two chemicals a pharmaceutical company needs to have on hand and $U(x, y)$ the gain from manufacturing a product whose synthesis requires different amounts of the chemicals depending on the process used. If G_1 costs a dollars per kilogram, G_2 costs b dollars per kilogram, and the total amount allocated for the purchase of G_1 and G_2 together is c dollars, then the company's managers want to maximize $U(x, y)$ given that $ax + by = c$. Thus, they need to solve a typical Lagrange multiplier problem.

  Suppose that

$$U(x, y) = xy + 2x$$

and that the equation $ax + by = c$ simplifies to

$$2x + y = 30.$$

Find the maximum value of U and the corresponding values of x and y subject to this latter constraint.

32. Locating a radio telescope You are in charge of erecting a radio telescope on a newly discovered planet. To minimize interference, you want to place it where the magnetic field of the planet is weakest. The planet is spherical, with a radius of 6 units. Based on a coordinate system whose origin is at the center of the planet, the strength of the magnetic field is given by $M(x, y, z) = 6x - y^2 + xz + 60$. Where should you locate the radio telescope?

Extreme Values Subject to Two Constraints

33. Maximize the function $f(x, y, z) = x^2 + 2y - z^2$ subject to the constraints $2x - y = 0$ and $y + z = 0$.

34. Minimize the function $f(x, y, z) = x^2 + y^2 + z^2$ subject to the constraints $x + 2y + 3z = 6$ and $x + 3y + 9z = 9$.

35. Minimum distance to the origin Find the point closest to the origin on the line of intersection of the planes $y + 2z = 12$ and $x + y = 6$.

36. Maximum value on line of intersection Find the maximum value that $f(x, y, z) = x^2 + 2y - z^2$ can have on the line of intersection of the planes $2x - y = 0$ and $y + z = 0$.

37. Extrema on a curve of intersection Find the extreme values of $f(x, y, z) = x^2yz + 1$ on the intersection of the plane $z = 1$ with the sphere $x^2 + y^2 + z^2 = 10$.

38. a. Maximum on line of intersection Find the maximum value of $w = xyz$ on the line of intersection of the two planes $x + y + z = 40$ and $x + y - z = 0$.

  b. Give a geometric argument to support your claim that you have found a maximum, and not a minimum, value of w.

39. Extrema on a circle of intersection Find the extreme values of the function $f(x, y, z) = xy + z^2$ on the circle in which the plane $y - x = 0$ intersects the sphere $x^2 + y^2 + z^2 = 4$.

40. Minimum distance to the origin Find the point closest to the origin on the curve of intersection of the plane $2y + 4z = 5$ and the cone $z^2 = 4x^2 + 4y^2$.

Theory and Examples

41. The condition $\nabla f = \lambda \nabla g$ is not sufficient Although $\nabla f = \lambda \nabla g$ is a necessary condition for the occurrence of an extreme value of $f(x, y)$ subject to the conditions $g(x, y) = 0$ and $\nabla g \neq \mathbf{0}$, it does not in itself guarantee that one exists. As a case in point, try using the method of Lagrange multipliers to find a maximum value of $f(x, y) = x + y$ subject to the constraint that $xy = 16$. The method will identify the two points $(4, 4)$ and $(-4, -4)$ as candidates for the location of extreme values. Yet the sum $(x + y)$ has no maximum value on the hyperbola $xy = 16$. The farther you go from the origin on this hyperbola in the first quadrant, the larger the sum $f(x, y) = x + y$ becomes.

42. A least squares plane The plane $z = Ax + By + C$ is to be "fitted" to the following points (x_k, y_k, z_k):

$$(0, 0, 0), \quad (0, 1, 1), \quad (1, 1, 1), \quad (1, 0, -1).$$

Find the values of A, B, and C that minimize

$$\sum_{k=1}^{4} (Ax_k + By_k + C - z_k)^2,$$

the sum of the squares of the deviations.

43. a. Maximum on a sphere Show that the maximum value of $a^2b^2c^2$ on a sphere of radius r centered at the origin of a Cartesian abc-coordinate system is $(r^2/3)^3$.

  b. Geometric and arithmetic means Using part (a), show that for nonnegative numbers a, b, and c,

$$(abc)^{1/3} \le \frac{a + b + c}{3};$$

  that is, the *geometric mean* of three nonnegative numbers is less than or equal to their *arithmetic mean*.

44. Sum of products Let $a_1, a_2, \ldots, a_n$ be n positive numbers. Find the maximum of $\sum_{i=1}^{n} a_i x_i$ subject to the constraint $\sum_{i=1}^{n} x_i^2 = 1$.

COMPUTER EXPLORATIONS

In Exercises 45–50, use a CAS to perform the following steps implementing the method of Lagrange multipliers for finding constrained extrema:

a. Form the function $h = f - \lambda_1 g_1 - \lambda_2 g_2$, where f is the function to optimize subject to the constraints $g_1 = 0$ and $g_2 = 0$.

b. Determine all the first partial derivatives of h, including the partials with respect to λ_1 and λ_2, and set them equal to 0.

c. Solve the system of equations found in part (b) for all the unknowns, including λ_1 and λ_2.

d. Evaluate f at each of the solution points found in part (c) and select the extreme value subject to the constraints asked for in the exercise.

45. Minimize $f(x, y, z) = xy + yz$ subject to the constraints $x^2 + y^2 - 2 = 0$ and $x^2 + z^2 - 2 = 0$.

46. Minimize $f(x, y, z) = xyz$ subject to the constraints $x^2 + y^2 - 1 = 0$ and $x - z = 0$.

47. Maximize $f(x, y, z) = x^2 + y^2 + z^2$ subject to the constraints $2y + 4z - 5 = 0$ and $4x^2 + 4y^2 - z^2 = 0$.

48. Minimize $f(x, y, z) = x^2 + y^2 + z^2$ subject to the constraints $x^2 - xy + y^2 - z^2 - 1 = 0$ and $x^2 + y^2 - 1 = 0$.

49. Minimize $f(x, y, z, w) = x^2 + y^2 + z^2 + w^2$ subject to the constraints $2x - y + z - w - 1 = 0$ and $x + y - z + w - 1 = 0$.

50. Determine the distance from the line $y = x + 1$ to the parabola $y^2 = x$. (*Hint:* Let (x, y) be a point on the line and (w, z) a point on the parabola. You want to minimize $(x - w)^2 + (y - z)^2$.)

14.9 Taylor's Formula for Two Variables

In this section we use Taylor's formula to derive the Second Derivative Test for local extreme values (Section 14.7) and the error formula for linearizations of functions of two independent variables (Section 14.6). The use of Taylor's formula in these derivations leads to an extension of the formula that provides polynomial approximations of all orders for functions of two independent variables.

Derivation of the Second Derivative Test

Let $f(x, y)$ have continuous partial derivatives in an open region R containing a point $P(a, b)$ where $f_x = f_y = 0$ (Figure 14.57). Let h and k be increments small enough to put the point $S(a + h, b + k)$ and the line segment joining it to P inside R. We parametrize the segment PS as

$$x = a + th, \qquad y = b + tk, \qquad 0 \le t \le 1.$$

If $F(t) = f(a + th, b + tk)$, the Chain Rule gives

$$F'(t) = f_x \frac{dx}{dt} + f_y \frac{dy}{dt} = hf_x + kf_y.$$

Since f_x and f_y are differentiable (they have continuous partial derivatives), F' is a differentiable function of t and

$$F'' = \frac{\partial F'}{\partial x}\frac{dx}{dt} + \frac{\partial F'}{\partial y}\frac{dy}{dt} = \frac{\partial}{\partial x}(hf_x + kf_y) \cdot h + \frac{\partial}{\partial y}(hf_x + kf_y) \cdot k$$

$$= h^2 f_{xx} + 2hk f_{xy} + k^2 f_{yy}. \qquad f_{xy} = f_{yx}$$

Since F and F' are continuous on $[0, 1]$ and F' is differentiable on $(0, 1)$, we can apply Taylor's formula with $n = 2$ and $a = 0$ to obtain

$$F(1) = F(0) + F'(0)(1 - 0) + F''(c)\frac{(1 - 0)^2}{2}$$

$$F(1) = F(0) + F'(0) + \frac{1}{2}F''(c) \tag{1}$$

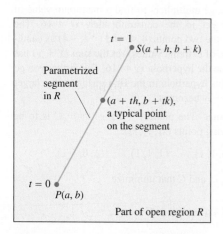

FIGURE 14.57 We begin the derivation of the Second Derivative Test at $P(a, b)$ by parametrizing a typical line segment from P to a point S nearby.

Within the figure:

$t = 1$ $S(a + h, b + k)$

Parametrized segment in R

$(a + th, b + tk)$, a typical point on the segment

$t = 0$ $P(a, b)$

Part of open region R

for some c between 0 and 1. Writing Equation (1) in terms of f gives

$$f(a + h, b + k) = f(a, b) + hf_x(a, b) + kf_y(a, b)$$

$$+ \frac{1}{2} \left(h^2 f_{xx} + 2hk f_{xy} + k^2 f_{yy} \right) \Big|_{(a+ch,\, b+ck)}. \tag{2}$$

Since $f_x(a, b) = f_y(a, b) = 0$, this reduces to

$$f(a + h, b + k) - f(a, b) = \frac{1}{2} \left(h^2 f_{xx} + 2hk f_{xy} + k^2 f_{yy} \right) \Big|_{(a+ch,\, b+ck)}. \tag{3}$$

The presence of an extremum of f at (a, b) is determined by the sign of $f(a + h, b + k) - f(a, b)$. By Equation (3), this is the same as the sign of

$$Q(c) = (h^2 f_{xx} + 2hk f_{xy} + k^2 f_{yy}) \big|_{(a+ch,\, b+ck)}.$$

Now, if $Q(0) \neq 0$, the sign of $Q(c)$ will be the same as the sign of $Q(0)$ for sufficiently small values of h and k. We can predict the sign of

$$Q(0) = h^2 f_{xx}(a, b) + 2hk f_{xy}(a, b) + k^2 f_{yy}(a, b) \tag{4}$$

from the signs of f_{xx} and $f_{xx}f_{yy} - f_{xy}^2$ at (a, b). Multiply both sides of Equation (4) by f_{xx} and rearrange the right-hand side to get

$$f_{xx}Q(0) = (hf_{xx} + kf_{xy})^2 + (f_{xx}f_{yy} - f_{xy}^2)k^2. \tag{5}$$

From Equation (5) we see that

1. If $f_{xx} < 0$ and $f_{xx}f_{yy} - f_{xy}^2 > 0$ at (a, b), then $Q(0) < 0$ for all sufficiently small nonzero values of h and k, and f has a *local maximum* value at (a, b).

2. If $f_{xx} > 0$ and $f_{xx}f_{yy} - f_{xy}^2 > 0$ at (a, b), then $Q(0) > 0$ for all sufficiently small nonzero values of h and k, and f has a *local minimum* value at (a, b).

3. If $f_{xx}f_{yy} - f_{xy}^2 < 0$ at (a, b), there are combinations of arbitrarily small nonzero values of h and k for which $Q(0) > 0$, and other values for which $Q(0) < 0$. Arbitrarily close to the point $P_0(a, b, f(a, b))$ on the surface $z = f(x, y)$ there are points above P_0 and points below P_0, so f has a *saddle point* at (a, b).

4. If $f_{xx}f_{yy} - f_{xy}^2 = 0$ at (a, b), another test is needed. The possibility that $Q(0)$ equals zero prevents us from drawing conclusions about the sign of $Q(c)$.

The Error Formula for Linear Approximations

We want to show that the difference $E(x, y)$, between the values of a function $f(x, y)$, and its linearization $L(x, y)$ at (x_0, y_0) satisfies the inequality

$$|E(x, y)| \leq \frac{1}{2} M(|x - x_0| + |y - y_0|)^2.$$

The function f is assumed to have continuous second partial derivatives throughout an open set containing a closed rectangular region R centered at (x_0, y_0). The number M is an upper bound for $|f_{xx}|, |f_{yy}|,$ and $|f_{xy}|$ on R.

The inequality we want comes from Equation (2). We substitute x_0 and y_0 for a and b, and $x - x_0$ and $y - y_0$ for h and k, respectively, and rearrange the result as

$$f(x, y) = \underbrace{f(x_0, y_0) + f_x(x_0, y_0)(x - x_0) + f_y(x_0, y_0)(y - y_0)}_{\text{linearization } L(x, y)}$$

$$+ \underbrace{\frac{1}{2} \left((x - x_0)^2 f_{xx} + 2(x - x_0)(y - y_0) f_{xy} + (y - y_0)^2 f_{yy} \right) \Big|_{(x_0+c(x-x_0),\, y_0+c(y-y_0))}}_{\text{error } E(x, y)}.$$

This equation reveals that

$$|E| \le \frac{1}{2} \left(|x - x_0|^2 |f_{xx}| + 2|x - x_0||y - y_0||f_{xy}| + |y - y_0|^2 |f_{yy}| \right).$$

Hence, if M is an upper bound for the values of $|f_{xx}|, |f_{xy}|$, and $|f_{yy}|$ on R,

$$|E| \le \frac{1}{2} \left(|x - x_0|^2 M + 2|x - x_0||y - y_0|M + |y - y_0|^2 M \right)$$

$$= \frac{1}{2} M(|x - x_0| + |y - y_0|)^2.$$

Taylor's Formula for Functions of Two Variables

The formulas derived earlier for F' and F'' can be obtained by applying to $f(x, y)$ the operators

$$\left(h\frac{\partial}{\partial x} + k\frac{\partial}{\partial y} \right) \qquad \text{and} \qquad \left(h\frac{\partial}{\partial x} + k\frac{\partial}{\partial y} \right)^2 = h^2 \frac{\partial^2}{\partial x^2} + 2hk \frac{\partial^2}{\partial x \partial y} + k^2 \frac{\partial^2}{\partial y^2}.$$

These are the first two instances of a more general formula,

$$F^{(n)}(t) = \frac{d^n}{dt^n} F(t) = \left(h\frac{\partial}{\partial x} + k\frac{\partial}{\partial y} \right)^n f(x, y), \tag{6}$$

which says that applying d^n/dt^n to $F(t)$ gives the same result as applying the operator

$$\left(h\frac{\partial}{\partial x} + k\frac{\partial}{\partial y} \right)^n$$

to $f(x, y)$ after expanding it by the Binomial Theorem.

If partial derivatives of f through order $n + 1$ are continuous throughout a rectangular region centered at (a, b), we may extend the Taylor formula for $F(t)$ to

$$F(t) = F(0) + F'(0)t + \frac{F''(0)}{2!} t^2 + \cdots + \frac{F^{(n)}(0)}{n!} t^{(n)} + \text{remainder},$$

and take $t = 1$ to obtain

$$F(1) = F(0) + F'(0) + \frac{F''(0)}{2!} + \cdots + \frac{F^{(n)}(0)}{n!} + \text{remainder}.$$

When we replace the first n derivatives on the right of this last series by their equivalent expressions from Equation (6) evaluated at $t = 0$ and add the appropriate remainder term, we arrive at the following formula.

Taylor's Formula for $f(x, y)$ at the Point (a, b)

Suppose $f(x, y)$ and its partial derivatives through order $n + 1$ are continuous throughout an open rectangular region R centered at a point (a, b). Then, throughout R,

$$f(a + h, b + k) = f(a, b) + (hf_x + kf_y)|_{(a, b)} + \frac{1}{2!} (h^2 f_{xx} + 2hk f_{xy} + k^2 f_{yy})|_{(a, b)}$$

$$+ \frac{1}{3!} (h^3 f_{xxx} + 3h^2 k f_{xxy} + 3hk^2 f_{xyy} + k^3 f_{yyy})|_{(a, b)} + \cdots + \frac{1}{n!} \left(h\frac{\partial}{\partial x} + k\frac{\partial}{\partial y} \right)^n f \bigg|_{(a, b)}$$

$$+ \frac{1}{(n + 1)!} \left(h\frac{\partial}{\partial x} + k\frac{\partial}{\partial y} \right)^{n+1} f \bigg|_{(a+ch, b+ck)}. \tag{7}$$

The first n derivative terms are evaluated at (a, b). The last term is evaluated at some point $(a + ch, b + ck)$ on the line segment joining (a, b) and $(a + h, b + k)$.

If $(a, b) = (0, 0)$ and we treat h and k as independent variables (denoting them now by x and y), then Equation (7) assumes the following form.

Taylor's Formula for $f(x, y)$ at the Origin

$$f(x, y) = f(0, 0) + xf_x + yf_y + \frac{1}{2!}(x^2 f_{xx} + 2xy f_{xy} + y^2 f_{yy})$$

$$+ \frac{1}{3!}(x^3 f_{xxx} + 3x^2 y f_{xxy} + 3xy^2 f_{xyy} + y^3 f_{yyy}) + \cdots + \frac{1}{n!}\left(x^n \frac{\partial^n f}{\partial x^n} + nx^{n-1}y \frac{\partial^n f}{\partial x^{n-1}\partial y} + \cdots + y^n \frac{\partial^n f}{\partial y^n}\right)$$

$$+ \frac{1}{(n+1)!}\left(x^{n+1}\frac{\partial^{n+1} f}{\partial x^{n+1}} + (n+1)x^n y \frac{\partial^{n+1} f}{\partial x^n \partial y} + \cdots + y^{n+1}\frac{\partial^{n+1} f}{\partial y^{n+1}}\right)\Bigg|_{(cx, cy)} \tag{8}$$

The first n derivative terms are evaluated at $(0, 0)$. The last term is evaluated at a point on the line segment joining the origin and (x, y).

Taylor's formula provides polynomial approximations of two-variable functions. The first n derivative terms give the polynomial; the last term gives the approximation error. The first three terms of Taylor's formula give the function's linearization. To improve on the linearization, we add higher-power terms.

EXAMPLE 1 Find a quadratic approximation to $f(x, y) = \sin x \sin y$ near the origin. How accurate is the approximation if $|x| \le 0.1$ and $|y| \le 0.1$?

Solution We take $n = 2$ in Equation (8):

$$f(x, y) = f(0, 0) + (xf_x + yf_y) + \frac{1}{2}(x^2 f_{xx} + 2xy f_{xy} + y^2 f_{yy})$$

$$+ \frac{1}{6}(x^3 f_{xxx} + 3x^2 y f_{xxy} + 3xy^2 f_{xyy} + y^3 f_{yyy})_{(cx, cy)}.$$

Calculating the values of the partial derivatives,

$$f(0, 0) = \sin x \sin y|_{(0,0)} = 0, \qquad f_{xx}(0, 0) = -\sin x \sin y|_{(0,0)} = 0,$$

$$f_x(0, 0) = \cos x \sin y|_{(0,0)} = 0, \qquad f_{xy}(0, 0) = \cos x \cos y|_{(0,0)} = 1,$$

$$f_y(0, 0) = \sin x \cos y|_{(0,0)} = 0, \qquad f_{yy}(0, 0) = -\sin x \sin y|_{(0,0)} = 0,$$

we have the result

$$\sin x \sin y \approx 0 + 0 + 0 + \frac{1}{2}(x^2(0) + 2xy(1) + y^2(0)), \qquad \text{or} \qquad \sin x \sin y \approx xy.$$

The error in the approximation is

$$E(x, y) = \frac{1}{6}(x^3 f_{xxx} + 3x^2 y f_{xxy} + 3xy^2 f_{xyy} + y^3 f_{yyy})|_{(cx, cy)}.$$

The third derivatives never exceed 1 in absolute value because they are products of sines and cosines. Also, $|x| \le 0.1$ and $|y| \le 0.1$. Hence

$$|E(x, y)| \le \frac{1}{6}((0.1)^3 + 3(0.1)^3 + 3(0.1)^3 + (0.1)^3) = \frac{8}{6}(0.1)^3 \le 0.00134$$

(rounded up). The error will not exceed 0.00134 if $|x| \le 0.1$ and $|y| \le 0.1$. ∎

Exercises 14.9

Finding Quadratic and Cubic Approximations

In Exercises 1–10, use Taylor's formula for $f(x, y)$ at the origin to find quadratic and cubic approximations of f near the origin.

1. $f(x, y) = xe^y$

2. $f(x, y) = e^x \cos y$

3. $f(x, y) = y \sin x$

4. $f(x, y) = \sin x \cos y$

5. $f(x, y) = e^x \ln(1 + y)$

6. $f(x, y) = \ln(2x + y + 1)$

7. $f(x, y) = \sin(x^2 + y^2)$

8. $f(x, y) = \cos(x^2 + y^2)$

9. $f(x, y) = \dfrac{1}{1 - x - y}$

10. $f(x, y) = \dfrac{1}{1 - x - y + xy}$

11. Use Taylor's formula to find a quadratic approximation of $f(x, y) = \cos x \cos y$ at the origin. Estimate the error in the approximation if $|x| \le 0.1$ and $|y| \le 0.1$.

12. Use Taylor's formula to find a quadratic approximation of $e^x \sin y$ at the origin. Estimate the error in the approximation if $|x| \le 0.1$ and $|y| \le 0.1$.

14.10 | Partial Derivatives with Constrained Variables

In finding partial derivatives of functions like $w = f(x, y)$, we have assumed x and y to be independent. In many applications, however, this is not the case. For example, the internal energy U of a gas may be expressed as a function $U = f(P, V, T)$ of pressure P, volume V, and temperature T. If the individual molecules of the gas do not interact, however, P, V, and T obey (and are constrained by) the ideal gas law

$$PV = nRT \qquad (n \text{ and } R \text{ constant}),$$

and fail to be independent. In this section we learn how to find partial derivatives in situations like this, which occur in economics, engineering, and physics.*

Decide Which Variables Are Dependent and Which Are Independent

If the variables in a function $w = f(x, y, z)$ are constrained by a relation like the one imposed on x, y, and z by the equation $z = x^2 + y^2$, the geometric meanings and the numerical values of the partial derivatives of f will depend on which variables are chosen to be dependent and which are chosen to be independent. To see how this choice can affect the outcome, we consider the calculation of $\partial w/\partial x$ when $w = x^2 + y^2 + z^2$ and $z = x^2 + y^2$.

EXAMPLE 1   Find $\partial w/\partial x$ if $w = x^2 + y^2 + z^2$ and $z = x^2 + y^2$.

Solution   We are given two equations in the four unknowns x, y, z, and w. Like many such systems, this one can be solved for two of the unknowns (the dependent variables) in terms of the others (the independent variables). In being asked for $\partial w/\partial x$, we are told that w is to be a dependent variable and x an independent variable. The possible choices for the other variables come down to

Dependent	Independent
w, z	x, y
w, y	x, z

In either case, we can express w explicitly in terms of the selected independent variables. We do this by using the second equation $z = x^2 + y^2$ to eliminate the remaining dependent variable in the first equation.

---

*This section is based on notes written for MIT by Arthur P. Mattuck.

In the first case, the remaining dependent variable is z. We eliminate it from the first equation by replacing it by $x^2 + y^2$. The resulting expression for w is

$$w = x^2 + y^2 + z^2 = x^2 + y^2 + (x^2 + y^2)^2$$
$$= x^2 + y^2 + x^4 + 2x^2y^2 + y^4$$

and

$$\frac{\partial w}{\partial x} = 2x + 4x^3 + 4xy^2. \tag{1}$$

This is the formula for $\partial w / \partial x$ when x and y are the independent variables.

In the second case, where the independent variables are x and z and the remaining dependent variable is y, we eliminate the dependent variable y in the expression for w by replacing y^2 in the second equation by $z - x^2$. This gives

$$w = x^2 + y^2 + z^2 = x^2 + (z - x^2) + z^2 = z + z^2$$

and

$$\frac{\partial w}{\partial x} = 0. \tag{2}$$

This is the formula for $\partial w / \partial x$ when x and z are the independent variables.

The formulas for $\partial w / \partial x$ in Equations (1) and (2) are genuinely different. We cannot change either formula into the other by using the relation $z = x^2 + y^2$. There is not just one $\partial w / \partial x$, there are two, and we see that the original instruction to find $\partial w / \partial x$ was incomplete. *Which* $\partial w / \partial x$? we ask.

The geometric interpretations of Equations (1) and (2) help to explain why the equations differ. The function $w = x^2 + y^2 + z^2$ measures the square of the distance from the point (x, y, z) to the origin. The condition $z = x^2 + y^2$ says that the point (x, y, z) lies on the paraboloid of revolution shown in Figure 14.58. What does it mean to calculate $\partial w / \partial x$ at a point $P(x, y, z)$ that can move only on this surface? What is the value of $\partial w / \partial x$ when the coordinates of P are, say, $(1, 0, 1)$?

If we take x and y to be independent, then we find $\partial w / \partial x$ by holding y fixed (at $y = 0$ in this case) and letting x vary. Hence, P moves along the parabola $z = x^2$ in the xz-plane. As P moves on this parabola, w, which is the square of the distance from P to the origin, changes. We calculate $\partial w / \partial x$ in this case (our first solution above) to be

$$\frac{\partial w}{\partial x} = 2x + 4x^3 + 4xy^2.$$

At the point $P(1, 0, 1)$, the value of this derivative is

$$\frac{\partial w}{\partial x} = 2 + 4 + 0 = 6.$$

If we take x and z to be independent, then we find $\partial w / \partial x$ by holding z fixed while x varies. Since the z-coordinate of P is 1, varying x moves P along a circle in the plane $z = 1$. As P moves along this circle, its distance from the origin remains constant, and w, being the square of this distance, does not change. That is,

$$\frac{\partial w}{\partial x} = 0,$$

as we found in our second solution. ∎

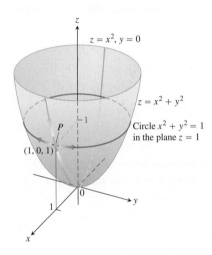

FIGURE 14.58 If P is constrained to lie on the paraboloid $z = x^2 + y^2$, the value of the partial derivative of $w = x^2 + y^2 + z^2$ with respect to x at P depends on the direction of motion (Example 1). (1) As x changes, with $y = 0$, P moves up or down the surface on the parabola $z = x^2$ in the xz-plane with $\partial w / \partial x = 2x + 4x^3$. (2) As x changes, with $z = 1$, P moves on the circle $x^2 + y^2 = 1$, $z = 1$, and $\partial w / \partial x = 0$.

How to Find $\partial w / \partial x$ When the Variables in $w = f(x, y, z)$ Are Constrained by Another Equation

As we saw in Example 1, a typical routine for finding $\partial w / \partial x$ when the variables in the function $w = f(x, y, z)$ are related by another equation has three steps. These steps apply to finding $\partial w / \partial y$ and $\partial w / \partial z$ as well.

> 1. *Decide* which variables are to be dependent and which are to be independent. (In practice, the decision is based on the physical or theoretical context of our work. In the exercises at the end of this section, we say which variables are which.)
> 2. *Eliminate* the other dependent variable(s) in the expression for w.
> 3. *Differentiate* as usual.

If we cannot carry out Step 2 after deciding which variables are dependent, we differentiate the equations as they are and try to solve for $\partial w / \partial x$ afterward. The next example shows how this is done.

EXAMPLE 2   Find $\partial w / \partial x$ at the point $(x, y, z) = (2, -1, 1)$ if

$$w = x^2 + y^2 + z^2, \qquad z^3 - xy + yz + y^3 = 1,$$

and x and y are the independent variables.

Solution   It is not convenient to eliminate z in the expression for w. We therefore differentiate both equations implicitly with respect to x, treating x and y as independent variables and w and z as dependent variables. This gives

$$\frac{\partial w}{\partial x} = 2x + 2z \frac{\partial z}{\partial x} \tag{3}$$

and

$$3z^2 \frac{\partial z}{\partial x} - y + y \frac{\partial z}{\partial x} + 0 = 0. \tag{4}$$

These equations may now be combined to express $\partial w / \partial x$ in terms of x, y, and z. We solve Equation (4) for $\partial z / \partial x$ to get

$$\frac{\partial z}{\partial x} = \frac{y}{y + 3z^2}$$

and substitute into Equation (3) to get

$$\frac{\partial w}{\partial x} = 2x + \frac{2yz}{y + 3z^2}.$$

The value of this derivative at $(x, y, z) = (2, -1, 1)$ is

$$\left(\frac{\partial w}{\partial x} \right)_{(2,-1,1)} = 2(2) + \frac{2(-1)(1)}{-1 + 3(1)^2} = 4 + \frac{-2}{2} = 3. \qquad \blacksquare$$

HISTORICAL BIOGRAPHY

Sonya Kovalevsky
(1850–1891)

Notation

To show what variables are assumed to be independent in calculating a derivative, we can use the following notation:

$$\left(\frac{\partial w}{\partial x} \right)_y \qquad \partial w / \partial x \text{ with } x \text{ and } y \text{ independent}$$

$$\left(\frac{\partial f}{\partial y} \right)_{x,t} \qquad \partial f / \partial y \text{ with } y, x \text{ and } t \text{ independent}$$

EXAMPLE 3   Find $(\partial w/\partial x)_{y,z}$ if $w = x^2 + y - z + \sin t$ and $x + y = t$.

Solution   With x, y, z independent, we have

$$t = x + y, \qquad w = x^2 + y - z + \sin(x + y)$$

$$\left(\frac{\partial w}{\partial x}\right)_{y,z} = 2x + 0 - 0 + \cos(x + y)\frac{\partial}{\partial x}(x + y)$$

$$= 2x + \cos(x + y). \qquad \blacksquare$$

Arrow Diagrams

In solving problems like the one in Example 3, it often helps to start with an arrow diagram that shows how the variables and functions are related. If

$$w = x^2 + y - z + \sin t \qquad \text{and} \qquad x + y = t$$

and we are asked to find $\partial w/\partial x$ when $x, y,$ and z are independent, the appropriate diagram is one like this:

$$\begin{pmatrix} x \\ y \\ z \end{pmatrix} \quad \rightarrow \quad \begin{pmatrix} x \\ y \\ z \\ t \end{pmatrix} \quad \rightarrow \quad w \tag{5}$$

Independent   Intermediate   Dependent
variables   variables   variable

To avoid confusion between the independent and intermediate variables with the same symbolic names in the diagram, it is helpful to rename the intermediate variables (so they are seen as *functions* of the independent variables). Thus, let $u = x$, $v = y$, and $s = z$ denote the renamed intermediate variables. With this notation, the arrow diagram becomes

$$\begin{pmatrix} x \\ y \\ z \end{pmatrix} \quad \rightarrow \quad \begin{pmatrix} u \\ v \\ s \\ t \end{pmatrix} \quad \rightarrow \quad w \tag{6}$$

Independent   Intermediate   Dependent
variables   variables and   variable
                 relations
                 $u = x$
                 $v = y$
                 $s = z$
                 $t = x + y$

The diagram shows the independent variables on the left, the intermediate variables and their relation to the independent variables in the middle, and the dependent variable on the right. The function w now becomes

$$w = u^2 + v - s + \sin t,$$

where

$$u = x, \qquad v = y, \qquad s = z, \qquad \text{and} \qquad t = x + y.$$

To find $\partial w/\partial x$, we apply the four-variable form of the Chain Rule to w, guided by the arrow diagram in Equation (6):

$$\frac{\partial w}{\partial x} = \frac{\partial w}{\partial u}\frac{\partial u}{\partial x} + \frac{\partial w}{\partial v}\frac{\partial v}{\partial x} + \frac{\partial w}{\partial s}\frac{\partial s}{\partial x} + \frac{\partial w}{\partial t}\frac{\partial t}{\partial x}$$

$$= (2u)(1) + (1)(0) + (-1)(0) + (\cos t)(1)$$

$$= 2u + \cos t$$

$$= 2x + \cos(x + y). \qquad \text{Substituting the original independent}$$
$$\text{variables } u = x \text{ and } t = x + y$$

Exercises 14.10

Finding Partial Derivatives with Constrained Variables

In Exercises 1–3, begin by drawing a diagram that shows the relations among the variables.

1. If $w = x^2 + y^2 + z^2$ and $z = x^2 + y^2$, find

    a. $\left(\dfrac{\partial w}{\partial y}\right)_z$ b. $\left(\dfrac{\partial w}{\partial z}\right)_x$ c. $\left(\dfrac{\partial w}{\partial z}\right)_y$.

2. If $w = x^2 + y - z + \sin t$ and $x + y = t$, find

    a. $\left(\dfrac{\partial w}{\partial y}\right)_{x,z}$ b. $\left(\dfrac{\partial w}{\partial y}\right)_{z,t}$ c. $\left(\dfrac{\partial w}{\partial z}\right)_{x,y}$

    d. $\left(\dfrac{\partial w}{\partial z}\right)_{y,t}$ e. $\left(\dfrac{\partial w}{\partial t}\right)_{x,z}$ f. $\left(\dfrac{\partial w}{\partial t}\right)_{y,z}$.

3. Let $U = f(P, V, T)$ be the internal energy of a gas that obeys the ideal gas law $PV = nRT$ (n and R constant). Find

    a. $\left(\dfrac{\partial U}{\partial P}\right)_V$ b. $\left(\dfrac{\partial U}{\partial T}\right)_V$.

4. Find

    a. $\left(\dfrac{\partial w}{\partial x}\right)_y$ b. $\left(\dfrac{\partial w}{\partial z}\right)_y$

    at the point $(x, y, z) = (0, 1, \pi)$ if

    $$w = x^2 + y^2 + z^2 \qquad \text{and} \qquad y \sin z + z \sin x = 0.$$

5. Find

    a. $\left(\dfrac{\partial w}{\partial y}\right)_x$ b. $\left(\dfrac{\partial w}{\partial y}\right)_z$

    at the point $(w, x, y, z) = (4, 2, 1, -1)$ if

    $$w = x^2 y^2 + yz - z^3 \qquad \text{and} \qquad x^2 + y^2 + z^2 = 6.$$

6. Find $(\partial u/\partial y)_x$ at the point $(u, v) = \left(\sqrt{2}, 1\right)$, if $x = u^2 + v^2$ and $y = uv$.

7. Suppose that $x^2 + y^2 = r^2$ and $x = r \cos \theta$, as in polar coordinates. Find

    $$\left(\frac{\partial x}{\partial r}\right)_\theta \qquad \text{and} \qquad \left(\frac{\partial r}{\partial x}\right)_y.$$

8. Suppose that

    $$w = x^2 - y^2 + 4z + t \qquad \text{and} \qquad x + 2z + t = 25.$$

    Show that the equations

    $$\frac{\partial w}{\partial x} = 2x - 1 \qquad \text{and} \qquad \frac{\partial w}{\partial x} = 2x - 2$$

each give $\partial w/\partial x$, depending on which variables are chosen to be dependent and which variables are chosen to be independent. Identify the independent variables in each case.

Theory and Examples

9. Establish the fact, widely used in hydrodynamics, that if $f(x, y, z) = 0$, then

    $$\left(\frac{\partial x}{\partial y}\right)_z \left(\frac{\partial y}{\partial z}\right)_x \left(\frac{\partial z}{\partial x}\right)_y = -1.$$

    (*Hint:* Express all the derivatives in terms of the formal partial derivatives $\partial f/\partial x$, $\partial f/\partial y$, and $\partial f/\partial z$.)

10. If $z = x + f(u)$, where $u = xy$, show that

    $$x\frac{\partial z}{\partial x} - y\frac{\partial z}{\partial y} = x.$$

11. Suppose that the equation $g(x, y, z) = 0$ determines z as a differentiable function of the independent variables x and y and that $g_z \neq 0$. Show that

    $$\left(\frac{\partial z}{\partial y}\right)_x = -\frac{\partial g/\partial y}{\partial g/\partial z}.$$

12. Suppose that $f(x, y, z, w) = 0$ and $g(x, y, z, w) = 0$ determine z and w as differentiable functions of the independent variables x and y, and suppose that

    $$\frac{\partial f}{\partial z}\frac{\partial g}{\partial w} - \frac{\partial f}{\partial w}\frac{\partial g}{\partial z} \neq 0.$$

    Show that

    $$\left(\frac{\partial z}{\partial x}\right)_y = -\frac{\dfrac{\partial f}{\partial x}\dfrac{\partial g}{\partial w} - \dfrac{\partial f}{\partial w}\dfrac{\partial g}{\partial x}}{\dfrac{\partial f}{\partial z}\dfrac{\partial g}{\partial w} - \dfrac{\partial f}{\partial w}\dfrac{\partial g}{\partial z}}$$

    and

    $$\left(\frac{\partial w}{\partial y}\right)_x = -\frac{\dfrac{\partial f}{\partial z}\dfrac{\partial g}{\partial y} - \dfrac{\partial f}{\partial y}\dfrac{\partial g}{\partial z}}{\dfrac{\partial f}{\partial z}\dfrac{\partial g}{\partial w} - \dfrac{\partial f}{\partial w}\dfrac{\partial g}{\partial z}}.$$

Chapter 14 Questions to Guide Your Review

1. What is a real-valued function of two independent variables? Three independent variables? Give examples.

2. What does it mean for sets in the plane or in space to be open? Closed? Give examples. Give examples of sets that are neither open nor closed.

3. How can you display the values of a function $f(x, y)$ of two independent variables graphically? How do you do the same for a function $f(x, y, z)$ of three independent variables?

4. What does it mean for a function $f(x, y)$ to have limit L as $(x, y) \to (x_0, y_0)$? What are the basic properties of limits of functions of two independent variables?

5. When is a function of two (three) independent variables continuous at a point in its domain? Give examples of functions that are continuous at some points but not others.

6. What can be said about algebraic combinations and composites of continuous functions?

7. Explain the two-path test for nonexistence of limits.

8. How are the partial derivatives $\partial f/\partial x$ and $\partial f/\partial y$ of a function $f(x, y)$ defined? How are they interpreted and calculated?

9. How does the relation between first partial derivatives and continuity of functions of two independent variables differ from the relation between first derivatives and continuity for real-valued functions of a single independent variable? Give an example.

10. What is the Mixed Derivative Theorem for mixed second-order partial derivatives? How can it help in calculating partial derivatives of second and higher orders? Give examples.

11. What does it mean for a function $f(x, y)$ to be differentiable? What does the Increment Theorem say about differentiability?

12. How can you sometimes decide from examining f_x and f_y that a function $f(x, y)$ is differentiable? What is the relation between the differentiability of f and the continuity of f at a point?

13. What is the general Chain Rule? What form does it take for functions of two independent variables? Three independent variables? Functions defined on surfaces? How do you diagram these different forms? Give examples. What pattern enables one to remember all the different forms?

14. What is the derivative of a function $f(x, y)$ at a point P_0 in the direction of a unit vector **u**? What rate does it describe? What geometric interpretation does it have? Give examples.

15. What is the gradient vector of a differentiable function $f(x, y)$? How is it related to the function's directional derivatives? State the analogous results for functions of three independent variables.

16. How do you find the tangent line at a point on a level curve of a differentiable function $f(x, y)$? How do you find the tangent plane and normal line at a point on a level surface of a differentiable function $f(x, y, z)$? Give examples.

17. How can you use directional derivatives to estimate change?

18. How do you linearize a function $f(x, y)$ of two independent variables at a point (x_0, y_0)? Why might you want to do this? How do you linearize a function of three independent variables?

19. What can you say about the accuracy of linear approximations of functions of two (three) independent variables?

20. If (x, y) moves from (x_0, y_0) to a point $(x_0 + dx, y_0 + dy)$ nearby, how can you estimate the resulting change in the value of a differentiable function $f(x, y)$? Give an example.

21. How do you define local maxima, local minima, and saddle points for a differentiable function $f(x, y)$? Give examples.

22. What derivative tests are available for determining the local extreme values of a function $f(x, y)$? How do they enable you to narrow your search for these values? Give examples.

23. How do you find the extrema of a continuous function $f(x, y)$ on a closed bounded region of the xy-plane? Give an example.

24. Describe the method of Lagrange multipliers and give examples.

25. How does Taylor's formula for a function $f(x, y)$ generate polynomial approximations and error estimates?

26. If $w = f(x, y, z)$, where the variables x, y, and z are constrained by an equation $g(x, y, z) = 0$, what is the meaning of the notation $(\partial w/\partial x)_y$? How can an arrow diagram help you calculate this partial derivative with constrained variables? Give examples.

Chapter 14 Practice Exercises

Domain, Range, and Level Curves

In Exercises 1–4, find the domain and range of the given function and identify its level curves. Sketch a typical level curve.

1. $f(x, y) = 9x^2 + y^2$
2. $f(x, y) = e^{x+y}$
3. $g(x, y) = 1/xy$
4. $g(x, y) = \sqrt{x^2 - y}$

In Exercises 5–8, find the domain and range of the given function and identify its level surfaces. Sketch a typical level surface.

5. $f(x, y, z) = x^2 + y^2 - z$
6. $g(x, y, z) = x^2 + 4y^2 + 9z^2$
7. $h(x, y, z) = \dfrac{1}{x^2 + y^2 + z^2}$
8. $k(x, y, z) = \dfrac{1}{x^2 + y^2 + z^2 + 1}$

Evaluating Limits

Find the limits in Exercises 9–14.

9. $\displaystyle \lim_{(x,y)\to(\pi,\, \ln 2)} e^y \cos x$

10. $\displaystyle \lim_{(x,y)\to(0,0)} \frac{2 + y}{x + \cos y}$

11. $\lim\limits_{(x,y)\to(1,1)} \dfrac{x - y}{x^2 - y^2}$

12. $\lim\limits_{(x,y)\to(1,1)} \dfrac{x^3 y^3 - 1}{xy - 1}$

13. $\lim\limits_{P\to(1,-1,e)} \ln|x + y + z|$

14. $\lim\limits_{P\to(1,-1,-1)} \tan^{-1}(x + y + z)$

By considering different paths of approach, show that the limits in Exercises 15 and 16 do not exist.

15. $\lim\limits_{\substack{(x,y)\to(0,0) \\ y\neq x^2}} \dfrac{y}{x^2 - y}$

16. $\lim\limits_{\substack{(x,y)\to(0,0) \\ xy\neq 0}} \dfrac{x^2 + y^2}{xy}$

17. Continuous extension Let $f(x, y) = (x^2 - y^2)/(x^2 + y^2)$ for $(x, y) \neq (0, 0)$. Is it possible to define $f(0, 0)$ in a way that makes f continuous at the origin? Why?

18. Continuous extension Let

$$f(x, y) = \begin{cases} \dfrac{\sin(x - y)}{|x| + |y|}, & |x| + |y| \neq 0 \\ 0, & (x, y) = (0, 0). \end{cases}$$

Is f continuous at the origin? Why?

Partial Derivatives

In Exercises 19–24, find the partial derivative of the function with respect to each variable.

19. $g(r, \theta) = r\cos\theta + r\sin\theta$

20. $f(x, y) = \dfrac{1}{2}\ln(x^2 + y^2) + \tan^{-1}\dfrac{y}{x}$

21. $f(R_1, R_2, R_3) = \dfrac{1}{R_1} + \dfrac{1}{R_2} + \dfrac{1}{R_3}$

22. $h(x, y, z) = \sin(2\pi x + y - 3z)$

23. $P(n, R, T, V) = \dfrac{nRT}{V}$ (the ideal gas law)

24. $f(r, l, T, w) = \dfrac{1}{2rl}\sqrt{\dfrac{T}{\pi w}}$

Second-Order Partials

Find the second-order partial derivatives of the functions in Exercises 25–28.

25. $g(x, y) = y + \dfrac{x}{y}$

26. $g(x, y) = e^x + y\sin x$

27. $f(x, y) = x + xy - 5x^3 + \ln(x^2 + 1)$

28. $f(x, y) = y^2 - 3xy + \cos y + 7e^y$

Chain Rule Calculations

29. Find dw/dt at $t = 0$ if $w = \sin(xy + \pi)$, $x = e^t$, and $y = \ln(t + 1)$.

30. Find dw/dt at $t = 1$ if $w = xe^y + y\sin z - \cos z$, $x = 2\sqrt{t}$, $y = t - 1 + \ln t$, and $z = \pi t$.

31. Find $\partial w/\partial r$ and $\partial w/\partial s$ when $r = \pi$ and $s = 0$ if $w = \sin(2x - y)$, $x = r + \sin s$, $y = rs$.

32. Find $\partial w/\partial u$ and $\partial w/\partial v$ when $u = v = 0$ if $w = \ln\sqrt{1 + x^2} - \tan^{-1}x$ and $x = 2e^u\cos v$.

33. Find the value of the derivative of $f(x, y, z) = xy + yz + xz$ with respect to t on the curve $x = \cos t$, $y = \sin t$, $z = \cos 2t$ at $t = 1$.

34. Show that if $w = f(s)$ is any differentiable function of s and if $s = y + 5x$, then

$$\frac{\partial w}{\partial x} - 5\frac{\partial w}{\partial y} = 0.$$

Implicit Differentiation

Assuming that the equations in Exercises 35 and 36 define y as a differentiable function of x, find the value of dy/dx at point P.

35. $1 - x - y^2 - \sin xy = 0$, $P(0, 1)$

36. $2xy + e^{x+y} - 2 = 0$, $P(0, \ln 2)$

Directional Derivatives

In Exercises 37–40, find the directions in which f increases and decreases most rapidly at P_0 and find the derivative of f in each direction. Also, find the derivative of f at P_0 in the direction of the vector $\mathbf{v}$.

37. $f(x, y) = \cos x\cos y$, $P_0(\pi/4, \pi/4)$, $\mathbf{v} = 3\mathbf{i} + 4\mathbf{j}$

38. $f(x, y) = x^2 e^{-2y}$, $P_0(1, 0)$, $\mathbf{v} = \mathbf{i} + \mathbf{j}$

39. $f(x, y, z) = \ln(2x + 3y + 6z)$, $P_0(-1, -1, 1)$, $\mathbf{v} = 2\mathbf{i} + 3\mathbf{j} + 6\mathbf{k}$

40. $f(x, y, z) = x^2 + 3xy - z^2 + 2y + z + 4$, $P_0(0, 0, 0)$, $\mathbf{v} = \mathbf{i} + \mathbf{j} + \mathbf{k}$

41. Derivative in velocity direction Find the derivative of $f(x, y, z) = xyz$ in the direction of the velocity vector of the helix

$$\mathbf{r}(t) = (\cos 3t)\mathbf{i} + (\sin 3t)\mathbf{j} + 3t\mathbf{k}$$

at $t = \pi/3$.

42. Maximum directional derivative What is the largest value that the directional derivative of $f(x, y, z) = xyz$ can have at the point $(1, 1, 1)$?

43. Directional derivatives with given values At the point $(1, 2)$, the function $f(x, y)$ has a derivative of 2 in the direction toward $(2, 2)$ and a derivative of -2 in the direction toward $(1, 1)$.
  a. Find $f_x(1, 2)$ and $f_y(1, 2)$.
  b. Find the derivative of f at $(1, 2)$ in the direction toward the point $(4, 6)$.

44. Which of the following statements are true if $f(x, y)$ is differentiable at (x_0, y_0)? Give reasons for your answers.
  a. If $\mathbf{u}$ is a unit vector, the derivative of f at (x_0, y_0) in the direction of $\mathbf{u}$ is $(f_x(x_0, y_0)\mathbf{i} + f_y(x_0, y_0)\mathbf{j}) \cdot \mathbf{u}$.
  b. The derivative of f at (x_0, y_0) in the direction of $\mathbf{u}$ is a vector.
  c. The directional derivative of f at (x_0, y_0) has its greatest value in the direction of ∇f.
  d. At (x_0, y_0), vector ∇f is normal to the curve $f(x, y) = f(x_0, y_0)$.

Gradients, Tangent Planes, and Normal Lines

In Exercises 45 and 46, sketch the surface $f(x, y, z) = c$ together with ∇f at the given points.

45. $x^2 + y + z^2 = 0$; $(0, -1, \pm 1)$, $(0, 0, 0)$

46. $y^2 + z^2 = 4$; $(2, \pm 2, 0)$, $(2, 0, \pm 2)$

In Exercises 47 and 48, find an equation for the plane tangent to the level surface $f(x, y, z) = c$ at the point P_0. Also, find parametric equations for the line that is normal to the surface at P_0.

47. $x^2 - y - 5z = 0$, $P_0(2, -1, 1)$

48. $x^2 + y^2 + z = 4$, $P_0(1, 1, 2)$

In Exercises 49 and 50, find an equation for the plane tangent to the surface $z = f(x, y)$ at the given point.

49. $z = \ln(x^2 + y^2)$, $(0, 1, 0)$

50. $z = 1/(x^2 + y^2)$, $(1, 1, 1/2)$

In Exercises 51 and 52, find equations for the lines that are tangent and normal to the level curve $f(x, y) = c$ at the point P_0. Then sketch the lines and level curve together with ∇f at P_0.

51. $y - \sin x = 1$, $P_0(\pi, 1)$ **52.** $\dfrac{y^2}{2} - \dfrac{x^2}{2} = \dfrac{3}{2}$, $P_0(1, 2)$

Tangent Lines to Curves

In Exercises 53 and 54, find parametric equations for the line that is tangent to the curve of intersection of the surfaces at the given point.

53. Surfaces: $x^2 + 2y + 2z = 4$, $y = 1$

  Point: $(1, 1, 1/2)$

54. Surfaces: $x + y^2 + z = 2$, $y = 1$

  Point: $(1/2, 1, 1/2)$

Linearizations

In Exercises 55 and 56, find the linearization $L(x, y)$ of the function $f(x, y)$ at the point P_0. Then find an upper bound for the magnitude of the error E in the approximation $f(x, y) \approx L(x, y)$ over the rectangle R.

55. $f(x, y) = \sin x \cos y$, $P_0(\pi/4, \pi/4)$

  R: $\left| x - \dfrac{\pi}{4} \right| \leq 0.1$, $\left| y - \dfrac{\pi}{4} \right| \leq 0.1$

56. $f(x, y) = xy - 3y^2 + 2$, $P_0(1, 1)$

  R: $|x - 1| \leq 0.1$, $|y - 1| \leq 0.2$

Find the linearizations of the functions in Exercises 57 and 58 at the given points.

57. $f(x, y, z) = xy + 2yz - 3xz$ at $(1, 0, 0)$ and $(1, 1, 0)$

58. $f(x, y, z) = \sqrt{2} \cos x \sin(y + z)$ at $(0, 0, \pi/4)$ and $(\pi/4, \pi/4, 0)$

Estimates and Sensitivity to Change

59. Measuring the volume of a pipeline   You plan to calculate the volume inside a stretch of pipeline that is about 36 in. in diameter and 1 mile long. With which measurement should you be more careful, the length or the diameter? Why?

60. Sensitivity to change   Is $f(x, y) = x^2 - xy + y^2 - 3$ more sensitive to changes in x or to changes in y when it is near the point $(1, 2)$? How do you know?

61. Change in an electrical circuit   Suppose that the current I (amperes) in an electrical circuit is related to the voltage V (volts) and the resistance R (ohms) by the equation $I = V/R$. If the voltage drops from 24 to 23 volts and the resistance drops from 100 to 80 ohms, will I increase or decrease? By about how much? Is the change in I more sensitive to change in the voltage or to change in the resistance? How do you know?

62. Maximum error in estimating the area of an ellipse   If $a = 10$ cm and $b = 16$ cm to the nearest millimeter, what should you expect the maximum percentage error to be in the calculated area $A = \pi ab$ of the ellipse $x^2/a^2 + y^2/b^2 = 1$?

63. Error in estimating a product   Let $y = uv$ and $z = u + v$, where u and v are positive independent variables.

  a. If u is measured with an error of 2% and v with an error of 3%, about what is the percentage error in the calculated value of y?

  b. Show that the percentage error in the calculated value of z is less than the percentage error in the value of y.

64. Cardiac index   To make different people comparable in studies of cardiac output, researchers divide the measured cardiac output by the body surface area to find the *cardiac index* C:

$$C = \frac{\text{cardiac output}}{\text{body surface area}}.$$

The body surface area B of a person with weight w and height h is approximated by the formula

$$B = 71.84 w^{0.425} h^{0.725},$$

which gives B in square centimeters when w is measured in kilograms and h in centimeters. You are about to calculate the cardiac index of a person 180 cm tall, weighing 70 kg, with cardiac output of 7 L/min. Which will have a greater effect on the calculation, a 1-kg error in measuring the weight or a 1-cm error in measuring the height?

Local Extrema

Test the functions in Exercises 65–70 for local maxima and minima and saddle points. Find each function's value at these points.

65. $f(x, y) = x^2 - xy + y^2 + 2x + 2y - 4$

66. $f(x, y) = 5x^2 + 4xy - 2y^2 + 4x - 4y$

67. $f(x, y) = 2x^3 + 3xy + 2y^3$

68. $f(x, y) = x^3 + y^3 - 3xy + 15$

69. $f(x, y) = x^3 + y^3 + 3x^2 - 3y^2$

70. $f(x, y) = x^4 - 8x^2 + 3y^2 - 6y$

Absolute Extrema

In Exercises 71–78, find the absolute maximum and minimum values of f on the region R.

71. $f(x, y) = x^2 + xy + y^2 - 3x + 3y$

  R: The triangular region cut from the first quadrant by the line $x + y = 4$

72. $f(x, y) = x^2 - y^2 - 2x + 4y + 1$

  R: The rectangular region in the first quadrant bounded by the coordinate axes and the lines $x = 4$ and $y = 2$

73. $f(x, y) = y^2 - xy - 3y + 2x$

  R: The square region enclosed by the lines $x = \pm 2$ and $y = \pm 2$

74. $f(x, y) = 2x + 2y - x^2 - y^2$

  R: The square region bounded by the coordinate axes and the lines $x = 2, y = 2$ in the first quadrant

75. $f(x, y) = x^2 - y^2 - 2x + 4y$

  R: The triangular region bounded below by the x-axis, above by the line $y = x + 2$, and on the right by the line $x = 2$

76. $f(x, y) = 4xy - x^4 - y^4 + 16$

R: The triangular region bounded below by the line $y = -2$, above by the line $y = x$, and on the right by the line $x = 2$

77. $f(x, y) = x^3 + y^3 + 3x^2 - 3y^2$

R: The square region enclosed by the lines $x = \pm 1$ and $y = \pm 1$

78. $f(x, y) = x^3 + 3xy + y^3 + 1$

R: The square region enclosed by the lines $x = \pm 1$ and $y = \pm 1$

Lagrange Multipliers

79. Extrema on a circle Find the extreme values of $f(x, y) = x^3 + y^2$ on the circle $x^2 + y^2 = 1$.

80. Extrema on a circle Find the extreme values of $f(x, y) = xy$ on the circle $x^2 + y^2 = 1$.

81. Extrema in a disk Find the extreme values of $f(x, y) = x^2 + 3y^2 + 2y$ on the unit disk $x^2 + y^2 \le 1$.

82. Extrema in a disk Find the extreme values of $f(x, y) = x^2 + y^2 - 3x - xy$ on the disk $x^2 + y^2 \le 9$.

83. Extrema on a sphere Find the extreme values of $f(x, y, z) = x - y + z$ on the unit sphere $x^2 + y^2 + z^2 = 1$.

84. Minimum distance to origin Find the points on the surface $x^2 - zy = 4$ closest to the origin.

85. Minimizing cost of a box A closed rectangular box is to have volume V cm^3. The cost of the material used in the box is a cents/cm^2 for top and bottom, b cents/cm^2 for front and back, and c cents/cm^2 for the remaining sides. What dimensions minimize the total cost of materials?

86. Least volume Find the plane $x/a + y/b + z/c = 1$ that passes through the point $(2, 1, 2)$ and cuts off the least volume from the first octant.

87. Extrema on curve of intersecting surfaces Find the extreme values of $f(x, y, z) = x(y + z)$ on the curve of intersection of the right circular cylinder $x^2 + y^2 = 1$ and the hyperbolic cylinder $xz = 1$.

88. Minimum distance to origin on curve of intersecting plane and cone Find the point closest to the origin on the curve of intersection of the plane $x + y + z = 1$ and the cone $z^2 = 2x^2 + 2y^2$.

Partial Derivatives with Constrained Variables

In Exercises 89 and 90, begin by drawing a diagram that shows the relations among the variables.

89. If $w = x^2 e^{yz}$ and $z = x^2 - y^2$ find

a. $\left(\dfrac{\partial w}{\partial y} \right)_z$ **b.** $\left(\dfrac{\partial w}{\partial z} \right)_x$ **c.** $\left(\dfrac{\partial w}{\partial z} \right)_y$.

90. Let $U = f(P, V, T)$ be the internal energy of a gas that obeys the ideal gas law $PV = nRT$ (n and R constant). Find

a. $\left(\dfrac{\partial U}{\partial T} \right)_P$ **b.** $\left(\dfrac{\partial U}{\partial V} \right)_T$.

Theory and Examples

91. Let $w = f(r, \theta)$, $r = \sqrt{x^2 + y^2}$, and $\theta = \tan^{-1}(y/x)$. Find $\partial w/\partial x$ and $\partial w/\partial y$ and express your answers in terms of r and θ.

92. Let $z = f(u, v)$, $u = ax + by$, and $v = ax - by$. Express z_x and z_y in terms of f_u, f_v, and the constants a and b.

93. If a and b are constants, $w = u^3 + \tanh u + \cos u$, and $u = ax + by$, show that

$$a \frac{\partial w}{\partial y} = b \frac{\partial w}{\partial x}.$$

94. Using the Chain Rule If $w = \ln(x^2 + y^2 + 2z)$, $x = r + s$, $y = r - s$, and $z = 2rs$, find w_r and w_s by the Chain Rule. Then check your answer another way.

95. Angle between vectors The equations $e^u \cos v - x = 0$ and $e^u \sin v - y = 0$ define u and v as differentiable functions of x and y. Show that the angle between the vectors

$$\frac{\partial u}{\partial x} \mathbf{i} + \frac{\partial u}{\partial y} \mathbf{j} \quad \text{and} \quad \frac{\partial v}{\partial x} \mathbf{i} + \frac{\partial v}{\partial y} \mathbf{j}$$

is constant.

96. Polar coordinates and second derivatives Introducing polar coordinates $x = r \cos \theta$ and $y = r \sin \theta$ changes $f(x, y)$ to $g(r, \theta)$. Find the value of $\partial^2 g/\partial \theta^2$ at the point $(r, \theta) = (2, \pi/2)$, given that

$$\frac{\partial f}{\partial x} = \frac{\partial f}{\partial y} = \frac{\partial^2 f}{\partial x^2} = \frac{\partial^2 f}{\partial y^2} = 1$$

at that point.

97. Normal line parallel to a plane Find the points on the surface

$$(y + z)^2 + (z - x)^2 = 16$$

where the normal line is parallel to the yz-plane.

98. Tangent plane parallel to xy-plane Find the points on the surface

$$xy + yz + zx - x - z^2 = 0$$

where the tangent plane is parallel to the xy-plane.

99. When gradient is parallel to position vector Suppose that $\nabla f(x, y, z)$ is always parallel to the position vector $x\mathbf{i} + y\mathbf{j} + z\mathbf{k}$. Show that $f(0, 0, a) = f(0, 0, -a)$ for any a.

100. One-sided directional derivative in all directions, but no gradient The *one-sided directional derivative of f at P* (x_0, y_0, z_0) *in the direction* $\mathbf{u} = u_1\mathbf{i} + u_2\mathbf{j} + u_3\mathbf{k}$ is the number

$$\lim_{s \to 0^+} \frac{f(x_0 + su_1, y_0 + su_2, z_0 + su_3) - f(x_0, y_0, z_0)}{s}.$$

Show that the one-sided directional derivative of

$$f(x, y, z) = \sqrt{x^2 + y^2 + z^2}$$

at the origin equals 1 in any direction but that f has no gradient vector at the origin.

101. Normal line through origin Show that the line normal to the surface $xy + z = 2$ at the point $(1, 1, 1)$ passes through the origin.

102. Tangent plane and normal line

a. Sketch the surface $x^2 - y^2 + z^2 = 4$.

b. Find a vector normal to the surface at $(2, -3, 3)$. Add the vector to your sketch.

c. Find equations for the tangent plane and normal line at $(2, -3, 3)$.

We are interested in what happens to these Riemann sums as the widths and heights of all the small rectangles in the partition of R approach zero. The **norm** of a partition P, written $\|P\|$, is the largest width or height of any rectangle in the partition. If $\|P\| = 0.1$ then all the rectangles in the partition of R have width at most 0.1 and height at most 0.1. Sometimes the Riemann sums converge as the norm of P goes to zero, written $\|P\| \to 0$. The resulting limit is then written as

$$\lim_{\|P\| \to 0} \sum_{k=1}^{n} f(x_k, y_k)\, \Delta A_k.$$

As $\|P\| \to 0$ and the rectangles get narrow and short, their number n increases, so we can also write this limit as

$$\lim_{n \to \infty} \sum_{k=1}^{n} f(x_k, y_k)\, \Delta A_k,$$

with the understanding that $\|P\| \to 0$, and hence $\Delta A_k \to 0$, as $n \to \infty$.

There are many choices involved in a limit of this kind. The collection of small rectangles is determined by the grid of vertical and horizontal lines that determine a rectangular partition of R. In each of the resulting small rectangles there is a choice of an arbitrary point (x_k, y_k) at which f is evaluated. These choices together determine a single Riemann sum. To form a limit, we repeat the whole process again and again, choosing partitions whose rectangle widths and heights both go to zero and whose number goes to infinity.

When a limit of the sums S_n exists, giving the same limiting value no matter what choices are made, then the function f is said to be **integrable** and the limit is called the **double integral** of f over R, written as

$$\iint\limits_{R} f(x, y)\, dA \qquad \text{or} \qquad \iint\limits_{R} f(x, y)\, dx\, dy.$$

It can be shown that if $f(x, y)$ is a continuous function throughout R, then f is integrable, as in the single-variable case discussed in Chapter 5. Many discontinuous functions are also integrable, including functions that are discontinuous only on a finite number of points or smooth curves. We leave the proof of these facts to a more advanced text.

Double Integrals as Volumes

When $f(x, y)$ is a positive function over a rectangular region R in the xy-plane, we may interpret the double integral of f over R as the volume of the 3-dimensional solid region over the xy-plane bounded below by R and above by the surface $z = f(x, y)$ (Figure 15.2). Each term $f(x_k, y_k)\Delta A_k$ in the sum $S_n = \sum f(x_k, y_k)\Delta A_k$ is the volume of a vertical rectangular box that approximates the volume of the portion of the solid that stands directly above the base ΔA_k. The sum S_n thus approximates what we want to call the total volume of the solid. We *define* this volume to be

$$\text{Volume} = \lim_{n \to \infty} S_n = \iint\limits_{R} f(x, y)\, dA,$$

where $\Delta A_k \to 0$ as $n \to \infty$.

As you might expect, this more general method of calculating volume agrees with the methods in Chapter 6, but we do not prove this here. Figure 15.3 shows Riemann sum approximations to the volume becoming more accurate as the number n of boxes increases.

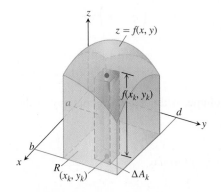

FIGURE 15.2 Approximating solids with rectangular boxes leads us to define the volumes of more general solids as double integrals. The volume of the solid shown here is the double integral of $f(x, y)$ over the base region R.

(a) $n = 16$ (b) $n = 64$ (c) $n = 256$

FIGURE 15.3 As n increases, the Riemann sum approximations approach the total volume of the solid shown in Figure 15.2.

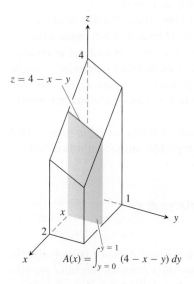

FIGURE 15.4 To obtain the cross-sectional area $A(x)$, we hold x fixed and integrate with respect to y.

Fubini's Theorem for Calculating Double Integrals

Suppose that we wish to calculate the volume under the plane $z = 4 - x - y$ over the rectangular region $R: 0 \leq x \leq 2, 0 \leq y \leq 1$ in the xy-plane. If we apply the method of slicing from Section 6.1, with slices perpendicular to the x-axis (Figure 15.4), then the volume is

$$\int_{x=0}^{x=2} A(x) \ dx, \tag{1}$$

where $A(x)$ is the cross-sectional area at x. For each value of x, we may calculate $A(x)$ as the integral

$$A(x) = \int_{y=0}^{y=1} (4 - x - y) \ dy, \tag{2}$$

which is the area under the curve $z = 4 - x - y$ in the plane of the cross-section at x. In calculating $A(x)$, x is held fixed and the integration takes place with respect to y. Combining Equations (1) and (2), we see that the volume of the entire solid is

$$
\begin{aligned}
\text{Volume} &= \int_{x=0}^{x=2} A(x) \ dx = \int_{x=0}^{x=2} \left(\int_{y=0}^{y=1} (4 - x - y) \, dy \right) dx \\
&= \int_{x=0}^{x=2} \left[4y - xy - \frac{y^2}{2} \right]_{y=0}^{y=1} dx = \int_{x=0}^{x=2} \left(\frac{7}{2} - x \right) dx \\
&= \left[\frac{7}{2}x - \frac{x^2}{2} \right]_{0}^{2} = 5.
\end{aligned}
\tag{3}
$$

If we just wanted to write a formula for the volume, without carrying out any of the integrations, we could write

$$\text{Volume} = \int_0^2 \int_0^1 (4 - x - y) \ dy \ dx.$$

The expression on the right, called an **iterated** or **repeated integral**, says that the volume is obtained by integrating $4 - x - y$ with respect to y from $y = 0$ to $y = 1$, holding x fixed, and then integrating the resulting expression in x with respect to x from $x = 0$ to $x = 2$. The limits of integration 0 and 1 are associated with y, so they are placed on the integral closest to dy. The other limits of integration, 0 and 2, are associated with the variable x, so they are placed on the outside integral symbol that is paired with dx.

What would have happened if we had calculated the volume by slicing with planes perpendicular to the y-axis (Figure 15.5)? As a function of y, the typical cross-sectional area is

$$A(y) = \int_{x=0}^{x=2} (4 - x - y)\, dx = \left[4x - \frac{x^2}{2} - xy \right]_{x=0}^{x=2} = 6 - 2y. \tag{4}$$

The volume of the entire solid is therefore

$$\text{Volume} = \int_{y=0}^{y=1} A(y)\, dy = \int_{y=0}^{y=1} (6 - 2y)\, dy = \left[6y - y^2 \right]_0^1 = 5,$$

in agreement with our earlier calculation.

Again, we may give a formula for the volume as an iterated integral by writing

$$\text{Volume} = \int_0^1 \int_0^2 (4 - x - y)\, dx\, dy.$$

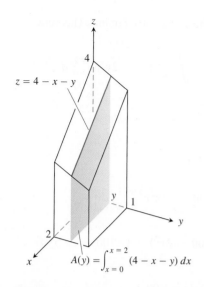

$$z = 4 - x - y$$

$$A(y) = \int_{x=0}^{x=2} (4 - x - y)\, dx$$

FIGURE 15.5 To obtain the cross-sectional area $A(y)$, we hold y fixed and integrate with respect to x.

The expression on the right says we can find the volume by integrating $4 - x - y$ with respect to x from $x = 0$ to $x = 2$ as in Equation (4) and integrating the result with respect to y from $y = 0$ to $y = 1$. In this iterated integral, the order of integration is first x and then y, the reverse of the order in Equation (3).

What do these two volume calculations with iterated integrals have to do with the double integral

$$\iint_R (4 - x - y)\, dA$$

over the rectangle $R \colon 0 \le x \le 2,\, 0 \le y \le 1$? The answer is that both iterated integrals give the value of the double integral. This is what we would reasonably expect, since the double integral measures the volume of the same region as the two iterated integrals. A theorem published in 1907 by Guido Fubini says that the double integral of any continuous function over a rectangle can be calculated as an iterated integral in either order of integration. (Fubini proved his theorem in greater generality, but this is what it says in our setting.)

HISTORICAL BIOGRAPHY

Guido Fubini
(1879–1943)

THEOREM 1—Fubini's Theorem (First Form)   If $f(x, y)$ is continuous throughout the rectangular region $R \colon a \le x \le b,\, c \le y \le d$, then

$$\iint_R f(x, y)\, dA = \int_c^d \int_a^b f(x, y)\, dx\, dy = \int_a^b \int_c^d f(x, y)\, dy\, dx.$$

Fubini's Theorem says that double integrals over rectangles can be calculated as iterated integrals. Thus, we can evaluate a double integral by integrating with respect to one variable at a time.

Fubini's Theorem also says that we may calculate the double integral by integrating in *either* order, a genuine convenience. When we calculate a volume by slicing, we may use either planes perpendicular to the x-axis or planes perpendicular to the y-axis.

EXAMPLE 1   Calculate $\iint_R f(x, y)\, dA$ for

$$f(x, y) = 100 - 6x^2y \quad \text{and} \quad R \colon \ 0 \le x \le 2,\ -1 \le y \le 1.$$

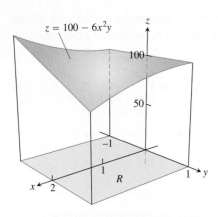

FIGURE 15.6 The double integral $\iint_R f(x, y)\, dA$ gives the volume under this surface over the rectangular region R (Example 1).

Solution Figure 15.6 displays the volume beneath the surface. By Fubini's Theorem,

$$\iint_R f(x, y)\, dA = \int_{-1}^{1}\int_{0}^{2} (100 - 6x^2y)\, dx\, dy = \int_{-1}^{1}\left[100x - 2x^3y \right]_{x=0}^{x=2} dy$$

$$= \int_{-1}^{1} (200 - 16y)\, dy = \left[200y - 8y^2 \right]_{-1}^{1} = 400.$$

Reversing the order of integration gives the same answer:

$$\int_{0}^{2}\int_{-1}^{1} (100 - 6x^2y)\, dy\, dx = \int_{0}^{2}\left[100y - 3x^2y^2 \right]_{y=-1}^{y=1} dx$$

$$= \int_{0}^{2}\left[(100 - 3x^2) - (-100 - 3x^2) \right] dx$$

$$= \int_{0}^{2} 200\, dx = 400. \qquad \blacksquare$$

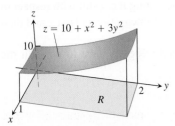

FIGURE 15.7 The double integral $\iint_R f(x, y)\, dA$ gives the volume under this surface over the rectangular region R (Example 2).

EXAMPLE 2 Find the volume of the region bounded above by the ellipitical paraboloid $z = 10 + x^2 + 3y^2$ and below by the rectangle $R: 0 \le x \le 1, 0 \le y \le 2$.

Solution The surface and volume are shown in Figure 15.7. The volume is given by the double integral

$$V = \iint_R (10 + x^2 + 3y^2)\, dA = \int_{0}^{1}\int_{0}^{2} (10 + x^2 + 3y^2)\, dy\, dx$$

$$= \int_{0}^{1}\left[10y + x^2y + y^3 \right]_{y=0}^{y=2} dx$$

$$= \int_{0}^{1} (20 + 2x^2 + 8)\, dx = \left[20x + \frac{2}{3}x^3 + 8x \right]_{0}^{1} = \frac{86}{3}. \qquad \blacksquare$$

Exercises 15.1

Evaluating Iterated Integrals
In Exercises 1–12, evaluate the iterated integral.

1. $\int_{1}^{2}\int_{0}^{4} 2xy\, dy\, dx$

2. $\int_{0}^{2}\int_{-1}^{1} (x - y)\, dy\, dx$

3. $\int_{-1}^{0}\int_{-1}^{1} (x + y + 1)\, dx\, dy$

4. $\int_{0}^{1}\int_{0}^{1}\left(1 - \frac{x^2 + y^2}{2} \right) dx\, dy$

5. $\int_{0}^{3}\int_{0}^{2} (4 - y^2)\, dy\, dx$

6. $\int_{0}^{3}\int_{-2}^{0} (x^2y - 2xy)\, dy\, dx$

7. $\int_{0}^{1}\int_{0}^{1} \frac{y}{1 + xy}\, dx\, dy$

8. $\int_{1}^{4}\int_{0}^{4}\left(\frac{x}{2} + \sqrt{y} \right) dx\, dy$

9. $\int_{0}^{\ln 2}\int_{1}^{\ln 5} e^{2x+y}\, dy\, dx$

10. $\int_{0}^{1}\int_{1}^{2} xye^x\, dy\, dx$

11. $\int_{-1}^{2}\int_{0}^{\pi/2} y \sin x\, dx\, dy$

12. $\int_{\pi}^{2\pi}\int_{0}^{\pi} (\sin x + \cos y)\, dx\, dy$

Evaluating Double Integrals over Rectangles
In Exercises 13–20, evaluate the double integral over the given region R.

13. $\iint_R (6y^2 - 2x)\, dA, \qquad R: \ 0 \le x \le 1, \ \ 0 \le y \le 2$

14. $\iint_R \left(\frac{\sqrt{x}}{y^2} \right) dA, \qquad R: \ 0 \le x \le 4, \ \ 1 \le y \le 2$

15. $\iint_R xy \cos y\, dA, \qquad R: \ -1 \le x \le 1, \ \ 0 \le y \le \pi$

16. $\displaystyle\iint_R y\sin(x+y)\,dA,$ $R:\ -\pi \le x \le 0,\ \ 0 \le y \le \pi$

17. $\displaystyle\iint_R e^{x-y}\,dA,$ $R:\ 0 \le x \le \ln 2,\ \ 0 \le y \le \ln 2$

18. $\displaystyle\iint_R xye^{xy^2}\,dA,$ $R:\ 0 \le x \le 2,\ \ 0 \le y \le 1$

19. $\displaystyle\iint_R \frac{xy^3}{x^2+1}\,dA,$ $R:\ 0 \le x \le 1,\ \ 0 \le y \le 2$

20. $\displaystyle\iint_R \frac{y}{x^2y^2+1}\,dA,$ $R:\ 0 \le x \le 1,\ \ 0 \le y \le 1$

In Exercises 21 and 22, integrate f over the given region.

21. Square $f(x,y) = 1/(xy)$ over the square $1 \le x \le 2, 1 \le y \le 2$

22. Rectangle $f(x,y) = y\cos xy$ over the rectangle $0 \le x \le \pi$, $0 \le y \le 1$

Volume Beneath a Surface $z = f(x, y)$

23. Find the volume of the region bounded above by the paraboloid $z = x^2 + y^2$ and below by the square $R: -1 \le x \le 1$, $-1 \le y \le 1$.

24. Find the volume of the region bounded above by the elliptical paraboloid $z = 16 - x^2 - y^2$ and below by the square $R: 0 \le x \le 2, 0 \le y \le 2$.

25. Find the volume of the region bounded above by the plane $z = 2 - x - y$ and below by the square $R: 0 \le x \le 1$, $0 \le y \le 1$.

26. Find the volume of the region bounded above by the plane $z = y/2$ and below by the rectangle $R: 0 \le x \le 4, 0 \le y \le 2$.

27. Find the volume of the region bounded above by the surface $z = 2\sin x \cos y$ and below by the rectangle $R: 0 \le x \le \pi/2$, $0 \le y \le \pi/4$.

28. Find the volume of the region bounded above by the surface $z = 4 - y^2$ and below by the rectangle $R: 0 \le x \le 1$, $0 \le y \le 2$.

15.2 Double Integrals over General Regions

In this section we define and evaluate double integrals over bounded regions in the plane which are more general than rectangles. These double integrals are also evaluated as iterated integrals, with the main practical problem being that of determining the limits of integration. Since the region of integration may have boundaries other than line segments parallel to the coordinate axes, the limits of integration often involve variables, not just constants.

Double Integrals over Bounded, Nonrectangular Regions

FIGURE 15.8 A rectangular grid partitioning a bounded nonrectangular region into rectangular cells.

To define the double integral of a function $f(x, y)$ over a bounded, nonrectangular region R, such as the one in Figure 15.8, we again begin by covering R with a grid of small rectangular cells whose union contains all points of R. This time, however, we cannot exactly fill R with a finite number of rectangles lying inside R, since its boundary is curved, and some of the small rectangles in the grid lie partly outside R. A partition of R is formed by taking the rectangles that lie completely inside it, not using any that are either partly or completely outside. For commonly arising regions, more and more of R is included as the norm of a partition (the largest width or height of any rectangle used) approaches zero.

Once we have a partition of R, we number the rectangles in some order from 1 to n and let ΔA_k be the area of the kth rectangle. We then choose a point (x_k, y_k) in the kth rectangle and form the Riemann sum

$$S_n = \sum_{k=1}^{n} f(x_k, y_k)\,\Delta A_k.$$

As the norm of the partition forming S_n goes to zero, $\|P\| \to 0$, the width and height of each enclosed rectangle goes to zero and their number goes to infinity. If $f(x, y)$ is a continuous function, then these Riemann sums converge to a limiting value, not dependent on any of the choices we made. This limit is called the **double integral** of $f(x, y)$ over R:

$$\lim_{\|P\| \to 0} \sum_{k=1}^{n} f(x_k, y_k)\,\Delta A_k = \iint_R f(x, y)\,dA.$$

The nature of the boundary of R introduces issues not found in integrals over an interval. When R has a curved boundary, the n rectangles of a partition lie inside R but do not cover all of R. In order for a partition to approximate R well, the parts of R covered by small rectangles lying partly outside R must become negligible as the norm of the partition approaches zero. This property of being nearly filled in by a partition of small norm is satisfied by all the regions that we will encounter. There is no problem with boundaries made from polygons, circles, ellipses, and from continuous graphs over an interval, joined end to end. A curve with a "fractal" type of shape would be problematic, but such curves arise rarely in most applications. A careful discussion of which type of regions R can be used for computing double integrals is left to a more advanced text.

Volumes

If $f(x, y)$ is positive and continuous over R, we define the volume of the solid region between R and the surface $z = f(x, y)$ to be $\iint_R f(x, y)\, dA$, as before (Figure 15.9).

If R is a region like the one shown in the xy-plane in Figure 15.10, bounded "above" and "below" by the curves $y = g_2(x)$ and $y = g_1(x)$ and on the sides by the lines $x = a$, $x = b$, we may again calculate the volume by the method of slicing. We first calculate the cross-sectional area

$$A(x) = \int_{y=g_1(x)}^{y=g_2(x)} f(x, y)\, dy$$

and then integrate $A(x)$ from $x = a$ to $x = b$ to get the volume as an iterated integral:

$$V = \int_a^b A(x)\, dx = \int_a^b \int_{g_1(x)}^{g_2(x)} f(x, y)\, dy\, dx. \tag{1}$$

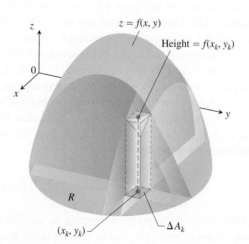

$$\text{Volume} = \lim \sum f(x_k, y_k)\, \Delta A_k = \iint_R f(x, y)\, dA$$

FIGURE 15.9 We define the volumes of solids with curved bases as a limit of approximating rectangular boxes.

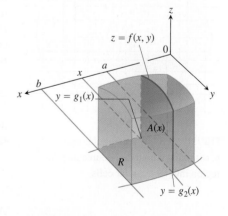

FIGURE 15.10 The area of the vertical slice shown here is $A(x)$. To calculate the volume of the solid, we integrate this area from $x = a$ to $x = b$:

$$\int_a^b A(x)\, dx = \int_a^b \int_{g_1(x)}^{g_2(x)} f(x, y)\, dy\, dx.$$

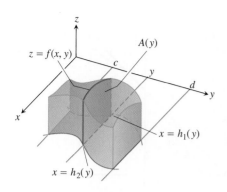

FIGURE 15.11 The volume of the solid shown here is

$$\int_c^d A(y)\, dy = \int_c^d \int_{h_1(y)}^{h_2(y)} f(x, y)\, dx\, dy.$$

For a given solid, Theorem 2 says we can calculate the volume as in Figure 15.10, or in the way shown here. Both calculations have the same result.

Similarly, if R is a region like the one shown in Figure 15.11, bounded by the curves $x = h_2(y)$ and $x = h_1(y)$ and the lines $y = c$ and $y = d$, then the volume calculated by slicing is given by the iterated integral

$$\text{Volume} = \int_c^d \int_{h_1(y)}^{h_2(y)} f(x, y)\, dx\, dy. \tag{2}$$

That the iterated integrals in Equations (1) and (2) both give the volume that we defined to be the double integral of f over R is a consequence of the following stronger form of Fubini's Theorem.

THEOREM 2—Fubini's Theorem (Stronger Form)   Let $f(x, y)$ be continuous on a region R.

1. If R is defined by $a \le x \le b$, $g_1(x) \le y \le g_2(x)$, with g_1 and g_2 continuous on $[a, b]$, then

$$\iint\limits_R f(x, y)\, dA = \int_a^b \int_{g_1(x)}^{g_2(x)} f(x, y)\, dy\, dx.$$

2. If R is defined by $c \le y \le d$, $h_1(y) \le x \le h_2(y)$, with h_1 and h_2 continuous on $[c, d]$, then

$$\iint\limits_R f(x, y)\, dA = \int_c^d \int_{h_1(y)}^{h_2(y)} f(x, y)\, dx\, dy.$$

EXAMPLE 1   Find the volume of the prism whose base is the triangle in the xy-plane bounded by the x-axis and the lines $y = x$ and $x = 1$ and whose top lies in the plane

$$z = f(x, y) = 3 - x - y.$$

Solution   See Figure 15.12. For any x between 0 and 1, y may vary from $y = 0$ to $y = x$ (Figure 15.12b). Hence,

$$V = \int_0^1 \int_0^x (3 - x - y)\, dy\, dx = \int_0^1 \left[3y - xy - \frac{y^2}{2} \right]_{y=0}^{y=x} dx$$

$$= \int_0^1 \left(3x - \frac{3x^2}{2} \right) dx = \left[\frac{3x^2}{2} - \frac{x^3}{2} \right]_{x=0}^{x=1} = 1.$$

When the order of integration is reversed (Figure 15.12c), the integral for the volume is

$$V = \int_0^1 \int_y^1 (3 - x - y)\, dx\, dy = \int_0^1 \left[3x - \frac{x^2}{2} - xy \right]_{x=y}^{x=1} dy$$

$$= \int_0^1 \left(3 - \frac{1}{2} - y - 3y + \frac{y^2}{2} + y^2 \right) dy$$

$$= \int_0^1 \left(\frac{5}{2} - 4y + \frac{3}{2}y^2 \right) dy = \left[\frac{5}{2}y - 2y^2 + \frac{y^3}{2} \right]_{y=0}^{y=1} = 1.$$

The two integrals are equal, as they should be.   ∎

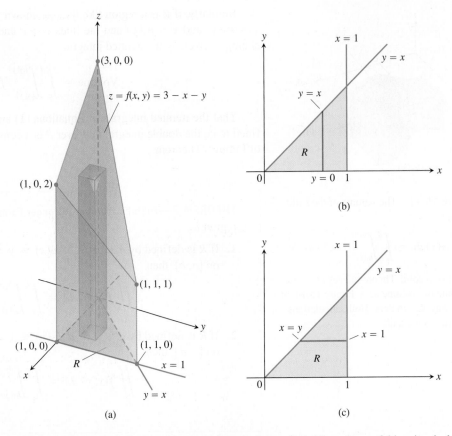

FIGURE 15.12 (a) Prism with a triangular base in the xy-plane. The volume of this prism is defined as a double integral over R. To evaluate it as an iterated integral, we may integrate first with respect to y and then with respect to x, or the other way around (Example 1). (b) Integration limits of

$$\int_{x=0}^{x=1} \int_{y=0}^{y=x} f(x, y) \, dy \, dx.$$

If we integrate first with respect to y, we integrate along a vertical line through R and then integrate from left to right to include all the vertical lines in R. (c) Integration limits of

$$\int_{y=0}^{y=1} \int_{x=y}^{x=1} f(x, y) \, dx \, dy.$$

If we integrate first with respect to x, we integrate along a horizontal line through R and then integrate from bottom to top to include all the horizontal lines in R.

Although Fubini's Theorem assures us that a double integral may be calculated as an iterated integral in either order of integration, the value of one integral may be easier to find than the value of the other. The next example shows how this can happen.

EXAMPLE 2 Calculate

$$\iint\limits_{R} \frac{\sin x}{x} \, dA,$$

where R is the triangle in the xy-plane bounded by the x-axis, the line $y = x$, and the line $x = 1$.

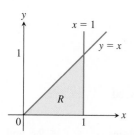

FIGURE 15.13 The region of integration in Example 2.

(a)

(b)

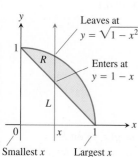

(c)

FIGURE 15.14 Finding the limits of integration when integrating first with respect to y and then with respect to x.

Solution The region of integration is shown in Figure 15.13. If we integrate first with respect to y and then with respect to x, we find

$$\int_0^1 \left(\int_0^x \frac{\sin x}{x}\, dy \right) dx = \int_0^1 \left(y\, \frac{\sin x}{x} \right]_{y=0}^{y=x} dx = \int_0^1 \sin x\, dx$$
$$= -\cos\,(1) + 1 \approx 0.46.$$

If we reverse the order of integration and attempt to calculate

$$\int_0^1 \int_y^1 \frac{\sin x}{x}\, dx\, dy,$$

we run into a problem because $\int ((\sin x)/x)\, dx$ cannot be expressed in terms of elementary functions (there is no simple antiderivative).

There is no general rule for predicting which order of integration will be the good one in circumstances like these. If the order you first choose doesn't work, try the other. Sometimes neither order will work, and then we need to use numerical approximations. ■

Finding Limits of Integration

We now give a procedure for finding limits of integration that applies for many regions in the plane. Regions that are more complicated, and for which this procedure fails, can often be split up into pieces on which the procedure works.

Using Vertical Cross-sections When faced with evaluating $\iint_R f(x, y)\, dA$, integrating first with respect to y and then with respect to x, do the following three steps:

1. *Sketch.* Sketch the region of integration and label the bounding curves (Figure 15.14a).

2. *Find the y-limits of integration.* Imagine a vertical line L cutting through R in the direction of increasing y. Mark the y-values where L enters and leaves. These are the y-limits of integration and are usually functions of x (instead of constants) (Figure 15.14b).

3. *Find the x-limits of integration.* Choose x-limits that include all the vertical lines through R. The integral shown here (see Figure 15.14c) is

$$\iint_R f(x, y)\, dA = \int_{x=0}^{x=1} \int_{y=1-x}^{y=\sqrt{1-x^2}} f(x, y)\, dy\, dx.$$

Using Horizontal Cross-sections To evaluate the same double integral as an iterated integral with the order of integration reversed, use horizontal lines instead of vertical lines in Steps 2 and 3 (see Figure 15.15). The integral is

$$\iint_R f(x, y)\, dA = \int_0^1 \int_{1-y}^{\sqrt{1-y^2}} f(x, y)\, dx\, dy.$$

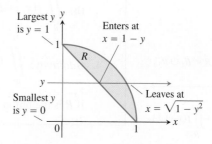

FIGURE 15.15 Finding the limits of integration when integrating first with respect to x and then with respect to y.

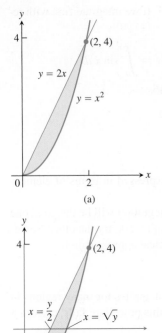

(a)

(b)

FIGURE 15.16 Region of integration for Example 3.

EXAMPLE 3   Sketch the region of integration for the integral

$$\int_0^2 \int_{x^2}^{2x} (4x + 2) \, dy \, dx$$

and write an equivalent integral with the order of integration reversed.

Solution   The region of integration is given by the inequalities $x^2 \le y \le 2x$ and $0 \le x \le 2$. It is therefore the region bounded by the curves $y = x^2$ and $y = 2x$ between $x = 0$ and $x = 2$ (Figure 15.16a).

To find limits for integrating in the reverse order, we imagine a horizontal line passing from left to right through the region. It enters at $x = y/2$ and leaves at $x = \sqrt{y}$. To include all such lines, we let y run from $y = 0$ to $y = 4$ (Figure 15.16b). The integral is

$$\int_0^4 \int_{y/2}^{\sqrt{y}} (4x + 2) \, dx \, dy.$$

The common value of these integrals is 8.   ■

Properties of Double Integrals

Like single integrals, double integrals of continuous functions have algebraic properties that are useful in computations and applications.

If $f(x, y)$ and $g(x, y)$ are continuous on the bounded region R, then the following properties hold.

1. *Constant Multiple:* $\displaystyle\iint\limits_R cf(x, y) \, dA = c \iint\limits_R f(x, y) \, dA$ (any number c)

2. *Sum and Difference:*

$$\iint\limits_R (f(x, y) \pm g(x, y)) \, dA = \iint\limits_R f(x, y) \, dA \pm \iint\limits_R g(x, y) \, dA$$

3. *Domination:*

   (a) $\displaystyle\iint\limits_R f(x, y) \, dA \ge 0$ if $f(x, y) \ge 0$ on R

   (b) $\displaystyle\iint\limits_R f(x, y) \, dA \ge \iint\limits_R g(x, y) \, dA$ if $f(x, y) \ge g(x, y)$ on R

4. *Additivity:* $\displaystyle\iint\limits_R f(x, y) \, dA = \iint\limits_{R_1} f(x, y) \, dA + \iint\limits_{R_2} f(x, y) \, dA$

   if R is the union of two nonoverlapping regions R_1 and R_2

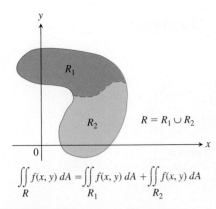

$$\iint\limits_R f(x, y) \, dA = \iint\limits_{R_1} f(x, y) \, dA + \iint\limits_{R_2} f(x, y) \, dA$$

FIGURE 15.17 The Additivity Property for rectangular regions holds for regions bounded by smooth curves.

Property 4 assumes that the region of integration R is decomposed into nonoverlapping regions R_1 and R_2 with boundaries consisting of a finite number of line segments or smooth curves. Figure 15.17 illustrates an example of this property.

The idea behind these properties is that integrals behave like sums. If the function $f(x, y)$ is replaced by its constant multiple $cf(x, y)$, then a Riemann sum for f

$$S_n = \sum_{k=1}^{n} f(x_k, y_k) \, \Delta A_k$$

is replaced by a Riemann sum for cf

$$\sum_{k=1}^{n} cf(x_k, y_k) \, \Delta A_k = c \sum_{k=1}^{n} f(x_k, y_k) \, \Delta A_k = cS_n.$$

Taking limits as $n \to \infty$ shows that $c \lim_{n\to\infty} S_n = c \iint_R f \, dA$ and $\lim_{n\to\infty} cS_n = \iint_R cf \, dA$ are equal. It follows that the constant multiple property carries over from sums to double integrals.

The other properties are also easy to verify for Riemann sums, and carry over to double integrals for the same reason. While this discussion gives the idea, an actual proof that these properties hold requires a more careful analysis of how Riemann sums converge.

(a)

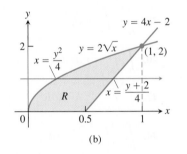

(b)

FIGURE 15.18   (a) The solid "wedgelike" region whose volume is found in Example 4. (b) The region of integration R showing the order $dx \, dy$.

EXAMPLE 4   Find the volume of the wedgelike solid that lies beneath the surface $z = 16 - x^2 - y^2$ and above the region R bounded by the curve $y = 2\sqrt{x}$, the line $y = 4x - 2$, and the x-axis.

Solution   Figure 15.18a shows the surface and the "wedgelike" solid whose volume we want to calculate. Figure 15.18b shows the region of integration in the xy-plane. If we integrate in the order $dy \, dx$ (first with respect to y and then with respect to x), two integrations will be required because y varies from $y = 0$ to $y = 2\sqrt{x}$ for $0 \le x \le 0.5$, and then varies from $y = 4x - 2$ to $y = 2\sqrt{x}$ for $0.5 \le x \le 1$. So we choose to integrate in the order $dx \, dy$, which requires only one double integral whose limits of integration are indicated in Figure 15.18b. The volume is then calculated as the iterated integral:

$$\iint_R (16 - x^2 - y^2) \, dA$$

$$= \int_0^2 \int_{y^2/4}^{(y+2)/4} (16 - x^2 - y^2) \, dx \, dy$$

$$= \int_0^2 \left[16x - \frac{x^3}{3} - xy^2 \right]_{x=y^2/4}^{x=(y+2)/4} dx$$

$$= \int_0^2 \left[4(y + 2) - \frac{(y + 2)^3}{3 \cdot 64} - \frac{(y + 2)y^2}{4} - 4y^2 + \frac{y^6}{3 \cdot 64} + \frac{y^4}{4} \right] dy$$

$$= \left[\frac{191y}{24} + \frac{63y^2}{32} - \frac{145y^3}{96} - \frac{49y^4}{768} + \frac{y^5}{20} + \frac{y^7}{1344} \right]_0^2 = \frac{20803}{1680} \approx 12.4. \qquad \blacksquare$$

Exercises 15.2

Sketching Regions of Integration
In Exercises 1–8, sketch the described regions of integration.
1. $0 \le x \le 3, \quad 0 \le y \le 2x$
2. $-1 \le x \le 2, \quad x - 1 \le y \le x^2$
3. $-2 \le y \le 2, \quad y^2 \le x \le 4$
4. $0 \le y \le 1, \quad y \le x \le 2y$

5. $0 \le x \le 1, \quad e^x \le y \le e$
6. $1 \le x \le e^2, \quad 0 \le y \le \ln x$
7. $0 \le y \le 1, \quad 0 \le x \le \sin^{-1} y$
8. $0 \le y \le 8, \quad \frac{1}{4} y \le x \le y^{1/3}$

Finding Limits of Integration

In Exercises 9–18, write an iterated integral for $\iint_R dA$ over the described region R using (a) vertical cross-sections, (b) horizontal cross-sections.

9.

10.

11.

12.

13. Bounded by $y = \sqrt{x}, y = 0,$ and $x = 9$

14. Bounded by $y = \tan x, x = 0,$ and $y = 1$

15. Bounded by $y = e^{-x}, y = 1,$ and $x = \ln 3$

16. Bounded by $y = 0, x = 0, y = 1,$ and $y = \ln x$

17. Bounded by $y = 3 - 2x, y = x,$ and $x = 0$

18. Bounded by $y = x^2$ and $y = x + 2$

Finding Regions of Integration and Double Integrals

In Exercises 19–24, sketch the region of integration and evaluate the integral.

19. $\displaystyle\int_0^\pi \int_0^x x \sin y \, dy \, dx$

20. $\displaystyle\int_0^\pi \int_0^{\sin x} y \, dy \, dx$

21. $\displaystyle\int_1^{\ln 8} \int_0^{\ln y} e^{x+y} \, dx \, dy$

22. $\displaystyle\int_1^2 \int_y^{y^2} dx \, dy$

23. $\displaystyle\int_0^1 \int_0^{y^2} 3y^3 e^{xy} \, dx \, dy$

24. $\displaystyle\int_1^4 \int_0^{\sqrt{x}} \frac{3}{2} e^{y/\sqrt{x}} \, dy \, dx$

In Exercises 25–28, integrate f over the given region.

25. Quadrilateral $f(x, y) = x/y$ over the region in the first quadrant bounded by the lines $y = x, y = 2x, x = 1,$ and $x = 2$

26. Triangle $f(x, y) = x^2 + y^2$ over the triangular region with vertices $(0, 0), (1, 0),$ and $(0, 1)$

27. Triangle $f(u, v) = v - \sqrt{u}$ over the triangular region cut from the first quadrant of the uv-plane by the line $u + v = 1$

28. Curved region $f(s, t) = e^s \ln t$ over the region in the first quadrant of the st-plane that lies above the curve $s = \ln t$ from $t = 1$ to $t = 2$

Each of Exercises 29–32 gives an integral over a region in a Cartesian coordinate plane. Sketch the region and evaluate the integral.

29. $\displaystyle\int_{-2}^0 \int_v^{-v} 2 \, dp \, dv$ (the pv-plane)

30. $\displaystyle\int_0^1 \int_0^{\sqrt{1-s^2}} 8t \, dt \, ds$ (the st-plane)

31. $\displaystyle\int_{-\pi/3}^{\pi/3} \int_0^{\sec t} 3 \cos t \, du \, dt$ (the tu-plane)

32. $\displaystyle\int_0^{3/2} \int_1^{4-2u} \frac{4 - 2u}{v^2} \, dv \, du$ (the uv-plane)

Reversing the Order of Integration

In Exercises 33–46, sketch the region of integration and write an equivalent double integral with the order of integration reversed.

33. $\displaystyle\int_0^1 \int_2^{4-2x} dy \, dx$

34. $\displaystyle\int_0^2 \int_{y-2}^0 dx \, dy$

35. $\displaystyle\int_0^1 \int_y^{\sqrt{y}} dx \, dy$

36. $\displaystyle\int_0^1 \int_{1-x}^{1-x^2} dy \, dx$

37. $\displaystyle\int_0^1 \int_1^{e^x} dy \, dx$

38. $\displaystyle\int_0^{\ln 2} \int_{e^y}^2 dx \, dy$

39. $\displaystyle\int_0^{3/2} \int_0^{9-4x^2} 16x \, dy \, dx$

40. $\displaystyle\int_0^2 \int_0^{4-y^2} y \, dx \, dy$

41. $\displaystyle\int_0^1 \int_{-\sqrt{1-y^2}}^{\sqrt{1-y^2}} 3y \, dx \, dy$

42. $\displaystyle\int_0^2 \int_{-\sqrt{4-x^2}}^{\sqrt{4-x^2}} 6x \, dy \, dx$

43. $\displaystyle\int_1^e \int_0^{\ln x} xy \, dy \, dx$

44. $\displaystyle\int_0^{\pi/6} \int_{\sin x}^{1/2} xy^2 \, dy \, dx$

45. $\displaystyle\int_0^3 \int_1^{e^y} (x + y) \, dx \, dy$

46. $\displaystyle\int_0^{\sqrt{3}} \int_0^{\tan^{-1} y} \sqrt{xy} \, dx \, dy$

In Exercises 47–56, sketch the region of integration, reverse the order of integration, and evaluate the integral.

47. $\displaystyle\int_0^\pi \int_x^\pi \frac{\sin y}{y} \, dy \, dx$

48. $\displaystyle\int_0^2 \int_x^2 2y^2 \sin xy \, dy \, dx$

49. $\displaystyle\int_0^1 \int_y^1 x^2 e^{xy} \, dx \, dy$

50. $\displaystyle\int_0^2 \int_0^{4-x^2} \frac{xe^{2y}}{4 - y} \, dy \, dx$

51. $\displaystyle\int_0^{2\sqrt{\ln 3}} \int_{y/2}^{\sqrt{\ln 3}} e^{x^2} \, dx \, dy$

52. $\displaystyle\int_0^3 \int_{\sqrt{x/3}}^1 e^{y^3} \, dy \, dx$

53. $\displaystyle\int_0^{1/16} \int_{y^{1/4}}^{1/2} \cos (16\pi x^5) \, dx \, dy$

54. $\displaystyle\int_0^8 \int_{\sqrt[3]{x}}^2 \frac{dy \, dx}{y^4 + 1}$

55. Square region $\iint_R (y - 2x^2) \, dA$ where R is the region bounded by the square $|x| + |y| = 1$

56. Triangular region $\iint_R xy \, dA$ where R is the region bounded by the lines $y = x, y = 2x,$ and $x + y = 2$

Volume Beneath a Surface $z = f(x, y)$

57. Find the volume of the region bounded above by the paraboloid $z = x^2 + y^2$ and below by the triangle enclosed by the lines $y = x, x = 0,$ and $x + y = 2$ in the xy-plane.

58. Find the volume of the solid that is bounded above by the cylinder $z = x^2$ and below by the region enclosed by the parabola $y = 2 - x^2$ and the line $y = x$ in the xy-plane.

59. Find the volume of the solid whose base is the region in the xy-plane that is bounded by the parabola $y = 4 - x^2$ and the line $y = 3x$, while the top of the solid is bounded by the plane $z = x + 4$.

60. Find the volume of the solid in the first octant bounded by the coordinate planes, the cylinder $x^2 + y^2 = 4$, and the plane $z + y = 3$.

61. Find the volume of the solid in the first octant bounded by the coordinate planes, the plane $x = 3$, and the parabolic cylinder $z = 4 - y^2$.

62. Find the volume of the solid cut from the first octant by the surface $z = 4 - x^2 - y$.

63. Find the volume of the wedge cut from the first octant by the cylinder $z = 12 - 3y^2$ and the plane $x + y = 2$.

64. Find the volume of the solid cut from the square column $|x| + |y| \leq 1$ by the planes $z = 0$ and $3x + z = 3$.

65. Find the volume of the solid that is bounded on the front and back by the planes $x = 2$ and $x = 1$, on the sides by the cylinders $y = \pm 1/x$, and above and below by the planes $z = x + 1$ and $z = 0$.

66. Find the volume of the solid bounded on the front and back by the planes $x = \pm\pi/3$, on the sides by the cylinders $y = \pm\sec x$, above by the cylinder $z = 1 + y^2$, and below by the xy-plane.

In Exercises 67 and 68, sketch the region of integration and the solid whose volume is given by the double integral.

67. $\displaystyle\int_0^3 \int_0^{2-2x/3} \left(1 - \frac{1}{3}x - \frac{1}{2}y\right) dy\, dx$

68. $\displaystyle\int_0^4 \int_{-\sqrt{16-y^2}}^{\sqrt{16-y^2}} \sqrt{25 - x^2 - y^2}\, dx\, dy$

Integrals over Unbounded Regions

Improper double integrals can often be computed similarly to improper integrals of one variable. The first iteration of the following improper integrals is conducted just as if they were proper integrals. One then evaluates an improper integral of a single variable by taking appropriate limits, as in Section 8.7. Evaluate the improper integrals in Exercises 69–72 as iterated integrals.

69. $\displaystyle\int_1^\infty \int_{e^{-x}}^1 \frac{1}{x^3 y}\, dy\, dx$

70. $\displaystyle\int_{-1}^1 \int_{-1/\sqrt{1-x^2}}^{1/\sqrt{1-x^2}} (2y + 1)\, dy\, dx$

71. $\displaystyle\int_{-\infty}^\infty \int_{-\infty}^\infty \frac{1}{(x^2 + 1)(y^2 + 1)}\, dx\, dy$

72. $\displaystyle\int_0^\infty \int_0^\infty xe^{-(x+2y)}\, dx\, dy$

Approximating Integrals with Finite Sums

In Exercises 73 and 74, approximate the double integral of $f(x, y)$ over the region R partitioned by the given vertical lines $x = a$ and horizontal lines $y = c$. In each subrectangle, use (x_k, y_k) as indicated for your approximation.

$$\iint_R f(x, y)\, dA \approx \sum_{k=1}^n f(x_k, y_k)\, \Delta A_k$$

73. $f(x, y) = x + y$ over the region R bounded above by the semicircle $y = \sqrt{1 - x^2}$ and below by the x-axis, using the partition $x = -1, -1/2, 0, 1/4, 1/2, 1$ and $y = 0, 1/2, 1$ with (x_k, y_k) the lower left corner in the kth subrectangle (provided the subrectangle lies within R)

74. $f(x, y) = x + 2y$ over the region R inside the circle $(x - 2)^2 + (y - 3)^2 = 1$ using the partition $x = 1, 3/2, 2, 5/2, 3$ and $y = 2, 5/2, 3, 7/2, 4$ with (x_k, y_k) the center (centroid) in the kth subrectangle (provided the subrectangle lies within R)

Theory and Examples

75. Circular sector Integrate $f(x, y) = \sqrt{4 - x^2}$ over the smaller sector cut from the disk $x^2 + y^2 \leq 4$ by the rays $\theta = \pi/6$ and $\theta = \pi/2$.

76. Unbounded region Integrate $f(x, y) = 1/[(x^2 - x)(y - 1)^{2/3}]$ over the infinite rectangle $2 \leq x < \infty, 0 \leq y \leq 2$.

77. Noncircular cylinder A solid right (noncircular) cylinder has its base R in the xy-plane and is bounded above by the paraboloid $z = x^2 + y^2$. The cylinder's volume is

$$V = \int_0^1 \int_0^y (x^2 + y^2)\, dx\, dy + \int_1^2 \int_0^{2-y} (x^2 + y^2)\, dx\, dy.$$

Sketch the base region R and express the cylinder's volume as a single iterated integral with the order of integration reversed. Then evaluate the integral to find the volume.

78. Converting to a double integral Evaluate the integral

$$\int_0^2 (\tan^{-1}\pi x - \tan^{-1} x)\, dx.$$

(*Hint:* Write the integrand as an integral.)

79. Maximizing a double integral What region R in the xy-plane maximizes the value of

$$\iint_R (4 - x^2 - 2y^2)\, dA?$$

Give reasons for your answer.

80. Minimizing a double integral What region R in the xy-plane minimizes the value of

$$\iint_R (x^2 + y^2 - 9)\, dA?$$

Give reasons for your answer.

81. Is it possible to evaluate the integral of a continuous function $f(x, y)$ over a rectangular region in the xy-plane and get different answers depending on the order of integration? Give reasons for your answer.

82 How would you evaluate the double integral of a continuous function $f(x, y)$ over the region R in the xy-plane enclosed by the triangle with vertices $(0, 1)$, $(2, 0)$, and $(1, 2)$? Give reasons for your answer.

83. Unbounded region Prove that

$$\int_{-\infty}^\infty \int_{-\infty}^\infty e^{-x^2-y^2}\, dx\, dy = \lim_{b\to\infty} \int_{-b}^b \int_{-b}^b e^{-x^2-y^2}\, dx\, dy$$

$$= 4\left(\int_0^\infty e^{-x^2}\, dx\right)^2.$$

84. Improper double integral   Evaluate the improper integral

$$\int_0^1 \int_0^3 \frac{x^2}{(y-1)^{2/3}} \, dy \, dx.$$

COMPUTER EXPLORATIONS

Use a CAS double-integral evaluator to estimate the values of the integrals in Exercises 85–88.

85. $\displaystyle\int_1^3 \int_1^x \frac{1}{xy} \, dy \, dx$ **86.** $\displaystyle\int_0^1 \int_0^1 e^{-(x^2+y^2)} \, dy \, dx$

87. $\displaystyle\int_0^1 \int_0^1 \tan^{-1} xy \, dy \, dx$ **88.** $\displaystyle\int_{-1}^1 \int_0^{\sqrt{1-x^2}} 3\sqrt{1-x^2-y^2} \, dy \, dx$

Use a CAS double-integral evaluator to find the integrals in Exercises 89–94. Then reverse the order of integration and evaluate, again with a CAS.

89. $\displaystyle\int_0^1 \int_{2y}^4 e^{x^2} \, dx \, dy$ **90.** $\displaystyle\int_0^3 \int_{x^2}^9 x \cos(y^2) \, dy \, dx$

91. $\displaystyle\int_0^2 \int_{y^3}^{4\sqrt{2y}} (x^2 y - xy^2) \, dx \, dy$ **92.** $\displaystyle\int_0^2 \int_0^{4-y^2} e^{xy} \, dx \, dy$

93. $\displaystyle\int_1^2 \int_0^{x^2} \frac{1}{x+y} \, dy \, dx$ **94.** $\displaystyle\int_1^2 \int_{y^3}^8 \frac{1}{\sqrt{x^2+y^2}} \, dx \, dy$

15.3   Area by Double Integration

In this section we show how to use double integrals to calculate the areas of bounded regions in the plane, and to find the average value of a function of two variables.

Areas of Bounded Regions in the Plane

If we take $f(x, y) = 1$ in the definition of the double integral over a region R in the preceding section, the Riemann sums reduce to

$$S_n = \sum_{k=1}^n f(x_k, y_k) \, \Delta A_k = \sum_{k=1}^n \Delta A_k. \tag{1}$$

This is simply the sum of the areas of the small rectangles in the partition of R, and approximates what we would like to call the area of R. As the norm of a partition of R approaches zero, the height and width of all rectangles in the partition approach zero, and the coverage of R becomes increasingly complete (Figure 15.8). We define the area of R to be the limit

$$\lim_{\|P\| \to 0} \sum_{k=1}^n \Delta A_k = \iint_R dA. \tag{2}$$

DEFINITION   The **area** of a closed, bounded plane region R is

$$A = \iint_R dA.$$

As with the other definitions in this chapter, the definition here applies to a greater variety of regions than does the earlier single-variable definition of area, but it agrees with the earlier definition on regions to which they both apply. To evaluate the integral in the definition of area, we integrate the constant function $f(x, y) = 1$ over R.

EXAMPLE 1   Find the area of the region R bounded by $y = x$ and $y = x^2$ in the first quadrant.

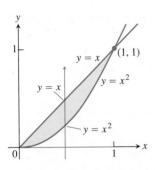

FIGURE 15.19 The region in Example 1.

Solution We sketch the region (Figure 15.19), noting where the two curves intersect at the origin and $(1, 1)$, and calculate the area as

$$A = \int_0^1 \int_{x^2}^{x} dy\, dx = \int_0^1 \Big[y\Big]_{x^2}^{x} dx$$

$$= \int_0^1 (x - x^2)\, dx = \Big[\frac{x^2}{2} - \frac{x^3}{3}\Big]_0^1 = \frac{1}{6}.$$

Notice that the single-variable integral $\int_0^1 (x - x^2)\, dx$, obtained from evaluating the inside iterated integral, is the integral for the area between these two curves using the method of Section 5.6. ∎

EXAMPLE 2 Find the area of the region R enclosed by the parabola $y = x^2$ and the line $y = x + 2$.

Solution If we divide R into the regions R_1 and R_2 shown in Figure 15.20a, we may calculate the area as

$$A = \iint_{R_1} dA + \iint_{R_2} dA = \int_0^1 \int_{-\sqrt{y}}^{\sqrt{y}} dx\, dy + \int_1^4 \int_{y-2}^{\sqrt{y}} dx\, dy.$$

On the other hand, reversing the order of integration (Figure 15.20b) gives

$$A = \int_{-1}^{2} \int_{x^2}^{x+2} dy\, dx.$$

This second result, which requires only one integral, is simpler and is the only one we would bother to write down in practice. The area is

$$A = \int_{-1}^{2} \Big[y\Big]_{x^2}^{x+2} dx = \int_{-1}^{2} (x + 2 - x^2)\, dx = \Big[\frac{x^2}{2} + 2x - \frac{x^3}{3}\Big]_{-1}^{2} = \frac{9}{2}. \quad ∎$$

(a)

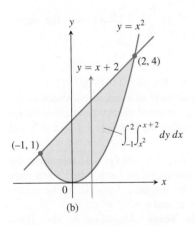

(b)

FIGURE 15.20 Calculating this area takes (a) two double integrals if the first integration is with respect to x, but (b) only one if the first integration is with respect to y (Example 2).

Average Value

The average value of an integrable function of one variable on a closed interval is the integral of the function over the interval divided by the length of the interval. For an integrable function of two variables defined on a bounded region in the plane, the average value is the integral over the region divided by the area of the region. This can be visualized by thinking of the function as giving the height at one instant of some water sloshing around in a tank whose vertical walls lie over the boundary of the region. The average height of the water in the tank can be found by letting the water settle down to a constant height. The height is then equal to the volume of water in the tank divided by the area of R. We are led to define the average value of an integrable function f over a region R as follows:

> **Average value** of f over $R = \dfrac{1}{\text{area of } R} \iint_R f\, dA.$ (3)

If f is the temperature of a thin plate covering R, then the double integral of f over R divided by the area of R is the plate's average temperature. If $f(x, y)$ is the distance from the point (x, y) to a fixed point P, then the average value of f over R is the average distance of points in R from P.

EXAMPLE 3 Find the average value of $f(x, y) = x \cos xy$ over the rectangle $R: 0 \leq x \leq \pi, \ 0 \leq y \leq 1$.

Solution The value of the integral of f over R is

$$\int_0^\pi \int_0^1 x \cos xy \, dy \, dx = \int_0^\pi \left[\sin xy \right]_{y=0}^{y=1} dx \qquad \int x \cos xy \, dy = \sin xy + C$$

$$= \int_0^\pi (\sin x - 0) \, dx = -\cos x \Big]_0^\pi = 1 + 1 = 2.$$

The area of R is π. The average value of f over R is $2/\pi$. ∎

Exercises 15.3

Area by Double Integrals

In Exercises 1–12, sketch the region bounded by the given lines and curves. Then express the region's area as an iterated double integral and evaluate the integral.

1. The coordinate axes and the line $x + y = 2$
2. The lines $x = 0, y = 2x$, and $y = 4$
3. The parabola $x = -y^2$ and the line $y = x + 2$
4. The parabola $x = y - y^2$ and the line $y = -x$
5. The curve $y = e^x$ and the lines $y = 0, x = 0$, and $x = \ln 2$
6. The curves $y = \ln x$ and $y = 2 \ln x$ and the line $x = e$, in the first quadrant
7. The parabolas $x = y^2$ and $x = 2y - y^2$
8. The parabolas $x = y^2 - 1$ and $x = 2y^2 - 2$
9. The lines $y = x, y = x/3$, and $y = 2$
10. The lines $y = 1 - x$ and $y = 2$ and the curve $y = e^x$
11. The lines $y = 2x, y = x/2$, and $y = 3 - x$
12. The lines $y = x - 2$ and $y = -x$ and the curve $y = \sqrt{x}$

Identifying the Region of Integration

The integrals and sums of integrals in Exercises 13–18 give the areas of regions in the xy-plane. Sketch each region, label each bounding curve with its equation, and give the coordinates of the points where the curves intersect. Then find the area of the region.

13. $\int_0^6 \int_{y^2/3}^{2y} dx \, dy$

14. $\int_0^3 \int_{-x}^{x(2-x)} dy \, dx$

15. $\int_0^{\pi/4} \int_{\sin x}^{\cos x} dy \, dx$

16. $\int_{-1}^2 \int_{y^2}^{y+2} dx \, dy$

17. $\int_{-1}^0 \int_{-2x}^{1-x} dy \, dx + \int_0^2 \int_{-x/2}^{1-x} dy \, dx$

18. $\int_0^2 \int_{x^2-4}^0 dy \, dx + \int_0^4 \int_0^{\sqrt{x}} dy \, dx$

Finding Average Values

19. Find the average value of $f(x, y) = \sin(x + y)$ over
    a. the rectangle $0 \leq x \leq \pi, \ 0 \leq y \leq \pi$.
    b. the rectangle $0 \leq x \leq \pi, \ 0 \leq y \leq \pi/2$.

20. Which do you think will be larger, the average value of $f(x, y) = xy$ over the square $0 \leq x \leq 1, 0 \leq y \leq 1$, or the average value of f over the quarter circle $x^2 + y^2 \leq 1$ in the first quadrant? Calculate them to find out.

21. Find the average height of the paraboloid $z = x^2 + y^2$ over the square $0 \leq x \leq 2, 0 \leq y \leq 2$.

22. Find the average value of $f(x, y) = 1/(xy)$ over the square $\ln 2 \leq x \leq 2 \ln 2, \ln 2 \leq y \leq 2 \ln 2$.

Theory and Examples

23. **Bacterium population** If $f(x, y) = (10{,}000e^y)/(1 + |x|/2)$ represents the "population density" of a certain bacterium on the xy-plane, where x and y are measured in centimeters, find the total population of bacteria within the rectangle $-5 \leq x \leq 5$ and $-2 \leq y \leq 0$.

24. **Regional population** If $f(x, y) = 100(y + 1)$ represents the population density of a planar region on Earth, where x and y are measured in miles, find the number of people in the region bounded by the curves $x = y^2$ and $x = 2y - y^2$.

25. **Average temperature in Texas** According to the *Texas Almanac*, Texas has 254 counties and a National Weather Service station in each county. Assume that at time t_0, each of the 254 weather stations recorded the local temperature. Find a formula that would give a reasonable approximation of the average temperature in Texas at time t_0. Your answer should involve information that you would expect to be readily available in the *Texas Almanac*.

26. If $y = f(x)$ is a nonnegative continuous function over the closed interval $a \leq x \leq b$, show that the double integral definition of area for the closed plane region bounded by the graph of f, the vertical lines $x = a$ and $x = b$, and the x-axis agrees with the definition for area beneath the curve in Section 5.3.

$$\iint\limits_{R} (9 - x^2 - y^2)\, dA = \int_{0}^{2\pi} \int_{0}^{1} (9 - r^2)\, r\, dr\, d\theta$$

$$= \int_{0}^{2\pi} \int_{0}^{1} (9r - r^3)\, dr\, d\theta$$

$$= \int_{0}^{2\pi} \left[\frac{9}{2} r^2 - \frac{1}{4} r^4 \right]_{r=0}^{r=1} d\theta$$

$$= \frac{17}{4} \int_{0}^{2\pi} d\theta = \frac{17\pi}{2}.$$

EXAMPLE 6   Using polar integration, find the area of the region R in the xy-plane enclosed by the circle $x^2 + y^2 + 4$, above the line $y = 1$, and below the line $y = \sqrt{3}x$.

Solution   A sketch of the region R is shown in Figure 15.28. First we note that the line $y = \sqrt{3}x$ has slope $\sqrt{3} = \tan\theta$, so $\theta = \pi/3$. Next we observe that the line $y = 1$ intersects the circle $x^2 + y^2 = 4$ when $x^2 + 1 = 4$, or $x = \sqrt{3}$. Moreover, the radial line from the origin through the point $(\sqrt{3}, 1)$ has slope $1/\sqrt{3} = \tan\theta$, giving its angle of inclination as $\theta = \pi/6$. This information is shown in Figure 15.28.

Now, for the region R, as θ varies from $\pi/6$ to $\pi/3$, the polar coordinate r varies from the horizontal line $y = 1$ to the circle $x^2 + y^2 = 4$. Substituting $r \sin\theta$ for y in the equation for the horizontal line, we have $r \sin\theta = 1$, or $r = \csc\theta$, which is the polar equation of the line. The polar equation for the circle is $r = 2$. So in polar coordinates, for $\pi/6 \le \theta \le \pi/3$, r varies from $r = \csc\theta$ to $r = 2$. It follows that the iterated integral for the area then gives

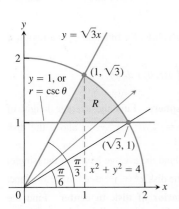

FIGURE 15.28   The region R in Example 6.

$$\iint\limits_{R} dA = \int_{\pi/6}^{\pi/3} \int_{\csc\theta}^{2} r\, dr\, d\theta$$

$$= \int_{\pi/6}^{\pi/3} \left[\frac{1}{2} r^2 \right]_{r=\csc\theta}^{r=2} d\theta$$

$$= \int_{\pi/6}^{\pi/3} \frac{1}{2} \left[4 - \csc^2\theta \right] d\theta$$

$$= \frac{1}{2} \left[4\theta + \cot\theta \right]_{\pi/6}^{\pi/3}$$

$$= \frac{1}{2} \left(\frac{4\pi}{3} + \frac{1}{\sqrt{3}} \right) - \frac{1}{2} \left(\frac{4\pi}{6} + \sqrt{3} \right) = \frac{\pi - \sqrt{3}}{3}.$$

Exercises 15.4

Regions in Polar Coordinates

In Exercises 1–8, describe the given region in polar coordinates.

1.

2.

3.

4.

5.

6.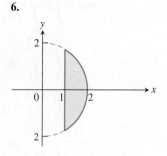

7. The region enclosed by the circle $x^2 + y^2 = 2x$.

8. The region enclosed by the semicircle $x^2 + y^2 = 2y, y \geq 0$.

Evaluating Polar Integrals

In Exercises 9–22, change the Cartesian integral into an equivalent polar integral. Then evaluate the polar integral.

9. $\displaystyle\int_{-1}^{1}\int_{0}^{\sqrt{1-x^2}} dy \, dx$

10. $\displaystyle\int_{0}^{1}\int_{0}^{\sqrt{1-y^2}} (x^2 + y^2) \, dx \, dy$

11. $\displaystyle\int_{0}^{2}\int_{0}^{\sqrt{4-y^2}} (x^2 + y^2) \, dx \, dy$

12. $\displaystyle\int_{-a}^{a}\int_{-\sqrt{a^2-x^2}}^{\sqrt{a^2-x^2}} dy \, dx$

13. $\displaystyle\int_{0}^{6}\int_{0}^{y} x \, dx \, dy$

14. $\displaystyle\int_{0}^{2}\int_{0}^{x} y \, dy \, dx$

15. $\displaystyle\int_{1}^{\sqrt{3}}\int_{1}^{x} dy \, dx$

16. $\displaystyle\int_{\sqrt{2}}^{2}\int_{\sqrt{4-y^2}}^{y} dx \, dy$

17. $\displaystyle\int_{-1}^{0}\int_{-\sqrt{1-x^2}}^{0} \frac{2}{1 + \sqrt{x^2 + y^2}} \, dy \, dx$

18. $\displaystyle\int_{-1}^{1}\int_{-\sqrt{1-x^2}}^{\sqrt{1-x^2}} \frac{2}{(1 + x^2 + y^2)^2} \, dy \, dx$

19. $\displaystyle\int_{0}^{\ln 2}\int_{0}^{\sqrt{(\ln 2)^2-y^2}} e^{\sqrt{x^2+y^2}} \, dx \, dy$

20. $\displaystyle\int_{-1}^{1}\int_{-\sqrt{1-y^2}}^{\sqrt{1-y^2}} \ln (x^2 + y^2 + 1) \, dx \, dy$

21. $\displaystyle\int_{0}^{1}\int_{x}^{\sqrt{2-x^2}} (x + 2y) \, dy \, dx$

22. $\displaystyle\int_{1}^{2}\int_{0}^{\sqrt{2x-x^2}} \frac{1}{(x^2 + y^2)^2} \, dy \, dx$

In Exercises 23–26, sketch the region of integration and convert each polar integral or sum of integrals to a Cartesian integral or sum of integrals. Do not evaluate the integrals.

23. $\displaystyle\int_{0}^{\pi/2}\int_{0}^{1} r^3 \sin\theta \cos\theta \, dr \, d\theta$

24. $\displaystyle\int_{\pi/6}^{\pi/2}\int_{1}^{\csc\theta} r^2 \cos\theta \, dr \, d\theta$

25. $\displaystyle\int_{0}^{\pi/4}\int_{0}^{2\sec\theta} r^5 \sin^2\theta \, dr \, d\theta$

26. $\displaystyle\int_{0}^{\tan^{-1}\frac{4}{3}}\int_{0}^{3\sec\theta} r^7 \, dr \, d\theta + \int_{\tan^{-1}\frac{4}{3}}^{\pi/2}\int_{0}^{4\csc\theta} r^7 \, dr \, d\theta$

Area in Polar Coordinates

27. Find the area of the region cut from the first quadrant by the curve $r = 2(2 - \sin 2\theta)^{1/2}$.

28. **Cardioid overlapping a circle** Find the area of the region that lies inside the cardioid $r = 1 + \cos\theta$ and outside the circle $r = 1$.

29. **One leaf of a rose** Find the area enclosed by one leaf of the rose $r = 12\cos 3\theta$.

30. **Snail shell** Find the area of the region enclosed by the positive x-axis and spiral $r = 4\theta/3, 0 \leq \theta \leq 2\pi$. The region looks like a snail shell.

31. **Cardioid in the first quadrant** Find the area of the region cut from the first quadrant by the cardioid $r = 1 + \sin\theta$.

32. **Overlapping cardioids** Find the area of the region common to the interiors of the cardioids $r = 1 + \cos\theta$ and $r = 1 - \cos\theta$.

Average values

In polar coordinates, the **average value** of a function over a region R (Section 15.3) is given by

$$\frac{1}{\text{Area}(R)} \iint\limits_{R} f(r, \theta) \, r \, dr \, d\theta.$$

33. **Average height of a hemisphere** Find the average height of the hemispherical surface $z = \sqrt{a^2 - x^2 - y^2}$ above the disk $x^2 + y^2 \leq a^2$ in the xy-plane.

34. **Average height of a cone** Find the average height of the (single) cone $z = \sqrt{x^2 + y^2}$ above the disk $x^2 + y^2 \leq a^2$ in the xy-plane.

35. **Average distance from interior of disk to center** Find the average distance from a point $P(x, y)$ in the disk $x^2 + y^2 \leq a^2$ to the origin.

36. **Average distance squared from a point in a disk to a point in its boundary** Find the average value of the *square* of the distance from the point $P(x, y)$ in the disk $x^2 + y^2 \leq 1$ to the boundary point $A(1, 0)$.

Theory and Examples

37. **Converting to a polar integral** Integrate $f(x, y) = [\ln (x^2 + y^2)]/\sqrt{x^2 + y^2}$ over the region $1 \leq x^2 + y^2 \leq e$.

38. **Converting to a polar integral** Integrate $f(x, y) = [\ln (x^2 + y^2)]/(x^2 + y^2)$ over the region $1 \leq x^2 + y^2 \leq e^2$.

39. **Volume of noncircular right cylinder** The region that lies inside the cardioid $r = 1 + \cos\theta$ and outside the circle $r = 1$ is the base of a solid right cylinder. The top of the cylinder lies in the plane $z = x$. Find the cylinder's volume.

40. **Volume of noncircular right cylinder** The region enclosed by the lemniscate $r^2 = 2\cos 2\theta$ is the base of a solid right cylinder whose top is bounded by the sphere $z = \sqrt{2 - r^2}$. Find the cylinder's volume.

41. **Converting to polar integrals**

a. The usual way to evaluate the improper integral $I = \int_{0}^{\infty} e^{-x^2} \, dx$ is first to calculate its square:

$$I^2 = \left(\int_{0}^{\infty} e^{-x^2} \, dx\right)\left(\int_{0}^{\infty} e^{-y^2} \, dy\right) = \int_{0}^{\infty}\int_{0}^{\infty} e^{-(x^2+y^2)} \, dx \, dy.$$

Evaluate the last integral using polar coordinates and solve the resulting equation for I.

b. Evaluate

$$\lim_{x \to \infty} \operatorname{erf}(x) = \lim_{x \to \infty} \int_0^x \frac{2e^{-t^2}}{\sqrt{\pi}} \, dt.$$

42. Converting to a polar integral Evaluate the integral

$$\int_0^\infty \int_0^\infty \frac{1}{(1 + x^2 + y^2)^2} \, dx \, dy.$$

43. Existence Integrate the function $f(x, y) = 1/(1 - x^2 - y^2)$ over the disk $x^2 + y^2 \leq 3/4$. Does the integral of $f(x, y)$ over the disk $x^2 + y^2 \leq 1$ exist? Give reasons for your answer.

44. Area formula in polar coordinates Use the double integral in polar coordinates to derive the formula

$$A = \int_\alpha^\beta \frac{1}{2} r^2 \, d\theta$$

for the area of the fan-shaped region between the origin and polar curve $r = f(\theta)$, $\alpha \leq \theta \leq \beta$.

45. Average distance to a given point inside a disk Let P_0 be a point inside a circle of radius a and let h denote the distance from P_0 to the center of the circle. Let d denote the distance from an arbitrary point P to P_0. Find the average value of d^2 over the region enclosed by the circle. (*Hint:* Simplify your work by placing the center of the circle at the origin and P_0 on the x-axis.)

46. Area Suppose that the area of a region in the polar coordinate plane is

$$A = \int_{\pi/4}^{3\pi/4} \int_{\csc \theta}^{2 \sin \theta} r \, dr \, d\theta.$$

Sketch the region and find its area.

COMPUTER EXPLORATIONS

In Exercises 47–50, use a CAS to change the Cartesian integrals into an equivalent polar integral and evaluate the polar integral. Perform the following steps in each exercise.

a. Plot the Cartesian region of integration in the xy-plane.

b. Change each boundary curve of the Cartesian region in part (a) to its polar representation by solving its Cartesian equation for r and θ.

c. Using the results in part (b), plot the polar region of integration in the $r\theta$-plane.

d. Change the integrand from Cartesian to polar coordinates. Determine the limits of integration from your plot in part (c) and evaluate the polar integral using the CAS integration utility.

47. $\displaystyle\int_0^1 \int_x^1 \frac{y}{x^2 + y^2} \, dy \, dx$ **48.** $\displaystyle\int_0^1 \int_0^{x/2} \frac{x}{x^2 + y^2} \, dy \, dx$

49. $\displaystyle\int_0^1 \int_{-y/3}^{y/3} \frac{y}{\sqrt{x^2 + y^2}} \, dx \, dy$ **50.** $\displaystyle\int_0^1 \int_y^{2-y} \sqrt{x + y} \, dx \, dy$

15.5 | Triple Integrals in Rectangular Coordinates

Just as double integrals allow us to deal with more general situations than could be handled by single integrals, triple integrals enable us to solve still more general problems. We use triple integrals to calculate the volumes of three-dimensional shapes and the average value of a function over a three-dimensional region. Triple integrals also arise in the study of vector fields and fluid flow in three dimensions, as we will see in Chapter 16.

Triple Integrals

If $F(x, y, z)$ is a function defined on a closed, bounded region D in space, such as the region occupied by a solid ball or a lump of clay, then the integral of F over D may be defined in the following way. We partition a rectangular boxlike region containing D into rectangular cells by planes parallel to the coordinate axes (Figure 15.29). We number the cells that lie completely inside D from 1 to n in some order, the kth cell having dimensions Δx_k by Δy_k by Δz_k and volume $\Delta V_k = \Delta x_k \Delta y_k \Delta z_k$. We choose a point (x_k, y_k, z_k) in each cell and form the sum

$$S_n = \sum_{k=1}^n F(x_k, y_k, z_k) \, \Delta V_k. \tag{1}$$

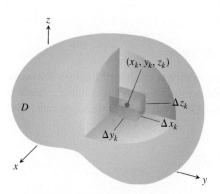

FIGURE 15.29 Partitioning a solid with rectangular cells of volume ΔV_k.

We are interested in what happens as D is partitioned by smaller and smaller cells, so that Δx_k, Δy_k, Δz_k and the norm of the partition $\|P\|$, the largest value among Δx_k, Δy_k, Δz_k, all approach zero. When a single limiting value is attained, no matter how the partitions and points (x_k, y_k, z_k) are chosen, we say that F is **integrable** over D. As before, it can be shown that when F is continuous and the bounding surface of D is formed from finitely many smooth surfaces joined together along finitely many smooth curves, then F is integrable. As $\|P\| \to 0$ and the number of cells n goes to ∞, the sums S_n approach a limit. We call this limit the **triple integral of F over D** and write

$$\lim_{n \to \infty} S_n = \iiint\limits_{D} F(x, y, z)\, dV \quad \text{or} \quad \lim_{\|P\| \to 0} S_n = \iiint\limits_{D} F(x, y, z)\, dx\, dy\, dz.$$

The regions D over which continuous functions are integrable are those having "reasonably smooth" boundaries.

Volume of a Region in Space

If F is the constant function whose value is 1, then the sums in Equation (1) reduce to

$$S_n = \sum F(x_k, y_k, z_k)\, \Delta V_k = \sum 1 \cdot \Delta V_k = \sum \Delta V_k.$$

As Δx_k, Δy_k, and Δz_k approach zero, the cells ΔV_k become smaller and more numerous and fill up more and more of D. We therefore define the volume of D to be the triple integral

$$\lim_{n \to \infty} \sum_{k=1}^{n} \Delta V_k = \iiint\limits_{D} dV.$$

> **DEFINITION** The **volume** of a closed, bounded region D in space is
>
> $$V = \iiint\limits_{D} dV.$$

This definition is in agreement with our previous definitions of volume, although we omit the verification of this fact. As we see in a moment, this integral enables us to calculate the volumes of solids enclosed by curved surfaces.

Finding Limits of Integration in the Order *dz dy dx*

We evaluate a triple integral by applying a three-dimensional version of Fubini's Theorem (Section 15.2) to evaluate it by three repeated single integrations. As with double integrals, there is a geometric procedure for finding the limits of integration for these single integrals.

To evaluate

$$\iiint\limits_{D} F(x, y, z)\, dV$$

over a region D, integrate first with respect to z, then with respect to y, and finally with respect to x. (You might choose a different order of integration, but the procedure is similar, as we illustrate in Example 2.)

1. *Sketch.* Sketch the region D along with its "shadow" R (vertical projection) in the xy-plane. Label the upper and lower bounding surfaces of D and the upper and lower bounding curves of R.

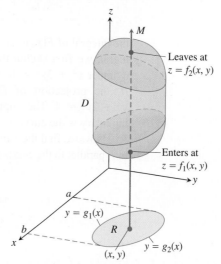

2. *Find the z-limits of integration.* Draw a line M passing through a typical point (x, y) in R parallel to the z-axis. As z increases, M enters D at $z = f_1(x, y)$ and leaves at $z = f_2(x, y)$. These are the z-limits of integration.

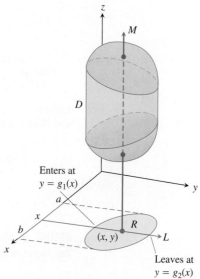

3. *Find the y-limits of integration.* Draw a line L through (x, y) parallel to the y-axis. As y increases, L enters R at $y = g_1(x)$ and leaves at $y = g_2(x)$. These are the y-limits of integration.

4. *Find the x-limits of integration.* Choose x-limits that include all lines through R parallel to the y-axis ($x = a$ and $x = b$ in the preceding figure). These are the x-limits of integration. The integral is

$$\int_{x=a}^{x=b} \int_{y=g_1(x)}^{y=g_2(x)} \int_{z=f_1(x, y)}^{z=f_2(x, y)} F(x, y, z)\, dz\, dy\, dx.$$

Follow similar procedures if you change the order of integration. The "shadow" of region D lies in the plane of the last two variables with respect to which the iterated integration takes place.

The preceding procedure applies whenever a solid region D is bounded above and below by a surface, and when the "shadow" region R is bounded by a lower and upper curve. It does not apply to regions with complicated holes through them, although sometimes such regions can be subdivided into simpler regions for which the procedure does apply.

EXAMPLE 1   Find the volume of the region D enclosed by the surfaces $z = x^2 + 3y^2$ and $z = 8 - x^2 - y^2$.

Solution   The volume is

$$V = \iiint_D dz\, dy\, dx,$$

the integral of $F(x, y, z) = 1$ over D. To find the limits of integration for evaluating the integral, we first sketch the region. The surfaces (Figure 15.30) intersect on the elliptical cylinder $x^2 + 3y^2 = 8 - x^2 - y^2$ or $x^2 + 2y^2 = 4$, $z > 0$. The boundary of the region R, the projection of D onto the xy-plane, is an ellipse with the same equation: $x^2 + 2y^2 = 4$. The "upper" boundary of R is the curve $y = \sqrt{(4 - x^2)/2}$. The lower boundary is the curve $y = -\sqrt{(4 - x^2)/2}$.

Now we find the z-limits of integration. The line M passing through a typical point (x, y) in R parallel to the z-axis enters D at $z = x^2 + 3y^2$ and leaves at $z = 8 - x^2 - y^2$.

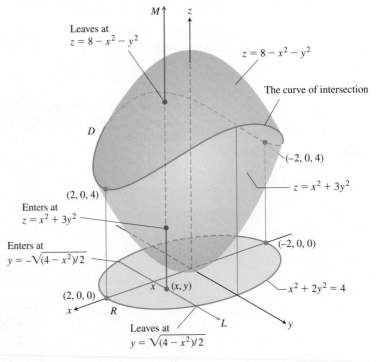

FIGURE 15.30   The volume of the region enclosed by two paraboloids, calculated in Example 1.

Next we find the y-limits of integration. The line L through (x, y) parallel to the y-axis enters R at $y = -\sqrt{(4 - x^2)/2}$ and leaves at $y = \sqrt{(4 - x^2)/2}$.

Finally we find the x-limits of integration. As L sweeps across R, the value of x varies from $x = -2$ at $(-2, 0, 0)$ to $x = 2$ at $(2, 0, 0)$. The volume of D is

$$
V = \iiint\limits_{D} dz\, dy\, dx
$$

$$
= \int_{-2}^{2} \int_{-\sqrt{(4-x^2)/2}}^{\sqrt{(4-x^2)/2}} \int_{x^2+3y^2}^{8-x^2-y^2} dz\, dy\, dx
$$

$$
= \int_{-2}^{2} \int_{-\sqrt{(4-x^2)/2}}^{\sqrt{(4-x^2)/2}} (8 - 2x^2 - 4y^2)\, dy\, dx
$$

$$
= \int_{-2}^{2} \left[(8 - 2x^2)y - \frac{4}{3}y^3 \right]_{y=-\sqrt{(4-x^2)/2}}^{y=\sqrt{(4-x^2)/2}} dx
$$

$$
= \int_{-2}^{2} \left(2(8 - 2x^2)\sqrt{\frac{4 - x^2}{2}} - \frac{8}{3}\left(\frac{4 - x^2}{2}\right)^{3/2} \right) dx
$$

$$
= \int_{-2}^{2} \left[8\left(\frac{4 - x^2}{2}\right)^{3/2} - \frac{8}{3}\left(\frac{4 - x^2}{2}\right)^{3/2} \right] dx = \frac{4\sqrt{2}}{3}\int_{-2}^{2} (4 - x^2)^{3/2}\, dx
$$

$$
= 8\pi\sqrt{2}. \qquad \text{\small After integration with the substitution } x = 2 \sin u \qquad \blacksquare
$$

In the next example, we project D onto the xz-plane instead of the xy-plane, to show how to use a different order of integration.

EXAMPLE 2 Set up the limits of integration for evaluating the triple integral of a function $F(x, y, z)$ over the tetrahedron D with vertices $(0, 0, 0)$, $(1, 1, 0)$, $(0, 1, 0)$, and $(0, 1, 1)$. Use the order of integration $dy\, dz\, dx$.

Solution We sketch D along with its "shadow" R in the xz-plane (Figure 15.31). The upper (right-hand) bounding surface of D lies in the plane $y = 1$. The lower (left-hand) bounding surface lies in the plane $y = x + z$. The upper boundary of R is the line $z = 1 - x$. The lower boundary is the line $z = 0$.

First we find the y-limits of integration. The line through a typical point (x, z) in R parallel to the y-axis enters D at $y = x + z$ and leaves at $y = 1$.

Next we find the z-limits of integration. The line L through (x, z) parallel to the z-axis enters R at $z = 0$ and leaves at $z = 1 - x$.

Finally we find the x-limits of integration. As L sweeps across R, the value of x varies from $x = 0$ to $x = 1$. The integral is

$$
\int_{0}^{1} \int_{0}^{1-x} \int_{x+z}^{1} F(x, y, z)\, dy\, dz\, dx. \qquad \blacksquare
$$

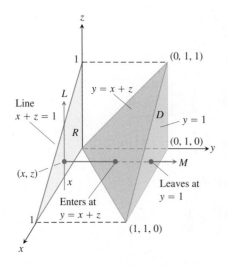

FIGURE 15.31 Finding the limits of integration for evaluating the triple integral of a function defined over the tetrahedron D (Examples 2 and 3).

EXAMPLE 3 Integrate $F(x, y, z) = 1$ over the tetrahedron D in Example 2 in the order $dz\, dy\, dx$, and then integrate in the order $dy\, dz\, dx$.

Solution First we find the z-limits of integration. A line M parallel to the z-axis through a typical point (x, y) in the xy-plane "shadow" enters the tetrahedron at $z = 0$ and exits through the upper plane where $z = y - x$ (Figure 15.32).

Next we find the y-limits of integration. On the xy-plane, where $z = 0$, the sloped side of the tetrahedron crosses the plane along the line $y = x$. A line L through (x, y) parallel to the y-axis enters the shadow in the xy-plane at $y = x$ and exits at $y = 1$ (Figure 15.32).

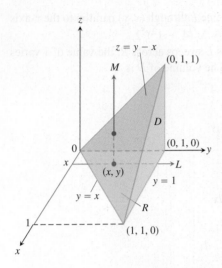

FIGURE 15.32 The tetrahedron in Example 3 showing how the limits of integration are found for the order $dz\,dy\,dx$.

Finally we find the x-limits of integration. As the line L parallel to the y-axis in the previous step sweeps out the shadow, the value of x varies from $x = 0$ to $x = 1$ at the point $(1, 1, 0)$ (see Figure 15.32). The integral is

$$\int_0^1 \int_x^1 \int_0^{y-x} F(x, y, z)\, dz\, dy\, dx.$$

For example, if $F(x, y, z) = 1$, we would find the volume of the tetrahedron to be

$$
\begin{aligned}
V &= \int_0^1 \int_x^1 \int_0^{y-x} dz\, dy\, dx \\
&= \int_0^1 \int_x^1 (y - x)\, dy\, dx \\
&= \int_0^1 \left[\frac{1}{2} y^2 - xy\right]_{y=x}^{y=1} dx \\
&= \int_0^1 \left(\frac{1}{2} - x + \frac{1}{2} x^2\right) dx \\
&= \left[\frac{1}{2} x - \frac{1}{2} x^2 + \frac{1}{6} x^3\right]_0^1 \\
&= \frac{1}{6}.
\end{aligned}
$$

We get the same result by integrating with the order $dy\,dz\,dx$. From Example 2,

$$
\begin{aligned}
V &= \int_0^1 \int_0^{1-x} \int_{x+z}^1 dy\, dz\, dx \\
&= \int_0^1 \int_0^{1-x} (1 - x - z)\, dz\, dx \\
&= \int_0^1 \left[(1 - x)z - \frac{1}{2} z^2\right]_{z=0}^{z=1-x} dx \\
&= \int_0^1 \left[(1 - x)^2 - \frac{1}{2}(1 - x)^2\right] dx \\
&= \frac{1}{2} \int_0^1 (1 - x)^2\, dx \\
&= -\frac{1}{6}(1 - x)^3 \Big]_0^1 = \frac{1}{6}.
\end{aligned}
$$

∎

Average Value of a Function in Space

The average value of a function F over a region D in space is defined by the formula

$$\textbf{Average value of } F \textbf{ over } D = \frac{1}{\text{volume of } D} \iiint_D F\, dV. \tag{2}$$

For example, if $F(x, y, z) = \sqrt{x^2 + y^2 + z^2}$, then the average value of F over D is the average distance of points in D from the origin. If $F(x, y, z)$ is the temperature at (x, y, z) on a solid that occupies a region D in space, then the average value of F over D is the average temperature of the solid.

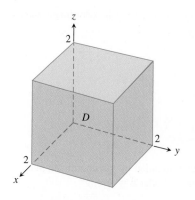

FIGURE 15.33 The region of integration in Example 4.

EXAMPLE 4 Find the average value of $F(x, y, z) = xyz$ throughout the cubical region D bounded by the coordinate planes and the planes $x = 2$, $y = 2$, and $z = 2$ in the first octant.

Solution We sketch the cube with enough detail to show the limits of integration (Figure 15.33). We then use Equation (2) to calculate the average value of F over the cube.

The volume of the region D is $(2)(2)(2) = 8$. The value of the integral of F over the cube is

$$\int_0^2 \int_0^2 \int_0^2 xyz \, dx \, dy \, dz = \int_0^2 \int_0^2 \left[\frac{x^2}{2} yz\right]_{x=0}^{x=2} dy \, dz = \int_0^2 \int_0^2 2yz \, dy \, dz$$

$$= \int_0^2 \left[y^2 z\right]_{y=0}^{y=2} dz = \int_0^2 4z \, dz = \left[2z^2\right]_0^2 = 8.$$

With these values, Equation (2) gives

$$\text{Average value of } xyz \text{ over the cube} = \frac{1}{\text{volume}} \iiint\limits_{\text{cube}} xyz \, dV = \left(\frac{1}{8}\right)(8) = 1.$$

In evaluating the integral, we chose the order $dx \, dy \, dz$, but any of the other five possible orders would have done as well. ∎

Properties of Triple Integrals

Triple integrals have the same algebraic properties as double and single integrals. Simply replace the double integrals in the four properties given in Section 15.2, page 864, with triple integrals.

Exercises 15.5

Triple Integrals in Different Iteration Orders

1. Evaluate the integral in Example 2 taking $F(x, y, z) = 1$ to find the volume of the tetrahedron in the order $dz \, dx \, dy$.

2. **Volume of rectangular solid** Write six different iterated triple integrals for the volume of the rectangular solid in the first octant bounded by the coordinate planes and the planes $x = 1$, $y = 2$, and $z = 3$. Evaluate one of the integrals.

3. **Volume of tetrahedron** Write six different iterated triple integrals for the volume of the tetrahedron cut from the first octant by the plane $6x + 3y + 2z = 6$. Evaluate one of the integrals.

4. **Volume of solid** Write six different iterated triple integrals for the volume of the region in the first octant enclosed by the cylinder $x^2 + z^2 = 4$ and the plane $y = 3$. Evaluate one of the integrals.

5. **Volume enclosed by paraboloids** Let D be the region bounded by the paraboloids $z = 8 - x^2 - y^2$ and $z = x^2 + y^2$. Write six different triple iterated integrals for the volume of D. Evaluate one of the integrals.

6. **Volume inside paraboloid beneath a plane** Let D be the region bounded by the paraboloid $z = x^2 + y^2$ and the plane $z = 2y$. Write triple iterated integrals in the order $dz \, dx \, dy$ and $dz \, dy \, dx$ that give the volume of D. Do not evaluate either integral.

Evaluating Triple Iterated Integrals

Evaluate the integrals in Exercises 7–20.

7. $\displaystyle\int_0^1 \int_0^1 \int_0^1 (x^2 + y^2 + z^2) \, dz \, dy \, dx$

8. $\displaystyle\int_0^{\sqrt{2}} \int_0^{3y} \int_{x^2+3y^2}^{8-x^2-y^2} dz \, dx \, dy$

9. $\displaystyle\int_1^e \int_1^{e^2} \int_1^{e^3} \frac{1}{xyz} \, dx \, dy \, dz$

10. $\displaystyle\int_0^1 \int_0^{3-3x} \int_0^{3-3x-y} dz \, dy \, dx$

11. $\displaystyle\int_0^{\pi/6} \int_0^1 \int_{-2}^3 y \sin z \, dx \, dy \, dz$

12. $\displaystyle\int_{-1}^1 \int_0^1 \int_0^2 (x + y + z) \, dy \, dx \, dz$

13. $\displaystyle\int_0^3 \int_0^{\sqrt{9-x^2}} \int_0^{\sqrt{9-x^2}} dz \, dy \, dx$

14. $\displaystyle\int_0^2 \int_{-\sqrt{4-y^2}}^{\sqrt{4-y^2}} \int_0^{2x+y} dz \, dx \, dy$

15. $\displaystyle\int_0^1 \int_0^{2-x} \int_0^{2-x-y} dz \, dy \, dx$

16. $\displaystyle\int_0^1 \int_0^{1-x^2} \int_3^{4-x^2-y} x \, dz \, dy \, dx$

17. $\displaystyle\int_0^\pi \int_0^\pi \int_0^\pi \cos(u + v + w) \, du \, dv \, dw$ (uvw-space)

18. $\displaystyle\int_0^1 \int_1^{\sqrt{e}} \int_1^e se^s \ln r \frac{(\ln t)^2}{t} \, dt \, dr \, ds$ (rst-space)

19. $\int_0^{\pi/4} \int_0^{\ln \sec v} \int_{-\infty}^{2t} e^x \, dx \, dt \, dv$ (*tvx*-space)

20. $\int_0^7 \int_0^2 \int_0^{\sqrt{4-q^2}} \dfrac{q}{r+1} \, dp \, dq \, dr$ (*pqr*-space)

Finding Equivalent Iterated Integrals

21. Here is the region of integration of the integral

$$\int_{-1}^{1} \int_{x^2}^{1} \int_0^{1-y} dz \, dy \, dx.$$

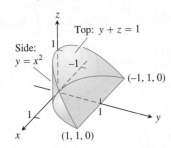

Rewrite the integral as an equivalent iterated integral in the order

a. *dy dz dx*   **b.** *dy dx dz*

c. *dx dy dz*   **d.** *dx dz dy*

e. *dz dx dy.*

22. Here is the region of integration of the integral

$$\int_0^1 \int_{-1}^0 \int_0^{y^2} dz \, dy \, dx.$$

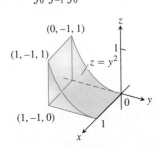

Rewrite the integral as an equivalent iterated integral in the order

a. *dy dz dx*   **b.** *dy dx dz*

c. *dx dy dz*   **d.** *dx dz dy*

e. *dz dx dy.*

Finding Volumes Using Triple Integrals

Find the volumes of the regions in Exercises 23–36.

23. The region between the cylinder $z = y^2$ and the *xy*-plane that is bounded by the planes $x = 0, x = 1, y = -1, y = 1$

24. The region in the first octant bounded by the coordinate planes and the planes $x + z = 1, y + 2z = 2$

25. The region in the first octant bounded by the coordinate planes, the plane $y + z = 2$, and the cylinder $x = 4 - y^2$

26. The wedge cut from the cylinder $x^2 + y^2 = 1$ by the planes $z = -y$ and $z = 0$

27. The tetrahedron in the first octant bounded by the coordinate planes and the plane passing through $(1, 0, 0), (0, 2, 0),$ and $(0, 0, 3)$

28. The region in the first octant bounded by the coordinate planes, the plane $y = 1 - x$, and the surface $z = \cos(\pi x/2)$, $0 \le x \le 1$

29. The region common to the interiors of the cylinders $x^2 + y^2 = 1$ and $x^2 + z^2 = 1$, one-eighth of which is shown in the accompanying figure

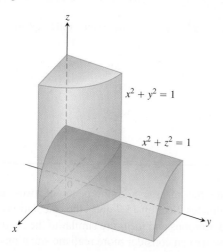

30. The region in the first octant bounded by the coordinate planes and the surface $z = 4 - x^2 - y$

31. The region in the first octant bounded by the coordinate planes, the plane $x + y = 4$, and the cylinder $y^2 + 4z^2 = 16$

32. The region cut from the cylinder $x^2 + y^2 = 4$ by the plane $z = 0$ and the plane $x + z = 3$

33. The region between the planes $x + y + 2z = 2$ and $2x + 2y + z = 4$ in the first octant

34. The finite region bounded by the planes $z = x$, $x + z = 8$, $z = y$, $y = 8$, and $z = 0$

35. The region cut from the solid elliptical cylinder $x^2 + 4y^2 \leq 4$ by the xy-plane and the plane $z = x + 2$

36. The region bounded in back by the plane $x = 0$, on the front and sides by the parabolic cylinder $x = 1 - y^2$, on the top by the paraboloid $z = x^2 + y^2$, and on the bottom by the xy-plane

Average Values

In Exercises 37–40, find the average value of $F(x, y, z)$ over the given region.

37. $F(x, y, z) = x^2 + 9$ over the cube in the first octant bounded by the coordinate planes and the planes $x = 2$, $y = 2$, and $z = 2$

38. $F(x, y, z) = x + y - z$ over the rectangular solid in the first octant bounded by the coordinate planes and the planes $x = 1$, $y = 1$, and $z = 2$

39. $F(x, y, z) = x^2 + y^2 + z^2$ over the cube in the first octant bounded by the coordinate planes and the planes $x = 1$, $y = 1$, and $z = 1$

40. $F(x, y, z) = xyz$ over the cube in the first octant bounded by the coordinate planes and the planes $x = 2$, $y = 2$, and $z = 2$

Changing the Order of Integration

Evaluate the integrals in Exercises 41–44 by changing the order of integration in an appropriate way.

41. $\displaystyle\int_0^4 \int_0^1 \int_{2y}^2 \frac{4 \cos(x^2)}{2\sqrt{z}}\, dx\, dy\, dz$

42. $\displaystyle\int_0^1 \int_0^1 \int_{x^2}^1 12xze^{zy^2}\, dy\, dx\, dz$

43. $\displaystyle\int_0^1 \int_{\sqrt[3]{z}}^1 \int_0^{\ln 3} \frac{\pi e^{2x} \sin \pi y^2}{y^2}\, dx\, dy\, dz$

44. $\displaystyle\int_0^2 \int_0^{4-x^2} \int_0^x \frac{\sin 2z}{4 - z}\, dy\, dz\, dx$

Theory and Examples

45. Finding an upper limit of an iterated integral   Solve for a:

$$\int_0^1 \int_0^{4-a-x^2} \int_a^{4-x^2-y} dz\, dy\, dx = \frac{4}{15}.$$

46. Ellipsoid   For what value of c is the volume of the ellipsoid $x^2 + (y/2)^2 + (z/c)^2 = 1$ equal to 8π?

47. Minimizing a triple integral   What domain D in space minimizes the value of the integral

$$\iiint_D (4x^2 + 4y^2 + z^2 - 4)\, dV?$$

Give reasons for your answer.

48. Maximizing a triple integral   What domain D in space maximizes the value of the integral

$$\iiint_D (1 - x^2 - y^2 - z^2)\, dV?$$

Give reasons for your answer.

COMPUTER EXPLORATIONS

In Exercises 49–52, use a CAS integration utility to evaluate the triple integral of the given function over the specified solid region.

49. $F(x, y, z) = x^2 y^2 z$ over the solid cylinder bounded by $x^2 + y^2 = 1$ and the planes $z = 0$ and $z = 1$

50. $F(x, y, z) = |xyz|$ over the solid bounded below by the paraboloid $z = x^2 + y^2$ and above by the plane $z = 1$

51. $F(x, y, z) = \dfrac{z}{(x^2 + y^2 + z^2)^{3/2}}$ over the solid bounded below by the cone $z = \sqrt{x^2 + y^2}$ and above by the plane $z = 1$

52. $F(x, y, z) = x^4 + y^2 + z^2$ over the solid sphere $x^2 + y^2 + z^2 \le 1$

15.6 | Moments and Centers of Mass

This section shows how to calculate the masses and moments of two- and three-dimensional objects in Cartesian coordinates. Section 15.7 gives the calculations for cylindrical and spherical coordinates. The definitions and ideas are similar to the single-variable case we studied in Section 6.6, but now we can consider more realistic situations.

Masses and First Moments

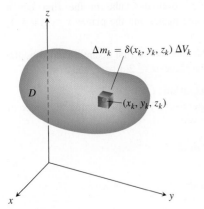

FIGURE 15.34 To define an object's mass, we first imagine it to be partitioned into a finite number of mass elements Δm_k.

If $\delta(x, y, z)$ is the density (mass per unit volume) of an object occupying a region D in space, the integral of δ over D gives the **mass** of the object. To see why, imagine partitioning the object into n mass elements like the one in Figure 15.34. The object's mass is the limit

$$M = \lim_{n \to \infty} \sum_{k=1}^{n} \Delta m_k = \lim_{n \to \infty} \sum_{k=1}^{n} \delta(x_k, y_k, z_k) \, \Delta V_k = \iiint_D \delta(x, y, z) \, dV.$$

The *first moment* of a solid region D about a coordinate plane is defined as the triple integral over D of the distance from a point (x, y, z) in D to the plane multiplied by the density of the solid at that point. For instance, the first moment about the yz-plane is the integral

$$M_{yz} = \iiint_D x\delta(x, y, z) \, dV.$$

The *center of mass* is found from the first moments. For instance, the x-coordinate of the center of mass is $\bar{x} = M_{yz}/M$.

For a two-dimensional object, such as a thin, flat plate, we calculate first moments about the coordinate axes by simply dropping the z-coordinate. So the first moment about the y-axis is the double integral over the region R forming the plate of the distance from the axis multiplied by the density, or

$$M_y = \iint_R x\delta(x, y) \, dA.$$

Table 15.1 summarizes the formulas.

FIGURE 15.35 Finding the center of mass of a solid (Example 1).

EXAMPLE 1   Find the center of mass of a solid of constant density δ bounded below by the disk $R: x^2 + y^2 \le 4$ in the plane $z = 0$ and above by the paraboloid $z = 4 - x^2 - y^2$ (Figure 15.35).

TABLE 15.1 Mass and first moment formulas

THREE-DIMENSIONAL SOLID

Mass: $M = \iiint\limits_{D} \delta \, dV$ $\qquad \delta = \delta(x, y, z)$ is the density at (x, y, z).

First moments about the coordinate planes:

$$M_{yz} = \iiint\limits_{D} x \, \delta \, dV, \qquad M_{xz} = \iiint\limits_{D} y \, \delta \, dV, \qquad M_{xy} = \iiint\limits_{D} z \, \delta \, dV$$

Center of mass:

$$\bar{x} = \frac{M_{yz}}{M}, \qquad \bar{y} = \frac{M_{xz}}{M}, \qquad \bar{z} = \frac{M_{xy}}{M}$$

TWO-DIMENSIONAL PLATE

Mass: $M = \iint\limits_{R} \delta \, dA$ $\qquad \delta = \delta(x, y)$ is the density at (x, y).

First moments: $M_y = \iint\limits_{R} x \, \delta \, dA, \qquad M_x = \iint\limits_{R} y \, \delta \, dA$

Center of mass: $\bar{x} = \frac{M_y}{M}, \qquad \bar{y} = \frac{M_x}{M}$

Solution By symmetry $\bar{x} = \bar{y} = 0$. To find $\bar{z}$, we first calculate

$$M_{xy} = \iiint\limits_{R}\int_{z=0}^{z=4-x^2-y^2} z \, \delta \, dz \, dy \, dx = \iint\limits_{R} \left[\frac{z^2}{2}\right]_{z=0}^{z=4-x^2-y^2} \delta \, dy \, dx$$

$$= \frac{\delta}{2} \iint\limits_{R} (4 - x^2 - y^2)^2 \, dy \, dx$$

$$= \frac{\delta}{2} \int_0^{2\pi} \int_0^2 (4 - r^2)^2 \, r \, dr \, d\theta \qquad \text{Polar coordinates simplify the integration.}$$

$$= \frac{\delta}{2} \int_0^{2\pi} \left[-\frac{1}{6}(4 - r^2)^3\right]_{r=0}^{r=2} d\theta = \frac{16\delta}{3} \int_0^{2\pi} d\theta = \frac{32\pi\delta}{3}.$$

A similar calculation gives the mass

$$M = \iiint\limits_{R}\int_0^{4-x^2-y^2} \delta \, dz \, dy \, dx = 8\pi\delta.$$

Therefore $\bar{z} = (M_{xy}/M) = 4/3$ and the center of mass is $(\bar{x}, \bar{y}, \bar{z}) = (0, 0, 4/3)$. ∎

When the density of a solid object or plate is constant (as in Example 1), the center of mass is called the **centroid** of the object. To find a centroid, we set δ equal to 1 and proceed to find $\bar{x}$, $\bar{y}$, and $\bar{z}$ as before, by dividing first moments by masses. These calculations are also valid for two-dimensional objects.

EXAMPLE 2 Find the centroid of the region in the first quadrant that is bounded above by the line $y = x$ and below by the parabola $y = x^2$.

FIGURE 15.36 The centroid of this region is found in Example 2.

Solution We sketch the region and include enough detail to determine the limits of integration (Figure 15.36). We then set δ equal to 1 and evaluate the appropriate formulas from Table 15.1:

$$M = \int_0^1 \int_{x^2}^x 1 \, dy \, dx = \int_0^1 \left[y \right]_{y=x^2}^{y=x} dx = \int_0^1 (x - x^2) \, dx = \left[\frac{x^2}{2} - \frac{x^3}{3} \right]_0^1 = \frac{1}{6}$$

$$M_x = \int_0^1 \int_{x^2}^x y \, dy \, dx = \int_0^1 \left[\frac{y^2}{2} \right]_{y=x^2}^{y=x} dx$$

$$= \int_0^1 \left(\frac{x^2}{2} - \frac{x^4}{2} \right) dx = \left[\frac{x^3}{6} - \frac{x^5}{10} \right]_0^1 = \frac{1}{15}$$

$$M_y = \int_0^1 \int_{x^2}^x x \, dy \, dx = \int_0^1 \left[xy \right]_{y=x^2}^{y=x} dx = \int_0^1 (x^2 - x^3) \, dx = \left[\frac{x^3}{3} - \frac{x^4}{4} \right]_0^1 = \frac{1}{12}.$$

From these values of M, M_x, and M_y, we find

$$\bar{x} = \frac{M_y}{M} = \frac{1/12}{1/6} = \frac{1}{2} \quad \text{and} \quad \bar{y} = \frac{M_x}{M} = \frac{1/15}{1/6} = \frac{2}{5}.$$

The centroid is the point $(1/2, 2/5)$. ∎

Moments of Inertia

An object's first moments (Table 15.1) tell us about balance and about the torque the object experiences about different axes in a gravitational field. If the object is a rotating shaft, however, we are more likely to be interested in how much energy is stored in the shaft or about how much energy is generated by a shaft rotating at a particular angular velocity. This is where the second moment or moment of inertia comes in.

Think of partitioning the shaft into small blocks of mass Δm_k and let r_k denote the distance from the kth block's center of mass to the axis of rotation (Figure 15.37). If the shaft rotates at a constant angular velocity of $\omega = d\theta/dt$ radians per second, the block's center of mass will trace its orbit at a linear speed of

$$v_k = \frac{d}{dt}(r_k\theta) = r_k \frac{d\theta}{dt} = r_k\omega.$$

The block's kinetic energy will be approximately

$$\frac{1}{2}\Delta m_k v_k^2 = \frac{1}{2} \Delta m_k (r_k\omega)^2 = \frac{1}{2} \omega^2 r_k^2 \, \Delta m_k.$$

The kinetic energy of the shaft will be approximately

$$\sum \frac{1}{2} \omega^2 r_k^2 \, \Delta m_k.$$

The integral approached by these sums as the shaft is partitioned into smaller and smaller blocks gives the shaft's kinetic energy:

$$\text{KE}_{\text{shaft}} = \int \frac{1}{2} \omega^2 r^2 \, dm = \frac{1}{2} \omega^2 \int r^2 \, dm. \qquad (1)$$

The factor

$$I = \int r^2 \, dm$$

is the *moment of inertia* of the shaft about its axis of rotation, and we see from Equation (1) that the shaft's kinetic energy is

$$\text{KE}_{\text{shaft}} = \frac{1}{2} I\omega^2.$$

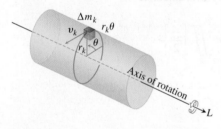

FIGURE 15.37 To find an integral for the amount of energy stored in a rotating shaft, we first imagine the shaft to be partitioned into small blocks. Each block has its own kinetic energy. We add the contributions of the individual blocks to find the kinetic energy of the shaft.

15.8 Substitutions in Multiple Integrals

The goal of this section is to introduce you to the ideas involved in coordinate transformations. You will see how to evaluate multiple integrals by substitution in order to replace complicated integrals by ones that are easier to evaluate. Substitutions accomplish this by simplifying the integrand, the limits of integration, or both. A thorough discussion of multivariable transformations and substitutions, and the *Jacobian*, is best left to a more advanced course following a study of linear algebra.

Substitutions in Double Integrals

The polar coordinate substitution of Section 15.4 is a special case of a more general substitution method for double integrals, a method that pictures changes in variables as transformations of regions.

Suppose that a region G in the uv-plane is transformed one-to-one into the region R in the xy-plane by equations of the form

$$x = g(u, v), \qquad y = h(u, v),$$

as suggested in Figure 15.53. We call R the **image** of G under the transformation, and G the **preimage** of R. Any function $f(x, y)$ defined on R can be thought of as a function $f(g(u, v), h(u, v))$ defined on G as well. How is the integral of $f(x, y)$ over R related to the integral of $f(g(u, v), h(u, v))$ over G?

The answer is: If g, h, and f have continuous partial derivatives and $J(u, v)$ (to be discussed in a moment) is zero only at isolated points, if at all, then

$$\iint\limits_{R} f(x, y)\, dx\, dy = \iint\limits_{G} f(g(u, v), h(u, v))\, |J(u, v)|\, du\, dv. \tag{1}$$

The factor $J(u, v)$, whose absolute value appears in Equation (1), is the *Jacobian* of the coordinate transformation, named after German mathematician Carl Jacobi. It measures how much the transformation is expanding or contracting the area around a point in G as G is transformed into R.

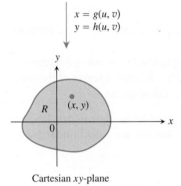

FIGURE 15.53 The equations $x = g(u, v)$ and $y = h(u, v)$ allow us to change an integral over a region R in the xy-plane into an integral over a region G in the uv-plane by using Equation (1).

DEFINITION   The **Jacobian determinant** or **Jacobian** of the coordinate transformation $x = g(u, v), y = h(u, v)$ is

$$J(u, v) = \begin{vmatrix} \dfrac{\partial x}{\partial u} & \dfrac{\partial x}{\partial v} \\[2ex] \dfrac{\partial y}{\partial u} & \dfrac{\partial y}{\partial v} \end{vmatrix} = \frac{\partial x}{\partial u}\frac{\partial y}{\partial v} - \frac{\partial y}{\partial u}\frac{\partial x}{\partial v}. \tag{2}$$

The Jacobian can also be denoted by

$$J(u, v) = \frac{\partial(x, y)}{\partial(u, v)}$$

to help us remember how the determinant in Equation (2) is constructed from the partial derivatives of x and y. The derivation of Equation (1) is intricate and properly belongs to a course in advanced calculus. We do not give the derivation here.

HISTORICAL BIOGRAPHY

Carl Gustav Jacob Jacobi
(1804–1851)

EXAMPLE 1   Find the Jacobian for the polar coordinate transformation $x = r \cos \theta$, $y = r \sin \theta$, and use Equation (1) to write the Cartesian integral $\iint_{R} f(x, y)\, dx\, dy$ as a polar integral.

Cartesian $r\theta$-plane

$$x = r \cos \theta$$
$$y = r \sin \theta$$

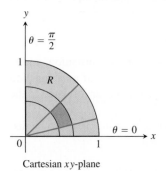

Cartesian xy-plane

FIGURE 15.54 The equations $x = r \cos \theta, y = r \sin \theta$ transform G into R.

Solution Figure 15.54 shows how the equations $x = r \cos \theta, y = r \sin \theta$ transform the rectangle $G: 0 \le r \le 1, 0 \le \theta \le \pi/2$, into the quarter circle R bounded by $x^2 + y^2 = 1$ in the first quadrant of the xy-plane.

For polar coordinates, we have r and θ in place of u and v. With $x = r \cos \theta$ and $y = r \sin \theta$, the Jacobian is

$$J(r, \theta) = \begin{vmatrix} \dfrac{\partial x}{\partial r} & \dfrac{\partial x}{\partial \theta} \\ \dfrac{\partial y}{\partial r} & \dfrac{\partial y}{\partial \theta} \end{vmatrix} = \begin{vmatrix} \cos \theta & -r \sin \theta \\ \sin \theta & r \cos \theta \end{vmatrix} = r(\cos^2 \theta + \sin^2 \theta) = r.$$

Since we assume $r \ge 0$ when integrating in polar coordinates, $|J(r, \theta)| = |r| = r$, so that Equation (1) gives

$$\iint\limits_{R} f(x, y) \, dx \, dy = \iint\limits_{G} f(r \cos \theta, r \sin \theta) \, r \, dr \, d\theta. \tag{3}$$

This is the same formula we derived independently using a geometric argument for polar area in Section 15.4.

Notice that the integral on the right-hand side of Equation (3) is not the integral of $f(r \cos \theta, r \sin \theta)$ over a region in the polar coordinate plane. It is the integral of the product of $f(r \cos \theta, r \sin \theta)$ and r over a region G in the *Cartesian $r\theta$-plane*. ∎

Here is an example of a substitution in which the image of a rectangle under the coordinate transformation is a trapezoid. Transformations like this one are called **linear transformations.**

EXAMPLE 2 Evaluate

$$\int_0^4 \int_{x=y/2}^{x=(y/2)+1} \frac{2x - y}{2} \, dx \, dy$$

by applying the transformation

$$u = \frac{2x - y}{2}, \qquad v = \frac{y}{2} \tag{4}$$

and integrating over an appropriate region in the uv-plane.

Solution We sketch the region R of integration in the xy-plane and identify its boundaries (Figure 15.55).

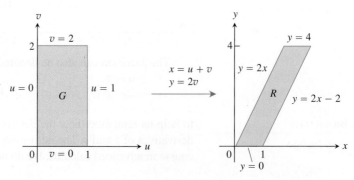

FIGURE 15.55 The equations $x = u + v$ and $y = 2v$ transform G into R. Reversing the transformation by the equations $u = (2x - y)/2$ and $v = y/2$ transforms R into G (Example 2).

To apply Equation (1), we need to find the corresponding uv-region G and the Jacobian of the transformation. To find them, we first solve Equations (4) for x and y in terms of u and v. From those equations it is easy to see that

$$x = u + v, \qquad y = 2v. \tag{5}$$

We then find the boundaries of G by substituting these expressions into the equations for the boundaries of R (Figure 15.55).

xy-equations for the boundary of R	Corresponding uv-equations for the boundary of G	Simplified uv-equations
$x = y/2$	$u + v = 2v/2 = v$	$u = 0$
$x = (y/2) + 1$	$u + v = (2v/2) + 1 = v + 1$	$u = 1$
$y = 0$	$2v = 0$	$v = 0$
$y = 4$	$2v = 4$	$v = 2$

The Jacobian of the transformation (again from Equations (5)) is

$$J(u, v) = \begin{vmatrix} \dfrac{\partial x}{\partial u} & \dfrac{\partial x}{\partial v} \\[2mm] \dfrac{\partial y}{\partial u} & \dfrac{\partial y}{\partial v} \end{vmatrix} = \begin{vmatrix} \dfrac{\partial}{\partial u}(u + v) & \dfrac{\partial}{\partial v}(u + v) \\[2mm] \dfrac{\partial}{\partial u}(2v) & \dfrac{\partial}{\partial v}(2v) \end{vmatrix} = \begin{vmatrix} 1 & 1 \\ 0 & 2 \end{vmatrix} = 2.$$

We now have everything we need to apply Equation (1):

$$\int_0^4 \int_{x=y/2}^{x=(y/2)+1} \frac{2x - y}{2}\, dx\, dy = \int_{v=0}^{v=2} \int_{u=0}^{u=1} u\,|J(u, v)|\, du\, dv$$

$$= \int_0^2 \int_0^1 (u)(2)\, du\, dv = \int_0^2 \left[u^2 \right]_0^1 dv = \int_0^2 dv = 2. \quad \blacksquare$$

EXAMPLE 3 Evaluate

$$\int_0^1 \int_0^{1-x} \sqrt{x + y}\,(y - 2x)^2\, dy\, dx.$$

Solution We sketch the region R of integration in the xy-plane and identify its boundaries (Figure 15.56). The integrand suggests the transformation $u = x + y$ and $v = y - 2x$. Routine algebra produces x and y as functions of u and v:

$$x = \frac{u}{3} - \frac{v}{3}, \qquad y = \frac{2u}{3} + \frac{v}{3}. \tag{6}$$

From Equations (6), we can find the boundaries of the uv-region G (Figure 15.56).

xy-equations for the boundary of R	Corresponding uv-equations for the boundary of G	Simplified uv-equations
$x + y = 1$	$\left(\dfrac{u}{3} - \dfrac{v}{3}\right) + \left(\dfrac{2u}{3} + \dfrac{v}{3}\right) = 1$	$u = 1$
$x = 0$	$\dfrac{u}{3} - \dfrac{v}{3} = 0$	$v = u$
$y = 0$	$\dfrac{2u}{3} + \dfrac{v}{3} = 0$	$v = -2u$

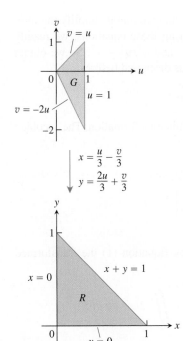

FIGURE 15.56 The equations $x = (u/3) - (v/3)$ and $y = (2u/3) + (v/3)$ transform G into R. Reversing the transformation by the equations $u = x + y$ and $v = y - 2x$ transforms R into G (Example 3).

The Jacobian of the transformation in Equations (6) is

$$J(u, v) = \begin{vmatrix} \dfrac{\partial x}{\partial u} & \dfrac{\partial x}{\partial v} \\[2mm] \dfrac{\partial y}{\partial u} & \dfrac{\partial y}{\partial v} \end{vmatrix} = \begin{vmatrix} \dfrac{1}{3} & -\dfrac{1}{3} \\[2mm] \dfrac{2}{3} & \dfrac{1}{3} \end{vmatrix} = \dfrac{1}{3}.$$

Applying Equation (1), we evaluate the integral:

$$\int_0^1 \int_0^{1-x} \sqrt{x+y}\,(y-2x)^2\,dy\,dx = \int_{u=0}^{u=1} \int_{v=-2u}^{v=u} u^{1/2} v^2 |J(u,v)|\,dv\,du$$

$$= \int_0^1 \int_{-2u}^{u} u^{1/2} v^2 \left(\frac{1}{3}\right) dv\,du = \frac{1}{3}\int_0^1 u^{1/2} \left[\frac{1}{3}v^3\right]_{v=-2u}^{v=u} du$$

$$= \frac{1}{9}\int_0^1 u^{1/2}(u^3 + 8u^3)\,du = \int_0^1 u^{7/2}\,du = \frac{2}{9}u^{9/2}\bigg]_0^1 = \frac{2}{9}. \qquad \blacksquare$$

In the next example we illustrate a nonlinear transformation of coordinates resulting from simplifying the form of the integrand. Like the polar coordinates' transformation, nonlinear transformations can map a straight line boundary of a region into a curved boundary (or vice versa with the inverse transformation). In general, nonlinear transformations are more complex to analyze than linear ones, and a complete treatment is left to a more advanced course.

EXAMPLE 4   Evaluate the integral

$$\int_1^2 \int_{1/y}^{y} \sqrt{\frac{y}{x}}\, e^{\sqrt{xy}}\,dx\,dy.$$

Solution   The square root terms in the integrand suggest that we might simplify the integration by substituting $u = \sqrt{xy}$ and $v = \sqrt{y/x}$. Squaring these equations, we readily have $u^2 = xy$ and $v^2 = y/x$, which imply that $u^2 v^2 = y^2$ and $u^2/v^2 = x^2$. So we obtain the transformation (in the same ordering of the variables as discussed before)

$$x = \frac{u}{v} \qquad \text{and} \qquad y = uv.$$

Let's first see what happens to the integrand itself under this transformation. The Jacobian of the transformation is

$$J(u, v) = \begin{vmatrix} \dfrac{\partial x}{\partial u} & \dfrac{\partial x}{\partial v} \\[2mm] \dfrac{\partial y}{\partial u} & \dfrac{\partial y}{\partial v} \end{vmatrix} = \begin{vmatrix} \dfrac{1}{v} & \dfrac{-u}{v^2} \\[2mm] v & u \end{vmatrix} = \frac{2u}{v}.$$

If G is the region of integration in the uv-plane, then by Equation (1) the transformed double integral under the substitution is

$$\iint_R \sqrt{\frac{y}{x}}\, e^{\sqrt{xy}}\,dx\,dy = \iint_G v e^u \frac{2u}{v}\,du\,dv = \iint_G 2u e^u\,du\,dv.$$

The transformed integrand function is easier to integrate than the original one, so we proceed to determine the limits of integration for the transformed integral.

The region of integration R of the original integral in the xy-plane is shown in Figure 15.57. From the substitution equations $u = \sqrt{xy}$ and $v = \sqrt{y/x}$, we see that the image of the left-hand boundary $xy = 1$ for R is the vertical line segment $u = 1, 2 \geq v \geq 1$, in G (see Figure 15.58). Likewise, the right-hand boundary $y = x$ of R maps to the horizontal line segment $v = 1, 1 \leq u \leq 2$, in G. Finally, the horizontal top boundary $y = 2$ of R

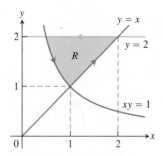

FIGURE 15.57   The region of integration R in Example 4.

FIGURE 15.58   The boundaries of the region G correspond to those of region R in Figure 15.57. Notice as we move counterclockwise around the region R, we also move counterclockwise around the region G. The inverse transformation equations $u = \sqrt{xy}, v = \sqrt{y/x}$ produce the region G from the region R.

maps to $uv = 2, 1 \leq v \leq 2$, in G. As we move counterclockwise around the boundary of the region R, we also move counterclockwise around the boundary of G, as shown in Figure 15.58. Knowing the region of integration G in the uv-plane, we can now write equivalent iterated integrals:

$$\int_1^2 \int_{1/y}^y \sqrt{\frac{y}{x}}\, e^{\sqrt{xy}}\, dx\, dy = \int_1^2 \int_1^{2/u} 2ue^u\, dv\, du.$$ Note the order of integration.

We now evaluate the transformed integral on the right-hand side,

$$\int_1^2 \int_1^{2/u} 2ue^u\, dv\, du = 2\int_1^2 vue^u\Big]_{v=1}^{v=2/u}\, du$$

$$= 2\int_1^2 (2e^u - ue^u)\, du$$

$$= 2\int_1^2 (2 - u)e^u\, du$$

$$= 2\Big[(2 - u)e^u + e^u\Big]_{u=1}^{u=2}$$ Integrate by parts.

$$= 2(e^2 - (e + e)) = 2e(e - 2). \qquad \blacksquare$$

Substitutions in Triple Integrals

The cylindrical and spherical coordinate substitutions in Section 15.7 are special cases of a substitution method that pictures changes of variables in triple integrals as transformations of three-dimensional regions. The method is like the method for double integrals except that now we work in three dimensions instead of two.

Suppose that a region G in uvw-space is transformed one-to-one into the region D in xyz-space by differentiable equations of the form

$$x = g(u, v, w), \qquad y = h(u, v, w), \qquad z = k(u, v, w),$$

as suggested in Figure 15.59. Then any function $F(x, y, z)$ defined on D can be thought of as a function

$$F(g(u, v, w), h(u, v, w), k(u, v, w)) = H(u, v, w)$$

defined on G. If g, h, and k have continuous first partial derivatives, then the integral of $F(x, y, z)$ over D is related to the integral of $H(u, v, w)$ over G by the equation

$$\iiint_D F(x, y, z)\, dx\, dy\, dz = \iiint_G H(u, v, w)\,|J(u, v, w)|\, du\, dv\, dw. \qquad (7)$$

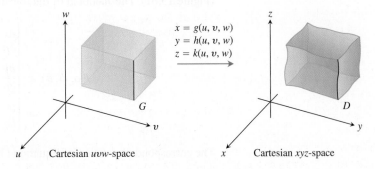

FIGURE 15.59 The equations $x = g(u, v, w)$, $y = h(u, v, w)$, and $z = k(u, v, w)$ allow us to change an integral over a region D in Cartesian xyz-space into an integral over a region G in Cartesian uyw-space using Equation (7).

Cube with sides parallel to the coordinate axes

G

r — Cartesian $r\theta z$-space

$x = r \cos \theta$
$y = r \sin \theta$
$z = z$

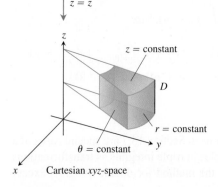

$z = $ constant

D

$r = $ constant

$\theta = $ constant

Cartesian xyz-space

FIGURE 15.60 The equations $x = r \cos \theta, y = r \sin \theta$, and $z = z$ transform the cube G into a cylindrical wedge D.

The factor $J(u, v, w)$, whose absolute value appears in this equation, is the **Jacobian determinant**

$$J(u, v, w) = \begin{vmatrix} \dfrac{\partial x}{\partial u} & \dfrac{\partial x}{\partial v} & \dfrac{\partial x}{\partial w} \\[2mm] \dfrac{\partial y}{\partial u} & \dfrac{\partial y}{\partial v} & \dfrac{\partial y}{\partial w} \\[2mm] \dfrac{\partial z}{\partial u} & \dfrac{\partial z}{\partial v} & \dfrac{\partial z}{\partial w} \end{vmatrix} = \frac{\partial(x, y, z)}{\partial(u, v, w)}.$$

This determinant measures how much the volume near a point in G is being expanded or contracted by the transformation from (u, v, w) to (x, y, z) coordinates. As in the two-dimensional case, the derivation of the change-of-variable formula in Equation (7) is omitted.

For cylindrical coordinates, r, θ, and z take the place of u, v, and w. The transformation from Cartesian $r\theta z$-space to Cartesian xyz-space is given by the equations

$$x = r \cos \theta, \qquad y = r \sin \theta, \qquad z = z$$

(Figure 15.60). The Jacobian of the transformation is

$$J(r, \theta, z) = \begin{vmatrix} \dfrac{\partial x}{\partial r} & \dfrac{\partial x}{\partial \theta} & \dfrac{\partial x}{\partial z} \\[2mm] \dfrac{\partial y}{\partial r} & \dfrac{\partial y}{\partial \theta} & \dfrac{\partial y}{\partial z} \\[2mm] \dfrac{\partial z}{\partial r} & \dfrac{\partial z}{\partial \theta} & \dfrac{\partial z}{\partial z} \end{vmatrix} = \begin{vmatrix} \cos \theta & -r \sin \theta & 0 \\ \sin \theta & r \cos \theta & 0 \\ 0 & 0 & 1 \end{vmatrix}$$

$$= r \cos^2 \theta + r \sin^2 \theta = r.$$

The corresponding version of Equation (7) is

$$\iiint\limits_{D} F(x, y, z) \, dx \, dy \, dz = \iiint\limits_{G} H(r, \theta, z) |r| \, dr \, d\theta \, dz.$$

We can drop the absolute value signs whenever $r \geq 0$.

For spherical coordinates, ρ, ϕ, and θ take the place of u, v, and w. The transformation from Cartesian $\rho\phi\theta$-space to Cartesian xyz-space is given by

$$x = \rho \sin \phi \cos \theta, \qquad y = \rho \sin \phi \sin \theta, \qquad z = \rho \cos \phi$$

(Figure 15.61). The Jacobian of the transformation (see Exercise 19) is

$$J(\rho, \phi, \theta) = \begin{vmatrix} \dfrac{\partial x}{\partial \rho} & \dfrac{\partial x}{\partial \phi} & \dfrac{\partial x}{\partial \theta} \\[2mm] \dfrac{\partial y}{\partial \rho} & \dfrac{\partial y}{\partial \phi} & \dfrac{\partial y}{\partial \theta} \\[2mm] \dfrac{\partial z}{\partial \rho} & \dfrac{\partial z}{\partial \phi} & \dfrac{\partial z}{\partial \theta} \end{vmatrix} = \rho^2 \sin \phi.$$

The corresponding version of Equation (7) is

$$\iiint\limits_{D} F(x, y, z) \, dx \, dy \, dz = \iiint\limits_{G} H(\rho, \phi, \theta) |\rho^2 \sin \phi| \, d\rho \, d\phi \, d\theta.$$

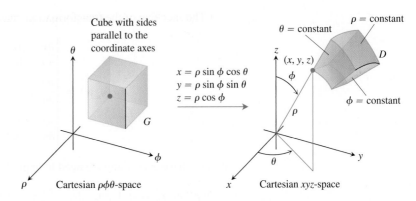

FIGURE 15.61 The equations $x = \rho \sin \phi \cos \theta, y = \rho \sin \phi \sin \theta$, and $z = \rho \cos \phi$ transform the cube G into the spherical wedge D.

We can drop the absolute value signs because $\sin \phi$ is never negative for $0 \le \phi \le \pi$. Note that this is the same result we obtained in Section 15.7.

Here is an example of another substitution. Although we could evaluate the integral in this example directly, we have chosen it to illustrate the substitution method in a simple (and fairly intuitive) setting.

EXAMPLE 5   Evaluate

$$\int_0^3 \int_0^4 \int_{x=y/2}^{x=(y/2)+1} \left(\frac{2x - y}{2} + \frac{z}{3} \right) dx \, dy \, dz$$

by applying the transformation

$$u = (2x - y)/2, \qquad v = y/2, \qquad w = z/3 \tag{8}$$

and integrating over an appropriate region in uvw-space.

Solution   We sketch the region D of integration in xyz-space and identify its boundaries (Figure 15.62). In this case, the bounding surfaces are planes.

To apply Equation (7), we need to find the corresponding uvw-region G and the Jacobian of the transformation. To find them, we first solve Equations (8) for x, y, and z in terms of u, v, and w. Routine algebra gives

$$x = u + v, \qquad y = 2v, \qquad z = 3w. \tag{9}$$

We then find the boundaries of G by substituting these expressions into the equations for the boundaries of D:

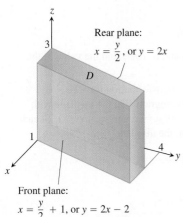

FIGURE 15.62 The equations $x = u + v, y = 2v$, and $z = 3w$ transform G into D. Reversing the transformation by the equations $u = (2x - y)/2, v = y/2$, and $w = z/3$ transforms D into G (Example 5).

xyz-equations for the boundary of D	Corresponding uvw-equations for the boundary of G	Simplified uvw-equations
$x = y/2$	$u + v = 2v/2 = v$	$u = 0$
$x = (y/2) + 1$	$u + v = (2v/2) + 1 = v + 1$	$u = 1$
$y = 0$	$2v = 0$	$v = 0$
$y = 4$	$2v = 4$	$v = 2$
$z = 0$	$3w = 0$	$w = 0$
$z = 3$	$3w = 3$	$w = 1$

The Jacobian of the transformation, again from Equations (9), is

$$J(u, v, w) = \begin{vmatrix} \dfrac{\partial x}{\partial u} & \dfrac{\partial x}{\partial v} & \dfrac{\partial x}{\partial w} \\[6pt] \dfrac{\partial y}{\partial u} & \dfrac{\partial y}{\partial v} & \dfrac{\partial y}{\partial w} \\[6pt] \dfrac{\partial z}{\partial u} & \dfrac{\partial z}{\partial v} & \dfrac{\partial z}{\partial w} \end{vmatrix} = \begin{vmatrix} 1 & 1 & 0 \\ 0 & 2 & 0 \\ 0 & 0 & 3 \end{vmatrix} = 6.$$

We now have everything we need to apply Equation (7):

$$\int_0^3 \int_0^4 \int_{x=y/2}^{x=(y/2)+1} \left(\frac{2x - y}{2} + \frac{z}{3} \right) dx \, dy \, dz$$

$$= \int_0^1 \int_0^2 \int_0^1 (u + w) |J(u, v, w)| \, du \, dv \, dw$$

$$= \int_0^1 \int_0^2 \int_0^1 (u + w)(6) \, du \, dv \, dw = 6 \int_0^1 \int_0^2 \left[\frac{u^2}{2} + uw \right]_0^1 dv \, dw$$

$$= 6 \int_0^1 \int_0^2 \left(\frac{1}{2} + w \right) dv \, dw = 6 \int_0^1 \left[\frac{v}{2} + vw \right]_0^2 dw = 6 \int_0^1 (1 + 2w) \, dw$$

$$= 6 \left[w + w^2 \right]_0^1 = 6(2) = 12. \qquad \blacksquare$$

Exercises 15.8

Jacobians and Transformed Regions in the Plane

1. a. Solve the system

$$u = x - y, \qquad v = 2x + y$$

for x and y in terms of u and v. Then find the value of the Jacobian $\partial(x, y)/\partial(u, v)$.

b. Find the image under the transformation $u = x - y$, $v = 2x + y$ of the triangular region with vertices $(0, 0)$, $(1, 1)$, and $(1, -2)$ in the xy-plane. Sketch the transformed region in the uv-plane.

2. a. Solve the system

$$u = x + 2y, \qquad v = x - y$$

for x and y in terms of u and v. Then find the value of the Jacobian $\partial(x, y)/\partial(u, v)$.

b. Find the image under the transformation $u = x + 2y$, $v = x - y$ of the triangular region in the xy-plane bounded by the lines $y = 0$, $y = x$, and $x + 2y = 2$. Sketch the transformed region in the uv-plane.

3. a. Solve the system

$$u = 3x + 2y, \qquad v = x + 4y$$

for x and y in terms of u and v. Then find the value of the Jacobian $\partial(x, y)/\partial(u, v)$.

b. Find the image under the transformation $u = 3x + 2y$, $v = x + 4y$ of the triangular region in the xy-plane bounded

by the x-axis, the y-axis, and the line $x + y = 1$. Sketch the transformed region in the uv-plane.

4. a. Solve the system

$$u = 2x - 3y, \qquad v = -x + y$$

for x and y in terms of u and v. Then find the value of the Jacobian $\partial(x, y)/\partial(u, v)$.

b. Find the image under the transformation $u = 2x - 3y$, $v = -x + y$ of the parallelogram R in the xy-plane with boundaries $x = -3$, $x = 0$, $y = x$, and $y = x + 1$. Sketch the transformed region in the uv-plane.

Substitutions in Double Integrals

5. Evaluate the integral

$$\int_0^4 \int_{x=y/2}^{x=(y/2)+1} \frac{2x - y}{2} \, dx \, dy$$

from Example 1 directly by integration with respect to x and y to confirm that its value is 2.

6. Use the transformation in Exercise 1 to evaluate the integral

$$\iint_R (2x^2 - xy - y^2) \, dx \, dy$$

for the region R in the first quadrant bounded by the lines $y = -2x + 4$, $y = -2x + 7$, $y = x - 2$, and $y = x + 1$.

7. Use the transformation in Exercise 3 to evaluate the integral

$$\iint_R (3x^2 + 14xy + 8y^2)\, dx\, dy$$

for the region R in the first quadrant bounded by the lines $y = -(3/2)x + 1$, $y = -(3/2)x + 3$, $y = -(1/4)x$, and $y = -(1/4)x + 1$.

8. Use the transformation and parallelogram R in Exercise 4 to evaluate the integral

$$\iint_R 2(x - y)\, dx\, dy.$$

9. Let R be the region in the first quadrant of the xy-plane bounded by the hyperbolas $xy = 1, xy = 9$ and the lines $y = x, y = 4x$. Use the transformation $x = u/v, y = uv$ with $u > 0$ and $v > 0$ to rewrite

$$\iint_R \left(\sqrt{\frac{y}{x}} + \sqrt{xy} \right) dx\, dy$$

as an integral over an appropriate region G in the uv-plane. Then evaluate the uv-integral over G.

10. a. Find the Jacobian of the transformation $x = u, y = uv$ and sketch the region $G: 1 \le u \le 2, 1 \le uv \le 2$, in the uv-plane.

b. Then use Equation (1) to transform the integral

$$\int_1^2 \int_1^2 \frac{y}{x}\, dy\, dx$$

into an integral over G, and evaluate both integrals.

11. Polar moment of inertia of an elliptical plate A thin plate of constant density covers the region bounded by the ellipse $x^2/a^2 + y^2/b^2 = 1$, $a > 0$, $b > 0$, in the xy-plane. Find the first moment of the plate about the origin. (*Hint:* Use the transformation $x = ar\cos\theta, y = br\sin\theta$.)

12. The area of an ellipse The area πab of the ellipse $x^2/a^2 + y^2/b^2 = 1$ can be found by integrating the function $f(x, y) = 1$ over the region bounded by the ellipse in the xy-plane. Evaluating the integral directly requires a trigonometric substitution. An easier way to evaluate the integral is to use the transformation $x = au, y = bv$ and evaluate the transformed integral over the disk $G: u^2 + v^2 \le 1$ in the uv-plane. Find the area this way.

13. Use the transformation in Exercise 2 to evaluate the integral

$$\int_0^{2/3} \int_y^{2-2y} (x + 2y)e^{(y-x)}\, dx\, dy$$

by first writing it as an integral over a region G in the uv-plane.

14. Use the transformation $x = u + (1/2)v, y = v$ to evaluate the integral

$$\int_0^2 \int_{y/2}^{(y+4)/2} y^3(2x - y)e^{(2x-y)^2}\, dx\, dy$$

by first writing it as an integral over a region G in the uv-plane.

15. Use the transformation $x = u/v, y = uv$ to evaluate the integral sum

$$\int_1^2 \int_{1/y}^{y} (x^2 + y^2)\, dx\, dy + \int_2^4 \int_{y/4}^{4/y} (x^2 + y^2)\, dx\, dy.$$

16. Use the transformation $x = u^2 - v^2, y = 2uv$ to evaluate the integral

$$\int_0^1 \int_0^{2\sqrt{1-x}} \sqrt{x^2 + y^2}\, dy\, dx.$$

(*Hint:* Show that the image of the triangular region G with vertices $(0, 0)$, $(1, 0)$, $(1, 1)$ in the uv-plane is the region of integration R in the xy-plane defined by the limits of integration.)

Finding Jacobians

17. Find the Jacobian $\partial(x, y)/\partial(u, v)$ of the transformation

a. $x = u\cos v, \quad y = u\sin v$

b. $x = u\sin v, \quad y = u\cos v$.

18. Find the Jacobian $\partial(x, y, z)/\partial(u, v, w)$ of the transformation

a. $x = u\cos v, \quad y = u\sin v, \quad z = w$

b. $x = 2u - 1, \quad y = 3v - 4, \quad z = (1/2)(w - 4)$.

19. Evaluate the appropriate determinant to show that the Jacobian of the transformation from Cartesian $\rho\phi\theta$-space to Cartesian xyz-space is $\rho^2 \sin\phi$.

20. Substitutions in single integrals How can substitutions in single definite integrals be viewed as transformations of regions? What is the Jacobian in such a case? Illustrate with an example.

Substitutions in Triple Integrals

21. Evaluate the integral in Example 5 by integrating with respect to x, y, and z.

22. Volume of an ellipsoid Find the volume of the ellipsoid

$$\frac{x^2}{a^2} + \frac{y^2}{b^2} + \frac{z^2}{c^2} = 1.$$

(*Hint:* Let $x = au, y = bv$, and $z = cw$. Then find the volume of an appropriate region in uvw-space.)

23. Evaluate

$$\iiint |xyz|\, dx\, dy\, dz$$

over the solid ellipsoid

$$\frac{x^2}{a^2} + \frac{y^2}{b^2} + \frac{z^2}{c^2} \le 1.$$

(*Hint:* Let $x = au, y = bv$, and $z = cw$. Then integrate over an appropriate region in uvw-space.)

24. Let D be the region in xyz-space defined by the inequalities

$$1 \le x \le 2, \quad 0 \le xy \le 2, \quad 0 \le z \le 1.$$

Evaluate

$$\iiint_D (x^2y + 3xyz)\, dx\, dy\, dz$$

by applying the transformation

$$u = x, \quad v = xy, \quad w = 3z$$

and integrating over an appropriate region G in uvw-space.

25. Centroid of a solid semiellipsoid   Assuming the result that the centroid of a solid hemisphere lies on the axis of symmetry three-eighths of the way from the base toward the top, show, by transforming the appropriate integrals, that the center of mass of a solid semiellipsoid $(x^2/a^2) + (y^2/b^2) + (z^2/c^2) \leq 1$, $z \geq 0$, lies on the z-axis three-eighths of the way from the base toward the top. (You can do this without evaluating any of the integrals.)

26. Cylindrical shells   In Section 6.2, we learned how to find the volume of a solid of revolution using the shell method; namely, if the region between the curve $y = f(x)$ and the x-axis from a to b $(0 < a < b)$ is revolved about the y-axis, the volume of the resulting solid is $\int_a^b 2\pi x f(x)\, dx$. Prove that finding volumes by using triple integrals gives the same result. (*Hint:* Use cylindrical coordinates with the roles of y and z changed.)

Chapter 15   Questions to Guide Your Review

1. Define the double integral of a function of two variables over a bounded region in the coordinate plane.

2. How are double integrals evaluated as iterated integrals? Does the order of integration matter? How are the limits of integration determined? Give examples.

3. How are double integrals used to calculate areas and average values. Give examples.

4. How can you change a double integral in rectangular coordinates into a double integral in polar coordinates? Why might it be worthwhile to do so? Give an example.

5. Define the triple integral of a function $f(x, y, z)$ over a bounded region in space.

6. How are triple integrals in rectangular coordinates evaluated? How are the limits of integration determined? Give an example.

7. How are double and triple integrals in rectangular coordinates used to calculate volumes, average values, masses, moments, and centers of mass? Give examples.

8. How are triple integrals defined in cylindrical and spherical coordinates? Why might one prefer working in one of these coordinate systems to working in rectangular coordinates?

9. How are triple integrals in cylindrical and spherical coordinates evaluated? How are the limits of integration found? Give examples.

10. How are substitutions in double integrals pictured as transformations of two-dimensional regions? Give a sample calculation.

11. How are substitutions in triple integrals pictured as transformations of three-dimensional regions? Give a sample calculation.

Chapter 15   Practice Exercises

Evaluating Double Iterated Integrals

In Exercises 1–4, sketch the region of integration and evaluate the double integral.

1. $\displaystyle\int_1^{10}\int_0^{1/y} y e^{xy}\, dx\, dy$

2. $\displaystyle\int_0^1\int_0^{x^3} e^{y/x}\, dy\, dx$

3. $\displaystyle\int_0^{3/2}\int_{-\sqrt{9-4t^2}}^{\sqrt{9-4t^2}} t\, ds\, dt$

4. $\displaystyle\int_0^1\int_{\sqrt{y}}^{2-\sqrt{y}} xy\, dx\, dy$

In Exercises 5–8, sketch the region of integration and write an equivalent integral with the order of integration reversed. Then evaluate both integrals.

5. $\displaystyle\int_0^4\int_{-\sqrt{4-y}}^{(y-4)/2} dx\, dy$

6. $\displaystyle\int_0^1\int_{x^2}^{x} \sqrt{x}\, dy\, dx$

7. $\displaystyle\int_0^{3/2}\int_{-\sqrt{9-4y^2}}^{\sqrt{9-4y^2}} y\, dx\, dy$

8. $\displaystyle\int_0^2\int_0^{4-x^2} 2x\, dy\, dx$

Evaluate the integrals in Exercises 9–12.

9. $\displaystyle\int_0^1\int_{2y}^2 4\cos(x^2)\, dx\, dy$

10. $\displaystyle\int_0^2\int_{y/2}^1 e^{x^2}\, dx\, dy$

11. $\displaystyle\int_0^8\int_{\sqrt[3]{x}}^2 \frac{dy\, dx}{y^4 + 1}$

12. $\displaystyle\int_0^1\int_{\sqrt[3]{y}}^1 \frac{2\pi \sin \pi x^2}{x^2}\, dx\, dy$

Areas and Volumes Using Double Integrals

13. **Area between line and parabola**   Find the area of the region enclosed by the line $y = 2x + 4$ and the parabola $y = 4 - x^2$ in the xy-plane.

14. **Area bounded by lines and parabola**   Find the area of the "triangular" region in the xy-plane that is bounded on the right by the parabola $y = x^2$, on the left by the line $x + y = 2$, and above by the line $y = 4$.

15. **Volume of the region under a paraboloid**   Find the volume under the paraboloid $z = x^2 + y^2$ above the triangle enclosed by the lines $y = x$, $x = 0$, and $x + y = 2$ in the xy-plane.

16. **Volume of the region under parabolic cylinder**   Find the volume under the parabolic cylinder $z = x^2$ above the region enclosed by the parabola $y = 6 - x^2$ and the line $y = x$ in the xy-plane.

Average Values

Find the average value of $f(x, y) = xy$ over the regions in Exercises 17 and 18.

17. The square bounded by the lines $x = 1$, $y = 1$ in the first quadrant

18. The quarter circle $x^2 + y^2 \leq 1$ in the first quadrant

Polar Coordinates

Evaluate the integrals in Exercises 19 and 20 by changing to polar coordinates.

19. $\displaystyle\int_{-1}^{1}\int_{-\sqrt{1-x^2}}^{\sqrt{1-x^2}} \frac{2\,dy\,dx}{(1 + x^2 + y^2)^2}$

20. $\displaystyle\int_{-1}^{1}\int_{-\sqrt{1-y^2}}^{\sqrt{1-y^2}} \ln(x^2 + y^2 + 1)\,dx\,dy$

21. Integrating over lemniscate Integrate the function $f(x, y) = 1/(1 + x^2 + y^2)^2$ over the region enclosed by one loop of the lemniscate $(x^2 + y^2)^2 - (x^2 - y^2) = 0$.

22. Integrate $f(x, y) = 1/(1 + x^2 + y^2)^2$ over

   a. Triangular region The triangle with vertices $(0, 0)$, $(1, 0)$, and $\left(1, \sqrt{3}\right)$.

   b. First quadrant The first quadrant of the xy-plane.

Evaluating Triple Iterated Integrals

Evaluate the integrals in Exercises 23–26.

23. $\displaystyle\int_{0}^{\pi}\int_{0}^{\pi}\int_{0}^{\pi} \cos(x + y + z)\,dx\,dy\,dz$

24. $\displaystyle\int_{\ln 6}^{\ln 7}\int_{0}^{\ln 2}\int_{\ln 4}^{\ln 5} e^{(x+y+z)}\,dz\,dy\,dx$

25. $\displaystyle\int_{0}^{1}\int_{0}^{x^2}\int_{0}^{x+y} (2x - y - z)\,dz\,dy\,dx$

26. $\displaystyle\int_{1}^{e}\int_{1}^{x}\int_{0}^{z} \frac{2y}{z^3}\,dy\,dz\,dx$

Volumes and Average Values Using Triple Integrals

27. Volume Find the volume of the wedge-shaped region enclosed on the side by the cylinder $x = -\cos y$, $-\pi/2 \le y \le \pi/2$, on the top by the plane $z = -2x$, and below by the xy-plane.

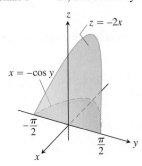

28. Volume Find the volume of the solid that is bounded above by the cylinder $z = 4 - x^2$, on the sides by the cylinder $x^2 + y^2 = 4$, and below by the xy-plane.

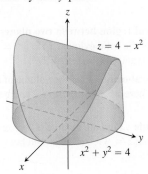

29. Average value Find the average value of $f(x, y, z) = 30xz\sqrt{x^2 + y}$ over the rectangular solid in the first octant bounded by the coordinate planes and the planes $x = 1$, $y = 3$, $z = 1$.

30. Average value Find the average value of ρ over the solid sphere $\rho \le a$ (spherical coordinates).

Cylindrical and Spherical Coordinates

31. Cylindrical to rectangular coordinates Convert

$$\int_{0}^{2\pi}\int_{0}^{\sqrt{2}}\int_{r}^{\sqrt{4-r^2}} 3\,dz\,r\,dr\,d\theta, \qquad r \ge 0$$

to **(a)** rectangular coordinates with the order of integration $dz\,dx\,dy$ and **(b)** spherical coordinates. Then **(c)** evaluate one of the integrals.

32. Rectangular to cylindrical coordinates **(a)** Convert to cylindrical coordinates. Then **(b)** evaluate the new integral.

$$\int_{0}^{1}\int_{-\sqrt{1-x^2}}^{\sqrt{1-x^2}}\int_{-(x^2+y^2)}^{(x^2+y^2)} 21xy^2\,dz\,dy\,dx$$

33. Rectangular to spherical coordinates **(a)** Convert to spherical coordinates. Then **(b)** evaluate the new integral.

$$\int_{-1}^{1}\int_{-\sqrt{1-x^2}}^{\sqrt{1-x^2}}\int_{\sqrt{x^2+y^2}}^{1} dz\,dy\,dx$$

34. Rectangular, cylindrical, and spherical coordinates Write an iterated triple integral for the integral of $f(x, y, z) = 6 + 4y$ over the region in the first octant bounded by the cone $z = \sqrt{x^2 + y^2}$, the cylinder $x^2 + y^2 = 1$, and the coordinate planes in **(a)** rectangular coordinates, **(b)** cylindrical coordinates, and **(c)** spherical coordinates. Then **(d)** find the integral of f by evaluating one of the triple integrals.

35. Cylindrical to rectangular coordinates Set up an integral in rectangular coordinates equivalent to the integral

$$\int_{0}^{\pi/2}\int_{1}^{\sqrt{3}}\int_{1}^{\sqrt{4-r^2}} r^3(\sin\theta\cos\theta)z^2\,dz\,dr\,d\theta.$$

Arrange the order of integration to be z first, then y, then x.

36. Rectangular to cylindrical coordinates The volume of a solid is

$$\int_{0}^{2}\int_{0}^{\sqrt{2x-x^2}}\int_{-\sqrt{4-x^2-y^2}}^{\sqrt{4-x^2-y^2}} dz\,dy\,dx.$$

   a. Describe the solid by giving equations for the surfaces that form its boundary.

   b. Convert the integral to cylindrical coordinates but do not evaluate the integral.

37. Spherical versus cylindrical coordinates Triple integrals involving spherical shapes do not always require spherical coordinates for convenient evaluation. Some calculations may be accomplished more easily with cylindrical coordinates. As a case in point, find the volume of the region bounded above by the sphere $x^2 + y^2 + z^2 = 8$ and below by the plane $z = 2$ by using **(a)** cylindrical coordinates and **(b)** spherical coordinates.

Masses and Moments

38. Finding I_z in spherical coordinates Find the moment of inertia about the z-axis of a solid of constant density $\delta = 1$ that is bounded above by the sphere $\rho = 2$ and below by the cone $\phi = \pi/3$ (spherical coordinates).

39. Moment of inertia of a "thick" sphere Find the moment of inertia of a solid of constant density δ bounded by two concentric spheres of radii a and b $(a < b)$ about a diameter.

40. Moment of inertia of an apple Find the moment of inertia about the z-axis of a solid of density $\delta = 1$ enclosed by the spherical coordinate surface $\rho = 1 - \cos \phi$. The solid is the red curve rotated about the z-axis in the accompanying figure.

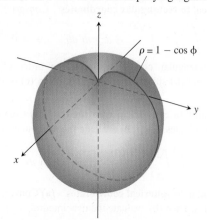

$\rho = 1 - \cos \phi$

41. Centroid Find the centroid of the "triangular" region bounded by the lines $x = 2$, $y = 2$ and the hyperbola $xy = 2$ in the xy-plane.

42. Centroid Find the centroid of the region between the parabola $x + y^2 - 2y = 0$ and the line $x + 2y = 0$ in the xy-plane.

43. Polar moment Find the polar moment of inertia about the origin of a thin triangular plate of constant density $\delta = 3$ bounded by the y-axis and the lines $y = 2x$ and $y = 4$ in the xy-plane.

44. Polar moment Find the polar moment of inertia about the center of a thin rectangular sheet of constant density $\delta = 1$ bounded by the lines

  a. $x = \pm 2$, $y = \pm 1$ in the xy-plane

  b. $x = \pm a$, $y = \pm b$ in the xy-plane.

(Hint: Find I_x. Then use the formula for I_x to find I_y and add the two to find I_0).

45. Inertial moment Find the moment of inertia about the x-axis of a thin plate of constant density δ covering the triangle with vertices $(0, 0)$, $(3, 0)$, and $(3, 2)$ in the xy-plane.

46. Plate with variable density Find the center of mass and the moments of inertia about the coordinate axes of a thin plate bounded by the line $y = x$ and the parabola $y = x^2$ in the xy-plane if the density is $\delta(x, y) = x + 1$.

47. Plate with variable density Find the mass and first moments about the coordinate axes of a thin square plate bounded by the lines $x = \pm 1$, $y = \pm 1$ in the xy-plane if the density is $\delta(x, y) = x^2 + y^2 + 1/3$.

48. Triangles with same inertial moment Find the moment of inertia about the x-axis of a thin triangular plate of constant density δ whose base lies along the interval $[0, b]$ on the x-axis and whose vertex lies on the line $y = h$ above the x-axis. As you will see, it does not matter where on the line this vertex lies. All such triangles have the same moment of inertia about the x-axis.

49. Centroid Find the centroid of the region in the polar coordinate plane defined by the inequalities $0 \le r \le 3$, $-\pi/3 \le \theta \le \pi/3$.

50. Centroid Find the centroid of the region in the first quadrant bounded by the rays $\theta = 0$ and $\theta = \pi/2$ and the circles $r = 1$ and $r = 3$.

51. a. Centroid Find the centroid of the region in the polar coordinate plane that lies inside the cardioid $r = 1 + \cos \theta$ and outside the circle $r = 1$.

  b. Sketch the region and show the centroid in your sketch.

52. a. Centroid Find the centroid of the plane region defined by the polar coordinate inequalities $0 \le r \le a$, $-\alpha \le \theta \le \alpha$ $(0 < \alpha \le \pi)$. How does the centroid move as $\alpha \to \pi^-$?

  b. Sketch the region for $\alpha = 5\pi/6$ and show the centroid in your sketch.

Substitutions

53. Show that if $u = x - y$ and $v = y$, then

$$\int_0^\infty \int_0^x e^{-sx} f(x - y, y) \, dy \, dx = \int_0^\infty \int_0^\infty e^{-s(u+v)} f(u, v) \, du \, dv.$$

54. What relationship must hold between the constants a, b, and c to make

$$\int_{-\infty}^\infty \int_{-\infty}^\infty e^{-(ax^2 + 2bxy + cy^2)} \, dx \, dy = 1?$$

(Hint: Let $s = \alpha x + \beta y$ and $t = \gamma x + \delta y$, where $(\alpha\delta - \beta\gamma)^2 = ac - b^2$. Then $ax^2 + 2bxy + cy^2 = s^2 + t^2$.)

Chapter 15   Additional and Advanced Exercises

Volumes

1. Sand pile: double and triple integrals The base of a sand pile covers the region in the xy-plane that is bounded by the parabola $x^2 + y = 6$ and the line $y = x$. The height of the sand above the point (x, y) is x^2. Express the volume of sand as **(a)** a double integral, **(b)** a triple integral. Then **(c)** find the volume.

2. Water in a hemispherical bowl A hemispherical bowl of radius 5 cm is filled with water to within 3 cm of the top. Find the volume of water in the bowl.

3. Solid cylindrical region between two planes Find the volume of the portion of the solid cylinder $x^2 + y^2 \le 1$ that lies between the planes $z = 0$ and $x + y + z = 2$.

4. Sphere and paraboloid Find the volume of the region bounded above by the sphere $x^2 + y^2 + z^2 = 2$ and below by the paraboloid $z = x^2 + y^2$.

5. Two paraboloids Find the volume of the region bounded above by the paraboloid $z = 3 - x^2 - y^2$ and below by the paraboloid $z = 2x^2 + 2y^2$.

6. Spherical coordinates   Find the volume of the region enclosed by the spherical coordinate surface $\rho = 2 \sin \phi$ (see accompanying figure).

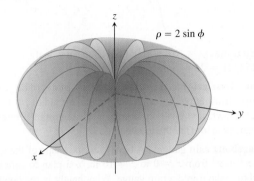

$\rho = 2 \sin \phi$

7. Hole in sphere   A circular cylindrical hole is bored through a solid sphere, the axis of the hole being a diameter of the sphere. The volume of the remaining solid is

$$V = 2 \int_0^{2\pi} \int_0^{\sqrt{3}} \int_1^{\sqrt{4-z^2}} r \, dr \, dz \, d\theta.$$

a. Find the radius of the hole and the radius of the sphere.

b. Evaluate the integral.

8. Sphere and cylinder   Find the volume of material cut from the solid sphere $r^2 + z^2 \le 9$ by the cylinder $r = 3 \sin \theta$.

9. Two paraboloids   Find the volume of the region enclosed by the surfaces $z = x^2 + y^2$ and $z = (x^2 + y^2 + 1)/2$.

10. Cylinder and surface $z = xy$ Find the volume of the region in the first octant that lies between the cylinders $r = 1$ and $r = 2$ and that is bounded below by the xy-plane and above by the surface $z = xy$.

Changing the Order of Integration

11. Evaluate the integral

$$\int_0^\infty \frac{e^{-ax} - e^{-bx}}{x} \, dx.$$

(*Hint:* Use the relation

$$\frac{e^{-ax} - e^{-bx}}{x} = \int_a^b e^{-xy} \, dy$$

to form a double integral and evaluate the integral by changing the order of integration.)

12. a. Polar coordinates   Show, by changing to polar coordinates, that

$$\int_0^{a \sin \beta} \int_{y \cot \beta}^{\sqrt{a^2-y^2}} \ln (x^2 + y^2) \, dx \, dy = a^2 \beta \left(\ln a - \frac{1}{2} \right),$$

where $a > 0$ and $0 < \beta < \pi/2$.

b. Rewrite the Cartesian integral with the order of integration reversed.

13. Reducing a double to a single integral   By changing the order of integration, show that the following double integral can be reduced to a single integral:

$$\int_0^x \int_0^u e^{m(x-t)} f(t) \, dt \, du = \int_0^x (x - t) e^{m(x-t)} f(t) \, dt.$$

Similarly, it can be shown that

$$\int_0^x \int_0^v \int_0^u e^{m(x-t)} f(t) \, dt \, du \, dv = \int_0^x \frac{(x - t)^2}{2} e^{m(x-t)} f(t) \, dt.$$

14. Transforming a double integral to obtain constant limits   Sometimes a multiple integral with variable limits can be changed into one with constant limits. By changing the order of integration, show that

$$\int_0^1 f(x) \left(\int_0^x g(x - y) f(y) \, dy \right) dx$$

$$= \int_0^1 f(y) \left(\int_y^1 g(x - y) f(x) \, dx \right) dy$$

$$= \frac{1}{2} \int_0^1 \int_0^1 g(|x - y|) f(x) f(y) \, dx \, dy.$$

Masses and Moments

15. Minimizing polar inertia   A thin plate of constant density is to occupy the triangular region in the first quadrant of the xy-plane having vertices $(0, 0)$, $(a, 0)$, and $(a, 1/a)$. What value of a will minimize the plate's polar moment of inertia about the origin?

16. Polar inertia of triangular plate   Find the polar moment of inertia about the origin of a thin triangular plate of constant density $\delta = 3$ bounded by the y-axis and the lines $y = 2x$ and $y = 4$ in the xy-plane.

17. Mass and polar inertia of a counterweight   The counterweight of a flywheel of constant density 1 has the form of the smaller segment cut from a circle of radius a by a chord at a distance b from the center ($b < a$). Find the mass of the counterweight and its polar moment of inertia about the center of the wheel.

18. Centroid of boomerang   Find the centroid of the boomerang-shaped region between the parabolas $y^2 = -4(x - 1)$ and $y^2 = -2(x - 2)$ in the xy-plane.

Theory and Examples

19. Evaluate

$$\int_0^a \int_0^b e^{\max (b^2 x^2, a^2 y^2)} \, dy \, dx,$$

where a and b are positive numbers and

$$\max (b^2 x^2, a^2 y^2) = \begin{cases} b^2 x^2 & \text{if } b^2 x^2 \ge a^2 y^2 \\ a^2 y^2 & \text{if } b^2 x^2 < a^2 y^2. \end{cases}$$

20. Show that

$$\iint \frac{\partial^2 F(x, y)}{\partial x \, \partial y} \, dx \, dy$$

over the rectangle $x_0 \le x \le x_1, y_0 \le y \le y_1$, is

$$F(x_1, y_1) - F(x_0, y_1) - F(x_1, y_0) + F(x_0, y_0).$$

21. Suppose that $f(x, y)$ can be written as a product $f(x, y) = F(x)G(y)$ of a function of x and a function of y. Then the integral of f over the rectangle $R: a \leq x \leq b, c \leq y \leq d$ can be evaluated as a product as well, by the formula

$$\iint_R f(x, y)\, dA = \left(\int_a^b F(x)\, dx \right)\left(\int_c^d G(y)\, dy \right). \qquad (1)$$

The argument is that

$$\iint_R f(x, y)\, dA = \int_c^d \left(\int_a^b F(x)G(y)\, dx \right) dy \qquad \text{(i)}$$

$$= \int_c^d \left(G(y) \int_a^b F(x)\, dx \right) dy \qquad \text{(ii)}$$

$$= \int_c^d \left(\int_a^b F(x)\, dx \right) G(y)\, dy \qquad \text{(iii)}$$

$$= \left(\int_a^b F(x)\, dx \right) \int_c^d G(y)\, dy. \qquad \text{(iv)}$$

a. Give reasons for steps (i) through (iv).

When it applies, Equation (1) can be a time-saver. Use it to evaluate the following integrals.

b. $\displaystyle \int_0^{\ln 2} \int_0^{\pi/2} e^x \cos y\, dy\, dx$ **c.** $\displaystyle \int_1^2 \int_{-1}^1 \frac{x}{y^2}\, dx\, dy$

22. Let $D_{\mathbf{u}} f$ denote the derivative of $f(x, y) = (x^2 + y^2)/2$ in the direction of the unit vector $\mathbf{u} = u_1\mathbf{i} + u_2\mathbf{j}$.

a. Finding average value Find the average value of $D_{\mathbf{u}} f$ over the triangular region cut from the first quadrant by the line $x + y = 1$.

b. Average value and centroid Show in general that the average value of $D_{\mathbf{u}} f$ over a region in the xy-plane is the value of $D_{\mathbf{u}} f$ at the centroid of the region.

23. The value of $\Gamma(1/2)$ The gamma function,

$$\Gamma(x) = \int_0^\infty t^{x-1} e^{-t}\, dt,$$

extends the factorial function from the nonnegative integers to other real values. Of particular interest in the theory of differential equations is the number

$$\Gamma\left(\frac{1}{2}\right) = \int_0^\infty t^{(1/2)-1} e^{-t}\, dt = \int_0^\infty \frac{e^{-t}}{\sqrt{t}}\, dt. \qquad (2)$$

a. If you have not yet done Exercise 41 in Section 15.4, do it now to show that

$$I = \int_0^\infty e^{-y^2}\, dy = \frac{\sqrt{\pi}}{2}.$$

b. Substitute $y = \sqrt{t}$ in Equation (2) to show that $\Gamma(1/2) = 2I = \sqrt{\pi}$.

24. Total electrical charge over circular plate The electrical charge distribution on a circular plate of radius R meters is $\sigma(r, \theta) = kr(1 - \sin \theta)$ coulomb/m^2 (k a constant). Integrate σ over the plate to find the total charge Q.

25. A parabolic rain gauge A bowl is in the shape of the graph of $z = x^2 + y^2$ from $z = 0$ to $z = 10$ in. You plan to calibrate the bowl to make it into a rain gauge. What height in the bowl would correspond to 1 in. of rain? 3 in. of rain?

26. Water in a satellite dish A parabolic satellite dish is 2 m wide and $1/2$ m deep. Its axis of symmetry is tilted 30 degrees from the vertical.

a. Set up, but do not evaluate, a triple integral in rectangular coordinates that gives the amount of water the satellite dish will hold. (*Hint:* Put your coordinate system so that the satellite dish is in "standard position" and the plane of the water level is slanted.) (*Caution:* The limits of integration are not "nice.")

b. What would be the smallest tilt of the satellite dish so that it holds no water?

27. An infinite half-cylinder Let D be the interior of the infinite right circular half-cylinder of radius 1 with its single-end face suspended 1 unit above the origin and its axis the ray from $(0, 0, 1)$ to ∞. Use cylindrical coordinates to evaluate

$$\iiint_D z(r^2 + z^2)^{-5/2}\, dV.$$

28. Hypervolume We have learned that $\int_a^b 1\, dx$ is the length of the interval $[a, b]$ on the number line (one-dimensional space), $\iint_R 1\, dA$ is the area of region R in the xy-plane (two-dimensional space), and $\iiint_D 1\, dV$ is the volume of the region D in three-dimensional space (xyz-space). We could continue: If Q is a region in 4-space ($xyzw$-space), then $\iiiint_Q 1\, dV$ is the "hypervolume" of Q. Use your generalizing abilities and a Cartesian coordinate system of 4-space to find the hypervolume inside the unit 3-dimensional sphere $x^2 + y^2 + z^2 + w^2 = 1$.

Chapter **15**   Technology Application Projects

Mathematica/Maple Module:

Take Your Chances: Try the Monte Carlo Technique for Numerical Integration in Three Dimensions
Use the Monte Carlo technique to integrate numerically in three dimensions.

Means and Moments and Exploring New Plotting Techniques, Part II
Use the method of moments in a form that makes use of geometric symmetry as well as multiple integration.

The formulas in Table 16.1 then give

$$M = \int_C \delta \, ds = \int_C (2 - z) \, ds = \int_0^\pi (2 - \sin t) \, dt = 2\pi - 2$$

$$M_{xy} = \int_C z\delta \, ds = \int_C z(2 - z) \, ds = \int_0^\pi (\sin t)(2 - \sin t) \, dt$$

$$= \int_0^\pi (2 \sin t - \sin^2 t) \, dt = \frac{8 - \pi}{2}$$

$$\bar{z} = \frac{M_{xy}}{M} = \frac{8 - \pi}{2} \cdot \frac{1}{2\pi - 2} = \frac{8 - \pi}{4\pi - 4} \approx 0.57.$$

With $\bar{z}$ to the nearest hundredth, the center of mass is $(0, 0, 0.57)$. ∎

Line Integrals in the Plane

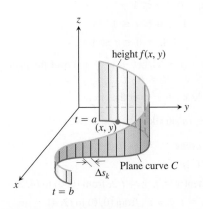

height $f(x, y)$

$t = a$

(x, y)

Δs_k

Plane curve C

$t = b$

FIGURE 16.5 The line integral $\int_C f \, ds$ gives the area of the portion of the cylindrical surface or "wall" beneath $z = f(x, y) \geq 0$.

There is an interesting geometric interpretation for line integrals in the plane. If C is a smooth curve in the xy-plane parametrized by $\mathbf{r}(t) = x(t)\mathbf{i} + y(t)\mathbf{j}$, $a \leq t \leq b$, we generate a cylindrical surface by moving a straight line along C orthogonal to the plane, holding the line parallel to the z-axis, as in Section 12.6. If $z = f(x, y)$ is a nonnegative continuous function over a region in the plane containing the curve C, then the graph of f is a surface that lies above the plane. The cylinder cuts through this surface, forming a curve on it that lies above the curve C and follows its winding nature. The part of the cylindrical surface that lies beneath the surface curve and above the xy-plane is like a "winding wall" or "fence" standing on the curve C and orthogonal to the plane. At any point (x, y) along the curve, the height of the wall is $f(x, y)$. We show the wall in Figure 16.5, where the "top" of the wall is the curve lying on the surface $z = f(x, y)$. (We do not display the surface formed by the graph of f in the figure, only the curve on it that is cut out by the cylinder.) From the definition

$$\int_C f \, ds = \lim_{n \to \infty} \sum_{k=1}^n f(x_k, y_k) \, \Delta s_k,$$

where $\Delta s_k \to 0$ as $n \to \infty$, we see that the line integral $\int_C f \, ds$ is the area of the wall shown in the figure.

Exercises 16.1

Graphs of Vector Equations
Match the vector equations in Exercises 1–8 with the graphs (a)–(h) given here.

a.

b.

c.

d.

e.

f.

g.

h.

1. $\mathbf{r}(t) = t\mathbf{i} + (1 - t)\mathbf{j}, \quad 0 \le t \le 1$

2. $\mathbf{r}(t) = \mathbf{i} + \mathbf{j} + t\mathbf{k}, \quad -1 \le t \le 1$

3. $\mathbf{r}(t) = (2 \cos t)\mathbf{i} + (2 \sin t)\mathbf{j}, \quad 0 \le t \le 2\pi$

4. $\mathbf{r}(t) = t\mathbf{i}, \quad -1 \le t \le 1$

5. $\mathbf{r}(t) = t\mathbf{i} + t\mathbf{j} + t\mathbf{k}, \quad 0 \le t \le 2$

6. $\mathbf{r}(t) = t\mathbf{j} + (2 - 2t)\mathbf{k}, \quad 0 \le t \le 1$

7. $\mathbf{r}(t) = (t^2 - 1)\mathbf{j} + 2t\mathbf{k}, \quad -1 \le t \le 1$

8. $\mathbf{r}(t) = (2 \cos t)\mathbf{i} + (2 \sin t)\mathbf{k}, \quad 0 \le t \le \pi$

Evaluating Line Integrals over Space Curves

9. Evaluate $\int_C (x + y) \, ds$ where C is the straight-line segment $x = t, y = (1 - t), z = 0$, from $(0, 1, 0)$ to $(1, 0, 0)$.

10. Evaluate $\int_C (x - y + z - 2) \, ds$ where C is the straight-line segment $x = t, y = (1 - t), z = 1$, from $(0, 1, 1)$ to $(1, 0, 1)$.

11. Evaluate $\int_C (xy + y + z) \, ds$ along the curve $\mathbf{r}(t) = 2t\mathbf{i} + t\mathbf{j} + (2 - 2t)\mathbf{k}, 0 \le t \le 1$.

12. Evaluate $\int_C \sqrt{x^2 + y^2} \, ds$ along the curve $\mathbf{r}(t) = (4 \cos t)\mathbf{i} + (4 \sin t)\mathbf{j} + 3t\mathbf{k}, -2\pi \le t \le 2\pi$.

13. Find the line integral of $f(x, y, z) = x + y + z$ over the straight-line segment from $(1, 2, 3)$ to $(0, -1, 1)$.

14. Find the line integral of $f(x, y, z) = \sqrt{3}/(x^2 + y^2 + z^2)$ over the curve $\mathbf{r}(t) = t\mathbf{i} + t\mathbf{j} + t\mathbf{k}, 1 \le t \le \infty$.

15. Integrate $f(x, y, z) = x + \sqrt{y} - z^2$ over the path from $(0, 0, 0)$ to $(1, 1, 1)$ (see accompanying figure) given by

$$C_1: \quad \mathbf{r}(t) = t\mathbf{i} + t^2\mathbf{j}, \quad 0 \le t \le 1$$

$$C_2: \quad \mathbf{r}(t) = \mathbf{i} + \mathbf{j} + t\mathbf{k}, \quad 0 \le t \le 1$$

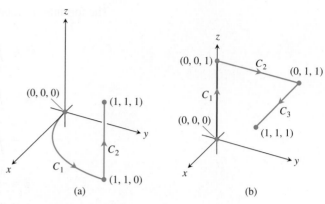

(a)   (b)

The paths of integration for Exercises 15 and 16.

16. Integrate $f(x, y, z) = x + \sqrt{y} - z^2$ over the path from $(0, 0, 0)$ to $(1, 1, 1)$ (see accompanying figure) given by

$$C_1: \quad \mathbf{r}(t) = t\mathbf{k}, \quad 0 \le t \le 1$$

$$C_2: \quad \mathbf{r}(t) = t\mathbf{j} + \mathbf{k}, \quad 0 \le t \le 1$$

$$C_3: \quad \mathbf{r}(t) = t\mathbf{i} + \mathbf{j} + \mathbf{k}, \quad 0 \le t \le 1$$

17. Integrate $f(x, y, z) = (x + y + z)/(x^2 + y^2 + z^2)$ over the path $\mathbf{r}(t) = t\mathbf{i} + t\mathbf{j} + t\mathbf{k}, 0 < a \le t \le b$.

18. Integrate $f(x, y, z) = -\sqrt{x^2 + z^2}$ over the circle

$$\mathbf{r}(t) = (a \cos t)\mathbf{j} + (a \sin t)\mathbf{k}, \quad 0 \le t \le 2\pi.$$

Line Integrals over Plane Curves

19. Evaluate $\int_C x \, ds$, where C is

   a. the straight-line segment $x = t, y = t/2$, from $(0, 0)$ to $(4, 2)$.

   b. the parabolic curve $x = t, y = t^2$, from $(0, 0)$ to $(2, 4)$.

20. Evaluate $\int_C \sqrt{x + 2y} \, ds$, where C is

   a. the straight-line segment $x = t, y = 4t$, from $(0, 0)$ to $(1, 4)$.

   b. $C_1 \cup C_2$; C_1 is the line segment from $(0, 0)$ to $(1, 0)$ and C_2 is the line segment from $(1, 0)$ to $(1, 2)$.

21. Find the line integral of $f(x, y) = ye^{x^2}$ along the curve $\mathbf{r}(t) = 4t\mathbf{i} - 3t\mathbf{j}, -1 \le t \le 2$.

22. Find the line integral of $f(x, y) = x - y + 3$ along the curve $\mathbf{r}(t) = (\cos t)\mathbf{i} + (\sin t)\mathbf{j}, 0 \le t \le 2\pi$.

23. Evaluate $\int_C \dfrac{x^2}{y^{4/3}} \, ds$, where C is the curve $x = t^2, y = t^3$, for $1 \le t \le 2$.

24. Find the line integral of $f(x, y) = \sqrt{y}/x$ along the curve $\mathbf{r}(t) = t^3\mathbf{i} + t^4\mathbf{j}, 1/2 \le t \le 1$.

25. Evaluate $\int_C (x + \sqrt{y}) \, ds$ where C is given in the accompanying figure.

26. Evaluate $\int_C \dfrac{1}{x^2 + y^2 + 1}\, ds$ where C is given in the accompanying figure.

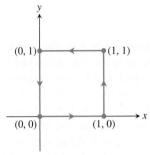

In Exercises 27–30, integrate f over the given curve.

27. $f(x, y) = x^3/y$, C: $y = x^2/2$, $0 \le x \le 2$

28. $f(x, y) = (x + y^2)/\sqrt{1 + x^2}$, C: $y = x^2/2$ from $(1, 1/2)$ to $(0, 0)$

29. $f(x, y) = x + y$, C: $x^2 + y^2 = 4$ in the first quadrant from $(2, 0)$ to $(0, 2)$

30. $f(x, y) = x^2 - y$, C: $x^2 + y^2 = 4$ in the first quadrant from $(0, 2)$ to $(\sqrt{2}, \sqrt{2})$

31. Find the area of one side of the "winding wall" standing orthogonally on the curve $y = x^2$, $0 \le x \le 2$, and beneath the curve on the surface $f(x, y) = x + \sqrt{y}$.

32. Find the area of one side of the "wall" standing orthogonally on the curve $2x + 3y = 6$, $0 \le x \le 6$, and beneath the curve on the surface $f(x, y) = 4 + 3x + 2y$.

Masses and Moments

33. Mass of a wire Find the mass of a wire that lies along the curve $\mathbf{r}(t) = (t^2 - 1)\mathbf{j} + 2t\mathbf{k}$, $0 \le t \le 1$, if the density is $\delta = (3/2)t$.

34. Center of mass of a curved wire A wire of density $\delta(x, y, z) = 15\sqrt{y + 2}$ lies along the curve $\mathbf{r}(t) = (t^2 - 1)\mathbf{j} + 2t\mathbf{k}$, $-1 \le t \le 1$. Find its center of mass. Then sketch the curve and center of mass together.

35. Mass of wire with variable density Find the mass of a thin wire lying along the curve $\mathbf{r}(t) = \sqrt{2}t\mathbf{i} + \sqrt{2}t\mathbf{j} + (4 - t^2)\mathbf{k}$, $0 \le t \le 1$, if the density is **(a)** $\delta = 3t$ and **(b)** $\delta = 1$.

36. Center of mass of wire with variable density Find the center of mass of a thin wire lying along the curve $\mathbf{r}(t) = t\mathbf{i} + 2t\mathbf{j} + (2/3)t^{3/2}\mathbf{k}$, $0 \le t \le 2$, if the density is $\delta = 3\sqrt{5 + t}$.

37. Moment of inertia of wire hoop A circular wire hoop of constant density δ lies along the circle $x^2 + y^2 = a^2$ in the xy-plane. Find the hoop's moment of inertia about the z-axis.

38. Inertia of a slender rod A slender rod of constant density lies along the line segment $\mathbf{r}(t) = t\mathbf{j} + (2 - 2t)\mathbf{k}$, $0 \le t \le 1$, in the

yz-plane. Find the moments of inertia of the rod about the three coordinate axes.

39. Two springs of constant density A spring of constant density δ lies along the helix

$$\mathbf{r}(t) = (\cos t)\mathbf{i} + (\sin t)\mathbf{j} + t\mathbf{k}, \qquad 0 \le t \le 2\pi.$$

a. Find I_z.

b. Suppose that you have another spring of constant density δ that is twice as long as the spring in part (a) and lies along the helix for $0 \le t \le 4\pi$. Do you expect I_z for the longer spring to be the same as that for the shorter one, or should it be different? Check your prediction by calculating I_z for the longer spring.

40. Wire of constant density A wire of constant density $\delta = 1$ lies along the curve

$$\mathbf{r}(t) = (t \cos t)\mathbf{i} + (t \sin t)\mathbf{j} + \left(2\sqrt{2}/3\right)t^{3/2}\mathbf{k}, \qquad 0 \le t \le 1.$$

Find $\bar{z}$ and I_z.

41. The arch in Example 3 Find I_x for the arch in Example 3.

42. Center of mass and moments of inertia for wire with variable density Find the center of mass and the moments of inertia about the coordinate axes of a thin wire lying along the curve

$$\mathbf{r}(t) = t\mathbf{i} + \frac{2\sqrt{2}}{3}t^{3/2}\mathbf{j} + \frac{t^2}{2}\mathbf{k}, \qquad 0 \le t \le 2,$$

if the density is $\delta = 1/(t + 1)$.

COMPUTER EXPLORATIONS

In Exercises 43–46, use a CAS to perform the following steps to evaluate the line integrals.

a. Find $ds = |\mathbf{v}(t)|\, dt$ for the path $\mathbf{r}(t) = g(t)\mathbf{i} + h(t)\mathbf{j} + k(t)\mathbf{k}$.

b. Express the integrand $f(g(t), h(t), k(t))|\mathbf{v}(t)|$ as a function of the parameter t.

c. Evaluate $\int_C f\, ds$ using Equation (2) in the text.

43. $f(x, y, z) = \sqrt{1 + 30x^2 + 10y}$; $\mathbf{r}(t) = t\mathbf{i} + t^2\mathbf{j} + 3t^2\mathbf{k}$, $0 \le t \le 2$

44. $f(x, y, z) = \sqrt{1 + x^3 + 5y^3}$; $\mathbf{r}(t) = t\mathbf{i} + \dfrac{1}{3}t^2\mathbf{j} + \sqrt{t}\mathbf{k}$, $0 \le t \le 2$

45. $f(x, y, z) = x\sqrt{y} - 3z^2$; $\mathbf{r}(t) = (\cos 2t)\mathbf{i} + (\sin 2t)\mathbf{j} + 5t\mathbf{k}$, $0 \le t \le 2\pi$

46. $f(x, y, z) = \left(1 + \dfrac{9}{4}z^{1/3}\right)^{1/4}$; $\mathbf{r}(t) = (\cos 2t)\mathbf{i} + (\sin 2t)\mathbf{j} + t^{5/2}\mathbf{k}$, $0 \le t \le 2\pi$

16.2 Vector Fields and Line Integrals: Work, Circulation, and Flux

Gravitational and electric forces have both a direction and a magnitude. They are represented by a vector at each point in their domain, producing a *vector field*. In this section we show how to compute the work done in moving an object through such a field by using a line integral involving the vector field. We also discuss velocity fields, such as the vector

FIGURE 16.6 Velocity vectors of a flow around an airfoil in a wind tunnel.

FIGURE 16.7 Streamlines in a contracting channel. The water speeds up as the channel narrows and the velocity vectors increase in length.

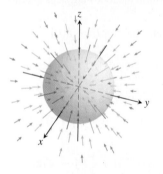

FIGURE 16.8 Vectors in a gravitational field point toward the center of mass that gives the source of the field.

field representing the velocity of a flowing fluid in its domain. A line integral can be used to find the rate at which the fluid flows along or across a curve within the domain.

Vector Fields

Suppose a region in the plane or in space is occupied by a moving fluid, such as air or water. The fluid is made up of a large number of particles, and at any instant of time, a particle has a velocity **v**. At different points of the region at a given (same) time, these velocities can vary. We can think of a velocity vector being attached to each point of the fluid representing the velocity of a particle at that point. Such a fluid flow is an example of a *vector field*. Figure 16.6 shows a velocity vector field obtained from air flowing around an airfoil in a wind tunnel. Figure 16.7 shows a vector field of velocity vectors along the streamlines of water moving through a contracting channel. Vector fields are also associated with forces such as gravitational attraction (Figure 16.8), and to magnetic fields, electric fields, and also purely mathematical fields.

Generally, a **vector field** is a function that assigns a vector to each point in its domain. A vector field on a three-dimensional domain in space might have a formula like

$$\mathbf{F}(x, y, z) = M(x, y, z)\mathbf{i} + N(x, y, z)\mathbf{j} + P(x, y, z)\mathbf{k}.$$

The field is **continuous** if the **component functions** M, N, and P are continuous; it is **differentiable** if each of the component functions is differentiable. The formula for a field of two-dimensional vectors could look like

$$\mathbf{F}(x, y) = M(x, y)\mathbf{i} + N(x, y)\mathbf{j}.$$

We encountered another type of vector field in Chapter 13. The tangent vectors **T** and normal vectors **N** for a curve in space both form vector fields along the curve. Along a curve $\mathbf{r}(t)$ they might have a component formula similar to the velocity field expression

$$\mathbf{v}(t) = f(t)\mathbf{i} + g(t)\mathbf{j} + h(t)\mathbf{k}.$$

If we attach the gradient vector ∇f of a scalar function $f(x, y, z)$ to each point of a level surface of the function, we obtain a three-dimensional field on the surface. If we attach the velocity vector to each point of a flowing fluid, we have a three-dimensional field defined on a region in space. These and other fields are illustrated in Figures 16.9–16.15. To sketch the fields, we picked a representative selection of domain points and drew the

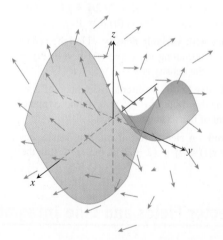

FIGURE 16.9 A surface, like a mesh net or parachute, in a vector field representing water or wind flow velocity vectors. The arrows show the direction and their lengths indicate speed.

$f(x, y, z) = c$

FIGURE 16.10   The field of gradient vectors ∇f on a surface $f(x, y, z) = c$.

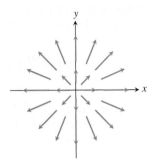

FIGURE 16.11   The radial field $\mathbf{F} = x\mathbf{i} + y\mathbf{j}$ of position vectors of points in the plane. Notice the convention that an arrow is drawn with its tail, not its head, at the point where $\mathbf{F}$ is evaluated.

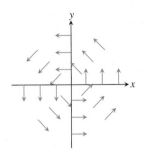

FIGURE 16.12   A "spin" field of rotating unit vectors

$$\mathbf{F} = (-y\mathbf{i} + x\mathbf{j})/(x^2 + y^2)^{1/2}$$

in the plane. The field is not defined at the origin.

vectors attached to them. The arrows are drawn with their tails, not their heads, attached to the points where the vector functions are evaluated.

Gradient Fields

The gradient vector of a differentiable scalar-valued function at a point gives the direction of greatest increase of the function. An important type of vector field is formed by all the

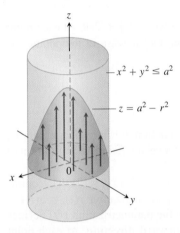

$x^2 + y^2 \le a^2$

$z = a^2 - r^2$

FIGURE 16.13   The flow of fluid in a long cylindrical pipe. The vectors $\mathbf{v} = (a^2 - r^2)\mathbf{k}$ inside the cylinder that have their bases in the xy-plane have their tips on the paraboloid $z = a^2 - r^2$.

FIGURE 16.14   The velocity vectors $\mathbf{v}(t)$ of a projectile's motion make a vector field along the trajectory.

WIND SPEED, M/S

0 2 4 6 8 10 12 14 16+

FIGURE 16.15   NASA's *Seasat* used radar to take 350,000 wind measurements over the world's oceans. The arrows show wind direction; their length and the color contouring indicate speed. Notice the heavy storm south of Greenland.

gradient vectors of the function (see Section 14.5). We define the **gradient field** of a differentiable function $f(x, y, z)$ to be the field of gradient vectors

$$\nabla f = \frac{\partial f}{\partial x}\mathbf{i} + \frac{\partial f}{\partial y}\mathbf{j} + \frac{\partial f}{\partial z}\mathbf{k}.$$

At each point (x, y, z), the gradient field gives a vector pointing in the direction of greatest increase of f, with magnitude being the value of the directional derivative in that direction. The gradient field is not always a force field or a velocity field.

EXAMPLE 1   Suppose that the temperature T at each point (x, y, z) in a region of space is given by

$$T = 100 - x^2 - y^2 - z^2,$$

and that $\mathbf{F}(x, y, z)$ is defined to be the gradient of T. Find the vector field $\mathbf{F}$.

Solution   The gradient field $\mathbf{F}$ is the field $\mathbf{F} = \nabla T = -2x\mathbf{i} - 2y\mathbf{j} - 2z\mathbf{k}$. At each point in space, the vector field $\mathbf{F}$ gives the direction for which the increase in temperature is greatest.   ∎

Line Integrals of Vector Fields

In Section 16.1 we defined the line integral of a scalar function $f(x, y, z)$ over a path C. We turn our attention now to the idea of a line integral of a vector field $\mathbf{F}$ along the curve C.

Assume that the vector field $\mathbf{F} = M(x, y, z)\mathbf{i} + N(x, y, z)\mathbf{j} + P(x, y, z)\mathbf{k}$ has continuous components, and that the curve C has a smooth parametrization $\mathbf{r}(t) = g(t)\mathbf{i} + h(t)\mathbf{j} + k(t)\mathbf{k}$, $a \leq t \leq b$. As discussed in Section 16.1, the parametrization $\mathbf{r}(t)$ defines a direction (or orientation) along C which we call the **forward direction**. At each point along the path C, the tangent vector $\mathbf{T} = d\mathbf{r}/ds = \mathbf{v}/|\mathbf{v}|$ is a unit vector tangent to the path and pointing in this forward direction. (The vector $\mathbf{v} = d\mathbf{r}/dt$ is the velocity vector tangent to C at the point, as discussed in Sections 13.1 and 13.3.) Intuitively, the line integral of the vector field is the line integral of the scalar tangential component of $\mathbf{F}$ along C. This tangential component is given by the dot product

$$\mathbf{F} \cdot \mathbf{T} = \mathbf{F} \cdot \frac{d\mathbf{r}}{ds},$$

so we have the following formal definition, where $f = \mathbf{F} \cdot \mathbf{T}$ in Equation (1) of Section 16.1.

DEFINITION   Let $\mathbf{F}$ be a vector field with continuous components defined along a smooth curve C parametrized by $\mathbf{r}(t)$, $a \leq t \leq b$. Then the **line integral of F along C** is

$$\int_C \mathbf{F} \cdot \mathbf{T} \, ds = \int_C \left(\mathbf{F} \cdot \frac{d\mathbf{r}}{ds}\right) ds = \int_C \mathbf{F} \cdot d\mathbf{r}.$$

We evaluate line integrals of vector fields in a way similar to how we evaluate line integrals of scalar functions (Section 16.1).

Evaluating the Line Integral of F = *M*i + *N*j + *P*k along
C: r(*t*) = *g*(*t*)i + *h*(*t*)j + *k*(*t*)k

1. Express the vector field **F** in terms of the parametrized curve *C* as **F**(**r**(*t*)) by substituting the components $x = g(t), y = h(t), z = k(t)$ of **r** into the scalar components $M(x, y, z), N(x, y, z), P(x, y, z)$ of **F**.

2. Find the derivative (velocity) vector $d\mathbf{r}/dt$.

3. Evaluate the line integral with respect to the parameter $t, a \le t \le b$, to obtain

$$\int_C \mathbf{F} \cdot d\mathbf{r} = \int_a^b \mathbf{F}(\mathbf{r}(t)) \cdot \frac{d\mathbf{r}}{dt}\, dt.$$

EXAMPLE 2   Evaluate $\int_C \mathbf{F} \cdot d\mathbf{r}$, where $\mathbf{F}(x, y, z) = z\,\mathbf{i} + xy\,\mathbf{j} - y^2\,\mathbf{k}$ along the curve C given by $\mathbf{r}(t) = t^2\mathbf{i} + t\mathbf{j} + \sqrt{t}\,\mathbf{k}, 0 \le t \le 1$.

Solution   We have

$$\mathbf{F}(\mathbf{r}(t)) = \sqrt{t}\,\mathbf{i} + t^3\mathbf{j} - t^2\mathbf{k}$$

and

$$\frac{d\mathbf{r}}{dt} = 2t\mathbf{i} + \mathbf{j} + \frac{1}{2\sqrt{t}}\,\mathbf{k}.$$

Thus,

$$\int_C \mathbf{F} \cdot d\mathbf{r} = \int_0^1 \mathbf{F}(\mathbf{r}(t)) \cdot \frac{d\mathbf{r}}{dt}\, dt$$

$$= \int_0^1 \left(2t^{3/2} + t^3 - \frac{1}{2}t^{3/2} \right) dt$$

$$= \left[\left(\frac{3}{2}\right)\left(\frac{2}{5}t^{5/2}\right) + \frac{1}{4}t^4 \right]_0^1 = \frac{17}{20}. \quad \blacksquare$$

Line Integrals With Respect to the *xyz* Coordinates

It is sometimes useful to write a line integral of a scalar function with respect to one of the coordinates, such as $\int_C M\,dx$. This integral is not the same as the arc length line integral $\int_C M\,ds$ we defined in Section 16.1. To define the new integral for the scalar function $M(x, y, z)$, we specify a vector field $\mathbf{F} = M(x, y, z)\mathbf{i}$ over the curve C parametrized by $\mathbf{r}(t) = g(t)\mathbf{i} + h(t)\mathbf{j} + k(t)\mathbf{k}, a \le t \le b$. With this notation we have $x = g(t)$ and $dx = g'(t)\,dt$. Then,

$$\mathbf{F} \cdot d\mathbf{r} = \mathbf{F} \cdot \frac{d\mathbf{r}}{dt}\,dt = M(x, y, z)g'(t)\,dt = M(x, y, z)\,dx.$$

So we *define* the line integral of M over C with respect to the coordinate x as

$$\int_C M(x, y, z)\,dx = \int_C \mathbf{F} \cdot d\mathbf{r}, \quad \text{where} \quad \mathbf{F} = M(x, y, z)\mathbf{i}.$$

In the same way, by defining $\mathbf{F} = N(x, y, z)\mathbf{j}$, or $\mathbf{F} = P(x, y, z)\mathbf{k}$, we obtain the integrals $\int_C N\,dy$ and $\int_C P\,dz$. Expressing everything in terms of the parameter t, we have the following formulas for these integrals:

$$\int_C M(x, y, z)\, dx = \int_a^b M(g(t), h(t), k(t))\, g'(t)\, dt \tag{1}$$

$$\int_C N(x, y, z)\, dy = \int_a^b N(g(t), h(t), k(t))\, h'(t)\, dt \tag{2}$$

$$\int_C P(x, y, z)\, dz = \int_a^b P(g(t), h(t), k(t))\, k'(t)\, dt \tag{3}$$

It often happens that these line integrals occur in combination, and we abbreviate the notation by writing

$$\int_C M(x, y, z)\, dx + \int_C N(x, y, z)\, dy + \int_C P(x, y, z)\, dz = \int_C M\, dx + N\, dy + P\, dz.$$

EXAMPLE 3   Evaluate the line integral $\int_C -y\, dx + z\, dy + 2x\, dz$, where C is the helix $\mathbf{r}(t) = (\cos t)\mathbf{i} + (\sin t)\mathbf{j} + t\mathbf{k}$, $0 \le t \le 2\pi$.

Solution   We express everything in terms of the parameter t, so $x = \cos t$, $y = \sin t$, $z = t$, and $dx = -\sin t\, dt$, $dy = \cos t\, dt$, $dz = dt$. Then,

$$\int_C -y\, dx + z\, dy + 2x\, dz = \int_0^{2\pi} [(-\sin t)(-\sin t) + t\cos t + 2\cos t]\, dt$$

$$= \int_0^{2\pi} [2\cos t + t\cos t + \sin^2 t]\, dt$$

$$= \left[2\sin t + (t\sin t + \cos t) + \left(\frac{t}{2} - \frac{\sin 2t}{4} \right) \right]_0^{2\pi}$$

$$= [0 + (0 + 1) + (\pi - 0)] - [0 + (0 + 1) + (0 - 0)]$$

$$= \pi. \qquad \blacksquare$$

Work Done by a Force over a Curve in Space

Suppose that the vector field $\mathbf{F} = M(x, y, z)\mathbf{i} + N(x, y, z)\mathbf{j} + P(x, y, z)\mathbf{k}$ represents a force throughout a region in space (it might be the force of gravity or an electromagnetic force of some kind) and that

$$\mathbf{r}(t) = g(t)\mathbf{i} + h(t)\mathbf{j} + k(t)\mathbf{k}, \qquad a \le t \le b,$$

is a smooth curve in the region. The formula for the work done by the force in moving an object along the curve is motivated by the same kind of reasoning we used in Chapter 6 to derive the formula $W = \int_a^b F(x)\, dx$ for the work done by a continuous force of magnitude $F(x)$ directed along an interval of the x-axis. For a curve C in space, we define the work done by a continuous force field $\mathbf{F}$ to move an object along C from a point A to another point B as follows.

We divide C into n subarcs $P_{k-1}P_k$ with lengths Δs_k, starting at A and ending at B. We choose any point (x_k, y_k, z_k) in the subarc $P_{k-1}P_k$ and let $\mathbf{T}(x_k, y_k, z_k)$ be the unit tangent vector at the chosen point. The work W_k done to move the object along the subarc $P_{k-1}P_k$ is approximated by the tangential component of the force $\mathbf{F}(x_k, y_k, z_k)$ times the arclength Δs_k approximating the distance the object moves along the subarc (see Figure 16.16).

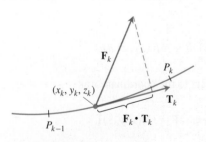

FIGURE 16.16   The work done along the subarc shown here is approximately $\mathbf{F}_k \cdot \mathbf{T}_k\, \Delta s_k$, where $\mathbf{F}_k = \mathbf{F}(x_k, y_k, z_k)$ and $\mathbf{T}_k = \mathbf{T}(x_k, y_k, z_k)$.

The total work done in moving the object from point A to point B is then approximated by summing the work done along each of the subarcs, so

$$W \approx \sum_{k=1}^{n} W_k \approx \sum_{k=1}^{n} \mathbf{F}(x_k, y_k, z_k) \cdot \mathbf{T}(x_k, y_k, z_k) \, \Delta s_k.$$

For any subdivision of C into n subarcs, and for any choice of the points (x_k, y_k, z_k) within each subarc, as $n \to \infty$ and $\Delta s_k \to 0$, these sums approach the line integral

$$\int_C \mathbf{F} \cdot \mathbf{T} \, ds.$$

This is just the line integral of $\mathbf{F}$ along C, which is defined to be the total work done.

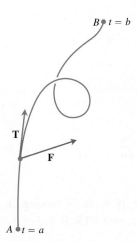

FIGURE 16.17 The work done by a force $\mathbf{F}$ is the line integral of the scalar component $\mathbf{F} \cdot \mathbf{T}$ over the smooth curve from A to B.

> **DEFINITION** Let C be a smooth curve parametrized by $\mathbf{r}(t)$, $a \leq t \leq b$, and $\mathbf{F}$ be a continuous force field over a region containing C. Then the **work** done in moving an object from the point $A = \mathbf{r}(a)$ to the point $B = \mathbf{r}(b)$ along C is
>
> $$W = \int_C \mathbf{F} \cdot \mathbf{T} \, ds = \int_a^b \mathbf{F}(\mathbf{r}(t)) \cdot \frac{d\mathbf{r}}{dt} \, dt. \tag{4}$$

The sign of the number we calculate with this integral depends on the direction in which the curve is traversed. If we reverse the direction of motion, then we reverse the direction of $\mathbf{T}$ in Figure 16.17 and change the sign of $\mathbf{F} \cdot \mathbf{T}$ and its integral.

Using the notations we have presented, we can express the work integral in a variety of ways, depending upon what seems most suitable or convenient for a particular discussion. Table 16.2 shows five ways we can write the work integral in Equation (4).

TABLE 16.2 Different ways to write the work integral for $\mathbf{F} = M\mathbf{i} + N\mathbf{j} + P\mathbf{k}$ over the curve $C : \mathbf{r}(t) = g(t)\mathbf{i} + h(t)\mathbf{j} + k(t)\mathbf{k}$, $a \leq t \leq b$

$\mathbf{W} = \displaystyle\int_C \mathbf{F} \cdot \mathbf{T} \, ds$	The definition
$= \displaystyle\int_C \mathbf{F} \cdot d\mathbf{r}$	Vector differential form
$= \displaystyle\int_a^b \mathbf{F} \cdot \frac{d\mathbf{r}}{dt} \, dt$	Parametric vector evaluation
$= \displaystyle\int_a^b \left(M \frac{dx}{dt} + N \frac{dy}{dt} + P \frac{dz}{dt} \right) dt$	Parametric scalar evaluation
$= \displaystyle\int_C M \, dx + N \, dy + P \, dz$	Scalar differential form

EXAMPLE 4 Find the work done by the force field $\mathbf{F} = (y - x^2)\mathbf{i} + (z - y^2)\mathbf{j} + (x - z^2)\mathbf{k}$ along the curve $\mathbf{r}(t) = t\mathbf{i} + t^2\mathbf{j} + t^3\mathbf{k}$, $0 \leq t \leq 1$, from $(0, 0, 0)$ to $(1, 1, 1)$ (Figure 16.18).

Solution First we evaluate $\mathbf{F}$ on the curve $\mathbf{r}(t)$:

$$\mathbf{F} = (y - x^2)\mathbf{i} + (z - y^2)\mathbf{j} + (x - z^2)\mathbf{k}$$
$$= \underbrace{(t^2 - t^2)}_{0}\mathbf{i} + (t^3 - t^4)\mathbf{j} + (t - t^6)\mathbf{k}. \quad \text{Substitute } x = t, \ y = t^2, z = t^3.$$

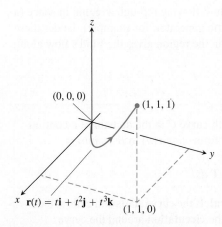

FIGURE 16.18 The curve in Example 4.

Then we find $d\mathbf{r}/dt$,

$$\frac{d\mathbf{r}}{dt} = \frac{d}{dt}(t\mathbf{i} + t^2\mathbf{j} + t^3\mathbf{k}) = \mathbf{i} + 2t\mathbf{j} + 3t^2\mathbf{k}.$$

Finally, we find $\mathbf{F} \cdot d\mathbf{r}/dt$ and integrate from $t = 0$ to $t = 1$:

$$\mathbf{F} \cdot \frac{d\mathbf{r}}{dt} = [(t^3 - t^4)\mathbf{j} + (t - t^6)\mathbf{k}] \cdot (\mathbf{i} + 2t\mathbf{j} + 3t^2\mathbf{k})$$

$$= (t^3 - t^4)(2t) + (t - t^6)(3t^2) = 2t^4 - 2t^5 + 3t^3 - 3t^8$$

so,

$$\text{Work} = \int_0^1 (2t^4 - 2t^5 + 3t^3 - 3t^8)\, dt$$

$$= \left[\frac{2}{5}t^5 - \frac{2}{6}t^6 + \frac{3}{4}t^4 - \frac{3}{9}t^9\right]_0^1 = \frac{29}{60}. \qquad \blacksquare$$

EXAMPLE 5　Find the work done by the force field $\mathbf{F} = x\mathbf{i} + y\mathbf{j} + z\mathbf{k}$ in moving an object along the curve C parametrized by $\mathbf{r}(t) = \cos(\pi t)\,\mathbf{i} + t^2\mathbf{j} + \sin(\pi t)\,\mathbf{k}, 0 \le t \le 1$.

Solution　We begin by writing $\mathbf{F}$ along C as a function of t,

$$\mathbf{F}(\mathbf{r}(t)) = \cos(\pi t)\,\mathbf{i} + t^2\mathbf{j} + \sin(\pi t)\,\mathbf{k}.$$

Next we compute $d\mathbf{r}/dt$,

$$\frac{d\mathbf{r}}{dt} = -\pi \sin(\pi t)\,\mathbf{i} + 2t\mathbf{j} + \pi \cos(\pi t)\,\mathbf{k}.$$

We then calculate the dot product,

$$\mathbf{F}(\mathbf{r}(t)) \cdot \frac{d\mathbf{r}}{dt} = -\pi \sin(\pi t)\cos(\pi t) + 2t^3 + \pi \sin(\pi t)\cos(\pi t) = 2t^3.$$

The work done is the line integral

$$\int_a^b \mathbf{F}(\mathbf{r}(t)) \cdot \frac{d\mathbf{r}}{dt}\, dt = \int_0^1 2t^3\, dt = \frac{t^4}{2}\bigg]_0^1 = \frac{1}{2}. \qquad \blacksquare$$

Flow Integrals and Circulation for Velocity Fields

Suppose that $\mathbf{F}$ represents the velocity field of a fluid flowing through a region in space (a tidal basin or the turbine chamber of a hydroelectric generator, for example). Under these circumstances, the integral of $\mathbf{F} \cdot \mathbf{T}$ along a curve in the region gives the fluid's flow along the curve.

DEFINITIONS　If $\mathbf{r}(t)$ parametrizes a smooth curve C in the domain of a continuous velocity field $\mathbf{F}$, the **flow** along the curve from $A = \mathbf{r}(a)$ to $B = \mathbf{r}(b)$ is

$$\text{Flow} = \int_C \mathbf{F} \cdot \mathbf{T}\, ds. \tag{5}$$

The integral in this case is called a **flow integral**. If the curve starts and ends at the same point, so that $A = B$, the flow is called the **circulation** around the curve.

The direction we travel along C matters. If we reverse the direction, then $\mathbf{T}$ is replaced by $-\mathbf{T}$ and the sign of the integral changes. We evaluate flow integrals the same way we evaluate work integrals.

EXAMPLE 6   A fluid's velocity field is $\mathbf{F} = x\mathbf{i} + z\mathbf{j} + y\mathbf{k}$. Find the flow along the helix $\mathbf{r}(t) = (\cos t)\mathbf{i} + (\sin t)\mathbf{j} + t\mathbf{k}, 0 \le t \le \pi/2$.

Solution   We evaluate $\mathbf{F}$ on the curve,

$$\mathbf{F} = x\mathbf{i} + z\mathbf{j} + y\mathbf{k} = (\cos t)\mathbf{i} + t\mathbf{j} + (\sin t)\mathbf{k} \qquad \text{Substitute } x = \cos t, z = t, y = \sin t.$$

and then find $d\mathbf{r}/dt$:

$$\frac{d\mathbf{r}}{dt} = (-\sin t)\mathbf{i} + (\cos t)\mathbf{j} + \mathbf{k}.$$

Then we integrate $\mathbf{F} \cdot (d\mathbf{r}/dt)$ from $t = 0$ to $t = \dfrac{\pi}{2}$:

$$\mathbf{F} \cdot \frac{d\mathbf{r}}{dt} = (\cos t)(-\sin t) + (t)(\cos t) + (\sin t)(1)$$
$$= -\sin t \cos t + t \cos t + \sin t$$

so,

$$\text{Flow} = \int_{t=a}^{t=b} \mathbf{F} \cdot \frac{d\mathbf{r}}{dt}\, dt = \int_0^{\pi/2} (-\sin t \cos t + t \cos t + \sin t)\, dt$$
$$= \left[\frac{\cos^2 t}{2} + t \sin t \right]_0^{\pi/2} = \left(0 + \frac{\pi}{2} \right) - \left(\frac{1}{2} + 0 \right) = \frac{\pi}{2} - \frac{1}{2}. \qquad \blacksquare$$

EXAMPLE 7   Find the circulation of the field $\mathbf{F} = (x - y)\mathbf{i} + x\mathbf{j}$ around the circle $\mathbf{r}(t) = (\cos t)\mathbf{i} + (\sin t)\mathbf{j}, 0 \le t \le 2\pi$ (Figure 16.19).

Solution   On the circle, $\mathbf{F} = (x - y)\mathbf{i} + x\mathbf{j} = (\cos t - \sin t)\mathbf{i} + (\cos t)\mathbf{j}$, and

$$\frac{d\mathbf{r}}{dt} = (-\sin t)\mathbf{i} + (\cos t)\mathbf{j}.$$

Then

$$\mathbf{F} \cdot \frac{d\mathbf{r}}{dt} = -\sin t \cos t + \underbrace{\sin^2 t + \cos^2 t}_{1}$$

gives

$$\text{Circulation} = \int_0^{2\pi} \mathbf{F} \cdot \frac{d\mathbf{r}}{dt}\, dt = \int_0^{2\pi} (1 - \sin t \cos t)\, dt$$
$$= \left[t - \frac{\sin^2 t}{2} \right]_0^{2\pi} = 2\pi.$$

As Figure 16.19 suggests, a fluid with this velocity field is circulating *counterclockwise* around the circle. $\qquad \blacksquare$

Flux Across a Simple Plane Curve

A curve in the xy-plane is **simple** if it does not cross itself (Figure 16.20). When a curve starts and ends at the same point, it is a **closed curve** or **loop**. To find the rate at which a fluid is entering or leaving a region enclosed by a smooth simple closed curve C in the xy-plane,

FIGURE 16.19   The vector field $\mathbf{F}$ and curve $\mathbf{r}(t)$ in Example 7.

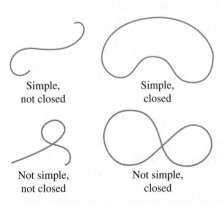

Simple, not closed

Simple, closed

Not simple, not closed

Not simple, closed

FIGURE 16.20   Distinguishing curves that are simple or closed. Closed curves are also called loops.

we calculate the line integral over C of $\mathbf{F} \cdot \mathbf{n}$, the scalar component of the fluid's velocity field in the direction of the curve's outward-pointing normal vector. The value of this integral is the *flux* of $\mathbf{F}$ across C. *Flux* is Latin for *flow*, but many flux calculations involve no motion at all. If $\mathbf{F}$ were an electric field or a magnetic field, for instance, the integral of $\mathbf{F} \cdot \mathbf{n}$ would still be called the flux of the field across C.

DEFINITION   If C is a smooth simple closed curve in the domain of a continuous vector field $\mathbf{F} = M(x, y)\mathbf{i} + N(x, y)\mathbf{j}$ in the plane, and if $\mathbf{n}$ is the outward-pointing unit normal vector on C, the **flux** of $\mathbf{F}$ across C is

$$\text{Flux of } \mathbf{F} \text{ across } C = \int_C \mathbf{F} \cdot \mathbf{n} \, ds. \tag{6}$$

Notice the difference between flux and circulation. The flux of $\mathbf{F}$ across C is the line integral with respect to arc length of $\mathbf{F} \cdot \mathbf{n}$, the scalar component of $\mathbf{F}$ in the direction of the outward normal. The circulation of $\mathbf{F}$ around C is the line integral with respect to arc length of $\mathbf{F} \cdot \mathbf{T}$, the scalar component of $\mathbf{F}$ in the direction of the unit tangent vector. Flux is the integral of the normal component of $\mathbf{F}$; circulation is the integral of the tangential component of $\mathbf{F}$.

To evaluate the integral for flux in Equation (6), we begin with a smooth parametrization

$$x = g(t), \qquad y = h(t), \qquad a \le t \le b,$$

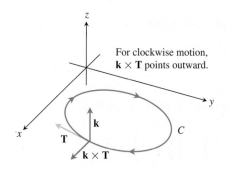

For clockwise motion, $\mathbf{k} \times \mathbf{T}$ points outward.

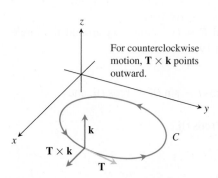

For counterclockwise motion, $\mathbf{T} \times \mathbf{k}$ points outward.

FIGURE 16.21 To find an outward unit normal vector for a smooth simple curve C in the xy-plane that is traversed counterclockwise as t increases, we take $\mathbf{n} = \mathbf{T} \times \mathbf{k}$. For clockwise motion, we take $\mathbf{n} = \mathbf{k} \times \mathbf{T}$.

that traces the curve C exactly once as t increases from a to b. We can find the outward unit normal vector $\mathbf{n}$ by crossing the curve's unit tangent vector $\mathbf{T}$ with the vector $\mathbf{k}$. But which order do we choose, $\mathbf{T} \times \mathbf{k}$ or $\mathbf{k} \times \mathbf{T}$? Which one points outward? It depends on which way C is traversed as t increases. If the motion is clockwise, $\mathbf{k} \times \mathbf{T}$ points outward; if the motion is counterclockwise, $\mathbf{T} \times \mathbf{k}$ points outward (Figure 16.21). The usual choice is $\mathbf{n} = \mathbf{T} \times \mathbf{k}$, the choice that assumes counterclockwise motion. Thus, although the value of the integral in Equation (6) does not depend on which way C is traversed, the formulas we are about to derive for computing $\mathbf{n}$ and evaluating the integral assume counterclockwise motion.

In terms of components,

$$\mathbf{n} = \mathbf{T} \times \mathbf{k} = \left(\frac{dx}{ds}\mathbf{i} + \frac{dy}{ds}\mathbf{j} \right) \times \mathbf{k} = \frac{dy}{ds}\mathbf{i} - \frac{dx}{ds}\mathbf{j}.$$

If $\mathbf{F} = M(x, y)\mathbf{i} + N(x, y)\mathbf{j}$, then

$$\mathbf{F} \cdot \mathbf{n} = M(x, y)\frac{dy}{ds} - N(x, y)\frac{dx}{ds}.$$

Hence,

$$\int_C \mathbf{F} \cdot \mathbf{n} \, ds = \int_C \left(M\frac{dy}{ds} - N\frac{dx}{ds} \right) ds = \oint_C M \, dy - N \, dx.$$

We put a directed circle $\circlearrowleft$ on the last integral as a reminder that the integration around the closed curve C is to be in the counterclockwise direction. To evaluate this integral, we express M, dy, N, and dx in terms of the parameter t and integrate from $t = a$ to $t = b$. We do not need to know $\mathbf{n}$ or ds explicitly to find the flux.

Calculating Flux Across a Smooth Closed Plane Curve

$$\text{(Flux of } \mathbf{F} = M\mathbf{i} + N\mathbf{j} \text{ across } C) = \oint_C M\, dy - N\, dx \qquad (7)$$

The integral can be evaluated from any smooth parametrization $x = g(t)$, $y = h(t)$, $a \le t \le b$, that traces C counterclockwise exactly once.

EXAMPLE 8   Find the flux of $\mathbf{F} = (x - y)\mathbf{i} + x\mathbf{j}$ across the circle $x^2 + y^2 = 1$ in the xy-plane. (The vector field and curve were shown previously in Figure 16.19.)

Solution   The parametrization $\mathbf{r}(t) = (\cos t)\mathbf{i} + (\sin t)\mathbf{j}$, $0 \le t \le 2\pi$, traces the circle counterclockwise exactly once. We can therefore use this parametrization in Equation (7). With

$$M = x - y = \cos t - \sin t, \qquad dy = d(\sin t) = \cos t\, dt$$

$$N = x = \cos t, \qquad dx = d(\cos t) = -\sin t\, dt,$$

we find

$$\text{Flux} = \int_C M\, dy - N\, dx = \int_0^{2\pi} (\cos^2 t - \sin t \cos t + \cos t \sin t)\, dt \qquad \text{Eq. (7)}$$

$$= \int_0^{2\pi} \cos^2 t\, dt = \int_0^{2\pi} \frac{1 + \cos 2t}{2}\, dt = \left[\frac{t}{2} + \frac{\sin 2t}{4} \right]_0^{2\pi} = \pi.$$

The flux of $\mathbf{F}$ across the circle is π. Since the answer is positive, the net flow across the curve is outward. A net inward flow would have given a negative flux.   ∎

Exercises 16.2

Vector Fields

Find the gradient fields of the functions in Exercises 1–4.

1. $f(x, y, z) = (x^2 + y^2 + z^2)^{-1/2}$

2. $f(x, y, z) = \ln\sqrt{x^2 + y^2 + z^2}$

3. $g(x, y, z) = e^z - \ln(x^2 + y^2)$

4. $g(x, y, z) = xy + yz + xz$

5. Give a formula $\mathbf{F} = M(x, y)\mathbf{i} + N(x, y)\mathbf{j}$ for the vector field in the plane that has the property that $\mathbf{F}$ points toward the origin with magnitude inversely proportional to the square of the distance from (x, y) to the origin. (The field is not defined at $(0, 0)$.)

6. Give a formula $\mathbf{F} = M(x, y)\mathbf{i} + N(x, y)\mathbf{j}$ for the vector field in the plane that has the properties that $\mathbf{F} = \mathbf{0}$ at $(0, 0)$ and that at any other point (a, b), $\mathbf{F}$ is tangent to the circle $x^2 + y^2 = a^2 + b^2$ and points in the clockwise direction with magnitude $|\mathbf{F}| = \sqrt{a^2 + b^2}$.

Line Integrals of Vector Fields

In Exercises 7–12, find the line integrals of $\mathbf{F}$ from $(0, 0, 0)$ to $(1, 1, 1)$ over each of the following paths in the accompanying figure.

a. The straight-line path C_1: $\mathbf{r}(t) = t\mathbf{i} + t\mathbf{j} + t\mathbf{k}$, $0 \le t \le 1$

b. The curved path C_2: $\mathbf{r}(t) = t\mathbf{i} + t^2\mathbf{j} + t^4\mathbf{k}$, $0 \le t \le 1$

c. The path $C_3 \cup C_4$ consisting of the line segment from $(0, 0, 0)$ to $(1, 1, 0)$ followed by the segment from $(1, 1, 0)$ to $(1, 1, 1)$

7. $\mathbf{F} = 3y\mathbf{i} + 2x\mathbf{j} + 4z\mathbf{k}$ 8. $\mathbf{F} = [1/(x^2 + 1)]\mathbf{j}$

9. $\mathbf{F} = \sqrt{z}\mathbf{i} - 2x\mathbf{j} + \sqrt{y}\mathbf{k}$ 10. $\mathbf{F} = xy\mathbf{i} + yz\mathbf{j} + xz\mathbf{k}$

11. $\mathbf{F} = (3x^2 - 3x)\mathbf{i} + 3z\mathbf{j} + \mathbf{k}$

12. $\mathbf{F} = (y + z)\mathbf{i} + (z + x)\mathbf{j} + (x + y)\mathbf{k}$

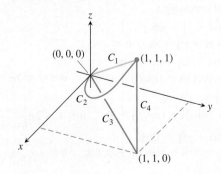

Line Integrals with Respect to x, y, and z

In Exercises 13–16, find the line integrals along the given path C.

13. $\displaystyle\int_C (x - y)\, dx$, where C: $x = t$, $y = 2t + 1$, for $0 \le t \le 3$

14. $\displaystyle\int_C \frac{x}{y}\, dy$, where C: $x = t$, $y = t^2$, for $1 \le t \le 2$

15. $\displaystyle\int_C (x^2 + y^2)\, dy$, where C is given in the accompanying figure.

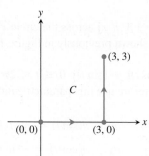

16. $\displaystyle\int_C \sqrt{x + y}\, dx$, where C is given in the accompanying figure.

17. Along the curve $\mathbf{r}(t) = t\mathbf{i} - \mathbf{j} + t^2\mathbf{k}$, $0 \le t \le 1$, evaluate each of the following integrals.

a. $\displaystyle\int_C (x + y - z)\, dx$

b. $\displaystyle\int_C (x + y - z)\, dy$

c. $\displaystyle\int_C (x + y - z)\, dz$

18. Along the curve $\mathbf{r}(t) = (\cos t)\mathbf{i} + (\sin t)\mathbf{j} - (\cos t)\mathbf{k}$, $0 \le t \le \pi$, evaluate each of the following integrals.

a. $\displaystyle\int_C xz\, dx$ **b.** $\displaystyle\int_C xz\, dy$ **c.** $\displaystyle\int_C xyz\, dz$

Work

In Exercises 19–22, find the work done by $\mathbf{F}$ over the curve in the direction of increasing t.

19. $\mathbf{F} = xy\mathbf{i} + y\mathbf{j} - yz\mathbf{k}$
$\mathbf{r}(t) = t\mathbf{i} + t^2\mathbf{j} + t\mathbf{k}$, $0 \le t \le 1$

20. $\mathbf{F} = 2y\mathbf{i} + 3x\mathbf{j} + (x + y)\mathbf{k}$
$\mathbf{r}(t) = (\cos t)\mathbf{i} + (\sin t)\mathbf{j} + (t/6)\mathbf{k}$, $0 \le t \le 2\pi$

21. $\mathbf{F} = z\mathbf{i} + x\mathbf{j} + y\mathbf{k}$
$\mathbf{r}(t) = (\sin t)\mathbf{i} + (\cos t)\mathbf{j} + t\mathbf{k}$, $0 \le t \le 2\pi$

22. $\mathbf{F} = 6z\mathbf{i} + y^2\mathbf{j} + 12x\mathbf{k}$
$\mathbf{r}(t) = (\sin t)\mathbf{i} + (\cos t)\mathbf{j} + (t/6)\mathbf{k}$, $0 \le t \le 2\pi$

Line Integrals in the Plane

23. Evaluate $\displaystyle\int_C xy\, dx + (x + y)\, dy$ along the curve $y = x^2$ from $(-1, 1)$ to $(2, 4)$.

24. Evaluate $\displaystyle\int_C (x - y)\, dx + (x + y)\, dy$ counterclockwise around the triangle with vertices $(0, 0)$, $(1, 0)$, and $(0, 1)$.

25. Evaluate $\displaystyle\int_C \mathbf{F} \cdot \mathbf{T}\, ds$ for the vector field $\mathbf{F} = x^2\mathbf{i} - y\mathbf{j}$ along the curve $x = y^2$ from $(4, 2)$ to $(1, -1)$.

26. Evaluate $\displaystyle\int_C \mathbf{F} \cdot d\mathbf{r}$ for the vector field $\mathbf{F} = y\mathbf{i} - x\mathbf{j}$ counterclockwise along the unit circle $x^2 + y^2 = 1$ from $(1, 0)$ to $(0, 1)$.

Work, Circulation, and Flux in the Plane

27. **Work** Find the work done by the force $\mathbf{F} = xy\mathbf{i} + (y - x)\mathbf{j}$ over the straight line from $(1, 1)$ to $(2, 3)$.

28. **Work** Find the work done by the gradient of $f(x, y) = (x + y)^2$ counterclockwise around the circle $x^2 + y^2 = 4$ from $(2, 0)$ to itself.

29. **Circulation and flux** Find the circulation and flux of the fields

$$\mathbf{F}_1 = x\mathbf{i} + y\mathbf{j} \quad \text{and} \quad \mathbf{F}_2 = -y\mathbf{i} + x\mathbf{j}$$

around and across each of the following curves.

a. The circle $\mathbf{r}(t) = (\cos t)\mathbf{i} + (\sin t)\mathbf{j}$, $0 \le t \le 2\pi$

b. The ellipse $\mathbf{r}(t) = (\cos t)\mathbf{i} + (4 \sin t)\mathbf{j}$, $0 \le t \le 2\pi$

30. **Flux across a circle** Find the flux of the fields

$$\mathbf{F}_1 = 2x\mathbf{i} - 3y\mathbf{j} \quad \text{and} \quad \mathbf{F}_2 = 2x\mathbf{i} + (x - y)\mathbf{j}$$

across the circle

$$\mathbf{r}(t) = (a \cos t)\mathbf{i} + (a \sin t)\mathbf{j}, \quad 0 \le t \le 2\pi.$$

In Exercises 31–34, find the circulation and flux of the field $\mathbf{F}$ around and across the closed semicircular path that consists of the semicircular arch $\mathbf{r}_1(t) = (a \cos t)\mathbf{i} + (a \sin t)\mathbf{j}$, $0 \le t \le \pi$, followed by the line segment $\mathbf{r}_2(t) = t\mathbf{i}$, $-a \le t \le a$.

31. $\mathbf{F} = x\mathbf{i} + y\mathbf{j}$ **32.** $\mathbf{F} = x^2\mathbf{i} + y^2\mathbf{j}$

33. $\mathbf{F} = -y\mathbf{i} + x\mathbf{j}$ **34.** $\mathbf{F} = -y^2\mathbf{i} + x^2\mathbf{j}$

35. **Flow integrals** Find the flow of the velocity field $\mathbf{F} = (x + y)\mathbf{i} - (x^2 + y^2)\mathbf{j}$ along each of the following paths from $(1, 0)$ to $(-1, 0)$ in the xy-plane.

a. The upper half of the circle $x^2 + y^2 = 1$

b. The line segment from $(1, 0)$ to $(-1, 0)$

c. The line segment from $(1, 0)$ to $(0, -1)$ followed by the line segment from $(0, -1)$ to $(-1, 0)$

36. **Flux across a triangle** Find the flux of the field $\mathbf{F}$ in Exercise 35 outward across the triangle with vertices $(1, 0)$, $(0, 1)$, $(-1, 0)$.

37. Find the flow of the velocity field $\mathbf{F} = y^2\mathbf{i} + 2xy\mathbf{j}$ along each of the following paths from $(0, 0)$ to $(2, 4)$.

c. Use any path from $(0, 0)$ to $(2, 4)$ different from parts (a) and (b).

38. Find the circulation of the field $\mathbf{F} = y\mathbf{i} + (x + 2y)\mathbf{j}$ around each of the following closed paths.

a.

b.

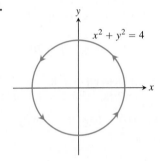

c. Use any closed path different from parts (a) and (b).

Vector Fields in the Plane

39. Spin field Draw the spin field

$$\mathbf{F} = -\frac{y}{\sqrt{x^2 + y^2}}\mathbf{i} + \frac{x}{\sqrt{x^2 + y^2}}\mathbf{j}$$

(see Figure 16.12) along with its horizontal and vertical components at a representative assortment of points on the circle $x^2 + y^2 = 4$.

40. Radial field Draw the radial field

$$\mathbf{F} = x\mathbf{i} + y\mathbf{j}$$

(see Figure 16.11) along with its horizontal and vertical components at a representative assortment of points on the circle $x^2 + y^2 = 1$.

41. A field of tangent vectors

a. Find a field $\mathbf{G} = P(x, y)\mathbf{i} + Q(x, y)\mathbf{j}$ in the xy-plane with the property that at any point $(a, b) \neq (0, 0)$, $\mathbf{G}$ is a vector of magnitude $\sqrt{a^2 + b^2}$ tangent to the circle $x^2 + y^2 = a^2 + b^2$ and pointing in the counterclockwise direction. (The field is undefined at $(0, 0)$.)

b. How is $\mathbf{G}$ related to the spin field $\mathbf{F}$ in Figure 16.12?

42. A field of tangent vectors

a. Find a field $\mathbf{G} = P(x, y)\mathbf{i} + Q(x, y)\mathbf{j}$ in the xy-plane with the property that at any point $(a, b) \neq (0, 0)$, $\mathbf{G}$ is a unit vector tangent to the circle $x^2 + y^2 = a^2 + b^2$ and pointing in the clockwise direction.

b. How is $\mathbf{G}$ related to the spin field $\mathbf{F}$ in Figure 16.12?

43. Unit vectors pointing toward the origin Find a field $\mathbf{F} = M(x, y)\mathbf{i} + N(x, y)\mathbf{j}$ in the xy-plane with the property that at each point $(x, y) \neq (0, 0)$, $\mathbf{F}$ is a unit vector pointing toward the origin. (The field is undefined at $(0, 0)$.)

44. Two "central" fields Find a field $\mathbf{F} = M(x, y)\mathbf{i} + N(x, y)\mathbf{j}$ in the xy-plane with the property that at each point $(x, y) \neq (0, 0)$, $\mathbf{F}$ points toward the origin and $|\mathbf{F}|$ is **(a)** the distance from (x, y) to the origin, **(b)** inversely proportional to the distance from (x, y) to the origin. (The field is undefined at $(0, 0)$.)

45. Work and area Suppose that $f(t)$ is differentiable and positive for $a \leq t \leq b$. Let C be the path $\mathbf{r}(t) = t\mathbf{i} + f(t)\mathbf{j}$, $a \leq t \leq b$, and $\mathbf{F} = y\mathbf{i}$. Is there any relation between the value of the work integral

$$\int_C \mathbf{F} \cdot d\mathbf{r}$$

and the area of the region bounded by the t-axis, the graph of f, and the lines $t = a$ and $t = b$? Give reasons for your answer.

46. Work done by a radial force with constant magnitude A particle moves along the smooth curve $y = f(x)$ from $(a, f(a))$ to $(b, f(b))$. The force moving the particle has constant magnitude k and always points away from the origin. Show that the work done by the force is

$$\int_C \mathbf{F} \cdot \mathbf{T}\, ds = k\left[(b^2 + (f(b))^2)^{1/2} - (a^2 + (f(a))^2)^{1/2}\right].$$

Flow Integrals in Space

In Exercises 47–50, $\mathbf{F}$ is the velocity field of a fluid flowing through a region in space. Find the flow along the given curve in the direction of increasing t.

47. $\mathbf{F} = -4xy\mathbf{i} + 8y\mathbf{j} + 2\mathbf{k}$
$\mathbf{r}(t) = t\mathbf{i} + t^2\mathbf{j} + \mathbf{k}, \quad 0 \leq t \leq 2$

48. $\mathbf{F} = x^2\mathbf{i} + yz\mathbf{j} + y^2\mathbf{k}$
$r(t) = 3t\mathbf{j} + 4t\mathbf{k}, \quad 0 \leq t \leq 1$

49. $\mathbf{F} = (x - z)\mathbf{i} + x\mathbf{k}$
$\mathbf{r}(t) = (\cos t)\mathbf{i} + (\sin t)\mathbf{k}, \quad 0 \leq t \leq \pi$

50. $\mathbf{F} = -y\mathbf{i} + x\mathbf{j} + 2\mathbf{k}$
$\mathbf{r}(t) = (-2 \cos t)\mathbf{i} + (2 \sin t)\mathbf{j} + 2t\mathbf{k}, \quad 0 \leq t \leq 2\pi$

51. Circulation Find the circulation of $\mathbf{F} = 2x\mathbf{i} + 2z\mathbf{j} + 2y\mathbf{k}$ around the closed path consisting of the following three curves traversed in the direction of increasing t.

$C_1: \quad \mathbf{r}(t) = (\cos t)\mathbf{i} + (\sin t)\mathbf{j} + t\mathbf{k}, \quad 0 \leq t \leq \pi/2$

$C_2: \quad \mathbf{r}(t) = \mathbf{j} + (\pi/2)(1 - t)\mathbf{k}, \quad 0 \leq t \leq 1$

$C_3: \quad \mathbf{r}(t) = t\mathbf{i} + (1 - t)\mathbf{j}, \quad 0 \leq t \leq 1$

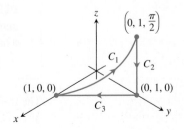

52. Zero circulation Let C be the ellipse in which the plane $2x + 3y - z = 0$ meets the cylinder $x^2 + y^2 = 12$. Show, without evaluating either line integral directly, that the circulation of the field $\mathbf{F} = x\mathbf{i} + y\mathbf{j} + z\mathbf{k}$ around C in either direction is zero.

53. Flow along a curve The field $\mathbf{F} = xy\mathbf{i} + y\mathbf{j} - yz\mathbf{k}$ is the velocity field of a flow in space. Find the flow from $(0, 0, 0)$ to $(1, 1, 1)$ along the curve of intersection of the cylinder $y = x^2$ and the plane $z = x$. (*Hint:* Use $t = x$ as the parameter.)

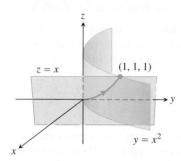

54. Flow of a gradient field Find the flow of the field $\mathbf{F} = \nabla(xy^2z^3)$:

a. Once around the curve C in Exercise 52, clockwise as viewed from above

b. Along the line segment from $(1, 1, 1)$ to $(2, 1, -1)$.

COMPUTER EXPLORATIONS

In Exercises 55–60, use a CAS to perform the following steps for finding the work done by force $\mathbf{F}$ over the given path:

a. Find $d\mathbf{r}$ for the path $\mathbf{r}(t) = g(t)\mathbf{i} + h(t)\mathbf{j} + k(t)\mathbf{k}$.

b. Evaluate the force $\mathbf{F}$ along the path.

c. Evaluate $\displaystyle\int_C \mathbf{F} \cdot d\mathbf{r}$.

55. $\mathbf{F} = xy^6\mathbf{i} + 3x(xy^5 + 2)\mathbf{j}$; $\mathbf{r}(t) = (2\cos t)\mathbf{i} + (\sin t)\mathbf{j}$,
$0 \le t \le 2\pi$

56. $\mathbf{F} = \dfrac{3}{1 + x^2}\mathbf{i} + \dfrac{2}{1 + y^2}\mathbf{j}$; $\mathbf{r}(t) = (\cos t)\mathbf{i} + (\sin t)\mathbf{j}$,
$0 \le t \le \pi$

57. $\mathbf{F} = (y + yz \cos xyz)\mathbf{i} + (x^2 + xz \cos xyz)\mathbf{j} + (z + xy \cos xyz)\mathbf{k}$; $\mathbf{r}(t) = (2\cos t)\mathbf{i} + (3\sin t)\mathbf{j} + \mathbf{k}$,
$0 \le t \le 2\pi$

58. $\mathbf{F} = 2xy\mathbf{i} - y^2\mathbf{j} + ze^x\mathbf{k}$; $\mathbf{r}(t) = -t\mathbf{i} + \sqrt{t}\mathbf{j} + 3t\mathbf{k}$,
$1 \le t \le 4$

59. $\mathbf{F} = (2y + \sin x)\mathbf{i} + (z^2 + (1/3)\cos y)\mathbf{j} + x^4\mathbf{k}$;
$\mathbf{r}(t) = (\sin t)\mathbf{i} + (\cos t)\mathbf{j} + (\sin 2t)\mathbf{k}$, $-\pi/2 \le t \le \pi/2$

60. $\mathbf{F} = (x^2y)\mathbf{i} + \dfrac{1}{3}x^3\mathbf{j} + xy\mathbf{k}$; $\mathbf{r}(t) = (\cos t)\mathbf{i} + (\sin t)\mathbf{j} + (2\sin^2 t - 1)\mathbf{k}$, $0 \le t \le 2\pi$

16.3 Path Independence, Conservative Fields, and Potential Functions

A **gravitational field G** is a vector field that represents the effect of gravity at a point in space due to the presence of a massive object. The gravitational force on a body of mass m placed in the field is given by $\mathbf{F} = m\mathbf{G}$. Similarly, an **electric field E** is a vector field in space that represents the effect of electric forces on a charged particle placed within it. The force on a body of charge q placed in the field is given by $\mathbf{F} = q\mathbf{E}$. In gravitational and electric fields, the amount of work it takes to move a mass or charge from one point to another depends on the initial and final positions of the object—not on which path is taken between these positions. In this section we study vector fields with this property and the calculation of work integrals associated with them.

Path Independence

If A and B are two points in an open region D in space, the line integral of $\mathbf{F}$ along C from A to B for a field $\mathbf{F}$ defined on D usually depends on the path C taken, as we saw in Section 16.1. For some special fields, however, the integral's value is the same for all paths from A to B.

> **DEFINITIONS** Let $\mathbf{F}$ be a vector field defined on an open region D in space, and suppose that for any two points A and B in D the line integral $\int_C \mathbf{F} \cdot d\mathbf{r}$ along a path C from A to B in D is the same over all paths from A to B. Then the integral $\int_C \mathbf{F} \cdot d\mathbf{r}$ is **path independent in D** and the field $\mathbf{F}$ is **conservative on D**.

The word *conservative* comes from physics, where it refers to fields in which the principle of conservation of energy holds. When a line integral is independent of the path C from point A to point B, we sometimes represent the integral by the symbol $\int_A^B$ rather than the usual line integral symbol $\int_C$. This substitution helps us remember the path-independence property.

Under differentiability conditions normally met in practice, we will show that a field **F** is conservative if and only if it is the gradient field of a scalar function f—that is, if and only if $\mathbf{F} = \nabla f$ for some f. The function f then has a special name.

DEFINITION   If **F** is a vector field defined on D and $\mathbf{F} = \nabla f$ for some scalar function f on D, then f is called a **potential function for F**.

A gravitational potential is a scalar function whose gradient field is a gravitational field, an electric potential is a scalar function whose gradient field is an electric field, and so on. As we will see, once we have found a potential function f for a field **F**, we can evaluate all the line integrals in the domain of **F** over any path between A and B by

$$\int_A^B \mathbf{F} \cdot d\mathbf{r} = \int_A^B \nabla f \cdot d\mathbf{r} = f(B) - f(A). \tag{1}$$

If you think of ∇f for functions of several variables as being something like the derivative f' for functions of a single variable, then you see that Equation (1) is the vector calculus analogue of the Fundamental Theorem of Calculus formula

$$\int_a^b f'(x)\, dx = f(b) - f(a).$$

Conservative fields have other remarkable properties. For example, saying that **F** is conservative on D is equivalent to saying that the integral of **F** around every closed path in D is zero. Certain conditions on the curves, fields, and domains must be satisfied for Equation (1) to be valid. We discuss these conditions next.

Assumptions on Curves, Vector Fields, and Domains

In order for the computations and results we derive below to be valid, we must assume certain properties for the curves, surfaces, domains, and vector fields we consider. We give these assumptions in the statements of theorems, and they also apply to the examples and exercises unless otherwise stated.

The curves we consider are **piecewise smooth**. Such curves are made up of finitely many smooth pieces connected end to end, as discussed in Section 13.1. We will treat vector fields **F** whose components have continuous first partial derivatives.

The domains D we consider are open regions in space, so every point in D is the center of an open ball that lies entirely in D (see Section 13.1). We also assume D to be **connected**. For an open region, this means that any two points in D can be joined by a smooth curve that lies in the region. Finally, we assume D is **simply connected**, which means that every loop in D can be contracted to a point in D without ever leaving D. The plane with a disk removed is a two-dimensional region that is *not* simply connected; a loop in the plane that goes around the disk cannot be contracted to a point without going into the "hole" left by the removed disk (see Figure 16.22c). Similarly, if we remove a line from space, the remaining region D is *not* simply connected. A curve encircling the line cannot be shrunk to a point while remaining inside D.

Simply connected

(a)

Simply connected

(b)

Not simply connected

(c)

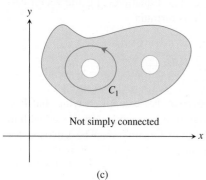

Not simply connected

(d)

FIGURE 16.22 Four connected regions. In (a) and (b), the regions are simply connected. In (c) and (d), the regions are not simply connected because the curves C_1 and C_2 cannot be contracted to a point inside the regions containing them.

Connectivity and simple connectivity are not the same, and neither property implies the other. Think of connected regions as being in "one piece" and simply connected regions as not having any "loop-catching holes." All of space itself is both connected and simply connected. Figure 16.22 illustrates some of these properties.

Caution Some of the results in this chapter can fail to hold if applied to situations where the conditions we've imposed do not hold. In particular, the component test for conservative fields, given later in this section, is not valid on domains that are not simply connected (see Example 5).

Line Integrals in Conservative Fields

Gradient fields **F** are obtained by differentiating a scalar function f. A theorem analogous to the Fundamental Theorem of Calculus gives a way to evaluate the line integrals of gradient fields.

THEOREM 1—Fundamental Theorem of Line Integrals Let C be a smooth curve joining the point A to the point B in the plane or in space and parametrized by $\mathbf{r}(t)$. Let f be a differentiable function with a continuous gradient vector $\mathbf{F} = \nabla f$ on a domain D containing C. Then

$$\int_C \mathbf{F} \cdot d\mathbf{r} = f(B) - f(A).$$

Like the Fundamental Theorem, Theorem 1 gives a way to evaluate line integrals without having to take limits of Riemann sums or finding the line integral by the procedure used in Section 16.2. Before proving Theorem 1, we give an example.

EXAMPLE 1 Suppose the force field $\mathbf{F} = \nabla f$ is the gradient of the function

$$f(x, y, z) = -\frac{1}{x^2 + y^2 + z^2}.$$

Find the work done by **F** in moving an object along a smooth curve C joining $(1, 0, 0)$ to $(0, 0, 2)$ that does not pass through the origin.

Solution An application of Theorem 1 shows that the work done by **F** along any smooth curve C joining the two points and not passing through the origin is

$$\int_C \mathbf{F} \cdot d\mathbf{r} = f(0, 0, 2) - f(1, 0, 0) = -\frac{1}{4} - (-1) = \frac{3}{4}. \qquad \blacksquare$$

The gravitational force due to a planet, and the electric force associated with a charged particle, can both be modeled by the field **F** given in Example 1 up to a constant that depends on the units of measurement.

Proof of Theorem 1 Suppose that A and B are two points in region D and that $C: \mathbf{r}(t) = g(t)\mathbf{i} + h(t)\mathbf{j} + k(t)\mathbf{k}, a \leq t \leq b$, is a smooth curve in D joining A to B.

We use the abbreviated form $\mathbf{r}(t) = x\mathbf{i} + y\mathbf{j} + z\mathbf{k}$ for the parametrization of the curve. Along the curve, f is a differentiable function of t and

$$\frac{df}{dt} = \frac{\partial f}{\partial x}\frac{dx}{dt} + \frac{\partial f}{\partial y}\frac{dy}{dt} + \frac{\partial f}{\partial z}\frac{dz}{dt} \qquad \text{Chain Rule in Section 14.4 with } x = g(t), y = h(t), z = k(t)$$

$$= \nabla f \cdot \left(\frac{dx}{dt}\mathbf{i} + \frac{dy}{dt}\mathbf{j} + \frac{dz}{dt}\mathbf{k}\right) = \nabla f \cdot \frac{d\mathbf{r}}{dt} = \mathbf{F} \cdot \frac{d\mathbf{r}}{dt}. \qquad \text{Because } \mathbf{F} = \nabla f$$

Therefore,

$$\int_C \mathbf{F} \cdot d\mathbf{r} = \int_{t=a}^{t=b} \mathbf{F} \cdot \frac{d\mathbf{r}}{dt}\, dt = \int_a^b \frac{df}{dt}\, dt \qquad \mathbf{r}(a) = A, \mathbf{r}(b) = B$$

$$= f(g(t), h(t), k(t)) \Big]_a^b = f(B) - f(A). \qquad \blacksquare$$

So we see from Theorem 1 that the line integral of a gradient field $\mathbf{F} = \nabla f$ is straightforward to compute once we know the function f. Many important vector fields arising in applications are indeed gradient fields. The next result, which follows from Theorem 1, shows that any conservative field is of this type.

> **THEOREM 2—Conservative Fields are Gradient Fields**   Let $\mathbf{F} = M\mathbf{i} + N\mathbf{j} + P\mathbf{k}$ be a vector field whose components are continuous throughout an open connected region D in space. Then $\mathbf{F}$ is conservative if and only if $\mathbf{F}$ is a gradient field ∇f for a differentiable function f.

Theorem 2 says that $\mathbf{F} = \nabla f$ if and only if for any two points A and B in the region D, the value of line integral $\int_C \mathbf{F} \cdot d\mathbf{r}$ is independent of the path C joining A to B in D.

Proof of Theorem 2   If $\mathbf{F}$ is a gradient field, then $\mathbf{F} = \nabla f$ for a differentiable function f, and Theorem 1 shows that $\int_C \mathbf{F} \cdot d\mathbf{r} = f(B) - f(A)$. The value of the line integral does not depend on C, but only on its endpoints A and B. So the line integral is path independent and $\mathbf{F}$ satisfies the definition of a conservative field.

On the other hand, suppose that $\mathbf{F}$ is a conservative vector field. We want to find a function f on D satisfying $\nabla f = \mathbf{F}$. First, pick a point A in D and set $f(A) = 0$. For any other point B in D define $f(B)$ to equal $\int_C \mathbf{F} \cdot d\mathbf{r}$, where C is *any* smooth path in D from A to B. The value of $f(B)$ does not depend on the choice of C, since $\mathbf{F}$ is conservative. To show that $\nabla f = F$ we need to demonstrate that $\partial f/\partial x = M$, $\partial f/\partial y = N$, and $\partial f/\partial z = P$.

Suppose that B has coordinates (x, y, z). By definition, the value of the function f at a nearby point B_0 located at (x_0, y, z) is $\int_{C_0} \mathbf{F} \cdot d\mathbf{r}$, where C_0 is any path from A to B_0. We take a path $C = C_0 \cup L$ from A to B formed by first traveling along C_0 to arrive at B_0 and then traveling along the line segment L from B_0 to B (Figure 16.23). When B_0 is close to B, the segment L lies in D and, since the value $f(B)$ is independent of the path from A to B,

$$f(x, y, z) = \int_{C_0} \mathbf{F} \cdot d\mathbf{r} + \int_L \mathbf{F} \cdot d\mathbf{r}.$$

Differentiating, we have

$$\frac{\partial}{\partial x} f(x, y, z) = \frac{\partial}{\partial x}\left(\int_{C_0} \mathbf{F} \cdot d\mathbf{r} + \int_L \mathbf{F} \cdot d\mathbf{r}\right).$$

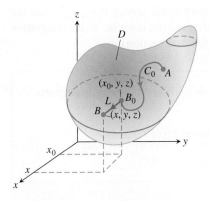

FIGURE 16.23 The function $f(x, y, z)$ in the proof of Theorem 2 is computed by a line integral $\int_{C_0} \mathbf{F} \cdot d\mathbf{r} = f(B_0)$ from A to B_0, plus a line integral $\int_L \mathbf{F} \cdot d\mathbf{r}$ along a line segment L parallel to the x-axis and joining B_0 to B located at (x, y, z). The value of f at A is $f(A) = 0$.

Only the last term on the right depends on x, so

$$\frac{\partial}{\partial x} f(x, y, z) = \frac{\partial}{\partial x} \int_L \mathbf{F} \cdot d\mathbf{r}.$$

Now parametrize L as $\mathbf{r}(t) = t\mathbf{i} + y\mathbf{j} + z\mathbf{k}$, $x_0 \le t \le x$. Then $d\mathbf{r}/dt = \mathbf{i}$, $\mathbf{F} \cdot d\mathbf{r}/dt = M$, and $\int_L \mathbf{F} \cdot d\mathbf{r} = \int_{x_0}^x M(t, y, z)\, dt$. Substitution gives

$$\frac{\partial}{\partial x} f(x, y, z) = \frac{\partial}{\partial x} \int_{x_0}^x M(t, y, z)\, dt = M(x, y, z)$$

by the Fundamental Theorem of Calculus. The partial derivatives $\partial f/\partial y = N$ and $\partial f/\partial z = P$ follow similarly, showing that $\mathbf{F} = \nabla f$. ∎

EXAMPLE 2 Find the work done by the conservative field

$$\mathbf{F} = yz\mathbf{i} + xz\mathbf{j} + xy\mathbf{k} = \nabla f, \quad \text{where} \quad f(x, y, z) = xyz,$$

along any smooth curve C joining the point $A(-1, 3, 9)$ to $B(1, 6, -4)$.

Solution With $f(x, y, z) = xyz$, we have

$$\int_C \mathbf{F} \cdot d\mathbf{r} = \int_A^B \nabla f \cdot d\mathbf{r} \qquad \text{\small $\mathbf{F} = \nabla f$ and path independence}$$

$$= f(B) - f(A) \qquad \text{\small Theorem 1}$$

$$= xyz\big|_{(1,6,-4)} - xyz\big|_{(-1,3,9)}$$

$$= (1)(6)(-4) - (-1)(3)(9)$$

$$= -24 + 27 = 3. \qquad ∎$$

A very useful property of line integrals in conservative fields comes into play when the path of integration is a closed curve, or loop. We often use the notation $\oint_C$ for integration around a closed path (discussed with more detail in the next section).

THEOREM 3—Loop Property of Conservative Fields The following statements are equivalent.

1. $\oint_C \mathbf{F} \cdot d\mathbf{r} = 0$ around every loop (that is, closed curve C) in D.

2. The field $\mathbf{F}$ is conservative on D.

Proof that Part 1 $\Rightarrow$ Part 2 We want to show that for any two points A and B in D, the integral of $\mathbf{F} \cdot d\mathbf{r}$ has the same value over any two paths C_1 and C_2 from A to B. We reverse the direction on C_2 to make a path $-C_2$ from B to A (Figure 16.24). Together, C_1 and $-C_2$ make a closed loop C, and by assumption,

$$\int_{C_1} \mathbf{F} \cdot d\mathbf{r} - \int_{C_2} \mathbf{F} \cdot d\mathbf{r} = \int_{C_1} \mathbf{F} \cdot d\mathbf{r} + \int_{-C_2} \mathbf{F} \cdot d\mathbf{r} = \int_C \mathbf{F} \cdot d\mathbf{r} = 0.$$

Thus, the integrals over C_1 and C_2 give the same value. Note that the definition of $\mathbf{F} \cdot d\mathbf{r}$ shows that changing the direction along a curve reverses the sign of the line integral.

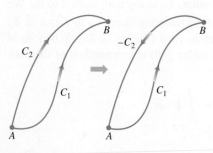

FIGURE 16.24 If we have two paths from A to B, one of them can be reversed to make a loop.

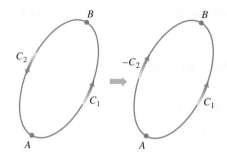

FIGURE 16.25 If A and B lie on a loop, we can reverse part of the loop to make two paths from A to B.

Proof that Part 2 $\Rightarrow$ Part 1   We want to show that the integral of $\mathbf{F} \cdot d\mathbf{r}$ is zero over any closed loop C. We pick two points A and B on C and use them to break C into two pieces: C_1 from A to B followed by C_2 from B back to A (Figure 16.25). Then

$$\oint_C \mathbf{F} \cdot d\mathbf{r} = \int_{C_1} \mathbf{F} \cdot d\mathbf{r} + \int_{C_2} \mathbf{F} \cdot d\mathbf{r} = \int_A^B \mathbf{F} \cdot d\mathbf{r} - \int_A^B \mathbf{F} \cdot d\mathbf{r} = 0. \qquad \blacksquare$$

The following diagram summarizes the results of Theorems 2 and 3.

$$\underset{\mathbf{F} = \nabla f \text{ on } D}{} \quad \overset{\text{Theorem 2}}{\Longleftrightarrow} \quad \underset{\substack{\mathbf{F} \text{ conservative} \\ \text{on } D}}{} \quad \overset{\text{Theorem 3}}{\Longleftrightarrow} \quad \underset{\substack{\displaystyle\oint_C \mathbf{F} \cdot d\mathbf{r} = 0 \\ \text{over any loop in } D}}{}$$

Two questions arise:

1. How do we know whether a given vector field $\mathbf{F}$ is conservative?
2. If $\mathbf{F}$ is in fact conservative, how do we find a potential function f (so that $\mathbf{F} = \nabla f$)?

Finding Potentials for Conservative Fields

The test for a vector field being conservative involves the equivalence of certain partial derivatives of the field components.

Component Test for Conservative Fields
Let $\mathbf{F} = M(x, y, z)\mathbf{i} + N(x, y, z)\mathbf{j} + P(x, y, z)\mathbf{k}$ be a field on a connected and simply connected domain whose component functions have continuous first partial derivatives. Then, $\mathbf{F}$ is conservative if and only if

$$\frac{\partial P}{\partial y} = \frac{\partial N}{\partial z}, \qquad \frac{\partial M}{\partial z} = \frac{\partial P}{\partial x}, \qquad \text{and} \qquad \frac{\partial N}{\partial x} = \frac{\partial M}{\partial y}. \qquad (2)$$

Proof that Equations (2) hold if F is conservative   There is a potential function f such that

$$\mathbf{F} = M\mathbf{i} + N\mathbf{j} + P\mathbf{k} = \frac{\partial f}{\partial x}\mathbf{i} + \frac{\partial f}{\partial y}\mathbf{j} + \frac{\partial f}{\partial z}\mathbf{k}.$$

Hence,

$$\begin{aligned}
\frac{\partial P}{\partial y} &= \frac{\partial}{\partial y}\left(\frac{\partial f}{\partial z}\right) = \frac{\partial^2 f}{\partial y \, \partial z} \\
&= \frac{\partial^2 f}{\partial z \, \partial y} \qquad \text{\small Mixed Derivative Theorem,} \\
&\qquad\qquad\qquad\quad \text{\small Section 14.3} \\
&= \frac{\partial}{\partial z}\left(\frac{\partial f}{\partial y}\right) = \frac{\partial N}{\partial z}.
\end{aligned}$$

The others in Equations (2) are proved similarly. $\blacksquare$

The second half of the proof, that Equations (2) imply that $\mathbf{F}$ is conservative, is a consequence of Stokes' Theorem, taken up in Section 16.7, and requires our assumption that the domain of $\mathbf{F}$ be simply connected.

Once we know that **F** is conservative, we usually want to find a potential function for **F**. This requires solving the equation $\nabla f = \mathbf{F}$ or

$$\frac{\partial f}{\partial x}\mathbf{i} + \frac{\partial f}{\partial y}\mathbf{j} + \frac{\partial f}{\partial z}\mathbf{k} = M\mathbf{i} + N\mathbf{j} + P\mathbf{k}$$

for f. We accomplish this by integrating the three equations

$$\frac{\partial f}{\partial x} = M, \qquad \frac{\partial f}{\partial y} = N, \qquad \frac{\partial f}{\partial z} = P,$$

as illustrated in the next example.

EXAMPLE 3   Show that $\mathbf{F} = (e^x \cos y + yz)\mathbf{i} + (xz - e^x \sin y)\mathbf{j} + (xy + z)\mathbf{k}$ is conservative over its natural domain and find a potential function for it.

Solution   The natural domain of **F** is all of space, which is connected and simply connected. We apply the test in Equations (2) to

$$M = e^x \cos y + yz, \qquad N = xz - e^x \sin y, \qquad P = xy + z$$

and calculate

$$\frac{\partial P}{\partial y} = x = \frac{\partial N}{\partial z}, \qquad \frac{\partial M}{\partial z} = y = \frac{\partial P}{\partial x}, \qquad \frac{\partial N}{\partial x} = -e^x \sin y + z = \frac{\partial M}{\partial y}.$$

The partial derivatives are continuous, so these equalities tell us that **F** is conservative, so there is a function f with $\nabla f = \mathbf{F}$ (Theorem 2).
    We find f by integrating the equations

$$\frac{\partial f}{\partial x} = e^x \cos y + yz, \qquad \frac{\partial f}{\partial y} = xz - e^x \sin y, \qquad \frac{\partial f}{\partial z} = xy + z. \qquad (3)$$

We integrate the first equation with respect to x, holding y and z fixed, to get

$$f(x, y, z) = e^x \cos y + xyz + g(y, z).$$

We write the constant of integration as a function of y and z because its value may depend on y and z, though not on x. We then calculate $\partial f/\partial y$ from this equation and match it with the expression for $\partial f/\partial y$ in Equations (3). This gives

$$-e^x \sin y + xz + \frac{\partial g}{\partial y} = xz - e^x \sin y,$$

so $\partial g/\partial y = 0$. Therefore, g is a function of z alone, and

$$f(x, y, z) = e^x \cos y + xyz + h(z).$$

We now calculate $\partial f/\partial z$ from this equation and match it to the formula for $\partial f/\partial z$ in Equations (3). This gives

$$xy + \frac{dh}{dz} = xy + z, \qquad \text{or} \qquad \frac{dh}{dz} = z,$$

so

$$h(z) = \frac{z^2}{2} + C.$$

Hence,

$$f(x, y, z) = e^x \cos y + xyz + \frac{z^2}{2} + C.$$

We have infinitely many potential functions of **F**, one for each value of C. ∎

EXAMPLE 4   Show that $\mathbf{F} = (2x - 3)\mathbf{i} - z\mathbf{j} + (\cos z)\mathbf{k}$ is not conservative.

Solution   We apply the component test in Equations (2) and find immediately that

$$\frac{\partial P}{\partial y} = \frac{\partial}{\partial y}(\cos z) = 0, \qquad \frac{\partial N}{\partial z} = \frac{\partial}{\partial z}(-z) = -1.$$

The two are unequal, so **F** is not conservative. No further testing is required. ∎

EXAMPLE 5   Show that the vector field

$$\mathbf{F} = \frac{-y}{x^2 + y^2}\mathbf{i} + \frac{x}{x^2 + y^2}\mathbf{j} + 0\mathbf{k}$$

satisfies the equations in the Component Test, but is not conservative over its natural domain. Explain why this is possible.

Solution   We have $M = -y/(x^2 + y^2)$, $N = x/(x^2 + y^2)$, and $P = 0$. If we apply the Component Test, we find

$$\frac{\partial P}{\partial y} = 0 = \frac{\partial N}{\partial z}, \qquad \frac{\partial P}{\partial x} = 0 = \frac{\partial M}{\partial z}, \qquad \text{and} \qquad \frac{\partial M}{\partial y} = \frac{y^2 - x^2}{(x^2 + y^2)^2} = \frac{\partial N}{\partial x}.$$

So it may appear that the field **F** passes the Component Test. However, the test assumes that the domain of **F** is simply connected, which is not the case. Since $x^2 + y^2$ cannot equal zero, the natural domain is the complement of the z-axis and contains loops that cannot be contracted to a point. One such loop is the unit circle C in the xy-plane. The circle is parametrized by $\mathbf{r}(t) = (\cos t)\mathbf{i} + (\sin t)\mathbf{j}$, $0 \le t \le 2\pi$. This loop wraps around the z-axis and cannot be contracted to a point while staying within the complement of the z-axis.

To show that **F** is not conservative, we compute the line integral $\oint_C \mathbf{F} \cdot d\mathbf{r}$ around the loop C. First we write the field in terms of the parameter t:

$$\mathbf{F} = \frac{-y}{x^2 + y^2}\mathbf{i} + \frac{x}{x^2 + y^2}\mathbf{j} = \frac{-\sin t}{\sin^2 t + \cos^2 t}\mathbf{i} + \frac{\cos t}{\sin^2 t + \cos^2 t}\mathbf{j} = (-\sin t)\mathbf{i} + (\cos t)\mathbf{j}.$$

Next we find $d\mathbf{r}/dt = (-\sin t)\mathbf{i} + (\cos t)\mathbf{j}$, and then calculate the line integral as

$$\oint_C \mathbf{F} \cdot d\mathbf{r} = \oint_C \mathbf{F} \cdot \frac{d\mathbf{r}}{dt}\, dt = \int_0^{2\pi} \left(\sin^2 t + \cos^2 t\right) dt = 2\pi.$$

Since the line integral of **F** around the loop C is not zero, the field **F** is not conservative, by Theorem 3. ∎

Example 5 shows that the Component Test does not apply when the domain of the field is not simply connected. However, if we change the domain in the example so that it is restricted to the ball of radius 1 centered at the point $(2, 2, 2)$, or to any similar ball-shaped region which does not contain a piece of the z-axis, then this new domain D *is* simply connected. Now the partial derivative Equations (2), as well as all the assumptions of the Component Test, are satisfied. In this new situation, the field **F** in Example 5 is conservative on D.

Just as we must be careful with a function when determining if it satisfies a property throughout its domain (like continuity or the intermediate value property), so must we also be careful with a vector field in determining the properties it may or may not have over its assigned domain.

Exact Differential Forms

It is often convenient to express work and circulation integrals in the differential form

$$\int_C M\,dx + N\,dy + P\,dz$$

discussed in Section 16.2. Such line integrals are relatively easy to evaluate if $M\,dx + N\,dy + P\,dz$ is the total differential of a function f and C is any path joining the two points from A to B. For then

$$\int_C M\,dx + N\,dy + P\,dz = \int_C \frac{\partial f}{\partial x}\,dx + \frac{\partial f}{\partial y}\,dy + \frac{\partial f}{\partial z}\,dz$$

$$= \int_A^B \nabla f \cdot d\mathbf{r} \qquad \nabla f \text{ is conservative.}$$

$$= f(B) - f(A). \qquad \text{Theorem 1}$$

Thus,

$$\int_A^B df = f(B) - f(A),$$

just as with differentiable functions of a single variable.

DEFINITIONS   Any expression $M(x, y, z)\,dx + N(x, y, z)\,dy + P(x, y, z)\,dz$ is a **differential form**. A differential form is **exact** on a domain D in space if

$$M\,dx + N\,dy + P\,dz = \frac{\partial f}{\partial x}\,dx + \frac{\partial f}{\partial y}\,dy + \frac{\partial f}{\partial z}\,dz = df$$

for some scalar function f throughout D.

Notice that if $M\,dx + N\,dy + P\,dz = df$ on D, then $\mathbf{F} = M\mathbf{i} + N\mathbf{j} + P\mathbf{k}$ is the gradient field of f on D. Conversely, if $\mathbf{F} = \nabla f$, then the form $M\,dx + N\,dy + P\,dz$ is exact. The test for the form's being exact is therefore the same as the test for $\mathbf{F}$ being conservative.

Component Test for Exactness of $M\,dx + N\,dy + P\,dz$

The differential form $M\,dx + N\,dy + P\,dz$ is exact on a connected and simply connected domain if and only if

$$\frac{\partial P}{\partial y} = \frac{\partial N}{\partial z}, \qquad \frac{\partial M}{\partial z} = \frac{\partial P}{\partial x}, \qquad \text{and} \qquad \frac{\partial N}{\partial x} = \frac{\partial M}{\partial y}.$$

This is equivalent to saying that the field $\mathbf{F} = M\mathbf{i} + N\mathbf{j} + P\mathbf{k}$ is conservative.

EXAMPLE 6   Show that $y\,dx + x\,dy + 4\,dz$ is exact and evaluate the integral

$$\int_{(1,1,1)}^{(2,3,-1)} y\,dx + x\,dy + 4\,dz$$

over any path from $(1, 1, 1)$ to $(2, 3, -1)$.

Solution   We let $M = y$, $N = x$, $P = 4$ and apply the Test for Exactness:

$$\frac{\partial P}{\partial y} = 0 = \frac{\partial N}{\partial z}, \qquad \frac{\partial M}{\partial z} = 0 = \frac{\partial P}{\partial x}, \qquad \frac{\partial N}{\partial x} = 1 = \frac{\partial M}{\partial y}.$$

These equalities tell us that $y\,dx + x\,dy + 4\,dz$ is exact, so

$$y\,dx + x\,dy + 4\,dz = df$$

for some function f, and the integral's value is $f(2, 3, -1) - f(1, 1, 1)$.

We find f up to a constant by integrating the equations

$$\frac{\partial f}{\partial x} = y, \qquad \frac{\partial f}{\partial y} = x, \qquad \frac{\partial f}{\partial z} = 4. \tag{4}$$

From the first equation we get

$$f(x, y, z) = xy + g(y, z).$$

The second equation tells us that

$$\frac{\partial f}{\partial y} = x + \frac{\partial g}{\partial y} = x, \qquad \text{or} \qquad \frac{\partial g}{\partial y} = 0.$$

Hence, g is a function of z alone, and

$$f(x, y, z) = xy + h(z).$$

The third of Equations (4) tells us that

$$\frac{\partial f}{\partial z} = 0 + \frac{dh}{dz} = 4, \qquad \text{or} \qquad h(z) = 4z + C.$$

Therefore,

$$f(x, y, z) = xy + 4z + C.$$

The value of the line integral is independent of the path taken from $(1, 1, 1)$ to $(2, 3, -1)$, and equals

$$f(2, 3, -1) - f(1, 1, 1) = 2 + C - (5 + C) = -3. \qquad \blacksquare$$

Exercises 16.3

Testing for Conservative Fields

Which fields in Exercises 1–6 are conservative, and which are not?

1. $\mathbf{F} = yz\mathbf{i} + xz\mathbf{j} + xy\mathbf{k}$

2. $\mathbf{F} = (y \sin z)\mathbf{i} + (x \sin z)\mathbf{j} + (xy \cos z)\mathbf{k}$

3. $\mathbf{F} = y\mathbf{i} + (x + z)\mathbf{j} - y\mathbf{k}$

4. $\mathbf{F} = -y\mathbf{i} + x\mathbf{j}$

5. $\mathbf{F} = (z + y)\mathbf{i} + z\mathbf{j} + (y + x)\mathbf{k}$

6. $\mathbf{F} = (e^x \cos y)\mathbf{i} - (e^x \sin y)\mathbf{j} + z\mathbf{k}$

Finding Potential Functions

In Exercises 7–12, find a potential function f for the field $\mathbf{F}$.

7. $\mathbf{F} = 2x\mathbf{i} + 3y\mathbf{j} + 4z\mathbf{k}$

8. $\mathbf{F} = (y + z)\mathbf{i} + (x + z)\mathbf{j} + (x + y)\mathbf{k}$

9. $\mathbf{F} = e^{y+2z}(\mathbf{i} + x\mathbf{j} + 2x\mathbf{k})$

10. $\mathbf{F} = (y \sin z)\mathbf{i} + (x \sin z)\mathbf{j} + (xy \cos z)\mathbf{k}$

11. $\mathbf{F} = (\ln x + \sec^2(x + y))\mathbf{i} +$
$$\left(\sec^2(x + y) + \frac{y}{y^2 + z^2}\right)\mathbf{j} + \frac{z}{y^2 + z^2}\mathbf{k}$$

12. $\mathbf{F} = \dfrac{y}{1 + x^2 y^2}\mathbf{i} + \left(\dfrac{x}{1 + x^2 y^2} + \dfrac{z}{\sqrt{1 - y^2 z^2}}\right)\mathbf{j} +$
$$\left(\dfrac{y}{\sqrt{1 - y^2 z^2}} + \dfrac{1}{z}\right)\mathbf{k}$$

Exact Differential Forms

In Exercises 13–17, show that the differential forms in the integrals are exact. Then evaluate the integrals.

13. $\displaystyle\int_{(0,0,0)}^{(2,3,-6)} 2x\,dx + 2y\,dy + 2z\,dz$

14. $\displaystyle\int_{(1,1,2)}^{(3,5,0)} yz\,dx + xz\,dy + xy\,dz$

15. $\displaystyle\int_{(0,0,0)}^{(1,2,3)} 2xy\,dx + (x^2 - z^2)\,dy - 2yz\,dz$

16. $\displaystyle\int_{(0,0,0)}^{(3,3,1)} 2x\,dx - y^2\,dy - \frac{4}{1+z^2}\,dz$

17. $\displaystyle\int_{(1,0,0)}^{(0,1,1)} \sin y \cos x\,dx + \cos y \sin x\,dy + dz$

Finding Potential Functions to Evaluate Line Integrals

Although they are not defined on all of space R^3, the fields associated with Exercises 18–22 are simply connected and the Component Test can be used to show they are conservative. Find a potential function for each field and evaluate the integrals as in Example 6.

18. $\displaystyle\int_{(0,2,1)}^{(1,\pi/2,2)} 2\cos y\,dx + \left(\frac{1}{y} - 2x\sin y\right)dy + \frac{1}{z}\,dz$

19. $\displaystyle\int_{(1,1,1)}^{(1,2,3)} 3x^2\,dx + \frac{z^2}{y}\,dy + 2z \ln y\,dz$

20. $\displaystyle\int_{(1,2,1)}^{(2,1,1)} (2x\ln y - yz)\,dx + \left(\frac{x^2}{y} - xz\right)dy - xy\,dz$

21. $\displaystyle\int_{(1,1,1)}^{(2,2,2)} \frac{1}{y}\,dx + \left(\frac{1}{z} - \frac{x}{y^2}\right)dy - \frac{y}{z^2}\,dz$

22. $\displaystyle\int_{(-1,-1,-1)}^{(2,2,2)} \frac{2x\,dx + 2y\,dy + 2z\,dz}{x^2 + y^2 + z^2}$

Applications and Examples

23. Revisiting Example 6   Evaluate the integral

$$\int_{(1,1,1)}^{(2,3,-1)} y\,dx + x\,dy + 4\,dz$$

from Example 6 by finding parametric equations for the line segment from $(1, 1, 1)$ to $(2, 3, -1)$ and evaluating the line integral of $\mathbf{F} = y\mathbf{i} + x\mathbf{j} + 4\mathbf{k}$ along the segment. Since $\mathbf{F}$ is conservative, the integral is independent of the path.

24. Evaluate

$$\int_C x^2\,dx + yz\,dy + (y^2/2)\,dz$$

along the line segment C joining $(0, 0, 0)$ to $(0, 3, 4)$.

Independence of path   Show that the values of the integrals in Exercises 25 and 26 do not depend on the path taken from A to B.

25. $\displaystyle\int_A^B z^2\,dx + 2y\,dy + 2xz\,dz$

26. $\displaystyle\int_A^B \frac{x\,dx + y\,dy + z\,dz}{\sqrt{x^2 + y^2 + z^2}}$

In Exercises 27 and 28, find a potential function for $\mathbf{F}$.

27. $\mathbf{F} = \dfrac{2x}{y}\mathbf{i} + \left(\dfrac{1-x^2}{y^2}\right)\mathbf{j}, \quad \{(x, y): y > 0\}$

28. $\mathbf{F} = (e^x \ln y)\mathbf{i} + \left(\dfrac{e^x}{y} + \sin z\right)\mathbf{j} + (y \cos z)\mathbf{k}$

29. Work along different paths   Find the work done by $\mathbf{F} = (x^2 + y)\mathbf{i} + (y^2 + x)\mathbf{j} + ze^z\mathbf{k}$ over the following paths from $(1, 0, 0)$ to $(1, 0, 1)$.

  a. The line segment $x = 1, y = 0, 0 \le z \le 1$

  b. The helix $\mathbf{r}(t) = (\cos t)\mathbf{i} + (\sin t)\mathbf{j} + (t/2\pi)\mathbf{k}, 0 \le t \le 2\pi$

  c. The x-axis from $(1, 0, 0)$ to $(0, 0, 0)$ followed by the parabola $z = x^2, y = 0$ from $(0, 0, 0)$ to $(1, 0, 1)$

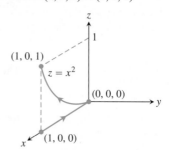

30. Work along different paths   Find the work done by $\mathbf{F} = e^{yz}\mathbf{i} + (xze^{yz} + z\cos y)\mathbf{j} + (xye^{yz} + \sin y)\mathbf{k}$ over the following paths from $(1, 0, 1)$ to $(1, \pi/2, 0)$.

  a. The line segment $x = 1, y = \pi t/2, z = 1 - t, 0 \le t \le 1$

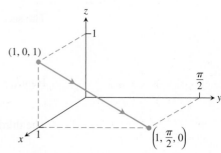

  b. The line segment from $(1, 0, 1)$ to the origin followed by the line segment from the origin to $(1, \pi/2, 0)$

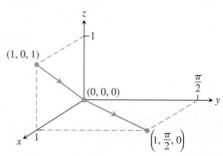

  c. The line segment from $(1, 0, 1)$ to $(1, 0, 0)$, followed by the x-axis from $(1, 0, 0)$ to the origin, followed by the parabola $y = \pi x^2/2, z = 0$ from there to $(1, \pi/2, 0)$

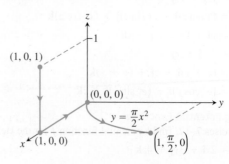

31. Evaluating a work integral two ways   Let $\mathbf{F} = \nabla(x^3 y^2)$ and let C be the path in the xy-plane from $(-1, 1)$ to $(1, 1)$ that consists of the line segment from $(-1, 1)$ to $(0, 0)$ followed by the line segment from $(0, 0)$ to $(1, 1)$. Evaluate $\int_C \mathbf{F} \cdot d\mathbf{r}$ in two ways.

  a. Find parametrizations for the segments that make up C and evaluate the integral.

  b. Use $f(x, y) = x^3 y^2$ as a potential function for $\mathbf{F}$.

32. Integral along different paths   Evaluate the line integral $\int_C 2x \cos y \, dx - x^2 \sin y \, dy$ along the following paths C in the xy-plane.

  a. The parabola $y = (x - 1)^2$ from $(1, 0)$ to $(0, 1)$

  b. The line segment from $(-1, \pi)$ to $(1, 0)$

  c. The x-axis from $(-1, 0)$ to $(1, 0)$

  d. The astroid $\mathbf{r}(t) = (\cos^3 t)\mathbf{i} + (\sin^3 t)\mathbf{j}$, $0 \le t \le 2\pi$, counterclockwise from $(1, 0)$ back to $(1, 0)$

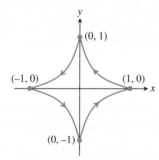

33. a. Exact differential form   How are the constants a, b, and c related if the following differential form is exact?
$$(ay^2 + 2czx) \, dx + y(bx + cz) \, dy + (ay^2 + cx^2) \, dz$$

  b. Gradient field   For what values of b and c will
$$\mathbf{F} = (y^2 + 2czx)\mathbf{i} + y(bx + cz)\mathbf{j} + (y^2 + cx^2)\mathbf{k}$$
  be a gradient field?

34. Gradient of a line integral   Suppose that $\mathbf{F} = \nabla f$ is a conservative vector field and
$$g(x, y, z) = \int_{(0,0,0)}^{(x,y,z)} \mathbf{F} \cdot d\mathbf{r}.$$
Show that $\nabla g = \mathbf{F}$.

35. Path of least work   You have been asked to find the path along which a force field $\mathbf{F}$ will perform the least work in moving a particle between two locations. A quick calculation on your part shows $\mathbf{F}$ to be conservative. How should you respond? Give reasons for your answer.

36. A revealing experiment   By experiment, you find that a force field $\mathbf{F}$ performs only half as much work in moving an object along path C_1 from A to B as it does in moving the object along path C_2 from A to B. What can you conclude about $\mathbf{F}$? Give reasons for your answer.

37. Work by a constant force   Show that the work done by a constant force field $\mathbf{F} = a\mathbf{i} + b\mathbf{j} + c\mathbf{k}$ in moving a particle along any path from A to B is $W = \mathbf{F} \cdot \overrightarrow{AB}$.

38. Gravitational field

  a. Find a potential function for the gravitational field
$$\mathbf{F} = -GmM \frac{x\mathbf{i} + y\mathbf{j} + z\mathbf{k}}{(x^2 + y^2 + z^2)^{3/2}}$$
  (G, m, and M are constants).

  b. Let P_1 and P_2 be points at distance s_1 and s_2 from the origin. Show that the work done by the gravitational field in part (a) in moving a particle from P_1 to P_2 is
$$GmM\left(\frac{1}{s_2} - \frac{1}{s_1}\right).$$

16.4 Green's Theorem in the Plane

If $\mathbf{F}$ is a conservative field, then we know $\mathbf{F} = \nabla f$ for a differentiable function f, and we can calculate the line integral of $\mathbf{F}$ over any path C joining point A to B as $\int_C \mathbf{F} \cdot d\mathbf{r} = f(B) - f(A)$. In this section we derive a method for computing a work or flux integral over a *closed* curve C in the plane when the field $\mathbf{F}$ is *not* conservative. This method, known as Green's Theorem, allows us to convert the line integral into a double integral over the region enclosed by C.

The discussion is given in terms of velocity fields of fluid flows (a fluid is a liquid or a gas) because they are easy to visualize. However, Green's Theorem applies to any vector field, independent of any particular interpretation of the field, provided the assumptions of the theorem are satisfied. We introduce two new ideas for Green's Theorem: *divergence* and *circulation density* around an axis perpendicular to the plane.

Divergence

Suppose that $\mathbf{F}(x, y) = M(x, y)\mathbf{i} + N(x, y)\mathbf{j}$ is the velocity field of a fluid flowing in the plane and that the first partial derivatives of M and N are continuous at each point of a region R. Let (x, y) be a point in R and let A be a small rectangle with one corner at (x, y) that, along with its interior, lies entirely in R. The sides of the rectangle, parallel to the coordinate axes, have lengths of Δx and Δy. Assume that the components M and N do not

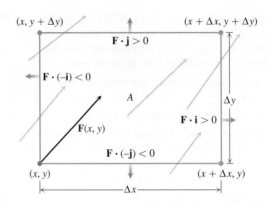

FIGURE 16.26 The rate at which the fluid leaves the rectangular region A across the bottom edge in the direction of the outward normal $-\mathbf{j}$ is approximately $\mathbf{F}(x, y) \cdot (-\mathbf{j}) \Delta x$, which is negative for the vector field $\mathbf{F}$ shown here. To approximate the flow rate at the point (x, y), we calculate the (approximate) flow rates across each edge in the directions of the red arrows, sum these rates, and then divide the sum by the area of A. Taking the limit as $\Delta x \to 0$ and $\Delta y \to 0$ gives the flow rate per unit area.

change sign throughout a small region containing the rectangle A. The rate at which fluid leaves the rectangle across the bottom edge is approximately (Figure 16.26)

$$\mathbf{F}(x, y) \cdot (-\mathbf{j}) \, \Delta x = -N(x, y)\Delta x.$$

This is the scalar component of the velocity at (x, y) in the direction of the outward normal times the length of the segment. If the velocity is in meters per second, for example, the flow rate will be in meters per second times meters or square meters per second. The rates at which the fluid crosses the other three sides in the directions of their outward normals can be estimated in a similar way. The flow rates may be positive or negative depending on the signs of the components of $\mathbf{F}$. We approximate the net flow rate across the rectangular boundary of A by summing the flow rates across the four edges as defined by the following dot products.

Fluid Flow Rates: Top: $\mathbf{F}(x, y + \Delta y) \cdot \mathbf{j} \, \Delta x = N(x, y + \Delta y)\Delta x$

Bottom: $\mathbf{F}(x, y) \cdot (-\mathbf{j}) \, \Delta x = -N(x, y)\Delta x$

Right: $\mathbf{F}(x + \Delta x, y) \cdot \mathbf{i} \, \Delta y = M(x + \Delta x, y)\Delta y$

Left: $\mathbf{F}(x, y) \cdot (-\mathbf{i}) \, \Delta y = -M(x, y)\Delta y.$

Summing opposite pairs gives

Top and bottom: $(N(x, y + \Delta y) - N(x, y))\Delta x \approx \left(\dfrac{\partial N}{\partial y} \Delta y\right)\Delta x$

Right and left: $(M(x + \Delta x, y) - M(x, y))\Delta y \approx \left(\dfrac{\partial M}{\partial x} \Delta x\right)\Delta y.$

Adding these last two equations gives the net effect of the flow rates, or the

Flux across rectangle boundary $\approx \left(\dfrac{\partial M}{\partial x} + \dfrac{\partial N}{\partial y}\right)\Delta x\Delta y.$

We now divide by $\Delta x\Delta y$ to estimate the total flux per unit area or *flux density* for the rectangle:

$$\frac{\text{Flux across rectangle boundary}}{\text{rectangle area}} \approx \left(\frac{\partial M}{\partial x} + \frac{\partial N}{\partial y}\right).$$

Source: div $\mathbf{F}\,(x_0, y_0) > 0$

A gas expanding
at the point (x_0, y_0).

Sink: div $\mathbf{F}\,(x_0, y_0) < 0$

A gas compressing
at the point (x_0, y_0).

FIGURE 16.27 If a gas is expanding at a point (x_0, y_0), the lines of flow have positive divergence; if the gas is compressing, the divergence is negative.

Finally, we let Δx and Δy approach zero to define the flux density of $\mathbf{F}$ at the point (x, y). In mathematics, we call the flux density the *divergence* of $\mathbf{F}$. The symbol for it is div $\mathbf{F}$, pronounced "divergence of $\mathbf{F}$" or "div $\mathbf{F}$."

DEFINITION The **divergence (flux density)** of a vector field $\mathbf{F} = M\mathbf{i} + N\mathbf{j}$ at the point (x, y) is

$$\text{div } \mathbf{F} = \frac{\partial M}{\partial x} + \frac{\partial N}{\partial y}. \tag{1}$$

A gas is compressible, unlike a liquid, and the divergence of its velocity field measures to what extent it is expanding or compressing at each point. Intuitively, if a gas is expanding at the point (x_0, y_0), the lines of flow would diverge there (hence the name) and, since the gas would be flowing out of a small rectangle about (x_0, y_0), the divergence of $\mathbf{F}$ at (x_0, y_0) would be positive. If the gas were compressing instead of expanding, the divergence would be negative (Figure 16.27).

EXAMPLE 1 The following vector fields represent the velocity of a gas flowing in the xy-plane. Find the divergence of each vector field and interpret its physical meaning. Figure 16.28 displays the vector fields.

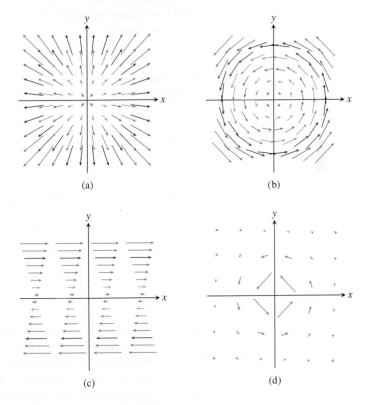

(a)   (b)

(c)   (d)

FIGURE 16.28 Velocity fields of a gas flowing in the plane (Example 1).

(a) *Uniform expansion or compression:* $\mathbf{F}(x, y) = cx\mathbf{i} + cy\mathbf{j}$

(b) *Uniform rotation:* $\mathbf{F}(x, y) = -cy\mathbf{i} + cx\mathbf{j}$

(c) *Shearing flow:* $\mathbf{F}(x, y) = y\mathbf{i}$

(d) *Whirlpool effect:* $\mathbf{F}(x, y) = \dfrac{-y}{x^2 + y^2}\mathbf{i} + \dfrac{x}{x^2 + y^2}\mathbf{j}$

Solution

(a) $\text{div } \mathbf{F} = \dfrac{\partial}{\partial x}(cx) + \dfrac{\partial}{\partial y}(cy) = 2c$: If $c > 0$, the gas is undergoing uniform expansion; if $c < 0$, it is undergoing uniform compression.

(b) $\text{div } \mathbf{F} = \dfrac{\partial}{\partial x}(-cy) + \dfrac{\partial}{\partial y}(cx) = 0$: The gas is neither expanding nor compressing.

(c) $\text{div } \mathbf{F} = \dfrac{\partial}{\partial x}(y) = 0$: The gas is neither expanding nor compressing.

(d) $\text{div } \mathbf{F} = \dfrac{\partial}{\partial x}\left(\dfrac{-y}{x^2 + y^2}\right) + \dfrac{\partial}{\partial y}\left(\dfrac{x}{x^2 + y^2}\right) = \dfrac{2xy}{(x^2 + y^2)^2} - \dfrac{2xy}{(x^2 + y^2)^2} = 0$: Again, the divergence is zero at all points in the domain of the velocity field.

∎

Cases (b), (c), and (d) of Figure 16.28 are plausible models for the two-dimensional flow of a liquid. In fluid dynamics, when the velocity field of a flowing liquid always has divergence equal to zero, as in those cases, the liquid is said to be **incompressible**.

Spin Around an Axis: The k-Component of Curl

The second idea we need for Green's Theorem has to do with measuring how a floating paddle wheel, with axis perpendicular to the plane, spins at a point in a fluid flowing in a plane region. This idea gives some sense of how the fluid is circulating around axes located at different points and perpendicular to the region. Physicists sometimes refer to this as the *circulation density* of a vector field $\mathbf{F}$ at a point. To obtain it, we return to the velocity field

$$\mathbf{F}(x, y) = M(x, y)\mathbf{i} + N(x, y)\mathbf{j}$$

and consider the rectangle A in Figure 16.29 (where we assume both components of $\mathbf{F}$ are positive).

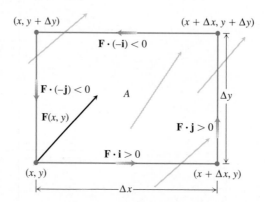

FIGURE 16.29 The rate at which a fluid flows along the bottom edge of a rectangular region A in the direction $\mathbf{i}$ is approximately $\mathbf{F}(x, y) \cdot \mathbf{i} \, \Delta x$, which is positive for the vector field $\mathbf{F}$ shown here. To approximate the rate of circulation at the point (x, y), we calculate the (approximate) flow rates along each edge in the directions of the red arrows, sum these rates, and then divide the sum by the area of A. Taking the limit as $\Delta x \to 0$ and $\Delta y \to 0$ gives the rate of the circulation per unit area.

The circulation rate of $\mathbf{F}$ around the boundary of A is the sum of flow rates along the sides in the tangential direction. For the bottom edge, the flow rate is approximately

$$\mathbf{F}(x, y) \cdot \mathbf{i} \, \Delta x = M(x, y)\Delta x.$$

From a routine calculation of the cross product we find

$$\mathbf{r}_u \times \mathbf{r}_v = \frac{F_x}{F_z}\mathbf{i} + \frac{F_y}{F_z}\mathbf{j} + \mathbf{k} \qquad F_z \neq 0$$

$$= \frac{1}{F_z}(F_x\mathbf{i} + F_y\mathbf{j} + F_z\mathbf{k})$$

$$= \frac{\nabla F}{F_z} = \frac{\nabla F}{\nabla F \cdot \mathbf{k}}$$

$$= \frac{\nabla F}{\nabla F \cdot \mathbf{p}}. \qquad \mathbf{p} = \mathbf{k}$$

Therefore, the surface area differential is given by

$$d\sigma = |\mathbf{r}_u \times \mathbf{r}_v|\, du\, dv = \frac{|\nabla F|}{|\nabla F \cdot \mathbf{p}|} dx\, dy. \qquad u = x \text{ and } v = y$$

We obtain similar calculations if instead the vector $\mathbf{p} = \mathbf{j}$ is normal to the xz-plane when $F_y \neq 0$ on S, or if $\mathbf{p} = \mathbf{i}$ is normal to the yz-plane when $F_x \neq 0$ on S. Combining these results with Equation (4) then gives the following general formula.

Formula for the Surface Area of an Implicit Surface

The area of the surface $F(x, y, z) = c$ over a closed and bounded plane region R is

$$\text{Surface area} = \iint\limits_R \frac{|\nabla F|}{|\nabla F \cdot \mathbf{p}|}\, dA, \qquad (7)$$

where $\mathbf{p} = \mathbf{i}, \mathbf{j}$, or $\mathbf{k}$ is normal to R and $\nabla F \cdot \mathbf{p} \neq 0$.

Thus, the area is the double integral over R of the magnitude of ∇F divided by the magnitude of the scalar component of ∇F normal to R.

We reached Equation (7) under the assumption that $\nabla F \cdot \mathbf{p} \neq 0$ throughout R and that ∇F is continuous. Whenever the integral exists, however, we define its value to be the area of the portion of the surface $F(x, y, z) = c$ that lies over R. (Recall that the projection is assumed to be one-to-one.)

EXAMPLE 7   Find the area of the surface cut from the bottom of the paraboloid $x^2 + y^2 - z = 0$ by the plane $z = 4$.

Solution   We sketch the surface S and the region R below it in the xy-plane (Figure 16.46). The surface S is part of the level surface $F(x, y, z) = x^2 + y^2 - z = 0$, and R is the disk $x^2 + y^2 \leq 4$ in the xy-plane. To get a unit vector normal to the plane of R, we can take $\mathbf{p} = \mathbf{k}$.

At any point (x, y, z) on the surface, we have

$$F(x, y, z) = x^2 + y^2 - z$$

$$\nabla F = 2x\mathbf{i} + 2y\mathbf{j} - \mathbf{k}$$

$$|\nabla F| = \sqrt{(2x)^2 + (2y)^2 + (-1)^2}$$

$$= \sqrt{4x^2 + 4y^2 + 1}$$

$$|\nabla F \cdot \mathbf{p}| = |\nabla F \cdot \mathbf{k}| = |-1| = 1.$$

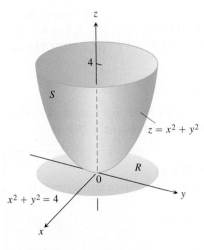

FIGURE 16.46 The area of this parabolic surface is calculated in Example 7.

In the region R, $dA = dx\,dy$. Therefore,

$$\text{Surface area} = \iint_R \frac{|\nabla F|}{|\nabla F \cdot \mathbf{p}|}\,dA \qquad\qquad \text{Eq. (7)}$$

$$= \iint_{x^2+y^2\le 4} \sqrt{4x^2 + 4y^2 + 1}\,dx\,dy$$

$$= \int_0^{2\pi}\int_0^2 \sqrt{4r^2 + 1}\,r\,dr\,d\theta \qquad\qquad \text{Polar coordinates}$$

$$= \int_0^{2\pi} \left[\frac{1}{12}(4r^2 + 1)^{3/2}\right]_0^2 d\theta$$

$$= \int_0^{2\pi} \frac{1}{12}(17^{3/2} - 1)\,d\theta = \frac{\pi}{6}\left(17\sqrt{17} - 1\right). \qquad\blacksquare$$

Example 7 illustrates how to find the surface area for a function $z = f(x, y)$ over a region R in the xy-plane. Actually, the surface area differential can be obtained in two ways, and we show this in the next example.

EXAMPLE 8   Derive the surface area differential $d\sigma$ of the surface $z = f(x, y)$ over a region R in the xy-plane **(a)** parametrically using Equation (5), and **(b)** implicitly, as in Equation (7).

Solution

(a) We parametrize the surface by taking $x = u$, $y = v$, and $z = f(x, y)$ over R. This gives the parametrization

$$\mathbf{r}(u, v) = u\mathbf{i} + v\mathbf{j} + f(u, v)\mathbf{k}$$

Computing the partial derivatives gives $\mathbf{r}_u + \mathbf{i} + f_u\mathbf{k}$, $\mathbf{r}_v = \mathbf{j} + f_v\mathbf{k}$ and

$$\mathbf{r}_u \times \mathbf{r}_v = -f_u\mathbf{i} - f_v\mathbf{j} + \mathbf{k}. \qquad \begin{vmatrix} \mathbf{i} & \mathbf{j} & \mathbf{k} \\ 1 & 0 & f_u \\ 0 & 1 & f_v \end{vmatrix}$$

Then $|\mathbf{r}_u \times \mathbf{r}_v|\,du\,dv = \sqrt{f_u^2 + f_v^2 + 1}\,du\,dv$. Substituting for u and v then gives the surface area differential

$$d\sigma = \sqrt{f_x^2 + f_y^2 + 1}\,dx\,dy.$$

(b) We define the implicit function $F(x, y, z) = f(x, y) - z$. Since (x, y) belongs to the region R, the unit normal to the plane of R is $\mathbf{p} = \mathbf{k}$. Then $\nabla F = f_x\mathbf{i} + f_y\mathbf{j} - \mathbf{k}$ so that $|\nabla F \cdot \mathbf{p}| = |-1| = 1$, $|\nabla F| = \sqrt{f_x^2 + f_y^2 + 1}$, and $|\nabla F|/|\nabla F \cdot \mathbf{p}| = |\nabla F|$. The surface area differential is again given by

$$d\sigma = \sqrt{f_x^2 + f_y^2 + 1}\,dx\,dy. \qquad\blacksquare$$

The surface area differential derived in Example 8 gives the following formula for calculating the surface area of the graph of a function defined explicitly as $z = f(x, y)$.

> **Formula for the Surface Area of a Graph $z = f(x, y)$**
> For a graph $z = f(x, y)$ over a region R in the xy-plane, the surface area formula is
>
> $$A = \iint_R \sqrt{f_x^2 + f_y^2 + 1}\,dx\,dy. \qquad\qquad (8)$$

Exercises 16.5

Finding Parametrizations

In Exercises 1–16, find a parametrization of the surface. (There are many correct ways to do these, so your answers may not be the same as those in the back of the book.)

1. The paraboloid $z = x^2 + y^2, z \leq 4$

2. The paraboloid $z = 9 - x^2 - y^2, z \geq 0$

3. **Cone frustum** The first-octant portion of the cone $z = \sqrt{x^2 + y^2}/2$ between the planes $z = 0$ and $z = 3$

4. **Cone frustum** The portion of the cone $z = 2\sqrt{x^2 + y^2}$ between the planes $z = 2$ and $z = 4$

5. **Spherical cap** The cap cut from the sphere $x^2 + y^2 + z^2 = 9$ by the cone $z = \sqrt{x^2 + y^2}$

6. **Spherical cap** The portion of the sphere $x^2 + y^2 + z^2 = 4$ in the first octant between the xy-plane and the cone $z = \sqrt{x^2 + y^2}$

7. **Spherical band** The portion of the sphere $x^2 + y^2 + z^2 = 3$ between the planes $z = \sqrt{3}/2$ and $z = -\sqrt{3}/2$

8. **Spherical cap** The upper portion cut from the sphere $x^2 + y^2 + z^2 = 8$ by the plane $z = -2$

9. **Parabolic cylinder between planes** The surface cut from the parabolic cylinder $z = 4 - y^2$ by the planes $x = 0$, $x = 2$, and $z = 0$

10. **Parabolic cylinder between planes** The surface cut from the parabolic cylinder $y = x^2$ by the planes $z = 0, z = 3$, and $y = 2$

11. **Circular cylinder band** The portion of the cylinder $y^2 + z^2 = 9$ between the planes $x = 0$ and $x = 3$

12. **Circular cylinder band** The portion of the cylinder $x^2 + z^2 = 4$ above the xy-plane between the planes $y = -2$ and $y = 2$

13. **Tilted plane inside cylinder** The portion of the plane $x + y + z = 1$

    a. Inside the cylinder $x^2 + y^2 = 9$

    b. Inside the cylinder $y^2 + z^2 = 9$

14. **Tilted plane inside cylinder** The portion of the plane $x - y + 2z = 2$

    a. Inside the cylinder $x^2 + z^2 = 3$

    b. Inside the cylinder $y^2 + z^2 = 2$

15. **Circular cylinder band** The portion of the cylinder $(x - 2)^2 + z^2 = 4$ between the planes $y = 0$ and $y = 3$

16. **Circular cylinder band** The portion of the cylinder $y^2 + (z - 5)^2 = 25$ between the planes $x = 0$ and $x = 10$

Surface Area of Parametrized Surfaces

In Exercises 17–26, use a parametrization to express the area of the surface as a double integral. Then evaluate the integral. (There are many correct ways to set up the integrals, so your integrals may not be the same as those in the back of the book. They should have the same values, however.)

17. **Tilted plane inside cylinder** The portion of the plane $y + 2z = 2$ inside the cylinder $x^2 + y^2 = 1$

18. **Plane inside cylinder** The portion of the plane $z = -x$ inside the cylinder $x^2 + y^2 = 4$

19. **Cone frustum** The portion of the cone $z = 2\sqrt{x^2 + y^2}$ between the planes $z = 2$ and $z = 6$

20. **Cone frustum** The portion of the cone $z = \sqrt{x^2 + y^2}/3$ between the planes $z = 1$ and $z = 4/3$

21. **Circular cylinder band** The portion of the cylinder $x^2 + y^2 = 1$ between the planes $z = 1$ and $z = 4$

22. **Circular cylinder band** The portion of the cylinder $x^2 + z^2 = 10$ between the planes $y = -1$ and $y = 1$

23. **Parabolic cap** The cap cut from the paraboloid $z = 2 - x^2 - y^2$ by the cone $z = \sqrt{x^2 + y^2}$

24. **Parabolic band** The portion of the paraboloid $z = x^2 + y^2$ between the planes $z = 1$ and $z = 4$

25. **Sawed-off sphere** The lower portion cut from the sphere $x^2 + y^2 + z^2 = 2$ by the cone $z = \sqrt{x^2 + y^2}$

26. **Spherical band** The portion of the sphere $x^2 + y^2 + z^2 = 4$ between the planes $z = -1$ and $z = \sqrt{3}$

Planes Tangent to Parametrized Surfaces

The tangent plane at a point $P_0(f(u_0, v_0), g(u_0, v_0), h(u_0, v_0))$ on a parametrized surface $\mathbf{r}(u, v) = f(u, v)\mathbf{i} + g(u, v)\mathbf{j} + h(u, v)\mathbf{k}$ is the plane through P_0 normal to the vector $\mathbf{r}_u(u_0, v_0) \times \mathbf{r}_v(u_0, v_0)$, the cross product of the tangent vectors $\mathbf{r}_u(u_0, v_0)$ and $\mathbf{r}_v(u_0, v_0)$ at P_0. In Exercises 27–30, find an equation for the plane tangent to the surface at P_0. Then find a Cartesian equation for the surface and sketch the surface and tangent plane together.

27. **Cone** The cone $\mathbf{r}(r, \theta) = (r \cos \theta)\mathbf{i} + (r \sin \theta)\mathbf{j} + r\mathbf{k}, r \geq 0$, $0 \leq \theta \leq 2\pi$ at the point $P_0(\sqrt{2}, \sqrt{2}, 2)$ corresponding to $(r, \theta) = (2, \pi/4)$

28. **Hemisphere** The hemisphere surface $\mathbf{r}(\phi, \theta) = (4 \sin \phi \cos \theta)\mathbf{i} + (4 \sin \phi \sin \theta)\mathbf{j} + (4 \cos \phi)\mathbf{k}, 0 \leq \phi \leq \pi/2, 0 \leq \theta \leq 2\pi$, at the point $P_0(\sqrt{2}, \sqrt{2}, 2\sqrt{3})$ corresponding to $(\phi, \theta) = (\pi/6, \pi/4)$

29. **Circular cylinder** The circular cylinder $\mathbf{r}(\theta, z) = (3 \sin 2\theta)\mathbf{i} + (6 \sin^2 \theta)\mathbf{j} + z\mathbf{k}, 0 \leq \theta \leq \pi$, at the point $P_0(3\sqrt{3}/2, 9/2, 0)$ corresponding to $(\theta, z) = (\pi/3, 0)$ (See Example 3.)

30. **Parabolic cylinder** The parabolic cylinder surface $\mathbf{r}(x, y) = x\mathbf{i} + y\mathbf{j} - x^2\mathbf{k}, -\infty < x < \infty, -\infty < y < \infty$, at the point $P_0(1, 2, -1)$ corresponding to $(x, y) = (1, 2)$

More Parametrizations of Surfaces

31. a. A *torus of revolution* (doughnut) is obtained by rotating a circle C in the xz-plane about the z-axis in space. (See the accompanying figure.) If C has radius $r > 0$ and center $(R, 0, 0)$, show that a parametrization of the torus is

$$\mathbf{r}(u, v) = ((R + r \cos u)\cos v)\mathbf{i}$$
$$+ ((R + r \cos u)\sin v)\mathbf{j} + (r \sin u)\mathbf{k},$$

where $0 \leq u \leq 2\pi$ and $0 \leq v \leq 2\pi$ are the angles in the figure.

b. Show that the surface area of the torus is $A = 4\pi^2 Rr$.

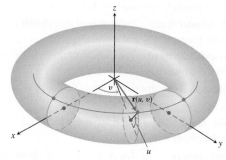

32. Parametrization of a surface of revolution Suppose that the parametrized curve C: $(f(u), g(u))$ is revolved about the x-axis, where $g(u) > 0$ for $a \le u \le b$.

a. Show that

$$\mathbf{r}(u, v) = f(u)\mathbf{i} + (g(u)\cos v)\mathbf{j} + (g(u)\sin v)\mathbf{k}$$

is a parametrization of the resulting surface of revolution, where $0 \le v \le 2\pi$ is the angle from the xy-plane to the point $\mathbf{r}(u, v)$ on the surface. (See the accompanying figure.) Notice that $f(u)$ measures distance *along* the axis of revolution and $g(u)$ measures distance *from* the axis of revolution.

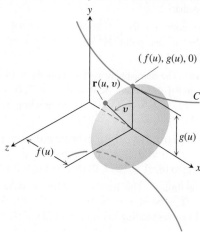

b. Find a parametrization for the surface obtained by revolving the curve $x = y^2$, $y \ge 0$, about the x-axis.

33. a. Parametrization of an ellipsoid The parametrization $x = a\cos\theta$, $y = b\sin\theta$, $0 \le \theta \le 2\pi$ gives the ellipse $(x^2/a^2) + (y^2/b^2) = 1$. Using the angles θ and ϕ in spherical coordinates, show that

$$\mathbf{r}(\theta, \phi) = (a\cos\theta\cos\phi)\mathbf{i} + (b\sin\theta\cos\phi)\mathbf{j} + (c\sin\phi)\mathbf{k}$$

is a parametrization of the ellipsoid $(x^2/a^2) + (y^2/b^2) + (z^2/c^2) = 1$.

b. Write an integral for the surface area of the ellipsoid, but do not evaluate the integral.

34. Hyperboloid of one sheet

a. Find a parametrization for the hyperboloid of one sheet $x^2 + y^2 - z^2 = 1$ in terms of the angle θ associated with the circle $x^2 + y^2 = r^2$ and the hyperbolic parameter u associated with the hyperbolic function $r^2 - z^2 = 1$. (*Hint:* $\cosh^2 u - \sinh^2 u = 1$.)

b. Generalize the result in part (a) to the hyperboloid $(x^2/a^2) + (y^2/b^2) - (z^2/c^2) = 1$.

35. (*Continuation of Exercise 34.*) Find a Cartesian equation for the plane tangent to the hyperboloid $x^2 + y^2 - z^2 = 25$ at the point $(x_0, y_0, 0)$, where $x_0^2 + y_0^2 = 25$.

36. Hyperboloid of two sheets Find a parametrization of the hyperboloid of two sheets $(z^2/c^2) - (x^2/a^2) - (y^2/b^2) = 1$.

Surface Area for Implicit and Explicit Forms

37. Find the area of the surface cut from the paraboloid $x^2 + y^2 - z = 0$ by the plane $z = 2$.

38. Find the area of the band cut from the paraboloid $x^2 + y^2 - z = 0$ by the planes $z = 2$ and $z = 6$.

39. Find the area of the region cut from the plane $x + 2y + 2z = 5$ by the cylinder whose walls are $x = y^2$ and $x = 2 - y^2$.

40. Find the area of the portion of the surface $x^2 - 2z = 0$ that lies above the triangle bounded by the lines $x = \sqrt{3}, y = 0$, and $y = x$ in the xy-plane.

41. Find the area of the surface $x^2 - 2y - 2z = 0$ that lies above the triangle bounded by the lines $x = 2, y = 0$, and $y = 3x$ in the xy-plane.

42. Find the area of the cap cut from the sphere $x^2 + y^2 + z^2 = 2$ by the cone $z = \sqrt{x^2 + y^2}$.

43. Find the area of the ellipse cut from the plane $z = cx$ (c a constant) by the cylinder $x^2 + y^2 = 1$.

44. Find the area of the upper portion of the cylinder $x^2 + z^2 = 1$ that lies between the planes $x = \pm 1/2$ and $y = \pm 1/2$.

45. Find the area of the portion of the paraboloid $x = 4 - y^2 - z^2$ that lies above the ring $1 \le y^2 + z^2 \le 4$ in the yz-plane.

46. Find the area of the surface cut from the paraboloid $x^2 + y + z^2 = 2$ by the plane $y = 0$.

47. Find the area of the surface $x^2 - 2\ln x + \sqrt{15}y - z = 0$ above the square R: $1 \le x \le 2, 0 \le y \le 1$, in the xy-plane.

48. Find the area of the surface $2x^{3/2} + 2y^{3/2} - 3z = 0$ above the square R: $0 \le x \le 1, 0 \le y \le 1$, in the xy-plane.

Find the area of the surfaces in Exercises 49–54.

49. The surface cut from the bottom of the paraboloid $z = x^2 + y^2$ by the plane $z = 3$

50. The surface cut from the "nose" of the paraboloid $x = 1 - y^2 - z^2$ by the yz-plane

51. The portion of the cone $z = \sqrt{x^2 + y^2}$ that lies over the region between the circle $x^2 + y^2 = 1$ and the ellipse $9x^2 + 4y^2 = 36$ in the xy-plane. (*Hint:* Use formulas from geometry to find the area of the region.)

52. The triangle cut from the plane $2x + 6y + 3z = 6$ by the bounding planes of the first octant. Calculate the area three ways, using different explicit forms.

53. The surface in the first octant cut from the cylinder $y = (2/3)z^{3/2}$ by the planes $x = 1$ and $y = 16/3$

54. The portion of the plane $y + z = 4$ that lies above the region cut from the first quadrant of the xz-plane by the parabola $x = 4 - z^2$

55. Use the parametrization

$$\mathbf{r}(x, z) = x\mathbf{i} + f(x, z)\mathbf{j} + z\mathbf{k}$$

and Equation (5) to derive a formula for $d\sigma$ associated with the explicit form $y = f(x, z)$.

56. Let S be the surface obtained by rotating the smooth curve $y = f(x)$, $a \leq x \leq b$, about the x-axis, where $f(x) \geq 0$.

    a. Show that the vector function

    $$\mathbf{r}(x, \theta) = x\mathbf{i} + f(x) \cos \theta \, \mathbf{j} + f(x) \sin \theta \, \mathbf{k}$$

    is a parametrization of S, where θ is the angle of rotation around the x-axis (see the accompanying figure).

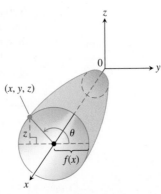

b. Use Equation (4) to show that the surface area of this surface of revolution is given by

$$A = \int_a^b 2\pi f(x) \sqrt{1 + [f'(x)]^2} \, dx.$$

16.6 Surface Integrals

To compute quantities such as the flow of liquid across a curved membrane or the upward force on a falling parachute, we need to integrate a function over a curved surface in space. This concept of a *surface integral* is an extension of the idea of a line integral for integrating over a curve.

Surface Integrals

Suppose that we have an electrical charge distributed over a surface S, and that the function $G(x, y, z)$ gives the *charge density* (charge per unit area) at each point on S. Then we can calculate the total charge on S as an integral in the following way.

Assume, as in Section 16.5, that the surface S is defined parametrically on a region R in the uv-plane,

$$\mathbf{r}(u, v) = f(u, v)\mathbf{i} + g(u, v)\mathbf{j} + h(u, v)\mathbf{k}, \qquad (u, v) \in R.$$

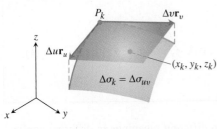

FIGURE 16.47 The area of the patch $\Delta\sigma_k$ is the area of the tangent parallelogram determined by the vectors $\Delta u \, \mathbf{r}_u$ and $\Delta v \, \mathbf{r}_v$. The point (x_k, y_k, z_k) lies on the surface patch, beneath the parallelogram shown here.

In Figure 16.47, we see how a subdivision of R (considered as a rectangle for simplicity) divides the surface S into corresponding curved surface elements, or patches, of area

$$\Delta\sigma_{uv} \approx |\mathbf{r}_u \times \mathbf{r}_v| \, du \, dv.$$

As we did for the subdivisions when defining double integrals in Section 15.2, we number the surface element patches in some order with their areas given by $\Delta\sigma_1, \Delta\sigma_2, \ldots, \Delta\sigma_n$. To form a Riemann sum over S, we choose a point (x_k, y_k, z_k) in the kth patch, multiply the value of the function G at that point by the area $\Delta\sigma_k$, and add together the products:

$$\sum_{k=1}^n G(x_k, y_k, z_k) \, \Delta\sigma_k.$$

Depending on how we pick (x_k, y_k, z_k) in the kth patch, we may get different values for this Riemann sum. Then we take the limit as the number of surface patches increases, their areas shrink to zero, and both $\Delta u \to 0$ and $\Delta v \to 0$. This limit, whenever it exists independent of all choices made, defines the **surface integral of G over the surface S** as

$$\iint_S G(x, y, z) \, d\sigma = \lim_{n \to \infty} \sum_{k=1}^n G(x_k, y_k, z_k) \, \Delta\sigma_k. \qquad (1)$$

Notice the analogy with the definition of the double integral (Section 15.2) and with the line integral (Section 16.1). If S is a piecewise smooth surface, and G is continuous over S, then the surface integral defined by Equation (1) can be shown to exist.

The formula for evaluating the surface integral depends on the manner in which S is described, parametrically, implicitly or explicitly, as discussed in Section 16.5.

Formulas for a Surface Integral

1. For a smooth surface S defined **parametrically** as $\mathbf{r}(u, v) = f(u, v)\mathbf{i} + g(u, v)\mathbf{j} + h(u, v)\mathbf{k}$, $(u, v) \in R$, and a continuous function $G(x, y, z)$ defined on S, the surface integral of G over S is given by the double integral over R,

$$\iint\limits_{S} G(x, y, z)\, d\sigma = \iint\limits_{R} G(f(u, v), g(u, v), h(u, v))\, |\mathbf{r}_u \times \mathbf{r}_v|\, du\, dv. \qquad (2)$$

2. For a surface S given **implicitly** by $F(x, y, z) = c$, where F is a continuously differentiable function, with S lying above its closed and bounded shadow region R in the coordinate plane beneath it, the surface integral of the continuous function G over S is given by the double integral over R,

$$\iint\limits_{S} G(x, y, z)\, d\sigma = \iint\limits_{R} G(x, y, z) \frac{|\nabla F|}{|\nabla F \cdot \mathbf{p}|}\, dA, \qquad (3)$$

where $\mathbf{p}$ is a unit vector normal to R and $\nabla F \cdot \mathbf{p} \neq 0$.

3. For a surface S given **explicitly** as the graph of $z = f(x, y)$, where f is a continuously differentiable function over a region R in the xy-plane, the surface integral of the continuous function G over S is given by the double integral over R,

$$\iint\limits_{S} G(x, y, z)\, d\sigma = \iint\limits_{R} G(x, y, f(x, y)) \sqrt{f_x^2 + f_y^2 + 1}\, dx\, dy. \qquad (4)$$

The surface integral in Equation (1) takes on different meanings in different applications. If G has the constant value 1, the integral gives the area of S. If G gives the mass density of a thin shell of material modeled by S, the integral gives the mass of the shell. If G gives the charge density of a thin shell, then the integral gives the total charge.

EXAMPLE 1 Integrate $G(x, y, z) = x^2$ over the cone $z = \sqrt{x^2 + y^2}$, $0 \leq z \leq 1$.

Solution Using Equation (2) and the calculations from Example 4 in Section 16.5, we have $|\mathbf{r}_r \times \mathbf{r}_\theta| = \sqrt{2}r$ and

$$\iint\limits_{S} x^2\, d\sigma = \int_0^{2\pi} \int_0^1 \left(r^2 \cos^2\theta\right)\left(\sqrt{2}r\right) dr\, d\theta \qquad x = r\cos\theta$$

$$= \sqrt{2} \int_0^{2\pi} \int_0^1 r^3 \cos^2\theta\, dr\, d\theta$$

$$= \frac{\sqrt{2}}{4} \int_0^{2\pi} \cos^2\theta\, d\theta = \frac{\sqrt{2}}{4} \left[\frac{\theta}{2} + \frac{1}{4}\sin 2\theta\right]_0^{2\pi} = \frac{\pi\sqrt{2}}{4}. \qquad \blacksquare$$

Surface integrals behave like other double integrals, the integral of the sum of two functions being the sum of their integrals and so on. The domain Additivity Property takes the form

$$\iint\limits_{S} G\, d\sigma = \iint\limits_{S_1} G\, d\sigma + \iint\limits_{S_2} G\, d\sigma + \cdots + \iint\limits_{S_n} G\, d\sigma.$$

When S is partitioned by smooth curves into a finite number of smooth patches with nonoverlapping interiors (i.e., if S is piecewise smooth), then the integral over S is the sum of the integrals over the patches. Thus, the integral of a function over the surface of a cube is the sum of the integrals over the faces of the cube. We integrate over a turtle shell of welded plates by integrating over one plate at a time and adding the results.

EXAMPLE 2   Integrate $G(x, y, z) = xyz$ over the surface of the cube cut from the first octant by the planes $x = 1, y = 1$, and $z = 1$ (Figure 16.48).

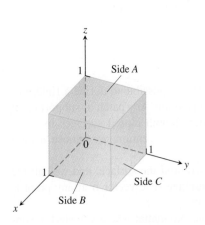

FIGURE 16.48   The cube in Example 2.

Solution   We integrate xyz over each of the six sides and add the results. Since $xyz = 0$ on the sides that lie in the coordinate planes, the integral over the surface of the cube reduces to

$$\iint\limits_{\substack{\text{Cube} \\ \text{surface}}} xyz\, d\sigma = \iint\limits_{\text{Side } A} xyz\, d\sigma + \iint\limits_{\text{Side } B} xyz\, d\sigma + \iint\limits_{\text{Side } C} xyz\, d\sigma.$$

Side A is the surface $f(x, y, z) = z = 1$ over the square region $R_{xy}\colon 0 \le x \le 1$, $0 \le y \le 1$, in the xy-plane. For this surface and region,

$$\mathbf{p} = \mathbf{k}, \qquad \nabla f = \mathbf{k}, \qquad |\nabla f| = 1, \qquad |\nabla f \cdot \mathbf{p}| = |\mathbf{k} \cdot \mathbf{k}| = 1$$

$$d\sigma = \frac{|\nabla f|}{|\nabla f \cdot \mathbf{p}|}\, dA = \frac{1}{1}\, dx\, dy = dx\, dy$$

$$xyz = xy(1) = xy$$

and

$$\iint\limits_{\text{Side } A} xyz\, d\sigma = \iint\limits_{R_{xy}} xy\, dx\, dy = \int_0^1 \int_0^1 xy\, dx\, dy = \int_0^1 \frac{y}{2}\, dy = \frac{1}{4}.$$

Symmetry tells us that the integrals of xyz over sides B and C are also $1/4$. Hence,

$$\iint\limits_{\substack{\text{Cube} \\ \text{surface}}} xyz\, d\sigma = \frac{1}{4} + \frac{1}{4} + \frac{1}{4} = \frac{3}{4}. \qquad \blacksquare$$

EXAMPLE 3   Integrate $G(x, y, z) = \sqrt{1 - x^2 - y^2}$ over the "football" surface S formed by rotating the curve $x = \cos z, y = 0, -\pi/2 \le z \le \pi/2$, around the z-axis.

Solution   The surface is displayed in Figure 16.44, and in Example 6 of Section 16.5 we found the parametrization

$$x = \cos u \cos v, \quad y = \cos u \sin v, \quad z = u, \quad -\frac{\pi}{2} \le u \le \frac{\pi}{2} \quad \text{and} \quad 0 \le v \le 2\pi,$$

where v represents the angle of rotation from the xz-plane about the z-axis. Substituting this parametrization into the expression for G gives

$$\sqrt{1 - x^2 - y^2} = \sqrt{1 - (\cos^2 u)(\cos^2 v + \sin^2 v)} = \sqrt{1 - \cos^2 u} = |\sin u|.$$

The surface area differential for the parametrization was found to be (Example 6, Section 16.5)

$$d\sigma = \cos u\, \sqrt{1 + \sin^2 u}\, du\, dv.$$

These calculations give the surface integral

$$\iint_S \sqrt{1 - x^2 - y^2}\, d\sigma = \int_0^{2\pi} \int_{-\pi/2}^{\pi/2} |\sin u| \cos u \sqrt{1 + \sin^2 u}\, du\, dv$$

$$= 2 \int_0^{2\pi} \int_0^{\pi/2} \sin u \cos u \sqrt{1 + \sin^2 u}\, du\, dv$$

$$= \int_0^{2\pi} \int_1^2 \sqrt{w}\, dw\, dv \qquad \begin{array}{l} w = 1 + \sin^2 u, \\ dw = 2 \sin u \cos u\, du \\ \text{When } u = 0, w = 1. \\ \text{When } u = \pi/2, w = 2. \end{array}$$

$$= 2\pi \cdot \frac{2}{3} w^{3/2} \Big]_1^2 = \frac{4\pi}{3} \left(2\sqrt{2} - 1 \right). \qquad \blacksquare$$

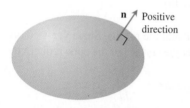

FIGURE 16.49 Smooth closed surfaces in space are orientable. The outward unit normal vector defines the positive direction at each point.

FIGURE 16.50 To make a Möbius band, take a rectangular strip of paper *abcd*, give the end *bc* a single twist, and paste the ends of the strip together to match *a* with *c* and *b* with *d*. The Möbius band is a nonorientable or one-sided surface.

Orientation

We call a smooth surface *S* **orientable** or **two-sided** if it is possible to define a field **n** of unit normal vectors on *S* that varies continuously with position. Any patch or subportion of an orientable surface is orientable. Spheres and other smooth closed surfaces in space (smooth surfaces that enclose solids) are orientable. By convention, we choose **n** on a closed surface to point outward.

Once **n** has been chosen, we say that we have **oriented** the surface, and we call the surface together with its normal field an **oriented surface**. The vector **n** at any point is called the **positive direction** at that point (Figure 16.49).

The Möbius band in Figure 16.50 is not orientable. No matter where you start to construct a continuous unit normal field (shown as the shaft of a thumbtack in the figure), moving the vector continuously around the surface in the manner shown will return it to the starting point with a direction opposite to the one it had when it started out. The vector at that point cannot point both ways and yet it must if the field is to be continuous. We conclude that no such field exists.

Surface Integral for Flux

Suppose that **F** is a continuous vector field defined over an oriented surface *S* and that **n** is the chosen unit normal field on the surface. We call the integral of **F** · **n** over *S* the flux of **F** across *S* in the positive direction. Thus, the flux is the integral over *S* of the scalar component of **F** in the direction of **n**.

> **DEFINITION** The **flux** of a three-dimensional vector field **F** across an oriented surface *S* in the direction of **n** is
>
> $$\text{Flux} = \iint_S \mathbf{F} \cdot \mathbf{n}\, d\sigma. \qquad (5)$$

The definition is analogous to the flux of a two-dimensional field **F** across a plane curve *C*. In the plane (Section 16.2), the flux is

$$\int_C \mathbf{F} \cdot \mathbf{n}\, ds,$$

the integral of the scalar component of **F** normal to the curve.

If **F** is the velocity field of a three-dimensional fluid flow, the flux of **F** across *S* is the net rate at which fluid is crossing *S* in the chosen positive direction. We discuss such flows in more detail in Section 16.7.

FIGURE 16.51 Finding the flux through the surface of a parabolic cylinder (Example 4).

EXAMPLE 4   Find the flux of $\mathbf{F} = yz\mathbf{i} + x\mathbf{j} - z^2\mathbf{k}$ through the parabolic cylinder $y = x^2, 0 \le x \le 1, 0 \le z \le 4$, in the direction $\mathbf{n}$ indicated in Figure 16.51.

Solution   On the surface we have $x = x, y = x^2$, and $z = z$, so we automatically have the parametrization $\mathbf{r}(x, z) = x\mathbf{i} + x^2\mathbf{j} + z\mathbf{k}, 0 \le x \le 1, 0 \le z \le 4$. The cross product of tangent vectors is

$$\mathbf{r}_x \times \mathbf{r}_z = \begin{vmatrix} \mathbf{i} & \mathbf{j} & \mathbf{k} \\ 1 & 2x & 0 \\ 0 & 0 & 1 \end{vmatrix} = 2x\mathbf{i} - \mathbf{j}.$$

The unit normal vectors pointing outward from the surface as indicated in Figure 16.51 are

$$\mathbf{n} = \frac{\mathbf{r}_x \times \mathbf{r}_z}{|\mathbf{r}_x \times \mathbf{r}_z|} = \frac{2x\mathbf{i} - \mathbf{j}}{\sqrt{4x^2 + 1}}.$$

On the surface, $y = x^2$, so the vector field there is

$$\mathbf{F} = yz\mathbf{i} + x\mathbf{j} - z^2\mathbf{k} = x^2 z\mathbf{i} + x\mathbf{j} - z^2\mathbf{k}.$$

Thus,

$$\mathbf{F} \cdot \mathbf{n} = \frac{1}{\sqrt{4x^2 + 1}}\left((x^2 z)(2x) + (x)(-1) + (-z^2)(0)\right)$$

$$= \frac{2x^3 z - x}{\sqrt{4x^2 + 1}}.$$

The flux of $\mathbf{F}$ outward through the surface is

$$\iint\limits_{S} \mathbf{F} \cdot \mathbf{n}\, d\sigma = \int_0^4 \int_0^1 \frac{2x^3 z - x}{\sqrt{4x^2 + 1}} |\mathbf{r}_x \times \mathbf{r}_z|\, dx\, dz$$

$$= \int_0^4 \int_0^1 \frac{2x^3 z - x}{\sqrt{4x^2 + 1}} \sqrt{4x^2 + 1}\, dx\, dz$$

$$= \int_0^4 \int_0^1 (2x^3 z - x)\, dx\, dz = \int_0^4 \left[\frac{1}{2}x^4 z - \frac{1}{2}x^2\right]_{x=0}^{x=1} dz$$

$$= \int_0^4 \frac{1}{2}(z - 1)\, dz = \frac{1}{4}(z - 1)^2\Big]_0^4$$

$$= \frac{1}{4}(9) - \frac{1}{4}(1) = 2. \qquad \blacksquare$$

If S is part of a level surface $g(x, y, z) = c$, then $\mathbf{n}$ may be taken to be one of the two fields

$$\mathbf{n} = \pm\frac{\nabla g}{|\nabla g|}, \tag{6}$$

depending on which one gives the preferred direction. The corresponding flux is

$$\text{Flux} = \iint\limits_{S} \mathbf{F} \cdot \mathbf{n}\, d\sigma$$

$$= \iint\limits_{R} \left(\mathbf{F} \cdot \frac{\pm\nabla g}{|\nabla g|}\right) \frac{|\nabla g|}{|\nabla g \cdot \mathbf{p}|}\, dA \qquad \text{Eqs. (6) and (3)}$$

$$= \iint\limits_{R} \mathbf{F} \cdot \frac{\pm\nabla g}{|\nabla g \cdot \mathbf{p}|}\, dA. \tag{7}$$

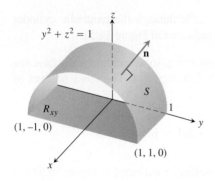

$y^2 + z^2 = 1$

$(1, -1, 0)$

R_{xy}

$(1, 1, 0)$

FIGURE 16.52 Calculating the flux of a vector field outward through the surface S. The area of the shadow region R_{xy} is 2 (Example 5).

EXAMPLE 5 Find the flux of $\mathbf{F} = yz\mathbf{j} + z^2\mathbf{k}$ outward through the surface S cut from the cylinder $y^2 + z^2 = 1$, $z \geq 0$, by the planes $x = 0$ and $x = 1$.

Solution The outward normal field on S (Figure 16.52) may be calculated from the gradient of $g(x, y, z) = y^2 + z^2$ to be

$$\mathbf{n} = +\frac{\nabla g}{|\nabla g|} = \frac{2y\mathbf{j} + 2z\mathbf{k}}{\sqrt{4y^2 + 4z^2}} = \frac{2y\mathbf{j} + 2z\mathbf{k}}{2\sqrt{1}} = y\mathbf{j} + z\mathbf{k}.$$

With $\mathbf{p} = \mathbf{k}$, we also have

$$d\sigma = \frac{|\nabla g|}{|\nabla g \cdot \mathbf{k}|} dA = \frac{2}{|2z|} dA = \frac{1}{z} dA.$$

We can drop the absolute value bars because $z \geq 0$ on S.

The value of $\mathbf{F} \cdot \mathbf{n}$ on the surface is

$$\begin{aligned}
\mathbf{F} \cdot \mathbf{n} &= (yz\mathbf{j} + z^2\mathbf{k}) \cdot (y\mathbf{j} + z\mathbf{k}) \\
&= y^2z + z^3 = z(y^2 + z^2) \qquad y^2 + z^2 = 1 \text{ on } S \\
&= z.
\end{aligned}$$

The surface projects onto the shadow region R_{xy}, which is the rectangle in the xy-plane shown in Figure 16.52. Therefore, the flux of $\mathbf{F}$ outward through S is

$$\iint_S \mathbf{F} \cdot \mathbf{n} \, d\sigma = \iint_S (z)\left(\frac{1}{z} dA\right) = \iint_{R_{xy}} dA = \text{area}(R_{xy}) = 2. \qquad \blacksquare$$

Moments and Masses of Thin Shells

Thin shells of material like bowls, metal drums, and domes are modeled with surfaces. Their moments and masses are calculated with the formulas in Table 16.3. The derivations are similar to those in Section 6.6. The formulas are like those for line integrals in Table 16.1, Section 16.1.

TABLE 16.3 Mass and moment formulas for very thin shells

Mass: $M = \iint_S \delta \, d\sigma$ $\qquad \delta = \delta(x, y, z) = $ density at (x, y, z) as mass per unit area

First moments about the coordinate planes:

$$M_{yz} = \iint_S x \delta \, d\sigma, \qquad M_{xz} = \iint_S y \delta \, d\sigma, \qquad M_{xy} = \iint_S z \delta \, d\sigma$$

Coordinates of center of mass:

$$\bar{x} = M_{yz}/M, \qquad \bar{y} = M_{xz}/M, \qquad \bar{z} = M_{xy}/M$$

Moments of inertia about coordinate axes:

$$I_x = \iint_S (y^2 + z^2) \delta \, d\sigma, \quad I_y = \iint_S (x^2 + z^2) \delta \, d\sigma, \quad I_z = \iint_S (x^2 + y^2) \delta \, d\sigma,$$

$$I_L = \iint_S r^2 \delta \, d\sigma \qquad r(x, y, z) = \text{distance from point } (x, y, z) \text{ to line } L$$

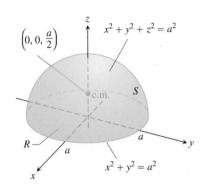

FIGURE 16.53 The center of mass of a thin hemispherical shell of constant density lies on the axis of symmetry halfway from the base to the top (Example 6).

EXAMPLE 6 Find the center of mass of a thin hemispherical shell of radius a and constant density δ.

Solution We model the shell with the hemisphere

$$f(x, y, z) = x^2 + y^2 + z^2 = a^2, \quad z \geq 0$$

(Figure 16.53). The symmetry of the surface about the z-axis tells us that $\bar{x} = \bar{y} = 0$. It remains only to find $\bar{z}$ from the formula $\bar{z} = M_{xy}/M$.

The mass of the shell is

$$M = \iint_S \delta \, d\sigma = \delta \iint_S d\sigma = (\delta)(\text{area of } S) = 2\pi a^2 \delta. \qquad \delta = \text{constant}$$

To evaluate the integral for M_{xy}, we take $\mathbf{p} = \mathbf{k}$ and calculate

$$|\nabla f| = |2x\mathbf{i} + 2y\mathbf{j} + 2z\mathbf{k}| = 2\sqrt{x^2 + y^2 + z^2} = 2a$$

$$|\nabla f \cdot \mathbf{p}| = |\nabla f \cdot \mathbf{k}| = |2z| = 2z$$

$$d\sigma = \frac{|\nabla f|}{|\nabla f \cdot \mathbf{p}|} \, dA = \frac{a}{z} \, dA.$$

Then

$$M_{xy} = \iint_S z\delta \, d\sigma = \delta \iint_R z\frac{a}{z} \, dA = \delta a \iint_R dA = \delta a(\pi a^2) = \delta \pi a^3$$

$$\bar{z} = \frac{M_{xy}}{M} = \frac{\pi a^3 \delta}{2\pi a^2 \delta} = \frac{a}{2}.$$

The shell's center of mass is the point $(0, 0, a/2)$. ∎

EXAMPLE 7 Find the center of mass of a thin shell of density $\delta = 1/z^2$ cut from the cone $z = \sqrt{x^2 + y^2}$ by the planes $z = 1$ and $z = 2$ (Figure 16.54).

Solution The symmetry of the surface about the z-axis tells us that $\bar{x} = \bar{y} = 0$. We find $\bar{z} = M_{xy}/M$. Working as in Example 4 of Section 16.5, we have

$$\mathbf{r}(r, \theta) = (r \cos \theta)\mathbf{i} + (r \sin \theta)\mathbf{j} + r\mathbf{k}, \quad 1 \leq r \leq 2, \quad 0 \leq \theta \leq 2\pi,$$

and

$$|\mathbf{r}_r \times \mathbf{r}_\theta| = \sqrt{2}\,r.$$

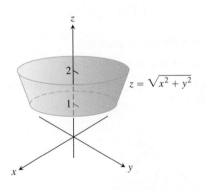

FIGURE 16.54 The cone frustum formed when the cone $z = \sqrt{x^2 + y^2}$ is cut by the planes $z = 1$ and $z = 2$ (Example 7).

Therefore,

$$M = \iint_S \delta \, d\sigma = \int_0^{2\pi} \int_1^2 \frac{1}{r^2} \sqrt{2}\,r \, dr \, d\theta$$

$$= \sqrt{2} \int_0^{2\pi} \left[\ln r\right]_1^2 \, d\theta = \sqrt{2} \int_0^{2\pi} \ln 2 \, d\theta$$

$$= 2\pi\sqrt{2} \ln 2,$$

$$M_{xy} = \iint\limits_{S} \delta z \, d\sigma = \int_0^{2\pi} \int_1^2 \frac{1}{r^2} \, r\sqrt{2}r \, dr \, d\theta$$

$$= \sqrt{2} \int_0^{2\pi} \int_1^2 dr \, d\theta$$

$$= \sqrt{2} \int_0^{2\pi} d\theta = 2\pi\sqrt{2},$$

$$\bar{z} = \frac{M_{xy}}{M} = \frac{2\pi\sqrt{2}}{2\pi\sqrt{2}\ln 2} = \frac{1}{\ln 2}.$$

The shell's center of mass is the point $(0, 0, 1/\ln 2)$. ∎

Exercises 16.6

Surface Integrals

In Exercises 1–8, integrate the given function over the given surface.

1. **Parabolic cylinder** $G(x, y, z) = x$, over the parabolic cylinder $y = x^2, 0 \le x \le 2, 0 \le z \le 3$

2. **Circular cylinder** $G(x, y, z) = z$, over the cylindrical surface $y^2 + z^2 = 4, z \ge 0, 1 \le x \le 4$

3. **Sphere** $G(x, y, z) = x^2$, over the unit sphere $x^2 + y^2 + z^2 = 1$

4. **Hemisphere** $G(x, y, z) = z^2$, over the hemisphere $x^2 + y^2 + z^2 = a^2, z \ge 0$

5. **Portion of plane** $F(x, y, z) = z$, over the portion of the plane $x + y + z = 4$ that lies above the square $0 \le x \le 1$, $0 \le y \le 1$, in the xy-plane

6. **Cone** $F(x, y, z) = z - x$, over the cone $z = \sqrt{x^2 + y^2}$, $0 \le z \le 1$

7. **Parabolic dome** $H(x, y, z) = x^2\sqrt{5 - 4z}$, over the parabolic dome $z = 1 - x^2 - y^2, z \ge 0$

8. **Spherical cap** $H(x, y, z) = yz$, over the part of the sphere $x^2 + y^2 + z^2 = 4$ that lies above the cone $z = \sqrt{x^2 + y^2}$

9. Integrate $G(x, y, z) = x + y + z$ over the surface of the cube cut from the first octant by the planes $x = a, y = a, z = a$.

10. Integrate $G(x, y, z) = y + z$ over the surface of the wedge in the first octant bounded by the coordinate planes and the planes $x = 2$ and $y + z = 1$.

11. Integrate $G(x, y, z) = xyz$ over the surface of the rectangular solid cut from the first octant by the planes $x = a, y = b$, and $z = c$.

12. Integrate $G(x, y, z) = xyz$ over the surface of the rectangular solid bounded by the planes $x = \pm a, y = \pm b$, and $z = \pm c$.

13. Integrate $G(x, y, z) = x + y + z$ over the portion of the plane $2x + 2y + z = 2$ that lies in the first octant.

14. Integrate $G(x, y, z) = x\sqrt{y^2 + 4}$ over the surface cut from the parabolic cylinder $y^2 + 4z = 16$ by the planes $x = 0$, $x = 1$, and $z = 0$.

15. Integrate $G(x, y, z) = z - x$ over the portion of the graph of $z = x + y^2$ above the triangle in the xy-plane having vertices $(0, 0, 0), (1, 1, 0)$, and $(0, 1, 0)$. (See accompanying figure.)

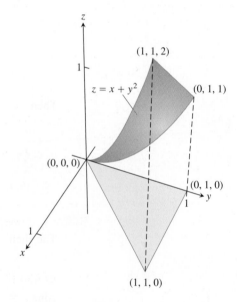

16. Integrate $G(x, y, z) = x$ over the surface given by

$$z = x^2 + y \quad \text{for} \quad 0 \le x \le 1, \quad -1 \le y \le 1.$$

17. Integrate $G(x, y, z) = xyz$ over the triangular surface with vertices $(1, 0, 0), (0, 2, 0)$, and $(0, 1, 1)$.

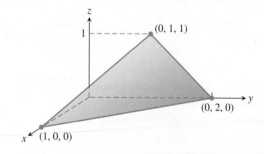

18. Integrate $G(x, y, z) = x - y - z$ over the portion of the plane $x + y = 1$ in the first octant between $z = 0$ and $z = 1$ (see the accompanying figure).

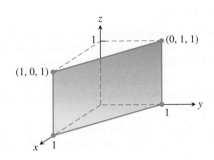

Finding Flux Across a Surface

In Exercises 19–28, use a parametrization to find the flux $\iint_S \mathbf{F}\cdot\mathbf{n}\,d\sigma$ across the surface in the given direction.

19. **Parabolic cylinder** $\mathbf{F} = z^2\mathbf{i} + x\mathbf{j} - 3z\mathbf{k}$ outward (normal away from the x-axis) through the surface cut from the parabolic cylinder $z = 4 - y^2$ by the planes $x = 0, x = 1$, and $z = 0$.

20. **Parabolic cylinder** $\mathbf{F} = x^2\mathbf{j} - xz\mathbf{k}$ outward (normal away from the yz-plane) through the surface cut from the parabolic cylinder $y = x^2, -1 \le x \le 1$, by the planes $z = 0$ and $z = 2$

21. **Sphere** $\mathbf{F} = z\mathbf{k}$ across the portion of the sphere $x^2 + y^2 + z^2 = a^2$ in the first octant in the direction away from the origin

22. **Sphere** $\mathbf{F} = x\mathbf{i} + y\mathbf{j} + z\mathbf{k}$ across the sphere $x^2 + y^2 + z^2 = a^2$ in the direction away from the origin

23. **Plane** $\mathbf{F} = 2xy\mathbf{i} + 2yz\mathbf{j} + 2xz\mathbf{k}$ upward across the portion of the plane $x + y + z = 2a$ that lies above the square $0 \le x \le a, 0 \le y \le a$, in the xy-plane

24. **Cylinder** $\mathbf{F} = x\mathbf{i} + y\mathbf{j} + z\mathbf{k}$ outward through the portion of the cylinder $x^2 + y^2 = 1$ cut by the planes $z = 0$ and $z = a$

25. **Cone** $\mathbf{F} = xy\mathbf{i} - z\mathbf{k}$ outward (normal away from the z-axis) through the cone $z = \sqrt{x^2 + y^2}, 0 \le z \le 1$

26. **Cone** $\mathbf{F} = y^2\mathbf{i} + xz\mathbf{j} - \mathbf{k}$ outward (normal away from the z-axis) through the cone $z = 2\sqrt{x^2 + y^2}, 0 \le z \le 2$

27. **Cone frustum** $\mathbf{F} = -x\mathbf{i} - y\mathbf{j} + z^2\mathbf{k}$ outward (normal away from the z-axis) through the portion of the cone $z = \sqrt{x^2 + y^2}$ between the planes $z = 1$ and $z = 2$

28. **Paraboloid** $\mathbf{F} = 4x\mathbf{i} + 4y\mathbf{j} + 2\mathbf{k}$ outward (normal away from the z-axis) through the surface cut from the bottom of the paraboloid $z = x^2 + y^2$ by the plane $z = 1$

In Exercises 29 and 30, find the flux of the field $\mathbf{F}$ across the portion of the given surface in the specified direction.

29. $\mathbf{F}(x, y, z) = -\mathbf{i} + 2\mathbf{j} + 3\mathbf{k}$
    S: rectangular surface $z = 0, \quad 0 \le x \le 2, \quad 0 \le y \le 3$, direction $\mathbf{k}$

30. $\mathbf{F}(x, y, z) = yx^2\mathbf{i} - 2\mathbf{j} + xz\mathbf{k}$
    S: rectangular surface $y = 0, \quad -1 \le x \le 2, \quad 2 \le z \le 7$, direction $-\mathbf{j}$

In Exercises 31–36, find the flux of the field $\mathbf{F}$ across the portion of the sphere $x^2 + y^2 + z^2 = a^2$ in the first octant in the direction away from the origin.

31. $\mathbf{F}(x, y, z) = z\mathbf{k}$

32. $\mathbf{F}(x, y, z) = -y\mathbf{i} + x\mathbf{j}$

33. $\mathbf{F}(x, y, z) = y\mathbf{i} - x\mathbf{j} + \mathbf{k}$

34. $\mathbf{F}(x, y, z) = zx\mathbf{i} + zy\mathbf{j} + z^2\mathbf{k}$

35. $\mathbf{F}(x, y, z) = x\mathbf{i} + y\mathbf{j} + z\mathbf{k}$

36. $\mathbf{F}(x, y, z) = \dfrac{x\mathbf{i} + y\mathbf{j} + z\mathbf{k}}{\sqrt{x^2 + y^2 + z^2}}$

37. Find the flux of the field $\mathbf{F}(x, y, z) = z^2\mathbf{i} + x\mathbf{j} - 3z\mathbf{k}$ outward through the surface cut from the parabolic cylinder $z = 4 - y^2$ by the planes $x = 0, x = 1$, and $z = 0$.

38. Find the flux of the field $\mathbf{F}(x, y, z) = 4x\mathbf{i} + 4y\mathbf{j} + 2\mathbf{k}$ outward (away from the z-axis) through the surface cut from the bottom of the paraboloid $z = x^2 + y^2$ by the plane $z = 1$.

39. Let S be the portion of the cylinder $y = e^x$ in the first octant that projects parallel to the x-axis onto the rectangle $R_{yz}: 1 \le y \le 2, 0 \le z \le 1$ in the yz-plane (see the accompanying figure). Let $\mathbf{n}$ be the unit vector normal to S that points away from the yz-plane. Find the flux of the field $\mathbf{F}(x, y, z) = -2\mathbf{i} + 2y\mathbf{j} + z\mathbf{k}$ across S in the direction of $\mathbf{n}$.

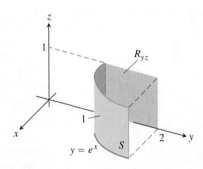

40. Let S be the portion of the cylinder $y = \ln x$ in the first octant whose projection parallel to the y-axis onto the xz-plane is the rectangle $R_{xz}: 1 \le x \le e, 0 \le z \le 1$. Let $\mathbf{n}$ be the unit vector normal to S that points away from the xz-plane. Find the flux of $\mathbf{F} = 2y\mathbf{j} + z\mathbf{k}$ through S in the direction of $\mathbf{n}$.

41. Find the outward flux of the field $\mathbf{F} = 2xy\mathbf{i} + 2yz\mathbf{j} + 2xz\mathbf{k}$ across the surface of the cube cut from the first octant by the planes $x = a, y = a, z = a$.

42. Find the outward flux of the field $\mathbf{F} = xz\mathbf{i} + yz\mathbf{j} + \mathbf{k}$ across the surface of the upper cap cut from the solid sphere $x^2 + y^2 + z^2 \le 25$ by the plane $z = 3$.

Moments and Masses

43. **Centroid** Find the centroid of the portion of the sphere $x^2 + y^2 + z^2 = a^2$ that lies in the first octant.

44. **Centroid** Find the centroid of the surface cut from the cylinder $y^2 + z^2 = 9, z \ge 0$, by the planes $x = 0$ and $x = 3$ (resembles the surface in Example 5).

45. **Thin shell of constant density** Find the center of mass and the moment of inertia about the z-axis of a thin shell of constant density δ cut from the cone $x^2 + y^2 - z^2 = 0$ by the planes $z = 1$ and $z = 2$.

46. **Conical surface of constant density** Find the moment of inertia about the z-axis of a thin shell of constant density δ cut from the cone $4x^2 + 4y^2 - z^2 = 0, z \ge 0$, by the circular cylinder $x^2 + y^2 = 2x$ (see the accompanying figure).

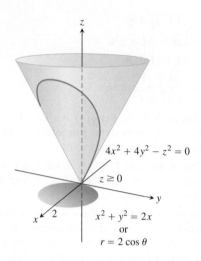

$4x^2 + 4y^2 - z^2 = 0$

$z \geq 0$

$x^2 + y^2 = 2x$

or

$r = 2\cos\theta$

47. Spherical shells

   a. Find the moment of inertia about a diameter of a thin spherical shell of radius a and constant density δ. (Work with a hemispherical shell and double the result.)

   b. Use the Parallel Axis Theorem (Exercises 15.6) and the result in part (a) to find the moment of inertia about a line tangent to the shell.

48. Conical Surface Find the centroid of the lateral surface of a solid cone of base radius a and height h (cone surface minus the base).

16.7 Stokes' Theorem

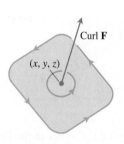

Curl **F**

(x, y, z)

FIGURE 16.55 The circulation vector at a point (x, y, z) in a plane in a three-dimensional fluid flow. Notice its right-hand relation to the rotating particles in the fluid.

As we saw in Section 16.4, the circulation density or curl component of a two-dimensional field $\mathbf{F} = M\mathbf{i} + N\mathbf{j}$ at a point (x, y) is described by the scalar quantity $(\partial N/\partial x - \partial M/\partial y)$. In three dimensions, circulation is described with a vector.

Suppose that $\mathbf{F}$ is the velocity field of a fluid flowing in space. Particles near the point (x, y, z) in the fluid tend to rotate around an axis through (x, y, z) that is parallel to a certain vector we are about to define. This vector points in the direction for which the rotation is counterclockwise when viewed looking down onto the plane of the circulation from the tip of the arrow representing the vector. This is the direction your right-hand thumb points when your fingers curl around the axis of rotation in the way consistent with the rotating motion of the particles in the fluid (see Figure 16.55). The length of the vector measures the rate of rotation. The vector is called the **curl vector** and for the vector field $\mathbf{F} = M\mathbf{i} + N\mathbf{j} + P\mathbf{k}$ it is defined to be

$$\text{curl } \mathbf{F} = \left(\frac{\partial P}{\partial y} - \frac{\partial N}{\partial z} \right)\mathbf{i} + \left(\frac{\partial M}{\partial z} - \frac{\partial P}{\partial x} \right)\mathbf{j} + \left(\frac{\partial N}{\partial x} - \frac{\partial M}{\partial y} \right)\mathbf{k}. \tag{1}$$

This information is a consequence of Stokes' Theorem, the generalization to space of the circulation-curl form of Green's Theorem and the subject of this section.

Notice that $(\text{curl } \mathbf{F}) \cdot \mathbf{k} = (\partial N/\partial x - \partial M/\partial y)$ is consistent with our definition in Section 16.4 when $\mathbf{F} = M(x, y)\mathbf{i} + N(x, y)\mathbf{j}$. The formula for curl $\mathbf{F}$ in Equation (1) is often written using the symbolic operator

$$\nabla = \mathbf{i} \frac{\partial}{\partial x} + \mathbf{j} \frac{\partial}{\partial y} + \mathbf{k} \frac{\partial}{\partial z}. \tag{2}$$

(The symbol ∇ is pronounced "del.") The curl of $\mathbf{F}$ is $\nabla \times \mathbf{F}$:

$$\nabla \times \mathbf{F} = \begin{vmatrix} \mathbf{i} & \mathbf{j} & \mathbf{k} \\ \dfrac{\partial}{\partial x} & \dfrac{\partial}{\partial y} & \dfrac{\partial}{\partial z} \\ M & N & P \end{vmatrix}$$

$$= \left(\frac{\partial P}{\partial y} - \frac{\partial N}{\partial z} \right)\mathbf{i} + \left(\frac{\partial M}{\partial z} - \frac{\partial P}{\partial x} \right)\mathbf{j} + \left(\frac{\partial N}{\partial x} - \frac{\partial M}{\partial y} \right)\mathbf{k}$$

$$= \text{curl } \mathbf{F}.$$

$$\text{curl } \mathbf{F} = \nabla \times \mathbf{F} \qquad (3)$$

EXAMPLE 1 Find the curl of $\mathbf{F} = (x^2 - z)\mathbf{i} + xe^z\mathbf{j} + xy\mathbf{k}$.

Solution We use Equation (3) and the determinant form, so

$$
\begin{aligned}
\text{curl } \mathbf{F} &= \nabla \times \mathbf{F} \\[4pt]
&= \begin{vmatrix}
\mathbf{i} & \mathbf{j} & \mathbf{k} \\[4pt]
\dfrac{\partial}{\partial x} & \dfrac{\partial}{\partial y} & \dfrac{\partial}{\partial z} \\[4pt]
x^2 - z & xe^z & xy
\end{vmatrix} \\[6pt]
&= \left(\frac{\partial}{\partial y}(xy) - \frac{\partial}{\partial z}(xe^z) \right)\mathbf{i} - \left(\frac{\partial}{\partial x}(xy) - \frac{\partial}{\partial z}(x^2 - z) \right)\mathbf{j} \\[6pt]
&\quad + \left(\frac{\partial}{\partial x}(xe^z) - \frac{\partial}{\partial y}(x^2 - z) \right)\mathbf{k} \\[6pt]
&= (x - xe^z)\mathbf{i} - (y + 1)\mathbf{j} + (e^z - 0)\mathbf{k} \\[4pt]
&= x(1 - e^z)\mathbf{i} - (y + 1)\mathbf{j} + e^z\mathbf{k} \qquad \blacksquare
\end{aligned}
$$

As we will see, the operator ∇ has a number of other applications. For instance, when applied to a scalar function $f(x, y, z)$, it gives the gradient of f:

$$\nabla f = \frac{\partial f}{\partial x}\mathbf{i} + \frac{\partial f}{\partial y}\mathbf{j} + \frac{\partial f}{\partial z}\mathbf{k}.$$

It is sometimes read as "del f" as well as "grad f."

Stokes' Theorem

Stokes' Theorem generalizes Green's Theorem to three dimensions. The circulation-curl form of Green's Theorem relates the counterclockwise circulation of a vector field around a simple closed curve C in the xy-plane to a double integral over the plane region R enclosed by C. Stokes' Theorem relates the circulation of a vector field around the boundary C of an oriented surface S in space (Figure 16.56) to a surface integral over the surface S. We require that the surface be **piecewise smooth**, which means that it is a finite union of smooth surfaces joining along smooth curves.

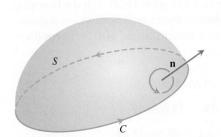

FIGURE 16.56 The orientation of the bounding curve C gives it a right-handed relation to the normal field $\mathbf{n}$. If the thumb of a right hand points along $\mathbf{n}$, the fingers curl in the direction of C.

THEOREM 6—Stokes' Theorem Let S be a piecewise smooth oriented surface having a piecewise smooth boundary curve C. Let $\mathbf{F} = M\mathbf{i} + N\mathbf{j} + P\mathbf{k}$ be a vector field whose components have continuous first partial derivatives on an open region containing S. Then the circulation of $\mathbf{F}$ around C in the direction counterclockwise with respect to the surface's unit normal vector $\mathbf{n}$ equals the integral of $\nabla \times \mathbf{F} \cdot \mathbf{n}$ over S.

$$\underbrace{\oint_C \mathbf{F} \cdot d\mathbf{r}}_{\substack{\text{Counterclockwise} \\ \text{circulation}}} = \underbrace{\iint_S \nabla \times \mathbf{F} \cdot \mathbf{n} \, d\sigma}_{\text{Curl integral}} \qquad (4)$$

Notice from Equation (4) that if two different oriented surfaces S_1 and S_2 have the same boundary C, their curl integrals are equal:

$$\iint_{S_1} \nabla \times \mathbf{F} \cdot \mathbf{n}_1 \, d\sigma = \iint_{S_2} \nabla \times \mathbf{F} \cdot \mathbf{n}_2 \, d\sigma.$$

Both curl integrals equal the counterclockwise circulation integral on the left side of Equation (4) as long as the unit normal vectors $\mathbf{n}_1$ and $\mathbf{n}_2$ correctly orient the surfaces.

If C is a curve in the xy-plane, oriented counterclockwise, and R is the region in the xy-plane bounded by C, then $d\sigma = dx\,dy$ and

$$(\nabla \times \mathbf{F}) \cdot \mathbf{n} = (\nabla \times \mathbf{F}) \cdot \mathbf{k} = \left(\frac{\partial N}{\partial x} - \frac{\partial M}{\partial y} \right).$$

Under these conditions, Stokes' equation becomes

$$\oint_C \mathbf{F} \cdot d\mathbf{r} = \iint_R \left(\frac{\partial N}{\partial x} - \frac{\partial M}{\partial y} \right) dx\,dy,$$

which is the circulation-curl form of the equation in Green's Theorem. Conversely, by reversing these steps we can rewrite the circulation-curl form of Green's Theorem for two-dimensional fields in del notation as

$$\oint_C \mathbf{F} \cdot d\mathbf{r} = \iint_R \nabla \times \mathbf{F} \cdot \mathbf{k} \, dA. \tag{5}$$

See Figure 16.57.

Green:

Stokes:

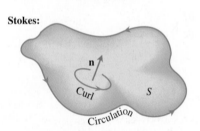

FIGURE 16.57 Comparison of Green's Theorem and Stokes' Theorem.

EXAMPLE 2 Evaluate Equation (4) for the hemisphere $S: x^2 + y^2 + z^2 = 9, z \geq 0$, its bounding circle $C: x^2 + y^2 = 9, z = 0$, and the field $\mathbf{F} = y\mathbf{i} - x\mathbf{j}$.

Solution The hemisphere looks much like the surface in Figure 16.56 with the bounding circle C in the xy-plane (see Figure 16.58). We calculate the counterclockwise circulation around C (as viewed from above) using the parametrization $\mathbf{r}(\theta) = (3\cos\theta)\mathbf{i} + (3\sin\theta)\mathbf{j}, 0 \leq \theta \leq 2\pi$:

$$d\mathbf{r} = (-3\sin\theta\,d\theta)\mathbf{i} + (3\cos\theta\,d\theta)\mathbf{j}$$

$$\mathbf{F} = y\mathbf{i} - x\mathbf{j} = (3\sin\theta)\mathbf{i} - (3\cos\theta)\mathbf{j}$$

$$\mathbf{F} \cdot d\mathbf{r} = -9\sin^2\theta\,d\theta - 9\cos^2\theta\,d\theta = -9\,d\theta$$

$$\oint_C \mathbf{F} \cdot d\mathbf{r} = \int_0^{2\pi} -9\,d\theta = -18\pi.$$

For the curl integral of $\mathbf{F}$, we have

$$\nabla \times \mathbf{F} = \left(\frac{\partial P}{\partial y} - \frac{\partial N}{\partial z} \right)\mathbf{i} + \left(\frac{\partial M}{\partial z} - \frac{\partial P}{\partial x} \right)\mathbf{j} + \left(\frac{\partial N}{\partial x} - \frac{\partial M}{\partial y} \right)\mathbf{k}$$

$$= (0 - 0)\mathbf{i} + (0 - 0)\mathbf{j} + (-1 - 1)\mathbf{k} = -2\mathbf{k}$$

$$\mathbf{n} = \frac{x\mathbf{i} + y\mathbf{j} + z\mathbf{k}}{\sqrt{x^2 + y^2 + z^2}} = \frac{x\mathbf{i} + y\mathbf{j} + z\mathbf{k}}{3} \qquad \text{Outer unit normal}$$

$$d\sigma = \frac{3}{z}\,dA \qquad \text{Section 16.6, Example 6, with } a = 3$$

$$\nabla \times \mathbf{F} \cdot \mathbf{n} \, d\sigma = -\frac{2z}{3} \frac{3}{z}\,dA = -2\,dA$$

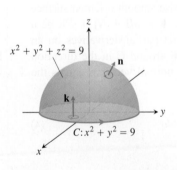

FIGURE 16.58 A hemisphere and a disk, each with boundary C (Examples 2 and 3).

and

$$\iint\limits_{S} \nabla \times \mathbf{F} \cdot \mathbf{n} \, d\sigma = \iint\limits_{x^2+y^2\leq 9} -2 \, dA = -18\pi.$$

The circulation around the circle equals the integral of the curl over the hemisphere, as it should. ∎

The surface integral in Stokes' Theorem can be computed using any surface having boundary curve *C*, provided the surface is properly oriented and lies within the domain of the field **F**. The next example illustrates this fact for the circulation around the curve *C* in Example 2.

EXAMPLE 3   Calculate the circulation around the bounding circle *C* in Example 2 using the disk of radius 3 centered at the origin in the *xy*-plane as the surface *S* (instead of the hemisphere). See Figure 16.58.

Solution   As in Example 2, $\nabla \times \mathbf{F} = -2\mathbf{k}$. For the surface being the described disk in the *xy*-plane, we have the normal vector $\mathbf{n} = \mathbf{k}$ so that

$$\nabla \times \mathbf{F} \cdot \mathbf{n} \, d\sigma = -2\mathbf{k} \cdot \mathbf{k} \, dA = -2 \, dA$$

and

$$\iint\limits_{S} \nabla \times \mathbf{F} \cdot \mathbf{n} \, d\sigma = \iint\limits_{x^2+y^2\leq 9} -2 \, dA = -18\pi,$$

a simpler calculation than before. ∎

EXAMPLE 4   Find the circulation of the field $\mathbf{F} = (x^2 - y)\mathbf{i} + 4z\mathbf{j} + x^2\mathbf{k}$ around the curve *C* in which the plane $z = 2$ meets the cone $z = \sqrt{x^2 + y^2}$, counterclockwise as viewed from above (Figure 16.59).

Solution   Stokes' Theorem enables us to find the circulation by integrating over the surface of the cone. Traversing *C* in the counterclockwise direction viewed from above corresponds to taking the *inner* normal **n** to the cone, the normal with a positive **k**-component.

We parametrize the cone as

$$\mathbf{r}(r, \theta) = (r\cos\theta)\mathbf{i} + (r\sin\theta)\mathbf{j} + r\mathbf{k}, \qquad 0 \leq r \leq 2, \quad 0 \leq \theta \leq 2\pi.$$

We then have

$$\mathbf{n} = \frac{\mathbf{r}_r \times \mathbf{r}_\theta}{|\mathbf{r}_r \times \mathbf{r}_\theta|} = \frac{-(r\cos\theta)\mathbf{i} - (r\sin\theta)\mathbf{j} + r\mathbf{k}}{r\sqrt{2}} \qquad \text{Section 16.5, Example 4}$$

$$= \frac{1}{\sqrt{2}}\left(-(\cos\theta)\mathbf{i} - (\sin\theta)\mathbf{j} + \mathbf{k}\right)$$

$$d\sigma = r\sqrt{2} \, dr \, d\theta \qquad \text{Section 16.5, Example 4}$$

$$\nabla \times \mathbf{F} = -4\mathbf{i} - 2x\mathbf{j} + \mathbf{k} \qquad \text{Example 1}$$

$$= -4\mathbf{i} - 2r\cos\theta\mathbf{j} + \mathbf{k}. \qquad x = r\cos\theta$$

Accordingly,

$$\nabla \times \mathbf{F} \cdot \mathbf{n} = \frac{1}{\sqrt{2}}\left(4\cos\theta + 2r\cos\theta\sin\theta + 1\right)$$

$$= \frac{1}{\sqrt{2}}\left(4\cos\theta + r\sin 2\theta + 1\right)$$

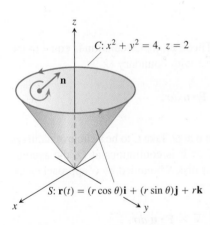

FIGURE 16.59 The curve *C* and cone *S* in Example 4.

and the circulation is

$$\oint_C \mathbf{F} \cdot d\mathbf{r} = \iint_S \nabla \times \mathbf{F} \cdot \mathbf{n} \, d\sigma \qquad \text{Stokes' Theorem, Eq. (4)}$$

$$= \int_0^{2\pi} \int_0^2 \frac{1}{\sqrt{2}} \left(4 \cos \theta + r \sin 2\theta + 1 \right) \left(r\sqrt{2} \, dr \, d\theta \right) = 4\pi. \qquad ∎$$

EXAMPLE 5   The cone used in Example 4 is not the easiest surface to use for calculating the circulation around the bounding circle C lying in the plane $z = 3$. If instead we use the flat disk of radius 3 centered on the z-axis and lying in the plane $z = 3$, then the normal vector to the surface S is $\mathbf{n} = \mathbf{k}$. Just as in the computation for Example 4, we still have $\nabla \times \mathbf{F} = -4\mathbf{i} - 2x\mathbf{j} + \mathbf{k}$. However, now we get $\nabla \times \mathbf{F} \cdot \mathbf{n} = 1$, so that

$$\iint_S \nabla \times \mathbf{F} \cdot \mathbf{n} \, d\sigma = \iint_{x^2 + y^2 \le 4} 1 \, dA = 4\pi. \qquad \text{The shadow is the disk of radius 2 in the } xy\text{-plane.}$$

This result agrees with the circulation value found in Example 4. $\qquad ∎$

Paddle Wheel Interpretation of $\nabla \times \mathbf{F}$

Suppose that $\mathbf{F}$ is the velocity field of a fluid moving in a region R in space containing the closed curve C. Then

$$\oint_C \mathbf{F} \cdot d\mathbf{r}$$

is the circulation of the fluid around C. By Stokes' Theorem, the circulation is equal to the flux of $\nabla \times \mathbf{F}$ through any suitably oriented surface S with boundary C:

$$\oint_C \mathbf{F} \cdot d\mathbf{r} = \iint_S \nabla \times \mathbf{F} \cdot \mathbf{n} \, d\sigma.$$

Suppose we fix a point Q in the region R and a direction $\mathbf{u}$ at Q. Take C to be a circle of radius ρ, with center at Q, whose plane is normal to $\mathbf{u}$. If $\nabla \times \mathbf{F}$ is continuous at Q, the average value of the $\mathbf{u}$-component of $\nabla \times \mathbf{F}$ over the circular disk S bounded by C approaches the $\mathbf{u}$-component of $\nabla \times \mathbf{F}$ at Q as the radius $\rho \to 0$:

$$(\nabla \times \mathbf{F} \cdot \mathbf{u})_Q = \lim_{\rho \to 0} \frac{1}{\pi \rho^2} \iint_S \nabla \times \mathbf{F} \cdot \mathbf{u} \, d\sigma.$$

If we apply Stokes' Theorem and replace the surface integral by a line integral over C, we get

$$(\nabla \times \mathbf{F} \cdot \mathbf{u})_Q = \lim_{\rho \to 0} \frac{1}{\pi \rho^2} \oint_C \mathbf{F} \cdot d\mathbf{r}. \qquad (6)$$

The left-hand side of Equation (6) has its maximum value when $\mathbf{u}$ is the direction of $\nabla \times \mathbf{F}$. When ρ is small, the limit on the right-hand side of Equation (6) is approximately

$$\frac{1}{\pi \rho^2} \oint_C \mathbf{F} \cdot d\mathbf{r},$$

which is the circulation around C divided by the area of the disk (circulation density). Suppose that a small paddle wheel of radius ρ is introduced into the fluid at Q, with its axle directed along $\mathbf{u}$ (Figure 16.60). The circulation of the fluid around C affects the rate

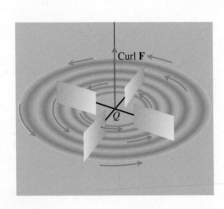

FIGURE 16.60 The paddle wheel interpretation of curl $\mathbf{F}$.

(c) Rotation about z-axis: $\mathbf{F}(x, y, z) = -y\mathbf{i} + x\mathbf{j}$

(d) Shearing along horizontal planes: $\mathbf{F}(x, y, z) = z\mathbf{j}$

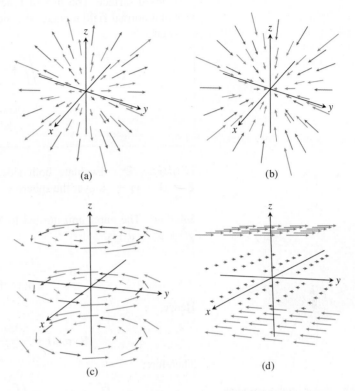

FIGURE 16.67   Velocity fields of a gas flowing in space (Example 1).

Solution

(a) $\operatorname{div} \mathbf{F} = \dfrac{\partial}{\partial x}(x) + \dfrac{\partial}{\partial y}(y) + \dfrac{\partial}{\partial z}(z) = 3$: The gas is undergoing uniform expansion at all points.

(b) $\operatorname{div} \mathbf{F} = \dfrac{\partial}{\partial x}(-x) + \dfrac{\partial}{\partial y}(-y) + \dfrac{\partial}{\partial z}(-z) = -3$: The gas is undergoing uniform compression at all points.

(c) $\operatorname{div} \mathbf{F} = \dfrac{\partial}{\partial x}(-y) + \dfrac{\partial}{\partial y}(x) = 0$: The gas is neither expanding nor compressing at any point.

(d) $\operatorname{div} \mathbf{F} = \dfrac{\partial}{\partial y}(z) = 0$: Again, the divergence is zero at all points in the domain of the velocity field, so the gas is neither expanding nor compressing at any point. ■

Divergence Theorem

The Divergence Theorem says that under suitable conditions, the outward flux of a vector field across a closed surface equals the triple integral of the divergence of the field over the region enclosed by the surface.

THEOREM 8—Divergence Theorem Let **F** be a vector field whose components have continuous first partial derivatives, and let S be a piecewise smooth oriented closed surface. The flux of **F** across S in the direction of the surface's outward unit normal field **n** equals the integral of $\nabla \cdot \mathbf{F}$ over the region D enclosed by the surface:

$$\iint_S \mathbf{F} \cdot \mathbf{n} \, d\sigma = \iiint_D \nabla \cdot \mathbf{F} \, dV. \qquad (2)$$

Outward   Divergence
flux   integral

EXAMPLE 2 Evaluate both sides of Equation (2) for the expanding vector field $\mathbf{F} = x\mathbf{i} + y\mathbf{j} + z\mathbf{k}$ over the sphere $x^2 + y^2 + z^2 = a^2$ (Figure 16.68).

Solution The outer unit normal to S, calculated from the gradient of $f(x, y, z) = x^2 + y^2 + z^2 - a^2$, is

$$\mathbf{n} = \frac{2(x\mathbf{i} + y\mathbf{j} + z\mathbf{k})}{\sqrt{4(x^2 + y^2 + z^2)}} = \frac{x\mathbf{i} + y\mathbf{j} + z\mathbf{k}}{a}. \quad x^2 + y^2 + z^2 = a^2 \text{ on } S$$

Hence,

$$\mathbf{F} \cdot \mathbf{n} \, d\sigma = \frac{x^2 + y^2 + z^2}{a} \, d\sigma = \frac{a^2}{a} \, d\sigma = a \, d\sigma.$$

Therefore,

$$\iint_S \mathbf{F} \cdot \mathbf{n} \, d\sigma = \iint_S a \, d\sigma = a \iint_S d\sigma = a(4\pi a^2) = 4\pi a^3. \quad \text{Area of } S \text{ is } 4\pi a^2.$$

The divergence of **F** is

$$\nabla \cdot \mathbf{F} = \frac{\partial}{\partial x}(x) + \frac{\partial}{\partial y}(y) + \frac{\partial}{\partial z}(z) = 3,$$

so

$$\iiint_D \nabla \cdot \mathbf{F} \, dV = \iiint_D 3 \, dV = 3\left(\frac{4}{3}\pi a^3\right) = 4\pi a^3. \qquad \blacksquare$$

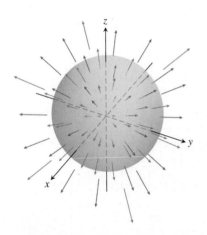

FIGURE 16.68 A uniformly expanding vector field and a sphere (Example 2).

EXAMPLE 3 Find the flux of $\mathbf{F} = xy\mathbf{i} + yz\mathbf{j} + xz\mathbf{k}$ outward through the surface of the cube cut from the first octant by the planes $x = 1$, $y = 1$, and $z = 1$.

Solution Instead of calculating the flux as a sum of six separate integrals, one for each face of the cube, we can calculate the flux by integrating the divergence

$$\nabla \cdot \mathbf{F} = \frac{\partial}{\partial x}(xy) + \frac{\partial}{\partial y}(yz) + \frac{\partial}{\partial z}(xz) = y + z + x$$

over the cube's interior:

$$\text{Flux} = \iint_{\substack{\text{Cube}\\\text{surface}}} \mathbf{F} \cdot \mathbf{n} \, d\sigma = \iiint_{\substack{\text{Cube}\\\text{interior}}} \nabla \cdot \mathbf{F} \, dV \qquad \text{The Divergence Theorem}$$

$$= \int_0^1 \int_0^1 \int_0^1 (x + y + z) \, dx \, dy \, dz = \frac{3}{2}. \qquad \text{Routine integration} \qquad \blacksquare$$

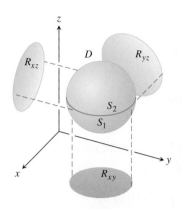

FIGURE 16.69 We prove the Divergence Theorem for the kind of three-dimensional region shown here.

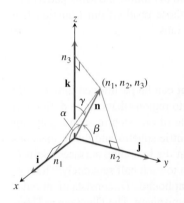

FIGURE 16.70 The components of $\mathbf{n}$ are the cosines of the angles α, β, and γ that it makes with $\mathbf{i}$, $\mathbf{j}$, and $\mathbf{k}$.

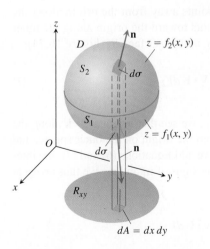

FIGURE 16.71 The region D enclosed by the surfaces S_1 and S_2 projects vertically onto R_{xy} in the xy-plane.

Proof of the Divergence Theorem for Special Regions

To prove the Divergence Theorem, we take the components of $\mathbf{F}$ to have continuous first partial derivatives. We first assume that D is a convex region with no holes or bubbles, such as a solid ball, cube, or ellipsoid, and that S is a piecewise smooth surface. In addition, we assume that any line perpendicular to the xy-plane at an interior point of the region R_{xy} that is the projection of D on the xy-plane intersects the surface S in exactly two points, producing surfaces

$$S_1: \quad z = f_1(x, y), \qquad (x, y) \text{ in } R_{xy}$$
$$S_2: \quad z = f_2(x, y), \qquad (x, y) \text{ in } R_{xy},$$

with $f_1 \leq f_2$. We make similar assumptions about the projection of D onto the other coordinate planes. See Figure 16.69.

The components of the unit normal vector $\mathbf{n} = n_1\mathbf{i} + n_2\mathbf{j} + n_3\mathbf{k}$ are the cosines of the angles α, β, and γ that $\mathbf{n}$ makes with $\mathbf{i}$, $\mathbf{j}$, and $\mathbf{k}$ (Figure 16.70). This is true because all the vectors involved are unit vectors. We have

$$n_1 = \mathbf{n} \cdot \mathbf{i} = |\mathbf{n}||\mathbf{i}| \cos \alpha = \cos \alpha$$
$$n_2 = \mathbf{n} \cdot \mathbf{j} = |\mathbf{n}||\mathbf{j}| \cos \beta = \cos \beta$$
$$n_3 = \mathbf{n} \cdot \mathbf{k} = |\mathbf{n}||\mathbf{k}| \cos \gamma = \cos \gamma.$$

Thus,

$$\mathbf{n} = (\cos \alpha)\mathbf{i} + (\cos \beta)\mathbf{j} + (\cos \gamma)\mathbf{k}$$

and

$$\mathbf{F} \cdot \mathbf{n} = M \cos \alpha + N \cos \beta + P \cos \gamma.$$

In component form, the Divergence Theorem states that

$$\iint_S \underbrace{(M \cos \alpha + N \cos \beta + P \cos \gamma)}_{\mathbf{F} \cdot \mathbf{n}} \, d\sigma = \iiint_D \underbrace{\left(\frac{\partial M}{\partial x} + \frac{\partial N}{\partial y} + \frac{\partial P}{\partial z} \right)}_{\text{div } \mathbf{F}} dx \, dy \, dz.$$

We prove the theorem by proving the three following equalities:

$$\iint_S M \cos \alpha \, d\sigma = \iiint_D \frac{\partial M}{\partial x} \, dx \, dy \, dz \tag{3}$$

$$\iint_S N \cos \beta \, d\sigma = \iiint_D \frac{\partial N}{\partial y} \, dx \, dy \, dz \tag{4}$$

$$\iint_S P \cos \gamma \, d\sigma = \iiint_D \frac{\partial P}{\partial z} \, dx \, dy \, dz \tag{5}$$

Proof of Equation (5) We prove Equation (5) by converting the surface integral on the left to a double integral over the projection R_{xy} of D on the xy-plane (Figure 16.71). The surface S consists of an upper part S_2 whose equation is $z = f_2(x, y)$ and a lower part S_1 whose equation is $z = f_1(x, y)$. On S_2, the outer normal $\mathbf{n}$ has a positive $\mathbf{k}$-component and

$$\cos \gamma \, d\sigma = dx \, dy \qquad \text{because} \qquad d\sigma = \frac{dA}{|\cos \gamma|} = \frac{dx \, dy}{\cos \gamma}.$$

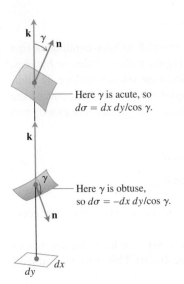

FIGURE 16.72 An enlarged view of the area patches in Figure 16.71. The relations $d\sigma = \pm dx\, dy/\cos\gamma$ come from Eq. (7) in Section 16.5.

See Figure 16.72. On S_1, the outer normal $\mathbf{n}$ has a negative $\mathbf{k}$-component and

$$\cos\gamma\, d\sigma = -dx\, dy.$$

Therefore,

$$\iint_S P\cos\gamma\, d\sigma = \iint_{S_2} P\cos\gamma\, d\sigma + \iint_{S_1} P\cos\gamma\, d\sigma$$

$$= \iint_{R_{xy}} P(x, y, f_2(x, y))\, dx\, dy - \iint_{R_{xy}} P(x, y, f_1(x, y))\, dx\, dy$$

$$= \iint_{R_{xy}} [P(x, y, f_2(x, y)) - P(x, y, f_1(x, y))]\, dx\, dy$$

$$= \iint_{R_{xy}} \left[\int_{f_1(x,y)}^{f_2(x,y)} \frac{\partial P}{\partial z}\, dz \right] dx\, dy = \iiint_D \frac{\partial P}{\partial z}\, dz\, dx\, dy.$$

This proves Equation (5). The proofs for Equations (3) and (4) follow the same pattern; or just permute x, y, z; M, N, P; α, β, γ, in order, and get those results from Equation (5). This proves the Divergence Theorem for these special regions. ∎

Divergence Theorem for Other Regions

The Divergence Theorem can be extended to regions that can be partitioned into a finite number of simple regions of the type just discussed and to regions that can be defined as limits of simpler regions in certain ways. For an example of one step in such a splitting process, suppose that D is the region between two concentric spheres and that $\mathbf{F}$ has continuously differentiable components throughout D and on the bounding surfaces. Split D by an equatorial plane and apply the Divergence Theorem to each half separately. The bottom half, D_1, is shown in Figure 16.73. The surface S_1 that bounds D_1 consists of an outer hemisphere, a plane washer-shaped base, and an inner hemisphere. The Divergence Theorem says that

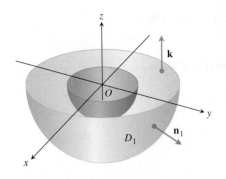

FIGURE 16.73 The lower half of the solid region between two concentric spheres.

$$\iint_{S_1} \mathbf{F} \cdot \mathbf{n}_1\, d\sigma_1 = \iiint_{D_1} \nabla \cdot \mathbf{F}\, dV_1. \tag{6}$$

The unit normal $\mathbf{n}_1$ that points outward from D_1 points away from the origin along the outer surface, equals $\mathbf{k}$ along the flat base, and points toward the origin along the inner surface. Next apply the Divergence Theorem to D_2, and its surface S_2 (Figure 16.74):

$$\iint_{S_2} \mathbf{F} \cdot \mathbf{n}_2\, d\sigma_2 = \iiint_{D_2} \nabla \cdot \mathbf{F}\, dV_2. \tag{7}$$

As we follow $\mathbf{n}_2$ over S_2, pointing outward from D_2, we see that $\mathbf{n}_2$ equals $-\mathbf{k}$ along the washer-shaped base in the xy-plane, points away from the origin on the outer sphere, and points toward the origin on the inner sphere. When we add Equations (6) and (7), the integrals over the flat base cancel because of the opposite signs of $\mathbf{n}_1$ and $\mathbf{n}_2$. We thus arrive at the result

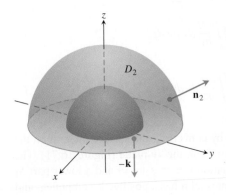

FIGURE 16.74 The upper half of the solid region between two concentric spheres.

$$\iint_S \mathbf{F} \cdot \mathbf{n}\, d\sigma = \iiint_D \nabla \cdot \mathbf{F}\, dV,$$

with D the region between the spheres, S the boundary of D consisting of two spheres, and $\mathbf{n}$ the unit normal to S directed outward from D.

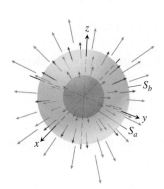

FIGURE 16.75 Two concentric spheres in an expanding vector field.

EXAMPLE 4   Find the net outward flux of the field

$$\mathbf{F} = \frac{x\mathbf{i} + y\mathbf{j} + z\mathbf{k}}{\rho^3}, \qquad \rho = \sqrt{x^2 + y^2 + z^2}$$

across the boundary of the region D: $0 < a^2 \le x^2 + y^2 + z^2 \le b^2$ (Figure 16.75).

Solution   The flux can be calculated by integrating $\nabla \cdot \mathbf{F}$ over D. We have

$$\frac{\partial \rho}{\partial x} = \frac{1}{2}(x^2 + y^2 + z^2)^{-1/2}(2x) = \frac{x}{\rho}$$

and

$$\frac{\partial M}{\partial x} = \frac{\partial}{\partial x}(x\rho^{-3}) = \rho^{-3} - 3x\rho^{-4}\frac{\partial \rho}{\partial x} = \frac{1}{\rho^3} - \frac{3x^2}{\rho^5}.$$

Similarly,

$$\frac{\partial N}{\partial y} = \frac{1}{\rho^3} - \frac{3y^2}{\rho^5} \qquad \text{and} \qquad \frac{\partial P}{\partial z} = \frac{1}{\rho^3} - \frac{3z^2}{\rho^5}.$$

Hence,

$$\text{div } \mathbf{F} = \frac{3}{\rho^3} - \frac{3}{\rho^5}(x^2 + y^2 + z^2) = \frac{3}{\rho^3} - \frac{3\rho^2}{\rho^5} = 0$$

and

$$\iiint_D \nabla \cdot \mathbf{F} \, dV = 0. \qquad \nabla \cdot \mathbf{F} = \text{div } \mathbf{F}$$

So the integral of $\nabla \cdot \mathbf{F}$ over D is zero and the net outward flux across the boundary of D is zero. There is more to learn from this example, though. The flux leaving D across the inner sphere S_a is the negative of the flux leaving D across the outer sphere S_b (because the sum of these fluxes is zero). Hence, the flux of $\mathbf{F}$ across S_a in the direction away from the origin equals the flux of $\mathbf{F}$ across S_b in the direction away from the origin. Thus, the flux of $\mathbf{F}$ across a sphere centered at the origin is independent of the radius of the sphere. What is this flux?

To find it, we evaluate the flux integral directly. The outward unit normal on the sphere of radius a is

$$\mathbf{n} = \frac{x\mathbf{i} + y\mathbf{j} + z\mathbf{k}}{\sqrt{x^2 + y^2 + z^2}} = \frac{x\mathbf{i} + y\mathbf{j} + z\mathbf{k}}{a}.$$

Hence, on the sphere,

$$\mathbf{F} \cdot \mathbf{n} = \frac{x\mathbf{i} + y\mathbf{j} + z\mathbf{k}}{a^3} \cdot \frac{x\mathbf{i} + y\mathbf{j} + z\mathbf{k}}{a} = \frac{x^2 + y^2 + z^2}{a^4} = \frac{a^2}{a^4} = \frac{1}{a^2}$$

and

$$\iint_{S_a} \mathbf{F} \cdot \mathbf{n} \, d\sigma = \frac{1}{a^2}\iint_{S_a} d\sigma = \frac{1}{a^2}(4\pi a^2) = 4\pi.$$

The outward flux of $\mathbf{F}$ across any sphere centered at the origin is 4π. ∎

Gauss's Law: One of the Four Great Laws of Electromagnetic Theory

There is still more to be learned from Example 4. In electromagnetic theory, the electric field created by a point charge q located at the origin is

$$\mathbf{E}(x, y, z) = \frac{1}{4\pi\epsilon_0} \frac{q}{|\mathbf{r}|^2} \left(\frac{\mathbf{r}}{|\mathbf{r}|}\right) = \frac{q}{4\pi\epsilon_0} \frac{\mathbf{r}}{|\mathbf{r}|^3} = \frac{q}{4\pi\epsilon_0} \frac{x\mathbf{i} + y\mathbf{j} + z\mathbf{k}}{\rho^3},$$

where ϵ_0 is a physical constant, $\mathbf{r}$ is the position vector of the point (x, y, z), and $\rho = |\mathbf{r}| = \sqrt{x^2 + y^2 + z^2}$. In the notation of Example 4,

$$\mathbf{E} = \frac{q}{4\pi\epsilon_0} \mathbf{F}.$$

The calculations in Example 4 show that the outward flux of $\mathbf{E}$ across any sphere centered at the origin is q/ϵ_0, but this result is not confined to spheres. The outward flux of $\mathbf{E}$ across any closed surface S that encloses the origin (and to which the Divergence Theorem applies) is also q/ϵ_0. To see why, we have only to imagine a large sphere S_a centered at the origin and enclosing the surface S (see Figure 16.76). Since

$$\nabla \cdot \mathbf{E} = \nabla \cdot \frac{q}{4\pi\epsilon_0} \mathbf{F} = \frac{q}{4\pi\epsilon_0} \nabla \cdot \mathbf{F} = 0$$

when $\rho > 0$, the integral of $\nabla \cdot \mathbf{E}$ over the region D between S and S_a is zero. Hence, by the Divergence Theorem,

$$\iint\limits_{\substack{\text{Boundary} \\ \text{of } D}} \mathbf{E} \cdot \mathbf{n} \, d\sigma = 0,$$

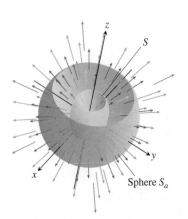

FIGURE 16.76 A sphere S_a surrounding another surface S. The tops of the surfaces are removed for visualization.

and the flux of $\mathbf{E}$ across S in the direction away from the origin must be the same as the flux of $\mathbf{E}$ across S_a in the direction away from the origin, which is q/ϵ_0. This statement, called *Gauss's Law*, also applies to charge distributions that are more general than the one assumed here, as you will see in nearly any physics text.

$$\text{Gauss's Law:} \quad \iint\limits_{S} \mathbf{E} \cdot \mathbf{n} \, d\sigma = \frac{q}{\epsilon_0}$$

Continuity Equation of Hydrodynamics

Let D be a region in space bounded by a closed oriented surface S. If $\mathbf{v}(x, y, z)$ is the velocity field of a fluid flowing smoothly through D, $\delta = \delta(t, x, y, z)$ is the fluid's density at (x, y, z) at time t, and $\mathbf{F} = \delta\mathbf{v}$, then the **continuity equation** of hydrodynamics states that

$$\nabla \cdot \mathbf{F} + \frac{\partial \delta}{\partial t} = 0.$$

If the functions involved have continuous first partial derivatives, the equation evolves naturally from the Divergence Theorem, as we now see.

First, the integral

$$\iint\limits_{S} \mathbf{F} \cdot \mathbf{n} \, d\sigma$$

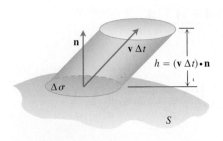

FIGURE 16.77 The fluid that flows upward through the patch $\Delta\sigma$ in a short time Δt fills a "cylinder" whose volume is approximately base $\times$ height $= \mathbf{v} \cdot \mathbf{n} \, \Delta\sigma \, \Delta t$.

is the rate at which mass leaves D across S (leaves because $\mathbf{n}$ is the outer normal). To see why, consider a patch of area $\Delta\sigma$ on the surface (Figure 16.77). In a short time interval Δt, the volume ΔV of fluid that flows across the patch is approximately equal to the volume of a cylinder with base area $\Delta\sigma$ and height $(\mathbf{v}\Delta t) \cdot \mathbf{n}$, where $\mathbf{v}$ is a velocity vector rooted at a point of the patch:

$$\Delta V \approx \mathbf{v} \cdot \mathbf{n} \, \Delta\sigma \, \Delta t.$$

The mass of this volume of fluid is about

$$\Delta m \approx \delta \mathbf{v} \cdot \mathbf{n} \, \Delta \sigma \, \Delta t,$$

so the rate at which mass is flowing out of D across the patch is about

$$\frac{\Delta m}{\Delta t} \approx \delta \mathbf{v} \cdot \mathbf{n} \, \Delta \sigma.$$

This leads to the approximation

$$\frac{\sum \Delta m}{\Delta t} \approx \sum \delta \mathbf{v} \cdot \mathbf{n} \, \Delta \sigma$$

as an estimate of the average rate at which mass flows across S. Finally, letting $\Delta \sigma \to 0$ and $\Delta t \to 0$ gives the instantaneous rate at which mass leaves D across S as

$$\frac{dm}{dt} = \iint_S \delta \mathbf{v} \cdot \mathbf{n} \, d\sigma,$$

which for our particular flow is

$$\frac{dm}{dt} = \iint_S \mathbf{F} \cdot \mathbf{n} \, d\sigma.$$

Now let B be a solid sphere centered at a point Q in the flow. The average value of $\nabla \cdot \mathbf{F}$ over B is

$$\frac{1}{\text{volume of } B} \iiint_B \nabla \cdot \mathbf{F} \, dV.$$

It is a consequence of the continuity of the divergence that $\nabla \cdot \mathbf{F}$ actually takes on this value at some point P in B. Thus,

$$(\nabla \cdot \mathbf{F})_P = \frac{1}{\text{volume of } B} \iiint_B \nabla \cdot \mathbf{F} \, dV = \frac{\displaystyle\iint_S \mathbf{F} \cdot \mathbf{n} \, d\sigma}{\text{volume of } B}$$

$$= \frac{\text{rate at which mass leaves } B \text{ across its surface } S}{\text{volume of } B}. \qquad (8)$$

The last term of the equation describes decrease in mass per unit volume.

Now let the radius of B approach zero while the center Q stays fixed. The left side of Equation (8) converges to $(\nabla \cdot \mathbf{F})_Q$, the right side to $(-\partial \delta / \partial t)_Q$. The equality of these two limits is the continuity equation

$$\nabla \cdot \mathbf{F} = -\frac{\partial \delta}{\partial t}.$$

The continuity equation "explains" $\nabla \cdot \mathbf{F}$: The divergence of $\mathbf{F}$ at a point is the rate at which the density of the fluid is decreasing there. The Divergence Theorem

$$\iint_S \mathbf{F} \cdot \mathbf{n} \, d\sigma = \iiint_D \nabla \cdot \mathbf{F} \, dV$$

now says that the net decrease in density of the fluid in region D is accounted for by the mass transported across the surface S. So, the theorem is a statement about conservation of mass (Exercise 31).

Unifying the Integral Theorems

If we think of a two-dimensional field $\mathbf{F} = M(x, y)\mathbf{i} + N(x, y)\mathbf{j}$ as a three-dimensional field whose $\mathbf{k}$-component is zero, then $\nabla \cdot \mathbf{F} = (\partial M/\partial x) + (\partial N/\partial y)$ and the normal form of Green's Theorem can be written as

$$\oint_C \mathbf{F} \cdot \mathbf{n} \, ds = \iint_R \left(\frac{\partial M}{\partial x} + \frac{\partial N}{\partial y} \right) dx \, dy = \iint_R \nabla \cdot \mathbf{F} \, dA.$$

Similarly, $\nabla \times \mathbf{F} \cdot \mathbf{k} = (\partial N/\partial x) - (\partial M/\partial y)$, so the tangential form of Green's Theorem can be written as

$$\oint_C \mathbf{F} \cdot d\mathbf{r} = \iint_R \left(\frac{\partial N}{\partial x} - \frac{\partial M}{\partial y} \right) dx \, dy = \iint_R \nabla \times \mathbf{F} \cdot \mathbf{k} \, dA.$$

With the equations of Green's Theorem now in del notation, we can see their relationships to the equations in Stokes' Theorem and the Divergence Theorem.

Green's Theorem and Its Generalization to Three Dimensions

Normal form of Green's Theorem: $\qquad \oint_C \mathbf{F} \cdot \mathbf{n} \, ds = \iint_R \nabla \cdot \mathbf{F} \, dA$

Divergence Theorem: $\qquad \iint_S \mathbf{F} \cdot \mathbf{n} \, d\sigma = \iiint_D \nabla \cdot \mathbf{F} \, dV$

Tangential form of Green's Theorem: $\qquad \oint_C \mathbf{F} \cdot d\mathbf{r} = \iint_R \nabla \times \mathbf{F} \cdot \mathbf{k} \, dA$

Stokes' Theorem: $\qquad \oint_C \mathbf{F} \cdot d\mathbf{r} = \iint_S \nabla \times \mathbf{F} \cdot \mathbf{n} \, d\sigma$

Notice how Stokes' Theorem generalizes the tangential (curl) form of Green's Theorem from a flat surface in the plane to a surface in three-dimensional space. In each case, the integral of the normal component of curl $\mathbf{F}$ over the interior of the surface equals the circulation of $\mathbf{F}$ around the boundary.

Likewise, the Divergence Theorem generalizes the normal (flux) form of Green's Theorem from a two-dimensional region in the plane to a three-dimensional region in space. In each case, the integral of $\nabla \cdot \mathbf{F}$ over the interior of the region equals the total flux of the field across the boundary.

There is still more to be learned here. All these results can be thought of as forms of a *single fundamental theorem*. Think back to the Fundamental Theorem of Calculus in Section 5.4. It says that if $f(x)$ is differentiable on (a, b) and continuous on $[a, b]$, then

$$\int_a^b \frac{df}{dx} \, dx = f(b) - f(a).$$

If we let $\mathbf{F} = f(x)\mathbf{i}$ throughout $[a, b]$, then $(df/dx) = \nabla \cdot \mathbf{F}$. If we define the unit vector field $\mathbf{n}$ normal to the boundary of $[a, b]$ to be $\mathbf{i}$ at b and $-\mathbf{i}$ at a (Figure 16.78), then

$$f(b) - f(a) = f(b)\mathbf{i} \cdot (\mathbf{i}) + f(a)\mathbf{i} \cdot (-\mathbf{i})$$
$$= \mathbf{F}(b) \cdot \mathbf{n} + \mathbf{F}(a) \cdot \mathbf{n}$$
$$= \text{total outward flux of } \mathbf{F} \text{ across the boundary of } [a, b].$$

FIGURE 16.78 The outward unit normals at the boundary of $[a, b]$ in one-dimensional space.

The Fundamental Theorem now says that

$$\mathbf{F}(b) \cdot \mathbf{n} + \mathbf{F}(a) \cdot \mathbf{n} = \int_{[a,b]} \nabla \cdot \mathbf{F} \, dx.$$

The Fundamental Theorem of Calculus, the normal form of Green's Theorem, and the Divergence Theorem all say that the integral of the differential operator $\nabla \cdot$ operating on a field $\mathbf{F}$ over a region equals the sum of the normal field components over the boundary of the region. (Here we are interpreting the line integral in Green's Theorem and the surface integral in the Divergence Theorem as "sums" over the boundary.)

Stokes' Theorem and the tangential form of Green's Theorem say that, when things are properly oriented, the integral of the normal component of the curl operating on a field equals the sum of the tangential field components on the boundary of the surface.

The beauty of these interpretations is the observance of a single unifying principle, which we might state as follows.

A Unifying Fundamental Theorem

The integral of a differential operator acting on a field over a region equals the sum of the field components appropriate to the operator over the boundary of the region.

Exercises 16.8

Calculating Divergence

In Exercises 1–4, find the divergence of the field.

1. The spin field in Figure 16.12

2. The radial field in Figure 16.11

3. The gravitational field in Figure 16.8 and Exercise 38a in Section 16.3

4. The velocity field in Figure 16.13

Calculating Flux Using the Divergence Theorem

In Exercises 5–16, use the Divergence Theorem to find the outward flux of $\mathbf{F}$ across the boundary of the region D.

5. **Cube** $\quad \mathbf{F} = (y - x)\mathbf{i} + (z - y)\mathbf{j} + (y - x)\mathbf{k}$

   D: The cube bounded by the planes $x = \pm 1$, $y = \pm 1$, and $z = \pm 1$

6. $\mathbf{F} = x^2\mathbf{i} + y^2\mathbf{j} + z^2\mathbf{k}$

   a. **Cube** $\quad D$: The cube cut from the first octant by the planes $x = 1, y = 1$, and $z = 1$

   b. **Cube** $\quad D$: The cube bounded by the planes $x = \pm 1$, $y = \pm 1$, and $z = \pm 1$

   c. **Cylindrical can** $\quad D$: The region cut from the solid cylinder $x^2 + y^2 \le 4$ by the planes $z = 0$ and $z = 1$

7. **Cylinder and paraboloid** $\quad \mathbf{F} = y\mathbf{i} + xy\mathbf{j} - z\mathbf{k}$

   D: The region inside the solid cylinder $x^2 + y^2 \le 4$ between the plane $z = 0$ and the paraboloid $z = x^2 + y^2$

8. **Sphere** $\quad \mathbf{F} = x^2\mathbf{i} + xz\mathbf{j} + 3z\mathbf{k}$

   D: The solid sphere $x^2 + y^2 + z^2 \le 4$

9. **Portion of sphere** $\quad \mathbf{F} = x^2\mathbf{i} - 2xy\mathbf{j} + 3xz\mathbf{k}$

   D: The region cut from the first octant by the sphere $x^2 + y^2 + z^2 = 4$

10. **Cylindrical can** $\quad \mathbf{F} = (6x^2 + 2xy)\mathbf{i} + (2y + x^2z)\mathbf{j} + 4x^2y^3\mathbf{k}$

    D: The region cut from the first octant by the cylinder $x^2 + y^2 = 4$ and the plane $z = 3$

11. **Wedge** $\quad \mathbf{F} = 2xz\mathbf{i} - xy\mathbf{j} - z^2\mathbf{k}$

    D: The wedge cut from the first octant by the plane $y + z = 4$ and the elliptical cylinder $4x^2 + y^2 = 16$

12. **Sphere** $\quad \mathbf{F} = x^3\mathbf{i} + y^3\mathbf{j} + z^3\mathbf{k}$

    D: The solid sphere $x^2 + y^2 + z^2 \le a^2$

13. **Thick sphere** $\quad \mathbf{F} = \sqrt{x^2 + y^2 + z^2}\,(x\mathbf{i} + y\mathbf{j} + z\mathbf{k})$

    D: The region $1 \le x^2 + y^2 + z^2 \le 2$

14. **Thick sphere** $\quad \mathbf{F} = (x\mathbf{i} + y\mathbf{j} + z\mathbf{k})/\sqrt{x^2 + y^2 + z^2}$

    D: The region $1 \le x^2 + y^2 + z^2 \le 4$

15. **Thick sphere** $\quad \mathbf{F} = (5x^3 + 12xy^2)\mathbf{i} + (y^3 + e^y \sin z)\mathbf{j} + (5z^3 + e^y \cos z)\mathbf{k}$

    D: The solid region between the spheres $x^2 + y^2 + z^2 = 1$ and $x^2 + y^2 + z^2 = 2$

16. **Thick cylinder** $\quad \mathbf{F} = \ln(x^2 + y^2)\mathbf{i} - \left(\dfrac{2z}{x} \tan^{-1}\dfrac{y}{x}\right)\mathbf{j} + z\sqrt{x^2 + y^2}\,\mathbf{k}$

    D: The thick-walled cylinder $1 \le x^2 + y^2 \le 2$, $\quad -1 \le z \le 2$

Properties of Curl and Divergence

17. div (curl G) is zero

a. Show that if the necessary partial derivatives of the components of the field $\mathbf{G} = M\mathbf{i} + N\mathbf{j} + P\mathbf{k}$ are continuous, then $\nabla \cdot \nabla \times \mathbf{G} = 0$.

b. What, if anything, can you conclude about the flux of the field $\nabla \times \mathbf{G}$ across a closed surface? Give reasons for your answer.

18. Let $\mathbf{F}_1$ and $\mathbf{F}_2$ be differentiable vector fields and let a and b be arbitrary real constants. Verify the following identities.

a. $\nabla \cdot (a\mathbf{F}_1 + b\mathbf{F}_2) = a\nabla \cdot \mathbf{F}_1 + b\nabla \cdot \mathbf{F}_2$

b. $\nabla \times (a\mathbf{F}_1 + b\mathbf{F}_2) = a\nabla \times \mathbf{F}_1 + b\nabla \times \mathbf{F}_2$

c. $\nabla \cdot (\mathbf{F}_1 \times \mathbf{F}_2) = \mathbf{F}_2 \cdot \nabla \times \mathbf{F}_1 - \mathbf{F}_1 \cdot \nabla \times \mathbf{F}_2$

19. Let $\mathbf{F}$ be a differentiable vector field and let $g(x, y, z)$ be a differentiable scalar function. Verify the following identities.

a. $\nabla \cdot (g\mathbf{F}) = g\nabla \cdot \mathbf{F} + \nabla g \cdot \mathbf{F}$

b. $\nabla \times (g\mathbf{F}) = g\nabla \times \mathbf{F} + \nabla g \times \mathbf{F}$

20. If $\mathbf{F} = M\mathbf{i} + N\mathbf{j} + P\mathbf{k}$ is a differentiable vector field, we define the notation $\mathbf{F} \cdot \nabla$ to mean

$$M\frac{\partial}{\partial x} + N\frac{\partial}{\partial y} + P\frac{\partial}{\partial z}.$$

For differentiable vector fields $\mathbf{F}_1$ and $\mathbf{F}_2$, verify the following identities.

a. $\nabla \times (\mathbf{F}_1 \times \mathbf{F}_2) = (\mathbf{F}_2 \cdot \nabla)\mathbf{F}_1 - (\mathbf{F}_1 \cdot \nabla)\mathbf{F}_2 + (\nabla \cdot \mathbf{F}_2)\mathbf{F}_1 - (\nabla \cdot \mathbf{F}_1)\mathbf{F}_2$

b. $\nabla(\mathbf{F}_1 \cdot \mathbf{F}_2) = (\mathbf{F}_1 \cdot \nabla)\mathbf{F}_2 + (\mathbf{F}_2 \cdot \nabla)\mathbf{F}_1 + \mathbf{F}_1 \times (\nabla \times \mathbf{F}_2) + \mathbf{F}_2 \times (\nabla \times \mathbf{F}_1)$

Theory and Examples

21. Let $\mathbf{F}$ be a field whose components have continuous first partial derivatives throughout a portion of space containing a region D bounded by a smooth closed surface S. If $|\mathbf{F}| \leq 1$, can any bound be placed on the size of

$$\iiint_D \nabla \cdot \mathbf{F} \, dV?$$

Give reasons for your answer.

22. The base of the closed cubelike surface shown here is the unit square in the xy-plane. The four sides lie in the planes $x = 0$, $x = 1$, $y = 0$, and $y = 1$. The top is an arbitrary smooth surface whose identity is unknown. Let $\mathbf{F} = x\mathbf{i} - 2y\mathbf{j} + (z + 3)\mathbf{k}$ and suppose the outward flux of $\mathbf{F}$ through Side A is 1 and through Side B is -3. Can you conclude anything about the outward flux through the top? Give reasons for your answer.

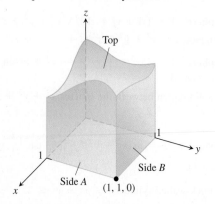

23. a. Show that the outward flux of the position vector field $\mathbf{F} = x\mathbf{i} + y\mathbf{j} + z\mathbf{k}$ through a smooth closed surface S is three times the volume of the region enclosed by the surface.

b. Let $\mathbf{n}$ be the outward unit normal vector field on S. Show that it is not possible for $\mathbf{F}$ to be orthogonal to $\mathbf{n}$ at every point of S.

24. Maximum flux Among all rectangular solids defined by the inequalities $0 \leq x \leq a, 0 \leq y \leq b, 0 \leq z \leq 1$, find the one for which the total flux of $\mathbf{F} = (-x^2 - 4xy)\mathbf{i} - 6yz\mathbf{j} + 12z\mathbf{k}$ outward through the six sides is greatest. What *is* the greatest flux?

25. Volume of a solid region Let $\mathbf{F} = x\mathbf{i} + y\mathbf{j} + z\mathbf{k}$ and suppose that the surface S and region D satisfy the hypotheses of the Divergence Theorem. Show that the volume of D is given by the formula

$$\text{Volume of } D = \frac{1}{3}\iint_S \mathbf{F} \cdot \mathbf{n} \, d\sigma.$$

26. Outward flux of a constant field Show that the outward flux of a constant vector field $\mathbf{F} = \mathbf{C}$ across any closed surface to which the Divergence Theorem applies is zero.

27. Harmonic functions A function $f(x, y, z)$ is said to be *harmonic* in a region D in space if it satisfies the Laplace equation

$$\nabla^2 f = \nabla \cdot \nabla f = \frac{\partial^2 f}{\partial x^2} + \frac{\partial^2 f}{\partial y^2} + \frac{\partial^2 f}{\partial z^2} = 0$$

throughout D.

a. Suppose that f is harmonic throughout a bounded region D enclosed by a smooth surface S and that $\mathbf{n}$ is the chosen unit normal vector on S. Show that the integral over S of $\nabla f \cdot \mathbf{n}$, the derivative of f in the direction of $\mathbf{n}$, is zero.

b. Show that if f is harmonic on D, then

$$\iint_S f \nabla f \cdot \mathbf{n} \, d\sigma = \iiint_D |\nabla f|^2 \, dV.$$

28. Outward flux of a gradient field Let S be the surface of the portion of the solid sphere $x^2 + y^2 + z^2 \leq a^2$ that lies in the first octant and let $f(x, y, z) = \ln\sqrt{x^2 + y^2 + z^2}$. Calculate

$$\iint_S \nabla f \cdot \mathbf{n} \, d\sigma.$$

($\nabla f \cdot \mathbf{n}$ is the derivative of f in the direction of outward normal $\mathbf{n}$.)

29. Green's first formula Suppose that f and g are scalar functions with continuous first- and second-order partial derivatives throughout a region D that is bounded by a closed piecewise smooth surface S. Show that

$$\iint_S f \nabla g \cdot \mathbf{n} \, d\sigma = \iiint_D (f \nabla^2 g + \nabla f \cdot \nabla g) \, dV. \qquad (9)$$

Equation (9) is **Green's first formula**. (*Hint:* Apply the Divergence Theorem to the field $\mathbf{F} = f \nabla g$.)

30. Green's second formula (*Continuation of Exercise 29.*) Interchange f and g in Equation (9) to obtain a similar formula. Then subtract this formula from Equation (9) to show that

$$\iint_S (f \nabla g - g \nabla f) \cdot \mathbf{n} \, d\sigma = \iiint_D (f \nabla^2 g - g\nabla^2 f) \, dV. \qquad (10)$$

This equation is **Green's second formula**.

31. **Conservation of mass** Let $\mathbf{v}(t, x, y, z)$ be a continuously differentiable vector field over the region D in space and let $p(t, x, y, z)$ be a continuously differentiable scalar function. The variable t represents the time domain. The Law of Conservation of Mass asserts that

$$\frac{d}{dt} \iiint_D p(t, x, y, z) \, dV = -\iint_S p\mathbf{v} \cdot \mathbf{n} \, d\sigma,$$

where S is the surface enclosing D.

a. Give a physical interpretation of the conservation of mass law if $\mathbf{v}$ is a velocity flow field and p represents the density of the fluid at point (x, y, z) at time t.

b. Use the Divergence Theorem and Leibniz's Rule,

$$\frac{d}{dt} \iiint_D p(t, x, y, z) \, dV = \iiint_D \frac{\partial p}{\partial t} \, dV,$$

to show that the Law of Conservation of Mass is equivalent to the continuity equation,

$$\nabla \cdot p\mathbf{v} + \frac{\partial p}{\partial t} = 0.$$

(In the first term $\nabla \cdot p\mathbf{v}$, the variable t is held fixed, and in the second term $\partial p / \partial t$, it is assumed that the point (x, y, z) in D is held fixed.)

32. **The heat diffusion equation** Let $T(t, x, y, z)$ be a function with continuous second derivatives giving the temperature at time t at the point (x, y, z) of a solid occupying a region D in space. If the solid's heat capacity and mass density are denoted by the constants c and ρ, respectively, the quantity $c\rho T$ is called the solid's **heat energy per unit volume**.

a. Explain why $-\nabla T$ points in the direction of heat flow.

b. Let $-k\nabla T$ denote the **energy flux vector**. (Here the constant k is called the **conductivity**.) Assuming the Law of Conservation of Mass with $-k\nabla T = \mathbf{v}$ and $c\rho T = p$ in Exercise 31, derive the diffusion (heat) equation

$$\frac{\partial T}{\partial t} = K \nabla^2 T,$$

where $K = k/(c\rho) > 0$ is the *diffusivity* constant. (Notice that if $T(t, x)$ represents the temperature at time t at position x in a uniform conducting rod with perfectly insulated sides, then $\nabla^2 T = \partial^2 T / \partial x^2$ and the diffusion equation reduces to the one-dimensional heat equation in Chapter 14's Additional Exercises.)

Chapter 16 Questions to Guide Your Review

1. What are line integrals? How are they evaluated? Give examples.

2. How can you use line integrals to find the centers of mass of springs? Explain.

3. What is a vector field? A gradient field? Give examples.

4. How do you calculate the work done by a force in moving a particle along a curve? Give an example.

5. What are flow, circulation, and flux?

6. What is special about path independent fields?

7. How can you tell when a field is conservative?

8. What is a potential function? Show by example how to find a potential function for a conservative field.

9. What is a differential form? What does it mean for such a form to be exact? How do you test for exactness? Give examples.

10. What is the divergence of a vector field? How can you interpret it?

11. What is the curl of a vector field? How can you interpret it?

12. What is Green's Theorem? How can you interpret it?

13. How do you calculate the area of a parametrized surface in space? Of an implicitly defined surface $F(x, y, z) = 0$? Of the surface which is the graph of $z = f(x, y)$? Give examples.

14. How do you integrate a function over a parametrized surface in space? Of surfaces that are defined implicitly or in explicit form? What can you calculate with surface integrals? Give examples.

15. What is an oriented surface? How do you calculate the flux of a three-dimensional vector field across an oriented surface? Give an example.

16. What is Stokes' Theorem? How can you interpret it?

17. Summarize the chapter's results on conservative fields.

18. What is the Divergence Theorem? How can you interpret it?

19. How does the Divergence Theorem generalize Green's Theorem?

20. How does Stokes' Theorem generalize Green's Theorem?

21. How can Green's Theorem, Stokes' Theorem, and the Divergence Theorem be thought of as forms of a single fundamental theorem?

Chapter 16 Practice Exercises

Evaluating Line Integrals

1. The accompanying figure shows two polygonal paths in space joining the origin to the point $(1, 1, 1)$. Integrate $f(x, y, z) = 2x - 3y^2 - 2z + 3$ over each path.

Path 1

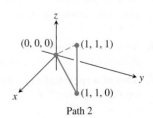

Path 2

2. The accompanying figure shows three polygonal paths joining the origin to the point (1, 1, 1). Integrate $f(x, y, z) = x^2 + y - z$ over each path.

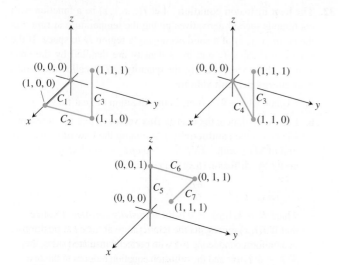

3. Integrate $f(x, y, z) = \sqrt{x^2 + z^2}$ over the circle

$$\mathbf{r}(t) = (a \cos t)\mathbf{j} + (a \sin t)\mathbf{k}, \qquad 0 \le t \le 2\pi.$$

4. Integrate $f(x, y, z) = \sqrt{x^2 + y^2}$ over the involute curve

$$\mathbf{r}(t) = (\cos t + t \sin t)\mathbf{i} + (\sin t - t \cos t)\mathbf{j}, \qquad 0 \le t \le \sqrt{3}.$$

Evaluate the integrals in Exercises 5 and 6.

5. $\displaystyle\int_{(-1,1,1)}^{(4,-3,0)} \frac{dx + dy + dz}{\sqrt{x + y + z}}$

6. $\displaystyle\int_{(1,1,1)}^{(10,3,3)} dx - \sqrt{\frac{z}{y}}\,dy - \sqrt{\frac{y}{z}}\,dz$

7. Integrate $\mathbf{F} = -(y \sin z)\mathbf{i} + (x \sin z)\mathbf{j} + (xy \cos z)\mathbf{k}$ around the circle cut from the sphere $x^2 + y^2 + z^2 = 5$ by the plane $z = -1$, clockwise as viewed from above.

8. Integrate $\mathbf{F} = 3x^2y\mathbf{i} + (x^3 + 1)\mathbf{j} + 9z^2\mathbf{k}$ around the circle cut from the sphere $x^2 + y^2 + z^2 = 9$ by the plane $x = 2$.

Evaluate the integrals in Exercises 9 and 10.

9. $\displaystyle\int_C 8x \sin y \, dx - 8y \cos x \, dy$

C is the square cut from the first quadrant by the lines $x = \pi/2$ and $y = \pi/2$.

10. $\displaystyle\int_C y^2 \, dx + x^2 \, dy$

C is the circle $x^2 + y^2 = 4$.

Finding and Evaluating Surface Integrals

11. Area of an elliptical region Find the area of the elliptical region cut from the plane $x + y + z = 1$ by the cylinder $x^2 + y^2 = 1$.

12. Area of a parabolic cap Find the area of the cap cut from the paraboloid $y^2 + z^2 = 3x$ by the plane $x = 1$.

13. Area of a spherical cap Find the area of the cap cut from the top of the sphere $x^2 + y^2 + z^2 = 1$ by the plane $z = \sqrt{2}/2$.

14. a. Hemisphere cut by cylinder Find the area of the surface cut from the hemisphere $x^2 + y^2 + z^2 = 4$, $z \ge 0$, by the cylinder $x^2 + y^2 = 2x$.

   b. Find the area of the portion of the cylinder that lies inside the hemisphere. (*Hint:* Project onto the xz-plane. Or evaluate the integral $\int h \, ds$, where h is the altitude of the cylinder and ds is the element of arc length on the circle $x^2 + y^2 = 2x$ in the xy-plane.)

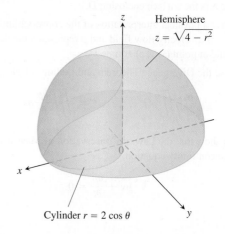

Cylinder $r = 2 \cos \theta$

15. Area of a triangle Find the area of the triangle in which the plane $(x/a) + (y/b) + (z/c) = 1$ $(a, b, c > 0)$ intersects the first octant. Check your answer with an appropriate vector calculation.

16. Parabolic cylinder cut by planes Integrate

   a. $g(x, y, z) = \dfrac{yz}{\sqrt{4y^2 + 1}}$

   b. $g(x, y, z) = \dfrac{z}{\sqrt{4y^2 + 1}}$

   over the surface cut from the parabolic cylinder $y^2 - z = 1$ by the planes $x = 0$, $x = 3$, and $z = 0$.

17. Circular cylinder cut by planes Integrate $g(x, y, z) = x^4y(y^2 + z^2)$ over the portion of the cylinder $y^2 + z^2 = 25$ that lies in the first octant between the planes $x = 0$ and $x = 1$ and above the plane $z = 3$.

18. Area of Wyoming The state of Wyoming is bounded by the meridians 111°3′ and 104°3′ west longitude and by the circles 41° and 45° north latitude. Assuming that Earth is a sphere of radius $R = 3959$ mi, find the area of Wyoming.

Parametrized Surfaces

Find parametrizations for the surfaces in Exercises 19–24. (There are many ways to do these, so your answers may not be the same as those in the back of the book.)

19. Spherical band The portion of the sphere $x^2 + y^2 + z^2 = 36$ between the planes $z = -3$ and $z = 3\sqrt{3}$

20. Parabolic cap The portion of the paraboloid $z = -(x^2 + y^2)/2$ above the plane $z = -2$

21. Cone The cone $z = 1 + \sqrt{x^2 + y^2}$, $z \le 3$

22. Plane above square The portion of the plane $4x + 2y + 4z = 12$ that lies above the square $0 \le x \le 2$, $0 \le y \le 2$ in the first quadrant

23. Portion of paraboloid The portion of the paraboloid $y = 2(x^2 + z^2)$, $y \leq 2$, that lies above the xy-plane

24. Portion of hemisphere The portion of the hemisphere $x^2 + y^2 + z^2 = 10$, $y \geq 0$, in the first octant

25. Surface area Find the area of the surface

$$\mathbf{r}(u, v) = (u + v)\mathbf{i} + (u - v)\mathbf{j} + v\mathbf{k},$$

$$0 \leq u \leq 1,\ 0 \leq v \leq 1.$$

26. Surface integral Integrate $f(x, y, z) = xy - z^2$ over the surface in Exercise 25.

27. Area of a helicoid Find the surface area of the helicoid

$\mathbf{r}(r, \theta) = (r \cos \theta)\mathbf{i} + (r \sin \theta)\mathbf{j} + \theta\mathbf{k}, \quad 0 \leq \theta \leq 2\pi, \quad 0 \leq r \leq 1$,
in the accompanying figure.

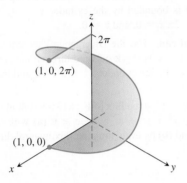

28. Surface integral Evaluate the integral $\iint_S \sqrt{x^2 + y^2 + 1}\, d\sigma$, where S is the helicoid in Exercise 27.

Conservative Fields

Which of the fields in Exercises 29–32 are conservative, and which are not?

29. $\mathbf{F} = x\mathbf{i} + y\mathbf{j} + z\mathbf{k}$

30. $\mathbf{F} = (x\mathbf{i} + y\mathbf{j} + z\mathbf{k})/(x^2 + y^2 + z^2)^{3/2}$

31. $\mathbf{F} = xe^y\mathbf{i} + ye^z\mathbf{j} + ze^x\mathbf{k}$

32. $\mathbf{F} = (\mathbf{i} + z\mathbf{j} + y\mathbf{k})/(x + yz)$

Find potential functions for the fields in Exercises 33 and 34.

33. $\mathbf{F} = 2\mathbf{i} + (2y + z)\mathbf{j} + (y + 1)\mathbf{k}$

34. $\mathbf{F} = (z \cos xz)\mathbf{i} + e^y\mathbf{j} + (x \cos xz)\mathbf{k}$

Work and Circulation

In Exercises 35 and 36, find the work done by each field along the paths from $(0, 0, 0)$ to $(1, 1, 1)$ in Exercise 1.

35. $\mathbf{F} = 2xy\mathbf{i} + \mathbf{j} + x^2\mathbf{k}$ **36.** $\mathbf{F} = 2xy\mathbf{i} + x^2\mathbf{j} + \mathbf{k}$

37. Finding work in two ways Find the work done by

$$\mathbf{F} = \frac{x\mathbf{i} + y\mathbf{j}}{(x^2 + y^2)^{3/2}}$$

over the plane curve $\mathbf{r}(t) = (e^t \cos t)\mathbf{i} + (e^t \sin t)\mathbf{j}$ from the point $(1, 0)$ to the point $(e^{2\pi}, 0)$ in two ways:

a. By using the parametrization of the curve to evaluate the work integral.

b. By evaluating a potential function for $\mathbf{F}$.

38. Flow along different paths Find the flow of the field $\mathbf{F} = \nabla(x^2ze^y)$

a. once around the ellipse C in which the plane $x + y + z = 1$ intersects the cylinder $x^2 + z^2 = 25$, clockwise as viewed from the positive y-axis.

b. along the curved boundary of the helicoid in Exercise 27 from $(1, 0, 0)$ to $(1, 0, 2\pi)$.

In Exercises 39 and 40, use the surface integral in Stokes' Theorem to find the circulation of the field $\mathbf{F}$ around the curve C in the indicated direction.

39. Circulation around an ellipse $\mathbf{F} = y^2\mathbf{i} - y\mathbf{j} + 3z^2\mathbf{k}$

C: The ellipse in which the plane $2x + 6y - 3z = 6$ meets the cylinder $x^2 + y^2 = 1$, counterclockwise as viewed from above

40. Circulation around a circle $\mathbf{F} = (x^2 + y)\mathbf{i} + (x + y)\mathbf{j} + (4y^2 - z)\mathbf{k}$

C: The circle in which the plane $z = -y$ meets the sphere $x^2 + y^2 + z^2 = 4$, counterclockwise as viewed from above

Masses and Moments

41. Wire with different densities Find the mass of a thin wire lying along the curve $\mathbf{r}(t) = \sqrt{2}t\mathbf{i} + \sqrt{2}t\mathbf{j} + (4 - t^2)\mathbf{k}$, $0 \leq t \leq 1$, if the density at t is **(a)** $\delta = 3t$ and **(b)** $\delta = 1$.

42. Wire with variable density Find the center of mass of a thin wire lying along the curve $\mathbf{r}(t) = t\mathbf{i} + 2t\mathbf{j} + (2/3)t^{3/2}\mathbf{k}$, $0 \leq t \leq 2$, if the density at t is $\delta = 3\sqrt{5 + t}$.

43. Wire with variable density Find the center of mass and the moments of inertia about the coordinate axes of a thin wire lying along the curve

$$\mathbf{r}(t) = t\mathbf{i} + \frac{2\sqrt{2}}{3}t^{3/2}\mathbf{j} + \frac{t^2}{2}\mathbf{k}, \qquad 0 \leq t \leq 2,$$

if the density at t is $\delta = 1/(t + 1)$.

44. Center of mass of an arch A slender metal arch lies along the semicircle $y = \sqrt{a^2 - x^2}$ in the xy-plane. The density at the point (x, y) on the arch is $\delta(x, y) = 2a - y$. Find the center of mass.

45. Wire with constant density A wire of constant density $\delta = 1$ lies along the curve $\mathbf{r}(t) = (e^t \cos t)\mathbf{i} + (e^t \sin t)\mathbf{j} + e^t\mathbf{k}$, $0 \leq t \leq \ln 2$. Find $\bar{z}$ and I_z.

46. Helical wire with constant density Find the mass and center of mass of a wire of constant density δ that lies along the helix $\mathbf{r}(t) = (2 \sin t)\mathbf{i} + (2 \cos t)\mathbf{j} + 3t\mathbf{k}, 0 \leq t \leq 2\pi$.

47. Inertia and center of mass of a shell Find I_z and the center of mass of a thin shell of density $\delta(x, y, z) = z$ cut from the upper portion of the sphere $x^2 + y^2 + z^2 = 25$ by the plane $z = 3$.

48. Moment of inertia of a cube Find the moment of inertia about the z-axis of the surface of the cube cut from the first octant by the planes $x = 1, y = 1$, and $z = 1$ if the density is $\delta = 1$.

Flux Across a Plane Curve or Surface

Use Green's Theorem to find the counterclockwise circulation and outward flux for the fields and curves in Exercises 49 and 50.

49. Square $\mathbf{F} = (2xy + x)\mathbf{i} + (xy - y)\mathbf{j}$

C: The square bounded by $x = 0, x = 1, y = 0, y = 1$

50. Triangle $F = (y - 6x^2)i + (x + y^2)j$

   C: The triangle made by the lines $y = 0$, $y = x$, and $x = 1$

51. Zero line integral Show that

$$\oint_C \ln x \sin y \, dy - \frac{\cos y}{x} \, dx = 0$$

for any closed curve C to which Green's Theorem applies.

52. a. Outward flux and area Show that the outward flux of the position vector field $F = xi + yj$ across any closed curve to which Green's Theorem applies is twice the area of the region enclosed by the curve.

   b. Let n be the outward unit normal vector to a closed curve to which Green's Theorem applies. Show that it is not possible for $F = xi + yj$ to be orthogonal to n at every point of C.

In Exercises 53–56, find the outward flux of F across the boundary of D.

53. Cube $F = 2xyi + 2yzj + 2xzk$

   D: The cube cut from the first octant by the planes $x = 1, y = 1, z = 1$

54. Spherical cap $F = xzi + yzj + k$

   D: The entire surface of the upper cap cut from the solid sphere $x^2 + y^2 + z^2 \le 25$ by the plane $z = 3$

55. Spherical cap $F = -2xi - 3yj + zk$

   D: The upper region cut from the solid sphere $x^2 + y^2 + z^2 \le 2$ by the paraboloid $z = x^2 + y^2$

56. Cone and cylinder $F = (6x + y)i - (x + z)j + 4yzk$

   D: The region in the first octant bounded by the cone $z = \sqrt{x^2 + y^2}$, the cylinder $x^2 + y^2 = 1$, and the coordinate planes

57. Hemisphere, cylinder, and plane Let S be the surface that is bounded on the left by the hemisphere $x^2 + y^2 + z^2 = a^2, y \le 0$, in the middle by the cylinder $x^2 + z^2 = a^2, 0 \le y \le a$, and on the right by the plane $y = a$. Find the flux of $F = yi + zj + xk$ outward across S.

58. Cylinder and planes Find the outward flux of the field $F = 3xz^2i + yj - z^3k$ across the surface of the solid in the first octant that is bounded by the cylinder $x^2 + 4y^2 = 16$ and the planes $y = 2z, x = 0$, and $z = 0$.

59. Cylindrical can Use the Divergence Theorem to find the flux of $F = xy^2i + x^2yj + yk$ outward through the surface of the region enclosed by the cylinder $x^2 + y^2 = 1$ and the planes $z = 1$ and $z = -1$.

60. Hemisphere Find the flux of $F = (3z + 1)k$ upward across the hemisphere $x^2 + y^2 + z^2 = a^2, z \ge 0$ **(a)** with the Divergence Theorem and **(b)** by evaluating the flux integral directly.

Chapter 16   Additional and Advanced Exercises

Finding Areas with Green's Theorem

Use the Green's Theorem area formula in Exercises 16.4 to find the areas of the regions enclosed by the curves in Exercises 1–4.

1. The limaçon $x = 2 \cos t - \cos 2t$, $y = 2 \sin t - \sin 2t$, $0 \le t \le 2\pi$

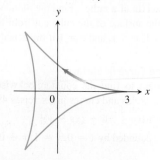

2. The deltoid $x = 2 \cos t + \cos 2t$, $y = 2 \sin t - \sin 2t$, $0 \le t \le 2\pi$

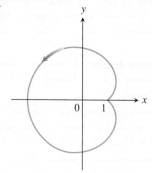

3. The eight curve $x = (1/2) \sin 2t$, $y = \sin t$, $0 \le t \le \pi$ (one loop)

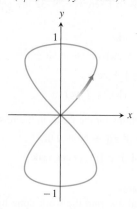

4. The teardrop $x = 2a \cos t - a \sin 2t$, $y = b \sin t$, $0 \le t \le 2\pi$

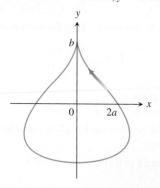

Theory and Applications

5. a. Give an example of a vector field $\mathbf{F}(x, y, z)$ that has value $\mathbf{0}$ at only one point and such that curl $\mathbf{F}$ is nonzero everywhere. Be sure to identify the point and compute the curl.

    b. Give an example of a vector field $\mathbf{F}(x, y, z)$ that has value $\mathbf{0}$ on precisely one line and such that curl $\mathbf{F}$ is nonzero everywhere. Be sure to identify the line and compute the curl.

    c. Give an example of a vector field $\mathbf{F}(x, y, z)$ that has value $\mathbf{0}$ on a surface and such that curl $\mathbf{F}$ is nonzero everywhere. Be sure to identify the surface and compute the curl.

6. Find all points (a, b, c) on the sphere $x^2 + y^2 + z^2 = R^2$ where the vector field $\mathbf{F} = yz^2\mathbf{i} + xz^2\mathbf{j} + 2xyz\mathbf{k}$ is normal to the surface and $\mathbf{F}(a, b, c) \neq \mathbf{0}$.

7. Find the mass of a spherical shell of radius R such that at each point (x, y, z) on the surface the mass density $\delta(x, y, z)$ is its distance to some fixed point (a, b, c) of the surface.

8. Find the mass of a helicoid

$$\mathbf{r}(r, \theta) = (r \cos \theta)\mathbf{i} + (r \sin \theta)\mathbf{j} + \theta\mathbf{k},$$

$0 \leq r \leq 1, 0 \leq \theta \leq 2\pi$, if the density function is $\delta(x, y, z) = 2\sqrt{x^2 + y^2}$. See Practice Exercise 27 for a figure.

9. Among all rectangular regions $0 \leq x \leq a, 0 \leq y \leq b$, find the one for which the total outward flux of $\mathbf{F} = (x^2 + 4xy)\mathbf{i} - 6y\mathbf{j}$ across the four sides is least. What *is* the least flux?

10. Find an equation for the plane through the origin such that the circulation of the flow field $\mathbf{F} = z\mathbf{i} + x\mathbf{j} + y\mathbf{k}$ around the circle of intersection of the plane with the sphere $x^2 + y^2 + z^2 = 4$ is a maximum.

11. A string lies along the circle $x^2 + y^2 = 4$ from $(2, 0)$ to $(0, 2)$ in the first quadrant. The density of the string is $\rho(x, y) = xy$.

    a. Partition the string into a finite number of subarcs to show that the work done by gravity to move the string straight down to the x-axis is given by

$$\text{Work} = \lim_{n \to \infty} \sum_{k=1}^{n} g\, x_k y_k^2 \Delta s_k = \int_C g\, xy^2\, ds,$$

    where g is the gravitational constant.

    b. Find the total work done by evaluating the line integral in part (a).

    c. Show that the total work done equals the work required to move the string's center of mass $(\bar{x}, \bar{y})$ straight down to the x-axis.

12. A thin sheet lies along the portion of the plane $x + y + z = 1$ in the first octant. The density of the sheet is $\delta(x, y, z) = xy$.

    a. Partition the sheet into a finite number of subpieces to show that the work done by gravity to move the sheet straight down to the xy-plane is given by

$$\text{Work} = \lim_{n \to \infty} \sum_{k=1}^{n} g\, x_k y_k z_k \, \Delta \sigma_k = \iint_S g\, xyz\, d\sigma,$$

    where g is the gravitational constant.

    b. Find the total work done by evaluating the surface integral in part (a).

    c. Show that the total work done equals the work required to move the sheet's center of mass $(\bar{x}, \bar{y}, \bar{z})$ straight down to the xy-plane.

13. Archimedes' principle If an object such as a ball is placed in a liquid, it will either sink to the bottom, float, or sink a certain distance and remain suspended in the liquid. Suppose a fluid has constant weight density w and that the fluid's surface coincides with the plane $z = 4$. A spherical ball remains suspended in the fluid and occupies the region $x^2 + y^2 + (z - 2)^2 \leq 1$.

    a. Show that the surface integral giving the magnitude of the total force on the ball due to the fluid's pressure is

$$\text{Force} = \lim_{n \to \infty} \sum_{k=1}^{n} w(4 - z_k) \, \Delta \sigma_k = \iint_S w(4 - z) \, d\sigma.$$

    b. Since the ball is not moving, it is being held up by the buoyant force of the liquid. Show that the magnitude of the buoyant force on the sphere is

$$\text{Buoyant force} = \iint_S w(z - 4)\mathbf{k} \cdot \mathbf{n} \, d\sigma,$$

    where $\mathbf{n}$ is the outer unit normal at (x, y, z). This illustrates Archimedes' principle that the magnitude of the buoyant force on a submerged solid equals the weight of the displaced fluid.

    c. Use the Divergence Theorem to find the magnitude of the buoyant force in part (b).

14. Fluid force on a curved surface A cone in the shape of the surface $z = \sqrt{x^2 + y^2}, 0 \leq z \leq 2$ is filled with a liquid of constant weight density w. Assuming the xy-plane is "ground level," show that the total force on the portion of the cone from $z = 1$ to $z = 2$ due to liquid pressure is the surface integral

$$F = \iint_S w(2 - z) \, d\sigma.$$

Evaluate the integral.

15. Faraday's Law If $\mathbf{E}(t, x, y, z)$ and $\mathbf{B}(t, x, y, z)$ represent the electric and magnetic fields at point (x, y, z) at time t, a basic principle of electromagnetic theory says that $\nabla \times \mathbf{E} = -\partial \mathbf{B}/\partial t$. In this expression $\nabla \times \mathbf{E}$ is computed with t held fixed and $\partial \mathbf{B}/\partial t$ is calculated with (x, y, z) fixed. Use Stokes' Theorem to derive Faraday's Law,

$$\oint_C \mathbf{E} \cdot d\mathbf{r} = -\frac{\partial}{\partial t} \iint_S \mathbf{B} \cdot \mathbf{n} \, d\sigma,$$

where C represents a wire loop through which current flows counterclockwise with respect to the surface's unit normal $\mathbf{n}$, giving rise to the voltage

$$\oint_C \mathbf{E} \cdot d\mathbf{r}$$

around C. The surface integral on the right side of the equation is called the *magnetic flux*, and S is any oriented surface with boundary C.

16. Let

$$\mathbf{F} = -\frac{GmM}{|\mathbf{r}|^3} \mathbf{r}$$

be the gravitational force field defined for $\mathbf{r} \neq \mathbf{0}$. Use Gauss's Law in Section 16.8 to show that there is no continuously differentiable vector field $\mathbf{H}$ satisfying $\mathbf{F} = \nabla \times \mathbf{H}$.

17. If $f(x, y, z)$ and $g(x, y, z)$ are continuously differentiable scalar functions defined over the oriented surface S with boundary curve C, prove that

$$\iint_S (\nabla f \times \nabla g) \cdot \mathbf{n}\, d\sigma = \oint_C f \nabla g \cdot d\mathbf{r}.$$

18. Suppose that $\nabla \cdot \mathbf{F}_1 = \nabla \cdot \mathbf{F}_2$ and $\nabla \times \mathbf{F}_1 = \nabla \times \mathbf{F}_2$ over a region D enclosed by the oriented surface S with outward unit normal $\mathbf{n}$ and that $\mathbf{F}_1 \cdot \mathbf{n} = \mathbf{F}_2 \cdot \mathbf{n}$ on S. Prove that $\mathbf{F}_1 = \mathbf{F}_2$ throughout D.

19. Prove or disprove that if $\nabla \cdot \mathbf{F} = 0$ and $\nabla \times \mathbf{F} = \mathbf{0}$, then $\mathbf{F} = \mathbf{0}$.

20. Let S be an oriented surface parametrized by $\mathbf{r}(u, v)$. Define the notation $d\boldsymbol{\sigma} = \mathbf{r}_u\, du \times \mathbf{r}_v\, dv$ so that $d\boldsymbol{\sigma}$ is a vector normal to the surface. Also, the magnitude $d\sigma = |d\boldsymbol{\sigma}|$ is the element of surface area (by Equation 5 in Section 16.5). Derive the identity

$$d\sigma = (EG - F^2)^{1/2}\, du\, dv$$

where

$$E = |\mathbf{r}_u|^2, \quad F = \mathbf{r}_u \cdot \mathbf{r}_v, \quad \text{and} \quad G = |\mathbf{r}_v|^2.$$

21. Show that the volume V of a region D in space enclosed by the oriented surface S with outward normal $\mathbf{n}$ satisfies the identity

$$V = \frac{1}{3} \iint_S \mathbf{r} \cdot \mathbf{n}\, d\sigma,$$

where $\mathbf{r}$ is the position vector of the point (x, y, z) in D.

Chapter 16 Technology Application Projects

Mathematica/Maple Module:

Work in Conservative and Nonconservative Force Fields
Explore integration over vector fields and experiment with conservative and nonconservative force functions along different paths in the field.

How Can You Visualize Green's Theorem?
Explore integration over vector fields and use parametrizations to compute line integrals. Both forms of Green's Theorem are explored.

Visualizing and Interpreting the Divergence Theorem
Verify the Divergence Theorem by formulating and evaluating certain divergence and surface integrals.

APPENDICES

A.1 Real Numbers and the Real Line

This section reviews real numbers, inequalities, intervals, and absolute values.

Real Numbers

Much of calculus is based on properties of the real number system. **Real numbers** are numbers that can be expressed as decimals, such as

$$-\frac{3}{4} = -0.75000\ldots$$

$$\frac{1}{3} = 0.33333\ldots$$

$$\sqrt{2} = 1.4142\ldots$$

The dots $\ldots$ in each case indicate that the sequence of decimal digits goes on forever. Every conceivable decimal expansion represents a real number, although some numbers have two representations. For instance, the infinite decimals $.999\ldots$ and $1.000\ldots$ represent the same real number 1. A similar statement holds for any number with an infinite tail of 9's.

The real numbers can be represented geometrically as points on a number line called the **real line**.

$$\xrightarrow{\qquad\qquad} $$

The number line showing: -2, $-1\ \frac{3}{4}$, $0\ \frac{1}{3}$, $1\ \sqrt{2}$, 2, $3\ \pi$, 4

The symbol $\mathbb{R}$ denotes either the real number system or, equivalently, the real line.

The properties of the real number system fall into three categories: algebraic properties, order properties, and completeness. The **algebraic properties** say that the real numbers can be added, subtracted, multiplied, and divided (except by 0) to produce more real numbers under the usual rules of arithmetic. *You can never divide by* 0.

The **order properties** of real numbers are given in Appendix 6. The useful rules at the left can be derived from them, where the symbol $\Rightarrow$ means "implies."

Notice the rules for multiplying an inequality by a number. Multiplying by a positive number preserves the inequality; multiplying by a negative number reverses the inequality. Also, reciprocation reverses the inequality for numbers of the same sign. For example, $2 < 5$ but $-2 > -5$ and $1/2 > 1/5$.

The **completeness property** of the real number system is deeper and harder to define precisely. However, the property is essential to the idea of a limit (Chapter 2). Roughly speaking, it says that there are enough real numbers to "complete" the real number line, in the sense that there are no "holes" or "gaps" in it. Many theorems of calculus would fail if the real number system were not complete. The topic is best saved for a more advanced course, but Appendix 6 hints about what is involved and how the real numbers are constructed.

Rules for inequalities

If a, b, and c are real numbers, then:

1. $a < b \Rightarrow a + c < b + c$
2. $a < b \Rightarrow a - c < b - c$
3. $a < b$ and $c > 0 \Rightarrow ac < bc$
4. $a < b$ and $c < 0 \Rightarrow bc < ac$
   Special case: $a < b \Rightarrow -b < -a$
5. $a > 0 \Rightarrow \dfrac{1}{a} > 0$
6. If a and b are both positive or both negative, then $a < b \Rightarrow \dfrac{1}{b} < \dfrac{1}{a}$

We distinguish three special subsets of real numbers.

1. The **natural numbers**, namely 1, 2, 3, 4, ...

2. The **integers**, namely 0, ±1, ±2, ±3, ...

3. The **rational numbers**, namely the numbers that can be expressed in the form of a fraction m/n, where m and n are integers and $n \neq 0$. Examples are

$$\frac{1}{3}, \quad -\frac{4}{9} = \frac{-4}{9} = \frac{4}{-9}, \quad \frac{200}{13}, \quad \text{and} \quad 57 = \frac{57}{1}.$$

The rational numbers are precisely the real numbers with decimal expansions that are either

(a) terminating (ending in an infinite string of zeros), for example,

$$\frac{3}{4} = 0.75000 \ldots = 0.75 \quad \text{or}$$

(b) eventually repeating (ending with a block of digits that repeats over and over), for example

$$\frac{23}{11} = 2.090909 \ldots = 2.\overline{09} \qquad \text{The bar indicates the block of repeating digits.}$$

A terminating decimal expansion is a special type of repeating decimal, since the ending zeros repeat.

The set of rational numbers has all the algebraic and order properties of the real numbers but lacks the completeness property. For example, there is no rational number whose square is 2; there is a "hole" in the rational line where $\sqrt{2}$ should be.

Real numbers that are not rational are called **irrational numbers**. They are characterized by having nonterminating and nonrepeating decimal expansions. Examples are π, $\sqrt{2}$, $\sqrt[3]{5}$, and $\log_{10} 3$. Since every decimal expansion represents a real number, it should be clear that there are infinitely many irrational numbers. Both rational and irrational numbers are found arbitrarily close to any point on the real line.

Set notation is very useful for specifying a particular subset of real numbers. A **set** is a collection of objects, and these objects are the **elements** of the set. If S is a set, the notation $a \in S$ means that a is an element of S, and $a \notin S$ means that a is not an element of S. If S and T are sets, then $S \cup T$ is their **union** and consists of all elements belonging either to S or T (or to both S and T). The **intersection** $S \cap T$ consists of all elements belonging to both S and T. The **empty set** $\varnothing$ is the set that contains no elements. For example, the intersection of the rational numbers and the irrational numbers is the empty set.

Some sets can be described by *listing* their elements in braces. For instance, the set A consisting of the natural numbers (or positive integers) less than 6 can be expressed as

$$A = \{1, 2, 3, 4, 5\}.$$

The entire set of integers is written as

$$\{0, \pm1, \pm2, \pm3, \ldots \}.$$

Another way to describe a set is to enclose in braces a rule that generates all the elements of the set. For instance, the set

$$A = \{x \mid x \text{ is an integer and } 0 < x < 6\}$$

is the set of positive integers less than 6.

Intervals

A subset of the real line is called an **interval** if it contains at least two numbers and contains all the real numbers lying between any two of its elements. For example, the set of all real numbers x such that $x > 6$ is an interval, as is the set of all x such that $-2 \leq x \leq 5$. The set of all nonzero real numbers is not an interval; since 0 is absent, the set fails to contain every real number between -1 and 1 (for example).

Geometrically, intervals correspond to rays and line segments on the real line, along with the real line itself. Intervals of numbers corresponding to line segments are **finite intervals**; intervals corresponding to rays and the real line are **infinite intervals**.

A finite interval is said to be **closed** if it contains both of its endpoints, **half-open** if it contains one endpoint but not the other, and **open** if it contains neither endpoint. The endpoints are also called **boundary points**; they make up the interval's **boundary**. The remaining points of the interval are **interior points** and together comprise the interval's **interior**. Infinite intervals are closed if they contain a finite endpoint, and open otherwise. The entire real line $\mathbb{R}$ is an infinite interval that is both open and closed. Table A.1 summarizes the various types of intervals.

TABLE A.1 Types of intervals

Notation	Set description	Type	Picture
(a, b)	$\{x \mid a < x < b\}$	Open	
$[a, b]$	$\{x \mid a \leq x \leq b\}$	Closed	
$[a, b)$	$\{x \mid a \leq x < b\}$	Half-open	
$(a, b]$	$\{x \mid a < x \leq b\}$	Half-open	
(a, ∞)	$\{x \mid x > a\}$	Open	
$[a, \infty)$	$\{x \mid x \geq a\}$	Closed	
$(-\infty, b)$	$\{x \mid x < b\}$	Open	
$(-\infty, b]$	$\{x \mid x \leq b\}$	Closed	
$(-\infty, \infty)$	$\mathbb{R}$ (set of all real numbers)	Both open and closed	

Solving Inequalities

The process of finding the interval or intervals of numbers that satisfy an inequality in x is called **solving** the inequality.

EXAMPLE 1   Solve the following inequalities and show their solution sets on the real line.

(a) $2x - 1 < x + 3$ **(b)** $-\dfrac{x}{3} < 2x + 1$ **(c)** $\dfrac{6}{x - 1} \geq 5$

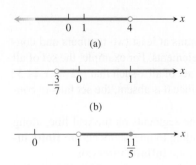

FIGURE A.1 Solution sets for the inequalities in Example 1.

Solution

(a)
$$2x - 1 < x + 3$$
$$2x < x + 4 \qquad \text{Add 1 to both sides.}$$
$$x < 4 \qquad \text{Subtract } x \text{ from both sides.}$$

The solution set is the open interval $(-\infty, 4)$ (Figure A.1a).

(b)
$$-\frac{x}{3} < 2x + 1$$
$$-x < 6x + 3 \qquad \text{Multiply both sides by 3.}$$
$$0 < 7x + 3 \qquad \text{Add } x \text{ to both sides.}$$
$$-3 < 7x \qquad \text{Subtract 3 from both sides.}$$
$$-\frac{3}{7} < x \qquad \text{Divide by 7.}$$

The solution set is the open interval $(-3/7, \infty)$ (Figure A.1b).

(c) The inequality $6/(x - 1) \geq 5$ can hold only if $x > 1$, because otherwise $6/(x - 1)$ is undefined or negative. Therefore, $(x - 1)$ is positive and the inequality will be preserved if we multiply both sides by $(x - 1)$, and we have

$$\frac{6}{x - 1} \geq 5$$
$$6 \geq 5x - 5 \qquad \text{Multiply both sides by } (x - 1).$$
$$11 \geq 5x \qquad \text{Add 5 to both sides.}$$
$$\frac{11}{5} \geq x. \qquad \text{Or } x \leq \frac{11}{5}.$$

The solution set is the half-open interval $(1, 11/5\,]$ (Figure A.1c). ∎

Absolute Value

The **absolute value** of a number x, denoted by $|x|$, is defined by the formula

$$|x| = \begin{cases} x, & x \geq 0 \\ -x, & x < 0. \end{cases}$$

EXAMPLE 2 $\quad |3| = 3, \quad |0| = 0, \quad |-5| = -(-5) = 5, \quad |-|a|| = |a|$ ∎

Geometrically, the absolute value of x is the distance from x to 0 on the real number line. Since distances are always positive or 0, we see that $|x| \geq 0$ for every real number x, and $|x| = 0$ if and only if $x = 0$. Also,

$$|x - y| = \text{the distance between } x \text{ and } y$$

on the real line (Figure A.2).

Since the symbol $\sqrt{a}$ always denotes the *nonnegative* square root of a, an alternate definition of $|x|$ is

$$|x| = \sqrt{x^2}.$$

It is important to remember that $\sqrt{a^2} = |a|$. Do not write $\sqrt{a^2} = a$ unless you already know that $a \geq 0$.

The absolute value has the following properties. (You are asked to prove these properties in the exercises.)

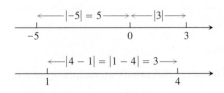

FIGURE A.2 Absolute values give distances between points on the number line.

Absolute Value Properties

1. $|-a| = |a|$ A number and its additive inverse or negative have the same absolute value.

2. $|ab| = |a||b|$ The absolute value of a product is the product of the absolute values.

3. $\left|\dfrac{a}{b}\right| = \dfrac{|a|}{|b|}$ The absolute value of a quotient is the quotient of the absolute values.

4. $|a + b| \le |a| + |b|$ The **triangle inequality**. The absolute value of the sum of two numbers is less than or equal to the sum of their absolute values.

FIGURE A.3 $|x| < a$ means x lies between $-a$ and a.

Absolute values and intervals

If a is any positive number, then

5. $|x| = a \iff x = \pm a$

6. $|x| < a \iff -a < x < a$

7. $|x| > a \iff x > a \text{ or } x < -a$

8. $|x| \le a \iff -a \le x \le a$

9. $|x| \ge a \iff x \ge a \text{ or } x \le -a$

Note that $|-a| \ne -|a|$. For example, $|-3| = 3$, whereas $-|3| = -3$. If a and b differ in sign, then $|a + b|$ is less than $|a| + |b|$. In all other cases, $|a + b|$ equals $|a| + |b|$. Absolute value bars in expressions like $|-3 + 5|$ work like parentheses: We do the arithmetic inside *before* taking the absolute value.

EXAMPLE 3

$$|-3 + 5| = |2| = 2 < |-3| + |5| = 8$$
$$|3 + 5| = |8| = |3| + |5|$$
$$|-3 - 5| = |-8| = 8 = |-3| + |-5|$$

The inequality $|x| < a$ says that the distance from x to 0 is less than the positive number a. This means that x must lie between $-a$ and a, as we can see from Figure A.3.

The statements in the table are all consequences of the definition of absolute value and are often helpful when solving equations or inequalities involving absolute values.

The symbol $\iff$ is often used by mathematicians to denote the "if and only if" logical relationship. It also means "implies and is implied by."

EXAMPLE 4   Solve the equation $|2x - 3| = 7$.

Solution   By Property 5, $2x - 3 = \pm 7$, so there are two possibilities:

$$2x - 3 = 7 \qquad 2x - 3 = -7 \qquad \text{Equivalent equations without absolute values}$$
$$2x = 10 \qquad\quad 2x = -4 \qquad\quad \text{Solve as usual.}$$
$$x = 5 \qquad\qquad x = -2$$

The solutions of $|2x - 3| = 7$ are $x = 5$ and $x = -2$.

EXAMPLE 5   Solve the inequality $\left|5 - \dfrac{2}{x}\right| < 1$.

Solution   We have

$$\left|5 - \frac{2}{x}\right| < 1 \iff -1 < 5 - \frac{2}{x} < 1 \qquad \text{Property 6}$$

$$\iff -6 < -\frac{2}{x} < -4 \qquad \text{Subtract 5.}$$

$$\iff 3 > \frac{1}{x} > 2 \qquad \text{Multiply by } -\frac{1}{2}.$$

$$\iff \frac{1}{3} < x < \frac{1}{2}. \qquad \text{Take reciprocals.}$$

Notice how the various rules for inequalities were used here. Multiplying by a negative number reverses the inequality. So does taking reciprocals in an inequality in which both sides are positive. The original inequality holds if and only if $(1/3) < x < (1/2)$. The solution set is the open interval $(1/3, 1/2)$. ∎

Exercises A.1

1. Express $1/9$ as a repeating decimal, using a bar to indicate the repeating digits. What are the decimal representations of $2/9$? $3/9$? $8/9$? $9/9$?

2. If $2 < x < 6$, which of the following statements about x are necessarily true, and which are not necessarily true?

   a. $0 < x < 4$

   b. $0 < x - 2 < 4$

   c. $1 < \frac{x}{2} < 3$

   d. $\frac{1}{6} < \frac{1}{x} < \frac{1}{2}$

   e. $1 < \frac{6}{x} < 3$

   f. $|x - 4| < 2$

   g. $-6 < -x < 2$

   h. $-6 < -x < -2$

In Exercises 3–6, solve the inequalities and show the solution sets on the real line.

3. $-2x > 4$

4. $5x - 3 \le 7 - 3x$

5. $2x - \frac{1}{2} \ge 7x + \frac{7}{6}$

6. $\frac{4}{5}(x - 2) < \frac{1}{3}(x - 6)$

Solve the equations in Exercises 7–9.

7. $|y| = 3$

8. $|2t + 5| = 4$

9. $|8 - 3s| = \frac{9}{2}$

Solve the inequalities in Exercises 10–17, expressing the solution sets as intervals or unions of intervals. Also, show each solution set on the real line.

10. $|x| < 2$

11. $|t - 1| \le 3$

12. $|3y - 7| < 4$

13. $\left|\frac{z}{5} - 1\right| \le 1$

14. $\left|3 - \frac{1}{x}\right| < \frac{1}{2}$

15. $|2s| \ge 4$

16. $|1 - x| > 1$

17. $\left|\frac{r + 1}{2}\right| \ge 1$

Solve the inequalities in Exercises 18–21. Express the solution sets as intervals or unions of intervals and show them on the real line. Use the result $\sqrt{a^2} = |a|$ as appropriate.

18. $x^2 < 2$

19. $4 < x^2 < 9$

20. $(x - 1)^2 < 4$

21. $x^2 - x < 0$

22. Do not fall into the trap of thinking $|-a| = a$. For what real numbers a is this equation true? For what real numbers is it false?

23. Solve the equation $|x - 1| = 1 - x$.

24. **A proof of the triangle inequality** Give the reason justifying each of the numbered steps in the following proof of the triangle inequality.

$$|a + b|^2 = (a + b)^2 \tag{1}$$
$$= a^2 + 2ab + b^2$$
$$\le a^2 + 2|a||b| + b^2 \tag{2}$$
$$= |a|^2 + 2|a||b| + |b|^2 \tag{3}$$
$$= (|a| + |b|)^2$$
$$|a + b| \le |a| + |b| \tag{4}$$

25. Prove that $|ab| = |a||b|$ for any numbers a and b.

26. If $|x| \le 3$ and $x > -1/2$, what can you say about x?

27. Graph the inequality $|x| + |y| \le 1$.

28. For any number a, prove that $|-a| = |a|$.

29. Let a be any positive number. Prove that $|x| > a$ if and only if $x > a$ or $x < -a$.

30. a. If b is any nonzero real number, prove that $|1/b| = 1/|b|$.

    b. Prove that $\left|\frac{a}{b}\right| = \frac{|a|}{|b|}$ for any numbers a and $b \ne 0$.

A.2 | Mathematical Induction

Many formulas, like

$$1 + 2 + \cdots + n = \frac{n(n + 1)}{2},$$

can be shown to hold for every positive integer n by applying an axiom called the *mathematical induction principle*. A proof that uses this axiom is called a *proof by mathematical induction* or a *proof by induction*.

The steps in proving a formula by induction are the following:

1. Check that the formula holds for $n = 1$.

2. Prove that if the formula holds for any positive integer $n = k$, then it also holds for the next integer, $n = k + 1$.

The induction axiom says that once these steps are completed, the formula holds for all positive integers n. By Step 1 it holds for $n = 1$. By Step 2 it holds for $n = 2$, and therefore by Step 2 also for $n = 3$, and by Step 2 again for $n = 4$, and so on. If the first domino falls, and the kth domino always knocks over the $(k + 1)$st when it falls, all the dominoes fall.

From another point of view, suppose we have a sequence of statements S_1, $S_2, \ldots, S_n, \ldots$, one for each positive integer. Suppose we can show that assuming any one of the statements to be true implies that the next statement in line is true. Suppose that we can also show that S_1 is true. Then we may conclude that the statements are true from S_1 on.

EXAMPLE 1   Use mathematical induction to prove that for every positive integer n,

$$1 + 2 + \cdots + n = \frac{n(n + 1)}{2}.$$

Solution   We accomplish the proof by carrying out the two steps above.

1. The formula holds for $n = 1$ because

$$1 = \frac{1(1 + 1)}{2}.$$

2. If the formula holds for $n = k$, does it also hold for $n = k + 1$? The answer is yes, as we now show. If

$$1 + 2 + \cdots + k = \frac{k(k + 1)}{2},$$

then

$$1 + 2 + \cdots + k + (k + 1) = \frac{k(k + 1)}{2} + (k + 1) = \frac{k^2 + k + 2k + 2}{2}$$

$$= \frac{(k + 1)(k + 2)}{2} = \frac{(k + 1)((k + 1) + 1)}{2}.$$

The last expression in this string of equalities is the expression $n(n + 1)/2$ for $n = (k + 1)$.

The mathematical induction principle now guarantees the original formula for all positive integers n. ∎

In Example 4 of Section 5.2 we gave another proof for the formula giving the sum of the first n integers. However, proof by mathematical induction is more general. It can be used to find the sums of the squares and cubes of the first n integers (Exercises 9 and 10). Here is another example.

EXAMPLE 2   Show by mathematical induction that for all positive integers n,

$$\frac{1}{2^1} + \frac{1}{2^2} + \cdots + \frac{1}{2^n} = 1 - \frac{1}{2^n}.$$

Solution   We accomplish the proof by carrying out the two steps of mathematical induction.

1. The formula holds for $n = 1$ because

$$\frac{1}{2^1} = 1 - \frac{1}{2^1}.$$

2. If

$$\frac{1}{2^1} + \frac{1}{2^2} + \cdots + \frac{1}{2^k} = 1 - \frac{1}{2^k},$$

then

$$\frac{1}{2^1} + \frac{1}{2^2} + \cdots + \frac{1}{2^k} + \frac{1}{2^{k+1}} = 1 - \frac{1}{2^k} + \frac{1}{2^{k+1}} = 1 - \frac{1 \cdot 2}{2^k \cdot 2} + \frac{1}{2^{k+1}}$$

$$= 1 - \frac{2}{2^{k+1}} + \frac{1}{2^{k+1}} = 1 - \frac{1}{2^{k+1}}.$$

Thus, the original formula holds for $n = (k + 1)$ whenever it holds for $n = k$.

With these steps verified, the mathematical induction principle now guarantees the formula for every positive integer n.   ∎

Other Starting Integers

Instead of starting at $n = 1$ some induction arguments start at another integer. The steps for such an argument are as follows.

1. Check that the formula holds for $n = n_1$ (the first appropriate integer).

2. Prove that if the formula holds for any integer $n = k \geq n_1$, then it also holds for $n = (k + 1)$.

Once these steps are completed, the mathematical induction principle guarantees the formula for all $n \geq n_1$.

EXAMPLE 3   Show that $n! > 3^n$ if n is large enough.

Solution   How large is large enough? We experiment:

n	1	2	3	4	5	6	7
$n!$	1	2	6	24	120	720	5040
3^n	3	9	27	81	243	729	2187

It looks as if $n! > 3^n$ for $n \geq 7$. To be sure, we apply mathematical induction. We take $n_1 = 7$ in Step 1 and complete Step 2.

Suppose $k! > 3^k$ for some $k \geq 7$. Then

$$(k + 1)! = (k + 1)(k!) > (k + 1)3^k > 7 \cdot 3^k > 3^{k+1}.$$

Thus, for $k \geq 7$,

$$k! > 3^k \quad \text{implies} \quad (k + 1)! > 3^{k+1}.$$

The mathematical induction principle now guarantees $n! \geq 3^n$ for all $n \geq 7$.   ∎

Proof of the Derivative Sum Rule for Sums of Finitely Many Functions

We prove the statement

$$\frac{d}{dx}(u_1 + u_2 + \cdots + u_n) = \frac{du_1}{dx} + \frac{du_2}{dx} + \cdots + \frac{du_n}{dx}$$

In Exercises 9–15, write an equation for each line described.

9. Passes through $(-1, 1)$ with slope -1

10. Passes through $(3, 4)$ and $(-2, 5)$

11. Has slope $-5/4$ and y-intercept 6

12. Passes through $(-12, -9)$ and has slope 0

13. Has y-intercept 4 and x-intercept -1

14. Passes through $(5, -1)$ and is parallel to the line $2x + 5y = 15$

15. Passes through $(4, 10)$ and is perpendicular to the line $6x - 3y = 5$.

In Exercises 16 and 17, find the line's x- and y-intercepts and use this information to graph the line.

16. $3x + 4y = 12$ 17. $\sqrt{2}x - \sqrt{3}y = \sqrt{6}$

18. Is there anything special about the relationship between the lines $Ax + By = C_1$ and $Bx - Ay = C_2 (A \neq 0, B \neq 0)$? Give reasons for your answer.

19. A particle starts at $A(-2, 3)$ and its coordinates change by increments $\Delta x = 5$, $\Delta y = -6$. Find its new position.

20. The coordinates of a particle change by $\Delta x = 5$ and $\Delta y = 6$ as it moves from $A(x, y)$ to $B(3, -3)$. Find x and y.

Circles

In Exercises 21–23, find an equation for the circle with the given center $C(h, k)$ and radius a. Then sketch the circle in the xy-plane. Include the circle's center in your sketch. Also, label the circle's x- and y-intercepts, if any, with their coordinate pairs.

21. $C(0, 2)$, $a = 2$ 22. $C(-1, 5)$, $a = \sqrt{10}$

23. $C(-\sqrt{3}, -2)$, $a = 2$

Graph the circles whose equations are given in Exercises 24–26. Label each circle's center and intercepts (if any) with their coordinate pairs.

24. $x^2 + y^2 + 4x - 4y + 4 = 0$

25. $x^2 + y^2 - 3y - 4 = 0$ 26. $x^2 + y^2 - 4x + 4y = 0$

Parabolas

Graph the parabolas in Exercises 27–30. Label the vertex, axis, and intercepts in each case.

27. $y = x^2 - 2x - 3$ 28. $y = -x^2 + 4x$

29. $y = -x^2 - 6x - 5$ 30. $y = \dfrac{1}{2}x^2 + x + 4$

Inequalities

Describe the regions defined by the inequalities and pairs of inequalities in Exercises 31–34.

31. $x^2 + y^2 > 7$ 32. $(x - 1)^2 + y^2 \leq 4$

33. $x^2 + y^2 > 1$, $x^2 + y^2 < 4$

34. $x^2 + y^2 + 6y < 0$, $y > -3$

35. Write an inequality that describes the points that lie inside the circle with center $(-2, 1)$ and radius $\sqrt{6}$.

36. Write a pair of inequalities that describe the points that lie inside or on the circle with center $(0, 0)$ and radius $\sqrt{2}$, and on or to the right of the vertical line through $(1, 0)$.

Theory and Examples

In Exercises 37–40, graph the two equations and find the points at which the graphs intersect.

37. $y = 2x$, $x^2 + y^2 = 1$ 38. $y - x = 1$, $y = x^2$

39. $y = -x^2$, $y = 2x^2 - 1$

40. $x^2 + y^2 = 1$, $(x - 1)^2 + y^2 = 1$

41. **Insulation** By measuring slopes in the figure, estimate the temperature change in degrees per inch for **(a)** the gypsum wallboard; **(b)** the fiberglass insulation; **(c)** the wood sheathing.

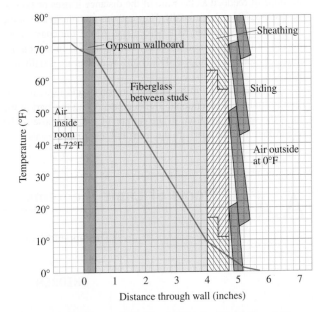

The temperature changes in the wall in Exercises 41 and 42.

42. **Insulation** According to the figure in Exercise 41, which of the materials is the best insulator? The poorest? Explain.

43. **Pressure under water** The pressure p experienced by a diver under water is related to the diver's depth d by an equation of the form $p = kd + 1$ (k a constant). At the surface, the pressure is 1 atmosphere. The pressure at 100 meters is about 10.94 atmospheres. Find the pressure at 50 meters.

44. **Reflected light** A ray of light comes in along the line $x + y = 1$ from the second quadrant and reflects off the x-axis (see the accompanying figure). The angle of incidence is equal to the angle of reflection. Write an equation for the line along which the departing light travels.

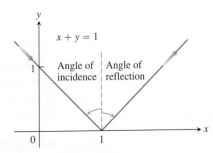

The path of the light ray in Exercise 44. Angles of incidence and reflection are measured from the perpendicular.

45. Fahrenheit vs. Celsius In the FC-plane, sketch the graph of the equation

$$C = \frac{5}{9}(F - 32)$$

linking Fahrenheit and Celsius temperatures. On the same graph sketch the line $C = F$. Is there a temperature at which a Celsius thermometer gives the same numerical reading as a Fahrenheit thermometer? If so, find it.

46. The Mt. Washington Cog Railway Civil engineers calculate the slope of roadbed as the ratio of the distance it rises or falls to the distance it runs horizontally. They call this ratio the **grade** of the roadbed, usually written as a percentage. Along the coast, commercial railroad grades are usually less than 2%. In the mountains, they may go as high as 4%. Highway grades are usually less than 5%.

The steepest part of the Mt. Washington Cog Railway in New Hampshire has an exceptional 37.1% grade. Along this part of the track, the seats in the front of the car are 14 ft above those in the rear. About how far apart are the front and rear rows of seats?

47. By calculating the lengths of its sides, show that the triangle with vertices at the points $A(1, 2)$, $B(5, 5)$, and $C(4, -2)$ is isosceles but not equilateral.

48. Show that the triangle with vertices $A(0, 0)$, $B\left(1, \sqrt{3}\right)$, and $C(2, 0)$ is equilateral.

49. Show that the points $A(2, -1)$, $B(1, 3)$, and $C(-3, 2)$ are vertices of a square, and find the fourth vertex.

50. Three different parallelograms have vertices at $(-1, 1)$, $(2, 0)$, and $(2, 3)$. Sketch them and find the coordinates of the fourth vertex of each.

51. For what value of k is the line $2x + ky = 3$ perpendicular to the line $4x + y = 1$? For what value of k are the lines parallel?

52. Midpoint of a line segment Show that the point with coordinates

$$\left(\frac{x_1 + x_2}{2}, \frac{y_1 + y_2}{2}\right)$$

is the midpoint of the line segment joining $P(x_1, y_1)$ to $Q(x_2, y_2)$.

A.4 | Proofs of Limit Theorems

This appendix proves Theorem 1, Parts 2–5, and Theorem 4 from Section 2.2.

THEOREM 1—Limit Laws   If L, M, c, and k are real numbers and

$$\lim_{x \to c} f(x) = L \quad \text{and} \quad \lim_{x \to c} g(x) = M, \quad \text{then}$$

1. *Sum Rule*: $\qquad\qquad\qquad\qquad \lim_{x \to c} (f(x) + g(x)) = L + M$

2. *Difference Rule*: $\qquad\qquad\quad \lim_{x \to c} (f(x) - g(x)) = L - M$

3. *Constant Multiple Rule*: $\qquad \lim_{x \to c} (k \cdot f(x)) = k \cdot L$

4. *Product Rule*: $\qquad\qquad\qquad \lim_{x \to c} (f(x) \cdot g(x)) = L \cdot M$

5. *Quotient Rule*: $\qquad\qquad\quad \lim_{x \to c} \dfrac{f(x)}{g(x)} = \dfrac{L}{M}, \qquad M \neq 0$

6. *Power Rule*: $\qquad\qquad\qquad \lim_{x \to c} [f(x)]^n = L^n$, n a positive integer

7. *Root Rule*: $\qquad\qquad\qquad\;\; \lim_{x \to c} \sqrt[n]{f(x)} = \sqrt[n]{L} = L^{1/n}$, n a positive integer

(If n is even, we assume that $\lim_{x \to c} f(x) = L > 0$.)

We proved the Sum Rule in Section 2.3 and the Power and Root Rules are proved in more advanced texts. We obtain the Difference Rule by replacing $g(x)$ by $-g(x)$ and M by $-M$ in the Sum Rule. The Constant Multiple Rule is the special case $g(x) = k$ of the Product Rule. This leaves only the Product and Quotient Rules.

Proof of the Limit Product Rule   We show that for any $\epsilon > 0$ there exists a $\delta > 0$ such that for all x in the intersection D of the domains of f and g,

$$0 < |x - c| < \delta \quad \Rightarrow \quad |f(x)g(x) - LM| < \epsilon.$$

Hence, all the *n*th roots of $z = re^{i\theta}$ are given by

$$\sqrt[n]{re^{i\theta}} = \sqrt[n]{r} \exp i\left(\frac{\theta}{n} + k\frac{2\pi}{n}\right), \qquad k = 0, \pm 1, \pm 2, \ldots . \qquad (15)$$

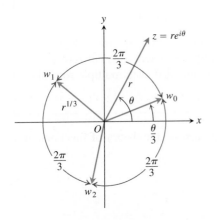

FIGURE A.28   The three cube roots of $z = re^{i\theta}$.

There might appear to be infinitely many different answers corresponding to the infinitely many possible values of *k*, but $k = n + m$ gives the same answer as $k = m$ in Equation (15). Thus, we need only take *n* consecutive values for *k* to obtain all the different *n*th roots of *z*. For convenience, we take

$$k = 0, 1, 2, \ldots, n - 1.$$

All the *n*th roots of $re^{i\theta}$ lie on a circle centered at the origin and having radius equal to the real, positive *n*th root of *r*. One of them has argument $\alpha = \theta/n$. The others are uniformly spaced around the circle, each being separated from its neighbors by an angle equal to $2\pi/n$. Figure A.28 illustrates the placement of the three cube roots, w_0, w_1, w_2, of the complex number $z = re^{i\theta}$.

EXAMPLE 5   Find the four fourth roots of -16.

Solution   As our first step, we plot the number -16 in an Argand diagram (Figure A.29) and determine its polar representation $re^{i\theta}$. Here, $z = -16$, $r = +16$, and $\theta = \pi$. One of the fourth roots of $16e^{i\pi}$ is $2e^{i\pi/4}$. We obtain others by successive additions of $2\pi/4 = \pi/2$ to the argument of this first one. Hence,

$$\sqrt[4]{16 \exp i\pi} = 2 \exp i\left(\frac{\pi}{4}, \frac{3\pi}{4}, \frac{5\pi}{4}, \frac{7\pi}{4}\right),$$

and the four roots are

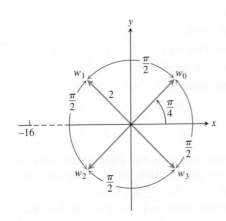

FIGURE A.29   The four fourth roots of -16.

$$w_0 = 2\left[\cos\frac{\pi}{4} + i\sin\frac{\pi}{4}\right] = \sqrt{2}(1 + i)$$

$$w_1 = 2\left[\cos\frac{3\pi}{4} + i\sin\frac{3\pi}{4}\right] = \sqrt{2}(-1 + i)$$

$$w_2 = 2\left[\cos\frac{5\pi}{4} + i\sin\frac{5\pi}{4}\right] = \sqrt{2}(-1 - i)$$

$$w_3 = 2\left[\cos\frac{7\pi}{4} + i\sin\frac{7\pi}{4}\right] = \sqrt{2}(1 - i). \qquad \blacksquare$$

The Fundamental Theorem of Algebra

One might say that the invention of $\sqrt{-1}$ is all well and good and leads to a number system that is richer than the real number system alone; but where will this process end? Are we also going to invent still more systems so as to obtain $\sqrt[4]{-1}$, $\sqrt[6]{-1}$, and so on? But it turns out this is not necessary. These numbers are already expressible in terms of the complex number system $a + ib$. In fact, the Fundamental Theorem of Algebra says that with the introduction of the complex numbers we now have enough numbers to factor every polynomial into a product of linear factors and so enough numbers to solve every possible polynomial equation.

> ### The Fundamental Theorem of Algebra
>
> Every polynomial equation of the form
>
> $$a_n z^n + a_{n-1} z^{n-1} + \cdots + a_1 z + a_0 = 0,$$
>
> in which the coefficients $a_0, a_1, \ldots, a_n$ are any complex numbers, whose degree n is greater than or equal to one, and whose leading coefficient a_n is not zero, has exactly n roots in the complex number system, provided each multiple root of multiplicity m is counted as m roots.

A proof of this theorem can be found in almost any text on the theory of functions of a complex variable.

Exercises A.7

Operations with Complex Numbers

1. **How computers multiply complex numbers** Find $(a, b) \cdot (c, d) = (ac - bd, ad + bc)$.

   a. $(2, 3) \cdot (4, -2)$ **b.** $(2, -1) \cdot (-2, 3)$

   c. $(-1, -2) \cdot (2, 1)$

   (This is how complex numbers are multiplied by computers.)

2. Solve the following equations for the real numbers, x and y.

   a. $(3 + 4i)^2 - 2(x - iy) = x + iy$

   b. $\left(\dfrac{1 + i}{1 - i}\right)^2 + \dfrac{1}{x + iy} = 1 + i$

   c. $(3 - 2i)(x + iy) = 2(x - 2iy) + 2i - 1$

Graphing and Geometry

3. How may the following complex numbers be obtained from $z = x + iy$ geometrically? Sketch.

   a. $\bar{z}$ **b.** $\overline{(-z)}$

   c. $-z$ **d.** $1/z$

4. Show that the distance between the two points z_1 and z_2 in an Argand diagram is $|z_1 - z_2|$.

In Exercises 5–10, graph the points $z = x + iy$ that satisfy the given conditions.

5. **a.** $|z| = 2$ **b.** $|z| < 2$ **c.** $|z| > 2$

6. $|z - 1| = 2$ 7. $|z + 1| = 1$

8. $|z + 1| = |z - 1|$ 9. $|z + i| = |z - 1|$

10. $|z + 1| \geq |z|$

Express the complex numbers in Exercises 11–14 in the form $re^{i\theta}$, with $r \geq 0$ and $-\pi < \theta \leq \pi$. Draw an Argand diagram for each calculation.

11. $\left(1 + \sqrt{-3}\right)^2$

12. $\dfrac{1 + i}{1 - i}$

13. $\dfrac{1 + i\sqrt{3}}{1 - i\sqrt{3}}$

14. $(2 + 3i)(1 - 2i)$

Powers and Roots

Use De Moivre's Theorem to express the trigonometric functions in Exercises 15 and 16 in terms of $\cos \theta$ and $\sin \theta$.

15. $\cos 4\theta$ 16. $\sin 4\theta$

17. Find the three cube roots of 1.

18. Find the two square roots of i.

19. Find the three cube roots of $-8i$.

20. Find the six sixth roots of 64.

21. Find the four solutions of the equation $z^4 - 2z^2 + 4 = 0$.

22. Find the six solutions of the equation $z^6 + 2z^3 + 2 = 0$.

23. Find all solutions of the equation $x^4 + 4x^2 + 16 = 0$.

24. Solve the equation $x^4 + 1 = 0$.

Theory and Examples

25. **Complex numbers and vectors in the plane** Show with an Argand diagram that the law for adding complex numbers is the same as the parallelogram law for adding vectors.

26. **Complex arithmetic with conjugates** Show that the conjugate of the sum (product, or quotient) of two complex numbers, z_1 and z_2, is the same as the sum (product, or quotient) of their conjugates.

27. **Complex roots of polynomials with real coefficients come in complex-conjugate pairs**

   a. Extend the results of Exercise 26 to show that $f(\bar{z}) = \overline{f(z)}$ if

   $$f(z) = a_n z^n + a_{n-1} z^{n-1} + \cdots + a_1 z + a_0$$

   is a polynomial with real coefficients $a_0, \ldots, a_n$.

   b. If z is a root of the equation $f(z) = 0$, where $f(z)$ is a polynomial with real coefficients as in part (a), show that the conjugate $\bar{z}$ is also a root of the equation. (*Hint:* Let $f(z) = u + iv = 0$; then both u and v are zero. Use the fact that $f(\bar{z}) = \overline{f(z)} = u - iv$.)

28. **Absolute value of a conjugate** Show that $|\bar{z}| = |z|$.

29. **When $z = \bar{z}$** If z and $\bar{z}$ are equal, what can you say about the location of the point z in the complex plane?

30. Real and imaginary parts Let $\text{Re}(z)$ denote the real part of z and $\text{Im}(z)$ the imaginary part. Show that the following relations hold for any complex numbers z, z_1, and z_2.

a. $z + \bar{z} = 2\text{Re}(z)$

b. $z - \bar{z} = 2i\text{Im}(z)$

c. $|\text{Re}(z)| \leq |z|$

d. $|z_1 + z_2|^2 = |z_1|^2 + |z_2|^2 + 2\text{Re}(z_1\bar{z}_2)$

e. $|z_1 + z_2| \leq |z_1| + |z_2|$

A.8 The Distributive Law for Vector Cross Products

In this appendix we prove the Distributive Law

$$\mathbf{u} \times (\mathbf{v} + \mathbf{w}) = \mathbf{u} \times \mathbf{v} + \mathbf{u} \times \mathbf{w},$$

which is Property 2 in Section 12.4.

Proof To derive the Distributive Law, we construct $\mathbf{u} \times \mathbf{v}$ a new way. We draw $\mathbf{u}$ and $\mathbf{v}$ from the common point O and construct a plane M perpendicular to $\mathbf{u}$ at O (Figure A.30). We then project $\mathbf{v}$ orthogonally onto M, yielding a vector $\mathbf{v}'$ with length $|\mathbf{v}|\sin\theta$. We rotate $\mathbf{v}'$ 90° about $\mathbf{u}$ in the positive sense to produce a vector $\mathbf{v}''$. Finally, we multiply $\mathbf{v}''$ by the length of $\mathbf{u}$. The resulting vector $|\mathbf{u}|\mathbf{v}''$ is equal to $\mathbf{u} \times \mathbf{v}$ since $\mathbf{v}''$ has the same direction as $\mathbf{u} \times \mathbf{v}$ by its construction (Figure A.30) and

$$|\mathbf{u}||\mathbf{v}''| = |\mathbf{u}||\mathbf{v}'| = |\mathbf{u}||\mathbf{v}|\sin\theta = |\mathbf{u} \times \mathbf{v}|.$$

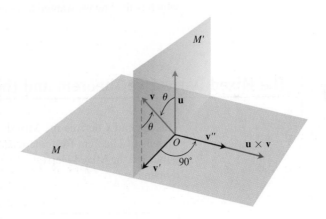

FIGURE A.30 As explained in the text, $\mathbf{u} \times \mathbf{v} = |\mathbf{u}|\mathbf{v}''$.

Now each of these three operations, namely,

1. projection onto M

2. rotation about $\mathbf{u}$ through 90°

3. multiplication by the scalar $|\mathbf{u}|$

when applied to a triangle whose plane is not parallel to $\mathbf{u}$, will produce another triangle. If we start with the triangle whose sides are $\mathbf{v}$, $\mathbf{w}$, and $\mathbf{v} + \mathbf{w}$ (Figure A.31) and apply these three steps, we successively obtain the following:

1. A triangle whose sides are $\mathbf{v}'$, $\mathbf{w}'$, and $(\mathbf{v} + \mathbf{w})'$ satisfying the vector equation

$$\mathbf{v}' + \mathbf{w}' = (\mathbf{v} + \mathbf{w})'$$

2. A triangle whose sides are $\mathbf{v}''$, $\mathbf{w}''$, and $(\mathbf{v} + \mathbf{w})''$ satisfying the vector equation

$$\mathbf{v}'' + \mathbf{w}'' = (\mathbf{v} + \mathbf{w})''$$

(the double prime on each vector has the same meaning as in Figure A.30)

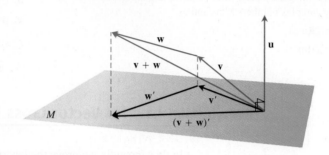

FIGURE A.31 The vectors, $\mathbf{v}$, $\mathbf{w}$, $\mathbf{v} + \mathbf{w}$, and their projections onto a plane perpendicular to $\mathbf{u}$.

3. A triangle whose sides are $|\mathbf{u}|\mathbf{v}''$, $|\mathbf{u}|\mathbf{w}''$, and $|\mathbf{u}|(\mathbf{v} + \mathbf{w})''$ satisfying the vector equation

$$|\mathbf{u}|\mathbf{v}'' + |\mathbf{u}|\mathbf{w}'' = |\mathbf{u}|(\mathbf{v} + \mathbf{w})''.$$

Substituting $|\mathbf{u}|\mathbf{v}'' = \mathbf{u} \times \mathbf{v}$, $|\mathbf{u}|\mathbf{w}'' = \mathbf{u} \times \mathbf{w}$, and $|\mathbf{u}|(\mathbf{v} + \mathbf{w})'' = \mathbf{u} \times (\mathbf{v} + \mathbf{w})$ from our discussion above into this last equation gives

$$\mathbf{u} \times \mathbf{v} + \mathbf{u} \times \mathbf{w} = \mathbf{u} \times (\mathbf{v} + \mathbf{w}),$$

which is the law we wanted to establish. ∎

A.9 The Mixed Derivative Theorem and the Increment Theorem

This appendix derives the Mixed Derivative Theorem (Theorem 2, Section 14.3) and the Increment Theorem for Functions of Two Variables (Theorem 3, Section 14.3). Euler first published the Mixed Derivative Theorem in 1734, in a series of papers he wrote on hydrodynamics.

THEOREM 2—The Mixed Derivative Theorem If $f(x, y)$ and its partial derivatives f_x, f_y, f_{xy}, and f_{yx} are defined throughout an open region containing a point (a, b) and are all continuous at (a, b), then

$$f_{xy}(a, b) = f_{yx}(a, b).$$

Proof The equality of $f_{xy}(a, b)$ and $f_{yx}(a, b)$ can be established by four applications of the Mean Value Theorem (Theorem 4, Section 4.2). By hypothesis, the point (a, b) lies in the interior of a rectangle R in the xy-plane on which f, f_x, f_y, f_{xy}, and f_{yx} are all defined. We let h and k be the numbers such that the point $(a + h, b + k)$ also lies in R, and we consider the difference

$$\Delta = F(a + h) - F(a), \tag{1}$$

where

$$F(x) = f(x, b + k) - f(x, b). \tag{2}$$

We apply the Mean Value Theorem to F, which is continuous because it is differentiable. Then Equation (1) becomes

$$\Delta = hF'(c_1), \tag{3}$$

where c_1 lies between a and $a + h$. From Equation (2),

$$F'(x) = f_x(x, b + k) - f_x(x, b),$$

so Equation (3) becomes

$$\Delta = h[f_x(c_1, b + k) - f_x(c_1, b)]. \tag{4}$$

Now we apply the Mean Value Theorem to the function $g(y) = f_x(c_1, y)$ and have

$$g(b + k) - g(b) = kg'(d_1),$$

or

$$f_x(c_1, b + k) - f_x(c_1, b) = kf_{xy}(c_1, d_1)$$

for some d_1 between b and $b + k$. By substituting this into Equation (4), we get

$$\Delta = hkf_{xy}(c_1, d_1) \tag{5}$$

for some point (c_1, d_1) in the rectangle R' whose vertices are the four points (a, b), $(a + h, b)$, $(a + h, b + k)$, and $(a, b + k)$. (See Figure A.32.)

By substituting from Equation (2) into Equation (1), we may also write

$$\begin{aligned}
\Delta &= f(a + h, b + k) - f(a + h, b) - f(a, b + k) + f(a, b) \\
&= [f(a + h, b + k) - f(a, b + k)] - [f(a + h, b) - f(a, b)] \\
&= \phi(b + k) - \phi(b),
\end{aligned} \tag{6}$$

where

$$\phi(y) = f(a + h, y) - f(a, y). \tag{7}$$

The Mean Value Theorem applied to Equation (6) now gives

$$\Delta = k\phi'(d_2) \tag{8}$$

for some d_2 between b and $b + k$. By Equation (7),

$$\phi'(y) = f_y(a + h, y) - f_y(a, y). \tag{9}$$

Substituting from Equation (9) into Equation (8) gives

$$\Delta = k[f_y(a + h, d_2) - f_y(a, d_2)].$$

Finally, we apply the Mean Value Theorem to the expression in brackets and get

$$\Delta = khf_{yx}(c_2, d_2) \tag{10}$$

for some c_2 between a and $a + h$.

Together, Equations (5) and (10) show that

$$f_{xy}(c_1, d_1) = f_{yx}(c_2, d_2), \tag{11}$$

where (c_1, d_1) and (c_2, d_2) both lie in the rectangle R' (Figure A.32). Equation (11) is not quite the result we want, since it says only that f_{xy} has the same value at (c_1, d_1) that f_{yx} has at (c_2, d_2). The numbers h and k in our discussion, however, may be made as small as we wish. The hypothesis that f_{xy} and f_{yx} are both continuous at (a, b) means that $f_{xy}(c_1, d_1) = f_{xy}(a, b) + \epsilon_1$ and $f_{yx}(c_2, d_2) = f_{yx}(a, b) + \epsilon_2$, where each of $\epsilon_1, \epsilon_2 \to 0$ as both $h, k \to 0$. Hence, if we let h and $k \to 0$, we have $f_{xy}(a, b) = f_{yx}(a, b)$.   ∎

The equality of $f_{xy}(a, b)$ and $f_{yx}(a, b)$ can be proved with hypotheses weaker than the ones we assumed. For example, it is enough for f, f_x, and f_y to exist in R and for f_{xy} to be continuous at (a, b). Then f_{yx} will exist at (a, b) and equal f_{xy} at that point.

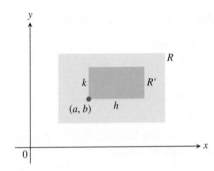

FIGURE A.32   The key to proving $f_{xy}(a, b) = f_{yx}(a, b)$ is that no matter how small R' is, f_{xy} and f_{yx} take on equal values somewhere inside R' (although not necessarily at the same point).

THEOREM 3—The Increment Theorem for Functions of Two Variables   Suppose that the first partial derivatives of $f(x, y)$ are defined throughout an open region R containing the point (x_0, y_0) and that f_x and f_y are continuous at (x_0, y_0). Then the change

$$\Delta z = f(x_0 + \Delta x, y_0 + \Delta y) - f(x_0, y_0)$$

in the value of f that results from moving from (x_0, y_0) to another point $(x_0 + \Delta x, y_0 + \Delta y)$ in R satisfies an equation of the form

$$\Delta z = f_x(x_0, y_0)\Delta x + f_y(x_0, y_0)\Delta y + \epsilon_1\Delta x + \epsilon_2\Delta y$$

in which each of $\epsilon_1, \epsilon_2 \to 0$ as both $\Delta x, \Delta y \to 0$.

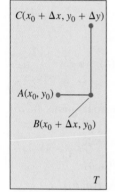

FIGURE A.33   The rectangular region T in the proof of the Increment Theorem. The figure is drawn for Δx and Δy positive, but either increment might be zero or negative.

Proof   We work within a rectangle T centered at $A(x_0, y_0)$ and lying within R, and we assume that Δx and Δy are already so small that the line segment joining A to $B(x_0 + \Delta x, y_0)$ and the line segment joining B to $C(x_0 + \Delta x, y_0 + \Delta y)$ lie in the interior of T (Figure A.33).

We may think of Δz as the sum $\Delta z = \Delta z_1 + \Delta z_2$ of two increments, where

$$\Delta z_1 = f(x_0 + \Delta x, y_0) - f(x_0, y_0)$$

is the change in the value of f from A to B and

$$\Delta z_2 = f(x_0 + \Delta x, y_0 + \Delta y) - f(x_0 + \Delta x, y_0)$$

is the change in the value of f from B to C (Figure A.34).

On the closed interval of x-values joining x_0 to $x_0 + \Delta x$, the function $F(x) = f(x, y_0)$ is a differentiable (and hence continuous) function of x, with derivative

$$F'(x) = f_x(x, y_0).$$

By the Mean Value Theorem (Theorem 4, Section 4.2), there is an x-value c between x_0 and $x_0 + \Delta x$ at which

$$F(x_0 + \Delta x) - F(x_0) = F'(c)\,\Delta x$$

or

$$f(x_0 + \Delta x, y_0) - f(x_0, y_0) = f_x(c, y_0)\,\Delta x$$

or

$$\Delta z_1 = f_x(c, y_0)\,\Delta x. \tag{12}$$

Similarly, $G(y) = f(x_0 + \Delta x, y)$ is a differentiable (and hence continuous) function of y on the closed y-interval joining y_0 and $y_0 + \Delta y$, with derivative

$$G'(y) = f_y(x_0 + \Delta x, y).$$

Hence, there is a y-value d between y_0 and $y_0 + \Delta y$ at which

$$G(y_0 + \Delta y) - G(y_0) = G'(d)\,\Delta y$$

or

$$f(x_0 + \Delta x, y_0 + \Delta y) - f(x_0 + \Delta x, y) = f_y(x_0 + \Delta x, d)\,\Delta y$$

or

$$\Delta z_2 = f_y(x_0 + \Delta x, d)\,\Delta y. \tag{13}$$

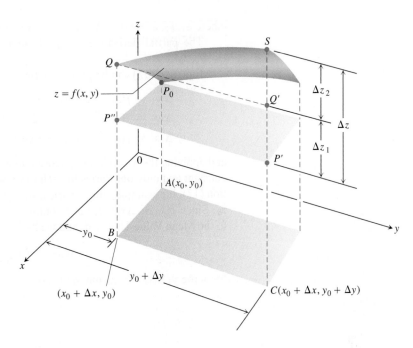

FIGURE A.34   Part of the surface $z = f(x, y)$ near $P_0(x_0, y_0, f(x_0, y_0))$. The points P_0, P', and P'' have the same height $z_0 = f(x_0, y_0)$ above the xy-plane. The change in z is $\Delta z = P'S$. The change

$$\Delta z_1 = f(x_0 + \Delta x, y_0) - f(x_0, y_0),$$

shown as $P''Q = P'Q'$, is caused by changing x from x_0 to $x_0 + \Delta x$ while holding y equal to y_0. Then, with x held equal to $x_0 + \Delta x$,

$$\Delta z_2 = f(x_0 + \Delta x, y_0 + \Delta y) - f(x_0 + \Delta x, y_0)$$

is the change in z caused by changing y_0 from $y_0 + \Delta y$, which is represented by $Q'S$. The total change in z is the sum of Δz_1 and Δz_2.

Now, as both Δx and $\Delta y \to 0$, we know that $c \to x_0$ and $d \to y_0$. Therefore, since f_x and f_y are continuous at (x_0, y_0), the quantities

$$\begin{align} \epsilon_1 &= f_x(c, y_0) - f_x(x_0, y_0), \\ \epsilon_2 &= f_y(x_0 + \Delta x, d) - f_y(x_0, y_0) \end{align} \tag{14}$$

both approach zero as both Δx and $\Delta y \to 0$.

Finally,

$$\begin{align} \Delta z &= \Delta z_1 + \Delta z_2 \\ &= f_x(c, y_0)\Delta x + f_y(x_0 + \Delta x, d)\Delta y && \text{From Eqs. (12) and (13)} \\ &= [f_x(x_0, y_0) + \epsilon_1]\Delta x + [f_y(x_0, y_0) + \epsilon_2]\Delta y && \text{From Eq. (14)} \\ &= f_x(x_0, y_0)\Delta x + f_y(x_0, y_0)\Delta y + \epsilon_1\Delta x + \epsilon_2\Delta y, \end{align}$$

where both ϵ_1 and $\epsilon_2 \to 0$ as both Δx and $\Delta y \to 0$, which is what we set out to prove.   ∎

Analogous results hold for functions of any finite number of independent variables. Suppose that the first partial derivatives of $w = f(x, y, z)$ are defined throughout an open region containing the point (x_0, y_0, z_0) and that f_x, f_y, and f_z are continuous at (x_0, y_0, z_0). Then

$$\begin{align} \Delta w &= f(x_0 + \Delta x, y_0 + \Delta y, z_0 + \Delta z) - f(x_0, y_0, z_0) \\ &= f_x\Delta x + f_y\Delta y + f_z\Delta z + \epsilon_1\Delta x + \epsilon_2\Delta y + \epsilon_3\Delta z, \end{align} \tag{15}$$

where $\epsilon_1, \epsilon_2, \epsilon_3 \rightarrow 0$ as $\Delta x, \Delta y$, and $\Delta z \rightarrow 0$.

The partial derivatives f_x, f_y, f_z in Equation (15) are to be evaluated at the point (x_0, y_0, z_0).

Equation (15) can be proved by treating Δw as the sum of three increments,

$$\Delta w_1 = f(x_0 + \Delta x, y_0, z_0) - f(x_0, y_0, z_0) \tag{16}$$

$$\Delta w_2 = f(x_0 + \Delta x, y_0 + \Delta y, z_0) - f(x_0 + \Delta x, y_0, z_0) \tag{17}$$

$$\Delta w_3 = f(x_0 + \Delta x, y_0 + \Delta y, z_0 + \Delta z) - f(x_0 + \Delta x, y_0 + \Delta y, z_0), \tag{18}$$

and applying the Mean Value Theorem to each of these separately. Two coordinates remain constant and only one varies in each of these partial increments $\Delta w_1, \Delta w_2, \Delta w_3$. In Equation (17), for example, only y varies, since x is held equal to $x_0 + \Delta x$ and z is held equal to z_0. Since $f(x_0 + \Delta x, y, z_0)$ is a continuous function of y with a derivative f_y, it is subject to the Mean Value Theorem, and we have

$$\Delta w_2 = f_y(x_0 + \Delta x, y_1, z_0) \, \Delta y$$

for some y_1 between y_0 and $y_0 + \Delta y$.

ANSWERS TO ODD-NUMBERED EXERCISES

CHAPTER 1

Section 1.1, pp. 11–13

1. $D: (-\infty, \infty),\quad R: [1, \infty)$ **3.** $D: [-2, \infty),\quad R: [0, \infty)$

5. $D: (-\infty, 3) \cup (3, \infty),\quad R: (-\infty, 0) \cup (0, \infty)$

7. **(a)** Not a function of x because some values of x have two values of y

  (b) A function of x because for every x there is only one possible y

9. $A = \dfrac{\sqrt{3}}{4}x^2,\quad p = 3x$ **11.** $x = \dfrac{d}{\sqrt{3}},\quad A = 2d^2,\quad V = \dfrac{d^3}{3\sqrt{3}}$

13. $L = \dfrac{\sqrt{20x^2 - 20x + 25}}{4}$

15. $(-\infty, \infty)$

17. $(-\infty, \infty)$

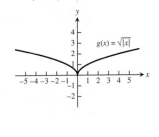

19. $(-\infty, 0) \cup (0, \infty)$

21. $(-\infty, -5) \cup (-5, -3] \cup [3, 5) \cup (5, \infty)$

23. **(a)** For each positive value of x, there are two values of y.   **(b)** For each value of $x \neq 0$, there are two values of y.

25.

27.

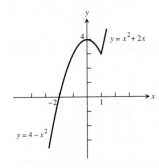

29. **(a)** $f(x) = \begin{cases} x, & 0 \le x \le 1 \\ -x + 2, & 1 < x \le 2 \end{cases}$

  (b) $f(x) = \begin{cases} 2, & 0 \le x < 1 \\ 0, & 1 \le x < 2 \\ 2, & 2 \le x < 3 \\ 0, & 3 \le x \le 4 \end{cases}$

31. **(a)** $f(x) = \begin{cases} -x, & -1 \le x < 0 \\ 1, & 0 < x \le 1 \\ -\frac{1}{2}x + \frac{3}{2}, & 1 < x < 3 \end{cases}$

  (b) $f(x) = \begin{cases} \frac{1}{2}x, & -2 \le x \le 0 \\ -2x + 2, & 0 < x \le 1 \\ -1, & 1 < x \le 3 \end{cases}$

33. **(a)** $0 \le x < 1$ **(b)** $-1 < x \le 0$ **35.** Yes

37. Symmetric about the origin   **39.** Symmetric about the origin

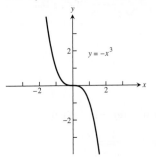

Dec. $-\infty < x < \infty$

Inc. $-\infty < x < 0$ and $0 < x < \infty$

41. Symmetric about the y-axis   **43.** Symmetric about the origin

Dec. $-\infty < x \le 0$; inc. $0 \le x < \infty$

Inc. $-\infty < x < \infty$

45. No symmetry

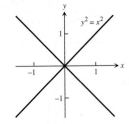

Dec. $0 \le x < \infty$

47. Even **49.** Even **51.** Odd **53.** Even

55. Neither **57.** Neither **59.** $t = 180$ **61.** $s = 2.4$

63. $V = x(14 - 2x)(22 - 2x)$

65. **(a)** h **(b)** f **(c)** g **67.** **(a)** $(-2, 0) \cup (4, \infty)$

71. $C = 5(2 + \sqrt{2})h$

Section 1.2, pp. 19–22

1. $D_f: -\infty < x < \infty,\quad D_g: x \geq 1,\quad R_f: -\infty < y < \infty,$
$R_g: y \geq 0,\quad D_{f+g} = D_{f \cdot g} = D_g,\quad R_{f+g}: y \geq 1,\quad R_{f \cdot g}: y \geq 0$

3. $D_f: -\infty < x < \infty,\quad D_g: -\infty < x < \infty,\quad R_f: y = 2,$
$R_g: y \geq 1,\quad D_{f/g}: -\infty < x < \infty,\quad R_{f/g}: 0 < y \leq 2,$
$D_{g/f}: -\infty < x < \infty,\quad R_{g/f}: y \geq 1/2$

5. (a) 2 **(b)** 22 **(c)** $x^2 + 2$ **(d)** $x^2 + 10x + 22$ **(e)** 5
(f) -2 **(g)** $x + 10$ **(h)** $x^4 - 6x^2 + 6$

7. $13 - 3x$ **9.** $\sqrt{\dfrac{5x + 1}{4x + 1}}$

11. (a) $f(g(x))$ **(b)** $j(g(x))$ **(c)** $g(g(x))$ **(d)** $j(j(x))$
(e) $g(h(f(x)))$ **(f)** $h(j(f(x)))$

13.

	$g(x)$	$f(x)$	$(f \circ g)(x)$
(a)	$x - 7$	$\sqrt{x}$	$\sqrt{x - 7}$
(b)	$x + 2$	$3x$	$3x + 6$
(c)	x^2	$\sqrt{x - 5}$	$\sqrt{x^2 - 5}$
(d)	$\dfrac{x}{x - 1}$	$\dfrac{x}{x - 1}$	x
(e)	$\dfrac{1}{x - 1}$	$1 + \dfrac{1}{x}$	x
(f)	$\dfrac{1}{x}$	$\dfrac{1}{x}$	x

15. (a) 1 **(b)** 2 **(c)** -2 **(d)** 0 **(e)** -1 **(f)** 0

17. (a) $f(g(x)) = \sqrt{\dfrac{1}{x} + 1},\ g(f(x)) = \dfrac{1}{\sqrt{x + 1}}$
(b) $D_{f \circ g} = (-\infty, -1] \cup (0, \infty),\ D_{g \circ f} = (-1, \infty)$
(c) $R_{f \circ g} = [0, 1) \cup (1, \infty),\ R_{g \circ f} = (0, \infty)$

19. $g(x) = \dfrac{2x}{x - 1}$

21. (a) $y = -(x + 7)^2$ **(b)** $y = -(x - 4)^2$

23. (a) Position 4 **(b)** Position 1 **(c)** Position 2
(d) Position 3

25. $(x + 2)^2 + (y + 3)^2 = 49$ **27.** $y + 1 = (x + 1)^3$

29. $y = \sqrt{x + 0.81}$ **31.** $y = 2x$

33. $y - 1 = \dfrac{1}{x - 1}$ **35.**

37. **39.**

41. **43.**

45. **47.**

49. **51.**

53.

55. (a) $D: [0, 2],\quad R: [2, 3]$ **(b)** $D: [0, 2],\quad R: [-1, 0]$

(c) $D : [0, 2], \quad R : [0, 2]$ **(d)** $D : [0, 2], \quad R : [-1, 0]$

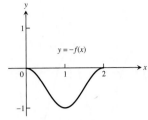

(e) $D : [-2, 0], \quad R : [0, 1]$ **(f)** $D : [1, 3], \quad R : [0, 1]$

(g) $D : [-2, 0], \quad R : [0, 1]$ **(h)** $D : [-1, 1], \quad R : [0, 1]$

57. $y = 3x^2 - 3$ **59.** $y = \dfrac{1}{2} + \dfrac{1}{2x^2}$ **61.** $y = \sqrt{4x + 1}$

63. $y = \sqrt{4 - \dfrac{x^2}{4}}$ **65.** $y = 1 - 27x^3$

67.   **69.**

71.   **73.**

75.   **77.**

79.   **81.**

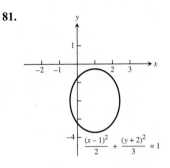

83. $\dfrac{(x + 4)^2}{16} + \dfrac{(y - 3)^2}{9} = 1$ Center: $(-4, 3)$

The major axis is the line segment between $(-8, 3)$ and $(0, 3)$.

85. (a) Odd   **(b)** Odd   **(c)** Odd   **(d)** Even   **(e)** Even
(f) Even   **(g)** Even   **(h)** Even   **(i)** Odd

Section 1.3, pp. 28–30

1. (a) 8π m   **(b)** $\dfrac{55\pi}{9}$ m   **3.** 8.4 in.

5.

θ	$-\pi$	$-2\pi/3$	0	$\pi/2$	$3\pi/4$
$\sin\theta$	0	$-\dfrac{\sqrt{3}}{2}$	0	1	$\dfrac{1}{\sqrt{2}}$
$\cos\theta$	-1	$-\dfrac{1}{2}$	1	0	$-\dfrac{1}{\sqrt{2}}$
$\tan\theta$	0	$\sqrt{3}$	0	UND	-1
$\cot\theta$	UND	$\dfrac{1}{\sqrt{3}}$	UND	0	-1
$\sec\theta$	-1	-2	1	UND	$-\sqrt{2}$
$\csc\theta$	UND	$-\dfrac{2}{\sqrt{3}}$	UND	1	$\sqrt{2}$

7. $\cos x = -4/5, \ \tan x = -3/4$

9. $\sin x = -\dfrac{\sqrt{8}}{3}, \ \tan x = -\sqrt{8}$

11. $\sin x = -\dfrac{1}{\sqrt{5}}, \ \cos x = -\dfrac{2}{\sqrt{5}}$

13. Period π

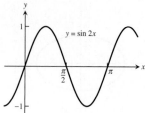

$y = \sin 2x$

15. Period 2

$y = \cos \pi x$

17. Period 6

$y = -\sin \frac{\pi x}{3}$

19. Period 2π

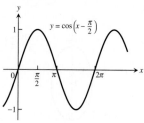

$y = \cos\left(x - \frac{\pi}{2}\right)$

21. Period 2π

$y = \sin\left(x - \frac{\pi}{4}\right) + 1$

23. Period $\pi/2$, symmetric about the origin

$s = \cot 2t$

25. Period 4, symmetric about the y-axis

$s = \sec \frac{\pi t}{2}$

29. $D : (-\infty, \infty)$, $R : y = -1, 0, 1$

$y = \lfloor \sin x \rfloor$ $y = \sin x$

39. $-\cos x$ **41.** $-\cos x$ **43.** $\dfrac{\sqrt{6} + \sqrt{2}}{4}$ **45.** $\dfrac{\sqrt{2} + \sqrt{6}}{4}$

47. $\dfrac{2 + \sqrt{2}}{4}$ **49.** $\dfrac{2 - \sqrt{3}}{4}$ **51.** $\dfrac{\pi}{3}, \dfrac{2\pi}{3}, \dfrac{4\pi}{3}, \dfrac{5\pi}{3}$

53. $\dfrac{\pi}{6}, \dfrac{\pi}{2}, \dfrac{5\pi}{6}, \dfrac{3\pi}{2}$ **59.** $\sqrt{7} \approx 2.65$ **63.** $a = 1.464$

65. $A = 2, B = 2\pi,$
$C = -\pi, D = -1$

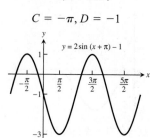

$y = 2\sin(x + \pi) - 1$

67. $A = -\dfrac{2}{\pi}, B = 4,$
$C = 0, D = \dfrac{1}{\pi}$

$y = -\dfrac{2}{\pi}\sin\left(\dfrac{\pi t}{2}\right) + \dfrac{1}{\pi}$

Section 1.4, p. 34

1. d   **3.** d
5. $[-3, 5]$ by $[-15, 40]$

$f(x) = x^4 - 4x^3 + 15$

7. $[-3, 6]$ by $[-250, 50]$

$f(x) = x^5 - 5x^4 + 10$

9. $[-3, 3]$ by $[-6, 6]$

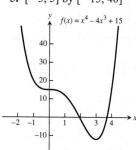

$f(x) = x\sqrt{9 - x^2}$

11. $[-2, 6]$ by $[-5, 4]$

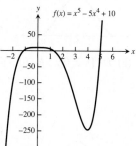

$y = 2x - 3x^{2/3}$

13. $[-2, 8]$ by $[-5, 10]$

$y = 5x^{2/5} - 2x$

15. $[-3, 3]$ by $[0, 10]$

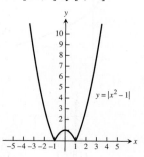

$y = |x^2 - 1|$

17. $[-10, 10]$ by $[-10, 10]$

$y = \dfrac{x + 3}{x + 2}$

19. $[-4, 4]$ by $[0, 3]$

$f(x) = \dfrac{x^2 + 2}{x^2 + 1}$

21. $[-10, 10]$ by $[-6, 6]$

23. $[-6, 10]$ by $[-6, 6]$

37.

39.

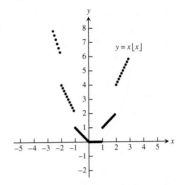

25. $\left[-\dfrac{\pi}{125}, \dfrac{\pi}{125}\right]$ by
$[-1.25, 1.25]$

27. $[-100\pi, 100\pi]$ by
$[-1.25, 1.25]$

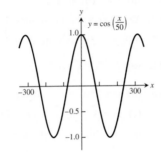

Section 1.5, pp. 39–40

1.

3.

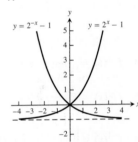

29. $\left[-\dfrac{\pi}{15}, \dfrac{\pi}{15}\right]$ by
$[-0.25, 0.25]$

31.

5.

7.

9.

33.

35.

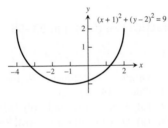

11. $16^{1/4} = 2$ **13.** $4^{1/2} = 2$ **15.** 5 **17.** $14^{\sqrt{3}}$ **19.** 4
21. $D: -\infty < x < \infty; R: 0 < y < 1/2$
23. $D: -\infty < t < \infty; R: 1 < y < \infty$
25. $x \approx 2.3219$ **27.** $x \approx -0.6309$ **29.** After 19 years
31. (a) $A(t) = 6.6\left(\dfrac{1}{2}\right)^{t/14}$ (b) About 38 days later

33. ≈ 11.433 years, or when interest is paid
35. $2^{48} \approx 2.815 \times 10^{14}$

Section 1.6, pp. 50–52

1. One-to-one   **3.** Not one-to-one   **5.** One-to-one
7. Not one-to-one   **9.** One-to-one
11. $D: (0, 1]$ $R: [0, \infty)$ **13.** $D: [-1, 1]$ $R: [-\pi/2, \pi/2]$

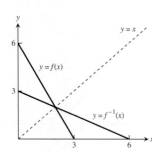

15. $D: [0, 6]$ $R: [0, 3]$ **17. (a)** Symmetric about the line $y = x$

19. $f^{-1}(x) = \sqrt{x} - 1$ **21.** $f^{-1}(x) = \sqrt[3]{x} + 1$
23. $f^{-1}(x) = \sqrt{x} - 1$
25. $f^{-1}(x) = \sqrt[5]{x}; D: -\infty < x < \infty; R: -\infty < y < \infty$
27. $f^{-1}(x) = 5\sqrt{x} - 1; D: -\infty < x < \infty;$
$R: -\infty < y < \infty$
29. $f^{-1}(x) = \dfrac{1}{\sqrt{x}}; D: x > 0; R: y > 0$
31. $f^{-1}(x) = \dfrac{2x + 3}{x - 1}; D: -\infty < x < \infty, x \ne 1;$
$R: -\infty < y < \infty, y \ne 2$
33. $f^{-1}(x) = 1 - \sqrt{x + 1}; D: -1 \le x < \infty;$
$R: -\infty < y \le 1$
35. (a) $f^{-1}(x) = \dfrac{1}{m}x$
  (b) The graph of f^{-1} is the line through the origin with slope $1/m$.
37. (a) $f^{-1}(x) = x - 1$

  (b) $f^{-1}(x) = x - b$. The graph of f^{-1} is a line parallel to the graph of f. The graphs of f and f^{-1} lie on opposite sides of the line $y = x$ and are equidistant from that line.
  (c) Their graphs will be parallel to one another and lie on opposite sides of the line $y = x$ equidistant from that line.
39. (a) $\ln 3 - 2\ln 2$ **(b)** $2(\ln 2 - \ln 3)$ **(c)** $-\ln 2$

  (d) $\dfrac{2}{3}\ln 3$ **(e)** $\ln 3 + \dfrac{1}{2}\ln 2$ **(f)** $\dfrac{1}{2}(3\ln 3 - \ln 2)$

41. (a) $\ln 5$ **(b)** $\ln(x - 3)$ **(c)** $\ln(t^2)$

43. (a) 7.2 **(b)** $\dfrac{1}{x^2}$ **(c)** $\dfrac{x}{y}$

45. (a) 1 **(b)** 1 **(c)** $-x^2 - y^2$
47. e^{2t+4} **49.** $e^{5t} + 40$ **51.** $y = 2xe^x + 1$
53. (a) $k = \ln 2$ **(b)** $k = (1/10)\ln 2$ **(c)** $k = 1000 \ln a$
55. (a) $t = -10 \ln 3$ **(b)** $t = -\dfrac{\ln 2}{k}$ **(c)** $t = \dfrac{\ln .4}{\ln .2}$
57. $4(\ln x)^2$
59. (a) 7 **(b)** $\sqrt{2}$ **(c)** 75 **(d)** 2 **(e)** 0.5 **(f)** -1
61. (a) $\sqrt{x}$ **(b)** x^2 **(c)** $\sin x$ **63. (a)** $\dfrac{\ln 3}{\ln 2}$ **(b)** 3 **(c)** 2
65. (a) $-\pi/6$ **(b)** $\pi/4$ **(c)** $-\pi/3$
67. (a) π **(b)** $\pi/2$
69. Yes, $g(x)$ is also one-to-one.
71. Yes, $f \circ g$ is also one-to-one.
73. (a) $f^{-1}(x) = \log_2\left(\dfrac{x}{100 - x}\right)$ **(b)** $f^{-1}(x) = \log_{1.1}\left(\dfrac{x}{50 - x}\right)$
75. (a) $y = \ln x - 3$ **(b)** $y = \ln(x - 1)$
  (c) $y = 3 + \ln(x + 1)$ **(d)** $y = \ln(x - 2) - 4$
  (e) $y = \ln(-x)$ **(f)** $y = e^x$
79. (a) Amount $= 8\left(\dfrac{1}{2}\right)^{t/12}$ **(b)** 36 hours

81. ≈ 44.081 years

Practice Exercises, pp. 53–55

1. $A = \pi r^2, C = 2\pi r, A = \dfrac{C^2}{4\pi}$ **3.** $x = \tan\theta, y = \tan^2\theta$
5. Origin   **7.** Neither   **9.** Even   **11.** Even
13. Odd   **15.** Neither
17. (a) Even   **(b)** Odd   **(c)** Odd   **(d)** Even   **(e)** Even
19. (a) Domain: all reals   **(b)** Range: $[-2, \infty)$
21. (a) Domain: $[-4, 4]$ **(b)** Range: $[0, 4]$
23. (a) Domain: all reals   **(b)** Range: $(-3, \infty)$
25. (a) Domain: all reals   **(b)** Range: $[-3, 1]$
27. (a) Domain: $(3, \infty)$ **(b)** Range: all reals
29. (a) Increasing   **(b)** Neither   **(c)** Decreasing
  (d) Increasing
31. (a) Domain: $[-4, 4]$ **(b)** Range: $[0, 2]$

33. $f(x) = \begin{cases} 1 - x, & 0 \le x < 1 \\ 2 - x, & 1 \le x \le 2 \end{cases}$

35. (a) 1 **(b)** $\dfrac{1}{\sqrt{2.5}} = \sqrt{\dfrac{2}{5}}$ **(c)** $x, x \ne 0$

  (d) $\dfrac{1}{\sqrt{1/\sqrt{x + 2} + 2}}$

37. (a) $(f \circ g)(x) = -x, x \ge -2, (g \circ f)(x) = \sqrt{4 - x^2}$
  (b) Domain $(f \circ g): [-2, \infty)$, domain $(g \circ f): [-2, 2]$
  (c) Range $(f \circ g): (-\infty, 2]$, range $(g \circ f): [0, 2]$

39.

41. Replace the portion for $x < 0$ with mirror image of the portion for $x > 0$ to make the new graph symmetric with respect to the y-axis.

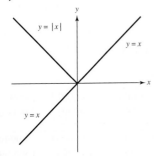

43. Reflects the portion for $y < 0$ across the x-axis
45. Reflects the portion for $y < 0$ across the x-axis
47. Adds the mirror image of the portion for $x > 0$ to make the new graph symmetric with respect to the y-axis

49. (a) $y = g(x - 3) + \dfrac{1}{2}$ **(b)** $y = g\left(x + \dfrac{2}{3}\right) - 2$

  (c) $y = g(-x)$ **(d)** $y = -g(x)$ **(e)** $y = 5g(x)$

  (f) $y = g(5x)$

51.

53.

55. Period π

57. Period 2

59.

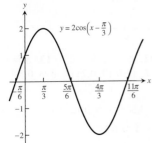

61. (a) $a = 1$ $b = \sqrt{3}$ **(b)** $a = 2\sqrt{3}/3$ $c = 4\sqrt{3}/3$

63. (a) $a = \dfrac{b}{\tan B}$ **(b)** $c = \dfrac{a}{\sin A}$

65. ≈ 16.98 m   **67. (b)** 4π
69. (a) Domain: $-\infty < x < \infty$ **(b)** Domain: $x > 0$
71. (a) Domain: $-3 \le x \le 3$ **(b)** Domain: $0 \le x \le 4$
73. $(f \circ g)(x) = \ln(4 - x^2)$ and domain: $-2 < x < 2$;
  $(g \circ f)(x) = 4 - (\ln x)^2$ and domain: $x > 0$;
  $(f \circ f)(x) = \ln(\ln x)$ and domain: $x > 1$;
  $(g \circ g)(x) = -x^4 + 8x^2 - 12$ and domain: $-\infty < x < \infty$.

79. (a) D: $(-\infty, \infty)$ R: $\left[\dfrac{-\pi}{2}, \dfrac{\pi}{2}\right]$ **(b)** D: $[-1, 1]$ R: $[-1, 1]$

81. (a) No   **(b)** Yes
83. (a) $f(g(x)) = \left(\sqrt[3]{x}\right)^3 = x$, $g(f(x)) = \sqrt[3]{x^3} = x$

  (b)

Additional and Advanced Exercises, pp. 55–57

  1. Yes. For instance: $f(x) = 1/x$ and $g(x) = 1/x$, or $f(x) = 2x$ and $g(x) = x/2$, or $f(x) = e^x$ and $g(x) = \ln x$.
  3. If $f(x)$ is odd, then $g(x) = f(x) - 2$ is not odd. Nor is $g(x)$ even, unless $f(x) = 0$ for all x. If f is even, then $g(x) = f(x) - 2$ is also even.

  5.

19. (a) Domain: all reals. Range: If $a > 0$, then (d, ∞); if $a < 0$, then $(-\infty, d)$.   **(b)** Domain: (c, ∞), range: all reals
21. (a) $y = 100{,}000 - 10000x$, $0 \le x \le 10$ **(b)** After 4.5 years
23. After $\dfrac{\ln(10/3)}{\ln 1.08} \approx 15.6439$ years. (If the bank only pays interest at the end of the year, it will take 16 years.)
25. $x = 2, x = 1$ **27.** $1/2$

CHAPTER 2

Section 2.1, pp. 63–65

  1. (a) 19   **(b)** 1

  3. (a) $-\dfrac{4}{\pi}$ **(b)** $-\dfrac{3\sqrt{3}}{\pi}$ **5.** 1

  7. (a) 4   **(b)** $y = 4x - 7$
  9. (a) 2   **(b)** $y = 2x - 7$
  11. (a) 12   **(b)** $y = 12x - 16$
  13. (a) -9 **(b)** $y = -9x - 2$

15. Your estimates may not completely agree with these.

(a)

PQ_1	PQ_2	PQ_3	PQ_4
43	46	49	50

The appropriate units are m/sec.

(b) ≈ 50 m/sec or 180 km/h

17. (a)

(b) $\approx \$56{,}000$/year
(c) $\approx \$42{,}000$/year

19. (a) $0.414213, 0.449489, (\sqrt{1+h}-1)/h$ **(b)** $g(x) = \sqrt{x}$

$1+h$	1.1	1.01	1.001	1.0001
$\sqrt{1+h}$	1.04880	1.004987	1.0004998	1.0000499
$(\sqrt{1+h}-1)/h$	0.4880	0.4987	0.4998	0.499

1.00001	1.000001
1.000005	1.0000005
0.5	0.5

(c) 0.5   **(d)** 0.5

21. (a) 15 mph, 3.3 mph, 10 mph   **(b)** 10 mph, 0 mph, 4 mph
(c) 20 mph when $t = 3.5$ hr

Section 2.2, pp. 73–76

1. (a) Does not exist. As x approaches 1 from the right, $g(x)$ approaches 0. As x approaches 1 from the left, $g(x)$ approaches 1. There is no single number L that all the values $g(x)$ get arbitrarily close to as $x \to 1$.
(b) 1   **(c)** 0   **(d)** 1/2
3. (a) True   **(b)** True   **(c)** False   **(d)** False
(e) False   **(f)** True   **(g)** True
5. As x approaches 0 from the left, $x/|x|$ approaches -1. As x approaches 0 from the right, $x/|x|$ approaches 1. There is no single number L that the function values all get arbitrarily close to as $x \to 0$.
7. Nothing can be said.   **9.** No; no; no   **11.** -9 **13.** -8
15. 5/8   **17.** 27   **19.** 16   **21.** 3/2   **23.** 1/10   **25.** -7
27. 3/2   **29.** $-1/2$ **31.** -1 **33.** 4/3   **35.** 1/6   **37.** 4
39. 1/2   **41.** 3/2   **43.** -1 **45.** 1   **47.** 1/3   **49.** $\sqrt{4-\pi}$
51. (a) Quotient Rule   **(b)** Difference and Power Rules
(c) Sum and Constant Multiple Rules
53. (a) -10 **(b)** -20 **(c)** -1 **(d)** 5/7
55. (a) 4   **(b)** -21 **(c)** -12 **(d)** $-7/3$
57. 2   **59.** 3   **61.** $1/(2\sqrt{7})$ **63.** $\sqrt{5}$
65. (a) The limit is 1.
67. (a) $f(x) = (x^2 - 9)/(x + 3)$

x	-3.1	-3.01	-3.001	-3.0001	-3.00001	-3.000001
$f(x)$	-6.1	-6.01	-6.001	-6.0001	-6.00001	-6.000001

x	-2.9	-2.99	-2.999	-2.9999	-2.99999	-2.999999
$f(x)$	-5.9	-5.99	-5.999	-5.9999	-5.99999	-5.999999

(c) $\lim\limits_{x \to -3} f(x) = -6$

69. (a) $G(x) = (x + 6)/(x^2 + 4x - 12)$

x	-5.9	-5.99	-5.999	-5.9999
$G(x)$	$-.126582$	$-.1251564$	$-.1250156$	$-.1250015$

-5.99999	-5.999999
$-.1250001$	$-.1250000$

x	-6.1	-6.01	-6.001	-6.0001
$G(x)$	$-.123456$	$-.124843$	$-.124984$	$-.124998$

-6.00001	-6.000001
$-.124999$	$-.124999$

(c) $\lim\limits_{x \to -6} G(x) = -1/8 = -0.125$

71. (a) $f(x) = (x^2 - 1)/(|x| - 1)$

x	-1.1	-1.01	-1.001	-1.0001	-1.00001	-1.000001
$f(x)$	2.1	2.01	2.001	2.0001	2.00001	2.000001

x	$-.9$	$-.99$	$-.999$	$-.9999$	$-.99999$	$-.999999$
$f(x)$	1.9	1.99	1.999	1.9999	1.99999	1.999999

(c) $\lim\limits_{x \to -1} f(x) = 2$

73. (a) $g(\theta) = (\sin \theta)/\theta$

θ	.1	.01	.001	.0001	.00001	.000001
$g(\theta)$	.998334	.999983	.999999	.999999	.999999	.999999

θ	$-.1$	$-.01$	$-.001$	$-.0001$	$-.00001$	$-.000001$
$g(\theta)$	.998334	.999983	.999999	.999999	.999999	.999999

$\lim\limits_{\theta \to 0} g(\theta) = 1$

75. (a) $f(x) = x^{1/(1-x)}$

x	.9	.99	.999	.9999	.99999	.999999
$f(x)$	.348678	.366032	.367695	.367861	.367877	.367879

x	1.1	1.01	1.001	1.0001	1.00001	1.000001
$f(x)$	.385543	.369711	.368063	.367897	.367881	.367878

$\lim\limits_{x \to 1} f(x) \approx 0.36788$

77. $c = 0, 1, -1$; the limit is 0 at $c = 0$, and 1 at $c = 1, -1$.
79. 7   **81. (a)** 5   **(b)** 5   **83. (a)** 0   **(b)** 0

Section 2.3, pp. 82–85

1. $\delta = 2$

3. $\delta = 1/2$

5. $\delta = 1/18$

7. $\delta = 0.1$ **9.** $\delta = 7/16$ **11.** $\delta = \sqrt{5} - 2$ **13.** $\delta = 0.36$
15. $(3.99, 4.01)$, $\delta = 0.01$ **17.** $(-0.19, 0.21)$, $\delta = 0.19$
19. $(3, 15)$, $\delta = 5$ **21.** $(10/3, 5)$, $\delta = 2/3$
23. $(-\sqrt{4.5}, -\sqrt{3.5})$, $\delta = \sqrt{4.5} - 2 \approx 0.12$
25. $(\sqrt{15}, \sqrt{17})$, $\delta = \sqrt{17} - 4 \approx 0.12$
27. $\left(2 - \dfrac{0.03}{m}, 2 + \dfrac{0.03}{m}\right)$, $\delta = \dfrac{0.03}{m}$
29. $\left(\dfrac{1}{2} - \dfrac{c}{m}, \dfrac{c}{m} + \dfrac{1}{2}\right)$, $\delta = \dfrac{c}{m}$ **31.** $L = -3$, $\delta = 0.01$

33. $L = 4,\quad \delta = 0.05$ **35.** $L = 4,\quad \delta = 0.75$
55. [3.384, 3.387]. To be safe, the left endpoint was rounded up and the right endpoint rounded down.
59. The limit does not exist as x approaches 3.

Section 2.4, pp. 90–92

1. **(a)** True **(b)** True **(c)** False **(d)** True **(e)** True
(f) True **(g)** False **(h)** False **(i)** False **(j)** False
(k) True **(l)** False
3. **(a)** 2, 1 **(b)** No, $\lim\limits_{x\to2^+} f(x) \neq \lim\limits_{x\to2^-} f(x)$
(c) 3, 3 **(d)** Yes, 3
5. **(a)** No **(b)** Yes, 0 **(c)** No
7. **(a)** **(b)** 1, 1 **(c)** Yes, 1

9. **(a)** $D : 0 \le x \le 2, R : 0 < y \le 1$ and $y = 2$
(b) $(0, 1) \cup (1, 2)$ **(c)** $x = 2$ **(d)** $x = 0$

11. $\sqrt{3}$ **13.** 1 **15.** $2/\sqrt{5}$ **17.** **(a)** 1 **(b)** -1
19. **(a)** 1 **(b)** 2/3 **21.** 1 **23.** 3/4 **25.** 2 **27.** 1/2
29. 2 **31.** 0 **33.** 1 **35.** 1/2 **37.** 0 **39.** 3/8
41. 3 **47.** $\delta = \epsilon^2,\ \lim\limits_{x\to5^+}\sqrt{x-5} = 0$
51. **(a)** 400 **(b)** 399 **(c)** The limit does not exist.

Section 2.5, pp. 101–103

1. No; discontinuous at $x = 2$; not defined at $x = 2$
3. Continuous **5.** **(a)** Yes **(b)** Yes **(c)** Yes **(d)** Yes
7. **(a)** No **(b)** No **9.** 0 **11.** 1, nonremovable; 0, removable
13. All x except $x = 2$ **15.** All x except $x = 3, x = 1$
17. All x **19.** All x except $x = 0$
21. All x except $n\pi/2$, n any integer
23. All x except $n\pi/2$, n an odd integer
25. All $x \ge -3/2$ **27.** All x **29.** All x
31. 0; continuous at $x = \pi$ **33.** 1; continuous at $y = 1$
35. $\sqrt{2}/2$; continuous at $t = 0$ **37.** 1; continuous at $x = 0$
39. $g(3) = 6$ **41.** $f(1) = 3/2$ **43.** $a = 4/3$ **45.** $a = -2, 3$
47. $a = 5/2, b = -1/2$ **71.** $x \approx 1.8794, -1.5321, -0.3473$
73. $x \approx 1.7549$ **75.** $x \approx 3.5156$ **77.** $x \approx 0.7391$

Section 2.6, pp. 114–116

1. **(a)** 0 **(b)** -2 **(c)** 2 **(d)** Does not exist **(e)** -1
(f) ∞ **(g)** Does not exist **(h)** 1 **(i)** 0

3. **(a)** -3 **(b)** -3 **5.** **(a)** 1/2 **(b)** 1/2 **7.** **(a)** $-5/3$
(b) $-5/3$ **9.** 0 **11.** -1 **13.** **(a)** 2/5 **(b)** 2/5
15. **(a)** 0 **(b)** 0 **17.** **(a)** 7 **(b)** 7 **19.** **(a)** 0 **(b)** 0
21. **(a)** $-2/3$ **(b)** $-2/3$ **23.** 2 **25.** ∞ **27.** 0 **29.** 1
31. ∞ **33.** 1 **35.** 1/2 **37.** ∞ **39.** $-\infty$ **41.** $-\infty$
43. ∞ **45.** **(a)** ∞ **(b)** $-\infty$ **47.** ∞ **49.** ∞ **51.** $-\infty$
53. **(a)** ∞ **(b)** $-\infty$ **(c)** $-\infty$ **(d)** ∞
55. **(a)** $-\infty$ **(b)** ∞ **(c)** 0 **(d)** 3/2
57. **(a)** $-\infty$ **(b)** 1/4 **(c)** 1/4 **(d)** 1/4 **(e)** It will be $-\infty$.
59. **(a)** $-\infty$ **(b)** ∞
61. **(a)** ∞ **(b)** ∞ **(c)** ∞ **(d)** ∞
63. **65.**

67. **69.** Here is one possibility.

71. Here is one possibility. **73.** Here is one possibility.

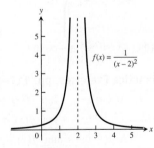

75. Here is one possibility. **79.** At most one

81. 0 **83.** $-3/4$ **85.** 5/2
93. **(a)** For every positive real number B there exists a corresponding number $\delta > 0$ such that for all x
$$x_0 - \delta < x < x_0 \implies f(x) > B.$$

(b) For every negative real number $-B$ there exists a corresponding number $\delta > 0$ such that for all x

$$x_0 < x < x_0 + \delta \implies f(x) < -B.$$

(c) For every negative real number $-B$ there exists a corresponding number $\delta > 0$ such that for all x

$$x_0 - \delta < x < x_0 \implies f(x) < -B.$$

99.

101.

103.

105.

107.

109. At ∞: ∞, at $-\infty$: 0

Practice Exercises, pp. 117–119

1. At $x = -1$: $\displaystyle\lim_{x \to -1^-} f(x) = \lim_{x \to -1^+} f(x) = 1$, so

$\displaystyle\lim_{x \to -1} f(x) = 1 = f(-1)$; continuous at $x = -1$

At $x = 0$: $\displaystyle\lim_{x \to 0^-} f(x) = \lim_{x \to 0^+} f(x) = 0$, so $\displaystyle\lim_{x \to 0} f(x) = 0$. However, $f(0) \neq 0$, so f is discontinuous at $x = 0$. The discontinuity can be removed by redefining $f(0)$ to be 0.

At $x = 1$: $\displaystyle\lim_{x \to 1^-} f(x) = -1$ and $\displaystyle\lim_{x \to 1^+} f(x) = 1$, so $\displaystyle\lim_{x \to 1} f(x)$ does not exist. The function is discontinuous at $x = 1$, and the discontinuity is not removable.

3. **(a)** -21 **(b)** 49 **(c)** 0 **(d)** 1 **(e)** 1 **(f)** 7
   (g) -7 **(h)** $-\dfrac{1}{7}$ **5.** 4

7. **(a)** $(-\infty, +\infty)$ **(b)** $[0, \infty)$ **(c)** $(-\infty, 0)$ and $(0, \infty)$
   (d) $(0, \infty)$

9. **(a)** Does not exist   **(b)** 0

11. $\dfrac{1}{2}$ **13.** $2x$ **15.** $-\dfrac{1}{4}$ **17.** $2/3$ **19.** $2/\pi$ **21.** 1

23. 4 **25.** $-\infty$ **27.** 0 **29.** 2 **31.** 0

33. No in both cases, because $\displaystyle\lim_{x \to 1} f(x)$ does not exist, and $\displaystyle\lim_{x \to -1} f(x)$ does not exist.

35. Yes, f does have a continuous extension, to $a = 1$ with $f(1) = 4/3$.

37. No   **41.** $2/5$ **43.** 0 **45.** $-\infty$ **47.** 0 **49.** 1 **51.** 1
53. $-\pi/2$ **55.** **(a)** $x = 3$ **(b)** $x = 1$ **(c)** $x = -4$

Additional and Advanced Exercises, pp. 119–121

3. 0; the left-hand limit was needed because the function is undefined for $v > c$.   **5.** $65 < t < 75$; within $5°F$

13. **(a)** B **(b)** A **(c)** A **(d)** A
21. **(a)** $\displaystyle\lim_{a \to 0} r_+(a) = 0.5$, $\displaystyle\lim_{a \to -1^+} r_+(a) = 1$
   (b) $\displaystyle\lim_{a \to 0} r_-(a)$ does not exist, $\displaystyle\lim_{a \to -1^+} r_-(a) = 1.$ç

25. 0 **27.** 1 **29.** 4 **31.** $y = 2x$ **33.** $y = x$, $y = -x$

CHAPTER 3

Section 3.1, pp. 125–126

1. P_1: $m_1 = 1$, P_2: $m_2 = 5$ **3.** P_1: $m_1 = 5/2$, P_2: $m_2 = -1/2$
5. $y = 2x + 5$ **7.** $y = x + 1$

9. $y = 12x + 16$

11. $m = 4$, $y - 5 = 4(x - 2)$
13. $m = -2$, $y - 3 = -2(x - 3)$

15. $m = 12, y - 8 = 12(t - 2)$

17. $m = \frac{1}{4}, y - 2 = \frac{1}{4}(x - 4)$

19. $m = -10$ **21.** $m = -1/4$ **23.** $(-2, -5)$

25. $y = -(x + 1), y = -(x - 3)$ **27.** 19.6 m/sec

29. 6π **33.** Yes   **35.** Yes   **37. (a)** Nowhere

39. (a) At $x = 0$ **41. (a)** Nowhere   **43. (a)** At $x = 1$

45. (a) At $x = 0$

Section 3.2, pp. 131–135

1. $-2x, 6, 0, -2$ **3.** $-\frac{2}{t^3}, 2, -\frac{1}{4}, -\frac{2}{3\sqrt{3}}$

5. $\frac{3}{2\sqrt{3\theta}}, \frac{3}{2\sqrt{3}}, \frac{1}{2}, \frac{3}{2\sqrt{2}}$ **7.** $6x^2$ **9.** $\frac{1}{(2t + 1)^2}$

11. $\frac{-1}{2(q + 1)\sqrt{q + 1}}$ **13.** $1 - \frac{9}{x^2}, 0$ **15.** $3t^2 - 2t, 5$

17. $\frac{-4}{(x - 2)\sqrt{x - 2}}, \quad y - 4 = -\frac{1}{2}(x - 6)$ **19.** 6

21. $1/8$ **23.** $\frac{-1}{(x + 2)^2}$ **25.** $\frac{-1}{(x - 1)^2}$ **27.** b   **29.** d

31. (a) $x = 0, 1, 4$ **33.**

(b)

35. (a) i) 1.5 °F/hr **ii)** 2.9 °F/hr
    iii) 0 °F/hr **iv)** -3.7 °F/hr

(b) 7.3 °F/hr at 12 P.M., -11 °F/hr at 6 P.M.

(c)

37. Since $\lim\limits_{h \to 0^+} \frac{f(0 + h) - f(0)}{h} = 1$

while $\lim\limits_{h \to 0^-} \frac{f(0 + h) - f(0)}{h} = 0$,

$f'(0) = \lim\limits_{h \to 0} \frac{f(0 + h) - f(0)}{h}$ does not exist and $f(x)$ is not differentiable at $x = 0$.

39. Since $\lim\limits_{h \to 0^+} \frac{f(1 + h) - f(1)}{h} = 2$ while

$\lim\limits_{h \to 0^-} \frac{f(1 + h) - f(1)}{h} = \frac{1}{2}, \quad f'(1) = \lim\limits_{h \to 0} \frac{f(1 + h) - f(1)}{h}$

does not exist and $f(x)$ is not differentiable at $x = 1$.

41. Since $f(x)$ is not continuous at $x = 0$, $f(x)$ is not differentiable at $x = 0$.

43. (a) $-3 \le x \le 2$ **(b)** None   **(c)** None

45. (a) $-3 \le x < 0, 0 < x \le 3$ **(b)** None   **(c)** $x = 0$

47. (a) $-1 \le x < 0, 0 < x \le 2$ **(b)** $x = 0$ **(c)** None

Section 3.3, pp. 143–145

1. $\frac{dy}{dx} = -2x, \frac{d^2y}{dx^2} = -2$

3. $\frac{ds}{dt} = 15t^2 - 15t^4, \frac{d^2s}{dt^2} = 30t - 60t^3$

5. $\frac{dy}{dx} = 4x^2 - 1 + 2e^x, \frac{d^2y}{dx^2} = 8x + 2e^x$

7. $\frac{dw}{dz} = -\frac{6}{z^3} + \frac{1}{z^2}, \frac{d^2w}{dz^2} = \frac{18}{z^4} - \frac{2}{z^3}$

9. $\frac{dy}{dx} = 12x - 10 + 10x^{-3}, \frac{d^2y}{dx^2} = 12 - 30x^{-4}$

11. $\frac{dr}{ds} = \frac{-2}{3s^3} + \frac{5}{2s^2}, \frac{d^2r}{ds^2} = \frac{2}{s^4} - \frac{5}{s^3}$

13. $y' = -5x^4 + 12x^2 - 2x - 3$

15. $y' = 3x^2 + 10x + 2 - \frac{1}{x^2}$ **17.** $y' = \frac{-19}{(3x - 2)^2}$

19. $g'(x) = \frac{x^2 + x + 4}{(x + 0.5)^2}$ **21.** $\frac{dv}{dt} = \frac{t^2 - 2t - 1}{(1 + t^2)^2}$

23. $f'(s) = \frac{1}{\sqrt{s}(\sqrt{s} + 1)^2}$ **25.** $v' = -\frac{1}{x^2} + 2x^{-3/2}$

27. $y' = \frac{-4x^3 - 3x^2 + 1}{(x^2 - 1)^2(x^2 + x + 1)^2}$ **29.** $y' = -2e^{-x} + 3e^{3x}$

31. $y' = 3x^2e^x + x^3e^x$ **33.** $y' = \frac{9}{4}x^{5/4} - 2e^{-2x}$

35. $\frac{ds}{dt} = 3t^{1/2}$ **37.** $y' = \frac{2}{7x^{5/7}} - ex^{e-1}$ **39.** $\frac{dr}{ds} = \frac{se^s - e^s}{s^2}$

41. $y' = 2x^3 - 3x - 1, y'' = 6x^2 - 3, y''' = 12x, y^{(4)} = 12,$
$y^{(n)} = 0$ for $n \ge 5$

43. $y' = 3x^2 + 4x - 8, y'' = 6x + 4, y''' = 6,$
$y^{(n)} = 0$ for $n \ge 4$

45. $y' = 2x - 7x^{-2}, y'' = 2 + 14x^{-3}$

47. $\frac{dr}{d\theta} = 3\theta^{-4}, \frac{d^2r}{d\theta^2} = -12\theta^{-5}$

49. $\frac{dw}{dz} = -z^{-2} - 1, \frac{d^2w}{dz^2} = 2z^{-3}$

51. $\frac{dw}{dz} = 6ze^{2z}(1 + z), \frac{d^2w}{dz^2} = 6e^{2z}(1 + 4z + 2z^2)$

53. (a) 13 **(b)** -7 **(c)** $7/25$ **(d)** 20

55. (a) $y = -\frac{x}{8} + \frac{5}{4}$ **(b)** $m = -4$ at $(0, 1)$
    (c) $y = 8x - 15, y = 8x + 17$

57. $y = 4x, y = 2$ **59.** $a = 1, b = 1, c = 0$

61. $(2, 4)$ **63.** $(0, 0), (4, 2)$ **65. (a)** $y = 2x + 2$ **(c)** $(2, 6)$

67. 50 **69.** $a = -3$

71. $P'(x) = na_nx^{n-1} + (n - 1)a_{n-1}x^{n-2} + \cdots + 2a_2x + a_1$

73. The Product Rule is then the Constant Multiple Rule, so the latter is a special case of the Product Rule.

75. (a) $\dfrac{d}{dx}(uvw) = uvw' + uv'w + u'vw$

(b) $\dfrac{d}{dx}(u_1 u_2 u_3 u_4) = u_1 u_2 u_3 u_4' + u_1 u_2 u_3' u_4 + u_1 u_2' u_3 u_4 + u_1' u_2 u_3 u_4$

(c) $\dfrac{d}{dx}(u_1 \cdots u_n) = u_1 u_2 \cdots u_{n-1} u_n' + u_1 u_2 \cdots u_{n-2} u_{n-1}' u_n + \cdots + u_1' u_2 \cdots u_n$

77. $\dfrac{dP}{dV} = -\dfrac{nRT}{(V - nb)^2} + \dfrac{2an^2}{V^3}$

Section 3.4, pp. 152–155

1. (a) -2 m, -1 m/sec
  (b) 3 m/sec, 1 m/sec; 2 m/sec², 2 m/sec²
  (c) Changes direction at $t = 3/2$ sec

3. (a) -9 m, -3 m/sec
  (b) 3 m/sec, 12 m/sec; 6 m/sec², -12 m/sec²
  (c) No change in direction

5. (a) -20 m, -5 m/sec
  (b) 45 m/sec, $(1/5)$ m/sec; 140 m/sec², $(4/25)$ m/sec²
  (c) No change in direction

7. (a) $a(1) = -6$ m/sec², $a(3) = 6$ m/sec²
  (b) $v(2) = 3$ m/sec   **(c)** 6 m

9. Mars: ≈ 7.5 sec, Jupiter: ≈ 1.2 sec   **11.** $g_s = 0.75$ m/sec²

13. (a) $v = -32t, |v| = 32t$ ft/sec, $a = -32$ ft/sec²
  (b) $t \approx 3.3$ sec   **(c)** $v \approx -107.0$ ft/sec

15. (a) $t = 2, t = 7$ **(b)** $3 \le t \le 6$
  (c)   **(d)**

![graph (c): |v| (m/sec) vs t (sec), labeled Speed; graph (d): a = dv/dt vs t]

17. (a) 190 ft/sec   **(b)** 2 sec   **(c)** 8 sec, 0 ft/sec
  (d) 10.8 sec, 90 ft/sec   **(e)** 2.8 sec
  (f) Greatest acceleration happens 2 sec after launch
  (g) Constant acceleration between 2 and 10.8 sec, -32 ft/sec²

19. (a) $\dfrac{4}{7}$ sec, 280 cm/sec   **(b)** 560 cm/sec, 980 cm/sec²
  (c) 29.75 flashes/sec

21. $C =$ position, $A =$ velocity, $B =$ acceleration

23. (a) \$110/machine   **(b)** \$80   **(c)** \$79.90

25. (a) $b'(0) = 10^4$ bacteria/h   **(b)** $b'(5) = 0$ bacteria/h
  (c) $b'(10) = -10^4$ bacteria/h

27. (a) $\dfrac{dy}{dt} = \dfrac{t}{12} - 1$

  (b) The largest value of $\dfrac{dy}{dt}$ is 0 m/h when $t = 12$ and the smallest value of $\dfrac{dy}{dt}$ is -1 m/h when $t = 0$.

(c)

![graph (c): y vs t, curve y = 6(1 - t/12)² and line dy/dt = t/12 - 1]

29. $t = 25$ sec $D = \dfrac{6250}{9}$ m

31.

![graph: s vs t, parabola s = 200t - 16t², line ds/dt = 200 - 32t, and d²s/dt² = -32]

(a) $v = 0$ when $t = 6.25$ sec
(b) $v > 0$ when $0 \le t < 6.25 \Rightarrow$ the object moves up; $v < 0$ when $6.25 < t \le 12.5 \Rightarrow$ the object moves down.
(c) The object changes direction at $t = 6.25$ sec.
(d) The object speeds up on $(6.25, 12.5]$ and slows down on $[0, 6.25)$.
(e) The object is moving fastest at the endpoints $t = 0$ and $t = 12.5$ when it is traveling 200 ft/sec. It's moving slowest at $t = 6.25$ when the speed is 0.
(f) When $t = 6.25$ the object is $s = 625$ m from the origin and farthest away.

33.

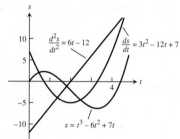

(a) $v = 0$ when $t = \dfrac{6 \pm \sqrt{15}}{3}$ sec

(b) $v < 0$ when $\dfrac{6 - \sqrt{15}}{3} < t < \dfrac{6 + \sqrt{15}}{3} \Rightarrow$ the object moves left; $v > 0$ when $0 \le t < \dfrac{6 - \sqrt{15}}{3}$ or $\dfrac{6 + \sqrt{15}}{3} < t \le 4 \Rightarrow$ the object moves right.

(c) The object changes direction at $t = \dfrac{6 \pm \sqrt{15}}{3}$ sec.

(d) The object speeds up on $\left(\dfrac{6 - \sqrt{15}}{3}, 2\right) \cup \left(\dfrac{6 + \sqrt{15}}{3}, 4\right]$ and slows down on $\left[0, \dfrac{6 - \sqrt{15}}{3}\right) \cup \left(2, \dfrac{6 + \sqrt{15}}{3}\right)$.

(e) The object is moving fastest at $t = 0$ and $t = 4$ when it is moving 7 units/sec and slowest at $t = \dfrac{6 \pm \sqrt{15}}{3}$ sec.

(f) When $t = \dfrac{6 + \sqrt{15}}{3}$ the object is at position $s \approx -6.303$ units and farthest from the origin.

Section 3.5, pp. 159–162

1. $-10 - 3 \sin x$ **3.** $2x \cos x - x^2 \sin x$

5. $-\csc x \cot x - \dfrac{2}{\sqrt{x}}$ **7.** $\sin x \sec^2 x + \sin x$ **9.** 0

11. $\dfrac{-\csc^2 x}{(1 + \cot x)^2}$ **13.** $4 \tan x \sec x - \csc^2 x$ **15.** $x^2 \cos x$

17. $3x^2 \sin x \cos x + x^3 \cos^2 x - x^3 \sin^2 x$

19. $\sec^2 t + e^{-t}$ **21.** $\dfrac{-2 \csc t \cot t}{(1 - \csc t)^2}$ **23.** $-\theta\,(\theta \cos \theta + 2 \sin \theta)$

25. $\sec \theta \csc \theta\,(\tan \theta - \cot \theta) = \sec^2 \theta - \csc^2 \theta$

27. $\sec^2 q$ **29.** $\sec^2 q$

31. $\dfrac{q^3 \cos q - q^2 \sin q - q \cos q - \sin q}{(q^2 - 1)^2}$

33. **(a)** $2 \csc^3 x - \csc x$ **(b)** $2 \sec^3 x - \sec x$

35.

37.

39. Yes, at $x = \pi$ **41.** No

43. $\left(-\dfrac{\pi}{4}, -1\right); \left(\dfrac{\pi}{4}, 1\right)$

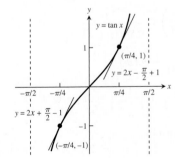

45. **(a)** $y = -x + \pi/2 + 2$ **(b)** $y = 4 - \sqrt{3}$

47. 0   **49.** $\sqrt{3}/2$ **51.** -1 **53.** 0

Section 3.6, pp. 167–170

1. $12x^3$ **3.** $3 \cos (3x + 1)$ **5.** $-\sin (\sin x) \cos x$

7. $10 \sec^2 (10x - 5)$

9. With $u = (2x + 1), y = u^5 : \dfrac{dy}{dx} = \dfrac{dy}{du}\dfrac{du}{dx} = 5u^4 \cdot 2 = 10(2x + 1)^4$

11. With $u = (1 - (x/7)), y = u^{-7} : \dfrac{dy}{dx} = \dfrac{dy}{du}\dfrac{du}{dx} = -7u^{-8} \cdot \left(-\dfrac{1}{7}\right) = \left(1 - \dfrac{x}{7}\right)^{-8}$

13. With $u = ((x^2/8) + x - (1/x)), y = u^4 : \dfrac{dy}{dx} = \dfrac{dy}{du}\dfrac{du}{dx} = 4u^3 \cdot \left(\dfrac{x}{4} + 1 + \dfrac{1}{x^2}\right) = 4\left(\dfrac{x^2}{8} + x - \dfrac{1}{x}\right)^3\left(\dfrac{x}{4} + 1 + \dfrac{1}{x^2}\right)$

15. With $u = \tan x, y = \sec u : \dfrac{dy}{dx} = \dfrac{dy}{du}\dfrac{du}{dx} = (\sec u \tan u)(\sec^2 x) = \sec (\tan x) \tan (\tan x) \sec^2 x$

17. With $u = \sin x, y = u^3 : \dfrac{dy}{dx} = \dfrac{dy}{du}\dfrac{du}{dx} = 3u^2 \cos x = 3 \sin^2 x\,(\cos x)$

19. $y = e^u, u = -5x, \dfrac{dy}{dx} = -5e^{-5x}$

21. $y = e^u, u = 5 - 7x, \dfrac{dy}{dx} = -7e^{(5 - 7x)}$

23. $-\dfrac{1}{2\sqrt{3 - t}}$ **25.** $\dfrac{4}{\pi}(\cos 3t - \sin 5t)$ **27.** $\dfrac{\csc \theta}{\cot \theta + \csc \theta}$

29. $2x \sin^4 x + 4x^2 \sin^3 x \cos x + \cos^{-2} x + 2x \cos^{-3} x \sin x$

31. $(3x - 2)^6 - \dfrac{1}{x^3\left(4 - \dfrac{1}{2x^2}\right)^2}$ **33.** $\dfrac{(4x + 3)^3 (4x + 7)}{(x + 1)^4}$

35. $(1 - x)e^{-x} + 3e^{3x}$ **37.** $\left(\dfrac{5}{2}x^2 - 3x + 3\right)e^{5x/2}$

39. $\sqrt{x}\,\sec^2 (2\sqrt{x}) + \tan (2\sqrt{x})$ **41.** $\dfrac{x \sec x \tan x + \sec x}{2\sqrt{7 + x \sec x}}$

43. $\dfrac{2 \sin \theta}{(1 + \cos \theta)^2}$ **45.** $-2 \sin (\theta^2) \sin 2\theta + 2\theta \cos (2\theta) \cos (\theta^2)$

47. $\left(\dfrac{t + 2}{2(t + 1)^{3/2}}\right) \cos\left(\dfrac{t}{\sqrt{t + 1}}\right)$ **49.** $2\theta e^{-\theta^2} \sin \left(e^{-\theta^2}\right)$

51. $2\pi \sin (\pi t - 2) \cos (\pi t - 2)$ **53.** $\dfrac{8 \sin (2t)}{(1 + \cos 2t)^5}$

55. $10t^{10} \tan^9 t \sec^2 t + 10t^9 \tan^{10} t$

57. $\dfrac{dy}{dt} = -2\pi \sin (\pi t - 1) \cdot \cos (\pi t - 1) \cdot e^{\cos^2 (\pi t - 1)}$

59. $\dfrac{-3t^6 (t^2 + 4)}{(t^3 - 4t)^4}$ **61.** $-2 \cos (\cos (2t - 5)) (\sin (2t - 5))$

55. $-\sqrt{2}$ m/sec, $\sqrt{2}$ m/sec, $\sqrt{2}$ m/sec^2, $\sqrt{2}$ m/sec^3

57. $c = 9$ **59.** $\sin x$

61. **(a)** **i)** 10 cm   **ii)** 5 cm   **iii)** $-5\sqrt{2} \approx -7.1$ cm

(b) **i)** 0 cm/sec   **ii)** $-5\sqrt{3} \approx -8.7$ cm/sec

iii) $-5\sqrt{2} \approx -7.1$ cm/sec

63. $\left(1 + \tan^4\left(\dfrac{t}{12}\right)\right)^2 \left(\tan^3\left(\dfrac{t}{12}\right)\sec^2\left(\dfrac{t}{12}\right)\right)$

65. $-\dfrac{t\sin(t^2)}{\sqrt{1 + \cos(t^2)}}$ **67.** $6\tan(\sin^3 t)\sec^2(\sin^3 t)\sin^2 t \cos t$

69. $3(2t^2 - 5)^3 (18t^2 - 5)$ **71.** $\dfrac{6}{x^3}\left(1 + \dfrac{1}{x}\right)\left(1 + \dfrac{2}{x}\right)$

73. $2\csc^2(3x - 1)\cot(3x - 1)$ **75.** $16(2x + 1)^2 (5x + 1)$

77. $2(2x^2 + 1)e^{x^2}$ **79.** 5/2   **81.** $-\pi/4$ **83.** 0   **85.** -5

87. **(a)** 2/3   **(b)** $2\pi + 5$ **(c)** $15 - 8\pi$ **(d)** 37/6   **(e)** -1

    (f) $\sqrt{2}/24$ **(g)** 5/32   **(h)** $-5/(3\sqrt{17})$ **89.** 5

91. **(a)** 1   **(b)** 1   **93.** $y = 1 - 4x$

95. **(a)** $y = \pi x + 2 - \pi$ **(b)** $\pi/2$

97. It multiplies the velocity, acceleration, and jerk by 2, 4, and 8, respectively.

99. $v(6) = \dfrac{2}{5}$ m/sec, $a(6) = -\dfrac{4}{125}$ m/sec^2

Section 3.7, pp. 174–175

1. $\dfrac{-2xy - y^2}{x^2 + 2xy}$ **3.** $\dfrac{1 - 2y}{2x + 2y - 1}$

5. $\dfrac{-2x^3 + 3x^2 y - xy^2 + x}{x^2 y - x^3 + y}$ **7.** $\dfrac{1}{y(x + 1)^2}$ **9.** $\cos^2 y$

11. $\dfrac{-\cos^2(xy) - y}{x}$ **13.** $\dfrac{-y^2}{y\sin\left(\dfrac{1}{y}\right) - \cos\left(\dfrac{1}{y}\right) + xy}$

15. $\dfrac{2e^{2x} - \cos(x + 3y)}{3\cos(x + 3y)}$ **17.** $-\dfrac{\sqrt{r}}{\sqrt{\theta}}$ **19.** $\dfrac{-r}{\theta}$

21. $y' = -\dfrac{x}{y}, y'' = \dfrac{-y^2 - x^2}{y^3}$

23. $\dfrac{dy}{dx} = \dfrac{xe^{x^2} + 1}{y}, \dfrac{d^2 y}{dx^2} = \dfrac{(2x^2 y^2 + y^2 - 2x)e^{x^2} - x^2 e^{2x^2} - 1}{y^3}$

25. $y' = \dfrac{\sqrt{y}}{\sqrt{y} + 1}, y'' = \dfrac{1}{2(\sqrt{y} + 1)^3}$

27. -2 **29.** $(-2, 1) : m = -1, (-2, -1) : m = 1$

31. **(a)** $y = \dfrac{7}{4}x - \dfrac{1}{2}$ **(b)** $y = -\dfrac{4}{7}x + \dfrac{29}{7}$

33. **(a)** $y = 3x + 6$ **(b)** $y = -\dfrac{1}{3}x + \dfrac{8}{3}$

35. **(a)** $y = \dfrac{6}{7}x + \dfrac{6}{7}$ **(b)** $y = -\dfrac{7}{6}x - \dfrac{7}{6}$

37. **(a)** $y = -\dfrac{\pi}{2}x + \pi$ **(b)** $y = \dfrac{2}{\pi}x - \dfrac{2}{\pi} + \dfrac{\pi}{2}$

39. **(a)** $y = 2\pi x - 2\pi$ **(b)** $y = -\dfrac{x}{2\pi} + \dfrac{1}{2\pi}$

41. Points: $(-\sqrt{7}, 0)$ and $(\sqrt{7}, 0)$, Slope: -2

43. $m = -1$ at $\left(\dfrac{\sqrt{3}}{4}, \dfrac{\sqrt{3}}{2}\right)$, $m = \sqrt{3}$ at $\left(\dfrac{\sqrt{3}}{4}, \dfrac{1}{2}\right)$

45. $(-3, 2) : m = -\dfrac{27}{8}; (-3, -2) : m = \dfrac{27}{8}; (3, 2) : m = \dfrac{27}{8};$

    $(3, -2) : m = -\dfrac{27}{8}$

47. $(3, -1)$

53. $\dfrac{dy}{dx} = -\dfrac{y^3 + 2xy}{x^2 + 3xy^2}, \dfrac{dx}{dy} = -\dfrac{x^2 + 3xy^2}{y^3 + 2xy}, \dfrac{dx}{dy} = \dfrac{1}{dy/dx}$

Section 3.8, pp. 184–185

1. **(a)** $f^{-1}(x) = \dfrac{x}{2} - \dfrac{3}{2}$ **3.** **(a)** $f^{-1}(x) = -\dfrac{x}{4} + \dfrac{5}{4}$

(b)   **(b)**

(c) 2, 1/2   **(c)** $-4, -1/4$

5. **(b)**

(c) Slope of f at $(1, 1) : 3$; slope of g at $(1, 1): 1/3$; slope of f at $(-1, -1) : 3$; slope of g at $(-1, -1): 1/3$

(d) $y = 0$ is tangent to $y = x^3$ at $x = 0$; $x = 0$ is tangent to $y = \sqrt[3]{x}$ at $x = 0$.

7. 1/9   **9.** 3   **11.** $1/x$ **13.** $2/t$ **15.** $-1/x$ **17.** $\dfrac{1}{\theta + 1}$

19. $3/x$ **21.** $2(\ln t) + (\ln t)^2$ **23.** $x^3 \ln x$ **25.** $\dfrac{1 - \ln t}{t^2}$

27. $\dfrac{1}{x(1 + \ln x)^2}$ **29.** $\dfrac{1}{x\ln x}$ **31.** $2\cos(\ln\theta)$

33. $-\dfrac{3x + 2}{2x(x + 1)}$ **35.** $\dfrac{2}{t(1 - \ln t)^2}$ **37.** $\dfrac{\tan(\ln\theta)}{\theta}$

39. $\dfrac{10x}{x^2 + 1} + \dfrac{1}{2(1 - x)}$

41. $\left(\dfrac{1}{2}\right)\sqrt{x(x + 1)}\left(\dfrac{1}{x} + \dfrac{1}{x + 1}\right) = \dfrac{2x + 1}{2\sqrt{x(x + 1)}}$

43. $\left(\dfrac{1}{2}\right)\sqrt{\dfrac{t}{t + 1}}\left(\dfrac{1}{t} - \dfrac{1}{t + 1}\right) = \dfrac{1}{2\sqrt{t}(t + 1)^{3/2}}$

45. $\sqrt{\theta + 3}(\sin\theta)\left(\dfrac{1}{2(\theta + 3)} + \cot\theta\right)$

47. $t(t + 1)(t + 2)\left[\dfrac{1}{t} + \dfrac{1}{t + 1} + \dfrac{1}{t + 2}\right] = 3t^2 + 6t + 2$

49. $\dfrac{\theta + 5}{\theta\cos\theta}\left[\dfrac{1}{\theta + 5} - \dfrac{1}{\theta} + \tan\theta\right]$

51. $\dfrac{x\sqrt{x^2 + 1}}{(x + 1)^{2/3}}\left[\dfrac{1}{x} + \dfrac{x}{x^2 + 1} - \dfrac{2}{3(x + 1)}\right]$

53. $\dfrac{1}{3}\sqrt[3]{\dfrac{x(x-2)}{x^2+1}}\left(\dfrac{1}{x}+\dfrac{1}{x-2}-\dfrac{2x}{x^2+1}\right)$ **55.** $-2\tan\theta$

57. $\dfrac{1-t}{t}$ **59.** $1/(1+e^{\theta})$ **61.** $e^{\cos t}(1-t\sin t)$

63. $\dfrac{ye^y\cos x}{1-ye^y\sin x}$ **65.** $\dfrac{dy}{dx}=\dfrac{y^2-xy\ln y}{x^2-xy\ln x}$ **67.** $2^x\ln x$

69. $\left(\dfrac{\ln 5}{2\sqrt{s}}\right)5^{\sqrt{s}}$ **71.** $\pi x^{(\pi-1)}$ **73.** $\dfrac{1}{\theta\ln 2}$ **75.** $\dfrac{3}{x\ln 4}$

77. $\dfrac{2(\ln r)}{r(\ln 2)(\ln 4)}$ **79.** $\dfrac{-2}{(x+1)(x-1)}$

81. $\sin(\log_7\theta)+\dfrac{1}{\ln 7}\cos(\log_7\theta)$ **83.** $\dfrac{1}{\ln 5}$

85. $\dfrac{1}{t}(\log_2 3)3^{\log_2 t}$ **87.** $\dfrac{1}{t}$ **89.** $(x+1)^x\left(\dfrac{x}{x+1}+\ln(x+1)\right)$

91. $(\sqrt{t})^t\left(\dfrac{\ln t}{2}+\dfrac{1}{2}\right)$ **93.** $(\sin x)^x(\ln\sin x+x\cot x)$

95. $(x^{\ln x})\left(\dfrac{\ln x^2}{x}\right)$

Section 3.9, pp. 191–192

1. (a) $\pi/4$ (b) $-\pi/3$ (c) $\pi/6$
3. (a) $-\pi/6$ (b) $\pi/4$ (c) $-\pi/3$
5. (a) $\pi/3$ (b) $3\pi/4$ (c) $\pi/6$
7. (a) $3\pi/4$ (b) $\pi/6$ (c) $2\pi/3$
9. $1/\sqrt{2}$ **11.** $-1/\sqrt{3}$ **13.** $\pi/2$ **15.** $\pi/2$ **17.** $\pi/2$

19. 0 **21.** $\dfrac{-2x}{\sqrt{1-x^4}}$ **23.** $\dfrac{\sqrt{2}}{\sqrt{1-2t^2}}$

25. $\dfrac{1}{|2s+1|\sqrt{s^2+s}}$ **27.** $\dfrac{-2x}{(x^2+1)\sqrt{x^4+2x^2}}$

29. $\dfrac{-1}{\sqrt{1-t^2}}$ **31.** $\dfrac{-1}{2\sqrt{t}(1+t)}$ **33.** $\dfrac{1}{(\tan^{-1}x)(1+x^2)}$

35. $\dfrac{-e^t}{|e^t|\sqrt{(e^t)^2-1}}=\dfrac{-1}{\sqrt{e^{2t}-1}}$ **37.** $\dfrac{-2s^2}{\sqrt{1-s^2}}$ **39.** 0

41. $\sin^{-1}x$
47. (a) Defined; there is an angle whose tangent is 2.
   (b) Not defined; there is no angle whose cosine is 2.
49. (a) Not defined; no angle has secant 0.
   (b) Not defined; no angle has sine $\sqrt{2}$.
59. (a) Domain: all real numbers except those having the form $\dfrac{\pi}{2}+k\pi$ where k is an integer; range: $-\pi/2<y<\pi/2$.
   (b) Domain: $-\infty<x<\infty$; range: $-\infty<y<\infty$
61. (a) Domain: $-\infty<x<\infty$; range: $0\le y\le\pi$
   (b) Domain: $-1\le x\le 1$; range: $-1\le y\le 1$
63. The graphs are identical.

Section 3.10, pp. 197–201

1. $\dfrac{dA}{dt}=2\pi r\dfrac{dr}{dt}$ **3.** 10 **5.** -6 **7.** $-3/2$
9. $31/13$ **11.** (a) $-180\ \text{m}^2/\text{min}$ (b) $-135\ \text{m}^3/\text{min}$
13. (a) $\dfrac{dV}{dt}=\pi r^2\dfrac{dh}{dt}$ (b) $\dfrac{dV}{dt}=2\pi hr\dfrac{dr}{dt}$
   (c) $\dfrac{dV}{dt}=\pi r^2\dfrac{dh}{dt}+2\pi hr\dfrac{dr}{dt}$

15. (a) 1 volt/sec (b) $-\dfrac{1}{3}$ amp/sec
   (c) $\dfrac{dR}{dt}=\dfrac{1}{I}\left(\dfrac{dV}{dt}-\dfrac{V}{I}\dfrac{dI}{dt}\right)$
   (d) $3/2$ ohms/sec, R is increasing.
17. (a) $\dfrac{ds}{dt}=\dfrac{x}{\sqrt{x^2+y^2}}\dfrac{dx}{dt}$
   (b) $\dfrac{ds}{dt}=\dfrac{x}{\sqrt{x^2+y^2}}\dfrac{dx}{dt}+\dfrac{y}{\sqrt{x^2+y^2}}\dfrac{dy}{dt}$ (c) $\dfrac{dx}{dt}=-\dfrac{y}{x}\dfrac{dy}{dt}$
19. (a) $\dfrac{dA}{dt}=\dfrac{1}{2}ab\cos\theta\dfrac{d\theta}{dt}$
   (b) $\dfrac{dA}{dt}=\dfrac{1}{2}ab\cos\theta\dfrac{d\theta}{dt}+\dfrac{1}{2}b\sin\theta\dfrac{da}{dt}$
   (c) $\dfrac{dA}{dt}=\dfrac{1}{2}ab\cos\theta\dfrac{d\theta}{dt}+\dfrac{1}{2}b\sin\theta\dfrac{da}{dt}+\dfrac{1}{2}a\sin\theta\dfrac{db}{dt}$
21. (a) $14\ \text{cm}^2/\text{sec}$, increasing (b) 0 cm/sec, constant
   (c) $-14/13$ cm/sec, decreasing
23. (a) -12 ft/sec (b) $-59.5\ \text{ft}^2/\text{sec}$ (c) -1 rad/sec
25. 20 ft/sec
27. (a) $\dfrac{dh}{dt}=11.19$ cm/min (b) $\dfrac{dr}{dt}=14.92$ cm/min
29. (a) $\dfrac{-1}{24\pi}$ m/min (b) $r=\sqrt{26y-y^2}$ m
   (c) $\dfrac{dr}{dt}=-\dfrac{5}{288\pi}$ m/min
31. 1 ft/min, $40\pi\ \text{ft}^2/\text{min}$ **33.** 11 ft/sec
35. Increasing at $466/1681\ \text{L/min}^2$
37. -5 m/sec **39.** -1500 ft/sec
41. $\dfrac{5}{72\pi}$ in./min, $\dfrac{10}{3}\ \text{in}^2/\text{min}$
43. (a) $-32/\sqrt{13}\approx-8.875$ ft/sec
   (b) $d\theta_1/dt=8/65$ rad/sec, $d\theta_2/dt=-8/65$ rad/sec
   (c) $d\theta_1/dt=1/6$ rad/sec, $d\theta_2/dt=-1/6$ rad/sec

Section 3.11, pp. 210–212

1. $L(x)=10x-13$ **3.** $L(x)=2$ **5.** $L(x)=x-\pi$
7. $2x$ **9.** $-x-5$ **11.** $\dfrac{1}{12}x+\dfrac{4}{3}$ **13.** $1-x$
15. $f(0)=1$. Also, $f'(x)=k(1+x)^{k-1}$, so $f'(0)=k$. This means the linearization at $x=0$ is $L(x)=1+kx$.
17. (a) 1.01 (b) 1.003
19. $\left(3x^2-\dfrac{3}{2\sqrt{x}}\right)dx$ **21.** $\dfrac{2-2x^2}{(1+x^2)^2}dx$
23. $\dfrac{1-y}{3\sqrt{y}+x}dx$ **25.** $\dfrac{5}{2\sqrt{x}}\cos(5\sqrt{x})dx$
27. $(4x^2)\sec^2\left(\dfrac{x^3}{3}\right)dx$
29. $\dfrac{3}{\sqrt{x}}(\csc(1-2\sqrt{x})\cot(1-2\sqrt{x}))dx$
31. $\dfrac{1}{2\sqrt{x}}\cdot e^{\sqrt{x}}dx$ **33.** $\dfrac{2x}{1+x^2}dx$ **35.** $\dfrac{2xe^{x^2}}{1+e^{2x^2}}dx$
37. $\dfrac{-1}{\sqrt{e^{-2x}-1}}dx$ **39.** (a) $.41$ (b) $.4$ (c) $.01$
41. (a) $.231$ (b) $.2$ (c) $.031$

43. (a) $-1/3$ **(b)** $-2/5$ **(c)** $1/15$ **45.** $dV = 4\pi r_0{}^2\, dr$
47. $dS = 12x_0\, dx$ **49.** $dV = 2\pi r_0 h\, dr$
51. (a) 0.08π m^2 **(b)** 2% **53.** $dV \approx 565.5$ in^3
55. (a) 2% **(b)** 4% **57.** $\frac{1}{3}$% **59.** 3%

61. The ratio equals 37.87, so a change in the acceleration of gravity on the moon has about 38 times the effect that a change of the same magnitude has on Earth.

65. (a) $L(x) = x\ln 2 + 1 \approx 0.69x + 1$
(b)

Practice Exercises, pp. 213–218

1. $5x^4 - 0.25x + 0.25$ **3.** $3x(x-2)$
5. $2(x+1)(2x^2 + 4x + 1)$
7. $3(\theta^2 + \sec\theta + 1)^2(2\theta + \sec\theta\tan\theta)$
9. $\dfrac{1}{2\sqrt{t}\left(1 + \sqrt{t}\right)^2}$ **11.** $2\sec^2 x\tan x$
13. $8\cos^3(1-2t)\sin(1-2t)$
15. $5(\sec t)(\sec t + \tan t)^5$
17. $\dfrac{\theta\cos\theta + \sin\theta}{\sqrt{2\theta}\sin\theta}$ **19.** $\dfrac{\cos\sqrt{2\theta}}{\sqrt{2\theta}}$
21. $x\csc\left(\dfrac{2}{x}\right) + \csc\left(\dfrac{2}{x}\right)\cot\left(\dfrac{2}{x}\right)$
23. $\dfrac{1}{2}x^{1/2}\sec(2x)^2\left[16\tan(2x)^2 - x^{-2}\right]$
25. $-10x\csc^2(x^2)$ **27.** $8x^3\sin(2x^2)\cos(2x^2) + 2x\sin^2(2x^2)$
29. $\dfrac{-(t+1)}{8t^3}$ **31.** $\dfrac{1-x}{(x+1)^3}$ **33.** $\dfrac{-1}{2x^2\left(1+\dfrac{1}{x}\right)^{1/2}}$
35. $\dfrac{-2\sin\theta}{(\cos\theta - 1)^2}$ **37.** $3\sqrt{2x+1}$ **39.** $-9\left[\dfrac{5x + \cos 2x}{(5x^2 + \sin 2x)^{5/2}}\right]$
41. $-2e^{-x/5}$ **43.** xe^{4x} **45.** $\dfrac{2\sin\theta\cos\theta}{\sin^2\theta} = 2\cot\theta$
47. $\dfrac{2}{(\ln 2)x}$ **49.** $-8^{-t}(\ln 8)$ **51.** $18x^{2.6}$
53. $(x+2)^{x+2}(\ln(x+2) + 1)$ **55.** $-\dfrac{1}{\sqrt{1-u^2}}$
57. $\dfrac{-1}{\sqrt{1-x^2}\cos^{-1}x}$ **59.** $\tan^{-1}(t) + \dfrac{t}{1+t^2} - \dfrac{1}{2t}$
61. $\dfrac{1-z}{\sqrt{z^2-1}} + \sec^{-1}z$ **63.** -1 **65.** $-\dfrac{y+2}{x+3}$
67. $\dfrac{-3x^2 - 4y + 2}{4x - 4y^{1/3}}$ **69.** $-\dfrac{y}{x}$ **71.** $\dfrac{1}{2y(x+1)^2}$
73. $-1/2$ **75.** y/x **77.** $-\dfrac{2e^{-\tan^{-1}x}}{1+x^2}$ **79.** $\dfrac{dp}{dq} = \dfrac{6q - 4p}{3p^2 + 4q}$
81. $\dfrac{dr}{ds} = (2r-1)(\tan 2s)$

83. (a) $\dfrac{d^2y}{dx^2} = \dfrac{-2xy^3 - 2x^4}{y^5}$ **(b)** $\dfrac{d^2y}{dx^2} = \dfrac{-2xy^2 - 1}{x^4y^3}$
85. (a) 7 **(b)** -2 **(c)** 5/12 **(d)** 1/4 **(e)** 12 **(f)** 9/2
(g) 3/4
87. 0 **89.** $\dfrac{3\sqrt{2e}^{\sqrt{3/2}}}{4}\cos\left(e^{\sqrt{3/2}}\right)$ **91.** $-\dfrac{1}{2}$ **93.** $\dfrac{-2}{(2t+1)^2}$
95. (a) **97. (a)**

$f(x) = \begin{cases} x^2, & -1 \le x < 0 \\ -x^2, & 0 \le x < 1 \end{cases}$

(b) Yes **(c)** Yes
(b) Yes **(c)** No

99. $\left(\dfrac{5}{2}, \dfrac{9}{4}\right)$ and $\left(\dfrac{3}{2}, -\dfrac{1}{4}\right)$ **101.** $(-1, 27)$ and $(2, 0)$
103. (a) $(-2, 16), (3, 11)$ **(b)** $(0, 20), (1, 7)$
105.

107. $\dfrac{1}{4}$ **109.** 4
111. Tangent: $y = -\dfrac{1}{4}x + \dfrac{9}{4}$, normal: $y = 4x - 2$
113. Tangent: $y = 2x - 4$, normal: $y = -\dfrac{1}{2}x + \dfrac{7}{2}$
115. Tangent: $y = -\dfrac{5}{4}x + 6$, normal: $y = \dfrac{4}{5}x - \dfrac{11}{5}$
117. $(1, 1)$: $m = -\dfrac{1}{2}$; $(1, -1)$: m not defined
119. B = graph of f, A = graph of f'
121.

123. (a) 0, 0 **(b)** 1700 rabbits, $\approx$1400 rabbits
125. -1 **127.** 1/2 **129.** 4 **131.** 1
133. To make g continuous at the origin, define $g(0) = 1$.
135. $\dfrac{2(x^2+1)}{\sqrt{\cos 2x}}\left[\dfrac{2x}{x^2+1} + \tan 2x\right]$
137. $5\left[\dfrac{(t+1)(t-1)}{(t-2)(t+3)}\right]^5\left[\dfrac{1}{t+1} + \dfrac{1}{t-1} - \dfrac{1}{t-2} - \dfrac{1}{t+3}\right]$

139. $\dfrac{1}{\sqrt{\theta}} (\sin \theta)^{\sqrt{\theta}} \left(\dfrac{\ln \sin \theta}{2} + \theta \cot \theta \right)$

141. (a) $\dfrac{dS}{dt} = (4\pi r + 2\pi h) \dfrac{dr}{dt}$ **(b)** $\dfrac{dS}{dt} = 2\pi r \dfrac{dh}{dt}$

   (c) $\dfrac{dS}{dt} = (4\pi r + 2\pi h) \dfrac{dr}{dt} + 2\pi r \dfrac{dh}{dt}$

   (d) $\dfrac{dr}{dt} = -\dfrac{r}{2r + h} \dfrac{dh}{dt}$

143. $-40 \text{ m}^2/\text{sec}$ **145.** 0.02 ohm/sec **147.** 22 m/sec

149. (a) $r = \dfrac{2}{5} h$ **(b)** $-\dfrac{125}{144\pi} \text{ ft/min}$

151. (a) $\dfrac{3}{5} \text{ km/sec or } 600 \text{ m/sec}$ **(b)** $\dfrac{18}{\pi} \text{ rpm}$

153. (a) $L(x) = 2x + \dfrac{\pi - 2}{2}$

   (b) $L(x) = -\sqrt{2}x + \dfrac{\sqrt{2}(4 - \pi)}{4}$

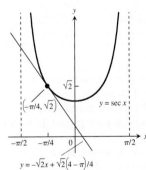

155. $L(x) = 1.5x + 0.5$ **157.** $dS = \dfrac{\pi r h_0}{\sqrt{r^2 + h_0^2}} dh$

159. (a) 4% **(b)** 8% **(c)** 12%

Additional and Advanced Exercises, pp. 218–220

1. (a) $\sin 2\theta = 2 \sin \theta \cos \theta$; $2 \cos 2\theta = 2 \sin \theta (-\sin \theta) + \cos \theta (2 \cos \theta)$; $2 \cos 2\theta = -2 \sin^2 \theta + 2 \cos^2 \theta$; $\cos 2\theta = \cos^2 \theta - \sin^2 \theta$

   (b) $\cos 2\theta = \cos^2 \theta - \sin^2 \theta$; $-2 \sin 2\theta = 2 \cos \theta (-\sin \theta) - 2 \sin \theta (\cos \theta)$; $\sin 2\theta = \cos \theta \sin \theta + \sin \theta \cos \theta$; $\sin 2\theta = 2 \sin \theta \cos \theta$

3. (a) $a = 1, b = 0, c = -\dfrac{1}{2}$ **(b)** $b = \cos a, c = \sin a$

5. $h = -4, k = \dfrac{9}{2}, a = \dfrac{5\sqrt{5}}{2}$

7. (a) $0.09y$ **(b)** Increasing at 1% per year

9. Answers will vary. Here is one possibility.

11. (a) $2 \text{ sec}, 64 \text{ ft/sec}$ **(b)** $12.31 \text{ sec}, 393.85 \text{ ft}$

15. (a) $m = -\dfrac{b}{\pi}$ **(b)** $m = -1, b = \pi$

17. (a) $a = \dfrac{3}{4}, b = \dfrac{9}{4}$ **19.** f odd $\Rightarrow f'$ is even

23. h' is defined but not continuous at $x = 0$; k' is defined *and* continuous at $x = 0$.

27. (a) 0.8156 ft **(b)** 0.00613 sec
   (c) It will lose about 8.83 min/day.

CHAPTER 4

Section 4.1, pp. 227–230

1. Absolute minimum at $x = c_2$; absolute maximum at $x = b$

3. Absolute maximum at $x = c$; no absolute minimum

5. Absolute minimum at $x = a$; absolute maximum at $x = c$

7. No absolute minimum; no absolute maximum

9. Absolute maximum at $(0, 5)$ **11.** (c)   **13.** (d)

15. Absolute minimum at $x = 0$; no absolute maximum

17. Absolute maximum at $x = 2$; no absolute minimum

19. Absolute maximum at $x = \pi/2$; absolute minimum at $x = 3\pi/2$

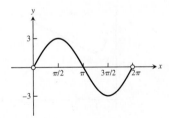

21. Absolute maximum: -3; absolute minimum: $-19/3$

23. Absolute maximum: 3; absolute minimum: -1

25. Absolute maximum: -0.25; absolute minimum: -4

27. Absolute maximum: 2; absolute minimum: -1

29. Absolute maximum: 2; absolute minimum: 0

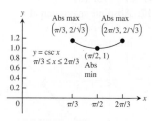

31. Absolute maximum: 1; absolute minimum: -1

33. Absolute maximum: $2/\sqrt{3}$; absolute minimum: 1

35. Absolute maximum: 2; absolute minimum: -1

37. Absolute maximum is $1/e$ at $x = 1$; absolute minimum is $-e$ at $x = -1$.

39. Absolute maximum value is $(1/4) + \ln 4$ at $x = 4$; absolute minimum value is 1 at $x = 1$; local maximum at $(1/2, 2 - \ln 2)$.

41. Increasing on $(0, 8)$, decreasing on $(-1, 0)$; absolute maximum: 16 at $x = 8$; absolute minimum: 0 at $x = 0$

43. Increasing on $(-32, 1)$; absolute maximum: 1 at $\theta = 1$; absolute minimum: -8 at $\theta = -32$

45. $x = 3$

47. $x = 1, x = 4$

49. $x = 1$

51. $x = 0$ and $x = 4$

53. Minimum value is 1 at $x = 2$.

55. Local maximum at $(-2, 17)$; local minimum at $\left(\frac{4}{3}, -\frac{41}{27}\right)$

57. Minimum value is 0 at $x = -1$ and $x = 1$.

59. There is a local minimum at $(0, 1)$.

61. Maximum value is $\frac{1}{2}$ at $x = 1$; minimum value is $-\frac{1}{2}$ at $x = -1$.

63. The minimum value is 2 at $x = 0$.

65. The minimum value is $-\frac{1}{e}$ at $x = \frac{1}{e}$.

67. The maximum value is $\frac{\pi}{2}$ at $x = 0$; an absolute minimum value is 0 at $x = 1$ and $x = -1$.

69.

Critical point or endpoint	Derivative	Extremum	Value
$x = -\frac{4}{5}$	0	Local max	$\frac{12}{25} 10^{1/3} \approx 1.034$
$x = 0$	Undefined	Local min	0

71.

Critical point or endpoint	Derivative	Extremum	Value
$x = -2$	Undefined	Local max	0
$x = -\sqrt{2}$	0	Minimum	-2
$x = \sqrt{2}$	0	Maximum	2
$x = 2$	Undefined	Local min	0

73.

Critical point or endpoint	Derivative	Extremum	Value
$x = 1$	Undefined	Minimum	2

75.

Critical point or endpoint	Derivative	Extremum	Value
$x = -1$	0	Maximum	5
$x = 1$	Undefined	Local min	1
$x = 3$	0	Maximum	5

77. (a) No

(b) The derivative is defined and nonzero for $x \neq 2$. Also, $f(2) = 0$ and $f(x) > 0$ for all $x \neq 2$.

(c) No, because $(-\infty, \infty)$ is not a closed interval.

(d) The answers are the same as parts (a) and (b) with 2 replaced by a.

79. Yes   **81.** g assumes a local maximum at $-c$.

83. (a) Maximum value is 144 at $x = 2$.

(b) The largest volume of the box is 144 cubic units, and it occurs when $x = 2$.

85. $\frac{v_0^2}{2g} + s_0$

87. Maximum value is 11 at $x = 5$; minimum value is 5 on the interval $[-3, 2]$; local maximum at $(-5, 9)$.

89. Maximum value is 5 on the interval $[3, \infty)$; minimum value is -5 on the interval $(-\infty, -2]$.

Section 4.2, pp. 236–238

1. $1/2$ **3.** 1   **5.** $\pm\sqrt{1 - \frac{4}{\pi^2}} \approx \pm 0.771$

7. $\frac{1}{3}(1 + \sqrt{7}) \approx 1.22, \frac{1}{3}(1 - \sqrt{7}) \approx -0.549$

9. Does not; f is not differentiable at the interior domain point $x = 0$.

11. Does   **13.** Does not; f is not differentiable at $x = -1$.

17. (a)

i)

ii)

iii)

iv)

29. Yes   **31. (a)** 4   **(b)** 3   **(c)** 3

33. (a) $\dfrac{x^2}{2} + C$ **(b)** $\dfrac{x^3}{3} + C$ **(c)** $\dfrac{x^4}{4} + C$

35. (a) $\dfrac{1}{x} + C$ **(b)** $x + \dfrac{1}{x} + C$ **(c)** $5x - \dfrac{1}{x} + C$

37. (a) $-\dfrac{1}{2}\cos 2t + C$ **(b)** $2\sin\dfrac{t}{2} + C$

 (c) $-\dfrac{1}{2}\cos 2t + 2\sin\dfrac{t}{2} + C$

39. $f(x) = x^2 - x$ **41.** $f(x) = 1 + \dfrac{e^{2x}}{2}$

43. $s = 4.9t^2 + 5t + 10$ **45.** $s = \dfrac{1 - \cos(\pi t)}{\pi}$

47. $s = e^t + 19t + 4$ **49.** $s = \sin(2t) - 3$

51. If $T(t)$ is the temperature of the thermometer at time t, then $T(0) = -19\,°C$ and $T(14) = 100\,°C$. From the Mean Value Theorem, there exists a $0 < t_0 < 14$ such that $\dfrac{T(14) - T(0)}{14 - 0} = 8.5\,°C/\sec = T'(t_0)$, the rate at which the temperature was changing at $t = t_0$ as measured by the rising mercury on the thermometer.

53. Because its average speed was approximately 7.667 knots, and by the Mean Value Theorem, it must have been going that speed at least once during the trip.

57. The conclusion of the Mean Value Theorem yields

$$\dfrac{\dfrac{1}{b} - \dfrac{1}{a}}{b - a} = -\dfrac{1}{c^2} \Rightarrow c^2\left(\dfrac{a - b}{ab}\right) = a - b \Rightarrow c = \sqrt{ab}.$$

61. $f(x)$ must be zero at least once between a and b by the Intermediate Value Theorem. Now suppose that $f(x)$ is zero twice between a and b. Then, by the Mean Value Theorem, $f'(x)$ would have to be zero at least once between the two zeros of $f(x)$, but this can't be true since we are given that $f'(x) \neq 0$ on this interval. Therefore, $f(x)$ is zero once and only once between a and b.

71. $1.09999 \leq f(0.1) \leq 1.1$

Section 4.3, pp. 241–243

1. (a) 0, 1
 (b) Increasing on $(-\infty, 0)$ and $(1, \infty)$; decreasing on $(0, 1)$
 (c) Local maximum at $x = 0$; local minimum at $x = 1$

3. (a) $-2, 1$
 (b) Increasing on $(-2, 1)$ and $(1, \infty)$; decreasing on $(-\infty, -2)$
 (c) No local maximum; local minimum at $x = -2$

5. (a) Critical point at $x = 1$
 (b) Decreasing on $(-\infty, 1)$, increasing on $(1, \infty)$
 (c) Local (and absolute) minimum at $x = 1$

7. (a) 0, 1
 (b) Increasing on $(-\infty, -2)$ and $(1, \infty)$; decreasing on $(-2, 0)$ and $(0, 1)$
 (c) Local minimum at $x = 1$

9. (a) $-2, 2$
 (b) Increasing on $(-\infty, -2)$ and $(2, \infty)$; decreasing on $(-2, 0)$ and $(0, 2)$
 (c) Local maximum at $x = -2$; local minimum at $x = 2$

11. (a) $-2, 0$
 (b) Increasing on $(-\infty, -2)$ and $(0, \infty)$; decreasing on $(-2, 0)$
 (c) Local maximum at $x = -2$; local minimum at $x = 0$

13. (a) $\dfrac{\pi}{2}, \dfrac{2\pi}{3}, \dfrac{4\pi}{3}$
 (b) Increasing on $\left(\dfrac{2\pi}{3}, \dfrac{4\pi}{3}\right)$; decreasing on $\left(0, \dfrac{\pi}{2}\right)$, $\left(\dfrac{\pi}{2}, \dfrac{2\pi}{3}\right)$, and $\left(\dfrac{4\pi}{3}, 2\pi\right)$
 (c) Local maximum at $x = 0$ and $x = \dfrac{4\pi}{3}$; local minimum at $x = \dfrac{2\pi}{3}$ and $x = 2\pi$

15. (a) Increasing on $(-2, 0)$ and $(2, 4)$; decreasing on $(-4, -2)$ and $(0, 2)$
 (b) Absolute maximum at $(-4, 2)$; local maximum at $(0, 1)$ and $(4, -1)$; absolute minimum at $(2, -3)$; local minimum at $(-2, 0)$

17. (a) Increasing on $(-4, -1)$, $(1/2, 2)$, and $(2, 4)$; decreasing on $(-1, 1/2)$
 (b) Absolute maximum at $(4, 3)$; local maximum at $(-1, 2)$ and $(2, 1)$; no absolute minimum; local minimum at $(-4, -1)$ and $(1/2, -1)$

19. (a) Increasing on $(-\infty, -1.5)$; decreasing on $(-1.5, \infty)$
 (b) Local maximum: 5.25 at $t = -1.5$; absolute maximum: 5.25 at $t = -1.5$

21. (a) Decreasing on $(-\infty, 0)$; increasing on $(0, 4/3)$; decreasing on $(4/3, \infty)$
 (b) Local minimum at $x = 0$ $(0, 0)$; local maximum at $x = 4/3$ $(4/3, 32/27)$; no absolute extrema

23. (a) Decreasing on $(-\infty, 0)$; increasing on $(0, 1/2)$; decreasing on $(1/2, \infty)$
 (b) Local minimum at $\theta = 0$ $(0, 0)$; local maximum at $\theta = 1/2$ $(1/2, 1/4)$; no absolute extrema

25. (a) Increasing on $(-\infty, \infty)$; never decreasing
 (b) No local extrema; no absolute extrema

27. (a) Increasing on $(-2, 0)$ and $(2, \infty)$; decreasing on $(-\infty, -2)$ and $(0, 2)$
 (b) Local maximum: 16 at $x = 0$; local minimum: 0 at $x = \pm 2$; no absolute maximum; absolute minimum: 0 at $x = \pm 2$

29. (a) Increasing on $(-\infty, -1)$; decreasing on $(-1, 0)$; increasing on $(0, 1)$; decreasing on $(1, \infty)$
 (b) Local maximum: 0.5 at $x = \pm 1$; local minimum: 0 at $x = 0$; absolute maximum: $1/2$ at $x = \pm 1$; no absolute minimum

31. (a) Increasing on $(10, \infty)$; decreasing on $(1, 10)$
 (b) Local maximum: 1 at $x = 1$; local minimum: -8 at $x = 10$; absolute minimum: -8 at $x = 10$

33. (a) Decreasing on $(-2\sqrt{2}, -2)$; increasing on $(-2, 2)$; decreasing on $(2, 2\sqrt{2})$
 (b) Local minima: $g(-2) = -4$, $g(2\sqrt{2}) = 0$; local maxima: $g(-2\sqrt{2}) = 0$, $g(2) = 4$; absolute maximum: 4 at $x = 2$; absolute minimum: -4 at $x = -2$

35. (a) Increasing on $(-\infty, 1)$; decreasing when $1 < x < 2$, decreasing when $2 < x < 3$; discontinuous at $x = 2$; increasing on $(3, \infty)$

(b) Local minimum at $x = 3$ $(3, 6)$; local maximum at $x = 1$ $(1, 2)$; no absolute extrema

37. (a) Increasing on $(-2, 0)$ and $(0, \infty)$; decreasing on $(-\infty, -2)$

(b) Local minimum: $-6\sqrt[3]{2}$ at $x = -2$; no absolute maximum; absolute minimum: $-6\sqrt[3]{2}$ at $x = -2$

39. (a) Increasing on $(-\infty, -2/\sqrt{7})$ and $(2/\sqrt{7}, \infty)$; decreasing on $(-2/\sqrt{7}, 0)$ and $(0, 2/\sqrt{7})$

(b) Local maximum: $24\sqrt[3]{2}/7^{7/6} \approx 3.12$ at $x = -2/\sqrt{7}$; local minimum: $-24\sqrt[3]{2}/7^{7/6} \approx -3.12$ at $x = 2/\sqrt{7}$; no absolute extrema

41. (a) Increasing on $((1/3) \ln (1/2), \infty)$, decreasing on $(-\infty, (1/3) \ln (1/2))$

(b) Local minimum is $\dfrac{3}{2^{2/3}}$ at $x = (1/3) \ln (1/2)$; no local maximum; absolute minimum is $\dfrac{3}{2^{2/3}}$ at $x = (1/3) \ln (1/2)$; no absolute maximum

43. (a) Increasing on (e^{-1}, ∞), decreasing on $(0, e^{-1})$

(b) A local minimum is $-e^{-1}$ at $x = e^{-1}$, no local maximum; absolute minimum is $-e^{-1}$ at $x = e^{-1}$, no absolute maximum

45. (a) Local maximum: 1 at $x = 1$; local minimum: 0 at $x = 2$

(b) Absolute maximum: 1 at $x = 1$; no absolute minimum

47. (a) Local maximum: 1 at $x = 1$; local minimum: 0 at $x = 2$

(b) No absolute maximum; absolute minimum: 0 at $x = 2$

49. (a) Local maxima: -9 at $t = -3$ and 16 at $t = 2$; local minimum: -16 at $t = -2$

(b) Absolute maximum: 16 at $t = 2$; no absolute minimum

51. (a) Local minimum: 0 at $x = 0$

(b) No absolute maximum; absolute minimum: 0 at $x = 0$

53. (a) Local maximum: 5 at $x = 0$; local minimum: 0 at $x = -5$ and $x = 5$

(b) Absolute maximum: 5 at $x = 0$; absolute minimum: 0 at $x = -5$ and $x = 5$

55. (a) Local maximum: 2 at $x = 0$; local minimum: $\dfrac{\sqrt{3}}{4\sqrt{3} - 6}$ at $x = 2 - \sqrt{3}$

(b) No absolute maximum; an absolute minimum at $x = 2 - \sqrt{3}$

57. (a) Local maximum: 1 at $x = \pi/4$; local maximum: 0 at $x = \pi$; local minimum: 0 at $x = 0$; local minimum: -1 at $x = 3\pi/4$

59. Local maximum: 2 at $x = \pi/6$; local maximum: $\sqrt{3}$ at $x = 2\pi$; local minimum: -2 at $x = 7\pi/6$; local minimum: $\sqrt{3}$ at $x = 0$

61. (a) Local minimum: $(\pi/3) - \sqrt{3}$ at $x = 2\pi/3$; local maximum: 0 at $x = 0$; local maximum: π at $x = 2\pi$

63. (a) Local minimum: 0 at $x = \pi/4$

65. Local maximum: 3 at $\theta = 0$; local minimum: -3 at $\theta = 2\pi$

67.

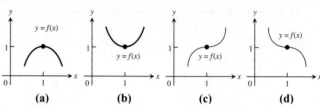

(a)   **(b)**   **(c)**   **(d)**

69. (a)   **(b)**

73. $a = -2, b = 4$

75. (a) Absolute minimum occurs at $x = \pi/3$ with $f(\pi/3) = -\ln 2$, and the absolute maximum occurs at $x = 0$ with $f(0) = 0$.

(b) Absolute minimum occurs at $x = 1/2$ and $x = 2$ with $f(1/2) = f(2) = \cos (\ln 2)$, and the absolute maximum occurs at $x = 1$ with $f(1) = 1$.

77. Minimum of $2 - 2 \ln 2 \approx 0.613706$ at $x = \ln 2$; maximum of 1 at $x = 0$

79. Absolute maximum value of $1/2e$ assumed at $x = 1/\sqrt{e}$

83. Increasing; $\dfrac{df^{-1}}{dx} = \dfrac{1}{9} x^{-2/3}$

85. Decreasing; $\dfrac{df^{-1}}{dx} = -\dfrac{1}{3} x^{-2/3}$

Section 4.4, pp. 251–254

1. Local maximum: $3/2$ at $x = -1$; local minimum: -3 at $x = 2$; point of inflection at $(1/2, -3/4)$; rising on $(-\infty, -1)$ and $(2, \infty)$; falling on $(-1, 2)$; concave up on $(1/2, \infty)$; concave down on $(-\infty, 1/2)$

3. Local maximum: $3/4$ at $x = 0$; local minimum: 0 at $x = \pm 1$; points of inflection at $\left(-\sqrt{3}, \dfrac{3\sqrt[3]{4}}{4}\right)$ and $\left(\sqrt{3}, \dfrac{3\sqrt[3]{4}}{4}\right)$; rising on $(-1, 0)$ and $(1, \infty)$; falling on $(-\infty, -1)$ and $(0, 1)$; concave up on $(-\infty, -\sqrt{3})$ and $(\sqrt{3}, \infty)$; concave down on $(-\sqrt{3}, \sqrt{3})$

5. Local maxima: $\dfrac{-2\pi}{3} + \dfrac{\sqrt{3}}{2}$ at $x = -2\pi/3$, $\dfrac{\pi}{3} + \dfrac{\sqrt{3}}{2}$ at $x = \pi/3$; local minima: $-\dfrac{\pi}{3} - \dfrac{\sqrt{3}}{2}$ at $x = -\pi/3$, $\dfrac{2\pi}{3} - \dfrac{\sqrt{3}}{2}$ at $x = 2\pi/3$; points of inflection at $(-\pi/2, -\pi/2), (0, 0)$, and $(\pi/2, \pi/2)$, rising on $(-\pi/3, \pi/3)$; falling on $(-2\pi/3, -\pi/3)$ and $(\pi/3, 2\pi/3)$, concave up on $(-\pi/2, 0)$ and $(\pi/2, 2\pi/3)$; concave down on $(-2\pi/3, -\pi/2)$ and $(0, \pi/2)$

7. Local maxima: 1 at $x = -\pi/2$ and $x = \pi/2$, 0 at $x = -2\pi$ and $x = 2\pi$; local minima: -1 at $x = -3\pi/2$ and $x = 3\pi/2$, 0 at $x = 0$; points of inflection at $(-\pi, 0)$ and $(\pi, 0)$; rising on $(-3\pi/2, -\pi/2), (0, \pi/2)$, and $(3\pi/2, 2\pi)$; falling on $(-2\pi, -3\pi/2), (-\pi/2, 0)$, and $(\pi/2, 3\pi/2)$; concave up on $(-2\pi, -\pi)$ and $(\pi, 2\pi)$; concave down on $(-\pi, 0)$ and $(0, \pi)$

9.

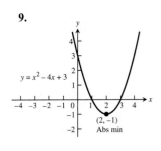

$y = x^2 - 4x + 3$
(2, −1) Abs min

11.

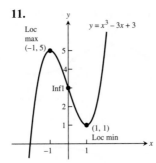

Loc max (−1, 5)
$y = x^3 - 3x + 3$
Infl
(1, 1) Loc min

13.

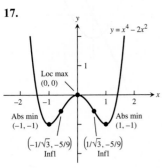

(2, 5) Loc max
Infl (1, 1)
(0, −3) Loc min
$y = -2x^3 + 6x^2 - 3$

15.

Infl (2, 1)
$y = (x-2)^3 + 1$

17.

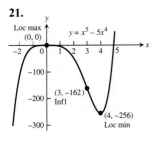

$y = x^4 - 2x^2$
Loc max (0, 0)
Abs min (−1, −1)
Abs min (1, −1)
$(-1/\sqrt{3}, -5/9)$ Infl
$(1/\sqrt{3}, -5/9)$ Infl

19.

(3, 27) Abs max
(2, 16) Infl
$y = 4x^3 - x^4$
Infl (0, 0)

21.

Loc max (0, 0)
$y = x^5 - 5x^4$
(3, −162) Infl
(4, −256) Loc min

23.

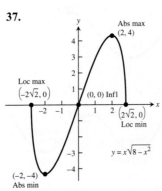

Abs max (2π, 2π)
$y = x + \sin x$
(π, π) Infl
Abs min

25.

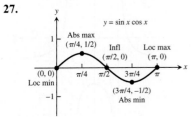

Loc max $(4\pi/3, 4\sqrt{3}\pi/3 + 1)$
$(2\pi, 2\sqrt{3}\pi - 2)$ Abs max
Infl $(5\pi/3, 5\sqrt{3}\pi/3 - 1)$ Loc min
$(3\pi/2, 3\sqrt{3}\pi/2)$
Infl $(\pi/2, \sqrt{3}\pi/2)$
(0, −2) Abs min
$y = \sqrt{3}x - 2\cos x$

27.

$y = \sin x \cos x$
Abs max (π/4, 1/2)
Infl (π/2, 0)
Loc max (π, 0)
(0, 0) Loc min
(3π/4, −1/2) Abs min

29.

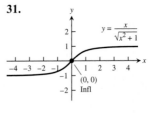

Vert tan at x = 0
$y = x^{1/5}$
(0, 0) Infl

31.

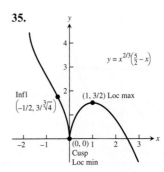

$y = \dfrac{x}{\sqrt{x^2+1}}$
(0, 0) Infl

33.

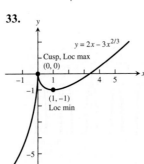

$y = 2x - 3x^{2/3}$
Cusp, Loc max (0, 0)
(1, −1) Loc min

35.

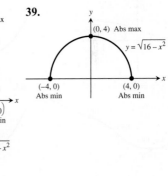

$y = x^{2/3}\left(\frac{5}{2} - x\right)$
Infl $(-1/2, 3/\sqrt[3]{4})$
(1, 3/2) Loc max
(0, 0) Cusp Loc min

37.

Abs max (2, 4)
Loc max $(-2\sqrt{2}, 0)$
(0, 0) Infl
$(2\sqrt{2}, 0)$ Loc min
$y = x\sqrt{8 - x^2}$
(−2, −4) Abs min

39.

(0, 4) Abs max
$y = \sqrt{16 - x^2}$
(−4, 0) Abs min
(4, 0) Abs min

41.

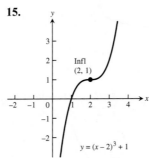

$y = \dfrac{x^2 - 3}{x - 2}$
(3, 6) Loc min
(1, 2) Loc max

43.

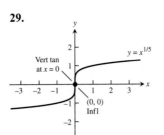

$y = \dfrac{8x}{x^2 + 4}$
(2, 2) Abs max $(2\sqrt{3}, \sqrt{3})$ Infl
$(-2\sqrt{3}, -\sqrt{3})$ Infl
(0, 0) Infl
(−2, −2) Abs min

45.

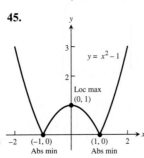

$y = x^2 - 1$
Loc max (0, 1)
(−1, 0) Abs min
(1, 0) Abs min

47.

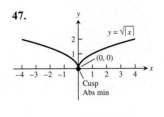

$y = \sqrt{|x|}$
(0, 0) Cusp Abs min

49.

51.

73. $y'' = -\sin t,\ 0 \le t \le 2\pi$

53.

55.

75. $y'' = -\dfrac{2}{3}(x+1)^{-5/3}$ **77.** $y'' = \dfrac{1}{3}x^{-2/3} + \dfrac{2}{3}x^{-5/3}$

57.

59. $y'' = 1 - 2x$

79. $y'' = \begin{cases} -2, & x < 0 \\ 2, & x > 0 \end{cases}$ **81.**

61. $y'' = 3(x-3)(x-1)$ **63.** $y'' = 3(x-2)(x+2)$

83.

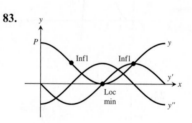

65. $y'' = 4(4-x)(5x^2 - 16x + 8)$

67. $y'' = 2\sec^2 x \tan x$ **69.** $y'' = -\dfrac{1}{2}\csc^2 \dfrac{\theta}{2},\ 0 < \theta < 2\pi$

85.   **87.**

71. $y'' = 2\tan\theta \sec^2\theta,\ -\dfrac{\pi}{2} < \theta < \dfrac{\pi}{2}$

89.   **91.**

93.

95.

97.

99.

101.

103.

Point	y'	y''
P	$-$	$+$
Q	$+$	0
R	$+$	$-$
S	0	$-$
T	$-$	$-$

105.

107. (a) Towards origin: $0 \le t < 2$ and $6 \le t \le 10$; away from origin: $2 \le t \le 6$ and $10 \le t \le 15$

    (b) $t = 2, t = 6, t = 10$ **(c)** $t = 5, t = 7, t = 13$

    (d) Positive: $5 \le t \le 7, 13 \le t \le 15$; negative: $0 \le t \le 5, 7 \le t \le 13$

109. ≈ 60 thousand units

111. Local minimum at $x = 2$; inflection points at $x = 1$ and $x = 5/3$

115. $b = -3$

119. $-1, 2$

121. $a = 1, b = 3, c = 9$

123. The zeros of $y' = 0$ and $y'' = 0$ are extrema and points of inflection, respectively. Inflection at $x = 3$, local maximum at $x = 0$, local minimum at $x = 4$.

125. The zeros of $y' = 0$ and $y'' = 0$ are extrema and points of inflection, respectively. Inflection at $x = -\sqrt[3]{2}$; local maximum at $x = -2$; local minimum at $x = 0$.

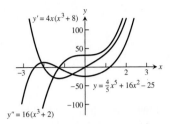

Section 4.5, pp. 261–262

1. $-1/4$ **3.** $5/7$ **5.** $1/2$ **7.** $1/4$ **9.** $-23/7$ **11.** $5/7$ **13.** 0

15. -16 **17.** -2 **19.** $1/4$ **21.** 2 **23.** 3 **25.** -1

27. $\ln 3$ **29.** $\dfrac{1}{\ln 2}$ **31.** $\ln 2$ **33.** 1 **35.** $1/2$ **37.** $\ln 2$

39. $-\infty$ **41.** $-1/2$ **43.** -1 **45.** 1 **47.** 0 **49.** 2

51. $1/e$ **53.** 1 **55.** $1/e$ **57.** $e^{1/2}$ **59.** 1 **61.** e^3

63. 0 **65.** 1 **67.** 3 **69.** 1 **71.** 0 **73.** ∞

75. (b) is correct.   **77.** (d) is correct.   **79.** $c = \dfrac{27}{10}$ **81. (b)** $\dfrac{-1}{2}$

83. -1 **87. (a)** $y = 1$ **(b)** $y = 0, y = \dfrac{3}{2}$

89. (a) We should assign the value 1 to $f(x) = (\sin x)^x$ to make it continuous at $x = 0$.

    (c) The maximum value of $f(x)$ is close to 1 near the point $x \approx 1.55$ (see the graph in part (a)).

Section 4.6, pp. 268–274

1. 16 in., 4 in. by 4 in.

3. (a) $(x, 1 - x)$ **(b)** $A(x) = 2x(1 - x)$

    (c) $\dfrac{1}{2}$ square units, 1 by $\dfrac{1}{2}$

5. $\dfrac{14}{3} \times \dfrac{35}{3} \times \dfrac{5}{3}$ in., $\dfrac{2450}{27}$ in^3 **7.** 80,000 m^2; 400 m by 200 m

9. (a) The optimum dimensions of the tank are 10 ft on the base edges and 5 ft deep.

    (b) Minimizing the surface area of the tank minimizes its weight for a given wall thickness. The thickness of the steel walls would likely be determined by other considerations such as structural requirements.

11. 9×18 in.   **13.** $\dfrac{\pi}{2}$ **15.** $h : r = 8 : \pi$

17. (a) $V(x) = 2x(24 - 2x)(18 - 2x)$ **(b)** Domain: $(0, 9)$

(c) Maximum volume ≈ 1309.95 in^3 when $x \approx 3.39$ in.
(d) $V'(x) = 24x^2 - 336x + 864$, so the critical point is at $x = 7 - \sqrt{13}$, which confirms the result in part (c).
(e) $x = 2$ in. or $x = 5$ in.

19. ≈ 2418.40 cm^3

21. (a) $h = 24$, $w = 18$
(b)

23. If r is the radius of the hemisphere, h the height of the cylinder, and V the volume, then $r = \left(\dfrac{3V}{8\pi}\right)^{1/3}$ and $h = \left(\dfrac{3V}{\pi}\right)^{1/3}$.

25. (b) $x = \dfrac{51}{8}$ **(c)** $L \approx 11$ in.

27. Radius $= \sqrt{2}$ m, height $= 1$ m, volume $= \dfrac{2\pi}{3}$ m^3

29. 1   **31.** $\dfrac{9b}{9 + \sqrt{3}\pi}$ m, triangle, $\dfrac{b\sqrt{3}\pi}{9 + \sqrt{3}\pi}$ m, circle

33. $\dfrac{3}{2} \times 2$ **35. (a)** 16   **(b)** -1

37. (a) $v(0) = 96$ ft/sec
(b) 256 ft at $t = 3$ sec
(c) Velocity when $s = 0$ is $v(7) = -128$ ft/sec.

39. ≈ 46.87 ft   **41. (a)** $6 \times 6\sqrt{3}$ in.

43. (a) $4\sqrt{3} \times 4\sqrt{6}$ in.

45. (a) $10\pi \approx 31.42$ cm/sec; when $t = 0.5$ sec, 1.5 sec, 2.5 sec, 3.5 sec; $s = 0$, acceleration is 0.
(b) 10 cm from rest position; speed is 0.

47. (a) $s = ((12 - 12t)^2 + 64t^2)^{1/2}$
(b) -12 knots, 8 knots
(c) No
(e) $4\sqrt{13}$. This limit is the square root of the sums of the squares of the individual speeds.

49. $x = \dfrac{a}{2}$, $v = \dfrac{ka^2}{4}$ **51.** $\dfrac{c}{2} + 50$

53. (a) $\sqrt{\dfrac{2km}{h}}$ **(b)** $\sqrt{\dfrac{2km}{h}}$

57. $4 \times 4 \times 3$ ft, \$288   **59.** $M = \dfrac{C}{2}$ **65. (a)** $y = -1$

67. (a) The minimum distance is $\dfrac{\sqrt{5}}{2}$.

(b) The minimum distance is from the point $(3/2, 0)$ to the point $(1, 1)$ on the graph of $y = \sqrt{x}$, and this occurs at the value $x = 1$, where $D(x)$, the distance squared, has its minimum value.

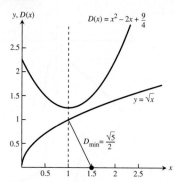

Section 4.7, pp. 277–279

1. $x_2 = -\dfrac{5}{3}, \dfrac{13}{21}$ **3.** $x_2 = -\dfrac{51}{31}, \dfrac{5763}{4945}$ **5.** $x_2 = \dfrac{2387}{2000}$

7. x_1, and all later approximations will equal x_0.

9.

$$y = \begin{cases} \sqrt{x}, & x \geq 0 \\ \sqrt{-x}, & x < 0 \end{cases}$$

11. The points of intersection of $y = x^3$ and $y = 3x + 1$ or $y = x^3 - 3x$ and $y = 1$ have the same x-values as the roots of part (i) or the solutions of part (iv).   **13.** 1.165561185

15. (a) Two   **(b)** 0.35003501505249 and -1.0261731615301

17. ± 1.3065629648764, ± 0.5411961001462 **19.** $x \approx 0.45$

21. 0.8192   **23.** 0, 0.53485   **25.** The root is 1.17951.

27. (a) For $x_0 = -2$ or $x_0 = -0.8$, $x_i \to -1$ as i gets large.
(b) For $x_0 = -0.5$ or $x_0 = 0.25$, $x_i \to 0$ as i gets large.
(c) For $x_0 = 0.8$ or $x_0 = 2$, $x_i \to 1$ as i gets large.
(d) For $x_0 = -\sqrt{21}/7$ or $x_0 = \sqrt{21}/7$, Newton's method does not converge. The values of x_i alternate between $-\sqrt{21}/7$ and $\sqrt{21}/7$ as i increases.

29. Answers will vary with machine speed.

Section 4.8, pp. 285–289

1. (a) x^2 **(b)** $\dfrac{x^3}{3}$ **(c)** $\dfrac{x^3}{3} - x^2 + x$

3. (a) x^{-3} **(b)** $-\dfrac{1}{3}x^{-3}$ **(c)** $-\dfrac{1}{3}x^{-3} + x^2 + 3x$

5. (a) $-\dfrac{1}{x}$ **(b)** $-\dfrac{5}{x}$ **(c)** $2x + \dfrac{5}{x}$

7. (a) $\sqrt{x^3}$ **(b)** $\sqrt{x}$ **(c)** $\dfrac{2\sqrt{x^3}}{3} + 2\sqrt{x}$

9. (a) $x^{2/3}$ **(b)** $x^{1/3}$ **(c)** $x^{-1/3}$

11. (a) $\ln x$ **(b)** $7 \ln x$ **(c)** $x - 5 \ln x$

13. (a) $\cos(\pi x)$ **(b)** $-3 \cos x$ **(c)** $-\dfrac{1}{\pi}\cos(\pi x) + \cos(3x)$

15. (a) $\tan x$ **(b)** $2\tan\left(\dfrac{x}{3}\right)$ **(c)** $-\dfrac{2}{3}\tan\left(\dfrac{3x}{2}\right)$

17. (a) $-\csc x$ **(b)** $\dfrac{1}{5}\csc(5x)$ **(c)** $2\csc\left(\dfrac{\pi x}{2}\right)$

19. (a) $\dfrac{1}{3}e^{3x}$ **(b)** $-e^{-x}$ **(c)** $2e^{x/2}$

21. (a) $\dfrac{1}{\ln 3}3^x$ **(b)** $\dfrac{-1}{\ln 2}2^{-x}$ **(c)** $\dfrac{1}{\ln(5/3)}\left(\dfrac{5}{3}\right)^x$

23. (a) $2\sin^{-1}x$ **(b)** $\dfrac{1}{2}\tan^{-1}x$ **(c)** $\dfrac{1}{2}\tan^{-1}2x$

25. $\dfrac{x^2}{2}+x+C$ **27.** $t^3+\dfrac{t^2}{4}+C$ **29.** $\dfrac{x^4}{2}-\dfrac{5x^2}{2}+7x+C$

31. $-\dfrac{1}{x}-\dfrac{x^3}{3}-\dfrac{x}{3}+C$ **33.** $\dfrac{3}{2}x^{2/3}+C$

35. $\dfrac{2}{3}x^{3/2}+\dfrac{3}{4}x^{4/3}+C$ **37.** $4y^2-\dfrac{8}{3}y^{3/4}+C$

39. $x^2+\dfrac{2}{x}+C$ **41.** $2\sqrt{t}-\dfrac{2}{\sqrt{t}}+C$ **43.** $-2\sin t+C$

45. $-21\cos\dfrac{\theta}{3}+C$ **47.** $3\cot x+C$ **49.** $-\dfrac{1}{2}\csc\theta+C$

51. $\dfrac{1}{3}e^{3x}-5e^{-x}+C$ **53.** $-e^{-x}+\dfrac{4^x}{\ln 4}+C$

55. $4\sec x-2\tan x+C$ **57.** $-\dfrac{1}{2}\cos 2x+\cot x+C$

59. $\dfrac{t}{2}+\dfrac{\sin 4t}{8}+C$ **61.** $\ln|x|-5\tan^{-1}x+C$

63. $\dfrac{3x^{(\sqrt{3}+1)}}{\sqrt{3}+1}+C$ **65.** $\tan\theta+C$ **67.** $-\cot x-x+C$

69. $-\cos\theta+\theta+C$

83. (a) Wrong: $\dfrac{d}{dx}\left(\dfrac{x^2}{2}\sin x+C\right)=\dfrac{2x}{2}\sin x+\dfrac{x^2}{2}\cos x=$
$x\sin x+\dfrac{x^2}{2}\cos x$

   (b) Wrong: $\dfrac{d}{dx}(-x\cos x+C)=-\cos x+x\sin x$

   (c) Right: $\dfrac{d}{dx}(-x\cos x+\sin x+C)=-\cos x+x\sin x+$
$\cos x=x\sin x$

85. (a) Wrong: $\dfrac{d}{dx}\left(\dfrac{(2x+1)^3}{3}+C\right)=\dfrac{3(2x+1)^2(2)}{3}=$
$2(2x+1)^2$

   (b) Wrong: $\dfrac{d}{dx}((2x+1)^3+C)=3(2x+1)^2(2)=$
$6(2x+1)^2$

   (c) Right: $\dfrac{d}{dx}((2x+1)^3+C)=6(2x+1)^2$

87. Right   **89. (b)**   **91.** $y=x^2-7x+10$

93. $y=-\dfrac{1}{x}+\dfrac{x^2}{2}-\dfrac{1}{2}$ **95.** $y=9x^{1/3}+4$

97. $s=t+\sin t+4$ **99.** $r=\cos(\pi\theta)-1$

101. $v=\dfrac{1}{2}\sec t+\dfrac{1}{2}$ **103.** $v=3\sec^{-1}t-\pi$

105. $y=x^2-x^3+4x+1$ **107.** $r=\dfrac{1}{t}+2t-2$

109. $y=x^3-4x^2+5$ **111.** $y=-\sin t+\cos t+t^3-1$

113. $y=2x^{3/2}-50$ **115.** $y=x-x^{4/3}+\dfrac{1}{2}$

117. $y=-\sin x-\cos x-2$

119. (a) (i) 33.2 units, **(ii)** 33.2 units, **(iii)** 33.2 units   **(b)** True
121. $t=88/k, k=16$

123. (a) $v=10t^{3/2}-6t^{1/2}$ **(b)** $s=4t^{5/2}-4t^{3/2}$
127. (a) $-\sqrt{x}+C$ **(b)** $x+C$ **(c)** $\sqrt{x}+C$
   (d) $-x+C$ **(e)** $x-\sqrt{x}+C$ **(f)** $-x-\sqrt{x}+C$

Practice Exercises, pp. 289–293

1. No   **3.** No minimum; absolute maximum: $f(1)=16$; critical
points: $x=1$ and $11/3$
5. Absolute minimum: $g(0)=1$; no absolute maximum; critical
point $x=0$
7. Absolute minimum: $2-2\ln 2$ at $x=2$; absolute maximum 1
at $x=1$
9. Yes, except at $x=0$ **11.** No   **15. (b)** one

17. (b) 0.8555 99677 2   **23.** Global minimum value of $\dfrac{1}{2}$ at $x=2$

25. (a) $t=0,6,12$ **(b)** $t=3,9$ **(c)** $6<t<12$
   (d) $0<t<6, 12<t<14$

27.   **29.**

31.   **33.**

35.   **37.**

39.   **41.**

43. (a) Local maximum at $x = 4$, local minimum at $x = -4$, inflection point at $x = 0$

(b)

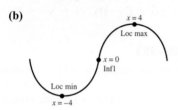

45. (a) Local maximum at $x = 0$, local minima at $x = -1$ and $x = 2$, inflection points at $x = (1 \pm \sqrt{7})/3$

(b)

47. (a) Local maximum at $x = -\sqrt{2}$, local minimum at $x = \sqrt{2}$, inflection points at $x = \pm 1$ and 0

(b)

53. **55.**

57. **59.**

61. 5 **63.** 0 **65.** 1 **67.** 3/7 **69.** 0 **71.** 1
73. $\ln 10$ **75.** $\ln 2$ **77.** 5 **79.** $-\infty$ **81.** 1 **83.** e^{bk}
85. (a) $0, 36$ **(b)** $18, 18$ **87.** 54 square units
89. height $= 2$, radius $= \sqrt{2}$

91. $x = 5 - \sqrt{5}$ hundred ≈ 276 tires,
  $y = 2(5 - \sqrt{5})$ hundred ≈ 553 tires
93. Dimensions: base is 6 in. by 12 in., height $= 2$ in.; maximum volume $= 144$ in.^{3}
95. $x_5 = 2.1958\,23345$ **97.** $\dfrac{x^4}{4} + \dfrac{5}{2}x^2 - 7x + C$
99. $2t^{3/2} - \dfrac{4}{t} + C$ **101.** $-\dfrac{1}{r+5} + C$ **103.** $(\theta^2 + 1)^{3/2} + C$
105. $\dfrac{1}{3}(1 + x^4)^{3/4} + C$ **107.** $10 \tan \dfrac{s}{10} + C$
109. $-\dfrac{1}{\sqrt{2}} \csc \sqrt{2}\,\theta + C$ **111.** $\dfrac{1}{2}x - \sin \dfrac{x}{2} + C$
113. $3 \ln x - \dfrac{x^2}{2} + C$ **115.** $\dfrac{1}{2}e^t + e^{-t} + C$
117. $\dfrac{\theta^{2-\pi}}{2 - \pi} + C$ **119.** $\dfrac{3}{2}\sec^{-1}|x| + C$
121. $y = x - \dfrac{1}{x} - 1$ **123.** $r = 4t^{5/2} + 4t^{3/2} - 8t$
125. Yes, $\sin^{-1}(x)$ and $-\cos^{-1}(x)$ differ by the constant $\pi/2$.
127. $1/\sqrt{2}$ units long by $1/\sqrt{e}$ units high, $A = 1/\sqrt{2e} \approx 0.43$ units^2
129. Absolute maximum $= 0$ at $x = e/2$, absolute minimum $= -0.5$ at $x = 0.5$
131. $x = \pm 1$ are the critical points; $y = 1$ is a horizontal asymptote in both directions; absolute minimum value of the function is $e^{-\sqrt{2}/2}$ at $x = -1$, and absolute maximum value is $e^{\sqrt{2}/2}$ at $x = 1$.
133. (a) Absolute maximum of $2/e$ at $x = e^2$, inflection point $(e^{8/3}, (8/3)e^{-4/3})$, concave up on $(e^{8/3}, \infty)$, concave down on $(0, e^{8/3})$
  (b) Absolute maximum of 1 at $x = 0$, inflection points $(\pm 1/\sqrt{2}, 1/\sqrt{e})$, concave up on $(-\infty, -1/\sqrt{2}) \cup (1/\sqrt{2}, \infty)$, concave down on $(-1/\sqrt{2}, 1/\sqrt{2})$
  (c) Absolute maximum of 1 at $x = 0$, inflection point $(1, 2/e)$, concave up on $(1, \infty)$, concave down on $(-\infty, 1)$

Additional and Advanced Exercises, pp. 293–296

1. The function is constant on the interval.
3. The extreme points will not be at the end of an open interval.
5. (a) A local minimum at $x = -1$, points of inflection at $x = 0$ and $x = 2$ **(b)** A local maximum at $x = 0$ and local minima at $x = -1$ and $x = 2$, points of inflection at $x = \dfrac{1 \pm \sqrt{7}}{3}$
9. No **11.** $a = 1, b = 0, c = 1$ **13.** Yes
15. Drill the hole at $y = h/2$.
17. $r = \dfrac{RH}{2(H - R)}$ for $H > 2R$, $r = R$ if $H \leq 2R$
19. (a) $\dfrac{10}{3}$ **(b)** $\dfrac{5}{3}$ **(c)** $\dfrac{1}{2}$ **(d)** 0 **(e)** $-\dfrac{1}{2}$ **(f)** 1 **(g)** $\dfrac{1}{2}$
  (h) 3
21. (a) $\dfrac{c-b}{2e}$ **(b)** $\dfrac{c+b}{2}$ **(c)** $\dfrac{b^2 - 2bc + c^2 + 4ae}{4e}$
  (d) $\dfrac{c+b+t}{2}$
23. $m_0 = 1 - \dfrac{1}{q}, m_1 = \dfrac{1}{q}$

25. $s = ce^{kt}$

27. (a) $k = -38.72$ **(b)** 25 ft

29. Yes, $y = x + C$ **31.** $v_0 = \dfrac{2\sqrt{2}}{3} b^{3/4}$

CHAPTER 5

Section 5.1, pp. 304–306

1. (a) 0.125   **(b)** 0.21875   **(c)** 0.625   **(d)** 0.46875
3. (a) 1.066667   **(b)** 1.283333   **(c)** 2.666667   **(d)** 2.083333
5. 0.3125, 0.328125   **7.** 1.5, 1.574603
9. (a) 87 in.   **(b)** 87 in.   **11. (a)** 3490 ft   **(b)** 3840 ft
13. (a) 74.65 ft/sec   **(b)** 45.28 ft/sec   **(c)** 146.59 ft

15. $\dfrac{31}{16}$ **17.** 1

19. (a) Upper = 758 gal, lower = 543 gal
(b) Upper = 2363 gal, lower = 1693 gal
(c) $\approx$ 31.4 h, $\approx$ 32.4 h

21. (a) 2   **(b)** $2\sqrt{2} \approx 2.828$

(c) $8 \sin\left(\dfrac{\pi}{8}\right) \approx 3.061$

(d) Each area is less than the area of the circle, π. As n increases, the polygon area approaches π.

Section 5.2, pp. 312–313

1. $\dfrac{6(1)}{1+1} + \dfrac{6(2)}{2+1} = 7$

3. $\cos(1)\pi + \cos(2)\pi + \cos(3)\pi + \cos(4)\pi = 0$

5. $\sin \pi - \sin\dfrac{\pi}{2} + \sin\dfrac{\pi}{3} = \dfrac{\sqrt{3}-2}{2}$ **7.** All of them   **9.** b

11. $\displaystyle\sum_{k=1}^{6} k$ **13.** $\displaystyle\sum_{k=1}^{4} \dfrac{1}{2^k}$ **15.** $\displaystyle\sum_{k=1}^{5} (-1)^{k+1}\dfrac{1}{k}$

17. (a) -15 **(b)** 1   **(c)** 1   **(d)** -11 **(e)** 16
19. (a) 55   **(b)** 385   **(c)** 3025
21. -56 **23.** -73 **25.** 240   **27.** 3376
29. (a) 21   **(b)** 3500   **(c)** 2620
31. (a) $4n$ **(b)** cn **(c)** $(n^2 - n)/2$
33. (a)   **(b)**

(c)

35. (a)   **(b)**

(c)

37. 1.2   **39.** $\dfrac{2}{3} - \dfrac{1}{2n} - \dfrac{1}{6n^2}, \ \dfrac{2}{3}$ **41.** $12 + \dfrac{27n+9}{2n^2}, \ 12$

43. $\dfrac{5}{6} + \dfrac{6n+1}{6n^2}, \ \dfrac{5}{6}$ **45.** $\dfrac{1}{2} + \dfrac{1}{n} + \dfrac{1}{2n^2}, \ \dfrac{1}{2}$

Section 5.3, pp. 321–325

1. $\displaystyle\int_0^2 x^2\, dx$ **3.** $\displaystyle\int_{-7}^5 (x^2 - 3x)\, dx$ **5.** $\displaystyle\int_2^3 \dfrac{1}{1-x}\, dx$

7. $\displaystyle\int_{-\pi/4}^0 \sec x\, dx$

9. (a) 0   **(b)** -8 **(c)** -12 **(d)** 10   **(e)** -2 **(f)** 16
11. (a) 5   **(b)** $5\sqrt{3}$ **(c)** -5 **(d)** -5
13. (a) 4   **(b)** -4 **15.** Area = 21 square units
17. Area = $9\pi/2$ square units   **19.** Area = 2.5 square units
21. Area = 3 square units   **23.** $b^2/4$ **25.** $b^2 - a^2$
27. (a) 2π **(b)** π **29.** 1/2   **31.** $3\pi^2/2$ **33.** 7/3
35. 1/24   **37.** $3a^2/2$ **39.** $b/3$ **41.** -14 **43.** -2
45. $-7/4$ **47.** 7   **49.** 0
51. Using n subintervals of length $\Delta x = b/n$ and right-endpoint values:

$$\text{Area} = \int_0^b 3x^2\, dx = b^3$$

53. Using n subintervals of length $\Delta x = b/n$ and right-endpoint values:

$$\text{Area} = \int_0^b 2x\, dx = b^2$$

55. $av(f) = 0$ **57.** $av(f) = -2$ **59.** $av(f) = 1$

61. (a) $av(g) = -1/2$ **(b)** $av(g) = 1$ **(c)** $av(g) = 1/4$

63. $c(b - a)$ **65.** $b^3/3 - a^3/3$ **67.** 9   **69.** $b^4/4 - a^4/4$

71. $a = 0$ and $b = 1$ maximize the integral.

73. Upper bound $= 1$, lower bound $= 1/2$

75. For example, $\displaystyle\int_0^1 \sin(x^2)\,dx \le \int_0^1 dx = 1$

77. $\displaystyle\int_a^b f(x)\,dx \ge \int_a^b 0\,dx = 0$ **79.** Upper bound $= 1/2$

Section 5.4, pp. 333–336

1. 6   **3.** $-10/3$ **5.** 8   **7.** 1   **9.** $2\sqrt{3}$ **11.** 0

13. $-\pi/4$ **15.** $1 - \dfrac{\pi}{4}$ **17.** $\dfrac{2 - \sqrt{2}}{4}$ **19.** $-8/3$

21. $-3/4$ **23.** $\sqrt{2} - \sqrt[4]{8} + 1$ **25.** -1 **27.** 16

29. $7/3$ **31.** $2\pi/3$ **33.** $\dfrac{1}{\pi}(4^\pi - 2^\pi)$ **35.** $\dfrac{1}{2}(e - 1)$

37. $\sqrt{26} - \sqrt{5}$ **39.** $(\cos\sqrt{x})\left(\dfrac{1}{2\sqrt{x}}\right)$ **41.** $4t^5$

43. $3x^2 e^{-x^3}$ **45.** $\sqrt{1 + x^2}$ **47.** $-\dfrac{1}{2}x^{-1/2}\sin x$ **49.** 0

51. 1   **53.** $2xe^{(1/2)x^2}$ **55.** 1   **57.** $28/3$ **59.** $1/2$ **61.** π

63. $\dfrac{\sqrt{2\pi}}{2}$ **65.** d, since $y' = \dfrac{1}{x}$ and $y(\pi) = \displaystyle\int_\pi^\pi \dfrac{1}{t}\,dt - 3 = -3$

67. b, since $y' = \sec x$ and $y(0) = \displaystyle\int_0^0 \sec t\,dt + 4 = 4$

69. $y = \displaystyle\int_2^x \sec t\,dt + 3$ **71.** $\dfrac{2}{3}bh$ **73.** \$9.00

75. a. $T(0) = 70°F$, $T(16) = 76°F$
   $T(25) = 85°F$
   b. $av(T) = 75°F$

77. $2x - 2$ **79.** $-3x + 5$

81. (a) True. Since f is continuous, g is differentiable by Part 1 of the Fundamental Theorem of Calculus.
   (b) True: g is continuous because it is differentiable.
   (c) True, since $g'(1) = f(1) = 0$.
   (d) False, since $g''(1) = f'(1) > 0$.
   (e) True, since $g'(1) = 0$ and $g''(1) = f'(1) > 0$.
   (f) False: $g''(x) = f'(x) > 0$, so g'' never changes sign.
   (g) True, since $g'(1) = f(1) = 0$ and $g'(x) = f(x)$ is an increasing function of x (because $f'(x) > 0$).

83. (a) $v = \dfrac{ds}{dt} = \dfrac{d}{dt}\displaystyle\int_0^t f(x)\,dx = f(t) \Rightarrow v(5) = f(5) = 2$ m/sec
   (b) $a = df/dt$ is negative since the slope of the tangent line at $t = 5$ is negative.
   (c) $s = \displaystyle\int_0^3 f(x)\,dx = \dfrac{1}{2}(3)(3) = \dfrac{9}{2}$ m since the integral is the area of the triangle formed by $y = f(x)$, the x-axis, and $x = 3$.
   (d) $t = 6$ since after $t = 6$ to $t = 9$, the region lies below the x-axis.
   (e) At $t = 4$ and $t = 7$, since there are horizontal tangents there.

(f) Toward the origin between $t = 6$ and $t = 9$ since the velocity is negative on this interval. Away from the origin between $t = 0$ and $t = 6$ since the velocity is positive there.

(g) Right or positive side, because the integral of f from 0 to 9 is positive, there being more area above the x-axis than below.

Section 5.5, pp. 342–344

1. $\dfrac{1}{6}(2x + 4)^6 + C$ **3.** $-\dfrac{1}{3}(x^2 + 5)^{-3} + C$

5. $\dfrac{1}{10}(3x^2 + 4x)^5 + C$ **7.** $-\dfrac{1}{3}\cos 3x + C$

9. $\dfrac{1}{2}\sec 2t + C$ **11.** $-6(1 - r^3)^{1/2} + C$

13. $\dfrac{1}{3}(x^{3/2} - 1) - \dfrac{1}{6}\sin(2x^{3/2} - 2) + C$

15. (a) $-\dfrac{1}{4}(\cot^2 2\theta) + C$ **(b)** $-\dfrac{1}{4}(\csc^2 2\theta) + C$

17. $-\dfrac{1}{3}(3 - 2s)^{3/2} + C$ **19.** $-\dfrac{2}{5}(1 - \theta^2)^{5/4} + C$

21. $(-2/(1 + \sqrt{x})) + C$ **23.** $\dfrac{1}{3}\tan(3x + 2) + C$

25. $\dfrac{1}{2}\sin^6\left(\dfrac{x}{3}\right) + C$ **27.** $\left(\dfrac{r^3}{18} - 1\right)^6 + C$

29. $-\dfrac{2}{3}\cos(x^{3/2} + 1) + C$ **31.** $\dfrac{1}{2\cos(2t + 1)} + C$

33. $-\sin\left(\dfrac{1}{t} - 1\right) + C$ **35.** $-\dfrac{\sin^2(1/\theta)}{2} + C$

37. $\dfrac{1}{16}(1 + t^4)^4 + C$ **39.** $\dfrac{2}{3}\left(2 - \dfrac{1}{x}\right)^{3/2} + C$

41. $\dfrac{2}{27}\left(1 - \dfrac{3}{x^3}\right)^{3/2} + C$

43. $\dfrac{1}{12}(x - 1)^{12} + \dfrac{1}{11}(x - 1)^{11} + C$

45. $-\dfrac{1}{8}(1 - x)^8 + \dfrac{4}{7}(1 - x)^7 - \dfrac{2}{3}(1 - x)^6 + C$

47. $\dfrac{1}{5}(x^2 + 1)^{5/2} - \dfrac{1}{3}(x^2 + 1)^{3/2} + C$ **49.** $\dfrac{-1}{4(x^2 - 4)^2} + C$

51. $e^{\sin x} + C$ **53.** $2\tan\left(e^{\sqrt{x}} + 1\right) + C$ **55.** $\ln|\ln x| + C$

57. $z - \ln(1 + e^z) + C$ **59.** $\dfrac{5}{6}\tan^{-1}\left(\dfrac{2r}{3}\right) + C$

61. $e^{\sin^{-1} x} + C$ **63.** $\dfrac{1}{3}(\sin^{-1} x)^3 + C$ **65.** $\ln|\tan^{-1} y| + C$

67. (a) $-\dfrac{6}{2 + \tan^3 x} + C$ **(b)** $-\dfrac{6}{2 + \tan^3 x} + C$
   (c) $-\dfrac{6}{2 + \tan^3 x} + C$

69. $\dfrac{1}{6}\sin\sqrt{3(2r - 1)^2 + 6} + C$ **71.** $s = \dfrac{1}{2}(3t^2 - 1)^4 - 5$

73. $s = 4t - 2\sin\left(2t + \dfrac{\pi}{6}\right) + 9$

75. $s = \sin\left(2t - \dfrac{\pi}{2}\right) + 100t + 1$ **77.** 6 m

Section 5.6, pp. 350–354

1. (a) 14/3   (b) 2/3   **3.** (a) 1/2   (b) −1/2
5. (a) 15/16   (b) 0   **7.** (a) 0   (b) 1/8   **9.** (a) 4   (b) 0
11. (a) 1/6   (b) 1/2   **13.** (a) 0   (b) 0   **15.** $2\sqrt{3}$
17. 3/4   **19.** $3^{5/2} - 1$ **21.** 3   **23.** $\pi/3$ **25.** e
27. ln 3   **29.** $(\ln 2)^2$ **31.** $\dfrac{1}{\ln 4}$ **33.** ln 2   **35.** ln 27   **37.** π
39. $\pi/12$ **41.** $2\pi/3$ **43.** $\sqrt{3} - 1$ **45.** $-\pi/12$
47. 16/3   **49.** $2^{5/2}$ **51.** $\pi/2$ **53.** 128/15   **55.** 4/3
57. 5/6   **59.** 38/3   **61.** 49/6   **63.** 32/3   **65.** 48/5
67. 8/3   **69.** 8   **71.** 5/3 (There are three intersection points.)
73. 18   **75.** 243/8   **77.** 8/3   **79.** 2   **81.** 104/15
83. 56/15   **85.** 4   **87.** $\dfrac{4}{3} - \dfrac{4}{\pi}$ **89.** $\pi/2$ **91.** 2   **93.** 1/2
95. 1   **97.** ln 16   **99.** 2   **101.** 2 ln 5
103. (a) $(\pm\sqrt{c}, c)$ (b) $c = 4^{2/3}$ (c) $c = 4^{2/3}$
105. 11/3   **107.** 3/4   **109.** Neither   **111.** $F(6) - F(2)$
113. (a) −3   (b) 3   **115.** $I = a/2$

Practice Exercises, pp. 354–357

1. (a) about 680 ft   (b)

3. (a) −1/2   (b) 31   (c) 13   (d) 0
5. $\displaystyle\int_1^5 (2x - 1)^{-1/2}\, dx = 2$ **7.** $\displaystyle\int_{-\pi}^0 \cos\frac{x}{2}\, dx = 2$
9. (a) 4   (b) 2   (c) −2   (d) -2π (e) 8/5
11. 8/3   **13.** 62   **15.** 1   **17.** 1/6   **19.** 18   **21.** 9/8
23. $\dfrac{\pi^2}{32} + \dfrac{\sqrt{2}}{2} - 1$ **25.** 4   **27.** $\dfrac{8\sqrt{2} - 7}{6}$
29. Min: −4, max: 0, area: 27/4   **31.** 6/5   **33.** 1
37. $y = \displaystyle\int_5^x \left(\frac{\sin t}{t}\right) dt - 3$ **39.** $y = \sin^{-1} x$
41. $y = \sec^{-1} x + \dfrac{2\pi}{3}, x > 1$ **43.** $-4(\cos x)^{1/2} + C$
45. $\theta^2 + \theta + \sin(2\theta + 1) + C$ **47.** $\dfrac{t^3}{3} + \dfrac{4}{t} + C$
49. $-\dfrac{1}{3}\cos(2t^{3/2}) + C$ **51.** $\tan(e^x - 7) + C$ **53.** $e^{\tan x} + C$
55. $\dfrac{-\ln 7}{3}$ **57.** ln (9/25)   **59.** $-\dfrac{1}{2}(\ln x)^{-2} + C$
61. $\dfrac{1}{2\ln 3}\left(3^{x^2}\right) + C$ **63.** $\dfrac{3}{2}\sin^{-1} 2(r - 1) + C$
65. $\dfrac{\sqrt{2}}{2}\tan^{-1}\left(\dfrac{x - 1}{\sqrt{2}}\right) + C$ **67.** $\dfrac{1}{4}\sec^{-1}\left|\dfrac{2x - 1}{2}\right| + C$
69. $e^{\sin^{-1}\sqrt{x}} + C$ **71.** $2\sqrt{\tan^{-1} y} + C$ **73.** 16
75. 2   **77.** 1   **79.** 8   **81.** $27\sqrt{3}/160$ **83.** $\pi/2$
85 $\sqrt{3}$ **87.** $6\sqrt{3} - 2\pi$ **89.** −1   **91.** 2   **93.** 1
95. $15/16 + \ln 2$ **97.** $e - 1$ **99.** 1/6   **101.** 9/14
103. $\dfrac{9\ln 2}{4}$ **105.** π **107.** $\pi/\sqrt{3}$ **109.** $\sec^{-1}|2y| + C$
111. $\pi/12$ **113.** (a) b (b) b

117. (a) $\dfrac{d}{dx}(x\ln x - x + C) = x\cdot\dfrac{1}{x} + \ln x - 1 + 0 = \ln x$
(b) $\dfrac{1}{e - 1}$
119. 25°F   **121.** $\sqrt{2 + \cos^3 x}$ **123.** $\dfrac{-6}{3 + x^4}$
125. $\dfrac{dy}{dx} = \dfrac{-2}{x}e^{\cos(2\ln x)}$ **127.** $\dfrac{dy}{dx} = \dfrac{1}{\sqrt{1 - x^2}\sqrt{1 - 2(\sin^{-1} x)^2}}$
129. Yes   **131.** $-\sqrt{1 + x^2}$
133. Cost $\approx$ \$10,899 using a lower sum estimate

Additional and Advanced Exercises, pp. 358–361

1. (a) Yes   (b) No   **5.** (a) 1/4   (b) $\sqrt[3]{12}$
7. $f(x) = \dfrac{x}{\sqrt{x^2 + 1}}$ **9.** $y = x^3 + 2x - 4$
11. 36/5   **13.** $\dfrac{1}{2} - \dfrac{2}{\pi}$

15. 13/3

17. 1/2   **19.** $\pi/2$ **21.** ln 2   **23.** 1/6   **25.** $\displaystyle\int_0^1 f(x)\, dx$
27. (b) πr^2
29. (a) 0   (b) −1   (c) $-\pi$ (d) $x = 1$
(e) $y = 2x + 2 - \pi$ (f) $x = -1, x = 2$ (g) $[-2\pi, 0]$
31. $2/x$ **33.** $\dfrac{\sin 4y}{\sqrt{y}} - \dfrac{\sin y}{2\sqrt{y}}$ **35.** $2x\ln|x| - x\ln\dfrac{|x|}{\sqrt{2}}$
37. $(\sin x)/x$ **39.** $x = 1$ **41.** $\dfrac{1}{\ln 2}, \dfrac{1}{2\ln 2}, 2:1$ **43.** 2/17

CHAPTER 6

Section 6.1, pp. 371–374

1. 16   **3.** 16/3   **5.** (a) $2\sqrt{3}$ (b) 8   **7.** (a) 60   (b) 36
9. 8π **11.** 10   **13.** (a) $s^2 h$ (b) $s^2 h$ **15.** $\dfrac{2\pi}{3}$
17. $4 - \pi$ **19.** $\dfrac{32\pi}{5}$ **21.** 36π **23.** π **25.** $\dfrac{\pi}{2}\left(1 - \dfrac{1}{e^2}\right)$
27. $\dfrac{\pi}{2}\ln 4$ **29.** $\pi\left(\dfrac{\pi}{2} + 2\sqrt{2} - \dfrac{11}{3}\right)$ **31.** 2π **33.** 2π
35. $4\pi\ln 4$ **37.** $\pi^2 - 2\pi$ **39.** $\dfrac{2\pi}{3}$ **41.** $\dfrac{117\pi}{5}$
43. $\pi(\pi - 2)$ **45.** $\dfrac{4\pi}{3}$ **47.** 8π **49.** $\dfrac{7\pi}{6}$

51. (a) 8π (b) $\dfrac{32\pi}{5}$ (c) $\dfrac{8\pi}{3}$ (d) $\dfrac{224\pi}{15}$

53. (a) $\dfrac{16\pi}{15}$ (b) $\dfrac{56\pi}{15}$ (c) $\dfrac{64\pi}{15}$ **55.** $V = 2a^2b\pi^2$

57. (a) $V = \dfrac{\pi h^2(3a - h)}{3}$ (b) $\dfrac{1}{120\pi}$ m/sec

61. $V = 3308$ cm^3 **63.** $\dfrac{4 - b + a}{2}$

Section 6.2, pp. 379–381

1. 6π **3.** 2π **5.** $14\pi/3$ **7.** 8π **9.** $5\pi/6$

11. $\dfrac{7\pi}{15}$ **13.** (b) 4π **15.** $\dfrac{16\pi}{15}(3\sqrt{2} + 5)$

17. $\dfrac{8\pi}{3}$ **19.** $\dfrac{4\pi}{3}$ **21.** $\dfrac{16\pi}{3}$

23. (a) 16π (b) 32π (c) 28π
   (d) 24π (e) 60π (f) 48π

25. (a) $\dfrac{27\pi}{2}$ (b) $\dfrac{27\pi}{2}$ (c) $\dfrac{72\pi}{5}$ (d) $\dfrac{108\pi}{5}$

27. (a) $\dfrac{6\pi}{5}$ (b) $\dfrac{4\pi}{5}$ (c) 2π (d) 2π

29. (a) About the x-axis: $V = \dfrac{2\pi}{15}$; about the y-axis: $V = \dfrac{\pi}{6}$

   (b) About the x-axis: $V = \dfrac{2\pi}{15}$; about the y-axis: $V = \dfrac{\pi}{6}$

31. (a) $\dfrac{5\pi}{3}$ (b) $\dfrac{4\pi}{3}$ (c) 2π (d) $\dfrac{2\pi}{3}$

33. (a) $\dfrac{4\pi}{15}$ (b) $\dfrac{7\pi}{30}$

35. (a) $\dfrac{24\pi}{5}$ (b) $\dfrac{48\pi}{5}$

37. (a) $\dfrac{9\pi}{16}$ (b) $\dfrac{9\pi}{16}$

39. Disk: 2 integrals; washer: 2 integrals; shell: 1 integral

41. (a) $\dfrac{256\pi}{3}$ (b) $\dfrac{244\pi}{3}$ **47.** $\pi\left(1 - \dfrac{1}{e}\right)$

Section 6.3, pp. 386–387

1. 12   **3.** $\dfrac{53}{6}$ **5.** $\dfrac{123}{32}$ **7.** $\dfrac{99}{8}$ **9.** 2

11. (a) $\displaystyle\int_{-1}^{2} \sqrt{1 + 4x^2}\,dx$ (c) ≈ 6.13

13. (a) $\displaystyle\int_{0}^{\pi} \sqrt{1 + \cos^2 y}\,dy$ (c) ≈ 3.82

15. (a) $\displaystyle\int_{-1}^{3} \sqrt{1 + (y + 1)^2}\,dy$ (c) ≈ 9.29

17. (a) $\displaystyle\int_{0}^{\pi/6} \sec x\,dx$ (c) ≈ 0.55

19. (a) $y = \sqrt{x}$ from $(1, 1)$ to $(4, 2)$
   (b) Only one. We know the derivative of the function and the value of the function at one value of x.

21. 1   **27.** Yes, $f(x) = \pm x + C$ where C is any real number.

31. $\dfrac{2}{27}(10^{3/2} - 1)$

Section 6.4, pp. 391–393

1. (a) $2\pi\displaystyle\int_{0}^{\pi/4} (\tan x)\sqrt{1 + \sec^4 x}\,dx$ (c) $S \approx 3.84$

(b)
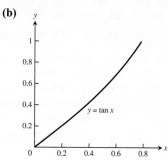

3. (a) $2\pi\displaystyle\int_{1}^{2} \dfrac{1}{y}\sqrt{1 + y^{-4}}\,dy$ (c) $S \approx 5.02$

(b)

5. (a) $2\pi\displaystyle\int_{1}^{4} (3 - x^{1/2})^2\sqrt{1 + (1 - 3x^{-1/2})^2}\,dx$ (c) $S \approx 63.37$

(b)

7. (a) $2\pi\displaystyle\int_{0}^{\pi/3}\left(\displaystyle\int_{0}^{y}\tan t\,dt\right)\sec y\,dy$ (c) $S \approx 2.08$

(b)
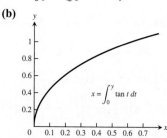

9. $4\pi\sqrt{5}$ **11.** $3\pi\sqrt{5}$ **13.** $98\pi/81$ **15.** 2π

17. $\pi(\sqrt{8} - 1)/9$ **19.** $35\pi\sqrt{5}/3$ **21.** $\pi\left(\dfrac{15}{16} + \ln 2\right)$

23. $253\pi/20$ **27.** Order 226.2 liters of each color.

Section 6.5, pp. 398–402

1. 400 N/m   **3.** 4 cm, 0.08 J
5. (a) 7238 lb/in.   (b) 905 in.-lb, 2714 in.-lb
7. 780 J   **9.** 72,900 ft-lb   **11.** 160 ft-lb
13. (a) 1,497,600 ft-lb   (b) 1 hr, 40 min
       (d) At 62.26 lb/ft^3: a) 1,494,240 ft-lb   b) 1 hr, 40 min
           At 62.59 lb/ft^3: a) 1,502,160 ft-lb   b) 1 hr, 40.1 min
15. 37,306 ft-lb   **17.** 7,238,299.47 ft-lb
19. 2446.25 ft-lb   **21.** 15,073,099.75 J
25. 85.1 ft-lb   **27.** 98.35 ft-lb   **29.** 91.32 in.-oz
31. 5.144×10^{10} J   **33.** 1684.8 lb
35. (a) 6364.8 lb   (b) 5990.4 lb   **37.** 1164.8 lb   **39.** 1309 lb
41. (a) 12,480 lb   (b) 8580 lb   (c) 9722.3 lb
43. (a) 93.33 lb   (b) 3 ft   **45.** $\dfrac{wb}{2}$
47. No. The tank will overflow because the movable end will have
       moved only $3\frac{1}{3}$ ft by the time the tank is full.

Section 6.6, pp. 411–413

1. $\bar{x} = 0, \bar{y} = 12/5$ **3.** $\bar{x} = 1, \bar{y} = -3/5$
5. $\bar{x} = 16/105, \bar{y} = 8/15$ **7.** $\bar{x} = 0, \bar{y} = \pi/8$
9. $\bar{x} \approx 1.44, \bar{y} \approx 0.36$
11. $\bar{x} = \dfrac{\ln 4}{\pi}, \bar{y} = 0$ **13.** $\bar{x} = 7, \bar{y} = \dfrac{\ln 16}{12}$
15. $\bar{x} = 3/2, \bar{y} = 1/2$
17. (a) $\dfrac{224\pi}{3}$ (b) $\bar{x} = 2, \bar{y} = 0$
      (c)

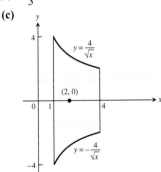

21. $\bar{x} = \bar{y} = 1/3$ **23.** $\bar{x} = a/3, \bar{y} = b/3$ **25.** $13\delta/6$
27. $\bar{x} = 0, \bar{y} = \dfrac{a\pi}{4}$
29. $\bar{x} = 1/2, \bar{y} = 4$ **31.** $\bar{x} = 6/5, \bar{y} = 8/7$
35. $V = 32\pi, S = 32\sqrt{2}\pi$
37. $4\pi^2$ **39.** $\bar{x} = 0, \bar{y} = \dfrac{2a}{\pi}$ **41.** $\bar{x} = 0, \bar{y} = \dfrac{4b}{3\pi}$
43. $\sqrt{2}\pi a^3(4 + 3\pi)/6$ **45.** $\bar{x} = \dfrac{a}{3}, \bar{y} = \dfrac{b}{3}$

Practice Exercises, pp. 413–415

1. $\dfrac{9\pi}{280}$ **3.** π^2 **5.** $\dfrac{72\pi}{35}$
7. (a) 2π (b) π (c) $12\pi/5$ (d) $26\pi/5$
9. (a) 8π (b) $1088\pi/15$ (c) $512\pi/15$
11. $\pi(3\sqrt{3} - \pi)/3$ **13.** π
15. $\dfrac{28\pi}{3}$ ft^3 **17.** $\dfrac{10}{3}$ **19.** $3 + \dfrac{1}{8}\ln 2$

21. $28\pi\sqrt{2}/3$ **23.** 4π **25.** 4640 J
27. 10 ft-lb, 30 ft-lb   **29.** 418,208.81 ft-lb
31. $22,500\pi$ ft-lb, 257 sec   **33.** $\bar{x} = 0, \bar{y} = 8/5$
35. $\bar{x} = 3/2, \bar{y} = 12/5$ **37.** $\bar{x} = 9/5, \bar{y} = 11/10$
39. 332.8 lb   **41.** 2196.48 lb

Additional and Advanced Exercises, pp. 415–416

1. $f(x) = \sqrt{\dfrac{2x - a}{\pi}}$ **3.** $f(x) = \sqrt{C^2 - 1}\,x + a$, where $C \geq 1$
5. $\dfrac{\pi}{30\sqrt{2}}$ **7.** 28/3   **9.** $\dfrac{4h\sqrt{3mh}}{3}$
11. $\bar{x} = 0, \bar{y} = \dfrac{n}{2n + 1}$, $(0, 1/2)$
15. (a) $\bar{x} = \bar{y} = 4(a^2 + ab + b^2)/(3\pi(a + b))$
       (b) $(2a/\pi, 2a/\pi)$
17. ≈ 2329.6 lb

CHAPTER 7

Section 7.1, pp. 425–427

1. $\ln\left(\dfrac{2}{3}\right)$ **3.** $\ln|y^2 - 25| + C$ **5.** $\ln|6 + 3\tan t| + C$
7. $\ln(1 + \sqrt{x}) + C$ **9.** 1   **11.** $2(\ln 2)^4$ **13.** 2
15. $2e^{\sqrt{r}} + C$ **17.** $-e^{-t^2} + C$ **19.** $-e^{1/x} + C$
21. $\dfrac{1}{\pi}e^{\sec \pi t} + C$ **23.** 1   **25.** $\ln(1 + e^r) + C$ **27.** $\dfrac{1}{2\ln 2}$
29. $\dfrac{1}{\ln 2}$ **31.** $\dfrac{6}{\ln 7}$ **33.** 32760   **35.** $3^{\sqrt{2}+1}$
37. $\dfrac{1}{\ln 10}\left(\dfrac{(\ln x)^2}{2}\right) + C$ **39.** $2(\ln 2)^2$ **41.** $\dfrac{3\ln 2}{2}$ **43.** $\ln 10$
45. $(\ln 10)\ln|\ln x| + C$ **47.** $y = 1 - \cos(e^t - 2)$
49. $y = 2(e^{-x} + x) - 1$ **51.** $y = x + \ln|x| + 2$ **53.** $\pi \ln 16$
55. $6 + \ln 2$ **57.** (b) 0.00469
69. (a) 1.89279   (b) -0.35621 (c) 0.94575   (d) -2.80735
       (e) 5.29595   (f) 0.97041   (g) -1.03972 (h) -1.61181

Section 7.2, pp. 433–435

9. $\dfrac{2}{3}y^{3/2} - x^{1/2} = C$ **11.** $e^y - e^x = C$
13. $-x + 2\tan\sqrt{y} = C$ **15.** $e^{-y} + 2e^{\sqrt{x}} = C$
17. $y = \sin(x^2 + C)$ **19.** $\dfrac{1}{3}\ln|y^3 - 2| = x^3 + C$
21. $4\ln(\sqrt{y} + 2) = e^{x^2} + C$
23. (a) -0.00001 (b) 10,536 years   (c) 82%
25. 54.88 g   **27.** 59.8 ft   **29.** 2.8147498×10^{14}
31. (a) 8 years   (b) 32.02 years
33. 15.28 years   **35.** 56,562 years
39. (a) 17.5 min   (b) 13.26 min
41. $-3°C$ **43.** About 6658 years   **45.** 54.44%

Section 7.3, pp. 441–444

1. $\cosh x = 5/4, \tanh x = -3/5, \coth x = -5/3,$
      $\operatorname{sech} x = 4/5, \operatorname{csch} x = -4/3$

3. $\sinh x = 8/15$, $\tanh x = 8/17$, $\coth x = 17/8$, $\operatorname{sech} x = 15/17$, $\operatorname{csch} x = 15/8$

5. $x + \dfrac{1}{x}$ **7.** e^{5x} **9.** e^{4x} **13.** $2\cosh\dfrac{x}{3}$

15. $\operatorname{sech}^2\sqrt{t} + \dfrac{\tanh\sqrt{t}}{\sqrt{t}}$ **17.** $\coth z$

19. $(\ln \operatorname{sech}\theta)(\operatorname{sech}\theta\tanh\theta)$ **21.** $\tanh^3 v$ **23.** 2

25. $\dfrac{1}{2\sqrt{x(1+x)}}$ **27.** $\dfrac{1}{1+\theta} - \tanh^{-1}\theta$

29. $\dfrac{1}{2\sqrt{t}} - \coth^{-1}\sqrt{t}$ **31.** $-\operatorname{sech}^{-1}x$ **33.** $\dfrac{\ln 2}{\sqrt{1+\left(\frac{1}{2}\right)^{2\theta}}}$

35. $|\sec x|$ **41.** $\dfrac{\cosh 2x}{2} + C$

43. $12\sinh\left(\dfrac{x}{2} - \ln 3\right) + C$ **45.** $7\ln|e^{x/7} + e^{-x/7}| + C$

47. $\tanh\left(x - \dfrac{1}{2}\right) + C$ **49.** $-2\operatorname{sech}\sqrt{t} + C$ **51.** $\ln\dfrac{5}{2}$

53. $\dfrac{3}{32} + \ln 2$ **55.** $e - e^{-1}$ **57.** $3/4$ **59.** $\dfrac{3}{8} + \ln\sqrt{2}$

61. $\ln(2/3)$ **63.** $\dfrac{-\ln 3}{2}$ **65.** $\ln 3$

67. (a) $\sinh^{-1}(\sqrt{3})$ (b) $\ln(\sqrt{3} + 2)$

69. (a) $\coth^{-1}(2) - \coth^{-1}(5/4)$ (b) $\left(\dfrac{1}{2}\right)\ln\left(\dfrac{1}{3}\right)$

71. (a) $-\operatorname{sech}^{-1}\left(\dfrac{12}{13}\right) + \operatorname{sech}^{-1}\left(\dfrac{4}{5}\right)$

(b) $-\ln\left(\dfrac{1+\sqrt{1-(12/13)^2}}{(12/13)}\right) + \ln\left(\dfrac{1+\sqrt{1-(4/5)^2}}{(4/5)}\right)$

$= -\ln\left(\dfrac{3}{2}\right) + \ln(2) = \ln(4/3)$

73. (a) 0 (b) 0

77. (b) $\sqrt{\dfrac{mg}{k}}$ (c) $80\sqrt{5} \approx 178.89$ ft/sec   **79.** 2π **81.** $\dfrac{6}{5}$

Section 7.4, pp. 448–449

1. (a) Slower   (b) Slower   (c) Slower   (d) Faster
(e) Slower   (f) Slower   (g) Same   (h) Slower

3. (a) Same   (b) Faster   (c) Same   (d) Same   (e) Slower
(f) Faster   (g) Slower   (h) Same

5. (a) Same   (b) Same   (c) Same   (d) Faster   (e) Faster
(f) Same   (g) Slower   (h) Faster

7. d, a, c, b

9. (a) False   (b) False   (c) True   (d) True   (e) True
(f) True   (g) False   (h) True

13. When the degree of f is less than or equal to the degree of g.

15. 1, 1

21. (b) $\ln(e^{17000000}) = 17{,}000{,}000 < (e^{17\times10^6})^{1/10^6}$
$= e^{17} \approx 24{,}154{,}952.75$
(c) $x \approx 3.4306311 \times 10^{15}$
(d) They cross at $x \approx 3.4306311 \times 10^{15}$.

23. (a) The algorithm that takes $O(n\log_2 n)$ steps

(b)

25. It could take one million for a sequential search; at most 20 steps for a binary search.

Practice Exercises, pp. 450–451

1. $-\cos e^x + C$ **3.** $\ln 8$ **5.** $2\ln 2$ **7.** $\dfrac{1}{2}(\ln(x-5))^2 + C$

9. $3\ln 7$ **11.** $2(\sqrt{2}-1)$ **13.** $y = \dfrac{\ln 2}{\ln(3/2)}$

15. $y = \ln x - \ln 3$ **17.** $y = \dfrac{1}{1-e^x}$

19. (a) Same rate   (b) Same rate   (c) Faster   (d) Faster
(e) Same rate   (f) Same rate

21. (a) True   (b) False   (c) False   (d) True   (e) True
(f) True

23. $1/3$ **25.** $1/e$ m/sec   **27.** $\ln 5x - \ln 3x = \ln(5/3)$

29. $1/2$ **31.** $y = \left(\tan^{-1}\left(\dfrac{x+C}{2}\right)\right)^2$

33. $y^2 = \sin^{-1}(2\tan x + C)$

35. $y = -2 + \ln(2 - e^{-x})$ **37.** $y = 4x - 4\sqrt{x} + 1$

39. 18,935 years

Additional and Advanced Exercises, p.p 451–452

1. (a) 1 (b) $\pi/2$ (c) π

3. $\tan^{-1}x + \tan^{-1}\left(\dfrac{1}{x}\right)$ is a constant and the constant is $\dfrac{\pi}{2}$ for $x > 0$; it is $-\dfrac{\pi}{2}$ for $x < 0$.

7. $\bar{x} = \dfrac{\ln 4}{\pi}$, $\bar{y} = 0$

CHAPTER 8

Section 8.1, pp. 459–461

1. $-2x\cos(x/2) + 4\sin(x/2) + C$

3. $t^2\sin t + 2t\cos t - 2\sin t + C$ **5.** $\ln 4 - \dfrac{3}{4}$

7. $xe^x - e^x + C$ **9.** $-(x^2 + 2x + 2)e^{-x} + C$

11. $y \tan^{-1}(y) - \ln \sqrt{1 + y^2} + C$

13. $x \tan x + \ln |\cos x| + C$

15. $(x^3 - 3x^2 + 6x - 6)e^x + C$

17. $(x^2 - 7x + 7)e^x + C$

19. $(x^5 - 5x^4 + 20x^3 - 60x^2 + 120x - 120)e^x + C$

21. $\dfrac{1}{2}(-e^{\theta} \cos \theta + e^{\theta} \sin \theta) + C$

23. $\dfrac{e^{2x}}{13}(3 \sin 3x + 2 \cos 3x) + C$

25. $\dfrac{2}{3}\left(\sqrt{3s + 9}\, e^{\sqrt{3s+9}} - e^{\sqrt{3s+9}}\right) + C$

27. $\dfrac{\pi\sqrt{3}}{3} - \ln(2) - \dfrac{\pi^2}{18}$

29. $\dfrac{1}{2}[-x \cos(\ln x) + x \sin(\ln x)] + C$

31. $\dfrac{1}{2} \ln |\sec x^2 + \tan x^2| + C$

33. $\dfrac{1}{2}x^2 (\ln x)^2 - \dfrac{1}{2}x^2 \ln x + \dfrac{1}{4}x^2 + C$

35. $-\dfrac{1}{x} \ln x - \dfrac{1}{x} + C$ **37.** $\dfrac{1}{4}e^{x^4} + C$

39. $\dfrac{1}{3}x^2 (x^2 + 1)^{3/2} - \dfrac{2}{15}(x^2 + 1)^{5/2} + C$

41. $-\dfrac{2}{5} \sin 3x \sin 2x - \dfrac{3}{5} \cos 3x \cos 2x + C$

43. $-\cos e^x + C$ **45.** $2\sqrt{x} \sin \sqrt{x} + 2 \cos \sqrt{x} + C$

47. $\dfrac{\pi^2 - 4}{8}$ **49.** $\dfrac{5\pi - 3\sqrt{3}}{9}$

51. (a) π (b) 3π (c) 5π (d) $(2n + 1)\pi$

53. $2\pi(1 - \ln 2)$ **55.** (a) $\pi(\pi - 2)$ (b) 2π

57. (a) 1 (b) $(e - 2)\pi$ (c) $\dfrac{\pi}{2}(e^2 + 9)$

(d) $\bar{x} = \dfrac{1}{4}(e^2 + 1), \bar{y} = \dfrac{1}{2}(e - 2)$

59. $\dfrac{1}{2\pi}(1 - e^{-2\pi})$ **61.** $u = x^n, dv = \cos x\, dx$

63. $u = x^n, dv = e^{ax}\, dx$ **67.** $x \sin^{-1} x + \cos(\sin^{-1} x) + C$

69. $x \sec^{-1} x - \ln |x + \sqrt{x^2 - 1}| + C$ **71.** Yes

73. (a) $x \sinh^{-1} x - \cosh(\sinh^{-1} x) + C$

(b) $x \sinh^{-1} x - (1 + x^2)^{1/2} + C$

Section 8.2, pp. 466–467

1. $\dfrac{1}{2} \sin 2x + C$ **3.** $-\dfrac{1}{4} \cos^4 x + C$ **5.** $\dfrac{1}{3} \cos^3 x - \cos x + C$

7. $-\cos x + \dfrac{2}{3} \cos^3 x - \dfrac{1}{5} \cos^5 x + C$ **9.** $\sin x - \dfrac{1}{3} \sin^3 x + C$

11. $\dfrac{1}{4} \sin^4 x - \dfrac{1}{6} \sin^6 x + C$ **13.** $\dfrac{1}{2}x + \dfrac{1}{4} \sin 2x + C$

15. $16/35$ **17.** 3π

19. $-4 \sin x \cos^3 x + 2 \cos x \sin x + 2x + C$

21. $-\cos^4 2\theta + C$ **23.** 4 **25.** 2

27. $\sqrt{\dfrac{3}{2}} - \dfrac{2}{3}$ **29.** $\dfrac{4}{5}\left(\dfrac{3}{2}\right)^{5/2} - \dfrac{18}{35} - \dfrac{2}{7}\left(\dfrac{3}{2}\right)^{7/2}$ **31.** $\sqrt{2}$

33. $\dfrac{1}{2} \tan^2 x + C$ **35.** $\dfrac{1}{3} \sec^3 x + C$ **37.** $\dfrac{1}{3} \tan^3 x + C$

39. $2\sqrt{3} + \ln(2 + \sqrt{3})$ **41.** $\dfrac{2}{3} \tan \theta + \dfrac{1}{3} \sec^2 \theta \tan \theta + C$

43. $4/3$ **45.** $2 \tan^2 x - 2 \ln(1 + \tan^2 x) + C$

47. $\dfrac{1}{4} \tan^4 x - \dfrac{1}{2} \tan^2 x + \ln |\sec x| + C$

49. $\dfrac{4}{3} - \ln \sqrt{3}$ **51.** $-\dfrac{1}{10} \cos 5x - \dfrac{1}{2} \cos x + C$ **53.** π

55. $\dfrac{1}{2} \sin x + \dfrac{1}{14} \sin 7x + C$

57. $\dfrac{1}{6} \sin 3\theta - \dfrac{1}{4} \sin \theta - \dfrac{1}{20} \sin 5\theta + C$

59. $-\dfrac{2}{5} \cos^5 \theta + C$ **61.** $\dfrac{1}{4} \cos \theta - \dfrac{1}{20} \cos 5\theta + C$

63. $\sec x - \ln |\csc x + \cot x| + C$ **65.** $\cos x + \sec x + C$

67. $\dfrac{1}{4}x^2 - \dfrac{1}{4}x \sin 2x - \dfrac{1}{8} \cos 2x + C$ **69.** $\ln(1 + \sqrt{2})$

71. $\pi^2/2$ **73.** $\bar{x} = \dfrac{4\pi}{3}, \bar{y} = \dfrac{8\pi^2 + 3}{12\pi}$

Section 8.3, pp. 470–471

1. $\ln |\sqrt{9 + x^2} + x| + C$ **3.** $\pi/4$ **5.** $\pi/6$

7. $\dfrac{25}{2} \sin^{-1}\left(\dfrac{t}{5}\right) + \dfrac{t\sqrt{25 - t^2}}{2} + C$

9. $\dfrac{1}{2} \ln \left|\dfrac{2x}{7} + \dfrac{\sqrt{4x^2 - 49}}{7}\right| + C$

11. $7\left[\dfrac{\sqrt{y^2 - 49}}{7} - \sec^{-1}\left(\dfrac{y}{7}\right)\right] + C$ **13.** $\dfrac{\sqrt{x^2 - 1}}{x} + C$

15. $-\sqrt{9 - x^2} + C$ **17.** $\dfrac{1}{3}(x^2 + 4)^{3/2} - 4\sqrt{x^2 + 4} + C$

19. $\dfrac{-2\sqrt{4 - w^2}}{w} + C$ **21.** $\dfrac{10}{3} \tan^{-1} \dfrac{5x}{6} + C$

23. $4\sqrt{3} - \dfrac{4\pi}{3}$ **25.** $-\dfrac{x}{\sqrt{x^2 - 1}} + C$

27. $-\dfrac{1}{5}\left(\dfrac{\sqrt{1 - x^2}}{x}\right)^5 + C$ **29.** $2 \tan^{-1} 2x + \dfrac{4x}{(4x^2 + 1)} + C$

31. $\dfrac{1}{2}x^2 + \dfrac{1}{2} \ln |x^2 - 1| + C$ **33.** $\dfrac{1}{3}\left(\dfrac{v}{\sqrt{1 - v^2}}\right)^3 + C$

35. $\ln 9 - \ln(1 + \sqrt{10})$ **37.** $\pi/6$ **39.** $\sec^{-1} |x| + C$

41. $\sqrt{x^2 - 1} + C$ **43.** $\dfrac{1}{2} \ln |\sqrt{1 + x^4} + x^2| + C$

45. $4 \sin^{-1} \dfrac{\sqrt{x}}{2} + \sqrt{x} \sqrt{4 - x} + C$

47. $\dfrac{1}{4} \sin^{-1} \sqrt{x} - \dfrac{1}{4} \sqrt{x} \sqrt{1 - x}(1 - 2x) + C$

49. $y = 2\left[\dfrac{\sqrt{x^2 - 4}}{2} - \sec^{-1}\left(\dfrac{x}{2}\right)\right]$

51. $y = \dfrac{3}{2} \tan^{-1}\left(\dfrac{x}{2}\right) - \dfrac{3\pi}{8}$ **53.** $3\pi/4$

55. (a) $\frac{1}{12}(\pi + 6\sqrt{3} - 12)$

(b) $\bar{x} = \dfrac{3\sqrt{3} - \pi}{4(\pi + 6\sqrt{3} - 12)}$, $\bar{y} = \dfrac{\pi^2 + 12\sqrt{3}\pi - 72}{12(\pi + 6\sqrt{3} - 12)}$

57. (a) $-\dfrac{1}{3}x^2(1 - x^2)^{3/2} - \dfrac{2}{15}(1 - x^2)^{5/2} + C$

(b) $-\dfrac{1}{3}(1 - x^2)^{3/2} + \dfrac{1}{5}(1 - x^2)^{5/2} + C$

(c) $\dfrac{1}{5}(1 - x^2)^{5/2} - \dfrac{1}{3}(1 - x^2)^{3/2} + C$

Section 8.4, pp. 479–480

1. $\dfrac{2}{x - 3} + \dfrac{3}{x - 2}$ **3.** $\dfrac{1}{x + 1} + \dfrac{3}{(x + 1)^2}$

5. $\dfrac{-2}{z} + \dfrac{-1}{z^2} + \dfrac{2}{z - 1}$ **7.** $1 + \dfrac{17}{t - 3} + \dfrac{-12}{t - 2}$

9. $\dfrac{1}{2}[\ln|1 + x| - \ln|1 - x|] + C$

11. $\dfrac{1}{7}\ln|(x + 6)^2(x - 1)^5| + C$ **13.** $(\ln 15)/2$

15. $-\dfrac{1}{2}\ln|t| + \dfrac{1}{6}\ln|t + 2| + \dfrac{1}{3}\ln|t - 1| + C$ **17.** $3\ln 2 - 2$

19. $\dfrac{1}{4}\ln\left|\dfrac{x + 1}{x - 1}\right| - \dfrac{x}{2(x^2 - 1)} + C$ **21.** $(\pi + 2\ln 2)/8$

23. $\tan^{-1}y - \dfrac{1}{y^2 + 1} + C$

25. $-(s - 1)^{-2} + (s - 1)^{-1} + \tan^{-1}s + C$

27. $\dfrac{2}{3}\ln|x - 1| + \dfrac{1}{6}\ln|x^2 + x + 1| - \sqrt{3}\tan^{-1}\left(\dfrac{2x + 1}{\sqrt{3}}\right) + C$

29. $\dfrac{1}{4}\ln\left|\dfrac{x - 1}{x + 1}\right| + \dfrac{1}{2}\tan^{-1}x + C$

31. $\dfrac{-1}{\theta^2 + 2\theta + 2} + \ln(\theta^2 + 2\theta + 2) - \tan^{-1}(\theta + 1) + C$

33. $x^2 + \ln\left|\dfrac{x - 1}{x}\right| + C$

35. $9x + 2\ln|x| + \dfrac{1}{x} + 7\ln|x - 1| + C$

37. $\dfrac{y^2}{2} - \ln|y| + \dfrac{1}{2}\ln(1 + y^2) + C$ **39.** $\ln\left(\dfrac{e^t + 1}{e^t + 2}\right) + C$

41. $\dfrac{1}{5}\ln\left|\dfrac{\sin y - 2}{\sin y + 3}\right| + C$

43. $\dfrac{(\tan^{-1}2x)^2}{4} - 3\ln|x - 2| + \dfrac{6}{x - 2} + C$

45. $\ln\left|\dfrac{\sqrt{x} - 1}{\sqrt{x} + 1}\right| + C$ **47.** $2\sqrt{1 + x} + \ln\left|\dfrac{\sqrt{x + 1} - 1}{\sqrt{x + 1} + 1}\right| + C$

49. $\dfrac{1}{4}\ln\left|\dfrac{x^4}{x^4 + 1}\right| + C$

51. $x = \ln|t - 2| - \ln|t - 1| + \ln 2$ **53.** $x = \dfrac{6t}{t + 2} - 1$

55. $3\pi\ln 25$ **57.** 1.10 **59. (a)** $x = \dfrac{1000e^{4t}}{499 + e^{4t}}$ **(b)** 1.55 days

Section 8.5, pp. 485–486

1. $\dfrac{2}{\sqrt{3}}\left(\tan^{-1}\sqrt{\dfrac{x - 3}{3}}\right) + C$

3. $\sqrt{x - 2}\left(\dfrac{2(x - 2)}{3} + 4\right) + C$

5. $\dfrac{(2x - 3)^{3/2}(x + 1)}{5} + C$

7. $\dfrac{-\sqrt{9 - 4x}}{x} - \dfrac{2}{3}\ln\left|\dfrac{\sqrt{9 - 4x} - 3}{\sqrt{9 - 4x} + 3}\right| + C$

9. $\dfrac{(x + 2)(2x - 6)\sqrt{4x - x^2}}{6} + 4\sin^{-1}\left(\dfrac{x - 2}{2}\right) + C$

11. $-\dfrac{1}{\sqrt{7}}\ln\left|\dfrac{\sqrt{7} + \sqrt{7 + x^2}}{x}\right| + C$

13. $\sqrt{4 - x^2} - 2\ln\left|\dfrac{2 + \sqrt{4 - x^2}}{x}\right| + C$

15. $\dfrac{e^{2t}}{13}(2\cos 3t + 3\sin 3t) + C$

17. $\dfrac{x^2}{2}\cos^{-1}x + \dfrac{1}{4}\sin^{-1}x - \dfrac{1}{4}x\sqrt{1 - x^2} + C$

19. $\dfrac{x^3}{3}\tan^{-1}x - \dfrac{x^2}{6} + \dfrac{1}{6}\ln(1 + x^2) + C$

21. $-\dfrac{\cos 5x}{10} - \dfrac{\cos x}{2} + C$ **23.** $8\left[\dfrac{\sin(7t/2)}{7} - \dfrac{\sin(9t/2)}{9}\right] + C$

25. $6\sin(\theta/12) + \dfrac{6}{7}\sin(7\theta/12) + C$

27. $\dfrac{1}{2}\ln(x^2 + 1) + \dfrac{x}{2(1 + x^2)} + \dfrac{1}{2}\tan^{-1}x + C$

29. $\left(x - \dfrac{1}{2}\right)\sin^{-1}\sqrt{x} + \dfrac{1}{2}\sqrt{x - x^2} + C$

31. $\sin^{-1}\sqrt{x} - \sqrt{x - x^2} + C$

33. $\sqrt{1 - \sin^2 t} - \ln\left|\dfrac{1 + \sqrt{1 - \sin^2 t}}{\sin t}\right| + C$

35. $\ln\left|\ln y + \sqrt{3 + (\ln y)^2}\right| + C$

37. $\ln\left|x + 1 + \sqrt{x^2 + 2x + 5}\right| + C$

39. $\dfrac{x + 2}{2}\sqrt{5 - 4x - x^2} + \dfrac{9}{2}\sin^{-1}\left(\dfrac{x + 2}{3}\right) + C$

41. $-\dfrac{\sin^4 2x \cos 2x}{10} - \dfrac{2\sin^2 2x \cos 2x}{15} - \dfrac{4\cos 2x}{15} + C$

43. $\dfrac{\sin^3 2\theta \cos^2 2\theta}{10} + \dfrac{\sin^3 2\theta}{15} + C$

45. $\tan^2 2x - 2\ln|\sec 2x| + C$

47. $\dfrac{(\sec \pi x)(\tan \pi x)}{\pi} + \dfrac{1}{\pi}\ln|\sec \pi x + \tan \pi x| + C$

49. $\dfrac{-\csc^3 x \cot x}{4} - \dfrac{3\csc x \cot x}{8} - \dfrac{3}{8}\ln|\csc x + \cot x| + C$

51. $\dfrac{1}{2}[\sec(e^t - 1)\tan(e^t - 1) +$
$\ln|\sec(e^t - 1) + \tan(e^t - 1)|] + C$

53. $\sqrt{2} + \ln(\sqrt{2} + 1)$ **55.** $\pi/3$

57. $2\pi\sqrt{3} + \pi\sqrt{2}\ln(\sqrt{2} + \sqrt{3})$

59. $\bar{x} = 4/3$, $\bar{y} = \ln\sqrt{2}$ **61.** 7.62 **63.** $\pi/8$ **67.** $\pi/4$

Section 8.6, pp. 493–495

1. I: (a) 1.5, 0   (b) 1.5, 0   (c) 0%
   II: (a) 1.5, 0   (b) 1.5, 0   (c) 0%
3. I: (a) 2.75, 0.08   (b) 2.67, 0.08   (c) $0.0312 \approx 3\%$
   II: (a) 2.67, 0   (b) 2.67, 0   (c) 0%
5. I: (a) 6.25, 0.5   (b) 6, 0.25   (c) $0.0417 \approx 4\%$
   II: (a) 6, 0   (b) 6, 0   (c) 0%
7. I: (a) 0.509, 0.03125   (b) 0.5, 0.009   (c) $0.018 \approx 2\%$
   II: (a) 0.5, 0.002604   (b) 0.5, 0.0004   (c) 0%
9. I: (a) 1.8961, 0.161   (b) 2, 0.1039   (c) $0.052 \approx 5\%$
   II: (a) 2.0045, 0.0066   (b) 2, 0.00454   (c) 0.2%
11. (a) 1   (b) 2   13. (a) 116   (b) 2
15. (a) 283   (b) 2   17. (a) 71   (b) 10
19. (a) 76   (b) 12   21. (a) 82   (b) 8
23. 15,990 ft^3 25. ≈ 10.63 ft
27. (a) ≈ 0.00021 (b) ≈ 1.37079 (c) $\approx 0.015\%$
31. (a) ≈ 5.870 (b) $|E_T| \le 0.0032$
33. 21.07 in.   35. 14.4

Section 8.7, pp. 505–507

1. $\pi/2$ 3. 2   5. 6   7. $\pi/2$ 9. $\ln 3$ 11. $\ln 4$ 13. 0
15. $\sqrt{3}$ 17. π 19. $\ln\left(1 + \dfrac{\pi}{2}\right)$ 21. -1 23. 1
25. $-1/4$ 27. $\pi/2$ 29. $\pi/3$ 31. 6   33. $\ln 2$
35. Diverges   37. Converges   39. Converges   41. Converges
43. Diverges   45. Converges   47. Converges   49. Diverges
51. Converges   53. Converges   55. Diverges   57. Converges
59. Diverges   61. Converges   63. Converges
65. (a) Converges when $p < 1$ (b) Converges when $p > 1$
67. 1   69. 2π 71. $\ln 2$ 73. (b) ≈ 0.88621
75. (a)

$$Si(x) = \int_0^x \frac{\sin t}{t}\, dt$$

$$y = \frac{\sin t}{t}$$

   (b) $\pi/2$
77. (a)

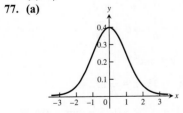

   (b) ≈ 0.683, ≈ 0.954, ≈ 0.997

Practice Exercises, pp. 507–509

1. $(x + 1)(\ln (x + 1)) - (x + 1) + C$
3. $x \tan^{-1}(3x) - \dfrac{1}{6}\ln(1 + 9x^2) + C$
5. $(x + 1)^2 e^x - 2(x + 1)e^x + 2e^x + C$

7. $\dfrac{2e^x \sin 2x}{5} + \dfrac{e^x \cos 2x}{5} + C$
9. $2\ln|x - 2| - \ln|x - 1| + C$
11. $\ln|x| - \ln|x + 1| + \dfrac{1}{x + 1} + C$
13. $-\dfrac{1}{3}\ln\left|\dfrac{\cos\theta - 1}{\cos\theta + 2}\right| + C$
15. $4\ln|x| - \dfrac{1}{2}\ln(x^2 + 1) + 4\tan^{-1}x + C$
17. $\dfrac{1}{16}\ln\left|\dfrac{(v - 2)^5(v + 2)}{v^6}\right| + C$
19. $\dfrac{1}{2}\tan^{-1}t - \dfrac{\sqrt{3}}{6}\tan^{-1}\dfrac{t}{\sqrt{3}} + C$
21. $\dfrac{x^2}{2} + \dfrac{4}{3}\ln|x + 2| + \dfrac{2}{3}\ln|x - 1| + C$
23. $\dfrac{x^2}{2} - \dfrac{9}{2}\ln|x + 3| + \dfrac{3}{2}\ln|x + 1| + C$
25. $\dfrac{1}{3}\ln\left|\dfrac{\sqrt{x + 1} - 1}{\sqrt{x + 1} + 1}\right| + C$ 27. $\ln|1 - e^{-s}| + C$
29. $-\sqrt{16 - y^2} + C$ 31. $-\dfrac{1}{2}\ln|4 - x^2| + C$
33. $\ln\dfrac{1}{\sqrt{9 - x^2}} + C$ 35. $\dfrac{1}{6}\ln\left|\dfrac{x + 3}{x - 3}\right| + C$
37. $-\dfrac{\cos^5 x}{5} + \dfrac{\cos^7 x}{7} + C$ 39. $\dfrac{\tan^5 x}{5} + C$
41. $\dfrac{\cos\theta}{2} - \dfrac{\cos 11\theta}{22} + C$ 43. $4\sqrt{1 - \cos(t/2)} + C$
45. At least 16   47. $T = \pi, S = \pi$ 49. 25°F
51. (a) ≈ 2.42 gal   (b) ≈ 24.83 mi/gal
53. $\pi/2$ 55. 6   57. $\ln 3$ 59. 2   61. $\pi/6$
63. Diverges   65. Diverges   67. Converges
69. $\dfrac{2x^{3/2}}{3} - x + 2\sqrt{x} - 2\ln\left(\sqrt{x} + 1\right) + C$
71. $\ln\left|\dfrac{\sqrt{x}}{\sqrt{x^2 + 1}}\right| - \dfrac{1}{2}\left(\dfrac{x}{\sqrt{x^2 + 1}}\right)^2 + C$
73. $-2\cot x - \ln|\csc x + \cot x| + \csc x + C$
75. $\dfrac{1}{12}\ln\left|\dfrac{3 + v}{3 - v}\right| + \dfrac{1}{6}\tan^{-1}\dfrac{v}{3} + C$
77. $\dfrac{\theta \sin(2\theta + 1)}{2} + \dfrac{\cos(2\theta + 1)}{4} + C$ 79. $\dfrac{1}{4}\sec^2\theta + C$
81. $2\left(\dfrac{(\sqrt{2 - x})^3}{3} - 2\sqrt{2 - x}\right) + C$ 83. $\tan^{-1}(y - 1) + C$
85. $\dfrac{1}{4}\ln|z| - \dfrac{1}{4z} - \dfrac{1}{4}\left[\dfrac{1}{2}\ln(z^2 + 4) + \dfrac{1}{2}\tan^{-1}\left(\dfrac{z}{2}\right)\right] + C$
87. $-\dfrac{1}{4}\sqrt{9 - 4t^2} + C$ 89. $\ln\left(\dfrac{e^t + 1}{e^t + 2}\right) + C$ 91. $1/4$
93. $\dfrac{2}{3}x^{3/2} + C$ 95. $-\dfrac{1}{5}\tan^{-1}(\cos 5t) + C$
97. $2\sqrt{r} - 2\ln(1 + \sqrt{r}) + C$
99. $\dfrac{1}{2}x^2 - \dfrac{1}{2}\ln(x^2 + 1) + C$

101. $\frac{2}{3} \ln |x + 1| + \frac{1}{6} \ln |x^2 - x + 1| +$

$\frac{1}{\sqrt{3}} \tan^{-1} \left(\frac{2x - 1}{\sqrt{3}} \right) + C$

103. $\frac{4}{7} \left(1 + \sqrt{x} \right)^{7/2} - \frac{8}{5} \left(1 + \sqrt{x} \right)^{5/2} + \frac{4}{3} \left(1 + \sqrt{x} \right)^{3/2} + C$

105. $2 \ln | \sqrt{x} + \sqrt{1 + x}| + C$

107. $\ln x - \ln |1 + \ln x| + C$

109. $\frac{1}{2} x^{\ln x} + C$ **111.** $\frac{1}{2} \ln \left| \frac{1 - \sqrt{1 - x^4}}{x^2} \right| + C$

113. (b) $\frac{\pi}{4}$ **115.** $x - \frac{1}{\sqrt{2}} \tan^{-1} \left(\sqrt{2} \tan x \right) + C$

Additional and Advanced Exercises, pp. 510–512

1. $x(\sin^{-1} x)^2 + 2(\sin^{-1} x)\sqrt{1 - x^2} - 2x + C$

3. $\frac{x^2 \sin^{-1} x}{2} + \frac{x\sqrt{1 - x^2} - \sin^{-1} x}{4} + C$

5. $\frac{1}{2} \left(\ln \left(t - \sqrt{1 - t^2} \right) - \sin^{-1} t \right) + C$

7. 0   **9.** $\ln(4) - 1$ **11.** 1   **13.** $32\pi/35$ **15.** 2π

17. (a) π **(b)** $\pi(2e - 5)$

19. (b) $\pi \left(\frac{8(\ln 2)^2}{3} - \frac{16(\ln 2)}{9} + \frac{16}{27} \right)$ **21.** $\left(\frac{e^2 + 1}{4}, \frac{e - 2}{2} \right)$

23. $\sqrt{1 + e^2} - \ln \left(\frac{\sqrt{1 + e^2}}{e} + \frac{1}{e} \right) - \sqrt{2} + \ln \left(1 + \sqrt{2} \right)$

25. $\frac{12\pi}{5}$ **27.** $a = \frac{1}{2}, -\frac{\ln 2}{4}$ **29.** $\frac{1}{2} < p \le 1$

33. $\frac{e^{2x}}{13} (3 \sin 3x + 2 \cos 3x) + C$

35. $\frac{\cos x \sin 3x - 3 \sin x \cos 3x}{8} + C$

37. $\frac{e^{ax}}{a^2 + b^2} (a \sin bx - b \cos bx) + C$ **39.** $x \ln (ax) - x + C$

41. $\frac{2}{1 - \tan (x/2)} + C$ **43.** 1   **45.** $\frac{\sqrt{3}\pi}{9}$

47. $\frac{1}{\sqrt{2}} \ln \left| \frac{\tan (t/2) + 1 - \sqrt{2}}{\tan (t/2) + 1 + \sqrt{2}} \right| + C$

49. $\ln \left| \frac{1 + \tan (\theta/2)}{1 - \tan (\theta/2)} \right| + C$

CHAPTER 9

Section 9.1, pp. 520–522

1. (d)   **3.** (a)

5.

7. $y' = x - y; y(1) = -1$ **9.** $y' = -(1 + y) \sin x; y(0) = 2$

11. $y(\text{exact}) = \frac{x}{2} - \frac{4}{x}$, $y_1 = -0.25$, $y_2 = 0.3$, $y_3 = 0.75$

13. $y(\text{exact}) = 3e^{x(x+2)}$, $y_1 = 4.2$, $y_2 = 6.216$, $y_3 = 9.697$

15. $y(\text{exact}) = e^{x^2} + 1$, $y_1 = 2.0$, $y_2 = 2.0202$, $y_3 = 2.0618$

17. $y \approx 2.48832$, exact value is e.

19. $y \approx -0.2272$, exact value is $1/\left(1 - 2\sqrt{5} \right) \approx -0.2880$.

23.

25.   **27.**

35. Euler's method gives $y \approx 3.45835$; the exact solution is $y = 1 + e \approx 3.71828$.

37. $y \approx 1.5000$; exact value is 1.5275.

Section 9.2, pp. 526–528

1. $y = \frac{e^x + C}{x}$, $x > 0$ **3.** $y = \frac{C - \cos x}{x^3}$, $x > 0$

5. $y = \frac{1}{2} - \frac{1}{x} + \frac{C}{x^2}$, $x > 0$ **7.** $y = \frac{1}{2} xe^{x/2} + Ce^{x/2}$

9. $y = x(\ln x)^2 + Cx$

11. $s = \frac{t^3}{3(t - 1)^4} - \frac{t}{(t - 1)^4} + \frac{C}{(t - 1)^4}$

13. $r = (\csc \theta)(\ln|\sec \theta| + C)$, $0 < \theta < \pi/2$

15. $y = \frac{3}{2} - \frac{1}{2} e^{-2t}$ **17.** $y = -\frac{1}{\theta} \cos \theta + \frac{\pi}{2\theta}$

19. $y = 6e^{x^2} - \frac{e^{x^2}}{x + 1}$ **21.** $y = y_0 e^{kt}$

23. (b) is correct, but **(a)** is not.   **25.** $t = \frac{L}{R} \ln 2$ sec

27. (a) $i = \frac{V}{R} - \frac{V}{R} e^{-3} = \frac{V}{R} (1 - e^{-3}) \approx 0.95 \frac{V}{R}$ amp   **(b)** 86%

29. $y = \frac{1}{1 + Ce^{-x}}$ **31.** $y^3 = 1 + Cx^{-3}$

Section 9.3, pp. 533–534

1. (a) 168.5 m   **(b)** 41.13 sec

3. $s(t) = 4.91\left(1 - e^{-(22.36/39.92)t} \right)$

5. $x^2 + y^2 = C$

7. $\ln|y| - \frac{1}{2}y^2 = \frac{1}{2}x^2 + C$

9. $y = \pm\sqrt{2x + C}$

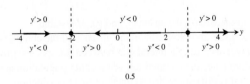

13. (a) 10 lb/min   **(b)** $(100 + t)$ gal   **(c)** $4\left(\dfrac{y}{100 + t}\right)$ lb/min

(d) $\dfrac{dy}{dt} = 10 - \dfrac{4y}{100 + t}, \quad y(0) = 50,$

$$y = 2(100 + t) - \frac{150}{\left(1 + \dfrac{t}{100}\right)^4}$$

(e) Concentration $= \dfrac{y(25)}{\text{amt. brine in tank}} = \dfrac{188.6}{125} \approx 1.5$ lb/gal

15. $y(27.8) \approx 14.8$ lb, $t \approx 27.8$ min

Section 9.4, pp. 540–541

1. $y' = (y + 2)(y - 3)$

(a) $y = -2$ is a stable equilibrium value and $y = 3$ is an unstable equilibrium.

(b) $y'' = 2(y + 2)\left(y - \dfrac{1}{2}\right)(y - 3)$

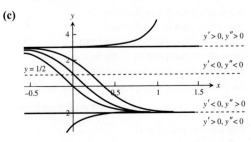

(c)

3. $y' = y^3 - y = (y + 1)y(y - 1)$

(a) $y = -1$ and $y = 1$ are unstable equilibria and $y = 0$ is a stable equilibrium.

(b) $y'' = (3y^2 - 1)y'$
$$= 3(y + 1)\left(y + 1/\sqrt{3}\right)y\left(y - 1/\sqrt{3}\right)(y - 1)$$

(c)

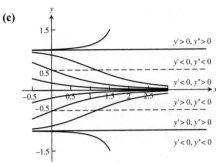

5. $y' = \sqrt{y}, y > 0$

(a) There are no equilibrium values.

(b) $y'' = \dfrac{1}{2}$

(c)

7. $y' = (y - 1)(y - 2)(y - 3)$

(a) $y = 1$ and $y = 3$ are unstable equilibria and $y = 2$ is a stable equilibrium.

(b) $y'' = (3y^2 - 12y + 11)(y - 1)(y - 2)(y - 3) =$

$$3(y - 1)\left(y - \frac{6 - \sqrt{3}}{3}\right)(y - 2)\left(y - \frac{6 + \sqrt{3}}{3}\right)(y - 3)$$

(c)

9. $\dfrac{dP}{dt} = 1 - 2P$ has a stable equilibrium at $P = \dfrac{1}{2}$;

$\dfrac{d^2P}{dt^2} = -2\dfrac{dP}{dt} = -2(1 - 2P).$

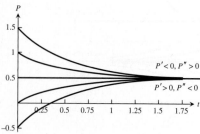

11. $\dfrac{dP}{dt} = 2P(P - 3)$ has a stable equilibrium at $P = 0$ and an unstable equilibrium at $P = 3$; $\dfrac{d^2P}{dt^2} = 2(2P - 3)\dfrac{dP}{dt} = 4P(2P - 3)(P - 3).$

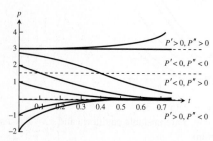

13. Before the catastrophe, the population exhibits logistic growth and $P(t)$ increases toward M_0, the stable equilibrium. After the catastrophe, the population declines logistically and $P(t)$ decreases toward M_1, the new stable equilibrium.

15. $\dfrac{dv}{dt} = g - \dfrac{k}{m}v^2, \quad g, k, m > 0$ and $v(t) \geq 0$

Equilibrium: $\dfrac{dv}{dt} = g - \dfrac{k}{m}v^2 = 0 \Rightarrow v = \sqrt{\dfrac{mg}{k}}$

Concavity: $\dfrac{d^2v}{dt^2} = -2\left(\dfrac{k}{m}v\right)\dfrac{dv}{dt} = -2\left(\dfrac{k}{m}v\right)\left(g - \dfrac{k}{m}v^2\right)$

(a)   **(b)**

(c) $v_{\text{terminal}} = \sqrt{\dfrac{160}{0.005}} = 178.9 \text{ ft/sec} = 122 \text{ mph}$

17. $F = F_p - F_r; ma = 50 - 5|v|; \dfrac{dv}{dt} = \dfrac{1}{m}(50 - 5|v|).$ The maximum velocity occurs when $\dfrac{dv}{dt} = 0$ or $v = 10 \text{ ft/sec}.$

19. Phase line:

If the switch is closed at $t = 0$, then $i(0) = 0$, and the graph of the solution looks like this:

As $t \to \infty$, $i(t) \to i_{\text{steady state}} = \dfrac{V}{R}.$

Section 9.5, pp. 545–547

1. Seasonal variations, nonconformity of the environments, effects of other interactions, unexpected disasters, etc.

3. This model assumes that the number of interactions is proportional to the product of x and y:

$\dfrac{dx}{dt} = (a - by)x, \quad a < 0,$

$\dfrac{dy}{dt} = m\left(1 - \dfrac{y}{M}\right)y - nxy = y\left(m - \dfrac{m}{M}y - nx\right).$

Rest points are $(0, 0)$, unstable, and $(0, M)$, stable.

5. (a) Logistic growth occurs in the absence of the competitor, and involves a simple interaction between the species: growth dominates the competition when either population is small, so it is difficult to drive either species to extinction.

(b) a: per capita growth rate for trout
    m: per capita growth rate for bass
    b: intensity of competition to the trout
    n: intensity of competition to the bass
    k_1: environmental carrying capacity for the trout
    k_2: environmental carrying capacity for the bass
    $\dfrac{a}{b}$: growth versus competition or net growth of trout
    $\dfrac{m}{n}$: relative survival of bass

(c) $\dfrac{dx}{dt} = 0$ when $x = 0$ or $y = \dfrac{a}{b} - \dfrac{a}{bk_1}x$,

$\dfrac{dy}{dt} = 0$ when $y = 0$ or $y = k_2 - \dfrac{k_2 n}{m}x$.

By picking $a/b > k_2$ and $m/n > k_1$, we insure that an equilibrium point exists inside the first quadrant.

Practice Exercises, pp. 547–548

1. $y = -\ln\left(C - \dfrac{2}{5}(x - 2)^{5/2} - \dfrac{4}{3}(x - 2)^{3/2}\right)$

3. $\tan y = -x \sin x - \cos x + C$ **5.** $(y + 1)e^{-y} = -\ln|x| + C$

7. $y = C\dfrac{x - 1}{x}$ **9.** $y = \dfrac{x^2}{4}e^{x/2} + Ce^{x/2}$

11. $y = \dfrac{x^2 - 2x + C}{2x^2}$ **13.** $y = \dfrac{e^{-x} + C}{1 + e^x}$ **15.** $xy + y^3 = C$

17. $y = \dfrac{2x^3 + 3x^2 + 6}{6(x + 1)^2}$ **19.** $y = \dfrac{1}{3}(1 - 4e^{-x^3})$

21. $y = e^{-x}(3x^3 - 3x^2)$

23.

x	y		x	y
0	0		1.1	1.6241
0.1	0.1000		1.2	1.8319
0.2	0.2095		1.3	2.0513
0.3	0.3285		1.4	2.2832
0.4	0.4568		1.5	2.5285
0.5	0.5946		1.6	2.7884
0.6	0.7418		1.7	3.0643
0.7	0.8986		1.8	3.3579
0.8	1.0649		1.9	3.6709
0.9	1.2411		2.0	4.0057
1.0	1.4273			

25. $y(3) \approx 0.8981$

27.

(a)

[-0.2, 4.5] by [-2.5, 0.5]

(b) Note that we choose a small interval of x-values because the y-values decrease very rapidly and our calculator cannot handle the calculations for $x \le -1$. (This occurs because the analytic solution is $y = -2 + \ln(2 - e^{-x})$, which has an asymptote at $x = -\ln 2 \approx -0.69$. Obviously, the Euler approximations are misleading for $x \le -0.7$.)

[-1, 0.2] by [-10, 2]

29. $y(\text{exact}) = \dfrac{1}{2}x^2 - \dfrac{3}{2}$; $y(2) \approx 0.4$; exact value is $\dfrac{1}{2}$.

31. $y(\text{exact}) = -e^{(x^2-1)/2}$; $y(2) \approx -3.4192$; exact value is $-e^{3/2} \approx -4.4817$.

33. (a) $y = -1$ is stable and $y = 1$ is unstable.

(b) $\dfrac{d^2y}{dx^2} = 2y\dfrac{dy}{dx} = 2y(y^2 - 1)$

(c)

Additional and Advanced Exercises, pp. 548–549

1. (a) $y = c + (y_0 - c)e^{-k(A/V)t}$
(b) Steady-state solution: $y_\infty = c$

5. $x^2(x^2 + 2y^2) = C$

7. $\ln|x| + e^{-y/x} = C$

9. $\ln|x| - \ln|\sec(y/x - 1) + \tan(y/x - 1)| = C$

CHAPTER 10

Section 10.1, pp. 559–562

1. $a_1 = 0, a_2 = -1/4, a_3 = -2/9, a_4 = -3/16$

3. $a_1 = 1, a_2 = -1/3, a_3 = 1/5, a_4 = -1/7$

5. $a_1 = 1/2, a_2 = 1/2, a_3 = 1/2, a_4 = 1/2$

7. $1, \dfrac{3}{2}, \dfrac{7}{4}, \dfrac{15}{8}, \dfrac{31}{16}, \dfrac{63}{32}, \dfrac{127}{64}, \dfrac{255}{128}, \dfrac{511}{256}, \dfrac{1023}{512}$

9. $2, 1, -\dfrac{1}{2}, -\dfrac{1}{4}, \dfrac{1}{8}, \dfrac{1}{16}, -\dfrac{1}{32}, -\dfrac{1}{64}, \dfrac{1}{128}, \dfrac{1}{256}$

11. $1, 1, 2, 3, 5, 8, 13, 21, 34, 55$ **13.** $a_n = (-1)^{n+1}, n \ge 1$

15. $a_n = (-1)^{n+1}(n)^2, n \ge 1$ **17.** $a_n = \dfrac{2^{n-1}}{3(n + 2)}, n \ge 1$

19. $a_n = n^2 - 1, n \ge 1$ **21.** $a_n = 4n - 3, n \ge 1$

23. $a_n = \dfrac{3n + 2}{n!}, n \ge 1$ **25.** $a_n = \dfrac{1 + (-1)^{n+1}}{2}, n \ge 1$

27. Converges, 2   **29.** Converges, -1 **31.** Converges, -5

33. Diverges   **35.** Diverges   **37.** Converges, 1/2

39. Converges, 0   **41.** Converges, $\sqrt{2}$ **43.** Converges, 1

45. Converges, 0   **47.** Converges, 0   **49.** Converges, 0

51. Converges, 1   **53.** Converges, e^7 **55.** Converges, 1

57. Converges, 1   **59.** Diverges   **61.** Converges, 4

63. Converges, 0   **65.** Diverges   **67.** Converges, e^{-1}

69. Converges, $e^{2/3}$ **71.** Converges, $x (x > 0)$

73. Converges, 0   **75.** Converges, 1   **77.** Converges, 1/2

79. Converges, 1   **81.** Converges, $\pi/2$ **83.** Converges, 0

85. Converges, 0   **87.** Converges, 1/2   **89.** Converges, 0

91. 8   **93.** 4   **95.** 5   **97.** $1 + \sqrt{2}$ **99.** $x_n = 2^{n-2}$

101. (a) $f(x) = x^2 - 2$, $1.414213562 \approx \sqrt{2}$
(b) $f(x) = \tan(x) - 1$, $0.7853981635 \approx \pi/4$
(c) $f(x) = e^x$, diverges

103. (b) 1   **111.** Nondecreasing, bounded

113. Not nondecreasing, bounded

115. Converges, nondecreasing sequence theorem

117. Converges, nondecreasing sequence theorem

119. Diverges, definition of divergence   **121.** Converges

123. Converges   **133.** (b) $\sqrt{3}$

Section 10.2, pp. 569–570

1. $s_n = \dfrac{2(1 - (1/3)^n)}{1 - (1/3)}, 3$ **3.** $s_n = \dfrac{1 - (-1/2)^n}{1 - (-1/2)}, 2/3$

5. $s_n = \dfrac{1}{2} - \dfrac{1}{n+2}, \dfrac{1}{2}$ **7.** $1 - \dfrac{1}{4} + \dfrac{1}{16} - \dfrac{1}{64} + \cdots, \dfrac{4}{5}$

9. $\dfrac{7}{4} + \dfrac{7}{16} + \dfrac{7}{64} + \cdots, \dfrac{7}{3}$

11. $(5 + 1) + \left(\dfrac{5}{2} + \dfrac{1}{3}\right) + \left(\dfrac{5}{4} + \dfrac{1}{9}\right) + \left(\dfrac{5}{8} + \dfrac{1}{27}\right) + \cdots, \dfrac{23}{2}$

13. $(1 + 1) + \left(\dfrac{1}{2} - \dfrac{1}{5}\right) + \left(\dfrac{1}{4} + \dfrac{1}{25}\right) + \left(\dfrac{1}{8} - \dfrac{1}{125}\right) + \cdots, \dfrac{17}{6}$

15. Converges, 5/3   **17.** Converges, 1/7   **19.** 23/99   **21.** 7/9

23. 1/15   **25.** 41333/33300   **27.** Diverges

29. Inconclusive   **31.** Diverges   **33.** Diverges

35. $s_n = 1 - \dfrac{1}{n+1}$; converges, 1   **37.** $s_n = \ln\sqrt{n+1}$; diverges

39. $s_n = \dfrac{\pi}{3} - \cos^{-1}\left(\dfrac{1}{n+2}\right)$; converges, $-\dfrac{\pi}{6}$

41. 1   **43.** 5   **45.** 1   **47.** $-\dfrac{1}{\ln 2}$ **49.** Converges, $2 + \sqrt{2}$

51. Converges, 1   **53.** Diverges   **55.** Converges, $\dfrac{e^2}{e^2 - 1}$

57. Converges, 2/9   **59.** Converges, 3/2   **61.** Diverges

63. Converges, 4   **65.** Diverges   **67.** Converges, $\dfrac{\pi}{\pi - e}$

69. $a = 1, r = -x$; converges to $1/(1 + x)$ for $|x| < 1$

71. $a = 3, r = (x - 1)/2$; converges to $6/(3 - x)$ for x in $(-1, 3)$

73. $|x| < \dfrac{1}{2}, \dfrac{1}{1 - 2x}$ **75.** $-2 < x < 0, \dfrac{1}{2 + x}$

77. $x \neq (2k + 1)\dfrac{\pi}{2}$, k an integer; $\dfrac{1}{1 - \sin x}$

79. (a) $\displaystyle\sum_{n=-2}^{\infty} \dfrac{1}{(n + 4)(n + 5)}$ (b) $\displaystyle\sum_{n=0}^{\infty} \dfrac{1}{(n + 2)(n + 3)}$

(c) $\displaystyle\sum_{n=5}^{\infty} \dfrac{1}{(n - 3)(n - 2)}$

89. (a) $r = 3/5$ (b) $r = -3/10$

91. $|r| < 1, \dfrac{1 + 2r}{1 - r^2}$ **93.** 8 m^2

Section 10.3, pp. 575–576

1. Converges   **3.** Converges   **5.** Converges   **7.** Diverges

9. Converges   **11.** Converges; geometric series, $r = \dfrac{1}{10} < 1$

13. Diverges; $\displaystyle\lim_{n\to\infty} \dfrac{n}{n+1} = 1 \neq 0$ **15.** Diverges; p-series, $p < 1$

17. Converges; geometric series, $r = \dfrac{1}{8} < 1$

19. Diverges; Integral Test

21. Converges; geometric series, $r = 2/3 < 1$

23. Diverges; Integral Test   **25.** Diverges; $\displaystyle\lim_{n\to\infty} \dfrac{2^n}{n+1} \neq 0$

27. Diverges; $\lim_{n\to\infty}\left(\sqrt{n}/\ln n\right) \neq 0$

29. Diverges; geometric series, $r = \dfrac{1}{\ln 2} > 1$

31. Converges; Integral Test   **33.** Diverges; nth-Term Test

35. Converges; Integral Test   **37.** Converges; Integral Test

39. Converges; Integral Test   **41.** $a = 1$

43. (a)

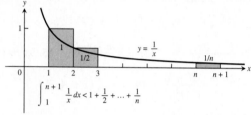

$$\int_1^{n+1} \dfrac{1}{x}\,dx < 1 + \dfrac{1}{2} + \cdots + \dfrac{1}{n}$$

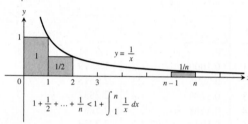

$$1 + \dfrac{1}{2} + \cdots + \dfrac{1}{n} < 1 + \int_1^n \dfrac{1}{x}\,dx$$

(b) ≈ 41.55

45. True   **47.** (b) $n \geq 251{,}415$

49. $s_8 = \displaystyle\sum_{n=1}^{8} \dfrac{1}{n^3} \approx 1.195$ **51.** 10^{60}

59. (a) $1.20166 \leq S \leq 1.20253$ (b) $S \approx 1.2021$, error < 0.0005

Section 10.4, pp. 580–581

1. Converges; compare with $\sum (1/n^2)$

3. Diverges; compare with $\sum\left(1/\sqrt{n}\right)$

5. Converges; compare with $\sum (1/n^{3/2})$

7. Converges; compare with $\displaystyle\sum \sqrt{\dfrac{n + 4n}{n^4 + 0}} = \sqrt{5}\sum \dfrac{1}{n^{3/2}}$

9. Converges   **11.** Diverges; limit comparison with $\sum (1/n)$

13. Diverges; limit comparison with $\sum\left(1/\sqrt{n}\right)$ **15.** Diverges

17. Diverges; limit comparison with $\sum\left(1/\sqrt{n}\right)$

19. Converges; compare with $\sum(1/2^n)$

21. Diverges; nth-Term Test

23. Converges; compare with $\sum (1/n^2)$

25. Converges; $\left(\dfrac{n}{3n + 1}\right)^n < \left(\dfrac{n}{3n}\right)^n = \left(\dfrac{1}{3}\right)^n$

27. Diverges; direct comparison with $\sum(1/n)$

29. Diverges; limit comparison with $\sum(1/n)$

31. Diverges; limit comparison with $\sum(1/n)$

33. Converges; compare with $\sum(1/n^{3/2})$

35. Converges; $\dfrac{1}{n2^n} \le \dfrac{1}{2^n}$ 37. Converges; $\dfrac{1}{3^{n-1}+1} < \dfrac{1}{3^{n-1}}$

39. Converges; comparison with $\sum (1/5n^2)$

41. Diverges; comparison with $\sum (1/n)$

43. Converges; comparison with $\sum \dfrac{1}{n(n-1)}$

 or limit comparison with $\sum (1/n^2)$

45. Diverges; limit comparison with $\sum (1/n)$

47. Converges; $\dfrac{\tan^{-1} n}{n^{1.1}} < \dfrac{\pi/2}{n^{1.1}}$

49. Converges; compare with $\sum (1/n^2)$

51. Diverges; limit comparison with $\sum (1/n)$

53. Converges; limit comparison with $\sum (1/n^2)$

63. Converges   65. Converges   67. Converges

Section 10.5, pp. 585–586

1. Converges   3. Diverges   5. Converges   7. Converges

9. Converges   11. Diverges   13. Converges   15. Converges

17. Converges; Ratio Test   19. Diverges; Ratio Test

21. Converges; Ratio Test

23. Converges; compare with $\sum (3/(1.25)^n)$

25. Diverges; $\displaystyle\lim_{n\to\infty} \left(1 - \dfrac{3}{n}\right)^n = e^{-3} \ne 0$

27. Converges; compare with $\sum (1/n^2)$

29. Diverges; compare with $\sum (1/(2n))$

31. Diverges; compare with $\sum (1/n)$ 33. Converges; Ratio Test

35. Converges; Ratio Test   37. Converges; Ratio Test

39. Converges; Root Test   41. Converges; compare with $\sum (1/n^2)$

43. Converges; Ratio Test   45. Converges; Ratio Test

47. Diverges; Ratio Test   49. Converges; Ratio Test

51. Converges; Ratio Test   53. Diverges; $a_n = \left(\dfrac{1}{3}\right)^{(1/n!)} \to 1$

55. Converges; Ratio Test   57. Diverges; Root Test

59. Converges; Root Test   61. Converges; Ratio Test   65. Yes

Section 10.6, pp. 591–592

1. Converges by Theorem 16

3. Converges; Alternating Series Test

5. Converges; Alternating Series Test

7. Diverges; $a_n \not\to 0$

9. Diverges; $a_n \not\to 0$

11. Converges; Alternating Series Test

13. Converges by Theorem 16

15. Converges absolutely. Series of absolute values is a convergent geometric series.

17. Converges conditionally; $1/\sqrt{n} \to 0$ but $\sum_{n=1}^{\infty} \dfrac{1}{\sqrt{n}}$ diverges.

19. Converges absolutely; compare with $\sum_{n=1}^{\infty} (1/n^2)$.

21. Converges conditionally; $1/(n+3) \to 0$ but $\sum_{n=1}^{\infty} \dfrac{1}{n+3}$ diverges (compare with $\sum_{n=1}^{\infty}(1/n)$).

23. Diverges; $\dfrac{3+n}{5+n} \to 1$

25. Converges conditionally; $\left(\dfrac{1}{n^2} + \dfrac{1}{n}\right) \to 0$ but $(1+n)/n^2 > 1/n$

27. Converges absolutely; Ratio Test

29. Converges absolutely by Integral Test   31. Diverges; $a_n \not\to 0$

33. Converges absolutely by Ratio Test

35. Converges absolutely since $\left|\dfrac{\cos n\pi}{n\sqrt{n}}\right| = \left|\dfrac{(-1)^{n+1}}{n^{3/2}}\right| = \dfrac{1}{n^{3/2}}$ (convergent p-series)

37. Converges absolutely by Root Test   39. Diverges; $a_n \to \infty$

41. Converges conditionally; $\sqrt{n+1} - \sqrt{n} = 1/(\sqrt{n} + \sqrt{n+1}) \to 0$, but series of absolute values diverges (compare with $\sum (1/\sqrt{n})$).

43. Diverges, $a_n \to 1/2 \ne 0$

45. Converges absolutely; $\operatorname{sech} n = \dfrac{2}{e^n + e^{-n}} = \dfrac{2e^n}{e^{2n}+1} < \dfrac{2e^n}{e^{2n}} = \dfrac{2}{e^n}$, a term from a convergent geometric series.

47. Converges conditionally; $\sum (-1)\dfrac{1}{2(n+1)}$ converges by Alternating Series Test; $\sum \dfrac{1}{2(n+1)}$ diverges by limit comparison with $\sum (1/n)$.

49. $|\text{Error}| < 0.2$ 51. $|\text{Error}| < 2 \times 10^{-11}$

53. $n \ge 31$ 55. $n \ge 4$ 57. 0.54030

59. (a) $a_n \ge a_{n+1}$ (b) $-1/2$

Section 10.7, pp. 600–602

1. (a) $1, -1 < x < 1$ (b) $-1 < x < 1$ (c) none

3. (a) $1/4, -1/2 < x < 0$ (b) $-1/2 < x < 0$ (c) none

5. (a) $10, -8 < x < 12$ (b) $-8 < x < 12$ (c) none

7. (a) $1, -1 < x < 1$ (b) $-1 < x < 1$ (c) none

9. (a) $3, -3 \le x \le 3$ (b) $-3 \le x \le 3$ (c) none

11. (a) ∞, for all x (b) for all x (c) none

13. (a) $1/2, -1/2 \le x < 1/2$ (b) $-1/2 < x < 1/2$ (c) $-1/2$

15. (a) $1, -1 \le x < 1$ (b) $-1 < x < 1$ (c) $x = -1$

17. (a) $5, -8 < x < 2$ (b) $-8 < x < 2$ (c) none

19. (a) $3, -3 < x < 3$ (b) $-3 < x < 3$ (c) none

21. (a) $1, -2 < x < 0$ (b) $-2 < x < 0$ (c) none

23. (a) $1, -1 < x < 1$ (b) $-1 < x < 1$ (c) none

25. (a) $0, x = 0$ (b) $x = 0$ (c) none

27. (a) $2, -4 < x \le 0$ (b) $-4 < x < 0$ (c) $x = 0$

29. (a) $1, -1 \le x \le 1$ (b) $-1 \le x \le 1$ (c) none

31. (a) $1/4, 1 \le x \le 3/2$ (b) $1 \le x \le 3/2$ (c) none

33. (a) ∞, for all x (b) for all x (c) none

35. (a) $1, -1 \le x < 1$ (b) $-1 < x < 1$ (c) -1

37. 3   39. 8   41. $-1/3 < x < 1/3, 1/(1-3x)$

43. $-1 < x < 3, 4/(3 + 2x - x^2)$

45. $0 < x < 16, 2/(4 - \sqrt{x})$

47. $-\sqrt{2} < x < \sqrt{2}, 3/(2 - x^2)$

49. $1 < x < 5, 2/(x-1), 1 < x < 5, -2/(x-1)^2$

51. (a) $\cos x = 1 - \dfrac{x^2}{2!} + \dfrac{x^4}{4!} - \dfrac{x^6}{6!} + \dfrac{x^8}{8!} - \dfrac{x^{10}}{10!} + \cdots$; converges for all x

 (b) Same answer as part (c)

 (c) $2x - \dfrac{2^3 x^3}{3!} + \dfrac{2^5 x^5}{5!} - \dfrac{2^7 x^7}{7!} + \dfrac{2^9 x^9}{9!} - \dfrac{2^{11} x^{11}}{11!} + \cdots$

53. (a) $\dfrac{x^2}{2} + \dfrac{x^4}{12} + \dfrac{x^6}{45} + \dfrac{17x^8}{2520} + \dfrac{31x^{10}}{14175}, \ -\dfrac{\pi}{2} < x < \dfrac{\pi}{2}$

(b) $1 + x^2 + \dfrac{2x^4}{3} + \dfrac{17x^6}{45} + \dfrac{62x^8}{315} + \cdots, \ -\dfrac{\pi}{2} < x < \dfrac{\pi}{2}$

Section 10.8, pp. 606–607

1. $P_0(x) = 1, P_1(x) = 1 + 2x, P_2(x) = 1 + 2x + 2x^2,$
$P_3(x) = 1 + 2x + 2x^2 + \dfrac{4}{3}x^3$

3. $P_0(x) = 0, P_1(x) = x - 1, P_2(x) = (x - 1) - \dfrac{1}{2}(x - 1)^2,$
$P_3(x) = (x - 1) - \dfrac{1}{2}(x - 1)^2 + \dfrac{1}{3}(x - 1)^3$

5. $P_0(x) = \dfrac{1}{2}, P_1(x) = \dfrac{1}{2} - \dfrac{1}{4}(x - 2),$
$P_2(x) = \dfrac{1}{2} - \dfrac{1}{4}(x - 2) + \dfrac{1}{8}(x - 2)^2,$
$P_3(x) = \dfrac{1}{2} - \dfrac{1}{4}(x - 2) + \dfrac{1}{8}(x - 2)^2 - \dfrac{1}{16}(x - 2)^3$

7. $P_0(x) = \dfrac{\sqrt{2}}{2}, P_1(x) = \dfrac{\sqrt{2}}{2} + \dfrac{\sqrt{2}}{2}\left(x - \dfrac{\pi}{4}\right),$
$P_2(x) = \dfrac{\sqrt{2}}{2} + \dfrac{\sqrt{2}}{2}\left(x - \dfrac{\pi}{4}\right) - \dfrac{\sqrt{2}}{4}\left(x - \dfrac{\pi}{4}\right)^2,$
$P_3(x) = \dfrac{\sqrt{2}}{2} + \dfrac{\sqrt{2}}{2}\left(x - \dfrac{\pi}{4}\right) - \dfrac{\sqrt{2}}{4}\left(x - \dfrac{\pi}{4}\right)^2$
$- \dfrac{\sqrt{2}}{12}\left(x - \dfrac{\pi}{4}\right)^3$

9. $P_0(x) = 2, P_1(x) = 2 + \dfrac{1}{4}(x - 4),$
$P_2(x) = 2 + \dfrac{1}{4}(x - 4) - \dfrac{1}{64}(x - 4)^2,$
$P_3(x) = 2 + \dfrac{1}{4}(x - 4) - \dfrac{1}{64}(x - 4)^2 + \dfrac{1}{512}(x - 4)^3$

11. $\displaystyle\sum_{n=0}^{\infty} \dfrac{(-x)^n}{n!} = 1 - x + \dfrac{x^2}{2!} - \dfrac{x^3}{3!} + \dfrac{x^4}{4!} - \cdots$

13. $\displaystyle\sum_{n=0}^{\infty} (-1)^n x^n = 1 - x + x^2 - x^3 + \cdots$

15. $\displaystyle\sum_{n=0}^{\infty} \dfrac{(-1)^n 3^{2n+1} x^{2n+1}}{(2n+1)!}$ **17.** $7\displaystyle\sum_{n=0}^{\infty} \dfrac{(-1)^n x^{2n}}{(2n)!}$ **19.** $\displaystyle\sum_{n=0}^{\infty} \dfrac{x^{2n}}{(2n)!}$

21. $x^4 - 2x^3 - 5x + 4$

23. $8 + 10(x - 2) + 6(x - 2)^2 + (x - 2)^3$

25. $21 - 36(x + 2) + 25(x + 2)^2 - 8(x + 2)^3 + (x + 2)^4$

27. $\displaystyle\sum_{n=0}^{\infty} (-1)^n (n + 1)(x - 1)^n$ **29.** $\displaystyle\sum_{n=0}^{\infty} \dfrac{e^2}{n!}(x - 2)^n$

31. $\displaystyle\sum_{n=0}^{\infty} (-1)^{n+1} \dfrac{2^{2n}}{(2n)!}\left(x - \dfrac{\pi}{4}\right)^{2n}$

33. $-1 - 2x - \dfrac{5}{2}x^2 - \cdots, \ -1 < x < 1$

35. $x^2 - \dfrac{1}{2}x^3 + \dfrac{1}{6}x^4 + \cdots, \ -1 < x < 1$

41. $L(x) = 0, Q(x) = -x^2/2$ **43.** $L(x) = 1, Q(x) = 1 + x^2/2$

45. $L(x) = x, Q(x) = x$

Section 10.9, pp. 613–614

1. $\displaystyle\sum_{n=0}^{\infty} \dfrac{(-5x)^n}{n!} = 1 - 5x + \dfrac{5^2 x^2}{2!} - \dfrac{5^3 x^3}{3!} + \cdots$

3. $\displaystyle\sum_{n=0}^{\infty} \dfrac{5(-1)^n(-x)^{2n+1}}{(2n+1)!} = \displaystyle\sum_{n=0}^{\infty} \dfrac{5(-1)^{n+1} x^{2n+1}}{(2n+1)!}$
$= -5x + \dfrac{5x^3}{3!} - \dfrac{5x^5}{5!} + \dfrac{5x^7}{7!} + \cdots$

5. $\displaystyle\sum_{n=0}^{\infty} \dfrac{(-1)^n(5x^2)^{2n}}{(2n)!} = 1 - \dfrac{25x^4}{2!} + \dfrac{625x^8}{4!} - \cdots$

7. $\displaystyle\sum_{n=1}^{\infty} (-1)^{n+1} \dfrac{x^{2n}}{n} = x^2 - \dfrac{x^4}{2} + \dfrac{x^6}{3} - \dfrac{x^8}{4} + \cdots$

9. $\displaystyle\sum_{n=0}^{\infty} (-1)^n \left(\dfrac{3}{4}\right)^n x^{3n} = 1 - \dfrac{3}{4}x^3 + \dfrac{3^2}{4^2}x^6 - \dfrac{3^3}{4^3}x^9 + \cdots$

11. $\displaystyle\sum_{n=0}^{\infty} \dfrac{x^{n+1}}{n!} = x + x^2 + \dfrac{x^3}{2!} + \dfrac{x^4}{3!} + \dfrac{x^5}{4!} + \cdots$

13. $\displaystyle\sum_{n=2}^{\infty} \dfrac{(-1)^n x^{2n}}{(2n)!} = \dfrac{x^4}{4!} - \dfrac{x^6}{6!} + \dfrac{x^8}{8!} - \dfrac{x^{10}}{10!} + \cdots$

15. $x - \dfrac{\pi^2 x^3}{2!} + \dfrac{\pi^4 x^5}{4!} - \dfrac{\pi^6 x^7}{6!} + \cdots = \displaystyle\sum_{n=0}^{\infty} \dfrac{(-1)^n \pi^{2n} x^{2n+1}}{(2n)!}$

17. $1 + \displaystyle\sum_{n=1}^{\infty} \dfrac{(-1)^n (2x)^{2n}}{2 \cdot (2n)!} =$
$1 - \dfrac{(2x)^2}{2 \cdot 2!} + \dfrac{(2x)^4}{2 \cdot 4!} - \dfrac{(2x)^6}{2 \cdot 6!} + \dfrac{(2x)^8}{2 \cdot 8!} - \cdots$

19. $x^2 \displaystyle\sum_{n=0}^{\infty} (2x)^n = x^2 + 2x^3 + 4x^4 + \cdots$

21. $\displaystyle\sum_{n=1}^{\infty} n x^{n-1} = 1 + 2x + 3x^2 + 4x^3 + \cdots$

23. $\displaystyle\sum_{n=1}^{\infty} (-1)^{n+1} \dfrac{x^{4n-1}}{2n-1} = x^3 - \dfrac{x^7}{3} + \dfrac{x^{11}}{5} - \dfrac{x^{15}}{7} + \cdots$

25. $\displaystyle\sum_{n=0}^{\infty} \left(\dfrac{1}{n!} + (-1)^n\right) x^n = 2 + \dfrac{3}{2}x^2 - \dfrac{5}{6}x^3 + \dfrac{25}{24}x^4 - \cdots$

27. $\displaystyle\sum_{n=1}^{\infty} \dfrac{(-1)^{n-1} x^{2n+1}}{3n} = \dfrac{x^3}{3} - \dfrac{x^5}{6} + \dfrac{x^7}{9} - \cdots$

29. $x + x^2 + \dfrac{x^3}{3} - \dfrac{x^5}{30} + \cdots$

31. $x^2 - \dfrac{2}{3}x^4 + \dfrac{23}{45}x^6 - \dfrac{44}{105}x^8 + \cdots$

33. $1 + x + \dfrac{1}{2}x^2 - \dfrac{1}{8}x^4 + \cdots$

35. $|\text{Error}| \le \dfrac{1}{10^4 \cdot 4!} < 4.2 \times 10^{-6}$

37. $|x| < (0.06)^{1/5} < 0.56968$

39. $|\text{Error}| < (10^{-3})^3/6 < 1.67 \times 10^{-10}, \ -10^{-3} < x < 0$

41. $|\text{Error}| < (3^{0.1})(0.1)^3/6 < 1.87 \times 10^{-4}$

49. (a) $Q(x) = 1 + kx + \dfrac{k(k-1)}{2}x^2$ **(b)** $0 \le x < 100^{-1/3}$

Section 10.10, pp. 620–622

1. $1 + \dfrac{x}{2} - \dfrac{x^2}{8} + \dfrac{x^3}{16}$ **3.** $1 + \dfrac{1}{2}x + \dfrac{3}{8}x^2 + \dfrac{5}{16}x^3 + \cdots$

5. $1 - x + \dfrac{3x^2}{4} - \dfrac{x^3}{2}$ **7.** $1 - \dfrac{x^3}{2} + \dfrac{3x^6}{8} - \dfrac{5x^9}{16}$

9. $1 + \dfrac{1}{2x} - \dfrac{1}{8x^2} + \dfrac{1}{16x^3}$

11. $(1 + x)^4 = 1 + 4x + 6x^2 + 4x^3 + x^4$

13. $(1 - 2x)^3 = 1 - 6x + 12x^2 - 8x^3$

15. 0.00267 **17.** 0.10000 **19.** 0.09994 **21.** 0.10000

23. $\dfrac{1}{13 \cdot 6!} \approx 0.00011$ **25.** $\dfrac{x^3}{3} - \dfrac{x^7}{7 \cdot 3!} + \dfrac{x^{11}}{11 \cdot 5!}$

27. (a) $\dfrac{x^2}{2} - \dfrac{x^4}{12}$

    (b) $\dfrac{x^2}{2} - \dfrac{x^4}{3 \cdot 4} + \dfrac{x^6}{5 \cdot 6} - \dfrac{x^8}{7 \cdot 8} + \cdots + (-1)^{15} \dfrac{x^{32}}{31 \cdot 32}$

29. $1/2$ 31. $-1/24$ 33. $1/3$ 35. -1 37. 2

39. $3/2$ 41. e 43. $\cos \dfrac{3}{4}$ 45. $\sqrt{3}/2$ 47. $\dfrac{x^3}{1 - x}$

49. $\dfrac{x^3}{1 + x^2}$ 51. $\dfrac{-1}{(1 + x)^2}$ 55. 500 terms   57. 4 terms

59. (a) $x + \dfrac{x^3}{6} + \dfrac{3x^5}{40} + \dfrac{5x^7}{112}$, radius of convergence $= 1$

    (b) $\dfrac{\pi}{2} - x - \dfrac{x^3}{6} - \dfrac{3x^5}{40} - \dfrac{5x^7}{112}$

61. $1 - 2x + 3x^2 - 4x^3 + \cdots$

67. (a) -1 (b) $(1/\sqrt{2})(1 + i)$ (c) $-i$

71. $x + x^2 + \dfrac{1}{3}x^3 - \dfrac{1}{30}x^5 + \cdots$, for all x

Practice Exercises, pp. 623–625

1. Converges to 1   3. Converges to -1 5. Diverges
7. Converges to 0   9. Converges to 1   11. Converges to e^{-5}
13. Converges to 3   15. Converges to $\ln 2$ 17. Diverges
19. $1/6$ 21. $3/2$ 23. $e/(e - 1)$ 25. Diverges
27. Converges conditionally   29. Converges conditionally
31. Converges absolutely   33. Converges absolutely
35. Converges absolutely   37. Converges absolutely
39. Converges absolutely
41. (a) $3, -7 \le x < -1$ (b) $-7 < x < -1$ (c) $x = -7$
43. (a) $1/3, 0 \le x \le 2/3$ (b) $0 \le x \le 2/3$ (c) None
45. (a) ∞, for all x (b) For all x (c) None
47. (a) $\sqrt{3}, -\sqrt{3} < x < \sqrt{3}$ (b) $-\sqrt{3} < x < \sqrt{3}$
    (c) None
49. (a) $e, -e < x < e$ (b) $-e < x < e$ (c) Empty set
51. $\dfrac{1}{1 + x}, \dfrac{1}{4}, \dfrac{4}{5}$ 53. $\sin x, \pi, 0$ 55. $e^x, \ln 2, 2$ 57. $\displaystyle\sum_{n=0}^{\infty} 2^n x^n$

59. $\displaystyle\sum_{n=0}^{\infty} \dfrac{(-1)^n \pi^{2n+1} x^{2n+1}}{(2n + 1)!}$ 61. $\displaystyle\sum_{n=0}^{\infty} \dfrac{(-1)^n x^{10n/3}}{(2n)!}$ 63. $\displaystyle\sum_{n=0}^{\infty} \dfrac{((\pi x)/2)^n}{n!}$

65. $2 - \dfrac{(x + 1)}{2 \cdot 1!} + \dfrac{3(x + 1)^2}{2^3 \cdot 2!} + \dfrac{9(x + 1)^3}{2^5 \cdot 3!} + \cdots$

67. $\dfrac{1}{4} - \dfrac{1}{4^2}(x - 3) + \dfrac{1}{4^3}(x - 3)^2 - \dfrac{1}{4^4}(x - 3)^3$

69. 0.4849171431 71. 0.4872223583 73. $7/2$ 75. $1/12$

77. -2 79. $r = -3, s = 9/2$ 81. $2/3$

83. $\ln\left(\dfrac{n + 1}{2n}\right)$; the series converges to $\ln\left(\dfrac{1}{2}\right)$.

85. (a) ∞ (b) $a = 1, b = 0$ 87. It converges.

Additional and Advanced Exercises, pp. 625–627

1. Converges; Comparison Test   3. Diverges; nth-Term Test
5. Converges; Comparison Test   7. Diverges; nth-Term Test

9. With $a = \pi/3$, $\cos x = \dfrac{1}{2} - \dfrac{\sqrt{3}}{2}(x - \pi/3) - \dfrac{1}{4}(x - \pi/3)^2$

    $+ \dfrac{\sqrt{3}}{12}(x - \pi/3)^3 + \cdots$

11. With $a = 0$, $e^x = 1 + x + \dfrac{x^2}{2!} + \dfrac{x^3}{3!} + \cdots$

13. With $a = 22\pi$, $\cos x = 1 - \dfrac{1}{2}(x - 22\pi)^2 + \dfrac{1}{4!}(x - 22\pi)^4$

    $- \dfrac{1}{6!}(x - 22\pi)^6 + \cdots$

15. Converges, limit $= b$ 17. $\pi/2$ 21. $b = \pm\dfrac{1}{5}$

23. $a = 2, L = -7/6$ 27. (b) Yes

31. (a) $\displaystyle\sum_{n=1}^{\infty} n x^{n-1}$ (b) 6 (c) $1/q$

33. (a) $R_n = C_0 e^{-kt_0}(1 - e^{-nkt_0})/(1 - e^{-kt_0})$,
    $R = C_0(e^{-kt_0})/(1 - e^{-kt_0}) = C_0/(e^{kt_0} - 1)$
    (b) $R_1 = 1/e \approx 0.368$,
    $R_{10} = R(1 - e^{-10}) \approx R(0.9999546) \approx 0.58195$;
    $R \approx 0.58198; 0 < (R - R_{10})/R < 0.0001$
    (c) 7

CHAPTER 11

Section 11.1, pp. 634–636

1.

3.

5.

7.

9.

11.

13.

15.

17.

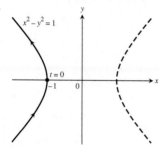

19. (a) $x = a \cos t$, $y = -a \sin t$, $0 \le t \le 2\pi$
  (b) $x = a \cos t$, $y = a \sin t$, $0 \le t \le 2\pi$
  (c) $x = a \cos t$, $y = -a \sin t$, $0 \le t \le 4\pi$
  (d) $x = a \cos t$, $y = a \sin t$, $0 \le t \le 4\pi$
21. Possible answer: $x = -1 + 5t$, $y = -3 + 4t$, $0 \le t \le 1$
23. Possible answer: $x = t^2 + 1$, $y = t$, $t \le 0$
25. Possible answer: $x = 2 - 3t$, $y = 3 - 4t$, $t \ge 0$
27. Possible answer: $x = 2 \cos t$, $y = 2\,|\sin t|$, $0 \le t \le 4\pi$
29. Possible answer: $x = \dfrac{-at}{\sqrt{1 + t^2}}$, $y = \dfrac{a}{\sqrt{1 + t^2}}$,
  $-\infty < t < \infty$
31. Possible answer: $x = \dfrac{4}{1 + 2\tan\theta}$, $y = \dfrac{4\tan\theta}{1 + 2\tan\theta}$,
  $0 \le \theta < \pi/2$ and $x = 0, y = 2$ if $\theta = \pi/2$
33. Possible answer: $x = 2 - \cos t$, $y = \sin t$, $0 \le t \le 2\pi$
35. $x = 2 \cot t$, $y = 2 \sin^2 t$, $0 < t < \pi$
37. $x = a \sin^2 t \tan t$, $y = a \sin^2 t$, $0 \le t < \pi/2$ **39.** $(1, 1)$

Section 11.2, pp. 643–645

1. $y = -x + 2\sqrt{2}$, $\dfrac{d^2 y}{dx^2} = -\sqrt{2}$

3. $y = -\dfrac{1}{2}x + 2\sqrt{2}$, $\dfrac{d^2 y}{dx^2} = -\dfrac{\sqrt{2}}{4}$

5. $y = x + \dfrac{1}{4}$, $\dfrac{d^2 y}{dx^2} = -2$ **7.** $y = 2x - \sqrt{3}$, $\dfrac{d^2 y}{dx^2} = -3\sqrt{3}$

9. $y = x - 4$, $\dfrac{d^2 y}{dx^2} = \dfrac{1}{2}$

11. $y = \sqrt{3}x - \dfrac{\pi\sqrt{3}}{3} + 2$, $\dfrac{d^2 y}{dx^2} = -4$

13. $y = 9x - 1$, $\dfrac{d^2 y}{dx^2} = 108$ **15.** $-\dfrac{3}{16}$ **17.** -6

19. 1 **21.** $3a^2\pi$ **23.** $ab\pi$ **25.** 4 **27.** 12

29. π^2 **31.** $8\pi^2$ **33.** $\dfrac{52\pi}{3}$ **35.** $3\pi\sqrt{5}$

37. $(\bar{x}, \bar{y}) = \left(\dfrac{12}{\pi} - \dfrac{24}{\pi^2}, \dfrac{24}{\pi^2} - 2\right)$

39. $(\bar{x}, \bar{y}) = \left(\dfrac{1}{3}, \pi - \dfrac{4}{3}\right)$ **41. (a)** π **(b)** π

43. (a) $x = 1$, $y = 0$, $\dfrac{dy}{dx} = \dfrac{1}{2}$ **(b)** $x = 0$, $y = 3$, $\dfrac{dy}{dx} = 0$

  (c) $x = \dfrac{\sqrt{3} - 1}{2}$, $y = \dfrac{3 - \sqrt{3}}{2}$, $\dfrac{dy}{dx} = \dfrac{2\sqrt{3} - 1}{\sqrt{3} - 2}$

45. $\left(\dfrac{\sqrt{2}}{2}, 1\right)$, $y = 2x$ at $t = 0$, $y = -2x$ at $t = \pi$

47. (a) $8a$ **(b)** $\dfrac{64\pi}{3}$

Section 11.3, pp. 648–649

1. a, e; b, g; c, h; d, f   **3.**

(a) $\left(2, \dfrac{\pi}{2} + 2n\pi\right)$ and $\left(-2, \dfrac{\pi}{2} + (2n + 1)\pi\right)$, n an integer
(b) $(2, 2n\pi)$ and $(-2, (2n + 1)\pi)$, n an integer
(c) $\left(2, \dfrac{3\pi}{2} + 2n\pi\right)$ and $\left(-2, \dfrac{3\pi}{2} + (2n + 1)\pi\right)$,
  n an integer
(d) $(2, (2n + 1)\pi)$ and $(-2, 2n\pi)$, n an integer
5. (a) $(3, 0)$ **(b)** $(-3, 0)$ **(c)** $\left(-1, \sqrt{3}\right)$ **(d)** $\left(1, \sqrt{3}\right)$
  (e) $(3, 0)$ **(f)** $\left(1, \sqrt{3}\right)$ **(g)** $(-3, 0)$ **(h)** $\left(-1, \sqrt{3}\right)$
7. (a) $\left(\sqrt{2}, \dfrac{\pi}{4}\right)$ **(b)** $(3, \pi)$

  (c) $\left(2, \dfrac{11\pi}{6}\right)$ **(d)** $\left(5, \pi - \tan^{-1}\dfrac{4}{3}\right)$

9. (a) $\left(-3\sqrt{2}, \dfrac{5\pi}{4}\right)$ **(b)** $(-1, 0)$

  (c) $\left(-2, \dfrac{5\pi}{3}\right)$ **(d)** $\left(-5, \pi - \tan^{-1}\dfrac{3}{4}\right)$

11.

13.

15.

17.

19.

21.

23.

25.

27. $x = 2$, vertical line through $(2, 0)$ **29.** $y = 0$, the x-axis
31. $y = 4$, horizontal line through $(0, 4)$
33. $x + y = 1$, line, $m = -1$, $b = 1$
35. $x^2 + y^2 = 1$, circle, $C(0, 0)$, radius 1
37. $y - 2x = 5$, line, $m = 2$, $b = 5$
39. $y^2 = x$, parabola, vertex $(0, 0)$, opens right
41. $y = e^x$, graph of natural exponential function
43. $x + y = \pm 1$, two straight lines of slope -1, y-intercepts $b = \pm 1$
45. $(x + 2)^2 + y^2 = 4$, circle, $C(-2, 0)$, radius 2
47. $x^2 + (y - 4)^2 = 16$, circle, $C(0, 4)$, radius 4
49. $(x - 1)^2 + (y - 1)^2 = 2$, circle, $C(1, 1)$, radius $\sqrt{2}$
51. $\sqrt{3}y + x = 4$ **53.** $r \cos \theta = 7$ **55.** $\theta = \pi/4$
57. $r = 2$ or $r = -2$ **59.** $4r^2 \cos^2 \theta + 9r^2 \sin^2 \theta = 36$
61. $r \sin^2 \theta = 4 \cos \theta$ **63.** $r = 4 \sin \theta$
65. $r^2 = 6r \cos \theta - 2r \sin \theta - 6$ **67.** $(0, \theta)$, where θ is any angle

Section 11.4, pp. 652–653

1. x-axis

3. y-axis

5. y-axis

7. x-axis, y-axis, origin

9. x-axis, y-axis, origin

11. y-axis, x-axis, origin

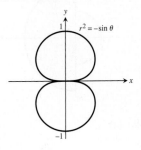

13. x-axis, y-axis, origin **15.** Origin
17. The slope at $(-1, \pi/2)$ is -1, at $(-1, -\pi/2)$ is 1.

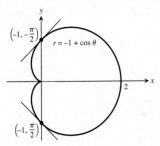

19. The slope at $(1, \pi/4)$ is -1, at $(-1, -\pi/4)$ is 1, at $(-1, 3\pi/4)$ is 1, at $(1, -3\pi/4)$ is -1.

21. (a)   **(b)**

23. (a)

(b)

25.

27.

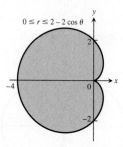

29. (a)

Section 11.5, pp. 656–657

1. $\dfrac{1}{6}\pi^3$ **3.** 18π **5.** $\dfrac{\pi}{8}$ **7.** 2 **9.** $\dfrac{\pi}{2} - 1$ **11.** $5\pi - 8$

13. $3\sqrt{3} - \pi$ **15.** $\dfrac{\pi}{3} + \dfrac{\sqrt{3}}{2}$ **17.** $\dfrac{8\pi}{3} + \sqrt{3}$

19. (a) $\dfrac{3}{2} - \dfrac{\pi}{4}$ **21.** $19/3$ **23.** 8

25. $3\left(\sqrt{2} + \ln\left(1 + \sqrt{2}\right)\right)$ **27.** $\dfrac{\pi}{8} + \dfrac{3}{8}$

Section 11.6, pp. 663–666

1. $y^2 = 8x$, $F(2, 0)$, directrix: $x = -2$

3. $x^2 = -6y$, $F(0, -3/2)$, directrix: $y = 3/2$

5. $\dfrac{x^2}{4} - \dfrac{y^2}{9} = 1$, $F\left(\pm\sqrt{13}, 0\right)$, $V(\pm 2, 0)$,

    asymptotes: $y = \pm\dfrac{3}{2}x$

7. $\dfrac{x^2}{2} + y^2 = 1$, $F(\pm 1, 0)$, $V\left(\pm\sqrt{2}, 0\right)$

9.

11.

13.

15.

17.

19.

21.

23.

25. $\dfrac{x^2}{4} + \dfrac{y^2}{2} = 1$

27. Asymptotes: $y = \pm x$

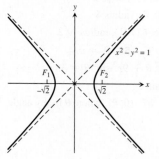

29. Asymptotes: $y = \pm x$

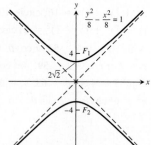

31. Asymptotes: $y = \pm 2x$

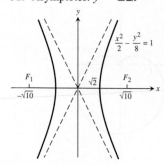

33. Asymptotes: $y = \pm x/2$

35. $y^2 - x^2 = 1$ **37.** $\dfrac{x^2}{9} - \dfrac{y^2}{16} = 1$

39. **(a)** Vertex: $(1, -2)$; focus: $(3, -2)$; directrix: $x = -1$
(b)

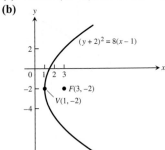

41. **(a)** Foci: $\left(4 \pm \sqrt{7}, 3\right)$; vertices: $(8, 3)$ and $(0, 3)$; center: $(4, 3)$
(b)

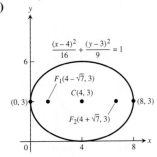

43. **(a)** Center: $(2, 0)$; foci: $(7, 0)$ and $(-3, 0)$; vertices: $(6, 0)$ and $(-2, 0)$; asymptotes: $y = \pm\dfrac{3}{4}(x - 2)$

(b)

45. $(y + 3)^2 = 4(x + 2)$, $V(-2, -3)$, $F(-1, -3)$, directrix: $x = -3$

47. $(x - 1)^2 = 8(y + 7)$, $V(1, -7)$, $F(1, -5)$, directrix: $y = -9$

49. $\dfrac{(x + 2)^2}{6} + \dfrac{(y + 1)^2}{9} = 1$, $F\left(-2, \pm\sqrt{3} - 1\right)$, $V(-2, \pm 3 - 1)$, $C(-2, -1)$

51. $\dfrac{(x - 2)^2}{3} + \dfrac{(y - 3)^2}{2} = 1$, $F(3, 3)$ and $F(1, 3)$, $V\left(\pm\sqrt{3} + 2, 3\right)$, $C(2, 3)$

53. $\dfrac{(x - 2)^2}{4} - \dfrac{(y - 2)^2}{5} = 1$, $C(2, 2)$, $F(5, 2)$ and $F(-1, 2)$, $V(4, 2)$ and $V(0, 2)$; asymptotes: $(y - 2) = \pm\dfrac{\sqrt{5}}{2}(x - 2)$

55. $(y + 1)^2 - (x + 1)^2 = 1$, $C(-1, -1)$, $F\left(-1, \sqrt{2} - 1\right)$ and $F\left(-1, -\sqrt{2} - 1\right)$, $V(-1, 0)$ and $V(-1, -2)$; asymptotes $(y + 1) = \pm(x + 1)$

57. $C(-2, 0)$, $a = 4$ **59.** $V(-1, 1)$, $F(-1, 0)$

61. Ellipse: $\dfrac{(x + 2)^2}{5} + y^2 = 1$, $C(-2, 0)$, $F(0, 0)$ and $F(-4, 0)$, $V\left(\sqrt{5} - 2, 0\right)$ and $V\left(-\sqrt{5} - 2, 0\right)$

63. Ellipse: $\dfrac{(x - 1)^2}{2} + (y - 1)^2 = 1$, $C(1, 1)$, $F(2, 1)$ and $F(0, 1)$, $V\left(\sqrt{2} + 1, 1\right)$ and $V\left(-\sqrt{2} + 1, 1\right)$

65. Hyperbola: $(x - 1)^2 - (y - 2)^2 = 1$, $C(1, 2)$, $F\left(1 + \sqrt{2}, 2\right)$ and $F\left(1 - \sqrt{2}, 2\right)$, $V(2, 2)$ and $V(0, 2)$;   asymptotes: $(y - 2) = \pm(x - 1)$

67. Hyperbola: $\dfrac{(y - 3)^2}{6} - \dfrac{x^2}{3} = 1$, $C(0, 3)$, $F(0, 6)$ and $F(0, 0)$, $V\left(0, \sqrt{6} + 3\right)$ and $V\left(0, -\sqrt{6} + 3\right)$; asymptotes: $y = \sqrt{2}x + 3$ or $y = -\sqrt{2}x + 3$

69. **(b)** $1:1$ **73.** Length $= 2\sqrt{2}$, width $= \sqrt{2}$, area $= 4$ **75.** 24π
77. $x = 0, y = 0: y = -2x; x = 0, y = 2: y = 2x + 2$; $x = 4, y = 0: y = 2x - 8$

79. $\bar{x} = 0$, $\bar{y} = \dfrac{16}{3\pi}$

Section 11.7, pp. 671–672

1. $e = \dfrac{3}{5}$, $F(\pm 3, 0)$; directrices are $x = \pm\dfrac{25}{3}$.

3. $e = \dfrac{1}{\sqrt{2}}$; $F(0, \pm 1)$; directrices are $y = \pm 2$.

5. $e = \dfrac{1}{\sqrt{3}}$; $F(0, \pm 1)$; directrices are $y = \pm 3$.

7. $e = \dfrac{\sqrt{3}}{3}$; $F\left(\pm\sqrt{3}, 0\right)$; directrices are $x = \pm 3\sqrt{3}$.

9. $\dfrac{x^2}{27} + \dfrac{y^2}{36} = 1$ **11.** $\dfrac{x^2}{4851} + \dfrac{y^2}{4900} = 1$

13. $\dfrac{x^2}{9} + \dfrac{y^2}{4} = 1$ **15.** $\dfrac{x^2}{64} + \dfrac{y^2}{48} = 1$

17. $e = \sqrt{2}$; $F\left(\pm\sqrt{2}, 0\right)$;

directrices are $x = \pm\dfrac{1}{\sqrt{2}}$.

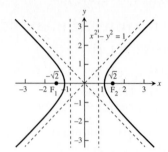

19. $e = \sqrt{2}$; $F(0, \pm4)$;

directrices are $y = \pm2$.

45. $y = 2 - x$

47. $y = \dfrac{\sqrt{3}}{3}x + 2\sqrt{3}$

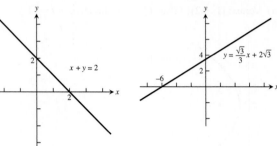

21. $e = \sqrt{5}$; $F\left(\pm\sqrt{10}, 0\right)$;

directrices are $x = \pm\dfrac{2}{\sqrt{10}}$.

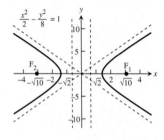

23. $e = \sqrt{5}$; $F\left(0, \pm\sqrt{10}\right)$;

directrices are $y = \pm\dfrac{2}{\sqrt{10}}$.

49. $r\cos\left(\theta - \dfrac{\pi}{4}\right) = 3$ **51.** $r\cos\left(\theta + \dfrac{\pi}{2}\right) = 5$

53.

55.

57. $r = 12\cos\theta$

59. $r = 10\sin\theta$

25. $y^2 - \dfrac{x^2}{8} = 1$ **27.** $x^2 - \dfrac{y^2}{8} = 1$ **29.** $r = \dfrac{2}{1 + \cos\theta}$

31. $r = \dfrac{30}{1 - 5\sin\theta}$ **33.** $r = \dfrac{1}{2 + \cos\theta}$ **35.** $r = \dfrac{10}{5 - \sin\theta}$

37.

39.

61. $r = -2\cos\theta$

63. $r = -\sin\theta$

41.

43.

65.

67.

69.

$$r = \frac{8}{4 + \cos \theta}$$

71.

$$r = \frac{1}{1 - \sin \theta}$$

73.

$$r = \frac{1}{1 + 2 \sin \theta}$$

75. (b)

Planet	Perihelion	Aphelion
Mercury	0.3075 AU	0.4667 AU
Venus	0.7184 AU	0.7282 AU
Earth	0.9833 AU	1.0167 AU
Mars	1.3817 AU	1.6663 AU
Jupiter	4.9512 AU	5.4548 AU
Saturn	9.0210 AU	10.0570 AU
Uranus	18.2977 AU	20.0623 AU
Neptune	29.8135 AU	30.3065 AU

Practice Exercises, pp.673–675

1.

3.

5.

7. $x = 3 \cos t, \quad y = 4 \sin t, \quad 0 \le t \le 2\pi$

9. $y = \frac{\sqrt{3}}{2} x + \frac{1}{4}, \frac{1}{4}$

11. (a) $y = \frac{\pm |x|^{3/2}}{8} - 1$ **(b)** $y = \frac{\pm \sqrt{1 - x^2}}{x}$

13. $\frac{10}{3}$ **15.** $\frac{285}{8}$ **17.** 10 **19.** $\frac{9\pi}{2}$ **21.** $\frac{76\pi}{3}$

23. $y = \frac{\sqrt{3}}{3} x - 4$ **25.** $x = 2$

27. $y = -\frac{3}{2}$

29. $x^2 + (y + 2)^2 = 4$

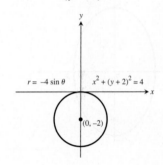

31. $\left(x - \sqrt{2} \right)^2 + y^2 = 2$

33. $r = -5 \sin \theta$

35. $r = 3 \cos \theta$

37.

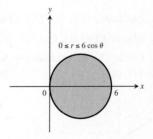

39. d **41.** l **43.** k **45.** i **47.** $\frac{9}{2} \pi$ **49.** $2 + \frac{\pi}{4}$

51. 8 **53.** $\pi - 3$

55. Focus is $(0, -1)$,
      directrix is $y = 1$.

57. Focus is $\left(\dfrac{3}{4}, 0\right)$,
      directrix is $x = -\dfrac{3}{4}$.

81. $r = \dfrac{4}{1 + 2\cos\theta}$ **83.** $r = \dfrac{2}{2 + \sin\theta}$

85. (a) 24π **(b)** 16π

Additional and Advanced Exercises, pp. 675–676

1. $x - \dfrac{7}{2} = \dfrac{y^2}{2}$

59. $e = \dfrac{3}{4}$

61. $e = 2$; the asymptotes
      are $y = \pm\sqrt{3}\,x$.

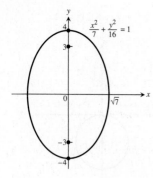

3. $3x^2 + 3y^2 - 8y + 4 = 0$ **5.** $F(0, \pm 1)$

7. (a) $\dfrac{(y-1)^2}{16} - \dfrac{x^2}{48} = 1$ **(b)** $\dfrac{\left(y + \dfrac{3}{4}\right)^2}{\left(\dfrac{25}{16}\right)} - \dfrac{x^2}{\left(\dfrac{75}{2}\right)} = 1$

11.

13.

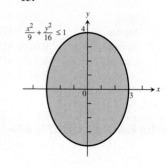

63. $(x - 2)^2 = -12(y - 3)$, $V(2, 3)$, $F(2, 0)$, directrix is $y = 6$.

65. $\dfrac{(x+3)^2}{9} + \dfrac{(y+5)^2}{25} = 1$, $C(-3, -5)$, $F(-3, -1)$ and
      $F(-3, -9)$, $V(-3, -10)$ and $V(-3, 0)$.

67. $\dfrac{\left(y - 2\sqrt{2}\right)^2}{8} - \dfrac{(x-2)^2}{2} = 1$, $C\left(2, 2\sqrt{2}\right)$,
      $F\left(2, 2\sqrt{2} \pm \sqrt{10}\right)$, $V\left(2, 4\sqrt{2}\right)$ and $V(2, 0)$, the asymptotes
      are $y = 2x - 4 + 2\sqrt{2}$ and $y = -2x + 4 + 2\sqrt{2}$.

69. Hyperbola: $C(2, 0)$, $V(0, 0)$ and $V(4, 0)$, the foci are
      $F\left(2 \pm \sqrt{5}, 0\right)$ and the asymptotes are $y = \pm\dfrac{x - 2}{2}$.

71. Parabola: $V(-3, 1)$, $F(-7, 1)$ and the directrix is $x = 1$.

73. Ellipse: $C(-3, 2)$, $F\left(-3 \pm \sqrt{7}, 2\right)$, $V(1, 2)$ and $V(-7, 2)$

75. Circle: $C(1, 1)$ and radius $= \sqrt{2}$

77. $V(1, 0)$

79. $V(2, \pi)$ and $V(6, \pi)$

15.

17. (a) $r = e^{2\theta}$ **(b)** $\dfrac{\sqrt{5}}{2}\left(e^{4\pi} - 1\right)$

19. $r = \dfrac{4}{1 + 2\cos\theta}$ **21.** $r = \dfrac{2}{2 + \sin\theta}$

23. $x = (a + b)\cos\theta - b\cos\left(\dfrac{a+b}{b}\theta\right)$,
      $y = (a + b)\sin\theta - b\sin\left(\dfrac{a+b}{b}\theta\right)$

27. $\dfrac{\pi}{2}$

CHAPTER 12

Section 12.1, pp. 681–682

1. The line through the point $(2, 3, 0)$ parallel to the z-axis
3. The x-axis   5. The circle $x^2 + y^2 = 4$ in the xy-plane
7. The circle $x^2 + z^2 = 4$ in the xz-plane
9. The circle $y^2 + z^2 = 1$ in the yz-plane
11. The circle $x^2 + y^2 = 16$ in the xy-plane
13. The ellipse formed by the intersection of the cylinder $x^2 + y^2 = 4$ and the plane $z = y$
15. The parabola $y = x^2$ in the xy-plane
17. (a) The first quadrant of the xy-plane
    (b) The fourth quadrant of the xy-plane
19. (a) The ball of radius 1 centered at the origin
    (b) All points more than 1 unit from the origin
21. (a) The ball of radius 2 centered at the origin with the interior of the ball of radius 1 centered at the origin removed
    (b) The solid upper hemisphere of radius 1 centered at the origin
23. (a) The region on or inside the parabola $y = x^2$ in the xy-plane and all points above this region
    (b) The region on or to the left of the parabola $x = y^2$ in the xy-plane and all points above it that are 2 units or less away from the xy-plane
25. (a) $x = 3$ (b) $y = -1$ (c) $z = -2$
27. (a) $z = 1$ (b) $x = 3$ (c) $y = -1$
29. (a) $x^2 + (y - 2)^2 = 4, z = 0$
    (b) $(y - 2)^2 + z^2 = 4, x = 0$ (c) $x^2 + z^2 = 4, y = 2$
31. (a) $y = 3, z = -1$ (b) $x = 1, z = -1$ (c) $x = 1, y = 3$
33. $x^2 + y^2 + z^2 = 25, z = 3$ 35. $0 \le z \le 1$ 37. $z \le 0$
39. (a) $(x - 1)^2 + (y - 1)^2 + (z - 1)^2 < 1$
    (b) $(x - 1)^2 + (y - 1)^2 + (z - 1)^2 > 1$
41. 3   43. 7   45. $2\sqrt{3}$ 47. $C(-2, 0, 2), a = 2\sqrt{2}$
49. $C(\sqrt{2}, \sqrt{2}, -\sqrt{2}), a = \sqrt{2}$
51. $(x - 1)^2 + (y - 2)^2 + (z - 3)^2 = 14$
53. $(x + 1)^2 + \left(y - \dfrac{1}{2}\right)^2 + \left(z + \dfrac{2}{3}\right)^2 = \dfrac{16}{81}$
55. $C(-2, 0, 2), a = \sqrt{8}$ 57. $C\left(-\dfrac{1}{4}, -\dfrac{1}{4}, -\dfrac{1}{4}\right), a = \dfrac{5\sqrt{3}}{4}$
59. (a) $\sqrt{y^2 + z^2}$ (b) $\sqrt{x^2 + z^2}$ (c) $\sqrt{x^2 + y^2}$
61. $\sqrt{17} + \sqrt{33} + 6$ 63. $y = 1$
65. (a) $(0, 3, -3)$ (b) $(0, 5, -5)$

Section 12.2, pp. 690–692

1. (a) $\langle 9, -6 \rangle$ (b) $3\sqrt{13}$ 3. (a) $\langle 1, 3 \rangle$ (b) $\sqrt{10}$
5. (a) $\langle 12, -19 \rangle$ (b) $\sqrt{505}$ 7. (a) $\left\langle \dfrac{1}{5}, \dfrac{14}{5} \right\rangle$ (b) $\dfrac{\sqrt{197}}{5}$
9. $\langle 1, -4 \rangle$ 11. $\langle -2, -3 \rangle$ 13. $\left\langle -\dfrac{1}{2}, \dfrac{\sqrt{3}}{2} \right\rangle$
15. $\left\langle -\dfrac{\sqrt{3}}{2}, -\dfrac{1}{2} \right\rangle$ 17. $-3\mathbf{i} + 2\mathbf{j} - \mathbf{k}$ 19. $-3\mathbf{i} + 16\mathbf{j}$
21. $3\mathbf{i} + 5\mathbf{j} - 8\mathbf{k}$

23. The vector $\mathbf{v}$ is horizontal and 1 in. long. The vectors $\mathbf{u}$ and $\mathbf{w}$ are $\dfrac{11}{16}$ in. long. $\mathbf{w}$ is vertical and $\mathbf{u}$ makes a 45° angle with the horizontal. All vectors must be drawn to scale.

(a)   (b)

(c)   (d)

25. $3\left(\dfrac{2}{3}\mathbf{i} + \dfrac{1}{3}\mathbf{j} - \dfrac{2}{3}\mathbf{k}\right)$ 27. $5(\mathbf{k})$
29. $\sqrt{\dfrac{1}{2}}\left(\dfrac{1}{\sqrt{3}}\mathbf{i} - \dfrac{1}{\sqrt{3}}\mathbf{j} - \dfrac{1}{\sqrt{3}}\mathbf{k}\right)$
31. (a) $2\mathbf{i}$ (b) $-\sqrt{3}\mathbf{k}$ (c) $\dfrac{3}{10}\mathbf{j} + \dfrac{2}{5}\mathbf{k}$ (d) $6\mathbf{i} - 2\mathbf{j} + 3\mathbf{k}$
33. $\dfrac{7}{13}(12\mathbf{i} - 5\mathbf{k})$
35. (a) $\dfrac{3}{5\sqrt{2}}\mathbf{i} + \dfrac{4}{5\sqrt{2}}\mathbf{j} - \dfrac{1}{\sqrt{2}}\mathbf{k}$ (b) $(1/2, 3, 5/2)$
37. (a) $-\dfrac{1}{\sqrt{3}}\mathbf{i} - \dfrac{1}{\sqrt{3}}\mathbf{j} - \dfrac{1}{\sqrt{3}}\mathbf{k}$ (b) $\left(\dfrac{5}{2}, \dfrac{7}{2}, \dfrac{9}{2}\right)$
39. $A(4, -3, 5)$ 41. $a = \dfrac{3}{2}, b = \dfrac{1}{2}$
43. $\approx \langle -338.095, 725.046 \rangle$
45. $|\mathbf{F}_1| = \dfrac{100 \cos 45°}{\sin 75°} \approx 73.205$ N,
    $|\mathbf{F}_2| = \dfrac{100 \cos 30°}{\sin 75°} \approx 89.658$ N,
    $\mathbf{F}_1 = \langle -|\mathbf{F}_1|\cos 30°, |\mathbf{F}_1|\sin 30° \rangle \approx \langle -63.397, 36.603 \rangle$,
    $\mathbf{F}_2 = \langle |\mathbf{F}_2|\cos 45°, |\mathbf{F}_2|\sin 45° \rangle \approx \langle 63.397, 63.397 \rangle$
47. $w = \dfrac{100 \sin 75°}{\cos 40°} \approx 126.093$ N,
    $|\mathbf{F}_1| = \dfrac{w \cos 35°}{\sin 75°} \approx 106.933$ N
49. (a) $(5 \cos 60°, 5 \sin 60°) = \left(\dfrac{5}{2}, \dfrac{5\sqrt{3}}{2}\right)$
    (b) $(5 \cos 60° + 10 \cos 315°, 5 \sin 60° + 10 \sin 315°) =$
    $\left(\dfrac{5 + 10\sqrt{2}}{2}, \dfrac{5\sqrt{3} - 10\sqrt{2}}{2}\right)$
51. (a) $\dfrac{3}{2}\mathbf{i} + \dfrac{3}{2}\mathbf{j} - 3\mathbf{k}$ (b) $\mathbf{i} + \mathbf{j} - 2\mathbf{k}$ (c) $(2, 2, 1)$

Section 12.3, pp. 698–700

1. (a) $-25, 5, 5$ (b) -1 (c) -5 (d) $-2\mathbf{i} + 4\mathbf{j} - \sqrt{5}\mathbf{k}$
3. (a) $25, 15, 5$ (b) $\dfrac{1}{3}$ (c) $\dfrac{5}{3}$ (d) $\dfrac{1}{9}(10\mathbf{i} + 11\mathbf{j} - 2\mathbf{k})$
5. (a) $2, \sqrt{34}, \sqrt{3}$ (b) $\dfrac{2}{\sqrt{3}\sqrt{34}}$ (c) $\dfrac{2}{\sqrt{34}}$
    (d) $\dfrac{1}{17}(5\mathbf{j} - 3\mathbf{k})$

7. (a) $10 + \sqrt{17}, \sqrt{26}, \sqrt{21}$ **(b)** $\dfrac{10 + \sqrt{17}}{\sqrt{546}}$

(c) $\dfrac{10 + \sqrt{17}}{\sqrt{26}}$ **(d)** $\dfrac{10 + \sqrt{17}}{26}(5\mathbf{i} + \mathbf{j})$

9. 0.75 rad **11.** 1.77 rad

13. Angle at $A = \cos^{-1}\left(\dfrac{1}{\sqrt{5}}\right) \approx 63.435$ degrees, angle at

$B = \cos^{-1}\left(\dfrac{3}{5}\right) \approx 53.130$ degrees, angle at

$C = \cos^{-1}\left(\dfrac{1}{\sqrt{5}}\right) \approx 63.435$ degrees.

23. Horizontal component: ≈ 1188 ft/sec, vertical component: ≈ 167 ft/sec

25. (a) Since $|\cos\theta| \le 1$, we have $|\mathbf{u}\cdot\mathbf{v}| = |\mathbf{u}||\mathbf{v}||\cos\theta| \le |\mathbf{u}||\mathbf{v}|(1) = |\mathbf{u}||\mathbf{v}|$.

(b) We have equality precisely when $|\cos\theta| = 1$ or when one or both of $\mathbf{u}$ and $\mathbf{v}$ are $\mathbf{0}$. In the case of nonzero vectors, we have equality when $\theta = 0$ or π, that is, when the vectors are parallel.

27. a

33. $x + 2y = 4$

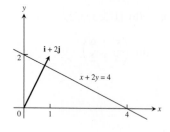

35. $-2x + y = -3$

37. $x + y = -1$

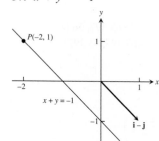

39. $2x - y = 0$

41. 5 J **43.** 3464 J **45.** $\dfrac{\pi}{4}$ **47.** $\dfrac{\pi}{6}$ **49.** 0.14

Section 12.4, pp. 704–706

1. $|\mathbf{u} \times \mathbf{v}| = 3$, direction is $\dfrac{2}{3}\mathbf{i} + \dfrac{1}{3}\mathbf{j} + \dfrac{2}{3}\mathbf{k}$; $|\mathbf{v} \times \mathbf{u}| = 3$, direction is $-\dfrac{2}{3}\mathbf{i} - \dfrac{1}{3}\mathbf{j} - \dfrac{2}{3}\mathbf{k}$

3. $|\mathbf{u} \times \mathbf{v}| = 0$, no direction; $|\mathbf{v} \times \mathbf{u}| = 0$, no direction

5. $|\mathbf{u} \times \mathbf{v}| = 6$, direction is $-\mathbf{k}$; $|\mathbf{v} \times \mathbf{u}| = 6$, direction is $\mathbf{k}$

7. $|\mathbf{u} \times \mathbf{v}| = 6\sqrt{5}$, direction is $\dfrac{1}{\sqrt{5}}\mathbf{i} - \dfrac{2}{\sqrt{5}}\mathbf{k}$; $|\mathbf{v} \times \mathbf{u}| = 6\sqrt{5}$, direction is $-\dfrac{1}{\sqrt{5}}\mathbf{i} + \dfrac{2}{\sqrt{5}}\mathbf{k}$

9.

11.

13.

15. (a) $2\sqrt{6}$ **(b)** $\pm\dfrac{1}{\sqrt{6}}(2\mathbf{i} + \mathbf{j} + \mathbf{k})$

17. (a) $\dfrac{\sqrt{2}}{2}$ **(b)** $\pm\dfrac{1}{\sqrt{2}}(\mathbf{i} - \mathbf{j})$

19. 8 **21.** 7 **23. (a)** None **(b)** $\mathbf{u}$ and $\mathbf{w}$ **25.** $10\sqrt{3}$ ft-lb

27. (a) True **(b)** Not always true **(c)** True **(d)** True **(e)** Not always true **(f)** True **(g)** True **(h)** True

29. (a) $\mathrm{proj}_\mathbf{v}\,\mathbf{u} = \dfrac{\mathbf{u}\cdot\mathbf{v}}{\mathbf{v}\cdot\mathbf{v}}\mathbf{v}$ **(b)** $\pm\mathbf{u} \times \mathbf{v}$ **(c)** $\pm(\mathbf{u} \times \mathbf{v}) \times \mathbf{w}$ **(d)** $|(\mathbf{u} \times \mathbf{v})\cdot\mathbf{w}|$ **(e)** $(\mathbf{u} \times \mathbf{v}) \times (\mathbf{u} \times \mathbf{w})$ **(f)** $|\mathbf{u}|\dfrac{\mathbf{v}}{|\mathbf{v}|}$

31. (a) Yes **(b)** No **(c)** Yes **(d)** No

33. No, $\mathbf{v}$ need not equal $\mathbf{w}$. For example, $\mathbf{i} + \mathbf{j} \ne -\mathbf{i} + \mathbf{j}$, but $\mathbf{i} \times (\mathbf{i} + \mathbf{j}) = \mathbf{i} \times \mathbf{i} + \mathbf{i} \times \mathbf{j} = \mathbf{0} + \mathbf{k} = \mathbf{k}$ and $\mathbf{i} \times (-\mathbf{i} + \mathbf{j}) = -\mathbf{i} \times \mathbf{i} + \mathbf{i} \times \mathbf{j} = \mathbf{0} + \mathbf{k} = \mathbf{k}$.

35. 2 **37.** 13 **39.** $\sqrt{129}$ **41.** $\dfrac{11}{2}$ **43.** $\dfrac{25}{2}$

45. $\dfrac{3}{2}$ **47.** $\dfrac{\sqrt{21}}{2}$

49. If $\mathbf{A} = a_1\mathbf{i} + a_2\mathbf{j}$ and $\mathbf{B} = b_1\mathbf{i} + b_2\mathbf{j}$, then

$$\mathbf{A} \times \mathbf{B} = \begin{vmatrix} \mathbf{i} & \mathbf{j} & \mathbf{k} \\ a_1 & a_2 & 0 \\ b_1 & b_2 & 0 \end{vmatrix} = \begin{vmatrix} a_1 & a_2 \\ b_1 & b_2 \end{vmatrix}\mathbf{k}$$

and the triangle's area is

$$\frac{1}{2}\left|\mathbf{A} \times \mathbf{B}\right| = \pm\frac{1}{2}\begin{vmatrix} a_1 & a_2 \\ b_1 & b_2 \end{vmatrix}.$$

The applicable sign is $(+)$ if the acute angle from $\mathbf{A}$ to $\mathbf{B}$ runs counterclockwise in the xy-plane, and $(-)$ if it runs clockwise.

Section 12.5, pp. 712–714

1. $x = 3 + t, \quad y = -4 + t, \quad z = -1 + t$

3. $x = -2 + 5t, \quad y = 5t, \quad z = 3 - 5t$

5. $x = 0, \quad y = 2t, \quad z = t$

7. $x = 1, \quad y = 1, \quad z = 1 + t$

9. $x = t, \quad y = -7 + 2t, \quad z = 2t$

11. $x = t, \quad y = 0, \quad z = 0$

13. $x = t, \quad y = t, \quad z = \dfrac{3}{2}t,$
$\quad 0 \le t \le 1$

15. $x = 1, \quad y = 1 + t,$
$\quad z = 0, \quad -1 \le t \le 0$

17. $x = 0, \quad y = 1 - 2t,$
$\quad z = 1, \quad 0 \le t \le 1$

19. $x = 2 - 2t, \quad y = 2t,$
$\quad z = 2 - 2t, \quad 0 \le t \le 1$

21. $3x - 2y - z = -3$ **23.** $7x - 5y - 4z = 6$
25. $x + 3y + 4z = 34$ **27.** $(1, 2, 3), -20x + 12y + z = 7$
29. $y + z = 3$ **31.** $x - y + z = 0$ **33.** $2\sqrt{30}$ **35.** 0
37. $\dfrac{9\sqrt{42}}{7}$ **39.** 3 **41.** $19/5$ **43.** $5/3$ **45.** $9/\sqrt{41}$
47. $\pi/4$ **49.** 1.38 rad   **51.** 0.82 rad   **53.** $\left(\dfrac{3}{2}, -\dfrac{3}{2}, \dfrac{1}{2}\right)$
55. $(1, 1, 0)$ **57.** $x = 1 - t, \quad y = 1 + t, \quad z = -1$
59. $x = 4, \quad y = 3 + 6t, \quad z = 1 + 3t$
61. $L1$ intersects $L2$; $L2$ is parallel to $L3$; $L1$ and $L3$ are skew.
63. $x = 2 + 2t, \quad y = -4 - t, \quad z = 7 + 3t; \quad x = -2 - t,$
$\quad y = -2 + (1/2)t, \quad z = 1 - (3/2)t$
65. $\left(0, -\dfrac{1}{2}, -\dfrac{3}{2}\right), (-1, 0, -3), (1, -1, 0)$
69. Many possible answers. One possibility: $x + y = 3$ and
$\quad 2y + z = 7$.
71. $(x/a) + (y/b) + (z/c) = 1$ describes all planes *except* those
through the origin or parallel to a coordinate axis.

Section 12.6, pp. 718–719

1. (d), ellipsoid   **3.** (a), cylinder   **5.** (l), hyperbolic paraboloid
7. (b), cylinder   **9.** (k), hyperbolic paraboloid   **11.** (h), cone
13.   **15.**

17.

19.

21.

23.

25.

27.

29.

31.

33.

35.

37.

$x^2 + y^2 - z^2 = 4$

39.

$x^2 + z^2 = 1$

41.

$z = -(x^2 + y^2)$

43.

$4y^2 + z^2 - 4x^2 = 4$

45. (a) $\dfrac{2\pi(9 - c^2)}{9}$ **(b)** 8π **(c)** $\dfrac{4\pi abc}{3}$

Practice Exercises, pp. 720–721

1. (a) $\langle -17, 32 \rangle$ **(b)** $\sqrt{1313}$

3. (a) $\langle 6, -8 \rangle$ **(b)** 10

5. $\left\langle -\dfrac{\sqrt{3}}{2}, -\dfrac{1}{2} \right\rangle$ [assuming counterclockwise]

7. $\left\langle \dfrac{8}{\sqrt{17}}, -\dfrac{2}{\sqrt{17}} \right\rangle$

9. Length $= 2$, direction is $\dfrac{1}{\sqrt{2}}\mathbf{i} + \dfrac{1}{\sqrt{2}}\mathbf{j}$.

11. $\mathbf{v}(\pi/2) = 2(-\mathbf{i})$

13. Length $= 7$, direction is $\dfrac{2}{7}\mathbf{i} - \dfrac{3}{7}\mathbf{j} + \dfrac{6}{7}\mathbf{k}$.

15. $\dfrac{8}{\sqrt{33}}\mathbf{i} - \dfrac{2}{\sqrt{33}}\mathbf{j} + \dfrac{8}{\sqrt{33}}\mathbf{k}$

17. $|\mathbf{v}| = \sqrt{2}, |\mathbf{u}| = 3, \mathbf{v} \cdot \mathbf{u} = \mathbf{u} \cdot \mathbf{v} = 3, \mathbf{v} \times \mathbf{u} = -2\mathbf{i} + 2\mathbf{j} - \mathbf{k}$,

$\mathbf{u} \times \mathbf{v} = 2\mathbf{i} - 2\mathbf{j} + \mathbf{k}, |\mathbf{v} \times \mathbf{u}| = 3, \theta = \cos^{-1}\left(\dfrac{1}{\sqrt{2}}\right) = \dfrac{\pi}{4}$,

$|\mathbf{u}|\cos\theta = \dfrac{3}{\sqrt{2}}, \text{proj}_\mathbf{v}\, \mathbf{u} = \dfrac{3}{2}(\mathbf{i} + \mathbf{j})$

19. $\dfrac{4}{3}(2\mathbf{i} + \mathbf{j} - \mathbf{k})$

21. $\mathbf{u} \times \mathbf{v} = \mathbf{k}$

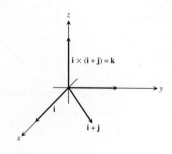

$\mathbf{i} \times (\mathbf{i} + \mathbf{j}) = \mathbf{k}$

23. $2\sqrt{7}$ **25. (a)** $\sqrt{14}$ **(b)** 1 **29.** $\sqrt{78}/3$

31. $x = 1 - 3t, \quad y = 2, \quad z = 3 + 7t$ **33.** $\sqrt{2}$

35. $2x + y + z = 5$

37. $-9x + y + 7z = 4$

39. $\left(0, -\dfrac{1}{2}, -\dfrac{3}{2}\right), (-1, 0, -3), (1, -1, 0)$ **41.** $\pi/3$

43. $x = -5 + 5t, \quad y = 3 - t, \quad z = -3t$

45. (b) $x = -12t, \quad y = 19/12 + 15t, \quad z = 1/6 + 6t$

47. Yes; $\mathbf{v}$ is parallel to the plane.

49. 3 **51.** $-3\mathbf{j} + 3\mathbf{k}$

53. $\dfrac{2}{\sqrt{35}}\left(5\mathbf{i} - \mathbf{j} - 3\mathbf{k}\right)$

55. $\left(\dfrac{11}{9}, \dfrac{26}{9}, -\dfrac{7}{9}\right)$

57. $(1, -2, -1); x = 1 - 5t, \quad y = -2 + 3t, \quad z = -1 + 4t$

59. $2x + 7y + 2z + 10 = 0$

61. (a) No **(b)** No **(c)** No **(d)** No **(e)** Yes

63. $11/\sqrt{107}$

65.

$x^2 + y^2 + z^2 = 4$

67.

$4x^2 + 4y^2 + z^2 = 4$

69.

$z = -(x^2 + y^2)$

71.

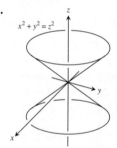

$x^2 + y^2 = z^2$

73.

$x^2 + y^2 - z^2 = 4$

75.

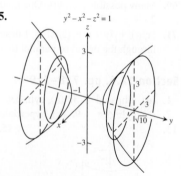

$y^2 - x^2 - z^2 = 1$

Additional and Advanced Exercises, pp. 722–724

1. $(26, 23, -1/3)$ **3.** $|\mathbf{F}| = 20$ lb

5. (a) $|\mathbf{F}_1| = 80$ lb, $|\mathbf{F}_2| = 60$ lb, $\mathbf{F}_1 = \langle -48, 64 \rangle$,

$\mathbf{F}_2 = \langle 48, 36 \rangle, \quad \alpha = \tan^{-1}\dfrac{4}{3}, \quad \beta = \tan^{-1}\dfrac{3}{4}$

(b) $|\mathbf{F}_1| = \dfrac{2400}{13} \approx 184.615$ lb, $|\mathbf{F}_2| = \dfrac{1000}{13} \approx 76.923$ lb,

$\mathbf{F}_1 = \left\langle \dfrac{-12,000}{169}, \dfrac{28,800}{169} \right\rangle \approx \langle -71.006, 170.414 \rangle,$

$\mathbf{F}_2 = \left\langle \dfrac{12,000}{169}, \dfrac{5000}{169} \right\rangle \approx \langle 71.006, 29.586 \rangle,$

$\alpha = \tan^{-1}\dfrac{12}{5}, \quad \beta = \tan^{-1}\dfrac{5}{12}$

9. (a) $\theta = \tan^{-1}\sqrt{2} \approx 54.74°$ **(b)** $\theta = \tan^{-1} 2\sqrt{2} \approx 70.53°$

13. (b) $\dfrac{6}{\sqrt{14}}$ **(c)** $2x - y + 2x = 8$

   (d) $x - 2y + z = 3 + 5\sqrt{6}$ and $x - 2y + z = 3 - 5\sqrt{6}$

15. $\dfrac{32}{41}\mathbf{i} + \dfrac{23}{41}\mathbf{j} - \dfrac{13}{41}\mathbf{k}$

17. (a) $0, 0$ **(b)** $-10\mathbf{i} - 2\mathbf{j} + 6\mathbf{k}, -9\mathbf{i} - 2\mathbf{j} + 7\mathbf{k}$
   (c) $-4\mathbf{i} - 6\mathbf{j} + 2\mathbf{k}, \mathbf{i} - 2\mathbf{j} - 4\mathbf{k}$
   (d) $-10\mathbf{i} - 10\mathbf{k}, -12\mathbf{i} - 4\mathbf{j} - 8\mathbf{k}$

19. The formula is always true.

CHAPTER 13

Section 13.1, pp. 731–733

1. $y = x^2 - 2x$, $\mathbf{v} = \mathbf{i} + 2\mathbf{j}$, $\mathbf{a} = 2\mathbf{j}$

3. $y = \dfrac{2}{9}x^2$, $\mathbf{v} = 3\mathbf{i} + 4\mathbf{j}$, $\mathbf{a} = 3\mathbf{i} + 8\mathbf{j}$

5. $t = \dfrac{\pi}{4}$: $\mathbf{v} = \dfrac{\sqrt{2}}{2}\mathbf{i} - \dfrac{\sqrt{2}}{2}\mathbf{j}$, $\mathbf{a} = \dfrac{-\sqrt{2}}{2}\mathbf{i} - \dfrac{\sqrt{2}}{2}\mathbf{j}$;
   $t = \pi/2$: $\mathbf{v} = -\mathbf{j}$, $\mathbf{a} = -\mathbf{i}$

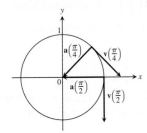

7. $t = \pi$: $\mathbf{v} = 2\mathbf{i}$, $\mathbf{a} = -\mathbf{j}$; $t = \dfrac{3\pi}{2}$: $\mathbf{v} = \mathbf{i} - \mathbf{j}$, $\mathbf{a} = -\mathbf{i}$

9. $\mathbf{v} = \mathbf{i} + 2t\mathbf{j} + 2\mathbf{k}$; $\mathbf{a} = 2\mathbf{j}$; speed: 3; direction: $\dfrac{1}{3}\mathbf{i} + \dfrac{2}{3}\mathbf{j} +$
   $\dfrac{2}{3}\mathbf{k}$; $\mathbf{v}(1) = 3\left(\dfrac{1}{3}\mathbf{i} + \dfrac{2}{3}\mathbf{j} + \dfrac{2}{3}\mathbf{k}\right)$

11. $\mathbf{v} = (-2\sin t)\mathbf{i} + (3\cos t)\mathbf{j} + 4\mathbf{k}$;
   $\mathbf{a} = (-2\cos t)\mathbf{i} - (3\sin t)\mathbf{j}$; speed: $2\sqrt{5}$;
   direction: $\left(-1/\sqrt{5}\right)\mathbf{i} + \left(2/\sqrt{5}\right)\mathbf{k}$;
   $\mathbf{v}(\pi/2) = 2\sqrt{5}\left[\left(-1/\sqrt{5}\right)\mathbf{i} + \left(2/\sqrt{5}\right)\mathbf{k}\right]$

13. $\mathbf{v} = \left(\dfrac{2}{t+1}\right)\mathbf{i} + 2t\mathbf{j} + t\mathbf{k}$; $\mathbf{a} = \left(\dfrac{-2}{(t+1)^2}\right)\mathbf{i} + 2\mathbf{j} + \mathbf{k}$;

   speed: $\sqrt{6}$; direction: $\dfrac{1}{\sqrt{6}}\mathbf{i} + \dfrac{2}{\sqrt{6}}\mathbf{j} + \dfrac{1}{\sqrt{6}}\mathbf{k}$;

   $\mathbf{v}(1) = \sqrt{6}\left(\dfrac{1}{\sqrt{6}}\mathbf{i} + \dfrac{2}{\sqrt{6}}\mathbf{j} + \dfrac{1}{\sqrt{6}}\mathbf{k}\right)$

15. $\pi/2$ **17.** $\pi/2$

19. $x = t$, $y = -1$, $z = 1 + t$ **21.** $x = t$, $y = \dfrac{1}{3}t$, $z = t$

23. (a) (i): It has constant speed 1   (ii): Yes
        (iii): Counterclockwise   (iv): Yes
   (b) (i): It has constant speed 2   (ii): Yes
        (iii): Counterclockwise   (iv): Yes
   (c) (i): It has constant speed 1   (ii): Yes
        (iii): Counterclockwise
        (iv): It starts at $(0, -1)$ instead of $(1, 0)$
   (d) (i): It has constant speed 1   (ii): Yes
        (iii): Clockwise   (iv): Yes
   (e) (i): It has variable speed   (ii): No
        (iii): Counterclockwise   (iv): Yes

25. $\mathbf{v} = 2\sqrt{5}\mathbf{i} + \sqrt{5}\mathbf{j}$

Section 13.2, pp. 738–742

1. $(1/4)\mathbf{i} + 7\mathbf{j} + (3/2)\mathbf{k}$ **3.** $\left(\dfrac{\pi + 2\sqrt{2}}{2}\right)\mathbf{j} + 2\mathbf{k}$

5. $(\ln 4)\mathbf{i} + (\ln 4)\mathbf{j} + (\ln 2)\mathbf{k}$

7. $\dfrac{e - 1}{2}\mathbf{i} + \dfrac{e - 1}{e}\mathbf{j} + \mathbf{k}$

9. $\mathbf{i} - \mathbf{j} + \dfrac{\pi}{4}\mathbf{k}$

11. $\mathbf{r}(t) = \left(\dfrac{-t^2}{2} + 1\right)\mathbf{i} + \left(\dfrac{-t^2}{2} + 2\right)\mathbf{j} + \left(\dfrac{-t^2}{2} + 3\right)\mathbf{k}$

13. $\mathbf{r}(t) = ((t + 1)^{3/2} - 1)\mathbf{i} + (-e^{-t} + 1)\mathbf{j} + (\ln(t + 1) + 1)\mathbf{k}$

15. $\mathbf{r}(t) = 8t\mathbf{i} + 8t\mathbf{j} + (-16t^2 + 100)\mathbf{k}$

17. $\mathbf{r}(t) = \left(\dfrac{3}{2}t^2 + \dfrac{6}{\sqrt{11}}t + 1\right)\mathbf{i} - \left(\dfrac{1}{2}t^2 + \dfrac{2}{\sqrt{11}}t - 2\right)\mathbf{j}$

   $+ \left(\dfrac{1}{2}t^2 + \dfrac{2}{\sqrt{11}}t + 3\right)\mathbf{k} = \left(\dfrac{1}{2}t^2 + \dfrac{2t}{\sqrt{11}}\right)(3\mathbf{i} - \mathbf{j} + \mathbf{k})$

   $+ (\mathbf{i} + 2\mathbf{j} + 3\mathbf{k})$

19. 50 sec

21. (a) 72.2 sec; 25,510 m   **(b)** 4020 m   **(c)** 6378 m

23. (a) $v_0 \approx 9.9$ m/sec   **(b)** $\alpha \approx 18.4°$ or $71.6°$

25. 39.3° or 50.7°

31. (b) $\mathbf{v}_0$ would bisect $\angle AOR$.

33. (a) (Assuming that "x" is zero at the point of impact)
        $\mathbf{r}(t) = (x(t))\mathbf{i} + (y(t))\mathbf{j}$, where $x(t) = (35 \cos 27°)t$ and
        $y(t) = 4 + (35 \sin 27°)t - 16t^2$.
   (b) At $t \approx 0.497$ sec, it reaches its maximum height of about
        7.945 ft.
   (c) Range ≈ 37.45 ft; flight time ≈ 1.201 sec
   (d) At $t \approx 0.254$ and $t \approx 0.740$ sec, when it is ≈ 29.554 and
        ≈ 14.396 ft from where it will land
   (e) Yes. It changes things because the ball won't clear the net.

35. 4.00 ft, 7.80 ft/sec

43. (a) $\mathbf{r}(t) = (x(t))\mathbf{i} + (y(t))\mathbf{j}$; where

$$x(t) = \left(\frac{1}{0.08}\right)(1 - e^{-0.08t})(152 \cos 20° - 17.6) \text{ and}$$

$$y(t) = 3 + \left(\frac{152}{0.08}\right)(1 - e^{-0.08t})(\sin 20°)$$

$$+ \left(\frac{32}{0.08^2}\right)(1 - 0.08t - e^{-0.08t})$$

(b) At $t \approx 1.527$ sec it reaches a maximum height of about 41.893 feet.

(c) Range ≈ 351.734 ft; flight time ≈ 3.181 sec

(d) At $t \approx 0.877$ and 2.190 sec, when it is about 106.028 and 251.530 ft from home plate

(e) No

Section 13.3, pp. 745–746

1. $\mathbf{T} = \left(-\frac{2}{3}\sin t\right)\mathbf{i} + \left(\frac{2}{3}\cos t\right)\mathbf{j} + \frac{\sqrt{5}}{3}\mathbf{k}, 3\pi$

3. $\mathbf{T} = \frac{1}{\sqrt{1 + t}}\mathbf{i} + \frac{\sqrt{t}}{\sqrt{1 + t}}\mathbf{k}, \frac{52}{3}$

5. $\mathbf{T} = -\cos t\mathbf{j} + \sin t\mathbf{k}, \frac{3}{2}$

7. $\mathbf{T} = \left(\frac{\cos t - t\sin t}{t + 1}\right)\mathbf{i} + \left(\frac{\sin t + t\cos t}{t + 1}\right)\mathbf{j}$

$$+ \left(\frac{\sqrt{2}t^{1/2}}{t + 1}\right)\mathbf{k}, \frac{\pi^2}{2} + \pi$$

9. $(0, 5, 24\pi)$ **11.** $s(t) = 5t, \quad L = \frac{5\pi}{2}$

13. $s(t) = \sqrt{3}e^t - \sqrt{3}, \quad L = \frac{3\sqrt{3}}{4}$ **15.** $\sqrt{2} + \ln(1 + \sqrt{2})$

17. (a) Cylinder is $x^2 + y^2 = 1$, plane is $x + z = 1$.

(b) and **(c)**

(d) $L = \int_0^{2\pi} \sqrt{1 + \sin^2 t}\, dt$ **(e)** $L \approx 7.64$

Section 13.4, pp. 751–752

1. $\mathbf{T} = (\cos t)\mathbf{i} - (\sin t)\mathbf{j}, \quad \mathbf{N} = (-\sin t)\mathbf{i} - (\cos t)\mathbf{j}, \quad \kappa = \cos t$

3. $\mathbf{T} = \frac{1}{\sqrt{1 + t^2}}\mathbf{i} - \frac{t}{\sqrt{1 + t^2}}\mathbf{j}, \quad \mathbf{N} = \frac{-t}{\sqrt{1 + t^2}}\mathbf{i} -$

$\frac{1}{\sqrt{1 + t^2}}\mathbf{j}, \quad \kappa = \frac{1}{2(\sqrt{1 + t^2})^3}$

5. (b) $\cos x$

7. (b) $\mathbf{N} = \frac{-2e^{2t}}{\sqrt{1 + 4e^{4t}}}\mathbf{i} + \frac{1}{\sqrt{1 + 4e^{4t}}}\mathbf{j}$

(c) $\mathbf{N} = -\frac{1}{2}\left(\sqrt{4 - t^2}\mathbf{i} + t\mathbf{j}\right)$

9. $\mathbf{T} = \frac{3\cos t}{5}\mathbf{i} - \frac{3\sin t}{5}\mathbf{j} + \frac{4}{5}\mathbf{k}, \quad \mathbf{N} = (-\sin t)\mathbf{i} - (\cos t)\mathbf{j},$

$\kappa = \frac{3}{25}$

11. $\mathbf{T} = \left(\frac{\cos t - \sin t}{\sqrt{2}}\right)\mathbf{i} + \left(\frac{\cos t + \sin t}{\sqrt{2}}\right)\mathbf{j},$

$\mathbf{N} = \left(\frac{-\cos t - \sin t}{\sqrt{2}}\right)\mathbf{i} + \left(\frac{-\sin t + \cos t}{\sqrt{2}}\right)\mathbf{j},$

$\kappa = \frac{1}{e^t\sqrt{2}}$

13. $\mathbf{T} = \frac{t}{\sqrt{t^2 + 1}}\mathbf{i} + \frac{1}{\sqrt{t^2 + 1}}\mathbf{j}, \quad \mathbf{N} = \frac{\mathbf{i}}{\sqrt{t^2 + 1}} - \frac{t\mathbf{j}}{\sqrt{t^2 + 1}},$

$\kappa = \frac{1}{t(t^2 + 1)^{3/2}}$

15. $\mathbf{T} = \left(\operatorname{sech}\frac{t}{a}\right)\mathbf{i} + \left(\tanh\frac{t}{a}\right)\mathbf{j},$

$\mathbf{N} = \left(-\tanh\frac{t}{a}\right)\mathbf{i} + \left(\operatorname{sech}\frac{t}{a}\right)\mathbf{j},$

$\kappa = \frac{1}{a}\operatorname{sech}^2\frac{t}{a}$

19. $1/(2b)$ **21.** $\left(x - \frac{\pi}{2}\right)^2 + y^2 = 1$

23. $\kappa(x) = 2/(1 + 4x^2)^{3/2}$ **25.** $\kappa(x) = |\sin x|/(1 + \cos^2 x)^{3/2}$

Section 13.5, pp. 756–757

1. $\mathbf{a} = |a|\mathbf{N}$ **3.** $\mathbf{a}(1) = \frac{4}{3}\mathbf{T} + \frac{2\sqrt{5}}{3}\mathbf{N}$ **5.** $\mathbf{a}(0) = 2\mathbf{N}$

7. $\mathbf{r}\left(\frac{\pi}{4}\right) = \frac{\sqrt{2}}{2}\mathbf{i} + \frac{\sqrt{2}}{2}\mathbf{j} - \mathbf{k}, \mathbf{T}\left(\frac{\pi}{4}\right) = -\frac{\sqrt{2}}{2}\mathbf{i} + \frac{\sqrt{2}}{2}\mathbf{j},$

$\mathbf{N}\left(\frac{\pi}{4}\right) = -\frac{\sqrt{2}}{2}\mathbf{i} - \frac{\sqrt{2}}{2}\mathbf{j}, \mathbf{B}\left(\frac{\pi}{4}\right) = \mathbf{k}$; osculating plane:

$z = -1$; normal plane: $-x + y = 0$; rectifying plane:

$x + y = \sqrt{2}$

9. $\mathbf{B} = \left(\frac{4}{5}\cos t\right)\mathbf{i} - \left(\frac{4}{5}\sin t\right)\mathbf{j} - \frac{3}{5}\mathbf{k}, \tau = -\frac{4}{25}$

11. $\mathbf{B} = \mathbf{k}, \tau = 0$ **13.** $\mathbf{B} = -\mathbf{k}, \tau = 0$ **15.** $\mathbf{B} = \mathbf{k}, \tau = 0$

17. Yes. If the car is moving on a curved path ($\kappa \neq 0$), then

$a_N = \kappa|\mathbf{v}|^2 \neq 0$ and $\mathbf{a} \neq \mathbf{0}$.

23. $\kappa = \frac{1}{t}, \rho = t$

29. Components of $\mathbf{v}$: $-1.8701, 0.7089, 1.0000$
Components of $\mathbf{a}$: $-1.6960, -2.0307, 0$
Speed: 2.2361; Components of $\mathbf{T}$: $-0.8364, 0.3170, 0.4472$
Components of $\mathbf{N}$: $-0.4143, -0.8998, -0.1369$
Components of $\mathbf{B}$: $0.3590, -0.2998, 0.8839$; Curvature: 0.5060
Torsion: 0.2813; Tangential component of acceleration: 0.7746
Normal component of acceleration: 2.5298

31. Components of $\mathbf{v}$: $2.0000, 0, -0.1629$
Components of $\mathbf{a}$: $0, -1.0000, -0.0086$; Speed: 2.0066
Components of $\mathbf{T}$: $0.9967, 0, -0.0812$
Components of $\mathbf{N}$: $-0.0007, -1.0000, -0.0086$
Components of $\mathbf{B}$: $-0.0812, 0.0086, 0.9967$;
Curvature: 0.2484
Torsion: 0.0411; Tangential component of
acceleration: 0.0007
Normal component of acceleration: 1.0000

Section 13.6, p. 760

1. $\mathbf{v} = (3a \sin \theta)\mathbf{u}_r + 3a(1 - \cos \theta)\mathbf{u}_\theta$
   $\mathbf{a} = 9a(2 \cos \theta - 1)\mathbf{u}_r + (18a \sin \theta)\mathbf{u}_\theta$
3. $\mathbf{v} = 2ae^{a\theta}\mathbf{u}_r + 2e^{a\theta}\mathbf{u}_\theta$
   $\mathbf{a} = 4e^{a\theta}(a^2 - 1)\mathbf{u}_r + 8ae^{a\theta}\mathbf{u}_\theta$
5. $\mathbf{v} = (-8 \sin 4t)\mathbf{u}_r + (4 \cos 4t)\mathbf{u}_\theta$
   $\mathbf{a} = (-40 \cos 4t)\mathbf{u}_r - (32 \sin 4t)\mathbf{u}_\theta$

Practice Exercises, pp. 761–762

1. $\dfrac{x^2}{16} + \dfrac{y^2}{2} = 1$

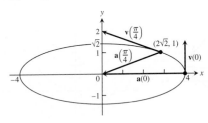

At $t = 0$: $a_T = 0$, $a_N = 4$, $\kappa = 2$;

At $t = \dfrac{\pi}{4}$: $a_T = \dfrac{7}{3}$, $a_N = \dfrac{4\sqrt{2}}{3}$, $\kappa = \dfrac{4\sqrt{2}}{27}$

3. $|\mathbf{v}|_{\max} = 1$ 5. $\kappa = 1/5$ 7. $dy/dt = -x$; clockwise
11. Shot put is on the ground, about 66 ft 3 in. from the stopboard.

15. Length $= \dfrac{\pi}{4}\sqrt{1 + \dfrac{\pi^2}{16}} + \ln\left(\dfrac{\pi}{4} + \sqrt{1 + \dfrac{\pi^2}{16}}\right)$

17. $\mathbf{T}(0) = \dfrac{2}{3}\mathbf{i} - \dfrac{2}{3}\mathbf{j} + \dfrac{1}{3}\mathbf{k}$; $\mathbf{N}(0) = \dfrac{1}{\sqrt{2}}\mathbf{i} + \dfrac{1}{\sqrt{2}}\mathbf{j}$;

    $\mathbf{B}(0) = -\dfrac{1}{3\sqrt{2}}\mathbf{i} + \dfrac{1}{3\sqrt{2}}\mathbf{j} + \dfrac{4}{3\sqrt{2}}\mathbf{k}$; $\kappa = \dfrac{\sqrt{2}}{3}$; $\tau = \dfrac{1}{6}$

19. $\mathbf{T}(\ln 2) = \dfrac{1}{\sqrt{17}}\mathbf{i} + \dfrac{4}{\sqrt{17}}\mathbf{j}$; $\mathbf{N}(\ln 2) = -\dfrac{4}{\sqrt{17}}\mathbf{i} + \dfrac{1}{\sqrt{17}}\mathbf{j}$;

    $\mathbf{B}(\ln 2) = \mathbf{k}$; $\kappa = \dfrac{8}{17\sqrt{17}}$; $\tau = 0$

21. $\mathbf{a}(0) = 10\mathbf{T} + 6\mathbf{N}$

23. $\mathbf{T} = \left(\dfrac{1}{\sqrt{2}}\cos t\right)\mathbf{i} - (\sin t)\mathbf{j} + \left(\dfrac{1}{\sqrt{2}}\cos t\right)\mathbf{k}$;

    $\mathbf{N} = \left(-\dfrac{1}{\sqrt{2}}\sin t\right)\mathbf{i} - (\cos t)\mathbf{j} - \left(\dfrac{1}{\sqrt{2}}\sin t\right)\mathbf{k}$;

    $\mathbf{B} = \dfrac{1}{\sqrt{2}}\mathbf{i} - \dfrac{1}{\sqrt{2}}\mathbf{k}$; $\kappa = \dfrac{1}{\sqrt{2}}$; $\tau = 0$

25. $\pi/3$ 27. $x = 1 + t$, $y = t$, $z = -t$ 31. $\kappa = 1/a$

Additional and Advanced Exercises, pp. 763–764

1. (a) $\left.\dfrac{d\theta}{dt}\right|_{\theta=2\pi} = 2\sqrt{\dfrac{\pi gb}{a^2 + b^2}}$

   (b) $\theta = \dfrac{gbt^2}{2(a^2 + b^2)}$, $z = \dfrac{gb^2t^2}{2(a^2 + b^2)}$

(c) $\mathbf{v}(t) = \dfrac{gbt}{\sqrt{a^2 + b^2}}\mathbf{T}$;

   $\dfrac{d^2\mathbf{r}}{dt^2} = \dfrac{bg}{\sqrt{a^2 + b^2}}\mathbf{T} + a\left(\dfrac{bgt}{a^2 + b^2}\right)^2\mathbf{N}$

   There is no component in the direction of $\mathbf{B}$.

5. (a) $\dfrac{dx}{dt} = \dot{r}\cos\theta - r\dot{\theta}\sin\theta$, $\dfrac{dy}{dt} = \dot{r}\sin\theta + r\dot{\theta}\cos\theta$

   (b) $\dfrac{dr}{dt} = \dot{x}\cos\theta + \dot{y}\sin\theta$, $r\dfrac{d\theta}{dt} = -\dot{x}\sin\theta + \dot{y}\cos\theta$

7. (a) $\mathbf{a}(1) = -9\mathbf{u}_r - 6\mathbf{u}_\theta$, $\mathbf{v}(1) = -\mathbf{u}_r + 3\mathbf{u}_\theta$ (b) 6.5 in.
9. (c) $\mathbf{v} = \dot{r}\mathbf{u}_r + r\dot{\theta}\mathbf{u}_\theta + \dot{z}\mathbf{k}$, $\mathbf{a} = (\ddot{r} - r\dot{\theta}^2)\mathbf{u}_r + (r\ddot{\theta} + 2\dot{r}\dot{\theta})\mathbf{u}_\theta + \ddot{z}\mathbf{k}$

CHAPTER 14

Section 14.1, pp. 771–773

1. (a) 0   (b) 0   (c) 58   (d) 33
3. (a) 4/5   (b) 8/5   (c) 3   (d) 0
5. Domain: all points (x, y) on or above line $y = x + 2$
7. Domain: all points (x, y) not lying on the graph of $y = x$ or $y = x^3$

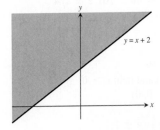

9. Domain: all points (x, y) satisfying $x^2 - 1 \le y \le x^2 + 1$

11. Domain: all points (x, y) for which $(x - 2)(x + 2)(y - 3)(y + 3) \ge 0$

13.

15.

17. (a) All points in the *xy*-plane   **(b)** All reals
   (c) The lines $y - x = c$ **(d)** No boundary points
   (e) Both open and closed   **(f)** Unbounded
19. (a) All points in the *xy*-plane   **(b)** $z \geq 0$
   (c) For $f(x, y) = 0$, the origin; for $f(x, y) \neq 0$, ellipses with the
   center $(0, 0)$, and major and minor axes along the *x*- and
   y-axes respectively
   (d) No boundary points   **(e)** Both open and closed
   (f) Unbounded
21. (a) All points in the *xy*-plane   **(b)** All reals
   (c) For $f(x, y) = 0$, the *x*- and *y*-axes; for $f(x, y) \neq 0$, hyperbolas with the *x*- and *y*-axes as asymptotes
   (d) No boundary points   **(e)** Both open and closed
   (f) Unbounded
23. (a) All (x, y) satisfying $x^2 + y^2 < 16$ **(b)** $z \geq 1/4$
   (c) Circles centered at the origin with radii $r < 4$
   (d) Boundary is the circle $x^2 + y^2 = 16$
   (e) Open   **(f)** Bounded
25. (a) $(x, y) \neq (0, 0)$ **(b)** All reals
   (c) The circles with center $(0, 0)$ and radii $r > 0$
   (d) Boundary is the single point $(0, 0)$
   (e) Open   **(f)** Unbounded
27. (a) All (x, y) satisfying $-1 \leq y - x \leq 1$
   (b) $-\pi/2 \leq z \leq \pi/2$
   (c) Straight lines of the form $y - x = c$ where $-1 \leq c \leq 1$
   (d) Boundary is two straight lines $y = 1 + x$ and $y = -1 + x$
   (e) Closed   **(f)** Unbounded
29. (a) Domain: all points (x, y) outside the circle $x^2 + y^2 = 1$
   (b) Range: all reals
   (c) Circles centered at the origin with radii $r > 1$
   (d) Boundary: $x^2 + y^2 = 1$
   (e) Open   **(f)** Unbounded
31. (f)   33. (a)   35. (d)
37. (a)

(b)

39. (a)

(b)

41. (a)

(b)

43. (a)

(b)

45. (a)

(b)

47. (a)

(b)

49. $x^2 + y^2 = 10$

51. $x + y^2 = 4$

53.

$f(x, y, z) = x^2 + y^2 + z^2 = 1$

55.

$f(x, y, z) = x + z = 1$

57.

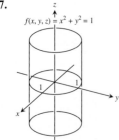

$f(x, y, z) = x^2 + y^2 = 1$

59.

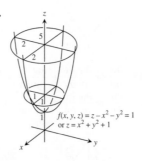

$f(x, y, z) = z - x^2 - y^2 = 1$
or $z = x^2 + y^2 + 1$

61. $\sqrt{x - y} - \ln z = 2$ **63.** $x^2 + y^2 + z^2 = 4$

65. Domain: all points (x, y)
satisfying $|x| < |y|$

67. Domain: all points (x, y)
satisfying $-1 \le x \le 1$ and
$-1 \le y \le 1$

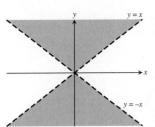

level curve: $y = 2x$

level curve:
$\sin^{-1} y - \sin^{-1} x = \dfrac{\pi}{2}$

Section 14.2, pp. 779–782

1. $5/2$ **3.** $2\sqrt{6}$ **5.** 1 **7.** $1/2$ **9.** 1 **11.** $1/4$ **13.** 0
15. -1 **17.** 2 **19.** $1/4$ **21.** 1 **23.** 3 **25.** $19/12$
27. 2 **29.** 3 **31. (a)** All (x, y) **(b)** All (x, y) except $(0, 0)$
33. (a) All (x, y) except where $x = 0$ or $y = 0$ **(b)** All (x, y)
35. (a) All (x, y, z) **(b)** All (x, y, z) except the interior of the
cylinder $x^2 + y^2 = 1$

37. (a) All (x, y, z) with $z \ne 0$ **(b)** All (x, y, z) with $x^2 + z^2 \ne 1$
39. (a) All points (x, y) satisfying $z > x^2 + y^2 + 1$
41. Consider paths along $y = x, x > 0$, and along $y = x, x < 0$.
43. Consider the paths $y = kx^2$, k a constant.
45. Consider the paths $y = mx$, m a constant, $m \ne -1$.
47. Consider the paths $y = kx^2$, k a constant, $k \ne 0$.
49. Consider the paths $x = 1$ and $y = x$.
51. (a) 1 **(b)** 0 **(c)** Does not exist
55. The limit is 1.   **57.** The limit is 0.
59. (a) $f(x, y)|_{y=mx} = \sin 2\theta$ where $\tan \theta = m$ **61.** 0
63. Does not exist   **65.** $\pi/2$ **67.** $f(0, 0) = \ln 3$
69. $\delta = 0.1$ **71.** $\delta = 0.005$ **73.** $\delta = 0.04$
75. $\delta = \sqrt{0.015}$ **77.** $\delta = 0.005$

Section 14.3, pp. 790–793

1. $\dfrac{\partial f}{\partial x} = 4x, \dfrac{\partial f}{\partial y} = -3$ **3.** $\dfrac{\partial f}{\partial x} = 2x(y + 2), \dfrac{\partial f}{\partial y} = x^2 - 1$

5. $\dfrac{\partial f}{\partial x} = 2y(xy - 1), \dfrac{\partial f}{\partial y} = 2x(xy - 1)$

7. $\dfrac{\partial f}{\partial x} = \dfrac{x}{\sqrt{x^2 + y^2}}, \dfrac{\partial f}{\partial y} = \dfrac{y}{\sqrt{x^2 + y^2}}$

9. $\dfrac{\partial f}{\partial x} = \dfrac{-1}{(x + y)^2}, \dfrac{\partial f}{\partial y} = \dfrac{-1}{(x + y)^2}$

11. $\dfrac{\partial f}{\partial x} = \dfrac{-y^2 - 1}{(xy - 1)^2}, \dfrac{\partial f}{\partial y} = \dfrac{-x^2 - 1}{(xy - 1)^2}$

13. $\dfrac{\partial f}{\partial x} = e^{x+y+1}, \dfrac{\partial f}{\partial y} = e^{x+y+1}$ **15.** $\dfrac{\partial f}{\partial x} = \dfrac{1}{x + y}, \dfrac{\partial f}{\partial y} = \dfrac{1}{x + y}$

17. $\dfrac{\partial f}{\partial x} = 2 \sin (x - 3y) \cos (x - 3y)$,

$\dfrac{\partial f}{\partial y} = -6 \sin (x - 3y) \cos (x - 3y)$

19. $\dfrac{\partial f}{\partial x} = yx^{y-1}, \dfrac{\partial f}{\partial y} = x^y \ln x$ **21.** $\dfrac{\partial f}{\partial x} = -g(x), \dfrac{\partial f}{\partial y} = g(y)$

23. $f_x = y^2, f_y = 2xy, f_z = -4z$
25. $f_x = 1, f_y = -y(y^2 + z^2)^{-1/2}, f_z = -z(y^2 + z^2)^{-1/2}$

27. $f_x = \dfrac{yz}{\sqrt{1 - x^2y^2z^2}}, f_y = \dfrac{xz}{\sqrt{1 - x^2y^2z^2}}, f_z = \dfrac{xy}{\sqrt{1 - x^2y^2z^2}}$

29. $f_x = \dfrac{1}{x + 2y + 3z}, f_y = \dfrac{2}{x + 2y + 3z}, f_z = \dfrac{3}{x + 2y + 3z}$

31. $f_x = -2xe^{-(x^2+y^2+z^2)}, f_y = -2ye^{-(x^2+y^2+z^2)}, f_z = -2ze^{-(x^2+y^2+z^2)}$

33. $f_x = \text{sech}^2(x + 2y + 3z), f_y = 2 \, \text{sech}^2(x + 2y + 3z)$,
$f_z = 3 \, \text{sech}^2(x + 2y + 3z)$

35. $\dfrac{\partial f}{\partial t} = -2\pi \sin (2\pi t - \alpha), \dfrac{\partial f}{\partial \alpha} = \sin (2\pi t - \alpha)$

37. $\dfrac{\partial h}{\partial \rho} = \sin \phi \cos \theta, \dfrac{\partial h}{\partial \phi} = \rho \cos \phi \cos \theta, \dfrac{\partial h}{\partial \theta} = -\rho \sin \phi \sin \theta$

39. $W_P(P, V, \delta, v, g) = V, W_V(P, V, \delta, v, g) = P + \dfrac{\delta v^2}{2g}$,

$W_\delta(P, V, \delta, v, g) = \dfrac{Vv^2}{2g}, W_v(P, V, \delta, v, g) = \dfrac{V\delta v}{g}$,

$W_g(P, V, \delta, v, g) = -\dfrac{V\delta v^2}{2g^2}$

41. $\dfrac{\partial f}{\partial x} = 1 + y, \dfrac{\partial f}{\partial y} = 1 + x, \dfrac{\partial^2 f}{\partial x^2} = 0, \dfrac{\partial^2 f}{\partial y^2} = 0,$

$\dfrac{\partial^2 f}{\partial y\,\partial x} = \dfrac{\partial^2 f}{\partial x\,\partial y} = 1$

43. $\dfrac{\partial g}{\partial x} = 2xy + y\cos x, \dfrac{\partial g}{\partial y} = x^2 - \sin y + \sin x,$

$\dfrac{\partial^2 g}{\partial x^2} = 2y - y\sin x, \dfrac{\partial^2 g}{\partial y^2} = -\cos y,$

$\dfrac{\partial^2 g}{\partial y\,\partial x} = \dfrac{\partial^2 g}{\partial x\,\partial y} = 2x + \cos x$

45. $\dfrac{\partial r}{\partial x} = \dfrac{1}{x + y}, \dfrac{\partial r}{\partial y} = \dfrac{1}{x + y}, \dfrac{\partial^2 r}{\partial x^2} = \dfrac{-1}{(x + y)^2}, \dfrac{\partial^2 r}{\partial y^2} = \dfrac{-1}{(x + y)^2},$

$\dfrac{\partial^2 r}{\partial y\,\partial x} = \dfrac{\partial^2 r}{\partial x\,\partial y} = \dfrac{-1}{(x + y)^2}$

47. $\dfrac{\partial w}{\partial x} = x^2 y\sec^2(xy) + 2x\tan(xy), \dfrac{\partial w}{\partial y} = x^3\sec^2(xy),$

$\dfrac{\partial^2 w}{\partial y\,\partial x} = \dfrac{\partial^2 w}{\partial x\,\partial y} = 2x^3 y\sec^2(xy)\tan(xy) + 3x^2\sec^2(xy)$

$\dfrac{\partial^2 w}{\partial x^2} = 4xy\sec^2(xy) + 2x^2 y^2\sec^2(xy)\tan(xy) + 2\tan(xy)$

$\dfrac{\partial^2 w}{\partial y^2} = 2x^4\sec^2(xy)\tan(xy)$

49. $\dfrac{\partial w}{\partial x} = \sin(x^2 y) + 2x^2 y\cos(x^2 y), \dfrac{\partial w}{\partial y} = x^3\cos(x^2 y),$

$\dfrac{\partial^2 w}{\partial y\,\partial x} = \dfrac{\partial^2 w}{\partial x\,\partial y} = 3x^2\cos(x^2 y) - 2x^4 y\sin(x^2 y)$

$\dfrac{\partial^2 w}{\partial x^2} = 6xy\cos(x^2 y) - 4x^3 y^2\sin(x^2 y)$

$\dfrac{\partial^2 w}{\partial y^2} = -x^5\sin(x^2 y)$

51. $\dfrac{\partial w}{\partial x} = \dfrac{2}{2x + 3y}, \dfrac{\partial w}{\partial y} = \dfrac{3}{2x + 3y}, \dfrac{\partial^2 w}{\partial y\,\partial x} = \dfrac{\partial^2 w}{\partial x\,\partial y} = \dfrac{-6}{(2x + 3y)^2}$

53. $\dfrac{\partial w}{\partial x} = y^2 + 2xy^3 + 3x^2 y^4, \dfrac{\partial w}{\partial y} = 2xy + 3x^2 y^2 + 4x^3 y^3,$

$\dfrac{\partial^2 w}{\partial y\,\partial x} = \dfrac{\partial^2 w}{\partial x\,\partial y} = 2y + 6xy^2 + 12x^2 y^3$

55. (a) x first   (b) y first   (c) x first
    (d) x first   (e) y first   (f) y first

57. $f_x(1, 2) = -13, f_y(1, 2) = -2$

59. $f_x(-2, 3) = 1/2, f_y(-2, 3) = 3/4$ **61.** (a) 3   (b) 2

63. 12   **65.** -2 **67.** $\dfrac{\partial A}{\partial a} = \dfrac{a}{bc\sin A}, \dfrac{\partial A}{\partial b} = \dfrac{c\cos A - b}{bc\sin A}$

69. $v_x = \dfrac{\ln v}{(\ln u)(\ln v) - 1}$

71. $f_x(x, y) = 0$ for all points (x, y),

$f_y(x, y) = \begin{cases} 3y^2, & y \geq 0 \\ -2y, & y < 0 \end{cases},$

$f_{xy}(x, y) = f_{yx}(x, y) = 0$ for all points (x, y)

89. Yes

Section 14.4, pp. 800–801

1. (a) $\dfrac{dw}{dt} = 0,$ (b) $\dfrac{dw}{dt}(\pi) = 0$

3. (a) $\dfrac{dw}{dt} = 1,$ (b) $\dfrac{dw}{dt}(3) = 1$

5. (a) $\dfrac{dw}{dt} = 4t\tan^{-1}t + 1,$ (b) $\dfrac{dw}{dt}(1) = \pi + 1$

7. (a) $\dfrac{\partial z}{\partial u} = 4\cos v\ln(u\sin v) + 4\cos v,$

$\dfrac{\partial z}{\partial v} = -4u\sin v\ln(u\sin v) + \dfrac{4u\cos^2 v}{\sin v}$

(b) $\dfrac{\partial z}{\partial u} = \sqrt{2}(\ln 2 + 2), \dfrac{\partial z}{\partial v} = -2\sqrt{2}(\ln 2 - 2)$

9. (a) $\dfrac{\partial w}{\partial u} = 2u + 4uv, \dfrac{\partial w}{\partial v} = -2v + 2u^2$

(b) $\dfrac{\partial w}{\partial u} = 3, \dfrac{\partial w}{\partial v} = -\dfrac{3}{2}$

11. (a) $\dfrac{\partial u}{\partial x} = 0, \dfrac{\partial u}{\partial y} = \dfrac{z}{(z - y)^2}, \dfrac{\partial u}{\partial z} = \dfrac{-y}{(z - y)^2}$

(b) $\dfrac{\partial u}{\partial x} = 0, \dfrac{\partial u}{\partial y} = 1, \dfrac{\partial u}{\partial z} = -2$

13. $\dfrac{dz}{dt} = \dfrac{\partial z}{\partial x}\dfrac{dx}{dt} + \dfrac{\partial z}{\partial y}\dfrac{dy}{dt}$

15. $\dfrac{\partial w}{\partial u} = \dfrac{\partial w}{\partial x}\dfrac{\partial x}{\partial u} + \dfrac{\partial w}{\partial y}\dfrac{\partial y}{\partial u} + \dfrac{\partial w}{\partial z}\dfrac{\partial z}{\partial u},$

$\dfrac{\partial w}{\partial v} = \dfrac{\partial w}{\partial x}\dfrac{\partial x}{\partial v} + \dfrac{\partial w}{\partial y}\dfrac{\partial y}{\partial v} + \dfrac{\partial w}{\partial z}\dfrac{\partial z}{\partial v}$

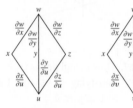

17. $\dfrac{\partial w}{\partial u} = \dfrac{\partial w}{\partial x}\dfrac{\partial x}{\partial u} + \dfrac{\partial w}{\partial y}\dfrac{\partial y}{\partial u}, \dfrac{\partial w}{\partial v} = \dfrac{\partial w}{\partial x}\dfrac{\partial x}{\partial v} + \dfrac{\partial w}{\partial y}\dfrac{\partial y}{\partial v}.$

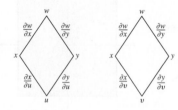

19. $\dfrac{\partial z}{\partial t} = \dfrac{\partial z}{\partial x}\dfrac{\partial x}{\partial t} + \dfrac{\partial z}{\partial y}\dfrac{\partial y}{\partial t}, \dfrac{\partial z}{\partial s} = \dfrac{\partial z}{\partial x}\dfrac{\partial x}{\partial s} + \dfrac{\partial z}{\partial y}\dfrac{\partial y}{\partial s}$

21. $\dfrac{\partial w}{\partial s} = \dfrac{dw}{du}\dfrac{\partial u}{\partial s}, \dfrac{\partial w}{\partial t} = \dfrac{dw}{du}\dfrac{\partial u}{\partial t}$

23. $\dfrac{\partial w}{\partial r} = \dfrac{\partial w}{\partial x}\dfrac{dx}{dr} + \dfrac{\partial w}{\partial y}\dfrac{dy}{dr} = \dfrac{\partial w}{\partial x}\dfrac{dx}{dr}$ since $\dfrac{dy}{dr} = 0$,

$\dfrac{\partial w}{\partial s} = \dfrac{\partial w}{\partial x}\dfrac{dx}{ds} + \dfrac{\partial w}{\partial y}\dfrac{dy}{ds} = \dfrac{\partial w}{\partial y}\dfrac{dy}{ds}$ since $\dfrac{dx}{ds} = 0$

25. $4/3$ **27.** $-4/5$ **29.** $\dfrac{\partial z}{\partial x} = \dfrac{1}{4}, \dfrac{\partial z}{\partial y} = -\dfrac{3}{4}$

31. $\dfrac{\partial z}{\partial x} = -1, \dfrac{\partial z}{\partial y} = -1$ **33.** 12 **35.** -7

37. $\dfrac{\partial z}{\partial u} = 2, \dfrac{\partial z}{\partial v} = 1$

39. $\dfrac{\partial w}{\partial t} = 2t\,e^{s^3+t^2}, \dfrac{\partial w}{\partial s} = 3s^2\,e^{s^3+t^2}$

41. -0.00005 amps/sec

47. $(\cos 1, \sin 1, 1)$ and $(\cos(-2), \sin(-2), -2)$

49. (a) Maximum at $\left(-\dfrac{\sqrt{2}}{2}, \dfrac{\sqrt{2}}{2}\right)$ and $\left(\dfrac{\sqrt{2}}{2}, -\dfrac{\sqrt{2}}{2}\right)$; minimum

at $\left(\dfrac{\sqrt{2}}{2}, \dfrac{\sqrt{2}}{2}\right)$ and $\left(-\dfrac{\sqrt{2}}{2}, -\dfrac{\sqrt{2}}{2}\right)$

(b) Max $= 6$, min $= 2$

51. $2x\sqrt{x^8 + x^3} + \displaystyle\int_0^{x^2} \dfrac{3x^2}{2\sqrt{t^4 + x^3}}\,dt$

Section 14.5, pp. 808–809

1.

3.

5.

7. $\nabla f = 3\mathbf{i} + 2\mathbf{j} - 4\mathbf{k}$ **9.** $\nabla f = -\dfrac{26}{27}\mathbf{i} + \dfrac{23}{54}\mathbf{j} - \dfrac{23}{54}\mathbf{k}$

11. -4 **13.** $21/13$ **15.** 3 **17.** 2

19. $\mathbf{u} = -\dfrac{1}{\sqrt{2}}\mathbf{i} + \dfrac{1}{\sqrt{2}}\mathbf{j}, (D_{\mathbf{u}}f)_{P_0} = \sqrt{2}; -\mathbf{u} = \dfrac{1}{\sqrt{2}}\mathbf{i} - \dfrac{1}{\sqrt{2}}\mathbf{j},$

$(D_{-\mathbf{u}}f)_{P_0} = -\sqrt{2}$

21. $\mathbf{u} = \dfrac{1}{3\sqrt{3}}\mathbf{i} - \dfrac{5}{3\sqrt{3}}\mathbf{j} - \dfrac{1}{3\sqrt{3}}\mathbf{k}, (D_{\mathbf{u}}f)_{P_0} = 3\sqrt{3};$

$-\mathbf{u} = -\dfrac{1}{3\sqrt{3}}\mathbf{i} + \dfrac{5}{3\sqrt{3}}\mathbf{j} + \dfrac{1}{3\sqrt{3}}\mathbf{k}, (D_{-\mathbf{u}}f)_{P_0} = -3\sqrt{3}$

23. $\mathbf{u} = \dfrac{1}{\sqrt{3}}(\mathbf{i} + \mathbf{j} + \mathbf{k}), (D_{\mathbf{u}}f)_{P_0} = 2\sqrt{3};$

$-\mathbf{u} = -\dfrac{1}{\sqrt{3}}(\mathbf{i} + \mathbf{j} + \mathbf{k}), (D_{-\mathbf{u}}f)_{P_0} = -2\sqrt{3}$

25.   **27.**

29. (a) $\mathbf{u} = \dfrac{3}{5}\mathbf{i} - \dfrac{4}{5}\mathbf{j}, D_{\mathbf{u}} f(1, -1) = 5$

(b) $\mathbf{u} = -\dfrac{3}{5}\mathbf{i} + \dfrac{4}{5}\mathbf{j}, D_{\mathbf{u}} f(1, -1) = -5$

(c) $\mathbf{u} = \dfrac{4}{5}\mathbf{i} + \dfrac{3}{5}\mathbf{j}, \mathbf{u} = -\dfrac{4}{5}\mathbf{i} - \dfrac{3}{5}\mathbf{j}$

(d) $\mathbf{u} = -\mathbf{j}, \mathbf{u} = \dfrac{24}{25}\mathbf{i} - \dfrac{7}{25}\mathbf{j}$

(e) $\mathbf{u} = -\mathbf{i}, \mathbf{u} = \dfrac{7}{25}\mathbf{i} + \dfrac{24}{25}\mathbf{j}$

31. $\mathbf{u} = \dfrac{7}{\sqrt{53}}\mathbf{i} - \dfrac{2}{\sqrt{53}}\mathbf{j}, -\mathbf{u} = -\dfrac{7}{\sqrt{53}}\mathbf{i} + \dfrac{2}{\sqrt{53}}\mathbf{j}$

33. No, the maximum rate of change is $\sqrt{185} < 14$.

35. $-7/\sqrt{5}$

Section 14.6, pp. 817–820

1. (a) $x + y + z = 3$

(b) $x = 1 + 2t, y = 1 + 2t, z = 1 + 2t$

3. (a) $2x - z - 2 = 0$ **(b)** $x = 2 - 4t, y = 0, z = 2 + 2t$

5. (a) $2x + 2y + z - 4 = 0$

(b) $x = 2t, y = 1 + 2t, z = 2 + t$

7. (a) $x + y + z - 1 = 0$ **(b)** $x = t, y = 1 + t, z = t$

9. $2x - z - 2 = 0$ **11.** $x - y + 2z - 1 = 0$

13. $x = 1, y = 1 + 2t, z = 1 - 2t$

15. $x = 1 - 2t, y = 1, z = \dfrac{1}{2} + 2t$

17. $x = 1 + 90t, y = 1 - 90t, z = 3$

19. $df = \dfrac{9}{11,830} \approx 0.0008$ **21.** $dg = 0$

23. (a) $\dfrac{\sqrt{3}}{2}\sin\sqrt{3} - \dfrac{1}{2}\cos\sqrt{3} \approx 0.935°C/\text{ft}$

(b) $\sqrt{3}\sin\sqrt{3} - \cos\sqrt{3} \approx 1.87°C/\text{sec}$

25. (a) $L(x, y) = 1$ **(b)** $L(x, y) = 2x + 2y - 1$

27. (a) $L(x, y) = 3x - 4y + 5$ **(b)** $L(x, y) = 3x - 4y + 5$

29. (a) $L(x, y) = 1 + x$ (b) $L(x, y) = -y + \dfrac{\pi}{2}$

31. (a) $W(20, 25) = 11°F$, $W(30, -10) = -39°F$,
    $W(15, 15) = 0°F$
  (b) $W(10, -40) \approx -65.5°F$, $W(50, -40) \approx -88°F$,
    $W(60, 30) \approx 10.2°F$
  (c) $L(v, T) \approx -0.36\,(v - 25) + 1.337(T - 5) - 17.4088$
  (d) i) $L(24, 6) \approx -15.7°F$
     ii) $L(27, 2) \approx -22.1°F$
     iii) $L(5, -10) \approx -30.2°F$

33. $L(x, y) = 7 + x - 6y$; 0.06 35. $L(x, y) = x + y + 1$; 0.08

37. $L(x, y) = 1 + x$; 0.0222

39. (a) $L(x, y, z) = 2x + 2y + 2z - 3$ (b) $L(x, y, z) = y + z$
  (c) $L(x, y, z) = 0$

41. (a) $L(x, y, z) = x$ (b) $L(x, y, z) = \dfrac{1}{\sqrt{2}}x + \dfrac{1}{\sqrt{2}}y$

  (c) $L(x, y, z) = \dfrac{1}{3}x + \dfrac{2}{3}y + \dfrac{2}{3}z$

43. (a) $L(x, y, z) = 2 + x$
  (b) $L(x, y, z) = x - y - z + \dfrac{\pi}{2} + 1$
  (c) $L(x, y, z) = x - y - z + \dfrac{\pi}{2} + 1$

45. $L(x, y, z) = 2x - 6y - 2z + 6$, 0.0024

47. $L(x, y, z) = x + y - z - 1$, 0.00135

49. Maximum error (estimate) ≤ 0.31 in magnitude

51. (a) $\pm 5\%$ (b) $\pm 7\%$

53. $\approx \pm 4.83\%$

55. Pay more attention to the smaller of the two dimensions. It will generate the larger partial derivative.

57. (a) 0.3%

59. f is most sensitive to a change in d.

61. Q is most sensitive to a change in h.

65. At $-\dfrac{\pi}{4}, -\dfrac{\pi}{2\sqrt{2}}$; at $0, 0$; at $\dfrac{\pi}{4}, \dfrac{\pi}{2\sqrt{2}}$

Section 14.7, pp. 826–829

1. $f(-3, 3) = -5$, local minimum   3. $f(-2, 1)$, saddle point

5. $f\left(3, \dfrac{3}{2}\right) = \dfrac{17}{2}$, local maximum

7. $f(2, -1) = -6$, local minimum   9. $f(1, 2)$, saddle point

11. $f\left(\dfrac{16}{7}, 0\right) = -\dfrac{16}{7}$, local maximum

13. $f(0, 0)$, saddle point; $f\left(-\dfrac{2}{3}, \dfrac{2}{3}\right) = \dfrac{170}{27}$, local maximum

15. $f(0, 0) = 0$, local minimum; $f(1, -1)$, saddle point

17. $f(0, \pm\sqrt{5})$, saddle points; $f(-2, -1) = 30$, local maximum; $f(2, 1) = -30$, local minimum

19. $f(0, 0)$, saddle point; $f(1, 1) = 2$, $f(-1, -1) = 2$, local maxima

21. $f(0, 0) = -1$, local maximum

23. $f(n\pi, 0)$, saddle points, for every integer n

25. $f(2, 0) = e^{-4}$, local minimum

27. $f(0, 0) = 0$, local minimum; $f(0, 2)$, saddle point

29. $f\left(\dfrac{1}{2}, 1\right) = \ln\left(\dfrac{1}{4}\right) - 3$, local maximum

31. Absolute maximum: 1 at $(0, 0)$; absolute minimum: -5 at $(1, 2)$

33. Absolute maximum: 4 at $(0, 2)$; absolute minimum: 0 at $(0, 0)$

35. Absolute maximum: 11 at $(0, -3)$; absolute minimum: -10 at $(4, -2)$

37. Absolute maximum: 4 at $(2, 0)$; absolute minimum: $\dfrac{3\sqrt{2}}{2}$ at $\left(3, -\dfrac{\pi}{4}\right)$, $\left(3, \dfrac{\pi}{4}\right)$, $\left(1, -\dfrac{\pi}{4}\right)$, and $\left(1, \dfrac{\pi}{4}\right)$

39. $a = -3$, $b = 2$

41. Hottest is $2\dfrac{1}{4}°$ at $\left(-\dfrac{1}{2}, \dfrac{\sqrt{3}}{2}\right)$ and $\left(-\dfrac{1}{2}, -\dfrac{\sqrt{3}}{2}\right)$; coldest is $-\dfrac{1}{4}°$ at $\left(\dfrac{1}{2}, 0\right)$.

43. (a) $f(0, 0)$, saddle point   (b) $f(1, 2)$, local minimum
  (c) $f(1, -2)$, local minimum; $f(-1, -2)$, saddle point

49. $\left(\dfrac{1}{6}, \dfrac{1}{3}, \dfrac{355}{36}\right)$ 51. $\left(\dfrac{9}{7}, \dfrac{6}{7}, \dfrac{3}{7}\right)$ 53. $3, 3, 3$ 55. 12

57. $\dfrac{4}{\sqrt{3}} \times \dfrac{4}{\sqrt{3}} \times \dfrac{4}{\sqrt{3}}$ 59. 2 ft $\times$ 2 ft $\times$ 1 ft

61. (a) On the semicircle, max $f = 2\sqrt{2}$ at $t = \pi/4$, min $f = -2$ at $t = \pi$. On the quarter circle, max $f = 2\sqrt{2}$ at $t = \pi/4$, min $f = 2$ at $t = 0, \pi/2$.
  (b) On the semicircle, max $g = 2$ at $t = \pi/4$, min $g = -2$ at $t = 3\pi/4$. On the quarter circle, max $g = 2$ at $t = \pi/4$, min $g = 0$ at $t = 0, \pi/2$.
  (c) On the semicircle, max $h = 8$ at $t = 0, \pi$; min $h = 4$ at $t = \pi/2$. On the quarter circle, max $h = 8$ at $t = 0$, min $h = 4$ at $t = \pi/2$.

63. i) min $f = -1/2$ at $t = -1/2$; no max   ii) max $f = 0$ at $t = -1, 0$; min $f = -1/2$ at $t = -1/2$ iii) max $f = 4$ at $t = 1$; min $f = 0$ at $t = 0$

67. $y = -\dfrac{20}{13}x + \dfrac{9}{13}$, $y|_{x=4} = -\dfrac{71}{13}$

Section 14.8, pp. 836–838

1. $\left(\pm\dfrac{1}{\sqrt{2}}, \dfrac{1}{2}\right)$, $\left(\pm\dfrac{1}{\sqrt{2}}, -\dfrac{1}{2}\right)$ 3. 39 5. $\left(3, \pm 3\sqrt{2}\right)$

7. (a) 8 (b) 64

9. $r = 2$ cm, $h = 4$ cm   11. Length $= 4\sqrt{2}$, width $= 3\sqrt{2}$

13. $f(0, 0) = 0$ is minimum, $f(2, 4) = 20$ is maximum.

15. Lowest $= 0°$, highest $= 125°$

17. $\left(\dfrac{3}{2}, 2, \dfrac{5}{2}\right)$ 19. 1 21. $(0, 0, 2)$, $(0, 0, -2)$

23. $f(1, -2, 5) = 30$ is maximum, $f(-1, 2, -5) = -30$ is minimum.

25. $3, 3, 3$ 27. $\dfrac{2}{\sqrt{3}}$ by $\dfrac{2}{\sqrt{3}}$ by $\dfrac{2}{\sqrt{3}}$ units

29. $(\pm 4/3, -4/3, -4/3)$ 31. $U(8, 14) = \$128$

33. $f(2/3, 4/3, -4/3) = \dfrac{4}{3}$ 35. $(2, 4, 4)$

37. Maximum is $1 + 6\sqrt{3}$ at $\left(\pm\sqrt{6}, \sqrt{3}, 1\right)$, minimum is $1 - 6\sqrt{3}$ at $\left(\pm\sqrt{6}, -\sqrt{3}, 1\right)$.

39. Maximum is 4 at $(0, 0, \pm 2)$, minimum is 2 at $\left(\pm\sqrt{2}, \pm\sqrt{2}, 0\right)$.

Section 14.9, p. 842

1. Quadratic: $x + xy$; cubic: $x + xy + \dfrac{1}{2}xy^2$

3. Quadratic: xy; cubic: xy

5. Quadratic: $y + \dfrac{1}{2}(2xy - y^2)$;

cubic: $y + \dfrac{1}{2}(2xy - y^2) + \dfrac{1}{6}(3x^2y - 3xy^2 + 2y^3)$

7. Quadratic: $\dfrac{1}{2}(2x^2 + 2y^2) = x^2 + y^2$; cubic: $x^2 + y^2$

9. Quadratic: $1 + (x + y) + (x + y)^2$;
cubic: $1 + (x + y) + (x + y)^2 + (x + y)^3$

11. Quadratic: $1 - \dfrac{1}{2}x^2 - \dfrac{1}{2}y^2$; $E(x, y) \le 0.00134$

Section 14.10, p. 846

1. (a) 0   **(b)** $1 + 2z$ **(c)** $1 + 2z$

3. (a) $\dfrac{\partial U}{\partial P} + \dfrac{\partial U}{\partial T}\left(\dfrac{V}{nR}\right)$ **(b)** $\dfrac{\partial U}{\partial P}\left(\dfrac{nR}{V}\right) + \dfrac{\partial U}{\partial T}$

5. (a) 5   **(b)** 5

7. $\left(\dfrac{\partial x}{\partial r}\right)_\theta = \cos\theta$

$\left(\dfrac{\partial r}{\partial x}\right)_y = \dfrac{x}{\sqrt{x^2 + y^2}}$

Practice Exercises, pp. 847–850

1. Domain: all points in the xy-plane; range: $z \ge 0$. Level curves are ellipses with major axis along the y-axis and minor axis along the x-axis.

3. Domain: all (x, y) such that $x \ne 0$ and $y \ne 0$; range: $z \ne 0$. Level curves are hyperbolas with the x- and y-axes as asymptotes.

5. Domain: all points in xyz-space; range: all real numbers. Level surfaces are paraboloids of revolution with the z-axis as axis.

7. Domain: all (x, y, z) such that $(x, y, z) \ne (0, 0, 0)$; range: positive real numbers. Level surfaces are spheres with center $(0, 0, 0)$ and radius $r > 0$.

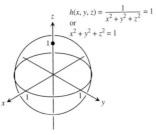

9. -2 **11.** $1/2$ **13.** 1   **15.** Let $y = kx^2, k \ne 1$

17. No; $\lim_{(x,y)\to(0,0)} f(x, y)$ does not exist.

19. $\dfrac{\partial g}{\partial r} = \cos\theta + \sin\theta, \dfrac{\partial g}{\partial \theta} = -r\sin\theta + r\cos\theta$

21. $\dfrac{\partial f}{\partial R_1} = -\dfrac{1}{R_1^2}, \dfrac{\partial f}{\partial R_2} = -\dfrac{1}{R_2^2}, \dfrac{\partial f}{\partial R_3} = -\dfrac{1}{R_3^2}$

23. $\dfrac{\partial P}{\partial n} = \dfrac{RT}{V}, \dfrac{\partial P}{\partial R} = \dfrac{nT}{V}, \dfrac{\partial P}{\partial T} = \dfrac{nR}{V}, \dfrac{\partial P}{\partial V} = -\dfrac{nRT}{V^2}$

25. $\dfrac{\partial^2 g}{\partial x^2} = 0, \dfrac{\partial^2 g}{\partial y^2} = \dfrac{2x}{y^3}, \dfrac{\partial^2 g}{\partial y\,\partial x} = \dfrac{\partial^2 g}{\partial x\,\partial y} = -\dfrac{1}{y^2}$

27. $\dfrac{\partial^2 f}{\partial x^2} = -30x + \dfrac{2 - 2x^2}{(x^2 + 1)^2}, \dfrac{\partial^2 f}{\partial y^2} = 0, \dfrac{\partial^2 f}{\partial y\,\partial x} = \dfrac{\partial^2 f}{\partial x\,\partial y} = 1$

29. $\left.\dfrac{dw}{dt}\right|_{t=0} = -1$

31. $\left.\dfrac{\partial w}{\partial r}\right|_{(r,s)=(\pi,0)} = 2, \left.\dfrac{\partial w}{\partial s}\right|_{(r,s)=(\pi,0)} = 2 - \pi$

33. $\left.\dfrac{df}{dt}\right|_{t=1} = -(\sin 1 + \cos 2)(\sin 1) + (\cos 1 + \cos 2)(\cos 1)$
$\qquad\qquad - 2(\sin 1 + \cos 1)(\sin 2)$

35. $\left.\dfrac{dy}{dx}\right|_{(x,y)=(0,1)} = -1$

37. Increases most rapidly in the direction $\mathbf{u} = -\dfrac{\sqrt{2}}{2}\mathbf{i} - \dfrac{\sqrt{2}}{2}\mathbf{j}$;

decreases most rapidly in the direction $-\mathbf{u} = \dfrac{\sqrt{2}}{2}\mathbf{i} + \dfrac{\sqrt{2}}{2}\mathbf{j}$;

$D_{\mathbf{u}}f = \dfrac{\sqrt{2}}{2}; D_{-\mathbf{u}}f = -\dfrac{\sqrt{2}}{2}; D_{\mathbf{u}_1}f = -\dfrac{7}{10}$ where $\mathbf{u}_1 = \dfrac{\mathbf{v}}{|\mathbf{v}|}$

39. Increases most rapidly in the direction $\mathbf{u} = \dfrac{2}{7}\mathbf{i} + \dfrac{3}{7}\mathbf{j} + \dfrac{6}{7}\mathbf{k}$;

decreases most rapidly in the direction $-\mathbf{u} = -\dfrac{2}{7}\mathbf{i} - \dfrac{3}{7}\mathbf{j} - \dfrac{6}{7}\mathbf{k}$;

$D_{\mathbf{u}}f = 7; D_{-\mathbf{u}}f = -7; D_{\mathbf{u}_1}f = 7$ where $\mathbf{u}_1 = \dfrac{\mathbf{v}}{|\mathbf{v}|}$

41. $\pi/\sqrt{2}$

43. (a) $f_x(1, 2) = f_y(1, 2) = 2$ **(b)** $14/5$

45.

47. Tangent: $4x - y - 5z = 4$; normal line:
$x = 2 + 4t, y = -1 - t, z = 1 - 5t$

49. $2y - z - 2 = 0$

51. Tangent: $x + y = \pi + 1$; normal line: $y = x - \pi + 1$

53. $x = 1 - 2t, y = 1, z = 1/2 + 2t$

55. Answers will depend on the upper bound used for
$|f_{xx}|, |f_{xy}|, |f_{yy}|$. With $M = \sqrt{2}/2, |E| \leq 0.0142$. With
$M = 1, |E| \leq 0.02$.

57. $L(x, y, z) = y - 3z, L(x, y, z) = x + y - z - 1$

59. Be more careful with the diameter.

61. $dI = 0.038$, % change in $I = 15.83\%$, more sensitive to voltage
change

63. (a) 5%   **65.** Local minimum of -8 at $(-2, -2)$

67. Saddle point at $(0, 0), f(0, 0) = 0$; local maximum of $1/4$ at
$(-1/2, -1/2)$

69. Saddle point at $(0, 0), f(0, 0) = 0$; local minimum of -4 at
$(0, 2)$; local maximum of 4 at $(-2, 0)$; saddle point at $(-2, 2)$,
$f(-2, 2) = 0$

71. Absolute maximum: 28 at $(0, 4)$; absolute minimum: $-9/4$ at
$(3/2, 0)$

73. Absolute maximum: 18 at $(2, -2)$; absolute minimum: $-17/4$ at
$(-2, 1/2)$

75. Absolute maximum: 8 at $(-2, 0)$; absolute minimum: -1 at $(1, 0)$

77. Absolute maximum: 4 at $(1, 0)$; absolute minimum: -4 at $(0, -1)$

79. Absolute maximum: 1 at $(0, \pm 1)$ and $(1, 0)$; absolute minimum:
-1 at $(-1, 0)$

81. Maximum: 5 at $(0, 1)$; minimum: $-1/3$ at $(0, -1/3)$

83. Maximum: $\sqrt{3}$ at $\left(\dfrac{1}{\sqrt{3}}, -\dfrac{1}{\sqrt{3}}, \dfrac{1}{\sqrt{3}}\right)$; minimum: $-\sqrt{3}$ at
$\left(-\dfrac{1}{\sqrt{3}}, \dfrac{1}{\sqrt{3}}, -\dfrac{1}{\sqrt{3}}\right)$

85. Width $= \left(\dfrac{c^2 V}{ab}\right)^{1/3}$, depth $= \left(\dfrac{b^2 V}{ac}\right)^{1/3}$, height $= \left(\dfrac{a^2 V}{bc}\right)^{1/3}$

87. Maximum: $\dfrac{3}{2}$ at $\left(\dfrac{1}{\sqrt{2}}, \dfrac{1}{\sqrt{2}}, \sqrt{2}\right)$ and $\left(-\dfrac{1}{\sqrt{2}}, -\dfrac{1}{\sqrt{2}}, -\sqrt{2}\right)$;

minimum: $\dfrac{1}{2}$ at $\left(-\dfrac{1}{\sqrt{2}}, \dfrac{1}{\sqrt{2}}, -\sqrt{2}\right)$ and $\left(\dfrac{1}{\sqrt{2}}, -\dfrac{1}{\sqrt{2}}, \sqrt{2}\right)$

89. (a) $(2y + x^2 z)e^{yz}$ **(b)** $x^2 e^{yz}\left(y - \dfrac{z}{2y}\right)$

(c) $(1 + x^2 y)e^{yz}$

91. $\dfrac{\partial w}{\partial x} = \cos \theta \dfrac{\partial w}{\partial r} - \dfrac{\sin \theta}{r} \dfrac{\partial w}{\partial \theta}, \dfrac{\partial w}{\partial y} = \sin \theta \dfrac{\partial w}{\partial r} + \dfrac{\cos \theta}{r} \dfrac{\partial w}{\partial \theta}$

97. $(t, -t \pm 4, t)$, t a real number

Additional and Advanced Exercises, pp. 851–852

1. $f_{xy}(0, 0) = -1, f_{yx}(0, 0) = 1$

7. (c) $\dfrac{r^2}{2} = \dfrac{1}{2}(x^2 + y^2 + z^2)$ **13.** $V = \dfrac{\sqrt{3}abc}{2}$

17. $f(x, y) = \dfrac{y}{2} + 4, g(x, y) = \dfrac{x}{2} + \dfrac{9}{2}$

19. $y = 2 \ln|\sin x| + \ln 2$

21. (a) $\dfrac{1}{\sqrt{53}}(2\mathbf{i} + 7\mathbf{j})$ **(b)** $\dfrac{-1}{\sqrt{29{,}097}}(98\mathbf{i} - 127\mathbf{j} + 58\mathbf{k})$

23. $w = e^{-c^2 \pi^2 t} \sin \pi x$

CHAPTER 15

Section 15.1, pp. 858–859

1. 24   **3.** 1   **5.** 16   **7.** $2 \ln 2 - 1$ **9.** $(3/2)(5 - e)$ **11.** $3/2$
13. 14   **15.** 0   **17.** $1/2$ **19.** $2 \ln 2$ **21.** $(\ln 2)^2$ **23.** $8/3$
25. 1   **27.** $\sqrt{2}$

Section 15.2, pp. 865–868

1.

3.

5.

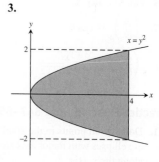

7.

9. (a) $0 \leq x \leq 2, x^3 \leq y \leq 8$
(b) $0 \leq y \leq 8, 0 \leq x \leq y^{1/3}$

11. (a) $0 \leq x \leq 3, x^2 \leq y \leq 3x$
(b) $0 \leq y \leq 9, \dfrac{y}{3} \leq x \leq \sqrt{y}$

13. (a) $0 \leq x \leq 9, 0 \leq y \leq \sqrt{x}$
(b) $0 \leq y \leq 3, y^2 \leq x \leq 9$

15. (a) $0 \leq x \leq \ln 3, e^{-x} \leq y \leq 1$
(b) $\dfrac{1}{3} \leq y \leq 1, -\ln y \leq x \leq \ln 3$

17. (a) $0 \leq x \leq 1, x \leq y \leq 3 - 2x$
(b) $0 \leq y \leq 1, 0 \leq x \leq y \cup 1 \leq y \leq 3, 0 \leq x \leq \dfrac{3 - y}{2}$

19. $\dfrac{\pi^2}{2} + 2$

21. $8 \ln 8 - 16 + e$

23. $e - 2$

25. $\dfrac{3}{2}\ln 2$ **27.** $-1/10$

29. 8

31. 2π

33. $\displaystyle\int_{2}^{4}\int_{0}^{(4-y)/2} dx\, dy$

35. $\displaystyle\int_{0}^{1}\int_{x^2}^{x} dy\, dx$

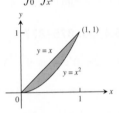

37. $\displaystyle\int_{1}^{e}\int_{\ln y}^{1} dx\, dy$

39. $\displaystyle\int_{0}^{9}\int_{0}^{(\sqrt{9-y})/2} 16x\, dx\, dy$

41. $\displaystyle\int_{-1}^{1}\int_{0}^{\sqrt{1-x^2}} 3y\, dy\, dx$

43. $\displaystyle\int_{0}^{1}\int_{e^y}^{e} xy\, dx\, dy$

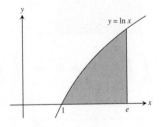

45. $\displaystyle\int_{1}^{e^3}\int_{\ln x}^{3} (x+y)\, dy\, dx$

47. 2

49. $\dfrac{e - 2}{2}$

51. 2

53. $1/(80\pi)$

55. $-2/3$

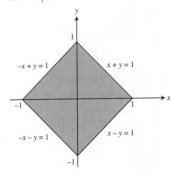

57. $4/3$ **59.** $625/12$ **61.** 16   **63.** 20   **65.** $2(1+\ln 2)$

67.

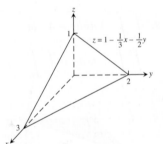

69. 1   **71.** π^2 **73.** $-\dfrac{3}{32}$ **75.** $\dfrac{20\sqrt{3}}{9}$

77. $\displaystyle\int_{0}^{1}\int_{x}^{2-x} (x^2 + y^2)\, dy\, dx = \dfrac{4}{3}$

79. R is the set of points (x, y) such that $x^2 + 2y^2 < 4$.

81. No, by Fubini's Theorem, the two orders of integration must give the same result.

85. 0.603   **87.** 0.233

Section 15.3, p. 870

1. $\int_0^2 \int_0^{2-x} dy\,dx = 2$ or   **3.** $\int_{-2}^1 \int_{y-2}^{-y^2} dx\,dy = \dfrac{9}{2}$

$\int_0^2 \int_0^{2-y} dx\,dy = 2$

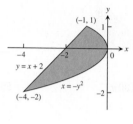

5. $\int_0^{\ln 2} \int_0^{e^x} dy\,dx = 1$ **7.** $\int_0^1 \int_{y^2}^{2y-y^2} dx\,dy = \dfrac{1}{3}$

9. $\int_0^2 \int_y^{3y} 1\,dx\,dy = 4$ or

$\int_0^2 \int_{x/3}^x 1\,dy\,dx + \int_2^6 \int_{x/3}^2 1\,dy\,dx = 4$

11. $\int_0^1 \int_{x/2}^{2x} 1\,dy\,dx + \int_1^2 \int_{x/2}^{3-x} 1\,dy\,dx = \dfrac{3}{2}$ or

$\int_0^1 \int_{y/2}^{2y} 1\,dx\,dy + \int_1^2 \int_{y/2}^{3-y} 1\,dx\,dy = \dfrac{3}{2}$

13. 12

15. $\sqrt{2} - 1$

17. $\dfrac{3}{2}$

19. (a) 0   **(b)** $4/\pi^2$ **21.** 8/3

23. $40{,}000(1 - e^{-2})\ln(7/2) \approx 43{,}329$

Section 15.4, pp. 875–877

1. $\dfrac{\pi}{2} \le \theta \le 2\pi, 0 \le r \le 9$ **3.** $\dfrac{\pi}{4} \le \theta \le \dfrac{3\pi}{4}, 0 \le r \le \csc\theta$

5. $0 \le \theta \le \dfrac{\pi}{6}, 1 \le r \le 2\sqrt{3}\sec\theta$;

$\dfrac{\pi}{6} \le \theta \le \dfrac{\pi}{2}, 1 \le r \le 2\csc\theta$

7. $-\dfrac{\pi}{2} \le \theta \le \dfrac{\pi}{2}, 0 \le r \le 2\cos\theta$ **9.** $\dfrac{\pi}{2}$

11. 2π **13.** 36   **15.** $2 - \sqrt{3}$ **17.** $(1 - \ln 2)\pi$

19. $(2\ln 2 - 1)(\pi/2)$ **21.** $\dfrac{2(1 + \sqrt{2})}{3}$

23.

$\int_0^1 \int_0^{\sqrt{1-x^2}} xy\,dy\,dx$ or $\int_0^1 \int_0^{\sqrt{1-y^2}} xy\,dx\,dy$

25.

$\int_0^2 \int_0^x y^2(x^2 + y^2)\,dy\,dx$ or $\int_0^2 \int_y^2 y^2(x^2 + y^2)\,dx\,dy$

27. $2(\pi - 1)$ **29.** 12π **31.** $(3\pi/8) + 1$ **33.** $\dfrac{2a}{3}$ **35.** $\dfrac{2a}{3}$

37. $2\pi\left(2 - \sqrt{e}\right)$ **39.** $\dfrac{4}{3} + \dfrac{5\pi}{8}$ **41. (a)** $\dfrac{\sqrt{\pi}}{2}$ **(b)** 1

43. $\pi \ln 4$, no   **45.** $\dfrac{1}{2}\left(a^2 + 2h^2\right)$

Section 15.5, pp. 883–886

1. $1/6$

3. $\displaystyle\int_0^1 \int_0^{2-2x} \int_0^{3-3x-3y/2} dz\,dy\,dx,\quad \int_0^2 \int_0^{1-y/2} \int_0^{3-3x-3y/2} dz\,dx\,dy,$

$\displaystyle\int_0^1 \int_0^{3-3x} \int_0^{2-2x-2z/3} dy\,dz\,dx,\quad \int_0^3 \int_0^{1-z/3} \int_0^{2-2x-2z/3} dy\,dx\,dz,$

$\displaystyle\int_0^2 \int_0^{3-3y/2} \int_0^{1-y/2-z/3} dx\,dz\,dy,$

$\displaystyle\int_0^3 \int_0^{2-2z/3} \int_0^{1-y/2-z/3} dx\,dy\,dz.$

The value of all six integrals is 1.

5. $\displaystyle\int_{-2}^2 \int_{-\sqrt{4-x^2}}^{\sqrt{4-x^2}} \int_{x^2+y^2}^{8-x^2-y^2} 1\,dz\,dx\,dy,$

$\displaystyle\int_{-2}^2 \int_{-\sqrt{4-y^2}}^{\sqrt{4-y^2}} \int_{x^2+y^2}^{8-x^2-y^2} 1\,dz\,dx\,dy,$

$\displaystyle\int_{-2}^2 \int_4^{8-y^2} \int_{-\sqrt{8-z-y^2}}^{\sqrt{8-z-y^2}} 1\,dx\,dz\,dy + \int_{-2}^2 \int_{y^2}^4 \int_{-\sqrt{z-y^2}}^{\sqrt{z-y^2}} 1\,dx\,dz\,dy,$

$\displaystyle\int_4^8 \int_{-\sqrt{8-z}}^{\sqrt{8-z}} \int_{-\sqrt{8-z-y^2}}^{\sqrt{8-z-y^2}} 1\,dx\,dy\,dz + \int_0^4 \int_{-\sqrt{z}}^{\sqrt{z}} \int_{-\sqrt{z-y^2}}^{\sqrt{z-y^2}} 1\,dx\,dy\,dz,$

$\displaystyle\int_{-2}^2 \int_4^{8-x^2} \int_{-\sqrt{8-z-x^2}}^{\sqrt{8-z-x^2}} 1\,dy\,dz\,dx + \int_{-2}^2 \int_{x^2}^4 \int_{-\sqrt{z-x^2}}^{\sqrt{z-x^2}} 1\,dy\,dz\,dx,$

$\displaystyle\int_4^8 \int_{-\sqrt{8-z}}^{\sqrt{8-z}} \int_{-\sqrt{8-z-x^2}}^{\sqrt{8-z-x^2}} 1\,dy\,dx\,dz + \int_0^4 \int_{-\sqrt{z}}^{\sqrt{z}} \int_{-\sqrt{z-x^2}}^{\sqrt{z-x^2}} 1\,dy\,dx\,dz.$

The value of all six integrals is 16π.

7. 1   **9.** 6   **11.** $\dfrac{5\left(2 - \sqrt{3}\right)}{4}$ **13.** 18

15. $7/6$ **17.** 0   **19.** $\dfrac{1}{2} - \dfrac{\pi}{8}$

21. (a) $\displaystyle\int_{-1}^1 \int_0^{1-x^2} \int_{x^2}^{1-z} dy\,dz\,dx$ **(b)** $\displaystyle\int_0^1 \int_{-\sqrt{1-z}}^{\sqrt{1-z}} \int_{x^2}^{1-z} dy\,dx\,dz$

(c) $\displaystyle\int_0^1 \int_0^{1-z} \int_{-\sqrt{y}}^{\sqrt{y}} dx\,dy\,dz$ **(d)** $\displaystyle\int_0^1 \int_0^{1-y} \int_{-\sqrt{y}}^{\sqrt{y}} dx\,dz\,dy$

(e) $\displaystyle\int_0^1 \int_{-\sqrt{y}}^{\sqrt{y}} \int_0^{1-y} dz\,dx\,dy$

23. $2/3$ **25.** $20/3$ **27.** 1   **29.** $16/3$ **31.** $8\pi - \dfrac{32}{3}$

33. 2   **35.** 4π **37.** $31/3$ **39.** 1   **41.** $2\sin 4$ **43.** 4

45. $a = 3$ or $a = 13/3$

47. The domain is the set of all points (x, y, z) such that $4x^2 + 4y^2 + z^2 \le 4$.

Section 15.6, pp. 891–893

1. $\bar{x} = 5/14, \bar{y} = 38/35$ **3.** $\bar{x} = 64/35, \bar{y} = 5/7$

5. $\bar{x} = \bar{y} = 4a/(3\pi)$ **7.** $I_x = I_y = 4\pi, I_0 = 8\pi$

9. $\bar{x} = -1, \bar{y} = 1/4$ **11.** $I_x = 64/105$

13. $\bar{x} = 3/8, \bar{y} = 17/16$

15. $\bar{x} = 11/3, \bar{y} = 14/27, I_y = 432$

17. $\bar{x} = 0, \bar{y} = 13/31, I_y = 7/5$

19. $\bar{x} = 0, \bar{y} = 7/10; I_x = 9/10, I_y = 3/10, I_0 = 6/5$

21. $I_x = \dfrac{M}{3}\left(b^2 + c^2\right), I_y = \dfrac{M}{3}\left(a^2 + c^2\right), I_z = \dfrac{M}{3}\left(a^2 + b^2\right)$

23. $\bar{x} = \bar{y} = 0, \bar{z} = 12/5, I_x = 7904/105 \approx 75.28,$

$I_y = 4832/63 \approx 76.70, I_z = 256/45 \approx 5.69$

25. (a) $\bar{x} = \bar{y} = 0, \bar{z} = 8/3$ **(b)** $c = 2\sqrt{2}$

27. $I_L = 1386$

29. (a) $4/3$ **(b)** $\bar{x} = 4/5, \bar{y} = \bar{z} = 2/5$

31. (a) $5/2$ **(b)** $\bar{x} = \bar{y} = \bar{z} = 8/15$ **(c)** $I_x = I_y = I_z = 11/6$

33. 3

37. (a) $I_{\text{c.m.}} = \dfrac{abc(a^2 + b^2)}{12}, R_{\text{c.m.}} = \sqrt{\dfrac{a^2 + b^2}{12}}$

(b) $I_L = \dfrac{abc(a^2 + 7b^2)}{3}, R_L = \sqrt{\dfrac{a^2 + 7b^2}{3}}$

Section 15.7, pp. 901–904

1. $\dfrac{4\pi\left(\sqrt{2} - 1\right)}{3}$ **3.** $\dfrac{17\pi}{5}$ **5.** $\pi\left(6\sqrt{2} - 8\right)$ **7.** $\dfrac{3\pi}{10}$

9. $\pi/3$

11. (a) $\displaystyle\int_0^{2\pi} \int_0^1 \int_0^{\sqrt{4-r^2}} r\,dz\,dr\,d\theta$

(b) $\displaystyle\int_0^{2\pi} \int_0^{\sqrt{3}} \int_0^1 r\,dr\,dz\,d\theta + \int_0^{2\pi} \int_{\sqrt{3}}^2 \int_0^{\sqrt{4-z^2}} r\,dr\,dz\,d\theta$

(c) $\displaystyle\int_0^1 \int_0^{\sqrt{4-r^2}} \int_0^{2\pi} r\,d\theta\,dz\,dr$

13. $\displaystyle\int_{-\pi/2}^{\pi/2} \int_0^{\cos\theta} \int_0^{3r^2} f(r, \theta, z)\,dz\,r\,dr\,d\theta$

15. $\displaystyle\int_0^\pi \int_0^{2\sin\theta} \int_0^{4-r\sin\theta} f(r, \theta, z)\,dz\,r\,dr\,d\theta$

17. $\displaystyle\int_{-\pi/2}^{\pi/2} \int_1^{1+\cos\theta} \int_0^4 f(r, \theta, z)\,dz\,r\,dr\,d\theta$

19. $\displaystyle\int_0^{\pi/4} \int_0^{\sec\theta} \int_0^{2-r\sin\theta} f(r, \theta, z)\,dz\,r\,dr\,d\theta$ **21.** π^2 **23.** $\pi/3$

25. 5π **27.** 2π **29.** $\left(\dfrac{8 - 5\sqrt{2}}{2}\right)\pi$

31. (a) $\displaystyle\int_0^{2\pi} \int_0^{\pi/6} \int_0^2 \rho^2 \sin\phi\,d\rho\,d\phi\,d\theta +$

$\displaystyle\int_0^{2\pi} \int_{\pi/6}^{\pi/2} \int_0^{\csc\phi} \rho^2 \sin\phi\,d\rho\,d\phi\,d\theta$

(b) $\displaystyle\int_0^{2\pi} \int_1^2 \int_{\pi/6}^{\sin^{-1}(1/\rho)} \rho^2 \sin\phi\,d\phi\,d\rho\,d\theta +$

$\displaystyle\int_0^{2\pi} \int_0^2 \int_0^{\pi/6} \rho^2 \sin\phi\,d\phi\,d\rho\,d\theta +$

$\displaystyle\int_0^{2\pi} \int_0^1 \int_0^{\pi/2} \rho^2 \sin\phi\,d\phi\,d\rho\,d\theta$

33. $\displaystyle\int_0^{2\pi} \int_0^{\pi/2} \int_{\cos\phi}^2 \rho^2 \sin\phi\,d\rho\,d\phi\,d\theta = \dfrac{31\pi}{6}$

35. $\int_0^{2\pi} \int_0^{\pi} \int_0^{1-\cos\phi} \rho^2 \sin\phi \, d\rho \, d\phi \, d\theta = \dfrac{8\pi}{3}$

37. $\int_0^{2\pi} \int_{\pi/4}^{\pi/2} \int_0^{2\cos\phi} \rho^2 \sin\phi \, d\rho \, d\phi \, d\theta = \dfrac{\pi}{3}$

39. (a) $8\int_0^{\pi/2} \int_0^{\pi/2} \int_0^{2} \rho^2 \sin\phi \, d\rho \, d\phi \, d\theta$

(b) $8\int_0^{\pi/2} \int_0^{2} \int_0^{\sqrt{4-r^2}} r \, dz \, dr \, d\theta$

(c) $8\int_0^{2} \int_0^{\sqrt{4-x^2}} \int_0^{\sqrt{4-x^2-y^2}} dz \, dy \, dx$

41. (a) $\int_0^{2\pi} \int_0^{\pi/3} \int_{\sec\phi}^{2} \rho^2 \sin\phi \, d\rho \, d\phi \, d\theta$

(b) $\int_0^{2\pi} \int_0^{\sqrt{3}} \int_1^{\sqrt{4-r^2}} r \, dz \, dr \, d\theta$

(c) $\int_{-\sqrt{3}}^{\sqrt{3}} \int_{-\sqrt{3-x^2}}^{\sqrt{3-x^2}} \int_1^{\sqrt{4-x^2-y^2}} dz \, dy \, dx$ (d) $5\pi/3$

43. $8\pi/3$ **45.** $9/4$ **47.** $\dfrac{3\pi-4}{18}$ **49.** $\dfrac{2\pi a^3}{3}$ **51.** $5\pi/3$

53. $\pi/2$ **55.** $\dfrac{4(2\sqrt{2}-1)\pi}{3}$ **57.** 16π **59.** $5\pi/2$

61. $\dfrac{4\pi(8-3\sqrt{3})}{3}$ **63.** $2/3$ **65.** $3/4$

67. $\bar{x} = \bar{y} = 0, \bar{z} = 3/8$ **69.** $(\bar{x}, \bar{y}, \bar{z}) = (0, 0, 3/8)$

71. $\bar{x} = \bar{y} = 0, \bar{z} = 5/6$ **73.** $I_x = \pi/4$ **75.** $\dfrac{a^4 h\pi}{10}$

77. (a) $(\bar{x}, \bar{y}, \bar{z}) = \left(0, 0, \dfrac{4}{5}\right), I_z = \dfrac{\pi}{12}$

(b) $(\bar{x}, \bar{y}, \bar{z}) = \left(0, 0, \dfrac{5}{6}\right), I_z = \dfrac{\pi}{14}$

81. $\dfrac{3M}{\pi R^3}$

85. The surface's equation $r = f(z)$ tells us that the point $(r, \theta, z) = (f(z), \theta, z)$ will lie on the surface for all θ. In particular, $(f(z), \theta + \pi, z)$ lies on the surface whenever $(f(z), \theta, z)$ lies on the surface, so the surface is symmetric with respect to the z-axis.

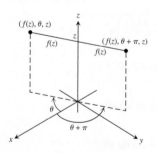

Section 15.8, pp. 912–914

1. (a) $x = \dfrac{u+v}{3}, y = \dfrac{v-2u}{3}; \dfrac{1}{3}$

(b) Triangular region with boundaries $u = 0, v = 0,$ and $u + v = 3$

3. (a) $x = \dfrac{1}{5}(2u - v), y = \dfrac{1}{10}(3v - u); \dfrac{1}{10}$

(b) Triangular region with boundaries $3v = u, v = 2u,$ and $3u + v = 10$

7. $64/5$ **9.** $\int_1^{2} \int_1^{3} (u+v)\dfrac{2u}{v} \, du \, dv = 8 + \dfrac{52}{3}\ln 2$

11. $\dfrac{\pi ab(a^2 + b^2)}{4}$ **13.** $\dfrac{1}{3}\left(1 + \dfrac{3}{e^2}\right) \approx 0.4687$ **15.** $\dfrac{225}{16}$

17. (a) $\begin{vmatrix} \cos v & -u\sin v \\ \sin v & u\cos v \end{vmatrix} = u\cos^2 v + u\sin^2 v = u$

(b) $\begin{vmatrix} \sin v & u\cos v \\ \cos v & -u\sin v \end{vmatrix} = -u\sin^2 v - u\cos^2 v = -u$

21. 12 **23.** $\dfrac{a^2 b^2 c^2}{6}$

Practice Exercises, pp. 914–916

1. $9e - 9$ **3.** $9/2$

5. $\int_{-2}^{0} \int_{2x+4}^{4-x^2} dy \, dx = \dfrac{4}{3}$ **7.** $\int_{-3}^{3} \int_0^{(1/2)\sqrt{9-x^2}} y \, dy \, dx = \dfrac{9}{2}$

9. $\sin 4$ **11.** $\dfrac{\ln 17}{4}$ **13.** $4/3$ **15.** $4/3$ **17.** $1/4$ **19.** π

21. $\dfrac{\pi - 2}{4}$ **23.** 0 **25.** $8/35$ **27.** $\pi/2$ **29.** $\dfrac{2(31 - 3^{5/2})}{3}$

31. (a) $\int_{-\sqrt{2}}^{\sqrt{2}} \int_{-\sqrt{2-y^2}}^{\sqrt{2-y^2}} \int_{\sqrt{x^2+y^2}}^{\sqrt{4-x^2-y^2}} 3 \, dz \, dx \, dy$

(b) $\int_0^{2\pi} \int_0^{\pi/4} \int_0^{2} 3\rho^2 \sin\phi \, d\rho \, d\phi \, d\theta$ (c) $2\pi(8 - 4\sqrt{2})$

33. $\int_0^{2\pi} \int_0^{\pi/4} \int_0^{\sec\phi} \rho^2 \sin\phi \, d\rho \, d\phi \, d\theta = \dfrac{\pi}{3}$

35. $\int_0^{1} \int_{\sqrt{1-x^2}}^{\sqrt{3-x^2}} \int_1^{\sqrt{4-x^2-y^2}} z^2 xy \, dz \, dy \, dx$

$+ \int_1^{\sqrt{3}} \int_0^{\sqrt{3-x^2}} \int_1^{\sqrt{4-x^2-y^2}} z^2 xy \, dz \, dy \, dx$

37. (a) $\dfrac{8\pi(4\sqrt{2} - 5)}{3}$ (b) $\dfrac{8\pi(4\sqrt{2} - 5)}{3}$

39. $I_z = \dfrac{8\pi\delta(b^5 - a^5)}{15}$

41. $\bar{x} = \bar{y} = \dfrac{1}{2 - \ln 4}$ **43.** $I_0 = 104$ **45.** $I_x = 2\delta$

47. $M = 4, M_x = 0, M_y = 0$ **49.** $\bar{x} = \dfrac{3\sqrt{3}}{\pi}, \bar{y} = 0$

51. **(a)** $\bar{x} = \dfrac{15\pi + 32}{6\pi + 48}, \bar{y} = 0$

(b)

Additional and Advanced Exercises, pp. 916–918

1. **(a)** $\displaystyle\int_{-3}^{2}\int_{x}^{6-x^2} x^2 \, dy \, dx$ **(b)** $\displaystyle\int_{-3}^{2}\int_{x}^{6-x^2}\int_{0}^{x^2} dz \, dy \, dx$

(c) $125/4$

3. 2π **5.** $3\pi/2$ **7.** **(a)** Hole radius $= 1$, sphere radius $= 2$

(b) $4\sqrt{3}\pi$ **9.** $\pi/4$ **11.** $\ln\left(\dfrac{b}{a}\right)$ **15.** $1/\sqrt[4]{3}$

17. Mass $= a^2 \cos^{-1}\left(\dfrac{b}{a}\right) - b\sqrt{a^2 - b^2}$,

$I_0 = \dfrac{a^4}{2}\cos^{-1}\left(\dfrac{b}{a}\right) - \dfrac{b^3}{2}\sqrt{a^2 - b^2} - \dfrac{b^3}{6}(a^2 - b^2)^{3/2}$

19. $\dfrac{1}{ab}\left(e^{a^2b^2} - 1\right)$ **21.** **(b)** 1 **(c)** 0

25. $h = \sqrt{20}$ in., $h = \sqrt{60}$ in.   **27.** $2\pi\left[\dfrac{1}{3} - \left(\dfrac{1}{3}\right)\dfrac{\sqrt{2}}{2}\right]$

CHAPTER 16

Section 16.1, pp. 923–925

1. Graph (c)   **3.** Graph (g)   **5.** Graph (d)   **7.** Graph (f)

9. $\sqrt{2}$ **11.** $\dfrac{13}{2}$ **13.** $3\sqrt{14}$ **15.** $\dfrac{1}{6}\left(5\sqrt{5} + 9\right)$

17. $\sqrt{3}\ln\left(\dfrac{b}{a}\right)$ **19.** **(a)** $4\sqrt{5}$ **(b)** $\dfrac{1}{12}(17^{3/2} - 1)$

21. $\dfrac{15}{32}\left(e^{16} - e^{64}\right)$ **23.** $\dfrac{1}{27}(40^{3/2} - 13^{3/2})$

25. $\dfrac{1}{6}(5^{3/2} + 7\sqrt{2} - 1)$ **27.** $\dfrac{10\sqrt{5} - 2}{3}$ **29.** 8

31. $\dfrac{1}{6}(17^{3/2} - 1)$ **33.** $2\sqrt{2} - 1$

35. **(a)** $4\sqrt{2} - 2$ **(b)** $\sqrt{2} + \ln\left(1 + \sqrt{2}\right)$ **37.** $I_z = 2\pi\delta a^3$

39. **(a)** $I_z = 2\pi\sqrt{2}\delta$ **(b)** $I_z = 4\pi\sqrt{2}\delta$ **41.** $I_x = 2\pi - 2$

Section 16.2, pp. 935–938

1. $\nabla f = -(x\mathbf{i} + y\mathbf{j} + z\mathbf{k})(x^2 + y^2 + z^2)^{-3/2}$

3. $\nabla g = -\left(\dfrac{2x}{x^2 + y^2}\right)\mathbf{i} - \left(\dfrac{2y}{x^2 + y^2}\right)\mathbf{j} + e^z\mathbf{k}$

5. $\mathbf{F} = -\dfrac{kx}{(x^2 + y^2)^{3/2}}\mathbf{i} - \dfrac{ky}{(x^2 + y^2)^{3/2}}\mathbf{j}$, any $k > 0$

7. **(a)** $9/2$ **(b)** $13/3$ **(c)** $9/2$

9. **(a)** $1/3$ **(b)** $-1/5$ **(c)** 0

11. **(a)** 2 **(b)** $3/2$ **(c)** $1/2$

13. $-15/2$ **15.** 36 **17.** **(a)** $-5/6$ **(b)** 0 **(c)** $-7/12$

19. $1/2$ **21.** $-\pi$ **23.** $69/4$ **25.** $-39/2$ **27.** $25/6$

29. **(a)** $\text{Circ}_1 = 0$, $\text{circ}_2 = 2\pi$, $\text{flux}_1 = 2\pi$, $\text{flux}_2 = 0$

(b) $\text{Circ}_1 = 0$, $\text{circ}_2 = 8\pi$, $\text{flux}_1 = 8\pi$, $\text{flux}_2 = 0$

31. $\text{Circ} = 0$, $\text{flux} = a^2\pi$ **33.** $\text{Circ} = a^2\pi$, $\text{flux} = 0$

35. **(a)** $-\dfrac{\pi}{2}$ **(b)** 0 **(c)** 1 **37.** **(a)** 32 **(b)** 32 **(c)** 32

39.

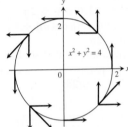

41. **(a)** $\mathbf{G} = -y\mathbf{i} + x\mathbf{j}$ **(b)** $\mathbf{G} = \sqrt{x^2 + y^2}\,\mathbf{F}$

43. $\mathbf{F} = -\dfrac{x\mathbf{i} + y\mathbf{j}}{\sqrt{x^2 + y^2}}$ **47.** 48 **49.** π **51.** 0 **53.** $\dfrac{1}{2}$

Section 16.3, pp. 947–949

1. Conservative   **3.** Not conservative   **5.** Not conservative

7. $f(x, y, z) = x^2 + \dfrac{3y^2}{2} + 2z^2 + C$

9. $f(x, y, z) = xe^{y+2z} + C$

11. $f(x, y, z) = x\ln x - x + \tan(x + y) + \dfrac{1}{2}\ln(y^2 + z^2) + C$

13. 49 **15.** -16 **17.** 1 **19.** $9\ln 2$ **21.** 0 **23.** -3

27. $\mathbf{F} = \nabla\left(\dfrac{x^2 - 1}{y}\right)$ **29.** **(a)** 1 **(b)** 1 **(c)** 1

31. **(a)** 2 **(b)** 2 **33.** **(a)** $c = b = 2a$ **(b)** $c = b = 2$

35. It does not matter what path you use. The work will be the same on any path because the field is conservative.

37. The force $\mathbf{F}$ is conservative because all partial derivatives of M, N, and P are zero. $f(x, y, z) = ax + by + cz + C$; $A = (xa, ya, za)$ and $B = (xb, yb, zb)$. Therefore, $\int \mathbf{F} \cdot d\mathbf{r} = f(B) - f(A) = a(xb - xa) + b(yb - ya) + c(zb - za) = \mathbf{F} \cdot \overrightarrow{AB}$.

Section 16.4, pp. 958–960

1. Flux $= 0$, $\text{circ} = 2\pi a^2$ **3.** Flux $= -\pi a^2$, $\text{circ} = 0$

5. Flux $= 2$, $\text{circ} = 0$ **7.** Flux $= -9$, $\text{circ} = 9$

9. Flux $= -11/60$, $\text{circ} = -7/60$ **11.** Flux $= 64/9$, $\text{circ} = 0$

13. Flux $= 1/2$, $\text{circ} = 1/2$ **15.** Flux $= 1/5$, $\text{circ} = -1/12$

17. 0 **19.** $2/33$ **21.** 0 **23.** -16π **25.** πa^2 **27.** $3\pi/8$

29. **(a)** 0 if C is traversed counterclockwise

(b) $(h - k)(\text{area of the region})$ **39.** **(a)** 0

Section 16.5, pp. 967–969

1. $\mathbf{r}(r, \theta) = (r \cos \theta)\mathbf{i} + (r \sin \theta)\mathbf{j} + r^2\mathbf{k}, 0 \le r \le 2$,
$0 \le \theta \le 2\pi$

3. $\mathbf{r}(r, \theta) = (r \cos \theta)\mathbf{i} + (r \sin \theta)\mathbf{j} + (r/2)\mathbf{k}, 0 \le r \le 6$,
$0 \le \theta \le \pi/2$

5. $\mathbf{r}(r, \theta) = (r \cos \theta)\mathbf{i} + (r \sin \theta)\mathbf{j} + \sqrt{9 - r^2}\,\mathbf{k}$,
$0 \le r \le 3\sqrt{2}/2, 0 \le \theta \le 2\pi$; Also: $\mathbf{r}(\phi, \theta) =$
$(3 \sin \phi \cos \theta)\mathbf{i} + (3 \sin \phi \sin \theta)\mathbf{j} + (3 \cos \phi)\mathbf{k}, 0 \le \phi \le \pi/4$,
$0 \le \theta \le 2\pi$

7. $\mathbf{r}(\phi, \theta) = \left(\sqrt{3} \sin \phi \cos \theta\right)\mathbf{i} + \left(\sqrt{3} \sin \phi \sin \theta\right)\mathbf{j} +$
$\left(\sqrt{3} \cos \phi\right)\mathbf{k}, \pi/3 \le \phi \le 2\pi/3, 0 \le \theta \le 2\pi$

9. $\mathbf{r}(x, y) = x\mathbf{i} + y\mathbf{j} + (4 - y^2)\mathbf{k}, 0 \le x \le 2, -2 \le y \le 2$

11. $\mathbf{r}(u, v) = u\mathbf{i} + (3 \cos v)\mathbf{j} + (3 \sin v)\mathbf{k}, 0 \le u \le 3$,
$0 \le v \le 2\pi$

13. (a) $\mathbf{r}(r, \theta) = (r \cos \theta)\mathbf{i} + (r \sin \theta)\mathbf{j} +$
$(1 - r \cos \theta - r \sin \theta)\mathbf{k}, 0 \le r \le 3, 0 \le \theta \le 2\pi$

(b) $\mathbf{r}(u, v) = (1 - u \cos v - u \sin v)\mathbf{i} + (u \cos v)\mathbf{j} +$
$(u \sin v)\mathbf{k}, 0 \le u \le 3, 0 \le v \le 2\pi$

15. $\mathbf{r}(u, v) = (4 \cos^2 v)\mathbf{i} + u\mathbf{j} + (4 \cos v \sin v)\mathbf{k}, 0 \le u \le 3$,
$-(\pi/2) \le v \le (\pi/2)$; another way: $\mathbf{r}(u, v) = (2 + 2 \cos v)\mathbf{i}$
$+ u\mathbf{j} + (2 \sin v)\mathbf{k}, 0 \le u \le 3, 0 \le v \le 2\pi$

17. $\int_0^{2\pi} \int_0^1 \frac{\sqrt{5}}{2} r \, dr \, d\theta = \frac{\pi\sqrt{5}}{2}$

19. $\int_0^{2\pi} \int_1^3 r\sqrt{5} \, dr \, d\theta = 8\pi\sqrt{5}$ **21.** $\int_0^{2\pi} \int_1^4 1 \, du \, dv = 6\pi$

23. $\int_0^{2\pi} \int_0^1 u\sqrt{4u^2 + 1} \, du \, dv = \frac{(5\sqrt{5} - 1)}{6}\pi$

25. $\int_0^{2\pi} \int_{\pi/4}^{\pi} 2 \sin \phi \, d\phi \, d\theta = \left(4 + 2\sqrt{2}\right)\pi$

27.   **29.**

33. (b) $A = \int_0^{2\pi} \int_0^{\pi} [a^2b^2 \sin^2 \phi \cos^2 \phi + b^2c^2 \cos^4 \phi \cos^2 \theta +$
$a^2c^2 \cos^4 \phi \sin^2 \theta]^{1/2} \, d\phi \, d\theta$

35. $x_0x + y_0y = 25$ **37.** $13\pi/3$ **39.** 4   **41.** $6\sqrt{6} - 2\sqrt{2}$

43. $\pi\sqrt{c^2 + 1}$ **45.** $\frac{\pi}{6}\left(17\sqrt{17} - 5\sqrt{5}\right)$ **47.** $3 + 2 \ln 2$

49. $\frac{\pi}{6}\left(13\sqrt{13} - 1\right)$ **51.** $5\pi\sqrt{2}$ **53.** $\frac{2}{3}\left(5\sqrt{5} - 1\right)$

Section 16.6, pp. 978–980

1. $\iint_S x \, d\sigma = \int_0^3 \int_0^2 u\sqrt{4u^2 + 1} \, du \, dv = \frac{17\sqrt{17} - 1}{4}$

3. $\iint_S x^2 \, d\sigma = \int_0^{2\pi} \int_0^{\pi} \sin^3 \phi \cos^2 \theta \, d\phi \, d\theta = \frac{4\pi}{3}$

5. $\iint_S z \, d\sigma = \int_0^1 \int_0^1 (4 - u - v)\sqrt{3} \, dv \, du = 3\sqrt{3}$
(for $x = u, y = v$)

7. $\iint_S x^2\sqrt{5 - 4z} \, d\sigma = \int_0^1 \int_0^{2\pi} u^2 \cos^2 v \cdot \sqrt{4u^2 + 1} \cdot$
$u\sqrt{4u^2 + 1} \, dv \, du = \int_0^1 \int_0^{2\pi} u^3(4u^2 + 1) \cos^2 v \, dv \, du = \frac{11\pi}{12}$

9. $9a^3$ **11.** $\frac{abc}{4}(ab + ac + bc)$ **13.** 2

15. $\frac{1}{30}\left(\sqrt{2} + 6\sqrt{6}\right)$ **17.** $\sqrt{6}/30$

19. -32 **21.** $\frac{\pi a^3}{6}$ **23.** $13a^4/6$ **25.** $2\pi/3$

27. $-73\pi/6$ **29.** 18   **31.** $\frac{\pi a^3}{6}$ **33.** $\frac{\pi a^2}{4}$

35. $\frac{\pi a^3}{2}$ **37.** -32 **39.** -4 **41.** $3a^4$

43. $\left(\frac{a}{2}, \frac{a}{2}, \frac{a}{2}\right)$ **45.** $(\bar{x}, \bar{y}, \bar{z}) = \left(0, 0, \frac{14}{9}\right), I_z = \frac{15\pi\sqrt{2}}{2}\delta$

47. (a) $\frac{8\pi}{3}a^4\delta$ **(b)** $\frac{20\pi}{3}a^4\delta$

Section 16.7, pp. 988–990

1. 4π **3.** $-5/6$ **5.** 0   **7.** -6π **9.** $2\pi a^2$ **13.** 12π
15. $-\pi/4$ **17.** -15π **25.** $16I_y + 16I_x$

Section 16.8, pp. 999–1001

1. 0   **3.** 0   **5.** -16 **7.** -8π **9.** 3π **11.** $-40/3$
13. 45π **15.** $12\pi\left(4\sqrt{2} - 1\right)$
21. The integral's value never exceeds the surface area of S.

Practice Exercises, pp. 1001–1004

1. Path 1: $2\sqrt{3}$; path 2: $1 + 3\sqrt{2}$ **3.** $4a^2$ **5.** 0

7. $8\pi \sin(1)$ **9.** 0   **11.** $\pi\sqrt{3}$ **13.** $2\pi\left(1 - \frac{1}{\sqrt{2}}\right)$

15. $\frac{abc}{2}\sqrt{\frac{1}{a^2} + \frac{1}{b^2} + \frac{1}{c^2}}$ **17.** 50

19. $\mathbf{r}(\phi, \theta) = (6 \sin \phi \cos \theta)\mathbf{i} + (6 \sin \phi \sin \theta)\mathbf{j} + (6 \cos \phi)\mathbf{k}$,
$\frac{\pi}{6} \le \phi \le \frac{2\pi}{3}, 0 \le \theta \le 2\pi$

21. $\mathbf{r}(r, \theta) = (r \cos \theta)\mathbf{i} + (r \sin \theta)\mathbf{j} + (1 + r)\mathbf{k}, 0 \le r \le 2$,
$0 \le \theta \le 2\pi$

23. $\mathbf{r}(u, v) = (u \cos v)\mathbf{i} + 2u^2\mathbf{j} + (u \sin v)\mathbf{k}, 0 \le u \le 1$,
$0 \le v \le \pi$

25. $\sqrt{6}$ **27.** $\pi\left[\sqrt{2} + \ln\left(1 + \sqrt{2}\right)\right]$ **29.** Conservative

31. Not conservative   **33.** $f(x, y, z) = y^2 + yz + 2x + z$

35. Path 1: 2; path 2: 8/3   **37. (a)** $1 - e^{-2\pi}$ **(b)** $1 - e^{-2\pi}$

39. 0   **41. (a)** $4\sqrt{2} - 2$ **(b)** $\sqrt{2} + \ln\left(1 + \sqrt{2}\right)$

43. $(\bar{x}, \bar{y}, \bar{z}) = \left(1, \frac{16}{15}, \frac{2}{3}\right); I_x = \frac{232}{45}, I_y = \frac{64}{15}, I_z = \frac{56}{9}$

45. $\bar{z} = \frac{3}{2}, I_z = \frac{7\sqrt{3}}{3}$ **47.** $(\bar{x}, \bar{y}, \bar{z}) = (0, 0, 49/12), I_z = 640\pi$

49. Flux: 3/2; circ: $-1/2$ **53.** 3   **55.** $\dfrac{2\pi}{3}\left(7 - 8\sqrt{2}\right)$

57. 0   **59.** π

Additional and Advanced Exercises, pp. 1004–1006

1. 6π **3.** 2/3

5. (a) $\mathbf{F}(x, y, z) = z\mathbf{i} + x\mathbf{j} + y\mathbf{k}$ (b) $\mathbf{F}(x, y, z) = z\mathbf{i} + y\mathbf{k}$
   (c) $\mathbf{F}(x, y, z) = z\mathbf{i}$

7. $\dfrac{16\pi R^3}{3}$ **9.** $a = 2, b = 1$. The minimum flux is -4.

11. (b) $\dfrac{16}{3}g$

   (c) Work $= \left(\displaystyle\int_C gxy \, ds\right) \bar{y} = g\displaystyle\int_C xy^2 \, ds = \dfrac{16}{3}g$

13. (c) $\dfrac{4}{3}\pi w$ **19.** False if $\mathbf{F} = y\mathbf{i} + x\mathbf{j}$

APPENDICES

Appendix 1, p. AP-6

1. $0.\overline{1}, 0.\overline{2}, 0.\overline{3}, 0.\overline{8}, 0.\overline{9}$ or 1

3. $x < -2$ **5.** $x \le -\dfrac{1}{3}$

7. $3, -3$ **9.** $7/6, 25/6$

11. $-2 \le t \le 4$ **13.** $0 \le z \le 10$

15. $(-\infty, -2] \cup [2, \infty)$ **17.** $(-\infty, -3] \cup [1, \infty)$

19. $(-3, -2) \cup (2, 3)$ **21.** $(0, 1)$ **23.** $(-\infty, 1]$

27. The graph of $|x| + |y| \le 1$ is the interior and boundary of the "diamond-shaped" region.

Appendix 3, pp. AP-16–AP-18

1. $2, -4; 2\sqrt{5}$ **3.** Unit circle

5. $m_\perp = -\dfrac{1}{3}$

7. (a) $x = -1$ (b) $y = 4/3$ **9.** $y = -x$

11. $y = -\dfrac{5}{4}x + 6$ **13.** $y = 4x + 4$ **15.** $y = -\dfrac{x}{2} + 12$

17. x-intercept $= \sqrt{3}, y$-intercept $= -\sqrt{2}$

19. $(3, -3)$

21. $x^2 + (y - 2)^2 = 4$ **23.** $\left(x + \sqrt{3}\right)^2 + (y + 2)^2 = 4$

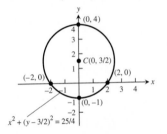

25. $x^2 + (y - 3/2)^2 = 25/4$ **27.**

29.

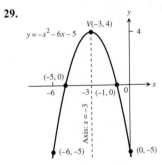

31. Exterior points of a circle of radius $\sqrt{7}$, centered at the origin

33. The washer between the circles $x^2 + y^2 = 1$ and $x^2 + y^2 = 4$ (points with distance from the origin between 1 and 2)

35. $(x + 2)^2 + (y - 1)^2 < 6$

37. $\left(\dfrac{1}{\sqrt{5}}, \dfrac{2}{\sqrt{5}}\right), \left(-\dfrac{1}{\sqrt{5}}, -\dfrac{2}{\sqrt{5}}\right)$

39. $\left(-\dfrac{1}{\sqrt{3}}, -\dfrac{1}{3}\right), \left(\dfrac{1}{\sqrt{3}}, -\dfrac{1}{3}\right)$

41. (a) ≈ -2.5 degrees/inch   (b) ≈ -16.1 degrees/inch
   (c) ≈ -8.3 degrees/inch   **43.** 5.97 atm

45. Yes: $C = F = -40°$

51. $k = -8,\quad k = 1/2$

Appendix 7, pp. AP-34–AP-35

1. **(a)** $(14, 8)$ **(b)** $(-1, 8)$ **(c)** $(0, -5)$

3. **(a)** By reflecting z across the real axis
  (b) By reflecting z across the imaginary axis
  (c) By reflecting z in the real axis and then multiplying the length of the vector by $1/|z|^2$

5. **(a)** Points on the circle $x^2 + y^2 = 4$
  (b) Points inside the circle $x^2 + y^2 = 4$
  (c) Points outside the circle $x^2 + y^2 = 4$

7. Points on a circle of radius 1, center $(-1, 0)$

9. Points on the line $y = -x$ **11.** $4e^{2\pi i/3}$ **13.** $1e^{2\pi i/3}$

15. $\cos^4\theta - 6\cos^2\theta\sin^2\theta + \sin^4\theta$ **17.** $1, -\dfrac{1}{2} \pm \dfrac{\sqrt{3}}{2}i$

19. $2i, -\sqrt{3} - i, \sqrt{3} - i$ **21.** $\dfrac{\sqrt{6}}{2} \pm \dfrac{\sqrt{2}}{2}i, -\dfrac{\sqrt{6}}{2} \pm \dfrac{\sqrt{2}}{2}i$

23. $1 \pm \sqrt{3}i, -1 \pm \sqrt{3}i$

INDEX

a, logarithms with base, 424
Abscissa, AP-10
Absolute change, 208
Absolute convergence, 588
Absolute Convergence Test, 589
Absolute extrema, finding, 226–227
Absolute (global) maximum, 222–224, 824–826
Absolute (global) minimum, 222–224, 824–826
Absolute value, AP-4–AP-6, AP-30
   properties of, AP-5
Absolute value function, derivative of, 166
   as piecewise-defined function, 5
Acceleration, 147, 149
   derivative of (jerk), 147, 148
   as derivative of velocity, 147, 148–149
   free fall and, 148
   normal component of, 752–756
   in polar coordinates, 757–760
   in space, 729
   tangential component of, 752–756
   velocity and position from, 234
Addition, of functions, 14–15
   of vectors, 685–686
Addition formulas, trigonometric, 26
Additivity, 921
   double integrals and, 864
Additivity Rule, for definite integrals, 317
Albert of Saxony, 577
Algebra, Fundamental Theorem of, AP-33–AP-34
Algebra operations, vector, 685–687
Algebra rules, for finite sums, 308–309
   for gradients, 807
   for natural logarithm, 44
Algebra systems, computer. *See* Computer algebra systems (CAS)
Algebraic functions, 10
Alternating series, 586
   harmonic, 586–587
Alternating Series Estimation Theorem, 588, 611
Alternating Series Test, 586
Angle convention, 23
Angle of elevation, 736
Angle of inclination, AP-11
Angles, 22–23
   between planes, 712

direction, 698
   in standard position, 23
   between vectors, 692–694
Angular velocity of rotation, 985
Antiderivative linearity rules, 281
Antiderivatives, 279–285
   definition of, 279
   and indefinite integrals, 284
   motion and, 282–283
   of vector function, 733
Antidifferentiation, 279
Applied optimization, 263–268
   of area of rectangle, 265
   examples from economics, 267–268
   examples from mathematics and physics, 265–267
   using least material, 264–265
   volume of box, 264
Approximations, center of, 202
   differential, error in, 206–207
   by differentials, 202
   error analysis of, 490–493
   linear, error formula for, 814, 839–840
   Newton's Method for roots, 275
   for roots and powers, 204
   by Simpson's Rule, 490–493
   standard linear, 202, 813
   tangent line, 202, 813
   by Taylor polynomials, 604
   trapezoidal, 487–488
   by Trapezoidal Rule, 492, 493
   using parabolas, 488–490
Arbitrary constant, 280
Arc length, 382–386
   along curve in space, 742–744
   differential formula for, 385–386
   of a function, 385, 744
   and line integrals, 919–920
Arc length differential, 641–642
Arc length formula, 384–385, 639, 742
Arc length parameter, 743
Arccosine function, 47–49
   identities involving, 49–50
Arcsecant, 186
Arcsine function, 47–49
   identities involving, 49–50
Arctangent, 186–187, 617–618

Area, 297–299
   of bounded regions in plane, 868–869
   cross-sectional, 363, 364
   under curve or graph, 319
   between curves, 347–349
      substitution and, 344–350
   as definite integral, 297
   definition of, 320
   by double integration, 868–870
   enclosed by astroid, 637
   estimation of, lower sum and, 299
      upper sum and, 298
   finite approximations for, 299
   under graph of nonnegative function, 304
   by Green's Theorem, 958
   infinite, 497
   optimizing, of rectangles, 216–217
   of parallelogram, 963
      vector cross product as, 701
   in polar coordinates, 873
   of smooth surface, 963–964
   surfaces and, 389, 642–643, 961–971
   of surfaces of revolution, 388–391, 642–643
   total, 298, 331–333
Area differential, 654
Argand diagrams, AP-29–AP-30
Argument, AP-30
Arrow diagram for a function, 2, 845–846
Associative laws, AP-23
Astroid, 638
   length of, 640
Asymptotes, 9
   of graphs, 103–114
   horizontal, 103, 105–108, 112
   of hyperbolas, 661–662
   oblique or slant, 108
   vertical, 103, 111–112
      definition of, 111
      finding equation for, 111–112
      integrands with, 499–500
Average rates of change, 60, 62
Average speed, 58–60
   over short time intervals, 59
Average value, 869–870
   of continuous functions, 320–321
      nonnegative, 303–304
   of multivariable functions, 869–870, 882–883

Average velocity, 146
a^x, definition of, 36, 423
    derivative of, 138, 179–181, 424
    inverse equations for, 424
    integral of, 424
    laws of exponents, 423
Axis(es), coordinate, AP-10
        moments of inertia about, 922, 976
    of ellipse, 659
    slicing and rotation about, volumes by,
        363–370
    spin around, 952–954

Base, a, logarithms with, 180–181, 424
    of cylinder, 363
    of exponential function, 37–38, 139, 423
Bernoulli, Daniel, 223
Bernoulli, Johann, 165, 254
Binary search, 447–448
Binomial series, 614–616
Binormal vector, 756
Birkhoff, George David, 339
Bolzano, Bernard, 147
Boundary points, 824–825, 826, AP-3
    for regions in plane, 767
    for regions in space, 769
Bounded intervals, 6
Bounded regions, 767
    absolute maxima and minima on, 824–826
    areas of, in plane, 868–870
Bounded sequences, 557–558
Box product, 702–704
Brachistochrones, 632–634
Branch diagram(s), for multivariable Chain
        rules, 794, 795, 796, 797, 798

Cable, hanging, 11, 443
Calculators, to estimate limits, 70–72
    graphing with, 30–34
Carbon-14 decay, 431–432
Cardioid, graphing of, 654
    length of, 655–656
    in polar coordinates, area enclosed by, 654
Carrying capacity, 539
Cartesian coordinate systems, 678–681
Cartesian coordinates, conversion to/from polar
        coordinates, 873–875
    in plane, AP-10
    related to cylindrical and spherical coordi-
        nates, 897
    related to cylindrical coordinates, 894
    related to polar coordinates, 641–648
    three-dimensional. *See* Three-dimensional
        coordinate systems
    triple integrals in, 877–883
Cartesian integrals, changing into polar inte-
        grals, 873–875
CAS. *See* Computer algebra systems
CAST rule, 24
Catenary, 11, 443
Cauchy, Augustin-Louis, 259
Cauchy's Mean Value Theorem, 259–260
Cavalieri, Bonaventura, 365
Cavalieri's principle, 365
Center of linear approximation, 202
Center of curvature, for plane curves, 749

Center of mass, 403
    centroid, 408–409
    coordinates of, 404, 922, 976
    moments and, 402–411, 886–891
    of solid, 887
    of thin flat plate, 404–407
    of thin shell, 977–978
    of wire or spring, 922
Centroids, 408–409, 887–888
    fluid forces and, 409
    Pappus's Theorems and, 410
Chain Rule, 162–167, 178, 180, 181, 188, 189,
        193, 194, 338, 422, 424, 440, 637, 748,
        757, 793–799, 804, 838, 846, 966
    derivative of composite function, 162–167
    for functions of three variables, 796–797
    for functions of two variables, 793–795
    for implicit differentiation, 666
    intuitive "proof" of, 163–164
    "outside-inside" rule and, 164
    with powers of function, 165–167
    proof of, 207–208
    repeated use of, 165
    Substitution Rule and, 336–341
    for two independent variables and three inter-
        mediate variables, 796–797
    for vector functions, 730, 731
Change, of base in a logarithm, 180, 424
    estimating, in special direction, 812
    exponential, 387–388
    rates of, 58–63, 123–124, 145–151
    sensitivity to, 151, 208–209
Charge, electrical, 996
Circle of curvature, for plane curves, 749–750
Circle, AP-14
Circles, length of, 640
    osculating, 749
    in plane, AP-13–AP-15
    polar equation for, 670–671
    standard equation for, 18, AP-14
Circulation, flux versus, 934
Circulation density, 952–954
Circulation for velocity fields, 932–933
Clairaut, Alexis, 788
Clairaut's Theorem, 788
Closed curve, 933
Closed region, 767, 769
Coefficients, binomial, 615
    determination for partial fractions, 478–479
    of polynomial, 8–9
    of power series, 593
    undetermined, 472
Combining functions, 14–22
Combining series, 567–568
Common functions, 7–11
Common logarithm function, 44
Commutativity laws, AP-23
Comparison tests, for convergence of improper
        integrals, 502–503
    for convergence of series, 576–579
Competitive-hunter model, 542–544
Complete ordered field, AP-23–AP-24
Completeness property of real numbers, 36,
        AP-23
Completing the square, AP-14–AP-15
Complex conjugate, AP-29

Complex numbers, AP-25–AP-34
    imaginary part of, AP-28
    real part of, AP-28
Component equation, for plane, 709
Component form of vectors, 683–685
Component functions, 725, 926
Component (scalar) of **u** in direction of **v**, 696
Component test, for conservative fields, 943, 945
    for exact differential form, 946
Composite functions, 15–16
    continuity of, 96, 779
    definition of, 15
    derivative of, 162–167
    limit of, 78
Compressing a graph, 16
Compression of a gas, uniform, 952
Computational formulas, for torsion, 755
Computer algebra systems (CAS), in evaluation
        of improper integrals, 500–501
    integral tables and, 481–482
    integrate command, 483
    integration with, 483–484
Computer graphing, 30–34
    of functions of two variables, 770
Computers, to estimate limits, 70–72
Concave down graph, 243
Concave up graph, 243, 244
Concavity, 243–244
    second derivative test for, 244
Conditional convergence, 588–589
Cones, elliptical, 715, 717
    parametrization of, 961
    surface area of, 964
Conics, eccentricity of, 666–668
    in polar coordinates, 657, 666–671
    polar equations of, 668–670
Connected region, 939
Connectedness, 100
Conservative fields, component test for, 943, 945
    finding potentials for, 943–946
    as gradient fields, 941
    line integrals in, 945
    loop property of, 942
    potentials for, 943–946
    and Stokes' theorem, 988
Constant, arbitrary, 280
    nonzero, 281
    rate, 428
    spring, 394
Constant-depth surface, fluid force on, 397
Constant force, 393
    work done by, 697
Constant Function Rule, 730
Constant functions, definition of, 7
    derivative of, 135
    limit of, 67
Constant Multiple Rules, for antiderivatives,
        281, 282, 284–285
    for derivatives, 136–137
    and divergent series, 567
    for finite sums, 308
    for gradients, 807
    for integrals, 317, 864
    for limits, 68
    for limits of functions of two variables, 775
    for limits of sequences, 554

for series, 567
Constrained maximum, 829–832
Constrained minimum, 829–832
Constrained variables, 842–846
Construction of reals, AP-24–AP-25
Continuity, 58–121 *See also* Discontinuity
  on an interval, 95
  of composites, 779
  differentiability and, 131, 790
  of function at a point, 93–95
  at interior point, 93
  of inverse functions, 96
  at left endpoint, 93
  for multivariable functions, 777–779
  partial derivatives and, 787
  of vector functions, 726–727
Continuity equation of hydrodynamics, 996–997
Continuity Test, 94
Continuous extension, 98–99
Continuous function theorem for sequences, 554
Continuous functions, 95–96
  absolute extrema of, 226–227, 779
  average value of, 320–321, 869, 882
  composite of, 97
  definition of, 95, 727
  differentiability and, 131
  extreme values of, on closed bounded sets,
    223, 779
  integrability of, 315
  Intermediate Value Theorem for, 99–101,
    266–267
  limits of, 92
  nonnegative, average value of, 303–304
  at point, 777
  properties of, 95
Continuous vector field, 926
Contour curve, 768
Convergence, 315
  absolute, 588
  conditional, 588–589
  of improper integrals, 496, 498–499
  interval of, 597
  of power series, 596–597
  radius of, 596–597
  of Riemann sums, 315
  of sequence, 551–553
  of series, 572
    geometric, 564
    power, 593–597
  of Taylor Series, 607–613
  tests for, 501–503, 590–591
Convergence Theorem for Power Series, 596
Coordinate axes, AP-10
  moments of inertia about, 922, 976
Coordinate conversion formulas, 901
Coordinate frame, left-handed, 678
  right-handed, 678
Coordinate pair, AP-10
Coordinate planes, 678
  first moments about, 922, 976
Coordinate systems, three-dimensional. *See*
  Three-dimensional coordinate systems
Coordinates, of center of mass, 404, 922, 966
  polar, integrals in, 871–872
  xyz, line integrals and, 929–930
Coplanar vectors, 687

Corner, 130
Cosecant, 23
Cosecant function, integral of, 421
  inverse of, 186, 190
Cosine(s), 23
  integrals of products of, 465–466
  integrals of products of powers of, 462–464
  law of, 26–27, 693
Cosine function, derivative of, 155–157
  graph of, 10
  integral of, 281
  inverse of, 48, 190
Costs, fixed, 150
  marginal, 150, 267
  variable, 150
Cotangent function, 23
  inverse of, 48, 186–187
  integral of, 421
Courant, Richard, 136
Critical point, 226, 248, 822, 826
Cross product, with determinants, 701–703
  proof of distributive law for, AP-35–AP-36
  properties of, 700–701
  right-hand rule for, 700
  of two vectors in space, 700–701
Cross Product Rule for derivatives of vector
    functions, 730–731
Cross-sections, 363–370
  horizontal, limits of integration and, 863–864
  vertical, limits of integration and, 863
Cube, integral over surface of, 973
Cube root function, 8
Cubic functions, 9
Curl, **k**-component of, 952–954
Curl vector, 980–981
Curvature, calculation of, 747, 756
  center of, 749
  of plane curves, 746–751
  radius of, 749
  in space, 750
Curved patch element, 963
Curves, area between, 347–349
    substitution and, 344–350
  area under, 319, 497
  assumptions for vector integral calculus,
    939–940
  closed, 933
  contour, 768
  generating for cylinder surface, 714
  graphing of, 347–348
  initial point of, 628
  level, 806–807
  negatively oriented, 954
  parametric, 628–629
  parametrically defined, length of, 638–640
  parametrized, 629, 642–643
  piecewise smooth, 728
  plane, curvature of, 746–751
    flux across, 933–935
    lengths of, 382–386, 639–642
    parametrizations of, 628–634, 725
  plates bounded by two, 407–408
  points of inflection of, 244–246, 249
  polar, graphing of, 642
    length of, 655–656
  positively oriented, 954

secant to, 60
sigmoid shape, 540
sketching, 243–251
slope of, definition of, 60–62, 122
  finding, 61, 123, 650
smooth, 3–4, 382–383, 638–639
  curvature of, 746–747
  length of, 742
  speed on, 744
  torsion of, 755
in space, 725–731
  arc length along, 742–744
  binormals to, 752
  formulas for, 756
  normals to, 748
  parametric equations for, 725
  vector equations for. *See* Vector functions
tangent line to, 122
tangents to, 58–63, 744, 806–807
terminal point of, 628
work done by force over, 930–932
$y = f(x)$, length of, 383
Cusp, 130
Cycloids, 632
Cylinder(s), 714–715
  base of, 363
  parabolic, flux through, 975
  parametrization of, 962
  slicing with, 374–376
  volume of, 363
Cylindrical coordinates, definition of, 893
  integration of, 895–897
  motion in, 757–758
  parametrization by, 962
  to rectangular coordinates, 894, 901
  from spherical coordinates, 901
  triple integrals in, 893–897
  volume differential in, 894
Cylindrical shells, 374–375
  volumes using, 374–379
Cylindrical solid, volume of, 363–364
Cylindrical surface, 388

De Moivre's Theorem, AP-32
Decay, exponential, 38–39, 428
Decay rate, radioactive, 39, 428
Decreasing function, 6, 238–239
Dedekind, Richard, 348, 568, AP-25
Definite integrals, 313–321
  applications of, 363–413
  definition of, 297, 313–315, 336
  evaluation of, by parts, 458
  existence of, 313–315
  Mean Value Theorem for, 325–328
  notation for, 314
  properties of, 316–318
  substitution in, 344–346
  of symmetric functions, 346–347
  of vector function, 734
Definite integration by parts, 458
Definite integration by substitution, 344
Degree, of polynomial, 9
Del (∇), 804, 984–986
Density, 404
  circulation, 952–954
  flux, 951

Dependent variable of function, 1, 765, 842–843
Derivative product rule, 140–141, 538
Derivative quotient rule, 141–142, 158–159, 587
Derivative rule for inverses, 177
Derivative sum rule, 137–138, AP-8
Derivative tests, for local extreme values, 224–225, 653–656, 821
Derivatives, of absolute value function, 166
   alternate formula for, 127
   applications of, 222–285
   calculation from definition, 127
   of composite function, 162–167, 730
   of constant function, 135
   constant multiple rule for, 136–137
   of cosine function, 155–157
   Cross Product Rule, 730
   definition of, 126
   difference rule for, 137–138
   directional. *See* Directional derivatives
   Dot Product Rule, 730
   in economics, 150–151
   of exponential functions, 138–139, 422–423
   as function, 122, 126–131
   functions from, graphical behavior of, 250–251
   General Power Rule for, 136, 182
   graphing of, 128–129
   higher-order, 142–143, 172
   of hyperbolic functions, 437
   of integral, 331
   of inverse functions, 176–184
   of inverse hyperbolic functions, 439–441
   of inverse trigonometric functions, 46–47, 190
   involving $\log_a x$, 181, 425, AP-27
   left-handed, 129–130
   of logarithms, 176–184
   notations for, 128
   *n*th, 143
   one-sided, 129–130
   partial. *See* Partial derivatives
   at point, 122–124, 130–131
   of power series, 598
   as rate of change, 145–151
   of reciprocal function, 127
   right-handed, 129–130
   second-order, 142–143
   of sine function, 155–156
   of square root function, 128
   symbols for, 143
   of tangent vector, 748
   third, 143
   of trigonometric functions, 155–159
   of vector function, 727–729
   as velocity, 146, 729
   of $y = \ln x$, 419
   of $y = \sec^{-1} u$, 188–189
   of $y = \sin^{-1} u$, 187–188
   of $y = \tan^{-1} u$, 188
Descartes, René, AP-10
Determinant(s), calculating the cross product, 704
   Jacobian, 905, 907, 908, 910
Difference quotient, 123, 127
   forms for, 127
   limit of, 124

Difference Rules, 135–143, 284
   for combining series, 567
   for derivatives, 137–138
   for gradient, 807
   for integrals, 317
   for limits, 68
   for limits of functions with two variables, 775
   for limits of sequences, 554, 566
   for vector functions, 730
Difference rules, for antiderivatives, 232, 235–236
Differentiability, 129–131, 782, 787, 789–790
Differentiable functions, 127, 636, 782, 790
   constant multiple rule of, 136
   continuous, 131, 638
   graph of, 205
   on interval, 129–130
   rules for, 136–143, 163, 730
   Taylor's formula for, 604
Differential approximation, error in, 206–207
Differential equations, autonomous, 535
   first-order, 514–549
      applications of, 528–533
      solutions, 429, 514–516
   initial value problems and, 282
   separable, 428–430
   system(s) of, 541–544
      trajectory of, 542
Differential forms, 946–947
Differential formula, short form of arc length, 385–386
Differentials, 201, 204–205, 814–816
   definition of, 204
   estimating with, 205–206
   surface area, for parametrized surface, 964
   total, 815, 816
Differentiation, 122–221
   of function, 127–128
   implicit, 170–174, 797–799
   and integration, as inverse processes, 331
   logarithmic, 181–184
   term-by-term for power series, 598
   of vector functions, rules for, 729–732
Differentiation rules, 135–143
Direct Comparison Test, 501, 502
Directed line segments, 683
Direction, along a path, 628–629, 919–920
   estimating change in, 812
   of vectors, 686
Direction cosines, 698
Directional derivatives, 802–808
   calculation of, 804–806
   defined, 803
   as dot product, 804
   estimating change with, 812
   and gradients, 804
   interpretation of, 803–804
   in plane, 802–803
   properties of, 805
Direction cosines, of ellipse, 667
   of hyperbola, 667
   of parabola, 667, 669
Dirichlet, Lejeune, 497
Discontinuity, 130
   in dy/dx, 384–385
   infinite, 94

jump, 94
   oscillating, 94
   point of, 93
   removable, 94
Discriminant (Hessian) of function, 823
Disk method, 366–368
Displacement, 146
   definition of, 302
   versus distance traveled, 302–303, 330
Display window, 30–33
Distance, in plane, AP-13–AP-15
   and spheres in space, 680–681
   in three-dimensional Cartesian coordinates, point to line, 708
      point to plane, 709–710, 711–712
      point to point, 680
Distance formula, 680, AP-14
Distance traveled, 299–301
   versus displacement, 302–303, 330
   estimating with finite sums, 301
   total, 302, 330
Distributive Law, AP-23
   proof of, AP-35–AP-36
   for vector cross products, 700
Divergence, of improper integrals, 496, 498–499
   *n*th-term test for, 566
   of sequence, 551–553
      to infinity, 553
      to negative infinity, 553
   of series, 563, 566
   tests for, 501–503, 590–591
   of vector field, 949–952, 990
Divergence Theorem, 992
   for other regions, 994–995
   for special regions, 993–994
Divergent sequence, 552
Divergent series, 563, 566
Domain, connected, 939
   of function, 1–3, 765, 766
   natural, 2
   of vector field, 926, 939
      connected, 939
      simply connected, 939
Dominant terms, 112–113
Domination, double integrals and, 864
Domination Rule for definite integrals, 317
Dot product, 692–697
   definition of, 693
   directional derivative as, 804
   of orthogonal vectors, 535
   properties of, 695–697
Dot Product Rule for vector functions, 730
Double-angle formulas, trigonometric, 26
Double integrals, Fubini's theorem for calculating, 856–858
   over bounded nonrectangular regions, 859–860
   over rectangles, 854–858
   in polar form, 871–875
   properties of, 864–865
   substitutions in, 905–909
   as volumes, 855, 856
Double integration, area by, 868–870
Dummy variable in integrals, 315

CREDITS

A Brief Table of Integrals

Basic Forms

1. $\displaystyle\int k\,dx = kx + C$ (any number k)

2. $\displaystyle\int x^n\,dx = \frac{x^{n+1}}{n+1} + C \quad (n \neq -1)$

3. $\displaystyle\int \frac{dx}{x} = \ln|x| + C$

4. $\displaystyle\int e^x\,dx = e^x + C$

5. $\displaystyle\int a^x\,dx = \frac{a^x}{\ln a} + C \quad (a > 0, a \neq 1)$

6. $\displaystyle\int \sin x\,dx = -\cos x + C$

7. $\displaystyle\int \cos x\,dx = \sin x + C$

8. $\displaystyle\int \sec^2 x\,dx = \tan x + C$

9. $\displaystyle\int \csc^2 x\,dx = -\cot x + C$

10. $\displaystyle\int \sec x \tan x\,dx = \sec x + C$

11. $\displaystyle\int \csc x \cot x\,dx = -\csc x + C$

12. $\displaystyle\int \tan x\,dx = \ln|\sec x| + C$

13. $\displaystyle\int \cot x\,dx = \ln|\sin x| + C$

14. $\displaystyle\int \sinh x\,dx = \cosh x + C$

15. $\displaystyle\int \cosh x\,dx = \sinh x + C$

16. $\displaystyle\int \frac{dx}{\sqrt{a^2 - x^2}} = \sin^{-1}\frac{x}{a} + C$

17. $\displaystyle\int \frac{dx}{a^2 + x^2} = \frac{1}{a}\tan^{-1}\frac{x}{a} + C$

18. $\displaystyle\int \frac{dx}{x\sqrt{x^2 - a^2}} = \frac{1}{a}\sec^{-1}\left|\frac{x}{a}\right| + C$

19. $\displaystyle\int \frac{dx}{\sqrt{a^2 + x^2}} = \sinh^{-1}\frac{x}{a} + C \quad (a > 0)$

20. $\displaystyle\int \frac{dx}{\sqrt{x^2 - a^2}} = \cosh^{-1}\frac{x}{a} + C \quad (x > a > 0)$

Forms Involving $ax + b$

21. $\displaystyle\int (ax+b)^n\,dx = \frac{(ax+b)^{n+1}}{a(n+1)} + C, \quad n \neq -1$

22. $\displaystyle\int x(ax+b)^n\,dx = \frac{(ax+b)^{n+1}}{a^2}\left[\frac{ax+b}{n+2} - \frac{b}{n+1}\right] + C, \quad n \neq -1, -2$

23. $\displaystyle\int (ax+b)^{-1}\,dx = \frac{1}{a}\ln|ax+b| + C$

24. $\displaystyle\int x(ax+b)^{-1}\,dx = \frac{x}{a} - \frac{b}{a^2}\ln|ax+b| + C$

25. $\displaystyle\int x(ax+b)^{-2}\,dx = \frac{1}{a^2}\left[\ln|ax+b| + \frac{b}{ax+b}\right] + C$

26. $\displaystyle\int \frac{dx}{x(ax+b)} = \frac{1}{b}\ln\left|\frac{x}{ax+b}\right| + C$

27. $\displaystyle\int \left(\sqrt{ax+b}\right)^n\,dx = \frac{2}{a}\frac{\left(\sqrt{ax+b}\right)^{n+2}}{n+2} + C, \quad n \neq -2$

28. $\displaystyle\int \frac{\sqrt{ax+b}}{x}\,dx = 2\sqrt{ax+b} + b\int \frac{dx}{x\sqrt{ax+b}}$

29. (a) $\int \frac{dx}{x\sqrt{ax+b}} = \frac{1}{\sqrt{b}} \ln\left|\frac{\sqrt{ax+b}-\sqrt{b}}{\sqrt{ax+b}+\sqrt{b}}\right| + C$ **(b)** $\int \frac{dx}{x\sqrt{ax-b}} = \frac{2}{\sqrt{b}} \tan^{-1}\sqrt{\frac{ax-b}{b}} + C$

30. $\int \frac{\sqrt{ax+b}}{x^2} dx = -\frac{\sqrt{ax+b}}{x} + \frac{a}{2}\int \frac{dx}{x\sqrt{ax+b}} + C$ **31.** $\int \frac{dx}{x^2\sqrt{ax+b}} = -\frac{\sqrt{ax+b}}{bx} - \frac{a}{2b}\int \frac{dx}{x\sqrt{ax+b}} + C$

Forms Involving $a^2 + x^2$

32. $\int \frac{dx}{a^2+x^2} = \frac{1}{a}\tan^{-1}\frac{x}{a} + C$ **33.** $\int \frac{dx}{(a^2+x^2)^2} = \frac{x}{2a^2(a^2+x^2)} + \frac{1}{2a^3}\tan^{-1}\frac{x}{a} + C$

34. $\int \frac{dx}{\sqrt{a^2+x^2}} = \sinh^{-1}\frac{x}{a} + C = \ln\left(x + \sqrt{a^2+x^2}\right) + C$

35. $\int \sqrt{a^2+x^2}\, dx = \frac{x}{2}\sqrt{a^2+x^2} + \frac{a^2}{2}\ln\left(x + \sqrt{a^2+x^2}\right) + C$

36. $\int x^2\sqrt{a^2+x^2}\, dx = \frac{x}{8}(a^2+2x^2)\sqrt{a^2+x^2} - \frac{a^4}{8}\ln\left(x + \sqrt{a^2+x^2}\right) + C$

37. $\int \frac{\sqrt{a^2+x^2}}{x} dx = \sqrt{a^2+x^2} - a\ln\left|\frac{a+\sqrt{a^2+x^2}}{x}\right| + C$

38. $\int \frac{\sqrt{a^2+x^2}}{x^2} dx = \ln\left(x + \sqrt{a^2+x^2}\right) - \frac{\sqrt{a^2+x^2}}{x} + C$

39. $\int \frac{x^2}{\sqrt{a^2+x^2}} dx = -\frac{a^2}{2}\ln\left(x + \sqrt{a^2+x^2}\right) + \frac{x\sqrt{a^2+x^2}}{2} + C$

40. $\int \frac{dx}{x\sqrt{a^2+x^2}} = -\frac{1}{a}\ln\left|\frac{a+\sqrt{a^2+x^2}}{x}\right| + C$ **41.** $\int \frac{dx}{x^2\sqrt{a^2+x^2}} = -\frac{\sqrt{a^2+x^2}}{a^2x} + C$

Forms Involving $a^2 - x^2$

42. $\int \frac{dx}{a^2-x^2} = \frac{1}{2a}\ln\left|\frac{x+a}{x-a}\right| + C$ **43.** $\int \frac{dx}{(a^2-x^2)^2} = \frac{x}{2a^2(a^2-x^2)} + \frac{1}{4a^3}\ln\left|\frac{x+a}{x-a}\right| + C$

44. $\int \frac{dx}{\sqrt{a^2-x^2}} = \sin^{-1}\frac{x}{a} + C$ **45.** $\int \sqrt{a^2-x^2}\, dx = \frac{x}{2}\sqrt{a^2-x^2} + \frac{a^2}{2}\sin^{-1}\frac{x}{a} + C$

46. $\int x^2\sqrt{a^2-x^2}\, dx = \frac{a^4}{8}\sin^{-1}\frac{x}{a} - \frac{1}{8}x\sqrt{a^2-x^2}(a^2-2x^2) + C$

47. $\int \frac{\sqrt{a^2-x^2}}{x} dx = \sqrt{a^2-x^2} - a\ln\left|\frac{a+\sqrt{a^2-x^2}}{x}\right| + C$ **48.** $\int \frac{\sqrt{a^2-x^2}}{x^2} dx = -\sin^{-1}\frac{x}{a} - \frac{\sqrt{a^2-x^2}}{x} + C$

49. $\int \frac{x^2}{\sqrt{a^2-x^2}} dx = \frac{a^2}{2}\sin^{-1}\frac{x}{a} - \frac{1}{2}x\sqrt{a^2-x^2} + C$ **50.** $\int \frac{dx}{x\sqrt{a^2-x^2}} = -\frac{1}{a}\ln\left|\frac{a+\sqrt{a^2-x^2}}{x}\right| + C$

51. $\int \frac{dx}{x^2\sqrt{a^2-x^2}} = -\frac{\sqrt{a^2-x^2}}{a^2x} + C$

Forms Involving $x^2 - a^2$

52. $\int \frac{dx}{\sqrt{x^2-a^2}} = \ln\left|x + \sqrt{x^2-a^2}\right| + C$

53. $\int \sqrt{x^2-a^2}\, dx = \frac{x}{2}\sqrt{x^2-a^2} - \frac{a^2}{2}\ln\left|x + \sqrt{x^2-a^2}\right| + C$

54. $\displaystyle\int \left(\sqrt{x^2 - a^2}\right)^n \, dx = \frac{x\left(\sqrt{x^2 - a^2}\right)^n}{n + 1} - \frac{na^2}{n + 1}\int \left(\sqrt{x^2 - a^2}\right)^{n-2} \, dx, \quad n \neq -1$

55. $\displaystyle\int \frac{dx}{\left(\sqrt{x^2 - a^2}\right)^n} = \frac{x\left(\sqrt{x^2 - a^2}\right)^{2-n}}{(2 - n)a^2} - \frac{n - 3}{(n - 2)a^2}\int \frac{dx}{\left(\sqrt{x^2 - a^2}\right)^{n-2}}, \quad n \neq 2$

56. $\displaystyle\int x\left(\sqrt{x^2 - a^2}\right)^n \, dx = \frac{\left(\sqrt{x^2 - a^2}\right)^{n+2}}{n + 2} + C, \quad n \neq -2$

57. $\displaystyle\int x^2\sqrt{x^2 - a^2} \, dx = \frac{x}{8}(2x^2 - a^2)\sqrt{x^2 - a^2} - \frac{a^4}{8}\ln\left|x + \sqrt{x^2 - a^2}\right| + C$

58. $\displaystyle\int \frac{\sqrt{x^2 - a^2}}{x} \, dx = \sqrt{x^2 - a^2} - a\sec^{-1}\left|\frac{x}{a}\right| + C$

59. $\displaystyle\int \frac{\sqrt{x^2 - a^2}}{x^2} \, dx = \ln\left|x + \sqrt{x^2 - a^2}\right| - \frac{\sqrt{x^2 - a^2}}{x} + C$

60. $\displaystyle\int \frac{x^2}{\sqrt{x^2 - a^2}} \, dx = \frac{a^2}{2}\ln\left|x + \sqrt{x^2 - a^2}\right| + \frac{x}{2}\sqrt{x^2 - a^2} + C$

61. $\displaystyle\int \frac{dx}{x\sqrt{x^2 - a^2}} = \frac{1}{a}\sec^{-1}\left|\frac{x}{a}\right| + C = \frac{1}{a}\cos^{-1}\left|\frac{a}{x}\right| + C$ 　　　**62.** $\displaystyle\int \frac{dx}{x^2\sqrt{x^2 - a^2}} = \frac{\sqrt{x^2 - a^2}}{a^2 x} + C$

Trigonometric Forms

63. $\displaystyle\int \sin ax \, dx = -\frac{1}{a}\cos ax + C$ 　　　　　　　**64.** $\displaystyle\int \cos ax \, dx = \frac{1}{a}\sin ax + C$

65. $\displaystyle\int \sin^2 ax \, dx = \frac{x}{2} - \frac{\sin 2ax}{4a} + C$ 　　　　　**66.** $\displaystyle\int \cos^2 ax \, dx = \frac{x}{2} + \frac{\sin 2ax}{4a} + C$

67. $\displaystyle\int \sin^n ax \, dx = -\frac{\sin^{n-1} ax \cos ax}{na} + \frac{n - 1}{n}\int \sin^{n-2} ax \, dx$

68. $\displaystyle\int \cos^n ax \, dx = \frac{\cos^{n-1} ax \sin ax}{na} + \frac{n - 1}{n}\int \cos^{n-2} ax \, dx$

69. (a) $\displaystyle\int \sin ax \cos bx \, dx = -\frac{\cos(a + b)x}{2(a + b)} - \frac{\cos(a - b)x}{2(a - b)} + C, \quad a^2 \neq b^2$

　　(b) $\displaystyle\int \sin ax \sin bx \, dx = \frac{\sin(a - b)x}{2(a - b)} - \frac{\sin(a + b)x}{2(a + b)} + C, \quad a^2 \neq b^2$

　　(c) $\displaystyle\int \cos ax \cos bx \, dx = \frac{\sin(a - b)x}{2(a - b)} + \frac{\sin(a + b)x}{2(a + b)} + C, \quad a^2 \neq b^2$

70. $\displaystyle\int \sin ax \cos ax \, dx = -\frac{\cos 2ax}{4a} + C$ 　　　　**71.** $\displaystyle\int \sin^n ax \cos ax \, dx = \frac{\sin^{n+1} ax}{(n + 1)a} + C, \quad n \neq -1$

72. $\displaystyle\int \frac{\cos ax}{\sin ax} \, dx = \frac{1}{a}\ln|\sin ax| + C$ 　　　　**73.** $\displaystyle\int \cos^n ax \sin ax \, dx = -\frac{\cos^{n+1} ax}{(n + 1)a} + C, \quad n \neq -1$

74. $\displaystyle\int \frac{\sin ax}{\cos ax} \, dx = -\frac{1}{a}\ln|\cos ax| + C$

75. $\displaystyle\int \sin^n ax \cos^m ax \, dx = -\frac{\sin^{n-1} ax \cos^{m+1} ax}{a(m + n)} + \frac{n - 1}{m + n}\int \sin^{n-2} ax \cos^m ax \, dx, \quad n \neq -m \quad \text{(reduces } \sin^n ax\text{)}$

76. $\displaystyle\int \sin^n ax \cos^m ax \, dx = \frac{\sin^{n+1} ax \cos^{m-1} ax}{a(m + n)} + \frac{m - 1}{m + n}\int \sin^n ax \cos^{m-2} ax \, dx, \quad m \neq -n \quad \text{(reduces } \cos^m ax\text{)}$

77. $\displaystyle\int \frac{dx}{b + c \sin ax} = \frac{-2}{a\sqrt{b^2 - c^2}} \tan^{-1}\left[\sqrt{\frac{b - c}{b + c}} \tan\left(\frac{\pi}{4} - \frac{ax}{2}\right)\right] + C, \quad b^2 > c^2$

78. $\displaystyle\int \frac{dx}{b + c \sin ax} = \frac{-1}{a\sqrt{c^2 - b^2}} \ln\left|\frac{c + b \sin ax + \sqrt{c^2 - b^2}\cos ax}{b + c \sin ax}\right| + C, \quad b^2 < c^2$

79. $\displaystyle\int \frac{dx}{1 + \sin ax} = -\frac{1}{a}\tan\left(\frac{\pi}{4} - \frac{ax}{2}\right) + C$ 80. $\displaystyle\int \frac{dx}{1 - \sin ax} = \frac{1}{a}\tan\left(\frac{\pi}{4} + \frac{ax}{2}\right) + C$

81. $\displaystyle\int \frac{dx}{b + c \cos ax} = \frac{2}{a\sqrt{b^2 - c^2}} \tan^{-1}\left[\sqrt{\frac{b - c}{b + c}} \tan\frac{ax}{2}\right] + C, \quad b^2 > c^2$

82. $\displaystyle\int \frac{dx}{b + c \cos ax} = \frac{1}{a\sqrt{c^2 - b^2}} \ln\left|\frac{c + b \cos ax + \sqrt{c^2 - b^2}\sin ax}{b + c \cos ax}\right| + C, \quad b^2 < c^2$

83. $\displaystyle\int \frac{dx}{1 + \cos ax} = \frac{1}{a}\tan\frac{ax}{2} + C$ 84. $\displaystyle\int \frac{dx}{1 - \cos ax} = -\frac{1}{a}\cot\frac{ax}{2} + C$

85. $\displaystyle\int x \sin ax\, dx = \frac{1}{a^2}\sin ax - \frac{x}{a}\cos ax + C$ 86. $\displaystyle\int x \cos ax\, dx = \frac{1}{a^2}\cos ax + \frac{x}{a}\sin ax + C$

87. $\displaystyle\int x^n \sin ax\, dx = -\frac{x^n}{a}\cos ax + \frac{n}{a}\int x^{n-1}\cos ax\, dx$ 88. $\displaystyle\int x^n \cos ax\, dx = \frac{x^n}{a}\sin ax - \frac{n}{a}\int x^{n-1}\sin ax\, dx$

89. $\displaystyle\int \tan ax\, dx = \frac{1}{a}\ln|\sec ax| + C$ 90. $\displaystyle\int \cot ax\, dx = \frac{1}{a}\ln|\sin ax| + C$

91. $\displaystyle\int \tan^2 ax\, dx = \frac{1}{a}\tan ax - x + C$ 92. $\displaystyle\int \cot^2 ax\, dx = -\frac{1}{a}\cot ax - x + C$

93. $\displaystyle\int \tan^n ax\, dx = \frac{\tan^{n-1} ax}{a(n - 1)} - \int \tan^{n-2} ax\, dx, \quad n \neq 1$ 94. $\displaystyle\int \cot^n ax\, dx = -\frac{\cot^{n-1} ax}{a(n - 1)} - \int \cot^{n-2} ax\, dx, \quad n \neq 1$

95. $\displaystyle\int \sec ax\, dx = \frac{1}{a}\ln|\sec ax + \tan ax| + C$ 96. $\displaystyle\int \csc ax\, dx = -\frac{1}{a}\ln|\csc ax + \cot ax| + C$

97. $\displaystyle\int \sec^2 ax\, dx = \frac{1}{a}\tan ax + C$ 98. $\displaystyle\int \csc^2 ax\, dx = -\frac{1}{a}\cot ax + C$

99. $\displaystyle\int \sec^n ax\, dx = \frac{\sec^{n-2} ax \tan ax}{a(n - 1)} + \frac{n - 2}{n - 1}\int \sec^{n-2} ax\, dx, \quad n \neq 1$

100. $\displaystyle\int \csc^n ax\, dx = -\frac{\csc^{n-2} ax \cot ax}{a(n - 1)} + \frac{n - 2}{n - 1}\int \csc^{n-2} ax\, dx, \quad n \neq 1$

101. $\displaystyle\int \sec^n ax \tan ax\, dx = \frac{\sec^n ax}{na} + C, \quad n \neq 0$ 102. $\displaystyle\int \csc^n ax \cot ax\, dx = -\frac{\csc^n ax}{na} + C, \quad n \neq 0$

Inverse Trigonometric Forms

103. $\displaystyle\int \sin^{-1} ax\, dx = x \sin^{-1} ax + \frac{1}{a}\sqrt{1 - a^2x^2} + C$ 104. $\displaystyle\int \cos^{-1} ax\, dx = x \cos^{-1} ax - \frac{1}{a}\sqrt{1 - a^2x^2} + C$

105. $\displaystyle\int \tan^{-1} ax\, dx = x \tan^{-1} ax - \frac{1}{2a}\ln(1 + a^2x^2) + C$

106. $\displaystyle\int x^n \sin^{-1} ax\, dx = \frac{x^{n+1}}{n + 1}\sin^{-1} ax - \frac{a}{n + 1}\int \frac{x^{n+1}\, dx}{\sqrt{1 - a^2x^2}}, \quad n \neq -1$

107. $\displaystyle\int x^n \cos^{-1} ax\, dx = \frac{x^{n+1}}{n + 1}\cos^{-1} ax + \frac{a}{n + 1}\int \frac{x^{n+1}\, dx}{\sqrt{1 - a^2x^2}}, \quad n \neq -1$

108. $\displaystyle\int x^n \tan^{-1} ax\, dx = \frac{x^{n+1}}{n + 1}\tan^{-1} ax - \frac{a}{n + 1}\int \frac{x^{n+1}\, dx}{1 + a^2x^2}, \quad n \neq -1$

Exponential and Logarithmic Forms

109. $\displaystyle\int e^{ax}\,dx = \frac{1}{a}e^{ax} + C$

110. $\displaystyle\int b^{ax}\,dx = \frac{1}{a}\frac{b^{ax}}{\ln b} + C, \quad b > 0, b \neq 1$

111. $\displaystyle\int xe^{ax}\,dx = \frac{e^{ax}}{a^2}(ax - 1) + C$

112. $\displaystyle\int x^n e^{ax}\,dx = \frac{1}{a}x^n e^{ax} - \frac{n}{a}\int x^{n-1}e^{ax}\,dx$

113. $\displaystyle\int x^n b^{ax}\,dx = \frac{x^n b^{ax}}{a\ln b} - \frac{n}{a\ln b}\int x^{n-1}b^{ax}\,dx, \quad b > 0, b \neq 1$

114. $\displaystyle\int e^{ax}\sin bx\,dx = \frac{e^{ax}}{a^2 + b^2}(a\sin bx - b\cos bx) + C$

115. $\displaystyle\int e^{ax}\cos bx\,dx = \frac{e^{ax}}{a^2 + b^2}(a\cos bx + b\sin bx) + C$

116. $\displaystyle\int \ln ax\,dx = x\ln ax - x + C$

117. $\displaystyle\int x^n(\ln ax)^m\,dx = \frac{x^{n+1}(\ln ax)^m}{n + 1} - \frac{m}{n + 1}\int x^n(\ln ax)^{m-1}\,dx, \quad n \neq -1$

118. $\displaystyle\int x^{-1}(\ln ax)^m\,dx = \frac{(\ln ax)^{m+1}}{m + 1} + C, \quad m \neq -1$

119. $\displaystyle\int \frac{dx}{x\ln ax} = \ln|\ln ax| + C$

Forms Involving $\sqrt{2ax - x^2}$, $a > 0$

120. $\displaystyle\int \frac{dx}{\sqrt{2ax - x^2}} = \sin^{-1}\left(\frac{x - a}{a}\right) + C$

121. $\displaystyle\int \sqrt{2ax - x^2}\,dx = \frac{x - a}{2}\sqrt{2ax - x^2} + \frac{a^2}{2}\sin^{-1}\left(\frac{x - a}{a}\right) + C$

122. $\displaystyle\int \left(\sqrt{2ax - x^2}\right)^n dx = \frac{(x - a)\left(\sqrt{2ax - x^2}\right)^n}{n + 1} + \frac{na^2}{n + 1}\int \left(\sqrt{2ax - x^2}\right)^{n-2} dx$

123. $\displaystyle\int \frac{dx}{\left(\sqrt{2ax - x^2}\right)^n} = \frac{(x - a)\left(\sqrt{2ax - x^2}\right)^{2-n}}{(n - 2)a^2} + \frac{n - 3}{(n - 2)a^2}\int \frac{dx}{\left(\sqrt{2ax - x^2}\right)^{n-2}}$

124. $\displaystyle\int x\sqrt{2ax - x^2}\,dx = \frac{(x + a)(2x - 3a)\sqrt{2ax - x^2}}{6} + \frac{a^3}{2}\sin^{-1}\left(\frac{x - a}{a}\right) + C$

125. $\displaystyle\int \frac{\sqrt{2ax - x^2}}{x}\,dx = \sqrt{2ax - x^2} + a\sin^{-1}\left(\frac{x - a}{a}\right) + C$

126. $\displaystyle\int \frac{\sqrt{2ax - x^2}}{x^2}\,dx = -2\sqrt{\frac{2a - x}{x}} - \sin^{-1}\left(\frac{x - a}{a}\right) + C$

127. $\displaystyle\int \frac{x\,dx}{\sqrt{2ax - x^2}} = a\sin^{-1}\left(\frac{x - a}{a}\right) - \sqrt{2ax - x^2} + C$

128. $\displaystyle\int \frac{dx}{x\sqrt{2ax - x^2}} = -\frac{1}{a}\sqrt{\frac{2a - x}{x}} + C$

Hyperbolic Forms

129. $\displaystyle\int \sinh ax\,dx = \frac{1}{a}\cosh ax + C$

130. $\displaystyle\int \cosh ax\,dx = \frac{1}{a}\sinh ax + C$

131. $\displaystyle\int \sinh^2 ax\,dx = \frac{\sinh 2ax}{4a} - \frac{x}{2} + C$

132. $\displaystyle\int \cosh^2 ax\,dx = \frac{\sinh 2ax}{4a} + \frac{x}{2} + C$

133. $\displaystyle\int \sinh^n ax\,dx = \frac{\sinh^{n-1}ax\cosh ax}{na} - \frac{n - 1}{n}\int \sinh^{n-2}ax\,dx, \quad n \neq 0$

134. $\displaystyle\int \cosh^n ax \, dx = \frac{\cosh^{n-1} ax \sinh ax}{na} + \frac{n-1}{n}\int \cosh^{n-2} ax \, dx, \quad n \neq 0$

135. $\displaystyle\int x \sinh ax \, dx = \frac{x}{a}\cosh ax - \frac{1}{a^2}\sinh ax + C$

136. $\displaystyle\int x \cosh ax \, dx = \frac{x}{a}\sinh ax - \frac{1}{a^2}\cosh ax + C$

137. $\displaystyle\int x^n \sinh ax \, dx = \frac{x^n}{a}\cosh ax - \frac{n}{a}\int x^{n-1}\cosh ax \, dx$

138. $\displaystyle\int x^n \cosh ax \, dx = \frac{x^n}{a}\sinh ax - \frac{n}{a}\int x^{n-1}\sinh ax \, dx$

139. $\displaystyle\int \tanh ax \, dx = \frac{1}{a}\ln(\cosh ax) + C$

140. $\displaystyle\int \coth ax \, dx = \frac{1}{a}\ln|\sinh ax| + C$

141. $\displaystyle\int \tanh^2 ax \, dx = x - \frac{1}{a}\tanh ax + C$

142. $\displaystyle\int \coth^2 ax \, dx = x - \frac{1}{a}\coth ax + C$

143. $\displaystyle\int \tanh^n ax \, dx = -\frac{\tanh^{n-1} ax}{(n-1)a} + \int \tanh^{n-2} ax \, dx, \quad n \neq 1$

144. $\displaystyle\int \coth^n ax \, dx = -\frac{\coth^{n-1} ax}{(n-1)a} + \int \coth^{n-2} ax \, dx, \quad n \neq 1$

145. $\displaystyle\int \operatorname{sech} ax \, dx = \frac{1}{a}\sin^{-1}(\tanh ax) + C$

146. $\displaystyle\int \operatorname{csch} ax \, dx = \frac{1}{a}\ln\left|\tanh \frac{ax}{2}\right| + C$

147. $\displaystyle\int \operatorname{sech}^2 ax \, dx = \frac{1}{a}\tanh ax + C$

148. $\displaystyle\int \operatorname{csch}^2 ax \, dx = -\frac{1}{a}\coth ax + C$

149. $\displaystyle\int \operatorname{sech}^n ax \, dx = \frac{\operatorname{sech}^{n-2} ax \tanh ax}{(n-1)a} + \frac{n-2}{n-1}\int \operatorname{sech}^{n-2} ax \, dx, \quad n \neq 1$

150. $\displaystyle\int \operatorname{csch}^n ax \, dx = -\frac{\operatorname{csch}^{n-2} ax \coth ax}{(n-1)a} - \frac{n-2}{n-1}\int \operatorname{csch}^{n-2} ax \, dx, \quad n \neq 1$

151. $\displaystyle\int \operatorname{sech}^n ax \tanh ax \, dx = -\frac{\operatorname{sech}^n ax}{na} + C, \quad n \neq 0$

152. $\displaystyle\int \operatorname{csch}^n ax \coth ax \, dx = -\frac{\operatorname{csch}^n ax}{na} + C, \quad n \neq 0$

153. $\displaystyle\int e^{ax} \sinh bx \, dx = \frac{e^{ax}}{2}\left[\frac{e^{bx}}{a+b} - \frac{e^{-bx}}{a-b}\right] + C, \quad a^2 \neq b^2$

154. $\displaystyle\int e^{ax} \cosh bx \, dx = \frac{e^{ax}}{2}\left[\frac{e^{bx}}{a+b} + \frac{e^{-bx}}{a-b}\right] + C, \quad a^2 \neq b^2$

Some Definite Integrals

155. $\displaystyle\int_0^\infty x^{n-1}e^{-x} \, dx = \Gamma(n) = (n-1)!, \quad n > 0$

156. $\displaystyle\int_0^\infty e^{-ax^2} \, dx = \frac{1}{2}\sqrt{\frac{\pi}{a}}, \quad a > 0$

157. $\displaystyle\int_0^{\pi/2} \sin^n x \, dx = \int_0^{\pi/2} \cos^n x \, dx = \begin{cases} \dfrac{1\cdot3\cdot5\cdot\,\cdots\,\cdot(n-1)}{2\cdot4\cdot6\cdot\,\cdots\,\cdot n}\cdot\dfrac{\pi}{2}, & \text{if } n \text{ is an even integer} \geq 2 \\[2mm] \dfrac{2\cdot4\cdot6\cdot\,\cdots\,\cdot(n-1)}{3\cdot5\cdot7\cdot\,\cdots\,\cdot n}, & \text{if } n \text{ is an odd integer} \geq 3 \end{cases}$

Trigonometry Formulas

1. Definitions and Fundamental Identities

Sine: $\sin\theta = \dfrac{y}{r} = \dfrac{1}{\csc\theta}$

Cosine: $\cos\theta = \dfrac{x}{r} = \dfrac{1}{\sec\theta}$

Tangent: $\tan\theta = \dfrac{y}{x} = \dfrac{1}{\cot\theta}$

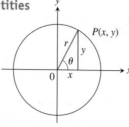

2. Identities

$$\sin(-\theta) = -\sin\theta, \quad \cos(-\theta) = \cos\theta$$

$$\sin^2\theta + \cos^2\theta = 1, \quad \sec^2\theta = 1 + \tan^2\theta, \quad \csc^2\theta = 1 + \cot^2\theta$$

$$\sin 2\theta = 2\sin\theta\cos\theta, \quad \cos 2\theta = \cos^2\theta - \sin^2\theta$$

$$\cos^2\theta = \frac{1 + \cos 2\theta}{2}, \quad \sin^2\theta = \frac{1 - \cos 2\theta}{2}$$

$$\sin(A + B) = \sin A\cos B + \cos A\sin B$$

$$\sin(A - B) = \sin A\cos B - \cos A\sin B$$

$$\cos(A + B) = \cos A\cos B - \sin A\sin B$$

$$\cos(A - B) = \cos A\cos B + \sin A\sin B$$

$$\tan(A + B) = \frac{\tan A + \tan B}{1 - \tan A\tan B}$$

$$\tan(A - B) = \frac{\tan A - \tan B}{1 + \tan A\tan B}$$

$$\sin\left(A - \frac{\pi}{2}\right) = -\cos A, \qquad \cos\left(A - \frac{\pi}{2}\right) = \sin A$$

$$\sin\left(A + \frac{\pi}{2}\right) = \cos A, \qquad \cos\left(A + \frac{\pi}{2}\right) = -\sin A$$

$$\sin A\sin B = \frac{1}{2}\cos(A - B) - \frac{1}{2}\cos(A + B)$$

$$\cos A\cos B = \frac{1}{2}\cos(A - B) + \frac{1}{2}\cos(A + B)$$

$$\sin A\cos B = \frac{1}{2}\sin(A - B) + \frac{1}{2}\sin(A + B)$$

$$\sin A + \sin B = 2\sin\frac{1}{2}(A + B)\cos\frac{1}{2}(A - B)$$

$$\sin A - \sin B = 2\cos\frac{1}{2}(A + B)\sin\frac{1}{2}(A - B)$$

$$\cos A + \cos B = 2\cos\frac{1}{2}(A + B)\cos\frac{1}{2}(A - B)$$

$$\cos A - \cos B = -2\sin\frac{1}{2}(A + B)\sin\frac{1}{2}(A - B)$$

Trigonometric Functions

Radian Measure

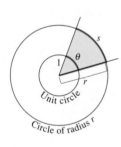

Degrees	Radians

$$\frac{s}{r} = \frac{\theta}{1} = \theta \quad \text{or} \quad \theta = \frac{s}{r},$$

$$180° = \pi \text{ radians}.$$

The angles of two common triangles, in degrees and radians.

y = sin x

Domain: $(-\infty, \infty)$
Range: $[-1, 1]$

y = cos x

Domain: $(-\infty, \infty)$
Range: $[-1, 1]$

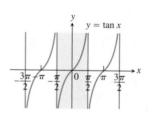

y = tan x

Domain: All real numbers except odd integer multiples of $\pi/2$
Range: $(-\infty, \infty)$

y = sec x

Domain: All real numbers except odd integer multiples of $\pi/2$
Range: $(-\infty, -1] \cup [1, \infty)$

y = csc x

Domain: $x \neq 0, \pm\pi, \pm 2\pi, \ldots$
Range: $(-\infty, -1] \cup [1, \infty)$

y = cot x

Domain: $x \neq 0, \pm\pi, \pm 2\pi, \ldots$
Range: $(-\infty, \infty)$